AF759599

Introduction to Aeroelasticity

Harijono Djojodihardjo

Introduction to Aeroelasticity

With Case-Studies

Springer

Harijono Djojodihardjo
The Institute for the Advancement
of Aerospace Science and Technology
"Persada Kriyareka Dirgantara"
Jakarta, Indonesia

ISBN 978-981-16-8077-9 ISBN 978-981-16-8078-6 (eBook)
https://doi.org/10.1007/978-981-16-8078-6

© The Editor(s) (if applicable) and The Author(s), under exclusive license to Springer Nature Singapore Pte Ltd. 2023
This work is subject to copyright. All rights are solely and exclusively licensed by the Publisher, whether the whole or part of the material is concerned, specifically the rights of translation, reprinting, reuse of illustrations, recitation, broadcasting, reproduction on microfilms or in any other physical way, and transmission or information storage and retrieval, electronic adaptation, computer software, or by similar or dissimilar methodology now known or hereafter developed.
The use of general descriptive names, registered names, trademarks, service marks, etc. in this publication does not imply, even in the absence of a specific statement, that such names are exempt from the relevant protective laws and regulations and therefore free for general use.
The publisher, the authors, and the editors are safe to assume that the advice and information in this book are believed to be true and accurate at the date of publication. Neither the publisher nor the authors or the editors give a warranty, expressed or implied, with respect to the material contained herein or for any errors or omissions that may have been made. The publisher remains neutral with regard to jurisdictional claims in published maps and institutional affiliations.

This Springer imprint is published by the registered company Springer Nature Singapore Pte Ltd.
The registered company address is: 152 Beach Road, #21-01/04 Gateway East, Singapore 189721, Singapore

Preface

This book is now finally prepared, after a long delay due to the author's many commitments during the author's services to the universities, government technology institutions and aircraft industry.

First of all, the author is deeply indebted to his mentor, who initiated his interest in aeroelasticity and structural dynamics, the legendary and aeroelasticity pioneer Prof. Holt Ashley of the Massachusetts Institute of Technology and Stanford University. His first memorable experience is obtained when he was assigned by Prof. Holt Ashley as his research assistant at the MIT Aeroelastic and Structures Laboratory way back in 1966 at the Department of Aeronautics and Astronautics there and at the same time working for his Sc.D. degree. He then was assigned to work on the concept Prof. Holt Ashley which was conceived during his sabbatical leave to India, and this is how the interest and lifelong education in this field started. Later on when Prof. Holt Ashley left for Stanford University, since the research was carried out at MIT, he enjoyed the dynamic and energetic supervision of Prof. Sheila E. Widnall and other members of the thesis committee including Prof. Rene H. Miller and Prof. Norman Ham, both from the Department of Aeronautics and Astronautics, and Prof. Justin E. Kerwin, from the Department of Naval Architecture and Marine Engineering. The author has also enjoyed sharing the views and vision of the late Prof. Martin Landahl. His thesis topic was "A *Numerical Method for the Calculation of Nonlinear, Unsteady, Lifting Potential Flow Problems,*" which was defended in October 1968.

Returning to Indonesia to his Alma Mater, the Institute of Technology Bandung (Institut Teknologi Bandung-ITB) after receiving his Sc.D. degree in February 1969, he resumed his academic duties there. Eventually, when the Study Program on Aeronautics and Astronautics was established at ITB, he taught unsteady aerodynamics, computational fluid dynamics and aeroelasticity. Prior to his retirement at ITB in 2005, he obtained an opportunity to teach Aeroelasticity at Universiti Sains Malaysia (University of Science Malaysia-USM) in 2004, which he continued delivering until 2008.

He was assigned in 1982 to head the Dynamics and Load Sub-Directorate at the Indonesian Aircraft Industry PT IPTN (which was established in 1976), to deal with dynamics, load, aeroelasticity and weight and balance for the CN-235 aircraft program and certification. That was the start of many exciting aeroelastic and structures activities, including the preparation of type certification documents and various aeroelastic testing activities. Established cooperation with DLR, Airbus (then MBB), NLR and Boeing addressed many aeroelastic areas, including ground vibration test, aeroelastic model testing and flight flutter testing, which included association with many knowledgeable aeroelasticians, including Prof.ir. R. Zwaan, Prof. Dr.-Ing. H. Försching and Prof. Dr.-Ing. H. Hönlinger. Later on, he enjoyed being associated with and inspired by Prof. Dr.-Ing. Boris Laschka in the wider area of fluid mechanics and aeroelasticity, for which the author has enjoyed being invited as a guest scholar at Technische Universität München in 1988 and 2002. These many gratifying experiences have developed the author's insight and love for aeroelasticity.

The present book has evolved from the author's teaching material at ITB and USM and his research at ITB, PT IPTN, USM and Universiti Putra Malaysia (UPM). Certainly, there are many excellent aeroelasticity books that have been published that addressed more comprehensive substance and much involved elaborations, in particular the classical books of Bisplinghoff, Ashley and Halfman's Aeroelasticity, Bisplinghoff and Ashley's Principles of Aeroelasticity, Fung's Introduction to the theory of aeroelasticity, Scanlan and Rosenbaum's Introduction to the study of aircraft vibration and flutter, Zwaan's Lecture on Aeroelasticity (circulated at the Delft University of Technology and Institut Teknologi Bandung), Dowell's Modern Course in Aeroelasticity, Hodges and Pierce's Introduction to Structural Dynamics and Aeroelasticity, and Wright and Cooper's Introduction to Aircraft Aeroelasticity and Loads, Hall and Kielb's Unsteady Aerodynamics, Aeroacoustics and Aeroelasticity of Turbomachines, Battoo's Beginners Guide to Literature in the Field of Aeroelasticity and Försching's Fundamental of Aeroelasticity (in German). The author has referred to these in the lectures as well as in this book.

Then the purpose of this book is to present an introductory version of the host of excellent classical and new publications but attempting to organize the essential parts of these in new structured narratives that are expected to facilitate the process of learning as well as utilizing the materials to solve new problems or develop new solution approaches.

Accordingly, in this book, only introductory materials for aeroelasticity are elaborated for students and novel practitioners, which has been presented with individual styles. Associated with these excellent books, this one may serve as an introductory before delving further into these excellent books.

To complement the introductory nature of the book, the author has incorporated some published material by the author and colleagues, which may illustrate and provide useful information for the readers. The examples and case studies incorporated in the book originating from the author's work and colleagues are intended to be instructive, since mostly are fundamental and solved using first principles. The author is indebted to colleagues Prof. Vladimir Zhuravlev from Moscow Aviation Institute, Prof. Polo Gasbarri from the University of Rome La Sapienza for their review, criticism and suggestions and former students Dr.ir. Bonifacius Bima Prananta (deceased), Dr.ir. Leonardo Gunawan, Ir. Mohamad Kusni M.Tek., both

Faculty Members at the Faculty of Mechanical and Aerospace Engineering, Institute of Technology Bandung, Dr.-Ing.Widjaja Kresna Sekar, now associated with a company serving Airbus, Dr.-Ing. Ismoyo Haryanto, Faculty Member of Diponegoro University, Ir. Yan Mursal M.Tek, an Aeroelastician just retiring from Pilatus (Switzerland), Dr.ir. Bambang Irawan Soemarwoto at NLR, the Netherlands, Ir. Agus Budiono Ph.D., a university instructor–professor, technopreneur and AI and control system expert, for their contributions in the joint work, and their review, criticism and suggestions.

In writing the book, many illustrations have been selected and presented. This has two merits:

- Pictures provide much more informations and inspirations than thousand words.
- Pictures will incite curiosity for further active learning efforts by the readers.

It is the hope of the author that this book will contribute to the further development of the field due to the involvements of more students and participating workers that may benefit from this book, as well as that the book will benefit from the readers' critical suggestions for improvement.

Jakarta, Indonesia
April 2020

Harijono Djojodihardjo

Acknowledgements

The author is indebted to colleagues Prof. Vladimir Zhuravlev from Moscow Aviation Institute, Prof. Polo Gasbarri from the University of Rome La Sapienza for their review, criticism and suggestions, Dr. Moh. Faisal Abd. Hamid at the Aerospace Engineering Department, Universiti Putra Malaysia, and Mr. Andi Eriawan from PT Dirgantara Indonesia for their assistance in providing information for many articles in the book, and former students Dr. ir. Bonifacius Bima Prananta (deceased), Dr. ir. Leonardo Gunawan, Ir. Mohamad Kusni M.Tek., both Faculty Members at the Faculty of Mechanical and Aerospace Engineering, Institute of Technology Bandung, Dr.-Ing. Widjaja Kresna Sekar, now associated with a company serving Airbus, Dr.-Ing. Ismoyo Haryanto, Faculty Member of Diponegoro University, Ir. Yan Mursal M.Tek., an Aeroelastician just retiring from Pilatus (Switzerland) (deceased), Dr. ir. Bambang Irawan Soemarwoto at NLR, the Netherlands, Ir. Agus Budiono Ph.D., a university instructor-Professor, technopreneur and AI and control system expert, for their contributions in the joint work, and their review, assistance, criticism and suggestions. The author would also like to thank Messrs. Irtan Safari, Yee Hong Ho, Yu Kok Hwa, Mahesa Akbar, Alif Syamim Syazwan Ramli, Mohamad Jafari and M. Anas Abd. Bari for their works that are incorporated in this book.

Furthermore the author wishes to thank Springer Nature and greatly appreciates the assistance of Messrs. Chandra Sekaran Arjunan and Ramesh Nath Premanth for their great care and dedication in preparing the publication of this book.

The author would like to welcome and will be grateful for any comments and suggestions for improving the book.

General Information

The book is intended for students, engineers and practitioners who are not too familiar with the subject of aeroelasticity or aeroelastomechanics, which is multidisciplinary. It is introductory in nature and attempts to draw the readers to the understanding, proficiency and enjoyment in applying fundamental (first) principles to this multidisciplinary topic. The presentation also attempts to unify the approach in order that the students, engineers and problem solvers can conceive, model and formulate the problem, identify the first principles to be applied, and devise solutions that can be based on experience, educated and systematic trial, and literature search.

Advanced and complex problems associated with the subjects will not be covered, as well as the elaboration of solutions method details covered in many commercial software. However, the book will elaborate on how the students and engineers can resort to fundamental principles in order to understand the basic and mainstream of the problem solution and to verify advanced and commercial software programs for basic physical understanding and plausibility, uncertainty and accuracy.

It will delve on the elements of aeroelasticity in order to train the students of the methodology, convenience and basic understanding of fundamental elements of mechanics, mathematical techniques, elasticity, aerodynamics and other related disciplines that may be relevant. It also attempts to introduce a paradigm for solving problems as they may first appear, without prior problem solver experience.

From the author's teaching and professional experience, an introductory or intermediate book that has been drawn from it is needed to facilitate students and practitioners to:

1. Understand and grasp the basic principles which can be used as mandatory tools for understanding and mapping solution approach to the problem at hand
2. Obtain some fundamental principles that can be applied to the problem
3. Attempt a skeleton approach that can be utilized in devising solution approach
4. Carry out mathematical formulation of the problem
5. Find appropriate solution for the mathematical problem formulated

6. To device systemic approach and utilize algorithm, modern method and appropriate computational software to facilitate and arrive at the solution of the problem.

In addition to the emphasis on fundamental principles, the book is written in a step-by-step fashion for the benefits of both students and instructors and novice practitioners.

By emphasizing on fundamental principles, the book focuses on the basic understanding of the concepts required in learning about aeroelasticity, from observation, reasoning and understanding fundamental physical principles. Fundamental and simple mathematics will be introduced to describe the features of aeroelastic problems and to devise simple concurrent physical and mathematical modeling. It will be accompanied by the introduction and understandings of the mechanisms that create the interactions that generate the aeroelastic phenomena considered. The students will also be led to the relation between observed phenomena, assumptions that may have to be adopted to arrive at physical and mathematical modeling, interpreting and verifying the results and the accompanied limitations, uncertainties and accuracies. The students will also be introduced to combine engineering problem-solving attitude and determination with simple mechanics problem-solving skills that coexist harmoniously with a useful mechanical intuition.

Author's Claim

While the data and information in this book are believed to be true and accurate at the date of publication, the author cannot accept any legal responsibility for any errors or omissions that may have been made.

Structured Approach for Learning and Solving Problems

A. Problem-Solving Paradigms

To Solve any Problem

- "Keep on Questioning":

"Forget Me Not's" Paradigm

For Understanding and Checking the Correctness of Mathematical Physics Equation(s)

- "THE DIMENSION" of each term in an equation, LHS and RHS, should be the same.
- "THE RANK" of each term (in view of matrix) in an equation, LHS and RHS, should be the same.
  - if one term is a scalar, each and every term in the equation should be scalar too;
  - if one term is a VECTOR, each and every term in the equation should be VECTOR too;
  - if one term is a MATRIX of rank n, each and every term in the equation should be a MATRIX of rank n too;
  - etc.

- "ORDER OF MAGNITUDE" considerations:

  To be physically meaningful and for solving convenience, the terms of the same order of magnitude should be grouped together in one equation.

Hence, if there are two groups of terms with two different orders of magnitude, the equations could be split into two equations.

This paradigm is beneficial for

- Mathematical derivations, such as the concept of limits and differentials
- To carry out practical solutions or first-order approximations, since engineers have to solve problems and explore further approximations.

To Facilitate Obtaining Solutions

- To solve any problems, it will be useful to start with "FIRST PRINCIPLES": These are physical and/or mathematics laws/theorems that are simple, logical or can be easily and rationally understood.
- To solve any problem, start by formulating each problem as understood, then formulate and built physical model (s), and start the solution process by utilizing "SIMPLIFICATION" and previous paradigms, in order to gain understanding from the solutions/results obtained. At the end the solution has to be assessed and the problem-solving process can be continued by progressive and more refined approximations and approach techniques gained from the experience.

B. Problem-Solving Schemes

Scheme 1

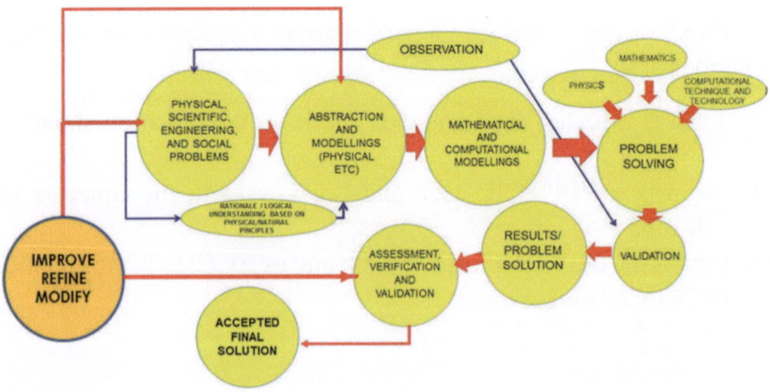

Fig. 1 Schematic of the analytical sequence for problem definition to problem solution – Scheme 1

Scheme 2

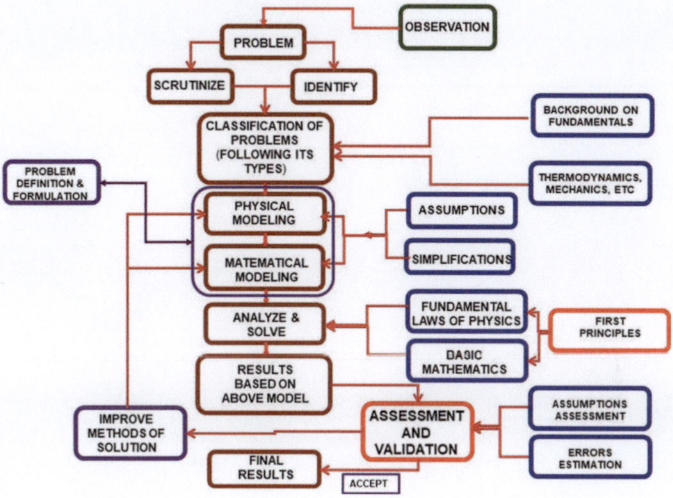

Fig. 2 Schematic of the analytical sequence for problem definition to problem solution – Scheme 2

Contents

Part I Foundation of Aeroelasticity

Part II Advanced Topics

About the Author

Prof. Harijono Djojodihardjo (Sarjana Teknik Mesin, Institut Teknologi Bandung, 1962; M.Sc. in Mechanical Engineering, University of Kentucky, 1964; Mech.E. 1965, S.M. in Naval Architecture and Marine Engineering, 1966 and Sc.D. in Aerodynamics and Gas Dynamics, 1968, all from Massachusetts Institute of Technology, Insinyur Professional Utama (IPU) certified by the Indonesian Institution of Engineers (PII) in 2009 and ASEAN Chartered Professional Engineer in 2010 by the ASEAN Chartered Professional Engineer, was born in Surabaya, Indonesia on 29 April 1940. He is an Academician of the International Academy of Astronautics. He has served as the Chairman of the Indonesian Space Agency, Deputy Chairman of the Indonesian Agency for the Assessment and Application of Technology (BPPT), Inspector General For Technology and Strategic Industry at the Office of the President, and as a Professor of Aerospace Engineering at Institut Teknologi Bandung, Universiti Sains Malaysia, and Universiti Putra Malaysia, with a combined teaching, research, industrial and management experience since 1962. He has served as Visiting Professor at Lehrstuhl fuer Fluidmechanik, Technische Universitaet Muenchen, April–October 2002, and Visiting Professor, International Cooperation Center for Engineering Education Development, Toyohashi University of Technology, November 2002–August 2003. He has also served as The Second Chairman and Rapporteur, the United Nations Committee on Peaceful Uses of Outer Space, for sessions in 2000, 2001 and 2002, and a member of the International Council for the Aeronautical Sciences (since 1990) and Technical Committee Member of the International Astronautical Federation (since 2004). He has published more than 200 publications including books, journal articles and conference presentations. Among the honors received, he was awarded two Medal of Merits from the President of the Republic of Indonesia, Bintang Jasa Utama and Bintang Mahaputra Utama, and an Engineering Award from the International Academy of Astronautics. Currently he is the Chairman of the Board, The Institute for the Advancement of Aerospace Science and Technology "Persada Kriyareka Dirgantara, Jakarta 15419, Indonesia, and Adjunct Professor at President University, Jababeka, Jakarta, Indonesia.

Part I
Foundation of Aeroelasticity

Chapter 1
Introduction and Overview

Abstract As an introductory chapter, it delves into the realm of aeroelasticity, which is also known as aeroelastomechanics; it will lead the reader from original query to observing the environment and identify of those phenomena that can be classified as aeroelastic phenomena, or something similar in day-to-day life. Any answer to this original question can lead one to explore one's imagination and satisfy one's understanding. Eventually, one could position oneself to get the best benefit and gain comprehensive understanding and significance of the subject. Historical background of aeroelasticity provides further perspectives, in particular to address aeroelastic phenomenon and eventually take necessary steps if necessary in responding to the demands of engineering solutions. Historical perspective is the study of a subject in light of its earliest phases and subsequent evolution. Such perspective may assist in obtaining an introductory or intermediate account that is needed to facilitate students and practitioners to understand and grasp the basic principles which can be used as mandatory tools for understanding and mapping solution approach to the problem at hand, as well as to obtain some fundamental principles that can be applied to the problem. Examples of aeroelastic problems and their system (block diagram) representation are introduced for illustration of the scope of aeroelasticity, such as some illustrative examples on the influence of aeroelastic phenomena on aircraft design for creative and proactive class discussions (Most Figures that appear in this chapter originated from NASA (or NACA)-based publication, including Garrick IE and Reed III WH in J Aircraft, 1981 [1], Garrick, I. E. and W. H. Reed, III. 2013. Document ID 19810045015, Conference Proceedings, NASA Langley Research Center Hampton, VA, United States, August 11, 2013 [2], and Dick SJ (ed) NASA'S First 50 Years Historical Perspectives, NASA SP-2010-4704:2010 [3]).

Keywords Aeroelasticity · Acousto-aeroelasticity · History of aeroelasticity · Lifting surface method · Singularity method · Unsteady aerodynamics

© The Author(s), under exclusive license to Springer Nature Singapore Pte Ltd. 2023

H. Djojodihardjo, *Introduction to Aeroelasticity*,
https://doi.org/10.1007/978-981-16-8078-6_1

1.1 Basic Introduction

What is aeroelasticity? The subject is also known as aeroelastomechanics. How should one attempt to answer this original question in a comprehensive way that can lead one to explore one's imagination and satisfy one's understanding?

How one should deal with aeroelasticity and position oneself in order to get the best benefit and gain comprehensive understanding and significance of the subject? With these questions in mind, then the following narrative essay may be of advantage. The essay will be divided into several questions

1. What is aeroelasticity?
2. How did it come about?
3. What are the pertinent practical issues indicative of aeroelasticity?
4. How does the discipline develop with time?
5. Typical phenomena indicative of aeroelasticity
6. What next as a consequence of aeroelasticity, to get a universal outlook?

More fundamental, to that end, there are three aspects that may be addressed:

1. A physical answer that may describe physical phenomena in nature and/or engineering that one may easily observe
2. A historical account when and how the phenomena entered into considerations, especially to be of particular concern and to be addressed for taking the phenomenon into considerations in one's life or how the phenomenon demands engineering solutions
3. To recognize how many different aspects or point of view, that later on may become scientific disciplines that have to be understood and considered.

1.1.1 Physical Answer Related to Observations

What is aeroelasticity? The answer should be descriptive, explanatory and prescriptive. The fundamental questions arising in mechanics are: What? Why? How? How much?

Day-to-day Aeroelastic Phenomena.

Observation from nature:

- Fluttering flag
- Motion of branches and rustles of leaves
- Others.

Why do leaves rustle? Why do trees and branches oscillate due to wind, and sometime they broke? In case of rustling leaves, one may argue that this occurs because the air flow over a leaf is not stable. Small speed differences in the air flow on either sides of the leaf create vortices that detach from the leaf surface, lowering the pressure on one side and causing it to flutter.

Why does flag flutter? This is a very good question, since the fluttering of a flag is a common scene observed by many. However, a very detailed scientific explanation is not that straight forward. In fact, the mechanism of flutter of a flag has been the subject of significant, if not continued, research.

Flag flutter is a flow instability which occurs due to coupled fluid structure interaction between the flow and the flag. This creates a net force on the flag which together with the flags structural forces cause flutter. When the strong winds blow on a particular day, the layers of air move with uneven velocity on both faces of the flag. Due to this, there will be a formation of uneven pressure on the faces of the flag. According to Bernoulli's theorem, the pressure difference exists above and below the flag, and it flutters.

1.2 Historical Background of Aeroelasticity [1–4]

Historical background of aeroelasticity provides a historical account and perspectives when and how the phenomena entered into considerations, particularly to be of particular concern and to be addressed for taking the phenomenon into considerations in one's life or how the phenomenon demands engineering solutions.

Historical perspective is the study of a subject in light of its earliest phases and subsequent evolution. Historical perspective differs from history in that the object of historical perspective is to sharpen one's vision of the present, not the past. Taking historical perspective means understanding the social, cultural, intellectual and emotional settings that shaped people's lives and actions in the past. Historical perspective related to aeroelasticity will be exemplified by some historical records in mechanized flight, or airplane.

Aeroelasticity did not play any significant role as it is now until early 1940s. At that time the aircraft speed was relatively very low, and the load requirements applied to aircraft with engines following the design specification criteria produced a relatively rigid structures so that aeroelastic phenomena did not appear, or various failure phenomena occurring at that time were not completely understood as caused by aeroelastic phenomena.

The history of mankind to attempt to fly may have started long before many myths in Egypt, China, Greece (such as Icarus) and India (such as the gods and Gatutkaca, a flying hero). The idea of a flying artifact has also been created by Leonardo da Vinci. In all of these attempts to fly using artificial means, like the present gliders and sailplanes, the requirement of lightweight and sufficiently strong structure is critical. Flying biosystems like insects and birds triggered many creative ideas. In the case of Icarus, the flying vehicle should also have thermal criterion. This eventually will lead to what is now known as aeroelasticity or even aerothermoelasticity.

An aeroelastic phenomenon that occurred dramatically in the early history of flight was divergence, which caused the death of a well-known aeronautical pioneer, Otto von Lillienthal. He was the first man to study gliding using a man-made glider. Karl Wilhelm Otto Lilienthal (May 23, 1848–August 10, 1896) attempted to glide

Fig. 1.1 Otto Lilienthal using his man-made glider. Each of both images depicts his successful attempts of mankind to fly. Courtesy of NASA [1, 2]

using an artificial wing which then experienced a divergence and caused the failure of the wing, although some may argue that during that flight on August 9, 1896, Lilienthal's glider stalled and he fell from a height of 17 m (56 ft.) (Fig. 1.1).

Otto Lilienthal is pioneer of aviation who became known as the "flying man." He was the first person to make well-documented, repeated, successful flights with gliders. The German inventor built several gliders in the 1890s and flew them more than 2000 times. His work with curved wings, based on extensive study of birds, inspired the two famous brothers, Wilbur and Orville Wright in the USA.

With the increase of aircraft speed, accompanied by load requirements that did not take into account such speed increase, the aerospace designers and engineers are faced with various problems now known as the aeroelastics.

The Langley Aerodrome was a pioneering but unsuccessful manned, powered flying machine designed at the close of the nineteenth century by *Smithsonian Institution* Secretary *Samuel Langley*. The U.S. Army paid $50,000 for the project in 1898 after Langley's successful flights with small-scale unmanned models two years earlier.

Although engineers were aware of the significance of aeroelastic problems, it was not only until the failure of the aircraft of Professor Samuel P. Langley that the effect of aeroelastic problem in the design of engined aircraft was made conspicuous. Professor Samuel P. Langley from the Smithsonian Institute was probably the first aircraft designer that encountered aeroelastic problem in engined aircraft. Aero-structural interaction occurring in 1903 was found on Langley Aerodrome which underwent Potomac River test. The wing failed during a catapult launch, and it broke while moving downward.

As Fig. 1.2 shows, the wings had a significant dihedral angle to provide lateral stability. In fact, Langley's design philosophy was to provide the airframe with inherent stability, independent of pilot input. As was done with the scale models, it was launched from a houseboat-mounted catapult. However, the Aerodrome quickly dove into the river, with no flight being achieved.

By reasoning and some analysis based on modern methods, it is quite probable that the failure of the wing of Langley's Aerodrome in 1903 was due to torsional divergence of the wing. An account on the causes of this accident was given by

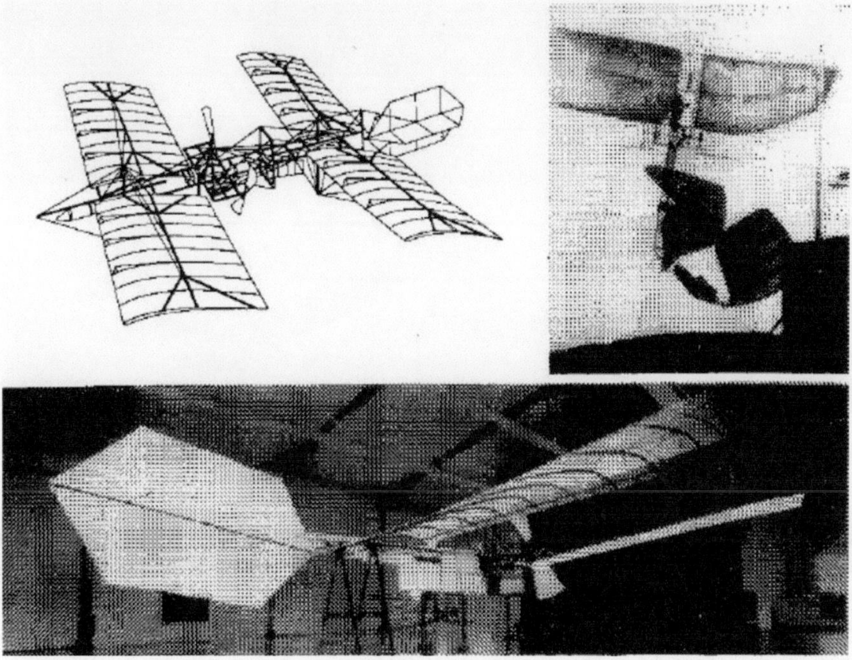

Fig. 1.2 Langley's Aerodrome, which plunged into Potomac River in 1903. Courtesy of NASA [1, 2, 5, 6]

Griffith Brewer from the Royal Aeronautical Society on the collapse of monoplane wings [2]. The failure of Langley occurred not long before the success of Wright brothers to fly a heavier-than-air aircraft for the first time.

Wright brothers may be inspired by the work of Otto Lilienthal. Wright brothers Bi-plane aircraft is depicted in Fig. 1.3. The success of Wright brothers bi-plane aircraft and the failure of Langley's monoplane may have caused the popularity of bi-pane aircrafts during the early stages of engined aircraft designs. At that time, the requirements for torsional rigidity have not been given due attention.

The aeroelastic problem faced on monoplanes has made aircraft designers resort to bi-panes. The monoplane used at that time did not have the required rigidity required to account for torsional rigidity, which then gave rise to flutter, the loss of control surface effectiveness and undesired deformation due to aerodynamic loading.

The development of cantilevered monoplane constituted a period at which aeroelastic research was initiated seriously, which was not carried out during the earlier period when the aeroelastic problems were dealt with in a cut-and-try method. Theory of wing load distribution and divergence was first introduced by Hans Reissner in 1926. Theory on the loss of lateral control and control surface reversal was published six years later by Cox and Pugsley in 1932. The mechanism of *flutter* in potential flow was well understood to be applicable for aircraft design in 1935, through research work carried out by Glauert, Frazer and Duncan, Kuessner dan Theodorsen [7, 8].

The second failure of Samuel Langley's prototype plane on the Potomac has been attributed to aeroelastic effects (specifically, torsional divergence). An early

Fig. 1.3 Wright brother's Bi-plane aircraft. Courtesy of NASA [2]

scientific work on the subject was George Bryan's *Theory of the Stability of a Rigid Aeroplane* published in 1906. Problems with torsional divergence plagued aircraft in the First World War and were solved largely by trial-and-error and ad hoc stiffening of the wing. The first recorded and documented case of flutter in an aircraft was that which occurred to a Handley Page O/400 bomber during a flight in 1916, when it suffered a violent tail oscillation, which caused extreme distortion of the rear fuselage and the elevators to move asymmetrically. Although the aircraft landed safely, in the subsequent investigation F. W. Lanchester was consulted. One of his recommendations was that left and right elevators should be rigidly connected by a stiff shaft, which was to subsequently become a design requirement. In addition, the National Physical Laboratory (NPL) was asked to investigate the phenomenon theoretically, which was subsequently carried out by Leonard Bairstow and Arthur Fage [9].

In 1926, Hans Reissner published a theory of wing divergence [8], leading to much further theoretical research on the subject. The term *aeroelasticity* itself was coined by Harold Roxbee Cox and Alfred Pugsley at the Royal Aircraft Establishment (RAE), Farnborough, in the early 1930s [9].

In the development of aeronautical engineering at Caltech, Theodore von Kármán started a course "Elasticity applied to Aeronautics [10]. " After teaching the course for one term, Kármán passed it over to Ernest Edwin Sechler, who developed aeroelasticity in that course and in publication of textbooks on the subject.

In 1947, Arthur Roderick Collar [11] defined aeroelasticity as "the study of the mutual interaction that takes place within the triangle of the inertial, elastic, and aerodynamic forces acting on structural members exposed to an airstream, and the influence of this study on design."

A more recent aeroelastic phenomenon that may occur and be observed by flyers is the glider limit cycle oscillation, may have noticeable aeroelastic phenomena, i.e. divergence, and may result in failure. Other modern pioneering ventures of combining environmental concern and energy-efficient airplane also pose some aeroelastic phenomena that need to be considered in the design. As the sport of soaring initially focused on exploiting ridge winds to maintain altitude, and the level of structural technology was unable to allow large spans, the low sink rates required were achieved by wings having large areas and fairly low aspect ratios [14].

There is a trade-off between low-induced drag for climb and low profile drag for cruise, which became a critical issue in the design of sailplane wings. Theoretical guidance for these designs was provided primarily by the lifting-line theory of Ludwig Prandtl and the minimum-induced drag, elliptical loading result of Max Munk. During this time, the need for greater spans and higher aspect ratios led to structural advancements in the primarily wooden airframes and the development of some very interesting wing geometries, such as the distinctive gull wings that were then popular.

The evolution of wing design through this period continued slowly until the introduction of new materials and laminar flow wing sections led to very rapid advancements beginning in the late 1950s. The use of glass-reinforced plastic structures, and later carbon-reinforced plastic, allowed designers to incorporate much larger aspect ratios than that had been possible earlier. These developments led to the adaptation of planforms having straight trailing edges and on to non-planar wing geometries and, now common practice, the use of winglets. These modern designs are best illustrated by Figs. 1.4, 1.5 and 1.6, mentioning only a few examples.

1.3 Scientific Aspects in Aeroelasticity

An introductory or intermediate book that has been drawn from teaching and professional experience should be helpful to facilitate students and practitioners alike to:

1. Understand and grasp the basic principles which can be used as mandatory tools for understanding and mapping solution approach to the problem at hand
2. Obtain some fundamental principles that can be applied to the problem
3. Attempt a skeleton approach that can be utilized in devising solution approach
4. Carry out mathematical formulation of the problem

Fig. 1.4 NASA HALE UAVs. Courtesy of NASA [12]

Fig. 1.5 NASA swift HALE test flight. Courtesy of NASA [13]

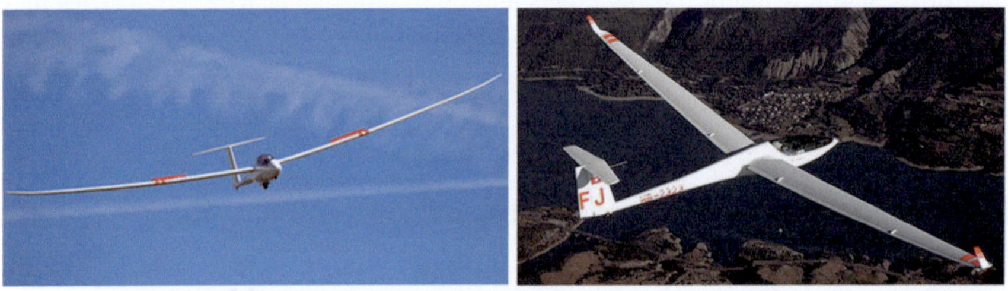

(a) (b)

Fig. 1.6 Glider that in the design should consider aeroelastic effects. **a** Glider with long wingspan (Courtesy of Gettyimages [15]); **b** Dg800 glider (Courtesy of Wikipedia [16])

5. Find appropriate solution for the mathematical problem formulated
6. To device systemic approach and utilize algorithm and modern method
7. Appropriate computational software to facilitate and arrive at the solution of the problem.

In addition to the emphasis of fundamental principles, the present book is written in a step-by-step fashion for the benefits of both students and instructors and beginners.

Aeroelasticity is the branch of physics and engineering that studies the interactions between the inertial, elastic and aerodynamic forces that occur when an elastic body is exposed to a fluid flow. The study of aeroelasticity may be broadly classified into two fields: *static aeroelasticity*, which deals with the static or steady-state response of an elastic body to a fluid flow; and *dynamic aeroelasticity*, which deals with the body's dynamic (typically vibrational) response.

Aircraft are prone to aeroelastic effects because they need to be lightweight and withstand large aerodynamic loads. Aircrafts are designed to avoid the following aeroelastic problems:

1. **Divergence** where the aerodynamic forces increase the angle of attack of a wing which further increases the force
2. **Control reversal** where control activation produces an opposite aerodynamic moment that reduces, or in extreme cases, reverses the control effectiveness
3. **Flutter** which is the uncontained vibration that can lead to the destruction of an aircraft.

Aeroelasticity problems can be prevented by adjusting the mass, stiffness or aerodynamics of structures which can be determined and verified through the use of calculations, *ground vibration tests* and *flight flutter trials.* Flutter of control surfaces is usually eliminated by the careful placement of *mass balances.*

The synthesis of aeroelasticity with thermodynamics is known as *aerothermoelasticity*, and its synthesis with control theory is known as *aeroservoelasticity*. Some known aeroelastic phenomena are briefly elaborated below in introductory sense.

What causes aeroelastic flutter?

Flutter is a dynamic instability of an elastic structure in a fluid flow, **caused** by positive feedback between the body's deflection and the force exerted by the fluid flow.

What is aircraft buffeting?

Buffeting is a vibration of the **aircraft** that may appear during maneuvers at cruising speed. Depending on the angle of attack, the flow may contain separations, which constitute an aerodynamic excitation.

What is divergence speed?

Divergence speed is the lowest **speed** at which the rate of change of aerodynamic deforming forces exceeds the rate of change of elastic restoring forces, or couples. It is the lowest **speed** at which control reversal caused by **divergence** can take place.

What is torsional aileron flutter?

Torsional aileron **flutter is a flutter** caused by the wing twisting under loads imposed on it by the movement of the ailerons. The center of gravity of the aileron is behind the hinge line; its inertia tends to make it lag behind, increasing the aileron lift, and so increasing the twisting motion.

What is wing divergence speed?

The moment increases as the square of the speed, but the stiffness remains constant. Therefore, at high speeds, the **wing** may be uncontrollable. The speed at which the **wing** stiffness can no longer counteract the twisting moment is called the "**wing divergence** speed."

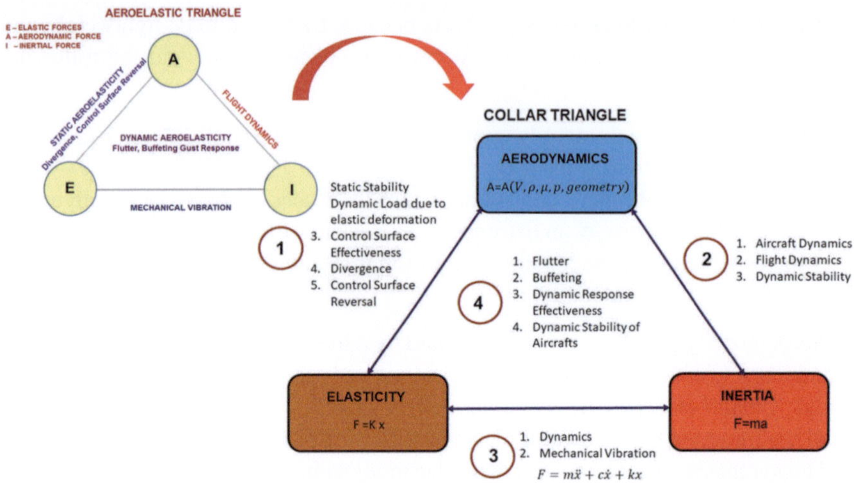

Fig. 1.7 Various depiction of aeroelasticity or aeroelastomechanics as defined by Collar, and known as Collar triangle

1.4 Definition of Aeroelasticity

Aeroelasticity is the study of the mutual interaction among inertial, aerodynamic and elastic forces acting on aircraft, buildings, surface vehicles and any elastic structure undergoing effects of inertia and airflow.

The interaction among these three forces can cause several undesired phenomena. Arthur Roderick Collar ingeniously suggested in 1946 that aeroelasticity could be usefully visualized as forming a triangle with vertices labeled with engineering disciplines: dynamics, fluid mechanics (aerodynamics) and structural mechanics, with related forces.

In 1947, he defined **aeroelasticity** as "the study of the mutual interaction that takes place within the triangle of the inertial, elastic, and aerodynamic forces acting on structural members exposed to an airstream, and the influence of this study on design." The Collar diagram (or **Collar Triangle**) is illustrated in Figs. 1.7 and 1.8.

Following Collar, one of the pioneers in aeroelasticity, as quoted by Bisplinghoff, Ashley dan Halfman and Försching, problems dealt with in aeroelasticity can be classified as the interactions between these various forces as depicted in the following diagram:

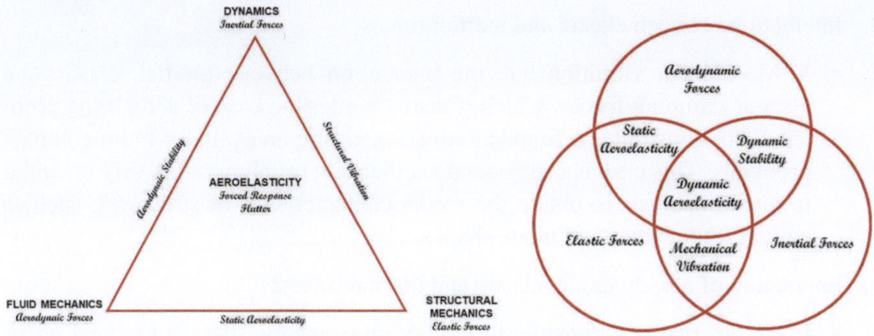

Fig. 1.8 Collar triangle of aeroelasticity element drawn in two versions: original and as three-ring aeroelastic interaction diagram

In the diagram, various aeroelastic problems can be classified into four different categories[1]:

1. Interactions between aerodynamic and elastic forces, i.e. forces due to the elastic effects of structures. These interactions can give rise to the following problems:
   - SSA Aeroelastic effects on the static stability of aircraft, i.e. the effect of elastic deformation of aircraft structure on the static stability of aircraft.
   - L Distribution of dynamic load, i.e. the influence of elastic deformation of aircraft structures on the pressure distribution or aerodynamic loading on aircraft structure.
   - C *Control surface effectiveness*, i.e. the influence of elastic deformation of aircraft structure on the controllability of aircraft.
   - D Divergence, that is the instability of lifting surface of the aircraft, in particular wing, during flight, at a speed known as the divergence speed, due to the deflection of the lifting surface (wing) due to aerodynamic forces without the ability of the aircraft structure elastic force (or stiffness) to balance it.
   - R Control surface reversal, i.e. a situation during flight at a speed known as *control reversal speed*, where the deflection (elastic deformation) of the control surface subject to the prevailing aerodynamic forces can no more be effective; in other words, in such situation, the deflection of the control surface does not increase the lift as it should be, but started to produce adverse effect.
2. Interaction between aerodynamic and inertial forces:
   - DS Aerodynamic stability of aircraft as a rigid body. This problem arises as a limiting situation, i.e. as a reference condition or a degenerate case of core aeroelastic problems.

[1] Synthesized from various classical references as lis contained in the list of references

3. Interaction between elastic and inertial forces:
   - V Mechanical vibration, i.e. the interaction between inertial, elastic and viscous damping forces, which is also a borderline case of aeroelastic problems, but constitutes a foundation of aeroelastic analysis for more complex problems. This case is a degenerate aeroelastic problem but is very essential in modal analysis to obtain the modal characteristics of structures, such as natural frequencies and mode shapes.
4. Interaction of aerodynamic, elastic and inertial forces:
   - F *Flutter*, that is a dynamic instability phenomenon that may prevail on an aircraft during flight, at a speed known as flutter speed, at which speed the motion-induced aerodynamic forces cannot be balanced by the elastic (and mechanical damping) and inertial forces.
   - B *Buffeting*, that is the transient vibration of aircraft structural components (such as tail plane) due to the aerodynamic forces generated by the wake or vortices behind a wing, nozzle or protrusion on the aircraft fuselage (such as landing gear) and other similar components of the aircraft. Buffeting may also appear due to flow separation near stall, or shock wave and boundary layer interaction at transonic speeds.
   - Z Dynamic response, that is the transient response of aircraft structural component due to sudden aerodynamic excitation during flight. These sudden aerodynamic excitations may be due to gust, air turbulence, landing, wake of preceding aircrafts during take-off and landing, sudden control surface or aircraft maneuvers, explosion in the vicinity of the aircraft and other similar forces.
   - DSA Dynamic stability of Aircraft. This is the influence of aeroelasticity on the dynamic stability of aircraft, due to the interaction among aerodynamic, elasticity and inertial forces.

With the progress of technology, the aeroelastic problems of aircraft are also developing. One concept of the scope of aeroelastic problems has been proposed by Bisplinghoff dan Ashley by considering aeroelasticity as the interaction between internal forces of the aircraft (or any other flight vehicle) as an open thermodynamic system with its environment (also in the thermodynamic sense).

The forces in the aircraft can appear as:

- Propulsive forces, to propel the aircraft
- Inertial forces, since the aircraft has a mass that is experiencing dynamic motion
- Elasto-plastic forces, since the aircraft structure is experiences structural/material deformation
- Control and guidance forces, which could be a reactive forces (aerodynamics, or forces due to its interaction with environmental forces), propulsion forces and other inertial forces.

Due to aircraft movement, the flight vehicle or aircraft environment is interacting with the aircraft (flight vehicle) and can give rise to:

- Aerodynamic forces, when the aircraft is moving in a fluid medium (air); the aerodynamic forces are a function of the motion, geometry and physical properties of the fluid.
- Gravitational forces, when the aircraft is moving in a gravitational field.
- Other environmental forces, such as:
  - Magneto-gas dynamic forces
  - Forces due to electromagnetic forces of the earth
  - Thermal radiation
  - Cosmic radiation
  - Meteorites or space debris
  - Gravitational forces of other celestial bodies.

1.5 Aeroelastic Problems in Engineering

Starting from the scope of aeroelasticity, the knowledge of aeroelasticity and the analytical method of aeroelasticity can be applied to aeronautical and other related engineering problems, such as[2]:

Vibration of elastic structures due to aerodynamic or wind forces, as occurring in towers, high-rise buildings, bridges, microwave or earth station antennae and the like:

- Vibration of elastic off-shore structures due to hydrodynamic forces, as found in *off-shore drilling platforms*
- Vibration in pipes or elastic conduits due to dynamic excitation of fluids flowing inside them, which may be found in oil pipes, rocket structures, blood vessels or arteries and so forth
- Vibration or aeroelastic phenomena occurring in propellers, compressor and turbine blades and wind turbines
- Vibration, dynamic response and control laws for large-space structures, due to the interactions of various forces in the structure and its environment.

Particularly in aeronautical engineering, aeroelastic problems are essential and need to be considered and solved, since the design stage up to the operation of aircrafts. In the design of aircrafts, with increasingly high operational velocities in the high subsonic, transonic and supersonic range, aeroelastic problems constitute the limiting factor that should be accounted for or considered, such as: divergence speed, flutter margin (which should be sufficiently higher than the operational speed of the aircraft).

---

[2] Synthesized from various classical references as lis contained in the list of references

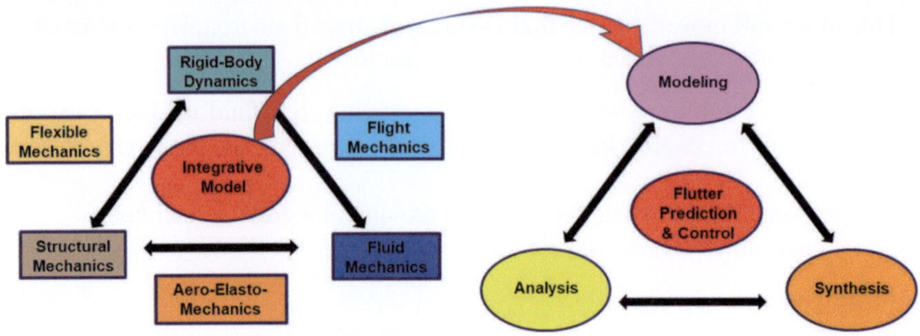

Fig. 1.9 From integrative modeling to flutter prediction control

With the increase of aircraft operational speeds, particularly in transonic region, the shock wave and boundary layer interaction will give rise to additional aerodynamic excitation on the aircraft structure. Dynamic response of aircraft due to atmospheric (environmental) disturbance, such as turbulence, gust, noise, landing dynamic impact, will influence the structural integrity, dynamic stability and aircraft endurance or fatigue life.

Similar problems are also met in many non-aeronautical engineering problems, particularly in the dynamic response phenomena and dynamic stability of structures. Figure 1.9 is adapted from FLEXOP [17], Flutter Free Flight Envelope Expansion Work Package [18], which considers flutter prediction, active suppression and the related mathematical modeling task that required Aeroelastic Tailoring Methods for Wing Design. Figure 1.9 shows two work packages, the first one (left) concerns the building up of the integrative aeroservoelastic model, while the second (right) employs three main tasks within the corresponding work package. These include modeling, analysis and synthesis, which depend on the availability of an integrative model produced in the first phase, i.e. the first work package shown on the left of Fig. 1.9. This task reconciles the rigid-body dynamics, the fluid mechanics and the structural mechanics. This task addresses the main goal of Flutter Analysis Methodologies, which is to develop techniques for flutter analyses in order to determine: flight envelope limits, flutter predictions and study flutter suppression.

1.6 Extended Concept: Hydro-elasticity, Aeroservoelasticity and Envaeroelastomechanics

These extended concepts are illustrated in Figs. 1.10a, b 1.11, 1.12 and 1.13. Engineering disciplines incorporated are indicated there.

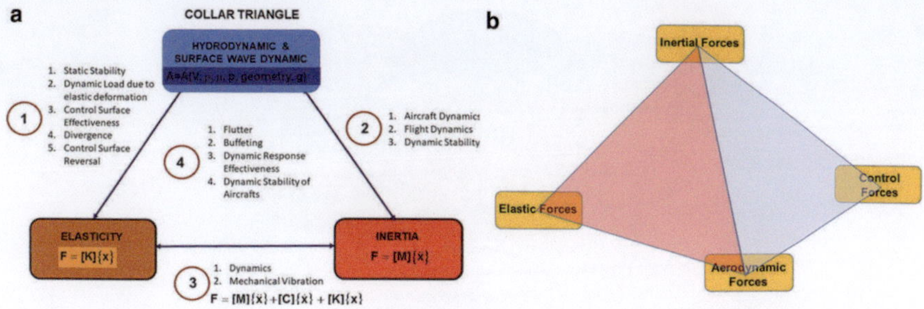

Fig. 1.10 **a** Hydro-elasticity, **b** Aeroservoelasticity

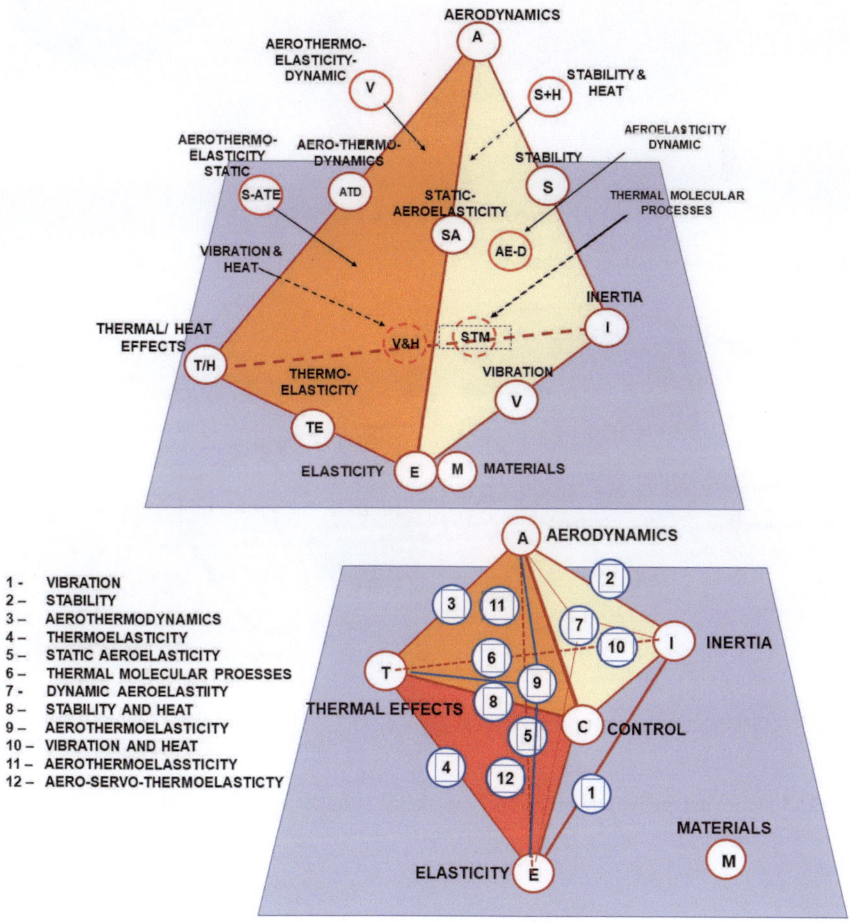

Fig. 1.11 Aerothermoelasticity and aeroservothermoelasticity tetrahedrons

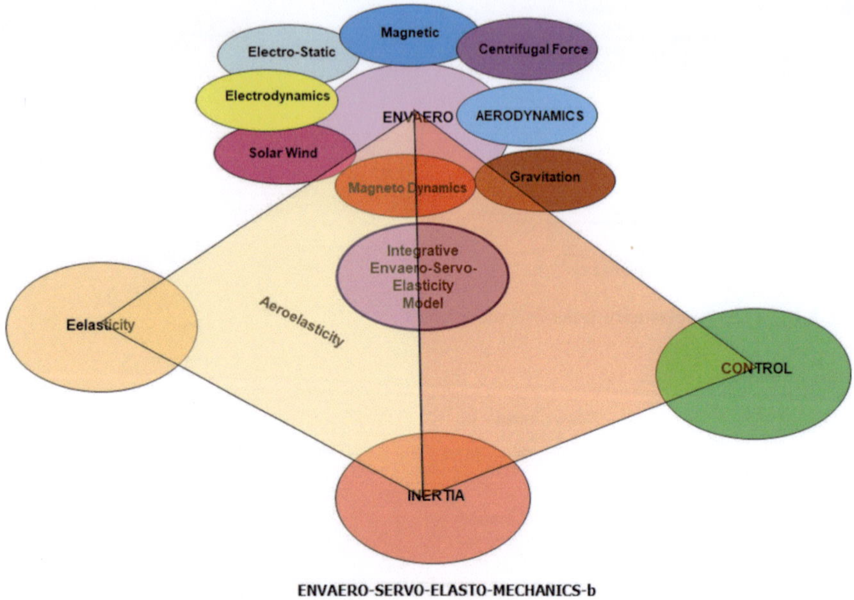

Fig. 1.12 Envaero-servoelastomechanics

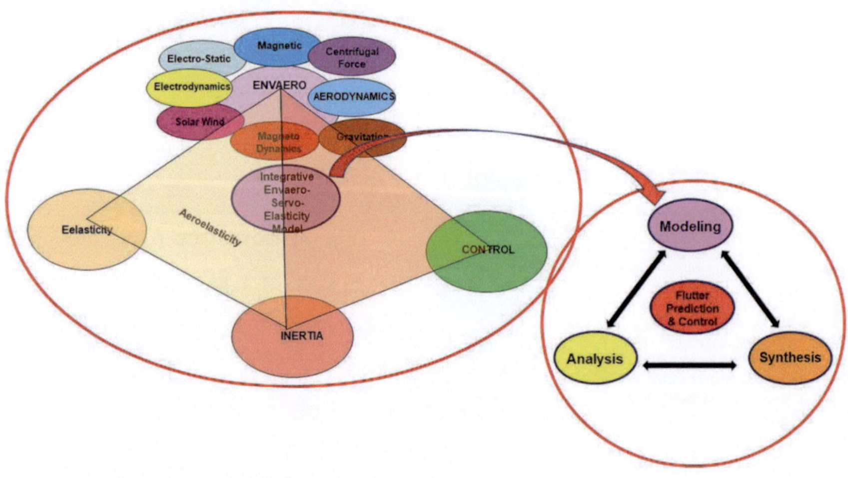

Fig. 1.13 From integrative envaeroelasticity modeling to flutter prediction control

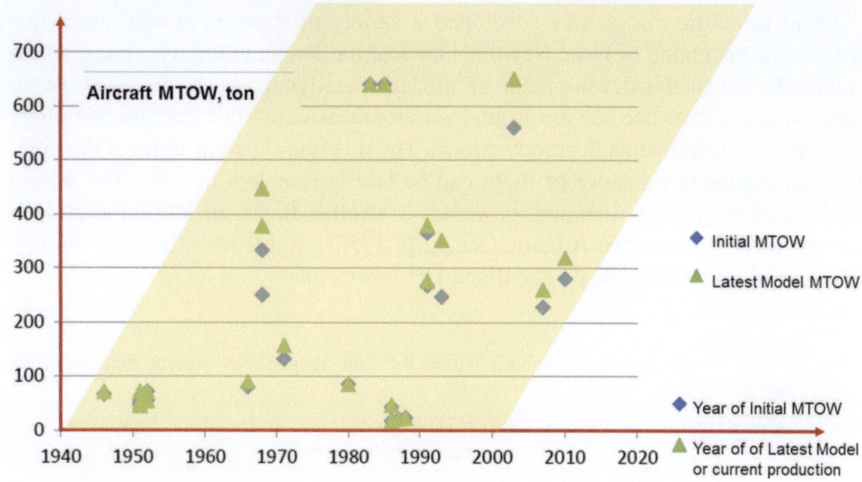

Fig. 1.14 Trend of Modern Aircrafts Maximum Take-Off Weight (MTOW) as a function of time.[3] List of aircrafts considered is given in Table 1.1

1.7 Extended Concept: Envaeroelastomechanics —Enviromental Forces Schematics

A—Aerodynamics	CF—Centrifugal force
G—Gravitational	M—Magnetic
SW—Solar wind	ES—Electrostatic
MD—Magnetodynamics	ED—Electrodynamics

This is an integral concept that anticipates and extends the realm initiated by aeroelasticity to incorporate other environmental forces on earth and in space.

1.8 Trend of Modern Aircrafts Development

Mankind has extended efforts in perusing the skies with man-made flying objects for over 2000 years. The history of aviation began with the invention of kites and gliders, before emerging to the multimillion-dollar aircraft industry of modern era.

The origin of the first man-made flying objects was kites circa 200 B.C. in China. During the seventeenth and eighteenth centuries, the discovery of hydrogen led to the first development of the hydrogen balloon, which carried people away at high altitudes and across several miles. In the nineteenth century, tethered balloons were used (Fig. 1.14).

[3] This figure is only schematic and qualitative. However Take-Off Weight data are obtained from authentic company sources or reliable websites.

Other scientific discoveries developed a variety of theories in mechanics that became the backbone of Isaac Newton's laws of motion and fluid dynamics, which eventually led to the development of modern aerodynamics. In the early twentieth century, gliders became the groundwork for massive aircraft, engine technology and further developments in aerodynamics. The timeline showing some of the highlights of the modern history of flight can be briefly sampled. In 1903, the Wright brothers make the first manned, powered, controlled flight. In 1919, the NC4 is the first plane to cross the Atlantic Ocean. In 1950s, flights are offered in the first commercial jet airliner, the de Havilland DH 106 Comet.

Table 1.1 List of selected aircrafts built within the time range 1945–present that are being incorporated in Fig. 1.14

No.	Aircraft name	MTOW [tons] (typical/selected representative model)	Year (year range)
1	Dassault Mercure [19]	66–71	1946–1958
2	ATR-42, ATR-72 [20]	ATR-42 16.6–18.6 ATR-72 23	ATR-42 1984–present ATR-72 1988–present
3	F-50 [21]	21	1987–1997
4	F-100 [21]	43–46	1986–1997
5	CN-235 [22]	17	1986–2022
6	Douglas DC-6, DC-6B [23]	45 (DC-6), 50 (DC-6B)	1951–1958
7	De Haviland DH.104, DH.106 Comet [24]	54–71	DH.006 1952–1997
8	Lockheed 1049 [25]	54	1951–1958
9	Airbus 320–200 [26]	85	1980–2022
10	Boeing 737/ Boeing 737-Max [27]	80–90	First design 1960s 1966–present
11	Boeing-787 [28]	228–260	2007–present
12	Airbus-300 [29]	132–157	1971–2017
13	Boeing-777 [30]	247 318–352 (777–300 ER)	1993–present
14	Airbus A-340 [31]	267–276(340-300X)-365–380	1991–2012
15	Boeing 747 [32]	250, 378 333–447(B747-8)	1968–2022
16	Airbus A-380 [33]	560–650	2003–2021
17	Airbus A-350 [34]	280–319-	2010–present
18	Antonov An-225 [35]	640	1985 (FF 1988)

1.9 Examples of Aeroelastic Problems and Their System (Block Diagram) Representation

Complete functional diagrams for an Oscillating Aircraft Structure[4] are illustrated in Figs. 1.15a–c and 1.16.

Possible aircraft excitations as represented by complete functional diagrams for an Oscillating Aircraft Structure in Figs. 1.15 and 1.16:

- Pilot input
- Atmospheric turbulence
- Buffet fluctuation (from other parts of A/C structure)
- Propeller slipstream
- Noise
- Landing impact
- Jettison of external store.

1.9.1 Two Major Categories of Aeroelastic Problems

(1) **Static and Dynamic Stability Boundaries of Structure Deformations**

- No aircraft excitation considered
- Motion-induced aerodynamic forces are able to amplify deformation possibly until structural failure (> $V_{critical}$)
- Aeroelastic instability
  - Deformation–vibration–flutter
  - Static deflection–divergence.

(2) **Static and Dynamic Responses**[5]

Aeroservoelastic Systems: Benefits, Problems and Opportunities (as illustrated in Fig. 1.17).

- Shape dynamic behavior of the flexible vehicle using active control: flight mechanics of the vehicle as a "rigid body"
- Optimum cruise shape
- Maneuver load/gust load alleviation
- Ride comfort (vibrations).

[4] For creative and proactive class discussions

[5] For creative and proactive class discussions

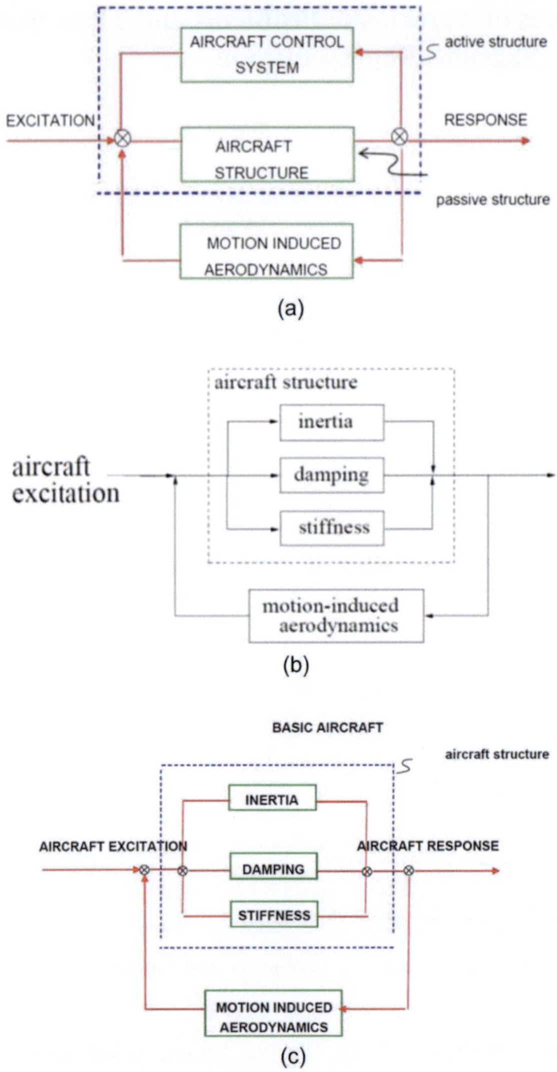

Fig. 1.15 Possible alternatives of complete functional diagrams and various block diagram representations for an oscillating aircraft structure

1.9.2 *Adverse Interactions: Flutter and Divergence*

- A control system designed for flight mechanics control, gust alleviation, ride comfort, etc., may interact with the dynamic aeroelastic structure to produce instabilities.

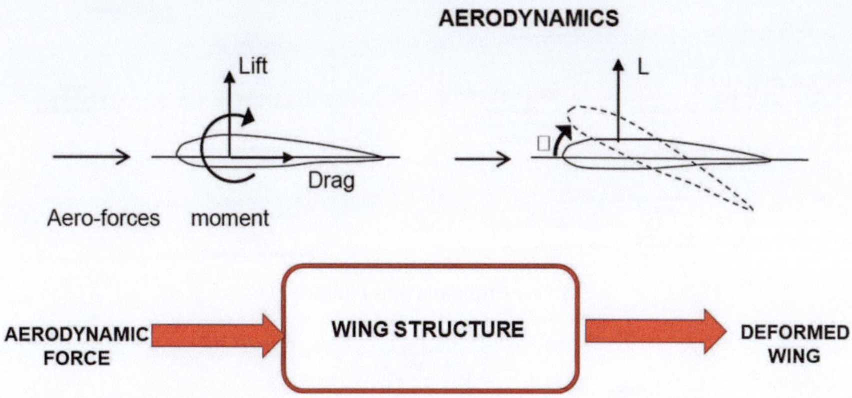

Fig. 1.16 Representations of an oscillating aircraft structure

- Find ways to decouple the active control system from the dynamics of the aeroelastic system.

  Opportunities—AFS as part of the integrated design from the START.

- Allow integrated optimization of the coupled structure/aerodynamic/control system from its early design stages, leading (potentially) to major weight savings and performance improvements.

  Technology state of the art.

- Gust alleviation systems are already certified on passenger airplanes as well as ride comfort augmentation and maneuver load control systems.
- Those aeroservoelastic systems operate in harmony with the aircraft flight control system (FCS).
- Active flutter suppression (AFS) has been thoroughly researched since the mid-1960s (when flight control systems began to become powerful and high bandwidth).
- Many academic/theoretical studies.
- Quite a number of wind tunnel tests using dynamically/aeroelastically scaled models of production or test aircraft with active controls.
- A few AFS flight tests of AFS-configured test vehicles
  - A B52 in the early 1970s
  - An F4F with external stores in the 1970s
  - NASA DAST UAV in the 1970s
  - Early 1980s, Lockheed/USAF
  - X56 UAV recently.

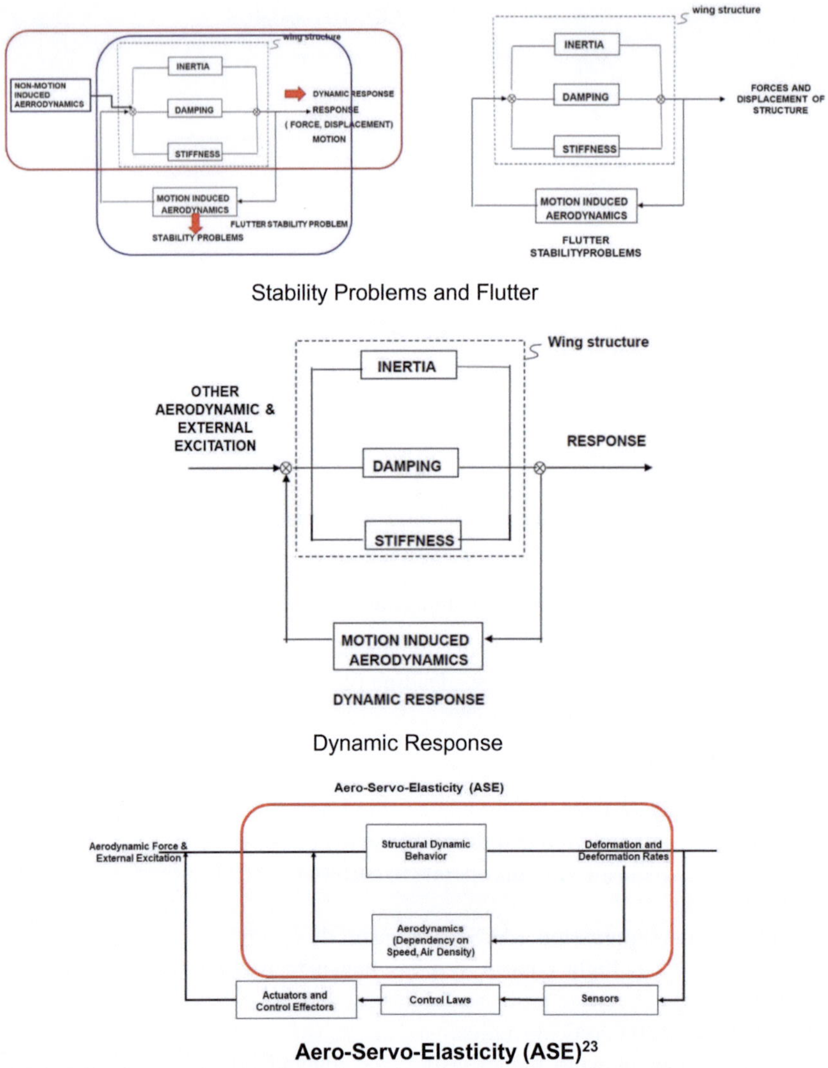

Fig. 1.17 Various block diagram representations of aeroelastic problems [36]

1.9.3 Basics of Aeroelasticity and Flutter Analysis

1. Basic aeroelastic stability problems
2. Principles of unsteady aerodynamics and modeling
3. Formulation of flutter stability equation
4. Flutter Analysis Method
5. "Robust" Flutter Analysis
6. Aeroelastomechanics outlook.

1.10 Some Illustrative Examples in Figures

1. The X-56A Multi-Utility Aeroelastic Demonstration (MAD) is an innovative modular unmanned air vehicle designed to test active flutter suppression and gust load alleviation, and breaking the flutter barrie in 2005, the Skunk Works® team began research in this area with a new design methodology that led to the Body Freedom Flutter (BFF) research program. Their work proved that flutter behavior can be accurately predicted and addressed through the creation of a new design paradigm of active control. This active control method produced a 75% increase in BFF speed, as proven in a series of flight tests.

For further information, reference can be made to Livne, The Active Flutter Suppression (AFS) Technology Evaluation Project(X-56A) (Fig. 1.18) [36–38].

2. Ikhana unmanned science and research aircraft system. A General Atomics Aeronautical Systems MQ-9 Predator B unmanned aircraft system (UAS) was acquired by NASA in November 2006 to support Earth science missions and advanced aeronautical technology development. The aircraft, named Ikhana, also acts as a test bed to develop capabilities and technologies to improve the utility of unmanned aircraft systems. Ikhana is a Native American Choctaw word meaning intelligent, conscious or aware. The name is descriptive of the research goals that NASA has established for the aircraft and its related systems [41, 42].
3. Boeing and the US FAA have come to a final agreement on the regulatory special condition required for the 747–8's outboard aileron modal suppression (OAMS) system designed to dampen out a structural vibration in the wing [37, 38].

Boeing 747–8s for outboard aileron modal suppression (OAMS) involves Boeing and the US FAA to come to a final agreement on the regulatory special condition required for the 747–8s outboard aileron modal suppression (OAMS) system designed to dampen out a structural vibration in the wing.

1.11 Influence of Aeroelastic Phenomena on Aircraft Design

Aeroelastic phenomena in modern high-speed aircraft have profound effects upon the design of structural members and somewhat lesser but nonetheless important effects upon mass distribution, lifting surface planforms and control system design. An example of the linear acceleration at the center of gravity due to a step-up gust for a transport aircraft for flutter suppression and gust alleviation using active control is exhibited in Fig. 1.19.

Flutter: Flutter is one of the most far-reaching effects on the design of high-speed aircrafts. Modern aircrafts are subject to many kinds of flutter phenomena. The classical type of flutter, which is very instructive for the understanding the phenomena, is associated with potential flow, i.e. smooth, non-separated flow, and involves the coupling of two or more degrees of freedom. The non-classical type of

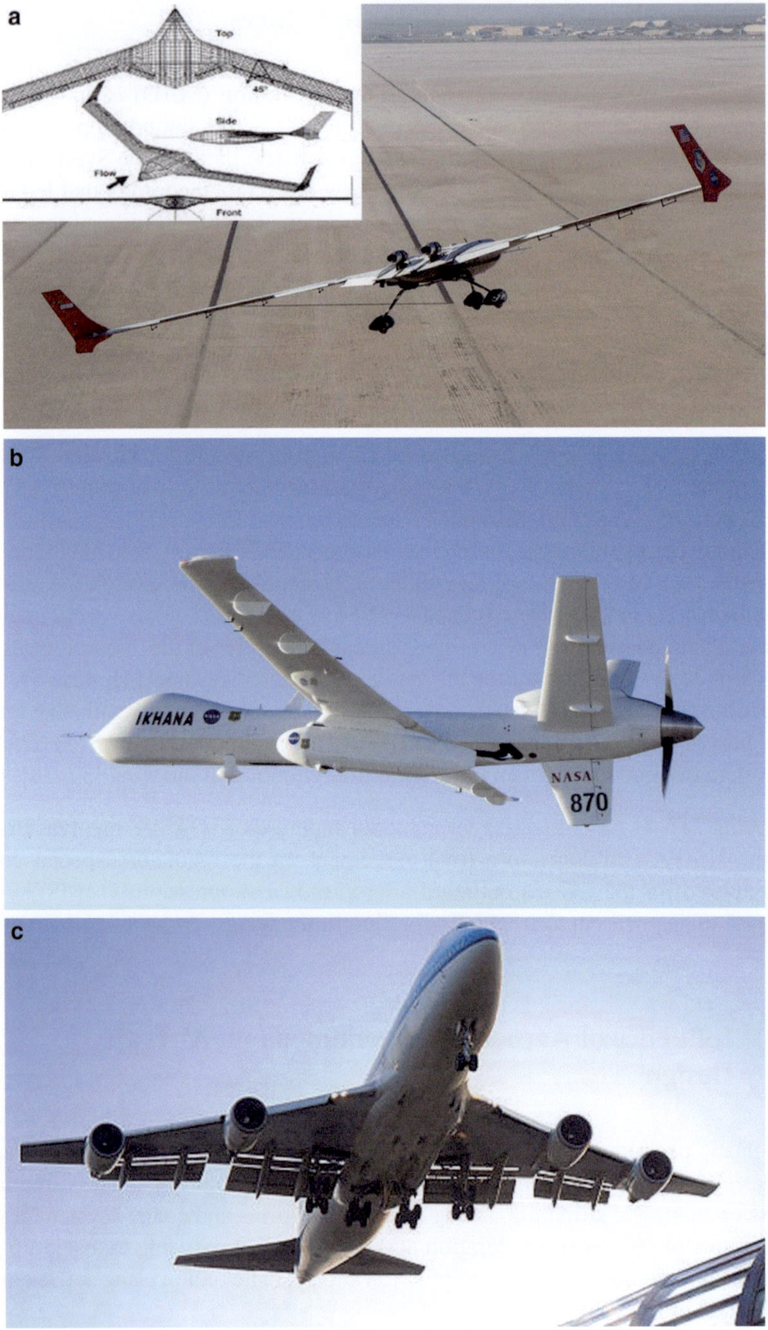
a
Top
45°
Side
Flow
Front
b
IKHANA
NASA
870
c

◀**Fig. 1.18** **a** NASA Lockheed Martin X-56A multi-utility aeroelastic demonstrator—CC (Courtesy of NASA [39, 40]). **b** NASA Ikhana unmanned science and research aircraft system (Courtesy of NASA [41, 42]). **c** Impression of Boeing 747 (Courtesy of istockphoto [43, 44]). Outboard aileron modal suppression (OAMS) has been tested on Boeing 747–8s [45]

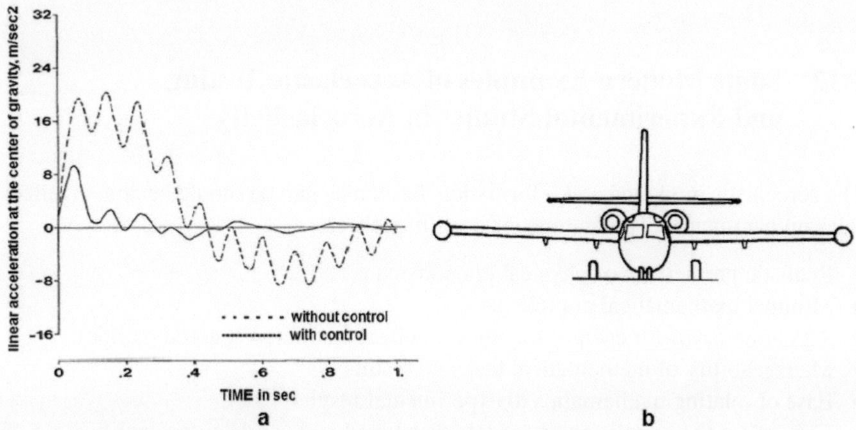

Fig. 1.19 **a** Variation with time of the linear acceleration at the center of gravity due to a step-up gust—Arava Transport with a 10 LE-TE active system spanning the whole of the wing; **b** General view of ARAVA Aircraft considered (Courtesy of NASA [46])

flutter, which is more difficult to analyze on a purely theoretical basis, may involve separated flow, periodic breakaway and reattachment of the flow, stalling conditions and various time-lag effects between the aerodynamic forces and the motion.

Buffeting. A serious buffeting phenomenon confronting designers is encountered by fighter aircrafts during pull-up to CLmax at high speed.

This often results in a rough transient vibrations in the tail due to aerodynamic impulses from the wake of the wing. The principal problems are those of reducing the severity of these vibrations and the provision of adequate strength.

1.11.1 Dynamic Loads Problem

This class of aeroelastic problem has its primary influence on structural design. In the prediction of design loads on an airplane structure in an accelerated condition, it is usually assumed that the airplane is perfectly rigid. Structural components designed by loads computed on this basis may fail due to dynamic overstress. External loads that are rapidly applied not only cause translation and rotation of the aircraft as a whole, but tend to excite vibrations of the structure. The additional inertial forces associated with these vibrations produce the dynamic overstress. As an illustration of the profound influence of aeroelasticity, consider the result of acceleration calculation on aircraft structure as depicted in Fig. 1.19 The figure illustrates the difference in the calculated result of the acceleration experienced by the fuselage and the wing tip of a typical swept wing flying at 460 mph equivalent airspeed at 11,000 ft. altitude when striking a gust.

1.12 Some Modern Examples of Aeroelastic Testing and Experimental Studies in Aeroelasticity

The aeroelastic modeling and information that can be gained should be able to reflect relevant parameters. The features of good models are:

- Realistic predictions of physical phenomena
- Minimal mathematical complexity
- Algebraic terms for each of the relevant effects (structures, aerodynamic)
- Manageability of mathematical task and results
- Ease of relating mathematics to experimental results
- Can reflect the simplicity of models developed and can lead to identification of physical phenomena and problem solving.

Figures 1.20 and 1.21 show two examples of flutter model test in wind tunnel for design validation and certification purposes. Figure 1.20 shows NASA TDT SUGAR (Subsonic Ultra-Green Aircraft Research) Truss-Braced Wing Wind Tunnel Model in the NASA Transonic Dynamics Tunnel (TDT). NASA TDT is a closed-circuit, continuous flow, variable pressure wind tunnel. The TDT is dedicated to identifying and solving aeroelasticity issues confronting fixed-wing aircraft and helicopter and tilt rotor configurations wing. Truss-braced wing (TBW) aircraft consists of two stiffening members, a strut and a jury added to each wing, which carry nearly all the compression or all the tension loads depending on load factors. The axial loads in the strut are transferred to the main wing and cause additional in-plane tensile or compressive loads in the inner wing. The jury connecting the strut and the main wing also restrains the wing detection under aerodynamic forces.

Figure 1.21 shows the wind tunnel flutter model test of N-250 Aeroelastic Model at the Indonesian Low Speed Tunnel at the Aerodynamic, Gas Dynamic and Vibration Laboratory of the Agency for the Assessment and Application of Technology BPPT (now BBTA3 BPPT).

1.13 Concluding Remarks

Despite its short coined terminology, Aeroelasticity is a very broad field, covering various disciplines like aerodynamics, structures, materials, control system and techniques. Aeroelastic phenomena that were first observed with concern for problem solving were aeroelastic instabilities, such as divergence and flutter. These have

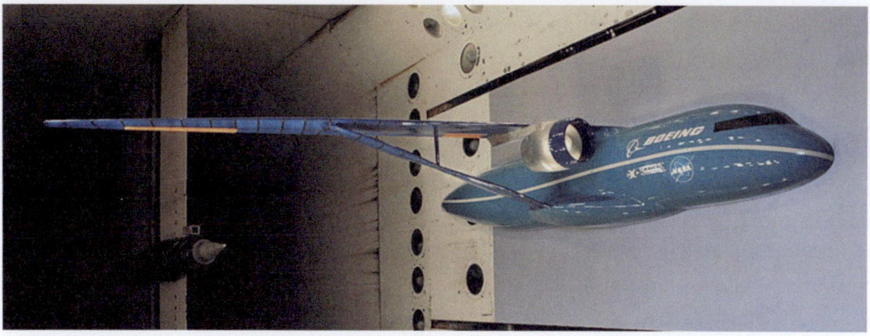

Fig. 1.20 NASA TDT SUGAR truss-braced wing wind tunnel model. Courtesy of NASA [47]

Fig. 1.21 N-250 aeroelastic model and experimental test set-up at ILST-LAGG [48][6]

imposed limitation on the design of aircraft for high-speed flights. The development of theories for aeroelastic analyses, which started with simplistic models of linear

[6] ILST-LAGG – Indonesin Low Speed Tunnel – The Laboratory of Aerodynamics, Gasdynamics and Vibration (the name adopted in 1980's. Now it has been changed).

modal analysis for structures and one-dimensional (1-D) quasi-steady aerodynamics, has progressively been developed to the point that various methods based on computational techniques in structures and fluids (such as the finite element method (FEM) and computational fluid dynamics (CFD)) and their coupling (such as CSD) are in current use. However, aeroelastic analysis tools and unsteady aerodynamics theories which were developed in the earlier part of the century are still in common usage, such as modal analysis, Theodorsen's unsteady aerodynamics, the Doublet Lattice Method and the V-g method. With the progress of high-powered computers and computational techniques solutions of many aeroelastic problems are made easier.

Traditionally, the most accurate aeroelasticity results have come from either an experimental investigation or more recently a complete numerical simulation by coupling finite element method and computational fluid dynamics analysis. Though such results are very accurate, can be obtained over a complete flight regime and can include "higher-order" phenomena and nonlinearities, they are also very expensive, especially so in the initial phase of design, when a number of design configurations may need to be analyzed. For a restricted problem, it is advantageous to base the analysis and to understand first principles in obtaining the first approximations to the problem and to check the plausibility of more sophisticated solutions. That also implies taking into account simplifications which do not compromise the quality of the results. Such approach could reduce the order of the problem while retaining high fidelity.

In particular, the use of first principles in physics, mathematics and engineering is mandatory to gain significant understanding and to initiate solution process, as well as assessing the plausibility of the application and results of existing methods and developing new methods.

The interaction between the first three of aerodynamics, structures, materials, control system and techniques can cause several undesirable phenomena: divergence (static aeroelastic phenomenon), flutter (dynamic aeroelastic phenomenon), limit cycle oscillations (nonlinear aeroelastic phenomenon) and vortex shedding, buffeting, galloping (unsteady aerodynamic phenomena). By understanding the physics of the problems, the first engineering attempts are how to avoid these phenomena that can lead to structural and functional failures. The means to solve the problems incorporate simplified aeroelastic analysis, aeroelastic design (divergence, flutter, control reversal), wind tunnel testing (aeroelastic scaling), ground vibration testing (complete modal analysis of aircraft structure) and flight flutter testing (demonstrate that flight envelope is flutter free). These will be the subject of further elaboration in subsequent chapters.

Appendices

The information, elaboration, discussion and questions outlined in a schematic fashion in the appendices can be used for teaching class problems or independent studies, en lieu of conventional problems.

Appendix 1: Studies in Aeroelasticity—For Creative and Proactive Class Discussions

Studies in aeroelasticity can be carried out for modeling, simulation and the concurrent design of aircraft.

In the modeling and simulation of aircrafts:

- Identify the physical problem parameters and assign symbols for convenience in the analysis
- Eliminate less important factors, which are carried out based on the judgment of the engineer
- Identify important physical constraints and boundary conditions
- Identify the problem physics and develop mathematical expressions describing the behavior
- Solve the problem and interpret the results.

Appendix 2: Problems and Issues for Creative and Proactive Class Discussions in Aeroelasticity

The figures displayed in this section are adapted from various literatures and are meant to incite curiosity and serve as practical examples found in practice.

a. **Aerodynamic concepts**

Lift and pitching moment coefficients

$$\left|L = qSC_L = qSC_{L_\alpha}\alpha\right|$$

$$M_{\text{ref}} = qScC_{M\text{ref}}$$

The Grumman X-29 was an American experimental aircraft that tested a forward-swept wing, canard control surfaces and other novel aircraft technologies. The X-29 was developed by Grumman, and the two built were flown by NASA and the United States Air Force. The aerodynamic instability of the X-29's airframe required the use of computerized fly-by-wire control. Composite materials were used to control the aeroelastic divergent twisting experienced by forward-swept wings and to reduce weight. The aircraft first flew in 1984, and two X-29s were flight tested through 1991 (Figs. 1.22 and 1.23).

b. **Historical Example—1911**

- The Bleriot XI monoplane
- Externally braced monoplane wings twist off at high speeds which should not happen for a safe plane.

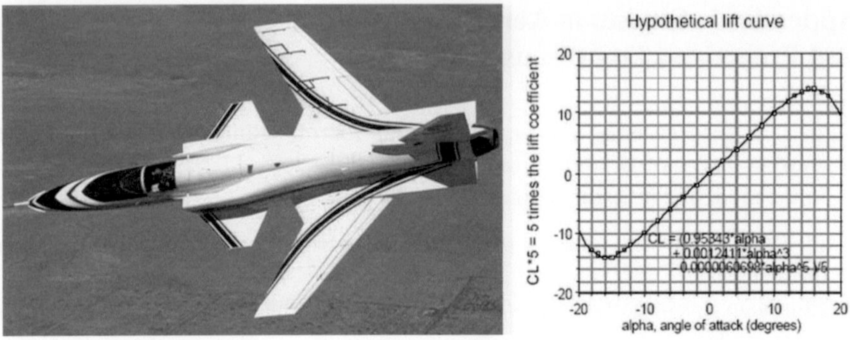

Fig. 1.22 Example of a fighter aircraft and hypothetical lift curve. Courtesy of NASA [49, 50]

- Suspected cause: torsional strength or flexibility does not meet the requirements.
- The phenomena demonstrated the interaction between wing lift and twist.

c. Load Factor

N—load factor is defined as

$$n = \frac{L}{W} = \frac{qSC_L}{W} = \frac{qSC_{L_\alpha}\alpha}{W}$$
$$A = A(V, \rho, \mu, p, \text{geometry})$$

$$F = m\ddot{x} + c\dot{x} + kx$$

d. Aero/Structural Interaction Model

Requirements:

- Simplicity
- Manageability
- Realistic (Fig. 1.24).

$$L = qSC_{L\alpha}(\alpha_0 + \theta)$$

1. Forces and moments on a typical section, a two-dimensional representation of half wing of an aircraft.

Further discussion and elaboration on such problem will be found in later chapters. The titles given on top of each schemes indicate, schematically, the procedures for obtaining what is indicated by the title of the particular scheme (Figs. 1.25, 1.26, 1.27, 1.28 and 1.29).

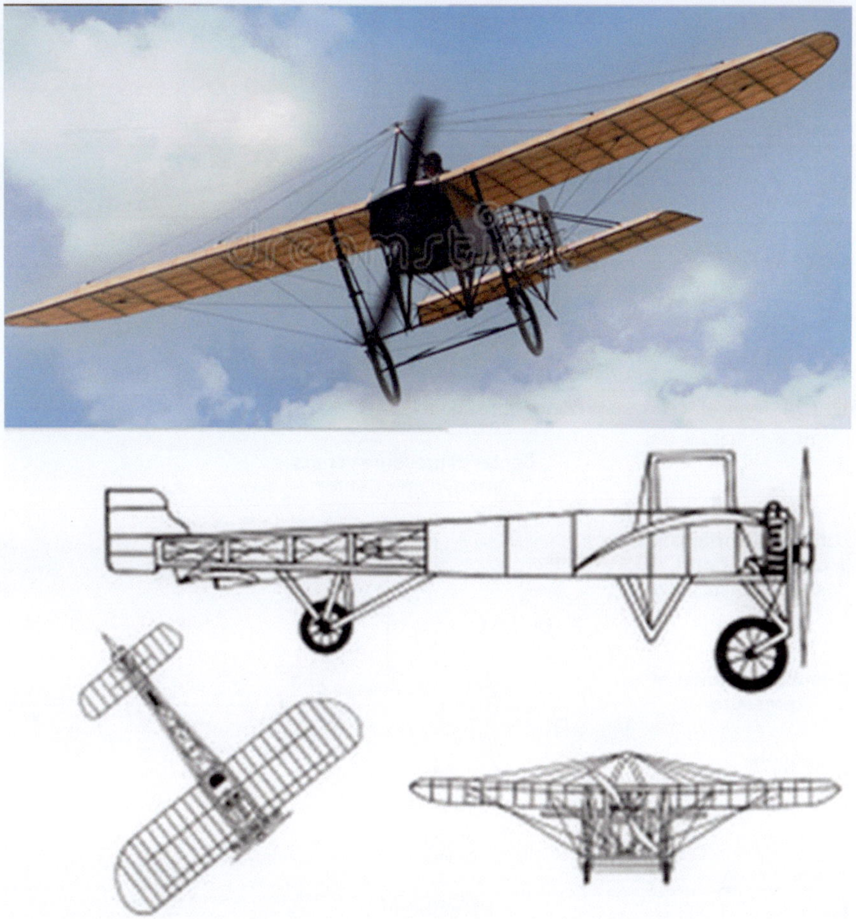

Fig. 1.23 Bleriot XI monoplane. Courtesy of Dreamline [51] and National Air and Space Museum, Smithsonian, Washington, D.C. [52]

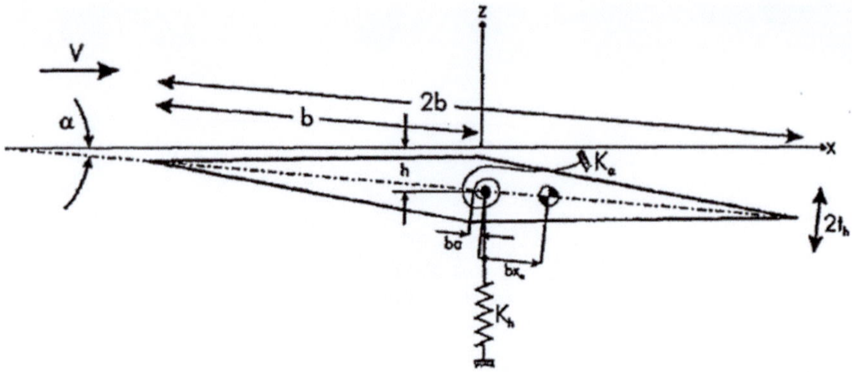

Fig. 1.24 Typical section, one degree of freedom

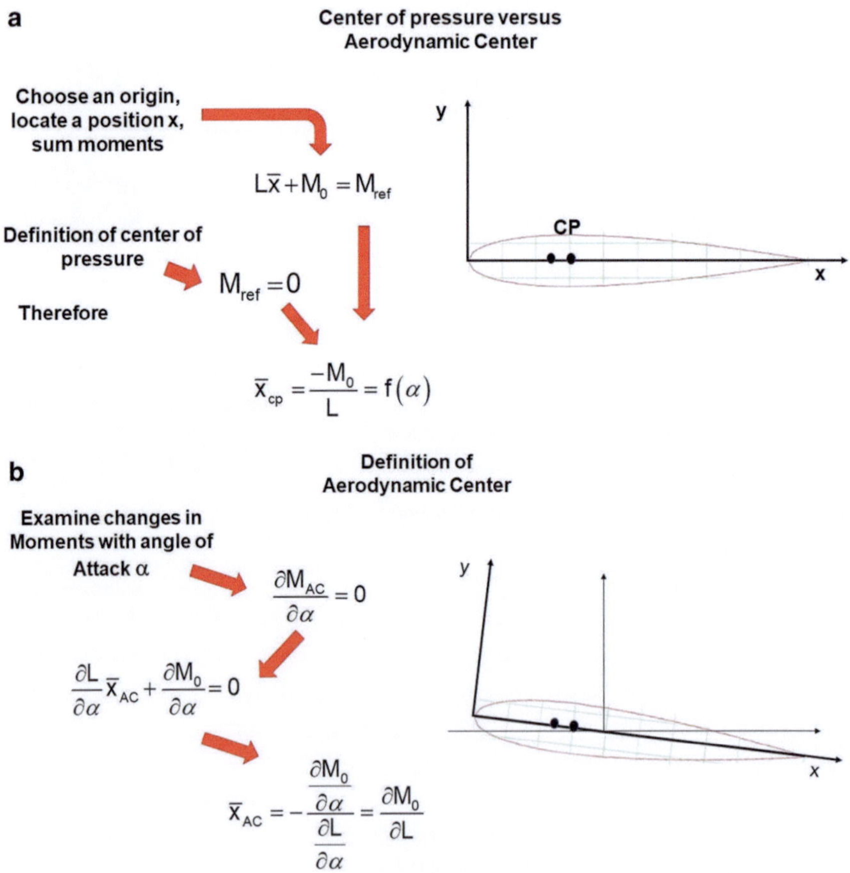

Fig. 1.25 **a** Center of pressure versus aerodynamic center. **b** Definition of aerodynamic center

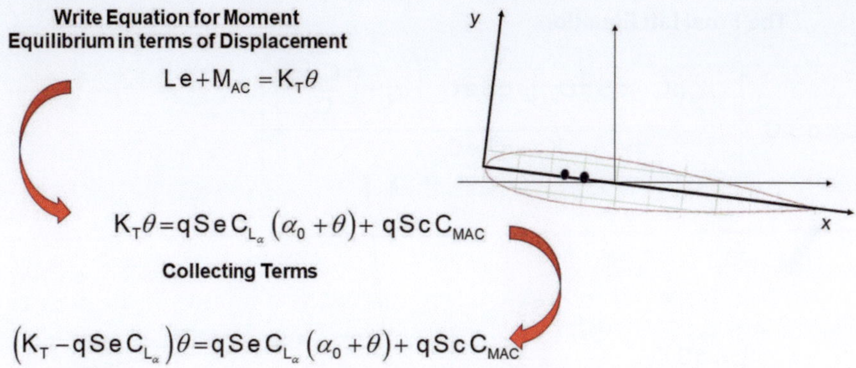

Fig. 1.26 Moment equilibrium in terms of displacement

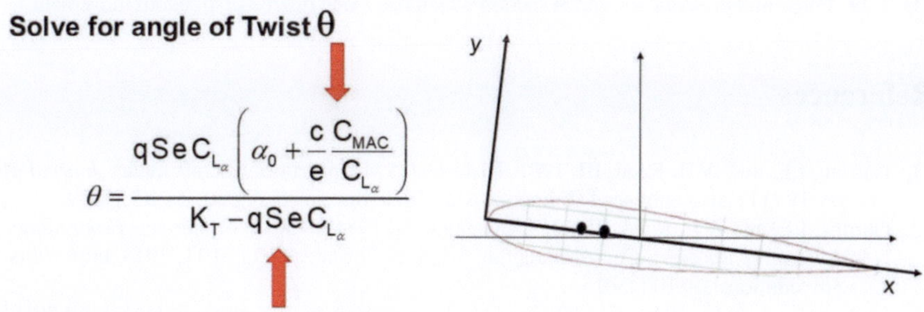

Fig. 1.27 Angle of twist

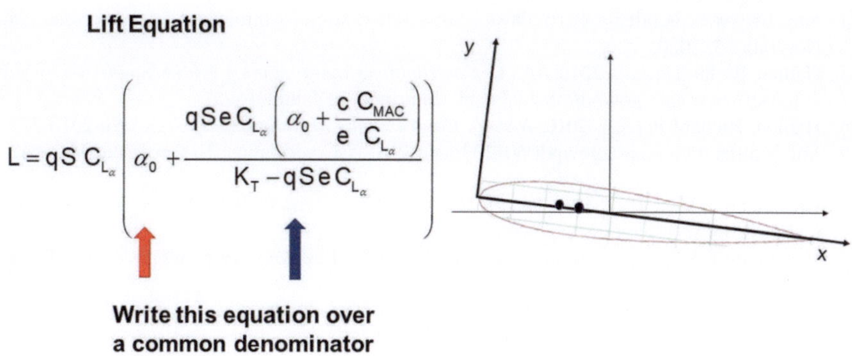

Fig. 1.28 Lift equation

The Final Lift Equation

$$L = qS\,C_{L_\alpha}\left(\frac{\alpha_0\left(K_T - qSeC_{L_\alpha}\right) + qSeC_{L_\alpha}\left(\alpha_0 + \frac{c}{e}\frac{C_{MAC}}{C_{L_\alpha}}\right)}{K_T - qSeC_{L_\alpha}}\right)$$

$$L = qS\,C_{L_\alpha}\left(\frac{\alpha_0 + \frac{qScC_{MAC}}{K_T}}{1 - \frac{qSeC_{L_\alpha}}{K_T}}\right)$$

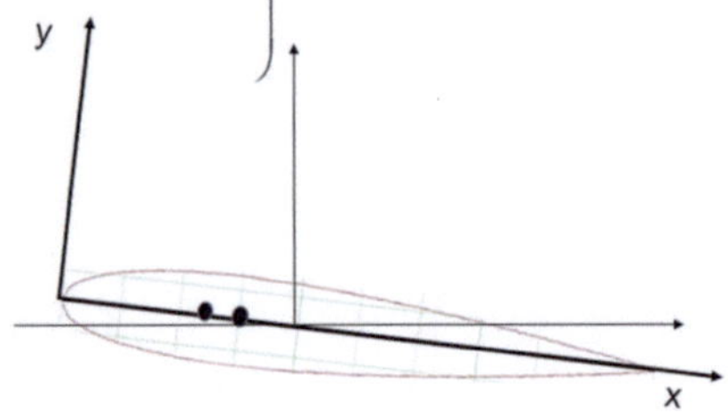

Fig. 1.29 Force and moment on typical section which has a one degree-of-freedom movement

References

1. Garrick, I.E. and W.H. Reed, III. 1981. Historical development of aircraft flutter. *Journal of Aircraft* 18 (11); also published Online: 22 May 2012. https://doi.org/10.2514/3.57579.
2. Garrick, I.E. and W.H. Reed, III. 2013. Document ID 19810045015, *Conference Proceedings*, NASA Langley Research Center Hampton, VA, United States, August 11, 2013. https://ntrs.nasa.gov/citations/19810045015.
3. Dick, Steve J. (ed.). 2010. *NASA'S First 50 Years Historical Perspectives*. NASA SP-2010-4704. https://www.nasa.gov/pdf/607087main_NASAsFirst50YearsHistoricalPerspectives-ebook.pdf, https://www.nasa.gov/sites/default/files/atoms/files/naca-nasa-aero-contributions-timeline.pdf. Accessed on November 5, 2020.
4. https://www.e-aircraftsupply.com/history-of-aviation-aircrafts-through-time/. Accessed on November 5, 2020.
5. Hallion, Richard P. (ed.). 2010. *NASA's Contributions to Aeronautics*. NASA/Sp-2010-570-Vol 1. https://www.nasa.gov/pdf/482993main_ContributionsVolume1.pdf.
6. Hallion, Richard P. (ed.). 2010. *NASA's Contributions to Aeronautics*. NASA/Sp-2010-570-Vol 2. https://www.nasa.gov/pdf/601333main_NASAsContributionsToAeronauticsVolume2-ebook.pdf.
7. Bisplinghoff, R.L., H. Ashley, and H. Halfman. 1996. *Aeroelasticity*. Dover Science. ISBN 0-486-69189-6.
8. Bisplinghoff, R.L., and H. Ashley. 1975. *Principles of Aeroelasticity*, copyright© 1967. Dover Publications.
9. Temple, G. 2004. Bairstow, Sir Leonard. In *Oxford Dictionary of National Biography* (online ed.), ed. Anita McConnell. Oxford University Press. https://doi.org/10.1093/ref:odnb/30543.
10. https://en.wikipedia.org/wiki/Aeroelasticity. Accessed on June 3, 2020.
11. Collar, A.R. 1946. *The Expanding Domain of Aeroelasticity*. Royal Aerronautical Society.
12. https://www.newscientist.com/article/dn3882-nasas-solar-wing-crashes-in-pacific/. Accessed on May 23, 2023.
13. https://www.nasa.gov/feature/ames/nasa-small-business-partnership-prepares-drone-for-30-day-science-flights. Accessed on May 23, 2023.
14. Maughmer, Mark D. 2003. *The Evolution of Sailplane Wing Design*. Tehnical Soaring, VOLUME XXVll -/11/y 2003, 14 July 2003. https://doi.org/10.2514/6.2003-2777.

15. https://www.gettyimages.com/detail/photo/austria-salzburg-glider-in-front-of-blue-sky-roy alty-free-image/509109577. Accessed on May 23, 2023.
16. https://en.wikipedia.org/wiki/File:Dg800.jpg. Accessed on May 23, 2023.
17. Wüstenhagen, M., T. Kier, M. Pusch, D. Ossmann, M.Y. Meddaikar, and A. Hermanutz. 2018. Aeroservoelastic stability of a 2D airfoil section equipped with a trailing edge flap. In *2018 Atmospheric Flight Mechanics Conference*, 3150 AIAA-1471524.
18. Iannelli, Andrea, Nandor Terkovics, Andres Marcos (UOB), Martin Leitner, Thiemo Kier, Gertjan Looye (DLR), and Roeland de Breuker (TUD). *Robust Analysis Techniques for Flexible Aircraft and Flutter.*
19. https://en.wikipedia.org/wiki/Dassault_Mercure. Accessed on June 3, 2020.
20. https://en.wikipedia.org/wiki/ATR_72. Accessed on April 12, 2021.
21. https://en.wikipedia.org/wiki/Fokker_100. Accessed on April 12, 2021
22. https://en.wikipedia.org/wiki/CASA/IPTN_CN-235. Accessed April 12, 2021.
23. https://en.wikipedia.org/wiki/Douglas_DC-6. Accessed April 12, 2021.
24. https://en.wikipedia.org/wiki/De_Havilland_Comet. Accessed April 12, 2021.
25. https://en.wikipedia.org/wiki/Lockheed_L-1049_Super_Constellation. Accessed April 12, 2021.
26. https://en.wikipedia.org/wiki/Airbus_A320. Accessed April 12, 2021.
27. https://en.wikipedia.org/wiki/Boeing_737_MAX. Accessed April 12, 2021.
28. https://www.boeing.com/commercial/787/. Accessed January 12, 2022.
29. https://en.wikipedia.org/wiki/Airbus_A300. Accessed January 12, 2022.
30. https://en.wikipedia.org/wiki/Boeing_777. Accessed January 12, 2022.
31. https://en.wikipedia.org/wiki/Airbus_A340. Accessed January 12, 2022.
32. https://en.wikipedia.org/wiki/Boeing_747. Accessed April 12, 2021.
33. https://en.wikipedia.org/wiki/Airbus_A380. Accessed January 12, 2022.
34. https://en.wikipedia.org/wiki/Airbus_A350. Accessed January 12, 2022.
35. https://en.wikipedia.org/wiki/Antonov_An-225. Accessed April 12, 2022.
36. Livne, Eli. 2014. *The Active Flutter Suppression (AFS) Technology Evaluation Project.*
37. Livne, Eli. 2014. *The Active Flutter Suppression (AFS) Technology Evaluation Project*, AMTAS Autumn Conference, Nov 12 https://depts.washington.edu/amtas/events/amtas_14fall/Livne.pdf, Accessed May 23, 2023.
38. Livne, Eli. and David R. Westlund, *The Active Flutter Suppression (AFS) Technology Evaluation Project*, https://depts.washington.edu/amtas/events /amtas_13fall/Livne-Flutter.pdf
39. https://www.nasa.gov/centers/armstrong/features/X-56A-suppresses-flutter-with-second-con troller.html. Accessed May 23, 2023.
40. https://www.nasa.gov/ sites/default/files/thumbnails/image/afrc2018-0272-53.jpg. Accessed May 23, 2023.
41. https://www.nasa.gov/pdf/470839main_ikhana_monograph.pdf. Accessed May 23, 2023.
42. NASA Armstrong Fact Sheet: Ikhana Predator B Unmanned Science and Research Aircraft System, https://www.nasa.gov/centers/armstrong/news/FactSheets/FS-097-DFRC.html. Accessed May 23, 2023.
43. https://www.istockphoto.com/id/search/2/image?phrase=boeing+747. Accessed January 12, 2022.
44. https://www.istockphoto.com/id/foto/pesawat-lepas-landas-gm15396128717020332. Accessed January 12, 2022.
45. https://www.federalregister.gov/documents/2011/05/26/2011-13022/special-conditions-boe ing-model-747-8-8f-airplanes-interaction-of-systems-and-structures. Accessed January 12, 2022.
46. Nissim, E. 1974. *Flutter Suppression and Gust Alleviation Using Active Controls.* NASA CR-19740018311.
47. Scott, Robert C., Timothy J. Allen, Christie J. Funk, Mark A. Castelluccio, Bradley W. Sexton, Scott Claggett, John Dykman, David A. Coulson, Robert E. Bartels. *Aeroservoelastic Wind-Tunnel Test of the SUGAR Truss Braced Wing Wind-Tunnel Model.* https://ntrs.nasa.gov/api/citations/20150006021/downloads/20150006021.pdf. Accessed May 23, 2023.

48. Djojodihardjo, H. 2002. *Aeroelastics and Unsteady Separated Aerodynamics of Aircrafts: Review and Outlook, in the Memorial Volume on the occasion of the Emeritierung of Professor Boris Laschka*, Muenchen, October 2002.
49. https://www.nasa.gov/centers/armstrong/history/experimental_aircraft/x-29.html. Accessed May 23, 2023.
50. https://www.nasa.gov/sites/default/files/images/344297main_EC91-491-6_full.jpg. Accessed May 23, 2023.
51. https://www.nasa.gov/centers/armstrong/history/experimental_aircraft/x-29.html
52. Tom D, Crouch, Curator Emeritus, National Air and Space Museum, Smithsonian, Washington, D.C https://www.britannica.com/biography/Louis-Bleriot. Accessed May 23, 2023.

Chapter 2
Fundamental Concepts from Theory of Elasticity

Abstract The present chapter provides a review and fundamentals of elasticity as one of the building blocks in aeroelasticity as well as aircraft structures, which may have overlapping significance in aerospace engineering. The presentation is intended to give the students, engineers and readers a comprehensive account on the principles and fundamentals of elasticity that contribute to the physical properties of engineering structures, their strength and integrity. In revealing the fundamentals of elasticity, mathematical representation is utilized, which requires some basic understanding of applied mathematics. The presentation starts with the mathematical foundation of aeroelasticity based on the equilibrium and compatibility equations for structures that can deform elastically. To obtain information on the elastic response which is an internal characteristic of the system, stress and strain relationships in the system are required and elaborated. The formulation of equation of motion for deformable bodies and the relationship between stress and deformation is elaborated in detail as required. The use of energy methods of analysis in deflection calculations and some structural problems for which energy methods are employed are elaborated. The deformation of aircraft structures under distributed forces and the influence functions approach is addressed and exemplified. Applications to aircraft structures such as for wings and bodies are elaborated comprehensively for fundamental understanding.

Keywords Aeroelasticity · Elasticity · Aerospace structure · Engineering mathematics · Materials and structures · Mechanics

2.1 Equilibrium and Compatibility Equations for Elastically Deformable Structures

Mathematical foundation of aeroelasticity is based on equilibrium and compatibility conditions of elastic structures that are subjected to external forces and displacement boundary conditions that are compatible.

© The Author(s), under exclusive license to Springer Nature Singapore Pte Ltd. 2023

H. Djojodihardjo, *Introduction to Aeroelasticity*,
https://doi.org/10.1007/978-981-16-8078-6_2

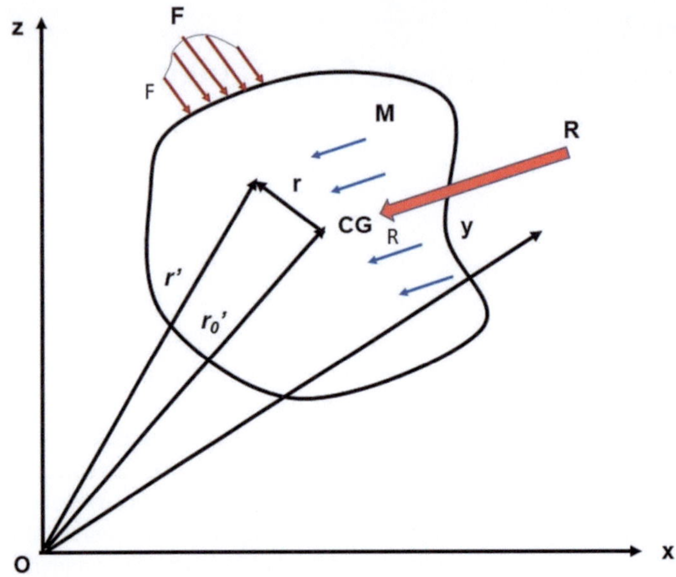

Fig. 2.1 Element of a three-dimensional structure that is free to move and experience forces and deformation in space

2.1.1 Equilibrium Equation

We will consider a three-dimensional structure that is free to move and experience deformation in space. The structure can experience a small displacement in the *x*–*y*-*z* coordinate system, which can be chosen arbitrarily and fixed in the inertial coordinate system.[1] Let *r'* be a position vector in the *x*-*y*-*z* coordinate system, as depicted in Fig. 2.1.

Let **F** be a surface force[2] per unit area and $\overline{R}$ a body force[3] per unit volume; then following Newton's second law of motion, the equilibrium equation for force can be written as (the bold symbol represents vector; see Fig. 2.1):

$$\iiint_V \mathbf{R}\mathrm{d}V + \iint_S \mathbf{F}\mathrm{d}S = \frac{\mathrm{d}}{\mathrm{d}t}\iiint_V \rho\frac{\mathrm{d}\mathbf{r}'}{\mathrm{d}t}\mathrm{d}V \tag{2.1}$$

and for moment

[1] An inertial coordinate system has its origin in absolute rest.

[2] A surface force *F* is a force exerted on the body by contact.

[3] A body force *R* is acting on individual particle point in the body; sometimes it is referred to as a force acting through a distance, without "touching" the subject (particle point). Body force is essentially a result of a potential field.

$$\iiint\limits_V \mathbf{r}' \times \mathbf{R}\mathrm{d}V + \iint\limits_S \mathbf{r}' \times \mathbf{F}\mathrm{d}S = \frac{\mathrm{d}}{\mathrm{d}t}\iiint\limits_V \left(\mathbf{r}' \times \rho\frac{\mathrm{d}\mathbf{r}'}{\mathrm{d}t}\right)\mathrm{d}V \tag{2.2}$$

where ρ is the density, i.e. mass per unit volume.

The center of gravity of the body should meet (and as a consequence of the definition of the center of gravity), and which can readily be proven the following requirement

$$Mr_0' = \iiint\limits_V \rho r' \mathrm{d}V \tag{2.3}$$

where *M is* total mass of the body

$\mathbf{r}'_0$ position vector within the *x-y-z* coordinate system to the center of gravity of the body.

Let:

$$\mathbf{r}' = \mathbf{r}_0' + \mathbf{r} \tag{2.4}$$

where $\mathbf{r}$ is the position vector of a particle point within the x–y-z coordinate system. Substitution of Eq. (2.4) into Eq. (2.1) and the use of Eq. (2.3) yield:

$$\iiint \mathbf{R}\mathrm{d}V + \iint \mathbf{F}\mathrm{d}S = \frac{\mathrm{d}}{\mathrm{d}t}\iiint \rho\frac{\mathrm{d}}{\mathrm{d}t}(\mathbf{r}_0' + \mathbf{r})\mathrm{d}V \tag{2.5}$$

which can be written as:

$$\mathbf{P} = \frac{\mathrm{d}}{\mathrm{d}t}\iiint \rho\frac{\mathrm{d}\mathbf{r}_0'}{\mathrm{d}t}\mathrm{d}V + \frac{\mathrm{d}}{\mathrm{d}t}\iiint \rho\frac{\mathrm{d}\mathbf{r}}{\mathrm{d}t}\mathrm{d}V \tag{2.6}$$

where

$$\mathbf{P} = \iiint \mathbf{R}\mathrm{d}V + \iint \mathbf{F}\mathrm{d}S \tag{2.7}$$

is the resultant vector of the forces working on the system, i.e. the total force on the body with mass M.

Since CG is the center of gravity of the body, it follows that:

$$\iiint \mathbf{r}\rho dV = 0 \tag{2.8}$$

$$\frac{\mathrm{d}}{\mathrm{d}t}\iiint \rho\frac{\mathrm{d}\mathbf{r}}{\mathrm{d}t}\mathrm{d}V = \frac{\mathrm{d}}{\mathrm{d}t}\iiint \frac{\mathrm{d}\mathbf{r}}{\mathrm{d}t}\rho\mathrm{d}V = \frac{\mathrm{d}}{\mathrm{d}t}\left(\frac{\mathrm{d}}{\mathrm{d}t}\left(\iiint \mathbf{r}\rho\mathrm{d}V\right)\right) = 0 \tag{2.9}$$

then it follows that the use of (2.7) and (2.9) reduces Eq. (2.6) to

$$\mathbf{P} = \frac{\mathrm{d}}{\mathrm{d}t}\frac{\mathrm{d}\mathbf{r}_0'}{\mathrm{d}t}\iiint \rho \mathrm{d}V \tag{2.10}$$

$$M = \iiint \rho \mathrm{d}V \tag{2.11}$$

then (2.10) reduces to:

$$\mathbf{P} = \frac{\mathrm{d}}{\mathrm{d}t}\left[M\frac{\mathrm{d}\mathbf{r}_0'}{\mathrm{d}t}\right] \tag{2.12}$$

which is the manifestation of Newton's second law of motion to a flexible body written in terms of its mass and the time rate of change of the position of its center of gravity. Equation (2.12) can be written as:

$$\frac{\mathrm{d}\mathbf{G}}{\mathrm{d}t} = \mathbf{P} \tag{2.13}$$

where **P** is the resultant vector of the forces working on the system, and $\overline{G}$ is the momentum vector, defined as:

$$\mathbf{G} = M\frac{\mathrm{d}\mathbf{r}_0'}{\mathrm{d}t} \tag{2.14}$$

Next, Eqs. (2.3) and (2.4) are substituted into Eq. (2.2), and then there is obtained: The LHS of Eq. (2.2):

$$\begin{aligned}\iiint\limits_V \mathbf{r}' \times \mathbf{R}\mathrm{d}V + \iint\limits_S \mathbf{r}' \times \mathbf{F}\mathrm{d}S &= \iiint\limits_V \mathbf{r}_0' \times \mathbf{R}\mathrm{d}V + \iiint\limits_V \mathbf{r} \times \mathbf{R}\mathrm{d}V + \iint\limits_S \mathbf{r}_0' \times \mathbf{F}\mathrm{d}S \\ &\quad + \iint\limits_S \mathbf{r} \times \mathbf{F}\mathrm{d}S \\ &= \iiint\limits_V \mathbf{r}_0' \times \mathbf{R}\mathrm{d}V + \iint\limits_S \mathbf{r}_0' \times \mathbf{F}\mathrm{d}S + \mathbf{L}\end{aligned} \tag{2.15}$$

where

$$\mathbf{L} = \iiint\limits_V \mathbf{r} \times \mathbf{R}\mathrm{d}V + \iint\limits_S \mathbf{r} \times \mathbf{F}\mathrm{d}S \tag{2.16}$$

is the moment acting on the system (with respect to the center of gravity.

The RHS (Right Hand Side) of Eq. (2.2) can now be mathematically manipulated as follows:

$$\frac{d}{dt}\iiint_V\left(\mathbf{r}'\times\rho\frac{\mathrm{d}\bar{r}}{\mathrm{d}t}\right)\mathrm{d}V=\frac{\mathrm{d}}{\mathrm{d}t}\iiint(\mathbf{r}_0'+\mathbf{r})\times\rho\left(\frac{\mathrm{d}(\mathbf{r}_0'+\mathbf{r})}{\mathrm{d}t}\right)\mathrm{d}V$$
$$=\frac{\mathrm{d}}{\mathrm{d}t}\iiint\mathbf{r}_0'\times\rho\frac{\mathrm{d}\mathbf{r}_0'}{\mathrm{d}t}\mathrm{d}V+\frac{\mathrm{d}}{\mathrm{d}t}\iiint\mathbf{r}\times\rho\frac{\mathrm{d}\mathbf{r}}{\mathrm{d}t}\mathrm{d}V$$
$$+\frac{\mathrm{d}}{\mathrm{d}t}\iiint\mathbf{r}_0'\times\rho\frac{\mathrm{d}\mathbf{r}}{\mathrm{d}t}\mathrm{d}V+\frac{\mathrm{d}}{\mathrm{d}t}\iiint\mathbf{r}\times\rho\frac{\mathrm{d}\rho_0'}{\mathrm{d}t}\mathrm{d}V$$
$$=\frac{\mathrm{d}\mathbf{H}}{\mathrm{d}t}+\frac{\mathrm{d}}{\mathrm{d}t}\iiint\mathbf{r}_0'\times\rho\frac{\mathrm{d}\mathbf{r}_0'}{\mathrm{d}t}\mathrm{d}V+\frac{\mathrm{d}}{\mathrm{d}t}\iiint\mathbf{r}\times\rho\frac{\mathrm{d}\mathbf{r}}{\mathrm{d}t}\mathrm{d}V$$
$$+\frac{d}{dt}\iiint\mathbf{r}\times\rho\frac{\mathrm{d}\mathbf{r}_0'}{\mathrm{d}t}\mathrm{d}V \tag{2.17}$$

where

$$\mathbf{H}\equiv\iiint\mathbf{r}\times\rho\frac{\mathrm{d}\mathbf{r}}{\mathrm{d}t}\mathrm{d}V \tag{2.18}$$

is defined as the angular momentum, or moment of momentum, of the system with respect to the center of gravity.

Using Eq. (2.1) in Eq. (2.15), there is obtained

$$\iiint_V\mathbf{r}_0'\times\mathbf{R}\mathrm{d}V+\iint_S\mathbf{r}_0'\times\mathbf{F}\mathrm{d}S=\mathbf{r}_0'\times\left[\iiint\mathbf{R}\mathrm{d}V+\iint\mathbf{F}\mathrm{d}S\right]$$
$$=\mathbf{r}_0'\times\frac{\mathrm{d}}{\mathrm{d}t}\iiint\rho\frac{\mathrm{d}\mathbf{r}'}{\mathrm{d}t}\mathrm{d}V \tag{2.19}$$

In Eq. (2.18)

$$\frac{\mathrm{d}}{\mathrm{d}t}\iiint\mathbf{r}_0'\times\rho\frac{\mathrm{d}\mathbf{r}_0'}{\mathrm{d}t}\mathrm{d}V+\frac{\mathrm{d}}{\mathrm{d}t}\iiint\mathbf{r}_0'\times\rho\frac{\mathrm{d}\mathbf{r}}{\mathrm{d}t}\mathrm{d}V$$
$$=\frac{d}{dt}\left\{\iiint\mathbf{r}_0'\times\rho\frac{\mathrm{d}\mathbf{r}_0'}{\mathrm{d}t}\mathrm{d}V+\iiint\mathbf{r}_0'\times\rho\frac{\mathrm{d}\mathbf{r}}{\mathrm{d}t}\mathrm{d}V\right\}$$
$$=\frac{d}{dt}\left\{\iiint\mathbf{r}_0'\times\rho\frac{\mathrm{d}\mathbf{r}'}{\mathrm{d}t}\mathrm{d}V\right\}$$
$$=\mathbf{r}_0'\times\frac{\mathrm{d}}{\mathrm{d}t}\iiint\rho\frac{\mathrm{d}\mathbf{r}'}{\mathrm{d}t}\mathrm{d}V+\frac{\mathrm{d}\mathbf{r}_0'}{\mathrm{d}t}\times\iiint\rho\frac{\mathrm{d}\mathbf{r}'}{\mathrm{d}t}\mathrm{d}V \tag{2.20}$$

Next, the last term in Eq. (2.17):

$$\frac{\mathrm{d}}{\mathrm{d}t}\iiint\mathbf{r}\times\rho\frac{\mathrm{d}\mathbf{r}_0'}{\mathrm{d}t}\mathrm{d}V=\frac{\mathrm{d}}{\mathrm{d}t}\left\{\left[\iiint\mathbf{r}\rho\mathrm{d}V\right]\times\frac{\mathrm{d}\mathbf{r}_0'}{\mathrm{d}t}\right\}=-\frac{\mathrm{d}}{\mathrm{d}t}\left\{\frac{\mathrm{d}\mathbf{r}_0'}{\mathrm{d}t}\times\iiint\mathbf{r}\rho\mathrm{d}V\right\} \tag{2.21}$$

Substituting Eq. (2.19) in Eq. (2.15), as well as substituting Eqs. (2.20) and (2.21) into Eq. (2.17), obtain

$$\mathbf{r}_0' \times \frac{\mathrm{d}}{\mathrm{d}t} \iiint \rho \frac{\mathrm{d}\mathbf{r}'}{\mathrm{d}t} \mathrm{d}V + \mathbf{L} = \frac{\mathrm{d}\mathbf{H}}{\mathrm{d}t} + \mathbf{r}_0' \times \frac{\mathrm{d}}{\mathrm{d}t} \iiint \rho \frac{\mathrm{d}\mathbf{r}'}{\mathrm{d}t} \mathrm{d}V + \frac{\mathrm{d}\mathbf{r}_0'}{\mathrm{d}t} \times \iiint \rho \frac{\mathrm{d}\mathbf{r}'}{\mathrm{d}t} \mathrm{d}V - \frac{\mathrm{d}}{\mathrm{d}t}\left(\frac{\mathrm{d}\mathbf{r}_0}{\mathrm{d}t} \times \iiint \mathbf{r}\rho \mathrm{d}V \right) \tag{2.22}$$

where

$$\frac{\mathrm{d}\mathbf{H}}{\mathrm{d}t} = \frac{\mathrm{d}}{\mathrm{d}t} \iiint \mathbf{r} \times \rho \frac{\mathrm{d}\mathbf{r}}{\mathrm{d}t} \mathrm{d}V \tag{2.23}$$

The integral part of the third term of the RHS of Eq. (2.22) is

$$\iiint \rho \frac{\mathrm{d}\mathbf{r}'}{\mathrm{d}t} \mathrm{d}V = \iiint \rho \frac{\mathrm{d}\mathbf{r}}{\mathrm{d}t} \mathrm{d}V + \iiint \rho \frac{\mathrm{d}\mathbf{r}_0'}{\mathrm{d}t} \mathrm{d}V = \iiint \rho \frac{\mathrm{d}\mathbf{r}}{\mathrm{d}t} \mathrm{d}V + \frac{\mathrm{d}\mathbf{r}_0'}{\mathrm{d}t} M \tag{2.24}$$

It follows that:

$$\frac{\mathrm{d}\mathbf{r}_0'}{\mathrm{d}t} \times \iiint \rho \frac{\mathrm{d}\mathbf{r}'}{\mathrm{d}t} \mathrm{d}V = \frac{\mathrm{d}\mathbf{r}_0'}{\mathrm{d}t} \times \iiint \rho \frac{\mathrm{d}\mathbf{r}}{\mathrm{d}t} \mathrm{d}V + \frac{\mathrm{d}\mathbf{r}_0'}{\mathrm{d}t} \times \frac{\mathrm{d}\mathbf{r}_0'}{\mathrm{d}t} M \tag{2.25}$$

Since Eq. (2.8) implies

$$\iiint \rho \frac{\mathrm{d}\mathbf{r}}{\mathrm{d}t} \mathrm{d}V = 0 \tag{2.26}$$

and

$$\frac{\mathrm{d}\mathbf{r}_0'}{\mathrm{d}t} \times \frac{\mathrm{d}\mathbf{r}_0'}{\mathrm{d}t} = 0 \tag{2.27}$$

then Eq. (2.25) is identically zero.

Similarly, Eq. (2.8) implies that the integral in the fourth term of Eq. (2.22) is

$$\frac{\mathrm{d}\mathbf{r}_0'}{\mathrm{d}t} \times \iiint \mathbf{r}\rho \mathrm{d}V = 0$$

Then Eq. (2.17) reduces to:

$$\mathbf{r}_0' \times \frac{\mathrm{d}}{\mathrm{d}t} \iiint \rho \frac{\mathrm{d}\mathbf{r}'}{\mathrm{d}t} \mathrm{d}V + \mathbf{L} = \frac{\mathrm{d}\mathbf{H}}{\mathrm{d}t} + \mathbf{r}_0' \times \frac{\mathrm{d}}{\mathrm{d}t} \iiint \rho \frac{\mathrm{d}\mathbf{r}'}{\mathrm{d}t} \mathrm{d}V \tag{2.28}$$

or:

$$\mathbf{L} = \frac{d\mathbf{H}}{dt} \tag{2.29}$$

Equation (2.5) implies that the motion of the center of gravity of the system, which is expressed in terms of position vector r_0', follows the motion of a single mass with the mass equal to the total mass of the system and is influenced by the resultant force of all external forces working on the system.

Similarly, Eq. (2.29), $\mathbf{L} = \frac{\mathrm{d}\mathbf{H}}{\mathrm{d}t}$, implies similar state of affairs for angular motion; i.e. that the rate of change of the resultant of the angular momentum with respect to the center of gravity is equal to the resultant of external moment with respect to the center of mass.

Equations (2.5) and (2.29) are the foundation of the calculation of basic motion of aeroelastic system but do not contain any information on the internal or elastic response.

2.1.2 Equilibrium Equation and Internal Stresses

To obtain information on the elastic response which is an internal characteristic of the system, stress and strain relationships in the system are required. Such relationship is elaborated subsequently.

The force vector **F** acting on the surface of the system can be distinguished into its components in the Cartesian coordinate system:

$$\mathbf{F} = \mathbf{i}F_x + \mathbf{j}F_y + \mathbf{k}F_z \tag{2.30}$$

where $\mathbf{i}, \mathbf{j},$ and $\mathbf{k}$ are unit vectors in the x-, y- and z-directions, respectively, and F_x, F_y and F_z are the components of F along these directions. Each force component is related to the stress component on the surface through the following relationships:

$$\begin{aligned} F_x &= \sigma_x \mathbf{n}\cdot\mathbf{i} + \tau_{xy}\mathbf{n}\cdot\mathbf{j} + \tau_{xz}\mathbf{n}\cdot\mathbf{k} \\ F_y &= \tau_{yx}\mathbf{n}\cdot\mathbf{i} + \sigma_y\mathbf{n}\cdot\mathbf{j} + \tau_{yz}\mathbf{n}\cdot\mathbf{k} \\ F_z &= \tau_{zx}\mathbf{n}\cdot\mathbf{i} + \tau_{zy}\mathbf{n}\cdot\mathbf{j} + \sigma_z\mathbf{n}\cdot\mathbf{k} \end{aligned} \tag{2.31a,b,c}$$

where **n** is a unit vector perpendicular to the surface in the outward direction of the surface. Here

$$\begin{aligned} \mathbf{n}\cdot\mathbf{i} &= \cos(x, n) \\ \mathbf{n}\cdot\mathbf{j} &= \cos(y, n) \\ \mathbf{n}\cdot\mathbf{k} &= \cos(z, n) \end{aligned} \tag{2.32a,b,c}$$

are the angle between the unit normal vector to the surface with the x-, y- and z-directions, respectively. Equations (2.31) and (2.32) can be written using vector notation as:

$$\mathbf{F} = \mathbf{n} \cdot \Phi \tag{2.33}$$

where the second-order stress tensor Φ can be expressed as:

$$\Phi = \begin{bmatrix} \sigma_x & \tau_{yx} & \tau_{zx} \\ \tau_{xy} & \sigma_y & \tau_{zy} \\ \tau_{xz} & \tau_{yz} & \sigma_z \end{bmatrix} \tag{2.34}$$

In vector form, Eq. (2.33) can be written as:

$$\begin{Bmatrix} F_x \\ F_y \\ F_z \end{Bmatrix} = \begin{bmatrix} \mathbf{n}\cdot\mathbf{i} & \mathbf{n}\cdot\mathbf{j} & \mathbf{n}\cdot\mathbf{k} \end{bmatrix} \begin{bmatrix} \sigma_x & \tau_{yx} & \tau_{zx} \\ \tau_{xy} & \sigma_y & \tau_{zy} \\ \tau_{xz} & \tau_{yz} & \sigma_z \end{bmatrix} \tag{2.35}$$

or

$$\begin{Bmatrix} F_x \\ F_y \\ F_z \end{Bmatrix} = \begin{Bmatrix} \sigma_x \mathbf{n}\cdot\mathbf{i} + \tau_{xy}\mathbf{n}\cdot\mathbf{j} + \tau_{xz}\mathbf{n}\cdot\mathbf{k} \\ \tau_{yx}\mathbf{n}\cdot\mathbf{i} + \sigma_y\mathbf{n}\cdot\mathbf{j} + \tau_{yz}\mathbf{n}\cdot\mathbf{k} \\ \tau_{zx}\mathbf{n}\cdot\mathbf{i} + \tau_{zy}\mathbf{n}\cdot\mathbf{j} + \sigma_z\mathbf{n}\cdot\mathbf{k} \end{Bmatrix} \tag{2.36}$$

Substituting Eq. (2.33) into the equation of dynamic equilibrium (2.1) and (2.2) gives:

$$\iiint \rho \frac{\mathrm{d}^2\mathbf{r}'}{\mathrm{d}t^2}\mathrm{d}V = \iiint\limits_V \mathbf{R}\mathrm{d}V + \iint\limits_S \mathbf{n}\cdot\Phi\mathrm{d}S \tag{2.37}$$

and

$$\iiint \left(\mathbf{r}'\times\rho\frac{\mathrm{d}^2\mathbf{r}'}{\mathrm{d}t^2}\right)\mathrm{d}V = \iiint\limits_V \mathbf{r}'\times\mathbf{R}\mathrm{d}V + \iint\limits_S \mathbf{r}'\times\mathbf{n}\cdot\Phi\mathrm{d}S \tag{2.38}$$

Equations (2.37) and (2.38) have also utilized Eqs. (2.3) and (2.4).

Equation (2.38) can be further worked out subsequently. From Eq. (2.2):

$$\frac{\mathrm{d}}{\mathrm{d}t}\iiint\limits_V \left(\mathbf{r}'\times\rho\frac{\mathrm{d}\mathbf{r}'}{\mathrm{d}t}\right)\mathrm{d}V = \iiint\limits_V \mathbf{r}'\times\mathbf{R}\mathrm{d}V + \iint\limits_S \mathbf{r}'\times\mathbf{F}\mathrm{d}S$$

Left-hand side (LHS:):

$$\frac{d}{dt}\iiint\limits_V \left(\mathbf{r}'\times\rho\frac{d\mathbf{r}'}{dt}\right)dV = \iiint\limits_V \frac{d\mathbf{r}'}{dt}\times\rho\frac{d\mathbf{r}'}{dt}dV + \iiint\limits_V \mathbf{r}'\times\rho\frac{d^2\mathbf{r}'}{dt^2}dV$$

$$= \iiint\limits_V \left(\frac{d\mathbf{r}'}{dt} \times \frac{d\mathbf{r}'}{dt}\right)\rho dV + \iiint\limits_V \mathbf{r}' \times \frac{d^2\mathbf{r}'}{dt^2}\rho dV$$

$$= \iiint\limits_V \mathbf{r}' \times \frac{d^2\mathbf{r}'}{dt^2}\rho dV = \iiint\limits_V (\mathbf{r}'_0 + \mathbf{r}) \times \frac{d^2(\mathbf{r}'_0 + r)}{dt^2}\rho dV$$

$$= \iiint\limits_V \mathbf{r}'_0 \times \frac{d^2\mathbf{r}'_0}{dt^2}\rho dV + \iiint\limits_V \mathbf{r}'_0 \times \frac{d^2\mathbf{r}}{dt^2}\rho dV$$

$$+ \iiint\limits_V \mathbf{r} \times \frac{d^2\mathbf{r}'_0}{dt^2}\rho dV + \iiint\limits_V \mathbf{r} \times \frac{d^2\mathbf{r}}{dt^2}\rho dV \tag{2.39}$$

Right-hand side (RHS):

$$\iiint\limits_V \mathbf{r}' \times \mathbf{R}\mathrm{d}V + \iint\limits_S \mathbf{r}' \times \mathbf{F}\mathrm{d}S = \iiint\limits_V \mathbf{r}'_0 \times \mathbf{R}\mathrm{d}V + \iint\limits_S \mathbf{r}'_0 \times \mathbf{F}\mathrm{d}S + \iiint\limits_V \mathbf{r} \times \mathbf{R}\mathrm{d}V$$

$$+ \iint\limits_S \mathbf{r} \times \mathbf{F}\mathrm{d}S$$

$$= \mathbf{r}'_0 \times \left[\iiint\limits_V \mathbf{R}\mathrm{d}V + \iint\limits_S \mathbf{F}\mathrm{d}S\right] + \iiint\limits_V r \times \mathbf{R}\mathrm{d}V$$

$$+ \iint\limits_S r \times \mathbf{F}\mathrm{d}S \tag{2.40}$$

From (2.1)

$$\frac{\mathrm{d}}{\mathrm{d}t}\iiint\limits_V \rho\frac{\mathrm{d}\mathbf{r}'}{\mathrm{d}t}\mathrm{d}V = \iiint\limits_V \mathbf{R}\mathrm{d}V + \iint\limits_S \mathbf{F}\mathrm{d}S \tag{2.1}$$

or

$$\iiint\limits_V \rho\frac{\mathrm{d}^2\mathbf{r}'}{\mathrm{d}t^2}\mathrm{d}V = \iiint\limits_V \mathbf{R}\mathrm{d}V + \iint\limits_S \mathbf{F}\mathrm{d}S \tag{2.41}$$

Hence

$$\mathbf{r}'_0 \times \iiint\limits_V \rho\frac{\mathrm{d}^2\mathbf{r}'}{\mathrm{d}t^2}\mathrm{d}V = \mathbf{r}'_0 \times \left[\iiint\limits_V \mathbf{R}\mathrm{d}V + \iint\limits_S \mathbf{F}\mathrm{d}S\right] \tag{2.42}$$

From (2.39) and (2.40), one gets

$$\iiint_V \mathbf{r}'_0 \times \frac{\mathrm{d}^2\mathbf{r}'_0}{\mathrm{d}t^2}\rho \mathrm{d}V + \iiint_V \mathbf{r}'_0 \times \frac{\mathrm{d}^2\mathbf{r}}{\mathrm{d}t^2}\rho \mathrm{d}V + \iiint_V \mathbf{r} \times \frac{\mathrm{d}^2\mathbf{r}'_0}{\mathrm{d}t^2}\rho \mathrm{d}V + \iiint_V \mathbf{r} \times \frac{\mathrm{d}^2\mathbf{r}}{\mathrm{d}t^2}\rho \mathrm{d}V$$

$$= \mathbf{r}'_0 \times \left[\iiint_V \mathbf{R}\mathrm{d}V + \iint_S \mathbf{F}\mathrm{d}S\right] + \iiint_V r \times \mathbf{R}\mathrm{d}V + \iint_S r \times \mathbf{F}\mathrm{d}S \tag{2.43}$$

and using Eq. (2.42), this reduces to

$$\iiint_V \left(\mathbf{r} \times \frac{\mathrm{d}^2\mathbf{r}'}{\mathrm{d}t^2}\right)\rho \mathrm{d}V = \iiint_V \mathbf{r} \times \mathbf{R}\mathrm{d}V + \iint_S \mathbf{r} \times \mathbf{F}\mathrm{d}S \tag{2.44}$$

Using Eq. (2.33) in (2.41) and (2.44), there is obtained

$$\iiint_V \rho \frac{\mathrm{d}^2\mathbf{r}'}{\mathrm{d}t^2}\mathrm{d}V = \iiint_V \mathbf{R}\mathrm{d}V + \iint_S \mathbf{n} \cdot \Phi \mathrm{d}S \tag{2.45a}$$

$$\iiint \left(\mathbf{r} \times \rho \frac{\mathrm{d}^2\mathbf{r}'}{\mathrm{d}t^2}\right)\mathrm{d}V = \iiint_V \mathbf{r} \times \mathbf{R}\mathrm{d}V + \iint_S \mathbf{r} \times \mathbf{n} \cdot \Phi \mathrm{d}S \tag{2.45b}$$

From divergence theorem

$$\iint \mathbf{n} \cdot \Phi \mathrm{d}S = \iiint \nabla \cdot \Phi \mathrm{d}V \tag{2.46}$$

Then the second term on RHS of Eq. (2.46) can be written as:

$$\iint \mathbf{r} \times \mathbf{n} \cdot \Phi \mathrm{d}S = -\iint \mathbf{n} \cdot \Phi \times \mathbf{r}\mathrm{d}S = -\iiint \nabla \cdot (\Phi \times \mathbf{r})\mathrm{d}V \tag{2.47}$$

Therefore Eq. (2.46) can be rewritten as:

$$\iiint \left(\mathbf{r} \times \rho \frac{\mathrm{d}^2\mathbf{r}'}{\mathrm{d}t^2}\right)\mathrm{d}V = \iiint_V \mathbf{r} \times \mathbf{R}\mathrm{d}V - \iiint \nabla \cdot (\Phi \times \mathbf{r})\mathrm{d}V \tag{2.48a}$$

or

$$\iiint \left(\mathbf{r} \times \rho \frac{\mathrm{d}^2\mathbf{r}'}{\mathrm{d}t^2}\right)\mathrm{d}V - \iiint_V \mathbf{r} \times \mathbf{R}\mathrm{d}V + \iiint \nabla \cdot (\Phi \times \mathbf{r})\mathrm{d}V = 0 \tag{2.48b}$$

Equations (2.45a, 2.45b) and (2.48a, 2.48b) can be further simplified into

$$\iiint_V \left[\rho\frac{d^2\mathbf{r}'}{dt^2} - \mathbf{R} - \nabla\cdot\Phi\right]dV = 0 \tag{2.49}$$

and

$$\iiint \left[\mathbf{r}\times\rho\frac{d^2\mathbf{r}'}{dt^2} - \mathbf{r}\times\mathbf{R} + \nabla\cdot(\Phi\times\mathbf{r})\right]dV = 0 \tag{2.50}$$

where

$$\nabla = \mathbf{i}\frac{\partial}{\partial x} + \mathbf{j}\frac{\partial}{\partial y} + \mathbf{k}\frac{\partial}{\partial z}$$

Since (2.49) and (2.50) are valid for any V, then the arguments should also be equal to zero. Hence:

$$\rho\frac{d^2\mathbf{r}'}{dt^2} - \mathbf{R} - \nabla\cdot\Phi = 0 \tag{2.51}$$

$$\mathbf{r}\times\rho\frac{d^2\mathbf{r}'}{dt^2} - \mathbf{r}\times\mathbf{R} + \nabla\cdot(\Phi\times\mathbf{r}) = 0 \tag{2.52}$$

Next, the acceleration and body force **R** can also be written in their components:

$$\frac{d^2\mathbf{r}'}{dt^2} = c_x\mathbf{i} + c_y\mathbf{j} + c_z\mathbf{k} \tag{2.53}$$

$$\mathbf{R} = X\mathbf{i} + Y\mathbf{j} + Z\mathbf{k} \tag{2.54}$$

where each component is parallel to the x-, y- and z-directions, respectively. Using Eqs. (2.31), (2.53) and (2.54) there is obtained:

$$\begin{aligned}
\rho a_x &= \frac{\partial\sigma_x}{\partial x} + \frac{\partial\tau_{yx}}{\partial y} + \frac{\partial\tau_{zx}}{\partial z} + X\\
\rho a_y &= \frac{\partial\tau_{xy}}{\partial x} + \frac{\partial\sigma_y}{\partial y} + \frac{\partial\tau_{zy}}{\partial z} + Y\\
\rho a_x &= \frac{\partial\tau_{xz}}{\partial x} + \frac{\partial\tau_{yz}}{\partial y} + \frac{\partial\sigma_z}{\partial z} + X
\end{aligned} \tag{2.55}$$

Using Eq. (2.52), it can be shown that the shear stress at two adjoining surfaces that are perpendicular to each other is the same. Hence:

$$\begin{aligned}\tau_{xy} &= \tau_{yx}\\ \tau_{xz} &= \tau_{zx}\\ \tau_{yz} &= \tau_{zy}\end{aligned} \tag{2.56}$$

2.1.3 Compatibility Equation and Equation of State

To derive the formulation of equation of motion for deformable bodies, the relationship between stress and deformation is required. Elastic deformation of a body can be expressed *u*, *v* and *w*, each along the *x*-, *y*- and *z*-axes, respectively. The resulting strain (deformation per unit length of original shape), by using small perturbation assumptions so that higher-order terms can be neglected, can be expressed as:

$$\begin{aligned}\varepsilon_x &= \frac{\partial u}{\partial x};\ \varepsilon_y = \frac{\partial v}{\partial y};\ \varepsilon_z = \frac{\partial w}{\partial z}\\ \gamma_{xy} &= \frac{\partial v}{\partial x} + \frac{\partial u}{\partial y};\ \gamma_{xz} = \frac{\partial w}{\partial x} + \frac{\partial u}{\partial z};\ \gamma_{yz} = \frac{\partial w}{\partial y} + \frac{\partial v}{\partial z}\end{aligned} \tag{2.57}$$

using conventional notation in the theory of aeroelasticity.

For homogeneous, isotropic and perfectly elastic body undergoing temperature change and deformation, the relationship (state equation between stress and strain) can be expressed as:

$$\begin{aligned}\varepsilon_x &= \alpha\Delta T + \frac{1}{E}\left[\sigma_x - \phi\left(\sigma_y + \sigma_z\right)\right]\\ \varepsilon_y &= \alpha\Delta T + \frac{1}{E}\left[\sigma_y - \phi(\sigma_x + \sigma_z)\right]\\ \varepsilon_z &= \alpha\Delta T + \frac{1}{E}\left[\sigma_z - \phi\left(\sigma_x + \sigma_y\right)\right]\\ \gamma_{xy} &= \frac{\tau_{xy}}{G};\ \gamma_{xz} = \frac{\tau_{xz}}{G};\ \gamma_{yz} = \frac{\tau_{yz}}{G}\end{aligned} \tag{2.58}$$

where

α thermal expansion coefficient of the material
ΔT temperature increment above the reference value.
E modulus of elasticity.
ν Poisson ratio.

The relationship among *E*, *G* and ν is given by:

$$G = \frac{E}{2(1 + \upsilon)}$$

In Eq. (2.24), shear strain is not influenced by temperature change. Reciprocal relationship associated with Eq. (2.24) is given by:

$$\sigma_x = \lambda e + 2G\varepsilon_x - \frac{\alpha E \Delta T}{1 - 2\upsilon}$$
$$\sigma_y = \lambda e + 2G\varepsilon_y - \frac{\alpha E \Delta T}{1 - 2\upsilon}$$
$$\sigma_z = \lambda e + 2G\varepsilon_z - \frac{\alpha E \Delta T}{1 - 2\upsilon}$$
$$\tau_{xy} = G\gamma_{xy}; \tau_{xz} = G\gamma_{xz}; \tau_{yz} = G\gamma_{yz} \tag{2.59}$$

where $e = \varepsilon_x + \varepsilon_y + \varepsilon_z$ is the volumetric expansion coefficient, and λ is known as the Lame elastic constant.[4,5,6]

2.1.4 Conditions for Solving the Equations

If there is an Apriori information on the temperature distribution, and if the appropriate boundary (2.8) and initial conditions are known, then the set of 18(eighteen Eqs. 2.21–2.24 or 2.26) define a complete basis for the calculation of the time change of stress, expansion and displacement of the body.

---

[4] http://web.mit.edu/16.20/homepage/3_Constitutive/Constitutive_files/module_3_no_solutions.pdf.

[5] https://encyclopediaofmath.org/wiki/Lam%C3%A9_constants.

[6] In continuum mechanics, the Lamé parameters (also called the Lamé coefficients, Lamé constants or Lamé moduli) are two material-dependent quantities denoted by λ and μ that arise in strain-stress relationships. In general, λ and μ are individually referred to as Lamé's first parameter and Lamé's second parameter, respectively. Other names are sometimes employed for one or both parameters, depending on context. For example, the parameter μ is referred to in fluid dynamics as the dynamic viscosity of a fluid (not the same units); whereas in the context of elasticity, μ is called the shear modulus and is sometimes denoted by G instead of μ. Typically, the notation G is seen paired with the use of Young's modulus, and the notation μ is paired with the use of λ. In homogeneous and isotropic materials, these define Hooke's law in 3D,

$\sigma = 2\mu e + \lambda\, tr(\varepsilon)$

where σ is the stress, ε the strain tensor, I the identity matrix and tr the trace function. Hooke's law may be written in terms of tensor components using index notation as

$\sigma_{ij} = 2\mu E_{ij} + \lambda \delta_{ij} E_{kk}$

where σ_{ij} is the stress tensor, E_{ij} the strain tensor and δ_{ij} the Kronecker delta. The two parameters together constitute a parameterization of the elastic moduli for homogeneous isotropic media, popular in mathematical literature, and are thus related to the other elastic moduli; for instance, the bulk modulus can be expressed as

$K = \lambda + \frac{2}{3}\mu$

Although the shear modulus, μ, must be positive, the Lamé's first parameter, λ, can be negative, in principle; however, for most materials it is also positive. The parameters are named after Gabriel Lamé. They have the same dimension as stress and are usually given in the pressure unit [Pa].

2.2 Thermodynamic Behavior of Elastic (Deformable Bodies) Under Dynamic and Thermal Loading

If on the body there are acting external forces and heating simultaneously, then both changes in kinetic and internal energy are taking place. If δT is the total kinetic energy change per unit volume, and δU_0 is the internal energy change per unit volume, both taking place when the body and surface forces, and heating are acting on the body for a time interval δt, then the first law of thermodynamics implies:

$$\delta T + \iiint_V \delta U_0 \mathrm{d}V = \delta W + \iiint_V \delta Q \mathrm{d}V \tag{2.60}$$

where δW is the work carried out by the body and surface forces within the time interval δt, and δQ is the mechanical energy equivalent of the heat energy per unit volume supplied to the body within the same time interval δt.

The kinetic energy is given by:

$$\tau = \frac{1}{2} \iiint \rho \left(\frac{\mathrm{d}\bar{r}}{\mathrm{d}t} \right) \cdot \left(\frac{\mathrm{d}\bar{r}}{\mathrm{d}t} \right) \mathrm{d}V \tag{2.61}$$

From the equivalence between work and energy:

$$\delta W = \delta T + \iiint (\sigma_x \delta \varepsilon_x + \sigma_y \delta \varepsilon_y + \sigma_z \delta \varepsilon_z + +\tau_{xy} \delta \gamma_{xy} + \tau_{xz} \delta \gamma_{xz} + \tau_{yz} \delta \gamma_{yz}) \mathrm{d}V \tag{2.62}$$

From Eqs. (2.61) and (2.62) there is obtained:

$$\delta U_0 = \delta Q + \sigma_x \delta \varepsilon_x + \sigma_y \delta \varepsilon_y + \sigma_z \delta \varepsilon_z + \tau_{xy} \delta \gamma_{xy} + \tau_{xz} \delta \gamma_{xz} + \tau_{yz} \delta \gamma_{yz} \tag{2.63}$$

The quantity δU_0 is a function that determines the configuration of the body per unit volume which is undergoing expansion due to temperature change. For any internal energy reference value of $U_0 = 0$, δU_0 is a function of internal energy.

The entropy change of a reversible system within a time interval δ*t* (or ΔT, as it may apply) is given by:

$$\delta S = \frac{\delta Q}{T_1} \tag{2.64}$$

where T_1 is the absolute temperature of an element volume of the body.

Since U_0 is a function of absolute expansion and temperature, using Eqs. (2.63) and (2.64) δ*S* can be obtained:

$$\begin{aligned}\delta S = &\frac{1}{T_1}\left[\frac{\partial U_0}{\partial T_1}\right]_\varepsilon \delta T_1 + \frac{1}{T_1}\left[\left(\frac{\partial U_0}{\partial \varepsilon_x}\right) - \sigma_x\right]\delta\varepsilon_x \\ &+ \frac{1}{T_1}\left[\left(\frac{\partial U_0}{\partial \varepsilon_y}\right) - \sigma_y\right]\delta\varepsilon_y + \frac{1}{T_1}\left[\left(\frac{\partial U_0}{\partial \varepsilon_z}\right) - \sigma_z\right]\delta\varepsilon_z \\ &+ \frac{1}{T_1}\left[\left(\frac{\partial U_0}{\partial \gamma_{xy}}\right) - \tau_{xy}\right]\delta\gamma_{xy} + \frac{1}{T_1}\left[\left(\frac{\partial U_0}{\partial \gamma_{xz}}\right) - \tau_{xz}\right]\delta\gamma_{xz} \\ &+ \frac{1}{T_1}\left[\left(\frac{\partial U_0}{\partial \gamma_{yz}}\right) - \tau_{yz}\right]\delta\gamma_{yz}\end{aligned} \quad (2.65)$$

The second law of thermodynamics implies that δS is a total differential of T_1 and strain; hence:

$$\begin{aligned}\frac{1}{T_1}\left[\frac{\partial U_0}{\partial \varepsilon_x}\right]_T &= \frac{\sigma_x}{T_1} - \left[\frac{\partial \sigma_x}{\partial T_1}\right]_\varepsilon \\ \frac{1}{T_1}\left[\frac{\partial U_0}{\partial \varepsilon_y}\right]_T &= \frac{\sigma_y}{T_1} - \left[\frac{\partial \sigma_y}{\partial T_1}\right]_\varepsilon \\ \frac{1}{T_1}\left[\frac{\partial U_0}{\partial \varepsilon_z}\right]_T &= \frac{\sigma_z}{T_1} - \left[\frac{\partial \sigma_z}{\partial T_1}\right]_\varepsilon\end{aligned} \quad (2.66)$$

$$\begin{aligned}\frac{1}{T_1}\left[\frac{\partial U_0}{\partial \gamma_{xy}}\right]_T &= \frac{\tau_{xy}}{T_1} - \left[\frac{\partial \tau_{xy}}{\partial T_1}\right]_\varepsilon \\ \frac{1}{T_1}\left[\frac{\partial U_0}{\partial \gamma_{xz}}\right]_T &= \frac{\tau_{xz}}{T_1} - \left[\frac{\partial \tau_{xz}}{\partial T_1}\right]_\varepsilon \\ \frac{1}{T_1}\left[\frac{\partial U_0}{\partial \gamma_{yz}}\right]_T &= \frac{\tau_{yz}}{T_1} - \left[\frac{\partial \tau_{yz}}{\partial T_1}\right]\end{aligned} \quad (2.67)$$

Define

$$C_\varepsilon = \left[\frac{\partial U_0}{\partial T_1}\right] \quad (2.68)$$

where C_ε is the heat capacity per unit volume with zero strain.

Then, from Eqs. (2.32) and (2.33) there is obtained:

$$\begin{aligned}\delta S = &C_\varepsilon \frac{\delta T}{T_1} - \left[\frac{\partial \sigma_x}{T_1}\right]\delta\varepsilon_x - \left[\frac{\partial \sigma_y}{T_1}\right]\delta\varepsilon_y - \left[\frac{\partial \sigma_z}{T_1}\right]\delta\varepsilon_z \\ &- \left[\frac{\partial \tau_{xy}}{T_1}\right]\delta\gamma_{xy} - \left[\frac{\partial \tau_{xz}}{T_1}\right]\delta\gamma_{xz} - \left[\frac{\partial \tau_{yz}}{T_1}\right]\delta\gamma_{yz}\end{aligned} \quad (2.69)$$

Combining Eqs. (2.65) and (2.69) leads to the expression for the heat supplied to the volume element within the time interval δt, i.e.:

$$\delta Q = C_\varepsilon T_1 - T_1\left[\frac{\partial\sigma_x}{T_1}\right]_\varepsilon \delta\varepsilon_x + \left[\frac{\partial\sigma_y}{T_1}\right]_\varepsilon \delta\varepsilon_y + \left[\frac{\partial\sigma_z}{T_1}\right]_\varepsilon \delta\varepsilon_z + \left[\frac{\partial\tau_{xy}}{T_1}\right]_\varepsilon \delta\gamma_{xy} + \left[\frac{\partial\tau_{xz}}{T_1}\right]_\varepsilon \delta\gamma_{xz} + \left[\frac{\partial\tau_{yz}}{T_1}\right]_\varepsilon \delta\gamma_{yz} \quad (2.70)$$

If the process is adiabatic, it follows from Eq. (2.30) that:

$$\delta U_0 = \sigma_x\delta\varepsilon_x + \sigma_y\delta\varepsilon_y + \sigma_z\delta\varepsilon_z + \tau_{xy}\delta\gamma_{xy} + \tau_{xz}\delta\gamma_{xz} + \tau_{yz}\delta\gamma_{yz} \quad (2.71)$$

The expression on the right-hand side is an exact differential; therefore the coefficient of each incremental term is an exact derivative of the function U_0:

$$\frac{\partial U_0}{\partial\varepsilon_x} = \sigma_x;\ \frac{\partial U_0}{\partial\varepsilon_y} = \sigma_y;\ \frac{\partial U_0}{\partial\varepsilon_z} = \sigma_z$$
$$\frac{\partial U_0}{\partial\gamma_{xy}} = \tau_{xy};\ \frac{\partial U_0}{\partial\gamma_{xz}} = \tau_{xz};\ \frac{\partial U_0}{\partial\gamma_{yz}} = \tau_{yz} \quad (2.72)$$

For isothermal expansion, $\delta T1$ in Eq. (2.36) = 0, and this result can be substituted in Eq. (2.30) to yield:

$$\delta U_0 + T_1\left\{\left(\frac{\partial\sigma_x}{T_1}\right)_\varepsilon \delta\varepsilon_x + \left(\frac{\partial\sigma_y}{T_1}\right)_\varepsilon \delta\varepsilon_y + \left(\frac{\partial\sigma_z}{T_1}\right)_\varepsilon \delta\varepsilon_z + \left(\frac{\partial\tau_{xy}}{T_1}\right)_\varepsilon \delta\gamma_{xy} + \left(\frac{\partial\tau_{xz}}{T_1}\right)_\varepsilon \delta\gamma_{xz} + \left(\frac{\partial\tau_{yz}}{T_1}\right)_\varepsilon \delta\gamma_{yz}\right\} = \sigma_x\delta\varepsilon_x + \sigma_y\delta\varepsilon_y + \sigma_z\delta\varepsilon_z + \tau_{xy}\delta\gamma_{xy} + \tau_{xz}\delta\gamma_{xz} + \tau_{yz}\delta\gamma_{yz} \quad (2.73)$$

or:

$$\delta F_0 \equiv \delta U_0 - T_1 dS = \sigma_x\delta\varepsilon_x + \sigma_y\delta\varepsilon_y + \sigma_z\delta\varepsilon_z + \tau_{xy}\delta\gamma_{xy} + \tau_{xz}\delta\gamma_{xz} + \tau_{yz}\delta\gamma_{yz} \quad (2.74)$$

where $F_0 = U_0 - T_1 S$ is the free energy per unit volume

$$\frac{\partial F_0}{\partial\varepsilon_x} = \sigma_x;\ \frac{\partial F_0}{\partial\varepsilon_y} = \sigma_y;\ \frac{\partial F_0}{\partial\varepsilon_z} = \sigma_z$$
$$\frac{\partial F_0}{\partial\gamma_{xy}} = \tau_{xy};\ \frac{\partial F_0}{\partial\gamma_{xz}} = \tau_{xz};\ \frac{\partial F_0}{\partial\gamma_{yz}} = \tau_{yz} \quad (2.75)$$

$$F_0 = \frac{1}{2}\left\{\lambda(\varepsilon_x+\varepsilon_y+\varepsilon_z)^2 + 2G\left(\varepsilon_x^2+\varepsilon_y^2+\varepsilon_z^2\right) + G\left(\gamma_{xy}^2+\gamma_{xz}^2+\gamma_{yz}^2\right) - \frac{2\alpha\varepsilon\Delta T}{1-2\upsilon}(\varepsilon_x+\varepsilon_y+\varepsilon_z)\right\} \quad (2.76)$$

2.3 Concepts from Strength of Materials

Elastic Axis

Locus of shear centers of the cross sections of a beam is called the elastic axis of that beam

Dynamic → aerodynamic, elastic and inertia forces
Static → no inertia forces.

From the theory of elasticity:

$$\frac{1}{R} = \frac{M}{EI}$$

where

$\frac{1}{R}$ change in curvature of the beam
E Young's modulus/Modulus of elasticity.
I Moment of inertia.
M Bending moment.

$$\sigma = \frac{My}{I}$$

where σ—bending stress
y—distance from neutral plane/axis (Figs. 2.2 and 2.3).

' per unit length.
L' $L' = \frac{1}{2}\rho V^2 C_L c = qcC_{L\alpha}(\theta + \alpha)$
or $L' = qc(C_{L\theta} + C_{L\alpha})$.

Shear Center

Point in the plane of the cross section through which the resultant shear force in the cross section must act to produce only bending, with no accompanying twist.

In many instances, to simplify and judging from the contex, the apostrophe sign is dropped for two-dimensional case (typical section).

Assume that h is the original length of spring, representing the bending elasticity of the wing (represented by the typical section). The weight of the wing is usually neglected.

Hence:

$$L' - W' = K_h \Delta h \tag{2.77}$$

Also

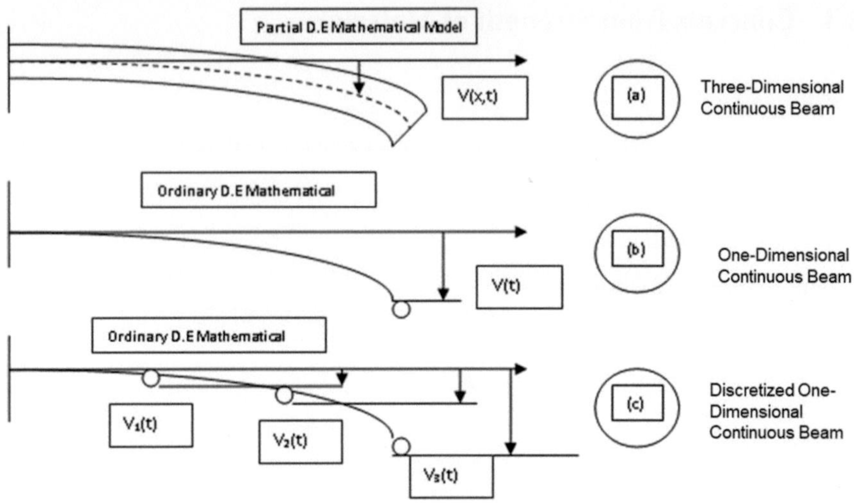

Fig. 2.2 Various beam physical and mathematical modeling

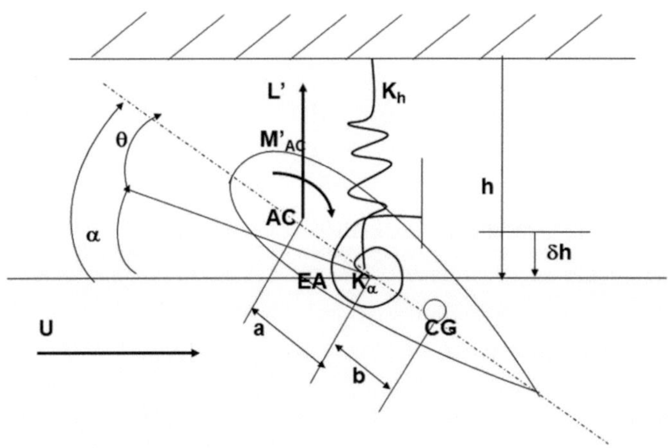

Fig. 2.3 Shear center

$$M'_{AC} = \frac{1}{2}\rho V^2 C_{MAC} c^2 \tag{2.78}$$

At the aerodynamic center AC, it is tacitly assumed that α, the twisting angle, has a positive for nose-up moment.

2.4 Fundamentals of Elasticity

$$q = \tau t \tag{2.79}$$

τt shear flow-shearing force per unit circumference
τ shearing stress (Fig. 2.4).

Equilibrium equation:

$$\left(\sigma_z + \frac{\partial \sigma_z}{\partial z}\mathrm{d}z\right)t\mathrm{d}s - \sigma_z t\mathrm{d}s + \left(\tau + \frac{\partial \tau}{\partial s}\mathrm{d}s\right)t\mathrm{d}z - \tau t\mathrm{d}z = 0 \tag{2.80}$$

Taking the limit $\mathrm{d}z \to 0$ then

$$\frac{\partial \sigma_z}{\partial z} + \frac{\partial \tau}{\partial s} = 0 \tag{2.81}$$

or multiplying by thickness t, if t is constant:

$$t\frac{\partial \sigma_z}{\partial z} + t\frac{\partial \tau}{\partial s} = t\frac{\partial \sigma_z}{\partial z} + \frac{\partial q}{\partial s} = 0 \tag{2.82}$$

where

$$q = \tau t \tag{2.83}$$

Example:

(a) Flat shear web

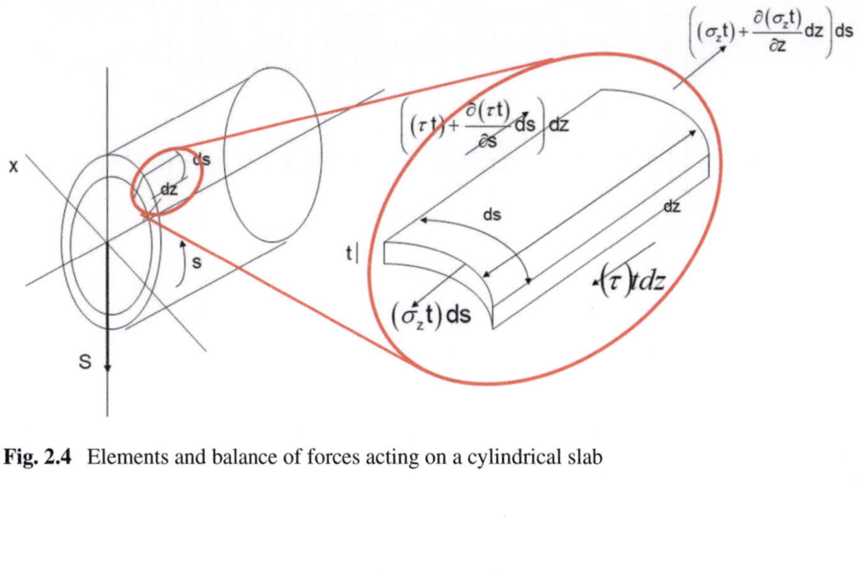

Fig. 2.4 Elements and balance of forces acting on a cylindrical slab

$$\tau t = q = \text{constant} \tag{2.84}$$

$$q = \frac{S}{h} \tag{2.85}$$

Pure shear

$$\sigma_z = 0$$

(b) Curved shear web.

See Figs. 2.5 and 2.6.

$$S = \int q \cos\theta ds \tag{2.86}$$

$$S = \int q dy = qh \tag{2.87}$$

$$q = \frac{S}{h} \tag{2.88}$$

where is the location of the resultant shear S?

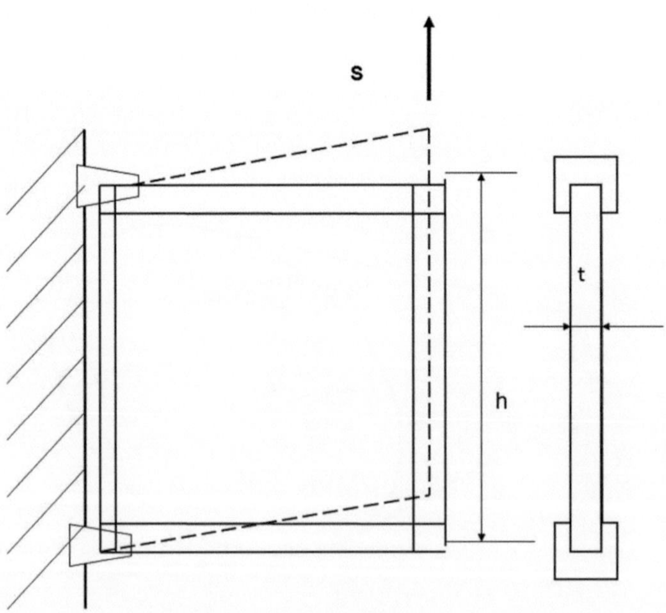

Fig. 2.5 Rectangular beam element subject to shear

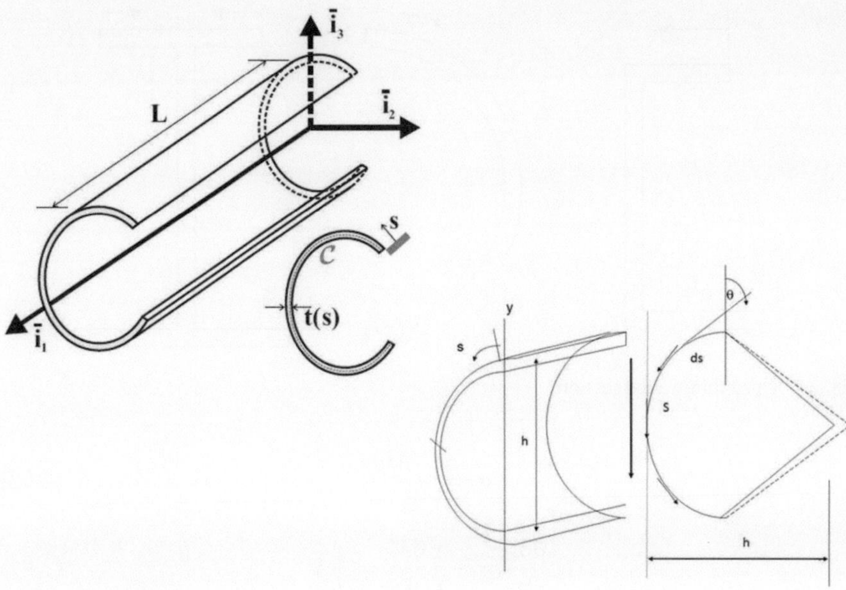

Fig. 2.6 Curved shear web

Equilibrium of moment:

$$M_o = \int qrds \tag{2.89}$$

$q =$ constant

$$M_o = q \int rds = qA \tag{2.90}$$

Both their moment should be equal too.
S times moment area ξ.
Hence

$$M_O = S\xi \tag{2.91}$$

$$\begin{aligned} \xi &= M_O/S - 2A/h \\ &= 2qA/S \end{aligned} \tag{2.92}$$

(c) Open thin-walled section.

See Fig. 2.7.

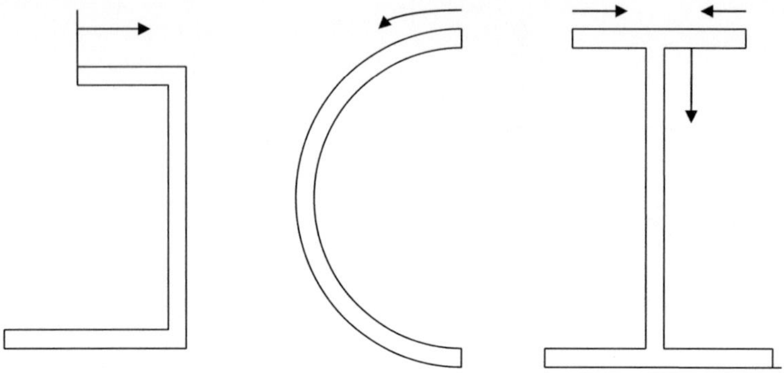

Fig. 2.7 Open thin-walled section

$$\sigma_z = -\frac{My}{I} \tag{2.93}$$

$$\frac{\partial \sigma_z}{\partial z} = -\frac{y}{I}\frac{\mathrm{d}M}{\mathrm{d}z} = \frac{y}{I}S \tag{2.94}$$

If shearing stress is assumed to be uniformly distributed over the wall thickness, the shear that can be obtained using integration

$$q = q_0 - \int_s^s t\frac{\partial \sigma_z}{\partial z}\mathrm{d}s \tag{2.95}$$

where

$$q_0 - q \text{ at } s_0.$$

Using (2.94)

$$q = q_0 - \frac{S}{I}\int_s^s ty\mathrm{d}s \tag{2.96}$$

The shear flow is determined when q_0 is known. For open thin-walled sections, the average value of shear stress across the wall thickness vanishes at the ends of section → following St. Venant theory of torsion.

Hence if $s0$ is taken as one of the ends, then $q_0 = 0$ there (Fig. 2.8).

$$\sum q_i = 0 \quad i = 1, 2, 3 \tag{2.97}$$

$$\Delta q = -\frac{S}{I}y\Delta A \tag{2.98}$$

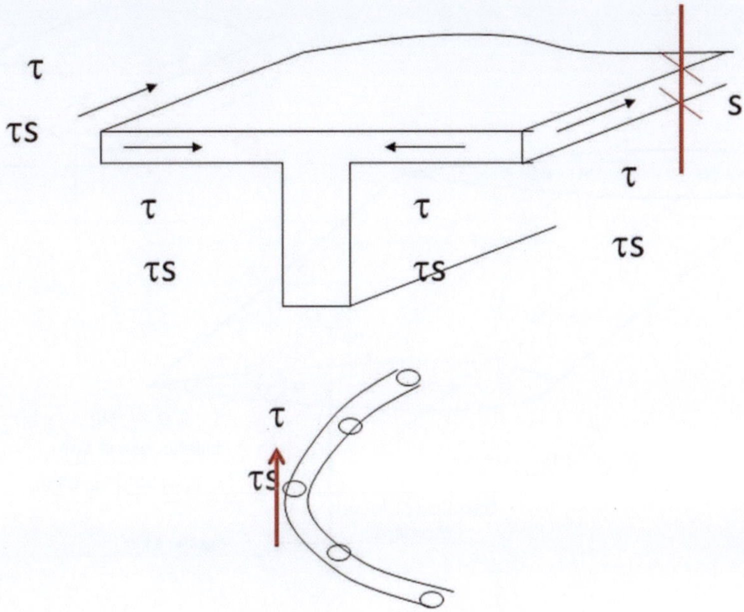

Fig. 2.8 Shear flow on an open thin-walled section

$$\xi = \frac{M_i}{S} = \frac{1}{S}\int qr\mathrm{d}s \tag{2.99}$$

Elastic Axis:

The locus of the shear centers of the cross sections of a beam is called the elastic axis of the beams (Fig. 2.9).

Flexural center, center of twist and flexural line are often used, to designate a point C or corresponding at which load can be applied without causing rotation of section ACB.

2.5 Elastic Properties of Structures

It will be assumed for convenience of analysis, that the aircraft structures considered are elastic. Certainly, such assumptions have limitations and consequences but will be an excellent start for further steps in problem solving efforts.

Consequently, for the loading situation illustrated below:

$$q = CQ \tag{2.100}$$

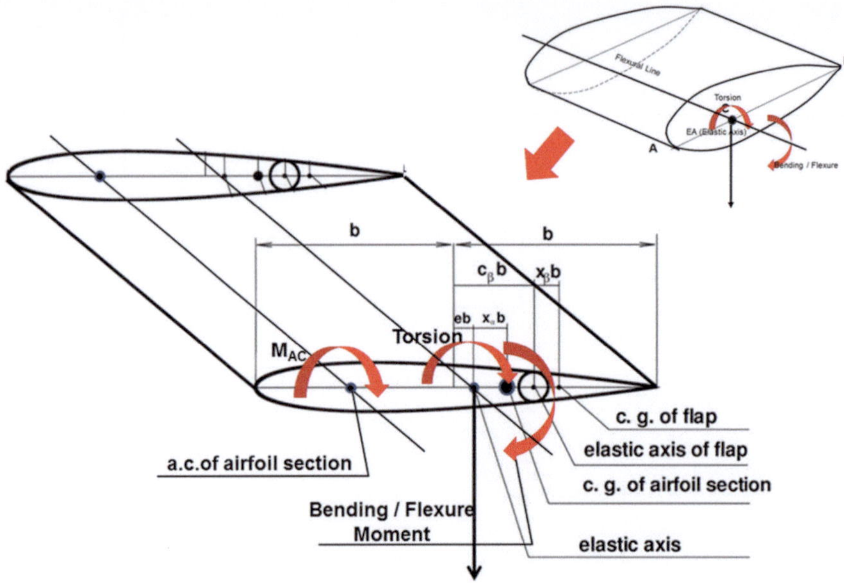

Fig. 2.9 Flexural center,[7] center of twist and flexural line

where q—deflection of the structure, and Q—applied force to the structure.

C—constant of proportionality (Fig. 2.10).

Work done by the external force during application is transformed completely into strain energy in the structure:

$$U = \frac{1}{2} q Q \tag{2.101}$$

Deformation at point i due to force at point j:

$$q_i = C_{ij} Q_j \tag{2.102}$$

Deformation at point i due to several forces at $j = 1, 2, \ldots, n$

$$q_i = \sum_{j=1}^{n} C_{ij} Q_j$$

$$j = 1, 2, \ldots, n \tag{2.103}$$

C_{ij} is called *flexibility influence coefficients.*

---

[7] With reference to a beam, the flexural center of any section is that point in the plane of the section through which a transverse load, applied at that section, must act if bending deflection only is to be produced, with no twist of the section. Also called shear center.

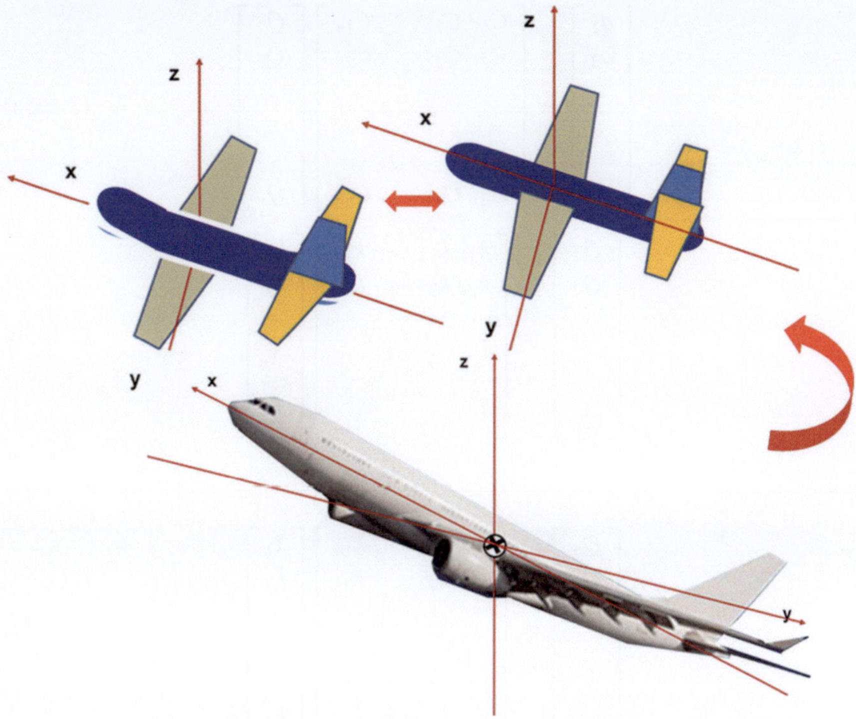

Fig. 2.10 Representation of an aircraft for basic aeroelastic analysis[8,9]

Conversely, the forces can be expressed as linear functions of the displacements by

$$Q_i = \sum_{j=1}^{n} k_{ij} q_j$$
$$j = 1, 2, \ldots, n \quad (2.104)$$

k_{ij} is called *stiffness influence coefficients.*

Equations (2.103) and (2.104) can be represented in matrix notation as

[8] Aircraft image Courtesy of stocksnap.

[9] https://stocksnap.io/search/image+of+widebody+aircraft+climbing; Shutter-stock.comm155701927.

$$\begin{bmatrix} q_1 \\ q_2 \\ \cdot \\ \cdot \\ \cdot \\ q_n \end{bmatrix} = \begin{bmatrix} C_{11} & C_{12} & \ldots & C_{1n} \\ C_{21} & C_{22} & \ldots & C_{2n} \\ \cdot & \cdot & \cdots & \cdot \\ \cdot & \cdot & \cdots & \cdot \\ \cdot & \cdot & \cdots & \cdot \\ C_{n1} & C_{n2} & \ldots & C_{nn} \end{bmatrix} \begin{bmatrix} Q_1 \\ Q_2 \\ \cdot \\ \cdot \\ \cdot \\ Q_n \end{bmatrix} \tag{2.105a}$$

$$\begin{bmatrix} Q_1 \\ Q_2 \\ \cdot \\ \cdot \\ \cdot \\ Q_n \end{bmatrix} = \begin{bmatrix} k_{11} & k_{12} & \cdots & k_{1n} \\ k_{21} & k_{22} & \cdots & k_{2n} \\ \cdot & \cdot & \cdots & \cdot \\ \cdot & \cdot & \cdots & \cdot \\ \cdot & \cdot & \cdots & \cdot \\ k_{n1} & k_{n2} & \cdots & k_{nn} \end{bmatrix} \begin{bmatrix} k_1 \\ k_2 \\ \cdot \\ \cdot \\ \cdot \\ k_n \end{bmatrix} \tag{2.105b}$$

and

$$\begin{bmatrix} Q_1 \\ Q_2 \\ \cdot \\ \cdot \\ \cdot \\ Q_n \end{bmatrix} = \begin{bmatrix} k_{11} & k_{12} & \cdots & k_{1n} \\ k_{21} & k_{22} & \cdots & k_{2n} \\ \cdot & \cdot & \cdots & \cdot \\ \cdot & \cdot & \cdots & \cdot \\ \cdot & \cdot & \cdots & \cdot \\ k_{n1} & k_{n2} & \cdots & k_{nn} \end{bmatrix} \begin{bmatrix} k_1 \\ k_2 \\ \cdot \\ \cdot \\ \cdot \\ k_n \end{bmatrix} \tag{2.105c}$$

In shorthand matrix notation:

$$\{\mathbf{q}\} = [\mathbf{C}]\{\mathbf{Q}\} \tag{2.106}$$

$$\{\mathbf{Q}\} = [\mathbf{k}]\{\mathbf{q}\} \tag{2.107}$$

is called the matrix of flexibility influence coefficient, while $[k]$ is called the matrix of stiffness influence coefficient and is related as:

$$[\mathbf{k}] = [\mathbf{C}]^{-1} \tag{2.108}$$

The process described by (2.108) is known as matrix inversion, and $[k]$ is said to be the reciprocal of $[C]$ (Figs. 2.11 and 2.12).

Bending deformation of beam:

$$M(y) = EIw''(y) \tag{2.109}$$

For a unit load at the tip, Eq. (2.11) becomes:

$$1 - y = EIw''(y) \tag{2.110}$$

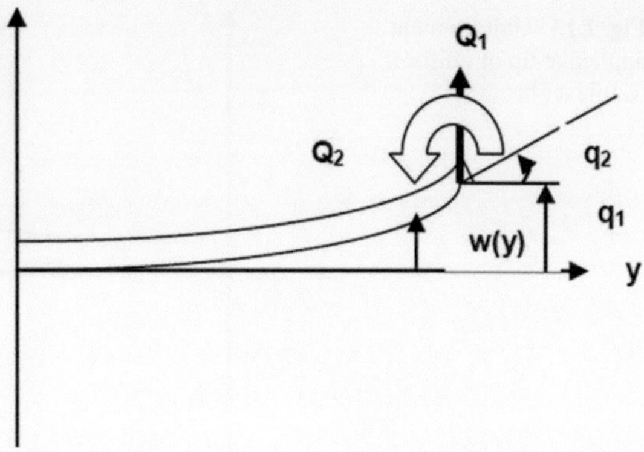

Fig. 2.11 .Force and moment applied at tip of uniform cantilever beam

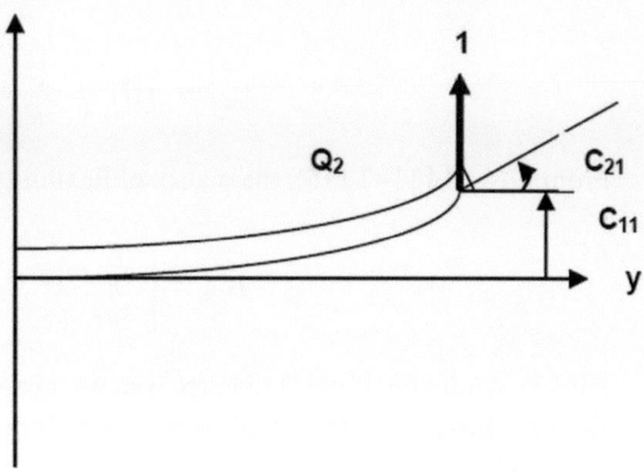

Fig. 2.12 Unit force applied at tip of uniform cantilever beam

Integrating and introducing the boundary condition $w(0) = w'(0) = 0$ at $y = 0$, obtain (Fig. 2.13)

$$C_{21} = w'(l) = \frac{1^2}{2EI} \tag{2.111}$$

$$C_{11} = w(l) = \frac{1^3}{3EI} \tag{2.112}$$

For a unit moment at the tip, Eq. (2.110) becomes:

$$1 = EIw''(y) \tag{2.113}$$

Integrating and introducing the boundary condition $w(0) = w'(0) = 0$ at $y = 0$, obtain

$$C_{22} = w'(l) = \frac{l}{EI} \tag{2.114}$$

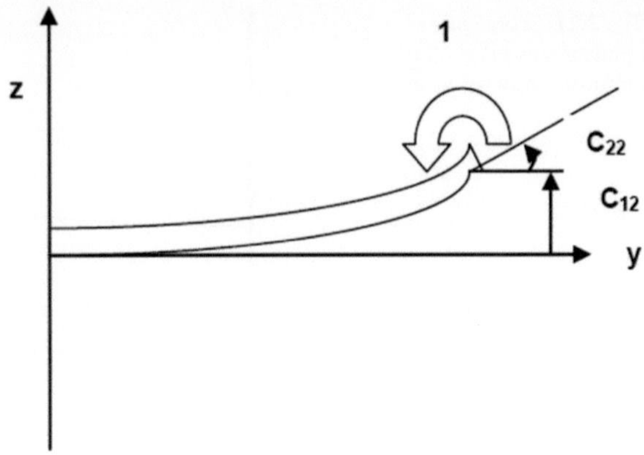

Fig. 2.13 Unit moment applied at tip of uniform cantilever beam

$$C_{12} = w(l) = \frac{l2}{2EI} \tag{2.115}$$

From Eqs. (2.111–2.115), the matrix of flexibility influence coefficient is

$$[C] = \begin{bmatrix} \frac{l^3}{3EI} & \frac{l^2}{2EI} \\ \frac{l^2}{2EI} & \frac{1}{EI} \end{bmatrix} \tag{2.116}$$

Direct calculation of the $[k]$ matrix is more involved.

Here we must solve the problem for computing k_{11} and k_{21} with the aid of the following Fig. 2.14.

The beam is given a unit displacement in the coordinate q_1, and zero displacement in q_2. The force Q_1 and the moment Q_2 required to hold the beam in this strained configuration are k_{11} and k_{21}, respectively.

Let $M(y) = k_{11}(l - y) + k_{21}$ and substitute it into (2.109), then

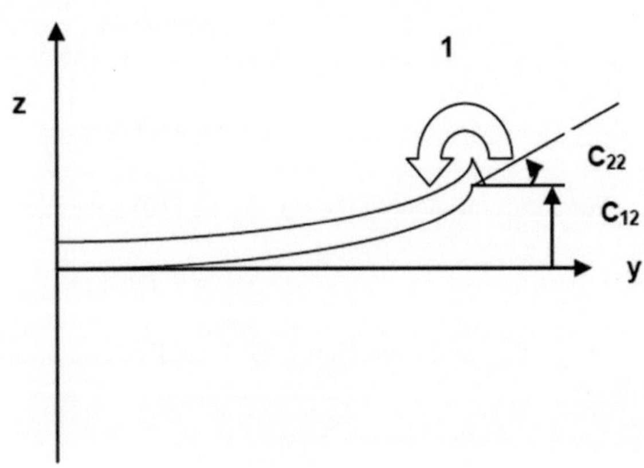

Fig. 2.14 Uniform cantilever beam with unit linear displacement at the tip

$$M(y) = EIw''(y) = k_{11}(1 - y) + k_{21} \tag{2.117}$$

Integrating and applying the boundary conditions:

$$w(0) = w'(0) = w'(l) = 0$$

and $w(l) = 1$, the following simultaneous equations are obtained

$$k_{11} + \frac{2}{1}k_{21} = 0 \tag{2.118a}$$

$$k_{11} + \frac{3}{2l}k_{21} = \frac{3EI}{l^3} \tag{2.118b}$$

It will give:

$$k_{11} = \frac{12EI}{l^3} \tag{2.119}$$

$$k_{21} = -\frac{6EI}{l^2} \tag{2.120}$$

To compute k_{22} and k_{12}, the following Fig. 2.15 should be used. Equation (2.117) for this case becomes:

$$M(y) = EIw''(y) = k_{12}(1 - y) + k_{22} \tag{2.121}$$

Integrating and applying the boundary conditions:

$$w(0) = w'(0) = w'(l) = 0$$

and $w'(l) = 1$, the following simultaneous equations are obtained

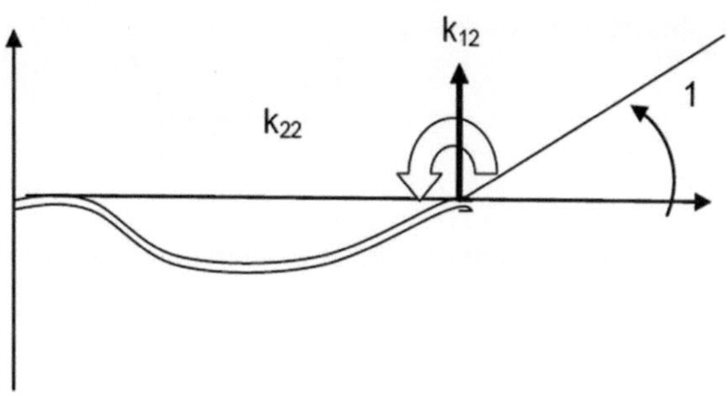

Fig. 2.15 .Uniform cantilever beam with unit linear displacement at the tip

$$k_{22} + \frac{1}{2}k_{12} = \frac{EI}{1} \tag{2.122}$$

$$k_{22} + \frac{21}{3}k_{12} = 0 \tag{2.123}$$

It will give:

$$k_{12} = \frac{6EI}{1} \tag{2.124}$$

$$k_{21} = \frac{4EI}{1} \tag{2.125}$$

Hence the matrix of stiffness influence coefficient becomes

$$[\mathbf{k}] = \begin{bmatrix} \frac{12EI}{1} & \frac{-6EI}{1^2} \\ \frac{-6EI}{1^2} & \frac{4EI}{1} \end{bmatrix} \tag{2.126}$$

It can be verified that the matrices (2.116) and (2.126) are reciprocal, which is left for exercise.

2.6 Strain Energy in Terms of Influence Coefficients

The total strain energy of a structure comprising n elements, using the principle of superposition due to its linear nature, can be expressed as, similar to (2.2):

$$U = \frac{1}{2}\sum_{i=1}^{n} q_i Q_i \tag{2.127}$$

Introducing stress–strain relationship (2.103) into (2.127), obtain

$$U = \frac{1}{2}\sum_{i=1}^{n}\sum_{j=1}^{n} C_{ij} Q_j Q_i \tag{2.128}$$

Similarly, by introducing stress–strain relationship (2.4) into (2.127), obtain

$$U = \frac{1}{2}\sum_{i=1}^{n}\sum_{j=1}^{n} k_{ij} q_i q_j \tag{2.129}$$

In matrix notation, these can be written as:

$$U = \frac{1}{2}\begin{bmatrix} Q_1 \; Q_2 \cdots Q_n \end{bmatrix}\begin{bmatrix} C_{11} & C_{12} & \cdots & C_{1n} \\ C_{21} & C_{22} & \cdots & C_{2n} \\ \cdot & \cdot & \cdots & \cdot \\ \cdot & \cdot & \cdots & \cdot \\ \cdot & \cdot & \cdots & \cdot \\ C_{n1} & C_{n2} & \cdots & C_{nn} \end{bmatrix}\begin{bmatrix} Q_1 \\ Q_2 \\ \cdot \\ \cdot \\ \cdot \\ Q_n \end{bmatrix} \tag{2.129}$$

and

$$U = \frac{1}{2}\begin{bmatrix} q_1 \; q_2 \ldots q_n \end{bmatrix}\begin{bmatrix} k_{11} & k_{12} & \ldots & k_{1n} \\ k_{21} & k_{22} & \ldots & k_{2n} \\ \cdot & \cdot & \cdots & \cdot \\ \cdot & \cdot & \cdots & \cdot \\ \cdot & \cdot & \cdots & \cdot \\ k_{n1} & k_{n2} & \ldots & k_{nn} \end{bmatrix}\begin{bmatrix} q_1 \\ q_2 \\ \cdot \\ \cdot \\ \cdot \\ q_n \end{bmatrix} \tag{2.130}$$

or in shorthand matrix notation:

$$U = \frac{1}{2}\{Q\}^T[C]\{Q\} \tag{2.131}$$

$$U = \frac{1}{2}\{q\}^T[k]\{q\} \tag{2.132}$$

2.7 Deformation Under Distributed Forces and Influence Functions

Consider, by way of illustration, a cantilever beam under distributed side load, $Z(y)$. The infinitesimal deflection $dw(y)$ at a point y, due to an infinitesimal element of the load $Z(\eta)d\,\eta$ at point η, can be expressed by

$$dw(y) = C(y, \eta)Z(\eta)d\eta \tag{2.133}$$

where $C(y, \eta)$ is a function giving the deflection at the point y due to a unit load at point η. The deflection at y due to the entre side load on the beam is obtained by integrating (2.36) along the length of the beam (here approximated by $y = l$):

$$w(y) = \int_0^l C(y, \eta)Z(\eta)d\eta \tag{2.133}$$

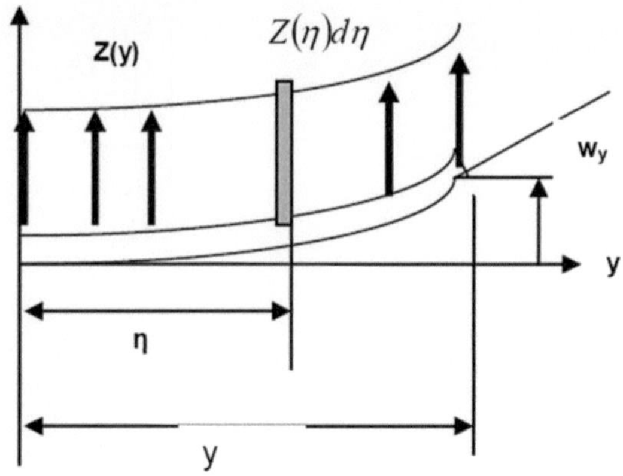

Fig. 2.16 Uniform cantilever beam under distributed side load

$C(y, \eta)$ is called the flexibility influence function or Green's function in one dimension, and it is a symmetrical function (Fig. 2.16):

$$C(y, \eta) = C(\eta, y) \tag{2.134}$$

If shear deformation is neglected, the influence function for the bending of a cantilever beam can be computed using Eq. (2.10). The bending moment on a cantilever beam due to a unit load applied at $y = \eta$ is given by:

$$M(y, \eta) = (\eta - y) \quad \text{for } y \leq \eta \tag{2.135}$$

$$M(y, \eta) = 0 \quad \text{for } y \geq \eta \tag{2.136}$$

Introducing (2.235) into (2.117):

$$EIC''(y, \mu) = (\eta - y) \quad \text{for } y \leq \eta \tag{2.137}$$

Integrating and introducing the boundary condition

$$C(0, \eta) = C'(0, \eta) = 0 \tag{2.138}$$

obtain the influence function of a uniform beam for this range as

$$C(y, \mu) = \frac{y^2}{6EI}(3\eta - y) \quad \text{for } y \leq \eta \tag{2.139}$$

Introducing (2.138) into (2.10):

$$EIC''(y, \mu) = (\eta - y) \quad for\ y \geq \eta \tag{2.140}$$

Integrating and evaluating the constants of integration by putting $y = \eta$ in Eq. (2.43), with the resultant requirements that:

$$C(\eta, \mu) = \frac{y^2}{3EI} \tag{2.141a}$$

$$C'(y, \mu) = \frac{2y}{6EI}(3\eta - y) - \frac{y^2}{6EI} = \frac{4\eta^2 - \eta^2}{6EI} = \frac{\eta^2}{2EI} \tag{2.141b}$$

obtain the influence function of a uniform beam for this range as

$$C(y, \mu) = \frac{y^2}{6EI}(3y - \eta) \quad \text{for } y \geq \eta \tag{2.142}$$

Notice that if y and η are interchanged either in (2.139) or (2.142), the other equation is obtained, thus verifying the reciprocal relationship (2.108) or (2.116) and (2.126).

Introducing Eqs. (2.139) and (2.142) into (2.133), obtain.

$$w(y) = \int_0^y \frac{\eta^2}{6EI}(3y - \eta)Z(\eta)d\eta + \int_y^l \frac{y^2}{6EI}(3\eta - y)Z(\eta)d\eta \tag{2.143}$$

As an example, take $Z(\eta)$ to be constant, i.e. $Z(\eta) = Z_0$.
Hence:

$$w(y) = \frac{Z_0 y^2}{6EI}\left(y^2 - 4yl + 6l^2\right) \tag{2.144}$$

2.8 Properties of Influence Functions

In general:

$$C(x, y; \xi, \eta) = C(\xi, \eta; x, y) \tag{2.145}$$

Equation (2.37)

$$w(y) = \int_0^l C(y, \eta)Z(\eta)\mathrm{d}\eta \tag{2.146}$$

Is an integral equation of the first kind, which expresses the deflection of the beam w in terms of the influence function $C(y, \eta)$ and the loading $Z(\eta)$. If the equation is considered to relate the deflection of the beam was known while the loading $Z(\eta)$ as unknown, it follows from the theory of integral equations that.

$$Z(y) = \int_0^l k(y, \eta)w(\eta)\mathrm{d}\eta \tag{2.147}$$

where $k(y, \eta)$ can be regarded as a hypothetical stiffness influence function. The process of inverting (2.37)–(2.51) is formidable, but in practical problems, this process can be carried out numerically and is much more tractable.

The strain energy of a continuously loaded one-dimensional system, analogous to (2.29)–(2.31), can be expressed as

$$U = \frac{1}{2}\int_0^l Z(y)w(y)\mathrm{d}y \tag{2.148}$$

Substituting (2.146), obtain the strain energy in terms of the flexibility influence function and the distributed load:

$$U = \frac{1}{2}\int_0^l Z(y)\int_0^l C(y, \eta)Z(\eta)\mathrm{d}\eta\mathrm{d}y \tag{2.149}$$

Alternatively, if (2.147) is substituted into (2.149), the strain energy can be expressed as

$$U = \frac{1}{2}\int_0^l w(y)\int_0^l k(y, \eta)w(\eta)\mathrm{d}\eta\mathrm{d}y \tag{2.150}$$

Note:

Properties of Influence Coefficients

In general:

$$C_{ij} = C_{ji} \tag{2.151}$$

which should be verified as an exercise.

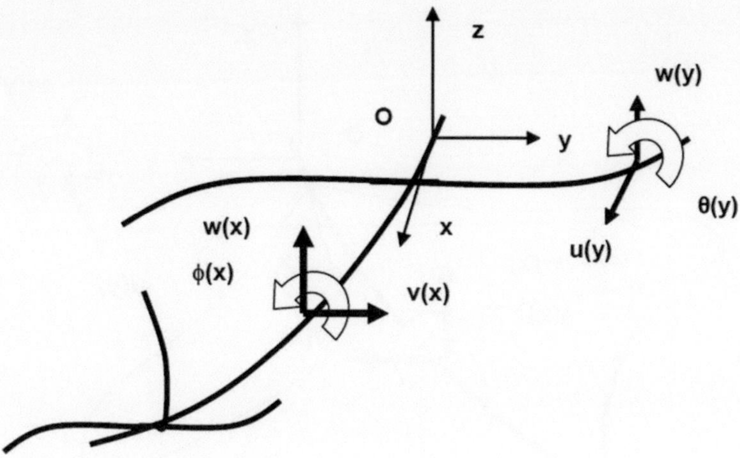

Fig. 2.17 Simplified elastic aircraft

2.9 The Simplified Elastic Airplane

Due to the complexity of aircraft structures, for purposes of analysis it would be very beneficial and instructive to introduce simplifying assumptions in order to calculate their elastic properties. In many cases, the wings, the fuselage and the horizontal and vertical tails can be regarded as beams rigid in cross sections perpendicular to their lengthwise direction. Then the aircraft becomes a collection of beams which can be represented by their intersecting elastic lines as depicted in the following Fig. 2.17.

For example, the displacement of a rigid cross section of the wing at a point y is described by the linear displacements $w(y)$ and $u(y)$, and by an angular displacement $\theta(y)$.

For very small aspect ratio wing, it may not be possible to assume rigid cross section, and chordwise deformation should be made possible. Then the wing can be represented by an elastic flat thin plate, as depicted in Fig. 2.18.

2.10 Deformations of Airplane Wings

Consider the deformation of an elastic wing under the influence of a distributed normal load $Z(x, y)$. The deformation can be expressed by a single function $w(x, y)$, defined by expanding expression (2.37) for two-dimensional loading, as follows:

$$w(x, y) = \iint_S C(x, y; \xi, \eta) Z(\xi, \eta) \mathrm{d}\xi \mathrm{d}\eta \qquad (2.152)$$

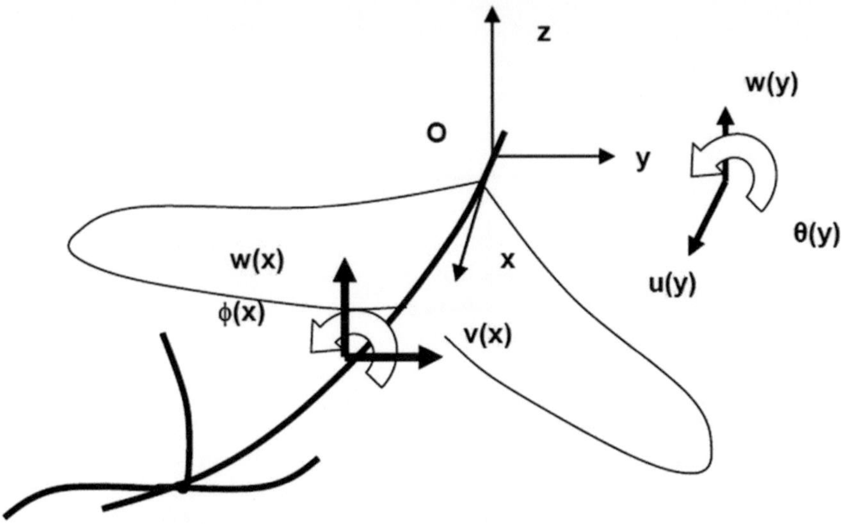

Fig. 2.18 Simplified elastic aircraft with low aspect ratio wing

If the wing is sufficiently slender so that the chordwise segments of the wing parallel to the x-axis can be assumed to be rigid, then the influence function can be written as:

$$C(x, y; \xi, \eta) = C^{xx}(y, \eta) - xC^{\theta z}(y, \eta) + \xi x C^{\theta\theta}(y, \eta) - \xi C^{z\theta}(y, \eta) \quad (2.153)$$

where the influence functions are defined by

$C^{pq}(y, \eta)$—linear or angular deflection in the p-direction at location y due to a unit force or torque in the q-direction at location η.

Example: $C^{z\theta}(y, \eta)$—linear deflection in the z-direction at location y due to a unit torque in the θ-direction at location η.

The deflection can be expressed in the form:

$$w(x, y) = w(y) - x\theta(y) \quad (2.154)$$

Substituting Eqs. (2.153) and (2.154) into Eq. (2.152), obtain.

$$w(y) = \int_0^l C^{zz}(y, \eta)Z(\eta)\mathrm{d}\eta + \int_0^l C^{z\theta}(y, \eta)t(\eta)\mathrm{d}\eta \quad (2.155)$$

$$\theta(y) = \int_0^l C^{\theta z}(y, \eta)Z(\eta)\mathrm{d}\eta + \int_0^l C^{\theta\theta}(y, \eta)t(\eta)\mathrm{d}\eta \quad (2.156)$$

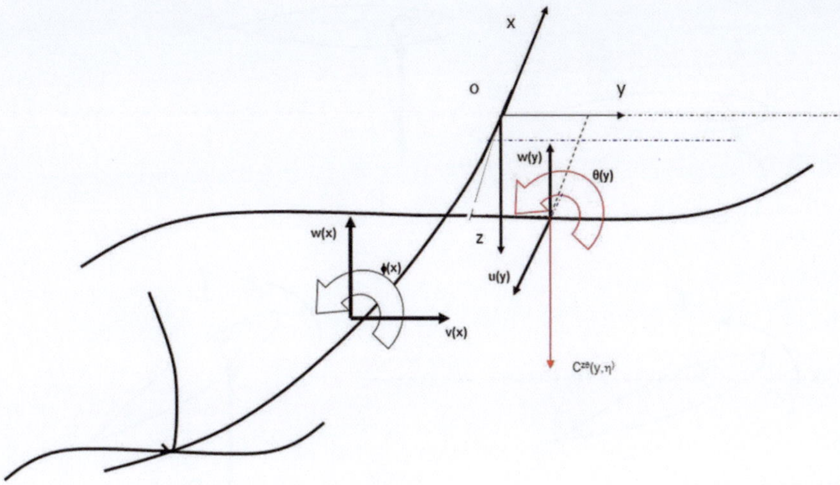

Fig. 2.19 Forces and moments at a simplified elastic aircraft

where?

$$Z(y) = \int_{chord} Z(\xi, \eta)\mathrm{d}\eta \quad (2.157)$$

$$t(y) = -\int_{chord} \xi Z(\xi, \eta)\mathrm{d}\eta \quad (2.158)$$

which is the running streamwise torque.[10]

A further simplification in Eqs. (2.157) and (2.158) can be made for unswept wings which are uniform, or which have a gradual change in section along the spanwise axis. Such wings can be approximately analyzed to allow the bending and torsion to be separable, so that there is no coupling between them. The concept of elastic axis has to be introduced. Then it is convenient to resolve the forces and moments on the wing as a beam into forces along and torques about the elastic axis, as illustrated in Figs. 2.19 and 2.20.

In this case, Eqs. (2.155) and (2.156) reduce to

$$\delta S = \frac{\delta Q}{T_1} w(y) = \int_0^l C^{zz}(y, \eta) Z(\eta)\mathrm{d}\eta \quad (2.159)$$

$$\delta S = \frac{\delta Q}{T_1} \theta(y) = \int_0^l C^{\theta\theta}(y, \eta) t(\eta)\mathrm{d}\eta \quad (2.160)$$

[10] The derivation of Eqs. (2.154–2.156) is elaborated in Appendix 2.A.

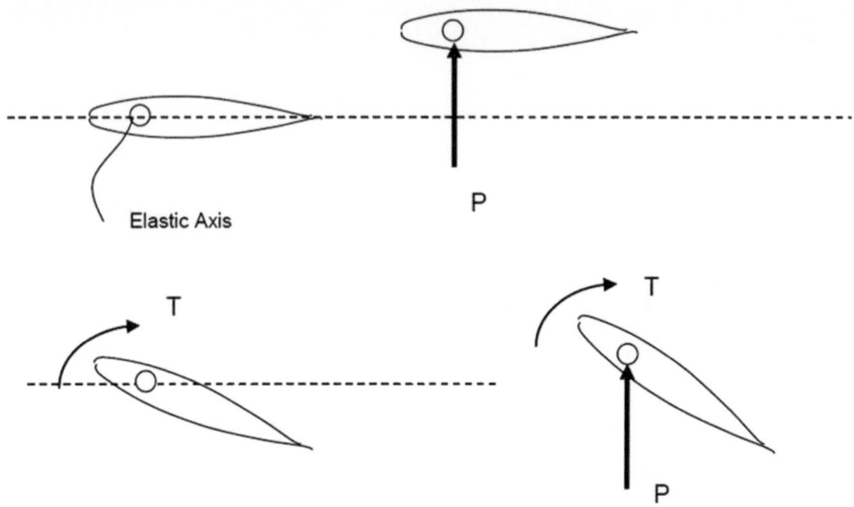

Fig. 2.20 Motion of typical section due to aerodynamic excitation

2.11 Energy Methods in Deflection Calculations

Energy methods are very useful in the determination of deformation under static and dynamic loads and in the calculation of influence functions and coefficients.

2.11.1 *Deflections Determination Using the Principle of Minimum Potential Energy*

The principle of minimum potential energy for the calculation of the displacements of conservative elastic system is based on the principle of virtual work.

The principle of virtual work, applied to deformable bodies, can be stated as follows:

> If a body is in equilibrium under the action of prescribed external forces, the virtual work done by these forces along a small displacement compatible with the geometric constraints (i.e. virtual displacement is equal to the strain energy.

The principle of virtual work can be stated as:

$$\delta W_e = \delta U \tag{2.161a}$$

where δW_e is the virtual work done by the external forces, and δU is the change of strain energy resulting from a small virtual displacement of the body.

The principle of minimum potential energy, applicable only to conservative systems, can be derived by transposing 2.161a:

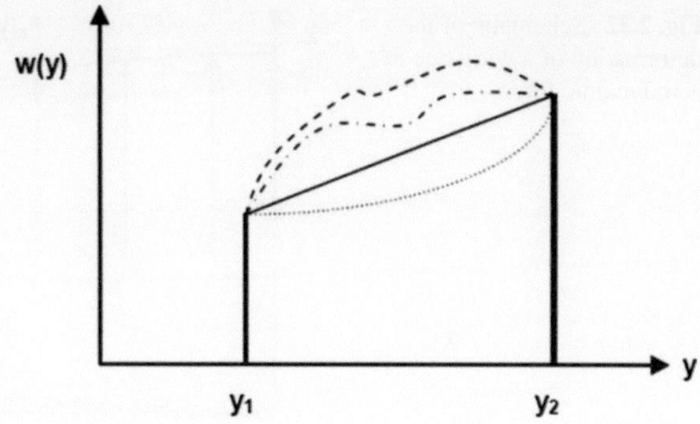

Fig. 2.21 Possible deformation configurations compatible with the geometric constraints

$$\delta W_e - \delta U = \delta(U - W_e) = 0 \tag{2.161b}$$

As expressed by Eq. 2.161b, the principle of minimum potential energy can be stated as (Fig. 2.21).

:

Among all possible deformation configurations compatible with the geometric constraints, the configuration which satisfies the equations of equilibrium is the one which minimizes the potential energy, $U - W_e$.

where $\delta(U - W_e) = 0$ implies minimization of the potential energy.

There are two principal uses of the principle of minimum potential energy in the deflection analysis:

1. The derivation of exact differential, integral or algebraic equations of equilibrium.
2. Approximate solutions to deflection problems of continuous systems which are difficult to solve by exact methods.

2.11.2 The Principle of Minimum Potential Energy Applied to Continuous Systems; Rayleigh–Ritz Method

Consider the application of the principle of minimum potential energy for determining the deflection of a wing idealized as a cantilever beam, under distributed load (vertical, along the z-axis, as depicted in the following Fig. 2.22.

Assume that the lateral deflection of the beam can be written as the sum of *independent* deflection functions $\boldsymbol{\gamma}_i(y)$, each of which satisfies the boundary conditions:

$$w(y) = \sum_{i=1}^{n} \gamma_i(y) q_i \tag{2.162}$$

where the q_i are generalized coordinates whose values are to be determined. For a cantilever beam, the condition of geometrical constraint on the functions $\gamma_i(y)$ is

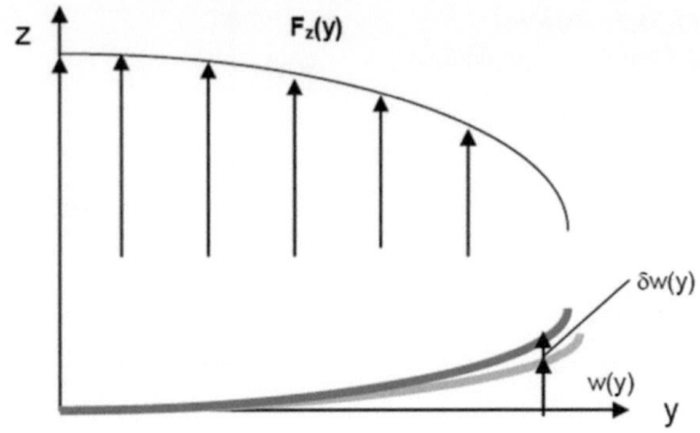

Fig. 2.22 Schematic of the deformation of a wing due to aerodynamic force

$$\gamma_i(0) = \gamma_i'(0) = 0 \tag{2.163}$$

The coordinates q^i are seen to fall within the definition of generalized coordinates, since they act as independent coordinates which represent possible displacements and which do not violate the conditions of geometric constraints.

Now let the beam be subject to an arbitrary virtual lateral displacement, shown also in the above figure, which has a magnitude:

$$\delta w(y) = \sum_{i=1}^{n} \gamma_i(y)\delta q_i \tag{2.164}$$

The corresponding work done by the external forces is given by:

$$\delta W_e = \int_0^l F_z(y)\delta w(y)\mathrm{d}y = \sum_{i=1}^{n} \left\{ \int_0^l F_z(y)\gamma_i(y)\mathrm{d}y \right\} \delta q_i \tag{2.165}$$

From strength of materials (or aircraft structures), the strain energy due to bending in a slender beam can be expressed in terms of lateral displacement by:

$$U = \frac{1}{2}\int_0^l EI\left(\frac{\mathrm{d}^2 w}{\mathrm{d}y^2}\right)^2 \mathrm{d}y \tag{2.166}$$

Substituting Eq. (2.62) into Eq. (2.66), there is obtained

$$U = \frac{1}{2}\sum_{i=1}^{n}\sum_{j=1}^{n} \left\{ \int_0^l EI\left(\frac{\mathrm{d}^2\gamma_i}{\mathrm{d}y^2}\right)\left(\frac{\mathrm{d}\gamma_j}{\mathrm{d}y^2}\right)\mathrm{d}y \right\} q_i q_j \tag{2.167}$$

The change in strain energy due to coordinate changes δ_i is then

$$\delta U = \frac{1}{2}\sum_{i=1}^{n}\sum_{j=1}^{n}\left\{\int_{0}^{l} EI\left(\frac{d^2\gamma_i}{dy^2}\right)\left(\frac{d\gamma_j}{dy^2}\right)dy\right\}q_j\delta q_i \tag{2.168}$$

Substituting Eqs. (2.165) and (2.168) into Eq. (2.161), principle of minimum potential energy, there is obtained

$$\sum_{i=1}^{n}\left[\sum_{j=1}^{n}\left\{\int_{0}^{l} EI\left(\frac{d^2\gamma_i}{dy^2}\right)\left(\frac{d^2\gamma_j}{dy^2}\right)dy\right\}q_j - \int_{0}^{l} F_z(y)\gamma_i(y)dy\right]\delta q_i = 0 \tag{2.169}$$

Since the δ_{qi} are independent arbitrary quantities, Eq. (2.69) can be satisfied only if the terms between square brackets are zero, i.e.

$$\left[\sum_{j=1}^{n}\left\{\int_{0}^{l} EI\left(\frac{d^2\gamma_i}{dy^2}\right)\left(\frac{d^2\gamma_j}{dy^2}\right)dy\right\}q_j - \int_{0}^{l} F_z(y)\gamma_i(y)dy\right] = 0$$
$$For\ i = 1, 2, 3, \ldots n \tag{2.170}$$

This is a set of simultaneous linear algebraic equations in the unknown generalized coordinates q_1, q_2, q_3, ..., q_n. The load distribution F_z is known, while γ_i, from structural dynamics, represent the deflection function compatible with the geometric constraints (Fig. 2.23).

Then in principle Eq. 2.170 can be solved for the unknown generalized coordinates $q_1, q_2, q_3, \ldots, q_n$. The final solution for the lateral deflection is obtained by substituting the values of generalized coordinates q_1, q_2, q_3, ..., q_n thus found into Eq. 2.162. If n is a finite quantity, this process yields an approximate solution for $w(y)$, the distribution of lateral deflection of the beam. The process of finding an approximate solution following this procedure is known as the Rayleigh–Ritz method.

Extending the procedure to a general three-dimensional body, let $F_x(x, y, z)$, $F_y(x, y, z)$, and $F_z(x, y, z)$, be the components of surface force $F_S(x, y, z)$ The work done

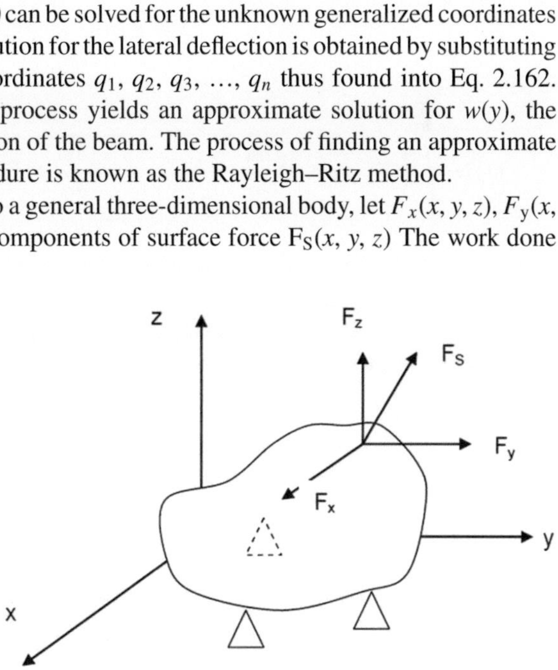

Fig. 2.23 General three-dimensional body free body diagram

by the surface forces during an arbitrary virtual displacement is then

$$\delta W_e = \int_0^l \big(F_x(x, y, z)\delta u(x, y, z) + F_y(x, y, z)\delta v + F_z(x, y, z)\delta w\big)\mathrm{d}S \quad (2.171)$$

The integration is over the surface.

Next let us assume that the displacement of the body can be expressed as a function of n discrete generalized coordinates q_i:

$$\begin{aligned} u &= u(x, y, z; q_1, q_2, \ldots, q_n) \\ v &= v(x, y, z; q_1, q_2, \ldots, q_n) \\ w &= w(x, y, z; q_1, q_2, \ldots, q_n) \end{aligned} \quad (2.172a,b,c)$$

In the beam application, (2.162) is an example of Eq. (2.172) for a specific case: Introduce (2.172) into (2.171) and obtain

$$\begin{aligned} \delta W_e &= \sum_{i=1}^{n} \left\{ \int_0^l \left(F_x(x, y, z)\frac{\partial u}{\partial q_i} + F_y(x, y, z)\frac{\partial v}{\partial q_i} + F_z(x, y, z)\frac{\partial w}{q_i} \right) S \right\} \delta q_i \\ &= \sum_{i=1}^{n} Q_i \delta q_i \end{aligned} \quad (2.173)$$

where

$$Q_i = \int_0^l \left(F_x(x, y, z)\frac{\partial u}{\partial q_i} + F_y(x, y, z)\frac{\partial v}{\partial q} + F_z(x, y, z)\frac{\partial w}{\partial q_i} \right)\mathrm{d}S \quad (2.174)$$

is the generalized force corresponding to the generalized coordinate q_i.

Then the strain energy can be expressed as a function of the generalized coordinates q_1, q_2, q_3, …, q_n and the change in strain energy due to an arbitrary virtual displacement of the generalized coordinates is

$$\delta U = \sum_{i=1}^{n} \frac{\partial U}{\partial q_i}\delta q_i \quad (2.175)$$

Substitution of Eqs. (2.74) and (2.75) into Eq. (2.61) yields

$$\sum_{i=1}^{n} \left\{ Q_i - \frac{\partial U}{\partial q_i} \right\} \delta q_i = 0 \quad (2.176)$$

Since δ_{qi} are independent and arbitrary quantities, this reduces to

$$\frac{\partial U}{\partial q_i} = Q_i \quad (i = 1, 2, \ldots n) \tag{2.177}$$

where Q_i is defined by (2.174).

Comparing with results from the beam:

$$\delta U = \frac{1}{2}\sum_{i=1}^{n}\sum_{j=1}^{n}\left\{\int_0^l EI\left(\frac{d^2\gamma_i}{dy^2}\right)\left(\frac{d\gamma_j}{dy^2}\right)dy\right\}q_j\delta q_i \tag{2.168}$$

$$\frac{\partial U}{\partial q_i} = Q_i \quad (i = 1, 2, \ldots n) \tag{2.177}$$

Then for the beam, this reduces to

$$\frac{\partial U}{\partial q_i} = \sum_{j=1}^{n}\left\{\int_0^l EI\left(\frac{d^2\gamma_i}{dy^2}\right)\left(\frac{d\gamma_j}{dy^2}\right)dy\right\}q_j \tag{2.178}$$

and comparing to (2.69):

$$\frac{\partial U}{\partial q_j} = \left[\sum_{j=1}^{n}\left\{\int_0^l EI\left(\frac{d^2\gamma_i}{dy^2}\right)\left(\frac{d^2\gamma_j}{dy^2}\right)dy\right\}q_j\right] = \int_0^l F_x(y)\gamma_j(y)dy = Q_j \tag{2.179}$$

Equation (2.177) is an equivalent form of the principle of minimum potential energy that is applicable to systems in which the space configuration can be described by a set of discrete generalized coordinates. When n is finite, the process of applying (2.177) to a continuous system, known as the Rayleigh–Ritz method, leads to approximate solutions.

2.11.3 Deflection by Castigliano's Theorem

Castigliano's theorem provides a method for computing deflections from the when the latter is expressed as a function of the applied loads. When Eq. (2.77) is differentiated with respect to Q_i, one obtains

$$\frac{\partial U}{\partial Q_i} = \frac{\partial}{\partial Q_i}\left(\frac{1}{2}\sum_{i=1}^{n}\sum_{1}^{n}C_{ij}Q_iQ_j\right) = \sum_{j=1}^{n}C_{ij}Q_j \tag{2.180}$$

Introducing stress–strain relationship (2.3) into (2.29), obtain

$$U = \frac{1}{2}\sum_{i=1}^{n}\sum_{j=1}^{n} C_{ij} Q_j Q_i \tag{2.180a}$$

Combining Eq. 2.180 with Eq. 2.3, one obtains the mathematical statement of Castigliano's theorem, Eq. 2.180a, and

$$q_i = \frac{\partial U}{\partial Q_i} \tag{2.180b}$$

Castigliano's theorem states that the partial derivative with respect to one of the loads of the strain energy, expressed as function of the applied loads, gives the deflection of the structure at the point of application, and in the direction of that load.

As an illustration of the application of the theorem for computing beam deflection, consider a cantilever beam subjected to an arbitrary side load. We would like to calculate the deflection at point a (Fig. 2.24).

Apply an additional dummy load P at point a. Let the bending moment distribution due to a one-unit load at a is given by m, then by the principle of superposition, the bending moment distribution due to a load P at a is given by mP. The strain energy due to normal strains in a slender beam can be expressed in terms of the applied bending moment distribution $\overline{M}$ which is given by:

$$U = \int_{0}^{l} \frac{\overline{M}^2}{2EI} dy \tag{2.181}$$

The strain energy due to the given side load plus the dummy load P is obtained by substituting

$$\overline{M} = M(y) + mP \tag{2.182}$$

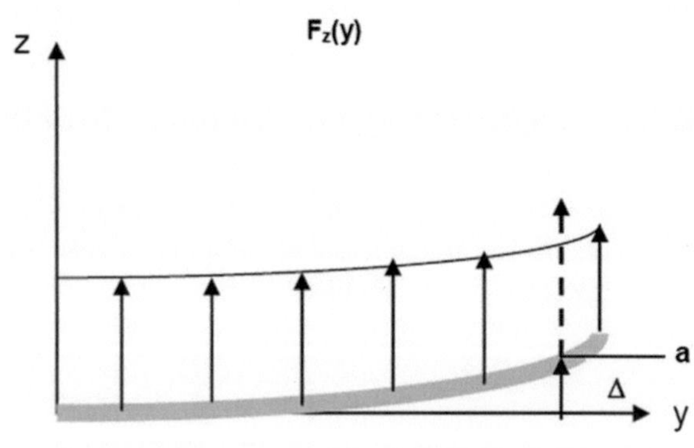

Fig. 2.24 Deflection of a wing along the wingspan due to aerodynamic forces

into Eq. (2.33).

$$U = \int_0^l \frac{[M(y) + mP]^2}{2EI} \mathrm{d}y \tag{2.183}$$

The deflection at a resulting from the given side load plus the dummy load P is obtained by applying Castigliano theorem:

$$\Delta = \frac{\partial U}{\partial P} = \int_0^l \frac{[M(y) + mP]m}{EI} \mathrm{d}y \tag{2.184}$$

Then the deflection at a due to the given side load alone is obtained from (2.132) by putting P to be identically zero, $P = 0$. Hence

$$\Delta = \frac{\partial U}{\partial P} = \int_0^l \frac{[M(y)]m}{EI} \mathrm{d}y \tag{2.185}$$

Equation (2.85) may be used to compute influence coefficients if written in the form:

$$C_{ij} = \int_0^l \frac{m_i m_j}{EI} \mathrm{d}y \tag{2.186}$$

where

m_i—bending moment distribution due to one unit load at i.
m_j—bending moment distribution due to one unit load at j.

2.12 Deformations of Slender Unswept Wings

To obtain a more realistic applications of the principles described previously, as well as understanding the basic procedure in obtaining the mandatory structural characteristics for structural dynamic and aeroelastic characteristic, the following example from BAH is elaborated.

Consider a typical beam in which the skin and shear webs are assumed to be thin: no bending rigidity, and the cross-sectional shape is preserved by closely spaced ribs which have infinite rigidity in their own plane but are completely free to warp out in their plane.

The beam may taper in the spanwise direction, and the properties of the cross section may vary with y. The reference axis is illustrated below (Fig. 2.25).

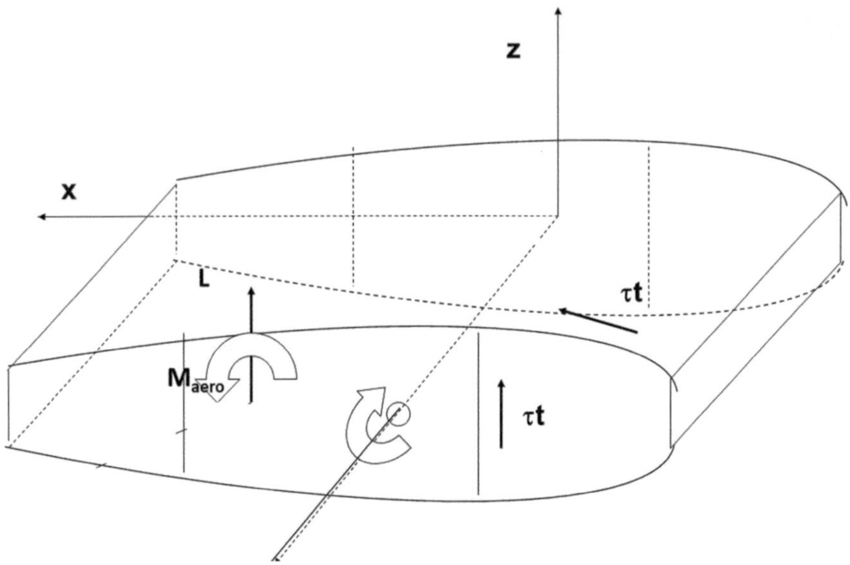

Fig. 2.25 Loaded cross section of a shell beam

It is convenient here to assume that the y-axis is chosen to pass through the centroid of the effective normal stress-carrying area and that the x- and z-axes are oriented so that the xy- and yz-planes pass through the principal bending axes.

A point on the cross section is located by the spanwise coordinate y and the tangential coordinate s. The tangential coordinate is measured positive in the counterclockwise direction for nthe peripheral skin and is positive in the positive direction of the z-axis for the interior webs.

A loaded cross section at a distance y from the origin is acted on by a moment $M(y)$ (with axis parallel to the x-direction), a shear $S(y)$, and a torque $T(y)$ (with axis parallel to the y-direction), with positive direction shown.

The point of application of $S(y)$ and $T(y)$ is at the *center of twist or shear center of the section.*

Normal stresses, assumed to be positive in the y-direction, are denoted by σ, and shear flows, assumed to be positive in the s-direction, are denoted by q.

By discussing slender beams, two simplifying assumptions are implied.

a. First, the beam is given complete freedom to warp when torque load is applied, which leads to St. Venant solution of the torsion problem: permits the simplification that bending and torsion are separate uncoupled actions.
b. Second, the plane sections are assumed to remain plane during bending, which allows the application of the well-known engineering bending theory.

2.12.1 Bending and Shearing Deformation

The wing is composed of skin elements in equilibrium under normal stresses σ and shear flows q. Since the ribs are assumed to be rigid in their own planes, and normal stresses in the chordwise direction can be neglected, the strain energy in the wing in terms of stresses is given by:

$$U = \frac{1}{2E}\int_0^l \int_{\text{cross section}} \overline{\sigma}^2 t \mathrm{d}s \mathrm{d}\lambda + \frac{1}{2G}\int_0^l \int_{\text{cross section}} \overline{q}^2 \frac{\mathrm{d}s}{t} \mathrm{d}\lambda \tag{2.187}$$

where t is the skin thickness, the integration on s is over the wing cross section, and the integration over λ is over the length of the wing beam.

Verification:

(a) Energy due to bending deformation, due to tension along the wing beam:

$$\sigma = \varepsilon E$$
$$\mathrm{d}Q = \sigma t \mathrm{d}s$$
$$\mathrm{d}q = \varepsilon \mathrm{d}\lambda$$

$$U = \frac{1}{2}\sum_{i=1}^{n} q_i Q_i \; or \; U = \frac{1}{2}\int_0^l dq \int_{\text{cross section}} dQ$$

$$U = \frac{1}{2}\int_0^l \varepsilon d\lambda \int_{\text{cross section}} \sigma t ds$$

$$U = \frac{1}{2E}\int_0^l \sigma d\lambda \int_{\text{cross section}} \sigma t ds = \frac{1}{2E}\int_0^l \sigma^2 \int_{\text{cross section}} t ds d\lambda$$

$$U = \frac{1}{2E}\int_0^l \int_{\text{cross section}} \overline{\sigma}^2 t ds d\lambda \qquad \text{(qed)}$$

(b) Energy due to torsional deformation, due to shear flow:

$$\tau = \gamma G$$

$$\mathrm{d}Q \equiv \mathrm{d}T = \tau t \mathrm{d}sh = \tau \mathrm{d}s.1$$

$$\mathrm{d}q = \gamma \mathrm{d}\gamma$$

$$U = \frac{1}{2}\int_0^l \mathrm{d}q \int_{\text{cross section}} \mathrm{d}Q \textit{ or } U = \frac{1}{2}\int_0^l \gamma \mathrm{d}\lambda \int_{\text{cross section}} \tau t \mathrm{d}s.1$$

$$U = \frac{1}{2}\int_0^l \gamma \mathrm{d}\lambda \int_{\text{cross section}} \tau t \mathrm{d}sh = \frac{1}{2G}\int_0^l \tau \int_{\text{cross section}} \tau t \mathrm{d}s.1\mathrm{d}\lambda \quad ((\text{i}))$$

$$U = \frac{1}{2G}\int_0^l \int_{\text{cross section}} \tau t \tau t \frac{\mathrm{d}s}{t} h \mathrm{d}\lambda = \frac{1}{2G}\int_0^l \int_{\text{cross section}} (\tau t)^2 \frac{\mathrm{d}s}{t} 1\mathrm{d}\lambda \quad ((\text{ii}))$$

Upon superposing (a) and (b), (2.187) is obtained[11]

$$U = \frac{1}{2E}\int_0^l \int_{\text{cross section}} \overline{\sigma}^2 t \mathrm{d}s \mathrm{d}\lambda + \frac{1}{2G}\int_0^l \int_{\text{cross section}} \overline{q}^2 \frac{\mathrm{d}s}{t} \mathrm{d}\lambda \tag{2.187}$$

Computation of the lateral deflection $w(y)$ of the wing due to bending moment distribution M [$\mathrm{ML^2T^{-2}}$], and shear distribution, S[$\mathrm{MLT^{-2}}$]. Note here that λ is the dummy running variable for integration, while y is the location of the point of interest.

To apply Castigliano theorem, an additional hypothetical external load P should be applied at the spanwise location $\lambda = y$. Then the total normal stress due to the given bending moment $M(\lambda)$ plus P is

$$\overline{\sigma}(\lambda, s) = \sigma(\lambda, s) + P\frac{\partial \overline{\sigma}}{\partial P} \tag{2.188}$$

[11] As a basic principle in assessing any mathematical relationship in physics, the correctness of the equation can be checked by dimensional analysis:

$$U = \frac{1}{2}\int_0^l \mathrm{d}q \int_{\text{cross section}} \mathrm{d}Q \rightarrow U = \frac{1}{2G}\int_0^l \int_{\text{cross section}} (\tau t)^2 \frac{\mathrm{d}s}{t} \mathrm{d}\lambda$$

$$U = \frac{1}{2}\int_0^l \mathrm{d}q \int_{\text{cross section}} \mathrm{d}Q\left[ML^2T^{-2}\right]$$

$$U\left[ML^2T^{-2}\right] = \frac{1}{2[ML^{-1}T^{-2}]}\int_0^l \int_{\text{cross section}} \left[MT^{-2}\right]^2 \frac{ds}{t} 1[L]d\lambda = \left[ML^2T^{-2}\right] \rightarrow \text{check.}$$

where $\sigma(\lambda, s)$ is the normal stress due to the given bending moment distribution $M(\lambda)$, and $\frac{\partial \overline{\sigma}}{\partial P}$ is the normal stress distribution due to the hypothetical force $P = 1$, located at $\lambda = y$.

Similarly, the total shear flow is denoted by (refer to Figs. 2.26 and 2.27)

$$\overline{q}(\lambda, s) = q(\lambda, s) + P\frac{\partial \overline{q}}{\partial P} \tag{2.189}$$

Substituting (2.188) and (2.189) into (2.187), obtain the total strain energy due to the given bending moment and shear distributions, plus the hypothetical load P:

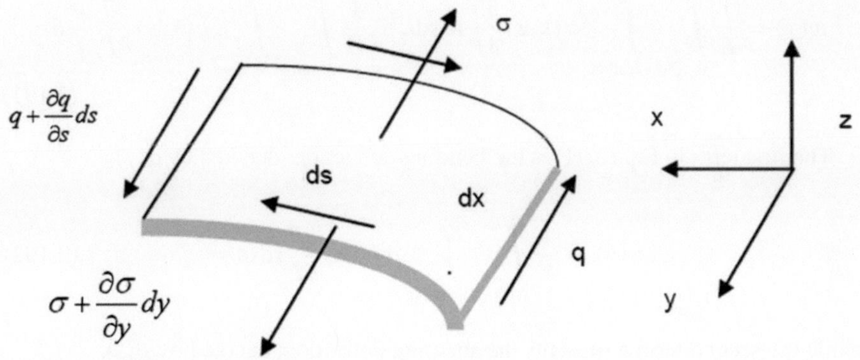

Fig. 2.26 Element of wing skin

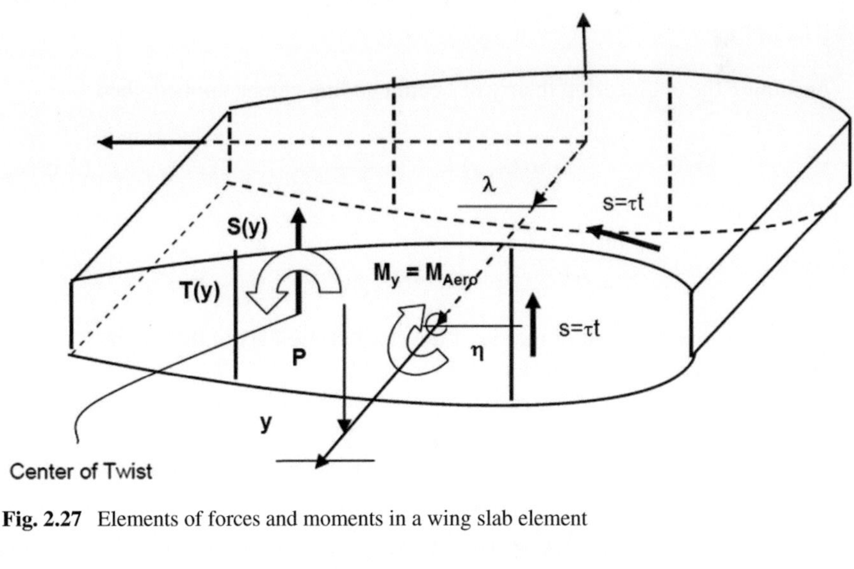

Fig. 2.27 Elements of forces and moments in a wing slab element

$$U = \frac{1}{2E}\int_0^l \int_{\text{cross section}} \left[\sigma(\lambda, s) + P\frac{\partial \overline{\sigma}}{\partial P}\right]^2 t ds d\lambda + \frac{1}{2G}\int_0^l \int_{\text{cross section}} \left[q(\lambda, s) + \frac{\partial \overline{q}}{\partial P}\right]^2 \frac{ds}{t} d\lambda \tag{2.190}$$

Differentiating with respect to P and setting $P = 0$ then gives the lateral deflection w(y):

$$w(y) = \frac{1}{E}\int_0^l \int_{\text{cross section}} \sigma(\lambda, s)\frac{\partial \overline{\sigma}}{\partial P} t ds d\lambda + \frac{1}{G}\int_0^l \int_{\text{cross section}} q(\lambda, s)\frac{\partial \overline{q}}{\partial P}\frac{ds}{t} d\lambda \tag{2.191}$$

The first term in Eq. (2.91) is the bending deflection, denoted by $\alpha(y)$:

$$\alpha(y) = \frac{1}{E}\int_0^l \int_{\text{cross section}} \sigma(\lambda, s)\frac{\partial \overline{\sigma}}{\partial P} t ds d\lambda \tag{2.192}$$

while the second term represents the shearing deflection, denoted by $\beta(y)$:

$$\beta(y) = \frac{1}{G}\int_0^l \int_{\text{cross section}} q(\lambda, s)\frac{\partial \overline{q}}{\partial P}\frac{ds}{t} d\lambda \tag{2.193}$$

Assuming the engineering theory of beam bending can be applied, then

$$\sigma(\lambda, s) = -\frac{M(\lambda)z}{I} \tag{2.194a}$$

$$\frac{\partial \overline{\sigma}(\lambda, s)}{\partial P} = -\begin{cases} -\frac{y-\lambda}{I} z & for\ y > \lambda \\ 0 & for\ y < \lambda \end{cases} \tag{2.194b}$$

where I is the area moment of inertia of the cross section about the x-axis.

Substituting (2.94) into (2.92), obtain

$$\alpha(y) = \frac{1}{E}\int_0^y \int_{\text{cross section}} \frac{M(\lambda)(y-\lambda)}{I}\frac{y-\lambda}{I} z^2 t ds d\lambda = \frac{1}{E}\int_0^y \frac{M(\lambda)(y-\lambda)}{I} d\lambda \int_{\text{cross section}} \frac{(z^2 t ds)}{I}$$

or

$$\alpha(y) = \int_0^y \frac{M(\lambda)(y-\lambda)}{EI} \mathrm{d}\lambda \tag{2.195}$$

Differentiation of Eq. (2.195) twice with respect to y will produce (recover the well-known relation between beam bending moment and curvature)[12]:

$$\alpha'' = \frac{M(\lambda)}{EI} \tag{2.196}$$

Next, the shear flow distribution due to $S = 1$ is denoted by $u(\lambda, s)$, then

$$q(\lambda, s) = Su(\lambda, s) \tag{2.197a}$$

$$\frac{\partial \overline{q}}{\partial P} = \begin{cases} u(\lambda, s), & \text{for } y > \lambda \\ 0 & \text{for } y < \lambda \end{cases} \tag{2.197b}$$

Substitution of (2.197a), (2.197b) into (2.193) gives

$$\beta(y) = \int_0^y \frac{S}{GK} \mathrm{d}\lambda \tag{2.198a}$$

where

$$K = \frac{1}{\int_{\text{cross section}} u^2 \frac{\mathrm{d}s}{t}} \tag{2.198b}$$

is a shearing constant of the beam cross section defined in terms of the shear flow distribution due to a unit shear force.

The quantity GK in Eq. (2.198a) is known as the shearing rigidity of the beam.

Differentiating Eq. (2.198a) with respect to y gives the relation between the shear and the first derivative of the shearing deflection[13]

---

[12] Note that α has the dimension of $[L]$; $\frac{M}{EI} = [ML^2T^{-2}][MT^{-2}L^{-1}]^{-1}[L^{-4}] = [L^{-1}]$, $\alpha'' = [L - 1]$.

[13] Note that β has the dimension of $[L]$; $u = [L - 1]$;
$q(\lambda, s) = Su(\lambda, s)\ [ML^2T^{-2}L^{-2}] = [MT^{-2}] = Su(\lambda, s) = [MLT^{-2}][L^{-1}]$
$G = [MLT^{-2}][L^{-2}][L] = [MT^{-2}]$, $K = [L^{-2}]^{-1} = [L^2]$.
$\beta' = \frac{S}{GK}; \frac{S}{GK} = [MLT^{-2}][MT^{-2}]^{-1}[L^2]^{-1} = [L^{-1}]$, $\beta = \frac{Sy}{GK} = \frac{[MLT^{-2}][L]}{[MT^{-2}][L^2]} = [-]$
$\beta(y) = \frac{1}{G}\int_0^l \int_{\text{cross section}} q(\lambda, s)\frac{\partial \bar{q}}{\partial P}\frac{\mathrm{d}s}{t}\mathrm{d}\lambda = \frac{1}{[MT^{-2}]}[MT^{-2}]\frac{[MT^{-2}]}{[MLT^{-2}]}[L] = [-]$

$$\beta' = \frac{S}{GK} \tag{2.199}$$

2.12.2 Influence Functions and Coefficients

Influence functions for both bending and shearing deformations can be computed from (2.191):

$$w(y) = \frac{1}{E}\int_0^l \int_{\text{cross section}} \sigma(\lambda, s)\frac{\partial \overline{\sigma}}{\partial P} t ds d\lambda + \frac{1}{G}\int_0^l \int_{\text{cross section}} q(\lambda, s)\frac{\partial \overline{q}}{\partial P}\frac{ds}{t} d\lambda \tag{2.191}$$

$C^{zz}(y, \eta)$ can be computed by applying the engineering theory of bending, by using (see Fig. 2.27):

$$\sigma(\lambda, s) = -\frac{\eta - \lambda}{\mathrm{I}} z \quad (0 < \lambda < y) \tag{2.200a}$$

$$\frac{\partial \overline{\sigma}}{\partial P} = -\frac{y - \lambda}{\mathrm{I}} z \quad (0 < \lambda < y) \tag{2.200b}$$

and

$$q(\lambda, s) = u(\lambda, s) \quad (0 < \lambda < y) \tag{2.200c}$$

$$\frac{\partial \overline{q}}{\partial P} = u(\lambda, s) \quad (0 < \lambda < y) \tag{2.200d}$$

and substitute them into (2.191).

For $\eta \geq y$, this gives the influence functions:

$$C^{zz}(y, \eta) = \int_0^y \frac{(\eta - \lambda)(y - \lambda)}{EI} \mathrm{d}\lambda \int_{\text{cross section}} \frac{z^2}{I} t \mathrm{d}s + \int_0^y \mathrm{d}\lambda \int_{\text{cross section}} \frac{(u(\lambda, s))^2}{G}\frac{\mathrm{d}s}{t}$$

or using (2.198b)

$$\alpha(y) = \frac{1}{E}\int_0^l \int_{\text{cross section}} \sigma(\lambda, s)\frac{\partial \overline{\sigma}}{\partial P} t ds d\lambda = \frac{1}{[MLT^{-2}L^{-2}]}[MLT^{-2}L^{-2}]\frac{[MLT^{-2}L^{-2}]}{[MLT^{-2}]}[L^3] = [L]$$

$$C^{zz}(y,\eta) = \int_0^y \frac{(\eta-\lambda)(y-\lambda)}{EI}\mathrm{d}\lambda + \int_0^y \frac{1}{GK}\mathrm{d}\lambda \quad for\ \eta \geq y \tag{2.201}$$

and for the case y ≥ η,

$$C^{zz}(y,\eta) = \int_0^\eta \frac{(\eta-\lambda)(y-\lambda)}{EI}\mathrm{d}\lambda + \int_0^\eta \frac{1}{GK}\mathrm{d}\lambda \quad for\ y \geq \eta \tag{2.202}$$

Influence coefficients for bending and shear can be obtained from Eqs. (2.101) and (2.102) by assigning explicit numerical values of interest for y and η.

2.12.3 Torsional Deformation and Influence Function

If during the application of torsional moments, the beam is free to warp, the strain energy is due entirely to shear stresses and is equal to (from 2.187)

$$U = \frac{1}{2E}\int_0^l \int_{\text{cross section}} \overline{\sigma}^2 t\,\mathrm{d}s\,\mathrm{d}\lambda + \frac{1}{2G}\int_0^l \int_{\text{cross section}} \overline{q}^2 \frac{\mathrm{d}s}{t}\mathrm{d}\lambda \tag{2.187}$$

then

$$U = \frac{1}{2G}\int_0^l \int_{\text{cross section}} \overline{q}^2 \frac{\mathrm{d}s}{t}\mathrm{d}\lambda \tag{2.203}$$

Applying Castigliano's theorem

$$U = \frac{1}{2}\sum_{i=1}^{n}\sum_{j=1}^{n} C_{ij} Q_j Q_i \tag{2.79}$$

$$q_i = \frac{\partial U}{\partial Q_i} \tag{2.80}$$

which states that "… the partial derivative with respect to one of the loads of the strain energy, expressed as function of the applied loads, gives the deflection of the structure at the point of application, and in the direction of that load …"

then for the bending problem, since the angle of twist of the beam due to a given distribution of applied torque, $T(\lambda)$, can be expressed as:

$$\theta(y) = \frac{\partial U}{\partial T} = \frac{1}{G}\int_0^l \int_{\text{cross section}} q\frac{\partial \overline{q}}{\partial T}\frac{ds}{t}d\lambda \tag{2.204a}$$

where $q(\lambda, s)$ is the shear distribution due to the applied torques, and $\frac{\partial \overline{q}}{\partial T}$ is the shear flow distribution due to a unit torque $T = 1$ applied at the spanwise station $\lambda = y$.

Let the shear flow distribution due to $T = 1$ be denoted by $v(\lambda, s)$, then

$$q(\lambda, s) = T(\lambda)v(\lambda, s) \tag{2.204b}$$

$$\frac{\partial \overline{q}}{\partial T} = \begin{cases} v(\lambda, s), & for\ y > \lambda \\ 0, & for\ y < \lambda \end{cases} \tag{2.204c}$$

Substituting Eqs. (2.204c) into (2.204a) gives

$$\begin{aligned}\theta(y) &= \frac{1}{G}\int_0^y \int_{\text{cross section}} T(\lambda)(v(\lambda, s))^2\frac{ds}{t}d\lambda \\ &= \int_0^y T(\lambda)\left(\frac{1}{G}\int_{\text{cross section}} \frac{v^2}{t}ds\right)d\lambda = \int_0^y \frac{T(\lambda)}{GJ}d\lambda\end{aligned} \tag{2.204d}$$

where

$$J = \frac{1}{\int_{\text{cross section}} \frac{v^2}{t}ds} \tag{2.204e}$$

is the torsion constant of the beam cross section.

The quantity GJ is known as the torsional stiffness of the beam (Fig. 2.28).

Differentiating Eq. (2.204d) with respect to y gives a relation between the applied torsional moment and the rate of twist (twist per unit length along the spanwise axis of the wing):

$$\theta' = \frac{T}{GJ} \tag{2.205}$$

The deflection problem associated with the torsional influence function is illustrated below.

A unit torque about the elastic axis is applied at a distance η from the origin, and the resulting angular displacement at y is $C^{\theta\theta}(y, \eta)$. The influence function $C^{\theta\theta}(y, \eta)$ can be calculated by substituting (2.204b) and (2.204c)

$$q(\lambda, s) = T(\lambda)v(\lambda, s) \quad for\ 0 < \lambda < \eta \tag{2.204b}$$

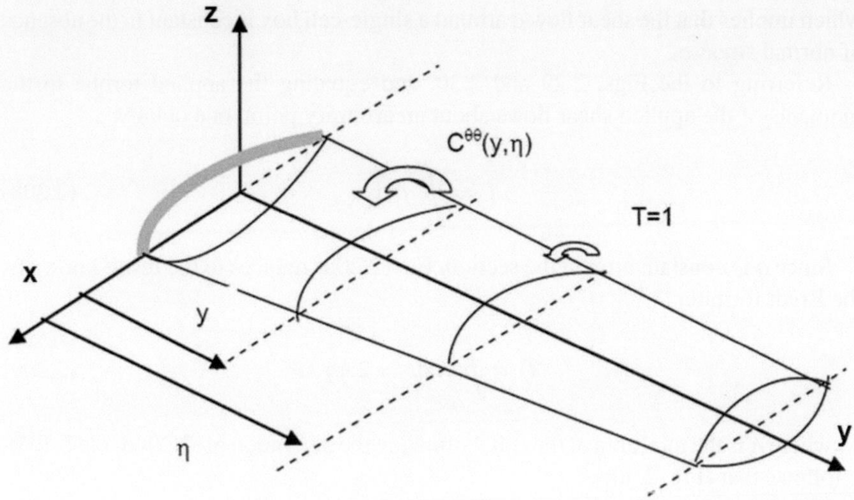

Fig. 2.28 Forces and moments along the span of a wing

$$\frac{\partial \overline{q}}{\partial T} = v(\lambda, s) \quad 0 < \lambda < y \tag{2.204c}$$

Into Eq. (2.204a).
Then, for $\eta \geq y$,

$$C^{\theta\theta}(y, \eta) = \int_0^y \frac{\mathrm{d}\lambda}{GJ}, \quad \eta \geq y \tag{2.205}$$

and for $y \geq \eta$

$$C^{\theta\theta}(y, \eta) = \int_0^\eta \frac{\mathrm{d}\lambda}{GJ}, \quad y \geq \eta, \tag{2.206}$$

The torsional constant is given by Eq. (2.204e) and can be evaluated at each section of the beam, providing the shear flow distribution $v(s, \lambda)$ due to a unit torque is known.

As an example, consider a single-cell beam. A torque T is applied to it. Assume that there is complete freedom to warp, and then no normal stress is induced. Referring to the free-body diagram of a skin element, it can be concluded that when $\sigma = 0$, then

$$\frac{\partial q}{\partial s} = 0 \tag{2.207}$$

which implies that the shear flow q around a single-cell box is constant in the absence of normal stress σ.

Referring to the Figs. 2.29 and 2.30, and equating the applied torque to the moments of the applied shear flows about an arbitrary point, one obtains

$$T = \oint qr\mathrm{d}s \tag{2.208}$$

Since q is constant around the section, Eq. (2.208) reduces to the result known as the Bredt formula:

$$T = \oint qr\mathrm{d}s = 2Aq \tag{2.209}$$

where A is the total area of the cell. Following the derivation of (2.204d) or (2.205), it follows that (Fig. 2.30)

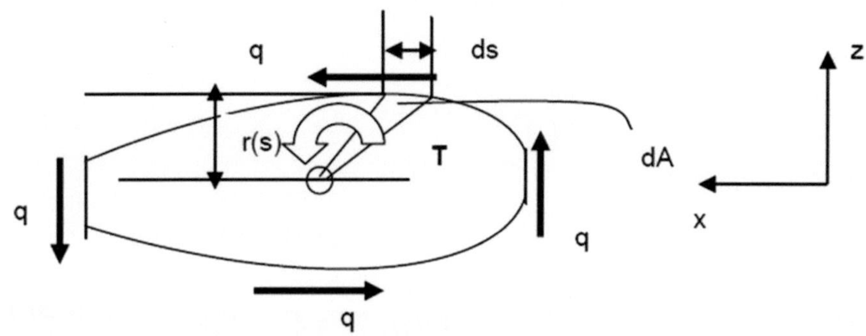

Fig. 2.29 Torque applied to a single-cell box

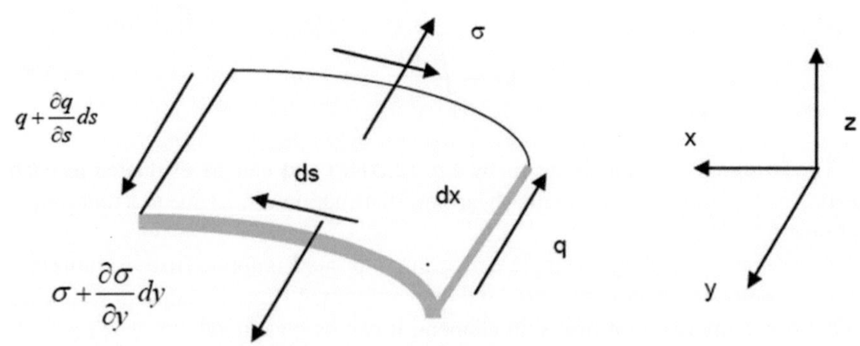

Fig. 2.30 Free-body diagram of an element of wing skin

$$2Aq = \int_0^y \frac{\mathrm{d}\lambda}{GJ}, q = v = \frac{1}{2A}\int_0^y \frac{\mathrm{d}\lambda}{GJ} \quad \eta \geq y \tag{2.210}$$

$$T = \oint qrds = 2Aq = 2Av \text{ and } v = \frac{T}{2A} \tag{2.211}$$

$$J = \frac{1}{\int\limits_{\text{cross section}} \frac{v^2}{t} ds} = \frac{1}{v^2 \oint \frac{ds}{t}} = \frac{4A^2}{T \oint \frac{ds}{t}} \tag{2.212}$$

Since the torsion constant is the value of 2.104d for T = one unit, the torsion constant of the beam cross section becomes

$$J = \frac{4A^2}{\oint \frac{ds}{t}} \tag{2.213a}$$

Hence:

$$GJ = \frac{4A^2}{\oint \frac{ds}{Gt}} \tag{2.213b}$$

which is known as **Bredt Batho** torsional stiffness theory for closed thin-walled cells.

2.12.4 Elastic Axis

The derivation of formulas in parts a and b has been carried out for computing bending and torsional flexibility influence coefficients of beams with straight elastic axis. The results can be used if the location of the elastic axis is known. The elastic axis can be located by drawing a spanwise line through the shear centers of the cross sections of the beam.

The shear center of each cross section is computed by establishing the points in the plane of the cross section at which the shear force can be applied to the section without producing a rate of twist (torsional moment of the section).

2.13 Example

Consider a typical wing of a jet transport, with the planform illustrated below. The elastic axis is straight and perpendicular to the root at 35% of the chord. The curves of bending rigidity EI, torsional rigidity GJ and shearing rigidity GK have been computed and are plotted. We would like to calculate a five-by-five matrices of bending and torsional influence coefficients at the wing station 90, 186, 268, 368 and 458.The wing planform of the typical jet transport considered is depicted in Fig. 2.31, while the bending, torsional and shear stiffness of the wing planform of typical jet transport have been computed as given in Table 2.1.

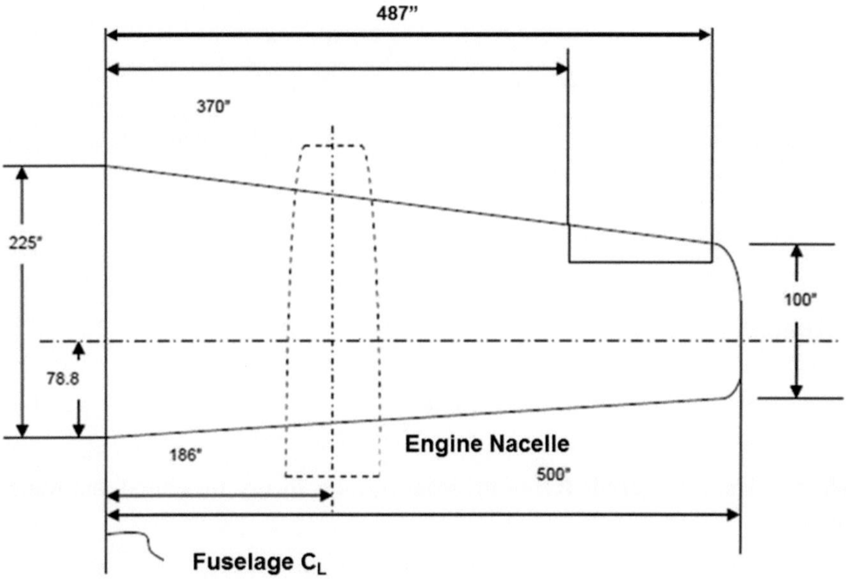

Fig. 2.31 Wing planform of typical jet transport

Table 2.1 Bending, torsional and shear stiffness of the wing planform of typical jet transport

Wing section	E-6	E-10	E 10
	GK	GJ	EI
90	26.5	2.6	
186	25	2.66	2.8
268	23	2.4	2.0
388	20	1.3	1.0
468	16	0.5	0.6

Solution:

Matrix of bending influence coefficients.

The bending influence coefficients are obtained by numerical evaluation of

$$C^{zz}(y,\eta)=\int_0^y \frac{(\eta-\lambda)(y-\lambda)}{EI}\mathrm{d}\lambda+\int_0^y \frac{1}{GK}\mathrm{d}\lambda \quad for \eta \geq y \tag{2.214}$$

and for the case y $\geq \eta$,

$$C^{zz}(y,\eta)=\int_0^\eta \frac{(\eta-\lambda)(y-\lambda)}{EI}\mathrm{d}\lambda+\int_0^\eta \frac{1}{GK}\mathrm{d}\lambda \; for\; y \geq \eta \tag{2.215}$$

Expanding (2.215) and letting $y = y_i$ and $\eta = y_j$, obtain an expression for the bending influence coefficient C_{ij}, i.e.

$$C_{ij}=y_i y_j\int_0^{y_j}\frac{1}{EI}\mathrm{d}\lambda-\left(y_i+y_j\right)\int_0^{y_j}\frac{\lambda \mathrm{d}\lambda}{EI}+\int_0^{y_j}\frac{\lambda^2\mathrm{d}\lambda}{EI}+\int_0^{y_j}\frac{1}{GK}\mathrm{d}\lambda,\; for\; y_i \geq y_j \tag{2.216a}$$

For $y_j = 90$, one can compute the first column of the bending influence coefficient matrix from

$$C_{ij}=90\, y_i\int_0^{90}\frac{1}{EI}\mathrm{d}\lambda-(y_i+90)\int_0^{90}\frac{\lambda \mathrm{d}\lambda}{EI}+\int_0^{90}\frac{\lambda^2\mathrm{d}\lambda}{EI}+\int_0^{90}\frac{1}{GK}\mathrm{d}\lambda\; for\; y_i \geq y_j \tag{2.216b}$$

where $y_j = 90$, 186, 268, 368 and 458 (Fig. 2.32).

The definite integrals can be evaluated by plotting 1/*EI*, λ/*EI*, λ^2/*EI* and 1/GK. Carrying out the integration in (b numerically using the function plotted above (approximated by curve fitting technique), the following matrix of bending influence coefficients is found to be:

$$\left[C^{zz}=\begin{bmatrix}0 & 0 & 0 & 0 & 0 & 0\\ 0 & 72.410 & 114.771 & 150.995 & 195.081 & 234.794\\ 0 & 114.771 & 253.306 & 461.299 & 714.949 & 943.234\\ 0 & 150.995 & 461.299 & 1247.53 & 1911.22 & 2508.54\\ 0 & 195.081 & 714.949 & 1911.22 & 2649.96 & 5237.42\\ 0 & 234.794 & 943.234 & 2508.54 & 5237.42 & 8434.02\end{bmatrix}\right]\cdot 10^{-7}\; in/lb \tag{2.217}$$

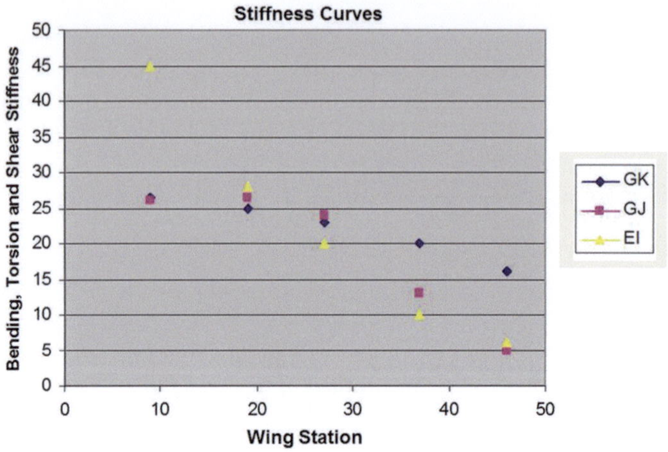

Fig. 2.32 Bending, torsional and shear stiffness of the wing planform of typical jet transport

where C^{zz} is the matrix of bending influence coefficients

Using similar procedure, matrix of torsional influence coefficients is obtained as follows:

$$\left[C^{\theta\theta} = \begin{bmatrix} 0 & 0 & 0 & 0 & 0 & 0 \\ 0 & 36 & 36 & 36 & 36 & 36 \\ 0 & 36 & 75.36 & 75.36 & 75.36 & 75.36 \\ 0 & 36 & 75.36 & 113.26 & 113.26 & 113.26 \\ 0 & 36 & 75.36 & 113.26 & 204.66 & 204.66 \\ 0 & 36 & 75.36 & 113.26 & 204.66 & 409.49 \end{bmatrix} \right] \cdot 10^{-10}\ rad/in - lb \quad (2.218)$$

where $C^{\theta\theta}$ is the matrix of bending influence coefficients.

2.14 Case Study: Vibration Analysis of a Cantilevered Beam with Spring Loading at the Tip as a Generic Elastic Structure[14]

Abstract

Although fundamental, the vibration of a cantilevered Euler–Bernoulli beam with spring attached at the tip is not found in the literature and is here solved analytically and numerically using finite element approach. The equation of motion of the beam is

[14] Reproduced from M Jafari, H Djojodihardjo, KA Ahmad, Vibration Analysis of A Cantilevered Beam with Spring Loading at the Tip as a Generic Elastic Structure, paper presented at AEROTECH III, UPM, 2015.

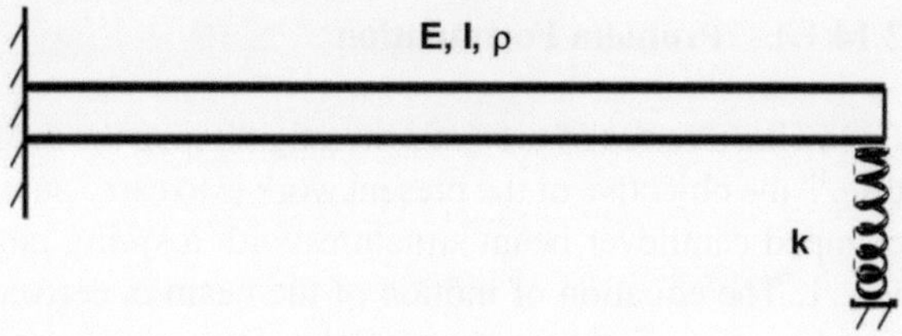

Fig. 2.33 Assumed model for the beam

obtained by using Hamilton's principle. Finite element method is utilized to write in-house program for the free vibration of the beam. Results show plausible agreements.

2.14.1 Introduction

Studies of vibrating beams and plates have wide variety of applications in mechanical engineering, aeronautical and civil structures.[15] For instructiveness, structures may be analyzed starting from simplified configurations. Three common models can be mentioned in beams theory[16]. Euler–Bernoulli beam is considered for thin and long span. In Euler–Bernoulli beam, plane sections remain plane and perpendicular to the beam axis, and shear stress and rotational inertia of the cross section are neglected (Fig. 2.33).

In Rayleigh theory the rotary inertia of the cross section of the beam is considered. Timoshenko beam is a more general theory and can be applied for a relatively thick beam, where shear deformations of the normal to the mid-surface of the beam are taken into account. The present research is carried out as a baseline approach for solving more complex problems in engineering and in the selection and utilization of commercial software. In-house computational routine is developed for various applications. The solution of a cantilevered beam with spring loading at the tip is elaborated and has the potential to be expanded in many variations. For example, an elastic robotic arm[17] can be modeled as a clamped beam with a mass and spring attached at the end.[18] Extension of the present approach could address flexible spacecraft and satellite structures.

---

[15] H. Djojodihardjo, H. and P. M. Ng (2008), Djojodihardjo, H. and A. Shokrani (2010), Djojodihardjo, H., Acta Astronautica, (2013), Djojodihardjo, H., (2013), Rossit, C. A. and Laura, P. A. (2001), and. Li, W. L., (2000).

[16] Rao, S. S. (2007).

[17] P. Gasbarri, R. Monti, C. De Angelisand M. Sabatini (2012).

[18] Maurizi, M. J., Rossi, R. E., Reyes, J. A. (1976).

2.14.1.1 Problem Formulation

Following a series of previous investigation on the analysis of impact resilient structure,[19] the objective of the present work is to carry out vibration analysis of an elastic clamped cantilever beam structures with a spring attached at the end, depicted in Fig. 1. The equation of motion of the beam is derived using Hamilton's principle. The boundary conditions of a Euler–Bernoulli beam are obtained via Hamilton's principle. The natural modes and natural frequencies of the beam are determined by using both numerical and analytical approaches. Analytical method is utilized to acquire the exact solution for the beam.[20] Hamilton approach[21] is applied to find the equation of motion and separation variables method to solve the partial differential equation of motion for natural frequency and mode shapes. Finite elements analysis via Galerkin method[22] is used to compare the numerical solutions to the exact ones.

In summary, the following simplifying assumptions are utilized in deriving the equations of motion. (1) For the Euler–Bernoulli beam, the rotation of cross sections of the beam is neglected in comparison to the translation. (2) The beam is assumed to have uniform elastic modulus cross section along the entire the beam. (3) The spring acting on the end of the beam is linear, so that Hook's law can be utilized. (4) The beam is assumed to have rectangular cross section.

In the Newtonian approach, the vibrational motion of the Euler–Bernoulli beam can be derived from Newton's second law, which is a vectorial equation of motion. In the utilization of Hamiltonian principle in the Lagrangian mechanics, the equation of motion of the beam is derived using Hamiltonian approach, which reduces to a scalar equation that is simpler to work with. Then the equation of motion of the system can be obtained as:

$$\delta \int_{t_1}^{t_2} L(x(t), \dot{x}(t), t)dt = 0, \ L = \Pi - T - W \tag{2.219}$$

where L is called Lagrangian. Π, T and W are potential energy, kinetic energy and work, respectively. Extension of the present analysis addresses piezoelectric actuator at the tip.[23]

---

[19] Djojodihardjo, H and Shokrani, A., (2010), Djojodihardjo, H. (2014).

[20] Rao, S. S. (2007).

[21] RAO, S. S. (2007), Tso, S. K. et al. (2003), [9] and Baruh, H. (1999), Meirovitch, L. (1986), Meirovith, L. (2001).

[22] Wrobel, L. C. and Aliabadi, M. H. (2002).

[23] Maurizi, M. J., Rossi, R. E., Reyes, J.A. (1976), Djojodihardjo, H. and Jafari, M. (2014).

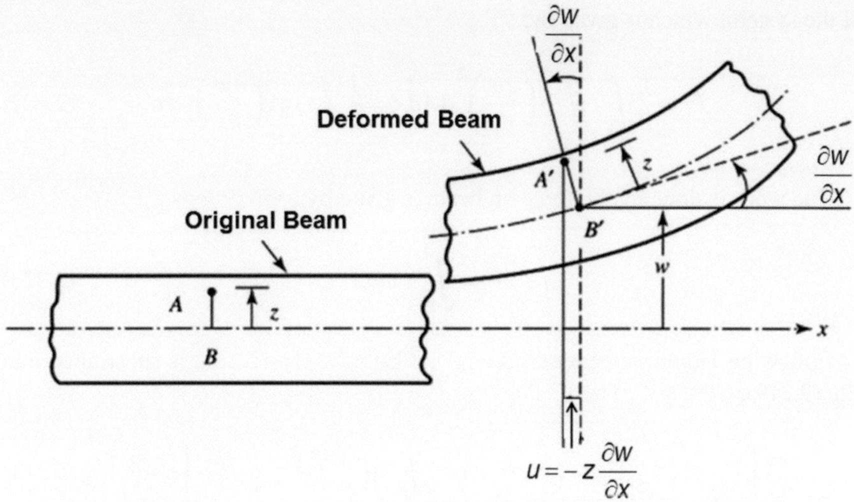

Fig. 2.34 Beam bending

2.14.2 *Detailed Vibration Analysis of Beam with Hinged Spring*

Hamilton's Principle

Figure 2.33 depicts the beam utilized in this investigation. In the Euler–Bernoulli beam, the shear displacement and the rotation of cross sections are neglected in comparison to the translation; hence the cross sections remain constant after deformation.[24] Then according to Fig. 2.34:

$$u = -z\frac{\partial w(x,t)}{\partial x}, \varepsilon_{xx} = \frac{\partial u}{\partial x} = -z\frac{\partial^2 w}{\partial x^2} \tag{2.220}$$

The strain energy (potential energy of the beam) due to the linear spring can be expressed as

$$\Pi = \frac{1}{2}\iiint_V \sigma_{xx}\varepsilon_{xx}\mathrm{d}V + \left(\frac{1}{2}kw^2(l,t)\right) = \frac{1}{2}\int_0^l z^2 E\left(\frac{\partial^2 w}{\partial x^2}\right)^2 \mathrm{d}x + \left(\frac{1}{2}kw^2(l,t)\right)$$
$$= \frac{1}{2}\int_0^l EI\left(\frac{\partial^2 w}{\partial x^2}\right)^2 \mathrm{d}x + \left(\frac{1}{2}kw^2(l,t)\right) \tag{2.221}$$

where I and E are the area of the moment of inertia and elasticity modulus, respectively. k is the stiffness of linear spring (spring constant), and T is the kinetic energy

---

[24] Rao, S. S. (2007).

of the system, which is given by:

$$T = \frac{1}{2}\int_0^l \iint\limits_A \rho\left(\frac{\partial w}{\partial t}\right)^2 dAdx = \frac{1}{2}\int_0^l \rho A\left(\frac{\partial w}{\partial t}\right)^2 dx \tag{2.222}$$

The work is done by the force on beam is given by:

$$W = \int_0^l fwdx \tag{2.223}$$

Following Hamilton general principle,[25] Eq. (2.221)–(2.223) is substituted into Eq. (2.219),

$$\delta\int_{t_1}^{t_2}\left\{\frac{1}{2}\int_0^l EI\left(\frac{\partial^2 w}{\partial x^2}\right)^2 dx + \left(\frac{1}{2}kw^2(l,t)\right) - \frac{1}{2}\int_0^l \rho A\left(\frac{\partial w}{\partial t}\right)^2 dx - \int_0^l fwdx\right\}dt = 0 \tag{2.224}$$

Using integration by parts from Eq. (2.224), the following equations can be obtained.

$$\frac{\partial^2}{\partial x^2}\left(EI\frac{\partial^2 w}{\partial x^2}\right) + \rho A\frac{\partial^2 w}{\partial t^2} = f(x,t) \tag{2.225}$$

For the transverse vibration of the beam, the boundary conditions for $x = 0, l$, are given by:

$$\left[-EI\frac{\partial^2 w}{\partial x^2}\delta\left(\frac{\partial w}{\partial x}\right)\right]_{x=0} = 0 \tag{2.226}$$

$$\left[EI\frac{\partial^2 w}{\partial x^2}\delta\left(\frac{\partial w}{\partial x}\right)\right]_{x=l} = 0 \tag{2.227}$$

$$\left[\frac{\partial}{\partial x}\left(EI\frac{\partial^2 w}{\partial x^2}\right)\delta w\right]_{x=0} = 0 \tag{2.228}$$

$$\left[\left(-\frac{\partial}{\partial x}\left(EI\frac{\partial^2 w}{\partial x^2}\right) + kw\right)\delta w\right]_{x=l} = 0 \tag{2.229}$$

Analytical Solution. To determine free vibration solution of the beam, the force vector should be set to zero,[26] $f(x, t) = 0$ For the uniform beam,

$$c^2\frac{\partial^4 w}{\partial x^4}(x,t) + \frac{\partial^2 w}{\partial t^2}(x,t) = 0 \tag{2.230}$$

[25] Baruh, H., (1999), Meirovitch, L (2001), Wrobel, L. C. and Aliabadi, M.H. (2002).

[26] Meirovitch, L. (2001), Kreyszig, E. (2010), Wrobel, L. C. And Aliabadi, M. H. (2002).

where

$$c = \sqrt{\frac{EI}{\rho A}} \tag{2.231}$$

The solution of Eq. (2.126) can be carried out using the methods of separation of variables.[27]

$$\begin{aligned} W(x) = {} & A(\cos\alpha x + \cosh\alpha x) + B(\cos\alpha x - \cosh\alpha x) \\ & + C(\sin\alpha x + \sinh\alpha x) + D(\sin\alpha x - \sinh\alpha x) \end{aligned} \tag{2.232}$$

where α is considered to be

$$\alpha^4 = \frac{\omega^2}{c^2} = \frac{\rho A\omega^2}{EI} \tag{2.233}$$

By applying boundary conditions Eqs. (2.226) and (2.227) for a cantilever beam, $A = C = 0$, Eq. (2.232) reduces to

$$W(x) = B(\cos\alpha x - \cosh\alpha x) + D(\sin\alpha x - \sinh\alpha x) \tag{2.234}$$

Applying boundary conditions of Eqs. (2.122) and (2.124), the following equations are obtained.

$$B(\cos\alpha l + \cosh\alpha l) + D(\sin\alpha l + \sinh\alpha l) = 0 \tag{2.235}$$

$$\begin{aligned} & B\left[EI\alpha^3(\sin\alpha l - \sinh\alpha l) - k(\cos\alpha l - \cosh\alpha l)\right] \\ & - D\left[-k(\sin\alpha l - \sinh\alpha l) - EI\alpha^3(\cos\alpha l + \cosh\alpha l)\right] = 0 \end{aligned} \tag{2.236}$$

For non-trivial solution of the constants B and D, Eqs. (2.235) and (2.236) can be written as:

$$\begin{vmatrix} \cos\alpha l + \cosh\alpha l & \sin\alpha l + \sinh\alpha l \\ EI\alpha^3(\sin\alpha l - \sinh\alpha l) - k(\cos\alpha l - \cosh\alpha l) & -k(\sin\alpha l - \sinh\alpha l) - EI\alpha^3(\cos\alpha l + \cosh\alpha l) \end{vmatrix} = 0 \tag{2.237}$$

Determining the determinant of Eq. (2.237) results in

$$1 + \frac{1}{\cos\alpha l\cosh\alpha l} - \frac{kl^3}{EI(\alpha l)^3}(\tanh\alpha l - \tan\alpha l) = 0 \tag{2.238}$$

After some elementary algebra, it can be shown that

---

[27] Kreyszig, E. (2010).

$$W(x) = B_n\left[\cos\alpha x - \cosh\alpha x + \left(-\frac{\cos\alpha l + \cosh\alpha l}{\sin\alpha l + \sinh\alpha l}\right)(\sin\alpha x - \sinh\alpha x)\right] \tag{2.239}$$

αl (hence α can be obtained from Eq. 2.238). By using Eq. (2.233), the natural frequencies of the system can be obtained. By substituting α and αl in Eq. (2.239). The natural mode of the system can be drawn.

2.14.3 Numerical Method-Finite Element Approach

There are many numerical methods that are available. The choice depends on the overall convenience and the efficiency and accuracy of the entire procedure. Among these, there are series solution (semi-analytic, matrix transfer method and the like.

The weighted residual method is applied to Eq. (2.224). There are numerous techniques for applying the weighted residual method; to that end, the Galerkin approach [14] is here utilized.

The differential Eq. (2.225) is multiplied by a test function ϕ(x). The test function is not required to satisfy the differential equation. The product of the test function and the differential equation is integrated over the domain. The integral is set equation to zero.

$$\int \phi(x)\left\{EI\frac{d^4W(x)}{dx^4} - \rho A\omega^2 W(x)\right\} = 0 \tag{2.240}$$

$$EI\int_0^l \left[\frac{d^2\phi(x)}{dx^2}\right]\left[\frac{d^2W(x)}{dx^2}\right]dx - \rho A\omega^2\int_0^l \phi(x)W(x)dx = 0 \tag{2.241}$$

$W(x)$ is assumed as

$$W(x) = L_1W_{j-1} + L_2\theta_{j-1} + L_3W_j + L_4\theta_j,(j-1)h \le x \le h \tag{2.242}$$

where W is displacement and θ is the rotation of the nodes. Now, a non-dimensional coordinate's ξ is introduced, according to Fig. 2.35. The essence of the Galerkin method is that the test function is chosen to be[28]

$$\phi(x) = W(x) \tag{2.243}$$

For a system of n elements, then

$$K_j - \omega^2 M_j = 0, j = 1, 2, ..., n \tag{2.244}$$

[28] Wrobel, L. C. and Aliabadi, M. H. (2002).

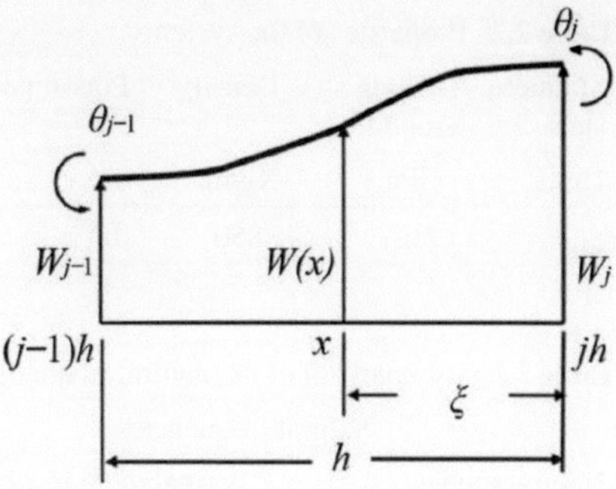

Fig. 2.35 Local coordinate system

where

$$K_j = \frac{EI}{h^3}\begin{bmatrix} 12 & 6h & -12 & 6h \\ 6h & 4h^2 & -6h & 2h^2 \\ -12 & -6h & 12 & -6h \\ 6h & 2h^2 & -6h & 4h^2 \end{bmatrix} \tag{2.245}$$

$$M_j = \frac{h\rho A}{420}\begin{bmatrix} 156 & 22h & 54 & -13h \\ 22h & 4h^2 & 13h & -3h^2 \\ 54 & 13h & 156 & -22h \\ -13h & -3h^2 & -22h & 4h^2 \end{bmatrix} \tag{2.246}$$

The natural frequencies of the beam can be obtained by determining the eigenvalue of Eq. (2.244). The eigenvector corresponds to the natural mode of the system.

A MATLAB program base on finite element method is prepared to evaluate the natural frequency and mode shapes of the beam. MATLAB helps to form the global mass and stiffness matrices for high numbers of elements, easily. Then, eigenvalue and eigenvector are achieved from mass and stiffness matrices according to Eq. (2.244). Natural frequencies of the system can be determined by taking the square root of eigenvalue.

2.14.4 Results and Discussions

Results. Hamiltonian approach is utilized to derive the beam's equation of motion. Natural frequencies and modes shapes of the beam are obtained from analytical and numerical methods. In following, the result of finite element method is compared with the exact solution. According to exact solution, the numerical results have very small error.

Here, the material of the beam is considered to be made of stainless steel. The properties and dimension of the beam are shown in Table 2.2.

Table 2.2 Properties of the system

Stainless steel	Young's modulus	Density	Poisson ratio	Length	Width	Thickness	Spring constant
Units	GPa	Kg/m^2	–	mm	mm	mm	N/m
	210	7850	0.3	450	20	3	10,000

Table 2.3 Comparison of the natural frequencies from the exact and numerical solutions

	Natural frequencies			
Natural modes	α	Analytical results Hz	Numerical results Hz	Numerical error %
1st	8.06796	46.4038120296	46.4038189479	1.1491 E−05
2nd	12.4237	110.035527965	110.032263817	2.9665 e−03
3rd	17.9451	229.571454472	229.569470466	8.8423 E−04

By solving Eq. (2.238) for α *and* αl and putting the result in Eq. (2.232), the natural frequencies for analytical approach can be found. The natural frequencies of the numerical approach are determined by Eq. (2.233).

The first three natural frequencies are determined by both approaches, because other frequencies can rarely occur. Referring to Table 2.3, the errors of numerical results are very small in comparison to the analytical ones. The first natural mode from numerical results has an excellent agreement with the exact answer and its error is nearly zero. For the next two modes, numerical results are obtained with good accuracy also.

Finite element modeling carried out in the present work follows earlier procedure elaborated by Djojodihardjo et al. [2–4]. By using MATLAB, natural mode or mode shape of the beam is drawn according to description in the previous sections. Two of the mode shapes obtained are illustrated in Fig. 2.36. The accuracy is considered satisfactory for the present purposes, although further refinement can be obtained by increasing the number of elements (as shown in Fig. 2.37).

Discussion. Maurizi et al. (1976) investigated the free vibration of a beam hinged at one end by a rotational spring and subjected to the restraining action of a translational spring at the other end.

The eigenfrequencies for the fundamental mode are presented for different values of the parameters $K_R L/EI$ and $K_T L^3/EI$, where K_R is the rotational spring constant, and K_T is the translational spring constant. They utilized the Euler–Bernoulli beam to find the natural frequency for the considered beam. After solving the equation of motion, one can obtain a general equation for natural frequencies.

$$\left(\frac{E^2 I^2}{K_R K_T L^4}\right) y_n^4 (\sin y_n \cosh y_n - \sinh y_n \cos y_n) - \left(\frac{EI}{K_T L^3}\right) y_n^3 (\cos y_n \cosh y_n + 1)$$
$$- 2\left(\frac{EI}{K_R L}\right) y_n \sin y_n \sinh y_n - (\sin y_n \cosh y_n - \sinh y_n \cos y_n) = 0 \qquad (2.247)$$

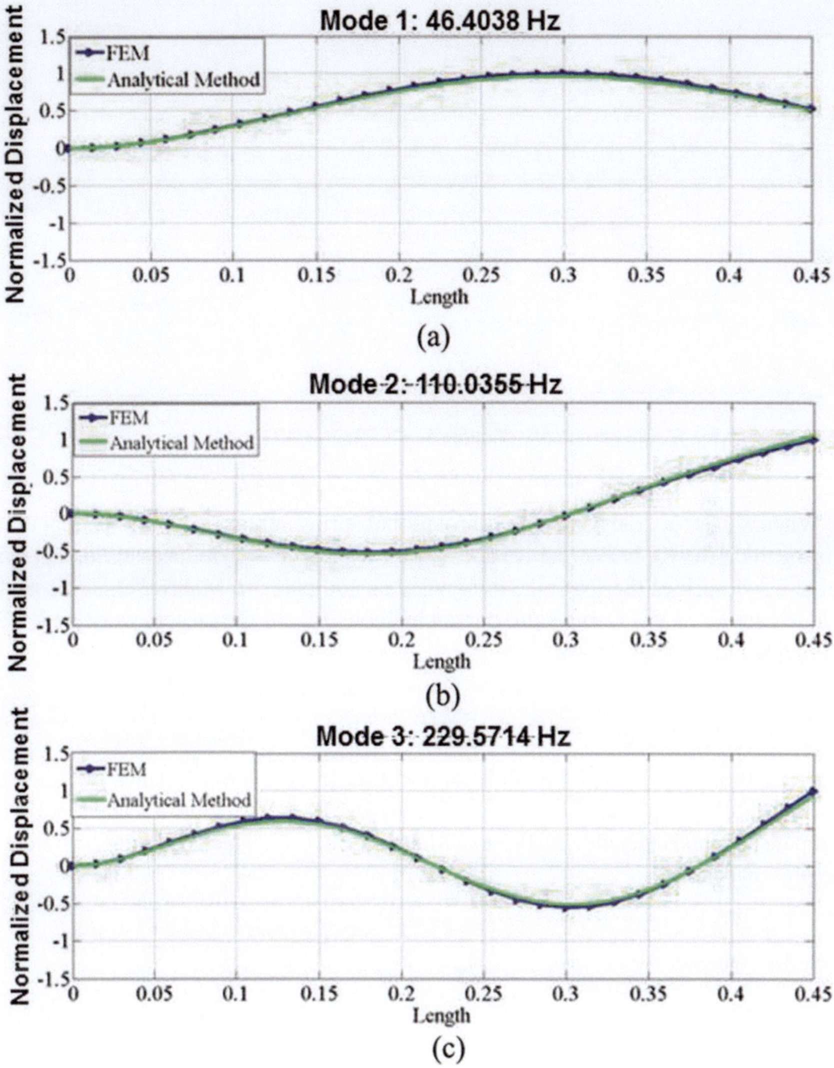

Fig. 2.36 **a** First natural mode, **b** second natural mode, **c** third mode; (1) Green Line: Analytical approach, (2) Blue line: Numerical approach

where y_n is eigenfrequencies of the system. The natural frequency of the system can be obtained as:

$$k_n^4 = \left(\frac{y_n}{L}\right)^4 = \omega_n^2\left(\frac{\rho A}{EI}\right)\ k_n^4 = \left(\frac{y_n}{L}\right)^4 = \omega_n^2\left(\frac{\rho A}{EI}\right) \tag{2.248}$$

$$y_n = k_n L \tag{2.249}$$

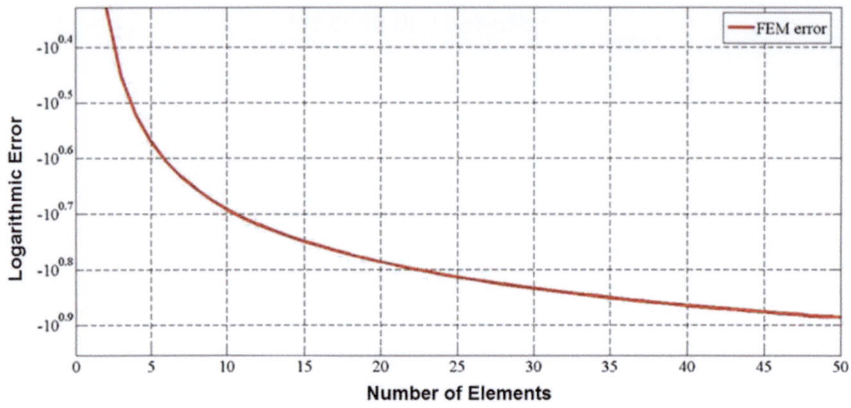

Fig. 2.37 FEM logarithmic error corresponding to the first natural frequency

When $K_R \to \infty$, the beam simple-rotational spring support becomes clamped. In this regard, if the K_R in Eq. (2.142) goes to infinity, Eq. (2.142) reduces to Eq. (2.133). Table 2.4 describes the first eigenfrequency as compares to Maurizi et al. research [15], when $K_R \to \infty$. Using values given in Table 2.1, for the present work $K_T L^3/EI$ is obtained to be 96.4286, while y_1 is

$$y_1 = k_1 L = \alpha_1 L \tag{2.250}$$

From Table 2.3, α_1 is 8.06796.It follows that, $y_1 = 3.630582$ Thus, the present $K_T L^3/EI$ is close to 100 in Table 2.4, and y_1 is near 3.640542. Table 2.4 and Fig. 2.38, compare the present result with those of Maurizi et al. [15], which can serve to validate the present results.

Table 2.4 Values of $y_1 =$ ($k_1 L$ as a function of $\mathbf{K_1 L/EI}$ **and** $\mathbf{K_T L^3/EI}$)

K_T_translational spring constant	K_R—torsional spring constant
0	1.886610
0.01	1.890076
0.1	2.010003
1.0	2.638925
10.0	3.640542
100.0	3.897801
∞	3.926602

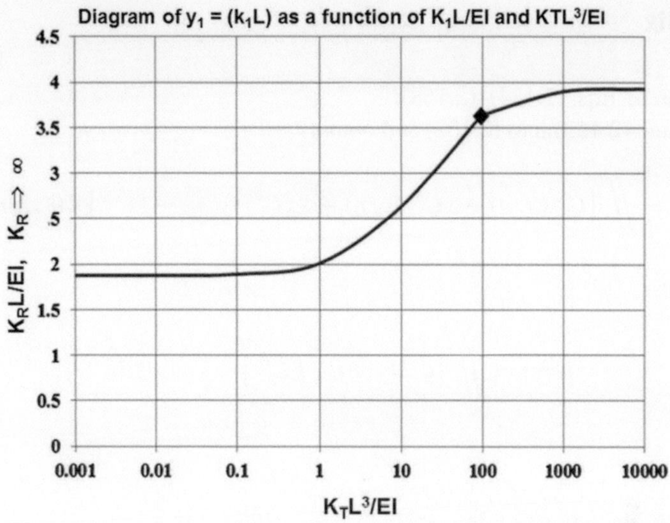

Fig. 2.38 First eigenfrequency and compared Maurizi's et al. result [15]

2.14.5 Conclusions

The equation of motion of Euler–Bernoulli beam is derived using Newtonian approach. The equation of motion and boundary conditions for the beam has also been derived using Hamiltonian approach. Modal analysis for the assumed beam is analytically and numerically evaluated. Finite element method via Galerkin approach is utilized for numerical analysis.

The analytical solution of the cantilevered beam with a spring at the end given here is the first contribution in the literature and has been validated through comparison with other result. The finite element results for the first three natural modes and natural frequencies show excellent agreements. The analytical work can be used as a baseline for more involved models, and the MATLAB finite element analysis code can be extended and applied to more complicated problems like plate modal analysis.

2.15 Concluding Remarks

The chapter has elaborated of a comprehensive aspect of elasticity and aircraft structures in the chapter which should give a good impression and understanding of the scope and fundamentals of elasticity. Integrated discussions with examples in aircraft structure have also been covered, which will provide the necessary background to understand, define and formulate solutions for aeroelastic problems.

Appendix

Derivation of Eqs. (2.154)–(2.158):

Substitute (2.153) into (2.152) and obtain

$$w(x, y) = \iint_S \left\{C^{zz}(y, \eta) - xC^{\theta z}(y, \eta) + \xi x C^{\theta\theta}(y, \eta) - \xi C^{z\theta}\right\} Z(\xi, \eta) d\xi d\eta \quad \text{(i)}$$

let

$$w(y) = \iint_S \left\{C^{zz}(y, \eta) - \xi C^{z\theta}\right\} Z(\xi, \eta) d\xi d\eta \quad \text{(ii)}$$

then

$$w(y) = \iint_S C^{zz}(y, \eta) Z(\xi, \eta) d\xi d\eta - \iint_S \xi C^{z\theta}(y, \eta) Z(\xi, \eta) d\xi d\eta \quad \text{(iii)}$$

$$w(y) = \int_0^l C^{zz}(y, \eta) \left\{ \int_{chord} Z(\xi, \eta) d\xi \right\} d\eta + \int_0^l C^{z\theta}(y, \eta) \left\{ - \int_{chord} \xi Z(\xi, \eta) d\xi \right\} d\eta \quad \text{(iv)}$$

and

$$w(y) = \int_0^l C^{zz}(y, \eta) Z(\eta) d\eta + \int_0^l C^{z\theta}(y, \eta) t(\eta) d\eta \quad \text{(v)}$$

Next let

$$\theta(y) = \iint_S \left\{C^{\theta z}(y, \eta) + \xi C^{\theta\theta}(y, \eta)\right\} Z(\xi, \eta) d\xi d\eta \quad \text{(vi)}$$

then

$$\theta(y) = \iint_S C^{\theta z} Z(\xi, \eta) d\xi d\eta + \iint_S C^{\theta\theta}(y, \mu) Z(\xi, \eta) d\xi d\eta \quad \text{(vii)}$$

or

$$\theta(y) = \int_0^l C^{\theta z}\left\{\int_{chord} Z(\xi,\eta)d\xi\right\}d\eta + \int_0^l C^{\theta\theta}(y,\mu)\left\{-\int_{chord} Z(\xi,\eta)d\xi\right\}d\eta \tag{viii}$$

$$\theta(y) = \int_0^l C^{\theta z}Z(\eta)d\eta + \int_0^l C^{\theta\theta}(y,\mu)t(\eta)d\eta \tag{ix}$$

where in v and ix:

$$Z(y) = \int_{chord} Z(\xi,\eta)d\eta \tag{2.157}$$

$$t(y) = -\int_{chord} \xi Z(\xi,\eta)d\eta \tag{2.158}$$

Comparing (i), (v) and (ix), then we can express (i) as (2.154)–(2.158).

In all these derivations, the effects of chordwise drag loads on the wing have been neglected. Equations (2.155) and (2.156) apply in general to the case of unswept or swept wings with structural discontinuities. The principal elastic effects of sweep and structural discontinuities are that the bending and torsional actions are coupled. Then one should note that the assumptions of rigidity along segments parallel to the x-axis become less valid with increasing angles of sweep. BAH mentioned that the error involved seems to be small for slender wings with angles of sweep up to about 45°.

References

1. Baruh, H. 1999. *Analytical Dynamics*. McGraw-Hill.
2. Fung, J.C. 1993. *An Introduction to the Theory of Aeroelasticity*. Dover Publications.
3. Bisplinghoff, R.L., H. Ashley, and R.L. Halfman. 1995. *Aeroelasticity*. Dover Publications, Inc., copyright ©1955 by Addison-Wesley Publishing Co., Inc., copyright © renewed 1983 by Bisplinghoff, R.L., Ashley, H., & Halfman, R.L.
4. Djojodihardjo, H., and P.M. Ng. 2008. Numerical simulation of impact loading on elastic structure and case studies. In *Paper IAC-08-C2.6.1, in Proceedings, 590th International Astronautical Congress*, 29 Sept–3 Oct 2008, Glasgow, United Kingdom.
5. Djojodihardjo, H., and A. Shokrani. 2010. Generic study and finite element analysis of impact loading on elastic panel structure. In *Paper IAC-10.C2.6.2, Proceedings, 61th International Astronautical Congress*, Sept–Oct 2010, Prague, The Czech Republic.
6. Djojodihardjo, H. 2013. Computational simulation for analysis and synthesis of impact resilient structure. *Acta Astronautica* 91 (Oct–Nov 2013): 283–301.
7. Djojodihardjo, H. 2013. Vibro-acoustic analysis of random vibration response of a flexible structure due to acoustic forcing. In *Beijing Conference, Space Technology & Systems Development*, Beijing, China.

8. Djojodihardjo, H., and M. Jafari. 2014. Vibration analysis of a cantilevered beam with piezo-electric actuator at the tip as a controllable elastic structure. In *To Be Presented at the International Astronautical Congress*. Toronto.
9. Gasbarri, P., R. Monti, C. De Angelis, M. Sabatini. 2012. Second order effects of the flexibility on the control of a spacecraft full-coupled model, IAA-AAS-DyCoSS1-10-08. In *Advances in Astronautical Sciences*, ed. P. Bainum, J.-M. Contant, A. Guerman. published by Univelt.
10. Kreyszig, E. 2010. *Advanced Engineering Mathematics*, 10th ed. Wiley.
11. Li, W.L. 2000. Free vibrations of beams with general boundary conditions. *Journal of Sound and Vibration.*
12. Maurizi, M.J., R.E. Rossi, and J.A. Reyes. 1976. Vibration frequencies for a uniform beam with one end spring-hinged and subjected to a translational restraint at the other end. *Journal of Sound and Vibration* 48 (4): 565–568.
13. Megson, T.H.G. 2007. *Aircraft Structures for Engineering Students.* Copyright ©2007, THG Megson, Elsevier.
14. Meirovitch, L. 1975. *Elements of Vibration Analysis.* McGraw-Hill, Inc.
15. Meirovitch, L. 1986. *Elements of Vibration Analysis.* McGraw-Hill.
16. Meirovitch, L. 2001. *Fundamentals of Vibration.* McGraw-Hill Higher Education.
17. Weaver, W., Jr., and P.R. Johnston. 1987. *Structural Dynamics by Finite Elements*, 07632. Prentice Hall, Inc.
18. Rao, S.S. 2007. *Vibration of Continuous System.* Wiley.
19. Rossit, C.A., and P.A.A. Laura. 2001. Free vibrations of a cantilever beam with a spring–mass system attached to the free end. *Ocean Engineering,* 28: 933–939.
20. Tso, S.K., T.W. Yang, W.L. Xu, and Sun, Z. Q. 2003. Vibration control for a flexible-link robot arm with deflection feedback. *International Journal of Non-Linear Mechanics.*
21. Wrobel, L.C., and M.H. Aliabadi. 2002. *The Boundary Element Method: Applications in Solids and Structures*, 1st ed. Wiley.

Chapter 3
Conservation Principles in Fluid Mechanics and Potential Flow Aerodynamics

Abstract The present chapter elaborates physical principles that govern theoretical, analytical and experimental fluid mechanics and aerodynamics, based on principles of mechanics, including those incorporated in thermodynamics that dictates conservation principles. Four generalh assumptions regarding the properties of the liquids and gases that form the subject of this book are made and retained throughout except cases where there are needs to consider more accurate considerations of more advanced and modern physics. These are assumptions that the fluid is a continuum, inviscid and adiabatic and either a perfect gas or a constant-density fluid (unless viscous fluids are considered). If there are discontinuities, such as shocks, compression and expansion wave or vortex sheets that are present, more physical aspects have to be incorporated in the considerations. For this purpose, these discontinuities will normally be treated as separate and serve as boundaries for continuous portions of the flow field. The laws of motion of the fluid are fundamental and can be found in any fundamental text on hydrodynamics or gas dynamics. The differential equations which apply the basic laws of physics to this situation are then elaborated. The concept of lifting flow over a cylinder will be utilized in the development of the theory of the lift generated by airfoils. Much of the materials here have been taken from the author's book "Mekanika Fluida" in Indonesian [1].

Keywords Aerodynamics · Conservation principles · Dragan lift · Fluid mechanics · Mechanics · Physics · Thermodynamics

---

Harijono Djojodihardjo, Mekanika Fluida (Fluid Mechanics, in Indonesia), Penerbit Erlangga, Jakarta, Indonesia, 1983.

© The Author(s), under exclusive license to Springer Nature Singapore Pte Ltd. 2023

H. Djojodihardjo, *Introduction to Aeroelasticity*,
https://doi.org/10.1007/978-981-16-8078-6_3

3.1 Potential Flow Fluid Dynamics; Conservation Principles

3.1.1 General Assumptions

In the present elaboration of Fluid Mechanics and Potential Flow Aerodynamics, four general assumptions regarding the properties of the liquids and gases that form the subject of the present chapter and book are made and maintained throughout, unless otherwise mentioned for some selected and special cases. These are as follows:

(1) The fluid is a continuum.
(2) The fluid is inviscid and adiabatic.
(3) The fluid either a perfect gas or a constant-density fluid.

If discontinuities, such as shocks, compression and expansion wave, or vortex sheets, are present, then each will be treated separately and may serve as boundaries for continuous portions of the flow field. The laws of motion of the fluid will be found derived in any fundamental text on Hydrodynamics, Fluid Mechanics or Gas dynamics. The differential equations which apply the basic laws of physics will be elaborated subsequently.

3.1.2 Thermodynamic Laws

Conservation Principle forms the basis of Thermodynamic and Fluid Sciences Laws. **Thermodynamics** has traditionally recognized **three** fundamental **laws**, simply named by an ordinal identification, the first **law**, the second **law** and the third **law**. The Zeroth **Law of Thermodynamics** states that if two **thermodynamic** systems are each in thermal equilibrium with a third, then they are in thermal equilibrium with each other. The First **Law of Thermodynamics**, also known as the **Law** of Conservation of Energy, states that energy cannot be created or destroyed in an isolated system. In any process, the total energy of the universe remains the same. The **Second Law of Thermodynamics** says that processes that involve the transfer or conversion of heat energy are irreversible. The **Second Law** also states that there is a natural tendency of any isolated system to degenerate into a more disordered state.

The **Third Law of Thermodynamics** states that the entropy of a system approaches a constant value as the temperature approaches absolute zero. The Fourth **Law of Thermodynamics** is the conservation of time. All of space, matter and energy are contained by time like a great bubble. Time is infinite and extends in all directions through a range of 1440 decimal degrees simultaneously and is flexible yet indestructible.

3.1.3 Thermodynamic Properties

Thermodynamic properties are divided into two broad types: intensive **properties** and **extensive properties**. An **extensive property** is any **property** that depends on the size (or extent) of the system under consideration.

Examples of extensive properties include Mass, M; Volume, V; amount of substance, n; energy, E; enthalpy, H; entropy, S; Gibbs energy, G; **heat** capacity, C_p; Helmholtz energy, A or F; and internal energy, U.

The ratio of two **extensive properties** of the same object or system is an **intensive property**. For example, the ratio of an object's mass and volume, which are two **extensive properties**, is density, which is an **intensive property**.

Temperature, pressure, specific volume and **density** are **examples** of intensive properties.

An extensive **property** of a system depends upon the total amount of material in the system. Mass, volume, internal energy, heat contents, free energy, entropy and heat capacity are all extensive **properties**.

Pressure does not depend upon the amount of substance because it is defined as force per unit area. **Pressure** is an **intensive property** (independent of amount).

3.1.4 Conservation Principle[1]

The results derived here is also known as Reynolds Transport Theorem [2–4]. Much of the work elaborated here has also previously appeared in reference [1]. The Conservation Principle in general is based on the fact that if we look at a system that possesses an extensive thermodynamic property, then the flux of that extensive property coming out of a system minus the flux of that extensive property entering the system should be the same as the rate of change of that extensive property in the system. Here, a system is defined as a finite mass of fluid with fixed identity, which is contained in a finite volume, which may change shape with time.

This principle will be demonstrated and derived as a general integral, which involves the use of conservation of mass or the equation of continuity.

We will look at a system with a volume V at the time instant $t = t_0$ and is bounded by a surface S_{t0}, as depicted in Fig. 3.1.

Then, after Δt elapses, at $t = t_0 + \Delta t$, the system occupies a new position, which is bounded by the surface S_t. Let us now consider a new variable B, which is an extensive property of the system ($B = bm$, where b is the corresponding intensive property and m is the mass of the system).

---

[1] The results derived here is also known as Reynolds Transport Theorem. See for example: Rutherford Aris, Vectors, Tensors and the Basic Equations of Fluid Mechanics, DOVER PUBLICATIONS, 1990, ISBN 0-486-66110-5 and Hyunse Yoon, Control Volume and Reynolds Transport Theorem, 10–11–2013, and Raj Nandkeolyar, Reynolds Transport Theorem and Navier-Stoke's Equation, http://www.nitjsr.ac.in/.

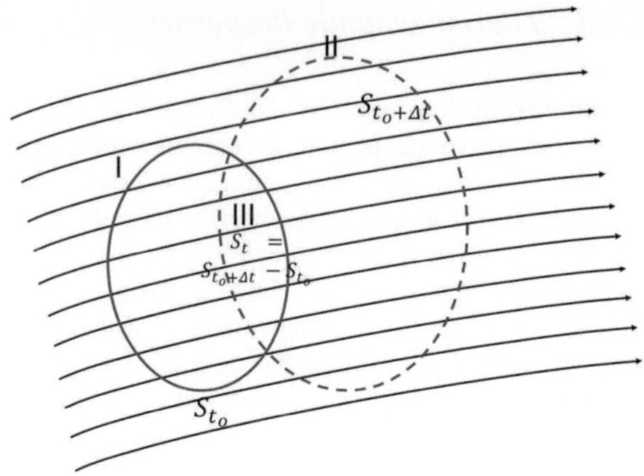

Fig. 3.1 The movement of fluid mass in a fluid flow and the definition of system and control volume by identifying volumes and surfaces in the flow for deriving the principle of mass conservation [1]

We note that a system S, which has a constant mass with fixed identity, is referred to as a **closed system**. However, during its movement from $t = t_0$, to $t = t_0 + \Delta t$, the surface of the system S has occupied different surfaces, S_{t0} and $S_t = S_{t0} + \Delta t$, respectively, which is due to the change of position of S during its infinitesimal movement.

Such changes takes place since the system S experiences local volume changes, i.e. at some places it loses some infinitesimal volumes, and at other places, gain some infinitesimal volumes, as a result of the movement of the System S, and accordingly, the movement of the bounding Surfaces S_{t0} and S_t.

Hence, one can write

$$\frac{DB}{Dt} = \frac{D}{Dt} \iiint \rho b \, \mathrm{d}V \tag{3.1}$$

where

$\iiint \rho b \, \mathrm{d}V$ is the mass of the system.

By referring to Fig. 3.1, the change of the extensive property B as represented by the Eq. (3.1) can be elaborated as:

$$\frac{DB}{Dt} = \lim_{\Delta t \to 0} \frac{\left(B^{I}_{t_0+\Delta t} + B^{II}_{t_0+\Delta t}\right) - \left(B^{I}_{t_0} + B^{II}_{t_0}\right)}{\Delta t} \tag{3.2a}$$

where

$B^{I}_{t_0+\Delta t}$ the amount of B in the volume bounded by the surface I at time $t_0 + \Delta t$
$B^{II}_{t_0+\Delta t}$ the amount of B in the volume bounded by the surface II at time $t_0 + \Delta t$
$B^{I}_{t_0}$ the amount of B in the volume bounded by the surface I at time t_0
$B^{II}_{t_0}$ the amount of B in the volume bounded by the surface II at time t_0.

The sum of the extensive property B in the first bracket represents the state at time $t_0 + \Delta t$ while the sum of the extensive property B in the second bracket represents the state at time t_0.

However, the sum of the extensive property B at time t_0 can be expressed by

$$\left(B^I_{t_0} + B^{II}_{t_0}\right)$$

Hence

$$\frac{DB}{Dt} = \lim_{\Delta t \to 0} \frac{\left(B^I_{t_0+\Delta t} + B^{II}_{t_0+\Delta t}\right) - \left(B^I_{t_0} + B^{II}_{t_0}\right)}{\Delta t} \frac{DB}{Dt}$$
$$= \lim_{\Delta t \to 0} \frac{\left(B^I_{t_0+\Delta t} - B^I_{t_0}\right)}{\Delta t} + \lim_{\Delta t \to 0} \frac{\left(B^{II}_{t_0+\Delta t} - B^{II}_{t_0}\right)}{\Delta t} \tag{3.2b}$$

It can be noted that:

$$\lim_{\Delta t \to 0} \frac{\left(B^I_{t_0+\Delta t} - B^I_{t_0}\right)}{\Delta t} = \lim_{\Delta t \to 0} \frac{\iiint_{\text{Volume bounded by the surface } S_0} (b\rho)\, \mathrm{d}V}{\Delta t} \tag{3.3}$$

To further elaborate the expression on the right-hand side of Eq. (3.3), one can look at a small plane element at the surface I that is adjacent to the surface II, as depicted in Fig. 3.2.

$$\lim_{\Delta t \to 0} \frac{B^I_{t_0+\Delta t}}{\Delta t} = \lim_{\Delta t \to 0} \sum_{\substack{\Delta A \\ \text{intersection of } B^I \text{ and } B^{II}}} \frac{\rho b \boldsymbol{V}_A \cdot \boldsymbol{n} \Delta A\, \Delta t}{\Delta t} \tag{3.4}$$

or

$$\frac{DB}{Dt} = \iint_{\text{Intersection of } B^I \text{ and } B^{II}} \rho b \boldsymbol{V}_A \cdot \boldsymbol{n}\, \mathrm{d}A = \iint_{\text{Intersection of } B^I \text{ and } B^{II}} \rho\, b\, \boldsymbol{V}_n\, \mathrm{d}A \tag{3.5}$$

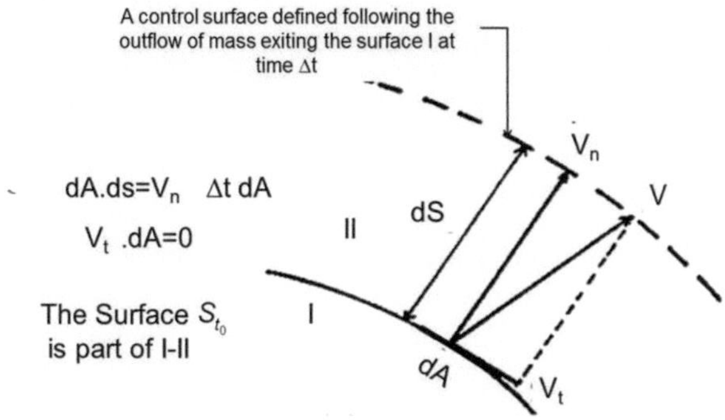

Fig. 3.2 The movement of a fluid mass with fixed identity across a small plane element at the surface I that is adjacent to the surface II

where

- V_n the velocity component at ΔA perpendicular the surface A outward from the fluid volume bounded by Surface B^I
- **V** the velocity vector at ΔA perpendicular the surface A outward from the fluid volume bounded by Surface B^I
- **n** the normal vector at ΔA perpendicular the surface A outward from the fluid volume bounded by Surface B^I
- dA infinitesimal area element at Surface A of B^I adjacent to B^{II}.

Similarly

$$\lim_{\Delta t \to 0} \frac{B_t^{II}}{\Delta t} == \iint\limits_{\text{Intersection of } B^I \text{ and } B^{II}} \rho\, b\, \boldsymbol{V}_A \cdot \boldsymbol{n}\, \mathrm{d}A \tag{3.6}$$

Therefore:

$$\frac{DB}{Dt} = \iiint\limits_{\text{Control Volume } V} \frac{\partial}{\partial t}(\rho\, b)\, \mathrm{d}V + \iint\limits_{S_{in}=\text{Left Control Surface}} (\rho\, b)\boldsymbol{V} \cdot \boldsymbol{n}\, \mathrm{d}A - \iint\limits_{S_{out}=\textit{Right Control Surface}} (\rho\, b)\boldsymbol{V} \cdot \boldsymbol{n}\, dA \tag{3.7}$$

or

$$\frac{DB}{Dt} = \iiint\limits_{\text{Control Volume } V} \frac{\partial}{\partial t}(\rho\, b)\, dV + \oiint\limits_{\text{Control Surface } S} (\rho\, b)\boldsymbol{V} \cdot \boldsymbol{n}\, dA \tag{3.8}$$

where the second and third terms of Eq. (3.7) have been unified in a closed surface integral covering the entire system, which in the limit of $\Delta t \to 0$, constitutes the control volume $V = \text{I} + \text{II} + \text{III}$. Equation (3.8) is known as **Reynolds Transport Equation** or **Reynolds Transport Theorem** [1–3].

It can readily be shown that for a multiply connected volume as depicted by Fig. 3.3, integral can be rewritten as

$$\frac{DB}{Dt} = \iint\limits_{\text{Control Volume } V} \frac{\partial}{\partial t}(\rho\, b)\; dV + \oiint\limits_{\text{Control Surface } S+\sigma} (\rho\, b)\boldsymbol{V} \cdot \boldsymbol{n}\, \mathrm{d}A \tag{3.9}$$

The derivation above applies for conservation principle for mass and energy, and any other thermodynamic property, and has been demonstrated using "System Approach" that can easily be extended to "Control Volume Approach [1]."

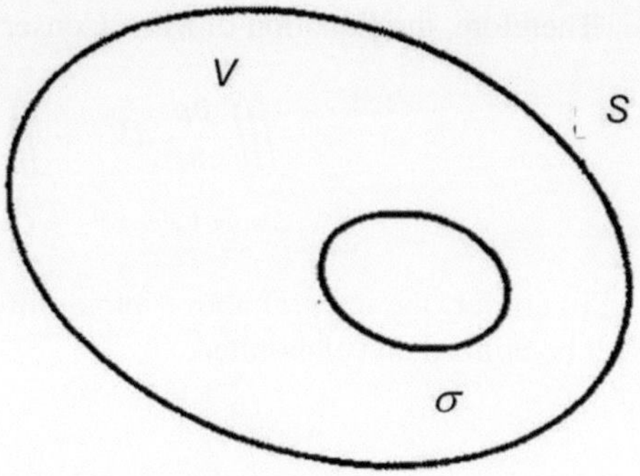

Fig. 3.3 Conservation principle for multiply-connected domain

3.2 Elaboration of Conservation and Compatibility Principle

3.2.1 Continuity Equation

To derive the equation of mass conservation (or the continuity equation) which is general in nature, two methods can be followed. The first uses a finite size control volume, while the second uses an infinitesimal control volume.

The first method is also known as the integral approach, which can follow closely the derivation and result obtained in the previous section, that is Eq. (3.9). For this purpose, since the extensive property of interest is the mass of the system, one can set B = m, or b = 1. Substituting this value into Eq. (3.9), there is obtained

$$\frac{Dm}{Dt} = \frac{D}{Dt}\iiint_{Control\ Volume\ V} \rho\, dV = \iiint_{Control\ Volume\ V} \frac{\partial \rho}{\partial t} dV + \oiint_{Control\ Surface\ S} \rho \boldsymbol{V} \cdot \boldsymbol{n}\, dA \tag{3.10a}$$

or

$$\frac{Dm}{Dt} = \frac{D}{Dt}\iiint_{Control\ Volume\ V} \rho\, \mathrm{d}V = \iiint_{Control\ Volume\ V} \frac{\partial \rho}{\partial t} \mathrm{d}V + \oiint_{Control\ Surface\ S} \rho\, \boldsymbol{V} \cdot \boldsymbol{n}\, \mathrm{d}A \tag{3.10b}$$

If there is no change of mass in the control volume, then

$$\frac{Dm}{Dt} = 0. \tag{3.10c}$$

Therefore, the Equation of Mass Conservation (or Continuity) can be written as:

$$\underset{Control\ Volume\ V}{\iiint} \frac{\partial \rho}{\partial t}\, dV + \underset{Control\ Surface\ S}{\oiint} \rho\, \mathbf{V} \cdot \boldsymbol{n}\, dA = 0 \tag{3.10}$$

To convert the integral above into a differential equation, some vector identities will be utilized and elaborated.

3.2.2 Gauss Theorem

$$\underset{Control\ Surface\ S}{\oiint} (\rho \boldsymbol{Q}) \cdot \boldsymbol{n}\, dA \equiv \underset{Control\ Volume\ V}{\iiint} \nabla \cdot (\rho \boldsymbol{Q}) dV \equiv \underset{Control\ Volume\ V}{\iiint} div(\rho \boldsymbol{Q}) dV \tag{3.11}$$

where Q is any vector variable and[2]

$$\nabla = \boldsymbol{i}\frac{\partial}{\partial x} + \boldsymbol{j}\frac{\partial}{\partial y} + \boldsymbol{k}\frac{\partial}{\partial z}$$

$$\boldsymbol{Q} = \boldsymbol{i}\, Q_x + \boldsymbol{j}\, Q_y + \boldsymbol{k} Q_z$$

so that

$$\nabla \cdot \boldsymbol{Q} = \frac{\partial Q_x}{\partial x} + \frac{\partial Q_y}{\partial y} + \frac{\partial Q_z}{\partial z}$$

Using this theorem, the vector variable **Q** can be replaced by velocity vector **V**. Then, Eqs. (3.10a–3.10d) can be converted to

$$\underset{Control\ Volume\ V}{\iiint} \frac{\partial \rho}{\partial t} \mathrm{d}V + \underset{Control\ Volume\ V}{\iiint} \nabla \cdot (\rho\, \boldsymbol{V}) \mathrm{d}V \tag{3.12a}$$

or:

$$\underset{Control\ Volume\ V}{\iiint} \left[\frac{\partial \rho}{\partial t} + \nabla \cdot (\rho\, \boldsymbol{V})\right] \mathrm{d}V = 0 \tag{3.12b}$$

---

[2] Note that V stands fo Volume and **V** stands for velocity vector.

Since the control volume is arbitrary, then Eq. (3.12b) implies that the integrand should be zero.

Therefore:

$$\left[\frac{\partial \rho}{\partial t} + \nabla \cdot (\rho \boldsymbol{V})\right] = 0 \tag{3.13}$$

This equation is known as the continuity equation in differential form, which has been derived using control volume approach, i.e. integral approach.

The second method to derive the continuity equation can be carried out by using a differential approach. For this purpose, an infinitesimal control volume $\Delta V = \Delta x\, \Delta y\, \Delta z$ is utilized, as depicted in Fig. 3.4. The Cartesian coordinate system O-xyz is also be utilized.

Following the principle of mass conservation (Eqs. 3.10a–3.10d), the rate of mass flow outward (mass outflow) from a control volume minus the rate of mass flow into (mass inflow) a control volume should be zero, if the mass in the control volume is constant. Otherwise, this difference should be the time rate of change of the mass in the control volume; positive if the mass outflow is less than the mass inflow.

The mass flowrate outward from the infinitesimal control volume $\Delta V = \Delta x\, \Delta y\, \Delta z$ is given by

in the x-direction

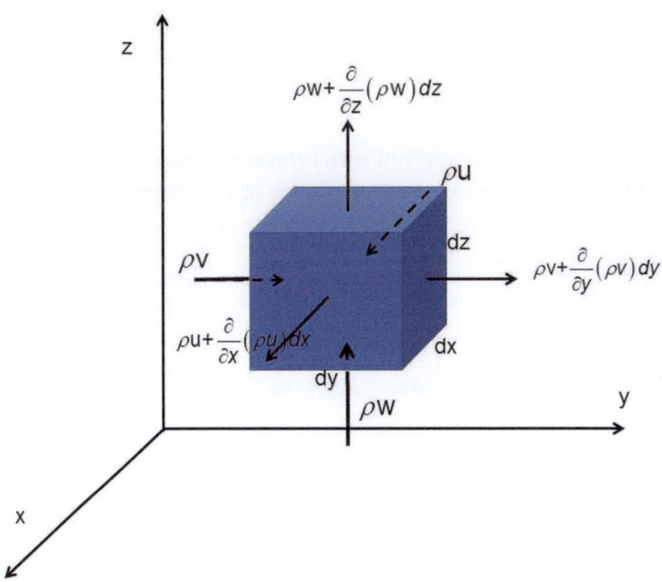

Fig. 3.4 Infinitesimal control volume $\Delta V = \Delta x\, \Delta y\, \Delta z$ for deriving continuity equation

$$\left(\rho u + \frac{\partial}{\partial x}(\rho u)\,\mathrm{d}x - \rho u\right)\mathrm{d}y\,\mathrm{d}z = \frac{\partial}{\partial t}(\rho u)\,\mathrm{d}x\,\mathrm{d}y\,\mathrm{d}z$$

in the y-direction

$$\left(\rho v + \frac{\partial}{\partial y}(\rho v)\,\mathrm{d}y - \rho v\right)\mathrm{d}z\,\mathrm{d}x = \frac{\partial}{\partial t}(\rho v)\,\mathrm{d}x\,\mathrm{d}y\,\mathrm{d}z$$

in the z-direction

$$\left(\rho w + \frac{\partial}{\partial z}(\rho w)\,\mathrm{d}z - \rho v\right)\mathrm{d}x\,\mathrm{d}y = \frac{\partial}{\partial t}(\rho w)\,\mathrm{d}x\,\mathrm{d}y\,\mathrm{d}z$$

This value should be the same as the rate of mass decrease in the infinitesimal control volume.

$$-\frac{\partial}{\partial t}(\rho \mathrm{d}x\,\mathrm{d}y\,\mathrm{d}z) = -\frac{\partial\,(\rho \mathrm{d}x\,\mathrm{d}y\,\mathrm{d}z)}{\partial t} = -\frac{\partial \rho}{\partial t}(\mathrm{d}x\,\mathrm{d}y\,\mathrm{d}z)$$

Therefore:

$$\frac{\partial}{\partial x}(\rho u)\,\mathrm{d}x\,\mathrm{d}y\,\mathrm{d}z + \frac{\partial}{\partial y}(\rho v)\mathrm{d}x\,\mathrm{d}y\,\mathrm{d}z + \frac{\partial}{\partial z}(\rho w)\,\mathrm{d}x\,\mathrm{d}y\,\mathrm{d}z = -\frac{\partial \rho}{\partial t}(\mathrm{d}x\,\mathrm{d}y\,\mathrm{d}z)$$

or

$$\frac{\partial \rho}{\partial t} + \nabla\cdot(\rho\,\boldsymbol{V}) = 0 \tag{3.13a}$$

Equation (3.13) **has been derived with reference to a person that is positioned in the absolutely stationary coordinate frame of reference O-x y z.**

Equation (3.13) can alternatively be written as

$$\frac{\partial \rho}{\partial t} + \nabla\cdot(\rho\,\boldsymbol{V}) = \frac{\partial \rho}{\partial t} + \rho\,\nabla\cdot\boldsymbol{V} + \boldsymbol{V}\cdot\nabla\rho$$
$$= \frac{\partial \rho}{\partial t} + \rho\left(\frac{\partial u}{\partial x} + \frac{\partial v}{\partial y} + \frac{\partial w}{\partial z}\right) + u\frac{\partial \rho}{\partial x} + v\frac{\partial \rho}{\partial y} + w\frac{\partial \rho}{\partial z} = 0 \tag{3.14}$$

or:

$$\frac{D\rho}{Dt} + \rho\,\nabla\cdot\boldsymbol{V} = \frac{D\rho}{Dt} + \rho\,\mathrm{div}\,\boldsymbol{V} = 0 \tag{3.15}$$

where:

$$\frac{D\rho}{Dt} = \left(\frac{\partial}{\partial t} + u\frac{\partial}{\partial x} + v\frac{\partial}{\partial y} + w\frac{\partial}{\partial z}\right)\rho \tag{3.16}$$

is the total or substantial derivative of ρ.

The first term in the total derivative bracket is $\left(\frac{\partial}{\partial t}\right)$, which is the local rate of change of the variable ρ at a spatial point within the fluid, while the other three terms represent the change of that variable ρ due the movement of the fluid, also referred to as the **convective terms**. In other words, the total derivative *Dp/Dt* represents the rate of change of the density of the fluid as a system that follows the movement of the fluid. Hence, Dp/Dt represents the rate of change of a collection of fluid mass of fixed identity and represents the rate of change that is observed by a person moving with the same velocity **V** of the fluid. **Hence,** Eq. (3.15) **is the continuity equation written on the basis of the observation of a person moving along with the fluid**.

If the fluid density is constant (the fluid is then referred to as incompressible), the continuity equation becomes

$$\frac{\partial u}{\partial x} + \frac{\partial v}{\partial y} + \frac{\partial w}{\partial z} = 0 \tag{3.17a}$$

or

$$\nabla \cdot \boldsymbol{V} = 0 \tag{3.17b}$$

Equations (3.17a), (3.17b) is always valid for velocities which are independent of time as well as dependent of time. A flow that is independent of time is called a stationary flow. For stationary flows, all partial derivatives with respect to time are zero.

3.3 Dynamics of Fluid Flow

3.3.1 Equation of Motion in Mechanics and Euler Equation in Fluid Dynamics

The Equation of Motion in Mechanics and Euler Equation in Fluid Dynamics originated from Newton's Second Law. For fluid, the derivation of the equation of motion can be carried out using two approaches, the differential approach applied to infinitesimal fluid element and the integral approach using a finite control volume.

In mechanics, the Newton's law of motion can be expressed as

$$\boldsymbol{F} = m\,\boldsymbol{a} \tag{3.18}$$

With components in the coordinate direction of x, y and z as:

$$F_x = m\,a_x$$
$$F_y = m\,a_y$$

$$F_z = m\, a_z$$

or

$$\boldsymbol{F} = \frac{D\,(m\; V)}{Dt} \tag{3.19}$$

where **F** is a force applied to the body, m the mass of the body, and **a** is the acceleration and **V** the velocity. The second approach, which is a fundamental form of an equation of motion that can be applied to fluids, relates the rate of change of momentum of a cluster of fluid mass to the summation of forces acting on the cluster of the fluid mass.

If the equation of motion is written in the form 18 or 19, then the coordinate frame of reference should not have any acceleration or rotation. In other words, the inertial system of coordinate (or absolutely stationary coordinate system) should be used. If another coordinate frame of reference which has an acceleration or a rotation, additional terms should be added, which will be elaborated later. In what follows here, an inertial coordinate system will be utilized.

3.3.2 *The Derivation of the Equation of Motion Using Differential Approach*

For this purpose, we will consider a fluid element with fixed identity and fixed mass, which is moving with the rest of the fluid. Let **u** and **a** denote the velocity and acceleration vector, respectively, at the center of mass of the fluid element. u, v and w denote the components of **u** along the x, y and z coordinate, respectively, while a_x, a_y and a_z denote the components of a along the x, y and z coordinate, respectively.

Then[3]:

$$a_x = \frac{du}{dt} = \frac{\partial u}{\partial t} + (\boldsymbol{V}\cdot\nabla)\,u = \frac{\partial u}{\partial t} + \boldsymbol{V}\cdot\left(\boldsymbol{i}\frac{\partial u}{\partial x} + \boldsymbol{j}\frac{\partial u}{\partial y} + \boldsymbol{k}\frac{\partial u}{\partial z}\right) \tag{3.20a}$$

$$a_x = \frac{du}{dt} = \frac{\partial u}{\partial t} + (\boldsymbol{V}\cdot\nabla)\,u = \frac{\partial u}{\partial t} + \left(\boldsymbol{V}\cdot\left(\boldsymbol{i}\frac{\partial}{\partial x} + \boldsymbol{j}\frac{\partial}{\partial y} + \boldsymbol{k}\frac{\partial}{\partial z}\right)\right)u \tag{3.20aa}$$

$$a_y = \frac{dv}{dt} = \frac{\partial v}{\partial t} + (\boldsymbol{V}\cdot\nabla)\,v = \frac{\partial v}{\partial t} + \boldsymbol{V}\cdot\left(\boldsymbol{i}\frac{\partial v}{\partial x} + \boldsymbol{j}\frac{\partial v}{\partial y} + \boldsymbol{k}\frac{\partial v}{\partial z}\right) \tag{3.20b}$$

$$a_y = \frac{dv}{dt} = \frac{\partial v}{\partial t} + (\boldsymbol{V}\cdot\nabla)\,v = \frac{\partial v}{\partial t} + \left(\boldsymbol{V}\cdot\left(\boldsymbol{i}\frac{\partial}{\partial x} + \boldsymbol{j}\frac{\partial}{\partial y} + \boldsymbol{k}\frac{\partial}{\partial z}\right)\right)v \tag{3.20bb}$$

[3] Here $\frac{D}{Dt} \equiv \frac{d}{dt}$ are used interchangeable.

$$a_z = \frac{dw}{dt} = \frac{\partial w}{\partial t} + (\boldsymbol{V} \cdot \nabla)\, w = \frac{\partial w}{\partial t} + \boldsymbol{V} \cdot \left(\boldsymbol{i}\,\frac{\partial w}{\partial x} + \boldsymbol{j}\,\frac{\partial w}{\partial y} + \boldsymbol{k}\,\frac{\partial w}{\partial z}\right) \tag{3.20c}$$

$$a_z = \frac{dw}{dt} = \frac{\partial w}{\partial t} + (\boldsymbol{V} \cdot \nabla)\, w = \frac{\partial w}{\partial t} + \left(\boldsymbol{V} \cdot \left(\boldsymbol{i}\,\frac{\partial}{\partial x} + \boldsymbol{j}\,\frac{\partial}{\partial y} + \boldsymbol{k}\,\frac{\partial}{\partial z}\right)\right) w \tag{3.20cc}$$

or

$$\boldsymbol{a} = \frac{d\boldsymbol{V}}{dt} = \frac{\partial \boldsymbol{V}}{\partial t} + (\boldsymbol{V} \cdot \nabla)\, \boldsymbol{V} = \frac{\partial \boldsymbol{V}}{\partial t} + \left(\boldsymbol{V} \cdot \left(i\,\frac{\partial}{\partial x} + j\,\frac{\partial}{\partial y} + k\,\frac{\partial}{\partial z}\right)\right) \boldsymbol{V} \tag{3.21}$$

or

$$\begin{aligned} a_x &= \frac{du}{dt} = \frac{\partial u}{\partial t} + (\boldsymbol{V} \cdot \nabla)\, u \\ &= \frac{\partial u}{\partial t} + \left((u\,\boldsymbol{i} + v\boldsymbol{j} + w\,\boldsymbol{k}) \cdot \left(\boldsymbol{i}\,\frac{\partial}{\partial x} + \boldsymbol{j}\,\frac{\partial}{\partial y} + \boldsymbol{k}\,\frac{\partial}{\partial z}\right)\right) u \\ &= \frac{\partial u}{\partial t} + \left(u\frac{\partial}{\partial x} + v\frac{\partial}{\partial y} + w\,\frac{\partial}{\partial z}\right) u \end{aligned} \tag{3.21a}$$

$$\begin{aligned} a_y &= \frac{dv}{dt} = \frac{\partial v}{\partial t} + (\boldsymbol{V} \cdot \nabla)\, v \\ &= \frac{\partial v}{\partial t} + \left((u\,\boldsymbol{i} + v\boldsymbol{j} + w\,\boldsymbol{k}) \cdot \left(\boldsymbol{i}\,\frac{\partial}{\partial x} + \boldsymbol{j}\,\frac{\partial}{\partial y} + \boldsymbol{k}\,\frac{\partial}{\partial z}\right)\right) v \\ &= \frac{\partial v}{\partial t} + \left(u\frac{\partial}{\partial x} + v\frac{\partial}{\partial y} + w\,\frac{\partial}{\partial z}\right) v \end{aligned} \tag{3.21b}$$

$$\begin{aligned} a_z &= \frac{dw}{dt} = \frac{\partial w}{\partial t} + (\boldsymbol{V} \cdot \nabla)\, w \\ &= \frac{\partial w}{\partial t} + \left((u\,\boldsymbol{i} + v\boldsymbol{j} + w\,\boldsymbol{k}) \cdot \left(\boldsymbol{i}\,\frac{\partial}{\partial x} + \boldsymbol{j}\,\frac{\partial}{\partial y} + \boldsymbol{k}\,\frac{\partial}{\partial z}\right)\right) w \\ &= \frac{\partial w}{\partial t} + \left(u\frac{\partial}{\partial x} + v\frac{\partial}{\partial y} + w\,\frac{\partial}{\partial z}\right) w \end{aligned} \tag{3.21c}$$

First, in what follows, the effect of friction is ignored. The forces that are acting on the fluid element can be differentiated into body force, which acts on each mass particle of the fluid element, and surface force, which acts on the surface of the fluid element. Examples of body force are gravitational, magnetic, electrostatic, Lorentz force and so forth.

The surface force is a force due to pressure distribution on the surface of the fluid mass element. The body force is proportional to the mass of the fluid element. The

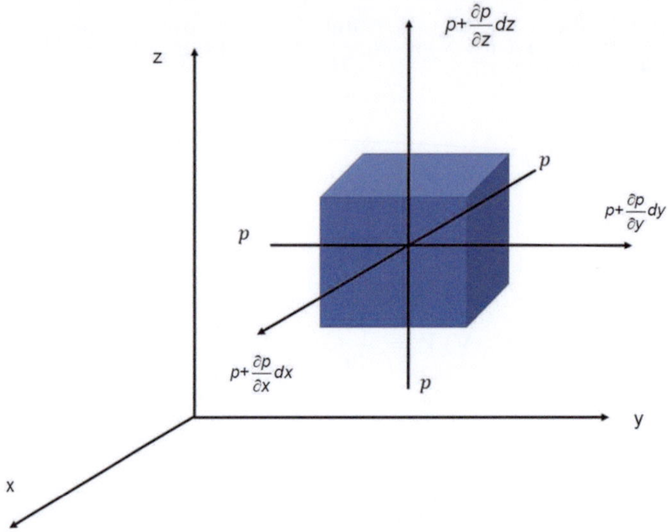

Fig. 3.5 Forces acting on a cubical fluid element without considering friction

components of the body force per unit mass of the fluid element are denoted by f_x, f_y and f_z (or f_x, f_y andn f_z). Figure 3.5 depicts these forces on a cubical fluid mass element.

Using Newton's law of motion, one can write

$$
\begin{aligned}
m\boldsymbol{a} &= \rho\, dx\, dy\, dz \frac{D\boldsymbol{V}}{Dt} = \rho\left(\frac{\partial u}{\partial t} + (\boldsymbol{V}\cdot\nabla)\,\boldsymbol{V}\right) dx\, dy\, dz \\
&= \rho\left(\frac{\partial \boldsymbol{V}}{\partial t} + \left(u\frac{\partial}{\partial x} + v\frac{\partial}{\partial y} + w\frac{\partial}{\partial z}\right)\boldsymbol{V}\right) dx\, dy\, dz \\
&= \left(-\frac{\partial p}{\partial x} + \rho \boldsymbol{f}\right) dx\, dy\, dz
\end{aligned}
\tag{3.22a}
$$

or

$$
m\boldsymbol{a} = \rho\, dx\, dy\, dz \frac{D\boldsymbol{V}}{Dt} = \boldsymbol{F} = (-\nabla p + \rho \boldsymbol{f}) dx\, dy\, dz \tag{3.22b}
$$

or:

$$
\frac{D\,\boldsymbol{V}}{Dt} = \frac{\partial\,\boldsymbol{V}}{\partial t} + (\boldsymbol{V}\cdot\nabla)\,\boldsymbol{V} = -\frac{1}{\rho}\nabla p + \boldsymbol{f} \tag{3.22c}
$$

Hence, in the x-direction

$$m\,a_x = \rho\,\mathrm{d}x\,\mathrm{d}y\,\mathrm{d}z\frac{Du}{Dt} = F_x = \left(-\frac{\partial p}{\partial x} + \rho f_x\right)\mathrm{d}x\,\mathrm{d}y\,\mathrm{d}z \tag{3.23a}$$

or,

$$\frac{Du}{Dt} = -\frac{1}{\rho}\frac{\partial p}{\partial x} + f_x \tag{3.23aa}$$

In the y-direction:

$$m\,a_y = \rho\,\mathrm{d}x\,\mathrm{d}y\,\mathrm{d}z\frac{Dv}{Dt} = F_y = \left(-\frac{\partial p}{\partial x} + \rho f_y\right)\mathrm{d}x\,\mathrm{d}y\,\mathrm{d}z \tag{3.23b}$$

or,

$$\frac{Dy}{Dt} = -\frac{1}{\rho}\frac{\partial p}{\partial y} + f_y \tag{3.23bb}$$

In the z-direction:

$$m\,a_z = \rho\,\mathrm{d}x\,\mathrm{d}y\,\mathrm{d}z\frac{Dw}{Dt} = F_z = \left(-\frac{\partial p}{\partial x} + \rho f_z\right)\mathrm{d}x\,\mathrm{d}y\,\mathrm{d}z \tag{3.23c}$$

or,

$$\frac{Dw}{Dt} = -\frac{1}{\rho}\frac{\partial p}{\partial z} + f_z \tag{3.23cc}$$

Equations (3.22a), (3.22b) and (3.22c), and (3.23a–3.23c) represent the equation of motion of a fluid mass element within an inertial coordinate frame of reference. There are no assumptions taken about the density of the fluid. Hence these equations are valid for compressible a well as incompressible fluid. This equation was first derived by Euler, hence the name Euler Equation.

3.3.3 The Derivation of the Equation of Motion Using Integral Approach

Equation (3.23a–3.23c) can also be derived using integral approach [1]. For a fluid with control volume V that is bounded by a surface S, its momentum is expressed by

$$\iiint \rho V\,\mathrm{d}\tau$$

where τ is a volume element within the control volume V. Hence, the rate of change of the momentum within the control volume V is given by:

$$\frac{\partial}{\partial t} \iiint \rho V \, d\tau \tag{3.24}$$

The integrand in the above integral is the product of the mass per unit volume and its velocity in the control volume V.

Following Newton's Second Law, then:

$$\frac{D}{Dt}(momentum\ within\ the\ Control\ Volume\ V) = Force \tag{3.25}$$

The resultant vector of the body force per unit mass in the control volume V can be defined as **f**; hence, the total force on the fluid mass cluster within the control volume V can be expressed as

$$\boldsymbol{F} = \iiint_V f \, d\tau \tag{3.26}$$

Next, the surface forces on the surface of the infinitesimal control volume $\Delta V = \Delta x,\ \Delta y,\ \Delta z$ as depicted in Fig. 3.5 will be considered. The i-th component of the surface forces on a surface element dA with normal $\mathbf{n}_j$ can be defined as

$$\sigma_{ij}\ n_j\ dA$$

where σ_{ij} is a stress tensor[4]; if there is no friction or viscosity, the stress tensor will only comprise three diagonal terms, i.e. σ_{ii}, since $\sigma_{ij} = 0$ if $i \neq j$. In general, $\sigma_{ii} \neq 0$. Therefore, the total surface forces acting on the fluid mass cluster in the infinitesimal control volume $\Delta V = \Delta x \Delta y \Delta z$ bounded by its surfaces can be given by[5]

$$\boldsymbol{P} = \oiint_S \sigma_i \boldsymbol{n}_i \, dS \tag{3.27}$$

If the working pressure in the fluid is given by

$$p = \frac{1}{3}\sigma_{ii} \tag{3.28}$$

---

[4] See Chap. 2.

[5] Note that it is an accepted convention that the prssure p is directed into the surface, i.e. in the direction opposite tote normal vector of the surface, while the stress σ (or σ_{ii} on the surface with i the coordinate along the normal vector of that surface) is positive outward along the direction of the normal vector of a surface.

Then

$$\boldsymbol{P} = \oiint\limits_{S} -\frac{1}{3}\sigma_{ii}\,\boldsymbol{n}_i\,dS = \oiint\limits_{S} -p\,\boldsymbol{n}\,dA \tag{3.29}$$

This means that the surface force is equal to the vector summation of the normal components of the pressure at each surface times its corresponding area.

By using Gauss theorem 3.11 on the Eq. (3.29), there is obtained

$$\boldsymbol{P} = \oiint\limits_{S} -\frac{1}{3}\sigma_{ii}\,\boldsymbol{n}_i\,dS = \oiint\limits_{S} -p\,\boldsymbol{n}\,dA \tag{3.30}$$

or

$$\boldsymbol{P} = \iiint\limits_{Control\ Volume\ V} -\frac{1}{3}\frac{\partial\sigma_{ij}}{\partial x_j}\,d\tau = \iiint\limits_{Control\ Volume\ V} -\frac{1}{3}\nabla\sigma\,d\tau = \iiint\limits_{Control\ Volume\ V} \nabla p\,d\tau \tag{3.30}$$

or

$$\boldsymbol{P} = \iiint\limits_{Control\ Volume\ V} -\frac{1}{3}\nabla\sigma\,d\tau \tag{3.31}$$

and

$$\boldsymbol{P} = \iiint\limits_{Control\ Volume\ V} \nabla p\,d\tau \tag{3.32}$$

Referring to Eq. (3.26) and utilizing Eqs. (3.27) up to (3.32), one obtains:

$$\frac{D}{Dt}\iiint\limits_{Control\ Volume\ V} \rho\,\boldsymbol{V}\,d\tau = \iiint\limits_{Control\ Volume\ V} \frac{\partial\sigma_{ij}}{\partial x_j}\,d\tau + \iiint\limits_{Control\ Volume\ V} \rho\boldsymbol{f}\,d\tau \tag{3.33}$$

To simplify the left-hand side, the Reynolds Transport Equation will be used. The Reynolds Transport Equation for $b \equiv \boldsymbol{V}$ reads:

$$\frac{DB}{Dt} = \frac{D}{Dt}\iiint\limits_{Control\ Volume\ V} \rho\,\boldsymbol{V}\,d\tau = \iiint\limits_{Control\ Volume\ V} \frac{\partial}{\partial t}(\rho\,\boldsymbol{V})\,dV + \oiint\limits_{Control\ Surface\ S} (\rho\,\boldsymbol{V})\,\boldsymbol{V}\cdot\boldsymbol{n}\,dA \tag{3.8c}$$

Hence

$$\frac{D}{Dt}\iiint\limits_{Control\ Volume\ V} \rho\, \boldsymbol{V}\ d\tau = \iiint\limits_{Control\ Volume\ V} \frac{\partial}{\partial t}(\rho\, \boldsymbol{V})\ d\tau + \oiint\limits_{Control\ Surface\ S} (\rho\, \boldsymbol{V}\,)\, \boldsymbol{V}\cdot \boldsymbol{n}\, dA \tag{3.33a}$$

Using Vector identity

$$\frac{D(\rho \boldsymbol{V})}{Dt} = \frac{\partial(\rho \boldsymbol{V})}{\partial t} + \boldsymbol{V}\cdot\nabla(\rho \boldsymbol{V}) \tag{3.33b}$$

Therefore

$$\frac{D}{Dt}\iiint\limits_{Control\ Volume\ V} \rho\, \boldsymbol{V}\ d\tau = \iiint\limits_{Control\ Volume\ V} \frac{\partial}{\partial t}(\rho\, \boldsymbol{V})\ dV + \iiint\limits_{Control\ Volume\ V} \boldsymbol{V}\cdot\nabla(\rho \boldsymbol{V}\,)\ dV \tag{3.33c}$$

Taking into consideration Gauss theorem[6]

$$\oiint\limits_{Control\ Surface\ S} (\rho\, \boldsymbol{Q})\cdot \boldsymbol{n}\, dA \equiv \iiint\limits_{Control\ Volume\ V} \nabla\cdot(\rho\, \boldsymbol{Q})dV \equiv \iiint\limits_{Control\ Volume\ V} \mathrm{div}(\rho\, \boldsymbol{Q})dV \tag{3.11}$$

or

$$\iiint\limits_{Control\ Volume\ V} \boldsymbol{V}\cdot\nabla(\rho \boldsymbol{V}\,)\ d\tau = \oiint\limits_{Control\ Surface\ S} (\rho\, \boldsymbol{V})\, \boldsymbol{V}\cdot \boldsymbol{n}\, dA \tag{3.34a}$$

or

$$\oiint\limits_{Control\ Surface\ S} (\rho\, \boldsymbol{V})\, \boldsymbol{V}\cdot \boldsymbol{n}\, dA = \iiint\limits_{Control\ Volume\ V} \boldsymbol{V}\cdot\nabla(\rho \boldsymbol{V}\,)\ d\tau \tag{3.34b}$$

we obtain

$$\frac{D}{Dt}\iiint\limits_{Control\ Volume\ V} \rho\, \boldsymbol{V}\ d\tau = \iiint\limits_{Control\ Volume\ V} \left(\frac{\partial(\rho\, \boldsymbol{V})}{\partial t} + \rho\, \boldsymbol{V}\,(\nabla\cdot \boldsymbol{V})\right) d\tau \tag{3.35}$$

This equation can be simplified further into:

---

[6] See also Lehman, Gauss' and Stokes' Theorems.

$$
\begin{aligned}
\frac{D}{Dt} \underset{Control\ Volume\ V}{\iiint} \rho\, \boldsymbol{V}\, d\tau
&= \underset{Control\ Volume\ V}{\iiint} \left(\frac{\partial(\rho\, \boldsymbol{V})}{\partial t} + \rho\, \boldsymbol{V}\, (\nabla \cdot \boldsymbol{V}) \right) d\tau \\
&= \underset{Control\ Volume\ V}{\iiint} \rho \frac{D(\boldsymbol{V})}{Dt} \\
&+ \underset{Control\ Volume\ V}{\iiint} \boldsymbol{V} \left\{ \frac{D\rho}{Dt} + \rho\, (\nabla \cdot \boldsymbol{V}) \right\} d\tau \\
&= \underset{Control\ Volume\ V}{\iiint} \rho \frac{D(\boldsymbol{V})}{Dt} \\
&+ \underset{Control\ Volume\ V}{\iiint} \boldsymbol{V} \left\{ \frac{\partial\rho}{\partial t} + (\boldsymbol{V} \cdot \nabla)\, \rho + \rho\, (\nabla \cdot \boldsymbol{V}) \right\} d\tau
\end{aligned}
\tag{3.36a}
$$

Next it is noted that the Equation of Mass Conservation (Continuity) is given by:

$$
\underset{Control\ Volume\ V}{\iiint} \left(\frac{D\rho}{Dt} + \rho\, (\nabla \cdot \boldsymbol{V}) \right) d\tau = 0 \tag{3.37a}
$$

or

$$
\begin{aligned}
\underset{Control\ Volume\ V}{\iiint} \left(\frac{D\rho}{Dt} + \rho\, (\nabla \cdot \boldsymbol{V}) \right) d\tau &= \underset{Control\ Volume\ V}{\iiint} \left\{ \frac{\partial\rho}{\partial t} \right. \\
&\left. + (\boldsymbol{V} \cdot \nabla)\, \rho + \rho\, (\nabla \cdot \boldsymbol{V}) \right\} d\tau \\
&= \underset{Control\ Volume\ V}{\iiint} \left(\frac{\partial\rho}{\partial t} + \nabla \cdot (\rho\, \boldsymbol{V}) \right) d\tau
\end{aligned}
\tag{3.37b}
$$

Hence:

$$
\frac{D}{Dt} \underset{Control\ Volume\ V}{\iiint} \rho\, \boldsymbol{V}\, d\tau = \underset{Control\ Volume\ V}{\iiint} \rho \frac{D\boldsymbol{V}}{Dt}\, d\tau \tag{3.38}
$$

Then Eq. (3.38) implies that the rate of change of momentum (of the mass) in the control volume V is equal to the volume integral of the rate of change of momentum of each fluid element in the control volume V.

Another way to obtain Eq. (3.38) for the rate of change of momentum (of the mass) in the control volume V can be obtained by applying the conservation principle for any extensive property B in the Eq. (3.8). Hence:

$$\frac{DB}{Dt} = \iiint_{Control\ Volume\ V} (\rho\, b)\mathrm{d}\tau + \iint_{Control\ Surface\ S} (\rho\, b)\, \boldsymbol{V} \cdot \boldsymbol{n}\, \mathrm{d}A \tag{3.38a}$$

or

$$\frac{D}{Dt} \iiint_{Control\ Volume\ V} (\rho\, b)\ \ \mathrm{d}\tau = \iiint_{Control\ Volume\ V} \frac{\partial}{\partial t}(\rho\, b)\, \mathrm{d}\tau + \oiint_{Control\ Surface\ S} (\rho\, b)\, \boldsymbol{V} \cdot \boldsymbol{n}\, \mathrm{d}A \tag{3.38b}$$

which, after substituting $b = \mathbf{V}$ becomes

$$\frac{D}{Dt}\left(\iiint_{Control\ Volume\ V} (\rho\, \boldsymbol{V})\ \ \mathrm{d}\tau \right) = \iiint_{Control\ Volume\ V} \frac{\partial}{\partial t}(\rho\, \boldsymbol{V})\ \mathrm{d}\tau + \oiint_{Control\ Surface\ S} (\rho\, \boldsymbol{V})\, \boldsymbol{V} \cdot \boldsymbol{n}\, \mathrm{d}A \tag{3.39}$$

Using Gauss theorem on the surface integral on the right-hand side yields

$$\frac{D}{Dt}\left(\iiint_{Control\ Volume\ V} (\rho\, \boldsymbol{V})\ \ \mathrm{d}\tau \right) = \iiint_{Control\ Volume\ V} \frac{\partial}{\partial t}(\rho\, \boldsymbol{V})\, \mathrm{d}\tau + \iiint_{Control\ Volume\ V} \nabla \cdot (\rho\, \boldsymbol{V}\ \boldsymbol{V})\, \mathrm{d}\psi \tag{3.40}$$

From vector identity, one obtains

$$\nabla \cdot (\rho\, \boldsymbol{V}\ \boldsymbol{V}) = \rho\ (\boldsymbol{V} \cdot \nabla)\, \boldsymbol{V} + \boldsymbol{V}\, \nabla \cdot (\rho \boldsymbol{V}) \tag{3.41}$$

Therefore

$$\frac{D}{Dt}\left(\iiint_{Control\ Volume\ V} (\rho\, \boldsymbol{V})\ d\tau \right)$$

$$= \left(\iiint\limits_{Control\ Volume\ V} \frac{D}{Dt} (\rho \boldsymbol{V})\, d\tau \right)$$

$$= \iiint\limits_{Control\ Volume\ V} \frac{\partial}{\partial t}(\rho \boldsymbol{V})\, d\tau$$

$$+ \iiint\limits_{Control\ Volume\ V} \nabla \cdot (\rho \boldsymbol{V}\ \boldsymbol{V})\, d\tau$$

$$= \iiint\limits_{Control\ Volume\ V} \frac{\partial}{\partial t}(\rho \boldsymbol{V})\, d\tau$$

$$+ \iiint\limits_{Control\ Volume\ V} \{\rho\, (\boldsymbol{V} \cdot \nabla)\, \boldsymbol{V} + \boldsymbol{V}\, \nabla \cdot (\rho \boldsymbol{V})\}\, d\tau$$

$$= \iiint\limits_{Control\ Volume\ V} \left(\rho \frac{\partial (\boldsymbol{V})}{\partial t} + \rho\, (\boldsymbol{V} \cdot \nabla) \boldsymbol{V} \right) d\tau$$

$$+ \iiint\limits_{Control\ Volume\ V} \boldsymbol{V} \underbrace{\left\{ \frac{\partial \rho}{\partial t} + \nabla \cdot (\rho\, \boldsymbol{V}) \right\}}_{=0} d\tau$$

$$= \iiint\limits_{Control\ Volume\ V} \rho \frac{D\,\boldsymbol{V}}{Dt}\, d\tau \tag{3.42}$$

which is the same as Eq. (3.38).

Combining Eqs. (3.33) and (3.38) one obtains

$$\iiint\limits_{Control\ Volume\ V} \rho\, \frac{D\boldsymbol{V}}{Dt}\, \mathrm{d}\tau = \iiint\limits_{Control\ Volume\ V} \frac{\partial \sigma_{ij}}{\partial x_i}\, \mathrm{d}\tau + \iiint\limits_{Control\ Volume\ V} \rho\, \boldsymbol{f}\, \mathrm{d}\tau \tag{3.43}$$

or

$$\iiint\limits_{Control\ Volume\ V} \left\{ \rho\, \frac{D\boldsymbol{V}}{Dt} - \frac{\partial \sigma_{ij}}{\partial x_i} - \rho\, \boldsymbol{f} \right\} \mathrm{d}\tau = 0 \tag{3.44}$$

Since this equation is written for an arbitrary volume V, then the integrand should be zero. Hence:

$$\rho\, \frac{D\boldsymbol{V}}{Dt} - \frac{\partial \sigma_{ij}}{\partial x_i} - \rho \boldsymbol{f} = 0 \tag{3.45}$$

or

$$\frac{D\boldsymbol{V}}{Dt} = \frac{\partial \boldsymbol{V}}{\partial t} + (\boldsymbol{V} \cdot \nabla)\,\boldsymbol{V} = \frac{1}{\rho}\frac{\partial \sigma_{ij}}{\partial x_i} + \boldsymbol{f} \text{ or } \rho\,\frac{D\boldsymbol{V}}{Dt}$$
$$= \rho\left(\frac{\partial \boldsymbol{V}}{\partial t} + (\boldsymbol{V} \cdot \nabla)\,\boldsymbol{V}\right) = \frac{\partial \sigma_{ij}}{\partial x_i} + \rho\boldsymbol{f} \qquad (3.45\text{a,b})$$

or

$$\frac{D\boldsymbol{V}}{Dt} = \frac{\partial \boldsymbol{V}}{\partial t} + (\boldsymbol{V} \cdot \nabla)\,\boldsymbol{V} = -\frac{\nabla p}{\rho} + \boldsymbol{f} \text{ or } \rho\frac{D\boldsymbol{V}}{Dt}$$
$$= \rho\left(\frac{\partial \boldsymbol{V}}{\partial t} + (\boldsymbol{V} \cdot \nabla)\,\boldsymbol{V}\right) = -\nabla p + \rho\boldsymbol{f} \qquad (3.45\text{c,d})$$

Hence by using infinitesimal fluid volume element, one obtains Eqs. 3.23a–3.23c again

$$\frac{D\boldsymbol{V}}{Dt} = \frac{\partial \boldsymbol{V}}{\partial t} + (\boldsymbol{V} \cdot \nabla)\,\boldsymbol{V} = -\frac{\nabla p}{\rho} + f \qquad (3.45\text{e})$$

Equations (3.45) are known as the ***Euler equation or the Momentum Conservation Equation*** in flow system (Fluid Mechanics).

Under the limitations of the present section, it can be easily shown [5][7] that the law of conservation of momentum (3.45e) can be written as

$$\boldsymbol{a} \equiv \frac{DV}{Dt} = \nabla\left[\Omega - \int \frac{dp}{\rho}\right] \qquad (3.45\text{f})$$

which is the Conservation Laws for a Barotropic Fluid in a Conservative Body Force Field. The term "barotropic" implies a unique pressure-density relation throughout the entire flow field; adiabatic-reversible or isentropic flow is the most important special case. As we shall see, (3.45f) can often be integrated to yield a useful relation among the quantities pressure, velocity, density and so forth, that applies throughout the entire flow.

Another consequence of barotropy is a simplification of Kelvin's theorem of the rate of change of circulation around a path C always composed of the same set of fluid particles. As can be followed in elementary textbooks on Fluid Mechanics, as a consequence of the equations of motion for inviscid fluid in a conservative body force field,

$$\frac{D\Gamma}{Dt} = \frac{D\left(\oint_C \boldsymbol{V} \cdot \mathrm{d}s\right)}{Dt} = -\oint_C \frac{\mathrm{d}p}{\rho} = \oint_C T\,\mathrm{d}\tilde{s} \qquad (3.45\text{g})$$

where

$\tilde{s}$ is the entropy.

[7] See Holt Ashley, Mårten Landahl, Aerodynamics of Wings and Bodies, Chap. 1.

The relationships as represented by the two last expressions represent the consequence of First Law of Thermodynamic.

$$\Gamma \equiv \oint_C \mathrm{V} \cdot \mathrm{ds} \tag{3.45h}$$

is the circulation or closed line integral of the tangential component of the velocity vector. Under the present limitations, we see that

$$\oint_C \frac{\mathrm{d}p}{\rho}$$

is the integral of a single-valued perfect differential and therefore must vanish. Hence we have the result

$$\frac{D\Gamma}{Dt} = 0 \tag{3.46}$$

for all such fluid paths, which means that the circulation is preserved. In particular, if the circulation around a path is initially zero, it will always remain so. The same result holds in a constant-density fluid where the quantity p in the denominator can be taken outside, leaving once more a perfect differential; this is true regardless of what assumptions are made about the thermodynamic behavior of the fluid. This result is known as Kelvin's theorem.

3.3.4 Other Forms of the Euler Equation for Fluid Motion

An alternative form of Eq. (3.45) can be obtained by using the following vector identity

$$\begin{aligned}\frac{1}{2}\nabla(\boldsymbol{V}\cdot\boldsymbol{V}) &= \frac{1}{2}\nabla(V)^2 = (\boldsymbol{V}\cdot\nabla)\,\boldsymbol{V} + \boldsymbol{V}\times(\nabla\times\boldsymbol{V})\\ &= (\boldsymbol{V}\cdot\nabla)\,\boldsymbol{V} + \boldsymbol{V}\times\mathrm{curl}\ \boldsymbol{V}\end{aligned} \tag{3.47}$$

Substituting this value in Eq. (3.45d) yields

$$\frac{\partial\boldsymbol{V}}{\partial t} + \frac{1}{2}\nabla(V)^2 + \frac{\nabla p}{\rho} - \boldsymbol{f} - \boldsymbol{V}\times\mathrm{curl}\ \boldsymbol{V} = 0 \tag{3.47a}$$

or

$$\frac{\partial\boldsymbol{V}}{\partial t} + \frac{1}{2}\nabla(V)^2 - \boldsymbol{V}\times\mathrm{curl}\ \boldsymbol{V} = -\frac{\nabla p}{\rho} + \boldsymbol{f} \tag{3.47b}$$

where

$$\text{curl } \boldsymbol{V} = \nabla \times \boldsymbol{V} \equiv \boldsymbol{\Omega} \tag{3.48}$$

In many occasions, the body force **f** can be derived from a certain potential, that is if **f** is conservative. In such cases, **f** can be written as

$$\mathbf{f} = -\nabla U \tag{3.49}$$

Then the Euler equation can be written following Lamb [5][8]

$$\frac{\partial \boldsymbol{V}}{\partial t} + \boldsymbol{V} \cdot \nabla \boldsymbol{V} = -\frac{\nabla p}{\rho} + \boldsymbol{f} \tag{3.50}$$

$$\frac{\partial \boldsymbol{V}}{\partial t} + \frac{1}{2}\nabla (V)^2 - \boldsymbol{V} \times \boldsymbol{\Omega} = -\frac{\nabla p}{\rho} - \nabla U \tag{3.51a}$$

or:

$$\frac{\partial \boldsymbol{V}}{\partial t} - \boldsymbol{V} \times \boldsymbol{\Omega} = -\nabla\left(\frac{V^2}{2} + U\right) - \frac{\nabla p}{\rho} \tag{3.51b}$$

which is universally valid provided **f** is a body force in a conservative potential field. The equation of motion can still be simplified, if the fluid can be assumed to be barotropic, i.e. $p = p(\rho)$.

Examples of barotropic fluids are

1. Compressible fluid moving adiabatically (hence $p \sim 1/v^{\gamma}$ or $p \sim \rho^{\gamma}$)
2. Compressible fluid moving isothermically (hence ($p \sim \rho$)
3. Incompressible fluid (ρ = constant).

for this special case, the term $\frac{1}{\rho}\nabla p$ can be simplified using the following procedure:

$$\mathrm{d}\boldsymbol{r} \cdot \frac{\nabla p}{\rho(p)} = \frac{\mathrm{d}p}{\rho(p)} = \mathrm{d}\int \frac{\mathrm{d}p}{\rho(p)} = \mathrm{d}\boldsymbol{r} \cdot \nabla \int \frac{\mathrm{d}p}{\rho(p)} \tag{3.52}$$

for any arbitrary d**r**. Therefore:

$$\frac{\nabla p}{\rho(p)} = \nabla \int \frac{\mathrm{d}p}{\rho(p)} \tag{3.53}$$

So that the "Lamb form" for barotropic fluid in a conservative potential force field becomes:

$$\frac{\partial \boldsymbol{V}}{\partial t} - \boldsymbol{V} \times \boldsymbol{\Omega} = -\nabla\left(\frac{V^2}{2} + U\right) - \int \frac{\mathrm{d}p}{\rho(p)} \tag{3.54}$$

---

[8] Reference [1].

3.3.5 Law of Conservation of Thermodynamic Energy (For Adiabatic Fluid)[9]

Following the Reynolds Transport Equation

$$\frac{DB}{Dt} = \iint \frac{\partial}{\partial t}(\rho\, b)\ \mathrm{d}V + \oiint (\rho\, b)\boldsymbol{V} \cdot \boldsymbol{n}\, \mathrm{d}A \tag{3.9}$$

Control Volume V *Control Surface S+σ*

and defining

$$b \equiv e + \frac{V^2}{2} \tag{3.55}$$

$$\boldsymbol{f} \equiv -\nabla U \tag{3.49}$$

Then we can write [6]

$$\frac{D}{Dt}\left[e + \frac{V^2}{2}\right] = \frac{\nabla \cdot (p\, V)}{\rho} + \boldsymbol{f} \cdot V \tag{3.56}$$

Here *e* is the internal energy per unit mass, and V represents the absolute magnitude of the velocity vector **V**.

Introducing the law of continuity and the definition of enthalpy,

$$h = e + \frac{p}{\rho} \tag{3.57}$$

we can modify (3.55) as[10]

$$\rho\frac{D}{Dt}\left[h + \frac{V^2}{2}\right] = \frac{\partial p}{\partial t} + \rho \boldsymbol{f} \cdot V \tag{3.58}$$

Newton's law can be used in combination with the Second Law of Thermodynamics to reduce the conservation of energy to the very simple form

$$\frac{Ds}{Dt} = 0 \tag{3.59}$$

where *s* is the entropy per unit mass. It must be emphasized that none of the forego 1 ng Eqs. (3.59) in particular, can be applied through a finite discontinuity in the flow

---

[9] See Holt Ashley, Mårten Landahl, Aerodynamics of Wings and Bodies, Chap. 1.

[10] The poof of formu; a (3.58) from (3.56) following the procedure outlined above will be left as an exercise.

field, such as a shock. It is an additional consequence of the second law that through an adiabatic shock s can only increase.

3.3.6 The Equation of Motion in a Non-inertial Coordinate System

If the control surface of the fluid is in motion, for convenience frame of reference coordinate System not moving with respect to the control surface can be chosen. The acceleration of a fluid element with respect to a moving coordinate system will be different to the absolute acceleration with respect to the Newtonian absolute frame of reference. Therefore, the equation of motion should be modified accordingly. A general example that can be demonstrated for a relative frame of coordinate system is experiencing a translatory or rotational motion.

First let us consider the differentiation of a unit vector **i** in a rotating frame of reference with respect to time in the absolute frame of reference. This can be expressed by

$$\left(\frac{\mathrm{d}\boldsymbol{i}}{\mathrm{d}t}\right)_a = \underbrace{\left(\frac{\mathrm{d}\boldsymbol{i}}{\mathrm{d}t}\right)_r}_{=0} + \Omega \times \boldsymbol{i} = \Omega \times \boldsymbol{i} \tag{3.55}$$

where

$\left(\frac{\mathrm{d}i}{\mathrm{d}t}\right)_a$ represent the rate of change (differentiation) of **i** with respect to time in the absolute frame of reference and

$\left(\frac{\mathrm{d}i}{\mathrm{d}t}\right)_r$ represent the rate of change (differentiation) of **i** with respect to time in the rotating (relative) frame of reference.

Since the unit vector **i** in the rotating frame of reference by definition is constant, its magnitude and direction are constant; hence $\left(\frac{\mathrm{d}i}{\mathrm{d}t}\right)_r \equiv 0$. Hence the last term in Eq. 3.55 results. This is depicted in Fig. 3.6. Here the unit vector **i** changes its direction since it is rotating with an angular velocity $\boldsymbol{\Omega}$.

Next, the time rate of change (differentiation) of any vector **R** in the moving (rotating, relative) coordinate frame of reference system in the absolute (Newtonian) frame of reference system will be derived. For this purpose, a relative frame of coordinate system that is rotating with an angular velocity $\boldsymbol{\Omega}$ about the absolute coordinate frame of reference system with the origin at the point O will be considered. The origin of the relative coordinate frame of reference system O' may also experience a translational motion in the absolute (or Newtonian) frame of coordinate system with an acceleration $\mathbf{a}_0$ as depicted in Fig. 3.7.

From Fig. 3.7, it follows that:

$$\boldsymbol{R} \equiv \boldsymbol{R}_a = \boldsymbol{R}_0 + \boldsymbol{R}_1 = \boldsymbol{R}_0 + \boldsymbol{R}_r \tag{3.56a}$$

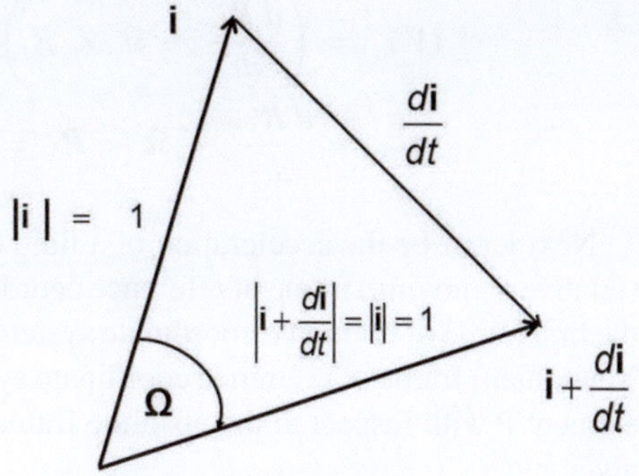

Fig. 3.6 Differentiation of a unit vector **i**

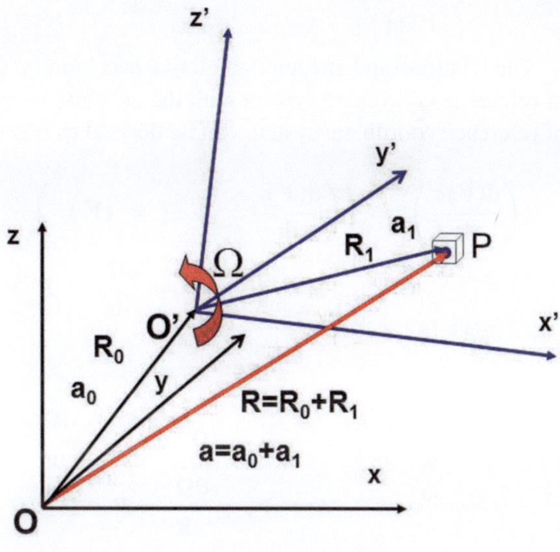

Fig. 3.7 Relative and absolute frame of reference coordinate systems

and if $\mathbf{R}_0 = 0$, then

$$\boldsymbol{R} \equiv \boldsymbol{R}_a = \boldsymbol{R}_1 = \boldsymbol{R}_r \tag{3.56b}$$

The velocity vector **V** that indicates the relative velocity of the fluid mass element P moving with the rotating frame of reference coordinate system with origin at O' with respect to the absolute frame of reference coordinate system with origin at O can then be given by

$$\left(\frac{\mathrm{d}\boldsymbol{R}}{\mathrm{d}t}\right)_a = \left(\frac{\mathrm{d}\boldsymbol{R}}{\mathrm{d}t} + \boldsymbol{\Omega} \times \boldsymbol{R}\right)_r = \left(\frac{\mathrm{d}\boldsymbol{R}}{\mathrm{d}t}\right)_r + \boldsymbol{\Omega} \times (\boldsymbol{R})_r \tag{3.57}$$

or

$$(\boldsymbol{V})_a = \left(\frac{d\,\boldsymbol{R}_1}{dt} + \boldsymbol{\Omega} \times \boldsymbol{R}\right)_r = (\boldsymbol{V})_r + \boldsymbol{\Omega} \times (\boldsymbol{R})_r$$
$$= \frac{d\,\boldsymbol{R}_1}{dt} + \boldsymbol{\Omega} \times \boldsymbol{R}_1 = \boldsymbol{V}_r + \boldsymbol{\Omega} \times \boldsymbol{R}_1 \tag{3.58}$$

Next let a1 be the acceleration of a fluid element *P* with respect to O' in the first (relative or moving) frame of reference coordinate system and a0 is the acceleration of the first frame of reference coordinate system with respect to the absolute (stationary, Newtonian) frame of reference coordinate system. Then, the acceleration of the fluid element P with respect to the absolute frame of reference coordinate system is then given by:

$$\boldsymbol{a} = \boldsymbol{a}_0 + \boldsymbol{a}_1 \tag{3.59}$$

The relationship between the relative acceleration $\mathbf{a}_1$ in the first or relative frame of reference coordinate system with the absolute acceleration in the absolute frame of reference coordinate system can be derived as follows, by referring to Fig. 3.7,[11]

$$\left(\frac{\mathrm{d}(\boldsymbol{V})_a}{\mathrm{d}t}\right)_a = \left(\left(\frac{\mathrm{d}(\boldsymbol{V})_r}{\mathrm{d}t}\right)_r + \boldsymbol{\Omega} \times (\boldsymbol{V})_r\right)_r$$
$$+ \left(\boldsymbol{\Omega} \times \left(\frac{\mathrm{d}\boldsymbol{R}}{\mathrm{d}t}\right)_r + \frac{\mathrm{d}\boldsymbol{\Omega}}{\mathrm{d}t} \times \boldsymbol{R} + \boldsymbol{\Omega} \times \boldsymbol{\Omega} \times \boldsymbol{R}\right) \tag{3.60a}$$

or

$$\boldsymbol{a} \equiv \frac{\mathrm{d}^2\boldsymbol{R}}{\mathrm{d}t^2} = \frac{\mathrm{d}^2\boldsymbol{R}_r}{\mathrm{d}t^2} + 2\boldsymbol{\Omega} \times \frac{\mathrm{d}\boldsymbol{R}_r}{\mathrm{d}t} + \frac{\mathrm{d}\boldsymbol{\Omega}}{\mathrm{d}t} \times \boldsymbol{R} + \boldsymbol{\Omega} \times \boldsymbol{\Omega} \times \boldsymbol{R}_r$$
$$\equiv \boldsymbol{a}_r + 2\boldsymbol{\Omega} \times \boldsymbol{V}_r + \frac{\mathrm{d}\boldsymbol{\Omega}}{\mathrm{d}t} \times \boldsymbol{R} + \boldsymbol{\Omega} \times \boldsymbol{\Omega} \times \boldsymbol{R}_r \tag{3.60b}$$

where $\boldsymbol{a}_r$ or $\frac{\mathrm{d}\boldsymbol{R}_r}{\mathrm{d}t}$ represents the rate of change of **R** as observed by an observer in the rotating frame of reference coordinate system. The second term, $2\boldsymbol{\Omega} \times \frac{d\boldsymbol{R}_r}{dt}$, represents the change of $\boldsymbol{V}_r$ due to the rotation of the rotating frame of reference coordinate system with respect to the absolute frame of reference coordinate system.

The terms on the right-hand side of Eq. (3.57) are the acceleration elements that can be identified as follows:

11
$$\left(\frac{d(\boldsymbol{V})_a}{dt}\right)_a = \left(\frac{d(\boldsymbol{V})_r}{dt}\right)_a + \left(\frac{d(\boldsymbol{\Omega} \times (\boldsymbol{R})_r)}{dt}\right)_a$$
$$= \left(\left(\frac{d(\boldsymbol{V})_r}{dt}\right)_r + \boldsymbol{\Omega} \times (\boldsymbol{V})_r\right)_r + \left(\left(\frac{d(\boldsymbol{\Omega} \times (\boldsymbol{R})_r)}{dt}\right) + \boldsymbol{\Omega} \times \boldsymbol{\Omega} \times \boldsymbol{R}\right)$$
$$= \left(\left(\frac{d(\boldsymbol{V})_r}{dt}\right)_r + \boldsymbol{\Omega} \times (\boldsymbol{V})_r\right)_r + \left(\boldsymbol{\Omega} \times \left(\frac{d\boldsymbol{R}}{dt}\right)_r + \frac{d\boldsymbol{\Omega}}{dt} \times \boldsymbol{R} + \boldsymbol{\Omega} \times \boldsymbol{\Omega} \times \boldsymbol{R}\right)$$
$$= \frac{d\boldsymbol{V}_r}{dt} + 2\boldsymbol{\Omega} \times \boldsymbol{V}_r + \frac{d\boldsymbol{\Omega}}{dt} \times \boldsymbol{R} + \boldsymbol{\Omega} \times \boldsymbol{\Omega} \times \boldsymbol{R}$$

1. $\mathbf{a}_r$ is the acceleration of the fluid element as observed by an observer in the rotating frame of reference coordinate system.
2. $2\boldsymbol{\Omega} \times \boldsymbol{V}_r$ is the acceleration due to the rotation of the rotating frame of reference coordinate system with respect to the absolute frame of reference coordinate system and is perpendicular both to $2\boldsymbol{\Omega} \times \boldsymbol{V}_r$ and $2\boldsymbol{\Omega} \times \boldsymbol{V}_r$ and is known as the Coriolis acceleration that contributes to the Coriolis force on the fluid element.
3. $\boldsymbol{\Omega} \times \boldsymbol{\Omega} \times \boldsymbol{R}_r$ is the centrifugal acceleration that gives rise to centrifugal force on the fluid element.
4. $\frac{d\boldsymbol{\Omega}}{dt} \times \boldsymbol{R}$ does not have any specific name; if the rotational velocity $\boldsymbol{\Omega}$ is constant, then this term reduces identically to zero. For example if one looks at the absolute acceleration of an object moving on the surface of the earth where the rotational velocity $\boldsymbol{\Omega}$ of the earth is considered to be constant.

3.3.7 Bernoulli's Equation

3.3.7.1 Simple Form of Bernoulli's Equation [1]

Although viscous effects will not be considered, the equation of motion of the fluid or is very complex and the general analytical solution is difficult to obtain, except for special cases. Therefore numerical method is often resorted to, and such approach has given rise to the progress of powerful approach known as computational fluid- or aerodynamics.

The difficulty in obtaining general analytical solution is due to the presence of non-linear terms, such as $u\frac{\partial u}{\partial x}$ and others.

However, with some limitation or constraints, the equation of motion can be integrated once. As the first example, we will consider a stationary two-dimensional incompressible flow, without viscous effect and body forces.

The Euler equation for this case is given by:

$$u\frac{\partial\ u}{\partial x} + v\frac{\partial\ u}{\partial y} = -\frac{1}{\rho}\frac{\partial p}{\partial x} \tag{3.61a}$$

$$u\frac{\partial\ v}{\partial x} + v\frac{\partial\ v}{\partial y} = -\frac{1}{\rho}\frac{\partial p}{\partial y} \tag{3.61b}$$

These pair of equations can be directly integrated along a streamline. For this purpose, the first equation is multiplied by dx and the second by dy, where dx and dy represents the projection of ds on the x- and y-axis, respectively.

Then there is obtained:

$$x \rightarrow \mathrm{d}x : u\frac{\partial u}{\partial x}\mathrm{d}x + v\frac{\partial u}{\partial y}\mathrm{d}x = -\frac{1}{\rho}\frac{\partial p}{\partial x}\mathrm{d}x \tag{3.62a}$$

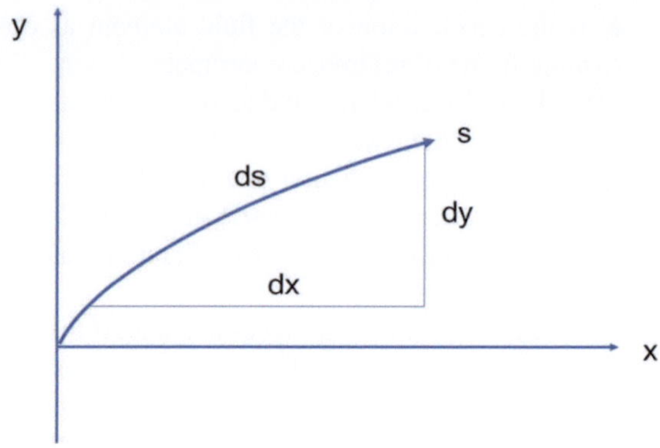

Fig. 3.8 Geometry of a piece of streamline ds with components dx and dy

$$y \to dy: \quad u\frac{\partial v}{\partial x}dy + v\frac{\partial v}{\partial y}dy = -\frac{1}{\rho}\frac{\partial p}{\partial y}dy \tag{3.62b}$$

From the definition of a streamline ds in terms of the components dx and dy in the coordinate frame of reference, as depicted in Fig. 3.8, are related as:

$$\frac{v}{u} = \frac{\mathrm{d}y}{\mathrm{d}x} \tag{3.63}$$

Hence

$$v \to u\frac{\mathrm{d}y}{\mathrm{d}x}: \ u\frac{\partial u}{\partial x}\mathrm{d}x + u\frac{\partial u}{\partial y}\mathrm{d}y = -\frac{1}{\rho}\frac{\partial p}{\partial x}\mathrm{d}x \tag{3.64a}$$

$$u \to v\frac{dx}{dy}: \ v\frac{\partial v}{\partial x}dx + v\frac{\partial v}{\partial y}dy = -\frac{1}{\rho}\frac{\partial p}{\partial y}dy \tag{3.64b}$$

Summing up Eqs. (3.64a) and (3.64b) yields

$$\frac{1}{2}\frac{\partial}{\partial x}\left(u^2 + v^2\right)\mathrm{d}x + \frac{1}{2}\frac{\partial}{\partial y}\left(u^2 + v^2\right)\mathrm{d}y = -\frac{1}{\rho}\left(\frac{\partial p}{\partial x}\mathrm{d}x + \frac{\partial p}{\partial y}\mathrm{d}y\right) \tag{3.65}$$

It can be seen that both sides of the Eq. (3.65) are perfect differential, while the velocity vector V is given by

$$V^2 = u^2 + v^2 \tag{3.66}$$

Therefore, Eq. 3.65 can be written as

$$d\left(\frac{V^2}{2} + \frac{p}{\rho}\right) = 0 \tag{3.67}$$

since ρ has been considered to be constant. Equation (3.67) can now be integrated to give

$$\frac{V^2}{2} + \frac{p}{\rho} = \text{constant} \tag{3.68a}$$

or:

$$\frac{V_1^2}{2} + \frac{p_1}{\rho} = \frac{V_2^2}{2} + \frac{p_2}{\rho} \tag{3.68b}$$

Here the subscripts 1 and 2 represent two points along one streamline. It can be easily shown that Eq. (3.68) is also valid for three dimensions, for which

$$V^2 = u^2 + v^2 + w^2 \tag{3.67}$$

The present result shows the direct relationship between pressure and velocity; if the velocities at these two points are known, the velocity at the point where the pressure is smaller will be larger.

Equation (3.68) is valid for one-dimensional flow or flow along a stream tube where the properties of the fluid is uniform across its cross section, or along a streamline. Equation (3.68) is the simplest form of the Bernoulli's equation or Bernoulli's integral.

3.3.7.2 Conservative Body Forces

We will now consider the presence in the Euler's equation of a body force, which has the property that it can be expressed as a gradient of a scalar function (or potential).

Hence

$$dU = \frac{\partial U}{\partial x} dx + \frac{\partial U}{\partial y} dy = f_x \, dx + f_y \, dy \tag{3.68}$$

$$f_x = \frac{\partial U}{\partial x};\ f_y = \frac{\partial U}{\partial y} \tag{3.68a,b}$$

where U is a function of x and y. Such force is called a ***conservative force***, and U is known as ***Force Potential***.

Since U is a function of x and y, then Eq. (3.68) follows.

If f_x and f_y are added to the Euler's Equation of Motion (3.45), and the above integration process for Bernoulli's equation are carried out, then there is obtained:

$$d\left(\frac{V^2}{2} + \frac{p}{\rho} - U\right) = 0 \tag{3.69}$$

or:

$$\frac{V_1^2}{2} + \frac{p_1}{\rho} - U_1 = \frac{V_2^2}{2} + \frac{p_2}{\rho} - U_2 \tag{3.70}$$

This equation is also valid for points along a streamline in a steady, non-viscous and incompressible flow.

3.3.7.3 Unsteady or Non-stationary Flow

If the unsteady terms are incorporated, then in the Eq. (3.70) one has to add the terms:

$$\text{in the + x direction } \frac{\partial u}{\partial t} \tag{3.71a}$$

$$\text{in the + y direction} \frac{\partial v}{\partial t} \tag{3.71b}$$

Both of these terms are the components of $\frac{\partial \mathbf{V}}{\partial t}$, where:

$$\frac{\partial \mathbf{V}}{\partial t}\mathrm{d}s = \frac{\partial u}{\partial t}\mathrm{d}x + \frac{\partial v}{\partial t}\mathrm{d}y \tag{3.72}$$

where **V** is the velocity vector and velocity vector V is given by

$$|\mathbf{V}|^2 = u^2 + v^2 \tag{3.73}$$

and ds is the streamline element.

Using the same procedure, one obtains:

$$\mathrm{d}\left[\int \frac{\partial|\mathbf{V}|}{\partial t}\mathrm{d}s + \frac{|\mathbf{V}|^2}{2} + \frac{p}{\rho} - U\right] = 0 \tag{3.74}$$

$$\int \frac{\partial\ |\mathbf{V}|}{\partial t} ds + \frac{|\mathbf{V}|^2}{2} + \frac{p}{\rho} - U \equiv B = \text{Constant} \tag{3.75}$$

The integration of the unsteady term is carried out along a certain (defined) streamline at a certain (defined) time span, starting from an arbitrarily chosen point of interest. Equation (3.74) shows that the quantity within the square bracket should be constant along the specified streamline. This constant is known as the Bernoulli constant B. However, since the integration is carried out along the specified streamline at certain time interval, B is still a function of time:

$$B = B(t) \tag{3.76}$$

By integrating Eq. (3.75) along two points 1 and 2, Eq. (3.74) or (3.75) becomes

$$\int_1^2 \frac{\partial V}{\partial t} ds + \frac{V_2^2}{2} + \frac{p_2}{\rho} - U_2 = \frac{V_1^2}{2} + \frac{p_1}{\rho} - U_1 \tag{3.77}$$

Equation (3.77) is valid for any integration between points 1 in 2 along the specified streamline for an incompressible flow under the influence of a conservative force field. The force potential that is usually of interest is that due to gravitation (hence the gravitational potential U[12]), the centrifugal force and the like. For the gravitational potential, the positive y value is chosen to be positive upwards and perpendicular to the surface of the earth (to be more precise, perpendicular to the earth gravitational potential surface, the geoid) [7].[13]

Hence:

$$f_x = 0 \tag{3.78a}$$

$$f_y = 0 \tag{3.78b}$$

and therefore:

$$U = gy \tag{3.79}$$

Using (3.54), then one obtains:

$$\int_1^2 \frac{\partial V}{\partial t} \mathrm{d}s + \frac{v_2^2}{2} + \frac{p_2}{\rho} - gy_2 = \frac{v_1^2}{2} + \frac{p_1}{\rho} - gy_1 \tag{3.80}$$

[12] Here the force potential is alternatively indicated by U or Ω, which should be understood from the physical context.

[13] **Geoid** is the equipotential gravity surface of the earth at mean sea level. At any point it is perpendicular to the direction of gravity. Courtesy of Wikipedia: https://en.wikipedia.org/wiki/Geoid.

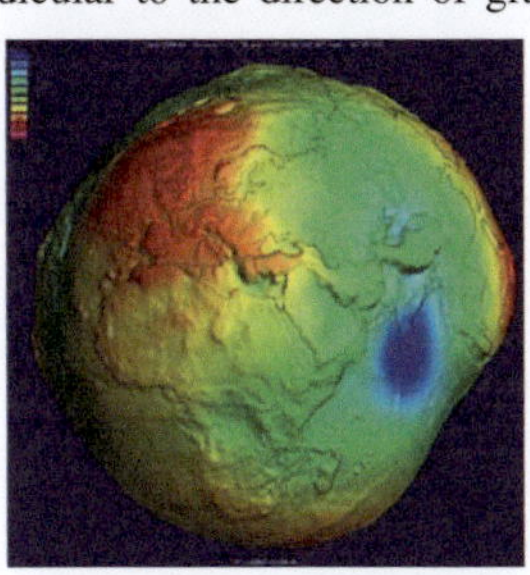

The Geoid

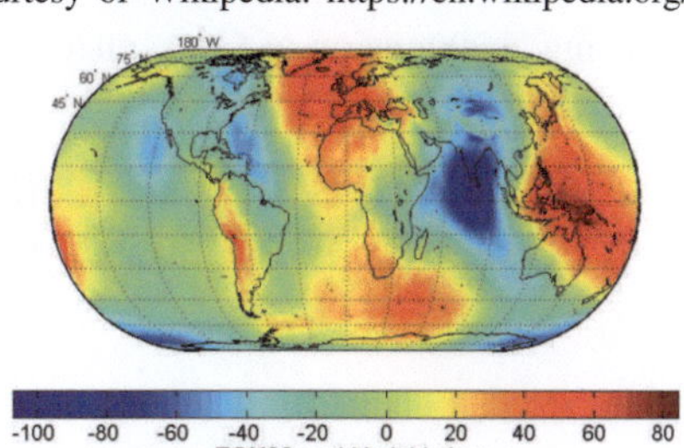

The Earth Gravitational Model

It should be noted that the integral of $\int_1^2 \frac{\partial V}{\partial t}\,\mathrm{d}s$ is not easy to evaluate, although the Eqs. (3.77) and (3.80) are valuable for many cases.

3.4 Momentum and Moment of Momentum Equations

3.4.1 Momentum Theorem

The momentum equation that is very general for a control volume has been earlier derived, which has the general form:

From (3.19) and (3.3) or (3.42)

$$\boldsymbol{F} = \frac{D(m\boldsymbol{V})}{Dt} = \frac{D}{Dt}\left(\iiint\limits_{Control\ Volume\ V} (\rho\,\boldsymbol{V})\,\mathrm{d}\tau\right) \tag{3.19b}$$

$$\frac{D}{Dt}\left(\iiint\limits_{Control\ Volume\ V} (\rho\boldsymbol{V})\mathrm{d}\tau\right) = \left(\iiint\limits_{Control\ Volume\ V} \rho\left(\frac{\partial \boldsymbol{V}}{\partial t} + (\boldsymbol{V}\cdot\nabla)\boldsymbol{V}\right)\mathrm{d}\tau\right) + \left(\iiint\limits_{Control\ Volume\ V} \boldsymbol{V}\underbrace{\left\{\frac{\partial\rho}{\partial t} + \nabla\cdot(\rho\boldsymbol{V})\right\}}_{=0}\mathrm{d}\tau\right) \tag{3.42b}$$

obtain

$$\boldsymbol{F} = \frac{D}{Dt}\iiint\limits_{V}(\rho\boldsymbol{V})\mathrm{d}\tau = \iiint\limits_{V}\frac{D}{Dt}(\rho\boldsymbol{V})\mathrm{d}\tau = \iiint\limits_{V}\left(\rho\frac{\partial(\boldsymbol{V})}{\partial t} + \rho(\boldsymbol{V}\cdot\nabla)\boldsymbol{V}\right)\mathrm{d}\tau$$

$$\boldsymbol{F} = \frac{D}{Dt}\int\limits_{V}\rho\mathbf{V}\,\mathrm{d}V = \int\limits_{V_0}\frac{\partial}{\partial t}(\rho\mathbf{V})\,\mathrm{d}V + \int\limits_{S_0}\rho\mathbf{V}(\mathbf{V}\cdot n)\mathrm{d}A \tag{3.81a}$$

Following an extension of Gauss theorem [8][14]

$$\iiint\limits_{Control\ Volume\ V}\boldsymbol{V}\cdot\nabla(\rho\boldsymbol{V})\,d\tau = \oiint\limits_{Control\ Surface\ S\ of\ V}(\rho\boldsymbol{V})\boldsymbol{V}\cdot\boldsymbol{n}dA \tag{3.81b}$$

Hence

[14] Lehman, Gauss' and Stokes' Theorems, www.lehman.edu.faculty.anchordoqui.VC-4.

$$F = \frac{D}{Dt}\iiint\limits_{V}(\rho V)\mathrm{d}\tau = \iiint\limits_{V}\frac{D}{Dt}(\rho V)\mathrm{d}\tau = \iiint\limits_{V}\frac{\partial}{\partial t}(\rho V)\mathrm{d}\tau + \iiint\limits_{V} V\cdot\nabla(\rho V)\mathrm{d}\tau$$

$$= \iiint\limits_{V}\frac{\partial}{\partial t}(\rho V)\mathrm{d}\tau + \oiint\limits_{\text{Control Surface } S \text{ of } V}(\rho V)V\cdot n\mathrm{d}A \quad (3.81c)$$

or

$$F = \frac{D}{Dt}\iiint\limits_{V_0}(\rho V)\mathrm{d}\tau = \iiint\limits_{V_0}\frac{D}{Dt}(\rho V)\mathrm{d}\tau$$

$$= \iiint\limits_{V_0}\frac{\partial}{\partial t}(\rho V)\mathrm{d}\tau + \oiint\limits_{\substack{\text{Control Surface } S_0 \text{ of}\\ \text{Control Volume } V_0}}(\rho V)V\cdot n\mathrm{d}A \quad (3.81)$$

where V_0 is the control volume, and S_0 is the closed surface of the control volume or the control surface. This equation is known as the Momentum theorem or Impulse theorem, which can be narrated as

> The rate of change per unit time of the momentum in a Control Volume and the flux (flow rate) of the momentum across the Control Surface is equal to the Force acting on the Control Volume.

3.4.2 *Angular Momentum or Moment of Momentum*

The Momentum theorem elaborated in preceding sections relates the force acting on a specified control volume with the total momentum flux across the surface of the control volume. In some cases, the moment of the momentum flux across the surface of the control volume should be considered. For this purpose, the angular momentum law from mechanics can be utilized, and adapted to the fluid flow.

From the second Newton's Law:

$$F = \frac{\mathrm{d}}{\mathrm{d}t}(mV) \quad (3.82)$$

where m, **F** and **V** are the mass, force on and velocity on a single mass particle. The moment produced by the force F with respect to a certain fixed point in space is given by

$$T = r \times F \quad (3.83)$$

where **r** is a radius vector from a fixed point in space to the point where **F** is acting. T is the product defined by Eq. (3.82).

Also, from the Second Newton's Law:

$$\boldsymbol{T} = \boldsymbol{r} \times \frac{\mathrm{d}}{\mathrm{d}t}(m\boldsymbol{V}) \tag{3.84}$$

One can now define a vector **H** as the product of a radius vector **r** to a point mass particle m as a linear momentum:

$$\boldsymbol{H} = \boldsymbol{r} \times \quad (m\boldsymbol{V}) \tag{3.85}$$

which is known as **angular momentum** or **moment of momentum**.

If H is differentiated with respect to time, then

$$\frac{\mathrm{d}\mathbf{H}}{\mathrm{d}t} = \frac{\mathrm{d}\mathbf{r}}{\mathrm{d}t} \times (m\mathbf{V}) + \mathbf{r} \times \frac{\mathrm{d}(m\mathbf{V})}{\mathrm{d}t} \tag{3.86}$$

However, $\frac{\mathrm{d}\mathbf{r}}{\mathrm{d}t} = \mathbf{V}$, and the cross product of a vector with itself is zero. Therefore

$$\frac{\mathrm{d}\boldsymbol{H}}{\mathrm{d}t} = \boldsymbol{r} \times \frac{\mathrm{d}(m\boldsymbol{V})}{\mathrm{d}t} \tag{3.87}$$

Equation (3.87) states that *the rate of change of the angular momentum H of a point mass particle with respect to time and with respect to a fixed point in space is equal to the moment of the rate of change of the particular moment acting on the point mass particle with respect to time and with respect to that particular fixed point in space.*

Now the above law derived for a particular point particle mass will be applied to the fluid mass in a control volume.

The moment acting on a particular control volume V with respect to a fixed point in space at a radius vector r can then be defined as an adaptation to Eq. (3.84):

$$\boldsymbol{T} = \frac{D}{Dt}\int_V \rho\boldsymbol{r} \times \boldsymbol{V}\, dV \tag{3.88}$$

which is obtained by integrating Eq. (3.84) for every mass particle in the control volume V. Proceeding following Reynolds Transport Relation and Gauss Divergence theorem,[15] there is obtained

$$\begin{aligned}
\boldsymbol{T} &= \frac{D}{Dt}\int\limits_{V} \rho\boldsymbol{r} \times \boldsymbol{V}\, dV \\
&= \iiint\limits_{Control\ Volume\ V} \left\{\frac{D(\rho\boldsymbol{r} \times \boldsymbol{V})}{Dt}\right\} + \rho(\boldsymbol{r} \times \boldsymbol{V})(\nabla\cdot\boldsymbol{V})\mathrm{d}\tau \\
&= \iiint\limits_{Control\ Volume\ V} \rho\frac{D(\boldsymbol{r} \times \boldsymbol{V})}{Dt}\mathrm{d}\tau \\
&+ \iiint\limits_{Control\ Volume\ V} (\boldsymbol{r} \times \boldsymbol{V})\left\{\frac{D\rho}{Dt} + \rho(\nabla\cdot\boldsymbol{V})\right\}\mathrm{d}\tau \\
&= \frac{\partial \boldsymbol{H}}{\partial t} + \oiint\limits_{Control\ Surface\ A\ of\ V} (\rho\boldsymbol{r} \times \boldsymbol{V})(\boldsymbol{V}\cdot\boldsymbol{n})\mathrm{d}A
\end{aligned} \tag{3.89}$$

where:

[15] Elaboration of the mathematical derivation

$$\begin{aligned}
\boldsymbol{T} &= \frac{D}{Dt}\int\limits_{V} \rho\boldsymbol{r} \times \boldsymbol{V}\, dV = \iiint\limits_{Control\ Volume\ V} \left\{\frac{D(\rho\boldsymbol{r} \times \boldsymbol{V})}{Dt}\right\} + \rho(\boldsymbol{r} \times \boldsymbol{V})(\nabla\cdot\boldsymbol{V})d\tau \\
&= \iiint\limits_{Control\ Volume\ V} \rho\frac{D(\boldsymbol{r} \times \boldsymbol{V})}{Dt}d\tau + \iiint\limits_{Control\ Volume\ V} (\boldsymbol{r} \times \boldsymbol{V})\left\{\frac{D\rho}{Dt} + \rho(\nabla\cdot\boldsymbol{V})\right\}d\tau \\
&= \iiint\limits_{Control\ Volume\ V} \frac{\partial}{\partial t}(\rho\boldsymbol{r} \times \boldsymbol{V})d\tau \\
&+ \iiint\limits_{Control\ Volume\ V} (\boldsymbol{r} \times \boldsymbol{V})\left\{\frac{\partial\rho}{\partial t} + (\nabla\cdot\boldsymbol{V})\rho + \rho(\nabla\cdot\boldsymbol{V})\right\}d\tau \\
&= \frac{\partial \boldsymbol{H}}{\partial t} + \iiint\limits_{Control\ Volume\ V} (\boldsymbol{r} \times \boldsymbol{V})\left\{\underbrace{\left(\frac{\partial\rho}{\partial t} + (\nabla\cdot\boldsymbol{V})\rho\right)}_{=0} + \rho(\nabla\cdot\boldsymbol{V})\right\}d\tau \\
&= \frac{\partial \boldsymbol{H}}{\partial t} + \oiint\limits_{Control\ Surface\ A\ of\ V} (\rho\boldsymbol{r} \times \boldsymbol{V})(\nabla\cdot\boldsymbol{V})dA \\
&= \frac{\partial \boldsymbol{H}}{\partial t} + \oiint\limits_{Control\ Surface\ A\ of\ V} (\rho\boldsymbol{r} \times \boldsymbol{V})(\boldsymbol{V}\cdot\boldsymbol{n})dA
\end{aligned}$$

$$\boldsymbol{H} = \iiint\limits_{Control\ Volume\ V} \rho \boldsymbol{r} \times \boldsymbol{V} \mathrm{d}V \tag{3.90}$$

is the angular momentum of the fluid within the control volume V_0 with respect to a specified fixed point in space, and S_0 is the area of the associated control surface. Equation (3.89) is the angular momentum theorem, which has similar form to the linear momentum theorem (3.81).

3.5 Energy Equations

3.5.1 The Derivation of Energy Equation Using Differential Method [1]

To derive the energy equation using differential method, we will consider the motion of a fluid element in an infinitesimal volume $\Delta x\ \Delta y\ \Delta z$ or an observer moving with the medium, the fluid element appears as a system. Therefore, the First Law of Thermodynamics for that system of fluid element is given by

$$\mathrm{d}E = \mathrm{d}Q - \mathrm{d}W \tag{3.91}$$

where E is the energy in the system, Q is the heat being transferred into the system, and W is the work carried out by the system. In order that E can be defined in the system, the system has to be in the state of equilibrium. For a fluid element that is in motion, the First Law of Thermodynamics as written in the form of Eq. (3.91) is valid as long as the process experienced in the fluid element as a system takes place statically. The energy that is contained in the fluid mass element as a system consists of *internal energy* and *mechanical energy*. The latter can be differentiated further into *kinetic* and *potential* energy.

If we want to express the First Law of Thermodynamics for the moving fluid element in rate quantities, then one can write

$$\frac{\mathrm{d}E}{\mathrm{d}t} = \frac{\mathrm{d}Q}{\mathrm{d}t} - \frac{\mathrm{d}W}{\mathrm{d}t} = \dot{Q} - \dot{W} \tag{3.92}$$

where

E the energy of the system of the fluid element
Q heat that has been transferred into the system of the fluid element
W work that is being performed by the system of the fluid element to its environment.

In other words:

{The rate of increase of the internal energy}
= {The rate of heat transferred into the system}
− {The rate of work carried out by the surface forces
(pressure and shear forces) on the environment} (3.92a)

The rate of change of the energy in the infinitesimal system fluid element is given by

$$\rho \Delta x \Delta y \Delta z \frac{DE}{Dt} \tag{3.93}$$

The heat transferred across the surface (boundary) of the system took place through conduction.

The amount of heat conducted in the x-direction is given by (see Fig. 3.9)

$$-\frac{\partial}{\partial}\left(\frac{q}{A}\right)_x \times (\Delta y \Delta z) \tag{3.94}$$

Similar equations can be written for the heat conducted in the direction of y and z.

Next the rate of work that is performed by the normal and tangential forces along the x-direction is given by (see Fig. 3.10)

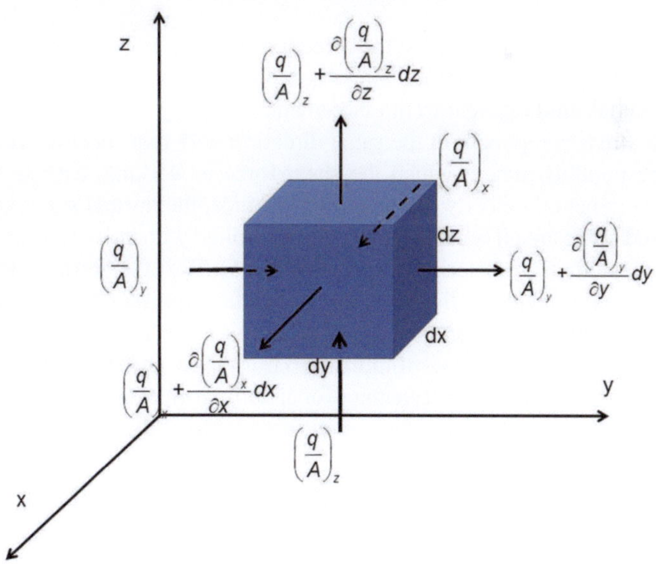

Fig. 3.9 Heat transferred into and out of an infinitesimal fluid element system

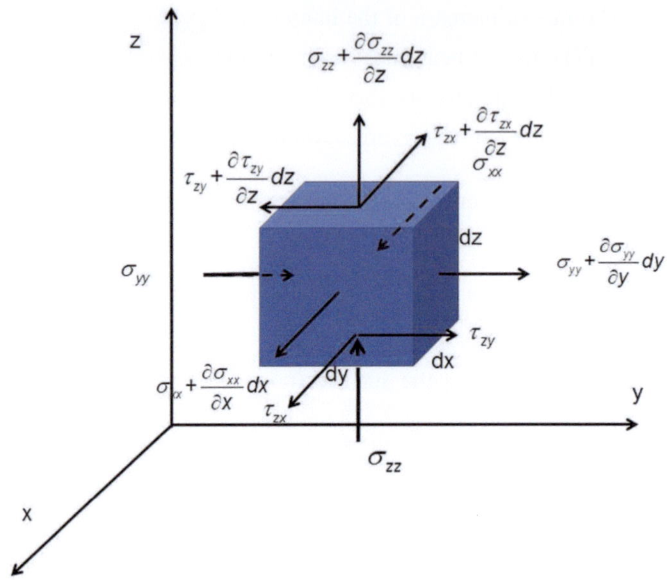

Fig. 3.10 Stress and force components at the surface of an infinitesimal fluid control volume

$$\sigma_x \frac{\partial u}{\partial x} \Delta x (\Delta y \Delta z) + \tau_{yz} \frac{\partial u}{\partial y} \Delta y (\Delta z \Delta x) + \tau_{zx} \frac{\partial u}{\partial z} \Delta z (\Delta x \Delta y) \tag{3.95}$$

Similar form can be written for the direction y and z. These expressions can be derived by multiplying the normal stress component times the area it is acting on the infinitesimal fluid element to the incremental velocity distance, and adding the other shear stress component in the same direction with that specific normal stress to the corresponding area at which the shear force is working, each to the corresponding incremental velocity distance. For instance, the normal stress component σ_x multiplied to the area it is acting, $\Delta y \Delta x$ is multiplied with its incremental velocity $\Delta u = \frac{\partial u}{\partial x} \Delta x$, yields the first term of expression (3.95). The other terms can be derived using the same rationale. By the same token, similar expressions like (3.95) can be derived for the work components in the y and z directions, respectively.

Collecting these terms and substituting into Eq. (3.92), and then dividing with Δx Δy Δz, the First Law of Thermodynamics for an infinitesimal system of fluid element without the generation of internal heat in differential form can then be written as

$$\begin{aligned} \rho \frac{De}{Dx} = &-\frac{\partial}{\partial x}\left(\frac{q}{A}\right)_x - \frac{\partial}{\partial y}\left(\frac{q}{A}\right)_y - \frac{\partial}{\partial z}\left(\frac{q}{A}\right)_z - \sigma_x \frac{\partial u}{\partial x} - \sigma_y \frac{\partial v}{\partial y} - \sigma_z \frac{\partial w}{\partial z} \\ &- \tau_{xy}\left(\frac{\partial u}{\partial y} + \frac{\partial v}{\partial x}\right) - \tau_{yz}\left(\frac{\partial v}{\partial z} + \frac{\partial w}{\partial y}\right) - \tau_{zx}\left(\frac{\partial w}{\partial x} + \frac{\partial u}{\partial z}\right) \end{aligned} \tag{3.96}$$

By referring to a textbook on Fluid Mechanics [1],[16] the fluid stress tensor components σ and τ can be related to fluid viscosity and deformation coefficients, and the First Law of Thermodynamics for an infinitesimal system of fluid element without the generation of internal heat in differential form can then be written further as

$$\rho\frac{De}{Dt} = -\frac{\partial}{\partial x}\left(\frac{q}{A}\right) - \frac{\partial}{\partial y}\left(\frac{q}{A}\right) - \frac{\partial}{\partial z}\left(\frac{q}{A}\right) - p\theta + n\Phi \tag{3.97}$$

where:

$$\theta = \nabla \cdot \boldsymbol{V} = \nabla \cdot \nabla\phi = \nabla^2\phi = \left(\frac{\partial\phi}{\partial x}\right)^2 + \left(\frac{\partial\phi}{\partial y}\right)^2 + \left(\frac{\partial\phi}{\partial z}\right)^2 \tag{3.98a}$$

and:

$$\begin{aligned}\Phi = 2&\left[\left(\frac{\partial u}{\partial x}\right)^2 + \left(\frac{\partial v}{\partial y}\right)^2 + \left(\frac{\partial w}{\partial z}\right)^2\right] \\ &+ \left[\left(\frac{\partial u}{\partial x} + \frac{\partial v}{\partial y}\right)^2 + \left(\frac{\partial w}{\partial x} + \frac{\partial u}{\partial z}\right)^2 + \left(\frac{\partial v}{\partial z} + \frac{\partial w}{\partial y}\right)^2\right] \\ &- \frac{2}{3}\theta^2\end{aligned} \tag{3.98b}$$

In vector notation:

$$\rho\frac{De}{Dt} = -\nabla.\left(\frac{\boldsymbol{q}}{A}\right) - \rho\nabla.\boldsymbol{V} + \mu\Phi \tag{3.99}$$

Here, θ expresses the divergence of the velocity vector V and ϕ the velocity potential, while $\mu\Phi$ denotes the dissipation function.

3.5.2 The Derivation of Energy Equation in Integral Form

The energy equation in Integral Form is found in many engineering applications and can be derived using a finite size Control Volume.

For this purpose, we utilize the integral form to express the Energy in the Control Volume V:

$$E = \oiiint_{Control\ Volume\ V} \rho e dV \tag{3.100}$$

[16] Such as H. Djojodihardjo, Mekanika Fluida (Fluid Mechanics), published by PT Erlangga, 1982.

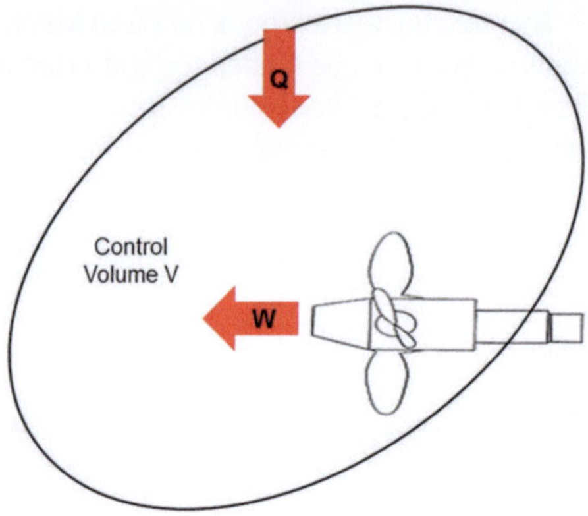

Fig. 3.11 Energy and work input into a system

The First Law of Thermodynamics for an finite Control Volume of fluid without the generation of internal heat in integral form, following Eq. (3.9), where B and ρb can be replaced by E and ρe, respectively, as in Eq. (3.100), can then be written as[17]

$$\frac{DE}{Dt} = \frac{D}{Dt} \oiiint\limits_{Control\ Volume\ V} \rho e d\tau = \oiiint\limits_{Control\ Volume\ V} \frac{\partial}{\partial t}(\rho e)d\tau + \oiint\limits_{Control\ Surface\ S\ of\ V} e(\rho \boldsymbol{V} \cdot \boldsymbol{n})dA \tag{3.101}$$

Then the time rate of work can be differentiated into rate of work due to pressure that is not utilized for useful work (W_p), time rate of dissipative work (W_q) and time rate of work that is being used for useful work W_s (shaft work). Refer also to Fig. 3.11, as a general case of Energy and Work input into a System.

Hence:

$$\dot{W} = \dot{W}_p + \dot{W}_f + \dot{W}_s \tag{3.102}$$

Next we will look at the rate of work due to pressure, which is generally acting at the surface of the Control Volume V (Control Surface)

$$W_p = \oiint\limits_{Control\ Durface\ S\ of\ V} p\boldsymbol{V} \cdot \boldsymbol{n}dA = \oiint\limits_{Control\ Surface\ S\ of\ V} \frac{p}{\rho}(\rho \boldsymbol{V} \cdot)\boldsymbol{n}dA \tag{3.103}$$

[17] It is customary to denote velocity with V and Volume with V also. The confusion can be avoided by undertanding the context. This situation occurs in this text also. However, to avoid such confusion, in this particular derivation, the variable τ is utilized. Again, this notation should not be confused with the variable stress.

The dissipative friction work is associated with the dissipative term, $\mu\sigma$, which is given by

$$\dot{W} = \underset{\text{Control Volume } V}{\oiiint} \rho\Phi d\tau \tag{3.104}$$

The detailed proof of this term is not elaborated here. What should be elaborated further is the useful work, which can be utilized for useful purposes outside the system; this is usually associated with shaft work in engineering systems. This work is due to the effect of pressure on an internal surface that is experiencing a movement (displaced along certain distances), such as the movement of an impeller due to pressure or the movement of a piston due to pressure. Such action can usually be represented by the movement of a weight along a certain height. In differential scale, this work is incorporated in the $p\nabla \cdot \mathbf{V}$ term.

By combining the useful work term in the surface integral on the right-hand side of the energy integral Eq. (3.101), this integral then becomes

$$\dot{Q} - \dot{W}_s - \dot{W}_f = \underset{\text{Control Volume } V}{\oiiint} \frac{\partial}{\partial t}(\rho e) d\tau + \underset{\text{Control Surface } S \text{ of } V}{\oiint} \left(e + \frac{p}{\rho}\right)(\rho \mathbf{V} \cdot \boldsymbol{n}) dA \tag{3.105}$$

where in general

$$e \equiv u + \frac{V^2}{2} + gz \tag{3.106}$$

and where

- e total energy per unit volume
- u internal energy per unit volume
- V velocity of the fluid flow
- g gravitational acceleration
- z elevation with respect to certain zero reference point in the chosen coordinate frame of reference.

In thermodynamics, $\left(u + \frac{p}{\rho}\right)$ is known as the enthalpy h. For unsteady flow, Eq. (3.105) can be reduced to

$$\dot{Q} - \dot{W}_s - \dot{W}_f = \underset{\text{Control Surface } S \text{ of } V}{\oiint} \left(u + \frac{p}{\rho} + \frac{V^2}{2} + gz\right)(\rho \mathbf{V} \cdot \boldsymbol{n}) dA \tag{3.107a}$$

or

$$\dot{Q} - \dot{W}_s - \dot{W}_f = \oiint_{Control\ Surface\ S\ of\ V} \left(h + \frac{V^2}{2} + gz\right)(\rho \boldsymbol{V} \cdot \boldsymbol{n})\mathrm{d}A \qquad (3.107b)$$

For one-dimensional flow, $\oint_{Control\ Surface\ S\ of\ V} (\rho \boldsymbol{V} \cdot \boldsymbol{n})\mathrm{d}A$ is equal to the mass flow rate. If Eq. (3.107a, 3.107b) is divided by the mass flow rate, and written for stations between the cross sections 1 and 2, there is obtained:

$$\dot{q} - \dot{w}_s - \dot{w}_f = \left(u_2 + \frac{p_2}{\rho} + \frac{V_2^2}{2} + gz_2\right) - \left(u_1 + \frac{p_1}{\rho} + \frac{V_1^2}{2} + gz_1\right) \qquad (3.108)$$

If we compare the Bernoulli Eq. (3.77) and the energy Eq. (3.108), their forms are similar. However, the energy equation is more general, since it is valid for compressible fluid and incorporating heat transfer and work component.

Consider the case where

$$\dot{w}_f = u_2 - u_1 \qquad (3.109a)$$

and

$$\dot{q} = 0 \qquad (3.109b)$$

In such a case, the work carried out to overcome the frictional forces has resulted in the increase of the internal energy in the fluid. For the fluid, such increase in the internal energy is always accompanied by an increase of the temperature. The relationship between the internal energy and temperature can usually be related by a linear equation, such as:

$$U_2 - U_1 = c(T_2 - T_1) \qquad (3.110)$$

where c is known as the specific heat of the fluid, and T is its temperature.

3.6 Some Geometric and Kinematic Properties of the Velocity Field

3.6.1 Gauss' Divergence Theorem[18]

Consider any volume V entirely within the field enclosed by a single closed bounding surface S as in Fig. (13.12).[19]

[18] This is another derivation of Gauss Divergence Theorem as previously elaborated in 2.2, in order to group it with other kinematic relationships.

[19] Note that V is the velocity vector of the fluid, while V can signify either scalar value of Velocity or Volume, as van b conjectured from the context.

$$\oiint_S \boldsymbol{V} \cdot \mathrm{d}S = \iiint_V (\nabla \cdot \boldsymbol{V})\,\mathrm{d}V \tag{3.111}$$

In Eq. (3.111), **n** is the unit normal vector directed outward from any infinitesimal or differential element of area dS on the surface S bounding the volume V. The derivation of this equation can be found in many books on Fluid Mechanics or Hydrodynamics. The theorem relates the tendency of the field lines to diverge or spread out within the volume V, to the net efflux of these lines from the boundary of V. It might therefore be described as an equation of continuity of field lines.

3.6.2 Stokes' Theorem on Rotation

Now we consider a closed curve C of the sort employed in 3.45 g and 3.46, except that the present result is instantaneous so that there is no question of a moving path composed of the same particles. Let S be any open surface which has the curve C as its boundary, as illustrated in Fig. 3.12b. The theorem refers to the circulation around the curve C can be expressed as [1, 8]

$$\Gamma \equiv \oint_C \boldsymbol{V} \cdot \mathrm{d}s = \iint_S \boldsymbol{n} \,.\, (\nabla \times \boldsymbol{V})\,\mathrm{d}S \tag{3.112}$$

Here

$$\nabla \times \boldsymbol{V} = \zeta \tag{3.113}$$

is the vorticity and can be shown to be equal to twice the angular velocity of a fluid particle about an axis through its own centroid. The theorem connects the spinning

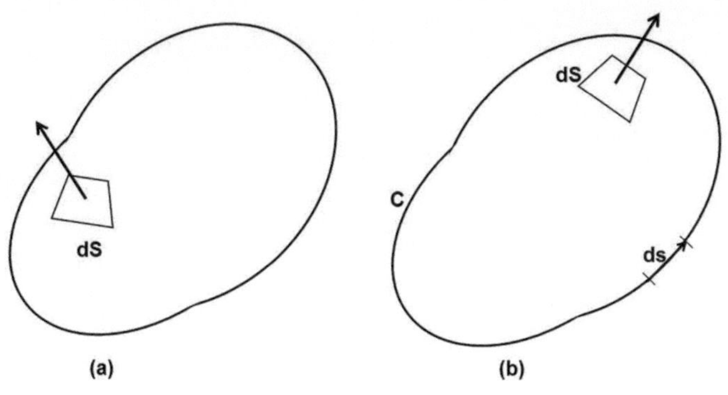

Fig. 3.12 Finite control volume and control surface in a flow field **a** finite control volume V surrounded by closed surface S in a flow field; **b** open surface S bounded by a closed curve C (Positive direction of n and s are as shown)

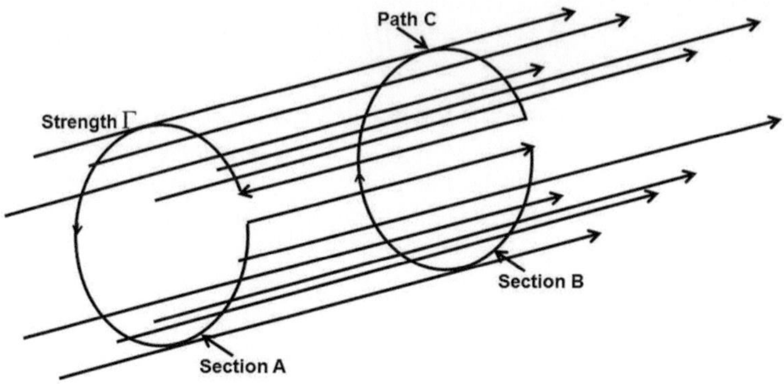

Fig. 3.13 Two cross sections of a vortex tube

tendency of the particles lying in surface S with the associated inclination of the fluid at the boundary of S to circulate in one direction or the other.

3.7 Vortex Theorems for the Ideal Fluid

In connection with the study of wing wakes, separation and related phenomena, it is of value to study the properties of the field vorticity vector $\boldsymbol{\zeta}$, the idea of a vortex line and a vortex tube. The arrows along such lines and tubes being are directed according to the right-hand rule of spin of fluid particles [6].[20]

Because $\boldsymbol{\zeta}$ is the curl of another vector, the field of vortex lines has certain properties that not all vector fields possess. Two of these are identified by the first two vortex theorems of Helmholtz. Although these theorems will be stated for the vorticity field, they are purely geometrical in nature and are unrelated in any way to the physics or dynamics of the fluid, or even to the requirement of continuity of mass.

3.7.1 First Vortex Theorem

The circulation around a given vortex tube, which implies the "strength" of the vortex, is the same everywhere along its length. This result can be proved in a variety of ways, such as by applying Stokes' theorem to a closed path in the surface of the vortex tube constructed as indicated in Fig. 3.13.

Stokes theorem on Rotation implies

[20] The present development follows closely that of Holt Ashley, Mårten Landahl, Aerodynamics of Wings and Bodies, Dover Books on Aeronautical Engineering, Revised Edition ISBN-13: 978-0486648996, ISBN-10: 0486648990.

$$\Gamma \equiv \oint_C \boldsymbol{V} \cdot \mathrm{d}s = \iint_S \boldsymbol{n} . (\nabla \times \boldsymbol{V}) \, \mathrm{d}S \tag{3.114}$$

Choosing for S the cylindrical surface lying in a wall of the tube, it is apparent that no vortex lines cross S. Therefore

$$\boldsymbol{n} . \boldsymbol{\zeta} \equiv \boldsymbol{n} . (\nabla \times \boldsymbol{V}) = 0 \tag{3.115}$$

Hence the circulation Γ around the whole of the curve C vanishes. By examining C, it is clear that

$$0 = \Gamma = \Gamma_B - \Gamma_A + \text{two pieces which cancels each other} \tag{3.116}$$

Hence $\Gamma_A = \Gamma_B$. Sections A and B can be chosen arbitrarily. Therefore, the circulation around the vortex is the same at all sections. I can also be observed that the circulation around the tube is always equals to

$$\iint \boldsymbol{n} . \boldsymbol{\zeta} \; dS$$

where the integral is taken over any surface which cuts through the tube but does not intersect any other vortex lines. Therefore this integral has the same value regardless of the orientation of the area used to cut through the tube, which implies that the number of vortex lines which make up the tube, or bundle, is everywhere the same.

3.7.2 Second Vortex Theorem

A vortex tube can never end in the fluid, but must close onto itself, end at a boundary, or go to infinity. Examples of the three kinds of behavior mentioned in this theorem are a smoke ring, a vortex bound to a two-dimensional airfoil spanning across from one wall to the other in a two-dimensional wind tunnel, and the downstream ends of horseshoe vortices representing the loading on a three-dimensional wing. This second theorem can be quite easily deduced from the continuity of circulation asserted by the first theorem; one simply notes that assuming an end for a vortex tube leads to a situation where the circulation is changing from one section to another along its length.

The first two vortex theorems are closely connected to the fact that the field of r is *solenoidal,* that is [6],

$$\nabla . \boldsymbol{\zeta} = 0 \tag{3.117}$$

When this result is inserted into Gauss' theorem, (3.111), it can be concluded that the number of vortex lines entering the closed surface should be equal to the number of vortex lines leaving that closed surface enclosing the volume V bounded by the two sections, A and B, and the cylindrical surface connecting them, whose generators are parallel to the vortex lines. Hence, flow streamlines cannot end, and the volume flux through any section is the same as that through any other section at a given instant of time. One may examine in the same light the field of tubes of the vector pV in a *steady* compressible flow.

3.7.3 Third Vortex Theorem

The third vortex theorem which is related to the dynamical properties of the fluid. Starting from the vector identity[21]

$$\boldsymbol{a} \equiv \frac{D\boldsymbol{V}}{Dt} = \frac{\partial \boldsymbol{V}}{\partial t} + (\boldsymbol{V}\cdot\nabla)\boldsymbol{V} = \frac{\partial \boldsymbol{V}}{\partial t} + \nabla\frac{(\boldsymbol{V}\cdot\boldsymbol{V})}{2} - \boldsymbol{V}\times\boldsymbol{\zeta} \tag{3.118}$$

The second step is to take the curl of (3.118) noting that the curl of a gradient should vanish,

$$\begin{aligned}\nabla\times\boldsymbol{a} &\equiv \nabla\times\frac{\partial \boldsymbol{V}}{\partial t} + \underbrace{\nabla\times\nabla\frac{(\boldsymbol{V}\cdot\boldsymbol{V})}{2}}_{=0} - \nabla\times(\boldsymbol{V}\times\boldsymbol{\zeta})\\ &\equiv \nabla\times\frac{\partial \boldsymbol{V}}{\partial t} - \nabla\times(\boldsymbol{V}\times\boldsymbol{\zeta})\end{aligned} \tag{3.119}$$

The operations $\nabla\times$ and $\partial\partial/t$ can be interchanged, so that the first term on the right becomes $\partial\varsigma\partial/t$. For any two vectors A and B, the following identity applies[22]

$$\nabla\times(\boldsymbol{A}\times\boldsymbol{B}) = (\boldsymbol{B}\cdot\nabla)\boldsymbol{A} - (\boldsymbol{A}\cdot\nabla)\boldsymbol{B} - \boldsymbol{B}(\nabla\cdot\boldsymbol{A}) + \boldsymbol{A}(\nabla\cdot\boldsymbol{B}) \tag{3.120}$$

Therefore, upon substituting the appropriate values, there is obtained

$$\begin{aligned}\nabla\times(\boldsymbol{V}\times\boldsymbol{\zeta}) &= (\boldsymbol{\zeta}\cdot\nabla)\boldsymbol{V} - (\boldsymbol{V}\cdot\nabla)\boldsymbol{\zeta} - \boldsymbol{\zeta}(\nabla\cdot\boldsymbol{V}) + \boldsymbol{A}(\nabla\cdot\boldsymbol{\zeta})\\ &= (\boldsymbol{\zeta}\cdot\nabla)\boldsymbol{V} - (\boldsymbol{V}\cdot\nabla)\boldsymbol{\zeta} - \boldsymbol{\zeta}(\nabla\cdot\boldsymbol{V}) + \boldsymbol{V}\left(\underbrace{\nabla\cdot(\nabla\times\boldsymbol{V})}_{=0}\right)\end{aligned} \tag{3.120a}$$

[21] Can be found in books on Hydrodynamics or Fluid Mechanics, such as Milne Thompson or Lamb.

[22] This vector identity can be found in books on Calculus and Applied Mathematics.

Substituting into (3.119,) we have

$$\nabla \times \boldsymbol{a} = \frac{D\boldsymbol{\zeta}}{Dt} - (\boldsymbol{\zeta} \cdot \nabla)\,\boldsymbol{V} + \boldsymbol{\zeta}(\nabla \cdot \boldsymbol{V}) \tag{3.120b}$$

Thus far, the development carried out are purely kinematical. Let us introduce the conservation of mass equation

$$\left[\frac{\partial \rho}{\partial t} + \nabla.(\rho \mathbf{V})\right] = 0 \tag{3.13a}$$

or

$$\frac{\partial \rho}{\partial t} + \nabla \cdot (\rho\,\boldsymbol{V}) = \frac{D\rho}{Dt} + \rho \nabla \cdot \boldsymbol{V} = 0 \tag{3.13b}$$

and obtain[23]

$$(\boldsymbol{\zeta} \cdot \nabla)\,\boldsymbol{V} = -\frac{\boldsymbol{\zeta}}{\rho}\frac{D\rho}{Dt} = \rho\,\boldsymbol{\zeta}\,\frac{D}{Dt}\left(\frac{1}{\rho}\right) \tag{3.121}$$

Substituting this term into the second term on the right-hand side of Eq. (3.121), and after some algebraic manipulation, there is obtained:

$$\frac{D}{Dt}\left(\frac{\boldsymbol{\zeta}}{\rho}\right) = \left(\frac{\boldsymbol{\zeta}}{\rho} \cdot \nabla\right)\boldsymbol{V} + \frac{\nabla \times \boldsymbol{a}}{\rho} \tag{3.122}$$

Taking into consideration that as indicated by

$$\boldsymbol{a} = \frac{D\boldsymbol{V}}{Dt} = \nabla\left[\Omega - \int \frac{dp}{\rho}\right]$$

a is a can be considered to be the gradient of another vector and its curl vanishes. Therefore, for inviscid, barotropic fluid in a conservative body force field, the foregoing result reduces to

$$\frac{D}{Dt}\left(\frac{\boldsymbol{\zeta}}{\rho}\right) = \left(\frac{\boldsymbol{\zeta}}{\rho} \cdot \nabla\right)\boldsymbol{V} \tag{3.123}$$

This last is what is usually known as the third vortex theorem of Helmholtz. In the continuum sense, it is an equation of conservation of angular momentum. If the specific entropy 8 is not uniform throughout the fluid, o~ can determine from a

---

23 $\frac{D\rho}{Dt} + \rho \nabla \cdot \boldsymbol{V} = 0 \frac{1}{\rho}\frac{D\rho}{Dt} = -\nabla \cdot \boldsymbol{V} \quad \nabla \cdot \boldsymbol{V} = \frac{D\rho}{Dt}\left(\frac{1}{\rho}\right)$.

$\boldsymbol{\zeta}(\nabla \cdot \boldsymbol{V}) = \rho\,\boldsymbol{\zeta}\frac{D\rho}{Dt}\left(\frac{1}{\rho}\right)$

combination of dynamical and thermodynamic considerations that

$$\nabla \times \boldsymbol{a} = \nabla \times (T \nabla \tilde{s}) \tag{3.124}$$

where $\tilde{s}$ stands for entropy. When inserted into (3.122), this demonstrates the role of entropy gradients in generating angular momentum, a result which is often associated with the name of Crocco. Further implication of this theorem is discussed.

3.8 Irrotational Flow and Velocity Potential

For further use and development in subsequent chapters, aerodynamic equations written in the velocity potential Φ will frequently be used.[24] For this purpose, the following section will elaborate such pertinent relationships in this section [6].[25]

One important consequence of fluids that can be considered to maintain irrotationality condition is the existence of a velocity potential. That is, the equation

$$\boldsymbol{\zeta} \equiv \nabla \times \boldsymbol{V} = 0 \tag{3.125}$$

is a necessary and sufficient condition for the existence of a potential Φ such that

$$\boldsymbol{V} \equiv \nabla \Phi \tag{3.126}$$

where Φ (x, y, z, t) or Φ $(\mathbf{r}, t)$ is the velocity potential in the entire flow field. Its existence allows the convenience in replacing a three-component vector $\mathbf{V}$ by

[24] Again, notation in Fluid echanics and Aerodynamics are often confusing, since several symbols are used for different variables or properties. However, such conusion can be clarified by understanding the physical context. Recall that earlier, related to Eq. (3.99).

$$\rho \frac{De}{Dt} = -\nabla \cdot \frac{q}{A} - \rho \nabla \cdot V + \mu \Phi \quad (3.99)$$

θ expresses the divergence of the velocity vector V, ϕ the velocity potential, while μ Φ denotes the dissipation function. In what follows, both Φ and ϕ will be used to signify velocity potential and is perturbatuon quantity.

[25] Some of the developments follow those in Holt Ashley, Mårten Landahl, Aerodynamics of Wings and Bodies.

a single scalar Φ as the principle dependent variable or unknown in theoretical investigations.[26]

As significant consequences of defining and assuring the existence of Φ, the following expressions can be established.

3.8.1 *The Bernoulli Equation for Irrotational Flow (Kelvin's Equation)*

This integral of the equations of fluid motion is derived by combining equations of motion (3.23a–3.23c) and Eq. (3.117) and assuming a distant acting force potential

$$\begin{aligned}\frac{D\boldsymbol{V}}{Dt} &= \frac{\partial \boldsymbol{V}}{\partial t} + \nabla\frac{(\boldsymbol{V}\cdot\boldsymbol{V})}{2} - \boldsymbol{V}\times\zeta \\ &= \frac{\partial \boldsymbol{V}}{\partial t} + \nabla\frac{(\boldsymbol{V}\cdot\boldsymbol{V})}{2} = \nabla\,\Omega - \nabla\int\frac{dp}{\rho}\end{aligned} \tag{3.127}$$

with the prevailing assumptions

$$\frac{\partial \boldsymbol{V}}{\partial t} = \frac{\partial}{\partial t}(\nabla\,\Phi) = \nabla\left(\frac{\partial\,\Phi}{\partial t}\right) \tag{3.128}$$

Therefore Eq. (3.137) can be rearranged into

$$\nabla\left[\frac{\partial\Phi}{\partial t} + \frac{(\boldsymbol{V}\cdot\boldsymbol{V})}{2} + \int\frac{dp}{\rho} - \Omega\right] = 0 \tag{3.129}$$

which, if we scrutinized it carefully is another form of the Bernoulli Eq. (3.77).[27]

The vanishing of the gradient as exhibited by Eq. (3.129) implies that, at most, the quantity involved will be a function of time throughout the entire field. Hence the least restricted form of this Bernoulli equation is

---

[26] Since
$$\begin{aligned}\nabla\times\boldsymbol{V} = \nabla\times\nabla\Phi &= \begin{vmatrix}\boldsymbol{i} & \boldsymbol{j} & \boldsymbol{k}\\ \frac{\partial}{\partial x} & \frac{\partial}{\partial y} & \frac{\partial}{\partial z}\\ \Phi_x & \Phi_y & \Phi_z\end{vmatrix}\\ &= \boldsymbol{i}\left(\frac{\partial\Phi_z}{\partial y} - \frac{\partial\Phi_y}{\partial z}\right) + \boldsymbol{j}\left(\frac{\partial\Phi_x}{\partial z} - \frac{\partial\Phi_z}{\partial x}\right) + \boldsymbol{k}\left(\frac{\partial\Phi_y}{\partial x} - \frac{\partial\Phi_x}{\partial y}\right)\\ &= \boldsymbol{i}\left(\frac{\partial^2\Phi}{\partial y\partial z} - \frac{\partial\Phi}{\partial z\partial y}\right) + \boldsymbol{j}\left(\frac{\partial^2\Phi}{\partial z\partial x} - \frac{\partial\Phi}{\partial x\partial z}\right)\\ &\quad + \boldsymbol{k}\left(\frac{\partial^2\Phi}{\partial x\partial y} - \frac{\partial\Phi}{\partial y\partial x}\right) = 0\end{aligned}$$

[27] $\int_1^2 \frac{\partial V}{\partial t}\,ds + \frac{V_2^2}{2} + \frac{p_2}{\rho} - U_2 = \frac{V_1^2}{2} + \frac{p_1}{\rho} - U_1 \quad (3.77)$

$$\frac{\partial \Phi}{\partial t} + \frac{(\boldsymbol{V} \cdot \boldsymbol{V})}{2} + \int \frac{\mathrm{d}p}{\rho} - \Omega = F(t) \tag{3.130}$$

If we consider there is a uniform stream U_∞ at remote points, or at infinity, then Φ will be constant and the pressure may be set equal to p_∞ and the force potential to Ω_∞ at some reference level

$$F(t) = \frac{U_\infty^2}{2} + \int^{p_\infty} \frac{\mathrm{d}p}{\rho} - \Omega_\infty = \text{constant} \tag{3.131}$$

Equation (3.130) can then be rewritten as

$$\frac{\partial \Phi}{\partial t} + \frac{1}{2}\left(V^2 - U_\infty^2\right) + \int^{p_\infty} \frac{\mathrm{d}p}{\rho} - \Omega_\infty = 0 \tag{3.132}$$

In isentropic flow with constant specific heat ratio γ, then after some algebra, Eq. (3.132) can be reorganized into a formula for the local pressure coefficient C_p:

$$\begin{aligned} C_p &= \frac{p - p_\infty}{\frac{1}{2}\rho U_\infty^2} \\ &= \frac{2}{\gamma M^2}\left\{\left[1 - \frac{\gamma - 1}{a_\infty^2}\left(\frac{\partial \Phi}{\partial t} + \frac{1}{2}\left(V^2 - U_\infty^2\right) + \Omega_\infty - \Omega\right)\right]^{\frac{\gamma}{\gamma-1}} - 1\right\} \end{aligned} \tag{3.133}$$

Here a is the speed of sound and $M \equiv \frac{U_\infty}{a_\infty}$ is Mach number. It may be convenient for some certain circumstances to recognize that

$$\int_{p_\infty}^{p} \frac{\mathrm{d}p}{\rho} = \int_{a_\infty^2}^{a^2} \frac{\mathrm{d}(a^2)}{\gamma - 1} = \frac{1}{\gamma - 1}\left[a^2 - a_\infty^2\right] \tag{3.134}$$

where

$$a^2 = \left(\frac{\partial p}{\partial \rho}\right)_s = \frac{dp}{d\rho} = \gamma R T = \frac{\gamma p}{\rho} \tag{3.135}$$

This substitution in 3.132 provides a convenient means of computing the local value of a or of the absolute temperature T,

$$a^2 - a_\infty^2 = -(\gamma - 1)\left\{\frac{\partial \Phi}{\partial t} + \frac{1}{2}\left(V^2 - U_\infty^2\right) + \Omega_\infty - \Omega\right\} \tag{3.136}$$

It should be noted that the term containing the body-force potential can usually be neglected in aeroelasticity and aeronautics.

3.8.2 The Partial Differential Equation for Φ

By substituting for ρ and V in the equation of continuity, the differential equation satisfied by the velocity potential can be written as:

$$\frac{1}{\rho}\frac{D\rho}{Dt} + \nabla \cdot \boldsymbol{V} = 0 \tag{3.137}$$

Writing the second term in terms of Φ yields

$$\nabla \cdot \boldsymbol{V} = \nabla \cdot (\nabla \Phi) = \nabla^2 \Phi \tag{3.138}$$

which is the well-known Laplace operator and for irrotational flow the governing partial differential equation for Φ reduces to:

$$\nabla^2 \Phi = \frac{\partial^2 \Phi}{\partial x^2} + \frac{\partial^2 \Phi}{\partial x^2} + \frac{\partial^2 \Phi}{\partial x^2} = 0 \tag{3.139}$$

in Cartesian coordinate.

To modify the first term of (3.136), the Bernoulli's equation is evaluated at infinity, where the flow has been assured to be uniform, with the special case of the fluid being at rest, $U_\infty = 0$. The body-force term then cancels out, yielding

$$\int_{p_\infty}^{p} \frac{dp}{\rho} = -\frac{\partial \Phi}{\partial t} - \frac{1}{2}\left(V^2 - U_\infty^2\right) \tag{3.140}$$

Utilizing Leibnitz rule for differentiation of a definite integral [9, 10][28]:

$$\frac{d}{dp}\int_{p_\infty}^{p} \frac{dp}{\rho} = -\frac{1}{\rho} \tag{3.141}$$

and applying the substantial derivative operator to (3.141), and utilizing (3.135), we obtain

[28] See for example, Differentiating an Integral-Leibniz' Rule, http://www.its.caltech.edu/~kcborder/Notes/LeibnizRule.pdf.

$$\frac{D}{Dt}\int_{p_\infty}^{p}\frac{dp}{\rho}=\left[\frac{d}{dp}\int_{p_\infty}^{p}\frac{dp}{\rho}\right]\frac{Dp}{Dt}$$
$$=\frac{1}{\rho}\frac{dp}{d\rho}\frac{Dp}{Dt}=\frac{a^2}{\rho}\frac{Dp}{Dt} \tag{3.142}$$

Dividing by a^2 and introducing the velocity potential Φ, we obtain

$$\frac{1}{\rho}\frac{Dp}{Dt}=-\frac{1}{a^2}\frac{D}{Dt}\left(\frac{\partial\Phi}{\partial t}+\frac{V^2}{2}\right)$$
$$=-\frac{1}{a^2}\left(\frac{\partial}{\partial t}+\boldsymbol{V}\cdot\nabla\right)\left(\frac{\partial\Phi}{\partial t}+\frac{V^2}{2}\right)$$
$$=-\frac{1}{a^2}\left(\frac{\partial^2\Phi}{\partial t^2}+2V\cdot\frac{\partial V}{\partial t}+\boldsymbol{V}\cdot\nabla\left(\frac{V^2}{2}\right)\right) \tag{3.143}$$

Finally, inserting (3.130) and (3.143) into (3.137) results in

$$\nabla^2\Phi-\frac{1}{a^2}\left(\frac{\partial^2\Phi}{\partial t^2}+\frac{\partial}{\partial t}(V^2)+\boldsymbol{V}\cdot\nabla\left(\frac{V^2}{2}\right)\right)=0 \tag{3.144}$$

By meticulous algebraic manipulations and resorting to basic physical properties of irrotational fluid, we have arrived at a celebrated governing equation and essentially the desired differential equation relating the velocity potential of the irrotational fluid to the velocity field and ideal gas relationship manifested in the speed of sound a^2. We then observed that:

1. If it is multiplied through by a^2, one sees that it is a differential equation second order and of third degree in the unknown dependent variable and its derivatives. It reduces to an ordinary wave equation in a situation where the speed of sound does not vary significantly from its ambient values, and where the squares of the velocity components can be neglected by comparison with a^2.
2. It is of interest that Garrick has pointed out that 3.144[29] can be reorganized into

$$\nabla^2\Phi=\frac{1}{a^2}\left(\frac{\partial}{\partial t}+\boldsymbol{V}\cdot\nabla\right)\left(\frac{\partial\Phi}{\partial t}+\boldsymbol{V}_c\cdot\nabla\Phi\right)=\frac{1}{a^2}\frac{D_c^2}{Dt^2}\Phi \tag{3.145}$$

where the subscript c on V_c and on the substantial derivative $\frac{D_c^2}{Dt^2}$ is intended to indicate that this velocity is treated as a constant during the second application of the operators $\frac{\partial}{\partial t}$ and (**V**. ∇). Then Eq. 3.145 appears as just a wave equation, with the propagation speed equal to the local value of *a, if* the prevailing process is observed relative to a coordinate system moving at the local fluid velocity *V*.

[29] Holt Ashley, Mårten Landahl, Aerodynamics of Wings and Bodies.

3.9 Problem Examples

3.9.1 Example 1: Transient Flow During Valve Opening

We would like to see the change of flow velocity of the fluid flow at Sect. 3.2 when the valve at Sect. 3.2 is opened. The pipe is connected to a container which is filled with the fluid at the level h above the valve elevation. The length of the pipe is L. The pressure of the ambient air is p_a, while its density is ρ_a. The fluid density is ρ.

Solution:

Euler's Equation along the streamline s can be written as:

$$\frac{1}{\rho}\frac{\partial p}{\partial s}\mathrm{d}s + g\frac{\partial p}{\partial s}\mathrm{d}s + V\frac{\partial V}{\partial s}\mathrm{d}s + V\frac{\partial V}{\partial t}\mathrm{d}t = 0 \tag{3.145}$$

If it is integrated between two points 1 and 2 at the streamline, then a Bernoulli integral is obtained:

$$\int_1^2 \frac{1}{\rho}\frac{\partial p}{\partial s}\mathrm{d}s + \int_1^2 g\frac{\partial p}{\partial s}\mathrm{d}s + \int_1^2 V\frac{\partial V}{\partial s}\mathrm{d}s + \int_1^2 V\frac{\partial V}{\partial t}\mathrm{d}s = 0 \tag{3.146}$$

now:

$$p_2 - p_1 = \rho g h \tag{3.147a}$$

and:

$$s_2 - s_1 = -h \tag{3.147b}$$

In general, $V_1 << V_2$, and can be neglected. Therefore (3.145) becomes:

$$\left(\frac{\rho_a}{\rho} - 1\right)gh + \frac{V_2^2}{2} + \int_1^2 \frac{\partial V}{\partial t}\mathrm{d}s = 0 \tag{3.148}$$

To evaluate $\int_1^2 \frac{\partial V}{\partial t}\mathrm{d}s$ Fig. 3.14 will be of great assistance. Looking carefully into Fig. 3.14, it can be noted that:

$$VA = V_2 A_2 \tag{3.149}$$

where V and A are velocity and cross-section at any point along the pipe L or the streamline, respectively, while V_2 and A_2 are velocity and cross-section at Sect. 3.2

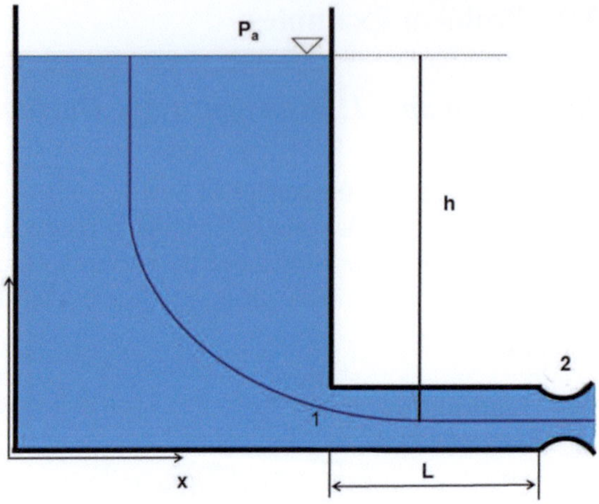

Fig. 3.14 Transient flow during valve opening

(i.e. the valve) shown there. If Eq. (3.128) is differentiated, one will obtain

$$\frac{\partial V}{\partial t} = \frac{1}{A}\left(V_2\frac{\partial A_2}{\partial t} + A_2\frac{\partial V_2}{\partial t}\right) \tag{3.150}$$

Since A_2 is the pipe cross-section at the valve, which will be stationary when the valve is opened suddenly, then

$$\frac{\partial A_2}{\partial t} \simeq 0 \tag{3.151}$$

Therefore

$$\frac{\partial V}{\partial t} = \frac{A_2}{A}\frac{\partial V_2}{\partial t} \tag{3.152a}$$

or

$$\int_1^2 \frac{\partial V}{\partial t}\mathrm{d}s = \frac{\partial V_2}{\partial t}\int_1^2 \frac{A_2}{A}\mathrm{d}s \tag{3.152b}$$

Let us assume

$$\int_1^2 \frac{A_2}{A}\mathrm{d}s \equiv L' \tag{3.153}$$

In general, the density of the air is much smaller than the density of the fluid, hence

$$\frac{\rho_a}{\rho} \ll 1 \tag{3.154}$$

and can be neglected in Eq. (3.148). In addition, when the steady condition is reached $\frac{\partial V}{\partial t} = 0$.

Hence Eq. (3.148) reduces to

$$\left(\frac{\rho_a}{\rho} - 1\right)gh + \frac{V_2^2}{2} + \int_1^2 \frac{\partial V}{\partial t}\mathrm{d}s \simeq -gh + \frac{V_2^2}{2} \simeq 0 \tag{3.155a}$$

or

$$V_{2\infty} \simeq \sqrt{2gh} \tag{3.155}$$

Hence Eq. (3.152a–3.152b) becomes

$$\frac{V_2^2 - V_{2\infty}^2}{2} + L'\frac{\partial V_2}{\partial t} \simeq 0 \tag{3.156}$$

This is the differential equation that governs the transient state at the time the valve is suddenly opened. The initial condition is:

$$\text{at t } = 0,\ \mathrm{V}_2 = 0 \tag{3.157}$$

Next the problem is to solve Eq. (3.156) subject to the initial condition (3.157).

For this purpose, the following approximation assumptions is made, where Eq. (3.156) is applied at station 2 during the transient state with initial condition (3.157):

$$\frac{V_2^2 - V_{2\infty}^2}{2} + L'\frac{\partial V_2}{\partial t} = \frac{V_2^2 - V_{2\infty}^2}{2} + L'\frac{\partial (V_2 + V_{2\infty})}{\partial t} = 0 \tag{3.158a}$$

or:

$$\frac{V_2^2 - V_{2\infty}^2}{2} + L'\frac{\partial V_2}{\partial t} = 0 \tag{3.158b}$$

since $V_{2\infty}$,, the transient velocity at Sect. 3.2 (the valve) at the end of the transient state can be assumed to be independent of time (constant). Division with $V_2 + V_{2\infty}$ and writing $\mathrm{d}V_2 = \mathrm{d}(V_2 + V_{2\infty})$ yields

$$\frac{V_2}{2L'}\mathrm{d}t + \mathrm{d}\ \ln(V_2 + V_{2\infty}) = \frac{V_{2\infty}}{2L'}\mathrm{d}t \tag{3.159}$$

Integrating between $t = 0$ to $t = \infty$ yields

$$\ln\left(1+\frac{V_2}{V_{2\infty}}\right)+\frac{1}{2L'}\int\limits_0^t V_2\mathrm{d}t=\frac{V_{2\infty}t}{2L'} \tag{3.160}$$

Next the Eq. (3.158b) is divided by $(V_2 - V_{2\infty})$ and $\mathrm{d}V_2$ is replaced by $\mathrm{d}(V_2 - V_{2\infty})$. Hence there is obtained

$$\frac{V_2^2-V_{2\infty}^2}{2(V_2-V_{2\infty})}+L'\frac{\mathrm{d}(V_2-V_{2\infty})}{(V_2-V_{2\infty})\mathrm{d}t}=0 \tag{3.161a}$$

or

$$\frac{V_2+V_{2\infty}}{2}+L'\frac{\mathrm{d}\ln(V_2-V_{2\infty})}{\mathrm{d}t}=0 \tag{3.162b}$$

or

$$\frac{V_2}{2L'}\mathrm{d}t+\mathrm{d}\ln(V_2-V_{2\infty})=\frac{V_{2\infty}}{2L'}\mathrm{d}t \tag{3.162c}$$

Integrating between $t = 0$ to $t = \infty$ yields

$$\ln\left(1-\frac{V_2}{V_{2\infty}}\right)+\frac{1}{2L'}\int\limits_0^t V_2\mathrm{d}t=\frac{V_{2\infty}t}{2L'} \tag{3.163}$$

Subtracting (3.163) from (3.160) yields

$$\ln\frac{\left(1+\frac{V_2}{V_{2\infty}}\right)}{\left(1-\frac{V_2}{V_{2\infty}}\right)}=-\frac{V_{2\infty}t}{L'} \tag{3.164}$$

Equation (3.164) gives the solution to the problem considered and is a transcendent equation which describes the velocity V_2 at the valve opening (or station 2) as a function of time, which is given in dimensionless variable $V_2/V_{2\infty}$ versus dimensionless variable $(V_{2\infty}\, t)/L'$. The state of affairs is depicted in Figs. 3.15 and 3.16.

Solution results of a similar but more complicated problem, which show the tank and pipeline circuit (Fig. 3.17) and pressure variation along the pipeline at two different instances (Figs. 3.18 and 3.19), are shown below for illustration [11].[30] For detail, the publication should be resorted to.

[30] Data and information based on Luis F. G. Pires, Renata C. C. Ladeia, Cláudio V. Barreto, Transient Flow Analysis of Fast Valve Closure in Short Pipelines, ASME IPC04-0367 2004.

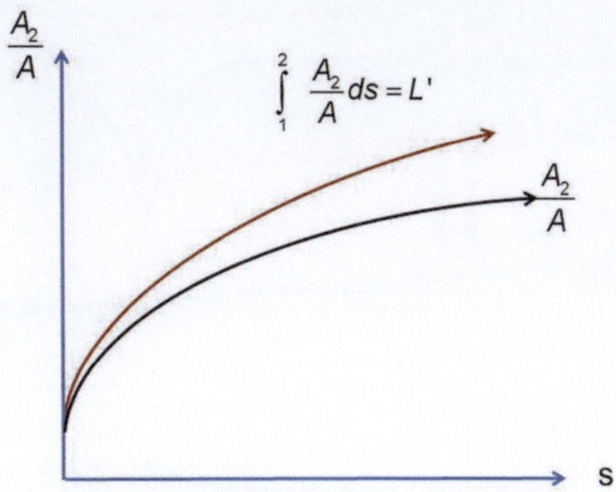

Fig. 3.15 Valve opening during transient flow

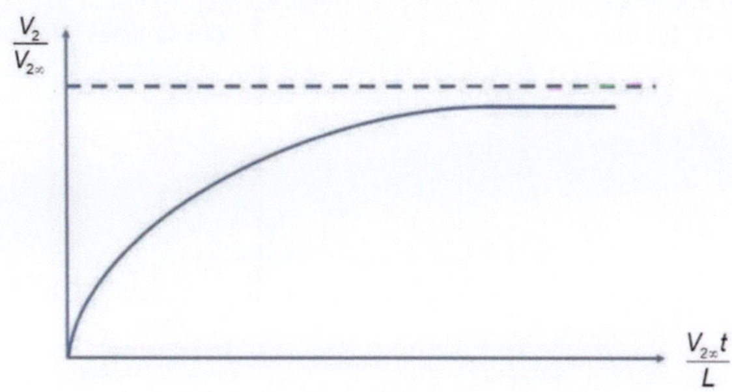

Fig. 3.16 Transient velocity at valve section during valve opening

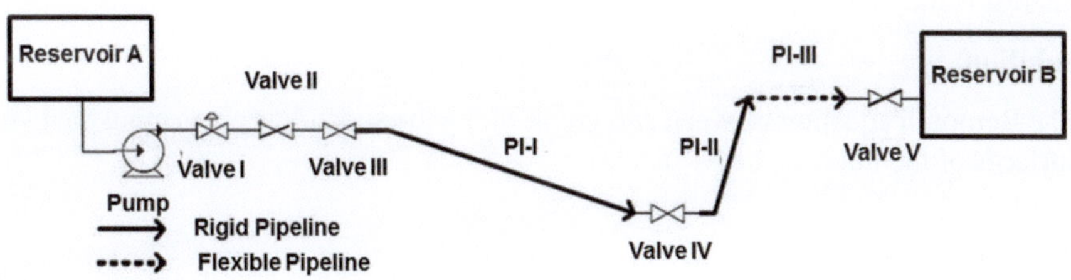

Fig. 3.17 Schematic of the modeled pipeline

3.9.2 *Example 2: Oscillating Fluid in a U-tube with Unequal Initial Column Height*

A fluid with density ρ fills the legs of a U-tube as illustrated in Fig. 3.20, with initial height l_1 and l_2 at the left and right sides, respectively. The cross-sectional areas of the two legs are equal. The distance between the two legs and the friction between

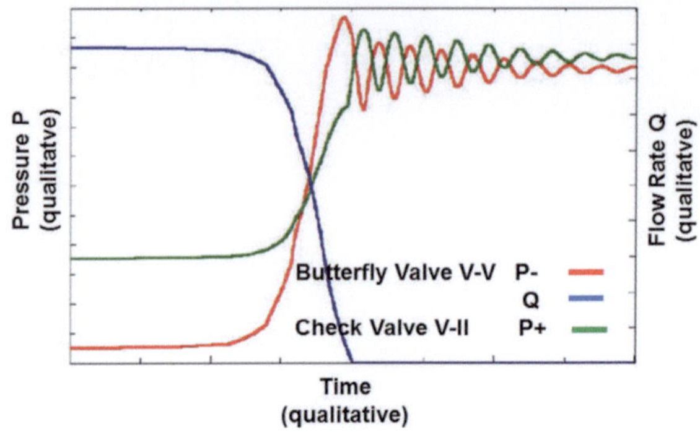

Fig. 3.18 Qualitative impression of pressure variation along the pipeline system of Fig. 3.17 at check valve V-II and block valves V-III and V-IV for a longer closure time period of the motorized block valve V-V

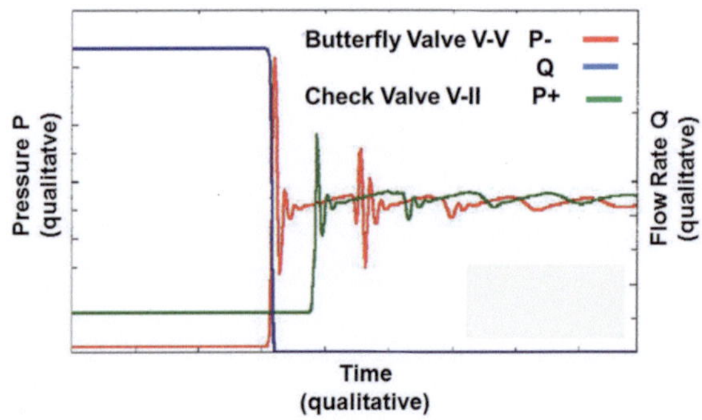

Fig. 3.19 Pressure transient at check valve V-II and block valves V-III and V-IV for an instantaneous closure of motorized block valve V-V

the fluid and the U-tube can be neglected for the present purpose. The problem of interest is the movement of the fluids.

Solution:

The Bernoulli equation between two points along the streamline connecting the two surfaces of the fluid can be written as

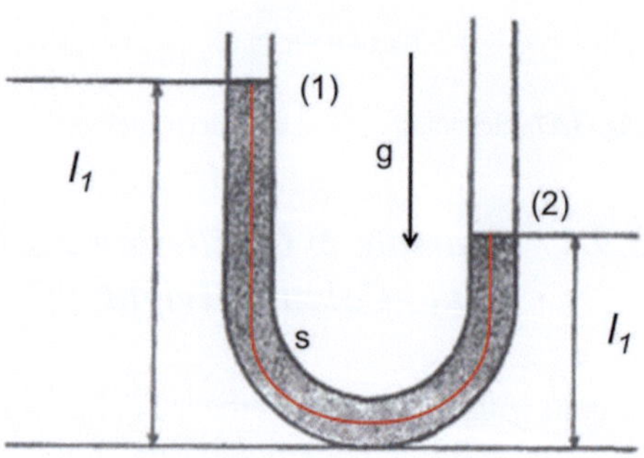

Fig. 3.20 A U-tube filled with fluid at initial time $t = 0$

$$\int_0^t \frac{\partial V}{\partial t} \mathrm{d}s + \frac{V_2^2}{2} + gl_2 = \frac{V_1^2}{2} + gl_1 \tag{3.165}$$

Assume that at each cross-section of the tube, the velocity of the fluid is uniform. Then Eq. (3.165) is valid for any streamline along the fluid.

From continuity:

$$|V_1| = |V_2| = V \tag{3.166}$$

Then the Bernoulli Eq. (3.145) becomes

$$\int_0^t \frac{\partial V}{\partial t} \mathrm{d}s + \frac{V_2^2}{2} + gl_2 = (l_2 + l_1)\frac{\partial V}{\partial t} + \frac{V_2^2}{2} + gl_2 = \frac{V_1^2}{2} + gl_1$$

or

$$(l_2 + l_1)\frac{\partial V}{\partial t} = g(l_1 - l)_2 \tag{3.167}$$

since $\frac{\partial V}{\partial t}$ is not a function of s (similar to V). Also, $l = (l_2 + l_1)$ and $V = \frac{\mathrm{d}l_2}{\mathrm{d}t}$. Substituting these values into Eq. (3.167) yields

$$\frac{\mathrm{d}^2 l}{\mathrm{d}t^2} = \frac{g(l_1 - l_2)}{l} \rightarrow \frac{\mathrm{d}^2 l}{\mathrm{d}t^2} = \frac{g(l - 2l_2)}{l} \rightarrow \frac{\mathrm{d}^2 l}{\mathrm{d}t^2} + \frac{2gl_2}{l} = g \tag{3.168}$$

This differential equation has a similar form the equation of harmonic oscillation with frequency

$$f = \frac{1}{2\pi}\sqrt{\frac{2g}{l}} \tag{3.169}$$

Therefore the fluid in the U-tube will oscillate with the frequency f given by Eq. (3.169).

3.9.3 *Example 3: Fluid Flow Through Converging Pipe*

Consider the flow of a fluid through a converging pipe as illustrated in Fig. 3.21. It is of interest to determine the difference of pressure at station 1 and 2 as a function of the volumetric flow rate Q and density ρ, where the flow rate is defined as

$$Q = VA \tag{3.170}$$

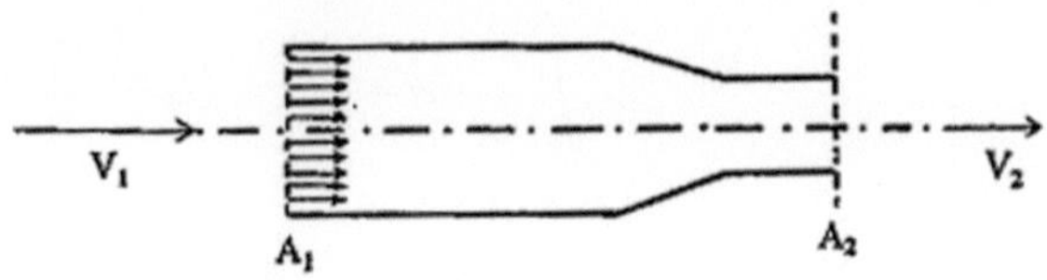

Fig. 3.21 Flow of a fluid through a converging pipe

It will be assumed that the velocity along the pipe is uniform at any cross-section. Similarly, the pressure at each cross-section will be uniform also. It is also assumed that the viscosity of the fluid is zero, which means that there will be no fluid friction.

Solution:

Since the density ρ is constant, the continuity equation will be

$$V_1A_1 = V_2A_2 = Q \tag{3.171}$$

The Bernoulli equation between the cross-sections 1 and 2 is then given by

$$\frac{V_1^2}{2} + \frac{p_1}{\rho} = \frac{V_2^2}{2} + \frac{p_2}{\rho} \tag{3.172}$$

Combining both equations to eliminate V_1, there is obtained

$$p_1 - p_2 = \frac{\rho V_1^2}{2}\left\{1 - \left(\frac{A_2^2}{A_1^2}\right)\right\} = \frac{\rho}{2}\frac{Q^2}{A_2^2}\left\{1 - \left(\frac{A_2^2}{A_1^2}\right)\right\} \tag{3.173}$$

Hence the difference in pressure is a function of the flow velocity, the density ρ and the cross-sectional area of the pipe. The flow conduit in the present example is known as the Venturi tube, which is utilized to measure the velocity of the fluid flow. For real fluids with viscosity, the mass flow rate ρQ should be multiplied by a correction factor, which in practice does not differ much from unity.

3.9.4 Example 4: Water Flow from an Orifice

Consider water with a density of ρ flowing through an orifice with a cross-sectional area A_0 at the bottom of a tank with cross sectional area A_t. The height of the water surface in the tank at each instant is denoted by y. The pressure of the air within the tank above the water surface is maintained at a pressure of P_t, while the pressure outside the tank at the orifice is the atmospheric pressure P_a. The influence of the gravitation will be considered, but the influence of the friction will be neglected. The velocity of the water jet at the orifice should be determined [1].

Solution:

First, one would like to find out possible stream lines starting from the surface of the water inside the tank flowing out through the orifice, as illustrated in Fig. 3.22.

a. First Approach

Choose one stream line from those depicted in Fig. 3.22, and write the Bernoulli equation between one point on the water surface in the tank, $A = A_t$ point 1 and another point 2 at the jet emanating from the orifice. Let the unsteady effects be neglected.

Hence:

$$\frac{V_1^2}{2} + \frac{p_a}{\rho} = \frac{V_1^2}{2} + \frac{p_1}{\rho} + gy \tag{3.174}$$

Note that the same equation will be obtained for any other streamline starting from the surface of the water in the tank and emanating through the orifice. It can be concluded that V_j will be uniform across the cross-section of the jet emanating through the orifice, as long as V_1 and y is the same.

From continuity:

$$V_1 A_t = V_j A_j \tag{3.175}$$

Eliminating V_1 in the Bernoulli equation using the continuity equation yields

$$V_j^2 = 2\left[gy + \frac{1}{\rho}(p_1 - p_a)\right]\frac{1}{1 - \left(\frac{A_j}{A_t}\right)^2} \tag{3.176}$$

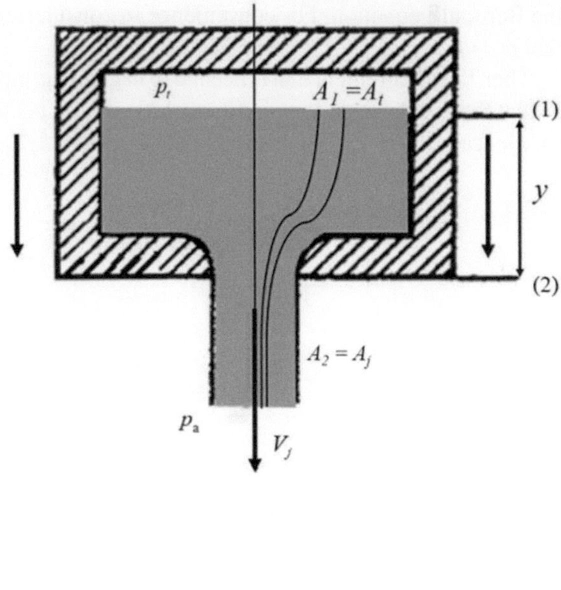

Fig. 3.22 Water flowing out from the tank through an orifice

Now the assumption that the flow is steady (or stationary) will be reviewed for its plausibility. When the flow *pf* the fluid through the orifice is taking place, y is decreasing; hence V_j is a function of time, $V_j = V_j(t)$. Then the stationary terms will be inspected, whether they can be neglected in the set of equations considered. However, it will be assumed that the flow will be steady at the orifice.

The integration of the Bernoulli equation along any streamline between two points 1 and 2 incorporating the unsteady terms, similar to Eq. (3.112), can be written as:

$$\int_1^2 \frac{1}{\rho}\frac{\partial p}{\partial s}\mathrm{d}s + \int_1^2 g\frac{\partial p}{\partial s}\mathrm{d}s + \int_1^2 V\frac{\partial V}{\partial s}\mathrm{d}s + \int_1^2 V\frac{\partial V}{\partial t}\mathrm{d}s = 0 \tag{3.177}$$

and considering the state of affairs, it simplifies to

$$\frac{V_j^2}{2} + \frac{p_a}{\rho} + \int_1^j V\frac{\partial V}{\partial t}\mathrm{d}t = \frac{V_1^2}{2} + \frac{p_1}{\rho} + gy \tag{3.178}$$

Since it is assumed that $\frac{\mathrm{d}V_j}{\mathrm{d}t} \cong 0$, this integral can be approximated by

$$\int_1^j V\frac{\partial V}{\partial t}\mathrm{d}t = \int_1^j V\frac{\partial V}{\partial t}\mathrm{d}s \approx \frac{\mathrm{d}V_1}{\mathrm{d}t}y = y\frac{\mathrm{d}V_j}{\mathrm{d}t}\frac{A_j}{A_t} \tag{3.179}$$

Here it has been tacitly assumed that the velocity at every cross-section is uniform and that the length of the streamline is close to y. In order that Eq. (3.140) is sufficiently accurate, the unsteady term (3.179) should be much smaller than the other terms in the Bernoulli equation. For convenience in considering the case, it will be assumed that $p_t = p_a$.

Then the requirement that is needed to neglect the unsteady term can be formulated as that the acceleration term in Eq. (3.179) should be less than the gravitational term gy. Hence

$$\left|\frac{A_j}{A_t}y\frac{\mathrm{d}V_j}{\mathrm{d}t}\right| \ll gy \tag{3.180a}$$

or

$$\left|\frac{A_j}{A_t}\frac{\mathrm{d}V_j}{\mathrm{d}t}\right| \ll g \tag{3.180b}$$

or

$$\left|\frac{1}{g}\frac{\mathrm{d}V_j}{\mathrm{d}t}\right| \ll \left|\frac{A_t}{A_j}\right| \tag{3.180c}$$

Therefore, the condition that the unsteady or acceleration term (3.179) can be ignored has been reduced to the condition that V_j can be approximated by its steady value, which means that the condition (3.144c) is met. Therefore the value of V_j can be approximated by its steady value

$$V_j^2 = 2gy\left[\frac{1}{1-\left(\frac{A_j}{A_t}\right)^2}\right] \tag{3.181}$$

So that

$$V_j\frac{\mathrm{d}V_j}{\mathrm{d}t} = \frac{g}{1-\left(\frac{A_j}{A_t}\right)^2}\frac{\mathrm{d}y}{\mathrm{d}t} \tag{3.182a}$$

or

$$V_j\frac{\mathrm{d}V_j}{\mathrm{d}t} \approx \frac{\mathrm{d}V_j}{\mathrm{d}t}\sqrt{2gh} = \frac{g}{1-\left(\frac{A_j}{A_t}\right)^2}\frac{\mathrm{d}y}{\mathrm{d}t} \tag{3.182b}$$

Now we will inspect, whether the stationary flow assumption is acceptable. When the flow start with the ejection of water through the orifice, y is decreasing with time. Therefore V_j is a function of time. We like to see, whether the unsteady terms can be neglected. We will make another assumption that the unsteadiness around the orifice area can be neglected. The integral of Eq. (3.179) is given by

$$\int_1^2 V_j\frac{\mathrm{d}V_j}{\mathrm{d}t}\mathrm{d}t \approx \frac{\mathrm{d}V_1}{\mathrm{d}t}y = y\frac{\mathrm{d}V_j}{\mathrm{d}t}\frac{A_j}{A_t} \tag{3.183}$$

In order that Eqs. (3.182a–3.182b) is sufficiently accurate, the unsteady term $\frac{\mathrm{d}V_j}{\mathrm{d}t}$ in the Eqs. (3.182a–3.182b) should be much less than the other acceleration term in the Bernoulli equation

$$\frac{V_1^2}{2} + \frac{p_a}{\rho} = \frac{V_1^2}{2} + \frac{p_1}{\rho} + gy \tag{3.184}$$

$$V_j^2 = 2\left[gy + \frac{1}{\rho}(p_1 - p_a)\right]\frac{1}{1-\left(\frac{A_j}{A_t}\right)^2} \tag{3.185}$$

In order that the acceleration term $y\frac{\mathrm{d}V_j}{\mathrm{d}t}$ is smaller than the gravitational term gy, then from Eqs. (3.182a–3.182b) and (3.183)

$$\left|\frac{A_j}{A_t}y\frac{\mathrm{d}V_j}{\mathrm{d}t}\right| \ll gy \tag{3.186a}$$

or

$$\left|\frac{A_j}{A_t}\frac{\mathrm{d}V_j}{\mathrm{d}t}\right| \ll g \tag{3.186b}$$

or

$$\left|\frac{\mathrm{d}V_j}{\mathrm{d}t}\right| \ll g\left|\frac{A_t}{A_j}\right| \tag{3.186c}$$

If $p_t = p_a$, there is already obtained

$$V_j^2 = \frac{2gy}{1-\left(\frac{A_j}{A_t}\right)^2} \tag{3.187a}$$

or

$$V_j = \sqrt{\frac{2gy}{1-\left(\frac{A_j}{A_t}\right)^2}} \tag{3.187b}$$

or

$$V_j = \sqrt{2gy\frac{1}{1-\left(\frac{A_j}{A_t}\right)^2}} = \sqrt{2gy}\sqrt{\frac{1}{1-\left(\frac{A_j}{A_t}\right)^2}} \tag{3.187c}$$

However:

$$\frac{\mathrm{d}y}{\mathrm{d}t} = -V_1 = -V_j\left(\frac{A_j}{A_t}\right) \tag{3.187}$$

Then[31]

31

$$V_j^2 = \frac{2gy}{1-\left(\frac{A_j}{A_t}\right)^2} \to 2V_j\frac{\mathrm{d}V_j}{\mathrm{d}t} = \frac{2g}{\left(1-\left(\frac{A_j}{A_t}\right)^2\right)}\frac{\mathrm{d}y}{\mathrm{d}t} = -V_j\frac{2g}{\left(1-\left(\frac{A_j}{A_t}\right)^2\right)}\left(\frac{A_j}{A_t}\right)$$

then $\frac{\mathrm{d}V_j}{\mathrm{d}t} = \frac{\left(\frac{A_j}{A_t}\right)}{\left(1-\left(\frac{A_j}{A_t}\right)^2\right)}g$

$$\frac{\mathrm{d}V_j}{\mathrm{d}t} = -\frac{\left(\frac{A_j}{A_t}\right)}{\left(1-\left(\frac{A_j}{A_t}\right)^2\right)}g \tag{3.188}$$

From (3.146), the non-stationary term can be neglected if

$$\left|\frac{\mathrm{d}V_j}{\mathrm{d}t}\right| \ll g\left|\frac{A_t}{A_j}\right| \quad \text{or} \quad \left|\frac{\left(\frac{A_j}{A_t}\right)}{\left(1-\left(\frac{A_j}{A_t}\right)^2\right)}\right| \ll \left|\frac{A_j}{A_t}\right| \tag{3.189a}$$

or

$$\left|\frac{\left(\frac{A_j}{A_t}\right)}{\left(1-\left(\frac{A_j}{A_t}\right)^2\right)}\right| \ll 1 \tag{3.189b}$$

If the difference in pressure due to acceleration terms has relative accuracy of 2%, then from Bernoulli equation the relative accuracy of the velocity will be about 1%. I order to have accuracy within this range, it follows from (3.189a) and (3.189b) that

$$\left|\frac{\left(\frac{A_j}{A_t}\right)}{\left(1-\left(\frac{A_j}{A_t}\right)^2\right)}\right| \ll 0.02 \tag{3.190a}$$

or

$$\left(\frac{A_j}{A_t}\right) \ll \frac{1}{7} \tag{3.190b}$$

It can be concluded that the non-stationary terms in this problem can be neglected and our assumption as such can be justified.

Referring to Fig. 3.22, one could write

$$\frac{\mathrm{d}y}{\mathrm{d}t} = -V_1 = -V_j\left(\frac{A_j}{A_t}\right)$$

or

$$V_1 = -\frac{\frac{\mathrm{d}y}{\mathrm{d}t}}{\left(\frac{A_j}{A_t}\right)} \tag{3.191}$$

From

$$V_j^2 = \frac{2gy}{1 - \left(\frac{A_j}{A_t}\right)} \tag{3.187a}$$

and if $p_t = p_a$, there is obtained

$$V_j \frac{\mathrm{d}V_j}{\mathrm{d}t} = g \frac{1}{\left(1 - \left(\frac{A_j}{A_t}\right)^2\right)} \frac{\mathrm{d}y}{\mathrm{d}t} \tag{3.192a}$$

or

$$\frac{\frac{\mathrm{d}y}{\mathrm{d}t}}{\left(\frac{A_j}{A_t}\right)} \frac{\mathrm{d}V_j}{\mathrm{d}t} + g \frac{1}{\left(1 - \left(\frac{A_j}{A_t}\right)^2\right)} \frac{\mathrm{d}y}{\mathrm{d}t} = 0 \tag{3.192b}$$

or

$$\left(\frac{1}{\left(\frac{A_j}{A_t}\right)} \frac{\mathrm{d}V_j}{\mathrm{d}t} + g \frac{1}{\left(1 - \left(\frac{A_j}{A_t}\right)^2\right)} \right) \frac{\mathrm{d}y}{\mathrm{d}t} = 0 \tag{3.192c}$$

since $.\frac{\mathrm{d}y}{\mathrm{d}t} \neq 0$. then

$$\left(\frac{1}{\left(\frac{A_j}{A_t}\right)} \frac{\mathrm{d}V_j}{\mathrm{d}t} + g \frac{1}{\left(1 - \left(\frac{A_j}{A_t}\right)^2\right)} \right) = 0 \tag{3.193}$$

Therefore

$$\frac{\mathrm{d}V_j}{\mathrm{d}t} = -\frac{\left(\frac{A_j}{A_t}\right)}{\left(1 - \left(\frac{A_j}{A_t}\right)^2\right)} g = 0 \tag{3.194}$$

or

$$V_j^{(t)} - V_j^{(0)} = \left(-\frac{\left(\frac{A_j}{A_t}\right)}{\left(1 - \left(\frac{A_j}{A_t}\right)^2\right)} \right) gt = 0 \tag{3.195a}$$

or

$$V_j = \left(\frac{\left(\frac{A_j}{A_t}\right)}{\left(1 - \left(\frac{A_j}{A_t}\right)^2\right)} \right) gt \tag{3.195b}$$

From (3.192b), the unsteady term can be ignored if

$$\left| \frac{\left(\frac{A_j}{A_t}\right)}{\left(1 - \left(\frac{A_j}{A_t}\right)^2\right)} \right| \ll 1 \tag{3.196}$$

and therefore the non-stationary terms in this problem can be neglected and our assumption as such can be justified.

Therefore the problem considered has the following answers:

1. $V_j = \sqrt{2gy \frac{1}{1 - \left(\frac{A_j}{A_t}\right)^2}} = \sqrt{2gy} \sqrt{\frac{1}{1 - \left(\frac{A_j}{A_t}\right)^2}}$
   which implies that V_j is a function of y
2. $V_j = \left(\frac{\left(\frac{A_j}{A_t}\right)^2}{\left(1 - \left(\frac{A_j}{A_t}\right)^2\right)} \right) gt$ which implies that V_j is a function of time t when the fluid starts flowing from the orifice.

Appendix: One-Dimensional Fluid Dynamics

Relation between Velocity and Cross-Sectional Area in One-Dimensional Gas Dynamics [1, 10][32]

In what follows, the conservation principle elaborated above is worked out for one-dimensional flow or a stream tube, or one-dimensional fluid dynamics. We should use conservation principle and may start with mass conservation principle. For this purpose we may relate what will be derived below with mass conservation equation which was derived using integral form, which results in Eq. (3.17a–3.17b), for example. The definition of *one-dimensional flow* or one-dimensional fluid dynamics applies to flow in a channel or tube, such as that illustrated in Fig. 3.23, which may be described by specifying the variation of the cross-sectional area along its axis, $A = A(x)$, and in which the flow properties are uniform over each cross-section, that is, $p = p(x)$, $\rho = \rho(x)$, etc. Similarly, the velocity u, which is normal to the

[32] Main Reference: Liepmann and Roshko [2].

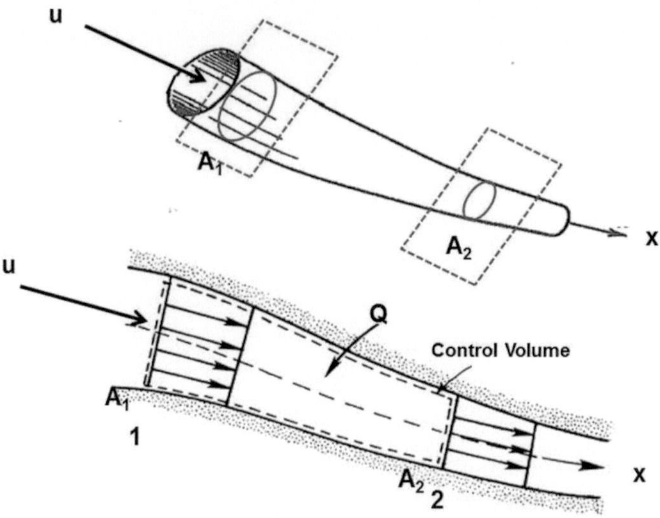

Fig. 3.23 One-dimensional flow in a stream tube

cross-section, should be uniform over each section, $u = u(x)$. These quantities may also be functions of time t, if the flow is *non-stationary non-steady.*

The Continuity Equation

If the flow in the tube of Fig. 3.23 is *steady,* the mass of fluid which passes a given section must eventually pass all the other sections farther downstream. This is simply a statement of the law of conservation of mass. At any two sections where conditions are uniform, the mass flows are equal, that is

$$\rho_1 u_1 A_1 = \rho_2 u_2 A_2 \tag{3.197}$$

This form of the continuity equation is very general, since it holds even if conditions *between* the sections are not uniform. If the flow *is* uniform at every section, the equation may be written as (refer also to Fig. 3.24)

$$\rho u A = \text{constant} = \frac{\mathrm{d}m}{\mathrm{d}t} = \dot{m} \tag{3.198}$$

and applied continuously along the tube. It may be differentiated, to give the differential form of the steady continuity equation,

$$\frac{\mathrm{d}}{\mathrm{d}x}(\rho u A) = 0 \tag{3.199}$$

Fig. 3.24 Flow through a tube segment

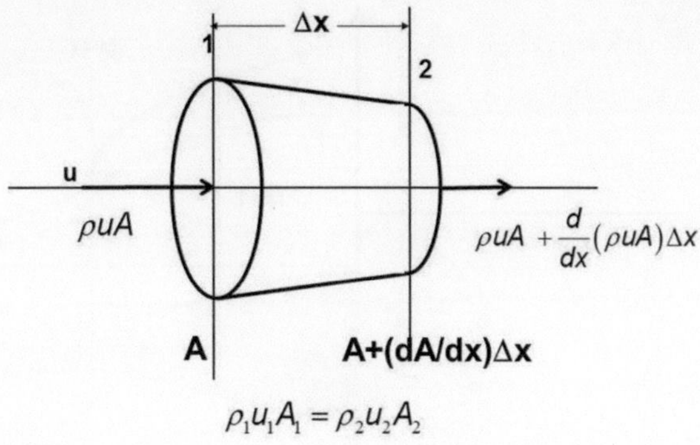

If the flow is unsteady, the continuity equation is obtained as follows. In Fig. 3.24 the mass contained between sections 1 and 2 is $Q = m = \rho A \Delta x$ and increases at the rate of $\frac{\partial}{\partial t}(\rho A \Delta x)$. This must be equal to the flow through section 1 minus the flow through section 2. Hence the net inflow is given by

$$-\frac{\mathrm{d}}{\mathrm{d}x}(\rho u A)\Delta x = \frac{\partial}{\partial t}(\rho A \Delta x) \tag{3.200}$$

The derivation for the Euler and Energy (and Bernoulli) equation can be carried out in a similar manner, and some are elaborated below.

The Area-Velocity Relation

To be able to demonstrate some of the effects of compressibility, one can consider the steady adiabatic flow in a stream tube of varying area, such as coming out from a reservoir as depicted in Figure A.7.4.1 as a "throat".

From the continuity Eq. (3.17) we have

$$\frac{\partial \rho}{\rho} + \frac{\partial u}{u \rho} + \frac{\partial A}{A} = 0 \tag{3.201}$$

For incompressible flow, $d\rho = 0$, Equation (3.201) reduces to the simple result that implies the increase of velocity is proportional to the decrease of area. The way in which this is modified by compressibility may be found by using Continuity Eq. (3.17) or its simplification for one-dimensional flow, in order to find the relations between density and velocity changes. We may rewrite it in the following form

$$u\mathrm{d}u = -\frac{\mathrm{d}\rho}{\rho} = -\frac{\mathrm{d}\rho}{\mathrm{d}\rho}\frac{\mathrm{d}\rho}{\rho} = -a^2\frac{\mathrm{d}\rho}{\rho} \tag{3.202}$$

Introducing the Mach number, this gives

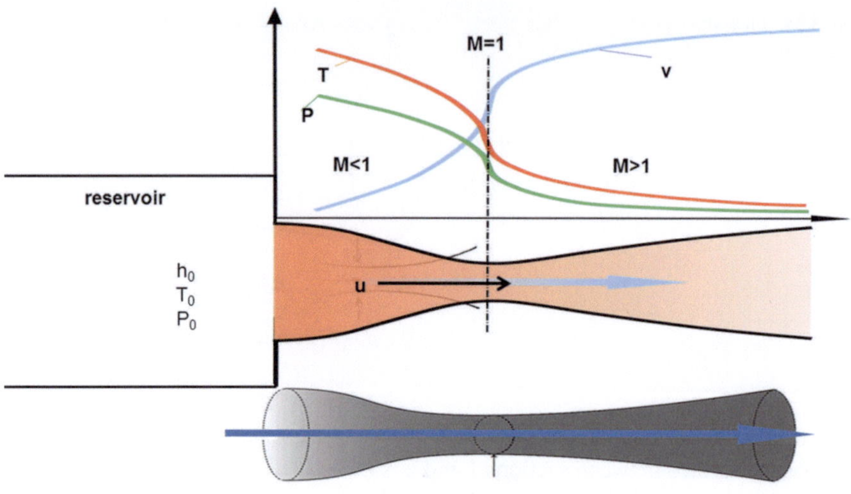

Fig. 3.25 Throat in a tube[33]

$$\frac{d\rho}{\rho} = -M^2 \frac{du}{u} \tag{3.203}$$

The role of M as a measure of compressibility is well illustrated here as also reflected by Fig. 3.25 At very low Mach numbers the density changes are so small, compared to the velocity changes, that they may be neglected in computing the flow field. In that case, it may be considered that $p = \text{const}$. Equivalent definitions of "incompressible flow" are $a = \infty$ or $M = 0$.

Substitution of this relation in Eq. (3.201) gives the area-velocity relations

$$\frac{du}{u} = \frac{\frac{dA}{A}}{1 - M^2} \tag{3.204}$$

which shows the following Mach number effects:

(1) At $M = 0$, a decrease in area gives a *proportional* increase in velocity.
(2) For values of M between 0 and 1, that is, at subsonic speeds, the relation is qualitatively the same as for incompressible flow, a decrease of area results in an increase in velocity, but the effect on the velocity is relatively greater, since the denominator is less than 1.
(3) At supersonic speeds, the denominator becomes negative and an increase of speed is produced by an *increase* of area. For one accustomed to thinking in terms of incompressible flow, this behavior may appear quite remarkable. It is due to the fact that at supersonic speeds the density decreases faster than the

[33] Created and adapted from various sources, among others de Laval nozzle—Wikipedia, https://en.wikipedia.org/wiki/De_Laval_nozzle.

velocity increases, so that the area must increase to maintain continuity of mass. This may be seen from Eq. A.7.4.2, which shows that for $M > 1$ the decrease in density is greater than the increase in velocity.

Then for $M = 1$, i.e. at *sonic* speed, $M = 1$, the area-velocity relation can be determined as follows. For this purpose, consider a tube in which the velocity increases continuously, from zero, and eventually becomes supersonic. The above discussion shows that the tube must converge in the subsonic and diverge in the supersonic portion. At $M = 1$ there must be a throat. This state of affairs is exhibited by Fig. 3.25. This can also be deduced from Eq. (3.205), which shows that, at $M = 1$, du/u can be finite only if dA $A = 0$. The same argument applies to the case where the velocity decreases continuously from supersonic to subsonic.

Energy Equation Considerations

From the discussions in this chapter, it was shown that in adiabatic flow where there is no heat input or output into the control volume or there is no hat exchange between the system and its environment, the energy equation for a perfect gas is

$$\frac{1}{2}u^2 + c_p T = c_p T_0 \tag{3.205}$$

With the expression $a^2 = \gamma RT$, where a = the speed of sound, and $c_p = \left(\frac{\gamma}{\gamma-1}\right)R$, one obtains:

$$\frac{u^2}{2} + \frac{a^2}{\gamma - 1} = \frac{a_0^2}{\gamma - 1} \tag{3.206}$$

Then, multiplying the last equation by $\frac{\gamma-1}{a^2}$ one obtains:

$$\frac{a_0^2}{a^2} = \frac{T_0}{T} = 1 + \frac{\gamma - 1}{2}M^2 \tag{3.207}$$

The isentropic relation

$$\frac{p}{p_0} = \frac{\rho}{\rho_0} = \left(\frac{T}{T_0}\right)^{\frac{\gamma}{\gamma-1}} \tag{3.208}$$

can be used to obtain

$$\frac{p}{p_0} = \left(1 + \frac{\gamma - 1}{2}M^2\right)^{\frac{\gamma}{\gamma-1}} \tag{3.209}$$

$$\frac{\rho}{\rho_0} = \left(1 + \frac{\gamma - 1}{2} M^2\right)^{\frac{\gamma}{\gamma - 1}} \tag{3.210}$$

In the Equations (3.206), (3.207) and (3.208) the values of T_0, dan a_0, are constant throughout the flow, so that they may be taken as those in the actual reservoir. In Eqs. (3.207) and (3.208) the values of p_0 and ρ_0 are the *local* "reservoir values." They are constant throughout only if the flow is isentropic.

Flow at Constant Area

Consider adiabatic, constant-area flow (Fig. 2.9a) through a non-equilibrium region (shown shaded). If sections 1 and 2 are outside this region, then the equations of continuity, momentum and energy are

$$\rho_1 u_1 = \rho_2 u_2 \tag{3.211}$$

$$p_1 + \frac{1}{2}\rho_1 u_1^2 = p_2 + \frac{1}{2}\rho_2 u_2^2 \tag{3.212}$$

$$h_1 + \frac{1}{2} u_1^2 = h_2 + \frac{1}{2} u_2^2 \tag{3.213}$$

The solution of these gives the relations that must exist between the flow parameters at the two sections; it will be worked out presently. There is no restriction on the size or details of the dissipation region so long as the reference sections are outside it. In particular, it may be idealized by the vanishingly thin region, shown in Fig. 2.9b, across which the flow parameters are said to "jump." The control sections 1 and 2 may then be brought arbitrarily close to it. Such a discontinuity is called a *shock wave*. Of course, a real fluid cannot have an actual discontinuity, and this is only an idealization of the very high gradients that actually occur in a shock wave, in

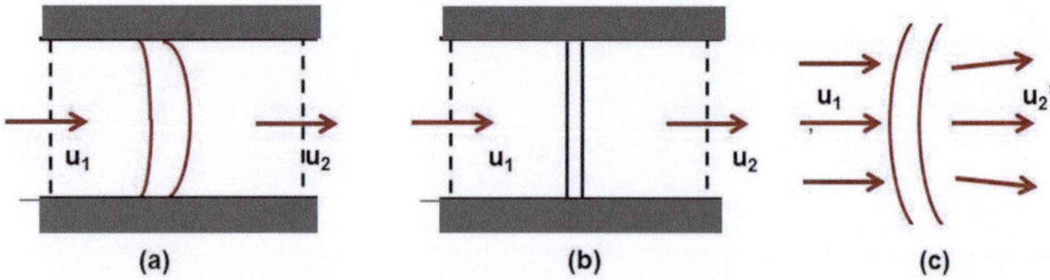

Fig. 3.26 A change of equilibrium conditions in constant area flow. **a** Uniform conditions on either side of a region of non-uniformity or dissipation; **b** normal shock wave; **c** shock wave normal to flow on streamline **a–b**

the transition from state 1 to 2. These severe gradients produce viscous stress and heat transfer, i.e. non-equilibrium conditions, inside the shock. Further detail can be followed in Liepmann and Roshko [10].[34]

The mechanism of shock-wave formation, as well as some details of conditions inside the dissipation region, will be discussed later. For application to most aerodynamic problems, it is sufficient to calculate the jumps in the equilibrium values and to represent the shock as a discontinuity. Since the reference sections may be brought arbitrarily near to the shock, the device of a constant area duct is no longer needed, that is, the results always apply *locally* to conditions on either side of a shock, provided it is normal to the streamline (Fig. 3.26c).

Of course, the shock relations may be applied to equilibrium sections of real constant-area ducts, such as the one shown in Fig. Fig. 3.26a, but it is necessary that the friction forces on the walls be negligible, since there are no friction terms in the momentum equation. An example is the constant-area supersonic diffuser, in which an adverse pressure gradient reduces the wall friction to negligible values. The diffusion occurs through a complicated, three-dimensional process involving interactions between shock waves and boundary layer. For equilibrium to be attained the diffuser must be *long,* in curious contrast to the normal shock, for which equilibrium is reached in a very short distance. For further detail, reference could be made to Liepmann and Roshko[35] and Shapiro.[36]

Problems

1. A tank has an orifice at its bottom. The cross section of the water jet emanating from the orifice is A_0, and the x coordinate is set at the orifice; hence $x_{orifice} = 0$. The water level in the tank measured from its bottom in h, which is maintained to be constant by continuous filling. Fluid friction and surface tension will be ignored, and the area of the orifice is much smaller than the area of the cross section of the tank A_0. Then find the tank area A as a function of A_0 and x (Fig. 3.27).

2. For a two-dimensional incompressible flow, if the *x* component of a velocity is

   a. $u = 3\ xy$
   b. $u = \ \times 2y$
   c. $u = 2x/y$

   Express the y component of such flow that will satisfy continuity.

3. A symmetrical sharp object is moving with a velocity *V* in the *X* direction in an incompressible fluid without friction with a density ρ. At the tip of the object, the fluid has also moved at the same velocity *V* with the same direction. What is the difference in the pressure of the fluid between the point right at the tip of the sharp object and with a point far ahead of that point, where its velocity is zero?

[34] Liepmann and Roshko [2].

[35] Liepmann and Roshko [2].

[36] Shapiro [3].

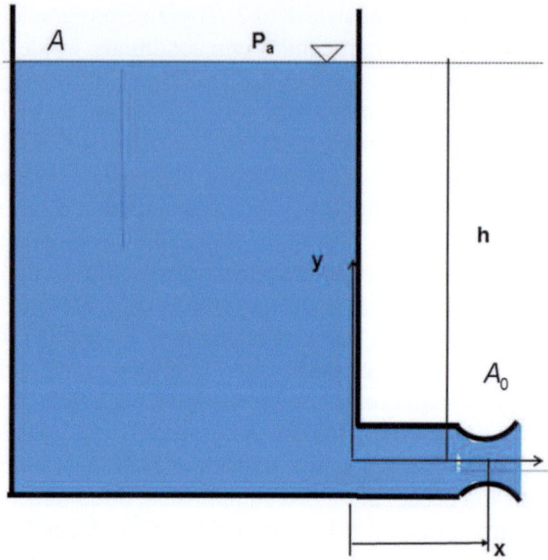

Fig. 3.27 Illustration for problem 1

Note: This flow is an unsteady (non-stationary) one. This problem can be solved using the unsteady form of the Bernoulli equation, and also by using the steady form of the Bernoulli equation using coordinate translation.

4. Oil is flowing in a series of pipes at a volumetric flow rate of 2 m^3/s. The pressure difference to overcome the friction is equivalent to 9 m per km of pipe length. A pump with 500 Hp and efficiency of 80% is being utilized at the pumping station. Assume that the density of the fluid is the same like the water density at room temperature. Then the pump station has to be located at every L km along the long pipe. Calculate L.
5. A large tank is filled with water until a height h above a small orifice. The area of the orifice is A_o. Assume that the water flows out of the orifice with a contraction ratio C_o. Find the force exerted on the tank due to the water jet emanating from the tank through the orifice.
6. A stationary rocket eject a hot gas at 6000 °C with a velocity of 2500 m/s. The corresponding mass flowrate is 1000 kgm/s. Assume that the exit pressure at the rocket nozzle P_e is the same as the atmospheric pressure P_a. Find the thrust of the rocket.
7. A turbojet in a wind tunnel sucks air at a speed of 150 m/s, while the density of the air flowing in the wind tunnel is $\rho = 0.003$ kgm/m^3. The air flow in the wind tunnel is assumed to be uniform across the cross-section of interest, and the area of the cross section of interest (test-section) is 0.1 m^2. The velocity of the gas jet exiting

$$V = 2V_0\left(1 - \left(\frac{r}{r_0}\right)^2\right)$$

from the turbojet is non-uniform and follow a distribution given by where r_o is the radius of the exit jet stream. The density of the jet is $\rho_j = 0.0015$ kg/m^3 and $V_0 = 600$ m/s.

a. Show that V_0 is the average velocity of the jet emanating from the turbo jet.
b. Determine the thrust acting on the turbo jet.
c. Determine the thrust acting on the turbo jet if the velocity of the jet emanating from the turbo jet is uniform with magnitude V_0.

References

1. Djojodihardjo, H. 1983. *Mekanika Fluida (Fluid Mechanics), in Indnesian.* Jakarta: Erlangga Publisher.
2. Aris, R. 1990.*Vectors, Tensors and the Basic Equations of Fluid Mechanics.* Dover Publications. ISBN 0-486-66110-5.
3. Yoon, H. *Control Volume and Reynolds Transport Theorem.* http://user.engineering.uiowa.edu/~fluids/posting/Lecture_Notes/Control%20Volume%20and%20Reynolds%20Transport%20Theorem_10-11-2013_Final.pdf. Accessed January 2020.
4. Nandkeolyar, R. *Reynolds Transport Theorem and Navier-Stoke's Equation.* http://www.nitjsr.ac.in/.
5. Lamb, Horace. 1932. *Hydrodynamics, Dover Books on Physics*, 6th Revised ed. Original Publisher, The University Press. ISBN 0521055156, 9780521055154; also Lamb, Horace. 1994. *Hydrodynamic.* Cambridge Unuversity Press.
6. Ashley, H., and M.T. Landahl. 1965. *Aerodynamics of Wings and Bodies*, 279. Dover Publications. ISBN-13: 978-0486648996, ISBN-10: 0486648990.
7. Wikipedia, Geoid. 2028. https://en.wikipedia.org/wiki/Geoid. Accessed 20 January 2028.
8. Lehman, *Gauss and Stokes Theorems.* https://www.lehman.edu/faculty/anchordoqui/VC-4.pdf.
9. Differentiating an Integral-Leibniz' Rule. http://www.its.caltech.edu/~kcborder/Notes/LeibnizRule.pdf. Accessed 20 January 2020.
10. Liepmann, H.W., and A. Roshko. 1957. *Elements of Gasdynamics.* Dover Books on Aeronautical Engineering Series.
11. Pires, L.F.G., R.C.C. Ladeia, and C.V. Barreto. 2004. *Transient Flow Analysis of Fast Valve Closure in Short Pipelines.* ASME IPC04-0367.

Chapter 4
Concepts of Typical Section

Abstract Typical section has been widely employed to analyze the aeroelasticity of the wing and the aircraft. Typical section typically represents the aeroelastic properties at about 70–75% of the half-wingspan (or the tail surface). Concept of typical section to represent the wing of aircraft has been introduced for multiple purposes, such as to give a fundamental and very instructive treatment of the problem of aeroelasticity, and accompanied by simple computation to provide quick estimate and general understanding of the problem of aircraft aeroelasticity, and can be employed in the early development of aeroelastic engineering approach. One advantage of the concept of typical section is that it can be applied for wings or aerodynamic surfaces of large aspect ratio. Relevant parameters of the wing, however, should be appropriately considered, such as the geometrical and stiffness parameters, the aerodynamic parameters and the kinematic and dynamic parameters commensurate with the aircraft flight. The aeroelastic differential equations of motion are then generated based on Newton's second Law of Motion, with progressive treatment of the wing configuration as on the essential element of the aircraft aeroelasticity [1, 2].

Keywords Aeroelasticity · Equation of motion · Kinematic and dynamics · Typical section

4.1 Typical Section—General

Typical Wing section [3, 4] has been utilized extensively in the study and analysis of aeroelasticity of aircraft. Typical section is a two-dimensional wing (airfoil) which typically represents the aeroelastic properties at about 70–75% of the half-wingspan (or the tail surface). It was introduced for the following purposes:

- It gives a fundamental and very instructive treatment of the problem of aeroelasticity
- It will give quick estimate, if not general understanding of the problem of aircraft aeroelasticity. It was generally used in the early development of aeroelasticity where only simple computation can be carried out.

© The Author(s), under exclusive license to Springer Nature Singapore Pte Ltd. 2023
H. Djojodihardjo, *Introduction to Aeroelasticity*,
https://doi.org/10.1007/978-981-16-8078-6_4

- One advantage of the concept of typical section is that it can be applied for wings or aerodynamic surfaces of large aspect ratio.

4.2 Typical Section—Definition

The definition introduced here follows that of Bisplinghoff et al. [3, 4] and Zwaan [5] in his class notes on the aeroelasticity of aircraft and represents many of those that have been utilized in standard textbooks (see references at the end of this chapter). It should be noted, however, that there are many different conventions. This concept of typical section is depicted in Fig. 4.1.

The following parameters need to be considered appropriately in the analysis of aircraft wing using the concept of typical section:

a. Geometrical and Stiffness Parameters:

b. Derived Parameters:

ω_h natural bending frequency of typical section, $\omega_h = (K_h/m)^{1/2}$
ω_α natural torsional frequency of typical section, $\omega_\alpha = (K_\alpha/I_\alpha)^{1/2}$

Notes:

m, S_α, I_α are calculated per unit span of the wing
EA is a locus of the *sectional shear center*.
Shear force at the *shear center* produces only translation.
Torsional Moment about the shear center produces only twist.

e. Aerodynamic Parameters:

AC Aerodynamic center

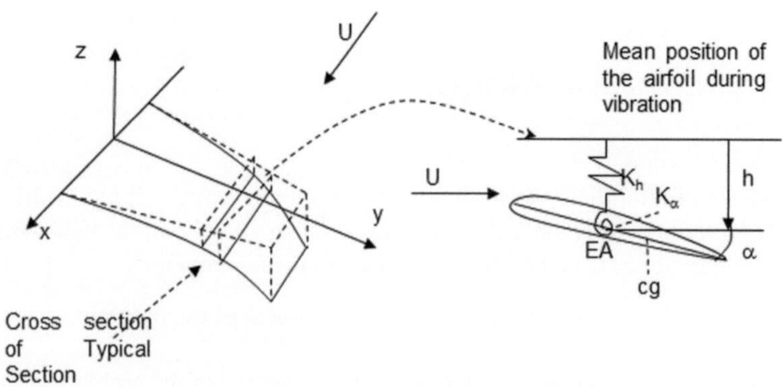

Fig. 4.1 Cross section of a typical: section and forces, Moments acting on it subject to flow-induced aerodynamic forces. The typical section is defined at about 07 wingspan representing the wing in a two-dimensional analysis [6]

U	Free-Stream Flow Velocity
α	angle of attack (between U vector and the chord line)
L	Lift; for typical section per unit span
M_{AC}	Pitching Moment about the Aerodynamic Center AC; for typical section per unit span
e	eccentricity factor
μ	mass ratio, $\mu = m/\pi \rho b^2$
ρ	the mass density of the typical section per unit span length [kg/m^3], which may be assumed to be uniform along the wingspan
b	half-chord of the typical section or the unswept wing beam model [m].

Notes:

Following the common notation used in textbooks on aeroelasticity, α can imply angle of attack or torsional deflection, depending on the context. To avoid confusion, indices can be used.

The parameters of typical section defined here are depicted in Figs. 4.2 and 4.3.

f. Control Surface Parameters:

δ deflection of the control surface with respect to the chord line

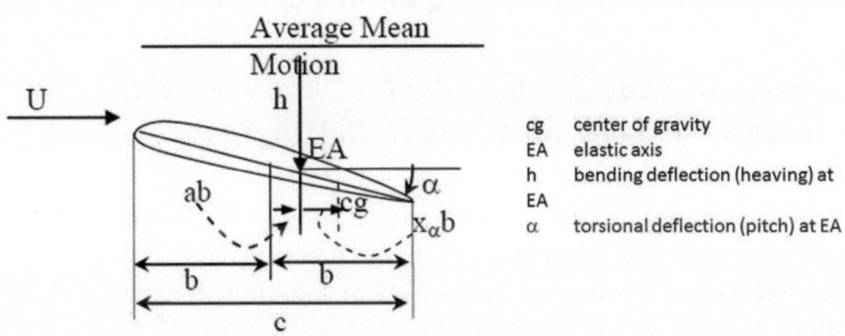

Fig. 4.2 Cross section of a typical; section with geometrical parameters [8, 9]

Fig. 4.3 Cross section of a typical; section with aerodynamic forces and moments

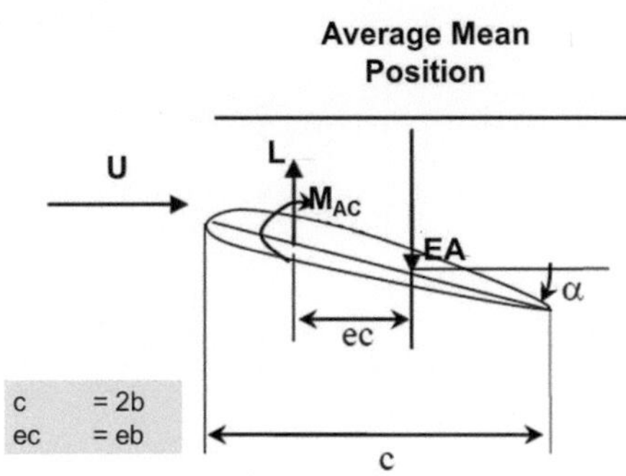

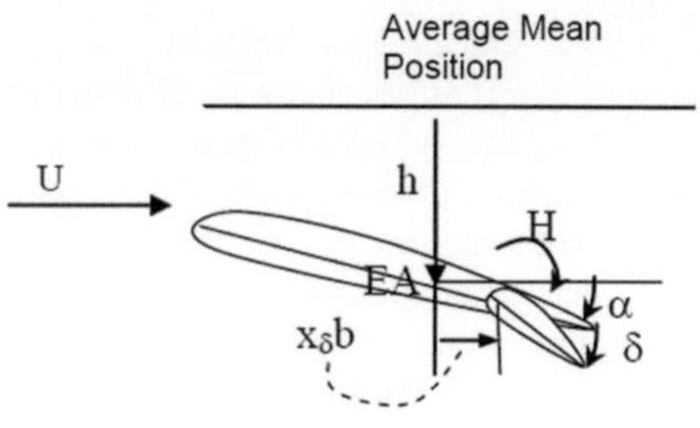

Fig. 4.4 Cross section of a typical section of a wing with a flap

S_δ static moment with respect to the hinge line of the control surface
I_δ mass moment of inertia with respect to the hinge line of the control surface
H hinge moment.

Some significant parameters of typical section with control surface are depicted in Fig. 4.4.

4.3 Unswept Uniform Wing Beam Model Aeroelastic Equations—Differential Equations of Motion

4.3.1 General: Wing Beam Under the Action of a Dynamic Transverse Load Along the Half-Wingspan

Consider the differential equation of motion of a wing as a beam under the action of a dynamic transverse load of intensity $F_z(y,t)$ as depicted in Fig. 4.5.

In the typical case shown here, the total lateral deflection $w(y,t)$ of the wing beam results from bending, which can further be represented by $\alpha(y,t)$.

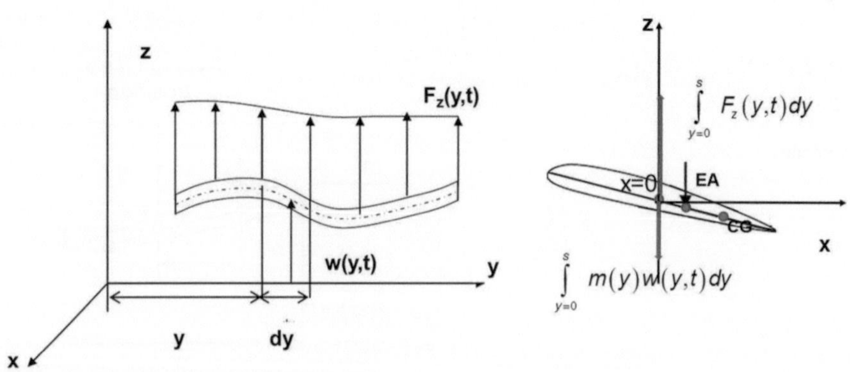

Fig. 4.5 Beam subjected to transverse load

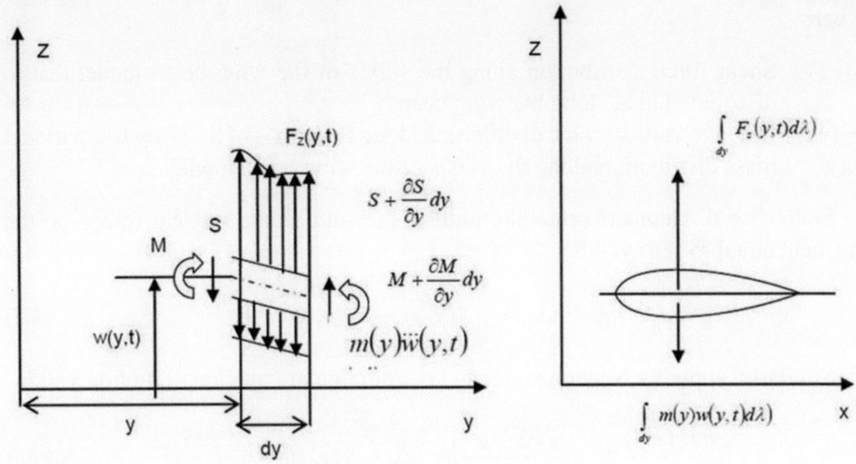

Fig. 4.6 Free-body diagram of beam element subjected to transverse load

$$w(y, t) = w_M(y, t) \equiv \alpha(y, t) \tag{4.1a}$$

and for the total lateral deflection $w(y,t)$ of the wing beam results from bending and shear, by

$$w(y, t) = w_M(y, t) + w_S(y, t) \equiv \alpha(y, t) + \beta(y, t) \tag{4.1b}$$

where[1]

$w_M(y,t)$, $\alpha(y,t)$ deflection due to bending moment along the y-axis of the wing beam model

$w_S(y,t)$, $\beta(y,t)$ deflection due to shear force distribution along the y-axis of the wing beam model.

Consider a free-body diagram for a beam segment dy taken from Fig. 4.5, as depicted by Fig. 4.6.

For the general dynamic equilibrium (for which the static equilibrium is the degenerate case for no time rate of change, i.e. on linear or rotational acceleration) Newton's second law can be applied.

Hence, if the beam experiences a lateral displacement acceleration $\ddot{w}(y, t)$ and rotational (angular with rotational axis in the x-direction) of $\ddot{\gamma}(y, t)$, where γ is the angular deformation, $\gamma(y, t) \equiv \frac{\partial \alpha(y,t)}{\partial y} = \alpha'$; hence, following Newton's second law.

$$S(y) + F_z(y, t) = m(y)\ddot{w}(y, t). \tag{4.2}$$

[1] Note that α and β here have been adopted for convenience, which should not be confused with α and β for angle of attack and torsional deflection. The interpretation of these symbols should be accompanied by understanding of the context.

where

$S(y,t)$ Shear force distribution along the y-axis of the wing beam model due to distributed load along the wing beam

$F_z(y,t)$ Discrete vertical force distribution along the y-axis of the wing beam model

$m(y)$ mass distribution along the y-axis of the wing beam model.

Following d'Alembert principle, putting the sum of the vertical forces on the segment equal to zero yields.

$$m(y)\ddot{w}(y,t) - S(y,t) - F_z(y,t) = 0 \tag{4.3}$$

Similarly, applying Newton's second law for angular (rotational) motion yields,

$$\frac{\partial M(y,t)}{\partial y}dy + \left(\frac{\partial S(y,t)}{\partial y}dy\right)dy + S(y,t)dy = \mu\frac{\partial\ddot{\alpha}}{\partial y}dy \tag{4.4a}$$

where μ^2 is the bending moment of inertia with respect to the neutral axis at section y (based on Mass per unit beam length with dimension $[ML]$).

Hence, neglecting terms of higher order in y, there is obtained

$$\frac{\partial M(y,t)}{\partial y} + S(y,t) = \mu\ddot{\gamma} = \mu\frac{\partial\ddot{\alpha}}{\partial y} \tag{4.4b}$$

or, following d'Alembert principle for rotational motion, by summing the moment on the segment about its center of gravity and equating it to zero gives

$$\frac{\partial M(y,t)}{\partial y} + S(y,t) - \mu\frac{\partial\ddot{\alpha}}{\partial y} = 0 \tag{4.4c}$$

where $w(y,t) \equiv \alpha(y,t)$ is lateral deflection due to bending and $\gamma(y,t) = \frac{\partial\alpha(y,t)}{\partial y}$ is the angular bending deformation along the unswept wing beam model. From the theory of beam bending and shear deflection (see, for example, Ref. [3]),

$$\frac{\partial^2\alpha}{\partial y^2} \equiv \alpha = \frac{M}{EI} \tag{4.5}$$

$$\frac{\partial\beta}{\partial y} = \beta' = \frac{S}{GA} \tag{4.6}$$

and the shear due to dynamic vertical load and bending (see, for example, Refs. [6–8])

[2] Again, for convenience, in the case of unswept uniform beam model here μ is the bending moment of inertia with respect to the neutral axis at section y, and not to be confused with mass ratio, $\mu = m/\pi\rho b^2$ for typical section.

$$S = \mu\left(\frac{\partial}{\partial y}\left(\frac{\partial^2\alpha}{\partial t^2}\right)\right) - \left(EI\frac{\partial^2\alpha}{\partial y^2}\right) \tag{4.7}$$

where α is the bending deflection and β the shear deflection.[3]

Substituting Eqs. 4.5 and 4.6 into Eqs. 4.4a and 4.4b, the following differential equations are obtained

$$m\frac{\partial^2 w}{\partial t^2} + \frac{\partial^2}{\partial y^2}\left(\frac{EI\partial^2\alpha(y,t)}{\partial y^2}\right) - \frac{\partial}{\partial y}\left(\mu\frac{\partial^2\alpha(y,t)}{\partial t^2}\right) = F_z(y,t) \tag{4.8}$$

or

$$m\ddot{w} + (EI\alpha) - (\mu\ddot{\alpha})' = F_z \tag{4.8a}$$

and

$$\frac{\partial\beta(y,t)}{\partial y} = \frac{1}{GA}\left[\mu\frac{\partial^2\alpha(y,t)}{\partial y^2} - \frac{\partial}{\partial y}\left(EI\left(\frac{\partial^2\alpha(y,t)}{\partial y^2}\right)\right)\right] \tag{4.9}$$

or

$$\beta' = \frac{1}{GA}\left[\mu\ddot{\alpha} - \left(EI\alpha''\right)'\right] \tag{4.9a}$$

In principle, Equations 4.7 and 4.8 can be solved simultaneously together with Eq. 4.1 for arbitrary variations in the elastic and inertial properties of the beam for all practical boundary conditions compatible for the associated physical (engineering) problem.

---

[3] In addition to the elasticity modulus E $[M/(LT^2)$, it should be noted that

For elastic materials, it is found that shear stress is proportional to the shear strain within elastic limit. The ratio is called modulus rigidity. It is denoted by the symbol "G" or "C".

$$G = \text{shear stress/shear strain} = t/ev = ML^{-1}T^{-2}$$

.

Bulk modulus (K): It is defined as the ratio of uniform stress intensity to the volumetric strain. It is denoted by the symbol K.

$$K = \text{stress intensity/volumetric strain} = s/ev = ML^{-1}T^{-2}.$$

4.3.2 *Differential Equation of Free Vibration of a Slender Beam*

Consider the free vibration of a slender beam, in which the cross-sectional dimensions are small in comparison with its length, and for which rotary inertial effects and transverse shear deformations may be neglected. The partial differential equation of lateral vibration for such simple case can be derived from Eqs. 4.1b, 4.7 and 4.8 by putting $G = \infty$ and $\mu = F_z(y,t) = 0$, which then yields (see, for example, Ref. [14])

$$m\ddot{w} + EI\frac{\partial^4 w}{\partial y^4} = 0 \tag{4.10a}$$

or

$$(EIw) + m\ddot{w} = 0 \tag{4.10b}$$

Solution of Eq. 4.10b is facilitated in many engineering problems by assuming that the spatial variation is independent of the time variation, which is a valid assumption for small spatial and time variation, which can be assumed to be linear. Such situation allows the assumption of synchronous motion, for which time rate of change is separable from the spatial changes; i.e. a ***separable partial differential equation*** can be assumed, with solution of the form (further detail can be found in Kreyszig [9]):

$$w(y, t) = W(y)T(t) \tag{4.11}$$

Substitution of Eqs. 4.11 into 4.10a, 4.10b gives:

$$-\frac{\ddot{T}}{T} = \frac{\left(EIW''\right)''}{mW} \tag{4.12}$$

Since y and T are independent variables, both quotients in Eq. 4.12 are independent of both y and t and can be equated to a separation constant, which for later practical applications, be defined as ϖ^2; Hence, one obtains ***two independent ordinary differential equations*** as follows:

$$\ddot{T} + \omega^2 T = 0 \tag{4.13}$$

$$\left(EIW''\right)'' - m\omega^2 W = 0 \tag{4.14}$$

Solutions to these equations provide the unknown functions $T(t)$, $W(y)$ and the constant ϖ^2. Since we are dealing with differential equations of higher order, i.e. 4.13 a second-order ordinary differential equation of T as a function of t, and 4.14 a fourth-order ordinary differential equation of W as a function of y, then appropriate initial and boundary conditions are required which should be part of realistic formulation of the associated engineering problem. For Eq. 4.13, two initial conditions are required to give the two constants of integration involved in the solution, and these involve specification of the initial displacement and velocity of the beam when $t = 0$. They may have the general form:

$$w(y, 0) = f_1(y);\ \dot{w}(y, 0) = f_2(y) \tag{4.15}$$

where $f_1(y)$ and $f_2(y)$ are arbitrary function of y.

Four boundary conditions are required for the fourth-order ordinary differential equation of W in order to solve for the four constants of integration involved associated with the realistic beam problem. These involve specifications of the support conditions (linear and angular deflection of the beam) on the two ends of the beam. The quantities ω^2 and $W(y)$ which satisfy Eq. 4.14 and the boundary conditions are called eigenvalues and eigenfunctions, respectively. There are infinite numbers of eigenvalues and eigenfunctions; for each eigenvalue, there is associated an eigenfunction. These have significant physical meaning in practical engineering problems. Every beam can vibrate in an infinite number of modes of vibration, and each mode has certain natural frequency. Each eigenfunction $W(y)$ represents the shape of a natural vibration mode (henceforth referred to as ***mode shape***), and the corresponding eigenvalue, ω^2, represents the square of ***the natural frequency of that mode*** [14]. At this point, we could look into the degenerate case, for which there is no time dependence. Then we refer to static aeroelasticity.

$$w(y, t) = w(y) = W(y) \tag{4.16}$$

$$\left(EIW''\right)'' - m\omega^2 W = 0 \tag{4.17}$$

$$L_{\text{sectional}} \Delta y = \int_{dy} F_z(\lambda) \mathrm{d}\lambda = \int_{\Delta y} L(\lambda) \mathrm{d}\lambda \tag{4.18}$$

4.3.3 Differential Equations of a Wing as a Slender Beam Under the Action of Torsional Load

The problem of the twisting motion of a slender beam can be treated in a manner analogous to the treatment of beams in bending, in terms of the derivation of the differential equations and, similarly, its integral equations.

The torsional deformations of a slender beam, according to St. Venant torsion theory, are given by

$$T(y) = GJ(y)\theta'(y) \tag{4.19}$$

where $T(y)$, GJ(y) and $\theta(y)$ are the applied torque, the torsional rigidity and the angle of twist, respectively.

A schematic of torsional deformation of a slender beam is depicted in Fig. 4.7.

The free-body diagram depicted in Fig. 4.8 illustrates the state of affairs associated with the equilibrium of torsional moments about the center of twist (elastic axis).

Equilibrium of torsional moments about the center of twist (elastic axis):

$$T'(y) + t(y,t) = 0 \tag{4.20}$$

where $t(y,t)$ is an applied external torsional moment per unit length and is taken positive in the positive θ direction, as depicted in Fig. 4.8. If the beam is undergoing free torsional vibration, then

$$t(y,t) = I_0\ddot{\theta}(y,t) \tag{4.21}$$

where I_0 is the moment of inertia per unit length about the center of twist of the beam. Differentiating Eq. 4.19 with respect to y and substituting Eq. 4.21, there is obtained

$$\left[GJ(y)\theta'(y)\right]' - I_0\ddot{\theta}(y,t) = 0 \tag{4.22}$$

which is the partial differential equation governing torsional free vibration of a beam.

Following similar assumptions and procedure like in free bending vibration of a beam, we can assume separability of the solution $\theta(y,t)$:

$$\theta(y,t) = \Theta(y)T(t) \tag{4.23}$$

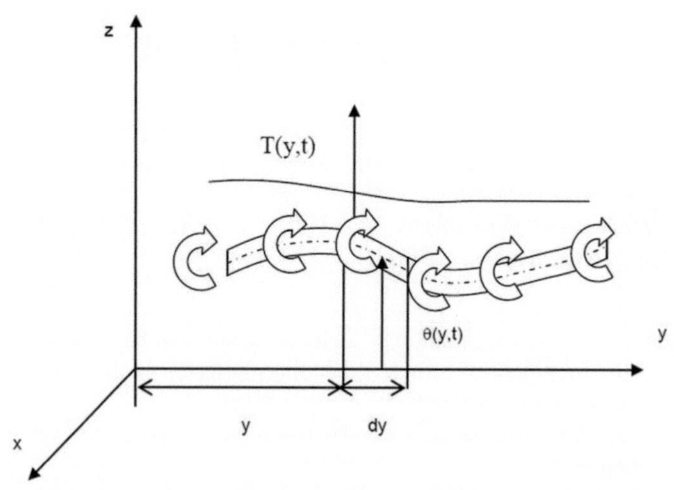

Fig. 4.7 Torsional deformation of a slender beam

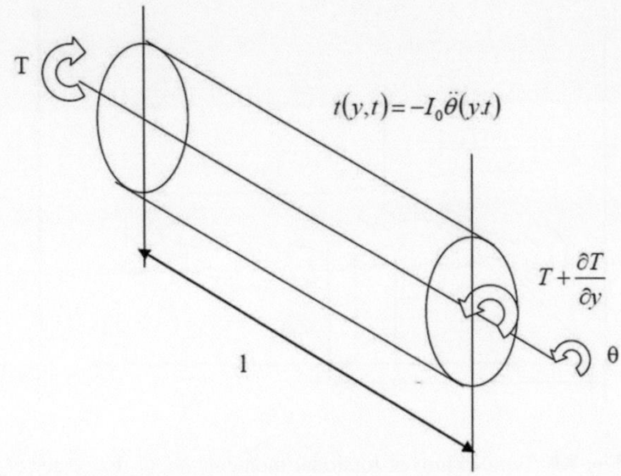

Fig. 4.8 Equilibrium of torsional moments about the center of twist (elastic axis)

Consequently, substituting 4.23 into 4.22 yields two partial differential equations for Θ and T:

$$\ddot{T} + \omega^2 T = 0 \tag{4.24}$$

$$\left(GJ\Theta'\right)' - I_0\omega^2\Theta = 0 \tag{4.25}$$

where again ω^2 is a separation constant, which signifies the eigenvalues (eigenfrequencies) of the free torsional vibration of the beam and also its torsional natural frequencies. Similar to the bending case, Eq. 4.24 must be supplemented by two initial conditions (of the initial torsional angle and torsional time rate of change and Eq. 4.25 by two boundary conditions of the points of support (each specifying the torsional angle).

At this point, again we could look into the degenerate case, for which there is no time dependence. Then we refer to static aeroelasticity.

Then the combined bending and torsion at the typical section, as schematically depicted in Fig. 4.9, produced total deflection at point (y, Δx) is given by:

$$w(y) = W(y) + \Theta(y)\Delta xx \tag{4.26}$$

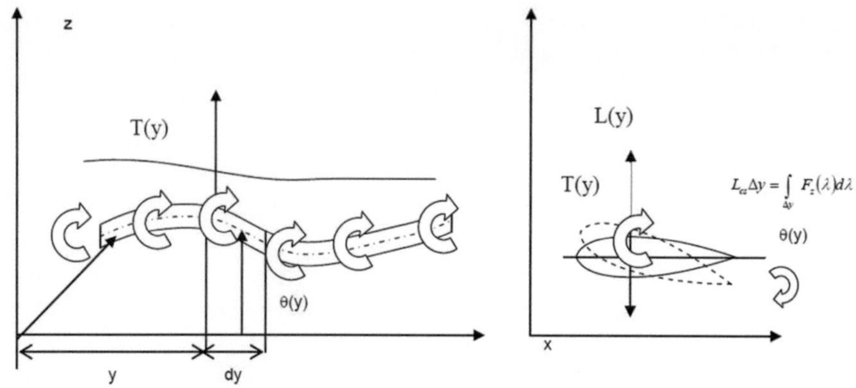

Fig. 4.9 Equilibrium of torsional moments about the center of twist and combined bending and torsion at the typical section

4.4 Aeroelastic Equations for a Simple Typical Wing Section Without Control Surface

The equation of motion of a typical wing section can be simply derived using Newton's equation of motion by the mechanical balance of the inertial, stiffness and the forces and moment acting in the typical section in the h, and α directions.

Such state of affairs is illustrated in Fig. 4.10. Using Newton's second law to the typical section above, the following sequence of equations can be written.

Motion in the $w(y)$ or h (heaving) direction:

Lift Force: L—lift per unit span, positive in the upward direction.

Inertial force: $m\ddot{h}$, positive in the downward direction.

Angular displacement of the typical section about its elastic axis (center of twist): $\Theta(y)$ which is here denoted by α. The rotational inertial force due to angular motion with respect to the elastic axis is given by

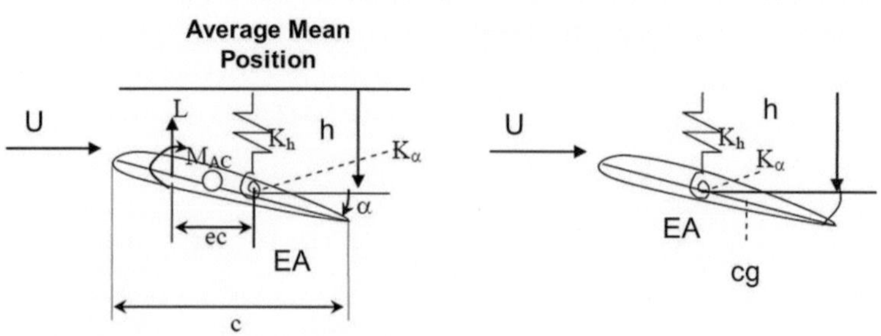

Fig. 4.10 Heaving (wing bending) and pitching (wing torsion) representation at the typical section

$$\int_{\text{typical section}} \ddot{\alpha} m \mathrm{d}x = S_\alpha \ddot{\alpha} \tag{4.27}$$

The bending resistance due to bending stiffness of the typical section is given by

$$\frac{\partial^2 h}{\partial y^2} = h = -\frac{M}{EI} \tag{4.28}$$

or if $M(y) = Py$

$$h(y) = -\frac{Py^3}{3EI} \tag{4.29}$$

Writing

$$F_{\text{bending stiffness}} = K_h.h \tag{4.30}$$

while

$$h(y)\frac{3EI}{y^3} = -P \tag{4.31}$$

then

$$K_h = -\frac{3EI}{y^3} \tag{4.32}$$

Hence Newton's law for bending deflection or heaving motion of the typical section can be written as

$$\text{Force} = \text{mass} \times \text{acceleration}$$

Force:

$$m\ddot{h} + S_\alpha \ddot{\alpha} + K_h h + L = 0 \tag{4.33a}$$

or

$$L + K_h h = -m\ddot{h} - S_\alpha \ddot{\alpha} \tag{4.33b}$$

Similarly, for the torsional motion, the torsional resistance due to torsional stiffness of the typical section is given by

$$\frac{\partial \Theta}{\partial y} = \frac{T(y)}{GJ(y)} \tag{4.34}$$

or

$$\Theta(y) = \frac{M_{EA} y}{GJ\,(y)} \tag{4.35}$$

Writing

$$T_{\text{torsional stiffness}} = K\alpha.\alpha \tag{4.36}$$

while

$$\Theta(y) = \alpha = \frac{M_{EA} y}{GJ\,(y)} \tag{4.37}$$

then

$$K_\alpha = -\frac{GJ\,(y)}{y} \tag{4.38}$$

The rotational inertial force with respect to the elastic axis due to heaving motion is given by

$$\int_{\text{typical section}} \ddot{h} m \mathrm{d}x = S_\alpha \ddot{\alpha} \tag{4.39}$$

Hence Newton's law for torsional deflection or pitching motion of the typical section can be written as

Torque due to aerodynamic minus stiffness effects = moment of inertia × angular acceleration.

Torque:

$$-M_{EA} + K_\alpha \alpha = -I_\alpha \ddot{\alpha} - S_\alpha \ddot{h} \tag{4.40a}$$

or

$$S_\alpha \ddot{h} + I_\alpha \ddot{\alpha} + K_\alpha \alpha - M_{EA} = 0 \tag{4.40b}$$

Alternatively, using the concept of generalized coordinates, Hamilton principle and the concept of Lagrangian, one may arrive at similar equations. The latter, however, is scalar-based and does not need the identification of each system and follows the principles of analytical dynamics.

$$m\ddot{h} + S_\alpha \ddot{\alpha} + K_h h + L = 0 \tag{4.41}$$

$$S_\alpha \ddot{h} + I_\alpha \ddot{\alpha} + K_\alpha \alpha - M_{EA} = 0 \tag{4.42}$$

where

$$M_{\mathrm{EA}} = \mathrm{Lec} + \mathrm{Mac}. \tag{4.43}$$

In matrix form, Eqs. 4.9 and 4.10a, 4.10b can be written as

$$\begin{bmatrix} m & S_\alpha \\ S_\alpha & I_\alpha \end{bmatrix} \begin{Bmatrix} \ddot{h} \\ \ddot{\alpha} \end{Bmatrix} + \begin{bmatrix} K_h & 0 \\ 0 & K_\alpha \end{bmatrix} \begin{Bmatrix} h \\ \alpha \end{Bmatrix} + \begin{Bmatrix} L \\ -M_{EA} \end{Bmatrix} = 0 \tag{4.44}$$

or

$$\begin{bmatrix} m & S_\alpha \\ S_\alpha & I_\alpha \end{bmatrix} \begin{Bmatrix} \ddot{h} \\ \ddot{\alpha} \end{Bmatrix} + \begin{bmatrix} K_h & 0 \\ 0 & K_\alpha \end{bmatrix} \begin{Bmatrix} h \\ \alpha \end{Bmatrix} = \begin{Bmatrix} -L \\ M_{EA} \end{Bmatrix} \tag{4.45}$$

4.5 Aeroelastic Equations for a Simple Typical Wing Section with Control Surface

4.5.1 General

The equation of motion of a typical wing section can be simply derived using Newton's equation of motion by the mechanical balance of the inertial, stiffness and the forces and moment acting in the typical section in the h, α and δ directions. Alternatively, using the concept of generalized coordinates, Hamilton principle and the concept of Lagrangian, one may arrive at similar equations. The latter, however, is scalar-based and does not need the identification of each system, and follows the principles of analytical dynamics.

$$m\ddot{h} + S_\alpha\ddot{\alpha} + S_\delta\ddot{\delta} + K_h h + L = 0 \tag{4.46}$$

$$S_\alpha\ddot{h} + I_\alpha\ddot{\alpha} + (I_\delta + x_\delta b S_\delta)\ddot{\delta} + K_\alpha\alpha - M_{EA} = 0 \tag{4.47}$$

$$S_\delta\ddot{h} + I_\delta\ddot{\delta} + (I_\delta + x_\delta b S_\delta)\ddot{\alpha} + K_\delta\delta - H = 0 \tag{4.48}$$

where

$$M_{\mathrm{EA}} = \mathrm{Lec} + \mathrm{Mac}. \tag{4.49}$$

H—Hinge Moment about the hinge point of the control surface.

At this point, it will be instructive to differentiate the aerodynamic forces acting on the typical section by the motion-induced aerodynamic force (denoted by index

M), which came about due to the vibration of the typical section, and the motion independent aerodynamic excitation (denoted by index D).

$$L = L^{(M)}(h, \alpha, \delta) + L^{(D)}(t) \tag{4.50}$$

$$M_{EA} = M_{EA}^{(M)}(h, \alpha, \delta) + M_{EA}^{(D)}(t) \tag{4.51}$$

$$H = H^{(M)}(h, \alpha, \delta) + H^{(D)}(t) \tag{4.52}$$

4.5.2 Flutter Equation

Conditions leading to Flutter are those characterized by self-induced aerodynamic excitation due to the vibration of the typical section, and no external aerodynamic excitation is working. In such case, then:

$$m\ddot{h} + S_\alpha\ddot{\alpha} + S_\delta\ddot{\delta} + K_h h + L^{(M)}(h, \alpha, \delta) = 0 \tag{4.53}$$

$$S_\alpha\ddot{h} + I_\alpha\ddot{\alpha} + (I_\delta + x_\delta b S_\delta)\ddot{\delta} + K_\alpha\alpha - M_{EA}^{(M)}(h, \alpha, \delta) = 0 \tag{4.54}$$

$$S_\delta\ddot{h} + I_\delta\ddot{\delta} + (I_\delta + x_\delta b S_\delta)\ddot{\alpha} + K_\delta\delta - H^{(M)}(h, \alpha, \delta) = 0 \tag{4.56}$$

The solution of the situation characterized by these set of equations gives rise to the dynamic stability of the deformation modes, which is dependent on the flight velocity and altitude.

4.5.3 Dynamic Response Equation

The most general case is given by the following sets of equations, at which the right-hand sides contain the external aerodynamic excitation.

$$m\ddot{h} + S_\alpha\ddot{\alpha} + S_\delta\ddot{\delta} + K_h h + L^{(M)}(h, \alpha, \delta) = -L^{(D)}(t) \tag{4.57}$$

$$S_\alpha\ddot{h} + I_\alpha\ddot{\alpha} + (I_\delta + x_\delta b S_\delta)\ddot{\delta} + K_\alpha\alpha - M_{EA}^{(M)}(h, \alpha, \delta) = M_{EA}^{(D)}(t) \tag{4.58}$$

$$S_\delta\ddot{h} + I_\delta\ddot{\delta} + (I_\delta + x_\delta b S_\delta)\ddot{\alpha} + K_\delta\delta - H^{(M)}(h, \alpha, \delta) = H^{(D)}(t) \tag{4.59}$$

The solution of these equations gives the dynamic response (h, α, δ) due to external aerodynamic excitation ($L^{(D)}$, ${M_{EA}}^{(D)}$, $H^{(D)}$) and is dependent also on the flight velocity and altitude. The flutter equation should be satisfied as the solution of the homogeneous equation.

4.5.4 Divergence Equation

The divergence equation can be viewed as the degenerate case of the above equations, in which time-dependent terms are omitted. For basic case concerns the induced deformation due to the motion-induced aerodynamic force.

$$K_h h + L^{(M)}(h, \alpha, \delta) = 0 \quad (4.60)$$

$$K_\alpha \alpha - M_{EA}^{(M)}(h, \alpha, \delta) = 0 \quad (4.61)$$

$$K_\delta \delta - H^{(M)}(h, \alpha, \delta) = 0 \quad (4.62)$$

The solution of these equations will give rise to static stability of the deformation modes (h, α, δ), which is dependent on the flight velocity and altitude.

4.5.5 Static Response Equation

Again, the time-dependent terms are neglected. Then

$$K_h h + L^{(M)}(h, \alpha, \delta) = -L^{(D)} \quad (4.63)$$

$$K_\alpha \alpha - M_{EA}^{(M)}(h, \alpha, \delta) = M_{EA}^{(D)} \quad (4.64)$$

$$K_\delta \delta - H^{(M)}(h, \alpha, \delta) = H^{(D)} \quad (4.65)$$

The solution of these equations will give rise to the static response (h, α, δ) due to the external aerodynamic excitation ($L^{(D)}$, ${M_{EA}}^{(D)}$, $H^{(D)}$) and is determined by flight velocity and altitude.

Notes: Since the divergence equation is the homogeneous part of the equation, the divergence condition should be satisfied.

4.5.6 Some Important Notes

In all the aeroelastic equations described above, some simplifications have taken place

- Only external aerodynamics have been considered.
- No structural damping has been considered.
- There are no aerodynamically induced forces due to longitudinal motion as well as there are no external aerodynamic excitation in the longitudinal direction.

However, these forces can readily be incorporated into these equations if required.

4.5.7 Linearity

The following assumptions have been considered:

- The mean (average) non-deflected position of the typical wing section representing the aircraft structure is considered to be the equilibrium position.
- The structural deflection in all these cases is considered to be small, in order that linear analysis of structures and aerodynamics can be utilized.

The linearization then makes possible the utilization of the following relationships. Aerodynamic forces:

$$L = L_{MEAN} + L(h, \alpha, \delta, t) \tag{4.66}$$

$$M = M_{MEAN} + M(h, \alpha, \delta, t) \tag{4.67}$$

$$H = H_{MEAN} + H(h, \alpha, \delta, t) \tag{4.68}$$

where $L(h, \alpha, \delta, t)$ is linear functions of (h, α, δ, t).

Therefore

$$\sin\alpha = \alpha + O(\alpha^3) \approx \alpha \tag{4.69}$$

$$\cos\alpha = 1 - O(\alpha^3) \approx 1 \tag{4.70}$$

In many cases, $L(h, \alpha, \delta, t)$ will be independent of L_{MEAN}.

As a consequence, aeroelastic analysis will be independent of performance study. One exception is in transonic flow, where such an assumption cannot be made, due to the occurring nonlinearities.

References

1. Chapter 3, Beams: Strain, Stress, Deflections, http://courses.washington.edu/me354a/.
2. Chapter_5-98, Stresses in Beam (Basic Topics), https://ocw.nthu.edu.tw/.
3. Chapter 9, Stresses: Beams in Bending, ocw.mit.edu-1-050-solid-mechanics.
4. Lecture Notes on Mechanics of Solids, https://nitsri.ac.in/. Mechanical Department.
5. Units of Elastic Constants. www.lkouniv.ac.in.
6. Bisplinghoff, R.L., H. Ashley, and R. L. Halfman. 1955. *Aeroelasticity*. Addison Wesley Publishing Company.
7. Bisplinghoff, R.L., and H. Ashley. 1962. *Principles of Aeroelasticity*. Wiley Inc.
8. Djojodihardjo, H. *Lecture Notes on Aeroelasticity at the Institute of Technology Bandung delivered in 1982 to 2005 (Author's repository).*
9. Djojodihardjo, H. *Lecture Notes on Aeroelasticity at the Universiti Sains Malaysia (University of Science Malaysia) in 2004 to 2008 (Author's repository).*
10. Dowell, E.H., H.C. Curtiss, R.H. Scanlan, and R. Sisto. 1980. *A Modern Course in Aeroelasticity*. Sijthoff & Noordhoff.
11. Dupen, B. 2016. *Applied Strength of Materials for Engineering Technology*, 10th ed. Purdue University. http://opus.ipfw.edu/mcetid_facpubs/48.
12. Fung, Y.C. 1955. *An Introduction to the Theory of Aeroelasticity*. Wiley Inc.
13. Kreyszig, E. 2011. *Advanced Engineering Mathematics*, 10th ed. Wiley. ISBN-13: 978-0470458365, ISBN-10: 0470458364.
14. Meirovitch, L. 1970. *Elements of Vibration Analysis*, 1st ed. McGraw-Hill.
15. Scanlan, R.H., and R. Rosenbaum. 1951. *Introduction to the Study of Aircraft Vibration and Flutter*. The Macmillan Company.
16. Weisshaar, T.A. 2001. *Class Notes on Aeroelasticity AAE 556—Aeroelasticity*. Purdue University. weisshaa@ecn.purdue.edu.
17. Zwaan, R. J. 1981. *Aeroelasticity of Aircraft, Lecture Notes, Special Lecture, Short Course offered at Institut Teknologi Bandung, Indonesia, Lecture Notes delivered at Special Meeting at ITB in August 1981* (Author's Repository; private collection).

Chapter 5
Static Aeroelasticity-Typical Section, One-Dimensional Model and Lifting Surface

Abstract Further discussions in the utilization of typical section to analyze aircraft wing aeroelasticity in this chapter cover selected basic topics in static aeroelasticity. Torsional divergence, aileron reversal, control surface effectiveness will first be discussed (The present chapter has been written as an update of the author's lecture notes on Aeroelasticity at Institut Teknologi Bandung (1982–2005) [1] and Universiti Sains MalaysIa (2004–2008) [2]. The theoretical development then proceeds with one-dimensional wing model. In formulating the governing equations for static aeroelasticity, divergence is viewed as an eigenvalue problem. In the analysis of the equilibrium equation for a wing with variable properties, Galerkin method is introduced and elaborated. A physically and mathematically more complex analysis of static aeroelasticity is introduced in analyzing the rolling of a straight wing. Here the problem is formulated in terms of an integral equation as compared to differential equation formulation approach, as well as the utilization of aerodynamic induction as compared to strip theory and Lumped element method as compared to modal approach or eigenfunction approach in previous problems. Then control surface reversal and rolling effectiveness, as well as two-dimensional aeroelastic models of lifting surface, are analyzed to give a comprehensive account of static aeroelasticity of aircraft wings and aerodynamic beam-like surfaces.

Keywords Aeroelasticity · Eigenvalue problem · Galerkin method · Lumped element method · Typical section · One-dimensional wing model · Lifting surface

5.1 Typical Wing Section

Typical section has been very convenient and instructive in modeling the aeroelastic characteristics of an aircraft wing. Following Bisplinghoff Ashley and Halfman [3–5][1], a typical section may represent the wing section at 0.7 span of a half-wing, i.e. one side of an aircraft wing. The typical section is then a two-dimensional representation of the wing at semi-span position with bending stiffness Kh and torsional stiffness $K\alpha$.

[1] Also Bisplinghoff and Ashley [2], Fung [5] (Dover).

© The Author(s), under exclusive license to Springer Nature Singapore Pte Ltd. 2023

H. Djojodihardjo, *Introduction to Aeroelasticity*,
https://doi.org/10.1007/978-981-16-8078-6_5

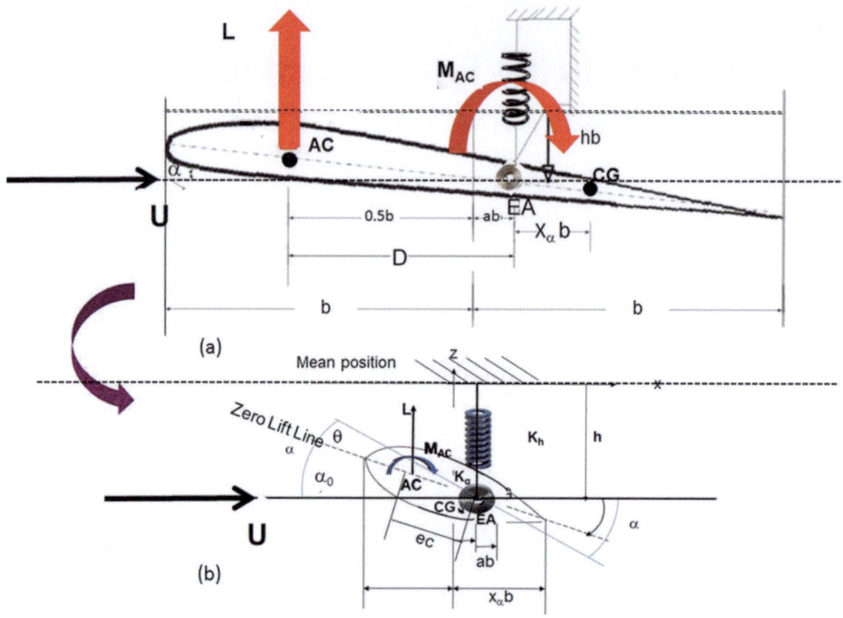

Fig. 5.1 Elements of a typical section and aerodynamic force and moment [1–3], [6] [2]

As a guidance, the typical values for the airfoil maximum thickness-to-chord ratio of majority of aircraft are about 6–18%. 1—For a low-speed aircraft with a high lift requirement (such as cargo aircraft), the typical wing (*t*/*c*)max is about 15–18%. 2—For a high-speed aircraft with a low lift requirement (such as high subsonic passenger aircraft), the typical wing (*t*/*c*)max is about 9–12%. 3—For the supersonic aircraft, the typical wing (*t*/*c*)max is about 3–9%.

Figure 5.1 depicts a basic typical section, which can be utilized to model the static aeroelastic characteristics of an aircraft wing. It is also referred to as a binary aeroelastic typical section model.

Typical section:

A two-dimensional representation of the wing at semi-span position with bending stiffness K_h and torsional stiffness K_α.

M = moment about elastic axis.

ac = aerodynamic center

$$\alpha = \alpha_0 + \alpha_e \tag{5.1}$$

[2] All Figures, unless otherwise elaborated with cited reference(s), have been taken and updated from the author's Lecture Notes on Aeroelasticity at the Institute of Technology Bandung delivered in 1982 to 2005 [5] and at the Universiti Sains Malaysia (University of Science Malaysia) in 2004 to 2008 [5].

$$M_y = M_{AC} + Le \tag{5.2}$$

M_y = moment about elastic axis/center.
M_{ac} = moment about aerodynamic center.
L = lift, not vertical force, positive up.
e = distance between aerodynamic center to elastic axis, positive aft.

From aerodynamics

$$L = C_L q S \tag{5.3}$$

$$q = \frac{1}{2}\rho V^2 \tag{5.4}$$

$$M_{AC} = C_{MAC} q Sc \tag{5.5}$$

$$C_L = C_{L0} + \frac{dC_L}{d\alpha}\alpha \tag{5.6}$$

C_{L0} = lift for zero angle of attack, α
C_{MAC} = moment coefficient about the aerodynamic center,
ρ = air density
V = air velocity
c = airfoil chord
S = wing area ($c \times l$)
l = wingspan (also denoted by s).

For a flat plate in two-dimensional flow with non-dimensional chord equal to unity, it follows that

$$\begin{aligned}
&\frac{\partial C_L}{\partial \alpha} = 2\pi \\
&C_{MAC,0} = C_{MAC,\alpha=0} = 0 \;\; C_{L,0} = C_{L,\alpha=0} = 0 \\
&\dot{M}_y = M_{AC} + Le \\
&M_y = eqS\left[\frac{\partial C_L}{\partial \alpha}(\alpha_0 + \alpha_e)\right] + qScC_{MAC,0}
\end{aligned} \tag{5.7a, b, c, d}$$

For symmetrical airfoil $C_{L,0} = 0$, $C_{MAC,0} = 0$

5.2 Torsional Divergence

5.2.1 General

The wing as represented by the typical section is restrained in twisting at the Elastic Axis, EA. For the particular problem considered, particularly for straight wing, bending deflection is not important (can be neglected with respect to the torsional deflection) [6].

The initial angle of attack of the typical sections is α_0. During the flight, a new angle of attack prevails at free-stream velocity U.

Hence

$$\alpha = \alpha_0 + \theta \tag{5.8}$$

θ is the angle of twist due to the load exerted by the dynamic pressure due to the flight velocity U.

Corresponding lift force:

$$L = q\,S\,C_L = q\,S\,\frac{\partial\,C_L}{\partial\,\alpha}\;\alpha\; = q\,S\,\frac{\partial\,C_L}{\partial\,\alpha}\;(\alpha\,+\,\theta) \tag{5.9}$$

and pitching moment coefficients

$$M_y = Lec + M_{\mathrm{AC}} = \left(C_L e + C_{M_{AC}}\right) cqS\left[\frac{\partial C_L}{\partial\alpha}(\alpha_0 + \theta)e + C_{M_{AC}}\right]cqS \tag{5.10}$$

where

$$q = \frac{1}{2}\rho U^2$$

is the dynamic pressure.

Due to the elastic forces on the wing structure, there exists a restoring torque on the wing structure and can be expressed by

$$T = K_\alpha\theta \tag{5.11}$$

where

K_α = elastic spring constant [moment/twist angle].

A stable equilibrium due to the torque exerted by the aerodynamic forces (5.9) and (5.10), and the restoring elastic torque of the wing structure is established and assured, if

$$M_{\mathrm{EA}} = T \tag{5.12}$$

or[3]

$$\left[\frac{\partial C_L}{\partial \alpha}(\alpha_0 + \theta)e + C_{M_{AC}}\right]cqS = \mathbf{K}_\alpha \theta \tag{5.13}$$

Therefore, the angle of twist for stable equilibrium, or the static aeroelastic response due to aerodynamic moment on the wing structure, is

$$\theta = \frac{cqS}{\mathbf{K}_\alpha} \frac{C_L \alpha_0 e + C_{M_{\mathrm{AC}}}}{1 + \frac{ecqS}{\mathbf{K}_\alpha} C_L} \tag{5.14}$$

Unconditional stability for $e \leq 0$, if the aerodynamic center AC is located downstream of the elastic axis. Example indicated is the weather vane.

For $e < 0$, which is the usual case, unstable twisting occurs for

$$1 + \frac{ecqS}{\mathbf{K}_\alpha} C_L \leq 0 \tag{5.15}$$

that is, the condition for torsional divergence.

The critical dynamic divergence pressure is then

$$q_D = \frac{\mathbf{K}_\alpha}{ecS\frac{\partial C_L}{\partial \alpha}} = \frac{\mathbf{K}_\alpha}{ecSC_{L_\alpha}} \tag{5.16}$$

and the critical divergence speed is

$$U_D = \sqrt{\frac{2\mathbf{K}_\alpha}{\rho ecSC_{L_\alpha}}} \tag{5.17}$$

Only positive dynamic pressure is meaningful. Then only if $e > 0$ will divergence occur.

The angle of twist becomes

$$\theta = \frac{\left(\frac{q}{q_D}\right)\alpha_0 + cSC_{M_{\mathrm{AC}}}}{1 - \left(\frac{q}{q_D}\right)} \tag{5.18a}$$

and if $C_{\mathrm{MAC}} = 0$, then

$$\theta = \frac{\left(\frac{q}{q_D}\right)\alpha_0}{1 - \left(\frac{q}{q_D}\right)} \tag{5.18b}$$

[3] The elastic deflection due to the torsion T or M_{EA} is denoted either by θ or α_e.

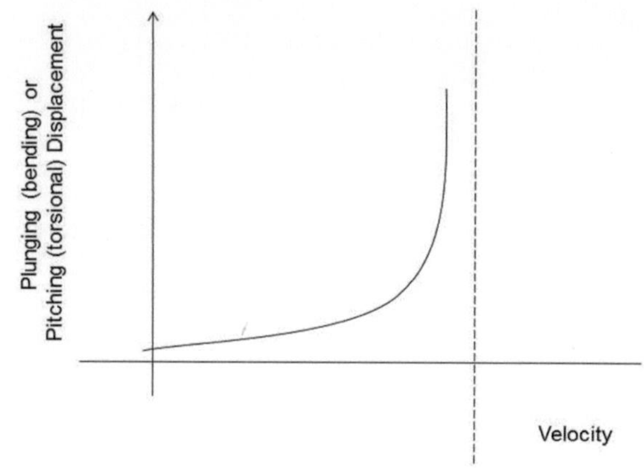

Fig. 5.2 Schematic of torsional divergence

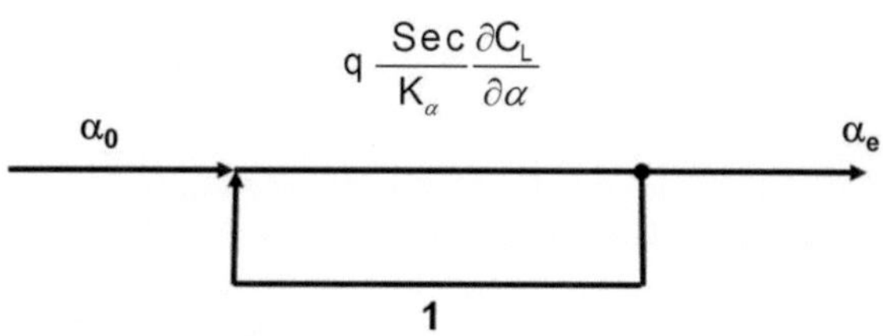

Fig. 5.3 Feedback representation at aeroelastic divergence

Hence,

Elastic twist does not become infinity large for any real airfoil due to the required infinity large aerodynamic moment.

The linear relationship between elastic twist and aerodynamic moment has a small range.

The relatively large elastic twist causes structure failure.

A schematic of torsional divergence is depicted in Fig. 5.2. Feedback representation at aeroelastic divergence is depicted in Fig. 5.3.

5.2.2 Physical Meaning of Torsional Divergence

For arbitrary $\Delta\theta$ beyond divergence speed, an increase of aerodynamic moment

$$\Delta M_{EA} = \frac{\partial C_L}{\partial \alpha} e\ c\ qS\ \Delta\theta \tag{5.19}$$

exceeds the increase of restoring torque:

$$\Delta \mathbf{T} = \mathbf{K}_\alpha \, \Delta\theta \tag{5.20}$$

Note that the torsional divergence indicates an eigenvalue problem characteristics.

Let us now compare the characteristic of the divergence problem with that of rigid wing. For a rigid wig, the restoring torque is given by

$$T^{(r)} = c\, qS \left[\frac{\partial C_L}{\partial \alpha} \alpha_0 \, e + C_{M_{AC}} \right] \tag{5.21}$$

The influence of twisting deformation

$$\frac{T}{T^{(r)}} = \frac{1}{1 - \frac{c\,q\,S}{K_\alpha} C_{L_\alpha}} = \frac{1}{1 - \frac{q}{q_D}} \tag{5.22}$$

Hence, T and consequently θ increase rapidly until structure collapses.

Torsional divergence is not important in modern aircraft with straight wings, but it is for forward-swept wings. Sometimes one has to design gliders with very long elastic wingspan and lightweight, and for such situation, one has to be very careful in avoiding the divergence speed.

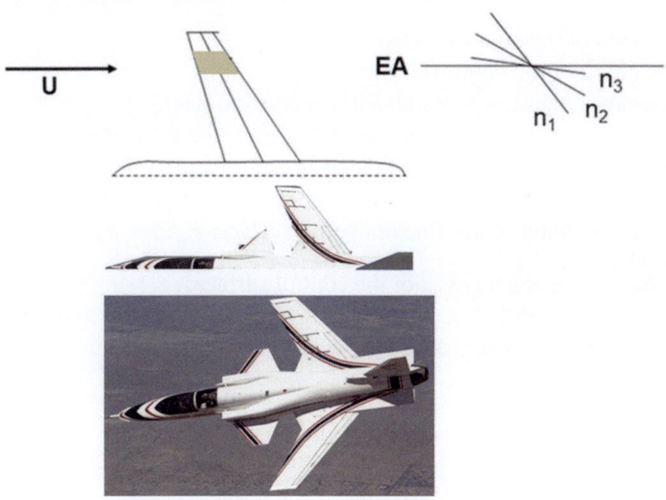

Fig. 5.4 Grumman X-29 and Schematic of a Swept Forward Wing, Courtesy of NASA [7]. Divergence in forward-swept wing can be avoided by application of composite materials, for example Grumman X-29, as illustrated in Fig. 5.4

5.2.3 Typical Section with Control Surface

Let:

$\alpha_0 = 0.$

$C_{\mathrm{MAC},0} = 0 \rightarrow \alpha = \alpha_c.$

Aerodynamic lift:

$$L = q\,S\,C_L = q\,S\left(\frac{\partial\,C_L}{\partial\,\alpha}\,\alpha + \frac{\partial\,C_L}{\partial\,\delta}\,\delta\right) \tag{5.23}$$

Aerodynamic moment:

$$M_{AC} = qS\,c\;C_{M_{AC}} = qS\,c\left(\frac{\partial\,C_{M_{AC}}}{\partial\,\delta}\,\delta\right) \tag{5.24}$$

Moment about hinge line:

$$H = q\,\overline{c}_H S_H\,C_H = q\,\overline{c}_H S_H\left(\frac{\partial\,C_H}{\partial\,\alpha}\,\alpha + \frac{\partial\,C_H}{\partial\,\delta}\,\delta\right) \tag{5.25}$$

where

S_H = the area of the control surface

$\overline{c}_H$ = the chord of the control surface

C_H = the non-dimensional aerodynamic hinge moment.

The basic purpose of a control surface is to change the lift or moment on the main lifting surface. We will examine the related aeroelastic effect for that purpose.

Elastic moment about elastic axis of main surface: $-K_\alpha\alpha$(+ nose-up).

Elastic moment about hinge line of control surface: $K_\delta(\delta - \delta_o)$ (+ tail down).

$\delta_e = \delta - \delta_0$ = the elastic twist of the control surface.

δ_0 = the difference between the angle of attack for zero aerodynamic control surface deflection and zero twist of the control surface elastic deformation.

The equation at static moment equilibrium

$$cqS\left(\frac{\partial\,C_L}{\partial\,\alpha}\alpha + \frac{\partial\,C_L}{\partial\delta}\delta\right) + qSc\frac{\partial\,C_{M_{AC}}}{\partial\delta}\delta - K_\alpha\alpha = 0 \tag{5.26}$$

$$qS_H\overline{c}_H\left(\frac{\partial\,C_H}{\partial\,\alpha}\alpha + \frac{\partial\,C_H}{\partial\delta}\delta\right) + K_\delta(\delta - \delta_0) = 0 \tag{5.27}$$

In addition to somewhat complicated form of the divergence condition, there is a new physical phenomenon associated with the control surface knowns as the control surface reversal, which is elaborated subsequently.

5.3 Aileron Reversal

5.3.1 General

Aileron has fixed deflection with respect to the wing. Due to the load on the deflected aileron, the wing is experiencing a twist. It would be of practical interest, whether such situation has a limiting condition, or instability. A schematic of aileron reversal is depicted in Fig. 5.5.

The lift force is:

$$L = qSC_L = qS\big(C_{L_\varepsilon}\theta + C_{L_\delta}\delta\big) \tag{5.28}$$

The pitching moment about the elastic axis EA is

$$M_{EA} = Lec + M_{AC} = \big[\big(\big(C_{L_\varepsilon}\theta + C_{L_{\delta_\delta}}\delta\big)e + C_{M_{AC}}\delta\big)\big]cqS \tag{5.29}$$

Restoring torque in wing structure:

$$\mathbf{T} = \mathbf{K}_\alpha\theta \tag{5.30}$$

Twisting is stable if:

$$\big[\big(C_{L_\varepsilon}\theta + C_{L_{\delta_\delta}}\delta\big)e + C_{M_{AC}}\delta\big]cqS = K_\alpha\theta \tag{5.31}$$

Angle of twist (static aeroelastic response) is:

$$\frac{\theta}{\delta} = \frac{cqS}{K_\alpha}\frac{C_{L_\delta}e + C_{M_{AC_\delta}}}{1 - \frac{cqSe}{K_\alpha}C_{L_\alpha}} \tag{5.32}$$

Note the occurrence of same condition for torsional divergence:

$$1 - \frac{cqSe}{K_\alpha}C_{L_\alpha} \tag{5.33}$$

or

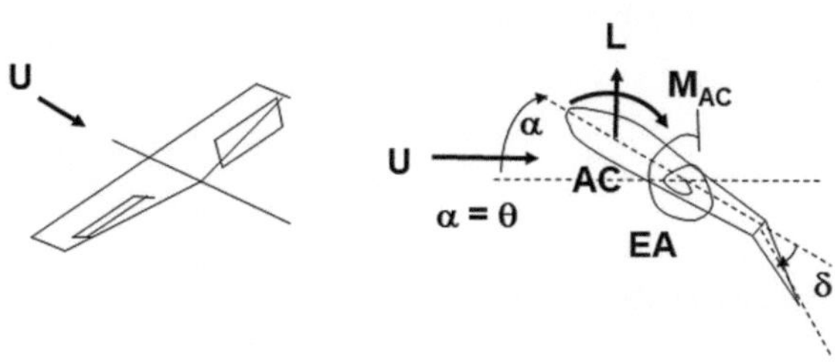

Fig. 5.5 Schematic of aileron reversal behavior

$$q_D = \frac{K_\alpha}{ecSC_{L_\alpha}} \tag{5.34}$$

5.3.2 Comparison with Rigid Wing

Consider sectional lift, being indicative of wing rolling power, from Eqs. 5.28 and 5.29.

$$L = \left[C_{L_\alpha} \frac{cqS}{K_\alpha} \frac{C_{L_\delta} e + C_{M_{AC_\delta}}}{1 - \frac{cqSe}{K_\alpha} C_{L_\alpha}} + C_{L_\delta} \right] qS\delta = \left[\frac{C_{L_\alpha} C_{M_{AC_\delta}} \frac{cqS}{K_\alpha} + C_{L_\delta}}{1 - \frac{cqSe}{K_\alpha} C_{L_\alpha}} \right] qS\delta \tag{5.35}$$

For rigid wing

$$L^{(r)} = C_{L_\delta} qS\delta \tag{5.36}$$

Aileron effectiveness: represented by

$$\frac{L}{L^{(r)}} = \left[\frac{\frac{C_{L\alpha}}{C_{L_\delta}} C_{M_{AC_\delta}} \frac{qSc}{K_\alpha} + 1}{1 - \frac{cqSe}{K_\alpha} C_{L_\alpha}} \right] \tag{5.37}$$

where C_{L_α} and C_{L_δ} are positive, $C_{M_{AC_\delta}}$ is negative.

Aileron reversal for

$$\frac{L}{L^{(r)}} \le 0.$$

Critical aileron reversal dynamic pressure:

$$q_r = -\frac{K_\alpha}{Sc} \frac{\frac{C_{L_\delta}}{C_{L\alpha}}}{C_{M_{AC_\delta}}} \tag{5.38}$$

Critical aileron reversal speed:

$$U_R = \sqrt{-\frac{2}{\rho} \frac{K_\alpha}{Sc} \frac{\frac{C_{L_\delta}}{C_{L\alpha}}}{C_{M_{AC_\delta}}}} \tag{5.39}$$

Aileron effectiveness becomes:

$$\frac{L}{L^{(r)}} = \frac{1 - \frac{q}{q_R}}{1 - \frac{q}{q_D}} = \frac{1 - \frac{q_D}{q_R}\frac{q}{q_D}}{1 - \frac{q}{q_D}} \tag{5.40}$$

The effectiveness of the aileron as a control surface is depicted schematically in Fig. 5.6, which declines with increasing airspeed. Other figures indicating its characteristics are schematically depicted in Figs. 5.7 and 5.8.

Note: Aileron reversal does not imply new instability.

Physical Meaning of Aileron Reversal:

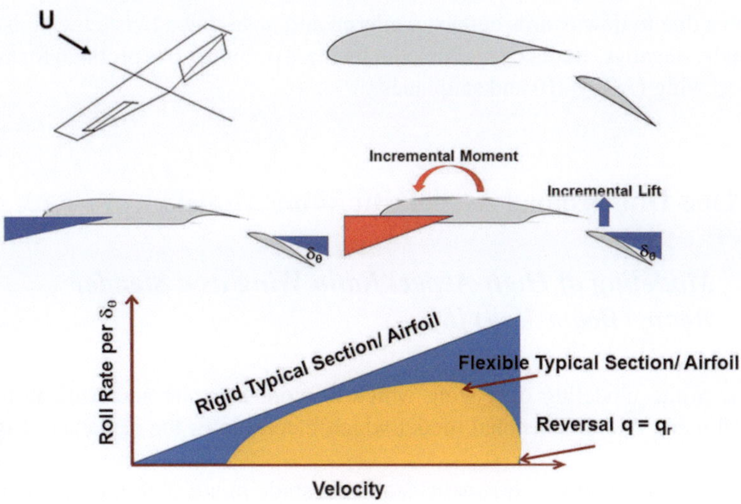

Fig. 5.6 Control effectiveness declines with increasing airspeed

Fig. 5.7 Schematic of aileron reversal behavior

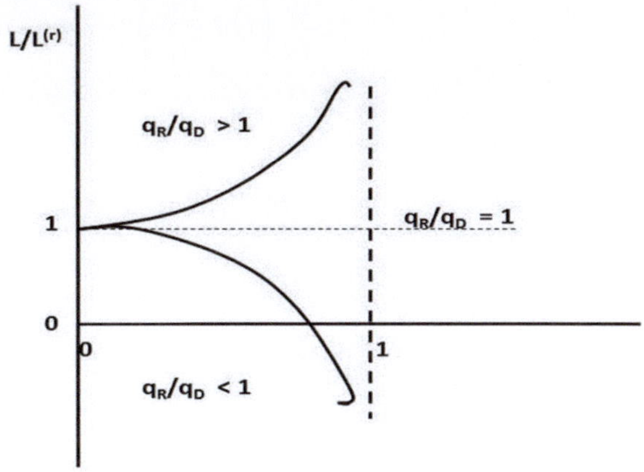

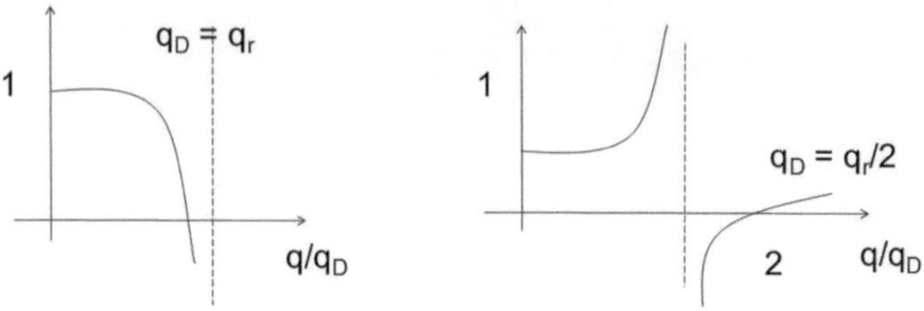

Fig. 5.8 Divergence and control reversal speed

Lift force due to downwards deflected aileron and nose-down twisted wing is unintentionally negative. Reduced aileron effectiveness poses serious problem for slender transport wing (AR $\geq$ 10) and sailplanes.

5.4 One-Dimensional Aeroelastic Wing Model

5.4.1 Modeling of High-Aspect Ratio Wing as a Slender Beam (Beam-Rod) [8]

More accurate modeling of a wing which is more accurate and realistic can be afforded using one-dimensional model which incorporates the elements of typical section.

Following the theory of prismatic beam, the static moment equilibrium equation for a slender beam at the *y* position along the beam (as shown in Fig. 5.9 can be written as

$$\frac{d}{dy}\left(GJ\,\frac{d\alpha_e}{dy}\right) + M_y = 0 \tag{5.41}$$

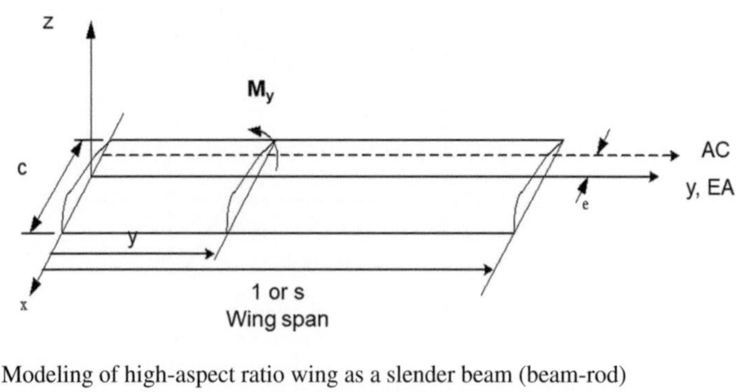

Fig. 5.9 Modeling of high-aspect ratio wing as a slender beam (beam-rod)

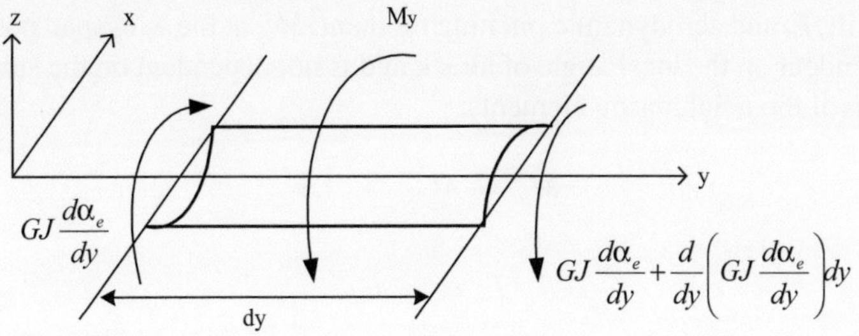

Fig. 5.10 Differential element of wing as slender beam

where

α_e torsional deflection of the wing (positive *nose-up*) with respect to the elastic axis, at *station y*.
M_y aerodynamic pitching moment (positive *nose-up*) with respect to the elastic axis, per unit span of the wing.
G Shear modulus.
J Polar moment of inertia of the wing section at the spanwise position y.
GJ Torsional stiffness.

for $\frac{d\alpha_e}{dy} > 0, GJ\frac{d\alpha_e}{dy} > 0.$
Derivation of the equilibrium equation at differential element dy:
Referring to Fig. 5.10, the equilibrium of moment at differential element dy can be expressed as:

$$-\left(GJ\frac{d\alpha_e}{dy}\right)+\left(GJ\frac{d\alpha_e}{dy}\right)+\frac{d}{dy}\left(GJ\frac{d\alpha_e}{dy}\right)+H.O.T.+M_y\,dy=0 \tag{5.42a}$$

where H.O.T means higher-order terms. In the limit of d$y \to 0$:

$$\frac{d}{dy}\left(GJ\frac{d\alpha_e}{dy}\right)+M_y\,dy=0 \tag{5.42b}$$

Equation (5.42a, 5.42b) is a second-order differential equation in y. There are two boundary conditions:

$$\alpha_e=0 \text{ at } y=0 \tag{5.43}$$

$$GJ\frac{d\alpha_e}{dy}=0 \text{ at } y=l \tag{5.44}$$

M_y is the aerodynamic pitching moment. The simplest computation of the aerodynamic pitching moment can be carried out using strip theory, that is by assuming that the Lift, L, and aerodynamic pitching moment, M_y, at the wingspan position y is only dependent on the local angle of attack and is not dependent on the surrounding conditions at the neighboring elements.

$$M_y = M_{AC} + L\,e \tag{5.45a}$$

$$L = q\,c\,C_L \tag{5.45b}$$

where[4,5]

$$C_L\,(y) = \frac{\partial\,C_L}{\partial\,\alpha}(\alpha_0\,(y) + \alpha_e\,(y)) \tag{5.45c}$$

$$M_{AC} = q\,S^2\,C_{M_{AC}} \tag{5.45d}$$

Using (5.45a, 5.45b, 5.45c, 5.45d) in (5.42a, 5.42b) and non-dimensionalization (for simplicity), and assuming prismatic beam (uniform along y), and defining:

$$\tilde{y} \equiv \frac{y}{l} \tag{5.46}$$

$$\lambda^2 = \frac{ql^2}{GJ}\,c\,\frac{\partial\,C_L}{\partial\,\alpha}\,e \tag{5.47}$$

$$K = -\frac{q\,c\,l^2}{GJ}\left(e\,\frac{\partial\,C_L}{\partial\,\alpha}\,\alpha_0 + C_{MAC}c\right) \tag{5.48}$$

then, (5.42a, 5.42b) becomes:

$$\frac{d^2\alpha_e}{d\tilde{y}^2} + \lambda^2\,\alpha_e = K \tag{5.49}$$

Using the boundary conditions (5.43) that has been non-dimensionalized,

$$\alpha_e = 0 \ at\ \tilde{y} = 0 \tag{5.50a}$$

---

[4] A more complete model can consider the influence of the angle of attack at position y on L and M, such as.

$$C_L\,(y) = \int\limits_{\eta=0}^{y} A(y-\eta)(\alpha_0\,(\eta) + \alpha_e\,(\eta))\,d\eta \tag{5.45ca}$$

.

[5] N the present book, $C_{M_{AC}}$ and C_{MAC} are alternatively used.

$$\frac{d\alpha_e}{d\tilde{y}} = 0 \; at \; \tilde{y} = l \tag{5.50b}$$

where

$\tilde{y}$ is non-dimensional y, such as by division with the wing semi-span s,

$$\tilde{y} = \frac{y}{s}$$

The general solutions of (5.49) are:

$$\alpha_e = A\sin(\lambda\tilde{y}) + B\cos(\lambda\tilde{y}) + \frac{K}{\lambda^2} \tag{5.51}$$

Using boundary condition (5.50a, 5.50b), one obtains:

$$B + \frac{K}{\lambda^2} = 0 \tag{5.52a}$$

$$\lambda[A\cos(\lambda) - B\sin(\lambda)] = 0 \tag{5.52b}$$

Solving (5.52a, 5.52b) one obtains:

$$A = -\frac{K}{\lambda^2}\tan(\lambda) \tag{5.53a}$$

$$B = -\frac{K}{\lambda^2} \tag{5.53b}$$

Substituting these into Eq. (5.51), we obtain the following relationship:

$$\alpha_e = \frac{K}{\lambda^2}\left[1 - \tan\lambda\sin(\lambda\tilde{y}) - \cos(\lambda\tilde{y})\right] \tag{5.54}$$

Divergence occurs when $\alpha_e \rightarrow \infty$[6] which implies that:

$$\tan\lambda \rightarrow \infty \text{ or } \cos\lambda \rightarrow 0$$

Hence,

$$\lambda = \lambda_m = (2m - 1)\frac{\pi}{2}, \quad \text{m} = 1, 2, 3, \ldots \tag{5.55}$$

as $\alpha_e \rightarrow \infty$

The lowest value is: $\lambda_1 = \frac{\pi}{2}$.

[6] Note that: $\lambda = 0$ is not a divergence condition. If (5.42a, 5.42b) is defined for $\lambda <<< 1$, then $\alpha_e = \frac{K}{\lambda^2}\left[1 - \lambda^2\tilde{y} - \left(1 - \frac{\lambda^2\tilde{y}}{2}\right) + \ldots\right]$ or $\alpha_e \rightarrow K\left[\frac{\tilde{y}}{2} - \tilde{y}\right]$ which has finite values when $\lambda \rightarrow 0$.

Using this value in (5.35), then

$$\left(\frac{\pi}{2}\right)^2 = \frac{ql^2}{GJ}\, c\, \frac{\partial C_L}{\partial \alpha}\, e \tag{5.56}$$

and the divergence dynamic pressure becomes:

$$q_D = \frac{\left(\frac{\pi}{2}\right)^2}{l^2 ec\left(\frac{\partial C_L}{\partial \alpha}\right)} \tag{5.57}$$

Note that $S = lc$ in (5.47) is equivalent to the cross-sectional area of the wing. Writing Eq. (5.48) as

$$K \equiv \frac{qcl^2}{GJ}\left[e\frac{\partial C_L}{\partial \alpha}\alpha_0 + C_{MAC_0}\, c\right] \equiv K_\alpha\, \alpha_0 + K_{C_{MAC}}\, c \tag{5.58}$$

then

$$\begin{aligned} K_\alpha &= \frac{q\,c\,l^2}{GJ}\, e\, \frac{\partial\, C_L}{\partial\, \alpha} = qlec\frac{l}{GJ}\frac{\partial C_L}{\partial \alpha} = \frac{\left(\frac{\pi}{2}\right)^2 \frac{GJ}{l}}{lec\frac{\partial C_L}{\partial \alpha}} lec\frac{l}{GJ}\frac{\partial C_L}{\partial \alpha} \\ &= \frac{\left(\frac{\pi}{2}\right)^2}{lec\frac{l}{GJ}\frac{\partial C_L}{\partial \alpha}} lec\frac{l}{GJ}\frac{\partial C_L}{\partial \alpha} \equiv \left(\frac{\pi}{2}\right)^2 \end{aligned} \tag{5.59}$$

or

$$K_\alpha = \left(\frac{\pi}{2}\right)^2 \tag{5.60c}$$

for $\lambda_1 = \frac{\pi}{2}$.

5.4.2 Static Aeroelasticity Using Eigenvalue and Eigenfunction Approach

5.4.2.1 First Case Study

Divergence can be viewed as an eigenvalue problem. Taking $\alpha_0 = 0$ and $C_{MAC_0} = 0$, so that from Eq. 5.36 $K = 0$, then (5.37) becomes:

$$\frac{\mathrm{d}^2\pi}{\mathrm{d}\,\tilde{y}^2} + \lambda^2\alpha = 0 \tag{5.61}$$

The prevailing boundary conditions become:

$$\alpha = 0 \text{ at } y = 0 \tag{5.62a}$$

$$\frac{d\alpha}{d\tilde{y}} = 0 \text{ at } y = 1 \tag{5.62b}$$

General solution of (5.61) is:

$$\alpha = A \sin \lambda \tilde{y} + B \cos \lambda \tilde{y} \tag{5.63}$$

Considering the boundary condition of (5.62a), (5.63) becomes:

$$B = 0 \tag{5.64a}$$

$$\lambda\,(A \sin \lambda \tilde{y} + B \cos \lambda \tilde{y}) = 0 \tag{5.64b}$$

Therefore the following solutions are obtained:

$$A = 0 \text{ (trivial solution)} \tag{5.65}$$

$$\lambda \cos \lambda = 0 \text{ and } A \neq 0 \text{ (exact solution)} \tag{5.66}$$

The final case or Eq. (5.66) is the divergence condition.

It can be shown from (5.49), that $\lambda = 0$ is not the divergence condition. The condition $\cos \lambda = 0$ results in eigenvalue:

$$\lambda = \lambda_m = (2m - 1)\frac{\pi}{2} \tag{5.67}$$

where $m = 1, 2, 3, 4 \ldots$ or integer.

From (5.48), each eigenvalue corresponds to eigenfunction $A \neq 0$, $B = 0$

$$\lambda_m \approx \sin \lambda_m \tilde{y} = \sin(2m - 1)\frac{\pi}{2}\,\tilde{y} \tag{5.68}$$

The eigenfunctions are attractive from various aspects:

It describes the torsion at the divergence dynamic pressure as described by Eqs. (5.52a, 5.52b).

It can be used to obtain solution in the form of a series expansion for each dynamic pressure.

Useful in forming approximate solution for wings with variable properties, for example with non-uniformity along the wingspan.

5.4.2.2 Second Case-Study

Suppose $\alpha_0 \neq 0$ and $C_{MAC_0} \neq 0$. Then (5.47) becomes:

$$\frac{\mathrm{d}^2\alpha_e}{\mathrm{d}\tilde{y}} + \lambda^2\,\alpha_e = K \tag{5.69}$$

Look for solution in the form of series expansion:

$$\alpha_e = \sum_n a_n\,\alpha_n\left(\widehat{y}\right) \tag{5.70}$$

$$K = \sum_n A_n\,\alpha_n\left(\widehat{y}\right) \tag{5.71}$$

where a_n and A_n are constants that still have to be determined. It can be shown that:

$$\int_0^l \alpha_n\left(\widehat{y}\right)\alpha_n\left(\widehat{y}\right)\mathrm{d}\tilde{y} = \frac{1}{2}\quad \text{for}\, m = n$$
$$\int_0^l \alpha_n\left(\widehat{y}\right)\alpha_n\left(\widehat{y}\right)\mathrm{d}\tilde{y} = 0\quad \text{for } m \neq n \tag{5.72}$$

This condition is known as the orthogonality condition. This condition can be used in the following development. First, A_n has to be determined. For that purpose multiply (5.72) with α_m and integrate from 0 to 1: $\int_0^l \ldots\ldots \mathrm{d}\tilde{y}$.

$$\int_0^1 K\,\alpha_m(\tilde{y})\,d\,\tilde{y} = \sum_n A_n \int_0^1 \alpha_n(\tilde{y})\,\alpha_m(\tilde{y})d\,\tilde{y} = \frac{1}{2}\,A_m \tag{5.73}$$

Using (5.72) obtain:

$$A_m = 2\int_0^l K\,\alpha_m\left(\widehat{y}\right)d\tilde{y} \tag{5.74}$$

Next a_n will be determined. Substitute (5.54) and (5.55) into (5.37) to obtain:

$$\sum_n\left[a_n\frac{d^2\alpha_n}{d\tilde{y}^2} + \lambda^2 a_n\alpha_n\right] = \sum_n A_n\,\alpha_n \tag{5.75}$$

Every eigenvalue λ_n satisfies (5.67):

$$\frac{\mathrm{d}^2\alpha_n}{\mathrm{d}\tilde{y}} + \lambda_n^2\alpha_n = K \tag{5.76}$$

Therefore, (5.75) can be written as:

$$\sum_n a_n[-\lambda_n^2 + \lambda^2]\alpha_n = \sum A_n\alpha_n \tag{5.77}$$

Multiply (5.77) with λ_n and integrate from 0 to 1, $\int_0^1 ......\mathrm{d}\tilde{y}$ and obtain:

$$|\lambda^2 - \lambda_m^2|\left(\frac{1}{2}a_m\right) = \frac{1}{2}A_m \tag{5.78}$$

Therefore

$$a_m = \frac{A_m}{|\lambda^2 - \lambda_m^2|} \tag{5.79}$$

From (5.77):

$$\alpha_e = \sum a_n\alpha_n = \sum_n \frac{A_n}{[\lambda^2 - \lambda_m^2]}\alpha_n(\tilde{y}) \tag{5.80}$$

where A_n is given by Eq. (5.74) and λ_n by Eq. (5.67).

5.5 Galerkin Method

The equilibrium equation for a wing with variable properties can be obtained by substituting (5.45a, 5.45b, 5.45c, 5.45d) in (5.42a, 5.42b).

In dimensional form, the results are:

$$\frac{\mathrm{d}}{\mathrm{d}y}\left(GJ\frac{\mathrm{d}}{\mathrm{d}y}\alpha_e\right) + eqc\frac{\partial C_L}{\partial\alpha}\alpha_e = -eqc\frac{\partial C_L}{\partial\alpha}\alpha_0 - qc^2C_{Mac_0} \tag{5.81}$$

and in non-dimensional form:

$$\frac{\mathrm{d}}{\mathrm{d}\tilde{y}}\left(\gamma\frac{\mathrm{d}\alpha_e}{\mathrm{d}\tilde{y}}\right) + \lambda^2\alpha_e\beta = K \tag{5.82}$$

where

$$\gamma \equiv \frac{GJ}{(GJ)_{\text{ref}}} \tag{5.83}$$

$$K = -\frac{qcl^2}{(GJ)_{\text{ref}}}\left[e\frac{\partial C_L}{\partial \alpha}\alpha_0 + C_{Mac_0}c\right] \tag{5.84}$$

$$\lambda^2 \equiv \frac{ql^2 c_{\text{ref}}}{(GJ)_{\text{ref}}}\left(\frac{\partial C_L}{\partial \alpha}\right)_{\text{ref}} e_{\text{ref}} \tag{5.85}$$

$$\beta = \frac{c}{c_{\text{ref}}}\frac{e}{e_{\text{ref}}}\frac{\left(\frac{\partial C_L}{\partial \alpha}\right)}{\left(\frac{\partial C_L}{\partial \alpha}\right)_{\text{ref}}} \tag{5.86}$$

As before, seek solution in the form:

$$\alpha_e = \sum_n a_n a_n(\tilde{y}) \tag{5.87}$$

$$K = \sum_n A_n a_n(\tilde{y}) \tag{5.88}$$

Substitute (5.68) and (5.69) into Eq. (5.82), multiply by α_n and integrate from 0 to 1, $\int_0^l$ ....d$\tilde{y}$, and obtain:

$$\sum_n a_n \left\{ \int_0^l \frac{\mathrm{d}}{\mathrm{d}\tilde{y}}\left(\gamma \frac{\mathrm{d}a_n}{\mathrm{d}\tilde{y}}\right) a_m \mathrm{d}\tilde{y} + \lambda^2 \int_0^l \beta\, \alpha_n \alpha_m \mathrm{d}\tilde{y} \right\} = \sum_n A_n \int_0^l \alpha_n \alpha_m \mathrm{d}\tilde{y} = \frac{A_m}{2} \tag{5.89}$$

The first and second terms cannot be simplified further unless the eigenfunction or "mode" that is used is the eigenfunction for wings with variable properties. Therefore, a_n is not related in a simple manner to A_n like in the case for wings with constant properties.

Equation (5.89) describes the algebraic equation for a_n. In matrix form,

$$[C_{mn}]\{a_n\} = \frac{1}{2}\{A_m\} \tag{5.90}$$

where

$$C_{mn} = \int_0^l \frac{\mathrm{d}}{\mathrm{d}\tilde{y}}\left(\lambda \frac{\mathrm{d}a_n}{\mathrm{d}\tilde{y}}\right) a_m \mathrm{d}\tilde{y} + \lambda^2 \int_0^l \beta\, \alpha_n \alpha_m \mathrm{d}\tilde{y} \tag{5.91}$$

By truncating the series (5.87) and (5.88) for a finite number of terms, in a formal way Eq. (5.71) has been solved to obtain α_n:

$$\{a_n\} = \frac{1}{2}[C_{mn}]^{-1}\{A_m\} \tag{5.92}$$

The divergent condition is the condition where he determinant of matrix C_{mn} becomes zero, therefore $a_n \to \infty$

$$\|[C_{mn}]\| = 0 \tag{5.93}$$

which is a polynomial in λ^2.

5.6 Rolling of a Straight Wing

5.6.1 General

A physically and mathematically more complex static aeroelastic problem is encountered in the rolling of a straight wing. The analysis will have the following characteristics:

- The problem is formulated in terms of an integral equation (compared to differential equation formulation in previous problems).
- Aerodynamic induction (compared to strip theory).

Lumped element method (compared to modal approach or eigenfunction approach).

Figure 5.11 depicts the schematic of the problem:

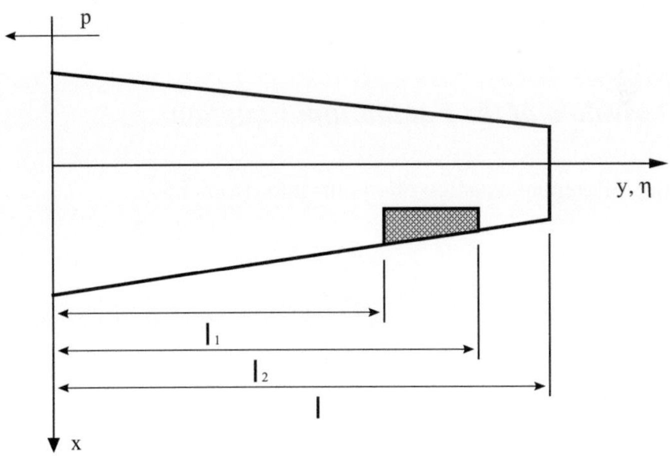

Fig. 5.11 Rolling of a straight wing

5.6.2 Integral Equation of Equilibrium

The integral equation of equilibrium can be written as:

$$\alpha(y) = \int_0^l C^{\alpha\alpha}(y, \eta) M_y(\eta) \mathrm{d}\eta \tag{5.94}$$

Physical significance of the influence coefficient $C^{\alpha\alpha}(y,\eta)$ and $C^{\alpha\alpha}(y,\gamma)$:
If a unit moment is applied at $y = \gamma$,

$$\begin{aligned} M_y(\eta) = \delta(\eta - \gamma) \text{ at } \eta \neq \gamma \text{ then } \delta(\eta - \gamma) = 0 \\ \text{at } \eta = \gamma \text{ then } \delta(\eta - \gamma) = \delta(0) = 1 \end{aligned} \tag{5.95}$$

Therefore (5.75) becomes:

$$\alpha(y) = \int_0^l C^{\alpha\alpha}(y, \eta)\delta(\eta - \gamma) d\eta = C^{\alpha\alpha}(y, \gamma) \tag{5.96}$$

Hence $C^{\alpha\alpha}(y, \gamma)$ is the twist at y due to the action of a unit twisting moment (torsion) at α. Similarly, $C^{\alpha\alpha}(y, \eta)$ is a unit twist at y due to the action of a unit twisting moment at α. $C^{\alpha\alpha}$ (y, η) or $C^{\alpha\alpha}(y, \gamma)$ is known as the structural influence function.

Note also that (5.75) implies that to obtain a total twist, a distributed torsion M_y has to be multiplied by $C^{\alpha\alpha}(y, \eta)$ and added (integrated) over the wingspan. $C^{\alpha\alpha}(y, \eta)$ can be measured.

5.6.3 Derivation of the Equilibrium Equation

Consider the differential equation of equilibrium (from 5.94):

$$\frac{\mathrm{d}}{\mathrm{d}y}\left(GJ\frac{\mathrm{d}\alpha}{\mathrm{d}y}\right) = -M_y \tag{5.97}$$

where

$$\alpha(0) = 0 \text{ and } \frac{\mathrm{d}\alpha}{\mathrm{d}y}(l) = 0 \tag{5.98a,b}$$

as the boundary condition.

For a unit torsion at $y = \eta$,

$$\frac{\mathrm{d}}{\mathrm{d}y} GJ \frac{\mathrm{d}C^{\alpha\alpha}}{\mathrm{d}y} = -\delta(y - \eta) \tag{5.99}$$

where

$$C^{\alpha\alpha}(0,\eta) = 0 \text{ and } \frac{\mathrm{d}C^{\alpha\alpha}}{\mathrm{d}y}(l,\eta) = 0 \tag{5.100}$$

Multiply (5.99) with $\alpha(y)$ and integrate over the wingspan:

$$\int_0^l \alpha(y)\frac{\mathrm{d}}{\mathrm{d}y}\left(GJ\frac{\mathrm{d}C^{\alpha\alpha}}{\mathrm{d}y}\right)\mathrm{d}y = -\int_0^l \delta(y-\eta)\alpha(y)\mathrm{d}y = -\alpha(\eta) \tag{5.101}$$

Carrying out integration by parts on the left-hand side of the equation obtain

$$\alpha GJ\frac{\mathrm{d}C^{\alpha\alpha}}{\mathrm{d}y}\bigg|_0^l - GJ\frac{\mathrm{d}\alpha}{\mathrm{d}y}C^{\alpha\alpha}\bigg|_0^l + \int_0^l C^{\alpha\alpha}\frac{\mathrm{d}}{\mathrm{d}y}\left(GJ\frac{\mathrm{d}\alpha}{\mathrm{d}y}\right)\mathrm{d}y = -\alpha(y) \tag{5.102}$$

Using the boundary conditions (5.98a) and (5.98b), the first two terms on the left-hand side reduce to zero. Hence the following expression results:

$$\alpha(\eta) = \int_0^l C^{\alpha\alpha}(y,\eta)M_y(y)\mathrm{d}y \tag{5.103}$$

Now interchange y and η:

$$\alpha(y) = \int_0^l C^{\alpha\alpha}(\eta, y)M_y(\eta)\mathrm{d}\eta \tag{5.104}$$

Equation (5.85) is identical to (5.84) if :

$$C^{\alpha\alpha}(\eta, y) = C^{\alpha\alpha}(y,\eta) \tag{5.105}$$

This will be proved below.

5.6.4 The Determination of $C^{\alpha\alpha}$

From Eq. (5.99) using (5.100), and after integration of (5.80) with respect to y from 0 to y_1, obtain

$$\begin{aligned} GJ(y_1)\frac{\mathrm{d}C^{\alpha\alpha}}{\mathrm{d}y}(y_1,\eta) - GJ(0)\frac{\mathrm{d}C^{\alpha\alpha}}{\mathrm{d}y}(0,\eta) &= -1 \text{ if } y_1 > \eta \\ &= 0 \text{ if } y_1 < \eta \equiv S(y_1,\eta) \end{aligned} \tag{5.106}$$

The function $S(y_1,\eta)$ is sketched below (Fig. 5.12),

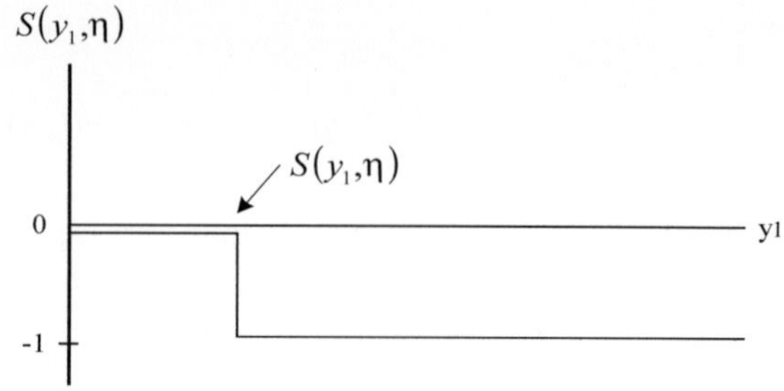

Fig. 5.12 Shear distribution along the wingspan

If (5.87) is divided by $GJ(y_1)$ and then integrated with respect to y_1 from *0* to y_2, and there is obtained

$$
\begin{aligned}
&C^{\alpha\alpha}(y_1,\eta) - C^{\alpha\alpha}(0,\eta) - GJ(0)\frac{\mathrm{d}C^{\alpha\alpha}}{\mathrm{d}y}(0,\eta)\int_0^{y_2}\frac{1}{GJ}\mathrm{d}y_1 \\
&= \int_0^{y_2}\frac{S(y_1,\eta)}{GJ(y_1)}dy_1 \\
&= \int_0^{y_2}\frac{1}{GJ(y_1)}dy_1 \quad \text{for } y_1 > \eta \\
&= 0 \quad \text{for } y_1 < \eta
\end{aligned}
\tag{5.107}
$$

From the boundary conditions, Eq. (5.98),

a. $C^{\alpha\alpha}(0,\eta) = 0$
b. $\frac{\mathrm{d}C^{\alpha\alpha}}{\mathrm{d}y}(l,\eta) = 0$

which can be used to evaluate the terms in (5.87) and (5.88) that have not yet been known. Evaluation of (5.87) at $y_1 = l$ yields

$$GJ(l)\frac{\mathrm{d}C^{\alpha\alpha}}{\mathrm{d}y}(l,\eta) - GJ(0)\frac{\mathrm{d}C^{\alpha\alpha}}{\mathrm{d}y}(0,\eta) = -1$$

if

$\frac{\mathrm{d}C^{\alpha\alpha}}{\mathrm{d}y}(l,\eta) = 0$ (from b) than $GJ(0)\frac{\mathrm{d}C^{\alpha\alpha}}{\mathrm{d}y}(0,\eta) = 1$.
Using (a) and (c), then (5.88) can be written as

$$
\begin{aligned}
C^{\alpha\alpha}(y_1,\eta) &= \int_0^{y_2}\frac{1}{GJ}\mathrm{d}y_1 - \int_\eta^{y_2}\frac{1}{GJ}\mathrm{d}y_1 \\
&= \int_0^{\eta}\frac{1}{GJ}\mathrm{d}y_1 \text{ for } y_2 > \eta \\
&= \int_0^{y_2}\frac{1}{GJ}\mathrm{d}y_1 \text{ for } y_2 < \eta
\end{aligned}
$$

By replacing the subscript y_2 by y, there is obtained:

$$C^{\alpha\alpha}(y_1, \eta) = \int_0^{\eta} \frac{1}{GJ} \mathrm{d}y_1 \text{ for } y > \eta$$
$$= \int_0^{y} \frac{1}{GJ} \mathrm{d}y_1 \text{ for } y < \eta \tag{5.108}$$

Therefore, from the above equations, it can be concluded that

$$C^{\alpha\alpha}(y, \eta) = C^{\alpha\alpha}(\eta, y) \tag{5.109}$$

This is known as the Maxwell's reciprocity theorem.

5.7 Aerodynamic Forces

First, the aerodynamic angle of attack should be identified, that is the angle between the chord of the airfoil and the relative velocity of the flow (Fig. 5.13).

From the figure, the total angle of attack due to wing twist and rolling is

$$\alpha_{\text{total}} = \alpha(y) - \frac{py}{U} \tag{5.110}$$

The control surface can be assumed to be rigid and produces the following rotation

$$\delta(y) = \delta_R \quad \text{for } l_1 < y < l_2$$

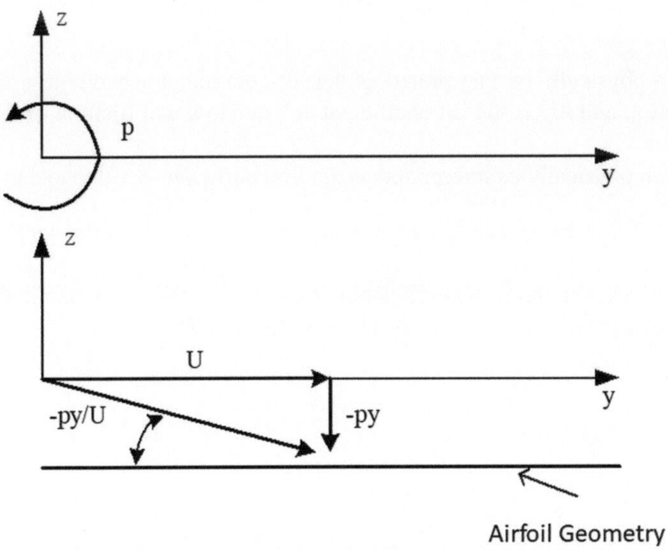

Fig. 5.13 Linearized aerodynamic approximation of an airfoil

$$= 0 \quad \text{otherwise} \tag{5.111}$$

Form aerodynamic theory or experiment,

$$C_L \equiv \frac{L}{qc} = \int_0^l A^{L\alpha}(y,\eta)\alpha_T(\eta)\frac{d\eta}{l} + \int_0^l A^{L\delta}(y,\eta)\delta(\eta)\frac{d\eta}{l} \tag{5.112}$$

$A^{L\alpha}$, $A^{L\delta}$ are influence function in non-dimensional form. Therefore, $A^{L\alpha}$ is the non-dimensional lift at y due to a unit angle of attack at η. Therefore Eq. (5.92) becomes:

$$C_L = \int_0^l A^{L\alpha}\alpha\frac{d\eta}{l} - \frac{pl}{U}\int_0^l A^{L\alpha}\frac{\eta}{l}\frac{d\eta}{l} + \delta_R\int_{l_1}^{l_2} A^{L\delta}\frac{d\eta}{l}$$

$$C_L = \int_0^l A^{L\alpha}\alpha\frac{d\eta}{l} + \frac{pl}{U}\frac{\partial C_L}{\partial\left(\frac{pl}{U}\right)} + \delta_R\frac{\partial C_L}{\partial \delta_R} \tag{5.113}$$

where

$$\frac{\partial C_L}{\partial\left(\frac{pl}{U}\right)}(y) \equiv -\int_0^l A^{L\alpha}\frac{\eta}{l}\frac{d\eta}{l}$$

and

$$\frac{\partial C_L}{\partial \delta_R} \equiv \int_{l_1}^{l_2} A^{L\delta}\frac{d\eta}{l}$$

$A^{L\alpha}$ can physically be interpreted as the lift coefficient at y due to a unit angle of attack at η, and $A^{L\delta}$ is the lift coefficient at y due to a unit rotation of the control surface at η.

$\frac{\partial C_L}{\partial\left(\frac{pl}{U}\right)}$ can physically be interpreted as the lift coefficient at y due to a unit rolling rate at η, and $\frac{\partial C_L}{\partial \delta_R}$ the lift coefficient at y due to a unit rotation of the control surface at η.

The aerodynamic moment coefficient with respect to the aerodynamic center a.c. at y due to the rotation of the control surface is given by:

$$C_{MAC} \equiv \frac{M_{AC}}{qc^2} = \frac{\partial C_{MAC}}{\partial \delta_R}\delta_R \tag{5.114}$$

where

$$\partial C_{MAC}\big/\partial\alpha_T \equiv 0$$

since it is defined with respect to the aerodynamic center. Therefore, the total moment with respect to the elastic axis becomes

$$\begin{aligned} M_y &= M_{AC} + Le \\ &= qc[C_{MAC}c + C_L e] \end{aligned} \tag{5.115}$$

Using (5.113) and (5.114), Eq. (5.115) becomes:

$$M_y = qc\left[c\frac{\partial C_{MAC}}{\partial \delta_R}\delta_R + e\left\{\int_0^l A^{L\alpha}\alpha\frac{\mathrm{d}\eta}{l} + \frac{\partial C_L}{\partial\left(\frac{pl}{U}\right)}\left(\frac{pl}{U}\right) + \frac{\partial C_L}{\partial \delta_R}\delta_R\right\}\right] \tag{5.116}$$

5.8 Aeroelastic Equilibrium Equation and Lumped Mass Method

$$\alpha(y_i) \cong \sum_{j=1}^{N} C^{\alpha\alpha}(y_i, \eta_j) M_y(\eta_j)\Delta\eta \quad i = 1, \ldots, \mathrm{N} \tag{5.117}$$

where $\Delta\eta$ is the segment width and N is the total number of segments. Similarly, Eq. (5.96) can be written as:

$$M_y = qc\left\{\left[c\frac{\partial C_{MAC}}{\partial \delta_R}\delta_R + \mathrm{e}\frac{\partial C_L}{\partial\left(\frac{pl}{U}\right)}\left(\frac{pl}{U}\right) + e\frac{\partial C_L}{\partial \delta_R}\delta_R\right] + \mathrm{e}\sum_{j=1}^{N} A^{L\alpha}(y_i, \eta_j)\alpha(\eta_j)\frac{\Delta\eta}{l}\right\} \quad i = 1, \ldots, \mathrm{N} \tag{5.118}$$

It is more convenient to utilize matrix notation in further manipulation of Eqs. (5.97) and (5.98). Hence

$$\{\alpha\} = \Delta\eta\left[C^{\alpha\alpha}\right]\{M_y\} \tag{5.119}$$

and

$$\begin{aligned} \{M_y\} = q\begin{bmatrix} \diagdown & & \\ & c^2 & \\ & & \diagdown \end{bmatrix}\left\{\frac{\partial C_{MAC}}{\partial \delta_R}\right\}\delta_R + q\begin{bmatrix} \diagdown & & \\ & ce & \\ & & \diagdown \end{bmatrix}\left\{\frac{\partial C_L}{\partial\left(\frac{pl}{U}\right)}\right\}\left(\frac{pl}{U}\right) \\ + q\begin{bmatrix} \diagdown & & \\ & ce & \\ & & \diagdown \end{bmatrix}\left\{\frac{\partial C_L}{\partial \delta_R}\right\}\delta_R + q\begin{bmatrix} \diagdown & & \\ & ce & \\ & & \diagdown \end{bmatrix}\left[A^{L\alpha}\right]\{\alpha\}\frac{\Delta\eta}{l} \end{aligned} \tag{5.120}$$

Substitution of (5.100) into (5.99) yields:

$$\left[\begin{bmatrix} \ddots & & \\ & 1 & \\ & & \ddots \end{bmatrix} - q\frac{(\Delta\eta)^2}{l}[E]\left[A^{L\alpha}\right]\right]\{\alpha\} = \{f\} \tag{5.121}$$

where $\{f\}$ is defined as

$$\{f\} = q[E]\left\|\left\{\left\{\frac{\partial C_L}{\partial\left(\frac{pl}{U}\right)}\right\}\left(\frac{pl}{U}\right)\right\}\Delta\eta + \frac{\partial C_L}{\partial\delta_R}\delta_R\right\| + q[F]\left\{\frac{\partial C_{MAC}}{\partial\delta_R}\right\}\delta_R\Delta\eta$$

$$[E] = \left[C^{\alpha\alpha}\right]\begin{bmatrix} \ddots & & \\ & ce & \\ & & \ddots \end{bmatrix}$$

$$[F] = \left[C^{\alpha\alpha}\right]\begin{bmatrix} \ddots & & \\ & c^2 & \\ & & \ddots \end{bmatrix}$$

Further, defining

$$[D] \equiv \begin{bmatrix} \ddots & & \\ & l & \\ & & \ddots \end{bmatrix} - q\frac{(\Delta\eta)^2}{l}[E]\left[A^{L\alpha}\right]$$

one may solve (5.101) as:

$$\{\alpha\} = [D]^{-1}\{f\} \tag{5.122}$$

Matrix D will have no inverse if

$$|D| = 0 \tag{5.123}$$

for which then,

$$\{\alpha\} \rightarrow \infty$$

This condition (which is an eigenvalue problem condition) is already known as the divergence. Equation (4.103) is the eigenvalue problem for the dynamic divergence pressure or divergence speed, q_D. Note also that Eq. (4.103) is a polynomial in q.

The smallest positive root (eigenvalue) of (4.103) indicates a physically meaningful quantity $q_{divergen}$. Since it may be more difficult or tedious to obtain the roots of g the polynomial, alternatively one may plot the value of $|D|$ with respect to q to obtain its roots, i.e. finding the zeroes of $|D|$. The roots are the values of the dynamic

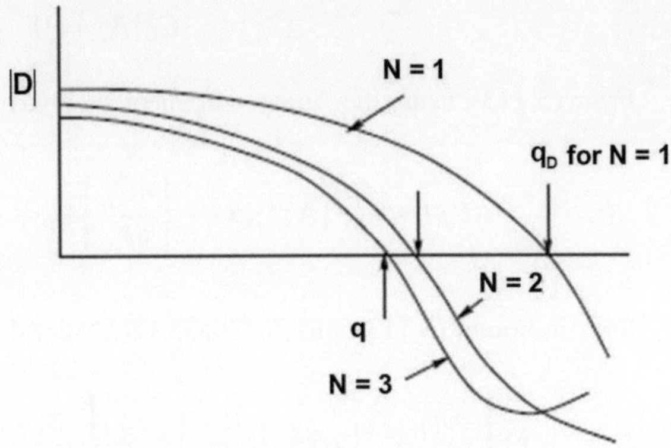

Fig. 5.14 Qualitative impression of the characteristic determinant versus dynamic pressure for rolling of a straight wing

pressure q's for which the values of $|D|$ are zero. The diagram for such scheme is qualitatively and schematically depicted in Fig. 5.14.

From these results, one can plot q_D (the smallest positive value of q for which $|D| = 0$) with respect to N. The exact values of q_D are obtained for $N \to \infty$. Usually, accurate values can be obtained for small values of N, say 10 or less. The divergence speeds calculated in the above fashion does not depend on the rolling of the wing, that is p is assumed to be known, for example $p = 0$.

5.9 Control Surface Reversal and Rolling Effectiveness

pl/U has been known before as it is given. In reality, pl/U is a function of δ_R and all parameters require the static equilibrium of the wing in roll. This is an additional degree of freedom. For the rolling equilibrium and stationary rolling speed p, rolling moment about the x-axis is zero.

$$M_{\text{Rolling}} \equiv 2\int_0^1 Ly\mathrm{d}y = 0 \tag{5.124}$$

Approximating 5.104:

$$\sum_i L_i y_i \Delta y = 0 \tag{5.125}$$

or in matrix notation:

$$2\lfloor y \rfloor \{L\} \Delta y = 0 \tag{5.126a}$$

or:

$$2q\lfloor cy\rfloor\{C_L\}\Delta y = 0 \tag{5.126b}$$

From (5.113), using the "lumped element" approximation and matrix notation:

$$\{C_L\} = \frac{\Delta\eta}{l}\left[A^{L\alpha}\right]\{\alpha\} + \left\{\frac{\partial C_L}{\partial\delta_R}\right\}\delta_R + \left\{\frac{\partial C_L}{\partial\left(\frac{pl}{U}\right)}\right\}\frac{pl}{U} \tag{5.127}$$

Substitution of (5.113) into (5.126a, 5.126b) gives:

$$\lfloor cy\rfloor\left\{\frac{\Delta\eta}{l}\left[A^{L\alpha}\right]\{\alpha\} + \left\{\frac{\partial C_L}{\partial\delta_R}\right\}\delta_R + \left\{\frac{\partial C_L}{\partial\left(\frac{pl}{U}\right)}\right\}\frac{pl}{U}\right\} = 0 \tag{5.128}$$

Note that Eq. (5.128) is a single algebraic Eq. (5.128) plus (5.119) and (5.120) are $2N + 1$ linear algebraic equation in $N(\alpha)$ plus $N(M_y)$ plus $1(p)$ unknown. As before, $\{M_y\}$ is normally eliminated by using (5.120) in (5.119) to obtain N Eqs. (5.101), plus 1, Eq. (5.128), equations in $N(\alpha)$ plus $1(p)$ unknowns. In either case, the divergence condition can be determined by setting the determinant of the coefficients to zero and determining the smallest positive eigenvalue, $q = q_D$.

For $q < q_D$, pl/U (and $\{\alpha\}$) may be determined from (5.121) and (5.128). Since the mathematical model is linear:

$$pl/U \sim \delta_R \tag{5.129}$$

and therefore, a convenient plot of the results is shown in Fig. 5.14

For $q \to q_D$, $pl \to \infty$ and $\alpha \to \infty$.

Another qualitatively different type of result may sometimes occur. This is indicated in Fig. 5.14:

$$\frac{\frac{pl}{U}}{\delta_R} \to 0 \quad \text{for } q \to q_R < q_D \tag{5.130}$$

Then rolling reversal is said to have occurred and the corresponding $q = q_R$ is called the reversal dynamic pressure. The basic phenomenon is the same as that encountered previously as control surface reversal.

5.10 Two-Dimensional Aeroelastic Model of Lifting Surface

In the following, we will consider structural modeling, aerodynamic modeling, and the combination of the two into an aeroelastic model, and its solution.

5.10.1 Two-Dimensional Structure—Integral Representation

The two-dimensional or plate analog to the one-dimensional or beam-rod model is:

$$w(x, y) = \iint C^{wp}(x, y; \xi, \eta) p(\xi, \eta) \mathrm{d}\xi . \mathrm{d}\eta \tag{5.131}$$

where

w vertical deflection at point x, y on plate
p pressure at point ξ, η on plate
C^{wp} deflection at point x, y due to unit pressure at ξ, η.

Note that w and p are taken as positive in the same direction. For the special case where

$$w(x, y) = h(y) + x\alpha(y) \tag{5.132}$$

and

$$C^{wp}(x, y; \xi, \eta) = C^{hF}(y, \eta) + xC^{\alpha F}(y, \eta) + \xi C^{hM}(y, \eta) + x\xi C^{\alpha M}(y, \eta) \tag{5.133}$$

and using the definition:
C^{hF} deflection of the y-axis due to a unit force F.
$C^{\alpha F}$ twist about the y-axis due to a unit force F, etc.
We can retrieve our beam-rod results. Note that (5.131) and (5.132) can be thought of as polynomial (Taylor Series) expansions of the deflection.
Substitution of (5.132), (5.133) into (5.131) yields:

$$\begin{aligned} h(y) + x\alpha(y) = &\left[\int C^{hF} \left(\int p(\xi, \eta) \mathrm{d}\xi \right) \mathrm{d}\eta + \int C^{hM} \left(\int \xi p(\xi, \eta) \mathrm{d}\xi \right) \mathrm{d}\eta \right] \\ &\times \left[\int C^{\alpha F} \left(\int p(\xi, \eta) \mathrm{d}\xi \right) \mathrm{d}\eta + \int C^{\alpha M} \left(\int \xi p(\xi, \eta) \mathrm{d}\xi \right) \mathrm{d}\eta \right] \end{aligned} \tag{5.134}$$

If y, η lie along an elastic axis, then $C^{hM} = C^{\alpha F} = 0$. Equating coefficients of like powers of x, we obtain:

$$h(y) = \int C^{hF}(y, \eta) F(\eta) \mathrm{d}\eta \tag{5.135a}$$

$$\alpha(y) = \int C^{\alpha M}(y, \eta) M(\eta) \mathrm{d}\eta \tag{5.135b}$$

where

$$F \equiv \int p.\mathrm{d}\xi \quad M \equiv \int p\xi \mathrm{d}\xi \tag{5.136}$$

Equation (5.135b) is our previous results. Since for static aeroelasticity problems, M is a function of α (and not a function of h). Equation (5.135b) can be solved independently of (5.135a). Subsequently (5.135b) may be solved to determine h if desired. Equation (5.135a) has no effect on divergence or control surface reversal, and hence it is justified in neglecting it in the foregoing discussions.

5.10.2 *Two-Dimensional Aerodynamic Surfaces—Integral Representation*

In a similar manner (for simplicity one only include deformation dependent aerodynamic forces to illustrate the method):

$$\frac{p(x, y)}{q} = \iint A^{pw}(x, y; \xi, \eta)\frac{\partial w}{\partial \xi}(\xi, \eta)\frac{\mathrm{d}\xi}{c_r}\frac{\mathrm{d}\eta}{l} \tag{5.137}$$

where

A^{pw} non-dimensional aerodynamic pressure at x, y due to unit $\frac{\partial w}{\partial \xi}$ at point (ξ, η).
c_r reference chord.
l reference span.

For g the special case:

$$w = h + x\alpha \tag{5.138}$$

and hence:

$$\frac{\partial w}{\partial x} = \alpha \tag{5.139}$$

Then our beam-rod aerodynamic results may be retrieved.
For example, we may compute the lift as:

$$L \equiv \int p\mathrm{d}x = qc_r \int_0^l A^{L\alpha} y(\xi, \eta)\alpha(\eta)\frac{\mathrm{d}\eta}{l} \tag{5.140}$$

where

$$A^{L\alpha} \equiv \iint A^{pw_x}(x, y; \xi, \eta)\alpha(\eta)\frac{\mathrm{d}\xi}{c_r}\frac{\mathrm{d}\eta}{c_r} \tag{5.141}$$

5.10.3 Solution by Lumped Mass Method

Approximating the integrals by sums and using matrix notation, (5.131) becomes:

$$\{w\} = \Delta\xi \, \Delta\eta \left[C^{wp}\right]\{p\} \tag{5.142}$$

and (5.137) becomes:

$$\{p\} = q \frac{\Delta\xi}{c_r} \frac{\Delta\eta}{l} \left[A^{pw_x}\right] \left(\frac{\partial w}{\partial \xi}\right) \tag{5.143}$$

Then:

$$\left(\frac{\partial w}{\partial \xi}\right)_i \cong \frac{w_{i+1} - w_{i-1}}{2\Delta\xi} \tag{5.144}$$

is the differential representation of the surface slope. Therefore[7]

$$\left(\frac{\partial w}{\partial \xi}\right) = \frac{1}{2\Delta\xi}[W]\{w\} = \frac{1}{2\Delta\xi} \begin{bmatrix} [W] & [0] & [0] & [0] \\ & [W] & [0] & [0] \\ & & [W] & [0] \\ & & & [W] \end{bmatrix} \{w\}^* \tag{5.145}$$

is the results obtained for four spanwise locations, where:

$$[W] = \begin{bmatrix} 0 & 1 & 0 & 0 & . \\ -1 & 0 & 1 & 0 & . \\ 0 & -1 & 0 & 1 & . \\ & & \cdots\cdots & & \\ \ldots & 0 & 0 & -1 & 0 \end{bmatrix} \tag{5.146}$$

number of chordwise locations

is a numerical weighting matrix. From (5.122), (5.123), and (5.124), we obtain an equation for w:

$$[D]\{w\} \equiv \left[[I] - q \frac{(\Delta\xi)^2}{c_r} \frac{(\Delta\eta)^2}{l} \frac{1}{2\Delta\xi} [C^{wp}][A^{pw_x}][W]\right]\{w\} = \{0\} \tag{5.147}$$

---

[7] For definiteness, assume a rectangular wing is divided into a finite number of small boxes. The weighting matrix $[W]$ is defined for loctions along the wing span and various chordwise boxes. The elements in the matrix, $\{\partial w/\partial \xi\}$ and $\{w\}$, are ordered following their location along the fixed spanwise location and then overball chordwise locations. This numerical scheme serves only as an illustration and need not necessarily be applied in practice.

where

[I] is the identity matrix.

For divergence to prevail:

$$|D| = 0 \tag{5.148}$$

which then allows the determination of q_D.

5.11 Closing Remarks

The present chapter focuses on static aeroelasticity and develop the static aeroelastic analysis by a simplified approach as fist approximation to understand the application of basic physical and mathematical principles using the typical section modeling. The analysis then is followed by one-dimensional aeroelastic wing model and progresses with more realistic models. Examples have been elaborated to allow the students and readers to follow logical procedures that will be the basis for dealing with more complex problems and realistic approaches.

References

1. Djojodihardjo, H. 2005. In *Lecture Notes on Aeroelasticity at the Intitute of Technology Bandung delivered in 1982 to 2005 (Author's repository).*
2. Djojodihardjo, H. 2008. In *Lecture Notes on Aeroelasticity at the Universiti Sains Malaysia (University of Science Malaysia) in 2004 t0 2008 (Author's repository).*
3. Bisplinghoff, R L., H. Ashley, and R.L. Halfman. 1955. *Aeroelasticity*. Addison-Wesley Publishing Company Inc.
4. Bisplinghoff, R.L., and H. Ashley. 1962. *Principles of Aeroelasticity*. Wiley Inc.
5. Fung, Y.C. 2002. *An iltroduction to the Theory of Aeroelasticity*. Courier Dover Publications.
6. Zwaan, R.J. 1981. *Aeroelasticity of Aircraft, Lecture Notes, Special Lecture, Short Course offered at Institut Teknologi Bandung, Indonesia, Lecture Notes delivered at Special Meeting at ITB in August 1981* (Author's Repository; private collection).
7. NASA. 2015. *NASA Armstrong Fact Sheet: X-29 Advanced Technology Demonstrator Aircraft*. https://www.nasa.gov/centers/armstrong/news/FactSheets/FS-008-DFRC.html.
8. Dowell, E.H., H.C. Curtiss, R.H. Scanlan, and R. Sisto. 1995. *A Modern Course in Aeroelasticity*. Kluwer Academic.

Chapter 6
Flutter Stability of a Typical Section-An Elementary Discussion

Abstract One of the aircraft mandatory requirements demands that aircraft flight is flutter-free. Flutter is a dynamic aeroelastic instability due to the interaction of inertial, elastic and aerodynamic forces. It is an undesirable phenomenon for safety requirements in aircraft since it causes divergent oscillations that may lead to structural damage or failure, performance and ride comfort degradation, or loss of control, in addition to comfort. Costly redesign is required if flutter is discovered at the aircraft certification stage. Emergence of flutter compromises not only the long-term durability of the wing structure but also the operational safety, flight performance and energy efficiency of the aircraft. Effectual means of flutter prevention are, therefore, mandatory in the certification of new flight vehicles. In the present chapter, flutter problems and solution in predicting their occurrence are discussed in a fundamental and elementary manner, to reveal the physical principles and their essential characteristics. The treatment is manifested by the use of simplified aerodynamic models and examples. However, this should not distract our attention to the relevant phenomenon to be revealed, which will be more complex with the presence of other physical elements that are here neglected and simplified mathematical approach adopted.

Keywords Flutter · Binary model · Control · Prediction · Suppression

6.1 Steady Aerodynamic Model

The angle of attack is equal to the incidence angle. The flow follows the angle of attack instantaneously as if at each instant the flow exists already all the time. Therefore the aerodynamic model is essentially a steady-state one.

Further simplifications:

- Ideal and incompressible flow
- Airfoil thickness neglected (flat plate).

In what follows, an airfoil with two degree of freedom is considered; these are the bending and torsional movement (i.e., binary system). Further simplification

© The Author(s), under exclusive license to Springer Nature Singapore Pte Ltd. 2023
H. Djojodihardjo, *Introduction to Aeroelasticity*,
https://doi.org/10.1007/978-981-16-8078-6_6

and assumption: *Only torsional motion induces aerodynamic forces*. The situation is depicted in Fig. 6.1.

$$L^{(M)}(t) = q\, S\, C_{L\alpha} \quad \alpha(t) \tag{6.1}$$

$$M_{\mathrm{AC}}^{(M)}(t) = 0 \tag{6.2}$$

For convenience, the superscript (M) will be omitted in the remaining development (Fig. 6.2).

$$M_{\mathrm{EA}} = 2\, L\, e\, b + M_{\mathrm{AC}} = 2\, q\, S\, e\, b\, C_{L\alpha} \quad a(t) \tag{6.3}$$

Flutter equations:

$$m\ddot{h} + S_{\alpha}\dot{\alpha} + K_h h + qSC_{L_\alpha}\alpha = 0 \tag{6.4}$$

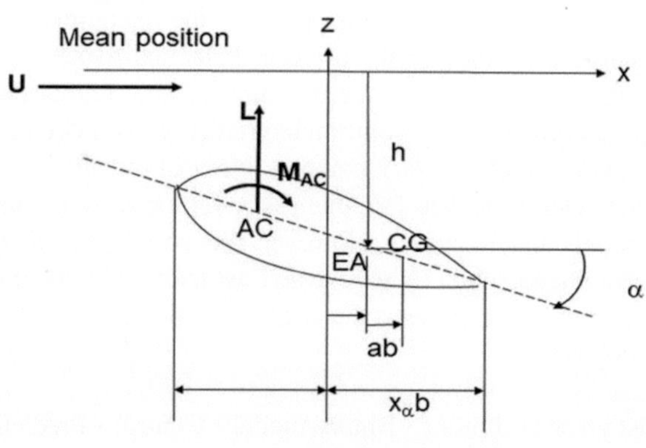

Fig. 6.1 An airfoil with two-degree of freedom: bending and torsion (binary system)

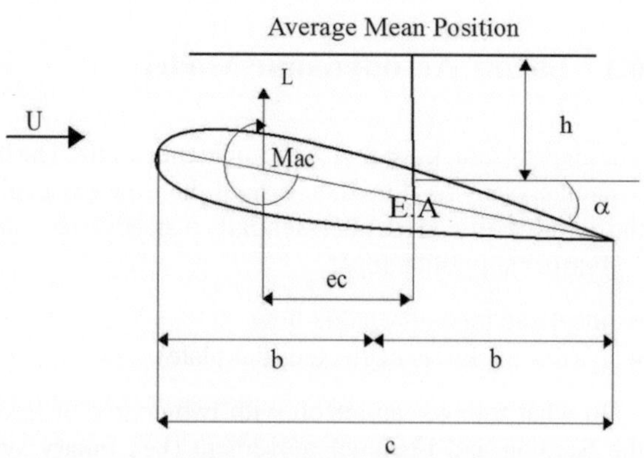

Fig. 6.2 Typical section representation of an airfoil with two-degree of freedom: bending and torsion

$$S_\alpha \ddot{h} + I_\alpha \ddot{\alpha} + K_\alpha \alpha - 2qSebC_{L_\alpha} \alpha = 0 \quad (6.5)$$

or in matrix notation:

$$[M]\{\ddot{x}\} + ([K] - q[A_0])\{x\} = 0 \quad (6.6)$$

where

$$[M] = \begin{bmatrix} m & S_\alpha \\ S_\alpha & I_\alpha \end{bmatrix} \text{ (inertia)} \quad (6.7)$$

$$K = \begin{bmatrix} K_h & 0 \\ 0 & K_\alpha \end{bmatrix} \text{ (structural stiffness)} \quad (6.8)$$

$$[A_0] = \begin{bmatrix} 0 & -SC_{L\alpha} \\ 0 & 2SebC_{L\alpha} \end{bmatrix} \text{ (aerodynamic stiffness)} \quad (6.9)$$

Essential for all flutter characteristics is *asymmetry of aerodynamic matrix*. (Compare with the skew-symmetric gyroscopic matrices of rotating mechanical systems).

Analysis of flutter stability can be carried out by assuming harmonic motion:

$$h = \hat{h}\, e^{\mathrm{pt}}, \quad (6.10a)$$

$$\alpha = \hat{\alpha}\, e^{\mathrm{pt}}, \quad (6.10b)$$

or

$$\{x\} = \{\hat{x}\}e^{\mathrm{pt}}, \quad (6.10c)$$

and then solve the resulting eigenvalue problem.

Substitution leads to

$$\left(mp^2 + K_h\right)\hat{h} + \left(S_\alpha p^2 + qSC_{L_\alpha}\right)\hat{\alpha} = 0 \quad (6.11)$$

$$S_\alpha p^2 \hat{h} + \left(I_\alpha p^2 + K_\alpha - 2qSebC_{L_\alpha}\right)\hat{\alpha} = 0 \quad (6.12)$$

and then solve the determinant equation

$$[M]p^2 + ([K] - q[A_0]) = 0 \quad (6.13a)$$

$$\left|[M]p^2 + ([K] - q[A_0])\right| = 0 \quad (6.13b)$$

The determinant of the set of Eqs. 6.11 and 6.12 is:

$$\left(I_\alpha m - S_\alpha^2\right)p^4 + \left(m\left(K_\alpha - 2qSebC_{L_\alpha}\right) + K_h I_\alpha + S_\alpha q S C_{L_\alpha}\right)p^2 + \left(K_h K_\alpha - 2qSebC_{L_\alpha} K_h\right) = 0 \tag{6.14a}$$

Reduction leads to the characteristic equation:

$$a_4 p^4 + a_2 p^2 + a_3 = 0 \tag{6.14b}$$

where

$$a_4 = \left(I_\alpha m - S_\alpha^2\right) \tag{6.15}$$

$$a_2 = \left(m\left(K_\alpha - 2qSebC_{L_\alpha}\right) + I_\alpha K_h + S_\alpha q S C_{L_\alpha}\right) = mK_\alpha + I_\alpha K_h - (2meb - S_\alpha)qSC_{L_\alpha} \tag{6.16}$$

$$a_0 = \left(K_\alpha K_h - 2qSebC_{L_\alpha} K_h\right) \tag{6.17}$$

Equation 6.14 is fourth-order polynomial (or second order in p^2) with 4 roots:

$$p_{1,2,3,4} = (\sigma + i\omega)_{1,2,3,4} = \pm\sqrt{\frac{1}{2a_4}\left(-a_2 \pm \sqrt{a_2^2 - 4a_4a_0}\right)} \tag{6.18}$$

Solutions are of the type:

$$\{x\} = \{\hat{x}\}e^{\sigma t}e^{i\omega t} \tag{6.19}$$

representing motion for which:

σ damping (decay).
ω frequency (rad/s).
$\{\hat{x}\}$ vibration mode.

The stability of motion is indicated by the value of σ

$\sigma > 0$ unstables
$\sigma = 0$ neutrals
$\sigma < 0$ stable

The following diagram and tables are reconstructed from the results of study carried out by Zwaan [1], Done [2, 3], showing;

- condition of coefficients
- type of solution and physical meaning

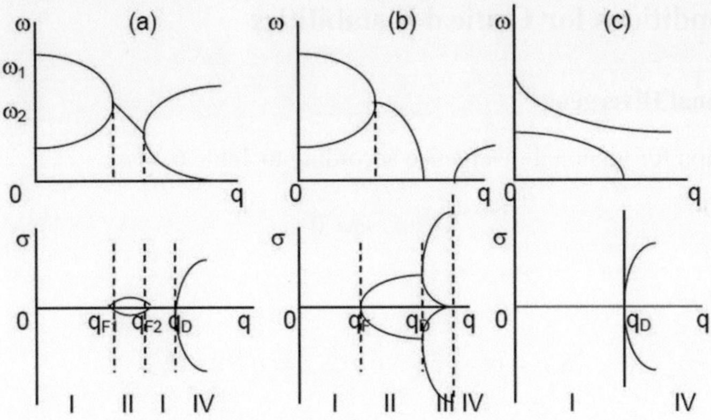

Fig. 6.3 Elementary examples of flutter diagram

- classification into stability categories.

All these stability categories are visualized by.

- roots in complex plane
- flutter diagrams (Fig. 6.3).

Flutter diagrams:

Usual presentation in practice embraces plots of roots:

- frequency ω versus speed or dynamic pressure
- damping σ versus speed or dynamic pressure.

Difference between flutter characteristics is caused by different coefficients a_4, a_2, a_0.

Three cases are shown:
At $q = 0$, solutions represent eigenfrequencies of the binary system.
At $q > 0$, there are three cases.

- Case a: - first instability is flutter at $q_{F!}$
  - stability recovered at q_{F2}
  - divergence at q_D
- Case b: - first instability is flutter at q_{F2}
  - transition to divergence at q_D
- Case c: - first instability is divergence at q_D

Solutions with negative frequencies have been deleted.
Only critical (=lowest) dynamic pressure or speeds have practical significance!

6.2 Conditions for Critical Instabilities

a. **Torsional Divergence**

Condition for torsional divergence according to Table 6.1 is:

$$a_0 <= 0 \tag{6.20}$$

From 6.17,

$$K_\alpha = \text{or} < 2q\,S\,e\,b\,C_{L\alpha} = q\,S\,e\,c\,C_{L\alpha} \tag{6.21}$$

Hence

$$q_D = \frac{K_\alpha}{ecSC_{L\alpha}} \tag{6.22}$$

which was already found in 5.16.

b. **Flutter**

Condition for flutter according to Table 6.1:

$$a_2^2 - a_4 a_0 < 0 \tag{6.23}$$

The conditions that will be studied here are to find out what locations of EA and CG will lead to flutter.

The condition represented by Eq. 6.23 can be written out in full using the typical section (airfoil) characteristics, i.e. using Eqs. 6.16 and 6.17:

Table 6.1 Types of instability of a typical airfoil, for steady aerodynamic model

$a_2^2 - 4a_4a_0$	> 0			< 0
a_0	> 0		< 0	$>< 0$
a_2	> 0	< 0	$>< 0$	$>< 0$
p2	$-\omega_1^2, -\omega_2^2$	σ_1^2, σ_2^2	$\sigma^2, -\omega^2$	$-g + ih$
p	$\pm i\omega_1, \pm i\omega_2$	$\pm i\sigma_1, \pm i\sigma_2$	$\pm \sigma, \pm i\omega$	$\pm \sigma, \pm i\omega$
Type of motion	Harmonic	Aperiodic	Aperiodic	Oscillatory
	2 pos freq	2 diverging	1 diverging	1 div.pos.freq
	2 neg freq	2 converging	1 converging	1 conv.pos.freq
			Harmonic	1 div.neg.freq
			1 pos freq	1 conv.neg.frq
Type of instability	Neutral	Divergent	Divergent	Flutter
Category	I	III	IV	II

For this purpose recall that:

$$\begin{aligned}
S_\alpha &= \mathrm{mx}_\alpha b \\
I_\alpha &= \mathrm{mr}_\alpha^2 b^2 \\
K_h &= m\omega_h^2 \\
K_\alpha &= I_\alpha \omega_\alpha^2 = \mathrm{mr}_\alpha^2 b^2 \omega_\alpha^2
\end{aligned} \tag{6.24}$$

and

$$a_4 = \mathrm{mb}^2 r_\alpha^2 \left(1 - \frac{x_\alpha^2}{r_\alpha^2}\right) \tag{6.25}$$

$$\begin{aligned}
a_2 &= m^2 b^2 r_\alpha^2 \omega_\alpha^2 \left[1 + \frac{\omega_h^2}{\omega_\alpha^2}\right] - m^2 b^2 r_\alpha^2 \omega_\alpha^2 (2e + x_\alpha) \frac{qSb}{K_\alpha} C_{L_\alpha} \\
&= m^2 b^2 r_\alpha^2 \omega_\alpha^2 \left[1 + \frac{\omega_h^2}{\omega_\alpha^2} - (2e + x_\alpha) Q\right]
\end{aligned} \tag{6.26}$$

where

$$Q \equiv \frac{qSb}{K_\alpha} C_{L_\alpha} \tag{6.27}$$

and

$$a_0 = m^2 b^2 r_\alpha^2 \omega_\alpha^2 \omega_h^2 [1 - 2eQ] \tag{6.28}$$

Substituting these values in 6.23 and canceling out

$$m^4 b^4 r_\alpha^4 \omega_\alpha^4$$

which is > 0, then obtain

$$\begin{aligned}
&\left[1 - \left(\frac{\omega_h}{\omega_\alpha}\right)^2 - \frac{qSb}{K_\alpha} C_{L_\alpha} (x_\alpha + 2e)\right]^2 \\
&\quad - 4\left(1 - \frac{x_\alpha^2}{r_\alpha^2}\right)\left(\frac{\omega_h}{\omega_\alpha}\right)^2 \left(1 - 2\frac{qSb}{K_\alpha} C_{L_\alpha} e\right) \le 0
\end{aligned} \tag{6.29}$$

Equation is quadratic in

$$Q = \frac{qSb}{K_\alpha} C_{L_\alpha}$$

then obtain:

$$\left[1+\left(\frac{\omega_h}{\omega_\alpha}\right)^2\right]^2 - 2(2e+x_\varepsilon)\left[1+\left(\frac{\omega_h}{\omega_\alpha}\right)^2\right]Q + (2e+x_\alpha)^2 Q^2$$
$$-4\left(1-\frac{x_\alpha^2}{r_\alpha^2}\right)\left(\frac{\omega_h}{\omega_\alpha}\right)^2 + 8e\left(1-\frac{x_\alpha^2}{r_\alpha^2}\right)\left(\frac{\omega_h}{\omega_\alpha}\right)^2 Q \le 0 \tag{6.30}$$

or

$$C_2 Q^2 + C_1 Q + C_0 \le 0 \tag{6.31}$$

where

$$C_2 = \left(2e - x_\alpha\right)^2 \tag{6.32}$$

$$C_1 = -2(2e+x_\alpha)\left[1+\frac{\omega_h^2}{\omega_\alpha^2}\right] + 8e\left(1-\frac{x_\alpha^2}{r_\alpha^2}\right)\left(\frac{\omega_h^2}{\omega_\alpha^2}\right)$$

or

$$C_1 = -2\left(2e + x_\alpha\left(\frac{\omega_h}{\omega_\alpha}\right)^2\left[x_\alpha - 2e + 4e\frac{x_\alpha^2}{r_\alpha^2}\right]\right) \tag{6.33}$$

and

$$C_0 = \left[1+\left(\frac{\omega_h}{\omega_\alpha}\right)^2\right]^2 - 4\left(\frac{\omega_h}{\omega_\alpha}\right)^2 + 4\left(\frac{x_\alpha^2}{r_\alpha^2}\right)\left(\frac{\omega_h}{\omega_\alpha}\right)^2$$

or

$$C_0 = \left[1-\left(\frac{\omega_h}{\omega_\alpha}\right)^2\right]^2 + 4\left(\frac{x_\alpha^2}{r_\alpha^2}\right)\left(\frac{\omega_h}{\omega_\alpha}\right)^2 \tag{6.34}$$

The roots of the quadratic Eq. 6.30 are given by

$$Q_1, Q_2 = -\frac{C_1}{2C_2} \pm \frac{1}{2C_2}\sqrt{C_1^2 - 4C_0C_2} \tag{6.35a}$$

or

$$Q_1, Q_2 = -\frac{1}{2\left[(2e - x_\alpha)^2\right]}\left[\begin{array}{l} -2\left(2e + x_\alpha + \left(\frac{\omega_h}{\omega_\alpha}\right)^2\left[x_\alpha - 2e + 4e\frac{x_\alpha^2}{r_\alpha^2}\right]\right) \\ \pm\sqrt{\begin{array}{l}\left[-2\left(2e + x_\alpha + \left(\frac{\omega_h}{\omega_\alpha}\right)^2\left[x_\alpha - 2e + 4e\frac{x_\alpha^2}{r_\alpha^2}\right]\right)\right]^2 \\ -4\left[\left[1 - \left(\frac{\omega_h}{\omega_\alpha}\right)^2\right]^2 + 4\left(\frac{x_\alpha^2}{r_\alpha^2}\right)\left(\frac{\omega_h}{\omega_\alpha}\right)^2\right] \\ \left[(2e - x_\alpha)^2\right]\end{array}}\end{array}\right] \quad (6.35)$$

To ensure that Q_1 and Q_2 have realistic values, so that flutter will be possible, the following conditions should be satisfied:

1. Q_1 and Q_2 should be real. Then the value of the terms within the square root sign should be positive:

$$x_\alpha\left(\frac{\omega_h}{\omega_\alpha}\right)^2\left(1 - \frac{x_\alpha^2}{r_\alpha^2}\right)\left[x_\alpha + 2e - 2e\left(\frac{\omega_h}{\omega_\alpha}\right)^2\left(1 + 2e\frac{x_\alpha}{r_\alpha^2}\right)\right] > 0 \quad (6.36)$$

or, since

$$\left(\frac{\omega_h}{\omega_\alpha}\right)^2\left(1 - \frac{x_\alpha^2}{r_\alpha^2}\right) \geq 0 \quad (6.37)$$

then

$$x_\alpha\left[x_\alpha + 2e - 2e\left(\frac{\omega_h}{\omega_\alpha}\right)^2\left(1 + 2e\frac{x_\alpha}{r_\alpha^2}\right)\right] > 0 \quad (6.38)$$

2. Q_1 and Q_2 should be non-negative.

Following the elementary rule,

$$Q_1 + Q_2 = -C_1/C_2 \quad (6.39)$$

Since $C_2 \geq 0$, this condition implies that

$$\left[x_\alpha + 2e + \left(\frac{\omega_h}{\omega_\alpha}\right)^2\left(x_\alpha - 2e + 4e\frac{x_\alpha^2}{r_\alpha^2}\right)\right] > 0 \quad (6.40)$$

These two conditions should be satisfied simultaneously in order for flutter to occur.

We then proceed to find out at the locations of EA and CG that will lead to flutter (Fig. 6.4).

Fig. 6.4 Moment and lift on a typical section for CG and EA position considerations

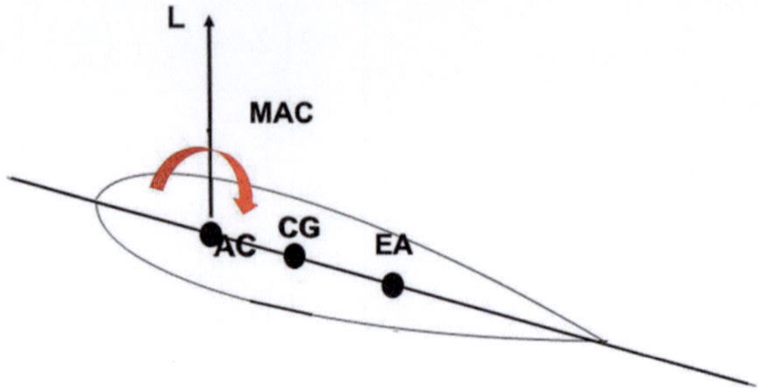

a. **CG ahead of EA (i.e. $x_\alpha < 0$)**

From condition 1:

$$x_\alpha\left[x_\alpha + 2e - 2e\left(\frac{\omega_h}{\omega_\alpha}\right)^2\left(1 + 2e\frac{x_\alpha}{r_\alpha^2}\right)\right] > 0 \tag{6.41}$$

or since $x_\alpha < 0$

$$\left[x_\alpha + 2e - 2e\left(\frac{\omega_h}{\omega_\alpha}\right)^2\left(1 + 2e\frac{x_\alpha}{r_\alpha^2}\right)\right] < 0 \tag{6.42}$$

$$x_\alpha + 2e - 2e\left(\frac{\omega_h}{\omega_\alpha}\right)^2 < 4e^2\left(\frac{\omega_h}{\omega_\alpha}\right)^2\frac{x_\alpha}{r_\alpha^2} \tag{6.43}$$

From condition 2:

$$x_\alpha + 2e + \left(\frac{\omega_h}{\omega_\alpha}\right)^2\left(x_\alpha - 2e + 4e\frac{x_\alpha^2}{r_\alpha^2}\right) > 0 \tag{6.44}$$

or

$$x_\alpha + 2e - 2e\left(\frac{\omega_h}{\omega_\alpha}\right)^2 + \left(\frac{\omega_h}{\omega_\alpha}\right)^2\left(x_\alpha + 4e\frac{x_\alpha^2}{r_\alpha^2}\right) \geq 0 \tag{6.45}$$

Substitution in condition 2:

$$4e^2\left(\frac{\omega_h}{\omega_\alpha}\right)^2\frac{x_\alpha}{r_\alpha^2} + \left(\frac{\omega_h}{\omega_\alpha}\right)^2\left(x_\alpha + 4e\frac{x_\alpha^2}{r_\alpha^2}\right) \geq 0 \tag{6.46}$$

or after division with $\left(\frac{\omega_h}{\omega_\alpha}\right)^2\frac{1}{r_\alpha^2}$ which is positive

$$4e^2x_\alpha + x_\alpha r_\alpha^2 + 4\mathrm{e}\mathrm{x}_\alpha^2 \geq 0 \tag{6.47}$$

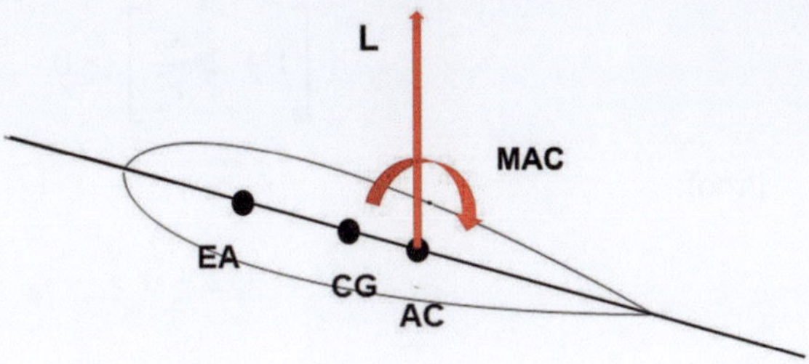

Fig. 6.5 Moment and lift on a typical section; AC behind EA

After division with x_α which is negative

$$4e^2 + r_\alpha^2 + 4ex_\alpha \le 0$$
$$4e^2 + x_\alpha^2 + 4ex_\alpha + r_\alpha^2 - x_\alpha^2 \le 0$$
$$(2e + x_\alpha)^2 + r_\alpha^2 - x_\alpha^2 \le 0 \tag{6.48}$$

since

$$(2e + x_\alpha)^2 \ge 0 \tag{6.49}$$

and

$$r_\alpha^2 + x_\alpha^2 \ge 0 \tag{6.50}$$

then conditions 1 and 2 cannot be satisfied. Hence no flutter is possible.

b. **The aerodynamic center AC is behind EA (i.e. e < 0)**

But ahead of the center of gravity CG ($2e + x_\alpha > 0$)

This location implies $x_\alpha > 0$ (Fig. 6.5).

From condition 1, then:

$$\left[x_\alpha + 2e - 2e\left(\frac{\omega_h}{\omega_\alpha}\right)^2\left(1 + 2e\frac{x_\alpha}{r_\alpha^2}\right)\right] \ge 0 \tag{6.51}$$

Since

$$[x_\alpha + 2e] \ge 0 \tag{6.52}$$

$$\left[2e\left(\frac{\omega_h}{\omega_\alpha}\right)^2\right] \ge 0 \tag{6.53}$$

and

$$\left[1+2e\frac{x_\alpha}{r_\alpha^2}\right]\geq 0 \tag{6.54}$$

Proof:

$$2e = 1/2 + a > -x_\alpha$$

Thus

$$1+2e\frac{x_\alpha}{r_\alpha^2} > 1-\frac{x_\alpha^2}{r_\alpha^2} > 0$$

as $x_\alpha < r_\alpha$

Hence condition 1 can be satisfied.

From condition 2:

$$\left[\underbrace{x_\alpha+2e}_{>0}+\underbrace{\left(\frac{\omega_h}{\omega_\alpha}\right)^2}_{>0}\left(\underbrace{2e+x_\alpha}_{>0}-4e\left(1-\frac{x_\alpha^2}{r_\alpha^2}\right)\right)\right]\geq 0 \tag{6.55}$$

Then condition 2 can also be satisfied.

Conclusion
Case b allows flutter to take place (flutter is possible).

Similar studies have been performed by Lind and Brenner [4], Zimmerman and Weissenburger [5] and, Rheinfurth and Swift [6].

6.3 Physical Explanation of Flutter Mechanism

Consider flutter solution for $q > q_F$.

Two aspects will be considered.

- Phase differences in flutter vibration modes
- Work done by aerodynamic forces

6.3.1 Phase Differences in Flutter Vibration Modes

From the eigenvalue Eqs. 6.16 and 6.17 are representing the equations of motion of the typical airfoil as a binary system

$$\left(mp^2+K_h\right)\hat{h}+\left(S_\alpha p^2+qSC_{L_\alpha}\right)\hat{\alpha}=0 \tag{6.11}$$

$$S_\alpha p^2 \hat{h} + \left(I_\alpha p^2 + K_\alpha - 2qSebC_{L_\alpha}\right)\hat{\alpha} = 0 \tag{6.12}$$

one could write the response of the airfoil expressed as the vibration modes $\hat{h}$ and $\hat{\alpha}$ as the amplitude of the heaving and pitching response:

$$\frac{\hat{h}}{b\hat{\alpha}} = \frac{\left(S_\alpha p^2 + qSC_{L_\alpha}\right)}{\left(mp^2 + K_h\right)b} \tag{6.56}$$

$$\frac{\hat{h}}{b\hat{\alpha}} = \frac{\left(I_\alpha p^2 + K_\alpha - 2qSebC_{L_\alpha}\right)}{S_\alpha p^2 b} \tag{6.57}$$

and the determinant

$$\begin{bmatrix} \left(mp^2 + K_h\right) & \left(S_\alpha p^2 + qSC_{L\alpha}\right) \\ \left(S_\alpha p^2\right) & \left(I_\alpha p^2 + K_\alpha - 2qSebC_{L_\alpha}\right) \end{bmatrix} = 0 \tag{6.58}$$

According to Table 6.1 (row 4 column 5) the following flutter solutions prevail

$$p^2 = g + ih \quad \text{(unstable)} \tag{6.59}$$

$$p^2 = -g - ih \quad \text{(stable)} \tag{6.60}$$

Consequences for flutter vibration modes will be investigated using vector diagrams, as sketched below (Figs. 6.6 and 6.7).

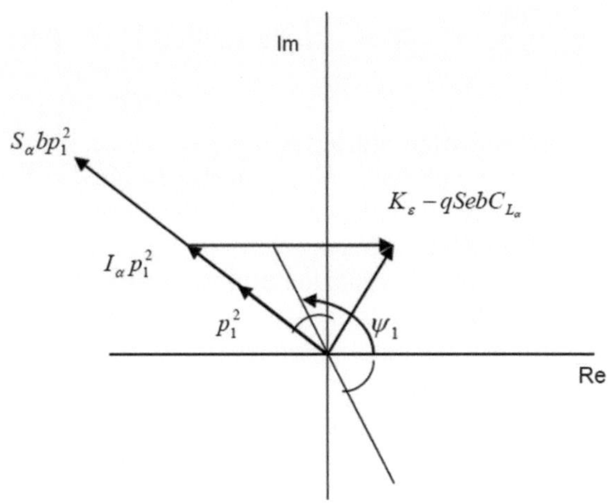

Fig. 6.6 A schematic of vector diagram for investigating the flutter vibration modes as Solution of 6.58

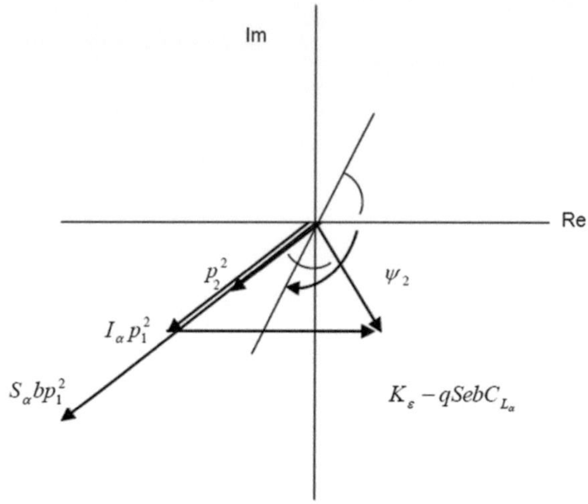

Fig. 6.7 A schematic of the representation of the elements of the determinant 6.58

$$h \text{ leads } \alpha, \quad \text{i.e.} \psi_1 > 0$$
$$h \text{ lags } \alpha, \quad \text{i.e.} \psi_1 < 0$$

6.3.2 Work Done by Aerodynamic Forces

In the present model of typical section, only lift can perform network over one cycle.

$$W = -\int_0^T \mathrm{Re}(L)\mathrm{dRe}(h) = -\int_0^T \mathrm{Re}(L)\mathrm{Re}\big(\dot{h}\big)\mathrm{d}t \tag{6.61}$$

L is positive upward, h is positive downward, so that $W > 0$ corresponds to positive work done by lift force. M_{EA} is in phase with α, so that zero work is done.

The following figures preset the time histories of $L(\alpha)$, h and $\dot{h}$, together with intervals in which aerodynamic work is positive (Fig. 6.8).

Alternative representation of time histories of lift force and motion during one cycle (Fig. 6.9).

Torsional motion α generates lift force $L(\alpha)$. If $L(\alpha)$ is directed in the same direction as $\dot{h}$, then positive work is done.

Flow acts at energy reservoir, when aerodynamic forces perform positive work. Energy flows from airflow into wing system, if work is negative, energy is extracted from wing system.

Remarks The above results are valid only for present simplified aerodynamic model. For more advanced models, the validity is judged from the trend.

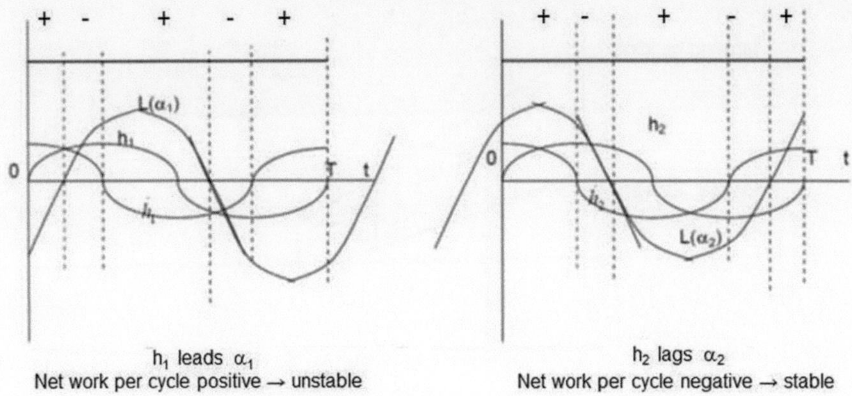

Fig. 6.8 Schematics of the solution diagram

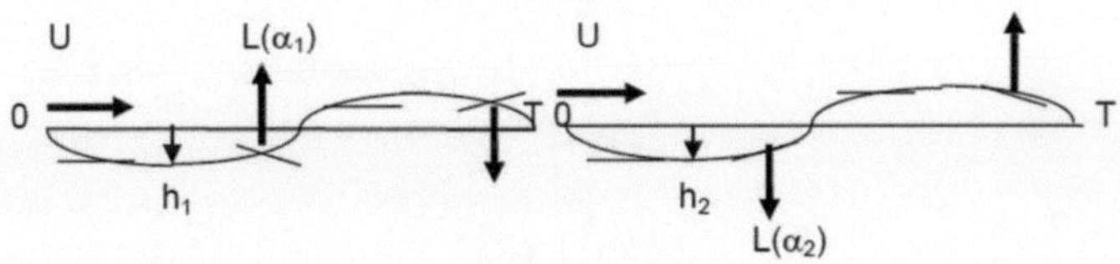

Fig. 6.9 Time histories of lift force and motion during one cycle

6.4 Example of Typical Section with Steady Aerodynamic Model

Consider a typical section of modern transport wing proposed by Isogai [7] (Fig. 6.10).

$$a = -2.0 \qquad C_{L_\alpha} = 2.0\pi$$
$$x_\alpha = 1.8 \qquad e = \text{-}0.75$$
$$r_\alpha^2 = 3.48 \qquad \mu = 100$$
$$\frac{\omega_h}{\omega_\alpha} = 1.0$$

The solution in the form of flutter diagram is shown in figures, in which $\frac{\omega_h}{\omega_\alpha}$, $\frac{\sigma}{\omega_\alpha}$ and $\frac{\hat{h}}{b\alpha}$ are shown as functions of $\frac{U}{b\omega_\alpha}$.

Flutter solutions with negative frequencies and with positive damping have been deleted. Flutter diagram of Isogai typical section using steady aerodynamic model is exhibited in Fig. 6.11 which qualitatively agrees with Zwaan's [1] result.

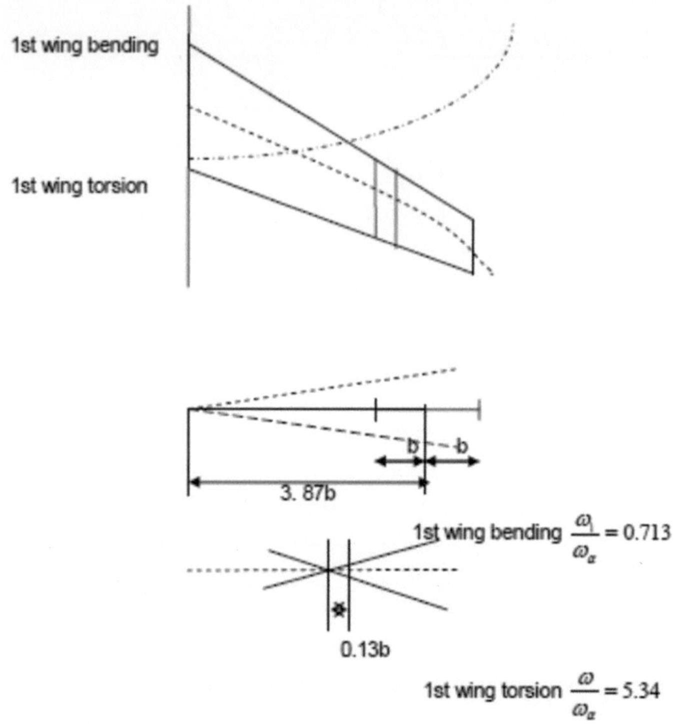

Fig. 6.10 State of affairs of wing deformation of Isogai Wing Example

6.5 Low-Frequency Refinement of Aerodynamic Model

First-order refinement of aerodynamic model can be achieved by replacing instantaneous incidence angle by instantaneous angle of attack α_{dyn}:

$$\alpha_{\text{dyn}}(t) = \frac{\dot{h}(t)}{U} + \alpha(t) \tag{6.62}$$

Note that the aerodynamic model utilized remains steady state, but the angle of attack has been extended to include angle of attack due to the heaving motion of the typical section.

Aerodynamic model is very common in the dynamic stability studies of complete aircraft.

Then the aerodynamic forces become:

$$L(t) = qSC_{L_\alpha}\left(\frac{\dot{h}(t)}{U} + \alpha(t)\right) \tag{6.63}$$

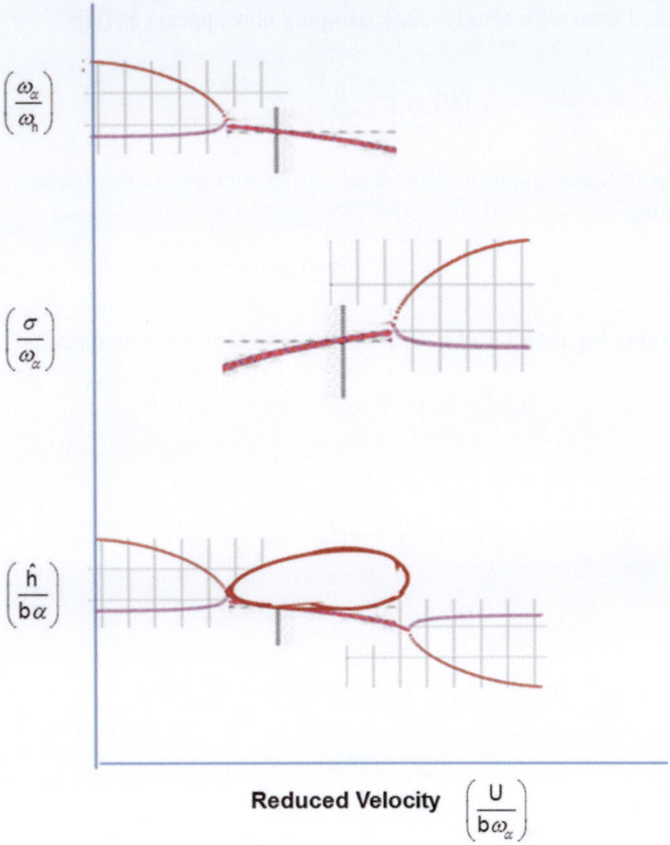

Fig. 6.11 Qualitative flutter diagram of typical section using steady aerodynamic model [1, 8, 9]

$$M_{\mathrm{EA}}(t) = 2qSebC_{L_\alpha}\left(\frac{\dot{h}(t)}{U} + \alpha(t)\right) \tag{6.64}$$

since

$$M_{\mathrm{AC}}(t) = 0 \tag{6.65}$$

and

$$\mathrm{ec} = 2\mathrm{eb} \tag{6.66}$$

Equation of motion becomes:

$$[M]\{\ddot{x}\} - [B]\{\dot{x}\} + ([K] - q[A_0])\{x\} = 0 \tag{6.67}$$

Additional term with aerodynamic damping now appears as [B]:

$$[B] = \begin{bmatrix} \frac{q}{U} SC_{L_\alpha} & 0 \\ -2\frac{q}{U} SebC_{L_\alpha} & 0 \end{bmatrix} \quad (6.68)$$

Analysis of flutter stability can now be carried out by assuming harmonic motion, i.e. by putting

$$\{x\} = \{\hat{x}\} e^{pt} \quad (6.69)$$

Substituting Eq. 6.69 into Eq. 6.67 leads to the characteristic equation:

$$a_4 p^4 + a_3 p^3 + a_2 p^2 + a_1 p + a_0 = 0 \quad (6.70)$$

where

$$a_4 = m I_\alpha - S_\alpha^2 \quad (6.71)$$

$$a_3 = \frac{q}{U} SC_{L\alpha}(2ebS_\alpha + I_\alpha) \quad (6.72)$$

$$a_2 = mK_\alpha + I_\alpha K_h - (2meb + S_\alpha) q SC_{L\alpha} \quad (6.73)$$

$$a_1 = \frac{q}{U} SC_{L_\alpha} K_\alpha \quad (6.74)$$

$$a_0 = K_h(K_\alpha - 2qSebC_{L\alpha}) \quad (6.75)$$

Equation 6.70 is a full fourth-order polynomial.

6.6 Example of Typical Section with Low-Frequency Aerodynamic Model

Consider the same typical section as in the steady aerodynamic case, using Isogai typical section. Following the same procedure as outlined in preceding section, the solution is obtained as exhibited by the flutter diagram of Isogai typical section using low-frequency aerodynamic model in Fig. 6.12. Flutter diagram of Isogai typical section using low-frequency refinement aerodynamic model as exhibited in Fig. 6.13.

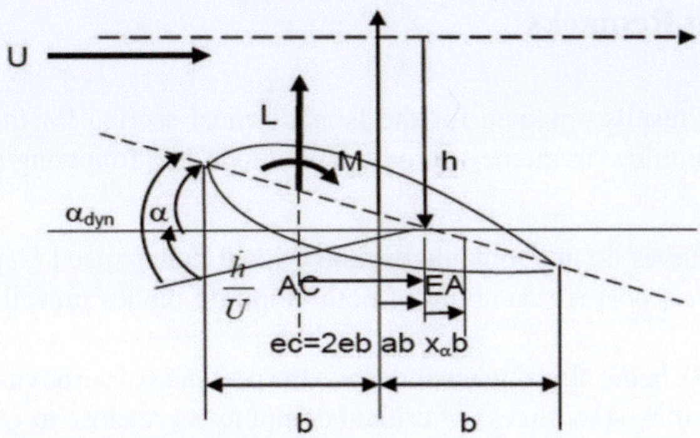

Fig. 6.12 Typical section aerodynamic model for low-frequency refinement

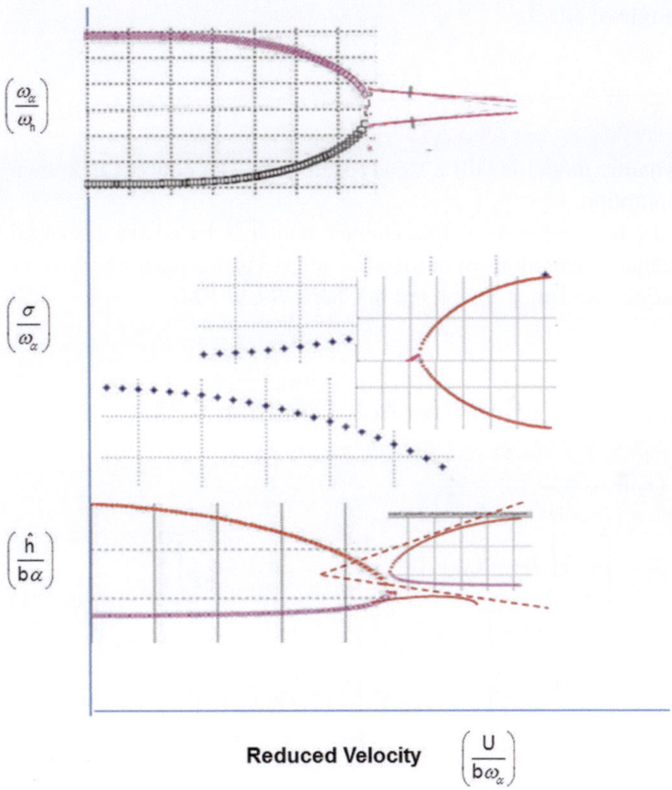

Fig. 6.13 Flutter diagram of Isogai typical section using low-frequency refinement of the aerodynamic model [1, 8, 9]

6.7 Closing Remarks

Comparing the results obtained for the Isogai typical section for the steady aerodynamic case and low-frequency aerodynamic mode, the following remarks are in order:

- Frequency curves do not coincide beyond critical flutter speed $U_F/b\omega_\alpha$
- Below $U_F/b\omega_\alpha$ positive damping of both vibration modes prevail (i.e., negative σ)
- $U_F/b\omega_\alpha \approx 20$, being slightly lower despite the presence of aerodynamic damping. This is caused by smoothness of critical damping curve close to $U_F/b\omega_\alpha$.

The last remark emphasizes the need for a complete aerodynamic modeling to trace possibly lower critical flutter speeds. One such possibility is completing dynamic angle of attack:

$$\alpha_{\text{dyn}}(t) = \frac{\dot{h}(t)}{U} + \alpha(t) - \frac{x - \text{ab}}{U}\dot{\alpha}(t) \tag{6.76}$$

Aerodynamic model is still a steady-state one. The concept is known as quasi-steady assumption.

The procedure outlined in this chapter which is based on the solution of the binary aeroelastic equations of motion 6.4 and 6.5 using quasi-steady or lower-order approximations is further elaborated in Chaps. 9 and 10.

Appendix: Verification/Proof oF Equation 6.35

$$Q_1, Q_2 = -\frac{1}{2\left[(2e - x_\alpha)^2\right]} \left[\begin{array}{l} -2\left(2e + x_\alpha + \left(\frac{\omega_h}{\omega_\alpha}\right)^2\left[x_\alpha - 2e + 4e\frac{x_\alpha^2}{r_\alpha^2}\right]\right) \\ \pm\sqrt{\dfrac{\left[-2\left(2e + x_\alpha + \left(\frac{\omega_h}{\omega_\alpha}\right)^2\left[x_\alpha - 2e + 4e\frac{x_\alpha^2}{r_\alpha^2}\right]\right)\right]^2 - 4\left[\left[1 - \left(\frac{\omega_h}{\omega_\alpha}\right)^2\right]^2 + 4\left(\frac{x_\alpha^2}{r_\alpha^2}\right)\left(\frac{\omega_h}{\omega_\alpha}\right)^2\right]}{\left[(2e - x_\alpha)^2\right]}} \end{array} \right] \tag{6.35a}$$

or

$$Q_1, Q_2 = -\frac{1}{\left[(2e - x_\alpha)^2\right]}\left[\begin{array}{c} -\left(2e + x_\alpha + \left(\frac{\omega_h}{\omega_\alpha}\right)^2\left[x_\alpha - 2e + 4e\frac{x_\alpha^2}{r_\alpha^2}\right]\right) \\ \pm\sqrt{\begin{array}{c}\left[-\frac{1}{2}\left(2e + x_\alpha + \left(\frac{\omega_h}{\omega_\alpha}\right)^2\left[x_\alpha - 2e + 4e\frac{x_\alpha^2}{r_\alpha^2}\right]\right)\right]^2 \\ -\left[\left[1 - \left(\frac{\omega_h}{\omega_\alpha}\right)^2\right]^2 + 4\left(\frac{x_\alpha^2}{r_\alpha^2}\right)\left(\frac{\omega_h}{\omega_\alpha}\right)^2\right]\left[(2e - x_\alpha)^2\right]\end{array}}\end{array}\right] \tag{6.35b}$$

$$Q_1, Q_2 = \frac{1}{2\left[(x_\alpha + 2e)^2\right]}\left[\begin{array}{l} \left\{\left[1 + \left(\frac{\omega_h}{\omega_\alpha}\right)^2\right](x_\alpha + 2e) - 4e\left(\frac{\omega_h}{\omega_\alpha}\right)^2\left(1 - \frac{x_\alpha^2}{r_\alpha^2}\right)\right\} \\ \pm\sqrt{x_\alpha\left(\frac{\omega_h}{\omega_\alpha}\right)^2\left(1 - \frac{x_\alpha^2}{r_\alpha^2}\right)} \\ \left[x_\alpha + 2e - 2e\left(\frac{\omega_h}{\omega_\alpha}\right)^2\left(1 + 2e\frac{x_\alpha}{r_\alpha^2}\right)\right]\end{array}\right] \tag{6.35c}$$

Verification

$$Q_1, Q_2 = -\frac{C_1}{2C_2} \pm \frac{1}{2C_2}\sqrt{C_1^2 - 4C_0C_2} = \frac{1}{C_2}\left[-\frac{C_1}{2} \pm \sqrt{\frac{C_1}{4}^2 - C_0C_2}\right] \tag{6.35d}$$

$$C_0 = \left[1 + \left(\frac{\omega_h}{\omega_\alpha}\right)^2\right]^2 - 4\left(\frac{\omega_h}{\omega_\alpha}\right)^2 + 4\left(\frac{x_\alpha^2}{r_\alpha^2}\right)\left(\frac{\omega_h}{\omega_\alpha}\right)^2$$

or

$$C_0 = \left[1 - \left(\frac{\omega_h}{\omega_\alpha}\right)^2\right]^2 + 4\left(\frac{x_\alpha^2}{r_\alpha^2}\right)\left(\frac{\omega_h}{\omega_\alpha}\right)^2 \tag{6.34}$$

$$C_1 = -2\left(2e + x_\alpha + \left(\frac{\omega_h}{\omega_\alpha}\right)^2\left[x_\alpha - 2e + 4e\frac{x_\alpha^2}{r_\alpha^2}\right]\right) \tag{6.33}$$

or

$$C_1 = -2(2e + x_\alpha)\left[1 + \frac{\omega_h^2}{\omega_\alpha^2}\right] + 8e\left(1 - \frac{x_\alpha^2}{r_\alpha^2}\right)\left(\frac{\omega_h^2}{\omega_\alpha^2}\right)$$

$$C_2 = \left(2e - x_\alpha\right)^2 \tag{6.32}$$

$$\begin{aligned}
-\frac{C_1}{2} &= \left(2e + x_\alpha + \left(\frac{\omega_h}{\omega_\alpha}\right)^2\left[x_\alpha - 2e + 4e\frac{x_\alpha^2}{r_\alpha^2}\right]\right)\\
&= \left\{(x_\alpha + 2e) + (x_\alpha - 2e)\left(\frac{\omega_h}{\omega_\alpha}\right)^2 + 4e\left(\frac{\omega_h}{\omega_\alpha}\right)^2\frac{x_\alpha^2}{r_\alpha^2}\right\}\\
&= \left\{(x_\alpha - 2e)\left(1 + \left(\frac{\omega_h}{\omega_\alpha}\right)^2\right) + 4e\left(\frac{\omega_h}{\omega_\alpha}\right)^2\frac{x_\alpha^2}{r_\alpha^2}\right\}
\end{aligned} \tag{6.33}$$

$$\begin{aligned}
-C_0C_2 &= -\left(2e - x_\alpha\right)^2\left\{\left[1 - \left(\frac{\omega_h}{\omega_\alpha}\right)^2\right]^2 + 4\left(\frac{x_\alpha^2}{r_\alpha^2}\right)\left(\frac{\omega_h}{\omega_\alpha}\right)^2\right\}\\
&= -\left\{\left(2e - x_\alpha\right)^2\left[1 - \left(\frac{\omega_h}{\omega_\alpha}\right)^2\right]^2 + 4\left(2e - x_\alpha\right)^2\left(\frac{x_\alpha^2}{r_\alpha^2}\right)\left(\frac{\omega_h}{\omega_\alpha}\right)^2\right\}
\end{aligned} \tag{6.34}$$

$$\begin{aligned}
\frac{C_1^2}{4} - C_0C_2 = &\left\{(x_\alpha - 2e)^2\left(1 + \left(\frac{\omega_h}{\omega_\alpha}\right)^2\right)^2\right.\\
&\left. + 8e(x_\alpha - 2e)\left(1 + \left(\frac{\omega_h}{\omega_\alpha}\right)^2\right)\left(\frac{\omega_h}{\omega_\alpha}\right)^2\frac{x_\alpha^2}{r_\alpha^2} + 16e^2\left(\frac{\omega_h}{\omega_\alpha}\right)^4\frac{x_\alpha^4}{r_\alpha^4}\right\}\\
&- \left(2e - x_\alpha\right)^2\left\{\left[1 - \left(\frac{\omega_h}{\omega_\alpha}\right)^2\right]^2 + 4\left(\frac{x_\alpha^2}{r_\alpha^2}\right)\left(\frac{\omega_h}{\omega_\alpha}\right)^2\right\}
\end{aligned}$$

or

$$\frac{C_1^2}{4} - C_0C_2 = \left\{\begin{array}{l}\left(1 + \left(\frac{\omega_h}{\omega_\alpha}\right)^2\right)^2(x_\alpha - 2e)^2 + 8e\left(\frac{\omega_h}{\omega_\alpha}\right)^2(x_\alpha - 2e)\left(1 + \left(\frac{\omega_h}{\omega_\alpha}\right)^2\right)\frac{x_\alpha^2}{r_\alpha^2}\\ -(x_\alpha - 2e)^2\left\{\left[1 - \left(\frac{\omega_h}{\omega_\alpha}\right)^2\right]^2 + 4\left(\frac{x_\alpha^2}{r_\alpha^2}\right)\left(\frac{\omega_h}{\omega_\alpha}\right)^2\right\} + 16e^2\left(\frac{\omega_h}{\omega_\alpha}\right)^4\frac{x_\alpha^4}{r_\alpha^4}\end{array}\right\}$$

Substituting C_1 (Eq. 6.33), C_2 (Eq. 6.32) and $\frac{C_1^2}{4} - C_0C_2$ (Eq. 6.34) into 6.35, there is obtained (Rederived Formula)

$$Q_1, Q_2 = \frac{1}{C_2}\left[-\frac{C_1}{2} \pm \sqrt{\frac{C_1{}^2}{4} - C_0C_2}\right]$$

$$= \frac{1}{(2e - x_\alpha)^2}\left[\begin{array}{l} \left\{(x_\alpha - 2e)\left(1 + \left(\frac{\omega_h}{\omega_\alpha}\right)^2\right) + 4e\left(\frac{\omega_h}{\omega_\alpha}\right)^2 \frac{x_\alpha^2}{r_\alpha^2}\right\} \\ \pm \sqrt{\left\{\begin{array}{l} \left(1 + \left(\frac{\omega_h}{\omega_\alpha}\right)^2\right)^2 (x_\alpha - 2e)^2 + 8e\left(\frac{\omega_h}{\omega_\alpha}\right)^2 (x_\alpha - 2e)\left(1 + \left(\frac{\omega_h}{\omega_\alpha}\right)^2\right)\frac{x_\alpha^2}{r_\alpha^2} \\ - (x_\alpha - 2e)^2\left\{\left[1 - \left(\frac{\omega_h}{\omega_\alpha}\right)^2\right]^2 + 4\left(\frac{x_\alpha^2}{r_\alpha^2}\right)\left(\frac{\omega_h}{\omega_\alpha}\right)^2\right\} \\ + 16e^2\left(\frac{\omega_h}{\omega_\alpha}\right)^4 \frac{x_\alpha^4}{r_\alpha^2} \end{array}\right\}} \end{array}\right] \tag{6.35}$$

which is an alternative form of the initially derived Eq. 6.35.

References

1. Zwaan, R.J. 1981. Aeroelasticity of aircraft, Lecture Notes, Special Lecture, Short Course offered at Institut Teknologi Bandung, Indonesia, August.
2. Done, G.T.S. 1968. The flutter and stability of undamped systems, ARC Reports and Memoranda No. 3553, HMSO.
3. Done, G.T.S. 1969. A study of binary flutter roots using a method of system synthesis, ARC Reports and Memoranda No. 3554, HMSO.
4. Lind, R., and M.J. Brenner. 1998. Robust flutter margin analysis that incorporates flight data. NASA-TP-1998–206543.
5. Zimmerman, Norman H., and Jason T. Weissenburger. 1964. Prediction of flutter onset speed based on flight testing at subcritical speeds. *AIAA Journal of Aircraft* 1 (4): 190–202.
6. Rheinfurth, Mario H., and Swift, Fredrick W. 1966. A new approach to the explanation of the flutter mechanism. NASA Technical Note TN D-3125.
7. Isogai, K. 1979. On the transonic-dip mechanism of flutter of a Sweptback Wing. *AIAA J* (TN).
8. Djojodihardjo, H. 2005. Lecture notes on aeroelasticity at the intitute of technology Bandung delivered in 1982 to 2005 (Author's repository).
9. Djojodihardjo, H. 2008. Lecture notes on aeroelasticity at the Universiti Sains Malaysia (University of Science Malaysia) in 2004 to 2008 (Author's repository).

Chapter 7
Introduction to Unsteady Aerodynamics

Abstract Fundamental fluid dynamics, which is founded on conservation principles related to mass, momentum and energy (first law of thermodynamics), as well as equation of state (which in most cases degenerates into the equation of ideal gas) is also one of the founding pillars of aeroelasticity, more so in the discipline of unsteady aerodynamics. Therefore this chapter elaborates the mandatory principles in great detail, including their mathematical formulation. The unsteady (or non-stationary) aerodynamics is an essential part of aeroelasticity and is usually the most difficult for the student and practitioner alike. Pioneering approach has been initiated by Wagner [1], Küssner [2] and Theodorsen [3] among others. Their work can be followed in their original work as well as classical books on aeroelasticity. The discussion presented here is developed from the fundamental theory of Fluid Mechanics Conservation Principles and Potential Flow Aerodynamics as elaborated in Chap. 3. Further discussions are elaborated in Chaps. 8 and 11 which incorporate some case studies. The present chapter is also introductory and focused on two-dimensional inviscid flow, where linearized principle can be applied. More advanced topics are elaborated in other chapters.

Keywords Aerodynamics · Fluid dynamics · Conservation principles · Unsteady aerodynamics

7.1 Basic Fluid Dynamic Equation[1]

In the development of unsteady aerodynamics utilized in aeroelasticity, one should bear in mind fundamental fluid dynamics, which is founded on conservation principles related to mass, momentum and energy (first law of thermodynamics), as well

[1] The present chapter has been prepared for the author's lectures from well-known books on aeroelasticity, in particular those of Bisplinghoff, R.L., Ashley, H. and Halfman, R.L., Aeroelasticity, Dover Publications, Inc., copyright ©1955 by Addison-Wesley Publishing Co., Inc., Bisplinghoff, R.L., and Ashley, H. [2], Dowell, Earl [9], and Wright and Cooper [26]. Materials related to Aerodynamics have also been based and elaborated on Ashley et al. [1]

© The Author(s), under exclusive license to Springer Nature Singapore Pte Ltd. 2023

H. Djojodihardjo, *Introduction to Aeroelasticity*,
https://doi.org/10.1007/978-981-16-8078-6_7

as equation of state (which in most cases degenerates into the equation of ideal gas). The basic assumptions concerning the nature of fluid are that it be inviscid and its thermodynamic processes be isentropic. Our attention at this stage is focused on the derivation of the equation of motion. The basic conservation laws have been elaborated in Chap. 3, which will be reviewed and utilized.

For this purpose, vector calculus is essential, in particular variations of what is known as Gauss theorem and some vector identities. These are:

(i)

$$\iint c\,\mathbf{n}\,dA = \iiint \nabla c\,\mathrm{d}V \tag{7.1}$$

(ii)

$$\iint \mathbf{b}\cdot\mathbf{n}\,\mathrm{d}A = \iiint \nabla\cdot\mathbf{b}\,\mathrm{d}V \tag{7.2}$$

(iii)

$$\iint a(\mathbf{b}\cdot\mathbf{n})\mathrm{d}A = \iiint [a(\nabla\cdot\mathbf{b}) + (\mathbf{b}\cdot\nabla)a]\mathrm{d}V \tag{7.3}$$

and vector identity.

(iv)

$$\nabla(\mathbf{a}\cdot\mathbf{a}) = 2(\mathbf{a}\cdot\nabla)\times\mathbf{a} + 2\mathbf{a}(\nabla\times\mathbf{a}) \tag{7.4}$$

where V is an arbitrary closed volume, A is surface area, $\vec{n}$ is the unit normal to the surface, positive outward, **a** and **b** are arbitrary vectors, and c is an arbitrary scalar.

7.2 Review of Fluid Dynamics

7.2.1 Conservation of Mass

Integral form

$$\iint \frac{\partial\rho}{\partial t} + \nabla\cdot(\rho\,\mathbf{q})\mathrm{d}V = 0 \tag{7.5}$$

Since V is arbitrary in (7.5), it follows that the integrand should also be zero. Then one obtains the differential form.

$$\frac{\partial \rho}{\partial t} + \nabla \cdot (\rho\, \mathbf{q}) = 0 \tag{7.6}$$

with alternative form

$$\frac{\partial \rho}{\partial t} + \rho \nabla \cdot \mathbf{q} + (\mathbf{q} \cdot \nabla)\rho = 0 \tag{7.7}$$

$$\frac{D\rho}{Dt} + \rho(\nabla \cdot \mathbf{q}) = 0 \tag{7.8}$$

where

$$\frac{D}{Dt} = \frac{\partial}{\partial t} + (\mathbf{q} \cdot \nabla).$$

7.2.2 Conservation of Momentum

$$\iiint \frac{\partial(\rho \mathbf{q})}{\partial t} \mathrm{d}V + \rho\, \mathbf{q}(\mathbf{q} \cdot \mathbf{n})\mathrm{d}A = \iint -p\mathbf{n}\, \mathrm{d}A \tag{7.9}$$

Using mathematical relations I and III to transform the area integrals and rearranging terms, obtain.

$$\iiint \left\{ \frac{\partial(\rho\, \mathbf{q})}{\partial t} + \rho\, \mathbf{q}(\nabla \cdot \mathbf{q}) + (\mathbf{q} \cdot \nabla)\rho \mathbf{q} + \nabla p \right\} \mathrm{d}V = 0 \tag{7.10}$$

Differential form (obtained from (7.10) since V is arbitrary)

$$\left\{ \frac{\partial(\rho\, \mathbf{q})}{\partial t} + \rho \mathbf{q}(\nabla \cdot \mathbf{q}) + (\mathbf{q} \cdot \nabla)\rho\, \mathbf{q} + \nabla p \right\} = 0 \tag{7.11}$$

or

$$\frac{\partial(\rho\, \mathbf{q})}{\partial t} + \rho \mathbf{q}(\nabla \cdot \mathbf{q}) + (\mathbf{q} \cdot \nabla)\rho\, \mathbf{q} = -\nabla\, p \tag{7.12}$$

Alternative form

$$\frac{D(\rho\, \mathbf{q})}{Dt} + \rho\, \mathbf{q}(\nabla \cdot \mathbf{q}) = -\nabla p \tag{7.13}$$

or

$$\rho \frac{D\, \mathbf{q}}{Dt} + \mathbf{q}\left[\rho\, \nabla \cdot \mathbf{q} + \frac{D\, \rho}{Dt}\right] = -\nabla\, p \tag{7.14}$$

where the terms in the bracket vanish from (7.8).

Hence

$$\rho \frac{D\mathbf{q}}{Dt} = -\nabla p \tag{7.14a}$$

Isentropic relation

$$\frac{p}{\rho^{\gamma}} = \text{constant} \tag{7.15}$$

7.2.3 Irrotational Flow, Kelvin's Theorem and Bernoulli's Equation

To solve the nonlinear, partial differential equations of basic fluid mechanics above, these equations have to be integrated. This can only be carried out only for very special cases, and generally, it is an impossible task, except by numerical approaches. However, there is one integration which may be performed and which is both interesting theoretically and useful for applications.

Consider the momentum Eq. (7.14) or (7.14a) which may be written as

$$\frac{D\mathbf{q}}{Dt} = -\frac{\nabla p}{\rho} \tag{7.16}$$

On the right-hand side, since $\rho = \rho(p)$, using Leibnitz' rule, one can write

$$\frac{\nabla p}{\rho} = \nabla \int_{p_{re}}^{p} \frac{dp_1}{\rho_1(p_1)} \tag{7.17}$$

where ρ_1, p_1 are dummy integration variables and p_{ref} is some constant reference pressure. On the left-hand side

$$\frac{D\mathbf{q}}{Dt} \equiv \frac{\partial \mathbf{q}}{\partial t} + (\mathbf{q} \cdot \nabla)\mathbf{q} \tag{7.18}$$

In Eq. (7.18) above, the second term may be written as

$$(\mathbf{q} \cdot \nabla)\mathbf{q} = \nabla \frac{(\mathbf{q} \cdot \mathbf{q})}{2} \tag{7.19}$$

from (iv) and the irrotationality condition being addressed.

Further, due to the assumption of irrotationality,

$$\nabla \times \mathbf{q} = 0 \tag{7.20}$$

which implies that $\mathbf{q}$ should be a gradient of a scalar potential,[2] say $\boldsymbol{\phi}$:

$$\mathbf{q} = \nabla\phi \tag{7.21}$$

where ϕ is the scalar velocity potential.

7.3 Differential Equations Based on Velocity Potential

7.3.1 Kelvin's Theorem and Velocity Potential

Recall from Sect. 3.12 in Chap. 3 on Irrotational Flow and Velocity Potential, that

$$\mathbf{Q} \equiv \nabla\ \Phi \quad \textbf{or} \quad \mathbf{V} \equiv \nabla\ \Phi$$

3.126[3]

or (7.21).

This is a consequence of the Kelvin's theorem, which states that a flow which is initially irrotational, $\nabla \times \mathbf{q} = 0$ will remain so at all subsequent time in the absence of dissipation, e.g. viscosity or shock waves.

Next we will consider the integration of the momentum Eq. (7.16), using relations (7.21). Hence

$$\frac{\partial}{\partial t}(\nabla\phi) + \nabla\left(\frac{\nabla\phi \cdot \nabla\phi}{2}\right) + \nabla\left(\int_{p_{ref}}^{p} \frac{d\,p_1}{\rho_1}\right) = 0 \tag{7.22}$$

or

$$\nabla\left[\frac{\partial\phi}{\partial t} + \frac{(\nabla\phi \cdot \nabla\phi)}{2} + \int_{p_{\text{ref}}}^{p} \frac{d\,p_1}{\rho_1}\right] = 0 \tag{7.23}$$

or

[2] Since from vector identity $\nabla\ \times\ (\nabla\phi)\ \equiv\ 0$

[3] As folowed in this book, V and Q or v and q are alternatively used to represent the velocity, both as scalar and as vector variables. Thus $\mathbf{Q} \equiv \nabla\ \Phi$, $\mathbf{V} \equiv \nabla\ \Phi$, $\mathbf{q} \equiv \nabla\ \phi$ and $\mathbf{v} \equiv \nabla\ \phi$ are alternatively found in this book.

$\mathbf{Q} \equiv \nabla\ \Phi$ or $\mathbf{V} \equiv \nabla\ \Phi$

or

$\mathbf{q} \equiv \nabla\ \phi$ or $\mathbf{v} \equiv \nabla\ \phi$

$$\frac{\partial \phi}{\partial t} + \frac{\nabla \phi \cdot \nabla \phi}{2} + \int_{p_{\text{ref}}}^{p} \frac{d\,p_1}{\rho_1} = F(t) \tag{7.24}$$

$F(t)$ can be evaluated by examining the fluid at some point where its state is known. For example, considering a flight vehicle flying at some constant velocity through an atmosphere. It is known that far away from the body (flight vehicle)

$$\mathbf{q} = U_\infty \mathbf{i} \tag{7.25a}$$

$$\phi = U_\infty\, x \tag{7.25b}$$

$$p = p_\infty \tag{7.25c}$$

If the lower limit is chosen to be

$$p_{\text{ref}} = p_\infty \tag{7.25d}$$

then (7.24) becomes

$$0 + \frac{U^2}{2} + 0 = F(t) \tag{7.26}$$

and F is found to be a constant independent of space *and* time. Using this finding in (7.23) it follows that

$$\frac{\partial \phi}{\partial t} + \frac{\nabla \phi \cdot \nabla \phi}{2} + \int_{p_{\text{ref}}}^{p} \frac{d\,p_1}{\rho_1} = \frac{U^2}{2} \tag{7.27}$$

This equation is usually referred to as Bernoulli's equation, although the derivation for unsteady flow is due to Kelvin. This equation can be referred to as "Kelvin-Bernoulli equation [4, 5]."

The practical merit of Kelvin-Bernoulli's equation is that it allows one to relate p to ϕ.

Using

$$\frac{p}{p_\infty} = \left(\frac{\rho}{\rho_\infty}\right)^{\gamma} \tag{7.28}$$

one may compute from (7.27) the following relationship for the pressure coefficient, as defined by

$$C_p = \frac{p - p_\infty}{\gamma M^2}$$
$$= \frac{2}{\gamma M^2}\left\{\left[1 + \frac{\gamma - 1}{2} M^2 \left(1 - \frac{\left(\mathbf{q}\cdot\mathbf{q} + 2\frac{\partial\phi}{\partial t}\right)}{U^2}\right)\right]^{\frac{\gamma}{\gamma-1}} - 1\right\} \tag{7.29}$$

where the Mach number M is defined as

$$M = \frac{U^2}{a^2} \tag{7.30}$$

and

$$a^2 = \frac{dp}{d\rho} = \frac{\gamma p}{\rho} \tag{7.31}$$

where a is the speed of sound.

7.3.2 *Derivation of Single Equation for Velocity Potential*

Most solutions are obtained by solving this equation. Let us begin with the equation for the conservation of mass

$$\frac{D\rho}{Dt} + \rho(\nabla\cdot\mathbf{q}) = 0 \tag{7.8}$$

or

$$\frac{1}{\rho}\frac{\partial\rho}{\partial t} + \frac{\mathbf{q}\cdot\nabla\rho}{\rho} + \nabla\cdot\mathbf{q} = 0 \tag{7.9}$$

Next consider the first term. Using Leibnitz' rule one may write

$$\frac{\partial\phi}{\partial t} + \frac{\nabla\phi\cdot\nabla\phi}{2} + \int\limits_{p_{\text{ref}}}^{p} \frac{dp_1}{\rho_1} = \frac{U^2}{2} \tag{7.27}$$

$$\frac{\partial}{\partial t}\left(\int\limits_{p_\infty}^{p}\frac{dp_1}{\rho_1}\right) = \frac{\partial\rho}{\partial t}\frac{\partial p}{\partial\rho}\frac{d}{dp}\left(\int\limits_{p_{\text{ref}}}^{p}\frac{dp_1}{\rho_1(p_1)}\right) = \frac{\partial\rho}{\partial t}a^2\frac{1}{\rho} \tag{7.33}$$

Hence

$$\frac{1}{\rho}\frac{\partial \rho}{\partial t} = \frac{1}{a^2}\frac{\partial}{\partial t}\left(\int_{p_\infty}^{p}\frac{dp_1}{\rho_1}\right) = -\frac{1}{a^2}\frac{\partial \rho}{\partial t}\left[\frac{\partial \phi}{\partial t} + \frac{\nabla\phi\cdot\nabla\phi}{2}\right] \tag{7.34}$$

from Kelvin-Bernoulli's Eq. (7.27).

In a similar fashion, the second equation may be written as

$$\mathbf{q}\cdot\frac{\nabla\rho}{\rho} = \frac{-\mathbf{q}\cdot\nabla\phi}{a^2}\left[\frac{\partial \phi}{\partial t} + \frac{\nabla\phi\cdot\nabla\phi}{2}\right] \tag{7.35}$$

Finally, the third term

$$\nabla\cdot\mathbf{q} = \nabla\cdot\nabla\phi = \nabla^2\phi \tag{7.36}$$

Collecting terms and rearranging, obtain

$$-\frac{1}{a^2}\left\{\begin{aligned}&\frac{\partial^2\phi}{\partial t^2} + \frac{\partial}{\partial t}\left(\frac{\nabla\phi\cdot\nabla\phi}{2}\right)\\ &+\nabla\phi\cdot\frac{\partial}{\partial t}(\nabla\phi) + \nabla\phi\cdot\nabla\left(\frac{\nabla\phi\cdot\nabla\phi}{2}\right)\end{aligned}\right\} + \nabla^2\phi = 0 \tag{7.37a}$$

or

$$\nabla^2\phi - \frac{1}{a^2}\left\{\frac{\partial}{\partial t}(\nabla\phi\cdot\nabla\phi) + \frac{\partial^2\phi}{\partial t^2} + \frac{\partial}{\partial t}\left(\frac{\nabla\phi\cdot\nabla\phi}{2}\right) + \nabla\phi\cdot\nabla\left(\frac{\nabla\phi\cdot\nabla\phi}{2}\right)\right\} = 0 \tag{7.37b}$$

Note that at this stage the aim of obtaining a single equation for the velocity potential ϕ is not completely accomplished, since in the above equation there are two unknowns, ϕ and a. A second independent relation between ϕ and a is required.

The simplest method of obtaining this relationship is to use

$$a^2 = \frac{dp}{d\rho} = \frac{\gamma p}{\rho} \tag{7.31}$$

and

$$\frac{p}{\rho^\gamma} = \text{constant} \tag{7.38}$$

in the Kelvin-Bernoulli's Eq. (7.27) to replace $\int_{p_{re}}^{p} \frac{1}{\rho_1(p_1)} \mathrm{d}p_1$. Then it can be shown[4] that one will arrive at Eq. (7.37b). It can also be verified that the following relationship holds

$$\frac{a^2 - a_\infty^2}{\lambda - 1} = \frac{U_\infty^2}{2} - \left(\frac{\partial \phi}{\partial t} + \frac{\nabla\phi \cdot \nabla\phi}{2}\right) \tag{7.39}$$

7.4 Small-Perturbation Theory

7.4.1 *Differential Equation*

Equations (7.37b) and (7.39) are often too difficult to solve. Therefore one seeks a simpler approximate method or theory to solve them. For that purposes, the usual approach is to linearize these equations using some plausible assumptions or approximations, which at the end need validation and will impose some limitation on their validity.

Following the usual linearization scheme, assume

$$a = a_\infty + \hat{a} \tag{7.40a}$$

$$p = p_\infty + \hat{p} \tag{7.40b}$$

$$\rho = \rho_\infty + \hat{\rho} \tag{7.40c}$$

$$\mathbf{q} = U_\infty \mathbf{i} + \hat{\mathbf{q}}$$

or

$$\nabla\phi = U_\infty \mathbf{i} + \nabla\hat{\phi} \tag{7.40d}$$

$$\phi = U_\infty x + \hat{\phi} \tag{7.40e}$$

[4] Equation (7.37b) can be obtained as follows:

$$\nabla^2\phi - \frac{1}{a^2}\left\{\frac{\partial}{\partial t}(\nabla\phi\cdot\nabla\phi) + \frac{\partial^2\phi}{\partial t^2} + \frac{\partial}{\partial t}\left(\frac{\nabla\phi\cdot\nabla\phi}{2}\right) + \nabla\phi\cdot\nabla\left(\frac{\nabla\phi\cdot\nabla\phi}{2}\right)\right\} = 0$$

from (7.27) $\frac{\partial\phi}{\partial t} + \frac{\nabla\phi\cdot\nabla\phi}{2} + \int_{p_{\mathrm{ref}}}^{p} \frac{1}{\rho_1} dp_1 = \frac{U_\infty^2}{2}$ using (7.31) $a^2 = \frac{dp}{d\rho} = \frac{\gamma p}{\rho}$ and (7.38) $\frac{p}{\rho^\gamma} =$ constant will be left as an excercise. The verification of Eq. (7.39) can also be carried out as an exercise.

Note that in the present case, the linearization is performed about a uniform flow with velocity U_∞. Using (7.40) in (7.37) and retaining only the lowest-order terms, one obtains:

First term:

$$\nabla^2\phi \rightarrow \nabla^2\hat{\phi} \tag{7.41}$$

Second term:

$$\begin{aligned}
&\frac{\partial}{\partial t}(\nabla\phi\cdot\nabla\phi)+\frac{\partial^2\phi}{\partial t^2}+\nabla\phi\cdot\nabla\left(\frac{\nabla\phi\cdot\nabla\phi}{2}\right)\\
&=2\left[U_\infty \boldsymbol{i}+\nabla\hat{\phi}\right]\cdot\frac{\partial}{\partial t}\left[U_\infty \boldsymbol{i}+\nabla\hat{\phi}\right]+\frac{\partial^2\hat{\phi}}{\partial t^2}\\
&+\left[U_\infty \boldsymbol{i}+\nabla\hat{\phi}\right]\cdot\nabla\left[\frac{U_\infty^2}{2}+U_\infty \boldsymbol{i}\cdot\nabla\hat{\phi}+\frac{1}{2}\nabla\hat{\phi}\cdot\nabla\hat{\phi}\right]\\
&=2U_\infty\frac{\partial^2\hat{\phi}}{\partial x\partial t}+\frac{\partial^2\hat{\phi}}{\partial t^2}+U_\infty^2\frac{\partial^2\hat{\phi}}{\partial x^2}+O\left(\hat{\phi}^2\right)
\end{aligned} \tag{7.42}$$

Thus the linear or small-perturbation equation becomes

$$\nabla^2\phi-\frac{1}{a^2}\left\{\frac{\partial^2\phi}{\partial t^2}+2U_\infty\frac{\partial^2\phi}{\partial x\partial t}+U_\infty^2\frac{\partial^2\phi}{\partial x^2}\right\}=0 \tag{7.43}$$

Note that we have replaced a by a_∞ which is correct to the lowest order. By examining

$$\frac{a^2-a_\infty^2}{\gamma-1}=\frac{U_\infty^2}{2}-\left(\frac{\partial\phi}{\partial t}+\frac{\nabla\phi\cdot\nabla\phi}{2}\right)=0 \tag{7.39}$$

one may show that

$$\hat{a}=-\frac{\gamma-1}{2}\frac{\left(\frac{\partial\phi}{\partial t}+U_\infty\frac{\partial\phi}{\partial x}\right)}{a_\infty} \tag{7.44}$$

In these expressions, the replacement of a by a_∞ is consistent as long as M is not too large, where $M=U_\infty/a_\infty$.

In a similar fashion, the relationship between pressure and velocity potential, (7.29), may be linearized to the relationship

$$C_p\approx\frac{\hat{p}}{\frac{\gamma}{2}p_\infty M^2}=-\frac{2}{U_\infty}\frac{\partial\hat{\phi}}{\partial x}-\frac{2}{U_\infty^2}\frac{\partial\hat{\phi}}{\partial t} \tag{7.45}$$

or

$$\hat{p} = -\rho_\infty \left[\frac{\partial \hat{\phi}}{\partial t} + U_\infty \frac{\partial \hat{\phi}}{\partial x} \right] \tag{7.46}$$

Also, noting that the acceleration potential ψ is defined by

$$\psi = \frac{P_\infty - P}{\rho} \tag{7.20}$$

then $\hat{P}$ may be related to acceleration potential

$$\frac{\hat{P}}{\rho_\infty} = -\left[\frac{\partial \hat{\psi}}{\partial t} + U_\infty \frac{\partial \hat{\psi}}{\partial x} \right] \tag{7.46b}$$

7.4.2 Boundary Conditions

The boundary conditions state that the normal velocity of the fluid at the body surface equals the normal velocity of the body.

Consider a body whose surface is described by

$$F(x, y, z, t) = 0 \tag{7.47a}$$

as depicted in Fig. 7.1.

At some time, t. At some later time, $t + \Delta t$, by

$$F(x + \Delta x, y + \Delta y, z + \Delta z, t + \Delta t) = 0 \tag{7.47b}$$

Now

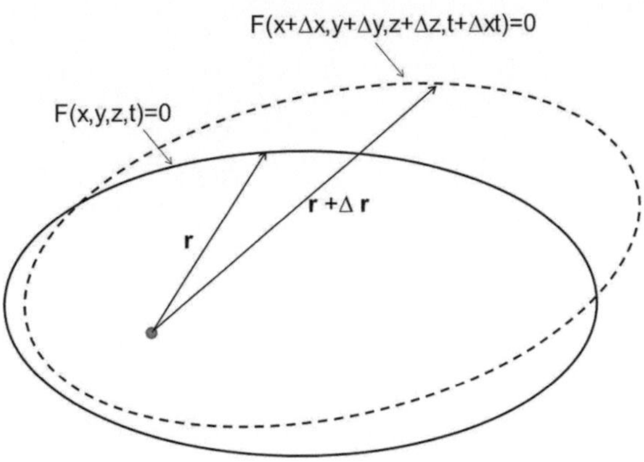

Fig. 7.1 Vector representation of body geometry spatial and time variation

$$\Delta F \equiv F(\mathbf{r} + \Delta\mathbf{r}, t + \Delta t) - F(\mathbf{r}, t) = 0 \tag{7.48}$$

Also

$$\begin{aligned}\Delta F &= \frac{\partial F}{\partial x}\Delta x + \frac{\partial F}{\partial y}\Delta y + \frac{\partial F}{\partial z}\Delta + \frac{\partial F}{\partial t}\Delta t \\ &= \nabla F \cdot \Delta\mathbf{r} + \frac{\partial F}{\partial t}\Delta t\end{aligned} \tag{7.49}$$

From (7.48) and (7.49)

$$\nabla F \cdot \Delta\mathbf{r} + \frac{\partial F}{\partial t}\Delta t = 0 \tag{7.50}$$

The unit normal at the surface $F(x,y,z,t) = 0$ is

$$\mathbf{n} = \frac{\nabla F}{|\nabla F|} \tag{7.51}$$

Also

$$\mathbf{V} \equiv \lim_{\Delta t \to 0} \frac{\Delta\mathbf{r}}{\Delta t} \equiv \text{body velocity} \equiv \mathbf{Q} \tag{7.52}$$

Hence the body normal velocity is

$$\mathbf{V} \cdot \mathbf{n} \equiv \frac{\Delta\mathbf{r}}{\Delta t} \cdot \frac{\nabla F}{|\nabla F|} \tag{7.53a}$$

and using from (7.50)

$$\mathbf{V} \cdot \mathbf{n} \equiv \frac{\Delta\mathbf{r}}{\Delta t} \cdot \frac{\nabla F}{|\nabla F|} = \frac{\Delta\mathbf{r} \cdot \nabla F}{\Delta t} \cdot \frac{1}{|\nabla F|} = -\frac{\partial F}{\partial t}\frac{1}{|\nabla F|} \tag{7.53b}$$

The boundary condition on the body is the normal fluid velocity that equals the normal body velocity on the body. Therefore, using (7.51) and (7.53)

$$\underbrace{\mathbf{V} \cdot \mathbf{n} = \mathbf{Q} \cdot \mathbf{n} = \mathbf{V} \cdot \frac{\nabla F}{|\nabla F|} = \mathbf{Q} \cdot \frac{\nabla F}{|\nabla F|}}_{\text{normal fluid velocity}} = \underbrace{-\frac{\partial F}{\partial t} \cdot \frac{1}{|\nabla F|}}_{\text{normal body velocity}} \tag{7.54}$$

or

$$\frac{\partial F}{\partial t} + \mathbf{V} \cdot \nabla F = 0 \tag{7.55a}$$

or

$$\left(\frac{\partial}{\partial t} + \mathbf{V} \cdot \nabla\right) F = 0 \tag{7.55b}$$

or

$$\frac{D}{Dt}\{F(x, y, z, t)\} = 0 \tag{7.55c}$$

on the body surface

$$F(x, y, z, t) = 0 \tag{7.55d}$$

Essentially that follows the rationale, that for $F(x,y,z,t)$ to be intact while moving through the fluid or being convected by the fluid, using Lagrangian approach, then Eqs. (7.55c) and (7.55d) follow [5].

7.5 Subsonic Flow

Subsonic flow is generally a more difficult problem area since all parts of the flow are disturbed due to the motion of the airfoil or aircraft. To counter this difficulty, an inverse method of solution has evolved, known as the "Kernel Approach [6]."

Following Dowell, the problem is formulated and solved in a formal way using Fourier transform. Also, the three-dimensional problem will be treated. The problem will be formulated in terms of pressure (or acceleration potential).

7.5.1 Derivation of the Integral Equation by Transform Methods and Solution by Collocation

Since there is no limited range of influence, in subsonic flow the Fourier transforms with respect to x and y will be used.[5] Simple harmonic time-dependent motion will be assumed following standard practice. Hence:

$$\phi = \overline{\phi}(x, y, z, t)e^{i\omega t} \tag{7.56}$$

Transformed:

$$\Phi^* = \int_{-\infty}^{\infty}\int \overline{\phi}(x, y, z, t)\exp(-i\alpha x - i\lambda y)\mathrm{d}x\mathrm{d}y \tag{7.57}$$

[5] X is in the flow direction and y in the spanwise direction.

Recalling Eq. (7.20).

$$\psi = \frac{P_\infty - P}{\rho} \tag{7.20}$$

and (7.46b),

$$\frac{\hat{p}}{\rho_\infty} = -\left[\frac{\partial\hat{\phi}}{\partial t} + U_\infty\frac{\partial\hat{\phi}}{\partial x}\right] \tag{7.46b}$$

then we may utilize the Bernoulli's equation

$$\hat{p} = -\rho_\infty\left[\frac{\partial\hat{\phi}}{\partial t} + U_\infty\frac{\partial\hat{\phi}}{\partial x}\right] \tag{7.58}$$

(which is closely related to the acceleration potential).

For convenience, the "hat" sign will be omitted from now on, remembering that the variables being considered are the perturbed ones.

Hence (7.58) may be transformed as

$$P^* = -\rho[i\omega + U_\infty i\alpha]\Phi^* \tag{7.59}$$

where

$$p = -\overline{p}(x, y, z, t)e^{i\omega t} \tag{7.60}$$

$$P^* = \int_{-\infty}^{\infty}\int \overline{p}(x, y, z, t)\exp(-i\alpha x - i\lambda y)\mathrm{d}x\mathrm{d}y \tag{7.61}$$

The transformed velocity potential Φ^* can be related to the transformed upwash W^*:

$$\Phi^*\big|_{z=0} = \frac{-W^*}{\left(\mu^2 + \gamma^2\right)^{\frac{1}{2}}} \tag{7.61}$$

Substituting (7.61) into (7.59)

$$P^* = \rho_\infty\frac{[i\omega + U_\infty i\alpha]}{\left(\mu^2 + \gamma^2\right)^{\frac{1}{2}}}W^*$$

or

$$\frac{\overline{W}^*}{U_\infty} = \frac{P^*}{\rho_\infty U_\infty^2} \frac{(\mu^2+\gamma^2)^{\frac{1}{2}}}{\left[\frac{i\omega}{U_\infty} + i\alpha\right]} \tag{7.62}$$

7.5.2 An Alternative Determination of the Kernel Function Using Green's Theorem

The transform methods are most efficient at least for formal derivations; however historically other approaches were first used. Many of these are now only of interest to history; however we should mention one other approach which is a powerful tool for non-steady aerodynamic problems. This is the use of Green's theorem.

For this purpose, one could start with the divergence theorem or Gauss's theorem

$$\iiint \nabla \cdot \mathbf{b}\, dV = -\frac{1}{4\pi} \iint \mathbf{b} \cdot \mathbf{n}\, dS \tag{7.63}$$

S surface area enclosing volume V
n outward normal
b arbitrary vector.

Assign $\mathbf{b} = \phi_1 \nabla\phi_2$ where ϕ_1, ϕ_2 are arbitrary scalars. Then (7.63) may be written as:

$$\iiint \nabla \cdot \phi_1 \nabla\phi_2\, dV = -\frac{1}{4\pi} \iint \mathbf{b} \cdot \phi_1 \nabla\phi_2 dS \tag{7.64}$$

Using the vector identity

$$\nabla \cdot c\,\mathbf{a} = c\nabla \cdot \mathbf{a} + \mathbf{a} \cdot \nabla c$$

where

c arbitrary scalar
$\boldsymbol{a}$ arbitrary vector

it then follows that

$$\nabla \cdot \phi_1 \nabla\phi_2 = \phi_1 \nabla^2 \cdot \phi_2 + \nabla\phi_2 \cdot \nabla\phi_1$$

and (7.64) becomes

$$\iiint [\phi_1 \nabla^2\phi_2 + \nabla\phi_2 \cdot \nabla\phi_1] dV = -\frac{1}{4\pi} \iint \mathbf{n} \cdot \phi_1 \nabla\phi_2\, dS \tag{7.65}$$

This is the first form of Green's theorem. Interchanging ϕ_1 and ϕ_2 in (7.65) and subtracting the result from (7.64) gives

$$\iiint \left[\phi_1 \nabla^2 \phi_2 - \nabla\phi_2 \cdot \nabla\phi_1\right] \mathrm{d}V = \iint \mathbf{n} \cdot (\phi_1 \nabla\phi_2 - \phi_2 \nabla\phi_1) \mathrm{d}S$$
$$= \iint \left(\phi_1 \frac{\partial\phi_2}{\partial n} - \phi_2 \frac{\partial\phi_1}{\partial n}\right) \mathrm{d}S \qquad (7.66)$$

This is the second (and generally more useful) form of Green's theorem. $\partial/\partial n$ denotes a derivative in the direction of the normal. Let us consider several special but informative cases.

After some algebra,[6] it then follows that

$$\phi_1(x, y, z) = -\frac{1}{4\pi} \iint \left(\phi_1 \frac{\partial}{\partial n} - \frac{\partial\phi_1}{\partial n}\right) \frac{1}{r} \, \mathrm{d}S \qquad (7.67)$$

The choice of $\phi_2 = 1/r$ may seem rather arbitrary. This can be motivated by noting that

$$\frac{\nabla^2 \phi_2}{4\pi} = -\delta(x - x_1)\delta(y - y_1)\delta(z - z_1) \qquad (7.68)$$

Hence a ϕ_2 which is the response to a delta function will be sought. This leads to the simplification of the volume integral.

7.5.3 Incompressible, Three-Dimensional Flow

To simplify matters we will first confine ourselves to $M = 0$. However, similar, but more complex calculations subsequently will be carried out for $M = 0.16$. For incompressible flow, the equation of motion is

$$\nabla^2 \phi = 0 \qquad (7.69a)$$

or

$$\nabla^2 p = 0 \qquad (7.69b)$$

where ϕ and p are (perturbation) velocity potential and pressure, respectively. Hence we may identify ϕ_1 in (7.67) with either variables as may be convenient. To confirm to convention in the aerodynamic theory literature, we will take the normal positive into the fluid and introduce a minus sign into (7.67) which now reads:

[6] Dowell, Earl [9].

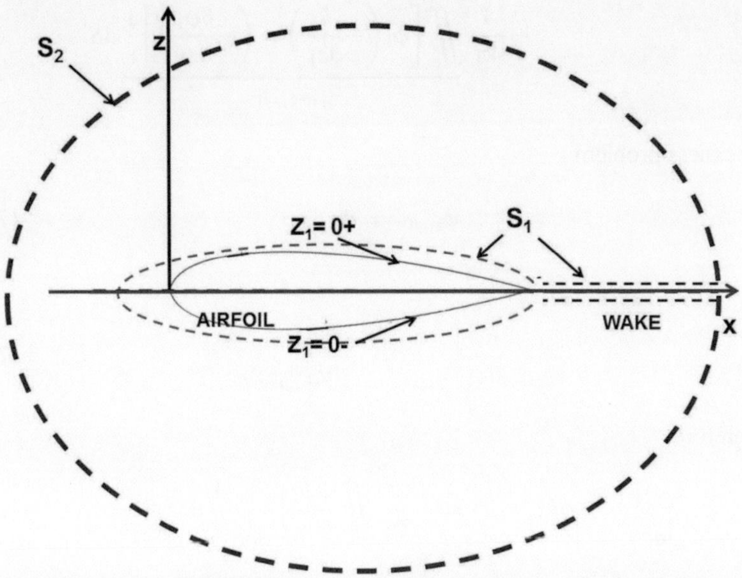

Fig. 7.2 Airfoil and flow field geometry

$$\phi_1(x, y, z) = \frac{1}{4\pi} \iint \left(\phi_1 \frac{\partial}{\partial n} - \frac{\partial \phi_1}{\partial n} \right) \frac{1}{r} \, \mathrm{d}S \tag{7.70}$$

For example for a planar airfoil surface

$$n \text{ on } S \text{ at } z1 = 0 + \text{is} + z1 \tag{7.71a}$$

$$n \text{ on } S \text{ at } z1 = 0 - \text{is} - z1 \tag{7.71b}$$

Note that *x, y, z* are any given point, while $x1$, $y1$, $z1$ are (dummy) integration variables, as depicted in Fig. 7.2. Here the area S can be identified as composed of two parts, the area of the airfoil plus wake, indicated by $S1$, and the area $S2$ of a sphere at infinity.

7.5.3.1 Thickness Problem (Non-lifting)

Let $\varphi_1 = \varphi$ be velocity potential. Because φ is bounded at $r \to \infty$, there is no contribution from $S2$. Hence

$$\phi(x, y, z) = -\frac{1}{4\pi} \underbrace{\iint \left[\phi_1 \frac{\partial}{\partial z_1} - \frac{\partial \phi_1}{\partial z_1} \right] \frac{1}{r} \, \mathrm{d}S}_{S_1 \, at \, z_1 = 0^+}$$

$$+\frac{1}{4\pi}\underbrace{\iint\left[\phi_1\left(-\frac{\partial}{\partial z_1}\right)-\left(-\frac{\partial\phi_1}{\partial z_1}\right)\right]\frac{1}{r}\,\mathrm{d}S}_{S_1\ at\ z_1=0^-} \tag{7.72}$$

For thickness problem

$$\phi_{z_1=0^+}=\phi_{z_1=0^-} \tag{7.73a}$$

and

$$\left.\frac{\partial\phi}{\partial z_1}\right|_{z_1=0^+}=-\left.\frac{\partial\phi}{\partial z_1}\right|_{z_1=0^-} \tag{7.73b}$$

Therefore

$$\phi(x,y,z)=-\frac{1}{2\pi}\iint\left.\frac{\partial\phi_1}{\partial z_1}\right|_{z_1=0^+}\frac{1}{r}\,\mathrm{d}S \tag{7.74}$$

and using the boundary condition

$$\phi(x,y,z)=-\frac{1}{2\pi}\iint w_a\frac{1}{r}\mathrm{d}S \tag{7.75}$$

where

$$w_a=\frac{\partial z_a}{\partial t}+U_\infty\frac{\partial z_a}{\partial x} \tag{7.76}$$

Note that this solution is valid for arbitrary time-dependent motion. Time only appears as a parameter in the solution $\phi\ (x,\ y,\ z)=\phi\ (x,\ y,\ z;\ t)$. This is a special consequence of $M\equiv 0$.

7.5.3.2 Lifting Problem

For the lifting problem it again will prove convenient to use pressure rather than velocity potential. Equation (4.3.22) becomes

$$p(x,y,z)=\frac{1}{4\pi}\iint\left[(p_{z=0^+}-p_{z=0^-})\frac{\partial}{\partial z_1}\left(\frac{1}{r}\right)-\left(\left.\frac{\partial p}{\partial z_1}\right|_{z=0^+}-\left.\frac{\partial p}{\partial z_1}\right|_{z=0^-}\right)\frac{1}{r}\right]\mathrm{d}S \tag{7.77}$$

Since

$$p_{z=0^+}=-p_{z=0^-} \tag{7.78a}$$

$$\left.\frac{\partial p}{\partial z_1}\right|_{z=0^+} - \left.\frac{\partial p}{\partial z_1}\right|_{z=0^-} = 0 \tag{7.78b}$$

and

$$\Delta p = p_{z=0^+} - p_{z=0^-} \tag{7.78c}$$

then

$$p(x, y, z) = \frac{1}{4\pi} \iint \Delta p \frac{\partial}{\partial z_1}\left(\frac{1}{r}\right) \mathrm{d}S \tag{7.79}$$

To be meaningful and to utilize the expression (7.79) for solving the problem at hand, p should be related to something that is being sought, that is, w. To simplify matters we shall specify harmonic motion,

$$p = \overline{p}\, e^{i\omega t} \tag{7.80a}$$

$$\phi = \overline{\phi}\, e^{i\omega t} \tag{7.80b}$$

From Bernoulli's equation

$$\overline{p} = \rho_\infty \left[i\omega\overline{\phi} + U_\infty \frac{\partial \overline{\phi}}{\partial x} \right] \tag{7.81}$$

Solving (7.81) by variation of parameters,[7]

$$\overline{\phi}(x, y, z) = -\int_{-\infty}^{x} \frac{\overline{p}(\lambda, y, z)}{\rho_\infty U_\infty} \exp\left[i\frac{\omega}{U_\infty}(\lambda - x) \right] \mathrm{d}\lambda \tag{7.82}$$

and using (7.79) one has

$$\overline{\phi}(x, y, z) = -\int_{-\infty}^{x} \exp\left[i\frac{\omega}{U_\infty}(\lambda - x) \right] \cdot \left\{ \frac{1}{4\pi} \iint \frac{\Delta \overline{p}(x_1, y_1, z_1 = 0)}{\rho_\infty U_\infty} \frac{\partial}{\partial z_1}\left(\frac{1}{r(\lambda)} \right) \mathrm{d}S \right\} \mathrm{d}\lambda \tag{7.83}$$

where

---

[7] See Kreyszig [14].

$$r(\lambda) = \sqrt{(\lambda - x_1)^2 + (y - y_1)^2 + (z - z_1)^2}$$
$$\mathrm{d}S = \mathrm{d}z_1 \mathrm{d}y_1$$

and let

$$\xi = \lambda - x_1 \quad \mathrm{d}\lambda = \mathrm{d}\xi \quad \lambda = \xi + x_1$$

After interchanging the order of integration with respect to ξ and S, obtain

$$\overline{\phi}(x, y, z) = -\frac{1}{4\pi} \iint \frac{\Delta\overline{p}(x_1, y_1, z_1 = 0)}{\rho_\infty U_\infty} \cdot \left\{ \int_{-\infty}^{x-x_1} \frac{\partial}{\partial z_1}\left(\frac{1}{r(\xi)}\right) \exp\left(i\frac{\omega}{U_\infty}(\xi - (x - x_1))\right) \mathrm{d}\xi \right\} \mathrm{d}S \tag{7.84}$$

After computing $\frac{\partial \overline{\phi}}{\partial z}$ and setting it equal $\frac{\partial \overline{\phi}}{\partial z} = \overline{w}_a$ from body boundary condition, on $z = 0$, it follows that

$$\overline{w}_a(x, y, z) \equiv \frac{\partial \overline{\phi}(x, y, z)}{\partial z} = -\frac{1}{4\pi} \iint \frac{\Delta\overline{p}(x_1, y_1, z_1 = 0)}{\rho_\infty U_\infty} \cdot \left\{ \frac{\partial}{\partial z} \int_{-\infty}^{x-x_1} \frac{\partial}{\partial z_1}\left(\frac{1}{r(\xi)}\right) \exp\left(i\frac{\omega}{U_\infty}(\xi - (x - x_1))\right) \mathrm{d}\xi \right\} \mathrm{d}S \tag{7.85}$$

At z = 0

$$\overline{w}_a(x, y, 0) \equiv \frac{\partial \overline{\phi}(x, y, 0)}{\partial z} = -\frac{1}{4\pi} \iint \frac{\Delta\overline{p}(x_1, y_1, z_1 = 0)}{\rho_\infty U_\infty} \cdot \left\{ \frac{\partial}{\partial z} \int_{-\infty}^{x-x_1} \frac{\partial}{\partial z_1}\left(\frac{1}{r(\xi)}\right) \exp\left(i\frac{\omega}{U_\infty}(\xi - (x - x_1))\right) \mathrm{d}\xi \right\} \mathrm{d}S \tag{7.86}$$

Note that

$$\frac{\partial}{\partial z}\left(\frac{1}{r}\right) = -\frac{\partial}{\partial z_1}\left(\frac{1}{r}\right) \tag{7.87}$$

Therefore

$$\frac{\overline{w}_a(x, y, 0)}{U_\infty} = \iint \frac{\Delta\overline{p}(x_1, y_1, z_1 = 0)}{\rho_\infty U_\infty^2} K(x - x_1, y - y_1, 0) \mathrm{d}x_1 \mathrm{d}y_1 \tag{7.88}$$

where

$$K(x-x_1, y-y_1, 0) \equiv \frac{1}{4\pi} \lim_{z\to 0} \frac{\partial^2}{\partial z^2} \left\{ \int_{-\infty}^{x-x_1} \frac{\exp\left\{ i \frac{\omega}{U_\infty} [\xi - (x - x_1)] \right\}}{r} \mathrm{d}\xi \right\} \tag{7.89a}$$

$$r \equiv \sqrt{\xi^2 + z^2 + (y - y_1)^2} \tag{7.89b}$$

The expression for the Kernel function may be simplified as

$$K(x-x_1, y-y_1, 0) \equiv \frac{\exp\left\{ -i \frac{\omega}{U_\infty} [(x - x_1)] \right\}}{4\pi} \left\{ \int_{-\infty}^{x-x_1} e^{i \frac{\omega \xi}{U_\infty}} \lim_{z\to 0} \frac{\partial^2}{\partial z^2} \left(\frac{1}{r} \right) \mathrm{d}\xi \right\} \tag{7.90}$$

since

$$\lim_{z\to 0} \frac{\partial^2}{\partial z^2} \left(\frac{1}{r} \right) = -\left[\xi^2 + (y - y_1)^2 \right]^{-\frac{3}{2}}$$

Then finally

$$K(x-x_1, y-y_1, 0) \equiv \frac{\exp\left\{ -i \frac{\omega}{U_\infty} [(x - x_1)] \right\}}{4\pi} \int_{-\infty}^{x-x_1} \frac{e^{+i \frac{\omega \xi}{U_\infty}}}{\left[\xi^2 + (y - y_1)^2 \right]^{-\frac{3}{2}}} \mathrm{d}\xi \tag{7.92}$$

The integral in (7.92) must be evaluated numerically.

7.5.4 Incompressible, Two-Dimensional Flow

A classical solution is due to Theodorsen[8] and others. Traditionally, the coordinate system origin is selected at mid-chord with $b \equiv$ half-chord. The governing differential equation for the velocity potential, ϕ, is

$$\nabla^2 \phi = 0. \tag{7.93}$$

Dowell [7] elaborates an alternative of the classical Theodorsen derivation by following Landahl derivation using integral equation and Fourier transform approach. This can be followed in Dowell's book [7]. The treatment of this problem is elaborated by Ashley and Landahl [5] using the method of matched asymptotic expansions.

---

[8] In addition to te original NACA Publication, the details of Theodorsen method are elaborated in [4]

7.6 Aerodynamic Lift and Moment for a Harmonically Oscillating Airfoil[9]

The present development follows that of Wright and Cooper to elaborate the unsteady aerodynamics of harmonically oscillating aerodynamic surface in incompressible flow, thus using potential flow aerodynamics, which was pioneered by Theodorsen [3].

The solution of the flow around the airfoil undergoing harmonic oscillations can be divided into two parts:

(a) *Circulatory terms*. Lift and moment terms occurring due to the vorticity in the flow (related to Theodorsen's function).
(b) *Non-circulatory terms*. "Apparent inertia" forces whose creation is not related to vorticity, i.e. as the airfoil moves, a cylindrical mass of air accelerates with the airfoil and introduces a reactive force and moment upon the airfoil. These terms are of minor importance for bending-torsion-type flutter of cantilever wings at low reduced frequencies, but are more important for flutter of control surfaces at higher reduced frequencies.

Consider a symmetric two-dimensional airfoil ($C_{M0} = 0$) of chord c, with the flexural axis positioned at distance $ab(= ac/2)$ aft of the mid-chord as shown in Fig. 7.3. The airfoil undergoes oscillatory harmonic motion in heave $z = z_0 e^{i\omega t}$ (positive downward) and pitch $\theta = \theta_0 e^{i\omega t}$ (positive nose-up). The classical solution for the lift and moment about the flexural axis, both expressed per unit span, may be written (Theodorsen 1935; Fung 1969; Bisplinghoff et al. 1996) as

$$L = \pi\,\rho\,b^2\left[\ddot{z} + V\,\dot{\theta} - b\,a\,\ddot{\theta}\right] + 2\pi\,\rho\,V\,b\,C(k)\left[\dot{z} + V\,\theta + b\left(\frac{1}{2} - a\right)\dot{\theta}\right] \tag{7.94}$$

$$\begin{aligned} M = {} & \pi\,\rho\,b^2\left[b\,a\,\ddot{z} - V\,b\left(\frac{1}{2} - a\right)\dot{\theta} - b^2\left(\frac{1}{8} + a^2\right)\ddot{\theta}\right] \\ & + 2\pi\,\rho\,V\,b^2\left(a + \frac{1}{2}\right)C(k)\left[\dot{z} + V\,\theta + b\left(\frac{1}{2} - a\right)\dot{\theta}\right] \end{aligned} \tag{7.95}$$

The derivation of these two equations can be found in Dowell[10] and other references in the literature. The first part of each expression shows the non-circulatory terms, and the second part shows the circulatory terms which are dependent upon the value of Theodorsen's function. There are terms dependent upon the displacement, velocity and acceleration of both heave and pitch motions, except for the heave displacement term (the vertical airfoil position does not affect the lift and moment). Here the two-dimensional lift curve slope has been taken as $\frac{\partial C_L}{\partial \alpha} = 2\pi$.

---

[9] Adapted from Wright and Cooper [26].
[10] Dowell, Earl [9].

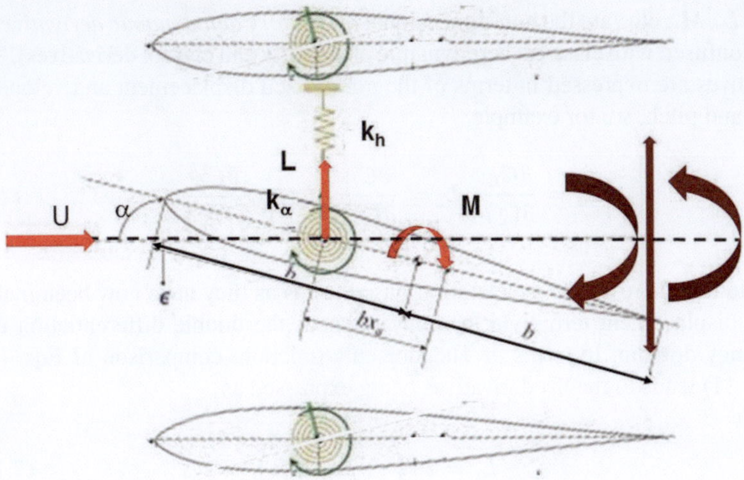

Fig. 7.3 Two-dimensional airfoil undergoing oscillating heave and pitch motion

7.7 Oscillatory Aerodynamic Derivatives

Taking the above expressions for the lift and moment about the flexural axis of the oscillating airfoil and substituting for the complex form of Theodorsen's function and the heave and pitch motions in complex algebra form (see Chap. 1), then Eqs. (7.94) and (7.95) become

$$L = \left\{ \begin{array}{l} \pi \rho b^2 \left[-\omega^2 z_0 + i\omega V \theta_0 + \omega^2 b a \theta_0 \right] \\ \quad + 2\pi \rho V b (F + iG) \left[i\omega z_0 + V \theta_0 + i\omega b \left(\frac{1}{2} - a \right) \theta_0 \right] \end{array} \right\} e^{i\omega t} \tag{7.96}$$

$$M = \left\{ \begin{array}{l} \pi \rho b^2 \left[-\omega^2 b a z_0 - i\omega V b \left(\frac{1}{2} - a \right) \theta_0 + b^2 \omega^2 \left(\frac{1}{8} + a^2 \right) \theta_0 \right] \\ \quad + 2\pi \rho V b^2 \left(a + \frac{1}{2} \right) (F + iG) \left[i\omega z_0 + V \theta_0 + i\omega b \left(\frac{1}{2} - a \right) \theta_0 \right] \end{array} \right\} e^{i\omega t} \tag{7.97}$$

These equations can then be written in the oscillatory derivative form

$$L = \rho V^2 b \left[(L_z + ikL_{\dot{z}}) \frac{z_0}{b} + (L_\theta + ikL_{\dot{\theta}}) \theta_0 \right] e^{i\omega t} \tag{7.98}$$

$$M = \rho V^2 b^2 \left[(M_z + ikM_{\dot{z}}) \frac{z_0}{b} + (M_\theta + ikM_{\dot{\theta}}) \theta_0 \right] e^{i\omega t} \tag{7.99}$$

where L_z, M_z, etc., are the non-dimensional *oscillatory aerodynamic derivatives* (not to be confused with classical aerodynamic or stability and control derivatives). These derivatives are expressed in terms of the normalized displacement and velocity for heave and pitch, so, for example,

$$L_z = \frac{\partial C_L}{\partial\left(\frac{z}{b}\right)} \quad L_{\dot{z}} = \frac{\partial C_L}{\partial\left(\frac{\dot{z}}{V}\right)} \quad L_{\dot{\theta}} = \frac{\partial C_L}{\partial\left(\frac{\dot{\theta}c}{V}\right)} \text{etc} \tag{7.100}$$

Note that there are no acceleration-based terms as they have now been included in the displacement terms via the conversion of the double differentiation to the frequency domain. In terms of Theodorsen's function, comparison of Eqs. (7.10) and (7.11) leads to the lift derivatives being expressed as

$$L_z = 2\pi\left(-\frac{k^2}{2} - Gk\right) \tag{7.101a}$$

$$L_{\dot{z}} = 2\pi F \tag{7.101b}$$

$$L_\theta = 2\pi\left[\frac{k^2 a}{2} + F - Gk\left(\frac{1}{2} - a\right)\right] \tag{7.101c}$$

$$L_{\dot{\theta}} = 2\pi\left[\frac{1}{2} + F\left(\frac{1}{2} - a\right) + \frac{G}{k}\right] \tag{7.101d}$$

and, from comparison of Eqs. (7.98) and (7.99), the relevant moment derivatives are

$$M_z = 2\pi\left(-\frac{k^2 a}{2} - k\left(a + \frac{1}{2}\right)Gk\right) \tag{7.102a}$$

$$M_{\dot{z}} = 2\pi\left(a + \frac{1}{2}\right)F \tag{7.102b}$$

$$M_\theta = 2\pi\left[\frac{k^2}{2}\left(\frac{1}{8} + a^2\right) + F\left(a + \frac{1}{2}\right) - kG\left(a + \frac{1}{2}\right)\left(\frac{1}{2} - a\right)\right] \tag{7.102c}$$

$$M_\theta = 2\pi\left[-\frac{k^2}{2}\left(\frac{1}{2} - a\right) + kF\left(a + \frac{1}{2}\right)\left(\frac{1}{2} - a\right) + \frac{G}{k}\left(a + \frac{1}{2}\right)\right] \tag{7.102d}$$

Apart from L_z and $L_{\dot{z}}$, the derivative values depend upon where the flexural axis is located on the chord.

The quasi-steady values of the aerodynamic derivatives ($k \to 0$, $F \to 1$, $G \to 0$) can be found as

$$L_z = 0 \quad L_{\dot{z}} = 2\pi \quad L_\theta = 2\pi \quad kL_{\dot{\theta}} = 0 \tag{7.103a,b,c,d}$$

$$M_{\dot{z}} = 0 \tag{7.103e}$$

$$M_{\dot{z}} = 2\pi\left(a + \frac{1}{2}\right) \tag{7.103f}$$

$$M_\theta = 2\pi\left(a + \frac{1}{2}\right) \tag{7.103g}$$

$$kM_\theta = 0 \tag{7.103h}$$

Note the singularity in the expressions for M_θ and $L_{\dot{\theta}}$ as $k \to 0$. However, since both $kL_{\dot{\theta}}$ and kM_θ tend to zero, then the contribution to the lift and moment from these derivatives is also zero a $k \to 0$. Therefore the concept of quasi-steady derivatives does not apply to the θ derivatives [8]. The other derivatives agree with the expressions found earlier for the quasi-steady forces and moments.

7.8 Aerodynamic Damping and Stiffness

Further insight into the effect of the unsteady aerodynamic forces can be obtained by considering

$$k = \frac{\omega b}{V} \quad z = z_0 e^{i\omega t} \quad \dot{z} = i\omega z_0 e^{i\omega t} \quad \theta = \theta_0 e^{i\omega t} \quad \text{and} \quad \dot{\theta} = i\omega\theta_0 e^{i\omega t} \tag{7.104a,b,c,d,e}$$

Substituting these expressions into the lift and moment Eqs. (7.12) gives

$$L = \rho V^2\left(L_z z + L_{\dot{z}}\frac{b\dot{z}}{V} + L_\theta b\theta + L_{\dot{\theta}}\frac{b^2\dot{\theta}}{V}\right) \tag{7.105a}$$

$$M = \rho V^2\left(M_z bz + M_{\dot{z}}\frac{b^2\dot{z}}{V} + M_\theta b^2\theta + M_{\dot{\theta}}\frac{b^3\dot{\theta}}{V}\right) \tag{7.105b}$$

and this can be written in the matrix form

$$\begin{Bmatrix} L \\ M \end{Bmatrix} = \rho V \begin{bmatrix} bL_{\dot{z}} & b^2 L_{\dot{\theta}} \\ b^2 M_{\dot{z}} & b^3 M_{\dot{\theta}} \end{bmatrix} \begin{Bmatrix} \dot{z} \\ \dot{\theta} \end{Bmatrix} + \rho V^2 \begin{bmatrix} L_z & bL_\theta \\ bM_z & b^2 M_\theta \end{bmatrix} \begin{Bmatrix} z \\ \theta \end{Bmatrix}$$
$$= \rho V \mathbf{B} \begin{Bmatrix} \dot{z} \\ \dot{\theta} \end{Bmatrix} + \rho V^2 \mathbf{C} \begin{Bmatrix} z \\ \theta \end{Bmatrix} \tag{7.106}$$

It can be seen that one term is proportional to the heave and pitch velocities, while the other term is proportional to the heave and pitch displacements. Thus, the aerodynamic forces acting on an airfoil undergoing oscillatory motion can be considered to behave in a similar way to that of damping and stiffness in a structure. Thus **B** and **C** are termed the aerodynamic damping and stiffness matrices, respectively.

A key difference to structural damping and stiffness matrices is that the aerodynamic matrices are non-symmetric, and this helps lead to the flutter aeroelastic instability; also, the damping and stiffness depend upon the flight condition, including the Mach number.

When applied to aeroelastic systems, as will be shown in the next chapter, the aerodynamic forces are considered together with the structural equations, and this leads to equations of motion in the classical form of

$$\mathbf{A}\ddot{\mathbf{q}} + (\rho V\mathbf{B} + \mathbf{D})\dot{\mathbf{q}}\,\rho V + \left(\rho V^2\mathbf{C} + \mathbf{E}\right)\mathbf{q} = 0 \tag{7.107}$$

where **A**, **B**, **C**, **D** *and* **E** are the structural inertia, aerodynamic damping, aerodynamic stiffness, structural damping and structural stiffness matrices, respectively, and $\boldsymbol{q}$ is the generalized coordinates (typically modal coordinates). It is important to note that the **B** *and* **C** matrices only apply for the reduced frequency for which they are defined; this can cause some difficulty for flutter calculations and will be discussed later.

Equation (7.20) is one of the most important equations in this book and describes the fundamental interaction between the flexible structure and the aerodynamic forces. Note that it is usual when considering aeroelastic systems to write the structural inertia, damping and stiffness matrices as **A**, **D** *and* **E**, respectively, rather than the **M**, **C** *and* **K** notation often used in classical structural dynamics.

7.9 Unsteady Aerodynamics of Thin Wing in Subsonic Flow, Derivation of Lift and Moment, Theodorsen Function[11,12]

There exist important phenomena, where unsteadiness cannot be overlooked, such as rapid maneuvers, response to atmospheric turbulence and flutter, which are familiar examples. We will carry out a short review of some significant results on time-dependent loading of wings, exact and linearized formulation, but focusing on linearized methods of solution. Transient and oscillating motion will be first introduced, for planar and non-planar configurations. In the present introductory chapter, attention will be focused to incompressible-flow case. However, extension

---

[11] The present development follows closely that of Theodorsen [3] and Ashley and Landahl [5].

[12] Theodor Theodorsen, General Theory Of Aerodynamic Instability And The Mechanism Of Flutter, NACA TR 496 and Holt Ashley, Mårten Landahl [1].

to compressible flow cases, which can be found in the literature, will be facilitated by such introduction.

Let irrotationality be assumed, under the limitations set forth in Chap. 3. The kinematics of the unsteady field are then fully described by a velocity potential Φ governed by the differential Eq. 3.144, from which the speed of sound is formally eliminated using

$$a^2 - a_\infty^2 = -(\gamma - 1)\left\{\frac{\partial \Phi}{\partial t} + \frac{1}{2}(V^2 - U_\infty^2) + \Omega_\infty - \Omega\right\} \tag{3.136}$$

and assigning $\Omega = 0$ here. Pressure distributions and generalized aerodynamic forces follow from

$$\begin{aligned} C_p &= \frac{p - p_\infty}{\frac{1}{2}\rho U_\infty^2} \\ &= \frac{2}{\gamma M^2}\left\{\left[1 - \frac{\gamma - 1}{a_\infty^2}\left(\frac{\partial \Phi}{\partial t} + \frac{1}{2}(V^2 - U_\infty^2) + \Omega_\infty - \Omega\right)\right]^{\frac{\gamma}{\gamma-1}} - 1\right\} \end{aligned} \tag{3.133}$$

The flow disturbances in a uniform stream can be represented by.

$$\Phi_0 = U_\infty x \tag{7.108}$$

that are generated by a thin lifting surface, Fig. 7.4, which is performing rapid, small displacements in a direction generally normal to its *x,y*-plane projection. Thus the wing might be vibrating elastically, undergoing sudden roll or pitch aerobatics, or an encounter with gusty air might give rise to a situation mathematically and physically analogous to vibrations.

With z_u and z_l, as given functions of position and time, then the boundary conditions which generalize the steady-state requirement of flow tangency at the surface (cf. 5.5) can be expressed by:

$$\Phi_z(x, y, z_u, t) = \frac{\partial z_u}{\partial x}\Phi_x(x, y, z_u, t) + \frac{\partial z_u}{\partial y}\Phi_y(x, y, z_u, t) + \frac{\partial z_u}{\partial t} \tag{7.109}$$

$$\Phi_z(x, y, z_u, t) = \frac{\partial z_l}{\partial x}\Phi_x(x, y, z_l, t) + \frac{\partial z_l}{\partial y}\Phi_y(x, y, z_l, t) + \frac{\partial z_l}{\partial t} \tag{7.110}$$

for *(x, y)* on *S*. There is the usual auxiliary condition of vanishing disturbances at points remote from the wing and its wake, but for *compressible* fluid this must be refined to ensure that such disturbances behave like outward-propagating waves. The Kutta-Joukowsky hypothesis of continuous pressure at subsonic trailing edges is also applied, although for high-frequency oscillation Abramson and Ransleben [9] observed that this hypothesis may not be valid.

Provided that there are no time-dependent variations of profile thickness, the upper and lower surface coordinates can be given by

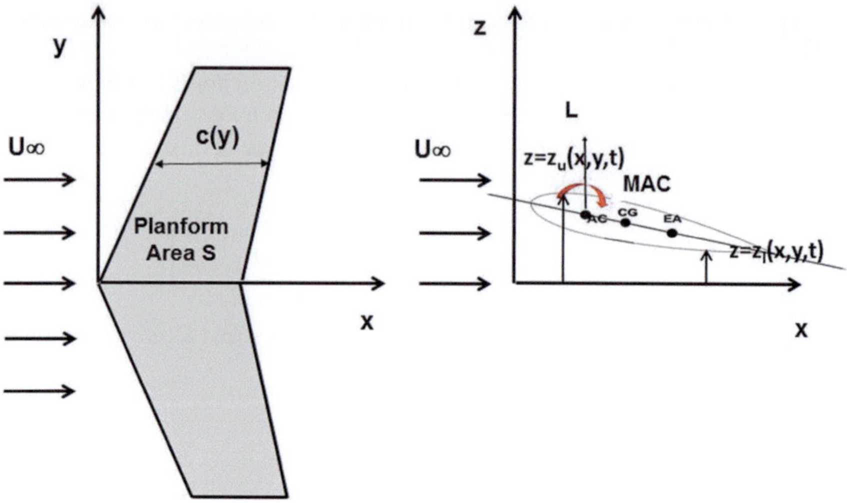

Fig. 7.4 Thin wing executing small motions normal to its mean plane

$$z_u = \varepsilon \overline{f}_u(x, y, t) = \tau \overline{g}(x, y) + \theta \overline{h}(x, y, t) \tag{7.110}$$

$$z_l = \varepsilon \overline{f}_l(x, y, t) = -\tau \overline{g}(x, y) + \theta \overline{h}(x, y, t) \tag{7.111}$$

Here ε is a dimensionless small parameter measuring the maximum crosswise extension of the wing, including the space occupied by its unsteady displacement. The angle of attack α can be thought of as encompassed by the θ-term; g and h are smooth functions as in steady motion; their x- and y-derivatives are everywhere of order unity; the t-derivative of h will be discussed below.

Ashley and Landahl matched asymptotic expansion (inner and outer expansion) [5] resulted in

$$\left.\begin{array}{l} \Phi^i_{2z} = \frac{\partial f_u}{\partial x} + \frac{1}{U_\infty}\frac{\partial f_u}{\partial t} \quad \text{at} \quad z = f_u \\ \Phi^i_{2z} = \frac{\partial f_l}{\partial x} + \frac{1}{U_\infty}\frac{\partial f_l}{\partial t} \quad \text{at} \quad z = f_l \end{array}\right\} \quad \text{for}(x, y) on\ S \tag{7.112}$$

Finally, a reduction of 3.133 and matching, to order ε, of Φ or Φ_x shows that

$$C_p = -2\varphi_x - \frac{2}{U_\infty}\varphi_t O\left(\varepsilon^2\right) \tag{7.113}$$

throughout the entire flow.[13]

[13] The small-perturbation Bernoulli equation again contains nonlinear terms when used in connection with unsteady motion of slender *bodies* rather than wings.

7.9.1 Two-Dimensional, Constant-Density Flow

The best-known of the classical solutions for unsteady loading for the oscillating thin airfoil at $M = 0$ shows that in this case of nearly constant density, a key distinction disappears between the steady and unsteady problems because the flow must satisfy a two-dimensional Laplace equation

$$\phi_{xx} + \phi_{zz} = 0 \tag{7.114}$$

The development will accordingly rely quite heavily on these results for a steadily lifting airfoil, which elaborates the required inversion for the oscillatory integral equation elaborated in Ashley and Landahl and to some extent in previous sections, while simultaneously enforcing Kutta's condition at the trailing edge.

With the lifting surface paralleling the *x,y*-plane between $x = 0$ and c, it can be assumed that is known over that area and given by (dimensionless)[14]

$$\varphi_x(x, 0, t) = \hat{\varphi}_x(x, 0, t)e^{i\omega t} \tag{7.115a}$$

and

$$w_0(x, t) = \hat{w}_0(x)e^{i\omega t} \tag{7.115b}$$

for $0 \leq x \leq c$.

The perturbation field has φ and u antisymmetrical in z, and allowance must be made for discontinuities in φ through the x,y-plane for $x > 0$. Hence, Eq. (7.114) and all other conditions can be satisfied by a vortex sheet similar to the one depicted in Fig. 7.2 and extended downstream by replacing the upper limit with infinity.

The following development is an extraction of parts of the derivation elaborated in Ashley and Landahl using the asymptotic (inner and outer) expansion method elaborated there, and adapted here to provide a coherent rationale, but omitting the required background which should be resorted to in Chaps. 5 and 13 there, and partly elaborated in Appendix of this Chapter. From the solution of the differential equation for the inner expansion of Φ, i.e.:

$$\Phi^i_{2_{\overline{zz}}^{\text{inner}}} = 0 \tag{7.116}$$

it follows that it requires a solution linear in $\overline{z}$; hence:

$$\Phi^i_2 = \overline{z}\left[\frac{\partial \overline{f}_u}{\partial x} + \frac{1}{U_\infty}\frac{\partial \overline{f}_u}{\partial t}\right] + \overline{g}_{2_u}(x, y, t) \tag{7.117}$$

[14] The ^ sign in $\hat{w}_0$ and others here indicates perturbantion quantities.

for $\overline{z} \geq \overline{f}_u$. Similar form applies for points below the lower surface. Extracting the linear terms from Eq. 3.144 and deriving the first-order outer differential equation gives

$$\left(1 - M^2\right)\Phi^o_{1_{xx}} + \Phi^o_{1_{yy}} + \Phi^o_{1_{zz}} - \frac{2\,M^2}{U_\infty}\Phi^o_{1_{zt}} - \frac{M^2}{U_\infty}\Phi^o_{1_{tt}} = 0 \tag{7.118}$$

After further derivation elaborated in Appendix A.7.2, there is obtained

$$\varphi_z = \theta\left[\frac{\partial \overline{h}}{\partial x} + \frac{1}{U_\infty}\frac{\partial \overline{h}}{\partial t}\right] \quad \text{at } z = 0 \tag{7.119}$$

Equations (7.115) and (7.119) are introduced through:

$$w_0(x,t) = -\frac{1}{2\pi}\underbrace{\int_0^\pi \frac{\gamma(x_1,t)}{x - x_1}\,\mathrm{d}x_1}_{\text{along curve } C} \quad \text{for} \quad 0 \leq x \leq c \tag{7.120}$$

where

$$\gamma = u(x, 0+, t) - u(x, 0-, t) = 2\,u(x, 0+, t) \tag{7.121}$$

For further utilization in successive development, an integrated vortex strength is defined as

$$\Gamma(x,t) = \int_0^x \gamma(x_1,t)\,\mathrm{d}x_1 = 2\int_0^x \varphi_x(x_1, 0+, t)\,\mathrm{d}x_1 = 2\varphi(x_1, 0+, t) \tag{7.122}$$

where $\Gamma(c, t)$ is the instantaneous circulation bound to the airfoil.

From the expression for C_p in (7.113) and the antisymmetry of C_p in z, it can be deduced that a discontinuity of pressure through the wake, which is actually physically impossible, can be avoided only if

$$\varphi_x + \frac{1}{U_\infty}\varphi_t = 0 \tag{7.123}$$

for $x > c$ along $z = 0+$. Equation (7.123) is a partial differential expression for $\varphi(x, 0+, t)$, which is solved subject to continuity of φ at the trailing edge by

$$\varphi(x, 0+, t) = \varphi\left(c, 0+, t - \frac{x - c}{U_\infty}\right) \tag{7.124}$$

Further relations for the wake can be derived from (7.122) and (7.124)

$$\Gamma(x,t) = \Gamma\left(c, t - \frac{x-c}{U_\infty}\right) \tag{7.125}$$

$$\gamma(x,t) = -\frac{1}{U_\infty}\Gamma_t\left(c, t - \frac{x-c}{U_\infty}\right) \tag{7.126}$$

Equation (7.126) can be interpreted as that wake vortex elements are convected downstream approximately at the flight speed U_∞ after being shed as countervortices from the trailing edge at a rate equal to the variation of bound circulation.

Introducing (7.126) into (7.120) and assuming that a linear, simple harmonic process has been going on indefinitely to replace all dependent variables with sinusoidal counterparts, one obtains

$$\hat{w}_0(x) = -\frac{1}{2\pi}\underbrace{\int_0^c \frac{\hat{\gamma}(x_1)}{x - x_1}\mathrm{d}x_1}_{\text{along curve } C} + \hat{g}(x) \tag{7.127}$$

after canceling the common factor $e^{i\omega t}$, where[15]

$$\overline{g}(x) = \frac{i\omega\,\overline{\Gamma}(c)}{2\pi\;U_\infty}\underbrace{\int_0^C \frac{e^{-i\left(\frac{\omega}{U_\infty}\right)(x_1-c)}}{x - x_1}\mathrm{d}x_1}_{\text{along curve } C} \tag{7.128}$$

Through some simple algebraic manipulations, an inversion of (7.127) is obtained

$$\gamma(x) = -\frac{1}{2\pi}\sqrt{\frac{c-x}{x}}\underbrace{\int_0^\infty \frac{w_0(x_1)}{x - x_1}\sqrt{\frac{x_1}{c - x_1}}\mathrm{d}x_1}_{\text{along curve C}} \tag{7.129}$$

If w_0 is replaced by $\hat{w}_0 - \hat{g}$, similar method can be utilized to invert (7.128), provided $\overline{\Gamma}(c)$ can be expressed in terms of known quantities. This requirement is met by multiplying (7.128) by $\sqrt{\frac{x_1}{c-x_1}}$ and integrating with respect to x along the chord. The integral on the left-hand side is easily evaluated for the most continuous function $\hat{w}_0(x)$. The integration of the two integrals appearing on the right is carried out as follows:

[15] If dimensionless x-variables are adopted in (7.127), based on reference length $l \equiv c/2$, it is clear how $k = \omega;c/2U_\infty$ will arise as one parameter of the problem.

(a)

$$
\begin{aligned}
I_1 &= -\frac{1}{2\pi}\int_0^c \sqrt{\frac{c-x}{x}} \int_{0\,\text{along curve } C}^{c} \frac{\hat{\gamma}(x_1)dx}{x-x_1}dx \\
&= -\frac{1}{2\pi}\int_0^c \hat{\gamma}(x_1) \int_{0\,\text{along curve } C}^{c} \sqrt{\frac{x}{c-x}}\frac{dx}{x-x_1}dx_1
\end{aligned}
\tag{7.130}
$$

The change of variable $x = (c/2)(1-\cos\theta)$ converts the inner integral here into one of the familiar results arising in Glauert's solution for the lifting line,

$$
\int_{0\,\text{along curve } C}^{c} \sqrt{\frac{x}{c-x}}\frac{dx}{x-x_1} = -\int_{0\,\text{along curve } C}^{\pi} \frac{(1-\cos\theta)}{\cos\theta-\cos\theta_1}d\theta = \pi \tag{7.131}
$$

for which

(b)

$$
I_1 = -\frac{1}{2}\int_{0\,\text{along curve } C}^{c} \hat{\gamma}(x_1)dx_1 = -\frac{1}{2}\hat{\Gamma}(c) \tag{7.132}
$$

$$
\begin{aligned}
I_2 &= \frac{1}{2\pi}\int_0^c \sqrt{\frac{x}{c-x}} \int_0^{\infty} \frac{e^{-i\frac{\omega}{U_\infty}(x-x_1)}}{x-x_1}\mathrm{d}x_1\mathrm{d}x \\
&= \frac{1}{2\pi}\int_0^{\infty} e^{-i\frac{\omega}{U_\infty}(x-x_1)} \int_0^c \sqrt{\frac{x}{c-x}}\frac{\mathrm{d}x}{x-x_1}\mathrm{d}x_1
\end{aligned}
\tag{7.133}
$$

For the inner integral here,[16]

$$
\int_0^c \sqrt{\frac{x}{c-x}}\frac{\mathrm{d}x}{x-x_1} = \pi\left[1-\sqrt{\frac{x_1}{x_1-c}}\right] \tag{7.134}
$$

The result

[16] See the details in Chaps. 5 and 13 in Ashley and Landahl [5].

$$I_2 = \frac{1}{2}\int\limits_0^{\infty} e^{-i\frac{\omega}{U_\infty}(x-x_1)}\left[1-\sqrt{\frac{x_1}{x_1-c}}\right]\mathrm{d}x_1 \tag{7.135}$$

is seen to be properly convergent because the integrand vanishes as $x_1 \Rightarrow \infty$. It is most easily evaluated by defining $\xi = (2x_1/c)-1$ to give

$$I_2 = \frac{ce^{ik}}{4}\int\limits_1^{\infty} e^{-ik\xi}\left[1-\sqrt{\frac{\xi+1}{\xi-1}}\right]\mathrm{d}\xi \tag{7.136}$$

After proper treatment of (7.136) as elaborated in Ashley and Landahl [5], there is obtained

$$I_2 = \frac{c}{2}\left[\frac{\pi}{4}e^{ik}\left(H_1^{(2)}(k)+iH_0^{(2)}(k)\right)+\frac{1}{2ik}\right] \tag{7.137}$$

where

$$H_0^{(2)}(k) \quad \text{and} \quad H_1^{(2)}(k) \tag{7.138}$$

are Hankel function of the second kind and order 0 and 1, respectively, (or $H_n^{(2)}(k)$ is Hankel function of the second kind and order n).

Substituting l_1 and l_2 given by Eqs. (7.136b) and (7.137) into the weighted integral of (7.127), we obtain

$$\hat{\Gamma}_c = \frac{4e^{-ik}\int_0^c\sqrt{\frac{x}{c-x}}\hat{w}_0(x)\mathrm{d}x}{\pi ik\left[H_1^{(2)}(k)+iH_0^{(2)}(k)\right]} \tag{7.139}$$

One could note that the appearance of $k = \omega c/2U_\infty$ as argument in (7.139) clarifies the choice of $c/2$ as the "natural" reference length for subsonic unsteady problems of this kind. Returning to (7.127), (7.128) has been used to express[17]

$$\hat{\gamma}(x) = \frac{2}{\pi}\sqrt{\frac{c-x}{x}}\underbrace{\int\limits_0^c \frac{\hat{w}_0(x_1)-\hat{g}(x)}{x-x_1}}_{\text{along curve } C}\sqrt{\frac{x_1}{c-x_1}}\mathrm{d}x_1 \tag{7.140}$$

Through (7.139), (7.138) and (7.123) we are now able to work out the pressure distribution, lift, moment, etc. due to any motion of the airfoil. Ashley and Landahl found a somewhat more efficient way of presenting these results than has appeared in the literature, which is elaborated below.

[17] See Ashley and Landahl [5] for detail.

Define an auxiliary function, connected to $\overline{w}_0$ as is $\overline{\Gamma}$ to $\overline{\gamma}$:

$$\hat{p}(x) = \int_0^x \hat{w}_0(x_1)\mathrm{d}x_1 = -\frac{1}{2\pi}\int_0^x \underbrace{\int_0^c \frac{\hat{\gamma}(x_2)}{x_1 - x_2}\mathrm{d}x_2\mathrm{d}x_1}_{\text{along curve } C} \tag{7.141}$$

The last integral here, being improper, is evaluated by replacing the infinite limit with R, inverting order, and letting $R \Rightarrow \infty$.

$$\int_0^x \underbrace{\int_0^R \frac{\hat{\gamma}(x_2)}{x_1 - x_2}\mathrm{d}x_2\mathrm{d}x_1}_{along\ curve\ C} = \int_0^R \hat{\gamma}(x_2)[\ln|x - x_2| - \ln x_2]\mathrm{d}x_2$$

$$= \left\{\hat{\Gamma}(x_2)\ln\left|\frac{x - x_2}{x_2}\right|\right\}\Bigg|_0^R + \underbrace{\int_0^R \frac{\hat{\Gamma}(x_2)}{x - x_2}\mathrm{d}x_2}_{\text{along curve } C} + \int_0^R \frac{\hat{\Gamma}(x_2)}{x_2}dx_2$$

$$= \underbrace{\int_0^\infty \frac{\hat{\Gamma}(x_1)}{x - x_1}\mathrm{d}x_1}_{\text{along curve } C} + 2\pi A \tag{7.142}$$

where

$$A = \frac{1}{2\pi}\int_0^\infty \frac{\hat{\Gamma}(x_1)}{x_1}\mathrm{d}x_1 \tag{7.143}$$

Note that the limit has been re-inserted before the last member of (7.142), and x_2 is replaced by $\times$ 1 as dummy variable. No principal value needs be taken in the integral defining A, because $\hat{\Gamma}$ vanishes at the leading edge as ${x_1}^{1/2}$, leaving only an integrable singularity.

Thus we obtain

$$\hat{p}(x) = -\frac{1}{2\pi}\underbrace{\int_0^\infty \frac{\hat{\Gamma}(x_1)}{x - x_1}\mathrm{d}x_1}_{\text{along curve } C} - A \tag{7.144}$$

Next one has to construct

$$\hat{w}_0 + i\frac{\omega}{U_\infty}\hat{p} = -\frac{1}{2\pi}\underbrace{\int\limits_0^\infty \frac{\hat{\gamma}(x_1) + i\omega\frac{\hat{\Gamma}(x_1)}{U_\infty}}{x - x_1}\mathrm{d}x_1}_{\text{along curve } C} - i\frac{\omega A}{U_\infty} \tag{7.145}$$

and to notice from (7.113), (7.120), (7.121) and (7.122) that the dimensionless pressure jump through the x-axis, positive in a sense to produce upward loading, is

$$\Delta C_p(x, t) = \Delta\hat{C}_p(x)e^{i\omega t} = \left[2\hat{\gamma}(x) + 2i\omega\frac{\hat{\Gamma}(x_1)}{U_\infty}\right]e^{i\omega t} \tag{7.146}$$

In view of the vanishing of ΔC_p for $x > c$, (7.145) becomes

$$\hat{w}_0 + i\omega\frac{\hat{p}}{U_\infty} + i\omega\frac{A}{U_\infty} = -\frac{1}{4\pi}\underbrace{\int\limits_0^\infty \frac{\Delta\hat{C}_p(x_1)}{x - x_1}\mathrm{d}x_1}_{\text{along curve } C} \tag{7.147}$$

Immediate application of the inversion (7.139) yields

$$\Delta\overline{C}_p(x_1) = \frac{4}{\pi}\sqrt{\frac{c-x}{x}}\underbrace{\int\limits_0^\infty \frac{\left[\overline{w}_0 + i\omega\frac{\overline{p}}{U_\infty}\right]}{x - x_1}\sqrt{\frac{x_1}{c - x_1}}\mathrm{d}x_1}_{\text{along curve } C} - 4i\omega\frac{A}{U_\infty}\sqrt{\frac{c-x}{x}} \tag{7.148}$$

An easy way of eliminating the constant A from (7.148) is to note that $\Delta\overline{C}_p$ and $2\,\overline{\gamma}$ approach one another at the leading edge $x = 0$, since $\overline{\Gamma}(0) = 0$. Multiplying through by $\sqrt{x}$ to cancel the singularity, we may therefore equate the following two limits. From (7.148):

$$\lim_{x\to 0}\left[\sqrt{\frac{x}{c}}\Delta\hat{C}_p(x)\right] = -\frac{4}{\pi}\underbrace{\int\limits_0^\infty \frac{\left[\hat{w}_0 + i\omega\frac{\hat{p}}{U_\infty}\right]}{\sqrt{\frac{x_1}{x_1(c-x_1)}}}\mathrm{d}x_1}_{\text{along curve } C} - 4i\omega\frac{A}{U_\infty} \tag{7.149}$$

and from (7.140):

$$\lim_{x\to 0}\left[\sqrt{\frac{x}{c}}2\hat{\gamma}(x)\right] = -\frac{4}{\pi}\underbrace{\int_0^c \frac{\hat{w}_0(x_1)-\hat{g}(x_1)}{\sqrt{(c-x_1)x_1}}\mathrm{d}x_1}_{\text{along curve } C} \tag{7.150}$$

Consequently

$$i\omega\frac{A}{U_\infty} = -\frac{1}{\pi}\left\{\int_0^c i\omega\frac{\hat{p}}{U_\infty}\frac{\mathrm{d}x_1}{\sqrt{x_1(c-x_1)}} + \int_0^c \hat{g}(x_1)\frac{\mathrm{d}x_1}{\sqrt{x_1(c-x_1)}}\right\} \tag{7.151}$$

After replacing $\overline{g}(x_1)$ through (7.128) and (7.139), we encounter the last of the two integrals evaluated in (7.137) and are ultimately led to

$$\begin{aligned}
i\omega\frac{A}{U_\infty} &= -\frac{1}{4\pi}\int_0^c i\frac{\omega}{U_\infty}\hat{p}_1(x_1)\frac{\mathrm{d}x_1}{\sqrt{x_1(c-x_1)}} \\
&\quad - \frac{2i}{\pi c}\frac{H_0^{(2)}(k)}{H_1^{(2)}(k)+iH_0^{(2)}(k)}\int_0^c \sqrt{\frac{x_1}{(c-x_1)}}\hat{w}_0(x_1)\mathrm{d}x_1 \\
&= -\frac{1}{4\pi}\int_0^c i\frac{\omega}{U_\infty}\hat{p}(x_1)\frac{\mathrm{d}x_1}{\sqrt{x_1(c-x_1)}} \\
&\quad + \frac{2}{\pi c}[C(k)-1]\int_0^c \sqrt{\frac{x_1}{(c-x_1)}}\hat{w}_0(x_1)\mathrm{d}x_1
\end{aligned} \tag{7.152}$$

where

$$C(k) = \frac{H_1^{(2)}(k)}{H_1^{(2)}(k)+iH_0^{(2)}(k)} \tag{7.153}$$

is the well-known Theodorsen's function. Finally (7.152) is used to eliminate A from (7.148) to obtain

$$\Delta\hat{C}_p(x_1) = \frac{4}{\pi}\sqrt{\frac{c-x}{x}}\underbrace{\int_0^\infty \hat{w}_0(x_1)\sqrt{\frac{x_1}{c-x_1}}\mathrm{d}x_1}_{\text{along curve } C}$$

$$+\frac{4}{\pi}\sqrt{x(c-x)}\underbrace{\int_0^\infty i\omega\frac{\hat{p}(x_1)}{U_\infty}}_{\text{along curve } C}\frac{dx_1}{(x-x_1)\sqrt{x_1(c-x_1)}}$$

$$+\frac{8}{\pi}[1-C(k)]\sqrt{\frac{c-x}{x}}\int_0^c \hat{w}_0(x_1)\sqrt{\frac{x_1}{c-x_1}}dx_1 \tag{7.154}$$

Starting from (7.154), the lift may be computed as the integral of the pressure

$$\frac{\overline{L}}{\frac{1}{2}\rho_\infty U_\infty^2\left(\frac{c}{2}\right)} \equiv \int_{-1}^{1}\Delta C_p dx^* = \left\{\begin{aligned} &-4C(k)\int_{-1}^{1}\frac{\hat{w}_0(\xi^*)}{U_\infty}\sqrt{\frac{1+\xi^*}{1-\xi^*}}d\xi^* \\ &-ik\int_{-1}^{1}\hat{w}_0\xi^*\sqrt{(1-\xi^*)}d\xi^* \end{aligned}\right\} \tag{7.155}$$

so that the amplitude of the oscillatory lift force per unit span can be obtained as

$$\frac{\overline{L}}{\frac{1}{2}\rho_\infty U_\infty^2\left(\frac{c}{2}\right)} = -4C(k)\frac{2}{c}\int_0^c \hat{w}_0(x_1)\sqrt{\frac{x_1}{c-x_1}}dx_1$$
$$-\frac{4ik}{\left(\frac{c}{2}\right)^2}\int_0^c \hat{w}_0(x_1)\sqrt{x_1(c-x_1)}dx_1 \tag{7.156}$$

Similarly for the moment about the point $x = \text{ba}$,

$$\frac{\overline{M}_y}{\frac{1}{2}\rho_\infty U_\infty^2\left(\frac{c}{2}\right)^2} \equiv \int_{-1}^{1}\overline{\Delta C_p}[x^*-a]dx^* \tag{7.157}$$

so that the amplitude of nose-up pitching moment per unit span, taken about an axis along the mid-chord line, is

$$\frac{\bar{M}}{\frac{1}{2}\rho_\infty U_\infty^2\left(\frac{c^2}{2}\right)} = \frac{2}{c}\int_0^c \hat{w}_0(x_1)\left[\sqrt{\frac{x_1}{c-x_1}}-\frac{4}{c}\sqrt{x_1(c-x_1)}\right]dx_1$$
$$-\int_0^c \frac{i\omega\hat{p}(x_1)}{U_\infty}\left[\frac{1}{\sqrt{x_1(c-x_1)}}-\frac{8}{c^2}\sqrt{x_1(c-x_1)}\right]dx_1$$

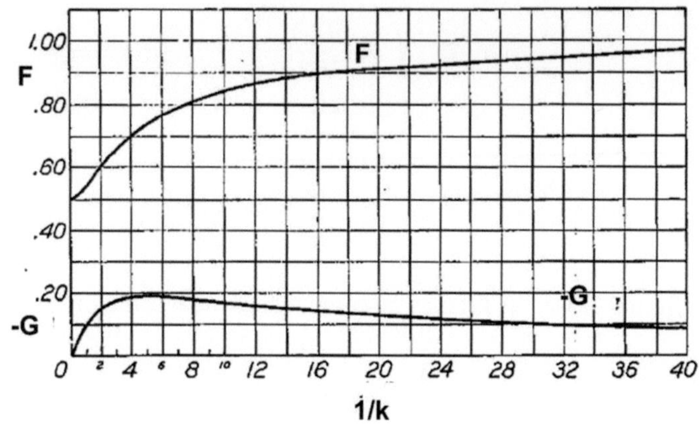

Fig. 7.5 Real and imaginary parts of Theodorsen function as functions of 1/k (Courtesy of NASA [3])

$$- C(k)\frac{2}{c}\int_{0}^{c} \hat{w}_0(x_1)\sqrt{\frac{x_1}{c - x_1}}\mathrm{d}x_1 \tag{7.158}$$

C(k) may be regarded as the lag in development of bound circulation due to the influence of the shed wake vortices. The so-called quasi-steady theory, which in one version corresponds to neglecting this wake effect, can be recovered by setting *C(k)* = 1 in (7.156) and (7.158). The Theodorsen function

$$C(k) = F(k) + iG(k) \tag{7.159}$$

is depicted in Figs. 7.5 and 7.6, where *F* is the real and *G* is the imaginary value, respectively.

7.10 Concluding Remarks

A brief but comprehensive account on the development of the fundamentals of fluid dynamics in the derivation in the mathematical expressions in unsteady aerodynamics has been elaborated in this chapter and would be useful in following their applications in further developments in the subsequent chapters.

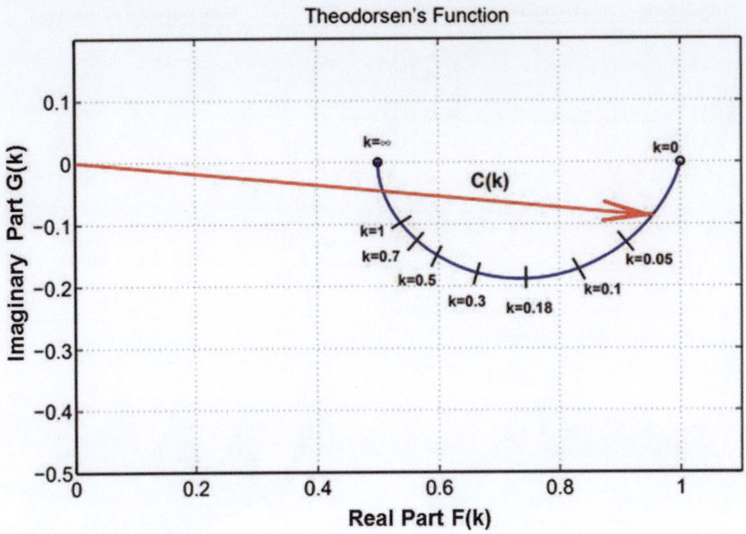

Fig. 7.6 Theodorsen function (Courtesy of NASA [3])

Appendix 1: Some Complementary Information Regarding Bessel and Hankel Functions Associated with Theodorsen Function C(k)

Some graphs of the Bessel function that form the Theodorsen function (7.153) and (7.159) are exhibited in Fig. 7.7, while the Hankel functions of order are defined in Eqs. 7.162 and 7.163.

The Bessel function of the first kind of order $\boldsymbol{\nu}$

$$\begin{aligned} J_\nu(x) &= \sum_{k=0}^{\infty} \frac{(-1)^k \left(\frac{x}{2}\right)^{\nu+2k}}{k!\Gamma(\nu+k+1)} \\ &= \frac{1}{\Gamma(1+\nu)}\left(\frac{x}{2}\right)^{\nu} \\ &\left\{1 - \frac{\left(\frac{x}{2}\right)^2}{1(1+\nu)}\left(1 - \frac{\left(\frac{x}{2}\right)^2}{2(1+\nu)}\left(1 - \frac{\left(\frac{x}{2}\right)^2}{3(1+\nu)}(1-\ldots)\right)\right)\right\} \end{aligned} \tag{7.160}$$

The Bessel function of the second kind of order $\boldsymbol{\nu}$

$$\begin{aligned} Y_\nu(x) &= \frac{2}{\pi} J_\nu(x)\left(\ln\frac{x}{2}\right) \\ &\quad - \frac{1}{\pi}\sum_{k=0}^{\nu-1} \frac{(\nu-k-1)!}{k!}\left(\frac{x}{2}\right)^{2k-\nu} \end{aligned}$$

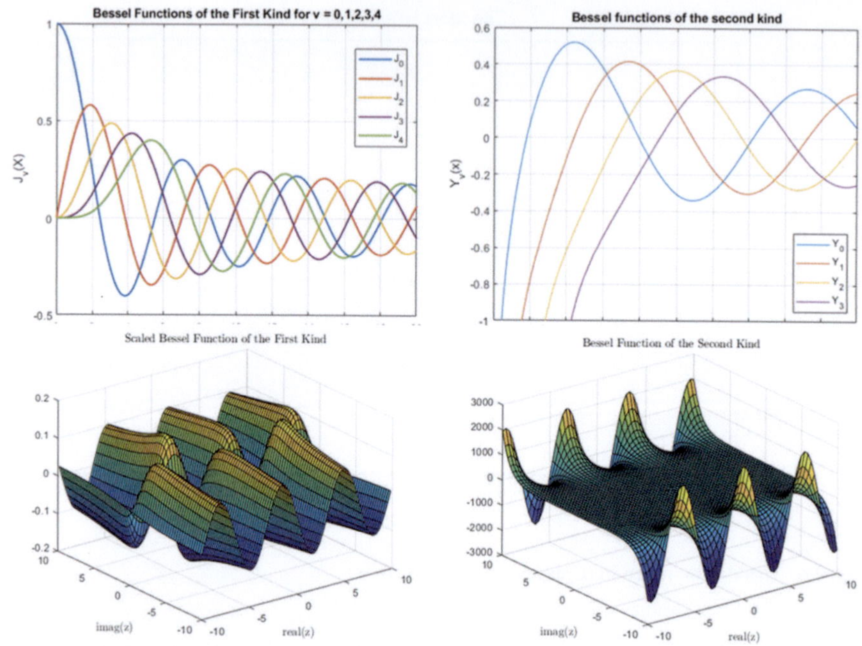

Fig. 7.7 Illustration and plot of the Bessel functions of the first kind and the Bessel functions of the second kind for $\nu = 1, 2, 3, 4$, calculated using MATLAB

$$+\frac{1}{\pi}\sum_{k=0}^{\infty}\frac{(-1)^{k-1}\left[\left(1+\frac{1}{2}+\ldots\ldots+\frac{1}{k}\right)+\left(1+\frac{1}{2}+\ldots\ldots+\frac{1}{k+\nu}\right)\right]}{k!(k+\nu)!}\left(\frac{x}{2}\right)^{2k+\nu} \tag{7.161}$$

$$H_\nu^{(1)}(x) = J_\nu(x) + iY_\nu(x) \tag{7.162}$$

$$H_\nu^{(2)}(x) = J_\nu(x) - iY_\nu(x) \tag{7.163}$$

Appendix 2: Some Excerpts from the Method of Asymptotic (Inner and Outer) Expansion for Two-Dimensional Unsteady Aerodynamics[18]

Expansion Procedure for the Equations of Motion

The method of asymptotic or inner and outer expansion reveals two paradigms; the first is to separate equalities in equations according to their order of magnitude, as articulated in the article at the beginning of this book; and the second is to capture physical significance related to two conditions and/or boundaries, not apparent in first or linearized approach, if applicable.

Now consider a two-dimensional flow in the *x*, z-plane around a thin airfoil located mainly along the x-axis with the free-stream velocity U_∞ in the direction of the positive x-axis. Referring to Figure 7.1, the location of the upper and lower airfoil surfaces can be given by

$$z_u = \varepsilon \overline{f}_u(x) = \tau \overline{g}(x) + \theta \overline{h}(x) - \alpha x \tag{7.164}$$

$$z_l = \varepsilon \overline{f}_l(x) = -\tau \overline{g}(x) + \theta \overline{h}(x) - \alpha x \tag{7.165}$$

where ε is a small dimensionless quantity measuring the maximum crosswise extension of the airfoil, τ its thickness ratio, α the angle of attack, and θ is a measure of the amount of camber. The functions $\overline{g}(x)$ and $\overline{h}(x)$ define the distribution of thickness and camber, respectively, along the chord. It will be assumed that $\overline{g}(x)$ and $\overline{h}(x)$ are both smooth and that $\overline{g}'(x)$ and $\overline{h}'(x)$ are of order of unity everywhere along the chord. A blunt leading edge is thus excluded. In the limit of $\varepsilon \Rightarrow 0$ the airfoil collapses to a segment along the x-axis and is assumed to be located between $x = 0$ and $x = c$.

With the use of asymptotic expansion method, the leading terms in a series expansion in ε of Φ will be sought to be used as an approximation for thin airfoils with small camber and angle of attack.

For two-dimensional steady flow the differential Eq. (7.118) for the velocity potential Φ simplifies to

$$\left(a^2 - \Phi_x^2\right)\Phi_{xx} + \left(a^2 - \Phi_z^2\right)\Phi_{zz} - 2\Phi_x\Phi_z\Phi_{xz} = 0 \tag{7.166}$$

where the velocity of sound is given by 3.135, which for the present case simplifies to

$$a^2 = a_\infty^2 - \frac{\gamma - 1}{2}\left(\Phi_x^2 + \Phi_z^2 - U_\infty^2\right) \tag{7.167}$$

---

[18] Adapted from Holt Ashley and Mårten Landahl [5].

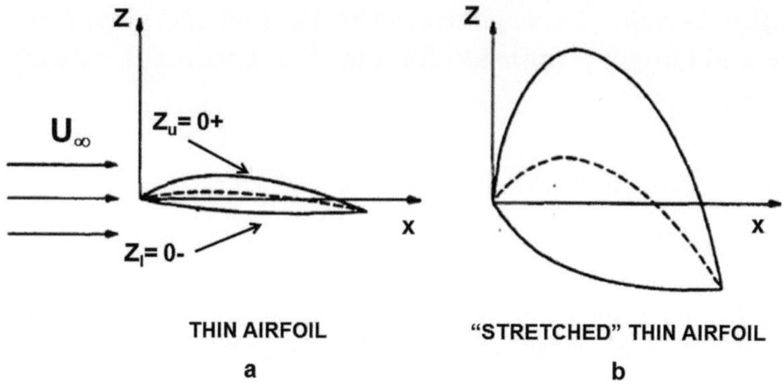

Fig. 7.8 Thin airfoil and airfoil in stretched coordinate system

Using the expression for Φ the pressure can be obtained using 3.144:

$$C_p = \frac{2}{\gamma M^2}\left\{\left[1 - \frac{\gamma - 1}{2\,a_\infty^2}\left(\Phi_x^2 + \Phi_z^2 - U_\infty^2\right)\right]^{\frac{\gamma}{\gamma-1}} - 1\right\} \tag{7.168}$$

See Fig. 7.8.

The boundary conditions are that the flow is undisturbed at infinity and tangential to the airfoil surface. Hence

$$\begin{aligned} \frac{\Phi_z}{\Phi_x} &= \varepsilon \frac{\mathrm{d}\overline{f}_u(x)}{\mathrm{d}x} \quad \text{on } z = \varepsilon \overline{f}_u \\ \frac{\Phi_z}{\Phi_x} &= \varepsilon \frac{\mathrm{d}\overline{f}_l(x)}{\mathrm{d}x} \quad \text{on } z = \varepsilon \overline{f}_l \end{aligned} \tag{7.169}$$

Additional boundary conditions required to make the solution unique for a subsonic flow are that the pressure is continuous at the trailing edge (Kutta-Joukowsky condition) and also everywhere outside the airfoil.

Now an outer expansion should first be sought of the form[19]

$$\Phi^o = U_\infty\left[\Phi_0^o(x, z) + \varepsilon\, \Phi_1^o(x, z) + \ldots\right] \tag{7.170}$$

The factor U_∞ is included for convenience; in this manner the first partial derivatives of the Φ_n-terms will be non-dimensional. Since the airfoil in the limit of $\varepsilon \Rightarrow 0$ collapses to a line parallel to the free stream, the zeroth-order term must represent parallel undisturbed flow. Thus

$$\Phi_0^o = x \tag{7.171}$$

[19] Notice that in Appendix 7.1, following Ashley and Landahl [5], by defining $\Phi_0^o = x$ in A.7.7 and not $\Phi_0^o = U_\infty x$ as conventionally done, like in A.7.28, Φ has the dimension of $[L]$.

By introducing the series (7.170) into (7.166) and (7.177), and using (7.171), after equating terms of order ε there is obtained

$$\left(1 - M^2\right)\Phi^o_{1xx} + \Phi^o_{1zz} = 0 \tag{7.172}$$

The only boundary condition available for this so far is that flow perturbations must vanish at large distances,

$$\Phi^o_{1xx}, \Phi^o_{1zz} \to 0 \quad \text{for} \quad \sqrt{x^2 + z^2} \to \infty \tag{7.173}$$

The remaining boundary conditions belong to the inner region and are to be obtained by matching. To enable the study of the flow in the immediate neighborhood of the airfoil in the limit of $\varepsilon \Rightarrow 0$, the inner solution is sought in the form

$$\Phi^i = U_\infty\left[\Phi^i_0(x, \overline{z}) + \varepsilon\, \Phi^i_1(x, \overline{z}) + \varepsilon^2 \Phi^i_2(x, \overline{z}) + \ldots\right] \tag{7.174}$$

where

$$\overline{z} = \frac{z}{\varepsilon} \tag{7.175}$$

since with such stretching the airflow shape then remains independent of ε (as depicted in Figure 7.1b and the width of the inner region becomes of order unity. The zeroth-order inner term is that of a parallel flow, that is, $\Phi^i_0 = x$, because the inner flow as well as the outer flow must be parallel in the limit $\varepsilon \Rightarrow 0$. Since the velocity component in the z direction, W, can be represented by $W = \Phi_z$, hence

$$W = \Phi_z = \frac{1}{\varepsilon}\Phi^i_{\overline{z}} = U_\infty\left[\Phi^i_{1\overline{z}}(x, \overline{z}) + \varepsilon\, \Phi^i_{2\overline{z}}(x, \overline{z}) + \ldots..\right] \tag{7.176}$$

then the zeroth-order inner term condition dictates that Φ^i_1 must be independent of $\overline{z}$, say $\Phi^i_1 = \overline{g}_1(x)$. In general this value should be different above and below the airfoil, since otherwise W would not vanish in the limit of $\varepsilon \Rightarrow 0$. Hence (7.174) may directly be simplified to

$$\Phi^i = U_\infty\left[x + \varepsilon\, \overline{g}_1(x) + \varepsilon^2\, \Phi^i_2(x, \overline{z}) + \ldots\right] \tag{7.177}$$

By substituting (7.173) into the differential equation (7.161) and the associated boundary condition (7.164) we find that

$$\Phi^i_{2\overline{z}}(x, \overline{z}) = 0 \tag{7.178}$$

$$\Phi^i_{2\overline{z}}(x, \overline{z}) = \frac{d\overline{f}_u}{dx} \quad \text{for} \quad \overline{z} = \overline{f}_u(x) \tag{7.179a}$$

$$\Phi^i_{2\overline{z}}(x,\overline{z}) = \frac{\mathrm{d}\overline{f}_l}{\mathrm{d}x} \quad \text{for} \quad \overline{z} = \overline{f}_l(x) \tag{7.179b}$$

Thus, the solution must be linear in $\overline{z}$,

$$\Phi^i_2(x,\overline{z}) = \overline{z}\frac{\mathrm{d}\overline{f}_u}{\mathrm{d}x} + \overline{g}_{2u}(x) \quad \text{for} \quad \overline{z} \geq \overline{f}_u \tag{7.180a}$$

$$\Phi^i_2(x,\overline{z}) = \overline{z}\frac{\mathrm{d}\overline{f}_l}{\mathrm{d}x} + \overline{g}_{2l}(x) \quad \text{for} \quad \overline{z} \geq \overline{f}_l \tag{7.180b}$$

These results mean that to lowest order the streamlines are parallel to the airfoil surface throughout the inner region.

The inner solution cannot give vanishing disturbances at infinity since this boundary condition belongs to the outer region.

Therefore the inner and outer solutions should be matched. This can be done in two ways; either one can use the limit matching principle for W or the asymptotic matching principle for Φ. From (7.180a, b) it follows that W^i is independent of $\overline{z}$ to lowest order, hence in the outer limit $\overline{z} \Rightarrow \infty$

$$W^{\text{io}} = \frac{1}{\varepsilon}\Phi^i_{\overline{z}}(x,\infty) = \varepsilon\frac{\mathrm{d}\overline{f}_u}{\mathrm{d}x} \tag{7.181}$$

Now

$$W^o = \varepsilon\,\Phi^o_{1\overline{z}} \tag{7.182}$$

Equating the inner limit ($z = 0+$) to outer limit (7.177), we obtain the following boundary condition:

$$\Phi^o_{1\overline{z}}\,(x, 0+) = \frac{\mathrm{d}\overline{f}_u}{\mathrm{d}x} \tag{7.183}$$

and in a similar manner

$$\Phi^o_{1\overline{z}}\,(x, 0-) = \frac{\mathrm{d}\overline{f}_l}{dx} \tag{7.184}$$

By matching the potential itself

$$\Phi^i = U_\infty\big[x + \varepsilon\,\overline{g}_1(x) + \varepsilon^2\Phi^i_2(x,\overline{z}) + \ldots\big] \tag{7.17}$$

$$\Phi^o = U_\infty\big[\Phi^o_0(x,z) + \varepsilon\,\Phi^o_1(x,z) + \ldots\big] \tag{7.10}$$

we find that

$$\overline{g}_{1u}(x) = \Phi_1^o(x, 0+) \tag{7.185}$$

To determine $\overline{g}_2$ it is necessary to go to a higher order in the outer solution.

To illustrate the use of the asymptotic matching principle, first express the two-term outer flow in inner variables,

$$\Phi^o = U_\infty\left[x + \varepsilon\,\Phi_1^o(x, \varepsilon\overline{z}) + \ldots\right] \tag{7.186}$$

and then take the three-term inner expansion of this, namely

$$\Phi^o = U_\infty\left[x + \varepsilon\,\Phi_1^o(x, 0+) + \varepsilon^2\overline{z}\,\Phi_1^o(x, 0+) + \ldots\right] \tag{7.187}$$

which upon re-expression in outer variables yields

$$\Phi^o = U_\infty\left[x + \varepsilon\,\Phi_1^o(x, 0+) + \varepsilon\,z\,\Phi_1^o(x, 0+) + \ldots\right] \tag{7.188}$$

The three-term inner expansion, expressed in outer variables, reads

$$\Phi^i = U_\infty\left[x + \varepsilon\,\overline{g}_{1u}(x) + \varepsilon\,z\frac{\mathrm{d}\overline{f}_u}{\mathrm{d}x} + \varepsilon^2\overline{g}_{2u}(x) + \ldots\right] \tag{7.189}$$

Thus equating of the two-term outer expansion of the three-term inner expansion

$$\Phi^i = U_\infty\left[x + \varepsilon\,\overline{g}_{1u}(x) + \varepsilon\,z\frac{\mathrm{d}\overline{f}_u}{\mathrm{d}x} + \ldots\right] \tag{7.190}$$

with the three-term inner expansion of the two-term outer expansion as given by

$$\overline{g}_{1u}(x) = \Phi_1^o(x, 0+) \tag{7.191a}$$

as before, (7.185), and from (7.188)

$$\Phi^i_{\overline{z}} = \frac{d\overline{f}_u}{dx} \tag{7.191b}$$

From the velocity components we may calculate the pressure coefficient by use of (7.168). Expanding in ε and using (7.181) we find that the pressure on the airfoil surface is given by

$$C_p = -2\,\varepsilon\,\Phi_{1x}^o(x, \pm) \tag{7.191}$$

where the plus sign is to be used for the upper surface and the minus sign for the lower one.

An examination of the expression (7.185) makes clear that such a simple inner solution cannot hold in regions where the flow changes rapidly in the x-direction as near the wing edges, or near discontinuities in airfoil surface slope. For a complete analysis of the entire flow field, these must be considered as separate inner flow regions to be matched locally to the outer flow. The singularities in the outer flow that are usually encountered at, for example, wing leading edges, do not occur in the real flow and should be interpreted rather as showing in what manner the perturbations due to the edge die off at large distances. For a discussion of edge effects on the basis of matched asymptotic expansions and for further detail the reader should refer to Ashley and Landahl [5] and Van Dyke [10].

Appendix 3: Applications of Two-Dimensional Unsteady Aerodynamics

Consider the flow disturbances in a uniform stream

$$\Phi_0^o = U_\infty x \tag{7.192}$$

that are generated by a thin lifting surface, as depicted in Fig. 7.4, which is performing rapid, small displacements, as a rigid body or elastically, in a direction generally normal to its x,y-plane projection, in a situation mathematically and physically analogous to vibrations.

Let z_u and z_l be given functions of position and time, representing the boundary conditions at the upper and lower surface, respectively, of the lifting surface, which generalize the steady-state requirement of flow tangency at the surface. Hence

$$\begin{aligned}\Phi_z(x, y, z_u, t) = & \frac{\partial z_u}{\partial x}\Phi_x(x, y, z_u, t) \\ & + \frac{\partial z_u}{\partial y}\Phi_y(x, y, z_u, t) + \frac{\partial z_u}{\partial t}\end{aligned} \tag{7.193a}$$

$$\Phi_z(x, y, z_l, t) = \frac{\partial z_l}{\partial x}\Phi_x(x, y, z_l, t) + \frac{\partial z_l}{\partial y}\Phi_y(x, y, z_l, t) + \frac{\partial z_l}{\partial t} \tag{7.193b}$$

for *(x, y)* on *S*. There is the usual auxiliary condition of vanishing disturbances at points remote from the wing and its wake, but for *compressible* fluid this must be refined to ensure that such disturbances behave like outward-propagating waves. The Kutta-Joukowsky hypothesis of continuous pressure at subsonic trailing edges is also applied.

Provided that there are no time-dependent variations of profile thickness, the upper and lower surface coordinates can be given by

$$z_u = \varepsilon \overline{f}_u(x, y, t) = \tau\overline{g}(x, y) + \theta\overline{h}(x, y, t) \tag{7.110}$$

$$z_l = \varepsilon\, \overline{f}_l(x, y, t) = -\tau \overline{g}(x, y) + \theta \overline{h}(x, y, t) \tag{7.111}$$

Here ε is a dimensionless small parameter measuring the maximum crosswise extension of the wing, including the space occupied by its unsteady displacement. The angle of attack α can be thought of as comprising the θ, $\overline{g}$ and $\overline{h}$ terms that are smooth functions as in steady motion; their x- and y-derivatives are everywhere of order unity.

The t-derivative of $\overline{g}$ and $\overline{h}$ will be elaborated as follows. Recognizing that in the limit $\varepsilon \Rightarrow 0$, the wing collapses to the x,y-plane, the perturbation vanishes, and the leading terms of these will be sought by the method of matched asymptotic expansions. Let the inner and outer series expansion be written as

$$\Phi^o = U_\infty\big[x + \varepsilon\, \Phi_1^o(x, y, z, t) + \ldots\big] \tag{7.194a}$$

$$\Phi^i = U_\infty\big[x + \varepsilon\, \Phi_1^i(x, y, \overline{z}, t) + \ldots\big] \tag{7.194b}$$

where

$$\overline{z} = \frac{z}{\varepsilon} \tag{7.195}$$

as in earlier developments. The presence of a uniform stream, which is a solution of 3.144, has already been recognized in the zeroth-order terms in (7.191a) and (7.191b).

When we insert (7.111) into (7.194), a new question arises as to the size of $\frac{\partial \overline{f}_u(x,y,t)}{\partial t}$ that is of $\frac{\partial \overline{h}(x,y,t)}{t}$. These derivatives may normally be expected to control the orders of magnitude of the time derivatives of Φ; hence of the terms that must be retained when (7.191a, b) are substituted into 3.144. To avoid the complexities of this issue, time and space rates of change are required to be of comparable magnitudes. For example, within the framework of linearized theory of a sinusoidal oscillation, these can be represented by

$$\overline{h}(x, y, t) = \hat{\overline{h}}(x, y) e^{i\omega t} \tag{7.196}$$

The combination of (7.196) with (7.193) followed by a non-dimensionalization of Φ_z and $\hat{\overline{h}}$ through division by U_∞ and typical length l, respectively, produces a term containing the factor

$$k = \frac{\omega l}{U_\infty} \tag{7.197}$$

Here k is known as the reduced frequency, and our present intention is to specify that $k = O(1)$.

With the foregoing limitation on sizes of time derivatives, we find that the development of small-perturbation unsteady flow theory parallels the steps elaborated in Appendix 7.2. Thus the condition that vertical velocity W must vanish as $\varepsilon \Rightarrow 0$ shows that Φ_1^i must be independent of $\overline{z}$, say

$$\Phi_1^i = \overline{g}_1(x, y, t) \tag{7.198}$$

By combining (7.194a, b) with (7.111), we conclude that Φ_2^i is the first term to possess a nonzero boundary condition,

$$\left.\begin{aligned} \Phi_{2_{\overline{z}}}^i &= \frac{\partial \overline{f}_u}{\partial x} + \frac{1}{U_\infty}\frac{\partial \overline{f}_u}{\partial t} \text{ at } \overline{z} = \overline{f}_u \\ \Phi_{2_{\overline{z}}}^i &= \frac{\partial \overline{f}_l}{\partial x} + \frac{1}{U_\infty}\frac{\partial \overline{f}_l}{\partial t} \text{ at } \overline{z} = \overline{f}_l \end{aligned}\right\} \text{ for } (x, y) \text{ on } S \tag{7.199}$$

The differential equation

$$\Phi_{2_{\overline{zz}}}^i(x, y, t) = 0 \tag{7.200}$$

requires a solution linear in $\overline{z}$; thus

$$\Phi_2^i(x, y, t) = \overline{z}\left[\frac{\partial \overline{f}_u(x, y, t)}{\partial x} + \frac{1}{U_\infty}\frac{\partial \overline{f}_u(x, y, t)}{\partial t}\right] + \overline{g}_{2_u}(x, y, t) \tag{7.201a}$$

For $z\ (x, y, t) \geq \overline{f}_u(x, y, t)$

$$\Phi_2^i(x, y, t) = \overline{z}\left[\frac{\partial \overline{f}_l(x, y, t)}{\partial x} + \frac{1}{U_\infty}\frac{\partial \overline{f}_l(x, y, t)}{\partial t}\right] + \overline{g}_{2_l}(x, y, t) \tag{7.201b}$$

For $z(x, y, z) \leq \overline{f}_l(x, y, t)$.

As in steady flow, the z-velocities are seen to remain unchanged along vertical lines through the inner field, and it will be shown to serve as a "cushion" that transmits both W and pressure directly from the outer field to the wing.

Using similar rationale, the first-order outer differential equation can be found to be

$$\left(1 - M^2\right)\Phi_{1_{xx}}^o + \Phi_{1_{yy}}^o + \Phi_{1_{zz}}^o - \frac{2M^2}{U_\infty}\Phi_{1_{xt}}^o - \frac{M^2}{U_\infty}\Phi_{1_{tt}}^o = 0 \tag{7.202}$$

By matching with W derived from (7.197), the following boundary conditions are obtained indirectly:

$$\left.\begin{aligned}\Phi^{o}_{1_z}(x, y, 0+, t) &= \frac{\partial \overline{f}_u}{\partial x} + \frac{1}{U_\infty}\frac{\partial \overline{f}_u}{\partial t}\\ \Phi^{o}_{1_z}(x, y, 0-, t) &= \frac{\partial \overline{f}_l}{\partial x} + \frac{1}{U_\infty}\frac{\partial \overline{f}_l}{\partial t}\end{aligned}\right\} \quad \text{for}(x, y) \ \text{on} \ S \tag{7.203}$$

Moreover, matching Φ itself identifies $\overline{g}_1$ with the potential Φ^o_1 at the inner limits $z = 0 \pm$.

The linear dependence of Φ^o_1 on $\overline{f}_u$ and $\overline{f}_l$ evident from (7.203) suggests that, in a small-perturbation solution which does not proceed beyond first order in ε, those portions of the flow that are symmetrical and antisymmetrical in z shoud be dealt with separately.

Adopt a perturbation potential, given by

$$\varepsilon\, \Phi^o_1 = \varphi(x, y, z, t) \tag{7.204}$$

and satisfying, together with proper conditions at infinity, the following system:

$$\left(1 - M^2\right)\varphi_{xx} + \varphi_{zz} + \varphi_{zz} - \frac{2M^2}{U_\infty}\varphi_{zt} - \frac{M^2}{U_\infty}\varphi_{tt} = 0 \tag{7.205}$$

$$\varphi_z = \theta\left[\frac{\partial \overline{h}}{\partial x} + \frac{1}{U_\infty}\frac{\partial \overline{h}}{\partial t}\right] \quad \text{at}\, z = 0 \tag{7.206}$$

for *(x, y)* on *S*. Corresponding to the pressure difference, φ has a discontinuity through *S*. The Kutta-Jotikowsky hypothesis also leads to unsteady discontinuities on the wake surface, which is approximated here by the part of the x,y-plane between the downstream wing-tip extensions.

Finally, a reduction of 3.133 and matching, to order ε, of Φ or Φ_x , shows that

$$C_p = -2\varphi_x - \frac{2}{U_\infty}\varphi_t + O\left(\varepsilon^2\right) \tag{7.207}$$

throughout the entire flow.

Appendix 4: Transonic Small-Disturbance Flow[20]

Introduction

The present section discusses a complementary topic in transonic small-disturbance flow, which concerns with thin airfoil and transonic flow. Thin airfoils can be depicted by Fig. 7.9.

[20] Elaboration in this section is based on those by Holt Ashley, Mårten Landahl [5] as the main reference.

A transonic flow is one in which local particle speeds both greater and less than sonic speed are found mixed together. Thus in the lower transonic range (ambient M slightly less than unity) there are one or more supersonic regions embedded in the subsonic flow, and similarly, in the upper transonic range the supersonic flow encloses one or more subsonic flow regions. Some typical transonic flow patterns are sketched in Fig. 7.10. Since in a transonic flow the body travels at nearly the same speed as the forward-going disturbances that it generates, one would expect that the flow perturbations are generally greater near $M = 1$ than in purely subsonic or supersonic flow. That this is indeed the case is demonstrated by experimental results, which show that the drag and lift coefficients are maximum in the transonic range. Associated with transonic flow is the notion that supersonic airplanes should experience and "overcome" the "sonic barrier," i.e. the sharp increase in drag experienced near M equal to unity.

Many of the special physical features and the associated analytical difficulties of a transonic flow may be qualitatively understood by considering the simplest case of one-dimensional fluid motion in a stream tube. After some manipulation, the combination of the Euler equation

$$U\frac{\mathrm{d}U}{\mathrm{d}x} = -\frac{1}{\rho}\frac{\mathrm{d}p}{\mathrm{d}x} \tag{7.208}$$

and the equation of continuity

$$\frac{\mathrm{d}}{\mathrm{d}x}(\rho U A) = 0 \tag{7.209}$$

where $A(x)$ is the stream-tube area yields

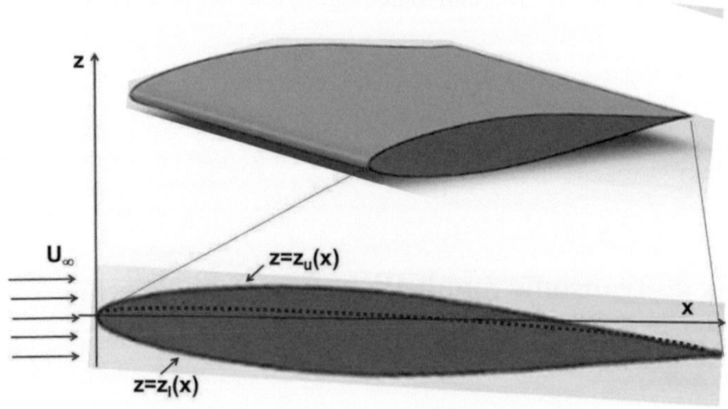

Fig. 7.9 Thin airfoil and a section of thin wing

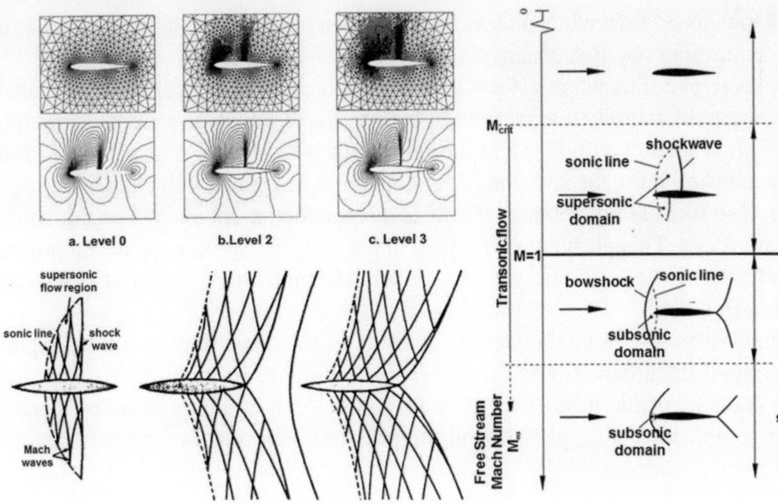

Fig. 7.10 Typical transonic flow patterns

$$\frac{dp}{\rho U^2} = \frac{dA}{A\left[1 - \left(\frac{U}{a}\right)^2\right]} = \frac{dA}{A\left[1 - M^2\right]} = \frac{dA}{A\beta^2} \tag{7.210}$$

where

$$\beta = \sqrt{\left[1 - M^2\right]} \tag{7.211}$$

This relation shows that for $U/a = 1$ the flow will resist any stream-tube area changes with an infinite force, i.e. it will effectively make the flow incompressible to gross changes in the stream-tube area. It should be noted, however, that it will not resist curvature changes or lateral displacement of a stream-tube pattern, as elaborated in Liepmann and Roshko [11] and Ashley and Landahl [5]. Therefore, the crossflow in planes normal to the free-stream direction will tend to be incompressible, as in the case of the flow near a slender body, so that much of the slender body analysis[21] applies in the transonic range to configurations that are not necessarily slender.

Due to the stream-tube area constraint, there will be a tendency for a stronger crossflow within the stream tube, and hence the effect of finite span will be maximum near $M = 1$. From Eq. (7.210), it also follows that in order to avoid large perturbation pressures and hence high drag one should avoid large (and sudden) cross-sectional area changes, which in essence is the statement of the Whitcomb transonic area rule [12, 13] as illustrated in Fig. 7.11 [12, 13] which is essential in fluid dynamics slender body in transonic flow. For the same reason one can see that the boundary layer can have a substantial influence on a transonic pressure distribution, since it provides a

[21] See, for example, Holt Ashley, Mårten Landah [5] and Liepmann and Roshko [16].

region of low-speed flow which is less "stiff" to area changes and hence can act as a "buffer" smoothing out area changes.

From such one-dimensional flow considerations, one practical difficulty also becomes apparent, namely that of wind tunnel testing at transonic speeds. Although a flow of $M = 1$ can be obtained in the minimum-area section of a nozzle with a moderate pressure ratio, the addition of a model, however small, will change the area distribution so that the flow no longer will correspond to an unbounded one of sonic free-stream speed. This problem was solved in the early 1950s with the development of the well-known slotted wall for transonic wind tunnels, such as that of NASA as depicted in Fig. 7.12.

The main difficulty in the theoretical analysis of transonic flow is that the equations for small-disturbance flow are basically nonlinear, in contrast to those for subsonic and supersonic flow, since even a small velocity change caused by a pressure change will have a large effect on the pressure-area relation. No satisfactory

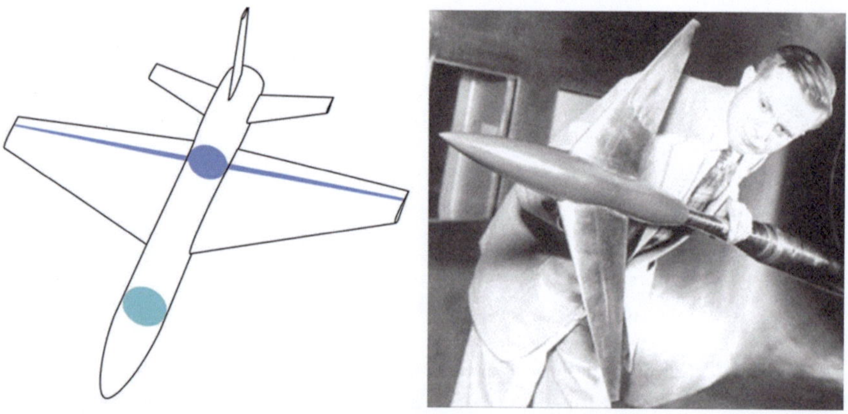

Fig. 7.11 Whitcomb transonic area rule. Although not congruent, the blue and light green shapes are roughly equal in area (Courtesy of NASA [12])

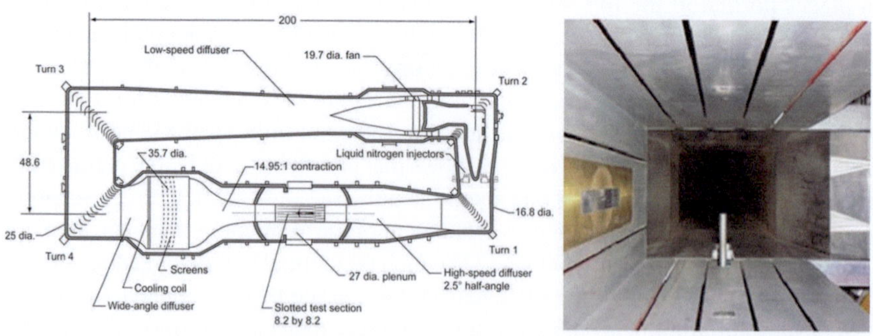

Fig. 7.12 Schematic NASA Transonic Wind Tunnel and picture of the slotted test section (Courtesy of NASA [14, 15])

general method exists for solving the transonic small-perturbation equations up to the present moment. For axisymmetric and three-dimensional flow, various approximate methods have been suggested. Some works on this topic are given in the references.[22]

In the case of two-dimensional flow it is possible, through the interchange of dependent and independent variables, to transform the nonlinear equations into linear ones in the hodograph plane.[23]

Small-Perturbation Flow Equations

For thin wings and finite, almost-plane wings, the derived equations of motion governing the sub- or supersonic flow around thin wings may be handled by simpler regular perturbation methods (Liepmann and Roshko, Chaps. 3 and 4) or by using the method of matched asymptotic expansions (Ashley and Landahl, Chaps. 5 and 7). Detailed elaboration to gain understanding of the method of matched asymptotic expansion is beyond the scope of this book, and the readers are suggested to resort to the above books. However, this technique will be followed to obtain the detail of the relevant variables associated with the utilization of the governing differential equation in the perturbed velocity potential relevant for unsteady aerodynamics and aeroelasticity.

The linearized small-perturbation techniques applied to the Bernoulli equation for irrotational flow (Kelvin's equation) for the velocity potential will be the starting point for the derivation of the small-perturbation flow equations for the velocity potential. Then the governing differential equations for the perturbation velocity potential are given by

$$\left(1 - M^2\right)\frac{\partial^2 \varphi}{\partial x^2} + \frac{\partial^2 \varphi}{\partial y^2} + \frac{\partial^2 \varphi}{\partial z^2} = 0 \tag{7.212}$$

and

$$C_p = -2\frac{\partial \varphi}{\partial x} \tag{7.213}$$

The boundary conditions for the perturbation velocity potential are given by (see Figure 7.7)

$$\left.\frac{\partial \varphi}{\partial z}\right|_{z=0+} = \left.\frac{\partial z_u}{\partial x}\right|_{z=0+} \quad \text{on } z = 0+ \text{ for } x,\, y \text{ on } S \tag{7.214a}$$

---

[22] Harijono Djojodihardjo [16]; M.G. Hall and M.C.F. Firmin [17]; Oscar Biblarz [18]; John T. Batina [19].

[23] See Holt Ashley, Mårten Landahl [5], and Ascher H, Shapiro [20]

$$\left.\frac{\partial \varphi}{\partial z}\right|_{z=0-} = \left.\frac{\partial z_l}{\partial x}\right|_{z=0-} \quad \text{on } z = 0- \ for\ x, y \text{ on } S \tag{7.214b}$$

The small-perturbation theory for sub- and supersonic flow breaks down at transonic speeds since from the linearized differential Eqs. (7.212) and (7.213) for the perturbation potential, which in the limit of $M \to 1$ becomes

$$\frac{\partial^2 \varphi}{\partial z^2} = 0 \tag{7.215}$$

for two-dimensional flow,

$$\frac{1}{r}\frac{\partial \varphi}{\partial r} + \frac{\partial^2 \varphi}{\partial r^2} = 0 \tag{7.216}$$

for axisymmetric flow.

Therefore, if the method of matched asymptotic expansions is applied, both the inner and outer flows will be described by the same differential equation, and it will in general not be possible to satisfy the boundary condition of vanishing perturbation velocities at infinity. For transonic flow it will hence be necessary to consider a different expansion that retains *at least one more* term in the equation *for* the first-order outer flow. Figure 7.10 exhibits typical transonic flow patterns **at various stages**.

In searching for such an expansion we may be guided by experiments. By testing airfoils, or bodies of revolution, of the same shape but different thickness ratios (affine bodies) in a sonic flow one will find that, as the thickness ratio is decreased, not only will the flow disturbances decrease, as would be expected, but also the disturbance pattern will persist to larger distances (see Fig. 7.13).

This would suggest that the significant portion of the outer flow will recede farther and farther away from the body as its thickness tends toward zero. In order to preserve, in the limit of vanishing body thickness, those portions of the outer flow field in which the condition of vanishing flow perturbations is to be applied we must therefore "compress" this (in the mathematical sense). Taking first the case of a two-dimensional airfoil with thickness but no lift, we shall therefore consider an expansion of the following form:

$$\Phi^i = U_\infty\left[x + \varepsilon \Phi_1^i(x, z) + \varepsilon^2 \Phi_2^i(x, z) + \ldots\ldots\right] \tag{7.217}$$

for the inner flow and

$$\Phi^o = U_\infty\left[x + \varepsilon\, \Phi_1^o(x, z) + \varepsilon^2 \Phi_2^o(x, z) + \ldots\ldots\right] \tag{7.218}$$

for the outer flow

Here ε is a small non-dimensional parameter measuring the perturbation level and

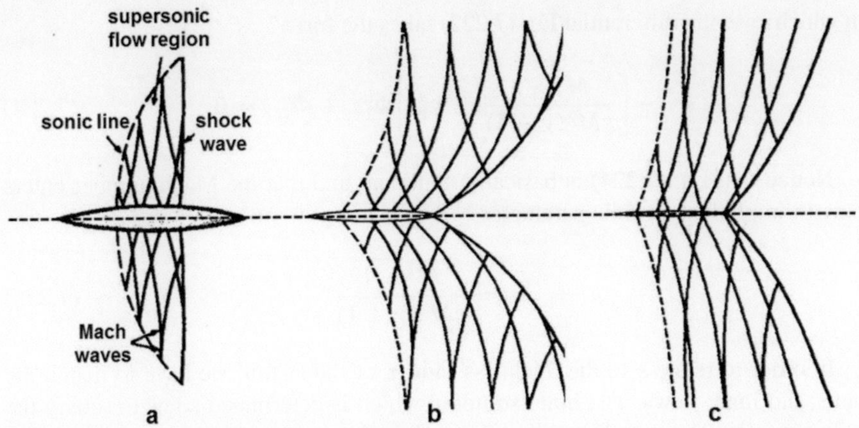

Fig. 7.13 Hypothetical flow patterns at $M = 1$ for related body geometries of two different thickness ratios (b and c)

$$\zeta = \delta z \tag{7.219}$$

with ζ being a function of ε so that $\delta - +0$ as $\varepsilon - +O$. Both ε and δ are related to the thickness ratio τ of the airfoil in a way that are elaborate in detail in Ashley and Landahl.

It is of interest that the result shows that the utilization of the inner and outer expansion will arrive at the following equations:

$$\begin{aligned}&\left[a^2 - U_\infty^2 - 2\varepsilon U_\infty^2 \Phi_{1x}^o - \varepsilon^2 U_\infty^2 \left(\Phi_{1x}^o\right)^2\right]\Phi_{1xx}^o \\ &+ \delta^2\left[a^2 - U_\infty^2 \varepsilon^2 \delta^2 \left(\Phi_{1\zeta}^o\right)^2\right]\Phi_{1\zeta\zeta}^o \\ &+ 2U_\infty^2\left(1 + +\varepsilon\Phi_{1x}^i\right)\varepsilon\delta^2\Phi_{1x\zeta}^o\Phi_{1\zeta}^o + \text{higher order trms} = 0\end{aligned} \tag{7.220}$$

where

$$a^2 = a_\infty^2 - \frac{(\gamma - 1)}{2}U_\infty^2\left[2\varepsilon\Phi_{1x}^o + \varepsilon^2\left(\Phi_{1x}^o\right)^2 + \varepsilon^2\delta^2\left(\Phi_{1\zeta}^o\right)^2\right] \tag{7.221}$$

By neglecting all terms of higher order in ε or δ we may simplify these to

$$\left[1 - M^2 - \varepsilon M^2(\gamma + 1)\Phi_{1x}^o\right]\Phi_{1xx}^o + \delta^2\Phi_{1\zeta\zeta}^o = 0 \tag{7.222}$$

In a transonic flow $1-M^2 = O(\varepsilon)$ so that the first term is of order ε, and a non-degenerate equation is obtained by setting $\delta^2 \sim \varepsilon$ For later purposes it is convenient to choose

$$\delta = M\sqrt{\varepsilon(\gamma + 1)} \tag{7.223}$$

in which case the differential Eq. (7.222) takes the form

$$-\left[\frac{M^2-1}{\varepsilon M^2(\gamma+1)}+\Phi^o_{1x}\right]\Phi^o_{1xx}+\Phi^o_{1\zeta\zeta}=0 \tag{7.224}$$

Notice that Eq. (7.224) is basically nonlinear and that the Mach number enters only through the transonic parameter

$$K_1=\frac{M^2-1}{\varepsilon M^2(\gamma+1)} \tag{7.225}$$

In order to relate ε to the thickness ratio τ of the airfoil, we have to match the outer and inner flows. The inner solution, which is determined so as to satisfy the tangency condition on the airfoil surface, gives that

$$\Phi^i_{1z}=\pm\frac{\tau}{\varepsilon}\frac{\mathrm{d}\overline{g}}{\mathrm{d}x} \tag{7.226}$$

where the upper sign is to be used for the region above the airfoil, that is, for $z\geq\tau g$, and the lower sign for the region below the airfoil. Since

$$\frac{\partial}{\mathrm{d}z}=\frac{\partial}{\mathrm{d}\zeta} \tag{7.227}$$

application of the limit matching principle will thus give

$$\Phi^o_{1\zeta}(x,\pm0)=\frac{\tau}{\delta\varepsilon}\frac{\mathrm{d}\overline{g}}{\mathrm{d}x} \tag{7.228}$$

For the expansion to be meaningful in the limit of $\varepsilon\to0$ we therefore set

$$\frac{\tau}{\delta\varepsilon}=1 \tag{7.229}$$

which upon use of $\delta=M\sqrt{\varepsilon(\gamma+1)}$ (7.A.16) gives

$$\varepsilon=\left(\frac{\tau}{M}\right)^{\frac{2}{3}}(\gamma+1)^{-\frac{1}{3}} \tag{7.230}$$

The disturbance magnitude is thus of order $\tau^{2/3}$, which is to be compared with τ in the sub- and supersonic cases. Further details are elaborated in Ashley and Landahl.[24]

Simple waves in the *x-t* plane. (a) Rightward-propagating wave, $s=F(x-a_1t)$; (b) leftward-propagating wave, $s=G(x+a_it)$.

---

[24] Holt Ashley, Mårten Landahl [1], 279 pages, ISBN-13: 978-0486648996, ISBN-10: 0486648990.

Appendix 5: Thin Airfoils in Supersonic Flow [25]

Thin Airfoils in Supersonic Flow

For $M > 1$ the differential equation governing rp may be written [5]

$$-B^2\frac{\partial^2\varphi}{\partial x^2}+\frac{\partial^2\varphi}{\partial z^2}=0 \qquad (7.231)$$

where $B=\sqrt{(M^2-1)}$. This equation is hyperbolic, which greatly simplifies the problem. A completely general solution of Eq. (7.231) can be easily shown to be

$$\varphi = F(x-Bz)+G(x+Bz) \qquad (7.232a)$$

which represents the right and left moving waves in one-dimensional wave motion in Liepmann and Roshko [11].[26]

There is a great similarity between Eq. (7.232) and the original complex representation (5.41) in Ashley and Landahl of φ in the incompressible case; in fact (7.232) may be obtained in a formal way from (5.41) simply by replacing z by $\pm\, iBz$. The lines

$$x - Bz = \text{constant} \qquad (7.233a)$$

$$x + Bz = \text{constant} \qquad (7.233b)$$

are the characteristics of the equation, see Fig. 7.12, which in the present context known as Mach lines. Disturbances in the flow propagate along the Mach lines.

In the first-order solution the actual Mach lines are approximated by those of the undisturbed stream. Since the disturbances must originate at the airfoil, it is evident that in the solution of (7.232), G must be zero for $z > 0$, whereas $F = 0$ for $z < 0$.

The solution satisfying (5.30) is thus

$$\varphi = -\frac{1}{B}z_u(x-B\,z) \quad \text{for } z>0 \qquad (7.234a)$$

$$\varphi = -\frac{1}{B}z_l(x+B\,z) \quad \text{for } z<0 \qquad (7.234a)$$

---

[25] The present section has resorted to Holt Ashley, Mårten Landahl [5] as main reference.

[26] Following Liepmann and Roshko [11], an alternation form is given by (see Fig. 7.14) .

$$u = f(x-a_1t)+g(x+a_1t) \qquad (7.232b)$$

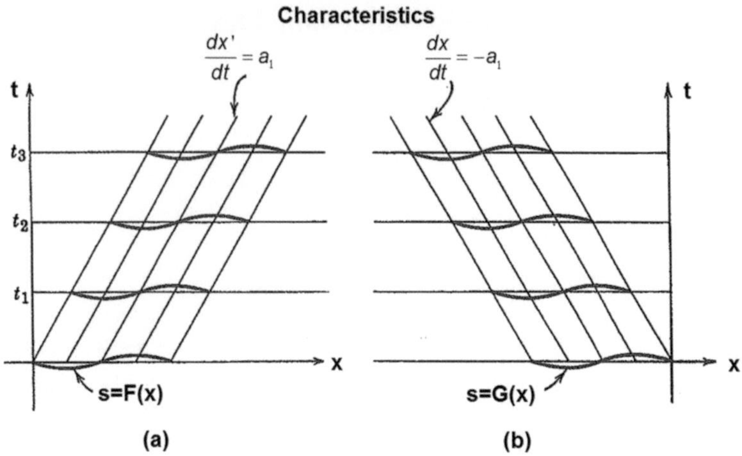

Fig. 7.14 Simple waves in the *x*-*t* plane. **a** Rightward-propagating wave, $s = F(x - a_1 t)$; **b** leftward-propagating wave, $s = G(x + a_i t)$

Then Eq. (7.233) will assume the following value

$$C_p = \frac{2}{B}\frac{\partial z_u}{\partial x} \tag{7.235}$$

(Ackeret formula). Comparisons of this simple result with experiments are shown in Fig. 7.13.[27] It is seen that the first-order theory tends to underestimate the pressure and in general is less accurate than that for incompressible flow. The deviation near the trailing edge is due to shock wave and boundary layer interaction which tends to make the higher pressure behind the oblique trailing-edge shocks leak upstream through the boundary layer (Fig. 7.14).

The much greater mathematical simplicity of supersonic flow problems than subsonic ones so strikingly demonstrated in the last two sections is primarily due to the absence of upstream influence for $M > 1$. Hence the flows on the upper and lower sides of the airfoil are independent, and there is no need to separate the flow into its thickness and lifting parts. Further details and discussion can be found in Holt Ashley, Mårten Landahl.[28]

An illustration on numerical values that compare the experimental results and theoretical ones is illustrated in Fig. 7.15, which exhibits the comparison of estimated and experimental pressure distributions for three airfoils differing only in maximum thickness, $M = 1.0$.

[27] Adapted from Holt Ashley, Mårten Landahl [1], 279 pages, ISBN-13: 978–0,486,648,996, ISBN-10: 0,486,648,990 as main reference.

[28] Holt Ashley, Mårten Landahl [1], 279 pages, ISBN-13: 978-0486648996, ISBN-10: 0486648990 as main reference.

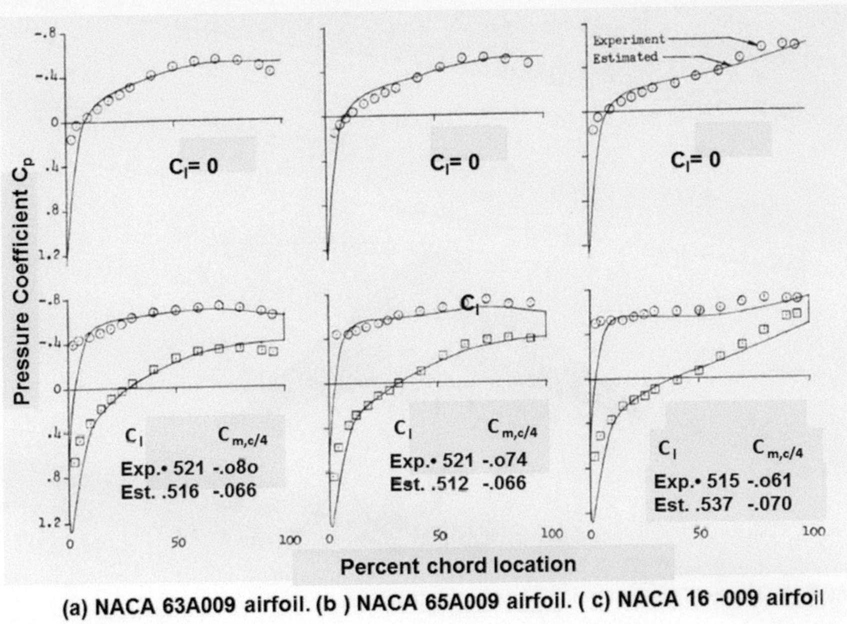

Fig. 7.15 Comparison of estimated and experimental pressure distributions for three airfoils differing only in maximum thickness, $M = 1.0$ (Courtesy of NASA [21]).

Appendix 6: Three-Dimensional Thin Wings in Steady Supersonic Flow

Introduction

In this section some developments in the theory for supersonic flow around three-dimensional wings will be reviewed. The linearized theory for thin wings as elaborated by Ashley and Landahl [5] will be used as the basis. Then the differential equation governing the perturbation velocity potential is similar to Eq. (7.212).

$$-B^2\frac{\partial^2\varphi}{\partial x^2}+\frac{\partial^2\varphi}{\partial y^2}+\frac{\partial^2\varphi}{\partial z^2}=0 \tag{7.236}$$

where $B^2 = M^2 - 1$. For three-dimensional thin wings at supersonic flow, the associated boundary conditions are given by

$$\left.\frac{\partial\varphi}{\partial z}\right|_{z=0+}=\left.\frac{\partial z_u}{\partial x}\right|_{z=0+} \quad \text{on } z=0+ \text{ for } x,\ y \text{ on } S \tag{7.237a}$$

$$\left.\frac{\partial\varphi}{\partial z}\right|_{z=0-}=\left.\frac{\partial z_l}{\partial x}\right|_{z=0-} \quad \text{on } z=0- \text{ for } x,\ y \text{ on } S \tag{7.237b}$$

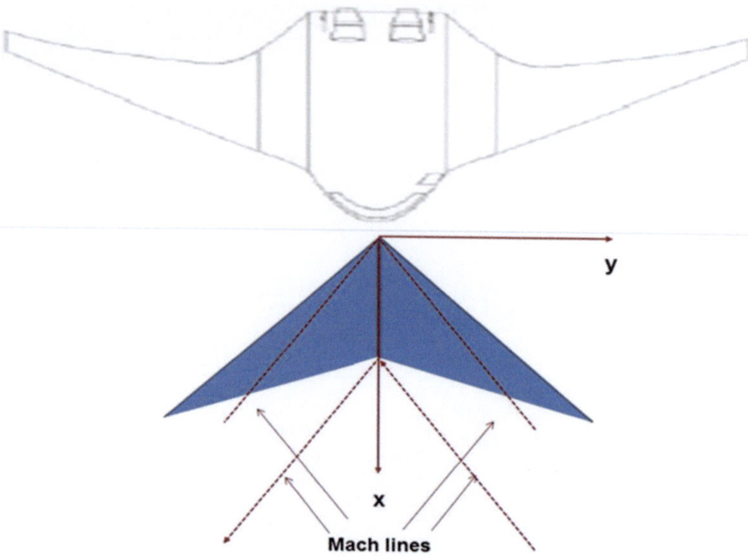

Fig. 7.16 Three-dimensional wing with simple platform

which results from the condition that the flow must be tangent to the wing surface. In addition one needs to prescribe that perturbations vanish ahead of the most upstream point of the wing.

The Kutta condition need not be specified in supersonic flow unless the wing has a trailing edge with a high sweep so that the velocity component normal to the edge is "subsonic"; such an edge will be said to be subsonic.

A wing may have both supersonic and subsonic leading and trailing edges. A wing with only supersonic leading and trailing edges is commonly referred to as having a "simple planform." An example of such a wing is given in Fig. 7.16.

Non-Lifting Wings

For a non-lifting wing $z_l = -z_u$ in equation. Therefore $w(x, y, 0+) = -w(x, y, 0-) \equiv w_0(x, y)$, for example, and the perturbation potential IP is thus symmetric in z. A convenient way to build up a symmetric potential is by covering the x,y-plane with a source distribution as was done in this chapter for the incompressible-flow case. The solution for a source of strength $!(x_1, y_1)\ \mathrm{d}x_1\ \mathrm{d}y_1$ at the point $x_1, y_1, 0$ can then be given by

$$d\varphi = -\frac{1}{2\pi}\frac{f(x_1, y_1)\mathrm{d}x_1\mathrm{d}y_1}{\sqrt{(x-x_1)^2 - B^2\left[(y-y_1)^2 + z^2\right]}} \tag{7.31}$$

After integration one obtains

$$\varphi(x, y, z) = -\frac{1}{2\pi}\iint\limits_{\Sigma} \frac{f(x_1, y_1)\mathrm{d}x_1\mathrm{d}y_1}{\sqrt{(x-x_1)^2 - B^2\left[(y-y_1)^2 + z^2\right]}} \tag{7.32}$$

where the region of integration Σ: is the portion of the x,y-plane intercepted by the upstream Mach cone from the field point *x, y, z.* This is illustrated in Fig. 7.17. Referring to Fig. 7.18 we see that the total volumetric outflow per unit area from the sources is *2Uoowo(x, Y),* each side contributing half of this value. Hence the nondimensional source strength is

$$f(x, y) = 2w_0(x, y) \tag{7.238}$$

By inserting (7.109) into (7.108) we obtain the following result, which was first derived by Puckett (1946):

$$\varphi(x, y, z) = -\frac{1}{\pi}\iint\limits_{\Sigma} \frac{w_0(x, y)\mathrm{d}x_1\mathrm{d}y_1}{\sqrt{(x-x_1)^2 - B^2\left[(y-y_1)^2 + z^2\right]}} \tag{7.239}$$

Next, the pressure coefficient is easily obtained from

$$C_p = -2\frac{\partial\varphi}{\partial x} = -2\varphi_x \tag{7.240}$$

When evaluating C_p on the wing surface $(z = \mathrm{O}\pm)$, it is convenient to integrate Eq. (7.239) first by parts with respect to x_1 and then perform the differentiation with respect to *x*. Noting the symmetry of the source solution in *x* and x_1, we then obtain

$$\begin{aligned} c_p(x, y, 0+) = &-\frac{2}{\pi}\int_A^B \frac{w_0(x_{\mathrm{LE}}, y_1)\mathrm{d}y_1}{\sqrt{(x-x_1)^2 - B^2\left[(y-y_1)^2\right]}} \\ &+ \iint\limits_{\Sigma} \frac{w_{0x_\Sigma}(x_1, y_1)\mathrm{d}x_1\mathrm{d}y_1}{\sqrt{(x-x_1)^2 - B^2\left[(y-y_1)^2\right]}} \end{aligned} \tag{7.241}$$

where *A* and *B* are the values of y_1 at the intersection of the upstream Mach lines from the point (x,y) with the leading edge of the wing (see Fig. 7.19): As a check of the above result we consider a two-dimensional wing for which *Wo* is independent of *y*. Thus the integration over y_1 can be carried out directly. Using standard integral tables we find that

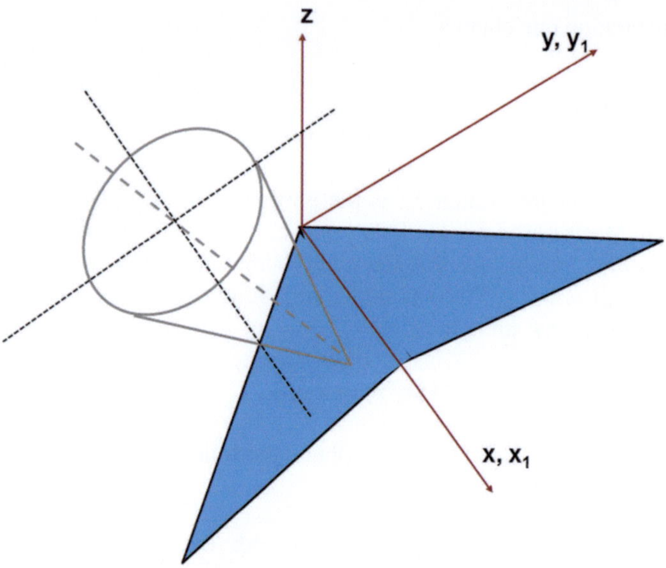

Fig. 7.17 Region of source distribution influencing the point (*x*,*y*,*x*)

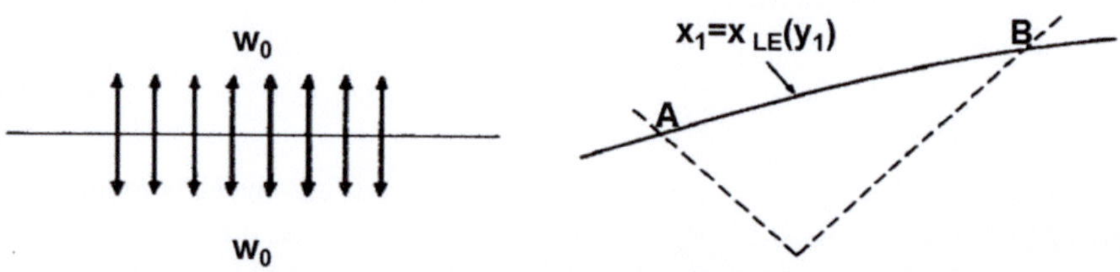

Fig. 7.18 Determination of source strength and integration limit

$$\int_{y-\frac{(x-x_1)}{B}}^{y+\frac{(x-x_1)}{B}} \frac{\mathrm{d}y_1}{\sqrt{(x-x_1)^2 - B^2\left[(y-y_1)^2\right]}} = \frac{\pi}{B} \tag{7.242}$$

Hence (7.A.36) gives

$$c_p(x, y, 0+) = \frac{2}{B}\left[w_0(0) + \int_0^x w_{0x1}\mathrm{d}x_1\right] = \frac{2w_0}{B} = \frac{2}{B}\frac{\partial z_u}{\partial x} \tag{7.243}$$

Lifting Wings of Simple Planform

For a wing with only supersonic edges, a wing with a "simple planform." There is no interaction between the flows on the upper and lower surfaces of the wing. Hence the flow above the wing, say, will be the same as on a symmetric wing with the same z_u, and (8.6) and (8.7) can therefore be directly applied. As an example, we give the result for an uncambered delta wing of sweep angle Λ (less than the sweep angle of the Mach lines) and angle of attack α. The lifting pressure difference between the lower and upper surfaces turns out to be [see, for example, Jones and Cohen (1960)]

$$\Delta C_p(x, y, 0+) = \frac{4\alpha}{\pi B} \frac{m}{\sqrt{m^2 - 1}} \mathrm{Re}\left\{ \cos^{-1} \frac{1 - mr}{m - \tau} + \cos^{-1} \frac{1 + mr}{m + \tau} \right\} \tag{7.244}$$

where

$$m = \frac{B}{\tan \Lambda} \tag{7.245a}$$

$$\tau = \frac{By}{x} \tag{7.245b}$$

and the vertex is located at the origin. It is seen that *t:Cp* is constant along rays through the wing vertex, as are all other physical properties of the flow. This is an example of so-called *conical flow* which will be discussed later. The pressure distribution given by (8.10) is shown in Fig. 8.6 for the case of $m = 1.2$. By integrating the lifting pressures over the wing surface one finds that the lift coefficient for a triangular wing becomes simply

$$C_p = -\frac{4\alpha}{B} \tag{7.246}$$

that is, the lift is independent of the sweep when the leading edges are supersonic. This result is identical to that for two-dimensional flow, as may be determined from (7.243).

For further development and derivation in obtaining the expressions for the disturbance potential and Cp for wing of simple planforms, the readers are referred to Ashley and Landahl's book. A summary of the results is presented below.

If w_0 ahead of the subsonic edge could somehow be calculated it would be possible to use

$$\varphi(x, y, z) = -\frac{1}{\pi} \iint_{\Sigma} \frac{w_0(x, y) dx_1 dy_1}{\sqrt{(x - x_1)^2 - B^2\left[(y - y_1)^2 + z^2\right]}} \tag{7.247}$$

Let us consider the following problem: a lifting surface with an angle of attack distribution *a(x, y)* has a leading edge that is partially supersonic, partially subsonic (see Fig. 7.18. The slope may be discontinuous, and the supersonic portion is assumed to end at O. The boundary conditions to be satisfied on the x,y-plane are

$$\varphi_{x=}(x, y, 0) = w_0 = -\alpha(x, y) \quad \text{on the wing} \tag{7.248a}$$

$$\varphi_{x=}(x, y, 0) = 0 \quad \text{ahead of the wing} \tag{7.248b}$$

The latter condition follows from the antisymmetry of φ_0 (Fig. 7.20).

The potential at a point P on the wing can be expressed by aid of (8.6) as follows:

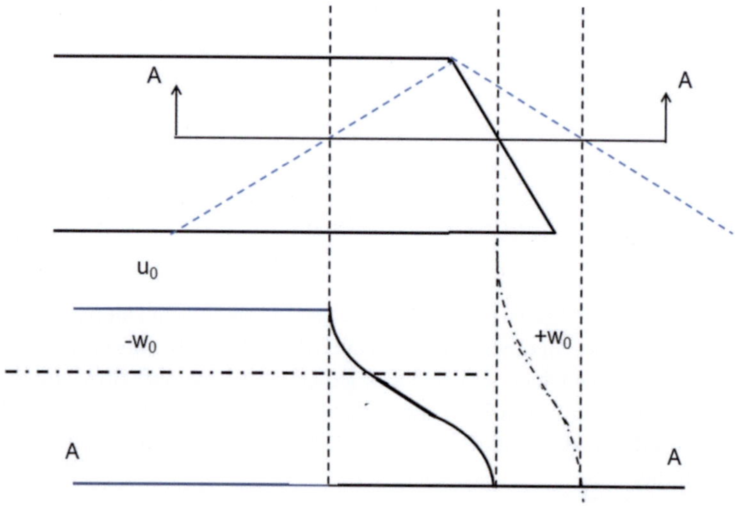

Fig. 7.19 Typical disturbance velocities uo and wo in the plane of a flat plate of planform shown

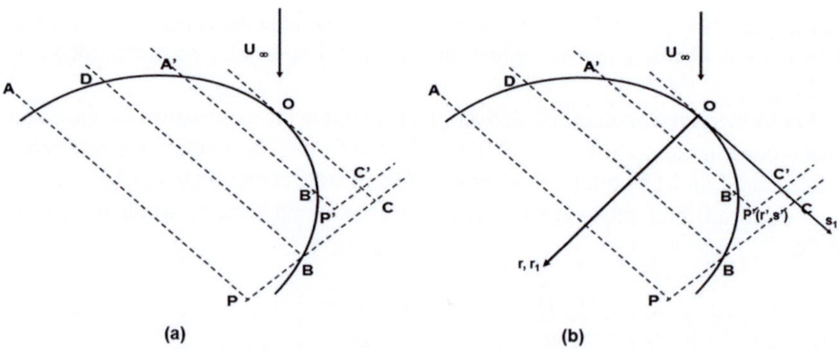

Fig. 7.20 **a** Wing planform with partially subsonic leading edge; **b** introduction of characteristic coordinates

$$\varphi(x, y, 0) = -\frac{1}{\pi} \iint\limits_{\text{OACP}} \frac{w_0(x_1, y_1)\mathrm{d}x_1\mathrm{d}y_1}{\sqrt{(x - x_1)^2 - B^2(y - y_1)^2}} \tag{7.249}$$

However, w_0 is not known in the region *OCB*. An integral equation for w_0 is obtained by use of Eq. (7.196) for a point in *OCB*. Then, because of Eq. (7.248b),

$$0 = -\frac{1}{\pi} \iint\limits_{A'OC'B'} \frac{w_0(x_1, y_1)\mathrm{d}x_1\mathrm{d}y_1}{\sqrt{(x - x_1)^2 - B^2(y - y_1)^2}} \tag{7.250}$$

To solve this integral equation it is convenient to introduce the characteristic coordinates rand 8 (see Fig. 8.9) defined by

$$r = (x - x_0) - B(y - y_0) \tag{7.251a}$$

$$s = (x - x_0) + B(y - y_0) \tag{7.251b}$$

where $x_{0,}$ y_0 are the coordinates of the point *O*. Similarly, as integration variables, we introduce

$$r_1 = (x_1 - x_0) - B(y_1 - y_0) \tag{7.252a}$$

$$s_1 = (x_1 - x_0) + B(y_1 - y_0) \tag{7.252b}$$

Thus

$$\mathrm{d}x_1\mathrm{d}y_. = \left|\frac{\partial(x_1, y_1)}{\partial(r_1, s_1)}\right|\mathrm{d}r_1\mathrm{d}s_1 = \frac{1}{2B}\mathrm{d}r_1\mathrm{d}s_1 \tag{7.253}$$

and the integral Eq. (7.250) takes the following form:

$$\int\limits_0^{r'} \int\limits_{s_A(r_1)}^{s'} \frac{w_0(r_1, s_1)\mathrm{d}r_1\mathrm{d}s_1}{\sqrt{(r' - r_1)(s' - s_1)}} = 0 \tag{7.254}$$

where $s_1 = s_A(r_1)$ is the equation of the supersonic portion of the leading edge (to the left of *O*). Equation (7.255) may be written in the following form:

$$\int\limits_0^{r'} \frac{F(r_1)\mathrm{d}r_1}{\sqrt{r' - r_1}} = 0 \tag{7.255}$$

where

$$F(r_1) = \int\limits_{s_A(r_1)}^{r'} \frac{w_0(r_1, s_1)\mathrm{d}s_1}{\sqrt{s' - s_1}} \tag{7.256}$$

Equations (7.255) and (7.256) are of the form

$$f(t) = \int\limits_{0}^{t} \frac{g(t_1)\mathrm{d}t_1}{\sqrt{t - t_1}} \tag{7.257}$$

This is well known as Abel's integral equation. Again, the solution of Abel's integral equation, in addition to be not necessary for the present problem, is beyond the scope of the present book. It is given in the appendix of Ashley and Landahl [5] book and can also be found in many classical mathematical books.[29]

References

1. Wagner, H. 1925. Uber die Entstehung des Dynamischen Auftriebes von Traflugeln. *Zeitscrift fur Angewandte Mathematik und Mechanik* 5 (1).
2. Kiissner, H.G., and G. von Gorup. 1960. Instationare Lineasierte Theorie der Flugelprofile endlicher Dicke in Inkompressibler Stromung, NR. 26. Mitteilungen Max-Planck Institute, Gottingen.
3. Theodorsen, Theodor. 1946. *General theory of aerodynamic instability and the mechanism of flutter*, NACA TR 496.
4. Bisplinghoff, R.L., H. Ashley, and R.L. Halfman, *Aeroelasticity*, Dover Publications, Inc., Copyright ©1955 by Addison-Wesley Publishing Co., Inc.
5. Ashley, Holt, Landahl, and T. Mårten. 1965. *Aerodynamics of wings and bodies.* Addison-Wesley Pubishing Co., ISBN-13: 978-0486648996, ISBN-10: 0486648990 which has been used as the main reference, 279.
6. Ballmann, J., R. Eppler, and W. Hackbusch., ed. 1988. Panel methods in fluid mechanics with emphasis on aerodynamics. *Proceedings of the Third GAMM-Seminar*, Kiel. Friedr. Vieweg and Sohn: Braunschweig (Notes on Numerical Fluid Mechanics, vol. 21). ISBN: 3-528-08095-7, January 16–18. https://doi.org/10.1007/978-3-663-13997-3.
7. Dowell, E.H., H.C. Curtiss, R.H. Scanlan, and R. Sisto. 1980. *A modern course in aeroelasticity.* Sijthoff and Noordhoff, Alphen aan den Rijn, The Netherlands, also.
8. Hancock, G.J., (ed.) 1980. *Specal course on unsteady aerodynamics.* AGARD-R-679.
9. Abramson, H.N., and G.E. Ransleben. 1963. *An experimental investigation of flutter of a fully submerged subcavitating hydrofoil.* Tech Rept No. 4, Contract Nonr - 3335(00), Southwest Research Institute, San Antonio-Houston, Texas.
10. van Dyke, Milton. 1975. *Perturbation methods in fluid mechanics.* Parabolic Press.
11. Liepmann, H.W., and A. Roshko. 1993. *Elements of gas dynamics.* Dover Publication.
12. Whitcomb, Richard T., Advanced transonic aerodynamic technology. https://ntrs.nasa.gov/search.jsp?R=197700034232020-03-22T12:09:03+00:00Z
13. Wallace, Lane E., The Whitcomb area rule-NACA aerodynamics research and innovation, https://history.nasa.gov/SP-4219/Chapter5.html.

[29] See, for example, Rudolf Gorenflo, Sergio Vessela, Abel Integral Equations, Springer Verlag, ISBN: 978-3-540-46949-0.

14. Penaranda, Frank E., and Freda, M. Shannon, (ed.). 1985. Aeronautical facilities catalogue, vol.1, Wind Tunnels, NASA RP-1132.
15. Rivers, M., J. Quest, R. Rudnik. 2015. Comparison of the NASA common research model european transonic wind tunnel test data to NASA test data. In *AIAA SciTech Forum*, 2015–1093. https://doi.org/10.2514/6.2015-1093.
16. Djojodihardjo, Harijono. 1972. Locally linearized solution of lifting transonic flow by method of parametric differentiation. *Proceedings I.T.B.* 6 (1).
17. Hall, M.G., and M.C.F. Firmin. 1974. Recent developments in methds for calculating transonic flows over wings. http://www.icas.org/ICAS_ARCHIVE/ICAS1974/Page%20134%20Hall_Firmin.pdf.
18. Biblarz, Oscar. 1975. An exact solution to the transonic equation. Naval PostGraduate Scool NPS-57Zi75041.
19. Batina, John T. 2005. Advanced small perturbation potential flow theory for unsteady aerodynamic and aeroelastic analyses. NASA TM-2005–213908.
20. Shapiro, A. H. 1953. *The dynamics and thermodynamics of compressible fluid flow*. vol. 1, 1st edn, Wiley.
21. Lindsay, Walter F., and Richard S. Dick., 1957. Two-dimensional chordwise load distributions. NACA RM-L51107.

Chapter 8
Unsteady Aerodynamics of Oscillating Objects with a Case Study

Abstract Unsteady aerodynamic theory and computational procedure are essential for calculating the aerodynamic forces in any aeroelastic problem. To gain a fundamental and comprehensive, and yet simple understanding, a two-dimensional physical model is elaborated. The model will be limited to an unsteady two-dimensional linearized model in an irrotational flow environment. Lifting and non-lifting configurations are illustrated. Illustrative examples and a case study are elaborated to obtain a good impression of problems and analytical approaches.

Keywords Unsteady aerodynamics · Binary model · Quasi-steady aerodynamics · Steady flow · Typical section

8.1 Introduction

In the present chapter, a more general unsteady aerodynamic model will be considered, but will be limited in the sense that it deals with linearized model, as well as irrotational flow. In the calculation of flutter boundaries, unsteady aerodynamic forces for harmonic or nearly harmonic motions are of utmost importance. Therefore, the discussion will be focused on the simplest unsteady aerodynamic model: harmonically oscillating thin airfoil in ideal inviscid incompressible flow.

Attention is given to physical aspects, and no rigorous mathematical treatment will be given. In spite of the simplicity of the model, the results can be used in "strip theory" approach during aircraft design.

8.2 Formulation of Unsteady Flow Problem

Consider an oscillating airfoil depicted in the following sketch.

© The Author(s), under exclusive license to Springer Nature Singapore Pte Ltd. 2023

H. Djojodihardjo, *Introduction to Aeroelasticity*,
https://doi.org/10.1007/978-981-16-8078-6_8

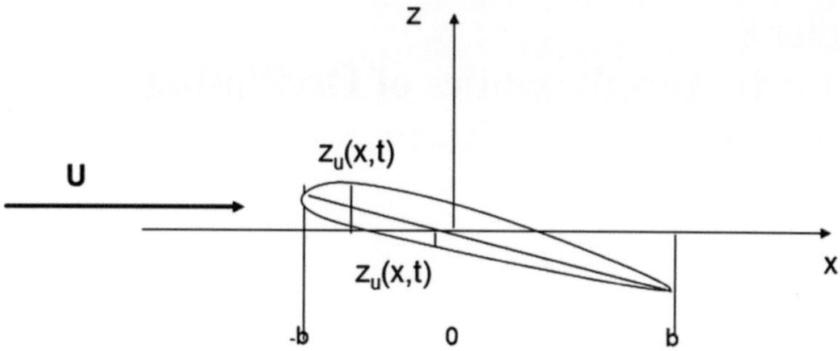

Fig. 8.1 Schematic of a linearized aerodynamic approach to harmonically oscillating thin airfoil in ideal inviscid incompressible flow

8.2.1 Irrotational Flow

The entire chapter will address irrotational flow, unless other flow conditions are specifically specified. Further, it tacitly assumed that (x, z) axes fixed to mean airfoil position.

8.2.2 Basic Flow Equation

Continuity[1]:

$$\operatorname{div} \cdot \mathbf{q} \equiv \nabla \cdot \mathbf{q} = 0 \tag{8.1}$$

where $q = (u, v, w)$ is the velocity vector

Dynamic equation:

$$\frac{\partial \mathbf{q}}{\partial t} + \frac{1}{\rho}\operatorname{grad} p \equiv \frac{\partial \mathbf{q}}{\partial t} + \frac{1}{\rho}\nabla p = 0 \tag{8.2}$$

where

p static pressure
ρ density.

Note

$\frac{d}{dt}$[2]is substantial (or total) derivative, i.e. the derivative when it is following a fluid particle

---

[1] Alternatively it can be written as div $\mathbf{V} \equiv \nabla \cdot \mathbf{V} = 0$; $\mathbf{V}$ and $\mathbf{q}$ are alternatively used as the symbol for velocity vector.

[2] Sometimes it is also written as $\frac{D}{Dt}$

$$\frac{d}{dt} \equiv \frac{\partial}{\partial t} + u\frac{\partial}{\partial x} + v\frac{\partial}{\partial y} + w\frac{\partial}{\partial z} = \frac{\partial}{\partial t} + \mathbf{q}\cdot\nabla = \frac{\partial}{\partial t} + \mathbf{q}\cdot \text{grad} \tag{8.3}$$

Next we introduce velocity potential Φ, which can be defined since the flow is irrotational (except at the airfoil wake).[3]

Note

$$\nabla \times \mathbf{q} = 0 \quad \text{or} \quad \nabla \otimes \mathbf{q} = 0 \tag{8.4}$$

implies that

$$\nabla \times (\nabla\Phi) = 0 \quad \text{thus} \quad \nabla \otimes (\nabla\Phi) = 0 \quad \text{or} \quad \nabla \otimes \text{grad}\,\Phi = 0 \tag{8.5}$$

where

$$\text{grad}\,\Phi = \mathbf{i}\frac{\partial\Phi}{\partial x} + \mathbf{j}\frac{\partial\Phi}{\partial y} + \mathbf{k}\frac{\partial\Phi}{\partial z} \tag{8.5a}$$

Since in three dimension

$$\nabla \otimes (\nabla\Phi) = \nabla \otimes grad\,\Phi = \begin{vmatrix} \mathbf{i} & \mathbf{j} & \mathbf{k} \\ \frac{\partial}{\partial x} & \frac{\partial}{\partial y} & \frac{\partial}{\partial z} \\ \frac{\partial\Phi}{\partial x} & \frac{\partial\Phi}{\partial x} & \frac{\partial\Phi}{\partial x} \end{vmatrix}$$

$$= \mathbf{i}\left(\frac{\partial^2\Phi}{\partial y\partial z} - \frac{\partial^2\Phi}{\partial z\partial y}\right) + \mathbf{j}\left(\frac{\partial^2\Phi}{\partial z\partial x} - \frac{\partial^2\Phi}{\partial x\partial z}\right) + \mathbf{k}\left(\frac{\partial^2\Phi}{\partial x\partial y} - \frac{\partial^2\Phi}{\partial y\partial x}\right) = 0 \tag{8.6}$$

Hence

$$\mathbf{q} = \text{grad}\,\Phi \tag{8.7}$$

In the two-dimensional unsteady aerodynamic theory, we are concerned only with the dimensions x and z (and u and w).

Next we will assume that moving airfoil induces small disturbances in the uniform free stream:

$$\Phi(x,z,t) = \underbrace{Ux}_{uniform\ free\ stream} + \underbrace{\varphi(x,z,t)}_{disturbance\ potential} \tag{8.8}$$

where

[3] In the present book, for cross product, the mathematical symbol $\times$ or $\otimes$ is used alternatively.

$$\begin{aligned} \varphi_x &= u \\ \varphi_z &= w \\ (u, w) &\ll U \end{aligned} \tag{8.8a}$$

This assumption enables linearization.
After substituting Φ, then we will obtain the following simplified equations:

$$\text{Continuity} \quad \frac{\partial^2 \varphi}{\partial x^2} + \frac{\partial^2 \varphi}{\partial z^2} = 0 \, \text{Laplace's equations} \tag{8.9}$$

and after linearization

$$\text{Dynamics} \quad p - p_\infty = -\rho\left(\frac{\partial \varphi}{\partial t} + U\frac{\partial \varphi}{\partial x}\right) \quad \text{Bernoulli's equation} \tag{8.10}$$

8.2.3 Boundary Conditions

The contour of the airfoil is given by

$$F(x, z, t) = 0 \tag{8.11}$$

or

$$\begin{aligned} &F_u = z - z_u\,(x, t) \quad \text{upper surface} \\ &(z_u, z_l) \ll b \end{aligned} \tag{8.12a}$$

$$F_l = z - z_l(x, t) \text{upper surface.} \tag{8.12b}$$

Boundary condition at contour:

$$\frac{\mathrm{d}F}{\mathrm{d}t} = \frac{\partial F}{\partial t} - (U + u)\frac{\partial F}{\partial x} + w\frac{\partial F}{\partial z} \tag{8.13}$$

That means that the fluid elements touch continuously the contour $F = 0$.
Introduction of F_u and F_l, and linearization yields

$$w_u\big(x, 0^+, t\big) = \frac{\partial z_u}{\partial t} + U\frac{\partial z_u}{\partial x} \tag{8.14a}$$

$$w_l\big(x, 0^-, t\big) = \frac{\partial z_l}{\partial t} + U\frac{\partial z_l}{\partial x} \tag{8.14b}$$

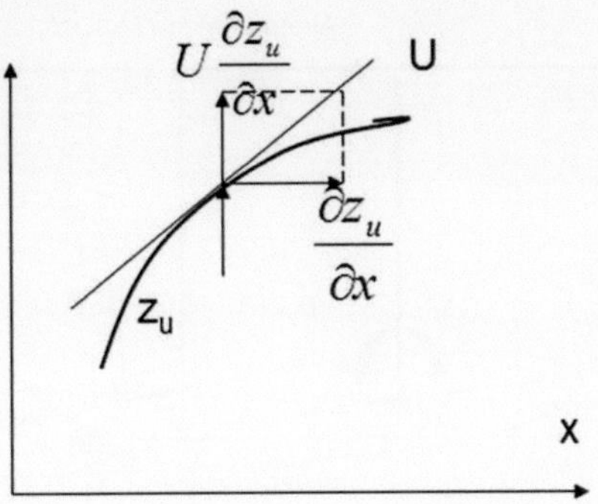

Fig. 8.2 Flow components at an element of an airfoil surface in linearized aerodynamics

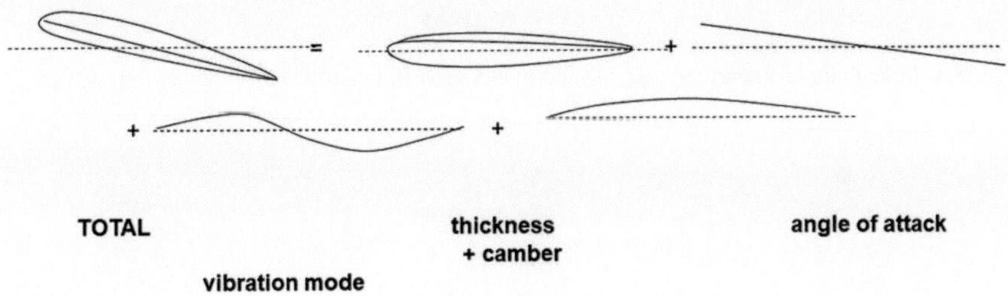

Fig. 8.3 Linear superposition of the components of an airfoil in linearized aerodynamics[4]

Physical meaning is illustrated in Figs. 8.1 and 8.2.

See Figs. 8.3, 8.4, 8.5 and 8.6.

Linearization implies further that the boundary conditions (BC) are imposed at x axis.

Separation of boundary condition into steady and unsteady parts

$$z_u(x,t) = z_t(x) + z_{\mathrm{as}}(x) + z_{\mathrm{ai}}(x,t) \tag{8.15a}$$

$$z_l(x,t) = -z_t(x) + z_{\mathrm{as}}(x) + z_{\mathrm{ai}}(x,t) \tag{8.15b}$$

8.2.4 Unsteady Boundary Condition

In what follows, the index *i* for the unsteady part will now on be omitted.

---

[4] *Note*

Boundary conditions remain imposed on the x-axis.

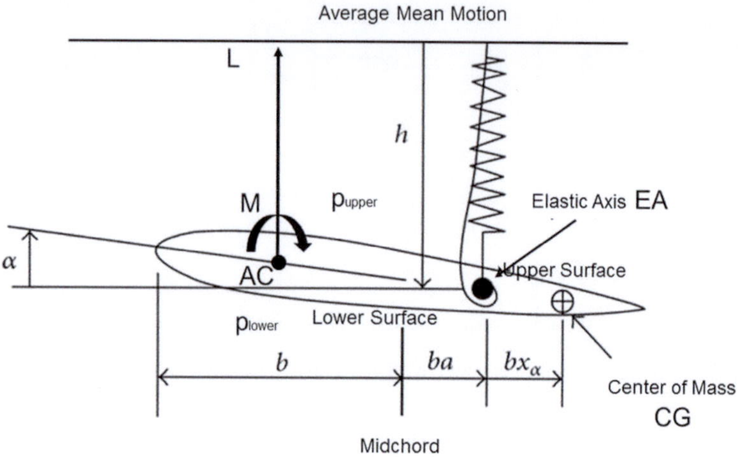

Fig. 8.4 Schematic of linearized typical section elements in linearized unsteady aerodynamics

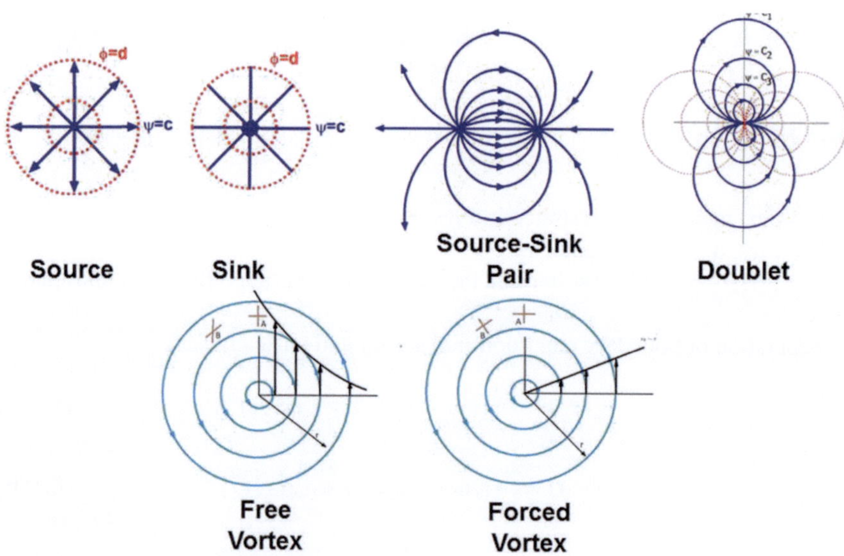

Fig. 8.5 Source, sink, source–sink pair, doublet, free and forced vortex

$$\left.\frac{\partial\varphi(x,z,t)}{\partial z}\right|_{z=0} = w_a(x,t) = \frac{\partial z_a}{\partial t} + U\frac{\partial z_a}{\partial x} \tag{8.16}$$

Proceeding to the unsteady flow problem:

The boundary condition at infinity:

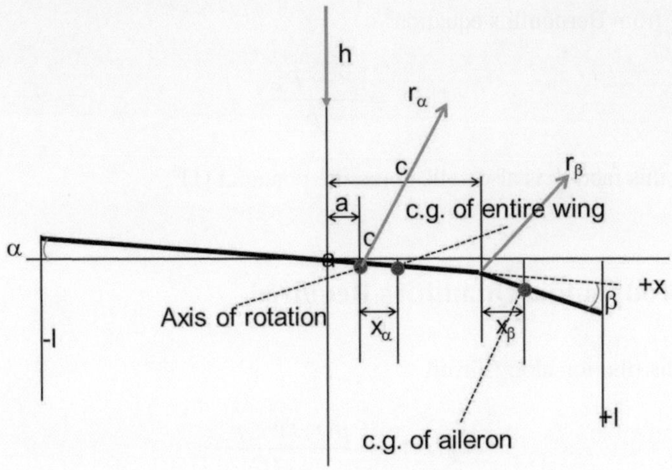

Fig. 8.6 Theodorsen parameters of the airfoil–aileron combination

$$(u, w) \to 0 \text{ for } (x, z) \to \infty \tag{8.17}$$

The Kutta condition at trailing edge should prevail (subsonic).

This implies that the velocity should be finite at the trailing edge.

The Kutta condition is equivalent to stating that the pressure, and hence φ, should be continuous.

Notice that these conditions are empirical and brings to some extent real flow effects to the (mathematical) solution.

8.3 Introduction to Acceleration Potential

Analogous to the velocity potential φ, where the relationship with the velocity is given by

$$\mathbf{V} \equiv \text{grad}\,\Phi \equiv \nabla\Phi \tag{8.18}$$

the acceleration potential can be defined as

$$a \equiv \text{grad}\,\psi \equiv \nabla\psi \tag{8.19a}$$

ψ also satisfies Laplace's equation:

$$\frac{\partial^2\psi}{\partial x^2} + \frac{\partial^2\psi}{\partial z^2} = 0 \tag{8.19b}$$

Hence, from Bernoulli's equation[5]

$$\psi = \frac{P - P_\infty}{\rho} \tag{8.20}$$

Due to this fact, ψ is also called pressure potential [1][6].

8.4 Aerodynamic Quantities Required

Pressure distribution along airfoil

$$C_p(x, t) = \frac{p(x, t) - p_\infty}{\frac{1}{2}\rho U^2} \tag{8.21}$$

Lift force (positive up)

$$L = \frac{1}{2}\rho U^2 \int_{-b}^{b} \left[p^-(x) - p^+(x)\right] dx \tag{8.22}$$

Moment about quarter-chord point (aerodynamic center) (positive trailing edge up)

$$M_{\frac{c}{4}} = \frac{1}{2}\rho U^2 \int_{-b}^{b} \left[p^-(x) - p^+(x)\right]\left(x + \frac{1}{2}b\right) dx \tag{8.23}$$

8.5 Methods of Solution for Harmonic Motions

Aerodynamic forces due to harmonic motions are essentials in analyzing and dealing with many aeroelastic problems.

For that purpose, we introduce

$$z_a(t) = \hat{z}_a\, e^{i\omega t} \quad C_p(t) = \hat{C}_p\, e^{i\omega t} \quad \varphi(t) = \hat{\varphi}\, e^{i\omega t} \quad L(t) = \hat{L}\, e^{i\omega t} \tag{8.24a,b,c,d,e,f}$$

$$\psi(t) = \hat{\psi}\, e^{i\omega t} \quad M(t) = \hat{M}_{c/4}\, e^{i\omega t}$$

---

[5] Elaboration on this notion is given in the Appendix 1.

[6] Further elabortion may be found in Ashley, H and Landahhl, [1] .

8.5.1 Boundary Value Problem

The boundary value problem then becomes

Laplace:

$$\frac{\partial^2 \varphi}{\partial x^2} + \frac{\partial^2 \varphi}{\partial z^2} = 0 \quad \text{or/and} \quad \frac{\partial^2 \psi}{\partial x^2} + \frac{\partial^2 \psi}{\partial z^2} = 0 \tag{8.25a,b}$$

Bernoulli:

$$\hat{p} - p_\infty = \rho\left(i\,\omega\,\hat{\varphi} + U\frac{\partial \hat{\varphi}}{\partial x}\right) = -\rho\,\hat{\psi} \tag{8.26}$$

Boundary conditions, airfoil:

$$\hat{w}_a = i\,\omega\,\hat{z}_a + U\frac{\partial \hat{z}_a}{\partial x} = 0 \tag{8.27a}$$

Boundary conditions, infinity:

$$(\hat{u}, \hat{w}) \to 0 \quad \text{for(x, z)} \to \infty \tag{8.27b}$$

Kutta conditions:

Finite velocity/continuous pressure at trailing edge.

8.5.2 Solution by Superposition of Elementary Solutions

Superposition of solution by adding up elementary solutions is made possible due to the linearity of governing equations and boundary conditions.

Elementary solutions satisfying Laplace's equation:

Source, Sink:

$$\varphi_s = \frac{\sigma}{2\pi}\ln r \quad r = \sqrt{x^2 + z^2} \tag{8.28}$$

Doublet with axis in z-direction:

$$\varphi_d = \frac{\mu}{2\pi}\frac{z}{x^2 + z^2} \tag{8.29}$$

Vortex:

$$\varphi_v = \frac{\Gamma\theta}{2\pi} \quad \theta = \text{arctg}\frac{z}{x} \tag{8.30}$$

8.6 Classical Methods

There are various methods that have been developed and of significant historical interest, among others by Theodorsen [2], Küssner [3], Schwarz [4] and Possio [5].

8.6.1 Velocity Potential Method [2][7]

Application of source–sink sheet along both sides of airfoil and wake.

Conformal mapping to circle.

Solution of circulatory and non-circulatory parts of velocity potential.

Theodorsen method is applicable only for incompressible flow.

Vortex Distribution Method [3, 4]
8

Application of vortex sheet along airfoil and wake.

Forming an integral equation for unknown vortex distribution.

Inversion to integral by Soehngen formula.

Kuessner and Schwarz method is applicable only for incompressible flow.

8.6.2 Integral Equation Method for Acceleration Potential [5]

Application of doublet-sheet along airfoil developed for subsonic compressible flow.

This method is discussed further below due to its potential for subsonic compressible flow and is applied for $M = 0$.

---

[7] Note 1:

Practical Aircraft Aeroelasticity.

Theodorsen Unsteady Aerrodynamics.

Three assumptions by Theodorsen.

- Attached flow and a small motion.
- Airfoil is assumed to be a flat plate.
- The wake is assumed to be flat.
- Theodorsen analysis assumes sinusoidal motion; the Equations of motion are only valid at zero airspeed or at the flutter condition.
- But they are also valid in the case of forced sinusoidal excitation.

Objective:

Calculate the response of an aeroelastic system with Theodorsen aerodynamics to any excitation force.

[8] For further detail, reference could be made to Bisplinghoff and Ashley [6]; Dowell, Curtiss, Scanlan., and Sisto [7]; Försching [8]; Platzer [9].

8.7 Methods of Solution for Harmonic Motions

8.7.1 *Integral Equation for Acceleration Potential*

The boundary value problem stated in Sect. 8.5.1 can be reduced to an integral equation

$$\frac{\hat{w}_a}{u}(x) = \int_{-b}^{b} \Delta\hat{C}_p(\xi)K(x-\xi;k)\mathrm{d}\xi \tag{8.31}$$

in which $\hat{w}_a(x)$ is prescribed downwash;

$$\frac{\hat{w}_a}{u}(x) = \frac{\partial \hat{z}_a}{\partial x} + i\frac{\omega}{U}\hat{z}_a = \frac{\partial \hat{z}_a}{\partial x} + ik\frac{\hat{z}_a}{b} \tag{8.32}$$

and

$\Delta\hat{C}_p(\xi)$ is unknown chordwise pressure distribution, to be solved.

$K(x-\xi;k)$ is a more complicated analytical Kernel function, with variable $(x-\xi)$ and k.

k is reduced frequency:

$$k = \frac{\omega b}{U}$$

Characteristics of $K(x, \xi; k)$:

- $K(x, \xi;k)$ represents upwash $\frac{\hat{w}_a}{u}$ at x due to unit pressure $\Delta\hat{C}_p(\xi)$ at ξ
- K(x, ξ;k) is integrable first order and possesses singularities for $\xi \to x$.

8.8 Standard Solution Methods of Integral Equation

The method is applicable to subsonic compressible flow. Incompressible flow is limiting case.

8.8.1 *Kernel Function Method*

- Pressure distribution $\Delta\hat{C}_p$ is expressed by finite sum of continuous pressure modes
- Pressure distribution should show correct characteristics at leading edge. For example, it is known from steady solutions:
- Leading edge characteristics: $\Delta\hat{C}_p(x) \sim \sqrt{\frac{1}{b+x}}$ leading edge peak/singularity

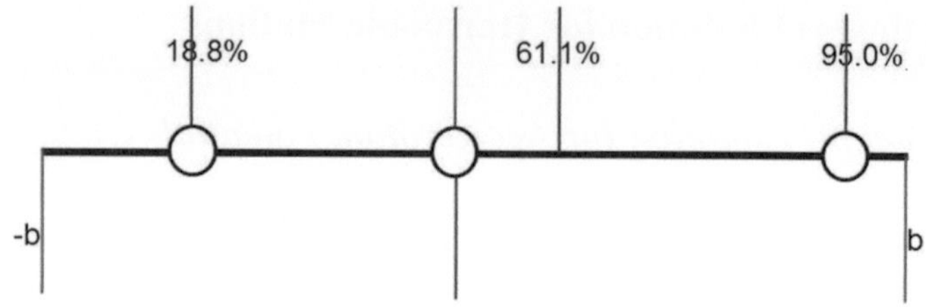

Fig. 8.7 Three collocation points

- Trailing edge characteristics $\Delta\hat{C}_p(x) \sim \sqrt{b-x}$: Kutta condition

  Appropriate pressure distribution

$$\Delta\hat{C}_p(\theta) = A_0 \cot g\left(\frac{\theta}{2}\right) + \sum_{n=1}^{N} A_n \sin(n\theta) \tag{8.33}$$

in which

$$\frac{x}{b} = -\cos\theta, \quad 0 \leq \theta \leq \pi \tag{8.34}$$

Substitute $\Delta\hat{C}_p$ in the integral equation, and let the equation be satisfied in a number of collocation points along the airfoil chord. The number of collocation points is equal to the number of pressure modes.

Optimum choice of collocation point locations

$$\theta_i = 2\pi \frac{i+2}{2N+3} \tag{8.35}$$

For three collocation points, for example (as depicted in Fig. 8.7):

- Integrations is carried out over (pressure mode) • K analytically or numerically; this leads to matrix equations:

$$\left\{\frac{\hat{w}_a}{U}\right\} = [d]\{A\} \tag{8.36}$$

$$\rightarrow \{A\} = [d]^{-1}\left\{\frac{\hat{w}_a}{U}\right\} \tag{8.37}$$

and solutions for $\Delta\hat{C}_p$, $\hat{\mathbf{L}}$ and $\hat{\mathbf{M}}$ $[d]$ is an $(N+1) \otimes (N+1)$ complex matrix of unsteady aerodynamic influence coefficients.

8.8.2 Doublet Lattice Method

- Chord is divided into $(N + 1)$ elements
- Loading on each element is represented by concentrated force at ¼ of element chord, i.e. $\Delta\hat{C}_p$ is represented by Dirac Delta functions:

$$\Delta\hat{C}_p(x) = \sum_{i=0}^{N} \Delta C_p(x_i)\delta(x - x_i) \quad \text{x}_\text{i} \text{ is loading point} \tag{8.38}$$

- Collocation points are chosen at ¾ of each element chord.
- Example: for $N = 2$ (depicted in Fig. 8.8)

- Element size needs not be uniform
- Integrations over (pressure function) • K analytically or numerically. This leads to matrix equation

$$\left\{\frac{\hat{w}_a}{U}\right\} = [D]\left\{\Delta\hat{C}_p\right\} \tag{8.39}$$

$$\rightarrow \left\{\Delta\hat{C}_p\right\} = [D]^{-1}\left\{\frac{\hat{w}_a}{U}\right\} \tag{8.40}$$

and solutions for $\hat{\mathbf{L}}$ and $\hat{\mathbf{M}}_{\frac{c}{4}}$

$[D]$ is an $(N + 1) \otimes (N + 1)$ complex matrix of unsteady aerodynamic influence coefficients.

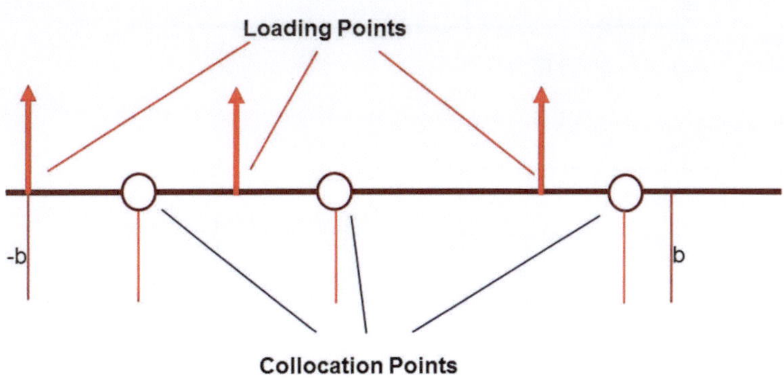

Fig. 8.8 Loading points and collocation points

8.8.3 *Relative Merits of Both Methods*

- Kernel function method provides continuous pressure distribution. For example, easy comparison with pressure data from windtunnel test can be carried out.
- Doublet lattices method is very flexible in applications. It is equally applicable in case of oscillating control.

8.9 Aerodynamic Loads Due to Oscillatory Translation

Translation

$$\hat{z}_p(x) = -\hat{h} \tag{8.41}$$

Collocation points and schematic of possible oscillations are depicted in Fig. 8.9. Pressure distribution:

Analytical solution:

$$A_0, A_1 \neq 0, A_2, \ldots\ldots = 0 \tag{8.42}$$

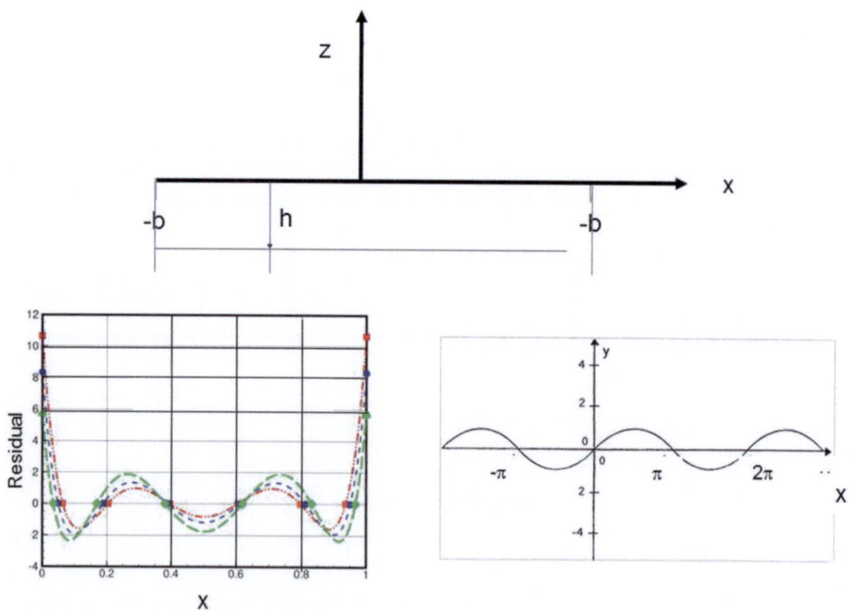

Fig. 8.9 Schematics of oscillatory translation

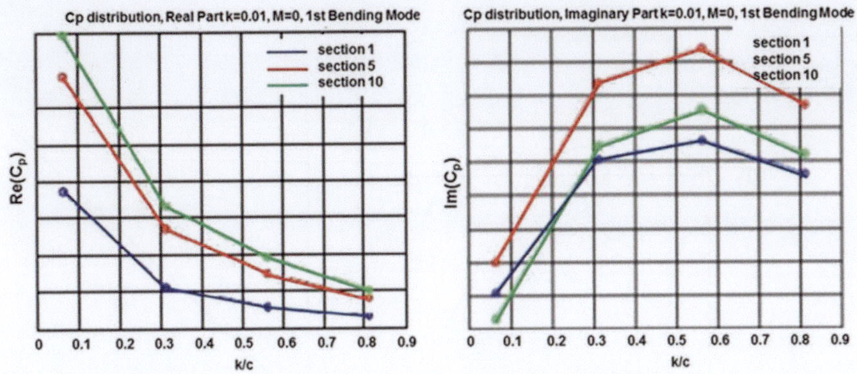

Fig. 8.10 Schematic plot of the real and imaginary parts of oscillatory pressure coefficients distribution along the typical section as a function of reduced frequency k[9]

$$\frac{\Delta\hat{C}_p}{\frac{\hat{h}}{b}} = 4\left[ikC(k)\cot\left(\frac{\theta}{2}\right) - k^2\sin\theta\right] = \frac{\Delta\hat{C}'_p}{\frac{\hat{h}}{b}} + i\frac{\Delta\hat{C}''_p}{\frac{\hat{h}}{b}} \tag{8.43}$$

$$\frac{\Delta\hat{C}_p}{\frac{\hat{h}}{b}} = 4\left[kG(k)\cot\left(\frac{\theta}{2}\right) - k^2\sin\theta\right] \tag{8.44}$$

$C(k)$ is the well-known (historical) Theodorsen function. Theodorsen function $C(k)$ can be expressed in terms of Hankel function $H_\nu{}^{(2)}$

$$C(k) = F(k) + iG(k) = \frac{H_1^{(2)}(k)}{H_1^{(2)}(k) + iH_0^{(2)}(k)} \tag{8.45}$$

$H_n^{(2)}(k)$ is Hankel function of 2nd kind and nth order. The Theodorsen functions $C(k)$, $F(k)$ and $G(k)$ are illustrated in Figs. 7.5 and 7.6, and Appendix 7.1 provides some complementary information regarding Bessel and Hankel functions associated with Theodorsen function $C(k)$.

A schematic plot of the real and imaginary parts of oscillatory pressure coefficients distribution along the typical section for various reduced frequencies k is illustrated in Fig. 8.10, while the schematic of computational results of the imaginary parts k_a' and the real part k_a " of the unsteady lift coefficient as a function of reduced frequency k is illustrated in Fig. 8.11. Figure 8.12 illustrates the schematic of computational results of the imaginary parts m_a' and the real part m_a " of the unsteady moment coefficient as a function of reduced frequency k.

Remarks $\mathrm{Re}\left[\frac{\Delta\hat{C}'_p}{\frac{\hat{h}}{b}}\right]$ occurs at most downward position of airfoil

[9] These figures are taken from Djojodihardjo et al [26].

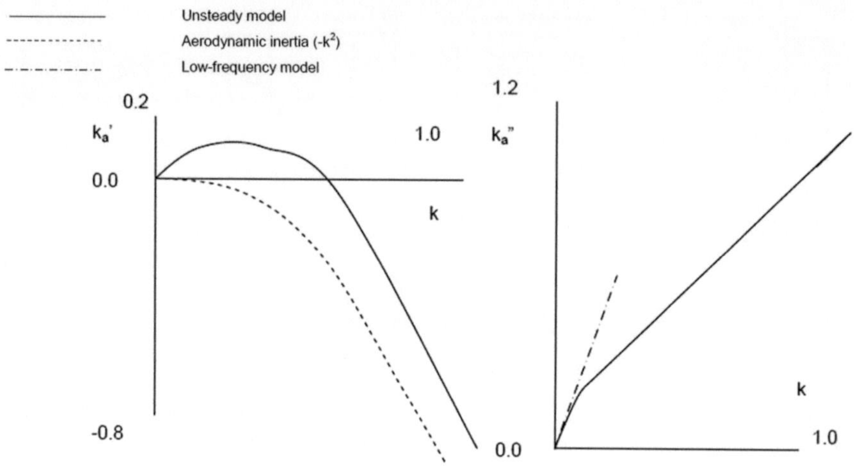

Fig. 8.11 Schematic of computational results of the imaginary parts k_a' and the real part k_a " of the unsteady lift coefficient as a function of reduced frequency k

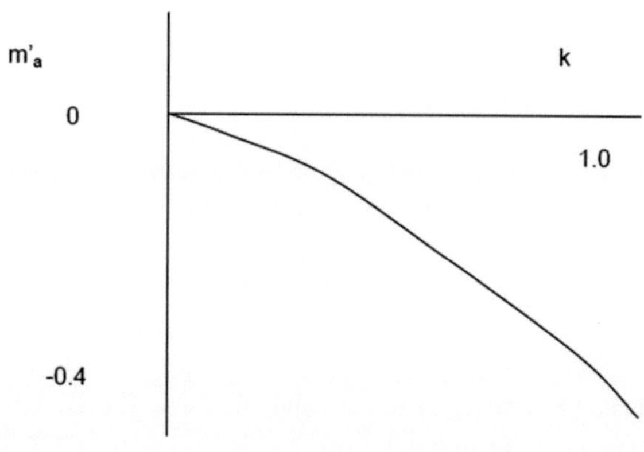

Fig. 8.12 Schematic of computational results of the imaginary parts m_a' and the real part m_a " of the unsteady moment coefficient as a function of reduced frequency k

$\mathrm{Im}\left[\frac{\Delta \hat{C}''_p}{\frac{\hat{h}}{b}}\right]$ occurs at mid-position when airfoil is moving down.
leading edge peak and Kutta condition at trailing edge prevail
Lift force:

$$\frac{\hat{L}}{\pi \rho U^2 b\left(\frac{\hat{h}}{b}\right)} = 2ikC(k) - k^2 = k'_a + ik''_a \tag{8.46}$$

where

$$k'_a = -2kG(k)$$
$$k''_a = 2kF(k) \tag{8.47a,b}$$

Moment about c/4

$$\frac{\hat{M}_{\frac{c}{4}}}{\pi\rho U^2 b^2\left(\frac{\hat{h}}{b}\right)} = -\frac{1}{2}k^2 = m'_a + im''_a \tag{8.48}$$

where

$$m'_a = -\frac{1}{2}k^2$$
$$cm''_a = 0 \tag{8.49a,b}$$

Remarks

1. The definition of unsteady aerodynamic coefficients k_a and m_a is introduced by Küssner.
2. Figures show that the significance of low-frequency model is restricted to $k <$ 0.05.

8.10 Aerodynamic Loads Due to Oscillatory Pitching

Physical Explanation of Unsteady Flow

The representation of unsteady flow by vortex sheet along airfoil and wake is depicted in Fig. 8.13.

Figure 8.14 illustrates the representation of unsteady flow by vortex sheet along airfol and wake. Reduced frequency k is the "unsteadiness" parameter, and is proportional to the ratio of the chord length $2b$ and the wave length L.

Airfoil oscillations induce variations in circulation according to Kelvin's theorem (conservation of circulation in a system of fluid which is originally at rest),

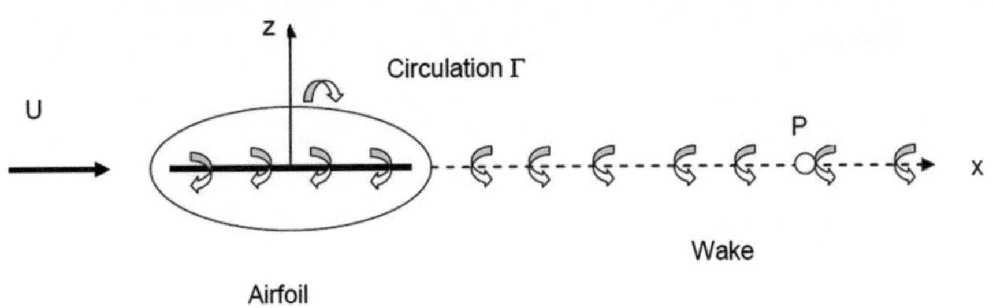

Fig. 8.13 Representation of unsteady flow by vortex sheet along the airfoil and wake

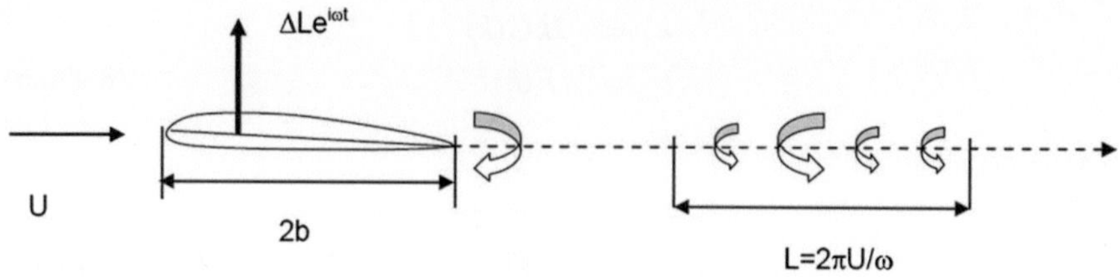

Fig. 8.14 Representation of unsteady flow by vortex sheet along airfol and wake

$$\left\{\frac{\mathrm{d}\Gamma}{\mathrm{d}t}\right\}_{\text{system of fluids}} = 0 \tag{8.50}$$

Each variation (changes) in circulation around the airfoil is accompanied by the formation of another vortex shed at the trailing edge with equal and opposite strength. Therefore, vortices are shed and drift downstream into the wake with convective stream velocity U.

Arrival at P after time (x-b)/U with phase lag relative to the generating motion is

$$\frac{\omega(x-b)}{U} = -\frac{k(x-b)}{b} \tag{8.51}$$

At P, vortex induces disturbance velocity at airfoil without further phase lag because of incompressibility assumption of the flow.

Reduced frequency k is the "unsteadiness" parameter and is proportional to the ratio of the chord length $2b$ and the wave length L

$$k = \frac{\omega b}{U} \tag{8.52}$$

All wake vortices represent information about previous airfoil motion; they form a "memory" of the flow.

8.11 Physical Interpretation of Unsteady Aerodynamic Forces

Aerodynamic forces can be expressed by the following approximation

$$\begin{aligned}\frac{\hat{L}}{\text{amplitude}} &= A_0 + A_1 k + A_2 k^2 \\ &= \underbrace{A_0}_{\text{Aerodyamic Stiffness}} + \underbrace{B_1}_{\text{Aerodyamic Damping}} \omega + \underbrace{B_2}_{\text{Inertia}} \omega^2\end{aligned} \tag{8.53}$$

It will be of interest to compare the development presented in the preceding section with respect to their analogy with mechanical quantities:

B_2 distinguishes the aerodynamic force in analogy with mechanical system as oscillation only for incompressible flow and is always positive.

$B_1 < 0$ may lead to unstable airfoil oscillation.

$A_0 < 0$ may lead to divergent airfoil motion.

8.12 Constant-Density Inviscid Flow Foundation for Unsteady Aerodynamic Applications

In what follows, some aerodynamic relationships that may be later utilized for elaborating practical problems which may be illustrated by case studies will be elaborated. The development follows closely those by Ashley and Landahl [1].

Recall that in Chap. 3, the governing equations for the desired differential equation relating the velocity potential of the irrotational fluid to the velocity field and ideal gas relationship manifested in the speed of sound a^2 are given by

$$\nabla^2\Phi - \frac{1}{a^2}\left[\frac{\partial^2\Phi}{\partial t^2} + \frac{\partial}{\partial t}\left(\frac{\nabla\Phi\cdot\nabla\Phi}{2}\right) + \nabla\Phi\cdot\nabla\left(\frac{\nabla\Phi\cdot\nabla\Phi}{2}\right)\right] = 0 \tag{8.54}$$

or

$$\nabla^2\Phi - \frac{1}{a^2}\left[\frac{\partial^2\Phi}{\partial t^2} + \frac{\partial}{\partial t}\left(\frac{V^2}{2}\right) + V\cdot\nabla\left(\frac{V^2}{2}\right)\right] = 0 \tag{8.55}$$

where

$$\mathbf{V} = \nabla\Phi \tag{8.56}$$

Writing the terms in **V** in terms of Φ yields

$$\nabla\cdot\mathbf{V} = \nabla\cdot(\nabla\Phi) = \nabla^2\Phi \tag{8.57}$$

where ∇^2 is the well-known Laplace operator and for irrotational flow the governing partial differential equation for Φ reduces to:

$$\nabla^2\Phi = \frac{\partial^2\Phi}{\partial x^2} + \frac{\partial^2\Phi}{\partial y^2} + \frac{\partial^2\Phi}{\partial z^2} \tag{8.58}$$

Cartesian Coordinate

Associated with this differential equation, the boundary conditions prescribe the values of the velocity potential or its normal derivative over the surfaces of a series of inner and outer boundaries. These conditions may be given in one of the following forms:

(1) The Neumann problem, in which the velocity component normal to the boundary surface $V_n = \mathbf{V} \cdot \mathbf{n} = \frac{\partial \Phi}{\partial n}$ is given.
(2) The Dirichlet problem, in which the value of Φ itself is given.
(3) The mixed (Poincare) problem, in which Φ is given over certain portions of the boundary and $\frac{\partial \Phi}{\partial n}$ is given over the remainder.

Among the very many concepts and practical solutions that might be considered worthy of presentation, we select some which are fundamental and instructive and which will be useful for subsequent work.

8.12.1 Green's Theorem

Let the boundary of the flow field remains at a finite distance. From Gauss' theorem for any vector Field A

$$\oiint_{S+\Sigma} \mathbf{A} \cdot \mathbf{n}\, \mathrm{d}S = -\iiint_{\mathbf{V}} \nabla \cdot \mathbf{A}\, \mathrm{d}\mathcal{V} \tag{8.59}$$

The minus sign here results from the reversal of direction of the normal vector to the boundary surface. Let Φ and Φ' be two continuous functions with finite, single valued.

first and second derivatives throughout the volume V, without specifying the physical significance of both. Let assume also that

$$\mathbf{A} = \Phi \nabla \Phi' \tag{8.60}$$

The integrand of the two sides of Eq. (8.2) may be expressed as

$$\mathbf{A} \cdot \mathbf{n} = \Phi\left[\nabla \Phi' \cdot \mathbf{n}\right] = \Phi \frac{\partial \Phi'}{\partial n} \tag{8.61}$$

$$\nabla \cdot \mathbf{A} = \nabla \cdot \left(\Phi \nabla \Phi'\right) = \Phi \nabla^2 \Phi' + \nabla \Phi \cdot \nabla \Phi' \tag{8.62}$$

Equations (8.61) and (8.62) are now substituted into Gauss theorem. After writing the result, we interchange the functions Φ and Φ', obtaining two alternative forms of the theorem:

$$\oiint_{S+\Sigma} \Phi \frac{\partial \Phi'}{\partial n} \mathrm{d}S = -\iiint_{\mathcal{V}} \left[\nabla \Phi \cdot \nabla \Phi' + \Phi \nabla^2 \Phi'\right] \mathrm{d}\mathbf{V} \tag{8.63}$$

$$\oiint_{S+\Sigma} \Phi' \frac{\partial \Phi}{\partial n} \mathrm{d}S = -\iiint_{\mathcal{V}} \left[\nabla\Phi' \cdot \nabla\Phi + \Phi'\nabla^2\Phi\right]\mathrm{d}\mathbf{V} \tag{8.64}$$

8.12.2 *Kinetic Energy*

To illustrate the application of Green's theorem, let Φ in (8.59) be the velocity potential of some flow at a certain instant of time and let $\Phi' = \Phi$. Then it follows that

$$\nabla^2\Phi \equiv \nabla^2\Phi' \tag{8.65}$$

Then we obtain a formula for the integral of the square of the fluid particle speed throughout the field

$$\oiint_{S+\Sigma} \Phi \frac{\partial \Phi}{\partial n} \mathrm{d}S = -\iiint_{\mathcal{V}} |\nabla\Phi|^2 \mathrm{d}\mathcal{V} = -\iiint_{\mathcal{V}} V^2 \mathrm{d}\mathcal{V} \tag{8.66}$$

If the last term of Eq. (8.66) is multiplied by one-half, the density of the fluid p and change its sign, then we obtain the total kinetic energy T of the fluid within the volume V. Hence

$$T = -\frac{\rho}{2} \oiint_{S+\Sigma} \Phi \frac{\partial \Phi}{\partial n} dS \tag{8.67}$$

Usually $\frac{\partial \Phi}{\partial n}$ is known from the boundary conditions. If the solid S is moving through an unlimited mass of fluid, with the fluid velocity $V = 0$ at infinity, it can easily be shown that the integral over Σ vanishes. Then Eq. (8.67) needs to only be integrated over the inner boundary S at the surface of the solid itself. Hence

$$T = -\frac{\rho}{2} \oiint_{S} \Phi \frac{\partial \Phi}{\partial n} dS \tag{8.68}$$

The Representation of Φ in Terms of Boundary Values

Let us consider a motion of a solid of volume V and surface S through a fluid and seek to find Φ_p at a certain point $P(x, y, z)$ arbitrarily located in the flow field. Let

$$\Phi' = \frac{1}{r} = \frac{1}{\sqrt{(x-x_1)^2 + (y-y_1)^2 + (z-z_1)^2}} \tag{8.69}$$

Here (x_1, y_1, z_1) is any other point, one on the boundary for instance. By direct substitution in the Laplace's equation, it can be shown that

$$\nabla^2\Phi' = \nabla^2\left(\frac{1}{r}\right) = 0 \tag{8.70}$$

everywhere except in the immediate vicinity of $r = 0$, where the Laplacian has an impulsive behavior corresponding to the local violation of the requirement of continuity.

8.13 Case Study—Simple Method to Calculathe the Oscillating Lift on a Circular Cylinder in Potential Flow [10]

Abstract
A simple potential flow model is presented to calculate the oscillating lift on a circular cylinder. In essence, the time-dependent lift acting on a cylinder is due to the presence of a vortex shed off the separation point on the cylinder. The vortex strength and its trajectory are predicted by employing the condition of zero force on the vortex and its feeding vortex sheet. The vortex breaks away from the cylinder and is convected by the fluid after the vortex strength reaches a maximum value.

At that instant, another vortex is shed off at the other side of the cylinder.

Oscillating lift with the right order of magnitude of available experimental results was predicted. For subcritical flow, good agreement is obtained between the predicted Strouhal number and that found experimentally.

8.13.1 Introduction

The oscillating lift acting on a circular cylinder has been a subject of continued interest, especially in the light of increased effort to understand and overcome wind excited vibration in the field of industrial aerodynamics. Several experimental investigations were performed by many investigators, among others Roshko [11, 19][10] Fung [13],[11] and Bishop and Hassan.[12] Alexander [14][13] has proposed a simple theory to calculate the oscillating lift acting on a circular cylinder. The present work presents an analytical model somewhat similar to that introduced by Alexander, but introduces

[10] References cited are listealphabetically at the Bibliography and References at the end of the chapter.

[11] Fung .

[12] Bishop and Hassan.

[13] Alexander [14].

additional conditions based on physical considerations. Alexander has preassumed the vortex trajectory and also assumed that the vortex shed off the cylinder breaks away from the cylinder or joins the vortex street at the instant when the lift force on the cylinder attained its maximum value, and hence suddenly the force on the cylinder due to the vortex vanishes. In the present theory, the above assumptions are removed, and instead, the vortex trajectory is computed employing the condition of zero force on the vortex and its feeding vortex sheet. The vortex breaks away from the cylinder and is convected by the fluid after its strength reaches a maximum value. The present model is capable of predicting the oscillating lift within the right order of magnitude of available experimental results. It should be noted, however, that to some extent most experimental results are affected by three-dimensional effects.

8.13.2 Analytical Model

Consider the flow around a circular cylinder, and assume that initially a simple vortex is shed off at one of the separation points on the cylinder, and then convected downstream. In this process, the vortex grows in strength. The feeding points—or separation points—will be defined from experimental evidence, or they may be arbitrarily defined a priori and verified a posteriori. It is assumed that as soon as the vortex attains its maximum strength, then it stops growing ln strength and breaks away from its feeding vortex sheet and starts to be convected downstream at von Karman's velocity and at constant strength. At the same time, another vortex is shed off on the other separation point. This assumption is based on the premise that the vortex cannot decrease ln strength, or unroll, and if such tendency seems to occur, the vortex will break away from its feeding vortex sheet and another vortex of opposite sense is shed. The initial path of the vortex shed off the cylinder is governed by the condition of zero force on the vortex and its feeding vortex sheet. The strength of the vortex at any instant is governed by the condition that the velocity at the separation point should always be equal to zero.

a. **Governing equations**

Figure 8.16 illustrates the analytical model of the flow about a cylinder at subcritical Reynolds number. At some instant after shedding, the vortex shed assumes a position z_0 and strength Γ. The complex potential at any position $z = x * iy$ for this situation is given by:

$$W(Z) = -U\left(z + \frac{a^2}{z}\right) - \frac{i\Gamma}{2\pi}\ln(z - z_0) + \frac{i\Gamma}{2\pi}\ln\left(\frac{a^2}{z} - \bar{z}_0\right) \tag{8.71}$$

The velocity field at any point $z(x,y)$ in the flow is given by

$$V = -\frac{dW(Z)}{dz} = -U\left(1 - \frac{a^2}{z^2}\right) + \frac{i\Gamma}{2\pi a}\left(\frac{a}{z - z_0} + \frac{a}{z}\frac{a}{a^2 - z_0 z}\right) \tag{8.72}$$

The u and v components are described by

$$u = U - \frac{Ua^2(x^2 - y^2)}{(x^2 + y^2)^2} + \frac{\Gamma}{2\pi}\frac{(y - y_0)}{(x - x_0)^2 + (y - y_0)^2} - \frac{a^2\Gamma}{2(x^2 + y^2)}\frac{\{y(xx_0 + yy_0 - a^2) + x(yx_0 + xy_0)\}}{\left\{(xx_0 + yy_0 - a^2)^2 + (yx_0 + xy_0)^2\right\}} \tag{8.73}$$

$$v = U - \frac{Ua^2xy}{(x^2 + y^2)^2} + \frac{\Gamma}{2\pi}\frac{(x - x_0)}{(x - x_0)^2 + (y - y_0)^2} - \frac{a^2\Gamma}{2(x^2 + y^2)}\frac{\{x(xx_0 + yy_0 - a^2) - y(yx_0 + xy_0)\}}{\left\{(xx_0 + yy_0 - a^2)^2 + (yx_0 + xy_0)^2\right\}} \tag{8.74}$$

Using Blasius theorem for the forces acting on the cylinder, one obtain

$$D - iL = \frac{1}{2}i\rho\oint_C \left(\frac{dW}{dz}\right)^2 dz \tag{8.75}$$

where the integration is carried out along a closed path around the cylinder, and D and L are the time-dependent drag and lift forces, respectively, acting on the cylinder.

Hence:

$$D = \frac{a^2x_0\rho\Gamma}{(x_0^2 + y_0^2)^2(x_0^2 + y_0^2 - a^2)}\left\{\frac{\Gamma}{2\pi}(x - x_0)^2 - 2y_0U(x_0^2 + y_0^2 - a^2)\right\} \tag{8.76}$$

$$L = \frac{a^2\rho\Gamma}{(x_0^2 + y_0^2)^2(x_0^2 + y_0^2 - a^2)}\left\{y_0\frac{\Gamma}{2\pi}(x - x_0)^2 - U(x_0^2 - y_0^2)(x_0^2 + y_0^2 - a^2)\right\} \tag{8.76}$$

Since:

$$C_D = \frac{D}{\rho U^2a}$$

and

$$C_L = \frac{L}{\rho U^2a}$$

then

$$C_D = \frac{a^2 x_0 \Gamma}{U^2 (x_0^2 + y_0^2)^2 (x_0^2 + y_0^2 - a^2)} \left\{ \frac{\Gamma}{2\pi} (x - x_0)^2 - 2 y_0 U (x_0^2 + y_0^2 - a^2) \right\} \tag{8.78}$$

$$C_L = \frac{a^2 \Gamma}{U^2 \left(x_0^2 + y_0^2\right)^2 \left(x_0^2 + y_0^2 - a^2\right)} \left\{ y_0 \frac{\Gamma}{2\pi} (x - x_0)^2 - U \left(x_0^2 - y_0^2\right) \left(x_0^2 + y_0^2 - a^2\right) \right\} \tag{8.79}$$

8.13.3 The Motion of the Vortex

The motion of the vortex is governed by the condition that the total force acting on the feeding vortex' sheet—which is assumed to have vanishing strength—and the point vortex is zero. Mathematically, there is a branch line between z_0 and z_1, the separation point, and the velocity potential 0 is discontinuous across this branch line. The discontinuity in ϕ is equal to Γ, the strength of the vortex. The state of affairs is schematically depicted in Fig. 8.15. Therefore there is a discontinuity in the pressure across this branch line with a magnitude of ρ Γ, and the total force on this vortex sheet is equal to

$$-i\rho\dot{\Gamma}(z_0 - z_1)$$

where z_1 is the location of the separation point.

The force acting on the concentrated vortex is equal to its circulation, times—ρ i, times the relative velocity between the fluid and the vortex at the point vortex, $V_0 - \dot{z}_0$, hence has a magnitude of $-i\rho\Gamma(V_0 - \dot{z}_0)$. To make the net force on the concentrated vortex and its feeding vortex sheet vanish, we must have:

$$-i\rho\dot{\Gamma}(z_0 - z_1) + i\rho\Gamma(V_0 - \dot{z}_0) = 0 \tag{8.80}$$

or

$$\dot{z}_0 + (z_0 - z_1)\frac{\dot{\Gamma}}{\Gamma} = V_0 \tag{8.81}$$

where V_0 has components u_0 and v_0 parallel to the x and y axes, respectively. From Eqs. (8.72) and (8.73), and employing a limit process, we obtain:

$$u_0 = U - Ua^2 \frac{(x_0^2 - y_0^2)}{(x_0^2 + y_0^2)^2} - \frac{\Gamma}{2\pi} \frac{a^2 y_0}{(x_0^2 + y_0^2)(x_0^2 + y_0^2 - a^2)} \tag{8.82}$$

$$v_0 = 2Ua^2 \frac{x_0 y_0}{(x_0^2 + y_0^2)^2} - \frac{\Gamma}{2\pi} \frac{a^2 y_0}{(x_0^2 + y_0^2)(x_0^2 + y_0^2 - a^2)} \tag{8.83}$$

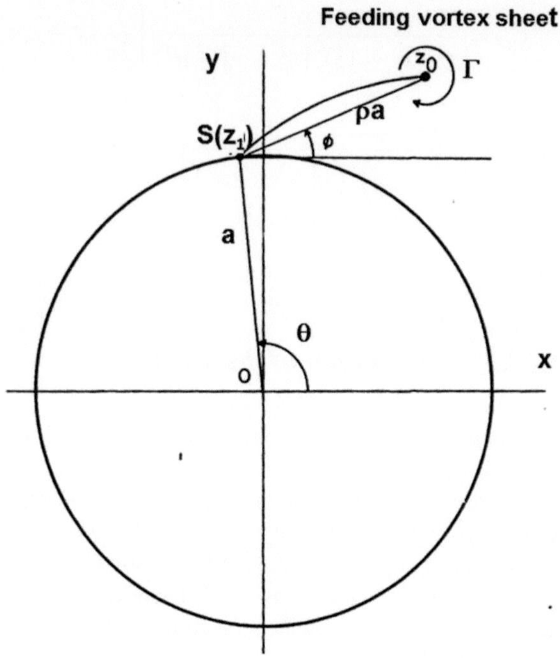

Fig. 8.15 Vortex shedding off traversing circular cylinder

The trajectory of the point vortex is a function of time and is described by:

$$x_0 = x_0(t) \tag{8.84a}$$

and

$$y_0 = y_0(t) \tag{8.84b}$$

Furthermore

$$u_0 = u_0(t) \tag{8.85a}$$

and

$$v_0 = v_0(t) \tag{8.85b}$$

From Eq. (8.77) we obtain

$$\frac{dx_0}{dt} + \frac{x_0 - x_1}{\Gamma}\frac{d\Gamma}{dt} = u_0 \tag{8.86a}$$

$$\frac{dy_0}{dt} + \frac{y_0 - y_1}{\Gamma}\frac{d\Gamma}{dt} = v_0 \tag{8.86b}$$

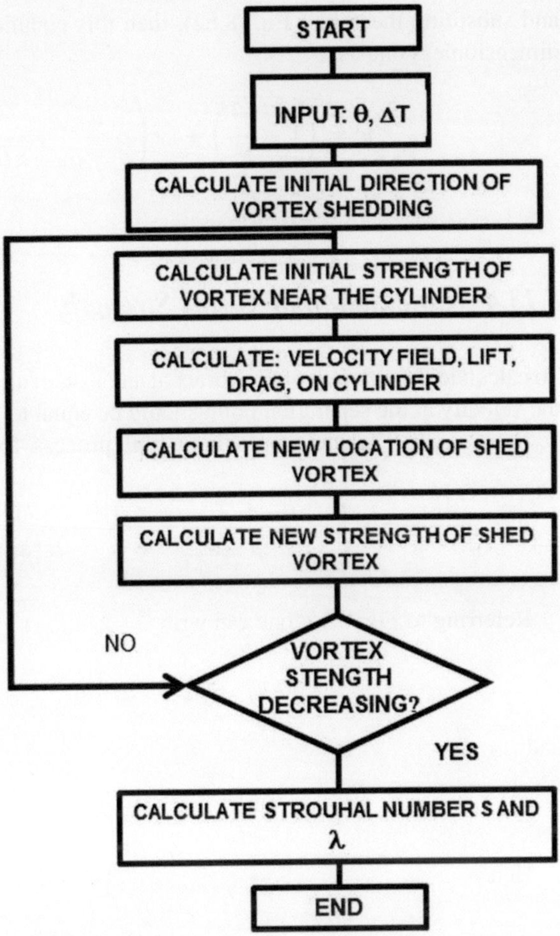

Fig. 8.16 Computational algorithm

If Γ at any instant is known, Eqs. (8.82a) and (8.82b) can be employed to predict the vortex trajectory ln a progressive fashion, since the initial position of the vortex. shed is known, i.e. at the separation point.

Defining

$$\gamma = \frac{\Gamma}{2\pi U a} \quad \text{as the dimensionless vortex strength}$$

and

$$w = \frac{V}{U} \quad \text{as the dimensionless velocity}$$

and substitute these into Eq. (8.82), then this equation can be transformed into a dimensionless one as

$$w = \left(1 - \frac{a^2}{z^2}\right) + i\gamma\left(\frac{a}{a - z_0} + \frac{a}{z\left(a^2 - \bar{z}_0 z\right)}\right) \tag{8.82a}$$

8.13.4 *Calculation of Vortex Strength*

To calculate the strength of the vortex at any instant along its path, the condition that the velocity at the separation point should be equal to zero is imposed.

From Eq. (8.72a) and employing a limit process, for $z = z_1$ there is obtained:

$$\gamma = \frac{\Gamma}{2\pi U a} = -\frac{i}{a}\frac{a^2 - z_1^2}{\frac{z_1^2}{z_1 - z_0} + \frac{z_1 a^2}{a^2 - \bar{z}_0 z}} = -\frac{i}{a}\frac{\left(a^2 - z_1^2\right)(z_1 - z_0)\left(a^2 - \bar{z}_0 z\right)}{2z_1^2 a^2 - z_1^3 \bar{z}_0 - \bar{z}_0 z_1 a^2} \tag{8.87}$$

Referring to Fig. 8.15, one can write[14]

$$z_0 - z_1 = a\rho e^{i\phi} \tag{8.88}$$

and

$$z_1 = a e^{i\phi} \tag{8.89}$$

Hence

$$z_0 = a\left(e^{i\phi} + \rho e^{i\phi}\right) \tag{8.90}$$

and

$$\bar{z}_0 = a\left(e^{-i\phi} + \rho e^{-i\phi}\right) \tag{8.91}$$

Substituting these values into Eq. (8.87) we obtain

$$\gamma = -\frac{i}{a}\frac{\left(a^2 - a^2 e^{2i\theta}\right)\left(-a\rho e^{i\theta}\right)\left[a^2 - a^2\left(1 - \rho e^{i(\phi - \theta)}\right)\right]}{2a^4 e^{2i\theta} - a^4 e^{3i\theta}\left(e^{-i\theta} + e^{-i\phi}\right) - a^4 e^{i\theta}\left(e^{i\theta} + \rho e^{-i\phi}\right)} \tag{8.92}$$

After some algebra, Eq. (8.92) can be reduced to

---

[14] Please note that in the formulae with ρ appearing as $\rho\, e^{i\,\phi}$ and the like, ρ is not fluid density, but is associatd with the distance between the separation point and the shed point vortex, as depicted in Fig. 9.1

$$\gamma = \rho \sin(\theta) \sec(\theta - \phi) \tag{8.93}$$

which is always real and positive, since from physical reasons,

$$\theta - \frac{\pi}{2} < \phi < \theta + \frac{\pi}{2}$$

Direction of the Vortex Shedding

To predict the direction of the initial vortex shedding, consider the flow situation shortly after a vortex is shed from the separation point $z1$. Substituting expression (8.87) into Eq. (8.82a), we obtain

$$v = 1 - \frac{1}{\left(e^{i\theta} + e^{-i\phi}\right)} - \frac{i\gamma}{\rho} \frac{1}{\left(e^{i\theta} + e^{-i\phi}\right)(1 + 2\cos(\theta - \phi))} \tag{8.94}$$

Substituting expression (8.93) into Eq. (8.94), and using Eq. (8.82a), we obtain

$$v = 1 - \frac{1}{\left(e^{i\theta} + e^{i\phi}\right)^2} - \frac{i \sin\theta \sec(\theta - \phi)}{\left(e^{i\theta} + e^{-i\phi}\right)(1 + 2\cos(\theta - \phi))} \tag{8.95}$$

Taylor series expansion of w with respect to ρ yields

$$w = w|_{\rho=0} + \rho \left(\frac{\partial w}{\partial \rho}\right)\Big|_{\rho=0} + \frac{\rho^2}{2!}\left(\frac{\partial^2 w}{\partial \rho^2}\right)\Big|_{\rho=0} + \ldots\ldots \tag{8.96}$$

Omitting terms of order ρ^2 and higher, and after some algebra, Eqs. (8.95) and (8.96) give

$$w = 1 - e^{-2i\theta} - ie^{-i\theta} \frac{\sin\theta \sec(\theta - \phi)}{(1 + 2\cos(\theta - \phi))} + ie^{i\phi - 2i\theta}\left[2e^{-i\theta} + \frac{i \sin\theta \sec(\theta - \phi)}{(1 + 2\cos(\theta - \phi))}\right] \tag{8.97}$$

Since the velocity at the separation point should be equal to zero, we obtain an additional condition that $w = 0$ for $\rho \to 0$. This condition yields:

$$1 - e^{-2i\theta} - ie^{-i\theta} \frac{\sin\theta \sec(\theta - \phi)}{(1 + 2\cos(\theta - \phi))} = 0 \tag{8.98}$$

Two relations between ϕ and θ result from Eq. (8.98), but only one relation is valid, which is

$$\phi = \theta - 0.677 \tag{8.99}$$

Equation (8.99) determines the direction of the initial vortex shedding.

8.13.5 Frequency of Vortex Shedding

At some distance downstream, the vortex being shed off the cylinder stops growing, breaks away from the feeding vortex sheet and joins the rows of von Karman vortices. This situation is assumed to take place when the vortex attained its maximum strength. At the same time, another vortex of opposite sense starts to be shed off at the other separation point, growing and traveling downstream in a similar fashion as the former vortex. Hence vortices are shed off alternately at either separation point and give rise to time dependent and oscillating lift and drag forces on the cylinder.

To calculate the frequency of vortex shedding, which in dimensionless form is characterized by the Strouhal number, we impose the condition that the vortices which join the von Karman vortex street attain the well-known von Karman vortex velocity. Hence:

$$u = U - \frac{\Gamma}{2l}\tanh\left(\frac{2\pi y}{l}\right) = \hat{U} \tag{8.100}$$

or

$$\frac{u}{U} = 1 - \gamma\frac{\pi a}{l}\tanh\left(\frac{2\pi y}{l}\right) \tag{8.101}$$

where

I the distance between two successive vortices.
y the lateral distance of the vortices from the symmetry axis.
U von Karman vortex velocity.

When the vortex attains its maximum strength, then

$$\dot{\gamma} \equiv \frac{d\gamma(t)}{dt} = 0 \tag{8.102}$$

and from Eqs. (8.80) and (8.81)[15] we obtain

$$\dot{z}_0 \equiv v_0 \tag{8.103}$$

and $v_0 = 0$ (which should be verified later).

15

$$-i\,\rho\,\dot{\Gamma}(z_0 - z_1) + i\,\rho\,\Gamma(V_0 - \dot{z}_0) = 0 \tag{8.80}$$

or

$$\dot{z}_0 + (z_0 - z_1)\frac{\dot{\Gamma}}{\Gamma} = V_0 \tag{8.81}$$

$\hat{U}$ is the computed u component of the vortex velocity at the instant when the vortex attains its maximum strength.

Equation (8.101) can be reduced further into the following form:

$$\tanh \alpha \equiv \frac{\beta}{\alpha} \tag{8.104}$$

where

$$\alpha \equiv \frac{2\pi y}{l} \tag{8.105}$$

$$\beta = \frac{2\pi\left(1 - \frac{\hat{U}}{U}\right)}{\frac{\gamma a}{y}} \tag{8.106}$$

Equation (8.106) can be solved graphically for calculating l. The Stroutral number can then be obtained from

$$S = \frac{nD}{U} = \frac{2\hat{U}a}{Ul} \tag{8.107}$$

8.13.6 Calculation of Vortex Trajectory

Equations (8.86a), (8.86b) and (8.93) comprise the set of equations that determine the strength and motion of the vortex. The initial condition is defined by assigning the location of the separation point, i.e. the initial position of the vortex, and computation of the direction of the initial vortex shedding, which is given by Eq. (8.98).

To compute the vortex trajectory, Eqs. (8.86a) and (8.86b) are integrated numerically, and the computation progresses in a step by step fashion. Integrating Eqs. (8.86a) and (8.86b) over the time increment Δt, and using linear approximation, we obtain:

$$x_0 + (x_0 - x_1)\ln\gamma - \ln\gamma \cdot x_0 = u_0 \cdot \Delta t \tag{8.108a}$$

$$y_0 + (y_0 - y_1)\ln\gamma - \ln\gamma \cdot y_0 = v_0 \cdot \Delta t \tag{8.108b}$$

where all variables have been rendered dimensionless through division with appropriate characteristic quantities.

The velocity components u_0 and v_0 of the vortex from Eqs. (8.103) and (8.104) are given by:

$$u_0 = 1 - \frac{\left(x_0^2 - y_0^2\right)}{\left(x_0^2 + y_0^2\right)^2} - \gamma \frac{y_0}{\left(x_0^2 + y_0^2\right)\left(x_0^2 + y_0^2 - 1\right)} \tag{8.109a}$$

$$v_0 = -\frac{2x_0 y_0}{\left(x_0^2 + y_0^2\right)^2} + \gamma \frac{y_0}{\left(x_0^2 + y_0^2\right)\left(x_0^2 + y_0^2 - 1\right)} \tag{8.109b}$$

Again, all variables in Eqs. (8.109a) and (8.109b) are dimensionless. The computation algorithm is shown in Fig. 8.16. The lift and drag forces acting on the cylinder due to the vortex throughout its trajectory is computed from Eqs. (8.106) and (8.107).

8.13.7 Result, Discussions and Concluding Remarks

Since the present theory does not explicitly incorporate viscosity, the locations of the separation points cannot be predicted, and recourse should be made to experimental evidence in order to determine these locations and then proceed with the calculation of the vortex trajectory [11–13]. Experimental results,indicate that the separation points, i.e. the point at which the boundary layer leaves the cylinder, occur at about $\theta_s = 95°$ for Re = 1.06×10^6 and between 50" and 80" for $1.08 \times 10^6 < \text{Re} < 2.12 \times 10^6$. Based on these results, some values of θ are assumed, and the vortex trajectory, lift and drag coefficients, vortex strength, vortex velocity, λ—the ratio between lateral separation and longitudinal separation of Karman vortices—and Strouhal number *S* are computed following the procedure outlined previously.

The results are tabulated in Table 8.1. Some values obtained experimentally and by Alexander [14] are also shown in Table 8.1 for comparison. For the case in which the separation point is located on the upstream side of the cylinder, say θ_s = I.74 rad, a maximum lift coefficient of 1.55 was obtained. Experimental results indicate considerable scatter, and time-dependent lift as high as 1.3 has been shown by Drescher (cited in [14]) and Macovsky (cited by Dowell et al [7]) and Bearman [12].

It should be noted that most experimental data are affected to some degree by three-dimensional effects, which results in lower measured llft coefficients than predicted two-dimensional values. Results obtained by Jordan and From [16] [16]employing numerical solution of the equations governing time-dependent, viscous, incompressible flow past a cylinder indicate peak to peak amplitude of the lift oscillation of 1.5 for Re = 400 and 1.9 for Re = 1000.

Their results are an adequate description of the flow past a circular cylinder for Re 400, but are not entirely valid for Re > 1000. The Strouhal number for this case was found to be 0.159, which is within the region of experimental results. Relf and Simmons [17, 18], Roshko [19] and Davis [20] obtained values of S between 1.9 and 2.4 for Reynolds number between 3×10^5 to 8×10^5, and S = 2.4−3.1 for Re 4.0

[16] Jordan and Fromm [16]

Table 8.1 Various computational results

θ	Δt	u_θ	v_θ	Γ_{max}	CL_{max}	CD_{max}	y_0	x_0	l	λ	s
1.74	0.03	0.404	0.001	1.049	3.95	0.07	1.265	0.214	5.05	0.501	0.16
1.74	0.02	0.57	0.02	1.041	2.80	−0.146	1.308	0.230	6.45	0.406	0.177
1.74	0.01	0.762	−0.04	1.024	1.55	−0.340	1.384	0.231	0.6	0.288	0.159
1.00	0.03	0.545	0.022	1.034	3.15	0.299	0.867	1.234	5.40	0.320	0.202
1.00	0.02	0.574	0.001	1.018	2.73	0.379	0.883	1.251	5.65	0.312	0.202
0.78	0.03	0.705	0.005	1.021	2.55	0.55	0.704	1.709	6.4	0.2	0.22
0.79	0.02	0.697	0.008	1.013	2.29	0.335	0.695	1.692	6.3	0.22	0.221
0.79	0.01	0.695	0.008	1.007	1.93	0.426	0.694	1.684	6.2	0.22	0.224
0.79		0.5	0		2.3	0		1.224	3.7	0.382	0.27
					1.3^{++}					0.2806**	0.18 0.23^{+}
						0.23*					

* From Alexander [14]; Alexander determines Sl = 1 and Sη = 0.191, where y0 = η. If S is taken to be 0.27, then η = 0.707, and hence l = 3.7, λ = 0.382; if S is taken to be 0.2, then η = 0.955, l = 5.0, λ = 0.28

** From Prandtl- & Tietjens [15]

+ From P.W. Bearman [12], at 105 < Re < 3.8 × 105

+ + From Drescher (cited by Alexander [14])

Peak-to-peak amplitude, from reference 12, for Re = 1.000

$\times$ 10^5–1.0 $\times$ 10^6, while experiment conducted by Delany and Sorensen [21] and Bearman [22] showed that S = 1.8–2.3 for Re = 1.0 $\times$ 10^5 to 3.8 $\times$ 10^5.

For supercritical region, i.e. for Re = 4 $\times$ 10^5, Bearman [22],[17,18] found that $S =$ 0.44–0.46, while Delany and Sorensen [21] found that S = 0.34 – 0.48 for Re 10^6. In addition, the present method results are associated with the value of λ of 0.288, which is in good agreement with experimental result 0.2806 reported in reference [23].

For Reynolds number greater than 10^6, the separation point occurs further downstream, i.e. at about 50° – 80°. Assuming separation points located between these values, there results maximum lift coefficient as high as 2.73 for $\theta_s = 1.0$ rad, with $C_L = 1.93$ for $\theta_{si} = 0.79$ rad. Experimental results indicate that S increases to about 0.24 before sudden reduction of drag coefficient occurs and increases further to 0.48 at supercritical region. The present method is able to predict the tendency of S to increase if the flow changes from subcritical to supercritical region, i.e. if the separation point moves downstream, but indicate lower values of S at both cases. Clearly the additional increase in S should be due to turbulence, which is not taken into account in the present theory. At supercritical region, λ varies between 0.22 and 0.3 for θ_s varying between 0.79 and 1.00.

The time-dependent drag obtained by the present method will be zero only when $v_0 = 0$, i.e. when the vortex joins von Karman vortex street. Since in general the vortex trajectory is not parallel to the free stream direction, the time-dependent drag is nonzero. Careful examination of Eq. (8.78) reveals that depending on the value of γ_0 and y_0 (lateral position of the vortex), the time-dependent drag coefficient may assume positive or negative value. Relatively large value of the time-dependent drag C_D is obtained when the vortex is close to the cylinder, and C_D decreases rapidly in inverse proportion to x_0 as x_0 increases.

The result indicates discrepancy in the lift coefficient in the order of 40% for forward separation point. Relatively better agreement is shown for after separation point, i.e. within 6%. These results also clearly indicate that the presence of small vertical velocity component of the vortex may give rise to significant changes in the values of time-dependent drag and vortex strength, in particular for forward separation point. Further evaluation of the theory should be accompanied by comparison with reliable experimental results, which merits further study.

Several computations performed with Δt varying from 0.01, 0.02' and 0.01 has indicated reasonable converging behavior of the solution, in particular for the calculation of y_0, C_{Lmax}, y_0, x_0, λ and S, as shown in Table 8.1. The variation of time-dependent lift coefficients $\theta_s = 1.74$ and $\theta_s = 0.79$ are shown in Fig. 8.17.

For comparison purposes for the merit of the present approach, a CFD simulation has been performed by Andira [24], as depicted in Fig. 8.18b and 8.18c, showing a simulation of the vortex pattern behind the circular cylinder following the case

---

[17] Bearman [3].

[18] Bearman [4].

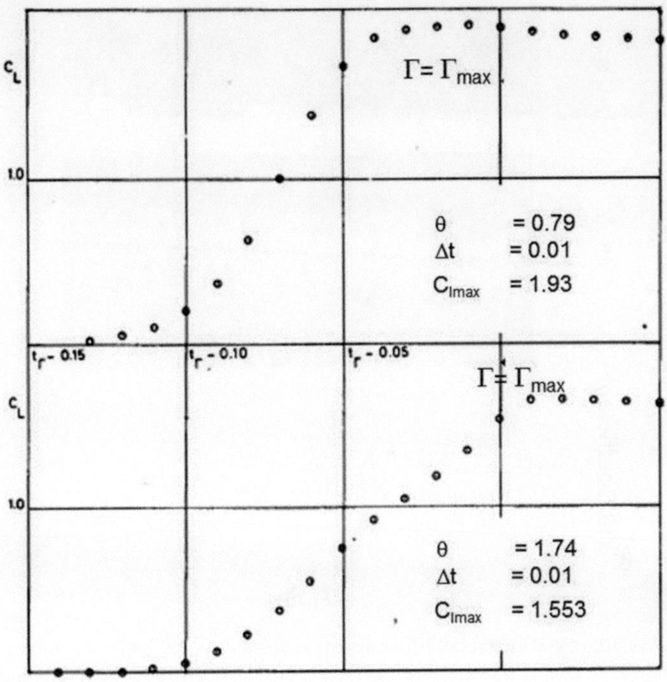

Fig. 8.17 Development of time-dependent lift due to a single vortex shed off the cylinder[19]

considered here for Reynolds number 1.0×10^6, using locally developed CFD Software CFDSOF. The data utilized are: Number of Cells: 49,286, Reynolds Number 1.0×10^6 Freestream Velocity: 10 m/s, Kinematic Viscosity: 1.0×10^5 m^2/s.

Turbulence Model: Menter SST k-ω, Governing Equation based on Large Eddy Simulation, Spatial Discretization Scheme: Second-Order Linear Upwind and Temporal Discretization Scheme: First-Order Euler Implicit. Figure 8.18 also shows the lift oscillation commensurate with the time-dependent flow pattern due to the von Karman vortex street.

The present simple model is in no way a substitute for viscous flow analysis, but has been shown to be useful in estimating the time dependent lift on a cylinder and in identifying factors contributing to the oscillating lift and drag, and the increase of Strouhal number.

In closing, after the presentation of the work at the Japanese International Symposium on Space Science and Technology in 1973, the author was informed of the classical work of Prof. Arthur Bryson [23], which has been frequently referred to by many other significant works by many authors, some of them are listed in the reference.

---

[19] Figures 8.18 and 8.19, and Table 8.1 have been reproduced by Andira [24]

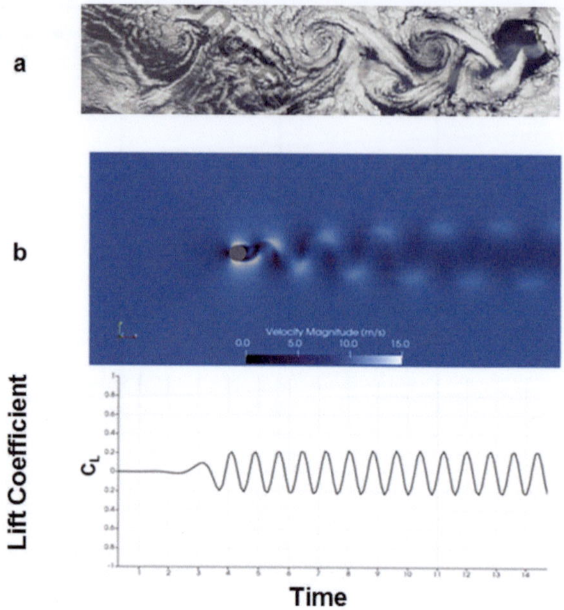

Fig. 8.18 **a**. Von Karman vortex street.(Courtesy of shutterstock.com[20]); **b**. Simulation of the vortex pattern behind the circular cylinder following the case considered here for Reynolds number 1.0×10^6 using Local Commercial Software CFDSOF [24]

8.14 List of Symbol

Used in the Case Study **Simple Method to Calculathe the Oscillating Lift on a Circular Cylinder In Potential Flow**.

a radius of the cylinder
C_L lift coefficient
C_D drag coefficient
D drag
i $\sqrt{-1}$
L lift
Re Reynolds number
S Strouhal number
t time
U free stream velocity
$\hat{U}$ von Karman vortex velocity
V velocity
V_0 velocity at point vortex at z_0
u_0 x-component of V_0

[20] www.shutterstock.com.875,918,670.

v_0 y-component of V_0
W complex potential$_0$
w V/U
x longitudinal coordinate
y lateral coordinate
z $x + iy$
Γ shed vortex strength
ϕ angular position of shed point vortex
γ dimensionless vortex strength
λ ratio between lateral to longitudinal distances of von Karman vortices
ρ distances between z_0 and z_1 expressed in units of a
θ angular position of separation point

Appendix 1: Acceleration or Pressure Potential[21]

From Bernoulli's equation

$$\frac{\partial \varphi}{\partial t} + U\frac{\partial \varphi}{\partial x} + \frac{p - p_\infty}{\rho} = 0$$

or

$$\frac{d\varphi}{dt} = \left(\frac{\partial \varphi}{\partial t} + U\frac{\partial \varphi}{\partial x}\right) = -\frac{p - p_\infty}{\rho} = -\frac{P - P_\infty}{\rho}$$

Hence from Bernoulli's equation, one may define

$$\psi = \frac{P - P_\infty}{\rho} \tag{8.20}$$

Due to this fact, ψ is also called pressure potential.[22]
Acceleration potential ψ should not be confused with stream function ψ.

Appendix 2: Velocity Potential and Stream Function

Velocity potential φ and stream function ψ in the two-dimensional Cartesian coordinate O-xz have been defined as

[21] Further details of Appendices 1 and 2, are elaborated in Djojodihardjo [25].

[22] Further elabortion may be found in Ashley, H and Landahhl [1]

$$u = \frac{\partial \phi}{\partial x} = \frac{\partial \psi}{\partial z}$$
$$v = \frac{\partial \phi}{\partial z} = -\frac{\partial \psi}{\partial x}$$

Both functions satisfy Laplace equation

$$\frac{\partial^2 \varphi}{\partial x^2} + \frac{\partial^2 \varphi}{\partial z^2} = 0 \quad \frac{\partial^2 \psi}{\partial x^2} + \frac{\partial^2 \psi}{\partial z^2} = 0$$

Both functions satisfy Cauchy–Riemann equations:

$$\frac{\partial \varphi}{\partial x} = \frac{\partial \psi}{\partial y} \quad \frac{\partial \varphi}{\partial y} = -\frac{\partial \psi}{\partial x}$$

and

$$\Phi(z) \equiv \Phi(x + iy) \equiv \varphi(x, y) + i\psi(x, y)$$

References

1. Ashley, H., and M.T. Landahl. 1965. *Aerodynamics of wings and bodies*. Addison-Wesley.
2. Theodorsen, T. 1934. *General theory of aerodynamic instability and the mechanism of flutter*. No. NACA TR . 4960.
3. Küssner, H.G. 1936. Zusammenfassender Bericht über den instationaren Auftrieb von Flügeln, *Luftfahrtforschung* 13 (12): 410–424.
4. Schwarz, L. 1940. Calculation of the pressure distribution of a wing harmonically oscillating in two-dimensional flow. *Luftfarhtforschung* 17 (11): 379.
5. Possio, C. 1938. L'azione aerodinamica su una superficie portante in mot0 oscillatorio. *Atti Reale Acc. Lincei* 26: 194–200
6. Bisplinghoff, R.L., and H. Ashley. 1962. *Principles of aeroelasticity*. Wiley.
7. Dowell, E.H., H.C. Curtiss, R.H. Scanlan, and R. Sisto. 1980. *A modern course in aeroelasticity*. Sijthoff and Noordhoff.
8. Försching, H.W. 1974. *Grundlagen der aeroelastik*. Springer.
9. Platzer, R.L., (ed.). 1960. AGARD manual on aeroelasncity in axial-flow turbomachines, vol. 2. Structural Dynamics and Aeroelasticity, AGARD 1960, vol. II, Chapter 2.
10. Djojodihardjo, H. 1973. Simple potential flow mode-calculation-oscillating lift-circular cylinder. *Journal of Mathematical and Fundamental Sciences* Institut Teknologi Bandung.
11. Roshko, A. 1961. Experi-nents of the flow past a circular cyllnder at very high reynolds number. *Journal of Fluid Mechanics* 10 (3).
12. Bearman, P.L.W. 1968. *The flow around a circular cylinder in the critical reynolds number regime*. NPL Aero Report 1257.
13. Fung, Y.C. 1960. Fluctuating lift and drag on a cylinder in a flovr at supercritical reynolds numbers. *Journal of Aerospace Sciences* 11: B01.
14. Alexander, A.J. 1968. A theory for the oscillating lift on a circular cylinde. *Symposiurn on Wind Effects on Buildings and Structures*. Loughborough University of Technology, Paper No. 19.

15. Prandtl, L., and O.C. Tietjens. 1957. *Applied hydro—and aeromechanics.* New York:Dover Publications, Inc.
16. Jordan, S.K. and J.E. Fromm. 1972. Oscillatory drag, lift and torque on a ciqcular cylinder in a uniform flow. *The Physics of Flujds* 15 (3).
17. Relf, E. F., and L.F.G. Simmons. 1954. The frequency of eddies generated by motion of circular cylinders through a Fluid. R. & M. 917, 1924, also Relf, E.F., and L.F.G. Simmons; On the frequency of the eddies generated by the motion of circular cylinders through a fluid, 509–511 | Published online: 08 Apr 2009; Roshko, A. On the Drag and Shedding Frequency of Two Dimensional Bluff Bodies. NACA TN 3169.
18. Relf, E.F., and L.F.G. Simmons. 2009. On the frequency of the eddies generated by the motion of circular cylinders through a fluid, 509–511.
19. Roshko, A. 1954. *On the drag and shedding frequency of two dimensional bluff bodies.* NACA TN 3169.
20. Davis, M.D. 1969. *An analytical study of separated flow about a circular cylinder*. MS Thesis, Naval PostGrad School.
21. Delany, N.K., and N.E. Sorensen. 1953. *Low speed drag of cylinders of various shapes.* NACA TN 3038.
22. Bearman, P.W. 1968. *Some effects of turbulence on the flow around bluff bodies.* NPL Aero Report 1264.
23. Bryson, A.E. 1959. Symmetric vortex separation on circular cylinders and cones. *Transactions ASME Series E: Journal of Applications and Mechanism* 26 (4): 643–648.
24. Andira, M.A. 2021. Personal communication.
25. Djojodihardjo, H. 1984. *Mekanika Fluida* (in Indonesian). Penerbit Erlangga.
26. Djojodihardjo, H. and Akbar et al. 2010. *Unified aerodynamic-acoustic formulation for aero-acoustic structure coupling.* Paper presented at. ICAS Congress, Nice, France.

Chapter 9
Flutter Calculation Methods

Abstract Proceeding with the elaboration of the foundation of dynamic aeroelasticity established in the earlier chapter, the present chapter then focuses on establishing the flutter calculation method, starting from basic illustrative cases to more complex problems. The chapter first elaborates a review of theoretical foundation for flutter stability for binary-bending-torsion flutter of typical section. Several methods of solving the aeroelastic stability of binary system will be introduced and discussed. Several case studies are elaborated for illustrative and instructional purposes.

Keywords Aeroelasticity · Binary system · Governing differential equations · Aeroelastic Stability · Flutter calculation methods

9.1 Review of Theoretical Foundation for Flutter Stability for Binary-Bending-Torsion Flutter of Typical Section

9.1.1 Quasi-Steady[1,2]

$$\begin{aligned} m\ddot{h} + S_\alpha\ddot{\alpha} + K_h h + qSC_{L_\alpha}\alpha = 0 \\ S_\alpha\ddot{h} + I_\alpha\ddot{\alpha} + K_\alpha\alpha - 2qSebC_{L_\alpha}\alpha = 0 \end{aligned} \tag{9.1}$$

In matrix notation:

$$[\mathbf{M}]\{\ddot{x}\} + ([\mathbf{K}] - q[A_0])\{x\} = 0 \tag{9.2}$$

with[3]

[1] The development elaborated here follows closely those of G.T.S. Done [1–3], and Zwaan [4].

[2] Examples worked out have also appeared in Djojodihardjo and Yee Hong Hoe [5], Djojodihardjo and Abd. Bari [6], Djojodihardjo and Ramli [7] and Djojodihardjo [8].

[3] Here I_α is given as the mass moment of inertia [ML^2], S_α is given as the mass static moment [ML], K_h is the spring constant, force per unit deflection, [MT^{-2}], and K_α is the torsional spring constant, moment per unit angular deflection [ML^{2}T^{-2}], with respect to the elastic axis of the typical section.

© The Author(s), under exclusive license to Springer Nature Singapore Pte Ltd. 2023
H. Djojodihardjo, *Introduction to Aeroelasticity*,
https://doi.org/10.1007/978-981-16-8078-6_9

$$[\mathbf{M}] = \begin{bmatrix} m & S_\alpha \\ S_\alpha & I_\alpha \end{bmatrix} \quad \text{– inertia matrix} \tag{9.3}$$

$$[\mathbf{K}] = \begin{bmatrix} K_h & 0 \\ 0 & K_\alpha \end{bmatrix} \quad \text{– structural stiffness} \tag{9.4}$$

$$[\mathbf{A}_0] = \begin{bmatrix} 0 & -SC_{L_\alpha} \\ 0 & 2SebC_{L_\alpha} \end{bmatrix} \quad \text{– aerodynamic stiffness} \tag{9.5}$$

representing the quasi-steady aerodynamics approximation.

Analysis for pure flutter stability:

$$h = \hat{h}e^{pt} \tag{9.6a}$$

$$\alpha = \hat{\alpha}e^{pt} \tag{9.6b}$$

or

$$\{\mathbf{x}\} = \{\hat{\mathbf{x}}\}e^{pt} \tag{9.7}$$

Solving the eigenvalue problem by substitution leads to

$$\begin{aligned} &(mp^2 + K_h)\hat{h} + (S_\alpha p^2 + qSC_{L_\alpha})\hat{\alpha} = 0 \\ &S_\alpha p^2\hat{h} + (I_\alpha p^2 + K_\alpha - 2qSebC_{L_\alpha})\hat{\alpha} = 0 \end{aligned} \tag{9.8}$$

Rewrite in non-dimensional form:

$$h = \hat{h}be^{pt} \tag{9.9a}$$

$$\alpha = \hat{\alpha}e^{pt} \tag{9.9b}$$

Hence one obtains

$$\begin{aligned} &(mp^2 + K_h)b\hat{h} + (S_\alpha p^2 + qSC_{L_\alpha})\hat{\alpha} = 0 \\ &S_\alpha p^2 b\hat{h} + (I_\alpha p^2 + K_\alpha - 2qSebC_{L_\alpha})\hat{\alpha} = 0 \end{aligned} \tag{9.10}$$

where

$$\tilde{h} = \frac{\hat{h}}{b} \quad \text{and} \quad h = \tilde{h}e^{pt} \tag{9.11}$$

or

$$\begin{bmatrix} (mp^2 + K_h)b & S_\alpha p^2 + qSC_{L_\alpha} \\ S_\alpha p^2 b & I_\alpha p^2 + K_\alpha - 2qSebC_{L_\alpha} \end{bmatrix} \begin{Bmatrix} \hat{h} \\ \hat{\alpha} \end{Bmatrix} = 0 \tag{9.12}$$

Rearrange to follow eigenvalue approach and K-method:

$$\begin{bmatrix} K_h b & qSC_{L_\alpha} \\ 0 & K_\alpha - 2qSebC_{L_\alpha} \end{bmatrix} \begin{Bmatrix} \hat{h} \\ \hat{\alpha} \end{Bmatrix} = \begin{bmatrix} -mb & -S_\alpha \\ -S_\alpha p^2 & -I_\alpha \end{bmatrix} p^2 \begin{Bmatrix} \hat{h} \\ \hat{\alpha} \end{Bmatrix} \tag{9.13}$$

or

$$[\mathbf{A}] \begin{Bmatrix} \hat{h} \\ \hat{\alpha} \end{Bmatrix} = [\mathbf{D}] p^2 \begin{Bmatrix} \hat{h} \\ \hat{\alpha} \end{Bmatrix} \tag{9.14}$$

For the eigenvalue problem in K-method:

$$[\mathbf{D}]^{-1}[\mathbf{A}] \begin{Bmatrix} \hat{h} \\ \hat{\alpha} \end{Bmatrix} = p^2 \begin{Bmatrix} \hat{h} \\ \hat{\alpha} \end{Bmatrix} = \lambda \begin{Bmatrix} \hat{h} \\ \hat{\alpha} \end{Bmatrix} \tag{9.15}$$

or

$$[\mathbf{A}(k)]\{\mathbf{x}\} = \lambda\{\mathbf{x}\} \tag{9.16}$$

where A depends on k since as seen from (9.12) and (9.13), it depends on q, where

$$q = \frac{1}{2}\rho U^2 = \frac{1}{2}\rho\left(\frac{\omega b}{k}\right)^2 \tag{9.17a}$$

Since

$$k = \left(\frac{\omega b}{U}\right) \quad \text{or} \quad U = \left(\frac{\omega b}{k}\right) \tag{9.17b}$$

In Eq. (9.16)

$$\lambda = -\begin{bmatrix} -\text{mb} & -S_\alpha \\ -S_\alpha p^2 & -I_\alpha \end{bmatrix} p^2 \tag{9.18}$$

What does this mean?

The determinant of Eq. 9.12 can be written as a polynomial equation in p^2

$$\begin{aligned}&(bmI_\alpha - S_\alpha b)p^4 + \left(I_\alpha K_h b + mb\left(K_\alpha - 2qSebC_{L_\alpha}\right) - S_\alpha qbSC_{L_\alpha}\right)p^2 \\ &\quad K_h b\left(K_\alpha - 2qSebC_{L_\alpha}\right) = 0\end{aligned} \tag{9.19}$$

or

$$A_4 p^4 + A_2 p^2 + A_0 = 0 \tag{9.20}$$

where

$$A_4 = \left(m I_\alpha - S_\alpha^2\right) b \tag{9.21}$$

$$\begin{aligned} A_2 &= (I_\alpha K_h + m K_\alpha) b - q b S C_{L_\alpha} (S_\alpha + 2emb) \\ &= (I_\alpha K_h + m K_\alpha) b - q m b^2 S C_{L_\alpha} \left(\frac{S_\alpha}{mb} + 2e \right) \end{aligned} \tag{9.22}$$

Since $S_\alpha = m x_\alpha b$

$$A_0 = K_h b \left(K_\alpha - 2 q b S C_{L_\alpha} \right) \tag{9.23}$$

Rewrite Eq. 9.12 as:

$$\begin{bmatrix} \left(p^2 + \frac{K_h}{m}\right) & \left(\frac{S_\alpha}{mb} p^2 + \frac{q S C_{L_\alpha}}{mb}\right) \\ p^2 & \left(\frac{I_\alpha}{S_\alpha b} p^2 + \frac{K_\alpha - 2 q S e b C_{L_\alpha}}{S_\alpha b}\right) \end{bmatrix} \begin{Bmatrix} \hat{h} \\ \hat{\alpha} \end{Bmatrix} = 0 \tag{9.24}$$

Then polynomial Eq. 9.19 and coefficients (9.21)–(9.23) become

$$\begin{aligned} &\left(\frac{I_\alpha}{S_\alpha b} - \frac{S_\alpha}{mb} \right) p^4 + \left(\frac{I_\alpha K_h}{m S_\alpha b} + \frac{K_\alpha - 2 q S e b C_{L_\alpha}}{S_\alpha b} - \frac{q S C_{L_\alpha}}{mb} \right) p^2 \\ &+ \frac{K_h}{m} \left(\frac{K_\alpha - 2 q S e b C_{L_\alpha}}{S_\alpha b} \right) = 0 \end{aligned} \tag{9.25}$$

and

$$P' = A_4' p^4 + A_2' p^2 + A_0' = 0 \tag{9.20b}$$

where

$$A_4' = \frac{I_\alpha}{S_\alpha b} - \frac{S_\alpha}{mb} \tag{9.26}$$

$$\begin{aligned} A_2' &= \left(\frac{I_\alpha K_h}{m S_\alpha b} + \frac{K_\alpha}{S_\alpha b} \right) - q S \left(\frac{2 e C_{L_\alpha}}{S_\alpha} + \frac{C_{L_\alpha}}{\mathrm{mb}} \right) \\ &= \frac{1}{b} \left(\frac{I_\alpha K_h}{m S_\alpha} + \frac{K_\alpha}{S_\alpha} \right) - q S C_{L_\alpha} \left(\frac{2e}{S_\alpha} + \frac{1}{\mathrm{mb}} \right) \\ &= \frac{1}{b S_\alpha} \left(\frac{I_\alpha K_h}{m} + K_\alpha \right) - \frac{q S C_{L_\alpha}}{\mathrm{mb}} \left(\frac{2e}{x_\alpha} + 1 \right) \end{aligned} \tag{9.27}$$

Since $S_\alpha = mx_\alpha b$

$$A_0' = \frac{K_h}{mbS_\alpha}\left(K_\alpha - 2qSebC_{L_\alpha}\right) \tag{9.28}$$

If we use

$$\mu = \frac{m}{\pi\rho b} \tag{9.29}$$

Equation 9.26–9.28 becomes:

$$A_4' = \frac{I_\alpha}{S_\alpha b} - \frac{S_\alpha}{\mathrm{mb}} \tag{9.30}$$

$$A_2' = \frac{1}{bS_\alpha}\left(\frac{I_\alpha K_h}{\mu\pi\rho b^2} + K_\alpha\right) - \frac{qSC_{L_\alpha}}{\mu\pi\rho b^2}\left(1 + \frac{2e}{x_\alpha}\right) \tag{9.31a}$$

$$A_0' = \frac{K_h}{\mathrm{mb}S_\alpha}\left(K_\alpha - 2qSebC_{L_\alpha}\right) \tag{9.31b}$$

Next solution of polynomial equation using MATLAB routine:

$$P' = \left[\,A_0'\ A_2\ A_4'\,\right] \tag{9.32}$$

$$p^2 = \mathrm{root}\left(P'\right) = \sqrt{P'} \tag{9.33}$$

$$(\sigma + i\omega) = p = \pm\sqrt{(p)^2} = \mathrm{Real}(p) + \mathrm{Im}(p) \tag{9.34}$$

9.1.2 GT Done-Type Analysis[4]

Obtain four roots of Eq. (9.33)

$$(\sigma_1 + i\omega_1) = p_1 = +\sqrt{(p)^2}\Big|_1 = +\mathrm{sqrt}\left(p^2\right)\big|_1 = \mathrm{Real}(p_1) + \mathrm{Imag}(p_1) \tag{9.35a}$$

$$(\sigma_2 + i\omega_2) = p_2 = -\sqrt{(p)^2}\Big|_1 = -\mathrm{sqrt}\left(p^2\right)\big|_1 = \mathrm{Real}(p_2) + \mathrm{Imag}(p_2) \tag{9.35b}$$

[4] References are made to the following publications listed in Reference: Done [1–3], Zwaan [4], and Djojodihardjo and Yee Hong Hoe [5].

$$(\sigma_3 + i\omega_3) = p_1 = +\sqrt{(p)^2}\Big|_2 = +\mathrm{sqrt}(p^2)\big|_2 = \mathrm{Real}(p_3) + \mathrm{Imag}(p_3) \quad (9.35c)$$

$$(\sigma_4 + i\omega_4) = p_4 = -\sqrt{(p)^2}\Big|_2 = -\mathrm{sqrt}(p^2)\big|_2 = \mathrm{Real}(p_4) + \mathrm{Imag}(p_4) \quad (9.35d)$$

Negative values ofω are taken out from the solution.

9.1.3 Preliminary Discussion—K-Method-Type Analysis[5,6,7]

The following discussion is preliminary, but comprehensive; further discussion on K-Method is elaborated in the following Sect. 9.2. Obtain two roots of Eq. 9.32

$$P' = \left[A_0' \; A_2 \; A_4' \right] \quad (9.32)$$

Let:

$$p^2 = \sqrt{P'} \quad (9.33)$$

Dimensional analysis:

$$P'^2 = \left[T^{-2}\right] \quad (9.32b)$$

For K-Method [4]

$$\lambda = \left(\frac{\omega_\alpha}{\omega}\right)^2 (1 + \mathrm{ig})[-] \quad (9.36)$$

To yield similar dimension for Eqs. (9.35) and (9.36), divide p^2 in Eq. (9.24) or (9.25) by ω_α^2 or define

$$\overline{p}^2 \equiv \frac{p'^2}{\omega_\alpha^2} \quad (9.37)$$

Then

$$\frac{1}{\overline{p}^2} = \frac{\omega_\alpha^2}{p'^2} \quad (9.38)$$

Solution of (9.25)

[5] Further detail of the analysis has been comprehensively presented by Zwaan [4].

[6] An alternative and different analysis is given by Hassig [9].

[7] Some notes on the incorrectess of the method was recently given by Patil [10].

$$p'^2 = \mathrm{root}(P') = \sqrt{P'} \tag{9.39}$$

or

$$\overline{p}^2 = \frac{p'^2}{\omega_\alpha^2} = \frac{1}{\omega_\alpha^2}\mathrm{root}(P') = \frac{1}{\omega_\alpha^2}\sqrt{P'} \tag{9.40}$$

Two roots:

$$\overline{p}_2^2 = \frac{p'^2}{\omega_\alpha^2} = \frac{1}{\omega_\alpha^2}\left(+\sqrt{P'}\right) = \frac{1}{\omega_\alpha^2}psq1 \tag{9.41}$$

$$\overline{p}_2^2 = \frac{p'^2}{\omega_\alpha^2} = \frac{1}{\omega_\alpha^2}\left(-\sqrt{P'}\right) = \frac{1}{\omega_\alpha^2}psq2 \tag{9.42}$$

$$\lambda = \left(\frac{\omega_\alpha}{\omega}\right)^2(1+\mathrm{ig}) = \left(\frac{\omega_\alpha}{p'}\right)^2 = \frac{\omega_\alpha^2}{psq1} = \mathrm{Real}(\lambda) + \mathrm{Imag}(\lambda) \tag{9.43}$$

$$\lambda_1 = \left(\frac{\omega_\alpha}{\omega}\right)^2(1+\mathrm{ig})\Big|_1 = \left(\frac{\omega_\alpha}{p_1'}\right)^2 = \frac{\omega_\alpha^2}{psq1} = \mathrm{Real}(\lambda_1) + \mathrm{Imag}(\lambda_1) \tag{9.43a}$$

$$\lambda_2 = \left(\frac{\omega_\alpha}{\omega}\right)^2(1+ig)\Big|_2 = \left(\frac{\omega_\alpha}{p_2'}\right)^2 = \frac{\omega_\alpha^2}{psq2} = \mathrm{Real}(\lambda_2) + Imag(\lambda_2) \tag{9.43b}$$

Hence

$$\omega_1 = \frac{\omega_\alpha}{\sqrt{\mathrm{Real}(\lambda_1)}} \tag{9.44}$$

$$\omega_2 = \frac{\omega_\alpha}{\sqrt{\mathrm{Real}(\lambda_2)}} \tag{9.45}$$

$$g_1 = \left(\frac{\omega_1}{\omega_\alpha}\right)^2 Imag(\lambda_1) \tag{9.46}$$

$$g_2 = \left(\frac{\omega_1}{\omega_\alpha}\right)^2 Imag(\lambda_2) \tag{9.47}$$

Further discussions for validation and comparison following this section are given in Appendix 1.[8]

[8] The numbering in Appendix 9.1 follows Sect. 9.1.3.

9.2 K-Method

Following closely the procedure elaborated in Zwaan [4], the Flutter equations expressed in generalized quantities are expressed by[9,10]

$$[\mathbf{M}]\{\ddot{q}\} + [\mathbf{D}]\{\dot{q}\} + [\mathbf{K}]\{q\} = \frac{1}{2}\left[A(t^*)\right]\{q\} \tag{9.66a}$$

or

$$[\mathbf{M}]\{\ddot{q}\} + [\mathbf{D}]\{\dot{q}\} + [\mathbf{K}]\{q\} = \frac{1}{2}\rho U^2\left[A\left(\frac{b}{U}s\right)\right]\{q\} \tag{9.66b}$$

where

$[\mathbf{M}]$	generalized mass matrix.
$[\mathbf{D}]$	generalized viscous damping matrix.
$[\mathbf{K}]$	generalized stiffness matrix.
$[A(t^*)] = \left[A\left(\frac{b}{U}s\right)\right]$	matrix of unsteady aerodynamic operators.
$\{q\}$	vector of generalized coordinates.

$$\{\hat{q}\} = \left(\frac{U}{b}\right)^2 p^2\{q\}$$

and

$$\{\hat{q}\} = \left(\frac{V}{b}\right)^2 p^2\{q\}$$

Generalized coordinates (degrees of freedom) represent usually vibration modes, up to a total number of N.

For typical section:

$$\{q\} \equiv \{q(h,\alpha)\}, \quad \{q\}^T \equiv \{q(h,\alpha)\}^T$$

---

[9] For later cross-reference, Eq. 9.66, as can be folowed in subsequest sections, is equivalent to.

$$\begin{bmatrix} K_h b & q S C_{L_\alpha} \\ 0 & K_\alpha - 2qSeb\,C_{L_\alpha} \end{bmatrix}\begin{Bmatrix} \hat{h} \\ \hat{\alpha} \end{Bmatrix} = \begin{bmatrix} -m\,b & -S_\alpha \\ -S_\alpha b & -I_\alpha \end{bmatrix} p^2 \begin{Bmatrix} \hat{h} \\ \hat{\alpha} \end{Bmatrix} \tag{9.13}$$

or

$$\left[\left(\frac{V}{c}\right)^2[M]p^2 + [K] - \left(\frac{1}{2}\rho V^2\right)[A(p)]\right]\{q\} = 0$$

.

[10] Here $[\mathbf{M}]$ is Mass Matrix and $[\mathbf{K}]$ stiffness matrix. M is Mach number.

Solution: substitution of

$$\{q\} \equiv \{\hat{q}\}e^{st}$$

$$s = (\sigma + i\omega)$$

$$\{q\} \equiv \{\hat{q}\}e^{(\sigma + i\omega)t}$$

into equ. of motion leads to a set of homogeneous equations

$$\left([\mathbf{M}]s^2 + [\mathbf{D}]s + [\mathbf{K}]\right)\{\hat{q}\} = \frac{1}{2}\rho U^2[A(p)]\{\hat{q}\} \tag{9.66c}$$

or

$$\left([\mathbf{M}]s^2 + [\mathbf{D}]s + [\mathbf{K}]\right)\{\hat{q}\} = \frac{1}{2}\rho U^2\left[A\left(\frac{b}{U}s\right)\right]\{\hat{q}\} \tag{9.66d}$$

where

$$p = \frac{b}{U}s = \frac{b}{U}(\sigma + i\omega)$$

which form eigenvalue problem for s.

- Assuming harmonic motion:

$$p = \frac{b}{U}s = \frac{b}{U}(\sigma + i\omega) \rightarrow Im(p) = \frac{b}{U}(i\omega) \rightarrow Im(s) = (i\omega)$$

- Assuming structural damping instead of viscous damping

  Flutter Eq. 9.66d becomes[11]

$$-\omega^2[\mathbf{M}]\{\hat{q}\} + (1 + ig)[\mathbf{K}]\{\hat{q}\} = \frac{1}{2}\rho U^2[A(M, k)]\{\hat{q}\} \tag{9.66e}$$

9.2.1 Reduction of Eigenvalue Problem

$$k = \frac{\omega b}{U} \quad A\left(\frac{b}{U}s\right) = A\left(\frac{\frac{\omega b}{U}s}{\omega}\right) = A\left(\frac{k}{\omega}i\omega\right)$$

[11] Here [**M**] is Mass Matrix and [**K**] stiffness matrix. M is Mach number.

$$\frac{(1+ig)}{\omega^2}[K]\{\hat{q}\} = \left([M] + \frac{1}{2}\rho U^2 \frac{1}{\omega^2}[A(M_\infty, k)]\right)\{\hat{q}\} \tag{9.67a}$$

$$[K]^{-1}\left([M] + \frac{1}{2}\rho U^2 \frac{1}{\omega^2}[A(M_\infty, k)]\right)\{\hat{q}\} = \frac{(1+ig)}{\omega^2}\{\hat{q}\} \tag{9.67b}$$

or, in short

$$[C(M_\infty, k)]\{\hat{q}\} = \lambda\{\hat{q}\} \tag{9.67c}$$

where

$$\lambda = \frac{(1+ig)}{\omega^2}$$

$$k = \frac{\omega b}{U}$$

G	artificial damping to enforce harmonic motion
C	is a complex variable
$\lambda_1, \lambda, \ldots\ldots \lambda_n$	are usually complex.

9.2.2 *Equivalent Modal Structural Damping*

The Flutter solution following K-Method is a stability analysis of Eq. 9.1 [11]. This equation cannot be solved directly because the Q_{aero} is dependent on the frequency parameter, ω. The solution technique "V-g" or the "K-Method" has been frequently used as elaborated in previous sections. In the K-Method the Q_{aer}o matrix is interpolated for a series of k values and the eigenvalue problem is solved for each interpolated aerodynamic matrix. The K-Method equation can be written as:

$$\left[K - \left\{M + \left(\frac{1}{2}\rho b^2\right)Q_{Aero}(M_\infty, k)\right\}\left(\frac{\omega^2}{(1+ig)}\right)\right]\{U\} = 0 \tag{9.66f}$$

This method solves for an equivalent modal structural damping that makes the motion oscillatory or neutrally stable. The value of "g" is the structural damping required for harmonic motion.

This accounts for and closely associated with the fact that the unsteady aerodynamic forces, Q_{aero}, are derived for harmonic motion. The solution is carried out sequentially by interpolating the unsteady aerodynamic matrix for different 1/k values, the eigenvalue is ω and g for each modal coordinate. Because a given 1/k was assumed, then the calculated root ω will also yield a corresponding velocity, V. Finally, the results are displayed in a V-G plot, as exemplified in Fig. 9.1 for combined aerodynamic and acoustic loading [12].

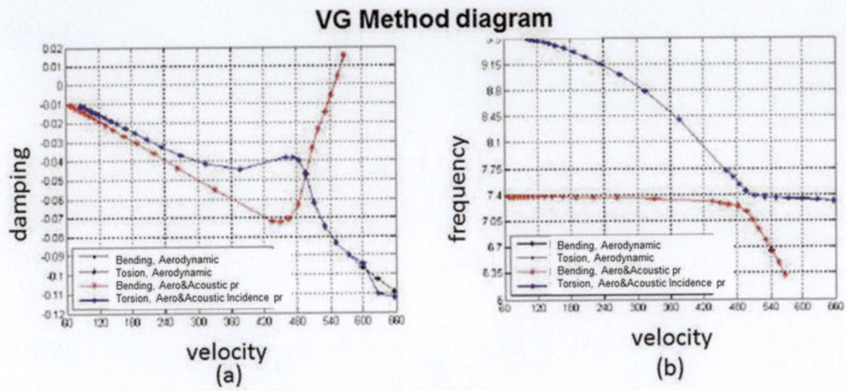

Fig. 9.1 **a** Damping and **b** frequency diagram for BAH wing calculated using V-g method written in MATLAB® for the acousto-aeroelastic problem (the total acoustic pressure already incorporates the scattering pressure)[12]

The flutter instability occurs when the damping, g, becomes positive. The K-Method is generally very fast, but it has a drawback in that sometimes the frequency and damping values "loop" around themselves and yield multi-value frequency and damping as a function of velocity. The K-Method solution is only valid when $g = 0$, and the structural motion is neutrally stable and matches the aerodynamic motion which is also neutrally stable. Neutrally stable is taken here to mean a non-converging or non-diverging oscillator motion. Generally, the K-Method is not a "Match Point" solution.

9.2.3 Agard Method [4, 13]

For a typical section

$$\frac{\hat{L}}{\pi\rho b^2} = \omega^2\left\{\frac{k_a}{k^2}\left[\frac{\hat{h}}{b} - (0.5+a)\hat{\alpha}\right] + \frac{k_b}{k^2}\hat{\alpha}\right\} \tag{9.68a}$$

$$\frac{\hat{M}_{AC}}{\pi\rho b^2} \approx \frac{\hat{M}_{c/4}}{\pi\rho b^2} = \omega^2\left\{\frac{m_a}{k^2}\left[\frac{\hat{h}}{b} - (0.5+a)\hat{\alpha}\right] + \frac{m_b}{k^2}\hat{\alpha}\right\} \tag{9.68b}$$

Eigenvalue problem format:

$$[A(k)]\{x\} = \lambda\{x\} \tag{9.69a}$$

or

$$[A(k)]\{\hat{q}\} = \lambda\{\hat{q}\} \tag{9.69b}$$

where

$$A_{11} = \left(\frac{\omega_\alpha}{\omega_h}\right)^2\left(\mu - \frac{k_a}{k^2}\right) \tag{9.70a}$$

$$A_{12} = \left(\frac{\omega_\alpha}{\omega_h}\right)^2\left(\mu x_\alpha - \frac{k_b}{k^2} + (0.5+a)\frac{k_a}{k^2}\right) \tag{9.70b}$$

$$A_{21} = \mu x_\alpha - \frac{m_a}{k^2} + (0.5+a)\frac{k_a}{k^2} \tag{9.70c}$$

$$A_{22} = \mu r_\alpha^2 \frac{m_b}{k^2} + (0.5+a)\left(\frac{m_a}{k^2} + \frac{k_a}{k^2}\right) - (0.5+a)^2\frac{k_a}{k^2} \tag{9.70d}$$

$$\{\hat{q}\}^T = \left\{\frac{\hat{h}}{b}, \hat{\alpha}\right\} \tag{9.70e}$$

$$\lambda = \left(\frac{\omega_\alpha}{\omega}\right)^2(1+ig) \tag{9.70c}$$

All elements in [A] depend on k rather than ω.
Theodorsen coefficients following Küssner:
Heaving:

$$k_a' = -2kG(k)$$

$$k_a'' = -2kF(k)$$

$$k_b' = -2kF(k) - 2kG(k) - \frac{k^2}{2}$$

$$k_b'' = 2kF(k) + 2kG(k) + k$$

$$m_a' = -\frac{k^2}{2}$$

$$m_a'' = 0$$

Pitching:

$$k_b' = -2kF(k) - 2kG(k) - \frac{k^2}{2}$$

$$k_b'' = 2kF(k) + 2kG(k) + k$$

$$m_a' = -\frac{3}{8}k^2$$

$$m_a'' = k$$

where $C(k)$ is the Theodorsen function defined by

$$C(k) = F(k) + iG(k) = \frac{H_1^{(2)}(k)}{H_1^{(2)}(k) + iH_0^{(2)}(k)}$$

$$F(k) = \mathrm{Re}(C(k)) = \mathrm{Re}\left(\frac{H_1^{(2)}(k)}{H_1^{(2)}(k) + iH_0^{(2)}(k)}\right)$$

$$G(k) = \mathrm{Im}(C(k)) = \mathrm{Im}\left(\frac{H_1^{(2)}(k)}{H_1^{(2)}(k) + iH_0^{(2)}(k)}\right)$$

Procedure:
For $M_\infty = 0$.

- Choose ρ or altitude and number of k values—$\rho = 1$ kg/m^3
- Calculate $[C]$ for each k

$$C(k) = F(k) + iG(k) = \frac{H_1^{(2)}(k)}{H_1^{(2)}(k) + iH_0^{(2)}(k)}$$

- Calculate eigenvalues $\lambda_1, \ldots.. \lambda_n$, only positive values are important
  Solve eigenvalue problem: $[A(k)]\{\hat{q}\} = \lambda\{\hat{q}\}$

$$\lambda = \left(\frac{\omega_\alpha}{\omega}\right)^2 (1 + ig)$$

- Calculate:

$$\lambda = \frac{1 + ig}{\omega^2} = \mathrm{Re}(\lambda) + \mathrm{Im}(\lambda) = \frac{1}{\omega^2} + \mathrm{Im}\left(\frac{g}{\omega^2}\right)$$

$$\omega = \sqrt{\frac{1}{\mathrm{Re}(\lambda)}}$$

$$g = \omega^2 \mathrm{Im}(\lambda)$$

- $\omega_1 = \frac{1}{\sqrt{\lambda_1}}$ $U_1 = \frac{\omega_1 b}{k_1}$,
- $\omega_2 = \frac{1}{\sqrt{\lambda_2}}$ $U_2 = \frac{\omega_2 b}{k_2}$,
- $\omega_3 = \frac{1}{\sqrt{\lambda_3}}$ $U_3 = \frac{\omega_3 b}{k_3}$,
  etc.

- Calculate U from

$$k = \frac{\omega b}{U}$$

$$U = \frac{\omega b_R}{k}$$

- Draw curves of ω versus U and versus U
  Flutter occurs when g crosses the U-axis, i.e. $g = 0$.

Solution of eigenvalue problem as example for this method is elaborated in Appendix 1, while an excerpt from the computational program is given in Appendix 2.

9.2.4 Low-Frequency Refinement [4]

$$\alpha_{dyn}(t) = \frac{\dot{h}(t)}{U} + \alpha(t)$$

Hence

$$L(t) = 2qSC_{L_\alpha}\left(\frac{\dot{h}(t)}{U} + \alpha(t)\right)$$

$$M_{EA}(t) = 2qSebC_{L_\alpha}\left(\frac{\dot{h}(t)}{U} + \alpha(t)\right)$$

Equation of motion becomes:

$$m\ddot{h} + S_\alpha\ddot{\alpha} + K_h h + 2qSC_{L_\alpha}\left(\frac{\dot{h}(t)}{U} + \alpha(t)\right) = 0$$

$$S_\alpha\ddot{h} + I_\alpha\ddot{\alpha} + K_\alpha\alpha - 2qSebC_{L_\alpha}\left(\frac{\dot{h}(t)}{U} + \alpha(t)\right) = 0$$

$$[M]\{\ddot{x}\} + [B]\{\dot{x}\} + ([K] - q[A])\{x\} = \{0\}$$

where an additional term appears:

The aerodynamic damping:

$$[B] = \begin{bmatrix} \frac{q}{U} SC_{L_\alpha} & 0 \\ -2\frac{q}{U} SebC_{L_\alpha} & 0 \end{bmatrix}$$

Assuming:

$$\{x\} = \{\hat{x}\}e^{pt}$$

Leads to characteristics equation:

$$a_4\, p^4 \;+\, a_3\, p^3 \;+\, a_2\, p^2 \;+\, a_1\, p \;+\, a_0 \;=\; 0$$

where

$$a_4 = mI_\alpha - S_\alpha^2$$

$$a_3 = \frac{q}{U} SC_{L_\alpha}(2ebS_\alpha + I_\alpha)$$

$$a_2 = mK_\alpha + I_\alpha K_\alpha - (2meb + S_\alpha)qSC_{L_\alpha}$$

$$a_1 = \frac{q}{U} SC_{L_\alpha} K_\alpha$$

$$a_0 = K_h\left(K_\alpha - 2qSebC_{L_\alpha}\right)$$

Another approach for this case is elaborated in Appendix 3.

9.3 Other Flutter Calculation Methods—The Expanding Domain of Aeroelasticity

During the past ten years the field of aeroelasticity and the developments in associated fields leading to the flutter solution has been for the most part regarded as discrete entities; now these disciplines are developing into component parts of an integrated whole—the dynamics of a deformable airplane [14].

The k method which has been comprehensively elaborated in previous sections, is regarded to be instructive, and constitutes a good introduction to the problem of flutter stability. Although it is capable to predict the correct flutter instability, however, the damping and frequency trends of the k method are known to be false.

There is some indications that in the future it may be necessary, or various enabling technologies have developed in such a way, especially when adequate computational aids become available, that many of methods of treatment presently used will be part of history, and a standard method of attack on the problem treated as a unified whole will be adopted. Therefore, in what follows, some flutter solution methods will not be elaborated in great detail, but sufficient to elaborate the principles, with a view to provide information that may trigger further creativity in the spirit just mentioned.

Recalling from 9.2 related to K-Method and following closely its development there, the Flutter equations expressed in generalized quantities can be expressed by

$$\begin{bmatrix} K_h b & qSC_{L\alpha} \\ 0 & K_\alpha - 2qSebC_{L\alpha} \end{bmatrix} \begin{Bmatrix} \overline{h} \\ \overline{\alpha} \end{Bmatrix} = \begin{bmatrix} -mb & -S_\alpha \\ -S_\alpha b & -I_\alpha \end{bmatrix} p^2 \begin{Bmatrix} \widehat{h} \\ \widehat{\alpha} \end{Bmatrix} \tag{9.13}$$

$$Det \begin{bmatrix} K_h\, b + mbp^2 & q\, S\, C_{L\alpha} + S_\alpha\, p^2 \\ S_\alpha\, bp^2 & (\, K_\alpha \; - 2\, q\, S\, e\, b\, C_{L\alpha}) + \, I_\alpha\, p^2 \end{bmatrix} \begin{Bmatrix} \widehat{h} \\ \widehat{\alpha} \end{Bmatrix} = 0 \tag{9.19a}$$

$$\begin{aligned} &\left(bmI_\alpha - S_\alpha^2 b\right)p^4 + (I_\alpha K_h b + mb(K_\alpha - 2qSebC_{L\alpha}) - S_\alpha bqSC_{L\alpha})p^2 \\ &\quad + K_h b(K_\alpha - 2qSebC_{L\alpha})\hat{\alpha} = 0 \end{aligned} \tag{9.19b}$$

or in a comprehensive formula[12]

$$[M]\,\{\ddot{q}\} \; + \; [D]\,\{\dot{q}\} + [K]\,\{q\} - \left(\frac{1}{2}\rho U^2\right)\left[A\left(\frac{b}{U}s\right)\right]\,\{q\} = 0 \tag{9.66b}$$

$$\left[\left(\frac{V}{c}\right)^2 [M]p^2 + [K] - \left(\frac{1}{2}\rho V^2\right)[A(p)]\right]\{q\} = 0 \tag{9.66g}$$

or

[12] Other formulae that can readily be derived.

$$[M]\{\ddot{q}\} + [D]\{\dot{q}\} + [K]\{q\} - \left(\frac{1}{2}\rho U^2\right)\left[A\left(\frac{b}{U}s\right)\right]\{q\} = 0 \tag{9.66b}$$

$$\left([M]s^2 + [D]s + [K]\right)\{\hat{q}\} = \frac{1}{2}\rho U^2 [A(p)]\{\hat{q}\} \tag{9.66c}$$

$$\left([M]s^2 + [D]s + [K]\right)\{\hat{q}\} = \frac{1}{2}\rho U^2\left[A\left(\frac{b}{U}s\right)\right]\{\hat{q}\} \tag{9.66d}$$

$$-\,\omega^2\,[\,M\,]\,\{\,\hat{q}\,\} + (1 + i\,g)[\,K\,]\,\{\,\hat{q}\,\} = \frac{1}{2}\rho U^2\,[\,A\,(M_\infty\,,\,k)\,]\,\{\,\hat{q}\,\} \tag{9.66e}$$

$$[M]\{\ddot{q}\} + [D]\{\dot{q}\} + [K]\{q\} - \left(\frac{1}{2}\rho U^2\right)[A(p)]\{q\} = 0 \tag{9.66f}$$

$$\left[K - \left\{M + \left(\frac{1}{2}\rho b^2\right) Q_{aero}(M, k)\right\}\left(\frac{\omega^2}{(1+ig)}\right)\right]\{U\} = 0 \quad (9.66h)$$

Since

$$p \equiv \frac{b}{U}s$$

$$\{q\} \equiv \{\hat{q}\}e^{st}$$

$$s \equiv \sigma + i\omega$$

where

$[M]$	generalized mass matrix.
$[D]$	generalized viscous damping matrix.
$[K]$	generalized stiffness matrix.
$[A(t*)]$	matrix of unsteady aerodynamic operators.
$[A(t*)] \equiv \left[A\left(\frac{b}{U}s\right)\right]$	matrix of unsteady aerodynamic operators.
$\{q\}$	vector of generalized coordinates

$$\{\hat{q}\} \equiv \left(\frac{U}{b}\right)^2 p^2\{q\}$$

$$\{\hat{q}\} \equiv \left[\left(\frac{V}{b}\right)^2 p^2\right]\{q\}$$

Generalized coordinates (degrees of freedom) represent usually vibration modes, up to total number N.

For typical section:

$$\{q\}^T \equiv \{q(h, \alpha)\}^T$$

Solution: substitution of

$$\{q\} \equiv \{\hat{q}\}\, e^{st}$$

$$s \equiv \sigma + i\omega$$

$$\{q\} \equiv \{\hat{q}\}e^{(\sigma+i\omega)t}$$

into equ. of motion leads to a set of homogeneous equations

$$\left([M]s^2 + [D]s + [K]\right)\{\hat{q}\} = \frac{1}{2}\rho U^2[A(p)]\{\hat{q}\} \tag{9.66c}$$

where

$$p \equiv \frac{b}{U}s \equiv \frac{b}{U}(\sigma + i\omega)$$

which form an eigenvalue problem.

9.3.1 Introduction to the P-K Method of Solution of the Flutter Solution—An Approximate True Damping Solution of the Flutter Equation by Determinant Iteration[13]

The flutter equation can be written in general form as

$$[M]\{\ddot{q}\} + [D]\{\dot{q}\} + [K]\{q\} - \left(\frac{1}{2}\rho U^2\right)\left[A\left(\frac{b}{U}s\right)\right]\{q\} = 0 \tag{9.66b}$$

Since p = bs/U = cs/V, these equations can be rewritten as

$$\left[\left(\frac{V}{c}\right)^2[M]p^2 + [K] - \left(\frac{1}{2}\rho V^2\right)[A(p)]\right]\{q\} = 0 \tag{9.66g}$$

$[K]$ defines the elastic characteristics by relating generalized forces (Qi) to generalized displacements $\{q\}$ following

$$[K][q\} = \{Q\} \tag{9.70}$$

M defines the inertia characteristics by relating the inertia forces Q to the generalized accelerations

$$\{Q_{\text{in}}\} = [M]\{q\} \tag{9.71}$$

$[A(p)]$ defines the unsteady aerodynamic forces through the equation

$$\{Q_{\text{aero}}\} = \left(\frac{1}{2}\rho V^2\right)[A(p)]\{q\} \tag{9.72}$$

---

[13] Hassig [9].

In Eqs. 9.66g and 9.72, p is the nondimensional differential operator $(c/V)(d/dt,)$. If the aerodynamic forces can be expressed as a sufficiently simple function of p, Eq. 9.1 defines a polynomial in p with real coefficients. For nonzero solutions for q the determinant formed by the matrix coefficients in these equations must be equal to zero. For a given value of the speed V, the determinant can then be solved directly for p. This leads to conjugate complex roots

$$p = \gamma k + ik \tag{9.73}$$

k defining a nondimensional reduced frequency $\omega c/V$ and γ defining a rate of decay

$$\gamma = 1/(2\pi)\ln(a_{n+1}/a_n) \tag{9.74}$$

where a_n and a_{n+1} are the amplitudes of successive cycles.

This method of solving the flutter equation will be called the p method. It can be used in the case of quasi-steady aerodynamics when $[A(p)]$ assumes the simple form

$$[A(p)] = [A_1]p + [A_2] \tag{9.75}$$

and also in the case of simple forms of unsteady aerodynamics.

Approximate formulation of unsteady aerodynamics that accounted for apparent mass, aerodynamic lag and aerodynamic coupling between different strips on the wing can be made which implies a relatively simple function p.

Then the flutter Eq. 9.66b can be expressed as

$$\begin{bmatrix} p^2[M] + \left(\frac{1}{V^2}\right)[K] + \frac{\sigma}{(p+b_1)}\left[p^2[A_2] + p[A_1] + [A_0]\right] \\ + \frac{\sigma}{(p+b_2)}\left[p^2[B_2] + p[B_1] + [B_0]\right] \end{bmatrix}\{q\} = 0 \tag{9.76}$$

The quantities b_1 and b_2 are taken from the assumed two-exponential lag function

$$1 - a_1 e^{-b_1 z} + a_2 e^{-b_2 z} \tag{9.77}$$

The matrices A_0, A_1, A_2 and B_0, B_1, B_2 contain the lift curve slope, aerodynamic influence coefficients, the aerodynamic-center and rotation center (corresponding to the chord point in two- dimensional theory) for each strip, and the damping in pitch.

More sophisticated formulations of the aerodynamics can be defined from the kernel function or doublet lattice approach (or the supersonic Mach box for supersonic case, which is beyond the scope of the present book), lead to aerodynamic matrices valid only for harmonic motion, $p = ik$. In that case, Eq. 9.66g takes the form

$$-\frac{1}{c}[M]k^2 + \left(\frac{1}{V^2}\right)[K] - \frac{1}{2}\rho[A(ik)]\ \{q\} = 0 \tag{9.78}$$

At chosen values of k, complex roots for $\left(\left(\frac{1}{V^2}\right), \lambda_r + i\lambda_i\right)$ are found and interpreted as

$$\lambda_r + i\lambda_i = \left(\frac{1}{V^2}\right)(1 + ig) \tag{9.79}$$

where p is the structural damping required for harmonic motion. This method of solving the flutter equation is here called the k method.

There exists a difference between the rate-of-decay damping, γ, found with the p method and the structural damping, g, found with the k-Method. For the k method p is plotted; for the P-Method the comparable quantity.

$$2\lambda = \frac{1}{\pi}\,\text{In}\left(\frac{a_{n+1}}{a_n}\right) \tag{9.80}$$

The p solution method may show flutter in the first mode, while the k method solution may show flutter in the second mode. Even the modal coupling could be different. However, the flutter speed obtained by both methods of solution may give identical results.

If g, ω, and V are found with the K Method, then,

$$2\lambda = \frac{1}{\pi}\,\text{In}\left(\frac{a_{n+1}}{a_n}\right) = g\left(1 - \frac{V}{\omega}\frac{d\omega}{dV}\right) \tag{9.81}$$

It should be noted that these methods are approximate and indirect. Indirect in the sense that they require the availability of the k-method solution.

The P-K Method

Equation 9.81 suggests an approximate method of finding a rate-of-decay type solution directly. Writing Eq. 9.66g in a form indicating that the aerodynamic matrix is only available for harmonic motion, and computing $[A(ik)]$ for an estimated value of k:

$$\left[\left(\frac{1}{V^2}\right)[M]p^2 + [K] - \left(\frac{1}{2}\rho V^2\right)[A(ik)]\right]\{q\} = 0 \tag{9.82}$$

one can solve for $p = \gamma k + ik$, compute $[A(ik)]$, solve Eq. 9.82 again which leads to $p = \gamma k_2 + ik_2$, etc., until the imaginary part of the solution equals the k value of the aerodynamics. Therefore this method of solution is here referred to as the P-K method. The rationale is based on notion that for sinusoidal motions with slowly increasing or decreasing amplitude, aerodynamics based on constant amplitude are a good approximation. Further detail can be followed in Hassig [9].

This P-K method of solving the flutter equation was designed to give a better approximation of the subcritical damping than the K-Method. It eliminates the looping sometime found in the K-Method. Both of these flutter solution methods,

P-K and K, in general yield the same flutter velocity, V_f, for $g = 0$ case. The equation of motion is written slightly different in the P-K Method as [11]

$$\left[Mp^2 + (1 + ig)K - \left(\frac{1}{2}\rho V^2\right) Q_{\text{aero}}(M, k)\right]\{U\} = 0 \tag{9.83}$$

where
$p = \text{eigenvalue} = \omega(\gamma + i)$.

γ = Transient Decay Rate
The eigenvalue p is represented by a damped sinusoidal motion, e^{pt} while the Q_{aero} term is expressed as undamped oscillatory motion, $e^{i\omega t}$. Then a mismatch between the assumed motion of the aerodynamic forces and the remaining structural forces is shown. Hassig's P-K Method is an approximate true damping method. The solution of Eq. 9.83 is carried out by matrix iterative techniques. The reduced frequency of the Q_{aero} term, k, is dependent of the circular frequency, ω, and is related by

$$\omega = \text{kV/b} = \text{Imaginary part of } p \tag{9.84}$$

For an assumed velocity, V, a set of eigenvalues are calculated for a given k value of the Q_{aero} term. If the calculated frequency, $\omega_{\text{Root,}}$ is not equal the assumed ω_{Aero} then the Q_{aero} term is interpolated to a new k value that is consistent with the $\omega_{\text{Root,}}$ value. This procedure of solving the eigenvalue problem is an iterative one. This iteration process is repeated until the structural root frequency, $\omega_{\text{Root,}}$, is equal to the assumed aerodynamic frequency, ω_{Aero}, for every root at every assumed velocity. This process involves the repeated interpolation of the Q_{aero} term for each velocity.

9.3.2 Non-Iterative P-K (NIPK) Method—A New Non-Iterative P-K Match Point Flutter Solution Method

Pitt [11] proposed a method that solves the P-K Eq. 9.83 in a non-iterative process. The non-iterative P-K Method starts as the traditional method with an assumed list of velocities. Once a velocity is chosen, the Q_{aero} term is no longer a function of k, but is a function of ω. For each hard k value the Q_{aero} term is now only a function of ω_{Aero}. The ω_{Aero} value is described as:

$$\omega = \text{kV/b for each hard } k \text{ values} \tag{9.85}$$

The eigenvalues, p, are found for Eq. 9.83 for an assumed ω_{Aero} value of the Q_{aero} term. The eigenvalues, p, yield a frequency and damping for each mode; which are called ω_{Root} and $\gamma_{\text{Root.}}$ Since Eq. 9.83 is solved for each hard k value, no interpolation of the Q_{aero} term is required. Therefore there is a series of ω_{Root} and γ_{Root} for each ω_{Aero} value of the Q_{aero} term. The key to this technique is to interpolate the ω_{Roots}

and γ_{Roots} so that $\omega_{\text{Aero}} = \omega_{\text{Root}}$, and to not interpolate the Q_{aero} term as a function of k.

It should be noted that the P-K flutter equation as written in Eq. 9.83 is not restricted to $k \neq 0$ reduced frequencies; $k = 0$ does not pose any special requirements. The k $= 0$ value will be used to predict the divergent speed which is associated with $\omega =$ 0. Further details and examples should be followed in Pitt [11].

9.3.3 An Analysis of the Flutter and Damping Characteristics Using the P-K Method of Flutter Solution [15]

In introducing the Concept of the Decay Rates it may be assumed that the wing typical section motion could be described through the superposition of normal modes as

$$\alpha(y,t) = \sum_{r=1}^{N} \alpha_r(y)\xi_r(t) \tag{9.86}$$

$$w(y,t) = \sum_{r=1}^{N} w_r(y)\xi_r(t) \tag{9.87}$$

Utilizing this representation, the equations of wing typical section bending and torsional motion are described by

$$M_r\ddot{\xi}_r + (1 + ig_r)\omega_r^2 M_r\xi_r = \Xi_r(t), \quad r = 1\, to\, N \tag{9.88}$$

where

$$\Xi_r(t) = -\int_0^R L(y,t)w_r(y)\,\mathrm{d}y + \int_0^R M_{EA}(y,t)\alpha_r(y)\,\mathrm{d}y \tag{9.89}$$

Let

$$\xi_r(t) = \overline{\xi}\exp(pt) \quad r = 1 \text{ to } N \tag{9.90}$$

where $p = (\gamma\omega + i\omega)$ and $\overline{\xi}_r$ is the complex amplitude of the r-th mode. Following its solution, every point on the lifting surface experiences a motion which can be expressed as

$$D\exp(\gamma\omega t)\cos(\omega t + \phi) \tag{9.91}$$

The amplitude of the oscillatory motion is D exp (ywt), where D is a function of the y location on the lifting surface. This amplitude decays exponentially when $y <$

0. u is the damped frequency of the vibration and \$ is the phase angle with respect to some reference time. It is interesting to note that at every point the motion decays with the same value of the logarithmic decrement.

Since $L(y, t)$ and M_{ea} (y, t) are linear functions of the normal coordinates, they will also take the form exp (pt), but could be with a phase shift.

$$L(y,t) = L_3(y)\frac{\sum_{r=1}^{N} w_r(y)\xi_r(t)}{b(y)} + L_4(y)\left(\sum_{r=1}^{N} \alpha_r(y)\xi_r(t)\right) = \overline{L}(y)\exp(\text{pt}) \quad (9.92)$$

Hence,

$$\overline{L}(y) = L_3(y)\frac{\sum_{r=1}^{N} w_r(y)\xi_r}{b(y)} + L_4(y)\left(\sum_{r=1}^{N} \alpha_r(y)\xi_r\right) \quad (9.93)$$

$$\overline{M}_{EA}(y) = M_3(y)\frac{\sum_{r=1}^{N} w_r(y)\xi_r}{b(y)} + M_4(y)\left(\sum_{r=1}^{N} \alpha_r(y)\xi_r\right) \quad (9.94)$$

where L_3, L_4, M_3 and M_4 are complex unsteady aerodynamic coefficients, which are functions of p/Ω and other parameters. These are represented by[14]

$$a_4 = mI_\alpha - S_\alpha^2$$

$$-\begin{Bmatrix} L_3(y) \\ L_4(y) \\ M_3(y) \\ M_4(y) \end{Bmatrix} = -\pi\rho_\infty\omega^2 b^4 \begin{Bmatrix} \frac{L_{hp}}{b} \\ \frac{[L_{\alpha p}-\varepsilon L_{hp}]}{b} \\ \frac{M_{hp}}{b} + \varepsilon L_{hp} \\ -\frac{M_{\alpha p}}{b} + \varepsilon\left(\frac{M_{hp}}{b} + L_{\alpha p}\right) + \varepsilon^2 L_{hp} \end{Bmatrix} \quad (9.95)$$

where the second subscript, p, denotes that these are p-type aerodynamic coefficients. An aerodynamic theory that is developed for a motion of the form exp (pt) is called p-type. A contrasting aerodynamic theory dealing with simple harmonic motion of the form exp $(i\omega t)$ is called k-type, where k is a conventional notation for the reduced frequency associated with ω. The P-K method uses k-type aerodynamics with modifications, to simulate a p-type motion. The generalized aerodynamic forces of Eq. 9.64 can be expressed by

---

[14] Modified from Viswanathan's original formul.

$$-\begin{Bmatrix} L_3(y) \\ L_4(y) \\ M_3(y) \\ M_4(y) \end{Bmatrix} = -\pi\rho_\infty\omega^2 b^4 \begin{Bmatrix} \frac{L_{\text{hp}}}{b} \\ \frac{[L_{\alpha p}-\varepsilon L_{\text{hp}}]}{b} \\ M_{\text{hp}} + \varepsilon L_{\text{hp}} \\ -M_{\alpha p} + \varepsilon\left(M_{\text{hp}} + L_{\alpha p}\right) + \varepsilon^2 L_{\text{hp}} \end{Bmatrix}$$

$$\Xi_r(t) = \overline{\Xi}\exp(\mathrm{pt}), \quad r = 1 \text{ to } N \tag{9.96}$$

where

$$\Xi_r(t) = \pi\rho_\infty\omega^2 b_{ref}^4 R \sum_{r=1}^{N} A_{\mathrm{rnp}}\overline{\xi}, \quad r = 1 \text{ to } N \tag{9.97}$$

and

$$A_{\mathrm{rnp}} = \int_0^1 \begin{bmatrix} \beta^2 L_{\mathrm{hp}} w_r * w_n * + \beta^3 (L_{\alpha p} - L_{\mathrm{hp}}\varepsilon)\alpha_r * w_n * \\ + \beta^3 (M_{\mathrm{hp}} - L_{\mathrm{hp}}\varepsilon)\alpha_r * w_n * \\ + \beta^4 (M_{\alpha p} - (L_{\mathrm{hp}} - M_{\mathrm{hp}})\varepsilon + L_{\mathrm{hp}}\varepsilon^2)\alpha_r * w_n * \end{bmatrix} d\eta \tag{9.98}$$

Equation 9.88 can now be written as

$$p^2 M_r^* \overline{\xi}_r + (1 + ig_r)\omega_r^2 M_r \overline{\xi}_r = \omega^2 \sum_{n=1}^{N} A_{rnp}\overline{\xi}_n, \quad r = 1\ to\ N \tag{9.99}$$

or in matrix form as

$$\begin{bmatrix} \searrow & & \\ & \left[\left(\frac{p}{\omega}\right)^2 + (1+ig_r)\left(\frac{\omega_r}{\omega}\right)^2 M_r^*\right] & \\ & & \searrow \end{bmatrix} \{\overline{\xi}\} = [A_p]\{\overline{\xi}\} \tag{9.100}$$

where A_{rnp} is the element on the r-th row and n-th column of matrix $[A_p]$. A_{rnp} can be evaluated after obtaining L_{hp}, $L_{\alpha p}$, M_{hp} and $M_{\alpha p}$ at several points along the radius by knowing the local Mach number, inflow ratio, frequency ratio, reduced frequency, decay rate, the number of wing typical sections and the phase angle between wing typical sections.

Inference of the Decay Rates from Forced Response

The decay rates of the aeroelastic modes provide a means of assessing the amount of damping present in the aeroelastic system. It was pointed out that at each rotor speed, ft, there in general would exist N aeroelastic modes, each individually characterized by a mode shape, frequency and decay rate. A knowledge of these quantities at sub-flutter rotor speeds provides a measure of the aeroelastic stability of wing typical section motion.

Almost all comprehensive unsteady aerodynamic theories are of the k-type. When a lifting wing typical section surface is acted upon by externally applied simple harmonic forces, the steady-state wing typical section response is also simple harmonic. The k-type aerodynamic theory will then predict the simple harmonic airload acting on the wing typical section. From the equations of motion, the resulting response can be obtained at all points of the surface for any prescribed simple harmonic forcing function. The response amplitudes will be functions of the frequency of excitation. As the forcing frequency is varied, the amplitude response can be evaluated as a function of the forcing frequency, and thus the frequency response function can be obtained.

For multi-degree-of-freedom systems, the above methods yield approximate results. If the damped natural frequencies of the various aeroelastic modes are far apart compared to the frequency width between their half power points, then the single degree of freedom assumption can be used to obtain an approximation to the decay rates.

For continuous systems with numerous degrees of freedom, the shape of the amplitude response curve, namely the plot of the response amplitude versus forcing frequency is dependent on the location of the response measurement and the point (or points) of application, of the forcing function. Consequently, for aeroelastic systems, these methods seem to be somewhat unreliable for predicting the damping. However, experimentally obtained frequency response characteristics are very useful in substantiating a k-type unsteady aerodynamic theory. Such a theory can be used in the P-K method to obtain the decay rates directly. Further detail can be followed in Viswanathan, S.P., An Analysis of the Flutter and Damping Characteristics, Ph.D Thesis, Gatech, 1977.

9.3.4 A Method for Efficient Flutter Analysis of Systems with Uncertain Modeling Parameters [16]

a. **Stability of the Nominal System** The nominal flutter equation can be written as

$$\left[p^2[\overline{M}] + \left(\frac{b}{u}\right)^2 [\overline{K}] - \left(\frac{1}{2}\rho b^2\right)[\overline{Q}(p)]\right]\{\eta\} = 0 \tag{9.101}$$

where M, K and Q represent the mass, stiffness and aerodynamic matrices respectively, $p = g + ik$ is the reduced Laplace variable, b is the aerodynamic reference length, u is the airspeed and ρ is the density of the air. The system (1) is formulated in the Laplace domain and can be seen as a nonlinear eigenvalue problem. The Laplace variable p represents an eigenvalue, and η represents the corresponding eigenvector. The system is said to be stable if all eigenvalues that make the flutter matrix singular have negative real part. This since these solutions correspond to a stable motion with decreasing amplitude in the time domain. If some eigenvalue has a real part $g \geq 0$ the system is said to be nominally unstable.

b. **Modal Projection** The mass and stiffness matrices are usually obtained from a finite element model of the aircraft structure and the dimension of the flutter problem can thus be fairly large. In order to reduce the size of the flutter problem the solution vector η can be approximated using a so-called modal basis. This does not only reduce the computational time, but it also gives the resulting modal mass and stiffness matrices a certain structure [17]. The modal basis consists of a number of eigenvectors to the structural eigenvalue problem.

$$\left[-\omega^2\left[\overline{M}\right]+\left[\overline{K}\right]\right]\{z\}=0 \tag{9.102}$$

The modal flutter equation is given by [17]

$$\left[p^2[I]+\left(\frac{b}{u}\right)^2[\Omega]-\left(\frac{1}{2}\rho b^2\right)\left[\overline{Q}(p)\right]\right]\{v\}=0 \tag{9.103}$$

Polar Formulation of the Modal Flutter Equation

Using $[Z]^T[M][Z]=[I]$ formulation, the stability of 9.103 can be analyzed by computation of eigenvalues of an ordinary linear system. For flutter analysis, a usual approximation is to let $Q(p)\approx Q(k)$ close to the imaginary axis [17]. If making a change of variables so that $p=$ reiθ then close to the imaginary axis k ≈ r and so $Q(p)$ can be approximated by $Q(r)$ in this area. Thus, 9.103 can be written

$$\left[r^2e^{i2\theta}[I]+\left(\frac{b}{u}\right)^2[\Omega]-\left(\frac{1}{2}\rho b^2\right)[Q(r)]\right]\{v\}=0 \tag{9.104}$$

when r ≠ 0, (4) can be reformulated as

$$\left[e^{i2\theta}[I]+\frac{1}{r^2}\left(\frac{b}{u}\right)^2\left([\Omega]-\left(\frac{1}{2}\rho u^2\right)\right)[Q(r)]\right]\{v\}=0 \tag{9.105}$$

The first term in 9.105 is a function of θ only and, since it is the identity matrix multiplied by a complex number, any v that is multiplied by it will become $e^{i2\theta}v$. Thus 9.105 is equal to

$$-e^{i2\theta}\{v\}=\left[\frac{1}{r^2}\left(\frac{b}{u}\right)^2\left([\Omega]-\left(\frac{1}{2}\rho u^2\right)\right)[Q(r)]\right]\{v\} \tag{9.106}$$

Letting

$$[A(u,r)]=\left[\frac{1}{r^2}\left(\frac{b}{u}\right)^2\left([\Omega]-\left(\frac{1}{2}\rho u^2\right)\right)[Q(r)]\right] \tag{9.107a}$$

then 9.105 can be written as

$$-e^{i2\theta}\{v\}=[A(u,r)]\{v\} \tag{9.107b}$$

Since $|-e^{i2\theta}| = 1$, this means that to have a solution, A in 9.105 has to have an eigenvalue λ with magnitude equal to one. Since A is a function of u, r and does not depend on θ, this implies that if there exists a λj with $|\lambda j| = 1$ for some u, r, then it is possible to find a solution for 9.105. Since $-e^{i2\theta}$ represents any complex number with magnitude equal to one, θ can be chosen so that $-e^{i2\theta} = \lambda_j$. For further detail, reference should be made to Lokatt's work.

9.3.5 Solution of the Uncertain Flutter Eigenvalue Problem Using μ-p Analysis [18]

To perform a robust flutter analysis for practical usefulness, a wind tunnel model with an external wing tip store was used as a test case. A parametric uncertainty description for the model, taking uncertainties in aerodynamic as well as mass properties into account has also been developed. To start with, a different sets of P-K eigenvalue solutions were visualized in the complex plane by sweeping the parameter space. Then, an analysis referred to as μ-p analysis is applied to calculate the boundaries of the same eigenvalue sets directly. The computed P-K and μ-p solutions show an excellent agreement for various uncertainty descriptions. Although there is observed a numerical issue for problems with multiple mass perturbations, the obtained results support the efficiency and practical usefulness of μ-p analysis.

The basic principle of the μ-p method is to exploit structured singular-value (or μ) analysis to investigate whether the system uncertainties can produce zero value of the flutter determinant for a given flutter eigenvalue *p*. This makes it possible to compute sets of feasible eigenvalues in the complex plane that can be used to predict damping bounds and perform robust flutter analysis. The μ-p method performed in this fashion extends the standard linear flutter analysis capability to take into account deterministic uncertainty and variation.

As a test-case, a delta-wing wind tunnel model was used. The model has a simple structural design and is made of glassfiber and carbon-fiber composite materials, with a semi-span of 0.88 m and mean chord of 0.70 m. The model was mounted vertically in the low-speed wind tunnel L2000 at the Royal Institute of Technology. The numerical analysis is based on a NASTRAN [19] model with shell elements for the wing, mass elements for the missile and aerodynamic panels for doublet-lattice aerodynamic loads. The relatively stiff missile was modeled as a rigid body attached to the wing tip. Twenty structural modes were used to establish a standard modal formulation of the flutter problem. To allow for subsequent μ-p analysis in MATLAB,[15] the required matrices and other data were extracted from NASTRAN [19] and stored in a format that could then be loaded into the MATLAB workspace. By taking aerodynamic uncertainty into account and performing successive model validation in the test, the robust flutter speed predicted by μ-p analysis differs by less than 1 m/s from the experimental value [18].

[15] MATLAB, Software Package [2].

The development of uncertainty modeling, a suitable aerodynamic uncertainty description for the present configuration is followed, which leads to an uncertain frequency-domain aerodynamic matrix in the form

$$Q(ik, \delta) = Q(ik) + \sum_{j=1}^{n_p} w_j Q_j(ik) \quad (9.108)$$

where δ is the vector of uncertainty parameters, n_p is the number of included patches, k is the reduced frequency, $Q_0(ik)$ is the nominal matrix, and $Q_j(ik)$ are aerodynamic perturbation matrices that determine the influence of the different patches.

By introducing uncertainty in the mass properties of the missile, a similar expression can be derived for the mass matrix. Uncertainty in three different parameters is considered: the mass m, the pitch moment of inertia J around the center of mass, and the location x of the center of mass in the streamwise direction (with respect to a certain grid point in the model). Similar to the aerodynamic case, the parameters are allowed to vary according to

$$m = (1 + w_m \delta_m) m_0 \quad (9.109)$$

$$J = (1 + w_J \delta_J) J_0 \quad (9.109b)$$

$$x = (1 + w_x \delta_x) x_0 \quad (9.109c)$$

where m_0, J_0 and x_0 are the nominal parameters, and $w_i > 0$ are real-valued bounds such that the real-valued uncertainty parameters $\delta_i \in (-1, 1)$. Based on Steiner's theorem for the mass moment of inertia of a rigid body, the uncertain mass matrix can now be written in the form

$$M(\delta) = M_0 + \hat{\delta}_J M_1 + \hat{\delta}_m M_2 + \hat{\delta}_x M_3 + \hat{\delta}_m \hat{\delta}_x M_4 + \hat{\delta}_x^2 M_5 + \hat{\delta}_m \hat{\delta}_x^2 M_6 \quad (9.110)$$

where $\hat{\delta}_i = w_i \delta_i$, M_0 is the nominal mass matrix and M_1-M_6 is a set of perturbation matrices. Equation 9.110 was derived by evaluating the consistent mass matrix for a set of different parameter values (m, J and x) and solving a linear system of equations for the perturbation matrices M1-M_6. The nominal modal basis was used for projection of the computed mass matrices to the modal subspace. The influence of mode shape variations due to the varying mass properties is thus neglected.

Using the uncertain aerodynamic and mass matrices defined in 9.108 and 9.110, respectively, the Robust Eigenvalue Analysis is first carried out using the resulting Laplace-domain equations of motion

$$[F(p, \delta)]\{\eta\} = \left[[M(\delta)]p^2 + \left(\frac{L}{V}\right)^2 [K_0] - \left(\frac{\rho L^2}{2}\right)[Q(ik, \delta)] \right]\{\eta\} \quad (9.111)$$

where $[K_0]$ is the nominal stiffness matrix; L is the aerodynamic reference length; V is the airspeed; ρ is the air density; $\{\eta\}$ is the vector of modal displacements; and $p = g + ik$ is the nondimensional Laplace variable with damping g and reduced frequency k. Equation 9.111 is an uncertain eigenvalue problem that depends on the unknown but bounded uncertainty parameters δ_k in the vector $\delta \in D$, where

$$D = \{\delta : |\delta_k| \leq 1 \quad \text{and}\, \delta_k \quad \text{real or complex}\} \tag{9.112}$$

The eigenvalues of the uncertain problem are defined as the values of p making the flutter matrix $F(p, \delta)$ singular for some $\delta \in D$. Consequently, p is an eigenvalue if $\det[F(p, \delta)] = 0$ for some $\delta \in D$. Clearly, if all eigenvalues of the uncertain problem have a negative real part, $g < 0$, the system is robustly stable subject to the uncertainty (no $\delta \in D$ can then destabilize the system).

The μ-p method exploits structured-singular-value analysis to investigate if some uncertainty $\delta \in D$ can make the flutter determinant $\det[F(p, \delta)] = 0$ for a given value of p. If this is true, then p is an eigenvalue according to the definition above. The first step is to pose 9.111 in a form that is suitable for μ-analysis. Next the structured singular value μ of a certain system matrix $P(p)$ subject to a block-structure of a corresponding uncertainty matrix Δ is computed. Then it can be shown that p is an eigenvalue of the uncertain eigenvalue problem if

$$\mu(p) = \mu[P(p)] \geq 1 \tag{9.113}$$

Consequently, by evaluating the real-valued scalar function $\mu(p)$ for a given value of p, it is possible to determine if p is an eigenvalue or not. This result can, for example, be exploited to compute the boundary of an eigenvalue set in the complex plane. First, the nominal eigenvalue is computed using a standard method. By searching in different directions in the complex plane, it is then possible to compute the points on the boundary where $\mu(p) = 1$ and visualize the result. This approach is used in the present study.

Further details of the analysis and results can be followed in the reference cited [18].

9.3.6 Modified P-K Method for Flutter Solution with Damping Iteration [20]

Flutter solution techniques that have been well developed in aeronautical engineering continue to be interesting topics for modern aeroelasticians. Many efforts are devoted to getting a "true damping" solution of the flutter equation. For example, the P-K method is widely used in routine flutter analysis for giving approximate true damping results, which are useful in flight flutter test programs. The promising g method for reliable damping prediction proposed by Chen [21] includes a first order of damping term in the flutter equation that is rigorously derived from the Laplace-domain aerodynamics. Further improvement for the g method can be found in the literature. In the present elaboration, the *g* method equation is reformulated and solved by a modified *P-K* method with damping iteration. Implications of the second order of damping term are also elaborated.

P-K Method and g Method for Solution of Flutter Equation

Assuming linear unsteady aerodynamics computed by panel method in the frequency domain or by linearized frequency-domain CFD results, the *P-K* method flutter equation presented by Hassig [9] reads

$$\left[\left(\frac{V}{L}\right)^2[M]p^2+\left(\frac{V}{L}\right)[B]p+[K]-\left(\frac{\rho V^2}{2}\right)[Q(ik)]\right]\{q\}=0 \tag{9.114}$$

where

$p=(V/L)s$ -the nondimensional Laplace parameter

$k=\omega\, L/V$-the reduced frequency,

and p can be expressed as

$$p=g+ik \tag{9.115}$$

Equation 9.114 was modified by Rodden and Johnson [19] as

$$\left[\left(\frac{V}{L}\right)^2[M]p^2+\left(\frac{V}{L}\right)[B]p+[K]-\left(\frac{\rho V^2}{2}\right)\frac{\left[Q^I(k)\right]}{k}p-\left(\frac{\rho V^2}{2}\right)\left[Q^R(ik)\right]\right]$$
$$\{q\}=0 \tag{9.116}$$

where $Q^R(k)$ and $Q^I(k)$ are the real part and imaginary part of $Q\_(ik)$, i.e.

$$[Q(k)]=\left[Q^R(ik)\right]+i\left[Q^I(k)\right] \tag{9.117}$$

Equation 9.116 can be rewritten in a state-space form as [20]:

$$[A(k)-pI_{2n}]\{u\}=0 \tag{9.118}$$

where $[A(k)]$ is the real matrix,

$$[A(k)]=\begin{bmatrix}0_n & I_n\\ [C_0] & [D_0]\end{bmatrix} \tag{9.119}$$

with

$$[C_0]=-\left(\frac{L}{V}\right)^2[M]^{-1}\left[[K]-\left(\frac{1}{2}\rho V^2\right)\left[Q^R(k)\right]\right] \tag{9.120a}$$

$$[D_0] = \left(\frac{L}{V}\right)^2 [M]^{-1}\left[\left(\frac{V}{L}\right)[B] - \left(\frac{1}{2}\rho V^2\right)\frac{[Q^I(k)]}{k}\right] \tag{9.120b}$$

and $\{u\} = \{\{q\}, p\{q\}\}^T$.

The P-K Method is then used to solve this nonlinear eigenvalue problem of Eq. 9.116.

However, by substituting Eq. 9.115 into the fourth term of Eq. 9.116, this equation can be rewritten as

$$\left[\left(\frac{V}{L}\right)^2 [M]p^2 + \left(\frac{V}{L}\right)[B]p + [K] - \left(\frac{\rho V^2}{2}\right)\frac{[Q^I(k)]}{k}g - \left(\frac{\rho V^2}{2}\right)[Q(ik)]\right]\{q\} = 0 \tag{9.121}$$

By comparing 9.121 with Eq. 9.114, it is found that there is an extra term, which is here identified as the added aerodynamic damping matrix:

$$-\left(\frac{1}{2}\rho V^2\right)\frac{[Q^I(k)]}{k}g$$

which was demonstrated to be valid only at small *k* or $g = 0$ for linearly varying *Q(ik)*. Based on the analytic property of the complex function $Q(p) = Q(g + ik)$ and using the Cauchy–Riemann equations, Chen [21] expanded *Q(p)* along the imaginary axis (i.e. $g = 0$) using the damping perturbation method for small g, which yields:

$$Q(p) = Q(ik) + gQ^I(ik) \tag{9.122}$$

where

$$Q^I(ik) = \frac{dQ(ik)}{d(ik)} = \frac{dQ(z)}{dz}\frac{dz}{d(ik)}$$

and $z \equiv ik$ can be calculated from *Q(ik)* by a central differencing scheme, except at $k = 0$. At $k = 0$, a forward differencing scheme is suggested by Chen [21].

The *g* method equation is formulated through replacing *Q(ik)* in Eq. 9.121 by *Q(p)* of Eq. 9.122 as expressed by Eq. 9.116.

Substituting Eq. 9.115 into Eq. 9.123 will yield a second-order linear system in terms of *g yields*,

$$\left[\left(\frac{V}{L}\right)^2 [M](g^2 - k^2) + \left(\frac{V}{L}\right)[B]g + [K] - \left(\frac{1}{2}\rho V^2\right)\frac{[Q^I(k)]}{k}g - \left(\frac{1}{2}\rho V^2\right)\mathrm{Re}([Q(ik)])\right]\{q\} = 0 \tag{9.123a}$$

and

$$\left[2\left(\frac{V}{L}\right)^2[M](gk)+\left(\frac{V}{L}\right)[B](k)+[K]-\left(\frac{1}{2}\rho V^2\right)\mathrm{Im}([Q(ik)])\right]\{q\}=0 \tag{9.123b}$$

Then a reduced-frequency-sweep technique is introduced by the g method to search for the condition Im $(g)=0$ and solve for the eigenvalues g.

Expanding *Q(p)* was further up to second order of damping perturbation yields

$$Q(p)\approx Q(ik)+gQ'(ik)+\frac{1}{2}g^2Q''(ik) \tag{9.124}$$

Equation 9.123 provides a reliable aerodynamic damping matrix that is valid in the complete k domain, and this eigenvalue problem in terms of g can be solved successfully by the g method.

P-K Method Equation with Damping Perturbation

To facilitate the eigenvalue analysis of Eq. 9.123 by the *P-K* method, by the pure imaginary unit number i of *Q(ik)* and *Q'(ik)* there are eliminated. These terms can be written in real part and imaginary part:

$$Q(ik)=Q^R(k)+iQ^I(ik) \tag{9.125}$$

$$Q'(ik)=Q'^R(ik)+iQ'^I(ik) \tag{9.126}$$

Solving for the imaginary unit number i in Eq. 9.115 gives

$$i=\frac{(p-g)}{k} \tag{9.127}$$

It should be noted that Eq. 9.127 is not valid at $k=0$. Fortunately, there is no need to introduce Eq. 9.127 for $k=0$, since the imaginary part of matrix *Q(ik)* will vanish as $k=0$ (which corresponds to stationary aerodynamics).

Substituting Eq. 9.127 into Eqs. 9.125 and 9.126 yields

$$Q(ik)=Q^R(k)+\left(\frac{(p-g)}{k}\right)Q^I(k) \tag{9.128}$$

$$Q'(ik)=Q'^R(k)+\left(\frac{(p-g)}{k}\right)Q'^I(k) \tag{9.129}$$

Replacing *Q(ik)* and *Q'(ik)* in Eq. 9.123 by Eqs. 9.128 and 9.129 yields the *P-K* method equation in terms of p,

$$\left(\frac{V}{L}\right)^2[M]p^2+\left(\frac{V}{L}\right)[B]p+[K]-\left(\frac{1}{2}\rho V^2\right)\left([Q'^R(k)]-[Q'^I(k)]\frac{(p-g)}{k}\right)g$$
$$\left(\frac{1}{2}\rho V^2\right)\left([Q'^R(k)]-[Q'^I(k)]\frac{(p-g)}{k}\right)\{q\}=0 \tag{9.130}$$

Equation 9.130 can be rewritten in a state-space form as:

$$[A(g.k)-pI_{2n}]\{u\}=0 \tag{9.131}$$

where $[A(g, k)]$ is the real matrix,

$$[A(g,k)]=\begin{bmatrix}0_n & I_n\\ [C] & [D]\end{bmatrix} \tag{9.132}$$

with

$$[C]=-\left(\frac{L}{V}\right)^2[M]^{-1}\left[[K]-\left(\frac{1}{2}\rho V^2\right)\left([Q^R(k)]+[Q'^R(k)]g\right.\right.$$
$$\left.\left.-\,[Q^I(k)]\frac{g}{k}-[Q'^I(k)]\frac{g^2}{k}\right)\right] \tag{9.133a}$$

$$[D]=-\left(\frac{L}{V}\right)^2[M]^{-1}\left[\left(\frac{V}{L}\right)[B]-\left(\frac{1}{2}\rho V^2\right)\right.$$
$$\left.\left([Q^I(k)]\frac{1}{k}-[Q'^I(k)]\frac{g}{k}\right)\right] \tag{9.133b}$$

and again $\{u\}=\{\{q\}, p\{q\}\}^T$.

By comparing Eqs. 9.121 and 9.132 to Eqs. 9.118 and 9.119, it can be conjectured that the new state matrix A depends not only on the reduced frequency k (imaginary part of p), but also on the damping term g (real part of p). Thus, modifications to the standard *P-K* method should be made.

Modifications to the Solution Algorithm of the Standard P-K Method

To adapt the standard *P-K* method to the dependence of matrix A on damping as stated above, it is suggested to include the damping iteration $g_{l+1}=p_l^R$ in addition to the reduced-frequency iteration $k_{l+1}=p_l^I$ in the solution algorithm of the standard *P-K* method. The damping convergence threshold should also be included:

$$|g_{l+1}-g_l|<\varepsilon \tag{9.134}$$

where ε is a user input with a default value of 0.001. Although these modifications were performed, the conventional reduced-frequency sequential process should still be carried out in the modified algorithm. The present *P-K* method will give the same number of roots as the number of structural modes used in the flutter analysis.

Further application to test case with detailed elaboration can be followed in the referenced work [20].

9.3.7 H Flutter Analysis Method—A Direct Harmonic Interpolation Method [22]

A novel flutter analysis method referred to as the H flutter analysis method automatically extents the aerodynamic forces data obtained for purely harmonic oscillatory motions to damped and diverging oscillatory motions by means of fitting free interpolation. The fitting free interpolation in the present reference is elaborated and verified for a two dimensional flat plate.

Usually flutter analysis is performed with two basic methods:

- **K-Method class**—the k method which predicts the correct flutter instability; although the damping and frequency trends of the k method are known to be false;
- **P-K Method class**—the pk method which in addition is capable to predict the damping and frequency trends fairly correctly near zero damping.

These methods are usually based on generalized aerodynamic forces obtained for purely harmonic oscillatory motions. The prediction of the damping and frequency trends can be further improved by methods belonging to the *pk* class such as:

- **g** the *g* Method, which improves the damping and frequency trends of the *pk* method automatically near-zero damping by taking into account the derivative of the generalized aerodynamic forces with respect to the damping at zero damping;
- **p** the P-Method, which improves the damping and frequency trends by taking into account the effect of non-zero damping by means of generalized aerodynamic forces which are approximately valid for the damping-frequency range of interest. It should be noted that methods which generate the aforementioned forces do hardly exist.

In general, analytical continuation of the generalized aerodynamic forces is applied with approximation errors as side-effect due to the fitting procedures associated with the generalized aerodynamic forces for purely harmonic oscillatory motion.

Also the more recent μ flutter analysis method can be grouped into the latter category since a fitting procedure is needed to transform the aerodynamics to the state space.

A novel flutter analysis method belonging to the *pk* class is introduced and described in this work. This so-called *H* flutter analysis method automatically extents the aerodynamic data obtained for purely harmonic oscillatory motions to damped and diverging oscillatory motions by means of a direct harmonic interpolation method thereby improving the prediction of damping and frequencies.

The latter procedure will be described and verified for a pitching flat plate. Results of a flutter analysis application will be presented for the well-known AGARD flutter test case.

The direct harmonic interpolation method or the interpolation/continuation method with respect to the generalized aerodynamic forces involves to acquisition of the generalized aerodynamic forces for nonzero damping. The generalized aerodynamic forces, which are computed for zero damping, have to be warped to the nonzero dampings space. Consequently, an interpolation is necessary which should provide the analytical continuation implicitly. Robust methods used are based on the class of spline techniques, which are also automatic and cardinal. Assuming the generalized aerodynamic forces GAF (0, k_m), with respect to purely harmonic oscillating motions are calculated for N distinct frequencies mk, the data is interpolated by:

$$GAF(g,k) = C^0 + C^g k + C^k k + \sum_{m=1}^{N} C_m E(g,k;0,k_m) \tag{9.135}$$

where C are the coefficients which are determined by satisfying the aforementioned equation at the N support points m and additional closure relations:

$$\mathrm{Im}(GAF(g,k)) = 0 \quad \forall g \tag{9.136}$$

$$\mathrm{Re}\sum_{m=1}^{N} C_m = 0 \tag{9.137}$$

$$\mathrm{Im}\sum_{m=1}^{N} C_m = 0 \tag{9.138}$$

$$\mathrm{Im}C^0 = 0 \tag{9.139}$$

$$\mathrm{Re}\left(C^g\right) = \mathrm{Im}\left(C^k\right) \tag{9.140}$$

$$\mathrm{Im}\left(C^g\right) = \mathrm{Re}\left(C^k\right) = 0 \tag{9.141}$$

The linear problem governed by Eqs. 9.135–9.141 is solved separately for the real and the imaginary parts. In particular Eq. 9.136 is satisfied for the real and the imaginary parts by assuming a symmetric and antisymmetric C_m distribution with respect to the g plane, respectively.

In Hounjet's work at the interpolation is required to be harmonic, in that the kernel function E satisfies the Laplace equation in a two-dimensional space spanned by the reduced damping g and the reduced frequency k, and can be expressed by:

$$\Delta E = \delta g \delta k \tag{9.143}$$

Two types of kernels were considered:

- **Discrete source kernel**

The Laplace kernel is consistently regularized according to an analogy with the determination of the auto-influence of a Laplace field panel:

$$E(g, k; 0, k_m) = \begin{cases} \frac{\ln r}{r} \\ E(0, k_m; 0, k_m)\left(1 - \frac{r}{R_c}\right) + E(G_c, Kc_m; 0, k_m)\left(\frac{r}{R_c}\right) \end{cases} \tag{9.145}$$

$$E(0, k_m; 0, k_m) = \frac{1}{2\pi} \ln(R_c) - \frac{1}{\pi} \tag{9.146}$$

$r = \sqrt{(k - k_m)^2 + g^2}$ is the distance to the source location k_m and R_C denotes the core size which is taken as the minimum distance between the support points:

$R_C = \sqrt{(K_C - k_m)^2 + G_C^2}$, G_C and K_C are locations on the cylinder with size R_C.

The singular kernel is first regularized by redefining the value of $E(0, k_m; 0, k_m)$ in the form of a weighted sum of neighboring values, and then by linear interpolation of E in the interval $r = 0$ and $r = R_C$. In redefining/regularizing the value of E at the origin, the following property is utilized:

$$\int_{S_C} \Delta E \; dk \; dg = 1 \tag{9.147}$$

where S_c denotes the cylinder with size R_c.

- **Continuous linear source kernel**

Alternatively, a distributed core which avoids the aforementioned regularization is utilized. A linear tent-like distribution of source singularities through successive frequency support points is chosen as the distributed core. Further detail can be found in the referenced paper by Hounjet.[16]

[16] Hounjet [22].

9.4 Concluding Remarks

A review of the classical bending-torsion flutter stability problem has been carried out following the analysis carried out by GTS Done [1–3] and further elaborated by Zwaan [4] that formulated the aeroelastic equation of motion for a binary system and typical section. These references are considered to be very instructive and to reveal several stability characteristics of the binary aeroelastic problems.

Two methods of approach are described: a. representation of flutter stability characteristics as the solution of the characteristic equation of the eigenvalue problem and b. the utilization of K-method in the said characteristic equation. These cases are illustrated by application of the analysis to BAH wing based typical section. Basic exercises demonstrate and illustrate the usefulness of the approach, which will be further elaborated with a case study in the following chapter.

Appendix 1: Discussions on K-Method, Following 9.1.3

Remarks 1: Rechecking Flutter Stability Equation Consistency

$$[\mathbf{A}]\begin{Bmatrix} \hat{h} \\ \hat{\alpha} \end{Bmatrix} = [\mathbf{D}]p^2\begin{Bmatrix} \hat{h} \\ \hat{\alpha} \end{Bmatrix} \tag{9.14}$$

Representing

$$\begin{bmatrix} K_h b & qSC_{L_\alpha} \\ 0 & K_\alpha - 2qSebC_{L_\alpha} \end{bmatrix}\begin{Bmatrix} \hat{h} \\ \hat{\alpha} \end{Bmatrix} = \begin{bmatrix} -mb & -S_\alpha \\ -S_\alpha p^2 & -I_\alpha \end{bmatrix}p^2\begin{Bmatrix} \hat{h} \\ \hat{\alpha} \end{Bmatrix} \tag{9.13}$$

where

$$[\mathbf{D}] = \begin{bmatrix} -mb & -S_\alpha \\ -S_\alpha b & -I_\alpha \end{bmatrix} \tag{9.48}$$

$$[\mathbf{A}] = \begin{bmatrix} K_h b & qSC_{L_\alpha} \\ 0 & K_\alpha - 2qSebC_{L_\alpha} \end{bmatrix} \tag{9.49}$$

Next: the determinant of [D]l be determined

$$Det[\mathbf{D}] = mbI_\alpha - S_\alpha^2 b = b\left(mI_\alpha - S_\alpha^2\right) \tag{9.50}$$

The inverse of the determinant [D] will be solved first by solving the simultaneous equation represented by Eq. 9.13 and utilizing the identity

$$[\mathbf{D}]^{-1}[\mathbf{D}] = \begin{bmatrix} 1 & 0 \\ 0 & 1 \end{bmatrix} = \begin{bmatrix} d_{11} & d_{12} \\ d_{21} & d_{22} \end{bmatrix} \begin{bmatrix} -mb & -S_\alpha \\ -S_\alpha b & -I_\alpha \end{bmatrix} \tag{9.51}$$

Hence

$$\begin{aligned} d_{11}mb + d_{12}S_\alpha b &= -1 \\ d_{11}S_\alpha \ + d_{12}I_\alpha &= \ \ 0 \\ d_{21}mb + d_{22}S_\alpha b &= \ \ 0 \\ d_{21}S_\alpha \ + d_{12}I_\alpha &= -1 \end{aligned} \tag{9.52}$$

so that

$$d_{11} = -\frac{I_\alpha}{S_\alpha^2 b - I_\alpha mb} = \frac{\frac{I_\alpha}{b}}{mI_\alpha - S_\alpha^2} \tag{9.53}$$

$$d_{12} = \frac{S_\alpha}{S_\alpha^2 b - I_\alpha mb} = \frac{\frac{S_\alpha}{b}}{mI_\alpha - S_\alpha^2} \tag{9.54}$$

$$d_{21} = \frac{S_\alpha}{mI_\alpha - S_\alpha^2} \tag{9.55}$$

$$d_{22} = \frac{-m}{mI_\alpha - S_\alpha^2} \tag{9.56}$$

Hence:

$$[\mathbf{D}]^{-1} = \begin{bmatrix} \frac{\frac{I_\alpha}{b}}{mI_\alpha - S_\alpha^2} & \frac{\frac{S_\alpha}{b}}{mI_\alpha - S_\alpha^2} \\ \frac{\frac{S_\alpha}{b}}{mI_\alpha - S_\alpha^2} & \frac{-m}{mI_\alpha - S_\alpha^2} \end{bmatrix} \tag{9.57}$$

Next the inverse of the determinant [**D**] will be solved by the use of Cramer's rule[17]

$$[\mathbf{D}]^{-1} = \frac{1}{\mathrm{Det}[\mathbf{D}]}\left[\mathrm{Adj}[\mathbf{D}]\right] \tag{9.58}$$

where

$\left[\mathrm{Adj}[\mathbf{D}]\right]$ is the adjoint of matrix [**D**], which is given by

$$\mathrm{Adj}[\mathbf{D}] = \left[\mathrm{Adj}[\mathbf{D}]\right] = \mathrm{Adj}\begin{bmatrix} -mb & -S_\alpha \\ -S_\alpha b & -I_\alpha \end{bmatrix} = \begin{bmatrix} C_{11} & C_{12} \\ C_{21} & C_{22} \end{bmatrix} \tag{9.59a}$$

where the minor M_{ij} and cofactor C_{ij} can be shown to be related by

[17] See Erwin Kreyszig [49], also Kyrchei [50].

$$\begin{bmatrix} C_{11} & C_{12} \\ C_{21} & C_{22} \end{bmatrix} = \begin{bmatrix} M_{11} & -M_{12} \\ -M_{21} & M_{22} \end{bmatrix} = \begin{bmatrix} -I_\alpha & S_\alpha b \\ S_\alpha & -mb \end{bmatrix} \quad (9.59b)$$

These yield

$$[\mathbf{D}]^{-1} = \frac{1}{\mathrm{Det}[\mathbf{D}]}[\mathrm{Adj}[\mathbf{D}]] = \frac{1}{b\left(mI_\alpha - S_\alpha^2\right)}\begin{bmatrix} -I_\alpha & S_\alpha b \\ S_\alpha & -mb \end{bmatrix} \quad (9.59c)$$

Comparing both the solution of the simultaneous equations and solution by Cramer's rule for $[\mathbf{D}^{-1}]$, the same solution is obtained as expected.

Remarks 2: Further Check and Validation

$$[\mathbf{D}]^{-1}[\mathbf{A}]\begin{Bmatrix} \hat{h} \\ \hat{\alpha} \end{Bmatrix} = p^2\begin{Bmatrix} \hat{h} \\ \hat{\alpha} \end{Bmatrix} = \lambda\begin{Bmatrix} \hat{h} \\ \hat{\alpha} \end{Bmatrix} \quad (9.15)$$

or

$$[\mathbf{A}(k)]\{\mathbf{x}\} = \lambda\{\mathbf{x}\} \quad (9.16)$$

or

$$\lambda = p^2 \quad (9.60)$$

There are two options in writing the baseline equation for K-Method:

$$\begin{bmatrix} K_h b & qSC_{L_\alpha} \\ 0 & K_\alpha - 2qSebC_{L_\alpha} \end{bmatrix}\begin{Bmatrix} \hat{h} \\ \hat{\alpha} \end{Bmatrix} = \begin{bmatrix} -mb & -S_\alpha \\ -S_\alpha p^2 & -I_\alpha \end{bmatrix} p^2 \begin{Bmatrix} \hat{h} \\ \hat{\alpha} \end{Bmatrix} \quad (9.61)$$

or

$$\begin{bmatrix} -mb & -S_\alpha \\ -S_\alpha p^2 & -I_\alpha \end{bmatrix}^{-1}\begin{bmatrix} K_h b & qSC_{L_\alpha} \\ 0 & K_\alpha - 2qSebC_{L_\alpha} \end{bmatrix}\begin{Bmatrix} \hat{h} \\ \hat{\alpha} \end{Bmatrix} = p^2\begin{Bmatrix} \hat{h} \\ \hat{\alpha} \end{Bmatrix} \quad (9.62)$$

which can be written as

$$\begin{bmatrix} \frac{-\frac{I_\alpha}{b}}{mI_\alpha - S_\alpha^2} & \frac{\frac{-S_\alpha}{b}}{mI_\alpha - S_\alpha^2} \\ \frac{-S_\alpha}{mI_\alpha - S_\alpha^2} & \frac{-m}{mI_\alpha - S_\alpha^2} \end{bmatrix}\begin{bmatrix} K_h b & qSC_{L_\alpha} \\ 0 & K_\alpha - 2qSebC_{L_\alpha} \end{bmatrix}\begin{Bmatrix} \hat{h} \\ \hat{\alpha} \end{Bmatrix} = p^2\begin{Bmatrix} \hat{h} \\ \hat{\alpha} \end{Bmatrix} \quad (9.63)$$

Or

$$\begin{bmatrix} \frac{-\frac{I_\alpha}{b}}{mI_\alpha - S_\alpha^2} & \left(\frac{-I_\alpha K_h q S C_{L\alpha}}{mI_\alpha - S_\alpha^2} + \frac{\frac{-S_\alpha}{b}\left(K_\alpha - 2qSebC_{L\alpha}\right)}{mI_\alpha - S_\alpha^2}\right) \\ \frac{-S_\alpha K_\alpha b}{mI_\alpha - S_\alpha^2} & \left(\frac{-S_\alpha K_h bqSC_{L\alpha}}{mI_\alpha - S_\alpha^2} + \frac{-m\left(K_\alpha - 2qSebC_{L\alpha}\right)}{mI_\alpha - S_\alpha^2}\right) - p^2 \end{bmatrix} \begin{Bmatrix} \hat{h} \\ \hat{\alpha} \end{Bmatrix} = p^2 \begin{Bmatrix} \hat{h} \\ \hat{\alpha} \end{Bmatrix} \tag{9.64}$$

But we have written Eq. 9.64 in its original form as

$$\begin{bmatrix} (mp^2 + K_h)b & S_\alpha p^2 + qSC_{L_\alpha} \\ S_\alpha p^2 b & I_\alpha p^2 + K_\alpha - 2qSebC_{L_\alpha} \end{bmatrix} \begin{Bmatrix} \hat{h} \\ \hat{\alpha} \end{Bmatrix} = 0 \tag{9.12}$$

which can be solved by writing Eq. 9.12 as a polynomial in p^2:

$$\begin{aligned} &(bmI_\alpha - S_\alpha b)p^4 + \left(I_\alpha K_h b + mb\left(K_\alpha - 2qSebC_{L_\alpha}\right) - S_\alpha qbSC_{L_\alpha}\right)p^2 \\ &K_h b\left(K_\alpha - 2qSebC_{L_\alpha}\right) = 0 \end{aligned} \tag{9.19}$$

Equation 9.64 can be rewritten as

$$\begin{bmatrix} \frac{-I_\alpha K_h}{(mI_\alpha - S_\alpha^2)} - p^2 & \left(\frac{-I_\alpha K_h q S C_{L\alpha}}{(mI_\alpha - S_\alpha^2)} + \frac{-\frac{S_\alpha}{b}\left(K_\alpha - 2qSebC_{L\alpha}\right)}{(mI_\alpha - S_\alpha^2)}\right) \\ \frac{-S_\alpha K_h b}{(mI_\alpha - S_\alpha^2)} & \left(\frac{-S_\alpha K_\alpha bqSC_{L\alpha}}{(mI_\alpha - S_\alpha^2)} + \frac{-m\left(K_\alpha - 2qSebC_{L\alpha}\right)}{(mI_\alpha - S_\alpha^2)}\right) - p^2 \end{bmatrix} \begin{Bmatrix} \hat{h} \\ \hat{\alpha} \end{Bmatrix} = 0 \tag{9.65}$$

which will end up in a more complicated coefficients than Eq. 9.19, although they should be similar!.

Remarks 3: Flutter Stability Analysis Similar to Done's Approach

In Flutter Stability Analysis similar to Done, the solution of

$$\begin{bmatrix} (mp^2 + K_h)b & S_\alpha p^2 + qSC_{L\alpha} \\ S_\alpha p^2 b & I_\alpha p^2 + K_\alpha - 2qSebC_{L\alpha} \end{bmatrix} \begin{Bmatrix} \hat{h} \\ \overline{\alpha} \end{Bmatrix} = 0 \tag{9.12}$$

is obtained by solving Eq. 9.12 as a polynomial equation in p^2 (see Eq. 9.19):

$$\begin{aligned} &\left(\frac{I_\alpha}{S_\alpha b} - \frac{S_\alpha b}{mb}\right)p^4 + \left(\frac{I_\alpha K_h}{mS_\alpha b} + \frac{K_\alpha - 2qSebC_{L_\alpha}}{S_\alpha b} - \frac{qSC_{L_\alpha}}{mb}\right)p^2 \\ &\frac{K_h}{m}\left(\frac{K_\alpha - 2qSebC_{L_\alpha}}{S_\alpha b}\right) = 0 \end{aligned} \tag{9.25}$$

or

$$P' \equiv A_4' p^4 + A_2' p^2 A_0' = 0 \tag{9.20b}$$

and

$$p^2 = \text{root}(P') = \sqrt{P'} \tag{9.33}$$

with dimension

$[T^{-2}]$.

For the eigenvalue problem in K-method:

$$[\mathbf{A}]\begin{Bmatrix} \hat{h} \\ \hat{\alpha} \end{Bmatrix} = [\mathbf{D}]p^2\begin{Bmatrix} \hat{h} \\ \hat{\alpha} \end{Bmatrix} \tag{9.14}$$

$$[\mathbf{D}]^{-1}[\mathbf{A}]\begin{Bmatrix} \hat{h} \\ \hat{\alpha} \end{Bmatrix} = p^2\begin{Bmatrix} \hat{h} \\ \hat{\alpha} \end{Bmatrix} = \lambda\begin{Bmatrix} \hat{h} \\ \hat{\alpha} \end{Bmatrix} \tag{9.15}$$

or

$$[\mathbf{A}(k)]\{\mathbf{x}\} = \lambda\{\mathbf{x}\} \tag{9.16}$$

K-method seeks solution of the form:

$$\lambda = \left(\frac{\omega_\alpha}{\omega}\right)^2 (1 + ig)[-] \tag{9.36}$$

where the dimension of $[A(k)] = [-]$

but not using:

$$\lambda' = \left(\frac{1}{\omega_\alpha}\right)^2 (g + i\omega)^2 = \left(\frac{\omega_\alpha}{p'}\right)^2 = \left(\frac{1}{\omega_\alpha}\right)^2$$

$$\left(g^2 - \omega^2 + 2ig\omega\right) = \text{Real}(\lambda) + \text{Imag}(\lambda)$$

where g could be the damping term and ω frequency.

$$\text{Real}(\lambda) = \left(\frac{1}{\omega_\alpha}\right)^2 \left(g^2 - \omega^2\right)$$

$$\text{mag}(\lambda) = 2ig\omega\left(\frac{1}{\omega_\alpha}\right)^2$$

which may not be convenient for obtaining ω and g, or

$$\lambda = \left(\frac{1}{\omega_\alpha}\right)^2 (g + i\omega)^2 = \text{Real}(\lambda) + \text{Imag}(\lambda)$$

This procedure may lead to false values of ω since one may expect that λ should be a function of ω but not $1/\omega$.[18]

Appendix 2: Solution of Eigenvalue Problem Example of 9.2.3

a. **Solution of Eigenvalue Problem**

The members of the matrix representing the coefficients of the equation of motion are given by

$$A_{11} = \left(\frac{\omega_\alpha}{\omega_h}\right)^2 \left(\mu - \frac{k_a}{k^2}\right)$$

$$A_{12} = \left(\frac{\omega_\alpha}{\omega_h}\right)^2 \left(\mu x_\alpha - \frac{k_b}{k^2} + (0.5 + a)\frac{k_a}{k^2}\right)$$

$$A_{21} = \mu x_\alpha - \frac{m_a}{k^2} + (0.5 + a)\frac{k_a}{k^2}$$

$$A_{22} = \mu r_\alpha^2 \frac{m_b}{k^2} + (0.5 + a)\left(\frac{m_a}{k^2} + \frac{k_b}{k^2}\right) - (0.5 + a)^2 \frac{k_a}{k^2}$$

Substitution into the eigenvalue problem

$$[A(k)]\{\hat{q}\} = \lambda\{\hat{q}\}$$

yields

$$\begin{bmatrix} \left(\frac{\omega_\alpha}{\omega_h}\right)^2 \left(\mu - \frac{k_a}{k^2}\right) & \left(\frac{\omega_\alpha}{\omega_h}\right)^2 \left(\mu x_\alpha - \frac{k_b}{k^2} + (0.5 + a)\frac{k_a}{k^2}\right) \\ \mu x_\alpha - \frac{m_a}{k^2} + (0.5 + a)\frac{k_a}{k^2} & \mu r_\alpha^2 \frac{m_b}{k^2} + (0.5 + a)\left(\frac{m_a}{k^2} + \frac{k_b}{k^2}\right) - (0.5 + a)^2 \frac{k_a}{k^2} \end{bmatrix}$$
$$\begin{Bmatrix} \frac{\hat{h}}{b} \\ \hat{\alpha} \end{Bmatrix} = \left(\frac{\omega_\alpha}{\omega}\right)^2 (1 + ig) \begin{Bmatrix} \frac{\hat{h}}{b} \\ \hat{\alpha} \end{Bmatrix}$$

The determinant of the eigenvalue problem:

$$\left(\left(\frac{\omega_\alpha}{\omega_h}\right)^2 \left(\mu - \frac{k_a}{k^2}\right) - \lambda^2\right)\left(\left(\mu r_\alpha^2 \frac{m_b}{k^2} + (0.5 + a)\left(\frac{m_a}{k^2} + \frac{k_b}{k^2}\right) - (0.5 + a)^2 \frac{k_a}{k^2}\right)\right.$$
$$\left. - \lambda^2\right) - \left(\mu x_\alpha - \frac{m_a}{k^2} + (0.5 + a)\frac{k_a}{k^2}\right)\left(\frac{\omega_\alpha}{\omega_h}\right)^2$$

---

[18] Verification of this case is left as an exercise.

$$\left(\mu x_\alpha - \frac{k_b}{k^2} + (0.5 + a)\frac{k_a}{k^2}\right) = 0$$

b. Excerpts From Computational Program

$$\left(A(1,1) - \lambda^2\right) * \left(A(2,2) - \lambda^2\right) = \lambda^4 - (A(1,1) + A(2,2))\lambda^2 + A(1,1) * A(2,2)$$

```
kx1 = 1;
kx2 = A(1,1) + A(2,2);
kx3 = A(1,1)*A(1,2);
p = [kx1 0 kx2 0 kx3];
lambda = roots(p);
if imag(lambda(1)) < 0.
    lambda1=0.0001;
    omega1=0.0001;
else if imag(lambda(2)) < 0.
    lambda2=0.0001;
    omega2=0.0001;
elseif imag(lambda(3)) < 0.
    lambda3=0.0001;
    omega3 = 0.0001;
elseif imag(lambda(4)) < 0.
    lambda4=0.0001;
    omega4 = 0.0001[19];
```

$\lambda^2 = \left(\frac{\omega_\alpha}{\omega}\right)^2\left((1 - g^2) + 2ig\right) = \mathrm{Re}(LSQ) + \mathrm{Im}(LSQ)$

$\lambda = \sqrt{\mathrm{Re}(LSQ)} = \mathrm{sqrt}(LSQ)$

$\lambda = \mathrm{Re}(LSQ) = \left(\frac{\omega_\alpha}{\omega}\right)^2(1 + ig)$

Only Re $(\lambda) \geq 0$ is significant.

Otherwise set $\lambda = 0$.

Next determine:

[19] $LSQ \equiv \lambda^2$

$$\lambda^2 = \left(\frac{\omega_\alpha}{\omega}\right)^2\left(\left(1 - g^2\right) + 2ig\right) = \mathrm{Re}(LSQ) + \mathrm{Im}(LSQ)$$

$$\lambda = \sqrt{\mathrm{Re}(LSQ)} = \mathrm{sqrt}(LSQ)$$

$$\lambda = \mathrm{Re}(LSQ) = \left(\frac{\omega_\alpha}{\omega}\right)^2(1 + ig)$$

$$\lambda = \left(\frac{\omega_\alpha}{\omega}\right)^2 (1 + ig)$$

$$\left(\frac{\omega}{\omega_\alpha}\right) = \frac{1}{\sqrt{\mathrm{Re}(\lambda)}}$$

$$\omega = \frac{\omega_\alpha}{\sqrt{\mathrm{Re}(\lambda)}}$$

where

g1 = imag(lambda(1))
omega1 = omegaalpha/sqrt(real(lambda(1)))
g2 = imag(lambda(2))
omega2 = omegaalpha/sqrt(real(lambda(2)))
g3 = imag(lambda(3))
omega3 = omegaalpha/sqrt(real(lambda(3)))
g4 = imag(lambda(4))
omega4 = omegaalpha/sqrt(real(lambda(4))).

Appendix 3: Another Approach on Example Problem 9.2.3

a. **General Discussion**

Equation of motion

$$m\ddot{h} + S_\alpha \ddot{\alpha} + K_h h + q\,S\,C_{L\alpha}\left(\frac{\dot{h}(t)}{U} + \alpha(t)\right) = 0$$

$$S_\alpha \ddot{h} + I_\alpha \ddot{\alpha} + K_\alpha \alpha - 2q\,S\,e\,b\,C_{L\alpha}\left(\frac{\dot{h}(t)}{U} + \alpha(t)\right) = 0$$

$$[M]\{\ddot{x}\} - [B]\{\dot{x}\} + ([K] - q[A])\{x\} = 0$$

$$\{x\} = \begin{Bmatrix} h \\ \alpha \end{Bmatrix} = \begin{Bmatrix} \tilde{h}b \\ \alpha \end{Bmatrix}$$

Assume harmonic motion

$$\{x\} = \{\hat{x}\}e^{\mathrm{pt}}$$

$$\{x\} = \begin{Bmatrix} h \\ \alpha \end{Bmatrix} = \begin{Bmatrix} \hat{h} \\ \hat{\alpha} \end{Bmatrix} e^{\mathrm{pt}} = \begin{Bmatrix} \tilde{h}b \\ \hat{\alpha} \end{Bmatrix} e^{\mathrm{pt}}$$

where

$$\hat{h} \equiv \tilde{h}b$$

and

$\tilde{h}$—dimensionless, just like α—in radian (dimensionless).

Hence

$$[B] = \begin{bmatrix} \frac{qb}{U} SC_{L\alpha} & 0 \\ \frac{2qb}{U} SebC_{L\alpha} & 0 \end{bmatrix}$$

$$[K] = \begin{bmatrix} K_h b & 0 \\ 0 & K_\alpha \end{bmatrix}$$

Therefore

$$[M]\{\ddot{x}\} - [B]\{\dot{x}\} + ([K] - q[A])\{x\} = 0$$

becomes

$$\begin{bmatrix} mb & S_\alpha \\ S_\alpha & I_\alpha \end{bmatrix} p^2 \begin{Bmatrix} \tilde{h} \\ \hat{\alpha} \end{Bmatrix} + \frac{qb}{U} \begin{bmatrix} SC_{L\alpha} & 0 \\ -2SebC_{L\alpha} & 0 \end{bmatrix} p \begin{Bmatrix} \tilde{h} \\ \hat{\alpha} \end{Bmatrix}$$
$$+ \left(\begin{bmatrix} K_h b & 0 \\ 0 & K_\alpha \end{bmatrix} + q \begin{bmatrix} 0 & S\,C_{L\alpha} \\ 0 & -2Seb\,C_{L\alpha} \end{bmatrix} \right) \begin{Bmatrix} \tilde{h} \\ \hat{\alpha} \end{Bmatrix} = 0$$

or

$$\begin{bmatrix} mbp^2 + \frac{2qb}{U} SC_{L\alpha} p + K_h b & S_\alpha p^2 + qSC_{L\alpha} \\ S_\alpha bp^2 - \frac{2qb}{U} SebC_{L\alpha} p & I_\alpha p^2 + K_\alpha - 2SebC_{L\alpha} \end{bmatrix} \begin{Bmatrix} \tilde{h} \\ \hat{\alpha} \end{Bmatrix} = 0$$

This leads to characteristics equation:

$$a_4 p^4 + a_3 p^3 + a_2 p_2 + a_1 p + a_0 = 0$$

where

$$a_4 = mbI_\alpha - S_\alpha^2 b$$

$$a_3 = \frac{q}{U} SC_{L_\alpha}(2ebS_\alpha + I_\alpha)$$

$$a_2 = mbK_\alpha + I_\alpha K_h b - \left(2meb^2 + S_\alpha b\right) qSC_{L_\alpha}$$

$$a_1 = -\frac{qb}{U} SC_{L_\alpha} K_\alpha$$

$$a_0 = K_h b\left(K_\alpha - 2qSebC_{L_\alpha}\right)$$

$$p = \sigma + i\omega = \text{Real}(r) + i \cdot \text{Imag}(r)$$

$$\overline{\sigma} = \frac{\sigma}{\omega_\alpha}$$

$$\overline{\omega} = \frac{\omega}{\omega_\alpha}$$

```
      P = [a4,a3,a2,a1,a0];
      r = roots(P)
sigma1 = real(r(1))/omegaalpha;
sigma2 = real(r(2))/omegaalpha;
sigma3 = real(r(3))/omegaalpha;
sigma4 = real(r(4))/omegaalpha;
omega1 = imag(r(1))/omegaalpha;
omega2 = imag(r(2))/omegaalpha;
omega3 = imag(r(3))/omegaalpha;
omega4 = imag(r(4))/omegaalpha;
```

b. **For K-Method**[20]

$$\lambda = \left(\frac{\omega_\alpha}{\omega}\right)^2 (1 + ig)[-] \tag{9.36}$$

To yield similar dimension for Eqs. 9.35 and 9.36, divide Eqs. 9.24 or 9.25 by ω_α^2 or define

$$\overline{p}^2 = \frac{p'^2}{\omega_\alpha^2} \tag{9.37}$$

Then

$$\frac{1}{\overline{p}^2} = \frac{\omega_\alpha^2}{p'^2} \tag{9.38}$$

Solution of (9.25)

$$p'^2 = \text{root}\left(P'\right) \tag{9.39}$$

or

[20] [−]—dimensionless.

$$\overline{p}^2 = \frac{p'^2}{\omega_\alpha^2} = \frac{1}{\omega_\alpha^2}\text{root}\big(P'\big) \tag{9.40}$$

Two roots:

$$\overline{p}_1^2 = \frac{p_1'^2}{\omega_\alpha^2} = \frac{1}{\omega_\alpha^2} psq1 \tag{9.41}$$

$$\overline{p}_2^2 = \frac{p_2'^2}{\omega_\alpha^2} = \frac{1}{\omega_\alpha^2} psq2 \tag{9.42}$$

$$\lambda = \left(\frac{\omega_\alpha}{\omega}\right)^2 (1 + ig) \equiv \left(\frac{\omega_\alpha}{p'}\right)^2 = \frac{\omega_\alpha^2}{psq} = \text{Real}(\lambda) + i \cdot \text{Imag}(\lambda) \tag{9.43}$$

Hence

$$\omega_1 = \frac{\omega_\alpha}{\sqrt{\text{Real}(\lambda_1)}} \tag{9.44}$$

$$\omega_2 = \frac{\omega_\alpha}{\sqrt{\text{Real}(\lambda 2)}} \tag{9.45}$$

$$g_1 = \left(\frac{\omega_1}{\omega_\alpha}\right)^2 \text{Imag}(\lambda_1) \tag{9.46}$$

$$g_2 = \left(\frac{\omega_2}{\omega_\alpha}\right)^2 \text{Imag}(\lambda_2) \tag{9.47}$$

Translating New Coefficient into Executable Program

$$a_4 = mbI_\alpha - S_\alpha^2 b$$

$$a_3 = \frac{q}{U} SC_{L_\alpha}(2ebS_\alpha + I_\alpha)$$

$$a_2 = mbK_\alpha + I_\alpha K_h b - \big(2meb^2 + S_\alpha b\big) qSC_{L_\alpha}$$

$$a_1 = -\frac{qb}{U} SC_{L_\alpha} K_\alpha$$

$$a_0 = K_h b\big(K_\alpha - 2qSebC_{L_\alpha}\big)$$

$$a_0 = K_h b(K_\alpha - 2qSebC_{L\alpha})$$

Notes

Statements for calculation using MATLAB.

```
a4 =b*( (I_alpha*m)-(S_alpha*S_alpha));
a3 = (q*b/V)*S*Cl_alpha*((2*e*b*S_alpha)+I_alpha);
a2 =b*((m*K_alpha)+(I_alpha*Kh)-((2*m*e*b)+S_alpha)*q*S*Cl_alpha);
a1 = (q*b/V)*S*Cl_alpha*K_alpha;
a0 = Kh*b*(K_alpha-(2*q*S*e*b*Cl_alpha));
P = [a4,a3,a2,a1,a0];
r = roots(P);
```

After some analysis, and noting that the characteristic equation is not a quadratic equation in λ, but a quartic in p2, then the equations should be reorganized. This can be carried out as elaborated in Appendix A (at the end of the book).

References

1. Done, G.T.S. 1963. *The effect of linear damping on flutter speed*. March, Aeronautical Research Council R&M No 3996.
2. Done, G.T.S. 1966. *The flutter and stability of undamped systems*. Aeronautical Research Council R&M 3553.
3. Done, G.T.S. 1967. *A study of binary flutter roots using a method of system synthesis*. Aeronautical Research Council, R&M 3554.
4. Zwaan, R.J. Aeroelasticity of aircraft, lecture notes, special lecture. Short Course offered at Institut Teknologi Bandung, Indonesia, Lecture Notes delivered at Special Meeting at ITB in August 1981 (private collection).
5. Djojodihardjo, H., and Yee Hong Hoe. 2007. Parametric study of the flutter characteristics of transport aircraft wings. *Proceedings of AEROTECH-II 2007 Conference on Aerospace Technology of XXI Century*, 20–21 June 2007, Kuala Lumpur.
6. Djojodihardjo, H., and M.A. Abd. Bari. 2014. Kinematic and unsteady aerodynamic modelling-flapping Bi-&Quad-Wing ornithopter—ICAS 2014–0433. *Congress of the International Council of the Aeronautical Sciences*.
7. Djojodihardjo, H., and A.S.S. Ramli. 2015. Analysis and computational study of the aerodynamics and aeroelasticity of flapping-IFASD..
8. Djojodihardjo, H. 2016. Analysis and computational study of the aerodynamics, aeroelasticity and flight dynamics of flapping wing ornithopter using linear approximation. In *AIAA SciTech Forum*. https://doi.org/10.2514/6.2016-2027.
9. Hassig, H.J. 1971. An approximate true damping solution of the flutter equation by determinant iteration. *Journal of Aircraft* 8 (11): 885–889. https://doi.org/10.2514/3.44311.
10. Patil, M.J., R.N. Yurkovich, and D.H. Hodges. 2004. Incorrectness of the k method for flutter calculations. *Journal of Aircraft*.
11. Dale, M. Pitt. 1999. A new non-iterative P-K match point flutter solution. 10.2514–6.1999–1353.
12. Djojodihardjo, H., M. Akbar, L. Gunawan, and M. Kusni. 2010. Unified aerodynamic-acoustic formulation for aero-acoustic structure coupling. Paper ICAS 2010–376, 27th Congress of the International Council of the Aeronautical Sciences, Nice, France.
13. Collar, A.R. 2016. The expanding domain of aeroelasticity. *The Aeronautical Journal* 50 (428), Published online by Cambridge University Press.
14. Max R. Platzer, (ed.). 1960. Agard manual on aeroelasncity in axial-flow turbomachines, vol. 2, Structural Dynamics and Aeroelasticity. AGARD-AG-298-Vol-2.

15. Viswanathan, S.P. 1977. *An analysis of the flutter and damping charactristics.* Ph.D. Thesis,GATECH.
16. Lokatt, Mikaela. *A method for efficient flutter analysis of aystems with uncertain modeling parameters.* M.Sc. Thesis, KTH, TRITA-MAT-E 2014:19 ISRN-KTH/MAT/E--14/19--SE.
17. Blades, Eric L., et al. 2011. A fluid-structure interaction simulation capability using the co-simulation engine (re Isogai Wing).
18. Borglund, D., and U. Ringertz. Solution of the uncertain flutter eigenvalue problem using μ-p analysis, IFASD-2009-048.
19. Rodden, W.P., and E.H. Johnson. 1994. MSC/NASTRAN aeroelastic analysis user's manual. *Version* 68: 237–257.
20. Gu, Yingsong and Yang, Zhichun. 2012. Modified P-K method for flutter solution with damping iteration. Technical Notes, *AIAA Journal*, AIAA-56983–579.
21. Chen, P.C. 2000. Damping perturbation method for flutter solution: The *g* method. *AIAA Journal* 38 (9): 1519–1524. https://doi.org/10.2514/2.1171.
22. Hounjet, Michael. 2008. *Flutter analysis method a direct harmonic interpolation method.* NLR-TP-2008–340.

Chapter 10
Dynamic Aeroelasticity of Typical Section with a Case Study

Abstract The study for the dynamics of aeroelastic system can be best illustrated by analyzing a generic aeroelastic system: the typical section. Such approach will be instructive as a device for exploring the mathematical tools and the physical content associated with the system. The present chapter elaborates dynamic aeroelasticity and flutter of typical section. Basic approach for the treatment of the dynamic aeroelasticity of a typical section will be discussed. Then a review will be made on the theoretical foundation of flutter stability for binary-bending-torsion flutter of typical section. The final section provides an example of the basic approach by elaboration to some detail of a case study on a parametric study on aeroelastic stability and flutter characteristics of aircraft wings as a case study. For the elaboration of dynamic aeroelasticity and flutter and review of the theoretical foundation for flutter stability for binary-bending-torsion flutter of typical section, the review presented here follows closely the elaboration in Zwaan (Zwaan [1]) and Mark R. Pletzer (Ed) (Max R Platzer, Ed., AGARD MANUAL on Aeroelasncity in Axial-Flow Turbomachines Volume 2, Structural Dynamics and Aeroelasticity, AGARD-AG-298-Vol-2.) for instructiveness. As instructive illustration, the case study is drawn from the author's work (Djojodihardjo, H and Yee, H H, Parametric Study of The Flutter Characteristics of Transport Aircraft Wings, Proceedings of AEROTECH-II 2007 Conference on Aerospace Technology of XXI Century, 20–21 June 2007, Kuala Lumpur; Djojodihardjo, H and Yee, H H, Parametric Study of The Flutter Characteristics of Transport Aircraft Wings, Proceedings of AEROTECH-II 2007 Conference on Aerospace Technology of XXI Century, 20–21 June 2007, Kuala Lumpur). For conceptual study purposes, it would be of advantage to look into a simple method that may give instructive results. In this connection, various modern transport aircrafts are classified and investigated in view of their structural dynamic and aeroelastic characteristics of their wing structure. The basic philosophy and formulation of flutter phenomena as reflected in the generic binary flutter problem is revisited and reflected to real aircraft for general understanding and conceptual design purposes.

Keywords Aerodynamics · Aeroelasticity · Aircraft Design · Unsteady aerodynamics

© The Author(s), under exclusive license to Springer Nature Singapore Pte Ltd. 2023

H. Djojodihardjo, *Introduction to Aeroelasticity*,
https://doi.org/10.1007/978-981-16-8078-6_10

10.1 Dynamic Aeroelasticity of Typical Section

As in previous chapters, the study for the dynamics of aeroelastic system can be best illustrated by analyzing a generic aeroelastic system: the typical section. Such approach will be instructive as a device for exploring the mathematical tools and the physical content associated with the system (Fig. 10.1).

Let us assume that the aerodynamic forces are already known, where $p(x, t)$ is the aerodynamic pressure, L is the resultant lift force and My is the resultant aerodynamic moment about the elastic axis.

The equation of motion is given as:

$$-\frac{\mathrm{d}}{\mathrm{d}t}(m\dot{h} + S_\alpha\dot{\alpha}) - K_h h - L = 0 \tag{10.1}$$

$$-\frac{\mathrm{d}}{\mathrm{d}t}(S_\alpha\dot{h} + I_\alpha\dot{\alpha}) - K_\alpha \alpha + M_y = 0 \tag{10.2}$$

or

$$m\ddot{h} + K_h h + S_\alpha\ddot{\alpha} = -L \tag{10.3}$$

$$S_\alpha\ddot{h} + I_\alpha\ddot{\alpha} + K_\alpha\alpha = M_y \tag{10.4}$$

where

$$L = \int p\,\mathrm{d}x \tag{10.5}$$

$$M = \int px\,\mathrm{d}x \tag{10.6}$$

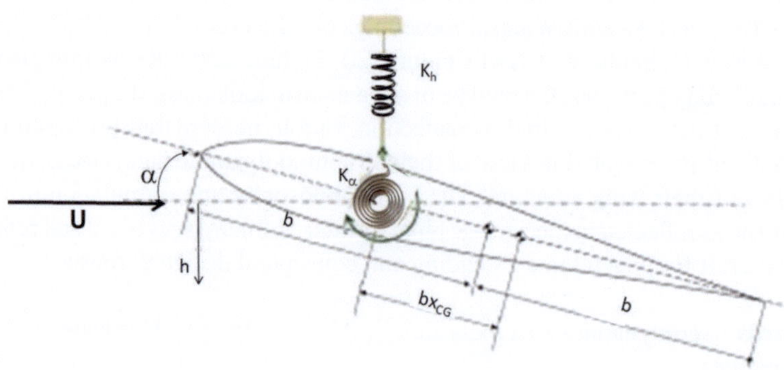

Fig. 10.1 Typical section geometry

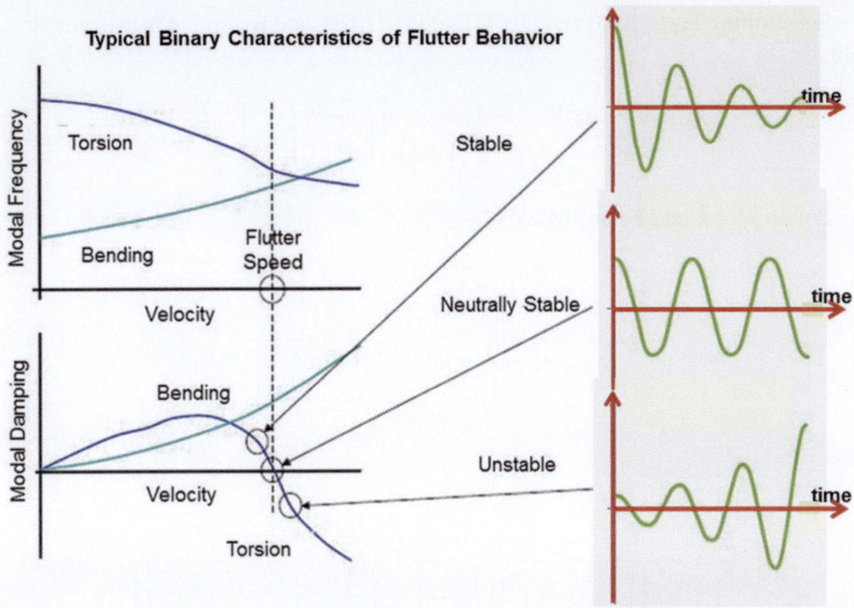

Fig. 10.2 Schematics of dominant binary characteristics of flutter speed

Define the uncoupled natural frequencies:

$$\omega_h^2 = \frac{K_h}{m} \tag{10.7a}$$

$$\omega_\alpha^2 = \frac{K_\alpha}{I_\alpha} \tag{10.7b}$$

which are the natural frequencies of the system for $S_\alpha = 0$ as will be seen subsequently (Fig. 10.2).

10.1.1 Sinusoidal Motion

The generic motion for unsteady situation can be given by sinusoidal motion as one of the simplest types of motion, as also done in previous chapters.

Let:

$$L = \overline{L}e^{i\omega t} \; M_y = \overline{M}_y e^{i\omega t} \tag{10.8}$$

$$h = \overline{h}e^{i\omega t} \alpha = \overline{\alpha}e^{i\omega t}$$

Substituting Eqs. (10.8) and (10.7) into Eq. (10.3), and writing in matrix notation, obtain:

$$\begin{bmatrix} m\left(-\omega^2+\omega_\alpha^2\right) & -S_\alpha\omega^2 \\ -S_\alpha\omega^2 & I_\alpha\left(-\omega^2+\omega_\alpha^2\right) \end{bmatrix} \begin{Bmatrix} \overline{h}e^{i\omega t} \\ \overline{\alpha}e^{i\omega t} \end{Bmatrix} = \begin{Bmatrix} -\overline{L}e^{i\omega t} \\ \overline{M}_y e^{i\omega t} \end{Bmatrix} \tag{10.9}$$

Solving for $\overline{h}$ and $\overline{\alpha}$, one obtains

$$\overline{h} = \frac{\begin{vmatrix} -\overline{L} & -S_\alpha\omega^2 \\ \overline{M}_y & I_\alpha\left(-\omega^2+\omega_\alpha^2\right) \end{vmatrix}}{\begin{vmatrix} m\left(-\omega^2+\omega_\alpha^2\right) & -S_\alpha\omega^2 \\ -S_\alpha\omega^2 & I_\alpha\left(-\omega^2+\omega_\alpha^2\right) \end{vmatrix}} \tag{10.10}$$

or

$$\frac{\overline{h}}{\overline{L}} = \frac{-\left[1-\left(\frac{\omega}{\omega_\alpha}\right)^2\right]+\frac{d}{b}\frac{x_\alpha}{r_\alpha^2}\left(\frac{\omega}{\omega_\alpha}\right)^2}{K_h\left\{\left[1-\left(\frac{\omega}{\omega_\alpha}\right)^2\right]\left[\left[1-\left(\frac{\omega}{\omega_\alpha}\right)^2\right]-\frac{x_\alpha^2}{r_\alpha^2}\left(\frac{\omega}{\omega_\alpha}\right)^2\left(\frac{\omega}{\omega_h}\right)^2\right]\right\}}$$
$$\equiv H_{hL}\left(\frac{\omega}{\omega_\alpha};\frac{\omega_h}{\omega_\alpha},\frac{d}{b},x_\alpha,r_\alpha\right) \tag{10.11}$$

where

$$d = \frac{\overline{M}_y}{\overline{L}} \tag{10.12}$$

where b is the reference length, usually selected as half-chord by tradition.

$$x_\alpha = \frac{S_\alpha}{\mathrm{mb}} = \frac{x_{c.g}}{b} \tag{10.13}$$

$$r_\alpha^2 = \frac{I_\alpha}{\mathrm{mb}} \tag{10.14}$$

A plot of H_{hL} is schematically (without specifying accurate numbers) shown in Fig. 10.3. Here $\left(\frac{\omega_1}{\omega_\alpha}\right)$ and $\left(\frac{\omega_2}{\omega_\alpha}\right)$ are the roots of the denominator, and they are the system "natural frequencies."

$$\left(\frac{\omega_1^2}{\omega_h\omega_\alpha}\right)\left(\frac{\omega_2^2}{\omega_h\omega_\alpha}\right) = \frac{\left[\frac{\omega_h}{\omega_\alpha}+\frac{\omega_\alpha}{\omega_h}\right]\pm\left\{\left[\frac{\omega_h}{\omega_\alpha}+\frac{\omega_\alpha}{\omega_h}\right]^2-4\left[1-\frac{x_\alpha^2}{r_\alpha^2}\right]\right\}}{2\left[1-\frac{x_\alpha^2}{r_\alpha^2}\right]} \tag{10.15}$$

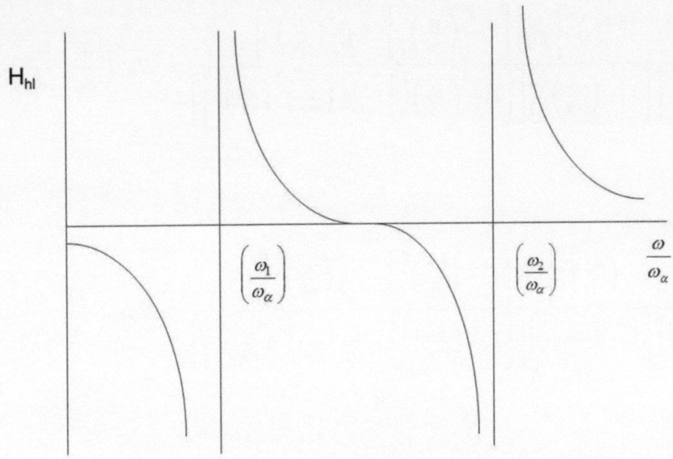

Fig. 10.3 Graph of transfer function H_{hl} as a function of $\frac{\omega}{\omega_\alpha}$

Similarly, following the same procedure, substituting (10.8) and (10.7) into (10.4), and writing in matrix notation, obtain:

$$\overline{\alpha} = \frac{\begin{vmatrix} m(-\omega^2+\omega_\alpha^2) & -\overline{L} \\ -S_\alpha\omega^2 & \overline{M}_y \end{vmatrix}}{\begin{vmatrix} m(-\omega^2+\omega_\alpha^2) & -S_\alpha\omega^2 \\ -S_\alpha\omega^2 & I_\alpha(-\omega^2+\omega_\alpha^2) \end{vmatrix}} \tag{10.16}$$

or

$$\frac{\overline{\alpha}}{\overline{L}} = \frac{\begin{vmatrix} m(-\omega^2+\omega_\alpha^2) & -1 \\ -S_\alpha\omega^2 & \frac{\overline{M}_y}{\overline{L}} \end{vmatrix}}{\begin{vmatrix} m(-\omega^2+\omega_\alpha^2) & -S_\alpha\omega^2 \\ -S_\alpha\omega^2 & I_\alpha(-\omega^2+\omega_\alpha^2) \end{vmatrix}}$$
$$= \frac{md\omega_h^2\left[1-\left(\frac{\omega}{\omega_\alpha}\right)^2\right] - mbx_\alpha\omega^2}{m^2b^2r_\alpha^2\omega_\alpha^2\omega_h^2\left\{\left[1-\left(\frac{\omega}{\omega_\alpha}\right)^2\right]\left[1-\left(\frac{\omega}{\omega_h}\right)^2\right] - \frac{x_\alpha^2}{r_\alpha^2}\left(\frac{\omega}{\omega_\alpha}\right)^2\left(\frac{\omega}{\omega_h}\right)^2\right\}} \tag{10.17}$$

Hence, for $H_{hL}\left(\frac{\omega}{\omega_\alpha}; \frac{\omega_h}{\omega_\alpha}, x_\alpha, r_\alpha\right)$ one may obtain

$$\frac{\overline{\alpha}}{\overline{L}} = \frac{\left(\frac{\omega_h}{\omega_\alpha}\right)^2\left\{\frac{d}{b^2r_\alpha^2}\left[1-\left(\frac{\omega}{\omega_\alpha}\right)^2\right]-\frac{x_\alpha}{br_\alpha^2}\left(\frac{\omega}{\omega_h}\right)^2\right\}}{K_h\left\{\left[1-\left(\frac{\omega}{\omega_\alpha}\right)^2\right]\left[\left[1-\left(\frac{\omega}{\omega_\alpha}\right)^2\right]-\frac{x_\alpha}{r_\alpha^2}\left(\frac{\omega}{\omega_\alpha}\right)^2\left(\frac{\omega}{\omega_h}\right)^2\right]\right\}} \equiv H_{\alpha L}\left(\frac{\omega}{\omega_\alpha};\frac{\omega_h}{\omega_\alpha},x_\alpha,r_\alpha\right) \tag{10.18a}$$

or

$$\frac{\overline{\alpha}}{\overline{Lb}} = \frac{\left(\frac{\omega_h}{\omega_\alpha}\right)^2\left\{\frac{d}{br_\alpha^2}\left[1-\left(\frac{\omega}{\omega_\alpha}\right)^2\right]-\frac{x_\alpha}{r_\alpha^2}\left(\frac{\omega}{\omega_h}\right)^2\right\}}{K_h\left\{\left[1-\left(\frac{\omega}{\omega_\alpha}\right)^2\right]\left[\left[1-\left(\frac{\omega}{\omega_\alpha}\right)^2\right]-\frac{x_\alpha}{r_\alpha^2}\left(\frac{\omega}{\omega_\alpha}\right)^2\left(\frac{\omega}{\omega_h}\right)^2\right]\right\}} \equiv H_{\alpha Lb}\left(\frac{\omega}{\omega_\alpha};\frac{\omega_h}{\omega_\alpha},x_\alpha,r_\alpha\right) \tag{10.18b}$$

or

$$\frac{\overline{\alpha}}{\overline{M}_y} = \frac{\left(\frac{\omega_h}{\omega_\alpha}\right)^2\left\{\frac{1}{b^2r_\alpha^2}\left[1-\left(\frac{\omega}{\omega_\alpha}\right)^2\right]-\frac{1}{db}\frac{x_\alpha}{r_\alpha^2}\left(\frac{\omega}{\omega_h}\right)^2\right\}}{K_h\left\{\left[1-\left(\frac{\omega}{\omega_\alpha}\right)^2\right]\left[\left[1-\left(\frac{\omega}{\omega_\alpha}\right)^2\right]-\frac{x_\alpha^2}{r_\alpha^2}\left(\frac{\omega}{\omega_\alpha}\right)^2\left(\frac{\omega}{\omega_h}\right)^2\right]\right\}} \equiv H_{\alpha L}\left(\frac{\omega}{\omega_\alpha};\frac{\omega_h}{\omega_\alpha},x_\alpha,r_\alpha\right) \tag{10.18c}$$

H_{hL} and $H_{\alpha F}$ are called the "transfer functions," and in this particular case of static aeroelasticity considered in the example, they are mechanical or structural transfer functions since they describe the motion of the structural system under specified loading.

In the numerator of Eq. (10.11), $\frac{\omega_3}{\omega_\alpha}$ is given by

$$\left(\frac{\omega_3}{\omega_h}\right)^2 = \frac{1}{1+\frac{d}{b}\frac{x_\alpha}{r_\alpha^2}} \tag{10.19}$$

is the root of the numerator of H_{hL}. Notice that this is not the root of $\mathrm{H}_{\alpha\mathrm{L}}$ since in general $\mathrm{H}_{\alpha\mathrm{L}}$ will vanish at a different frequency.

The infinite response which occurs at the natural frequencies ω_1 and ω_2 for both ω_α and ω_h, since it is given by the common denominator of Eqs. (10.11) and (10.16), is not an instability. It is the resonance of the system with infinite amplitude since here the system does not have any damping. In real physical system, some damping may prevail which will then limit the magnitude of the transfer functions at natural frequencies. However, the expression for the transfer function involving any damping (structural or aerodynamic) will be more involved.

10.1.2 Periodic Motion

The above analysis can be generalized to any periodic motion by expanding the motion into a Fourier (sinusoidal) series. For such a case, define:

$$T_0 \equiv \text{basic period}$$
$$\omega_0 \equiv 2\pi/T_0 \quad \text{fundamental frequency}$$

Then a periodic force, $L(t)$, may be written as

$$L(t) = \sum_{n=-\infty}^{\infty} L_n e^{+in\omega_0 t} \tag{10.20}$$

where

$$L_n = \frac{1}{T_0} \int_{-\frac{T_0}{2}}^{\frac{T_0}{2}} L(t) e^{+in\omega_0 t} \mathrm{d}t \tag{10.21}$$

Using (10.21) and (10.11), the response of the system in terms of h coordinate is then given by

$$h(t) = \sum_{n=-\infty}^{\infty} H_{hL}\left(\frac{\omega_0 n}{\omega_\alpha}\right) L_n e^{+in\omega_0 t} \tag{10.22}$$

10.1.3 Arbitrary Motion

By taking the limit as the basic period becomes infinitely long, i.e. $T_0 \to \infty$, results for non-periodic motion can be obtained.

Define:

$$\omega \equiv n\omega_0$$

$\Delta\omega \equiv \Delta n\omega_0 = 2\pi/\mathrm{T}_0$ frequency increment.

$L^*(\omega) \equiv \frac{L_n}{\Delta\omega} = \frac{L_n T_0}{2\pi}$ force per frequency increment.

Substituting these values into Eq. (10.20), and replacing the summation sign with integral sign, obtain

$$L(t) = \int_{-\infty}^{\infty} L^*(\omega)e^{+i\omega t}\,\mathrm{d}\omega \tag{10.23}$$

and it follows that Eqs. (10.21) and (10.22) become

$$L^*(\omega) = \frac{1}{2\pi}\int_{-\infty}^{\infty} L(t)e^{-i\omega t}\,\mathrm{d}t \tag{10.24}$$

$$h(t) = \int_{-\infty}^{\infty} H_{\mathrm{hl}}\left(\frac{\omega}{\omega_\alpha}\right)L^*(\omega)e^{i\omega t}\,\mathrm{d}\omega \tag{10.25}$$

An interesting alternate form of (10.25) can be obtained by substituting (10.24) into (10.25). Using dummy variable for indicating time variable within the integral sign, obtain:

$$h(t) = \int_{-\infty}^{\infty} H_{\mathrm{hl}}\left(\frac{\omega}{\omega_\alpha}\right)\left(\frac{1}{2\pi}\int_{-\infty}^{\infty} L(\tau)e^{-i\omega\tau}\,\mathrm{d}\tau\right)e^{i\omega t}\mathrm{d}\omega \tag{10.26a}$$

Interchanging the order of integration

$$\begin{aligned} h(t) &= \frac{1}{2\pi}\int_{-\infty}^{\infty}\int_{-\infty}^{\infty} H_{\mathrm{hl}}\left(\frac{\omega}{\omega_\alpha}\right)e^{i\omega t}\left(L(\tau)e^{-i\omega\tau}d\omega\right)\mathrm{d}\tau \\ &= \int_{-\infty}^{\infty} L(\tau)\left\{\frac{1}{2\pi}\int_{-\infty}^{\infty} H_{\mathrm{hl}}\left(\frac{\omega}{\omega_\alpha}\right)e^{i\omega(t-\tau)}\,\mathrm{d}\omega\right\}\mathrm{d}\tau \\ &= \int_{-\infty}^{\infty} I_{\mathrm{hL}}(t-\tau)L(\tau)\,\mathrm{d}\tau \end{aligned} \tag{10.26}$$

where it is defined that

$$I_{\mathrm{hL}}(t) \equiv \frac{1}{2\pi}\int_{-\infty}^{\infty} H_{\mathrm{hl}}\left(\frac{\omega}{\omega_\alpha}\right)e^{i\omega t}\,\mathrm{d}\omega \tag{10.27}$$

Comparing (10.25) and (10.27), one could note that I_{hL} is the response to $L^*(\omega) = 1/2\pi$, or from (10.23) and (10.24), $L(t) = \delta(\mathrm{t})$. Hence I is the response of the system to an impulse force, and correspondingly it is referred to as the "**impulse function.**"

Equations (10.23–10.25) are the "***Fourier Transform***" relations, and the third form of Eq. (10.26) is also called the "***Convolution Integral.***"

10.2 Parametric Study of Aeroelastic Stability and Flutter Characteristics of Aircraft Wings as a Case Study[1]

10.2.1 Introduction

With the progress of aircraft technology, materials and lighter aircraft structures, modern aircrafts are designed to be lighter and much more flexible. For the aircraft designers, manufacturers, as well as operators, the state of the art and trends of the structural dynamic characteristics of modern aircraft structures are of great interest and relevance. For instructional and conceptual study purposes, it would be of advantage to look into a simple method that may give indicative results. It is with such objectives that the present study is carried out. Here an analytical approach is followed to investigate the structural dynamic and aeroelastic characteristics of the wing structure of modern transport aircrafts and to obtain some correlation between some representative specifications to their structural dynamic and aeroelastic characteristics of their wing structure. For conceptual design and instructional purposes, it would be desirable to be able to identify certain pertinent correlation between structural characteristics and flutter perspectives. It is with such objectives in mind that we will look into a fundamental equation governing flutter, and for this purpose, the basic two-degree of freedom aeroelastic equation for typical section will be revisited. A typical section, which is usually taken at 70–75% of the wingspan from the root, has been considered to be representative of the aeroelastic behavior of the wing.[2] Two problems will be investigated, and these are the flutter stability analysis, and the solution of flutter equation using the classical VG or K-Method, chosen for its eloquence. Along this line a simple (non-exhaustive) parametric study will be carried out, to illustrate structural characteristics relevant for flutter behavior. For this purpose, we will look into three different aerodynamics approximation in the flutter equation, all for the incompressible case, to obtain the influence of such approximation to flutter behavior. For this purpose, the well-known Bisplinghoff–Ashley–Halfman (BAH) typical section,[3] modified as required, will be utilized, which also serves as benchmarking purposes.

---

[1] Originally presented as Djojodihardjo, H and Yee, H H, Parametric Study of The Flutter Characteristics of Transport Aircraft Wings, Proceedings of AEROTECH-II 2007 Conference on Aerospace Technology of XXI Century, 20-21 June 2007, Kuala Lumpur.

[2] Y. C. Fung, The Theory of Aeroelasticity, Dover Publication, 1955; Bisplinghoff, R.L., Ashley, H. and Halfman, R.L., Aeroelasticity, Dover Publications, Inc., copyright 1955 by Addison-Wesley Publishing Co., Inc., copyright © renewed 1983 by Bisplinghoff, R.L., Ashley, H. and Halfman, R.L.; and Zwaan, R.J., Aeroelasticity of Aircraft, Lecture Notes, Special Lecture, Short Course offered at Institut Teknologi Bandung, Indonesia, Lecture Notes delivered at Special Meeting at ITB in August 1981(private collection).

[3] BAH typical section or wing, where BAH refers to the typical section mentioned as an example in Bisplinghoff, R.L., Ashley, H. and Halfman, R.L., Aeroelasticity, Dover Publications, Inc., copyright ©1955 by Addison-Wesley Publishing Co., Inc., has been frequently used for benchmarking purposes in many literatures, although nowadays many others have also been utilized, especially for more updated situations.

For flutter stability analysis, the analysis well-elaborated in Zwaan,[4] using Done's aeroelastic stability analysis the generic bending-torsion binary flutter of typical section will be closely followed. For the solution of the flutter equation to obtain the onset of flutter, VG or K-Method as described in AGARD Manual on Aeroelasticity [4] will be utilized. Results of the analysis will be discussed in view of actual aircraft data. Hence, in addition for its usefulness for instructional purposes, this investigation will be useful for conceptual and preliminary design purposes for which simple but representative parametric studies could be based on such basic analytical tool.

10.2.2 *Typical Section Representation of 3D Wing for Aeroelastic Analysis*

For purposes of analysis which could reflect the influence of relevant parameters as found in actual aircraft wings, the use of typical section for the aeroelastic analysis of aircraft wing has been commonly adopted. Although such approach will not reflect the actual behavior of real aircraft wings, its usefulness can be drawn from the qualitative behavior as functions of relevant parameters (as elaborated subsequently) and later be assessed by comparison of the results obtained and to more elaborate computation (such as using NASTRAN©[5]) or actual data. For this purpose, the basic governing equation for the two-degree-of freedom (binary) flutter using incompressible and linearized aerodynamics can be utilized and given in the following form:

$$\begin{aligned} m\ddot{h} + S_\alpha\ddot{\alpha} + K_h h &= -L \\ S_\alpha\ddot{h} + I_\alpha\ddot{\alpha} + K_\alpha\alpha &= M \end{aligned} \qquad (10.28)$$

where m is the mass per unit span of the typical section, S_α is the static moment of the typical section with respect to the elastic axis, I_α is the polar moment of inertia of the typical section with respect to the elastic axis, K_h and K_α are the bending (heaving) and torsional spring stiffness, respectively, of the typical section. The typical section experiences movement in two degrees of freedom, i.e. h, heaving (bending) displacement in the vertical direction (positive downward), and α, pitching angular displacement (positive nose-up). L and M_{AC} are the aerodynamic lift and moment, respectively; both L and M_{AC} are acting on the aerodynamic center (L positive upward, M_{AC} positive nose-up) (Fig. 10.4).

---

[4] Zwaan, Op.Cit.

[5] Rodden,W.P., and Johnson, E.H., *MSC/NASTRAN Aeroelastic Analysis User's Manual*, Version 68, 1994, pp. 237-257.

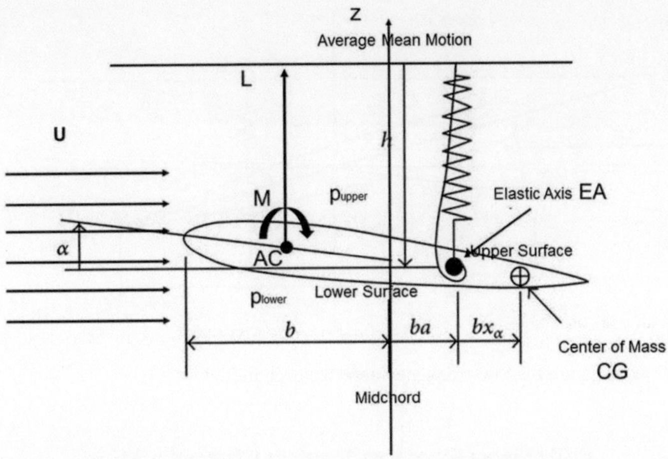

Fig. 10.4 Free-body diagram for a typical section

Next three different aerodynamic approximations, with increasing complexity, will be utilized; these are the steady state, low-frequency oscillation and classical unsteady (harmonic) aerodynamics as described by Theodorsen.[6]

10.2.2.1 Steady Aerodynamic Model

For the steady aerodynamic model, the aerodynamic lift L and moment M_{AC}, as well known in steady linearized aerodynamics, are given by:

$$L(t) = qSC_{L\alpha}\alpha(t) \tag{10.29}$$

and

$$M_{AC} = 0 \tag{10.30}$$

Since linearized aerodynamics is used, the airfoil essentially is regarded as a flat plate. With the assistance of schematic Figs. 10.5, 10.6 and 10.7, the aerodynamic moment with respect to the elastic axis can be shown to be given by

$$M_{EA} = 2Leb + M_{AC} = 2qSebC_{L\alpha}\alpha(t) \tag{10.31}$$

Hence Eq. (10.28) can be rewritten as

[6] Theodorsen, T., General Theory of Aerodynamic Instability and the Mechanism of Flutter, NACA Report 496, 1935.

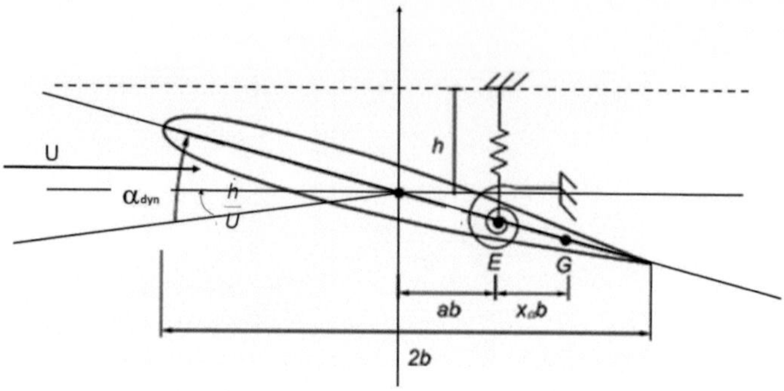

Fig. 10.5 Flow situation for low-frequency aerodynamic model

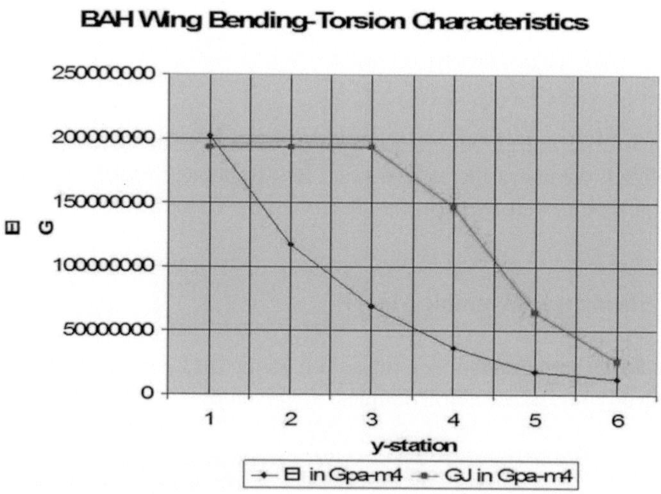

Fig. 10.6 Bending and torsional stiffness of BAH wing, in [GPa-m^4]

$$\begin{aligned} m\ddot{h} + S_\alpha \ddot{\alpha} + K_h h + qsC_{L\alpha}\alpha = 0 \\ S_\alpha \ddot{h} + I_\alpha \ddot{\alpha} + K_\alpha \alpha - 2qSebC_{L\alpha}\alpha = 0 \end{aligned} \tag{10.32}$$

which is known as the flutter equation. This is an eigenvalue or stability equation. In matrix notation, it can be rewritten as:

$$[M]\{\ddot{x}\} + ([K] - q[A_o])\{x\} = \{0\} \tag{10.33}$$

where

$$[M] = \begin{bmatrix} m & S_\alpha \\ S_\alpha & I_\alpha \end{bmatrix} \text{ (Inertia)} \tag{10.34a}$$

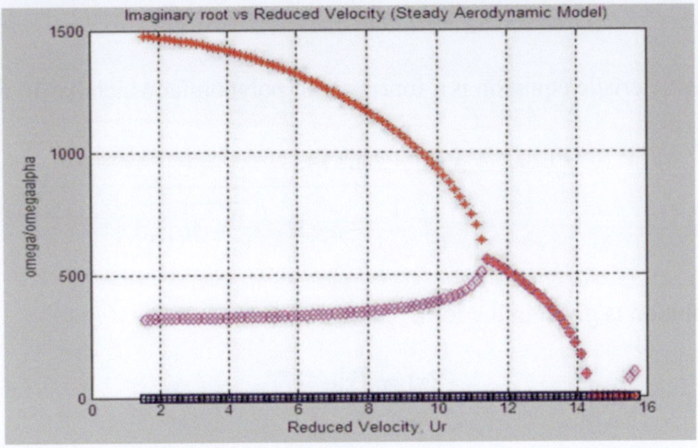

Fig. 10.7 Frequency-reduced velocity diagram

$$[K] = \begin{bmatrix} K_h & 0 \\ 0 & K_\alpha \end{bmatrix} \text{ (Structural stiffness)} \tag{10.34b}$$

$$[A_o] = \begin{bmatrix} 0 & -\mathrm{SC}_{L\alpha} \\ 0 & 2\mathrm{seb}C_{L\alpha} \end{bmatrix} \text{(Aerodynamic "Stiffness")} \tag{10.34c}$$

For convenience, following the practice in aeroelasticity, the analysis of flutter stability can be obtained by assuming a solution of the form:

$$\begin{aligned} h &= \hat{h}e^{\mathrm{pt}} \\ \alpha &= \hat{\alpha}e^{\mathrm{pt}} \end{aligned} \text{ and } \{x\} = \{\hat{x}\}e^{pt} \tag{10.35a,b,c}$$

Solving as eigenvalue problem, going through the algebra, will result in the flutter stability characteristic equation given by:

$$\begin{aligned} &I_\alpha m p^4 - S_\alpha^2 p^4 + p^2\left[mK_\alpha - 2qSebC_{L\alpha}m + I_\alpha K_h\right] \\ &- p^2 S_\alpha qSC_{L\alpha} + K_\alpha K_h - 2qSebC_{L\alpha}K_h = 0 \end{aligned} \tag{10.35}$$

which has a general form of:

$$a_4 p^4 + a_2 p^2 + a_0 = 0 \tag{10.36}$$

where

$$\begin{aligned} a_4 &= I_\alpha m - S_\alpha^2 \\ a_2 &= m(K_\alpha - 2qSebC_{L\alpha}) + I_\alpha K_h - S_\alpha qSC_{L\alpha} \\ &= mK_\alpha + I_\alpha K_h - (2meb + S_\alpha)qSC_{L\alpha} \end{aligned}$$

$$a_0 = K_h(K_\alpha - 2qSebC_{L\alpha}) \tag{10.37a,b,c}$$

The characteristic equation is a fourth-order polynomial which has four roots:

$$\begin{aligned} p_{1,2,3,4} &= (\sigma + i\omega)_{1,2,3,4} \\ &= \pm\sqrt{\frac{1}{2a_4}\left(-a_2 \pm \sqrt{a_2^2 - 4a_4a_0}\right)} \end{aligned} \tag{10.38}$$

and the solution is given in the form

$$\{x\} = \{\hat{x}\}e^{\sigma t}e^{i\omega t} \tag{10.39}$$

where

σ damping
ω frequency
$\{\hat{x}\}$ Vibration mode is representing the displacement vector.

Following Done (1968), as summarized by Zwaan,[7] the solution can be conveniently and comprehensively represented by damping and frequency diagrams as functions of either dynamic pressure q, or reduced frequency (or reduced velocity U_R) $k_R = U/(b.\omega)$, or velocity, as illustrated subsequently, or summarized in a table. The table allows the classification of solution according to the values of the coefficients a_0, a_2, a_4, and the stability categories that result. The stability of motion depends on the value of σ (aerodynamic damping). As can be concluded from Eq. (10.61) and summarized in Table 10.1, if $\sigma > 0$, then the displacement vector $\{x\}$ will oscillate with increasing amplitude in time, and the resulting motion will be unstable. If $\sigma = 0$, a neutrally stable oscillation will result. Only when $\sigma > 0$ the oscillation will subside in time.

10.2.2.2 Low-Frequency Refinement Aerodynamic Model

First-order refinement of the aerodynamic model is obtained by replacing the instantaneous incidence angle by the instantaneous angle of attack α_{dyn} contributed by the angle of incidence $\alpha(t)$ and the contribution of the heaving motion; hence

$$\alpha_{\text{dyn}}(t) = \frac{\dot{h}}{U} + \alpha(t) \tag{10.40}$$

and consequently the aerodynamic lift and moment become:

---

[7] Zwaan, R.J., Aeroelasticity of Aircraft, Lecture Notes, Special Lecture, Short Course offered at Institut Teknologi Bandung, Indonesia, Lecture Notes delivered at Special Meeting at ITB in August 1981(private collection.

Table 10.1 Flutter stability solution categories

a_2^2-$4a_4a_0$	> 0			< 0
a_0	> 0		< 0	> < 0
a_2	> 0	< 0	> < 0	> < 0
p2	$-\omega_1^2, -\omega_2^2$	σ_1^2, σ_2^2	$\sigma^2, -\omega^2$	$-g + ih$
p	$\pm i\omega_1, \pm i\omega_2$	$\pm i\sigma_1, \pm i\sigma_2$	$\pm \sigma, \pm i\omega$	$\pm \sigma, \pm i\omega$
Type of Motion	Harmonic	Aperiodic	Aperiodic	Oscillatory
	2 pos freq	2 diverging	1 diverging	1 div.pos.freq
	2 neg freq	2 converging	1 converging	1 conv.pos.freq
			Harmonic	1 div.neg.freq
			1 pos freq	1 conv.neg.frq
			1 neg freq	
Type of Instability	Neutral	Divergent	Divergent	Flutter
Category	I	III	IV	II

$$L(t) = qSC_{L\alpha}\left(\frac{\dot{h}}{U} + \alpha\right)$$

$$M_{EA}(t) = 2qSebC_{L\alpha}\left(\frac{\dot{h}}{U} + \alpha\right) \tag{10.41}$$

Hence the flutter equation becomes:

$$m\ddot{h} + S_\alpha\ddot{\alpha} + K_h h + qSC_{L\alpha}\left(\frac{\dot{h}}{U} + \alpha\right) = 0$$

$$S_\alpha\ddot{h} + I_\alpha\ddot{\alpha} + K_\alpha\alpha - 2qSebC_{L\alpha}\left(\frac{\dot{h}}{U} + \alpha\right) = 0 \tag{10.42}$$

or in matrix notation:

$$[M]\{\ddot{x}\} - [B]\{\dot{x}\} + ([K] - q[A_o])\{x\} = \{0\} \tag{10.43}$$

$$[B] = \begin{bmatrix} \frac{q}{U}SC_{L\alpha} & 0 \\ -2\frac{q}{U}SebC_{L\alpha} & 0 \end{bmatrix} \text{ (Aerodynamic damping)} \tag{10.44}$$

Substitution of appropriate equations leads to the flutter stability equation (eigenvalue problem):

$$\left(mp^2 + \frac{q}{U}SC_{L\alpha}p + K_h\right)\hat{h} + (S_\alpha p^2 + qSC_{L\alpha})\hat{\alpha} = 0$$

$$\left(S_\alpha p^2 - 2\frac{q}{U}SebC_{L\alpha}p\right)\hat{h} + (I_\alpha p^2 + K_\alpha - 2qSebC_{L\alpha})\hat{\alpha} = 0 \tag{10.45}$$

The determinant of the equation is given by:

$$\begin{vmatrix} mp^2 + \frac{q}{U}SC_{L\alpha}p + K_h & S_\alpha p^2 + qSC_{L\alpha} \\ S_\alpha p^2 - 2\frac{q}{U}SebC_{L\alpha}p & I_\alpha p^2 + K_\alpha - 2qSebC_{L\alpha} \end{vmatrix} = 0 \tag{10.46}$$

Going through the algebra, the following characteristic equation is obtained:

$$\begin{aligned} &p^4[mI_\alpha - S_\alpha^2] + p^3[\frac{q}{U}SC_{L\alpha}(2ebS_\alpha + I_\alpha)] \\ &\quad + p^2[mK_\alpha + I_\alpha K_h - (2meb + S_\alpha)qSC_{L\alpha}] \\ &\quad + p[\frac{q}{U}SC_{L\alpha}K_\alpha] + K_h(K_\alpha - 2qSebC_{L\alpha}) = 0 \end{aligned} \tag{10.47}$$

or

$$a_4p^4 + a_3p^3 + a_2p^2 + a_1p + a_0 = 0 \tag{10.48}$$

where

$$\begin{aligned} a_4 &= mI_\alpha - S_\alpha^2 \\ a_3 &= \frac{q}{U}SC_{L\alpha}(2ebS_\alpha + I_\alpha) \\ a_2 &= mK_\alpha + I_\alpha K_h - (2meb + S_\alpha)qSC_{L\alpha} \\ a_1 &= \frac{q}{U}SC_{L\alpha}K_\alpha \\ a_0 &= K_h(K_\alpha - sqSebC_{L\alpha}) \end{aligned} \tag{10.49a,b,c,d,e}$$

This equation is a fully fourth-order polynomial.

10.2.2.3 Theodorsen Unsteady Aerodynamic Model

Following AGARD Manual on Aeroelasticity, 1960, Vol.VI,[8] the unsteady aerodynamic forces for harmonic motions can be represented by

$$\begin{aligned} \frac{\hat{L}}{\pi\rho b^3} &= \omega^2\left\{\frac{k_a}{k^2}\left[\frac{\hat{h}}{b} - (0.5 + a)\hat{\alpha}\right] + \frac{k_b}{k^2}\hat{\alpha}\right\} \\ \frac{M_{AC}}{\pi\rho b^4} &= -\omega^2\left\{\frac{m_a}{k^2}\left[\frac{\hat{h}}{b} - (0.5 + a)\hat{\alpha}\right] + \frac{m_b}{k^2}\hat{\alpha}\right\} \end{aligned} \tag{10.50}$$

8 -, AGARD Manual on Aeroelasticity, Vol. VI, 1961.

where

$$
\begin{aligned}
k_a &= k_a' + ik_a'' \\
k_a' &= -2kG(k) \\
k_a'' &= 2kF(k) \\
k_b &= k_b' + ik_b'' \\
k_b' &= -2F(k) - 2kG(k) - \frac{k^2}{2} \\
k_b'' &= 2kF(k) + 2G(k) + k \\
m_a &= m_a' + im_a'' \\
m_a' &= -\frac{1}{2}k^2 \\
m_b'' &= 0 \\
m_b &= m_b' + im_b'' \\
m_b' &= -\frac{3}{8}k^2 \\
m_b'' &= k
\end{aligned} \tag{10.51a–l}
$$

were defined by Küssner,[9] while F(k) and G(k) are the real and imaginary parts of the Theodorsen function C(k). Substituting Eq. (10.50) into the right-hand side of Eq. (10.28) using Eq. (10.51), then the eigenvalue or flutter stability equation equivalence of Eq. (10.45)

$$
[A(k)]\{\widehat{q}\} = \lambda\{\widehat{q}\} \tag{10.52}
$$

where A is a 2 × 2 matrix whose elements are given by

$$
\begin{aligned}
A_{11} &= \left(\frac{\omega_\alpha}{\omega_h}\right)^2\left(\mu - \frac{k_a}{k^2}\right) \\
A_{12} &= \left(\frac{\omega_\alpha}{\omega_h}\right)^2\left(\mu x_\alpha - \frac{k_b}{k^2} + (0.5 + a)\frac{k_a}{k^2}\right) \\
A_{21} &= \mu x_\alpha - \frac{m_a}{k^2} + (0.5 + a)\frac{k_a}{k^2} \\
A_{22} &= \mu r_\alpha^2 \frac{m_b}{k^2} + (0.5 + a)\left(\frac{m_a}{k^2} + \frac{k_b}{k^2}\right) - (0.5 + a)^2\frac{k_a}{k^2}
\end{aligned} \tag{10.53a–d}
$$

Equation (10.52) is another polynomial of the eigenvalue λ.

---

[9] Küssner, H.G., Zusammenfassender Bericht ueber den iinstationaeren Auftrieb von Fluegeln, Luftfahrtforschung., Bd. 13, Nr. 12, December 1963.

At this point, the following remarks are in order. First, following the flutter stability analysis as carried out by Done and elaborated in Chap. 9 (Sect. 9.1.2), similar procedure for solving Eq. (10.75) can be followed to obtain the roots λ, which in general are complex. The square root of λ is also complex numbers, consisting of the real parts, which are indicative of the damping terms and imaginary parts, which are indicative of the frequencies of the flutter solution. A diagram indicating the values of the frequencies and damping terms of the flutter solution can be drawn, from which one may obtain significant information on the behavior of the flutter solution as a function of flow properties, represented by the airspeed U, dynamic pressure q or reduced frequency $k_R = U/\omega b$.

Second, following the classical procedure known as VG [2] or K-Method [3], the flutter equation can be solved for the prevailing modal frequencies (bending and torsion for the particular binary flutter problem) and the associated damping, by assuming a solution of the form

$$\lambda = \left(\frac{\omega_\alpha}{\omega}\right)^2 (1 + \mathrm{ig}) \tag{10.54}$$

Then the frequency and damping of the flutter solution are given by

$$\begin{aligned}
\lambda &= \frac{1 + ig}{\omega^2} = \mathrm{Re}(\lambda) + \mathrm{Im}(\lambda) = \frac{1}{\omega^2} + \mathrm{Im}\left(\frac{g}{\omega^2}\right) \\
\omega &= \sqrt{\frac{1}{\mathrm{Re}(\lambda)}} \\
g &= \omega^2 \mathrm{Im}(\lambda)
\end{aligned} \tag{10.55a,b,c}$$

It will subsequently be shown that the onset of flutter is given by the value of U (then here is the flutter velocity UF), for which one of the damping (g) curve crosses the zero-axis. Numerical results show that at this point, the two modal frequencies (bending and torsion) will approach each other, which is confirmed by experimental results.

10.2.3 Case Studies

Case Study 1: Parametric Study of Typical Section

Based on the theoretical foundation described above, numerical computation is carried out for the well-known BAH wing,[10] which have been modified to allow

[10] In all these case studies, BAH wing refers to that defined by Bisplinghoff, R.L., Ashley, H. and Halfman, R.L., Aeroelasticity, Dover Publications, Inc., copyright 1955 by Addison-Wesley Publishing Co., Inc., copyright © renewed 1983 by Bisplinghoff, R.L., Ashley, H. and Halfman, R.L., while Thedorsen's refers to the seminal work, Theodorsen, T., General Theory of Aerodynamic Instability and the Mechanism of Flutter, NACA Report 496, 1935.

Table 10.2 BAH wing typical section as baseline

$\mu = m/(\pi\,\rho\, b^2)$		100
M = Area$_{Airfoil}/\rho_{Airfoil}$ [m^{2}kg/m^3]		2331.050025
Span [m]	BAH[11] p-45	12.7
c [m]	193.75	4.92
c for typical section [m]	c	4.92125
b = c/2		2.46063
e		0.51
bR	5.468	1.6404
omega-alpha, ω_α [rad/sec]	BAH, p 181	22.36
omega-h, ω_h [rad/sec]		12.8
omega-alpha /omega-alpha ω_h/ω_α [-]		0.5724508
ρ [kg/m^3]		1.225
$C_{L\alpha}$	2π	6.285714286
$M = 100\pi\rho b^2$		2331.050025
y		9.525
E [Gpa]	[Gpa	0.69 E + 12
$I = bh^3/12$	[m^4]	0.003054925
G [Gpa]	[Gpa	0.207 E + 12

evaluation of some uncertain factors or for parametric study purposes, while the example discussed here refers an example of modern transport aircraft synthesized by the author. The results are elaborated subsequently.

10.2.3.1 Solution of Flutter Stability Solution of BAH Wing-based Typical Section for Steady

Aerodynamic Model. The structural properties of the BAH wing are summarized in Table 10.2.

The flutter stability diagram for BAH wing-based typical section is given in Fig. 10.8.

In all these diagrams, the stability of motion depends on the value of σ (aerodynamic damping), i.e. $\sigma > 0$ indicate unstable motion, $\sigma = 0$ neutrally stable motion, and $\sigma < 0$ stable motion.

[11] Bisplinghoff, R.L., Ashley, H. and Halfman, R.L.,(1955), Aeroelasticity, Dover Publications, Inc., copyright 1955 by Addison-Wesley Publishing Co., Inc.

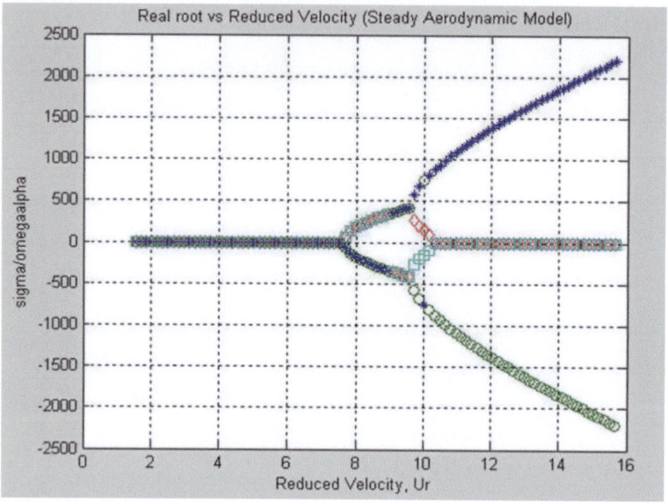

Fig. 10.8 Damping-reduced velocity diagram steady aerodynamic model

10.2.3.2 Solution of Flutter Stability Solution of BAH Wing-Based Typical Section for Low-Frequency Aerodynamic Model

The flutter stability diagram for BAH wing-based typical section is given in Fig. 10.9.

Again, in all these diagrams, the stability of motion depends on the value of σ (aerodynamic damping), i.e. $\sigma > 0$ indicate unstable motion, $\sigma = 0$ neutrally stable motion, and $\sigma < 0$ stable motion. Only positive frequencies are realistic.

10.2.3.3 Solution of Flutter Stability Solution of BAH Wing-Based Typical Section for Unsteady (Harmonic) Aerodynamic Model (Theodorsen's[12])

The flutter stability diagram for BAH wing-based typical section[13] is given in Fig. 10.10.

As in previous cases, in all these diagrams, the stability of motion depends on the value of σ (aerodynamic damping), i.e. $\sigma > 0$ indicate unstable motion, $\sigma =$

---

[12] Thedorsen's refers to the seminal work, Theodorsen, T., General Theory of Aerodynamic Instability and the Mechanism of Flutter, NACA Report 496, 1935.

[13] Reference has been made to Bisplinghoff, R.L., Ashley, H. and Halfman, R.L., Aeroelasticity, Dover Publications, Inc., copyright 1955 by Addison-Wesley Publishing Co., Inc., copyright © renewed 1983 by Bisplinghoff, R.L., Ashley, H. and Halfman, R.L., and Zwaan, R.J., Aeroelasticity of Aircraft, Lecture Notes, Special Lecture, Short Course offered at Institut Teknologi Bandung, Indonesia, Lecture Notes delivered at Special Meeting at ITB in August 1981(private collection).

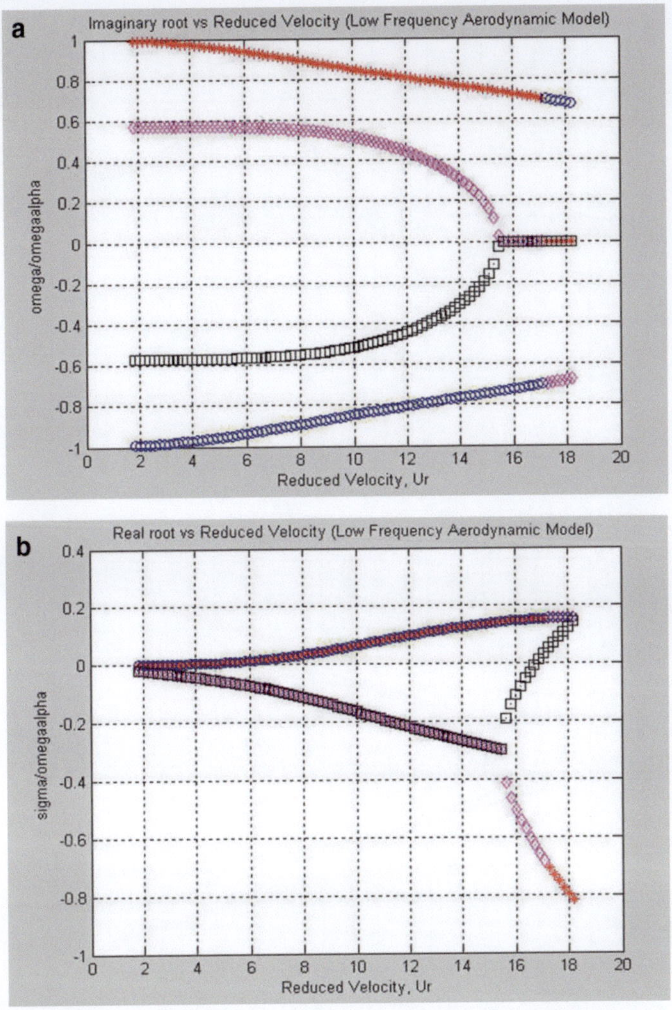

Fig. 10.9 **a** Frequency-reduced velocity diagram low-frequency aerodynamic model. **b** Damping-reduced velocity diagram low-frequency aerodynamic model

0 neutrally stable motion, and $\sigma < 0$ stable motion. Only positive frequencies are realistic.

Case Study 2: Parametric Study of Three-Dimensional Wing

In the parametric study for three-dimensional wings, simplifications are made to facilitate analysis. For illustration, four cases will be studied; these are the uniform

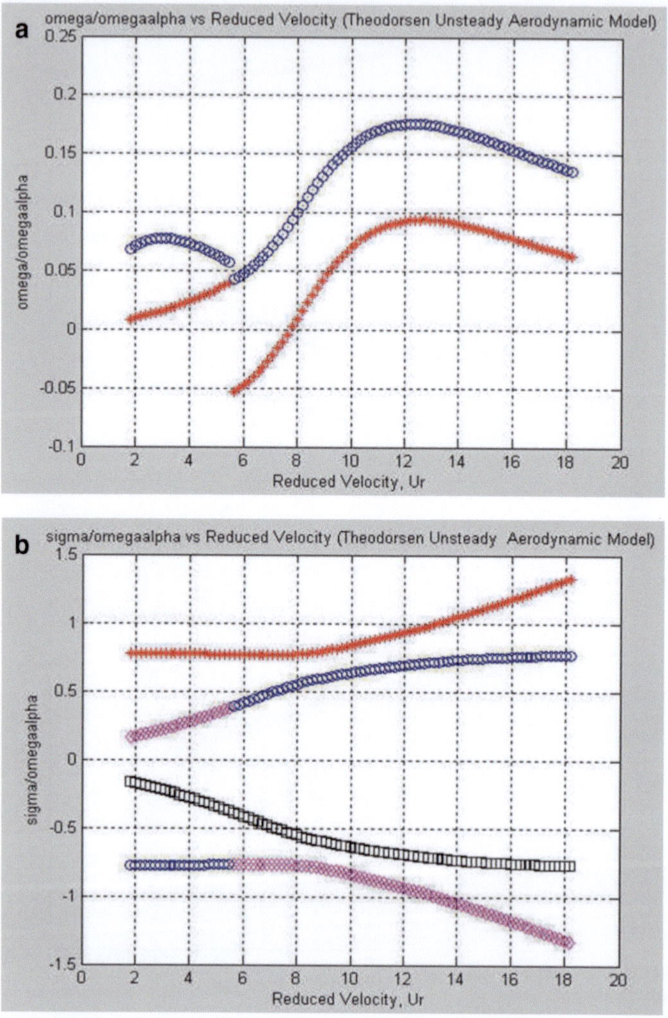

Fig. 10.10 **a** Frequency-reduced velocity diagram unsteady (Theodorsen) aerodynamic model. **b** Damping-reduced velocity diagram unsteady (Theodorsen) aerodynamic model

cantilever straight wing, uniform cantilever swept back tapered wing, AIRBUS A-300-like and Boeing 747-like wing.[14] The normal trend of the occurrence of instabilities will be investigated by calculating the flutter instability diagrams. Due to space limitation, only the first example is illustrated below.

[14] Synthesized and modified from information obtained from open publications by the AIRBUS and Boeing Companies; also https://www.militaryfactory.com/aircraft/compare-aircraft-results.php?form=form&aircraft1=250&aircraft2=247&Submit=Compare+Aircraft.

Table 10.3 Assumed properties of uniform cantilever straight wing

Assumed Properties of Uniform Cantilever Straight Wing			
Span [m]	12.7	Bending Spring Stiffness K_h [N/m]	8.89 E + 05
Chord c [m]	5.715		
Density, ρ [kg/m³]	1.225		
Lift Coefficient Slope $C_{L\alpha}$ [-]	0.1788	Torsional Spring Stiffness K_α [N]	3.3668 E + 07
Eccentricity Ec [-]	0.25		
Mass per unit span m [kg/m]	0.0161	Distance between elastic axis and Center of Gravity, x_α [m]	0.2
Radius of Gyration r_α [m]	0.5	Reduced Semi-Chord b_R [-]	1

10.2.4 Solution of Problems Addressed in Case Studies

10.2.4.1 Uniform Cantilever Straight Wing

The characteristics of the wing studied are summarized in Table 10.3.

The use of such properties is meant only for rough estimate, so that some practical figures can be obtained, without losing generalities, since further parametric study can be tailored using a range of specific realistic values or by having exact knowledge of the structural properties of a particular wing which may be indicated by a point in the calculated curve. The flutter stability diagram is exhibited in Fig. 10.11.

10.2.4.2 Uniform Cantilever Swept-Back Tapered Wing

The properties considered for the uniform cantilever swept-back tapered wing are given in Table 10.4.

10.2.5 Case Study: Boeing 747-Like Wing[15]

The assumed properties of the Boeing 747-like wing are given in Table 10.5.

---

[15] The example given here is very hypothetical, although the assumptions regarding the aeroelastic, geometry, material and others have been sought to be a close estimate to the reference airplane, the computational work only illustrates the efforts to have reasonable assumptions, and application of the method to a large airplane. No comparison and validation of results can be justified at this stage with this example.

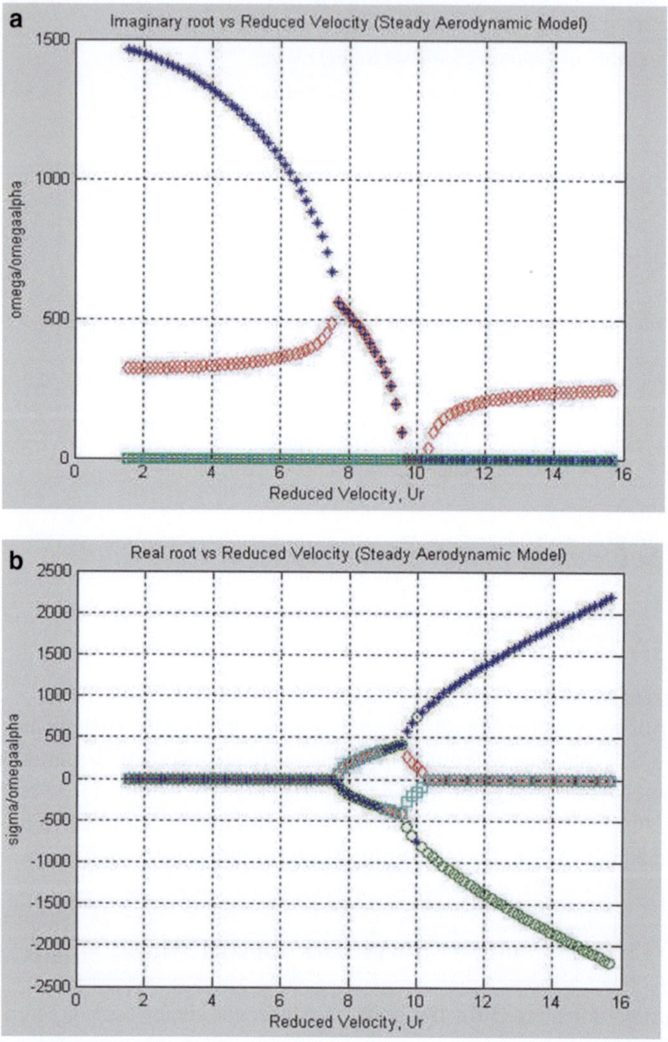

Fig. 10.11 **a** Frequency-reduced velocity diagram uniform cantilevered wing, steady aerodynamic model. **b** Damping-reduced velocity diagram uniform cantilevered wing, steady aerodynamic model. **c** Frequency-reduced velocity diagram uniform cantilevered wing, low-frequency aerodynamic model. **d** Damping-reduced velocity diagram uniform cantilevered wing, low-frequency aerodynamic model. e Frequency-reduced velocity diagram uniform cantilevered wing, unsteady (Theodorsen) aerodynamic model. f Damping-reduced velocity diagram uniform cantilevered wing, unsteady (Theodorsen) aerodynamic model

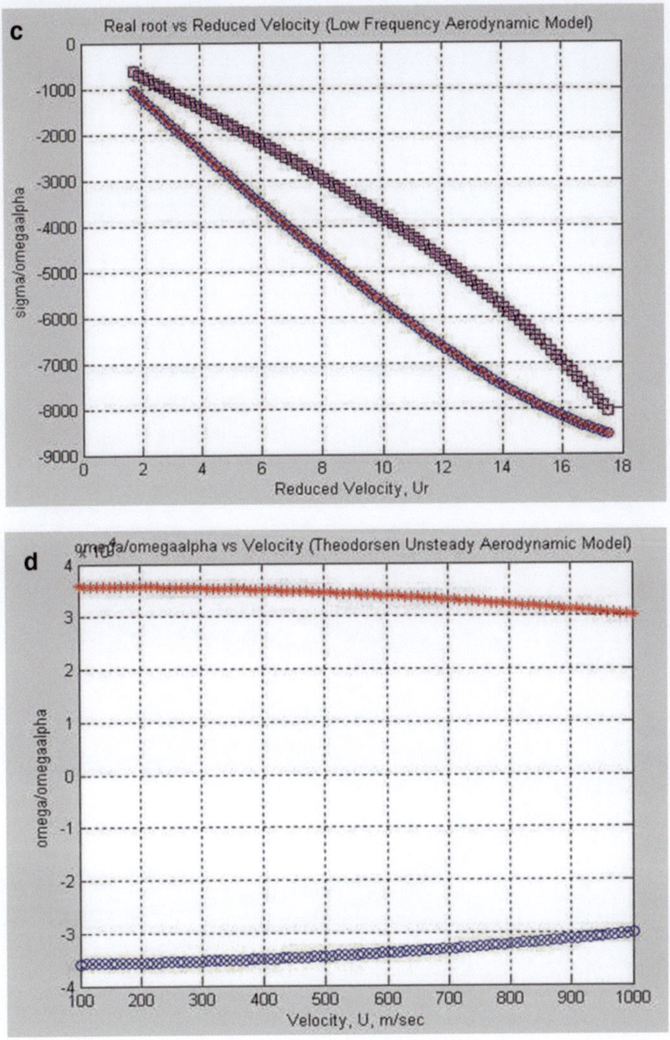

Fig. 10.11 (continued)

10.2.6 *Case Study 3: Determination of the Onset of Flutter for Typical Wing Section Using K-Method*

The solution of flutter equation for BAH wing-based typical section using Theodorsen unsteady aerodynamics is illustrated in Fig. 10.14.

Figure 10.8a, b show that the onset of flutter for the particular BAH wing-based typical section with specifications described in Table 10.2 occurs at a velocity of

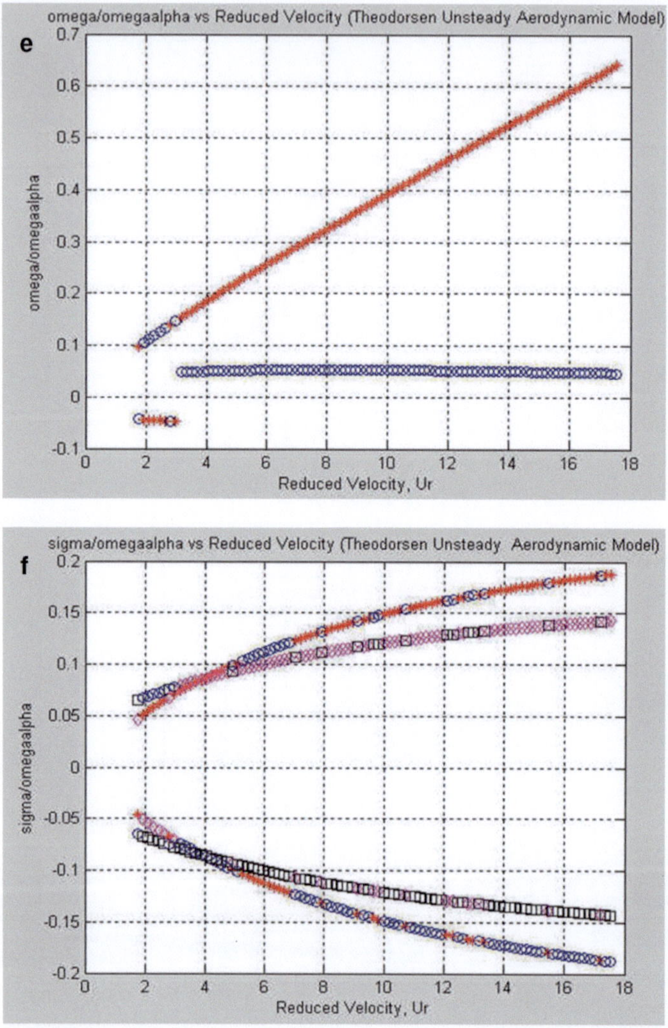

Fig. 10.11 (continued)

410 m/sec. Further parametric study can be carried out to vary the structural properties of the wing as represented by the typical section.

Table 10.4 Assumed properties of uniform cantilever swept-back tapered wing

Assumed Properties of Swept-Back Tapered Wing			
Span S [m]	45.5	Bending Spring Stiffness K_h [N/m]	6.309 e + 08
Chord c [m]	5.0995	Sweep Angle Λ [-]	21.9
Density ρ [kg/m^3]	1.225	Mass per unit span m [kg/m]	0.0161
$C_{L\alpha}$ [-]	2π	Torsional Spring Stiffness K_α [N]	2.141 e + 10
Eccentricity Ec [-]	0.25c	Reduced Semi-Chord b_R [-]	1
x_α [m]	0.2	Radius of Gyration r_α [m]	0.5

Table 10.5 Assumed properties of Boeing 747-like wing

Assumed Properties of Boeing 747-like Wing			
Span S [m]	30.30	Bending Spring Stiffness K_h [N/m]	9.214 e + 07
Chord c [m]	8.286	Torsional Spring Stiffness K_α [N]	1.628 E + 10
Density ρ [kg/m^3]	1.225	Reduced Semi-Chord b_R [-]	0.333
$C_{L\alpha}$ [-]	2π	x_α [m]	0.2
Eccentricity Ec [-]	0.25c		
Mass per unit span m [kg/m]	0.0155		
Radius of Gyration r_α [m]	0.5	Sweep Angle Λ [°]	37.5

10.2.7 Case Study 4: Parametric Study of Typical Section Subject to Changes in Its Sectional Characteristics

The binary flutter as formulated above can be used to study the influence of certain typical section parameters. For this purpose, one significant parameter, i.e. a, will be varied, and its influence on the flutter solution can be studied. Figure 10.15 exhibits the influence of varying a from −3.0 to −1.5 to the onset of flutter.

10.2.8 Discussion and Analysis

As tabulated in Table 10.1 for the stability of bending-torsion flutter equation for typical section subject to aerodynamic load modeled as steady aerodynamic, flutter may occur at situations described in the last column in that table. The calculated results are shown in Fig. 10.8 and following figures. Only the diverging positive frequency reflects the physically realizable state. Difference between flutter characteristics is caused by different coefficients a_4, a_2 and a_0. Following this table, flutter may be possible if a_2^2-$4a_4a_0 < 0$. For low-frequency aerodynamic model, Table 10.1 no more applies. Similar analysis can be carried out using the new flutter stability Eq. 21 for low-frequency aerodynamic model and Eq. 25 for unsteady incompressible flow. The Done-type diagrams given in Figs. 10.8, 10.9, 10.10, 10.11, 10.12, 10.13, 10.14 and 10.15 summarize all stability possibilities, including divergence. The analysis carried out using VG or K-method addresses the possibilities of the onset of flutter in a pair of diagrams which are more comprehensive.

10.2.9 Remarks on Case Studies Addressed

The classical bending-torsion flutter stability problem has been carried out as a case study to look into its utilization for instructional and conceptual study utilization. In particular, three simplified cases using linearized incompressible aerodynamics can be differentiated in the formulation of approximate flutter analysis with progressive sophistication. Two methods of approach are described: a. representation of flutter stability characteristics as the solution of the characteristic equation of the eigenvalue problem and b. the utilization of K-method in the said characteristic equation. These cases are illustrated by application of the analysis to BAH wing-based typical section. Further application of the method for parametric study of simplified three-dimensional wings and real aircraft wing exemplified by Boeing 747-like wing has also been made.

10.3 Concluding Remarks

The present chapter has elaborated a summary and review of the dynamics of aeroelastic system and for instructional purposes has resorted to the analysis of a generic

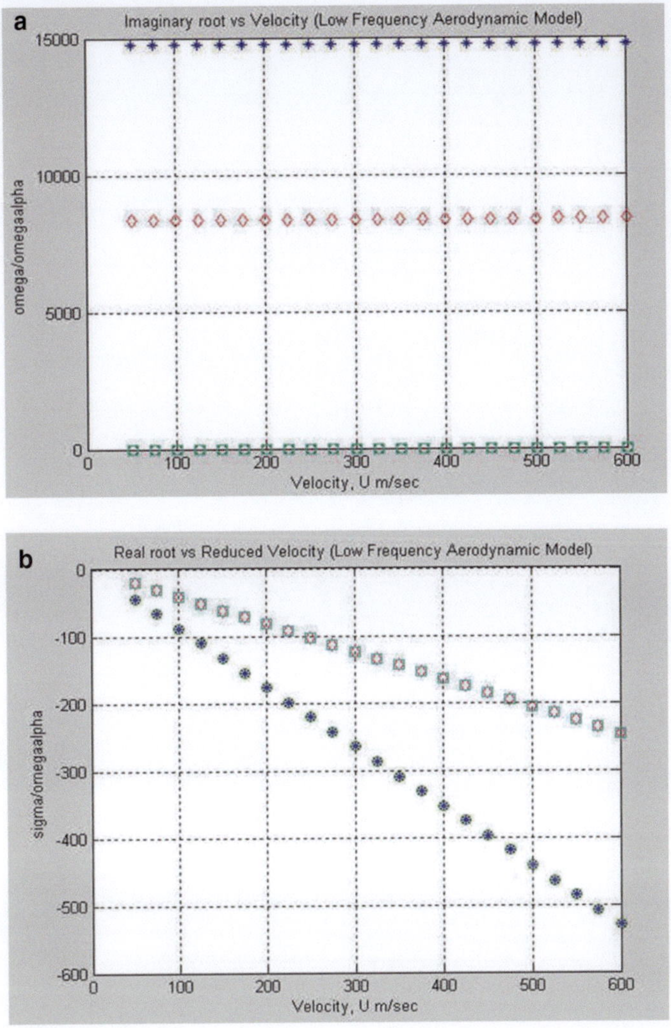

Fig. 10.12 **a** Frequency-velocity diagram, tapered swept-back wing, low-frequency aerodynamic model. **b** Damping-velocity diagram, tapered swept-back wing, low-frequency aerodynamic model. **c** Frequency-velocity diagram, tapered swept-back wing, unsteady (Theodorsen) aerodynamic model. **d** Damping-velocity diagram, tapered swept-back wing, unsteady (Theodorsen) aerodynamic model

aeroelastic system: the typical section. The elaborated discussions cover the utilization of the basic physics and mathematical tools which started from first principles: the equation of motion, elastic properties and differential equations as appropriate, following the Collar triangle paradigm.

The topics covered are:

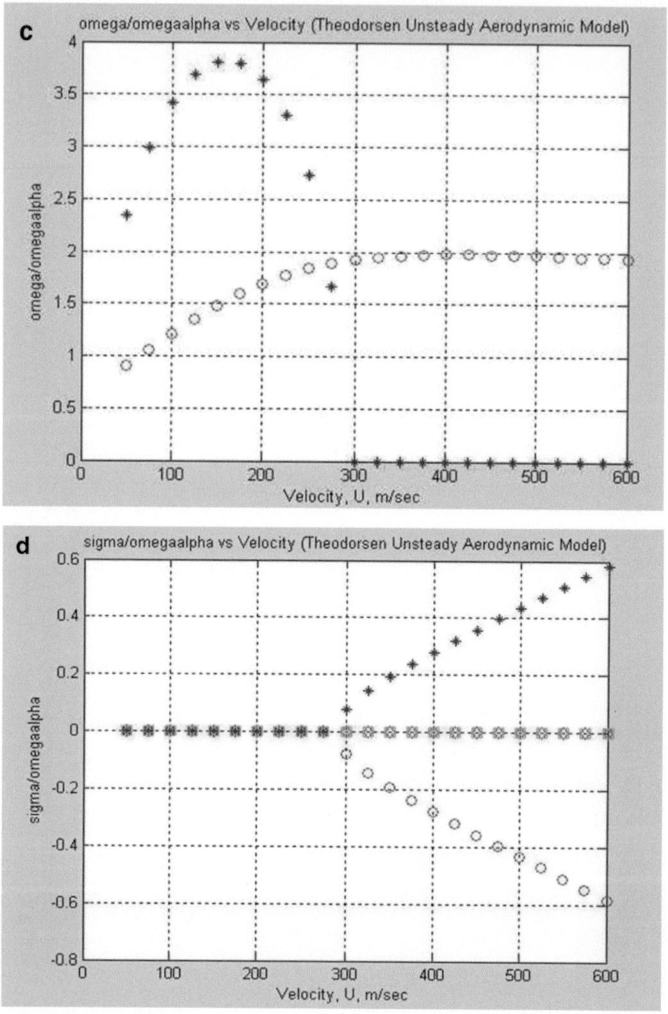

Fig. 10.12 (continued)

- Dynamic Aeroelasticity of Typical Section
- Review of Theoretical Foundation of Flutter Stability of Binary-Bending-Torsion Flutter of a Typical Section
- Parametric Study of Aeroelastic Stability and Flutter Characteristicsof Aircraft Wings a Case Study.

For the elaboration of dynamic aeroelasticity and flutter and review of the theoretical foundation for flutter stability for binary-bending-torsion flutter of typical section, a revisit to the classical bending-torsion flutter stability problem has been

a

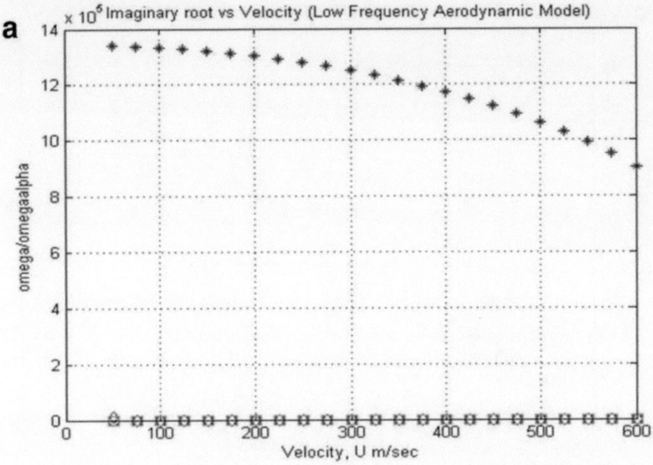

b

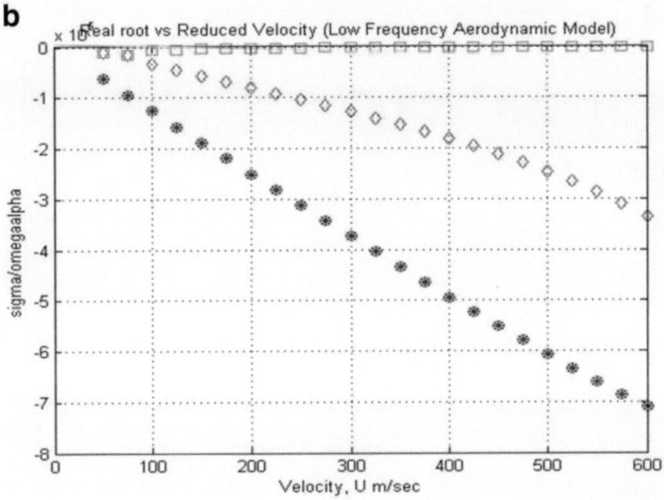

Fig. 10.13 **a** Frequency-velocity diagram, Boeing 747-like wing, low-frequency aerodynamic model. **b** Damping-velocity diagram, Boeing 747-like wing, low-frequency aerodynamic model. **c** Frequency-velocity diagram, Boeing 747-like wing, unsteady (Theodorsen) aerodynamic model. **d** Damping-velocity diagram, Boeing 747-like wing, unsteady (Theodorsen) aerodynamic model

carried out in the third part to look into its utilization for instructional and conceptual study utilization of binary aeroelastic approach for three simplified cases using linearized incompressible aerodynamics can be differentiated in the formulation of approximate flutter analysis with progressive sophistication [6–10].

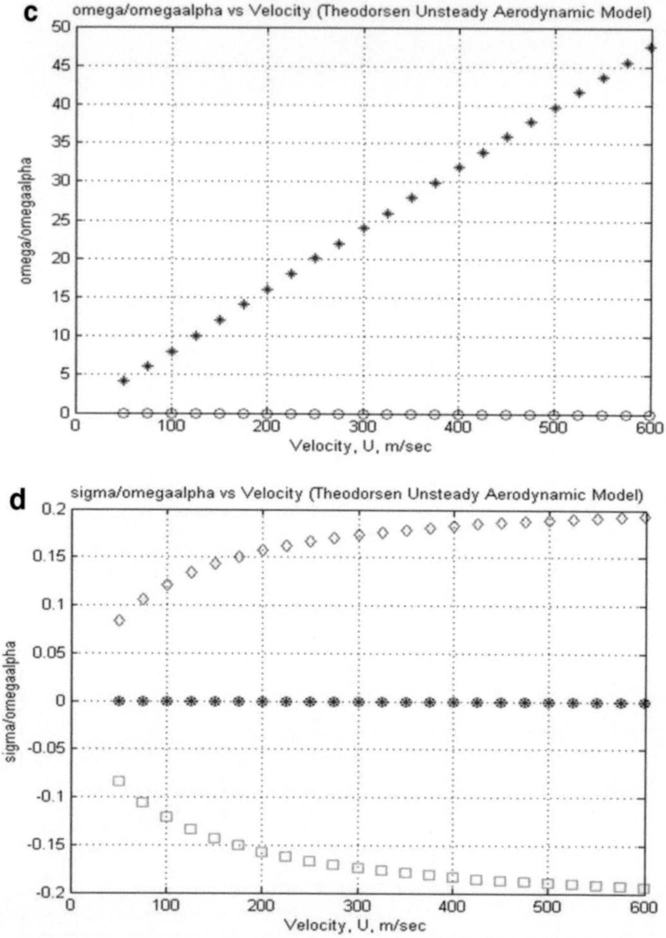

Fig. 10.13 (continued)

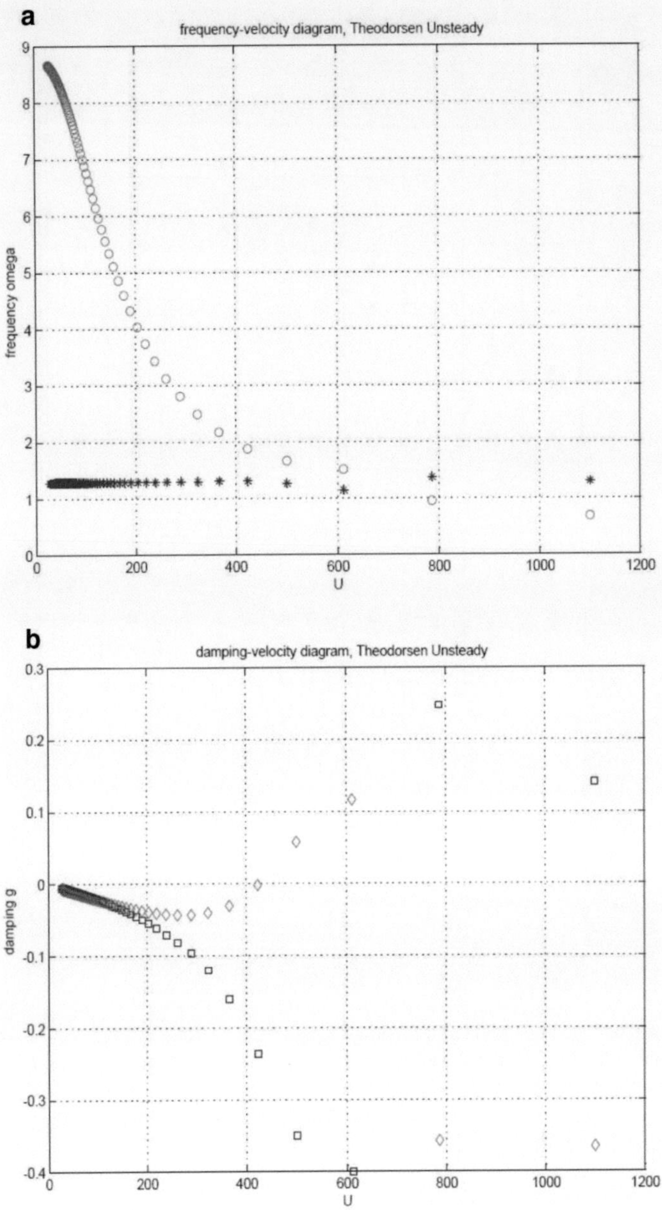

Fig. 10.14 **a** Bending and torsional frequencies of BAH. Typical section as a function of free-stream velocity U. **b** Bending and torsional artificial damping of BAH typical section as a function of free-stream velocity U

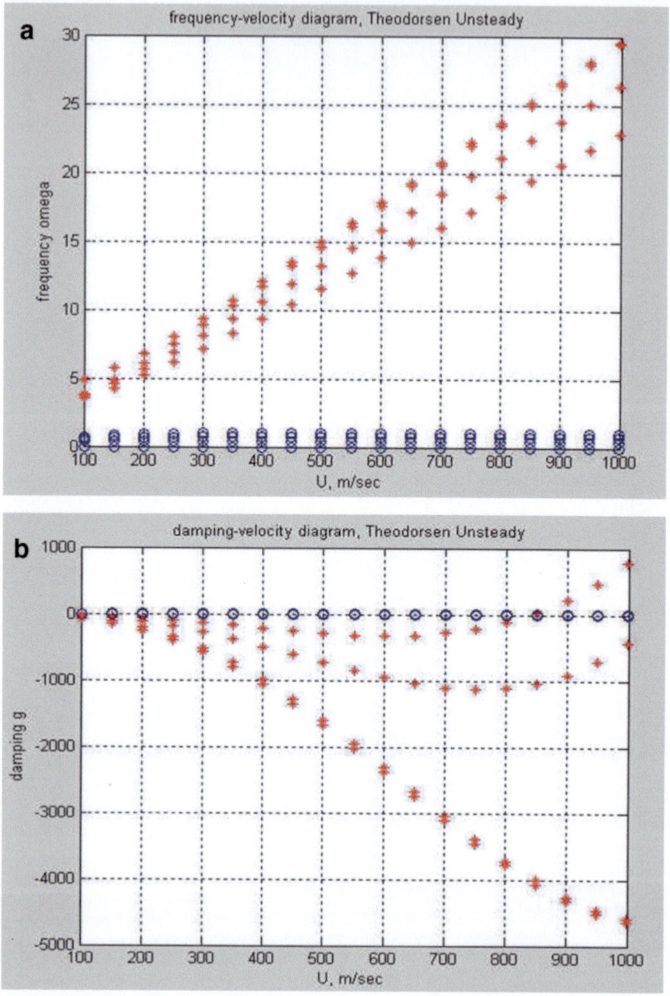

Fig. 10.15 **a** Changes in bending and torsional frequencies of BAH typical section as a function of free-stream velocity U if a is varied from −2.5 to −1.5. **b** Changes in bending and torsional artificial damping of BAH typical section as a function of free-stream velocity U if a is varied from −2.5 t −1.5

References

1. Zwaan, R.J. 1981.Aeroelasticity of aircraft, lecture notes, special lecture, Short Course offered at Institut Teknologi Bandung, Indonesia.
2. Platzer, Max R., (ed.). 1960. Agard manual on aeroelasncity in axial-flow turbomachines, vol. 2, Structural dynamics and aeroelasticity. AGARD-AG-298-Vol-2.
3. Djojodihardjo, H., and Hong-Hoe. Yee. 2007. Parametric atudy of the flutter characteristics of transport aircraft wings. In *Proceedings of AEROTECH-II 2007 Conference on Aerospace Technology of XXI Century*, 20–21 June, Kuala Lumpur.
4. Done, G.T.S. 1963. *The effect of linear damping on flutter speed*. March, Aeronautical Research Council R&M No 3996.
5. Done, G.T.S. 1966. *The flutter stability of undamped systems*. Aeronautical Research Council R&M 3553.
6. Done, G.T.S. 1967. *A study of binary flutter roots using a method of system synthesis*. Aeronautical Research Council, R&M 3554.
7. Rodden, W.P., and E.H. Johnson. 1994. MSC/NASTRAN aeroelastic analysis user's manual. *Version* 68, ISBN 10: 1585240060, ISBN 13: 9781585240067.
8. Theodorsen, T. 1935. *General theory of aerodynamic instability and the mechanism of flutter*. NACA Report 496.
9. Küssner, H.G. 1963. Zusammenfassender bericht ueber den iinstationaeren auftrieb von fluegeln. *Luftfahrtforschung., Bd* 13, Nr. 12.
10. Bisplinghoff, R.L., and H. Ashley. 1955. *Aeroelasticity*. Addison-Wesley.

References

Part II
Advanced Topics

Chapter 11
Unsteady Aerodynamics with Case Studies

Abstract Three-dimensional unsteady aerodynamics and computational approaches for continuation of Chap. 8 in unsteady Aerodynamics are elaborated in the present chapter (Reproduced from and based on several papers: [1–4]). Further it illustrates some of the contributions in lifting surfaces based on potential aerodynamic approaches and serves to provide and elaborate a good spectrum of problems and analytical approaches focusing on case studies in three-dimensional unsteady aerodynamic theory and computational methods. The first case study illustrates one of the first attempts in unsteady lifting potential flow solution method using geometric discretization and singularity distributions on the aerodynamic surface, which contributes to the present Boundary Element Method. The second case study addresses a more complex situation, that is the unsteady subsonic three-dimensional flow with separation bubble, and utilizes the singularity method. The third case study reflects the application of these previous techniques in addressing aircraft buffeting phenomena, a practical aeroelastic problem, using dynamic response approach The last case study elaborates the application of the three-dimensional unsteady aerodynamics lifting surface method to address combined aeroelasticity and acoustic excitation in a unified boundary element approach. These case studies are presented to provide an illustration of the application of lifting surface or boundary element method to solve unsteady aerodynamic problems.

Keywords Aeroelasticity · Acousto-aeroelasticity · Boundary element method · Lifting surface method · Singularity method · Unsteady aerodynamics

Nomenclature

A; A_{ij}	Area of the surface element; matrix of velocity, influence coefficient
B_i	Total apparent downwash velocity
C_{ij}	Velocity potential influence coefficient
C_L, C'_M, c_i, c_m	Lift and moment coefficients; sectional or two-dimensional lift and moment coefficients

© The Author(s), under exclusive license to Springer Nature Singapore Pte Ltd. 2023

H. Djojodihardjo, *Introduction to Aeroelasticity*,
https://doi.org/10.1007/978-981-16-8078-6_11

c	Half-chord
i, j, k	Unit vectors in the x, y and z directions
K_v, K_{v2}	Velocity kernel functions
$K_{Vx}, K_{Vy}, K_{Vxx}, K_{Vxy}, K_{Vyy}$	First- and second-order derivatives of K_v
M_x	Area moment about the x axis
n, ν; n	Unit normal vectors; normal coordinate
N, NC, NS	Total number of surface elements, number of elements along the chord and number of elements along the semi-span
Q	Velocity vector at a field point
R	$\|\mathbf{x} - \xi\|$ scalar distance between a field and a source point
$S(x,t)$	The surface of the wing
s	Distance travelled by the wing in half-chord; chord-wise coordinate along the surface
t; Δt	Dimensionless time or time; time increment
U(t)); U_0	Freestream velocity; a reference velocity
$W(x,t)$	The surface of the wake
x, ζ; x, y, z	Coordinate vector of a point; its components
α	Angle of incidence
Γ	Vortex strength, circulation
θ	Angle
σ; σ_x; σ_{xx}	Doublet strength; its derivatives
ϕ	Velocity potential

Subscripts

o	Centroid of a surface element
i, j, k	Dummy indices; refer to element i, j, or k
x, y, z	Refer to the x, y and z directions
l	Lower surface
u	Upper surface

Superscripts

s	Step doublet distribution or the surface of the wing
w	Surface of the wake

11.1 Introduction

We will consider a more general unsteady aerodynamic model, but will be limited in the sense that in deals with linearized model, as well as irrotational flow. In the calculation of flutter boundaries, unsteady aerodynamic forces for harmonic or nearly harmonic motions are of utmost importance. Therefore, in this section the discussion will be focused on the simplest unsteady aerodynamic model: Harmonically oscillating thin airfoil in ideal inviscid incompressible flow. Attention is given to physical aspects, and no rigorous mathematical treatment will be given. In spite of the simplicity of the model, the results can be used in "strip theory" approach during aircraft design.

11.2 Formulation of Unsteady Flow Problem

Consider an oscillating airfoil depicted in the following sketch, Fig. 11.1.

11.2.1 Irrotational Flow

(x, z) axes fixed to mean airfoil position.

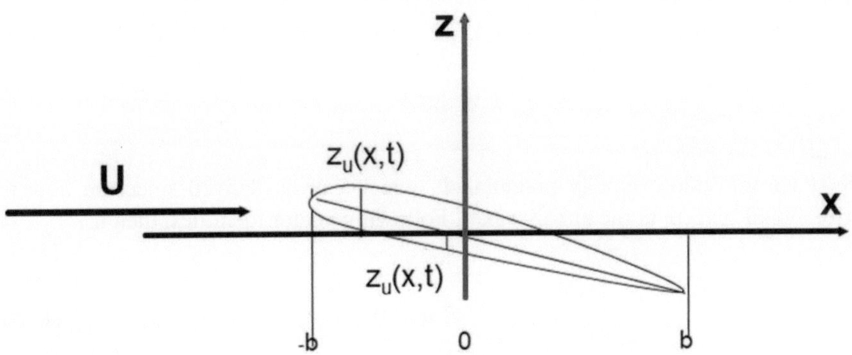

Fig. 11.1 Schematic of a linearized aerodynamic approach to harmonically oscillating thin airfoil in ideal inviscid incompressible flow

11.2.2 Basic Flow Equation[1]

Continuity:

$$div\ \mathbf{q} = 0 \tag{11.1}$$

where $\mathbf{q} = (u, w)$ is the velocity vector

Dynamic equation:

$$\frac{d\mathbf{q}}{dt} + \frac{1}{\rho} grad\ p = 0 \tag{11.2}$$

where p—static pressure

ρ—density.

Here:

$\frac{d}{dt}$ is substantial (or total) derivative.

The substantial (or total) derivative can be written when an observer is following a fluid particle

$$\begin{aligned} \frac{d}{dt} &= \frac{\partial}{\partial t} + u\frac{\partial}{\partial x} + v\frac{\partial}{\partial y} + w\frac{\partial}{\partial z} \quad \text{or} \\ \frac{D}{Dt} &= \frac{\partial}{\partial t} + \mathbf{q}\cdot\nabla \quad \text{or} \\ \frac{D}{Dt} &= \frac{\partial}{\partial t} + \mathbf{q}\cdot \text{grad} \end{aligned} \tag{11.3}$$

Next we introduce velocity potential Φ, which can be defined since the flow is irrotational (except at the airfoil wake). Following vector identities, then if

$$\nabla \otimes \mathbf{q} = 0 \tag{11.4}$$

It implies that

$$\nabla \otimes (\nabla\Phi) = 0 \tag{11.5}$$

[1] In the present book q and Q, as well as **q** and **Q**, are alternatively used to denote fluid velocity vector, which by context can be distinguished from V as volume. Similarly, the vector cross product is represented by either $\times$ or $\otimes$. Also, alternatively the substantial derivative $\frac{d}{dt}$ and $\frac{D}{Dt}$ are alternatively utilized.

Since in three dimensions

$$\nabla \otimes (\nabla \Phi) = \begin{vmatrix} \mathbf{i} & \mathbf{j} & \mathbf{k} \\ \frac{\partial}{\partial x} & \frac{\partial}{\partial y} & \frac{\partial}{\partial z} \\ \frac{\partial \Phi}{\partial x} & \frac{\partial \Phi}{\partial y} & \frac{\partial \Phi}{\partial z} \end{vmatrix}$$
$$= \mathbf{i}\left(\frac{\partial^2 \Phi}{\partial y \partial z} - \frac{\partial^2 \Phi}{\partial z \partial y}\right) + \mathbf{j}\left(\frac{\partial^2 \Phi}{\partial z \partial x} - \frac{\partial^2 \Phi}{\partial x \partial z}\right) + \mathbf{k}\left(\frac{\partial^2 \Phi}{\partial x \partial y} - \frac{\partial^2 \Phi}{\partial y \partial x}\right) = 0 \tag{11.6}$$

Hence:

$$\mathbf{q} = \text{grad}\,\Phi \tag{11.7}$$

In the two-dimensional unsteady aerodynamic theory, we are concerned only with the dimensions x and z (and u and w).[2]

Next we will assume that moving airfoil induces small disturbances in the uniform free stream:

$$\Phi(x, z, t) = \underset{\text{uniform free stream}}{U} x + \underset{\text{disturbance potential}}{\phi} (x, z, t) \tag{11.8}$$

where

$$\frac{d\mathbf{q}}{dt} + \frac{1}{\rho}\, grad\, p = 0$$
$$\phi_x = u$$
$$\phi_y = w$$
$$(u, w) \ll U$$

This assumption enables linearization.

After substituting Φ, then we will obtain the following simplified equations:

Continuity

$$\frac{\partial^2 \phi}{\partial^2 x} + \frac{\partial^2 \phi}{\partial^2 y} = 0 \text{ Laplace's equation} \tag{11.9}$$

and after linearization

[2] Or equivalent expression in polar or cylindrical coordinates, u_r for the radial component and u_θ; for the tangential component of the velocity.

Dynamics

$$p - p_\infty = \rho\left(\frac{\partial \phi}{\partial t} + U\frac{\partial \phi}{\partial x}\right) = 0 \text{ Bernoulli's equation} \tag{11.10}$$

11.2.3 Boundary Condition

The contour of the airfoil is given by

$$F(x, z, t) = 0 \tag{11.11}$$

or

$$F_u = z - z_u(x, t) \text{ upper surface} \tag{11.12a}$$

$(z_u, z_l) << b$

$$F_l = z - z_l(x, t) \text{ lower surface} \tag{11.12b}$$

Boundary condition at contour:

$$\frac{dF}{dt} = \frac{\partial F}{\partial t} + (U + u)\frac{\partial F}{\partial x} + w\frac{\partial F}{\partial z} \tag{11.13}$$

That means that the fluid elements touch continuously the contour $F = 0$. Introduction of F_u and F_l, and linearization yields

$$w_u\left(x, 0^+, t\right) = \frac{\partial z_u}{\partial t} + U\frac{\partial z_u}{\partial x} \tag{11.14a}$$

$$w_l\left(x, 0^+, t\right) = \frac{\partial z_l}{\partial t} + U\frac{\partial z_l}{\partial x} \tag{11.14b}$$

Physical meaning (Fig. 11.2):

Linearization implies further that the boundary conditions (BC) are imposed at x axis.

Separation of boundary condition into steady and unsteady parts (Fig. 11.3)

$$z_u(X, t) = z_t(X) + z_{as}(X) + z_{ai}(X, t) \tag{11.15a}$$

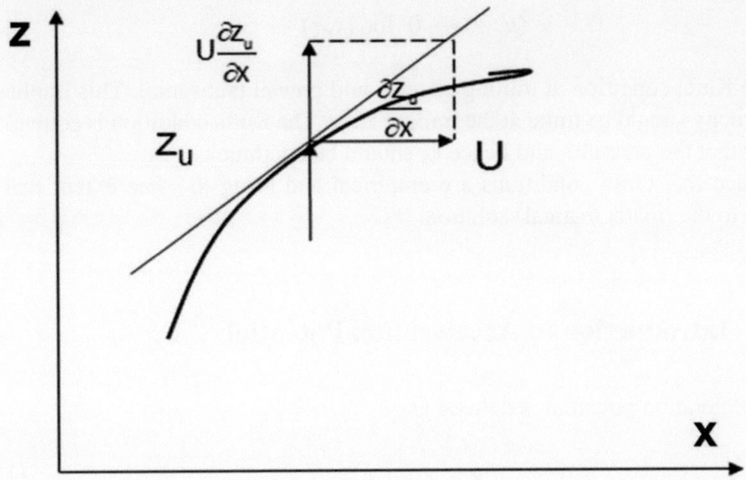

Fig. 11.2 Flow components at an element of an airfoil surface in linearized aerodynamics

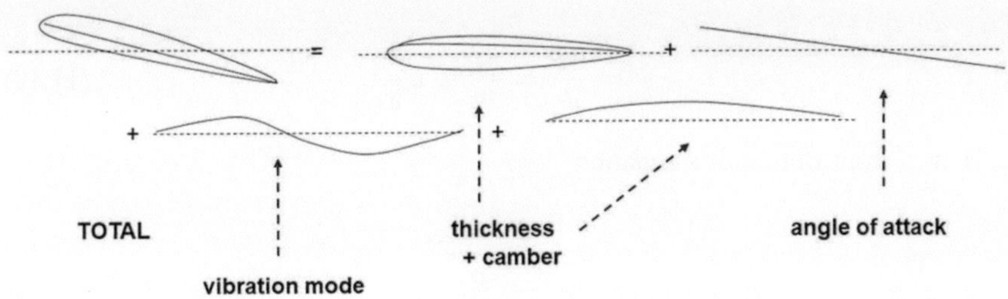

Fig. 11.3 Linear superposition of the components of an airfoil in linearized aerodynamics. *Note* Boundary conditions remain imposed on the x-axis

$$z_l(X, t) = -z_t(X) + z_{as}(X) + z_{ai}(X, t) \tag{11.15b}$$

11.2.4 Unsteady Boundary Condition

(the index i for the unsteady part will now on be omitted)

$$\left.\frac{\partial \varphi(x, z, t)}{\partial z}\right|_{z=0} = w_a(x, t) = \frac{\partial z_a}{\partial t} + U\frac{\partial z_a}{\partial x} \tag{11.16}$$

Next we proceed to the discussions on unsteady flow problem. The boundary condition at infinity:

$$(u, w) \to 0 \text{ for } (x, z) \to \infty \tag{11.17}$$

The Kutta condition at trailing edge should prevail (subsonic). This implies that the velocity should be finite at the trailing edge. The Kutta condition is equivalent to stating that the pressure, and hence ϕ, should be continuous.

Notice that these conditions are empirical and bring to some extent real flow effects to the (mathematical) solution.

11.3 Introduction to Acceleration Potential

The acceleration potential is defined as

$$\mathbf{a} \equiv grad\ \psi \equiv \nabla \psi \tag{11.18}$$

ψ satisfies Laplace's equation.

Relation with velocity potential

$$\psi = \frac{d\phi}{dt} = \frac{\partial \phi}{\partial t} + u\frac{\partial \phi}{\partial x} \tag{11.19}$$

Hence, from Bernoulli's equation

$$\psi = \frac{P - P_\infty}{\rho} \tag{11.20}$$

Due to this fact, ψ is also called pressure potential.

It should also be noted that when the equations of fluid motion can be simplified to the form[3]:

$$\mathbf{a} = \frac{D\mathbf{Q}}{Dt} = \frac{\partial \mathbf{Q}}{\partial t} + (\mathbf{Q} \cdot \nabla)\mathbf{Q} = \nabla\left[\Omega - \int \frac{dp}{\rho}\right] \tag{11.21}$$

It follows that

$$\nabla \times \mathbf{a} = 0 \tag{11.22}$$

Noting that, in irrotational flow,

$$\nabla \times \mathbf{Q} = 0 \tag{11.4}$$

[3] One could also write as $\mathbf{a} = \frac{D\mathbf{Q}}{Dt} = \frac{\partial \mathbf{Q}}{\partial t} + (\mathbf{Q} \cdot \nabla)\mathbf{Q} = \rho\, \mathbf{g} - \frac{\nabla p}{\rho}$. In addition, one could also use V instead of Q. Hence $\mathbf{a} = \frac{D\mathbf{V}}{Dt} = \frac{\partial \mathbf{V}}{\partial t} + (\mathbf{V} \cdot \nabla)\, \mathbf{V} = \nabla\left[\Omega - \int \frac{dp}{\rho}\right]$.

Then

$$\mathbf{Q} = \nabla\Phi \tag{11.5}$$

since $\nabla \times \mathbf{Q} = 0$ or $\nabla \times \nabla\Phi = 0$

Therefore one can conclude that as a consequence of (11.24), then

$$\mathbf{a} \equiv grad\ \psi \equiv \nabla\ \psi \tag{11.18}$$

Here $\psi(x, y, z, t)$ or $\psi(x, t)$ or $\psi(r, t)$ is a scalar function that is known as the acceleration potential. Consequently one could define

$$\psi = \Omega - \int \frac{1}{\rho} dp + G(t) \tag{11.23}$$

where $G(t)$ is a function of time that is usually non-essential.

11.4 Case Study I: Calculation of Nonlinear, Unsteady Lifting Potential Flow Problems[4]

Summary

A numerical method is developed for attacking lifting problems of a general three-dimensional wing executing an arbitrary motion in a potential flow. The formulation of the problem is exact in the sense that the boundary conditions are satisfied on the surface of the wing and the geometry of the time-dependent wake is predicted. The problem is then governed by a Fredholm integral equation of the first kind which relates the velocity potential doublet distribution to the normal velocity on the surface of the wing, subject to the Kutta condition of smooth flow off the trailing edge and the condition that the wake cannot sustain any pressure difference. The solution of the integral equation is obtained in a step-by-step fashion, and the appearing surface integrals are evaluated by employing a discrete set of approximate surface elements and assumption of the doublet distribution over the element which is based on a formal expansion of the velocity influence coefficient. Verification of the method is made by comparison with solutions of classical problems. To demonstrate its utility, the method is applied to obtain the airloads due to transient motion of wings with simple planforms. The nonlinear effects due to the wake, aspect ratio, and thickness are discussed. Results of these calculations indicate that the present method can be applied to treat more complicated geometries and motion to obtain accurate and efficient prediction of the airloads.

---

[4] Djojodihardjo, RH and Widnall, SE, Numerical Method for the Calculation Nonlinear Unsteady Lifting Potential Flow Problem AIAA 7,10,1969.

11.4.1 Introduction

The need for a more rigorous approach to the treatment of the three-dimensional flow problems involving complex motions and flow geometries derives from recent progress in V/STOL aircraft and hydrofoil craft technology. Many basic approximations inherent in classical aerodynamic theory such as the linearized boundary conditions, small perturbation velocities, and the concept of planar and rigid wake in the projected plane of the wing need careful examination when the downwash velocity associated with the generation of lift is large compared with the flight velocity. Experimental results by Greidanus, van de Vooren and Bergh [5], Ransleben and Abramsomn [6] and Spurck [7] indicate discrepancies in the forces and moments from those predicted by linearized theories, which are attributable to the thickness of the airfoil and the viscosity of the flow.

Before any theory that accounts for the effect of viscosity may be developed, it is necessary to have available methods for predicting the pressure distribution on bodies of arbitrary geometries which execute arbitrary motions.

Refinements of the linearized two-dimensional unsteady-potential flow theory have been suggested by Küssner and von Gorup [8], Hewson-Browne [9], van de Vooren and van de Vel [10], Giesing [11], and Chen and Wirtz [12]. Although rapid progress has been achieved in the linearized lifting surface theory for planar and non-planar configurations in steady and unsteady flows, as is well summarized in review papers by Ashley et al. [13] and Landahl and Stark [14] comparatively less attention has been devoted to the case of the deforming wake and thickness effects in the more significant three-dimensional situations. Attempts to account for the deforming unsteady wake have been made in the calculation of helicopter rotor airloads, such as the works of Miller [15, 16] which involve a combined analytical and numerical procedure involving lifting line and lifting surface approximations. The progress in high-speed digital computer technology has made possible a numerical approach with a more realistic flow model, an approach which may be termed exact in the sense that true boundary locations are accounted for and the exact solution is attained uniformly as some computation network is refined.

Such an approach has been suggested by Ashley [17] and Ashley et al. [13] to account for the unsteady distorted wake, thickness and flow geometry. The potentialities of purely numerical approaches have been demonstrated by several investigators. Hess and Smith [18] developed a very general method for calculating the non-lifting potential flow about arbitrary bodies. Giesing 7 developed a method for calculating the lifting two-dimensional flow about airfoils of arbitrary geometries, and Rubbert et al. [19] developed a method to calculate the steady lifting flow about a fan-in-wing configuration including boundary-layer effects.

Obviously a numerical method to solve the unsteady, nonlinear potential flow problems associated with a wing of arbitrary geometry executing arbitrary motion should now be developed. With the foregoing in mind, the objectives of the present work can be enunciated in three parts: (1) to perform an exact numerical analysis for calculating the unsteady airloads and for predicting the geometry of the wake, (2) to verify the method through comparison with classical solutions and (3) to apply the method to some unsteady (transient) problems and to investigate the nonlinear effects due to the deforming wake (hence the time history of the motion), the effects of thickness and the effects of three-dimensionality of the flowfield.

11.4.2 Formulation of the Problem

Consider a lifting system configuration consisting of a three-dimensional body of arbitrary geometry which executes an arbitrary translation in an incompressible ideal fluid, as depicted in Fig. 11.4a. Let the surface of the lifting body be represented by

$$S(\mathbf{x}, t) = 0 \tag{11.24}$$

and the wake following the body be defined by a surface of velocity discontinuity given by

$$W(\mathbf{x}, t) = 0 \tag{11.25}$$

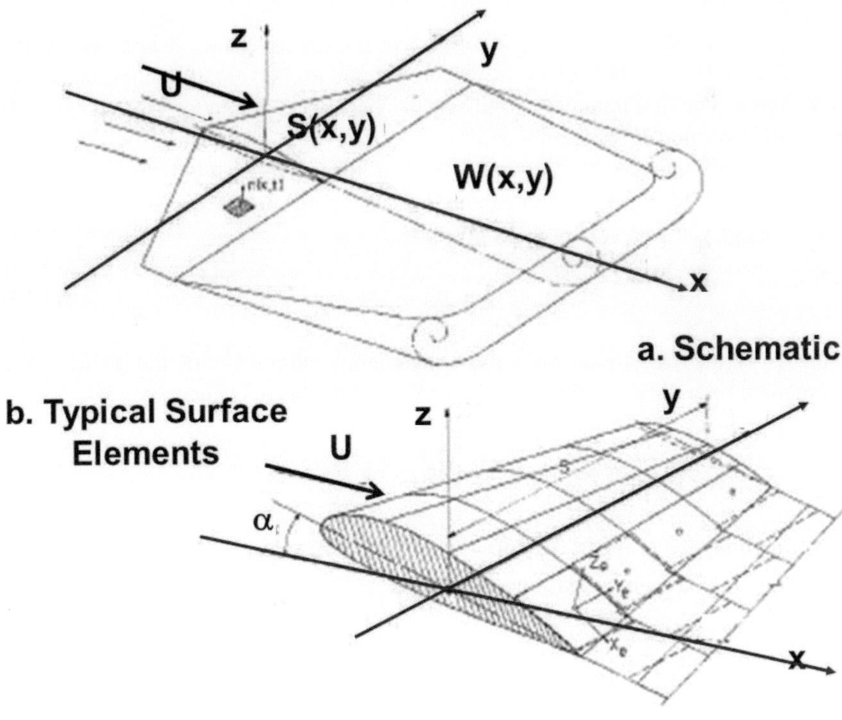

Fig. 11.4 Lifting system configuration

which is of zero thickness.

The velocity potential for this flow then satisfies Laplace's equation and is subject to the following boundary conditions: (1) quiescent fluid at infinity, (2) the kinematical flow tangency condition, which states that the fluid and the surface with which contact is maintained must have the same normal velocity, (3) the Kutta–Joukowski condition of smooth flow off the trailing edge and (4) the dynamical condition of zero pressure discontinuity across the wake sheet. These conditions are sufficient to ensure the existence of a unique velocity potential. The problem posed would be to find the velocity potential and the wake geometry for a specified wing in unsteady motion. For convenience, the coordinate frame of reference is fixed on the body. Further, all quantities are rendered dimensionless by division with appropriate characteristic quantities such as chord length 2c, reference velocity U_0, or a combination of both. By use of Green's theorem, it can be shown [17, 20] that the velocity potential on the surfaces S and W and the region external to them is given by

$$\begin{aligned}\phi(\mathbf{x}, t) &= \frac{1}{4\pi}\iint_S \sigma(\boldsymbol{\xi}, t)\frac{\partial}{\partial\nu(\boldsymbol{\xi}, t)}\left[\frac{1}{R(\mathbf{x}, \boldsymbol{\xi}, t)}\right]dS \\ &= \frac{1}{4\pi}\iint_S \Delta\phi_u(\boldsymbol{\xi}, t)\frac{\partial}{\partial\nu(\boldsymbol{\xi}, t)}\left[\frac{1}{R(\mathbf{x}, \boldsymbol{\xi}, t)}\right]dS\end{aligned} \tag{11.26}$$

where σ is the potential doublet strength over the entire surface S and $\Delta\phi u$ is the doublet strength over the surface W, which is equal to the velocity potential discontinuity there. The first boundary condition has been incorporated in Eq. (11.26). ν is the outward normal at the source point $(\boldsymbol{\xi}, t)$ and $R = |\mathbf{x} - \boldsymbol{\xi}|$ where $P(\mathbf{x}, t)$ is a field point. For P($\mathbf{x}$, t) on S or W, the first or second integral in Eq. (11.26), respectively, is singular, but proper evaluation can be made by a limiting process.

The velocity at any field point is given by

$$\mathbf{Q} = \mathbf{i}\mathrm{U} + \nabla\phi \tag{11.27}$$

where U is the freestream velocity, conveniently aligned with the x axis. The kinematical boundary condition on the surface of the body is given by

$$\frac{dS}{dt} = \frac{\partial S}{\partial t} + (U\mathbf{i} + \nabla\phi)\cdot\nabla S \text{ on } S(\mathbf{x}, t) = 0 \tag{11.28}$$

Noting that $\frac{\nabla S}{|\nabla S|} \equiv \mathbf{n}$ and $\frac{1}{|\nabla S|}\frac{\partial S}{\partial t} = \frac{\partial \mathbf{n}}{\partial t}$, the dimensionless velocity of S normal to itself,[5] Eq. (11.28) can be rearranged to yield

$$\frac{\partial\phi}{\partial n} = -U\mathbf{i}\cdot\mathbf{n} - \frac{\partial\mathbf{n}}{\partial t} \text{ on } S(\mathbf{x}, t) = 0 \tag{11.29}$$

---

5 $\frac{1}{|\nabla S|}\frac{\partial S}{\partial t} \equiv \frac{\partial\left(\frac{\nabla S}{|\nabla S|}\right)}{\partial t} \equiv \frac{\partial\mathbf{n}}{\partial t}$.

Substituting Eq. (11.26) into Eq. (11.29) and rearranging, we obtain

$$\frac{1}{4\pi}\iint_S \sigma(\boldsymbol{\xi}, t)\frac{\partial^2}{\partial n \partial \nu}\left[\frac{1}{R(\mathbf{x}, \boldsymbol{\xi}, t)}\right]dS = -U\mathbf{i}\cdot\mathbf{n} - \frac{\partial n}{\partial t}$$
$$- \frac{1}{4\pi}\iint_S \Delta\phi_u(\boldsymbol{\xi}, t)\frac{\partial^2}{\partial n \partial \nu}\left[\frac{1}{R(\mathbf{x}, \boldsymbol{\xi}, t)}\right]dS \tag{11.30}$$

which is a singular Fredholm integral equation of the first kind and comprises the ensuing equation for the velocity potential field of the flow. Since the integral involves a single kernel, the coefficient of the doublet strength of the integral, evaluation should be made in the principal value sense [21, 22].

11.4.3 Dynamical Condition Governing the Wake

From momentum consideration, the force acting on the lifting configuration consisting of *S* and *W* is given by

$$\mathbf{F} = \frac{D}{Dt}\left(\iint_S \phi\mathbf{n}dS + \iint_{W_u} \Delta\phi_w\mathbf{n}\,dS\right) \tag{11.31}$$

Since the wake cannot sustain any pressure difference across it,

$$\mathbf{F} = \frac{D}{Dt}\left(\iint_{W_u} \Delta\phi_w\mathbf{n}dS\right) = 0 \tag{11.32}$$

which can be further reduced to

$$\frac{D}{Dt}\{\Delta\phi_w\} = 0 \tag{11.33}$$

That is, once a fluid particle in the wake is imparted a doublet strength $\Delta\phi_w$, this value remains constant as it is convected downstream. This condition is equivalent to Kelvin's theorem.

11.4.4 The Kutta–Joukowski Condition and the Generation of the Wake

A general statement known as the extended Kutta–Joukowski condition implies that the rear dividing streamline leaves the airfoil at the trailing edge. Its tangent at the trailing edge, in general, passes through the interior of the airfoil [23]. The Kutta–Joukowski condition was originally applied in a two-dimensional steady flow in order to obtain a finite velocity at the trailing edge, and as a consequence, the flowfield is uniquely determined. This hypothesis implicitly accounts for the viscous effects otherwise neglected in inviscid theory. In the classical unsteady flow calculations, the Kutta–Joukowski condition is applied at any instant [24] or for each set of wake vortex and its image inside the circle (in the circle plane) [25].

In the present inviscid approach for a three-dimensional unsteady flow, the foregoing general statement is applied. Since the wake then emerges from the trailing edge, the dynamical condition 11.32 governing the wake also applies at the trailing edge. This condition can be shown [20] to be equivalent to the finiteness of the velocity at the trailing edge, provided no discontinuous motion of the airfoil takes place. As another consequence, $\Delta\phi$ is continuous through the trailing edge, as also mentioned in [13, 18]. For a thick body with a sharp trailing edge, $\Delta\phi$ at the trailing edge can be defined in the limiting sense as the difference between the values of the doublet strengths at the upper and lower surfaces there; thus

$$\sigma_{TE} = \lim_{x \to TE} (\sigma_u - \sigma_l) \tag{11.33}$$

The velocity of the wake shedding at the trailing edge is determined by

$$\mathbf{Q}_{\mathrm{TE}} = U\mathbf{i} + (\nabla\phi)_m \tag{11.34}$$

where

$$(\nabla\phi)_m = \frac{1}{2}\left[(\nabla\phi)_u + (\nabla\phi)_l\right] \tag{11.35}$$

11.4.5 Pressure, Forces and Moment

From the Bernoulli equation and the use of expression 11.27, the pressure coefficient can be written as

$$c_p = -2U\left(\frac{x}{U_0}\frac{\partial U}{\partial t} + \frac{\partial\phi}{\partial t}\right) - (2iU \cdot \nabla\phi + \nabla\phi \cdot \nabla\phi) \tag{11.36}$$

For a constant freestream velocity, $U = U_0$ and Eq. (11.36) is further simplified. The lift, drag and moment coefficients can be obtained by the usual procedure of integrating the vertical and horizontal components and the moments of these components with respect to the origin, respectively, over the surface of the wing and appropriately divided by the largest projected area of the wing or by the product of this area with the root chord of the wing, as appropriate.

An alternative formula can be derived to calculate the lift, employing the velocity potential distribution directly. Taking the vertical component of Eq. (11.31) and the use of expression 11.27, the lift on the wing can be shown to be

$$C_L = \frac{1}{U_\infty^2 A}\left(\frac{\partial}{\partial t}\iint\limits_S \phi(\mathbf{n}\cdot\mathbf{k})dS + U\iint\limits_S \frac{\partial\phi}{\partial s}(\mathbf{n}\cdot\mathbf{k})\mathrm{d}S\right) \tag{11.37}$$

which can be evaluated by numerical means upon knowing the values of ϕ at any point on the surface S. Since it is not necessary to calculate the pressure coefficients, rapid computation results. Similar formulas can be derived for the drag and moment coefficients. The first integral on the right-hand side of Eq. (11.36) represents the unsteady part, and the second has the same form as in steady flow.

To obtain the starting lift for impulsive motion, for which the accuracy of a numerical differentiation is difficult to assess, an alternative method is suggested below. For this purpose, we could write

$$\phi(\mathbf{x}, t) = PO[B(\mathbf{x}, t)] \tag{11.38}$$

in the absence of the wake. The notation PO is the potential operator as employed[6] [11]

$$C_{Lm} = \frac{1}{U_\infty^2 A}\frac{\partial}{\partial t}\iint\limits_S \phi dS = \frac{1}{U_\infty^2 A}\iint\limits_S \frac{\partial\phi}{\partial t}dS \tag{11.39}$$

Since the potential operator as implied in [26] is a linear one, then

$$\frac{\partial\phi(\mathbf{x}_j)}{\partial t} = \frac{\partial}{\partial t}\left[\mathrm{PO}\left[B(\mathbf{x}_j, t)\right]\right] = \mathrm{PO}\left\{\frac{\partial}{\partial t}\left[B(\mathbf{x}_j, t)\right]\right\} \tag{11.40}$$

which should be used in Eq. (11.39). Equation (11.40) can be evaluated according to a numerical scheme developed subsequently to evaluate ϕ in 11.38.

Hence this procedure is similar to solving the boundary value problem associated with the flow situation at the starting instant, where the boundary condition of flow tangency at the surface of the body should be replaced by its time derivative. For an impulsively started elliptical planform, this procedure yields a result that agrees with

[6] U_0 or U_∞ is chosen here as appropriate for physical plausibility and for convenience.

that of Jones [27]. This procedure can also be employed to calculate the unsteady term due to discontinuous motion.

11.4.6 Method of Solution

The problem of obtaining the velocity potential field of the ensuing flow has been reduced to an integral equation which relates the velocity potential doublet distribution on the surface of the lifting system configuration to the normal induced velocity variation on the surface of the wing, as expressed by Eq. (11.30), and which is subject to the dynamical condition 11.24 on the wake sheet and the Kutta–Joukowski condition of smooth flow off the trailing edge. The velocity potential field is then determined from Eq. (11.26). The integral Eq. (11.30) is nonlinear in the sense that solutions for various flows cannot be superimposed. The nonlinearity arises from the fact that the doublet distribution on the wake as well as the wake location is implicitly dependent on the doublet distribution on the surface of the wing, since the location of the wake sheet at any instant is a function of the previous velocity potential field. In addition, the normal component of the free stream velocity to the surface of the wing—the first term on the right-hand side of Eq. (11.30)—is exact rather than linearized as in most lifting surface theories.

To solve Eq. (11.30), two problems immediately arise: (1) the determination of the geometry of the wake $W(x, t)$ and (2) the solution of, the integral equation for known right-hand side terms. One possible approach is to employ a step-by-step procedure, so that at each step, with the aid of condition (9) and the Kutta–Joukowski condition, the terms on the right-hand side of 30 would be known. Solution of the integral equation then becomes a matter of numerical analysis.

11.4.7 Step-by-Step Procedure

At the initial instant of motion, no wake exists, and hence no wake term appears on the right-hand side of Eq. (11.30) and the remaining equation can be solved for $a(x, i)$. To solve for the situation at successive instants, an assumption could be made regarding the variation of the velocity field during a small time increment Δt. Then the location of a wake element emerging off the trailing edge can be predicted from the following set of equations:

$$\mathbf{x}_w(t_0, t) = \int_{t_0}^{t} \mathbf{Q}(\mathbf{x}, t_i) dt_k + \mathbf{x}_w(t_0, t_i) \tag{11.41}$$

$$\mathbf{Q}(\mathbf{x}, t) = U\mathbf{i} + \nabla_s \phi_{\text{wm}}(\mathbf{x}, t) + \mathbf{n}\frac{\partial \phi(\mathbf{x}, t)}{\partial n} \tag{11.42}$$

where ϕ_{wm} is the mean velocity potential on the wake sheet and the subscript s on the gradient operator indicates that the operation is performed along the wake surface, and where

$$\frac{\partial\phi(\mathbf{x},t)}{\partial n} = \frac{1}{4\pi}\oiint_S \sigma(\boldsymbol{\xi},t)\frac{\partial^2}{\partial n(\mathbf{x},t)\partial\nu(\boldsymbol{\xi},t)}\left[\frac{1}{R(\mathbf{x},\boldsymbol{\xi},t)}\right]dS$$
$$+\frac{1}{4\pi}\iint_W \Delta\phi_w(\boldsymbol{\xi},t)\frac{\partial^2}{\partial n(\mathbf{x},t)\partial\nu(\boldsymbol{\xi},t)}\left[\frac{1}{R(\mathbf{x},\boldsymbol{\xi},t)}\right]dS \tag{11.43}$$

Obviously, ϕ_{wm} can be defined and Eq. (11.43) can be used after the wake has emerged. The wake elements carry with them the values of $\Delta\phi_w$ imparted at the instant of shedding, which is numerically set equal to the velocity potential jump at the trailing edge at that instant (Eq. 11.33). Each value of $\Delta\phi_w$ is fixed for any wake element of fixed identity, according 11.43. Setting t_0 equal to the instant of shedding for each wake element, then

$$\mathbf{x}_w(t_0,t_i) \equiv \mathbf{x}_{\mathrm{TE}} \tag{11.44}$$

Hence, at each time step, all terms on the right-hand side of Eq. (11.30) are again known. By repeated use of such procedure, the transient loading of any motion can be obtained. The steady or periodic loading is obtained after all transient effects subside. Clearly, such a procedure is best suited for solving transient problems.

11.4.8 Numerical Method

It is convenient to cast the integral Eq. (11.30) into the following form:

$$[A_{ij}]\{\sigma_i\} = \{B_i\} - \left[A_{kj}^{(w)}\right]\{\Delta\phi_{wk}\} \tag{11.45}$$

incorporating a discrete set of surface elements to represent the lifting surface configuration. Here

$$A_{ij} = \frac{1}{4\pi}\iint_{el.i}\frac{\sigma(\boldsymbol{\xi},t)}{\sigma(\mathbf{x}_0,t)}\frac{\partial^2}{\partial n(\mathbf{x},t)\partial\nu(\boldsymbol{\xi},t)}\left[\frac{1}{R(\mathbf{x},\boldsymbol{\xi},t)}\right]dS \tag{11.46}$$

is the velocity influence coefficient, representing the velocity induced at a field point $P(x_j,t)$ due to doublet distribution over the surface element i and x0i is a representative point on the element. Further,

$$B_j = -U\mathbf{i}\cdot\mathbf{n}(\mathbf{x}_j, t) - \frac{\partial n(\mathbf{x}_j, t)}{\partial t} \tag{11.47}$$

The solution can be obtained by inverting the matrix of influence coefficient A; thus

$$\{\sigma\} = [A]^{-1}\left\{B_i'\right\} \tag{11.48}$$

where

$$\left\{B_i'\right\} = \{B_i\} - \left[A_{kj}^{(w)}\right]\{\Delta\phi_{wk}\} \tag{11.49}$$

The velocity potential ϕ is then given by

$$\{\phi_s\} = \left[C_{ij}\right]\left\{\sigma_j\right\} + \left[C_{ik}^{(w)}\right]\{\Delta\phi_{wk}\} \tag{11.50}$$

and

$$\{\phi_{\text{wm}}\} = \left[C_{ij}^{(\text{sw})}\right]\left\{\sigma_j\right\} + \left[C_{ik}^{(\text{ww})}\right]\{\Delta\phi_{\text{wk}}\} \tag{11.51}$$

where

$$C_{ij} = \frac{1}{4\pi}\iint\limits_{el.i} \frac{\sigma(\boldsymbol{\xi}, t)}{\sigma(\mathbf{x}_0, t)}\frac{\partial}{\partial \nu(\boldsymbol{\xi}, t)}\left[\frac{1}{R(\mathbf{x}, \boldsymbol{\xi}, t)}\right]dS \tag{11.52}$$

Combining Eqs. (11.48) and (11.50), the potential operator employed in the preceding section can be defined. Hence

$$\phi(\mathbf{x}_j, t) = \text{PO}\left(\left\{B(\mathbf{x}_j, t)\right\} - \left[A_{\text{kj}}^{(w)}\right]\{\Delta\phi_{\text{wk}}\}\right) + \left[C_{\text{kj}}^{(w)}\right]\{\Delta\phi_{\text{wk}}\}j \tag{11.53}$$

which reduces to 11.38 in the absence of the wake.

To evaluate the integrals 11.46 and 11.52, three processes are involved. These are the simplification of the geometry of the surface elements and the proper selection of the doublet distribution function, both for the purpose of analytically evaluating these integrals and efficient computation, and a suitable choice of control and source points required for best accuracy. The source point is defined as a representative point at a surface element, at which the value of the doublet distribution is sought, and a control point is defined as a point at which the flow tangency condition is to be satisfied. In some methods, these sets of points belong to a single set, generally referred as the collocation points, whereas in other methods, distinct sets of points are utilized, such as the familiar set in lifting line theory, in which the vortex line is specified on the quarter-chord points and the downwash velocities are evaluated at the three-quarter-chord points. Since the kernel functions of the integrals 11.46 and

11.52 are singular, care should be exercised in the selection of the doublet distribution functions and the control and source points.

One possible way to simplify the geometry of the surface elements is to employ planar quadrilaterals, using a procedure essentially similar to that of Ref. [26]. The characteristic dimension of such an approximate element should be much smaller than the local radius of curvature of the original surface element. Figure 11.4b illustrates a typical arrangement of surface elements. Further approximation by rectangular surface elements may also be possible and is in fact desirable in view of its symmetrical properties and simplified analytical evaluation of the integrals 11.46 and 11.52.

11.4.9 Systematic Expansion of the Velocity Influence Coefficient

To choose a proper doublet distribution function over each of the surface elements, attention is focused on the far-field and near-field behavior of the velocity influence coefficient and means for efficient, yet accurate, simplification.

A formal analysis can be performed by Taylor series expansion of σ and the velocity kernel function $\frac{\partial^2}{\partial n(\mathbf{x},t)\partial \nu(\boldsymbol{\xi},t)}\left[\frac{1}{R(\mathbf{x},\boldsymbol{\xi},t)}\right]$. Hence

$$\begin{aligned}\sigma(x) = \sigma(0) &+ x\left.\frac{\partial\sigma}{\partial x}\right|_{x=0} + y\left.\frac{\partial\sigma}{\partial y}\right|_{x=0} + \frac{1}{2}x^2\left.\frac{\partial^2\sigma}{\partial x^2}\right|_{x=0}\\ &+ \frac{1}{2}y^2\left.\frac{\partial^2\sigma}{\partial y^2}\right|_{x=0} + xy\left.\frac{\partial^2\sigma}{\partial x\partial y}\right|_{x=0} + \ldots\end{aligned} \tag{11.54}$$

and

$$\begin{aligned}K_\nu(\mathbf{x},\boldsymbol{\xi},t) = K_\nu(\mathbf{x},0,t) &+ x\left.\frac{\partial K_\nu}{\partial x}\right|_{x=0} + y\left.\frac{\partial K_\nu}{\partial y}\right|_{x=0} + \frac{1}{2}x^2\left.\frac{\partial^2 K_\nu}{\partial x^2}\right|_{x=0}\\ &+ \frac{1}{2}y^2\left.\frac{\partial^2 K_\nu}{\partial y^2}\right|_{x=0} + xy\left.\frac{\partial^2 K_\nu}{\partial x\partial y}\right|_{x=0} + \ldots\end{aligned} \tag{11.55}$$

Here we have employed an element coordinate system with its origin at the source point of the surface element being considered. For a field point in the far field, substituting 11.54 and 11.55 into 11.46 and retaining terms up to the second order yields the following expression:

$$A_{ij} = \frac{A}{4\pi}\left[\begin{array}{l} \left\{\begin{array}{l} 1 + \overline{\sigma}_x \dfrac{M_x}{A} + \overline{\sigma}_y \dfrac{M_y}{A} + \dfrac{1}{2}\overline{\sigma}_{xx}\dfrac{I_{xx}}{A} \\ + \overline{\sigma}_{xy}\dfrac{I_{xy}}{A} + \dfrac{1}{2}\overline{\sigma}_{yy}\dfrac{I_{yy}}{A} \end{array}\right\} K \\ K_x\left\{\dfrac{M_x}{A} + \overline{\sigma}_x\dfrac{I_{xx}}{A}\right\} + K_y\left\{\dfrac{M_y}{A} + \overline{\sigma}_y\dfrac{I_{yy}}{A}\right\} \\ + \dfrac{1}{2}K_{xx}\dfrac{I_{xx}}{A} + K_{xy}\dfrac{I_{xy}}{A} + \dfrac{1}{2}K_{yy}\dfrac{I_{yy}}{A} + \ldots \end{array}\right] \tag{11.56}$$

At the centroid of the surface element, $M_x = M_y = 0 = I_{xy}$ and to a second-order accuracy, we could set

$$A_{ij} = \frac{AK_v(x, 0, t)}{4\pi} \tag{11.57}$$

It is then suitable to locate the source point at the centroid of the surface element. Expression 11.32, in addition, can be interpreted as a multipole expansion, since the coefficients of $K_{v_x} = \frac{\partial}{\partial x}(K_v)$ and $K_{v_y} = \frac{\partial}{\partial y}(K_v)$ represent the strengths of quadrupoles at the source point, whereas the coefficients of $K_{v_{xx}} = \frac{\partial^2}{\partial x^2}(K_v)$ and $K_{v_{yy}} = \frac{\partial^2}{\partial y^2}(K_v)$ are octupoles, with appropriate orientations. The use of 11.32 or its approximation is dictated by the accuracy, as compared to a more exact formula derived for the near-field region. For the near-field region, we would employ 11.30 and then perform the integration 11.46 analytically.

Since additional unknowns represented by the derivatives of a are introduced in the process of matrix inversion 11.48, resort is made to a finite-difference version, so that these derivatives are expressed in terms of cr/s. For most applications, we could disregard the variation of a along one coordinate, so that in, general we could write Eq. (11.43) as

$$\sigma(x) = \sigma(0)\left(a + bx + cx^2 + \ldots\right) \tag{11.58}$$

and hence

$$A_{ij} = \frac{1}{4\pi}\iint\limits_{el.i} \left(a + bx + cx^2 + \ldots\right)K_v\left(x_j, x, t\right)dS \tag{11.59}$$

Since $K_v\left(x_j, x, t\right)$ is a singular function of the fifth power, the velocity influence integral 11.59 is singular for field points located in a region for which the doublet distribution function 11.58 or its slope (x-derivative) possesses a discontinuity in that region [20]. Obviously a control point should not be located at this region, and to avoid any such region, one has to represent the doublet distribution by a function that possesses no discontinuity up to its slope. In view of this argument, representation of cr by a function of order higher than two is not necessary. This argument is valid for representation of o- as a function of a single variable.

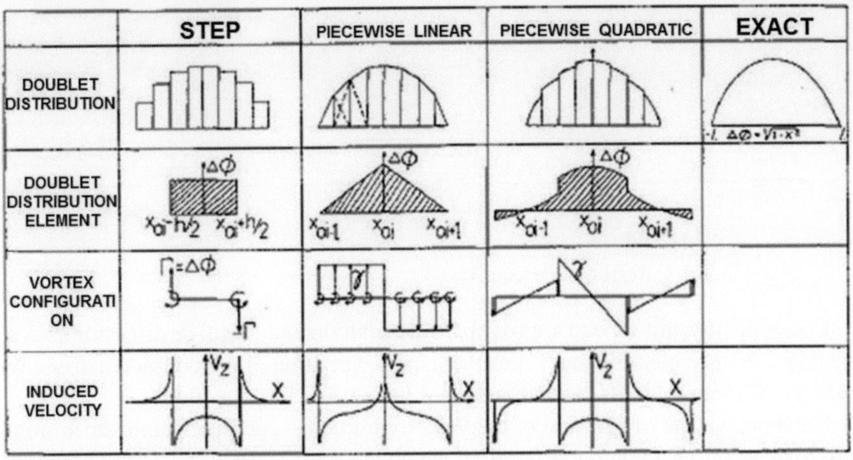

Fig. 11.5 Approximate doublet distribution functions[7]

Before proceeding to the three-dimensional problems, much could be gained from a study of the two-dimensional situations. Consider three finite-difference approximations of a based on 11.33, written in general forms as.

(1) $\sigma(x) = \alpha$ (step distribution, zeroth-order approximation)
(2) $\sigma(x) = \alpha + \beta x$ (piecewise linear distribution, first-order approximation)
(3) $\sigma(x) = \alpha + \beta x + \gamma \times 2$ (piecewise quadratic distribution, second-order approximation

where α, β and γ are coefficients numerically determined by finite-difference analysis and dependent on the values of σi 's in the neighborhood of each source point in question.

These distributions are illustrated in Fig. 11.5. The control points for the first and third distributions are located at the source points. No discontinuity in the slope of σ would result in the minimization of error, as can be verified by treating the near-field behavior in a way similar to that employed in the choice of the source point [20]. The piecewise linear distribution can be regarded as a superposition of elementary triangular distributions, as shown in Fig. 11.5. Since the vertices are points of slope discontinuity, and by locating the control point midway between the vertices unequal number of source and control points result (except for the steady-state case by imposing the Kutta–Joukowski condition at the trailing edge), the use of a piecewise linear distribution does not seem plausible and will not be considered further in the three-dimensional case.

The piecewise quadratic distribution is of interest. Such a distribution can be represented, between a region bounded by

---

7 Such approach is reminiscent of presently well-developed partition of unity distribution and shape functions in boundary element and finite element methods.

$$x_{0i} - \frac{1}{2}h \text{ and } x_{0i} + \frac{1}{2}h$$

By

$$\sigma = \frac{(x - x_{0i})(x - x_{0i+1})}{(x_{0i-1} - x_{0i})(x_{0i-1} - x_{0i+1})}\sigma_{i-1} + \frac{(x - x_{0i})(x - x_{0i+1})}{(x_{0i-1} - x_{0i})(x_{0i-1} - x_{0i+1})}\sigma_i + \frac{(x - x_{0i})(x - x_{0i+1})}{(x_{0i-1} - x_{0i})(x_{0i-1} - x_{0i+1})}\sigma_{i+1} \tag{11.60}$$

Consequently, this type of approximate distribution would still be discontinuous at the edges of each element, but for each element, variation of the doublet distribution over the element up to its second order has been taken into account.

Carrying out the integration of the velocity influence coefficient and substituting into expression 11.45, and rearranging, we obtain

$$\sum_{i=1}^{N} \Delta\phi_i\left(A_{i,j^o} + A_{i+1,j^r} + A_{i-1,j^l}\right) = B_j^{'} \tag{11.61}$$

where

$$A_{i,j^o} = \frac{1}{2\pi}\iint_{el\, i} \frac{(x - x_{0i+1})(x - x_{0i-1})}{(x_{0i} - x_{0i+1})(x_{0i} - x_{0i-1})} K_{v2}\left(\mathbf{x}, \mathbf{x}_j\right)dx \tag{11.62}$$

$$A_{i+1,j^r} = \frac{1}{2\pi}\iint_{el\, i+1} \frac{(x - x_{0i+1})(x - x_{0i+2})}{(x_{0i} - x_{0i+1})(x_{0i} - x_{0i+2})} K_{v2}\left(\mathbf{x}, \mathbf{x}_j\right)dx \tag{11.63}$$

$$A_{i-1,j^l} = \frac{1}{2\pi}\iint_{el\, i-1} \frac{(x - x_{0i-1})(x - x_{0i-2})}{(x_{0i} - x_{0i-1})(x_{0i} - x_{0i-2})} K_{v2}\left(\mathbf{x}, \mathbf{x}_j\right)dx \tag{11.64a}$$

$$K_{v2}\left(\mathbf{x}, \mathbf{x}_j\right) = \frac{\left[(x - x_i)^2 - z_j^2\right]n_{zj} - 2z_j(x - x_i)n_{xj}}{\left[(x - x_i)^2 - z_j^2\right]^2} \tag{11.64b}$$

which can readily be evaluated. A_{ij^0} corresponds to the doublet distribution extending within the element i, whereas A_{i+1,j^r} and A_{i-1,j^l} are contributed by the parts of the element of the doublet distribution associated with the source element i which extend on the right-hand side and the left-hand side, respectively, as illustrated in the fourth column, second row of Fig. 11.5.

In the three-dimensional situation, we would consider planar quadrilateral and rectangular surface elements. For a rectangular surface element characterized by the sides D_1 and D_2, the velocity influence coefficient normal to the surface element of the control point, for a step distribution, is given by

$$A_{ij} = -n_{zj}A_{1ij} - 3n_{xj}A_{2ij} - 3n_{yj}A_{3ij} \tag{11.65}$$

where

$$A_{1ij} = \frac{1}{4\pi}\int_{-\frac{D_1}{2}}^{\frac{D_1}{2}}\int_{-\frac{D_1}{2}}^{\frac{D_1}{2}}\left[\frac{3z_j^2}{\left[(x-x_j)^2+(y-y_j)^2+z_j^2\right]^{\frac{5}{2}}} - \frac{1}{\left[(x-x_j)^2+(y-y_j)^2+z_j^2\right]^{\frac{3}{2}}}\right]dxdy \tag{11.66}$$

$$A_{2ij} = \frac{1}{4\pi}\int_{-\frac{D_1}{2}}^{\frac{D_1}{2}}\int_{-\frac{D_1}{2}}^{\frac{D_1}{2}}\left[\frac{(y-y_j)z_j}{\left[(x-x_j)^2+(y-y_j)^2+z_j^2\right]^{\frac{5}{2}}}\right]dxdy \tag{11.67}$$

$$A_{3ij} = \frac{1}{4\pi}\int_{-\frac{D_1}{2}}^{\frac{D_1}{2}}\int_{-\frac{D_1}{2}}^{\frac{D_1}{2}}\left[\frac{(x-x_j)z_j}{\left[(x-x_j)^2+(y-y_j)^2+z_j^2\right]^{\frac{5}{2}}}\right]dxdy \tag{11.68}$$

Each of these terms represents the induced velocity in the z, x and y direction, respectively, of the surface element i. The step distribution is equivalent to the vortex lattice of rectangular shape (or of the shape of the polygonal surface element), and by the use of Biot–Savart law, Eq. (11.41) could again be obtained. For $R/D > 2.5$, where D is the largest of D_1 and D_2, a multipole expansion approximate formula has proved to give sufficient accuracy. Since the expression for the influence coefficient associated with the piecewise quadratic distribution, which was chosen to take place along the chordwise direction, is lengthy, it will not be reproduced here. Detailed description is outlined in Djojodihardjo [20].

11.4.10 Application

11.4.10.1 Application to the Starting Instant Problem

Figure 11.6 displays the doublet distribution over a flat plate of infinite aspect ratio, obtained by employing step and piecewise quadratic distribution functions. The spacing between the elements is uniform. The step doublet distribution approximation predicts a doublet distribution that overestimates the exact one; for 32 elements, an agreement within 3% is indicated, but agreement becomes worse near the leading and trailing edges.

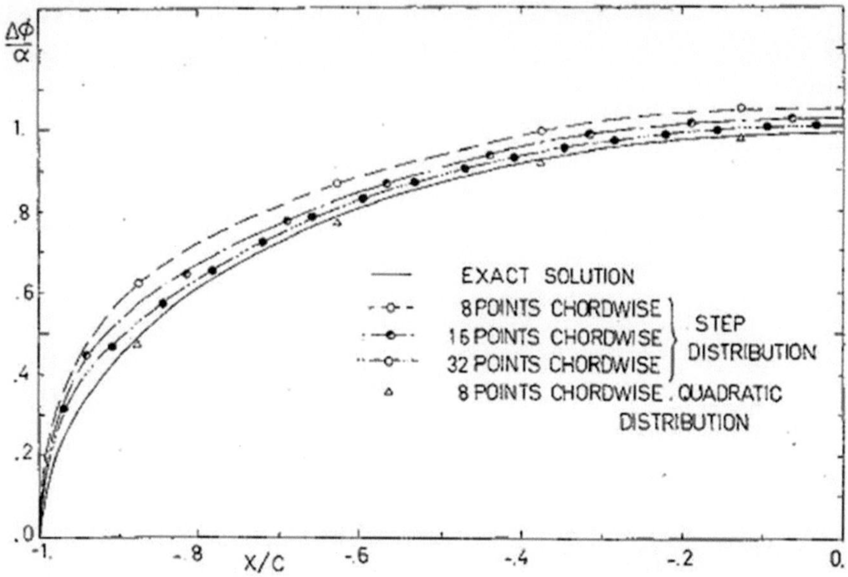

Fig. 11.6 Two-dimensional chordwise distribution of doublet at the instant of starting

Convergence of the solution can be observed with the increase of the number of elements, but near the edges, increase of the number of elements alone is not sufficient, since $\Delta\phi$ varies rapidly there. Although variable spacing to allow denser elements at regions of rapidly varying doublet distribution would improve the accuracy, the use of a more accurate representation at the edges would result in the best accuracy. Rapid convergence (in the sense of a small number of elements) could be obtained by the use of a piecewise quadratic distribution. For 8 elements an agreement within 1% is indicated, but the result underestimates the exact distribution. Such a behavior can be verified to be a consequence of the behavior of the self-induced velocity influence coefficient, which dominates the induced downwash at any field point.

Another point of interest is the ability of the doublet scheme to deal with the thickness problem. For this purpose, the method is applied to the non-lifting flow around a circular cylinder. The result indicates an excellent agreement with the exact solution, i.e. 0.5% for 64 elements, as shown in Fig. 11.7.

Figure 11.8a displays the doublet distribution over an elliptical planform using step and piecewise quadratic doublet distribution functions. Rectangular elements were employed, and care is taken to obtain proper representation of the tip sections. Again, better convergence is indicated by the piecewise quadratic distribution approximation. In these calculations, only 16 elements arranged as a network of 4 by 4 were used. The agreement with the exact solution is fair for the step distribution.

To investigate the convergence behavior, the method is applied to a rectangular planform, for which no surface element approximation is necessary. The results

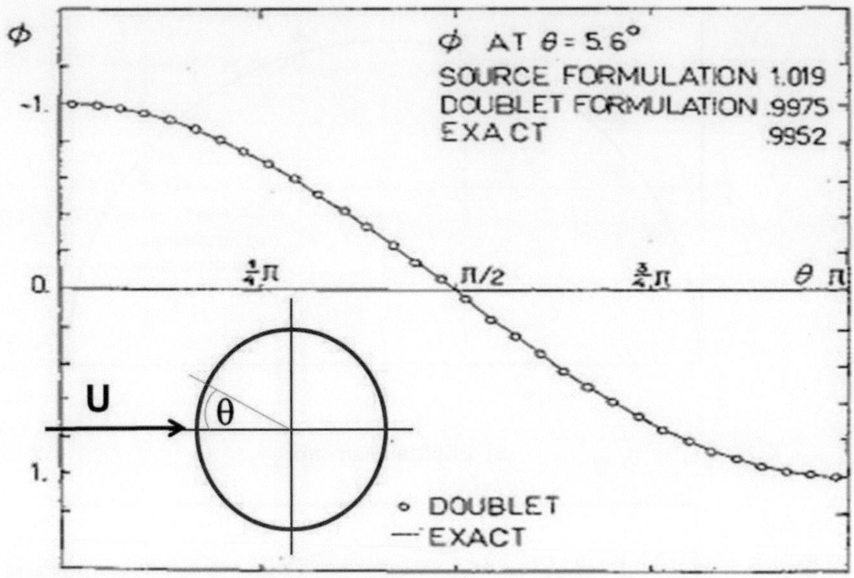

Fig. 11.7 Velocity potential distribution over the surface of a cylinder in non-lifting potential flow

shown in Fig. 11.8b indicate a converging behavior for the step distribution approximation with 64 elements arranged as a 16 by 4 network and for a piecewise quadratic distribution approximation with a 4 by 4 network. Although better convergence seems to be indicated by the use of a piecewise quadratic distribution approximation, in further application, the choice between the two distributions would be dictated by the computation time required for a comparable accuracy. A separate study to optimize the computation is necessary. Nevertheless, the step distribution has given a reasonably good prediction, except for the tip regions, for which the inaccuracy would be of a local nature. Improvement of the accuracy can be made by incorporating a better (higher order) approximation, as before, or by employing more elements at this region. The results just described lend support to the application of the method to the more complicated problems which involve the unsteady deforming wake.

11.4.10.2 Application to Impulsively Started Planforms and Comparison of Results

The transient loading on a wing impulsively started with a uniform velocity $U = U_0$ will be considered. To compare the accuracy provided by the present method, two-dimensional cases are first considered.

Figure 11.9 exhibits the indicial circulation experienced by a two-dimensional flat plate. Comparison is made with the prediction of Jones [27] which is based on a linearized formulation of the problem. In particular, Jones' calculation is based on

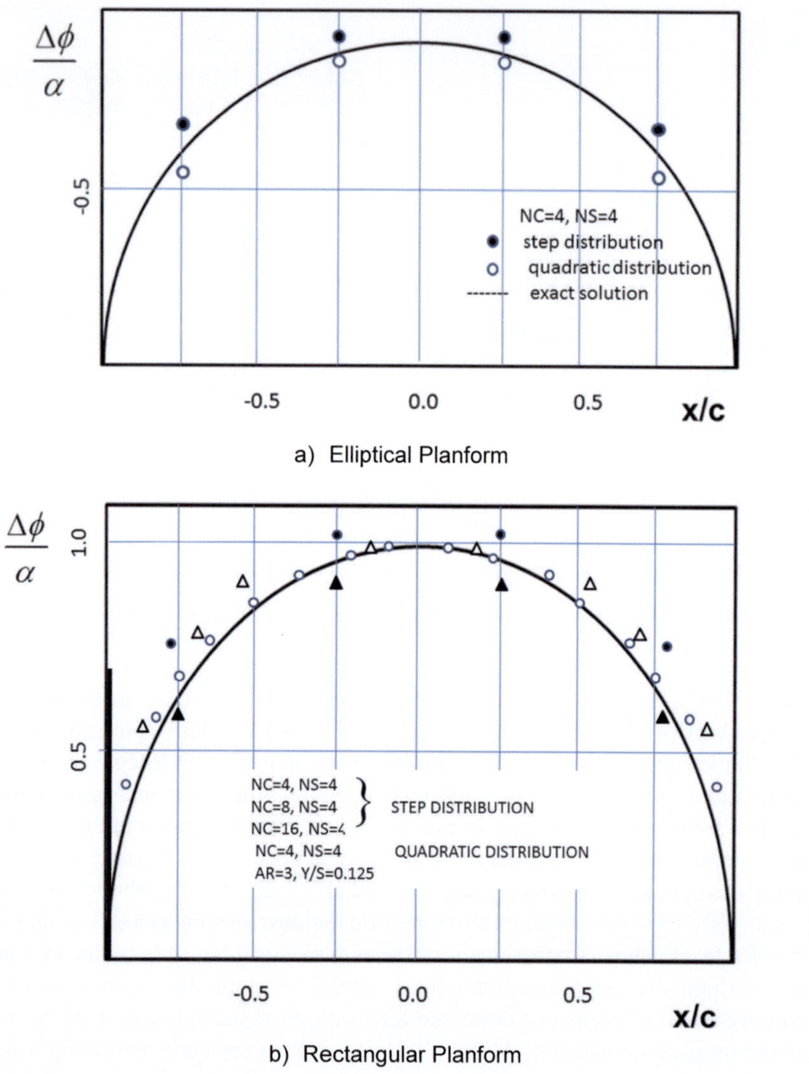

Fig. 11.8 Chordwise doublet distribution for a finite planform at the starting instant, AR = 3, at midspan section; **a** elliptical planform, **b** rectangular planform[8]

operational methods to solve two integral equations simultaneously; the first equation relates the induced downwash to the circulation of the airfoil along a straight path, and the second is an integral equation relating the circulation along the flight path to the rise of circulation following a sudden start of motion. The present computation incorporates a step doublet distribution over each element. In the linear region, say

[8] Reproduced qualitatively from Djojodihardjo et al. [1].

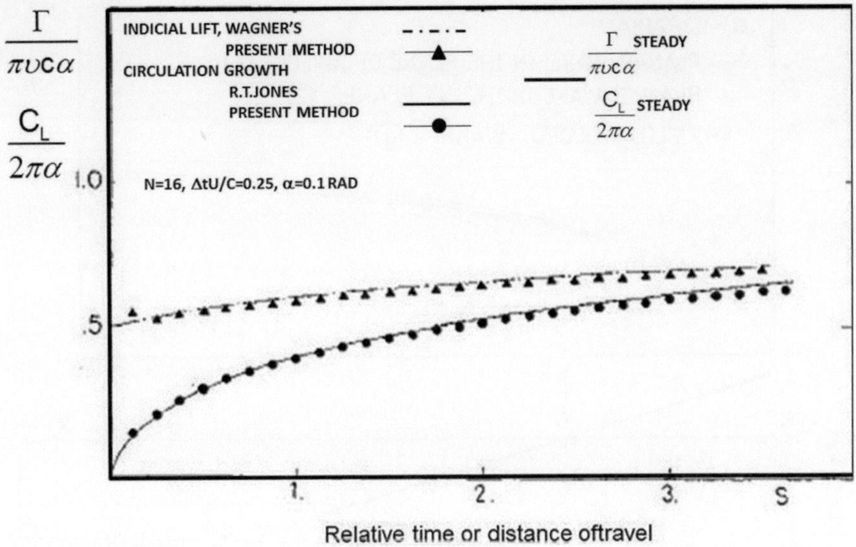

Fig. 11.9 Indicial lift and growth of circulation for an impulsively started two-dimensional flat plate

a. = 0.1 rad, the agreement is excellent. Similarly, the indicial lift (Wagner function, [24]) compares very well with standard result [27, 28] for a small angle of attack.

The computation of lift was based on the superposition of the steady part according to the alternate method discussed in Sec. II and the numerical integration of the unsteady pressure coefficient over the surface S. These two parts are both circulatory, except at the starting instant, for which the latter term is the added mass.

Several time steps were employed, the results indicating satisfactory convergence for $\frac{\Delta tU}{\frac{1}{2}c} = 0.125$ or less; the time step is commensurate with the spacing of the elements along the chord.

The nonlinearities introduced by the wake geometry can be observed from Fig. 11.10a. Calculation of the wake trajectory is obtained by incorporating the method to evaluate Eqs. (11.30) and (11.31) numerically, which may involve the use of a predictor–corrector scheme [11, 20] for improved accuracy. Three wake geometries were considered, as depicted in Fig. 11.10b, namely a freestream directed wake, a straight wake tangent to the limiting streamline at the trailing edge and a field-induced deforming wake with rolled up starting vortex. For an angle of incidence of 0.3 rad, the indicial circulation associated with these wake configurations differ within 4% at s, the distance travelled by the airfoil, equal to 1.25. As s increases, the indicial circulation for the field-induced deformed wake approaches that of the freestream directed wake.

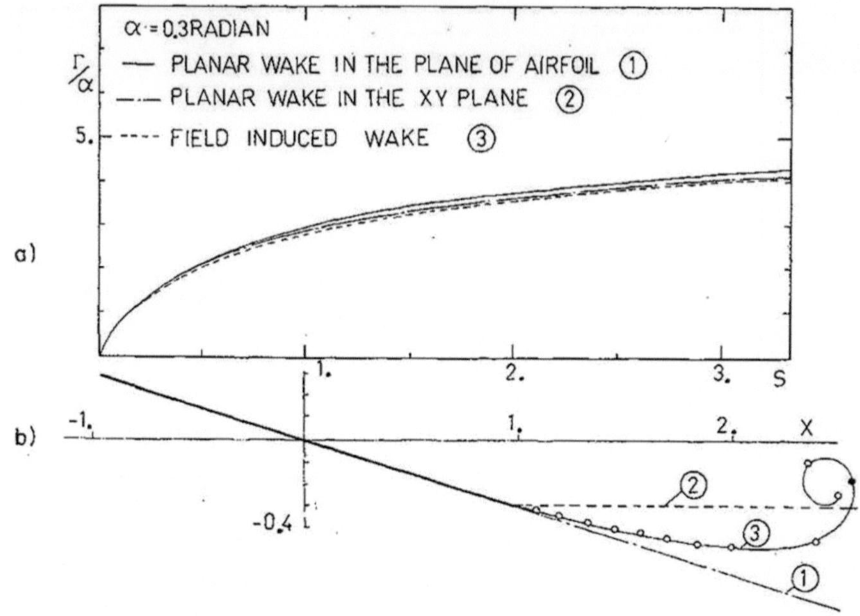

Fig. 11.10 Effect of the wake geometry on the growth of circulation of an impulsively started flat plate (2 dimensions)

To illustrate the capabilities of the present method, Fig. 11.11 shows the Küssner function (due to Von Karman and Sears [25]) predicted by the present method for a gust onset velocity equal to 0.1 U_0 striking a two-dimensional flat plate.

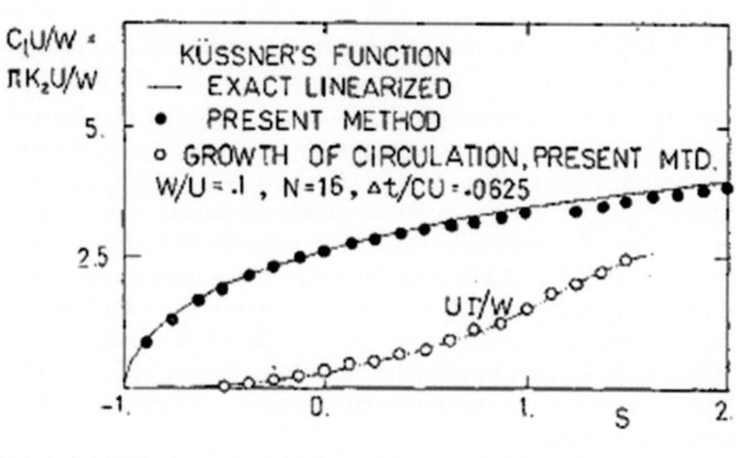

Fig. 11.11 Indicial lift and growth of circulation for a two-dimensional flat plate entering a sharp-edged gust

Figure 11.12 indicates the wake pattern behind an impulsively started, oscillating flat plate that exhibits a qualitative similarity with the visual observation of Bratt [29] for a steady-state oscillation and Giesing's [11] calculation for a similar transient oscillation.

The indicial sectional circulations for rectangular and elliptical planforms are shown in Fig. 11.10a and b, and comparison is made to the two-dimensional results. The more significant indicial lift function for a rectangular planform of aspect ratio 6 is shown in Fig. 11.13c and exhibits an excellent agreement with the linearized theory of Jones [30] for a small angle of incidence. These results again serve to verify the feasibility of the present method.

The effects of the aspect ratio and the geometry of the planform on the indicial loading can be studied from these results. The dependency on the angle of incidence is shown in Fig. 11.14, where the method has been applied to a rectangular planform of aspect ratio 6; the effect of thickness can be observed from Fig. 11.15, for rectangular planforms with NACA profiles of thickness ratios 0.12 and 0.15. Similar behavior is also given by Giesing [11].

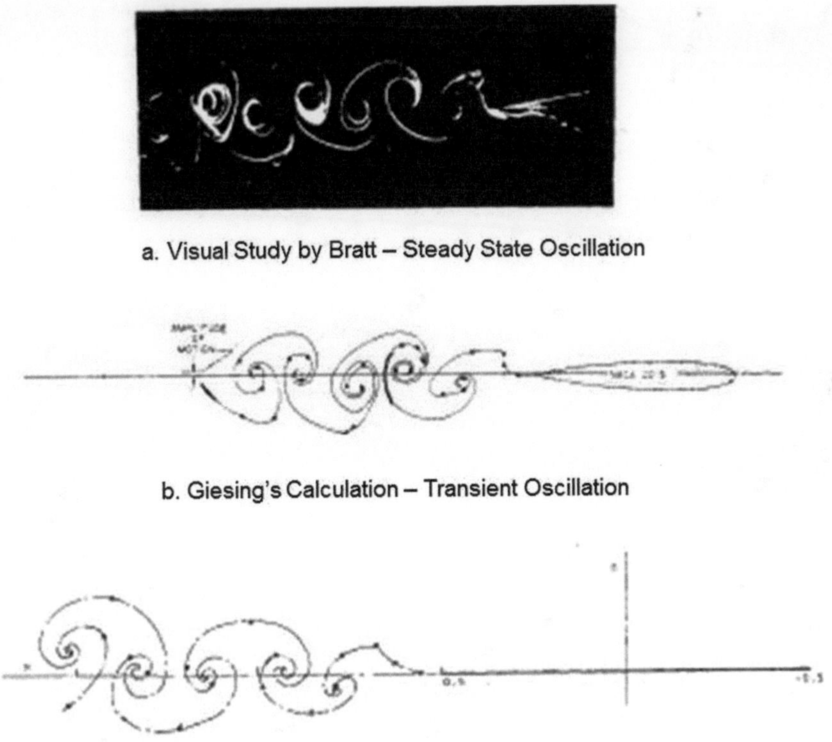

a. Visual Study by Bratt – Steady State Oscillation

b. Giesing's Calculation – Transient Oscillation

c. Present Method for Two-Dimensional Flat Plate – Transient Oscillation

Fig. 11.12 Wake pattern behind an oscillating foil

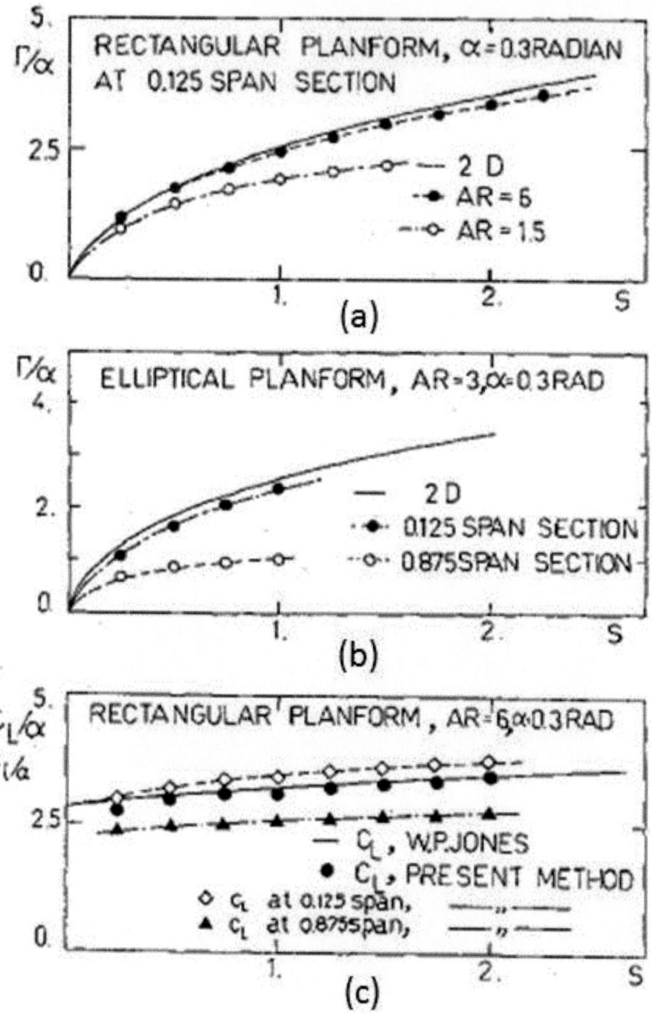

Fig. 11.13 Indicial loading of an impulsively started finite planform; **a** growth of circulation for a rectangular planform, **b** growth of circulation for an elliptical planform, **c** indicial lift for a rectangular planform

11.4.10.3 Concluding Remarks of Case-Study I

A numerical method has been developed which has been demonstrated in its potentialities to solve nonlinear transient lifting potential flow problems for wings of arbitrary planforms with thickness, although for comparison only simple planforms were considered. Employing a finite-difference version of the Taylor series expansion, it has been shown that step doublet distribution is sufficient to predict accurate airloads,

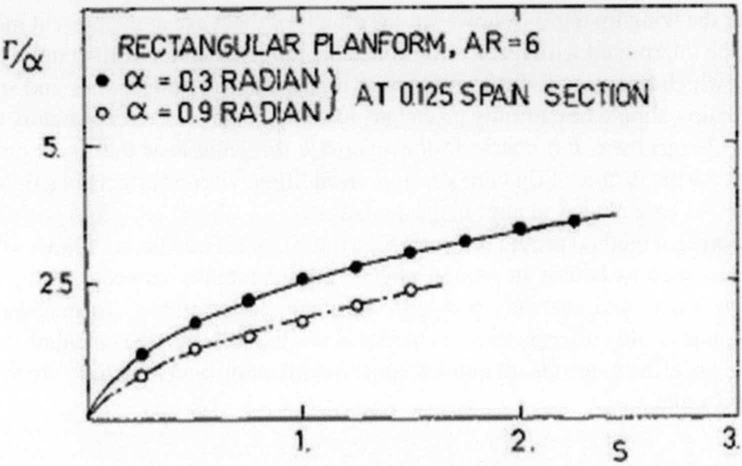

Fig. 11.14 Growth of circulation for an impulsively started rectangular planform at various angles of incidence

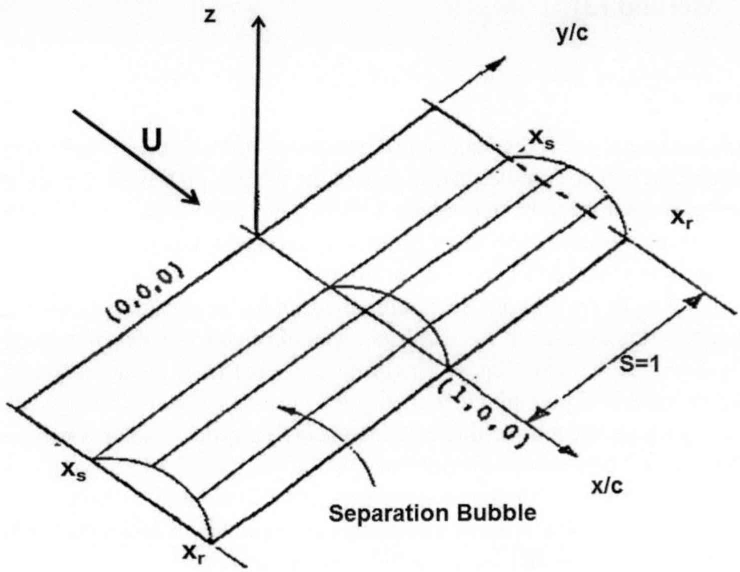

Fig. 11.15 Schematic of a wing with separation bubble; linearized situation

and improvement of the accuracy is also suggested according to the framework of the method outlined herein.

In particular, the present method finds its usefulness in accounting for the thickness effects and moderate angle of incidences beyond the range of linear theory. In addition, it is possible to predict the location of the unsteady and deforming wake.

Near the wing tip regions, however, the effects of the wake at the present moment should be interpreted with reservation, since the influence of the rolling up of the tip vortex, which becomes more significant at higher angle of incidences and smaller aspect ratios, should be carefully taken into account. In the present exploratory calculations, the grid was too coarse in the spanwise direction near the tip to properly account for the details of tip vortex rollup. In addition, viscous effects in a tip vortex core should be included in any such calculation.

The present method provides a basis for a more careful treatment of more difficult problems, such as bodies in ground effects, the interaction between a wing and a returning wake, and unsteady problems with aeroelastic effects. To treat periodic motion, one could either proceed as before along the step-by-step calculation, until all transient effects subside, or employ some modification which includes an initially assumed wake.

11.5 Case Study II: Calculation of 3D Unsteady Subsonic Flow with Separation Bubble Using Singularity Method [2]

Summary

A simple analytical method to calculate the unsteady airloads on an oscillating wing with separation bubbles on the suction side of the wing is presented. The oscillation of the wing is assumed to be harmonica. Linearized aerodynamics is assumed in the analysis. It is also assumed that the separation and reattachment points are known and fixed. Thus, the separation bubble does not move and the flow is separated during the whole cycle of oscillation. Such simplified model is considered beneficial for preliminary estimation of the unsteady aerodynamic characteristics of an oscillatory flow about wing with separation, such as those met in buffeting phenomenon. In order to apply the linearized potential flow theory, the flow is assumed to be inviscid and is entropic. Although such modelling is by no means a physical one, it is believed that it can serve as a mathematical convenience in obtaining the load acting in separated flow situation. Results of the method are presented. The validity of the method for fully attached flow case was insured by comparison with published results, while for separated flow cases, qualitative discussions are presented.

11.5.1 Introduction

The prediction of buffeting characteristics remains a challenging problem for aeroelasticians. The common difficulty found by engineers in this case is usually the calculation of unsteady aerodynamic load associated with flow separation which is nonlinear in nature. For preliminary design the purposes, the use of Navier–Stokes solver or even the method of inviscid-viscous strong interaction to analyze the unsteady aerodynamic load may not be economically justified. In many practical applications, engineers are concerned mainly with the prediction of the load itself rather than detail of the of the flow separation that occurs.

The flow aforementioned problem motivated many authors, such as Chi [31], Ericsson and Reding [32], Perumal and Sisto [33], Dowell [34], Huber et al. [35] and Djojodihardjo et al. [36] to work on this problem using a simplified model of potential flow theory The modelling is not aimed at the physics of the flow separation itself, but rather on the load that occurred during the course.

The results for two-dimensional cases seemed to have an encouraging agreement with the experimental data. Thus, although the approach proposed do not physically model the flow separation in great detail, it can reflect the essential mechanism present in such phenomena [37].

Since all the cases considered previously were two-dimensional, a three-dimensional method would be of interest for common practical applications. This case study extends the method earlier developed by Djojodihardjo et al. [36] for two-dimensional case and applied to wings of finite span. The kernel function method used to solve the unsteady linearized compressible potential flow without separation is extended to cover also the separated region. The boundary conditions comprise not only the slip condition on the body and no disturbance at infinity, but also a prescribed pressure at separated region. The three-dimensional kernel function theory of Laschka [38] is adapted and modified for this case. To satisfy all the boundary conditions, doublet distribution is placed on the entire lifting surface and source distribution on the separated region, following the success of such modeling in an earlier paper by Djojodihardjo and Widnall [1].

11.5.2 Problem Formulation

A wing modeled as at a steady undisturbed flow having semi-span equal to unity is considered. The wing is executing harmonic oscillations in pitching and heaving modes. At the upper side of the wing, separation bubbles may appear. This situation is sketched in Fig. 11.15.

11.5.3 Governing Equation

The equation governing unsteady linearized compressible potential flow can be obtained by assuming small disturbance with respect to the free stream condition. This equation is known as the Kelvin–Bernoulli equation, which for an acceleration potential formulation reads:

$$(1 - M^2)\psi_{xx} + \psi_{yy} + \psi_{zz} - \frac{2M}{a_\infty}\psi_{xt} + \frac{1}{a_\infty^2}\psi_{tt} = 0 \tag{11.69}$$

For a harmonic type of oscillation, this equation can be simplified further to

$$(1 - M^2)\psi_{xx} + \psi_{yy} + \psi_{zz} - \frac{i2kM}{a_\infty}\psi_x + \frac{k^2}{a_\infty^2}\psi = 0 \tag{11.70}$$

In the second equation, ψ is a complex quantity representing the amplitude of the acceleration potential and the phase lag with respect to the wing motion.

11.5.4 Boundary Conditions

a. On the attached Flow Region:

The kinematical boundary condition of inviscid flow model dictates that the flow has to be tangential to the surface. This condition can be written as

$$w(x, y; \tau) = \left(\frac{\partial}{\partial x} + \frac{\partial}{\partial \tau}\right) z(x, y; \tau) \tag{11.71}$$

where

$$w(x, y; \tau) = \overline{w}(x, y)e^{ik\tau} \tag{11.72}$$

$$z(x, y; \tau) = \overline{z}(x, y)e^{ik\tau} \tag{11.73}$$

And where k is the reduced frequency and τ is the non-dimensional time. For a heaving and pitching motion, the surface can be represented by:

$$z(x, y; \tau) = -\overline{h}(x, y)e^{ik\tau} \tag{11.74}$$

and

$$z(x, y; \tau) = -(x - x_c)\overline{\alpha}(x, y)e^{ik\tau} \tag{11.75}$$

respectively. Consequently, the boundary conditions become

(i) For heaving motion

$$\overline{w} = -ik\overline{h} \tag{11.76}$$

and

(ii) For pitching motion

$$\overline{w} = -[1 + ik(x - x_c)]\overline{\alpha} \tag{11.77}$$

b. On the separated Flow Region:

The separation is considered to take place only on the upper side of the wing. On this region some assumptions have to be made regarding the velocity and the pressure of the fluid, based on experimental evidence. The true situation may be obtained using viscous flow approach, which is presently beyond the scope of this work. It may be assumed that the velocity on the separated region is the same as that of the undisturbed flow.

Alternatively, it may be assumed a priori that the pressure difference in the separation bubble is equal to some value γ, which may be zero [31]:

$$\overline{C}_{p_u} = \gamma \tag{11.78}$$

11.5.5 Systems of Integral Equations

The system of integral equation for the stated problem is obtained from two sets of integral equations. The first set is the velocity (normal wash) equation with the kinematical boundary condition on the attached region, and the other is the pressure equation with the prescribed pressure boundary condition on the separated region. In each set of the integral equations, both singularities distributed on the attached and separated regions contribute to the appropriate boundary conditions. Therefore the system of integral equations comprise four components or four kernel functions. In this chapter the four kernel functions are derived.

11.5.6 Induced Velocity at the Attached Flow Region

The first kernel of the integral equation is obtained from the induced velocity at the attached flow region. Since Eq. (11.70) is linear, then its solution for the attached flow subject to the boundary conditions can be formed by the superposition of its fundamental solution. A fundamental solution of the Kelvin–Bernoulli equation for a lifting problem is the doublet flow. In the frequency domain approach of an acceleration potential formulation equation, the induced potential of a doublet point takes the form of

$$\psi_D(x, y; \tau) = -\frac{1}{4\pi} A e^{i\mu k M\left(x-x'\right)} \frac{\partial}{\partial z}\left(\frac{e^{i\mu k\tau}}{r}\right) \tag{11.79}$$

where

$$A = \frac{\Delta\overline{C}_p\left(x', y', z'\right)}{2} \tag{11.80}$$

The downwash at the attached flow region induced by the doublet distribution over the entire surface is then

$$\overline{w}_1(\mathbf{x}, t) = -\frac{1}{8\pi}\int_{-s}^{s}\int_{x_{le}}^{x_{te}} \Delta\overline{C}_p\left(x', y'\right) K_1\left(x, x'; y, y'\right) dx' dy' \tag{11.81}$$

Equation (11.15a, 11.15b) is the usual integral equation of the lifting surface approach for attached flow. The kernel of this integral equation is introduced by Küssner [39] as

$$K_1 = e^{-ik\left(x-x'\right)} \lim_{z\to 0}\frac{\partial^2}{\partial z^2}\int_{-x}^{x-x'}\left(\frac{e^{ik\left[\frac{\left(x-x'\right)}{\beta^2}-\mu\tau\right]}}{r}\right) d\left(x-x'\right) \tag{11.82}$$

or in a convenient form in the development here, as was formulated by Watkins et al. [40] and Watkins et al. [41], as:

$$K_1 = \frac{e^{ik\left(\xi-\xi'\right)}}{s^2\left(\eta-\eta'\right)}\left\{-k\left|\eta-\eta'\right|\left[\begin{array}{l} K_1\left(k\left|\eta-\eta'\right|\right) \\ +\frac{\pi \mathrm{i}}{2}\left(I_1 k\left|\eta-\eta'\right| - L_1 k\left|\eta-\eta'\right|\right) \\ -\mathrm{i}\int_0^{UBr}\frac{\tau}{\left(1+\tau^2\right)^{\frac{1}{2}}} e^{-\mathrm{i}k\left(\eta-\eta'\right)\tau} d\tau \end{array}\right]\right\} \tag{11.83}$$

where ξ and η are the non-dimensional coordinates in x and y directions respectively, and

$$r' = \left[\left(\xi-\xi'\right)^2 + \beta^2\left(\eta-\eta'\right)^2\right]^{\frac{1}{2}} \tag{11.84}$$

the upper boundary of the integral is[9]

[9] $U \cdot Br \equiv$ Upper Boundary.

$$U \cdot Br = \frac{1}{(\eta - \eta')\beta^2}\left[\left(\xi - \xi'\right) + Mr'\right] \tag{11.85}$$

11.5.7 Induced Velocity at the Separated Flow Region

The contribution of the separated region, on which a source distribution is placed, to the first set of integral equation is derived using velocity potential formulation. Since separation occurs only in the upper surface, acceleration potential is not appropriate for this situation; velocity potential formulation is more suitable. The Kelvin–Bernoulli equation for the velocity potential formulation takes the same form as Eq. (11.2), since a linear relationship between pressure and velocity potential is assumed. The velocity potential induced by a point source is

$$\phi_S = \frac{1}{4\pi}A^+ e^{i\mu kM\left(x-x'\right)}\left(\frac{e^{i\mu k\tau}}{r}\right) \tag{11.86}$$

At the lifting surface, a limiting process can be followed to determine the following relationship

$$\Delta\overline{w}\left(x', y', z'\right) = -A^+ \tag{11.87}$$

The downwash induced by the source distributed on the upper surface where the flow is separated then becomes

$$\overline{w}_1(\xi, \eta) = \int_{-1}^{1}\int_{\xi_s}^{x_r} \Delta\overline{w}\left(\xi', \eta'\right)K_3\left(\xi, \eta; \xi', \eta'\right)d\xi' d\eta' + \left.\frac{\Delta\overline{w}}{2}\right|_{\xi=\xi', \eta=\eta'} \tag{11.88}$$

Notice that in the above equation the singular point has been isolated and treated separately, taking into account Eq. (11.64a, 11.64b). The separated region for the time being is taken to be uniform in the spanwise direction; this condition can easily be modified by examining the experimental results. The kernel function of Eq. (11.88) is

$$K_3 = \frac{1}{4\pi}\lim_{z\to 0}\left(-\beta^2 z\Delta\overline{w}\left(\xi', \eta'\right)e^{-i\mu kM\left(\xi-\xi'\right)}\frac{e^{-i\mu kr'}\left(-i\mu kr' - 1\right)}{\left(r'\right)^3}\right) \tag{11.89}$$

It is obvious that the last equation is zero on the lifting surface since linearized aerodynamics has been utilized. The remaining equation is then

$$\overline{w}_1(\xi, \eta) = \left.\frac{\Delta\overline{w}}{2}\right|_{\xi=\xi', \eta=\eta'} \tag{11.90}$$

11.5.8 Induced Pressure at the Attached Flow Region

The third and fourth integral equations are pressure equations, with the pressure on the upper surface where the separation occurs as the boundary condition. The third equation is obtained from the pressure induced by the doublet distribution in the attached flow region.

From the definition of acceleration potential and the linearized Bernoulli equation the pressure coefficient is

$$\overline{C}_p(x, y) = -2\psi_D(x, y, z) \tag{11.91}$$

Taking Eqs. (11.67) and Eq. (11.56) into account the pressure coefficient on the upper surface of the wing is then

$$\overline{C}_{pu}(\xi, \eta) = \int_{-1}^{1}\int_{\xi_{le}}^{\xi_{te}} \Delta\overline{C}_p\left(\xi', \eta'\right)\, K_2\left(\xi, \eta; \xi', \eta'\right) d\xi' d\eta' - \left.\frac{\Delta\overline{C}_p}{2}\right|_{\xi=\xi', \eta=\eta'} \tag{11.92}$$

where, again, the singularity is isolated and treated independently and the separated region is distributed uniformly in the spanwise direction. The kernel function is worked out in the same manner as in two-dimensional case [37] to obtain

$$K_2 = \frac{1}{4\pi}\lim_{z\to 0}\left(\beta^2 z e^{-i\mu k\left[M\left(\xi-\xi'\right)-r'\right]}\frac{\left(-i\mu k r' - 1\right)}{\left(r'\right)^3}\right) \tag{11.93}$$

Careful examination will show that for a planar surface, Eq. (11.93) will go to zero when z approaches zero. The remaining pressure equation then becomes a very simple equation as

$$\overline{C}_{pu}(x, y) = -\left.\frac{\Delta\overline{C}_p}{2}\right|_{\xi=\xi', \eta=\eta'} \tag{11.94}$$

11.5.9 Induced Pressure at the Separated Flow Region

The kernel's last integral equation is the contribution of the velocity potential source to the pressure integral equation. The pressure coefficient is related to the velocity potential through the linearized Bernoulli equation as

$$\overline{C}_p(x, y, z) = -2\left(\frac{\partial\phi_s}{\partial\tau} + \frac{\partial\phi_s}{\partial x}\right) \tag{11.95}$$

Substituting the expression of induced velocity potential due to sources distributed on the separated region into Eq. (11.71), one obtains the pressure at the upper surface in the form

$$\overline{C}_{pu}(\xi,\eta) = \int_{-1}^{1}\int_{\xi_s}^{x_r} \Delta\overline{w}\left(\xi',\eta'\right)K_4\left(\xi,\eta;\xi',\eta'\right)d\xi' d\eta' \tag{11.96}$$

where the kernel

$$K_4 = -\frac{1}{2\pi}e^{-\mathrm{i}\mu k\left[M\left(\xi-\xi'\right)-r'\right]}\left(\frac{ik}{\beta^2 r} - \frac{\left(\xi-\xi'\right)\left(\mathrm{i}\mu k - \frac{1}{r'}\right)}{\left(r'\right)^2}\right) \tag{11.97}$$

Superposing all the components contributing to the integral equation as elaborated in this section, the system of integral equation is found to be

$$\overline{w}_1(\xi,\eta) = \int_{-1}^{1}\int_{\xi_s}^{x_r} \Delta\overline{C}_p\left(\xi',\eta'\right)K_1\left(\xi,\eta;\xi',\eta'\right)d\xi' d\eta' + \left.\frac{\Delta\overline{w}}{2}\right|_{\xi=\xi',\eta=\eta'} \tag{11.98}$$

$$\overline{C}_{pu}(\xi,\eta) = \int_{-1}^{1}\int_{\xi_s}^{\xi_r} \Delta\overline{w}\left(\xi',\eta'\right)K_4\left(\xi,\eta;\xi',\eta'\right)d\xi' d\eta' + \left.\frac{\Delta\overline{C}_p}{2}\right|_{\xi=\xi',\eta=\eta'} \tag{11.99}$$

This system of integral equation comprises two Fredholm integral equations of the second kind, since the unknowns also exist outside the integrals.

11.5.9.1 Method of Solution

Consistent to the approach identified at the beginning of this case study, the method employed for solving the system of integral equation is the kernel function method. The solution is assumed to stem in a function space which must satisfy the physical characteristics of the solution. Each component of the function space is valid throughout the domain (in this case the lifting surface); this is the main difference as compared to the finite element approach where the function space is only valid in an element. The choice of the series of functions is carried out taking into consideration four aspects, namely the completeness of the function space, the physical characteristic modelled, the convenience of the integration and the rapid convergence of the series.

In the present work, the unknowns, the distributions of pressure difference and downwash jump across the lifting surface are represented by the series expansion as

$$\Delta\overline{C}_p(X,y)=\sum_{r=0}^{R}a_r(y)h_r(X) \tag{11.100}$$

in which each function $h(X)$ should exhibit a well-known singularity behavior for $x\rightarrow 0$ of the type,

$$\lim_{\varepsilon\rightarrow 0}\varepsilon^{\frac{1}{2}}$$

This singularity models the suction force at the leading edge. At the trailing edge where Kutta condition applies, the result of thin airfoil theory dictates that the pressure difference should go to zero in a manner also like $\lim_{\varepsilon\rightarrow 0}\varepsilon^{\frac{1}{2}}$.

One set of functions which satisfies those physical characteristics in the chordwise direction is the one introduced by Glauert [42]

$$h_r(X)=\frac{2}{\pi}\frac{\cos(r\phi)+\cos((r+1)\phi)}{\sin\phi} \tag{11.101}$$

where

$$X=\frac{x-x_{LE}(\eta')}{c(\eta')}=\frac{1}{2}(1-\cos(\phi)) \tag{11.102}$$

The collocation points are defined as

$$\phi=\frac{2\pi(p+1)}{2P+3},\quad p=1,2,\ldots P \tag{11.103}$$

The coefficients of the series are not constants, but they will also be expanded in the spanwise direction, taking into account the physical characteristics of the flow.

Similarly for the downwash jump across the lifting surface, the same series expansion is used

$$\Delta\overline{w}(X,y)=\sum_{q=0}^{Q}d_q(y)h_q(X) \tag{11.104}$$

After substitution of the series expansion, the system of integral equations becomes

$$\bar{w}_1(\xi,\eta)=\sum_{r=0}^{R}1\int_{-1}^{1}a*_r\left(\eta'\right)\int_{\xi_{le}}^{\xi_{te}}h_r(X)K_1\left(\xi,\eta;\xi',\eta'\right)d\xi'd\eta' + \sum_{q=0}^{Q}\int_{-1}^{1}d*_q\left(\eta'\right)\frac{h_q(X)}{2}\bigg|_{\xi=\xi',\eta=\eta'}d\eta' \quad (11.105)$$

$$\overline{C}_{pu}(\xi,\eta)=\sum_{q=0}^{Q}\int_{-1}^{1}d*_q\left(\eta'\right)\int_{\xi_s}^{\xi_r}h_q(X)K_4\left(\xi,\eta;\xi',\eta'\right)d\xi'd\eta' + \sum_{r=0}^{R}\int_{-1}^{1}a*_r(\eta)\frac{h_r(X)}{2}\bigg|_{\xi=\xi',\eta=\eta'}d\eta' \quad (11.106)$$

The asterisks in the coefficients stand for the appropriate non-dimensionalization of the coefficients in the spanwise direction. After the integration in the chordwise direction, the above set of equations can be written as

$$\bar{w}_1(\xi,\eta)=\sum_{r=0}^{R}\int_{-1}^{1}a*_r\left(\eta'\right)f_{1r}\left(\xi,\eta;\eta'\right)d\eta' + \sum_{q=0}^{Q}\int_{-1}^{1}d*_q\left(\eta'\right)\frac{h_q(X)}{2}\bigg|_{\xi=\xi',\eta=\eta'}d\eta' \quad (11.107)$$

$$\bar{C}_{pu}(\xi,\eta)=\sum_{r=0}^{R}\int_{-1}^{1}a*_r(\eta)\frac{h_r(X)}{2}\bigg|_{\xi=\xi',\eta=\eta'}d\eta' + \sum_{q=0}^{Q}\int_{-1}^{1}d*_q\left(\eta'\right)f_{4q}\left(\xi,\eta;\eta'\right)d\eta' \quad (11.108)$$

The characteristic of the flow in the spanwise directions can be well represented by expanding the coefficients $a^*(\eta)$ and $d^*(\eta)$ in a Fourier sine series. Introducing a coordinate transformation, where $\eta=\cos(\theta)$, the series can be written as

$$a_r^*(\theta)=\sum_{\mu=0}^{M}A_\mu\sin(\mu\theta) \quad (11.109)$$

The integration in the spanwise direction is worked out using Multhopp method, which is closely related to the choice of the series. Substituting the series into the system of integral equation and carrying out the integration, one obtains

$$\overline{w}_1(\xi,\eta) = \sum_{r=0}^{R}\sum_{m=1}^{M} a^*_{\rm rm} c1_{\rm rm} f_{1r}(\xi,\eta;\theta_m) + \sum_{q=0}^{Q}\sum_{m=1}^{M} d_{\rm qm}\left(\eta'\right) c2_{\rm rm}\frac{h_{\rm qm}}{2} \tag{11.110}$$

$$\overline{C}_{pu}(\xi,\eta) = \sum_{r=0}^{R}\sum_{m=1}^{M} a^*_{\rm rm}\left(\eta'\right) c3_{\rm rm}\frac{h_{\rm rm}}{2} + \sum_{q=0}^{Q}\sum_{m=1}^{M} d^*_{\rm rm} c4_{rm} f_{4q}(\xi,\eta;\theta_m) \tag{11.111}$$

Here, c1, c2, c3, c4 are the quadrature constants of the Multhopp integration method.

The last set of equations indicates the presence of $M(R+1)$ number of unknowns, in the form of coefficients of expansions. The same number of collocation points, where the equation has to be satisfied, should be taken. The collocation points in the chordwise direction have been introduced before, namely

$$\phi = \frac{2\pi(p+1)}{2P+3}, \quad p = 1, 2, \ldots P \tag{11.112}$$

In the spanwise direction the collocation points are chosen to follow a cosine spacing

$$\theta = \frac{\upsilon\pi}{N+3}, \quad \upsilon = 1, 2, \ldots \boldsymbol{N} \tag{11.113}$$

To have the same number of collocation points P must be equal to R and N must be equal to M. The set of algebraic equations at the collocation points are

$$\overline{w}_{1\ pv} = \sum_{r=0}^{R}\sum_{m=1}^{M} a^*_{\rm rm} g_{1\ \rm rm\ pv} + \sum_{q=0}^{Q}\sum_{m=1}^{M} d^*_{\rm qm} g_{3\ \rm rm\ pv} \tag{11.114}$$

$$\overline{C}_{P_{u\,pv}} = \sum_{r=0}^{R}\sum_{m=1}^{M} a^*_{\rm rm} g_{2\ \rm rm\ pv} + \sum_{q=0}^{Q}\sum_{m=1}^{M} d^*_{\rm qm} g_{4\ \rm rm\ pv} \tag{11.115}$$

where p and υ run from O to P and from 1 to N respectively. The equation in the matrix form

$$\begin{bmatrix} g_1 & g_2 \\ g_2 & g_4 \end{bmatrix}\begin{Bmatrix} a* \\ d* \end{Bmatrix} = \begin{Bmatrix} \overline{w}_1 \\ \overline{C}_{P_u} \end{Bmatrix} \tag{11.116}$$

Solving this set of algebraic equations, the coefficients of the series expansion are obtained.

The pressure coefficient and the other aerodynamic coefficients can then be easily calculated from this result.

11.5.9.2 Pressure, Lift and Moment Coefficient

The pressure coefficient consists of the attached and separated part, which are calculated separately. Both of them give their contribution to the lift and moment coefficients.

$$\Delta\bar{C}_p(X,\eta) = \begin{cases} \Delta\bar{C}_{p_{at}}(X,\eta) & \text{for } 0 < x < x_{\text{sep}} \text{ and } x_{\text{reattach}} < x < c(\eta) \\ \Delta\bar{C}_{p_{\text{sep}}}(X,\eta) & \text{for } x_{\text{sep}} < x < x_{\text{reattach}} \end{cases} \tag{11.117}$$

The pressure difference on the attached region can directly be calculated from the series expansion as,

$$\Delta\overline{C}_{p_{att}}(X,\eta) = \sum_{r=0}^{R} a_r(\eta)\, h_r(X) \tag{11.118}$$

The pressure difference at the separated region may be calculated using the following procedure. The definition of pressure difference in the separated region is

$$\Delta\overline{C}_{p_{sep}}(X,\eta) = \overline{C}_{p_l}(X,\eta) - \overline{C}_{p_u}(X,\eta) \tag{11.119}$$

The lower pressure coefficient is calculated simply from

$$\overline{C}_{p_l}(X,\eta) = \sum_{r=0}^{R} a_r(\eta)\,\frac{h_r(X)}{2} \tag{11.120}$$

and from the pressure integral equation, the upper pressure is

$$\overline{C}_{p_u}(X,\eta) = -\sum_{r=0}^{R} a_r(\eta)\,\frac{h_r(X)}{2} + \sum_{q=0}^{Q}\sum_{m=1}^{M} d_{\text{qm}} g_{4\ \text{rm pv}} \tag{11.121}$$

Substituting the two expressions of the upper and lower pressure, the pressure difference is then lower pressure, the pressure difference is then

$$\Delta\overline{C}_p(X,\eta) = \sum_{r=0}^{R} a_r(\eta)\, h_r(X) - \sum_{q=0}^{Q}\sum_{m=1}^{M} d_{\text{qm}} g_{4\ \text{rm pv}} \tag{11.122}$$

The sectional lift coefficient may then be calculated as

$$\overline{C}_l(\eta) = \int_0^{x_{sep}} \Delta\overline{C}_{p_{att}}\left(X',\eta\right)dX' + \int_{x_{sep}}^{x_{reatt}} \Delta\overline{C}_{p_{sep}}\left(X',\eta\right)dX'$$

$$= a_0(\eta) - \sum_{q=0}^{Q}\sum_{m=1}^{M} d_{qm} \sum_{j=0}^{J} g_{1\ qm\ vp} \sin\phi_j \left(\frac{\phi_{reatt} - \phi_{sep}}{2(J+1)}\right) \tag{11.123}$$

The sectional moment coefficient is calculated in the same manner as the lift coefficients

$$\bar{C}_m(\eta) = -\int_0^{x_{sep}} \Delta\bar{C}_{p_{att}}\left(X', \eta\right)\left(X' - X_c\right)dX' - \int_{x_{sep}}^{x_{reatt}} \Delta\bar{C}_{p_{sep}}\left(X', \eta\right)\left(X' - X_c\right)dX'$$

$$= \frac{1}{4}[a_1(\eta) + (4X_c - 1)a_0(\eta)]$$

$$+ \sum_{q=0}^{Q}\sum_{m=1}^{M} d_{qm} \sum_{j=0}^{J} g_{4\ qm\ vp}\left(X_j - X_c\right) \sin\phi_j \left(\frac{\phi_{reatt} - \phi_{sep}}{2(J+1)}\right) \tag{11.124}$$

The wing lift coefficient is obtained by integrating the sectional lift coefficient in the spanwise direction

$$\overline{C}_L = \frac{1}{s}\int_{-s}^{s} C_l(y)c(y)dy = \frac{\pi}{SM}\sum_{m=1}^{M} C_l(\theta_m)c(\theta_m)\sin\theta_m \tag{11.125}$$

In the same manner the moment coefficient of the wing is calculated from

$$\overline{C}_L = \frac{1}{sc}\int_{-s}^{s} C_m(y)c^2(y)dy = \frac{\pi}{ScM}\sum_{m=1}^{M} C_m(\theta_m)c^2(\theta_m)\sin\theta_m \tag{11.126}$$

11.5.9.3 Results

Figures 11.16a and b show the sectional lift and moment coefficients, respectively, for pitching oscillations pivoted at $X_c = 0.5$ in a fully attached flow, for a lifting surface with an aspect ratio of 2, and for incompressible case. The results agree very well with those obtained by Laschka [38].

Figures 11.17a and b show sectional moment coefficient for fully attached flow, compressible, for pitching and heaving motion, respectively for $M = 08$. The moment coefficients of the pitching oscillation have been referred to the pivot point, while for heaving oscillation to the quarter chord. The agreement appears to be very good. The slight difference indicated in the results may be attributed to numerical errors in approximating the integral, since this difference behaves consistently. The distribution of pressure differences in the chordwise direction for pitching oscillations around midchord with reduced frequency of 2 for a wing with aspect ratio of 2 in

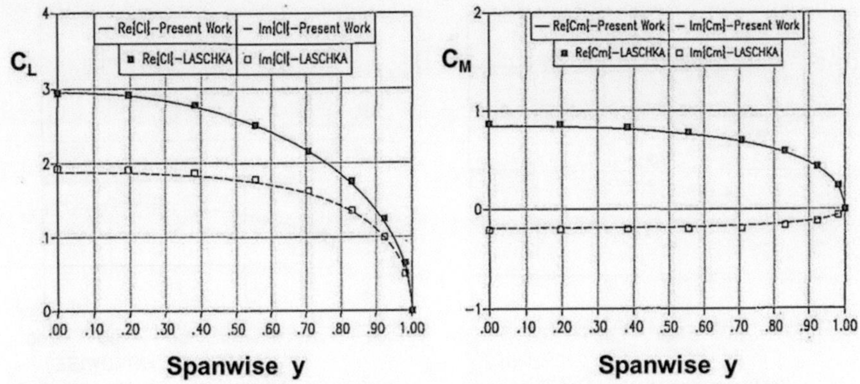

Fig. 11.16 Sectional lift coefficient; fully attached flow, pitching motion. $M = 0, X_c = 0.5\ k = 1$, AR $= 2$

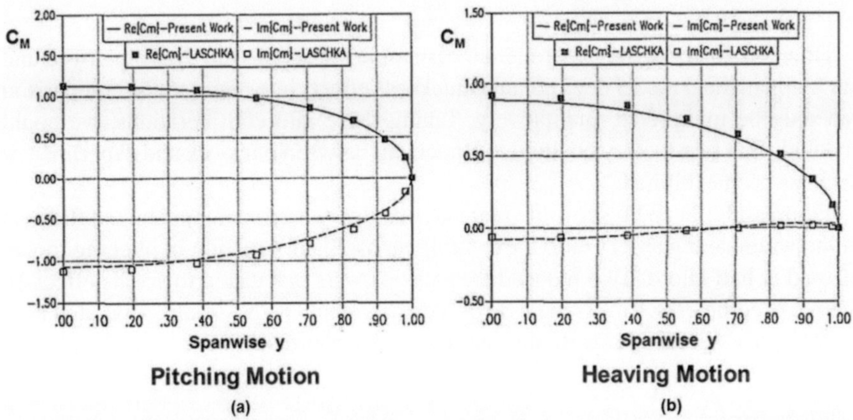

Fig. 11.17 Sectional moment coefficient; fully attached flow $M = 0.8, X_c = 0.5\ k = 1$, AR $= 2$

incompressible flow has also been checked,[10] and the results exhibit good agreement with those of Laschka's.

For the separated three-dimensional flow case to the best of the author's knowledge no theoretical results, under the similar assumptions, as well as experimental results, for a fixed separation point, are available. Therefore we would like to embark upon a qualitative comparisons to existing data. As a first step toward this end, computational results for steady state case at Mach number 0.147 for a wing with separation point at 35% chord and for fully separated wing are shown in Fig. 11.18a, while the fully attached flow case is shown in Fig. 11.18b. These results are compared to experimental ones obtained at IPTN,[11] both for a wing with aspect ratio of 4.4.

[10] The detail is given in the referenced paper, Djojodihardjo et al. [2].

[11] Now PT Dirgantara Indonesia.

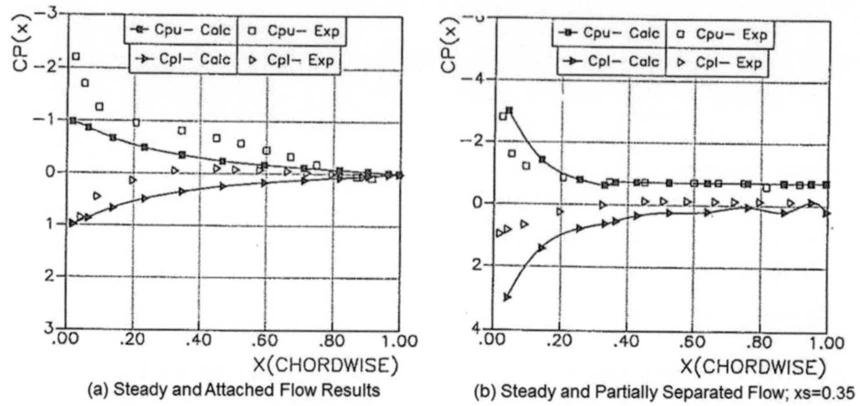

Fig. 11.18 Qualitative comparison for **a** steady and attached flow results; **b** steady and partially separated flow; $x_s = 0.35$ $M = 0.147$, $y = 0.2875$, AR $= 4.444$. Experiment for NACA 54A2015

However since for the experimental case finite thickness wing was employed and our mathematical model does not take thickness effects into account, the comparison can only be interpreted qualitatively. Taking these into considerations, we could conclude that satisfactory qualitative agreement between our work and experimental data has been exhibited.

Figures 11.19a and b show the real and imaginary sectional moment coefficients of the wing having aspect ratio 2 executing an oscillatory motion in pitching mode, pivoted at half chord. Two reduced frequencies were calculated to see its effect on moment coefficients. This is the case of incompressible with partially separated flow on the upper side; the separation is started at 60% chord.

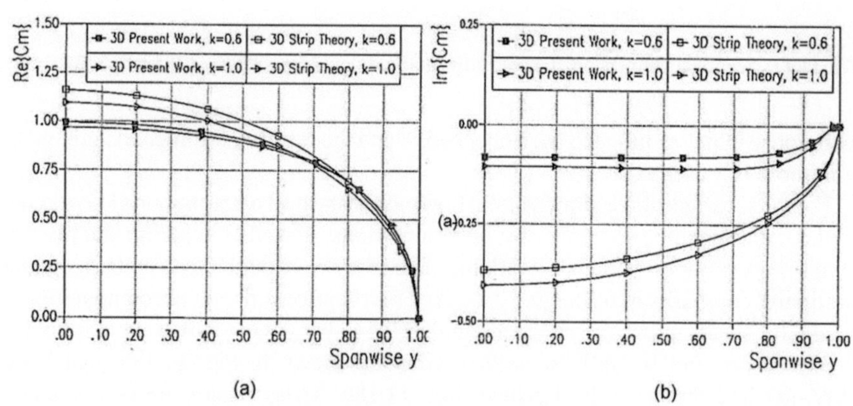

Fig. 11.19 Influence of reduced frequency on the sectional moment coefficient, for pitching motion; $x_s = 0.35$, $M = 0.0$, AR $= 2$, gamma $= 0.0$. **a** real part; **b** imaginary part

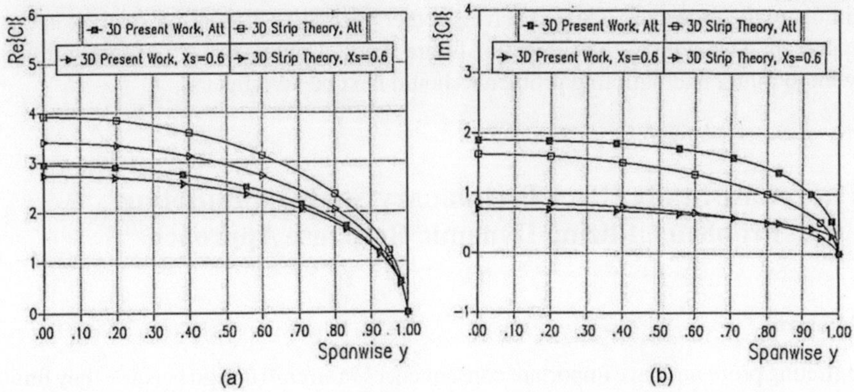

Fig. 11.20 Comparison of the real and imaginary values of lift coefficient of attached and separated flow for pitching motion; $x_s = 0.35$, $M = 0.0$, AR = 2, gamma = 0.0. **a** real part; **b** imaginary part

The results of the present method are compared to the three-dimensional ones of the strip theory corrected with elliptic downwash of Djojodihardjo and Sekar [43]. The two-dimensional theory used in the strip theory is the one developed by the authors [36]. The differences exhibited in Fig. 19a and b have been expected to appear, since the strip theory uses a simple prescribed downwash. Consistent differences shown between the two results indicate correct tendencies as the reduced frequency increases.

The effect of separation is shown in Figs. 11.20a and b for the same Mach number and vibration mode as the previous case. Two cases were calculated, namely fully attached and separated flow starting from 60% chord. The results again are compared to the strip theory with prescribed elliptic downwash. Both methods show a decrease in lift coefficients due to separation as expected. Conclusion similar to that of Figs. 11.19a and b can be drawn for Fig. 11.20a and b.

11.5.9.4 Concluding Remarks for Case Study II

Extension of the linearized theory for unsteady two-dimensional subsonic flow around oscillating airfoil with separated region has been carried out for finite wings. Although comparison with the results of other three-dimensional theoretical work cannot be made due to lack of data, comparison with strip theory based on two-dimensional results corrected with a simple prescribed downwash shows a correct tendency.

The approach taken here is linearized aerodynamics, and planar surface has been implicitly treated throughout. However, for non-planar surfaces, the work can be extended by approximating the surfaces by finite set of planar elements and by the use of modified series expansion taking into account the local angle of incidence. The kernel functions can then be evaluated numerically or exactly, wherever appropriate, similar to the procedure adopted here. Another alternative is to utilize doublet lattice approach. This will be the subject of future work. The applicability of the present work to flutter and buffeting problems should next be investigated.

11.6 Case Study III: A Preliminary Study on Buffeting Problem Utilizing Dynamic Response Approach[12]

Summary

Buffeting problems have important consequences in aircraft design because they limit cruise velocity and aircraft performance. For this purpose, a preliminary analytical study of dynamic response caused buffeting is necessary. In present case study, an analytical approach of buffeting is performed utilizing dynamic response technique based on assumption that there are no coupling between motion-induced unsteady aerodynamic force and independent-motion unsteady airload. The motion-induced unsteady airload is calculated based on a method developed by Djojodihardjo et al., and the buffeting excitation load is based on Powell and Bull's experiment. The dynamic response of the structure is determined according to methods developed by Forching and Zingel.

11.6.1 Introduction

The problem of dynamic response of an aircraft structure due to flow separation, known as buffeting, is an important problem in aeroelasticity when the aircraft is operated at high speed or high angle of attack. For civil transport aircrafts, buffeting phenomenon will limit their cruise speed and comfort.

Whereas, for fighter aircrafts, this will limit their maneuvering performance. In aircraft design, the problem is related with fatigue and fracture of aircraft structure [44].

The analytical approach to buffeting problem associated with subsonic flow was developed by many researchers, such as Forching [44], Mabey [13], Zingel [17], Becker [45] and others. Because of the complexity of the phenomenon, no purely analytical approach were used in their approach. All their predictions were based on semi-empirical method, where the quantity of unsteady aerodynamic forces associated with separation and attached flow are measured by wind tunnel experiment and then the dynamic response of the structure are calculated by using several methods such as the power spectrum method. A purely analytical approach to predict the quantity of unsteady airloads on buffeting problem was developed by Djojodihardjo

[12] Based on work carried out in Haryanto [4].

et. al., using Kernel function technique [43, 46, 47]. This technique is used in the present case study to predict the buffeting behavior.

11.6.2 Problem Formulation

Figure 11.21 illustrates a typical cross sectional of wing or horizontal tail on uniform flow with the separation region on the upper surface [46]. By assuming that there are no coupling between motion-induced unsteady airloads and independent-motion unsteady airloads, the problem of analytical buffeting prediction can be modeled as forced vibration system as illustrated in Fig. 11.22 (modified to suit the oresent case from Försching [48]). In this model the independentmotion unsteady airloads are associated with the driving excitation forces or buffeting forces, and the motion-induced unsteady airloads produce aerodynamics damping that together with structural damping will limit the dynamics aeroelastic response of the structure.

The motion-induced unsteady airloads can be determined by using lifting surface theory, and the buffeting forces are approximated base on Powell and Bull's experimental result [49].

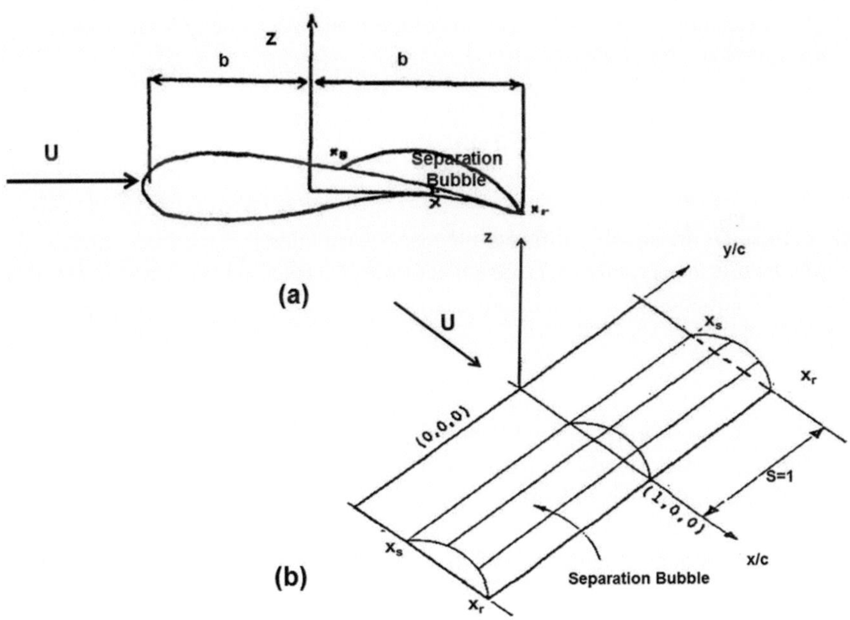

Fig. 11.21 Airfoil with separation bubble

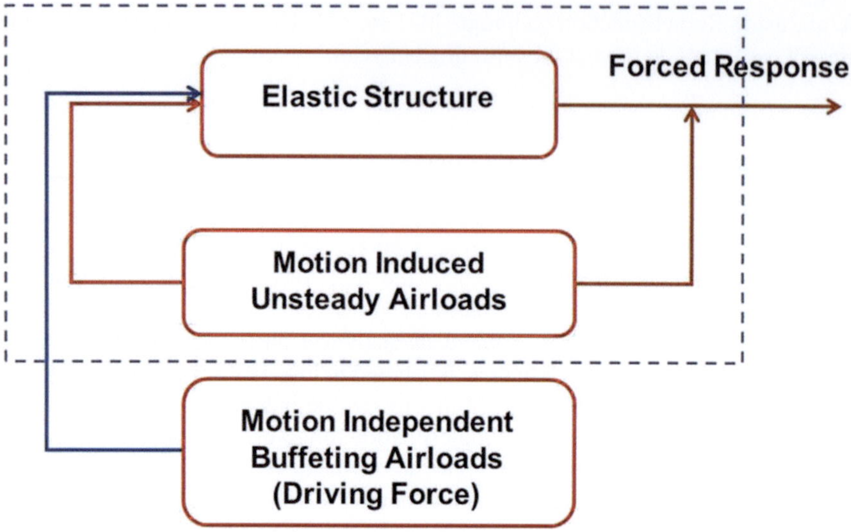

Fig. 11.22 Forced oscillation model of buffeting

Mathematical Formulation

From flexure beam theory, if the wing is treated as a slender beam in which the cross-sectional dimensions are small in comparison with the length, the equation of equilibrium can be written as [50]

$$\frac{\partial^2}{\partial y^2}\left[EI\frac{\partial^2 u(y,t)}{\partial y^2}\right] = -m\ddot{u}(y,t) \tag{11.127}$$

where El and m are bending stiffness and mass distribution per unit span respectively.

Meanwhile $u(y, t)$ and $\ddot{u}(y, t)$ are deflection and acceleration of structure element. Hence, the equation of motion of wing is:

$$\frac{\partial^2}{\partial y^2}\left[EI\frac{\partial^2 u(y,t)}{\partial y^2}\right] = -m\,\ddot{u}(y,t) + F^A(t) + F^F(t) \tag{11.128}$$

where $F^A(t)$ is motion-induced force and $F^F(t)$ indicate independent-motion forces.

In the application of oscillation model approach to predict the structural buffeting response aircraft, the principle of superposition may be applied. Hence, the deflection of the aircraft structure at any point P and any time t can be expressed in the form:

$$u(P,t) = \sum_{r=1}^{n} u_r(P)q_r(t) \tag{11.129}$$

such that the equation of motion can be written as:

$$M_r\ddot{q}_r(t) + K_r q_r(t) - \sum_{r=1}^{n} A_{rs} q_s(t) = Q_s^F(t) \tag{11.130}$$

where K_r and M_r are generalized stiffness and generalized mass of the structure and can be determined by following equation:

$$\begin{aligned} K_r &= \int_0^l \frac{\partial^2}{\partial y^2}\left[EI\frac{\partial^2 u_r(y)}{\partial y^2}\right] u_r(y)dy \\ &= \omega_r \int_0^l u_r(y)m(y)u_r^T(y)dy = -M_r\omega_r \end{aligned} \tag{11.131}$$

In Eq. (11.105) the generalized unsteady aerodynamic forces of the motion-induced unsteady pressure, $p_s^A(P,t)$, when the structure vibrates in the s-th mode terms, is given by:

$$A_{rs}(t) = \iint_S u_r(P)p_s^A(P,t)dS \tag{11.132}$$

The term on the right-hand side of Eq. (11.105) expresses the generalized aerodynamics forcing term associated with pressure fluctuation $p^F(P,\ t)$ due to flow separation, which can be written as

$$Q_r^F(t) = \iint_S p^F(P,t)u_r(P)dS \tag{11.133}$$

or in term of cross spectral densities of the generalized unsteady airloads:

$$S_{Q_{rs}}(\omega) = \iint_S \iint_S u_r^T(P)S_P\left(P,P',\omega\right)u_s\left(P'\right)dSdS' \tag{11.134}$$

where $S_p(P, P', \omega)$ is the diagonal matrix of cross-spectral density of the fluctuating pressure at two different points (P, P'). Hence, the power spectrum of deflection of the aircraft structure at point P can be determined by the following input–output relation [37]:

$$S_u(P,\omega) = \sum_{r=1}^{n}\sum_{s1}^{n} H_r^*(P,\omega)H_s^T(P,\omega)S_{Q_{rs}}(\omega) \tag{11.135}$$

where H_r (P, ω) is the transfer function of dynamic system mode r, and $H_r^*(P, \omega)$ is its conjugate form.

By assuming that both input and output are harmonic,

$$q_r(\omega) = \overline{q}_r(\omega)e^{i\omega t} \tag{11.136}$$

$$Q_r(\omega) = \overline{Q}_r(\omega)e^{i\omega t} \tag{11.137}$$

Equation (11.130) can be transformed into

$$M_r(\omega_r^2 - \omega^2)\overline{q}_r(\omega) - \sum_{r=1}^{n} A_{rs}\overline{q}_s(\omega) = \overline{Q}_r^F(\omega) \tag{11.138}$$

Solving the linear simultaneous equation of Eq. (11.10), the transfer function can be determined as follows:

$$H_{u,r}(P, \omega) = \frac{u_r(P)\overline{q}_r(\omega)}{\overline{Q}_r^F(\omega)} \tag{11.139}$$

The bending moment at point P of wing can be derived from the simple beam theory which will give:

$$M(P, t) = EI\frac{\partial^2(u(P, t))}{\partial y^2} \tag{11.140}$$

Combining Eqs. (11.12) and (11.11) yields transfer function for the bending moment of the structure and is given by

$$H_{M,r}(P, \omega) = \frac{\left[EI\frac{\partial^2(u_r(P,t))}{\partial y^2}\right]^2 \overline{q}_r(\omega)}{\overline{Q}_r^F(\omega)} \tag{11.141}$$

By neglecting the structure damping the mean-square values of bending moment can be formulated as [37, 51]:

$$\overline{M_{P,r}^2} = \frac{\left[EI\frac{\partial^2(u_r(P))}{\partial y^2}\right]^2}{8\pi M_r\omega_r^3\gamma_{\text{aero}}} S_{Q,r}(\omega_r) \tag{11.142}$$

where γ_{aero} is the aerodynamic damping factor. If it is assumed that the mean value of both the displacement and the bending moment are zero, then the values of their deviation standard are equal to the root-mean-square value, $\overline{M}_{rms}$.

11.6.3 *Unsteady Airloads*

11.6.3.1 Motion-Induced Unsteady Airloads

Because the motion-induced on the unsteady airloads on the buffeting problem are assumed to have linear relationship with respect to the system displacements, then their quantities can be obtained by using the lifting surface theory.

By assuming small disturbance with respect to the free stream condition, the equation of unsteady linearized compressible potential flow in perturbation velocity potential is given by [52]:

$$\left(1 - M^2\right)\varphi_{xx} + \varphi_{zz} - \frac{2ikM}{a_\infty}\varphi_x + \frac{k^2}{a_\infty^2}\varphi = 0 \tag{11.143}$$

where φ is a complex function representing the amplitude of perturbation potential and M is Mach number.

For inviscid flow case, the boundary condition on the attached flow region is given by

$$w(x, \tau) = \left(\frac{\partial}{\partial x} + \frac{\partial}{\partial y}\right) z(x, \tau) \tag{11.144}$$

where

$$w(x, \tau) = \overline{w}(x)e^{ik\tau} \tag{11.145}$$

$$z(x, \tau) = \overline{z}(x)e^{ik\tau} \tag{11.146}$$

where k is the reduced frequency and τ is non-dimensional time.

Pressure distribution on the airfoil surface can be determined by solving equation:

$$\overline{w}(x) = \oint_{chord} \Delta\overline{C}_p K(kx_0, M)dx \tag{11.147}$$

where $K(kx_o, M)$ is the Kernel function defined in Bisplinghoff et al. [52].

The separation in considered to take place only on the upper side of the wing. On this region some assumptions have to be made regarding the velocity and the pressure of the fluid, based on the experimental evidence. The true situation may be obtained using viscous flow approach, which is beyond the scope of this work.

It may be assumed that the velocity on the separated region is the same as that of undistributed flow. Alternatively, it may be assumed a priori that the pressure difference in separation region is equal to some value

$$\overline{C}_{p_u} = \overline{\gamma} \tag{11.148}$$

More detailed procedures to determine the pressure distribution on the separation flow region are elaborated by Djojodihardjo and Sekar [43], Djojodihardjo, Sekar, Prananta [2], Djojodihardjo and Sekar [46]. The pressure distribution on the wing surface then can be calculated by strip theory.

11.6.3.2 Motion Independent Unsteady Airloads

By assuming that the pressure fluctuations have a statistical regularity, then the magnitude of the motion independent unsteady airloads can be written in by means of statistical form. If the structure can be approximated to have a plate shape and the flow separations are turbulent, the pressure fluctuation can be modeled based on Bull and Powell's experiment.

Based on Bull and Powell's experimental joint structural acceptance function read as follows [26, 49, 53, 54].

$$J_{rs}(\omega) = \frac{1}{S^2 S_p(\omega)} \iint_S \iint_S u_r(P) S_P\left(P, P', \omega\right) u_s\left(P'\right) dS dS' \tag{11.149}$$

where S and $S_p(w)$ are wing surface area and auto-correlation power spectrum density of pressure fluctuation, respectively.

According to Crocco's experiment [53] the cross-correlation power spectral density of the pressure fluctuation is

$$S_p\left(P, P', \omega\right) = S_P(\omega) \exp\left[-\frac{\omega}{U_c}(0.1|\xi_1| + 0.55|\xi_2|)\right] \cos\left(\frac{\omega}{U_c}|\xi_1|\right) \tag{11.150}$$

where ξ_1 and ξ_2 are the distances between two points in the longitudinal and transversal direction, respectively, to flow, and U_c is the convection velocity which is often assumed to be equal to about 70% of free stream velocity. For two-dimensional case Eq. (11.124) can be reduced to the form

$$S_p\left(P, P', \omega\right) = S_P(\omega) \exp\left[-\frac{\omega}{U_c}(0.1|\xi_1|)\right] \cos\left(\frac{\omega}{U_c}|\xi_1|\right) \tag{11.151}$$

The auto-correlation power spectral density at $\omega = \omega_1$, can be approximated by Blevins [53]:

$$S_p(\omega) = 4K^2 q^2 c^3 \tag{11.152}$$

q and c are dynamic pressure and chord. K is a constant which can be determined based on experiment.

11.6.4 Discussions on the Computational Results

The wing structure which is used in this study is approximated by a rectangular semi-span wing which was used in Boyden and Johnson's experiment [42].

The computational result of the first natural bending mode of this model is about 270 Hz. This value is the same as Boyden and Johnson's experiment results. Separation points are not determined by computational method. Approximation as a first estimate to the locations has been attempted by establishing a model by capitalizing on the available experimental result of Boyden and Johnson.

Following this approach, approximation function of the locations of separation points has been constructed as a function of the angle of attack which are modeled as follows:

$$\frac{x_s}{b} = \begin{cases} 1.0 \quad \text{if}\,\alpha_s < \alpha \\ 2\dfrac{(\alpha - \alpha_s)}{\left(\alpha_f - \alpha_s\right)} - 1.0 \;\; \text{if}\,\alpha_s < \alpha < \alpha_f \\ 1.0 \quad \text{if}\,\alpha > \alpha_f \end{cases} \tag{11.153}$$

where α_s and α_f are the angle of attack associated with the onset of separation, and a fully separated situation, respectively, which are modeled expressed to be as follows:

$$\alpha_s = 2.357 \times 10^{-6}\ \text{Re} + 5.162 \tag{11.154a}$$

$$\alpha_s = 1.010 \times 10^{-6}\ \text{Re} + 12.212 \tag{11.154b}$$

where Re is Reynolds number.

Figure 11.23 illustrates non-dimensional form of the chordwise coordinate of the separation points which are assumed to be located in a straight line along the span as illustrated by Fig. 11.21b.

The real and the imaginary parts of the motion-induced unsteady aerodynamic load for various location of separation points are illustrated in Fig. 11.24 (real part) and Fig. 11.25 (real and imaginary). The pressure distribution on the separated region of the upper surface is assumed to be equal to zero. The independent-motion unsteady airload quantities in terms of the acceptance function[13] quantities for the first mode are illustrated in Fig. 11.26.

Based on experimental result of Boyden and Johnson [42], the K value of Eq. (11.152) is found to be:

$$K = \frac{\rho_\infty U_\infty}{\mu} \tag{11.155}$$

[13] For acceptance function, one may refer to Chen, XZ, Matsumoto, M, Kareem, A, Time Domain Flutter and Buffeting Response Analysis of Bridges, https://www3.nd.edu/~nathaz/journals/ (2000).

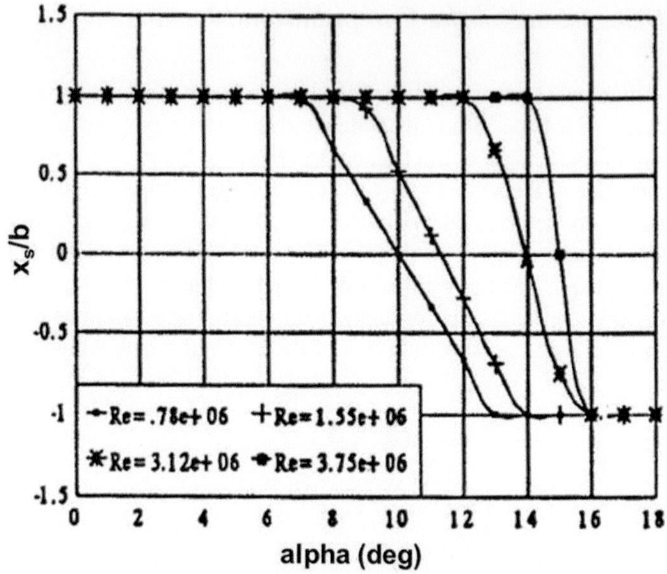

Fig. 11.23 Separation point model as a function of Reynolds number and angle of attack

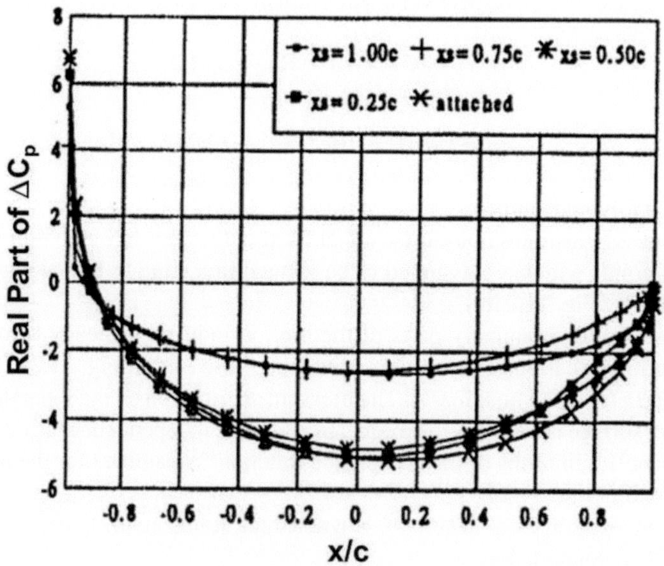

Fig. 11.24 Real part of pressure distribution of bending mode, $y/l = 0.5$, $t = 1.67$ and $M = 0.3$

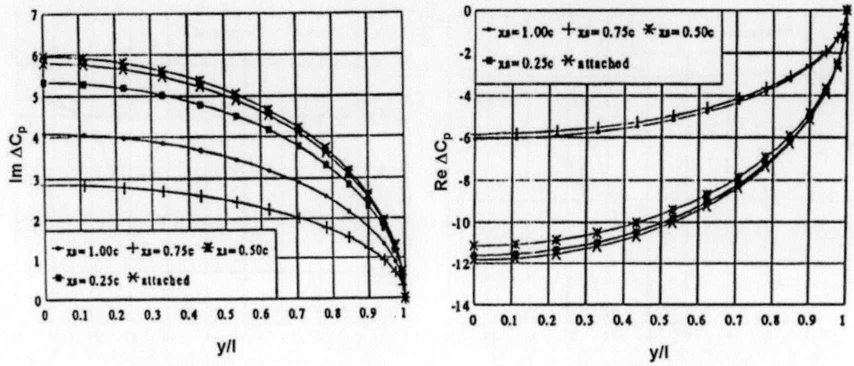

Fig. 11.25 Pressure distribution of bending mode, $x/c = 0.55$, $k = 1.67$ and $M = 0.3$; **a** imaginary part and **b** real part

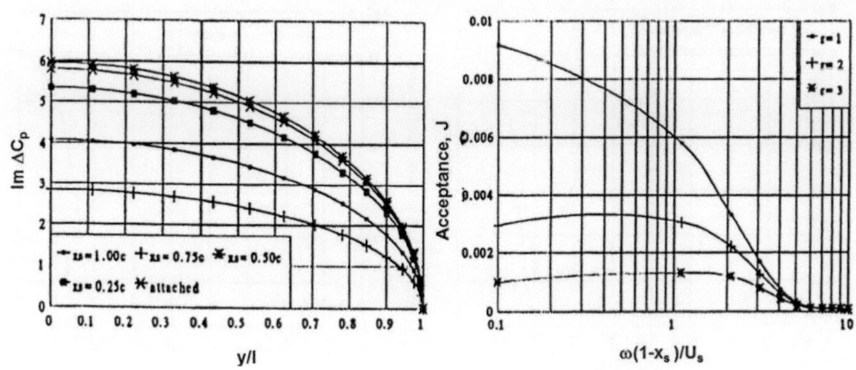

Fig. 11.26 **a** Imaginary part of pressure distribution of bending mode $x/c = 0.55$, $k = 1.67$ and $M = 0.3$; **b** acceptance function for independent-motion airloads

where ρ_∞, U_∞ and μ are air density, velocity and viscosity of the undisturbed flow, respectively. Utilizing this approach, the root-mean-square bending moment coefficients, $\frac{M_{rms}}{qcS}$, for various Reynolds number and angle of attack can be found, as illustrated in Figs. 11.27 and 11.28.

Compared to experimental results, the computational results give lower values of the standard deviation of the bending moment. These results may be attributed to the following:

(1) Before the onset of buffeting, the flow is assumed to be fully attached in the computation. Therefore no unsteady load results.
(2) At the onset of buffeting, the pressure in the separation bubble is assumed to be zero.

The results also showed that this method to some extent is capable to reveal the mechanism for buffeting.

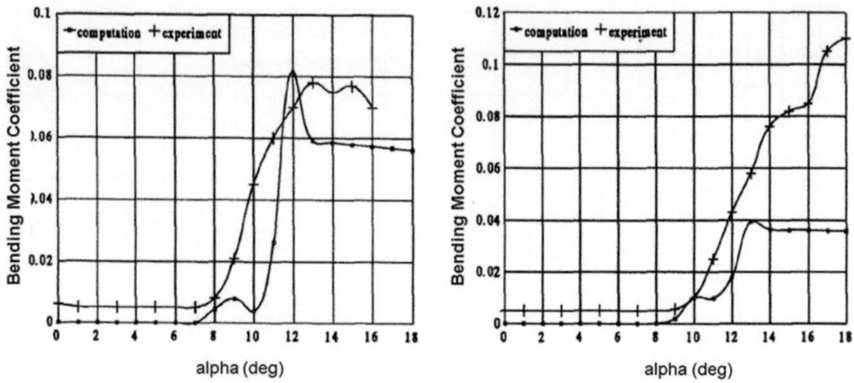

Fig. 11.27 Bending moment coefficient for **a** Re = 0.0.78 + 06, **b** Re = 0.1.55e + 06

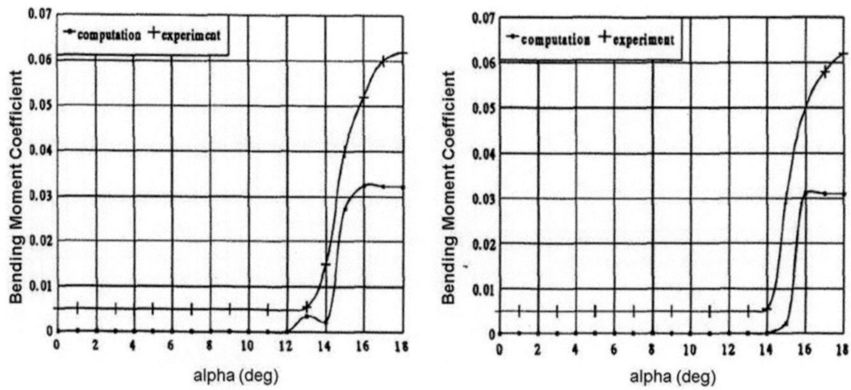

Fig. 11.28 Bending moment coefficient for **a** Re = 3.12e + 06, **b** Re = 3.75 + 06

11.6.5 Concluding Remarks for Case-Study III

A method for the dynamic response calculation wing with separated flow has been developed. The procedure is based on forced vibration model approach suggested by Försching and Zingel which utilizes power spectrum method.

In addition, the motion-dependent unsteady aerodynamic forces have been calculated following the method of Djojodihardjo et al., and the motion-independent unsteady aerodynamic forces have been estimated utilizing experimental result of Bull and Powell that has been adapted.

Results have shown the following:

(1) For simplified case, dynamic response behavior that qualitatively agrees with experimental data has been indicated.

(2) The method has indicated the validity and plausibility of a systematic approach for buffeting prediction, which utilizes theoretical approach for the motion-dependent aerodynamic forces and empirical approach for motion independent aerodynamic forces.

However, further work is required to establish the method for more general case.

11.7 Case Study IV: Unified Aerodynamic-Acoustic Formulation for Aero-Acoustic Structure Coupling[14]

Summary

As a follow-up on the author's work on coupled BE/FE Method and modeling of structural acoustic interaction, the present work utilizes BE formulation for both the acoustic and aerodynamic problem. For this purpose the unsteady aerodynamic problem which was earlier utilizes well-established lifting surface method is reformulated using BE approach. This scheme is presently outlined and worked out in the general framework of acousto-aeroelastic stability problem developed previously.

11.7.1 Introduction

Earlier work by the author and colleagues [55–60] addressed progressive stages of the development of coupled BE/FE Method and Modelling of structural–acoustic interaction. The computational approach based on the formulation of the boundary element computation of the acoustic load following the governing Helmholtz equation for acoustic disturbance propagation on the structure allows the incorporation of the direct or hydrodynamic part of the acoustic influence on the aeroelastic stability as a specific problem investigated.

Various components of the method have been validated and the method produced results in qualitative agreement with well established analytical and experimental results of Holström [61], Marquez et al. [62], and Huang [63]. To address the problem associated with fluid structure interaction, in particular the vibration of structures due to sound waves, aerodynamics and their combined effects, a generic approach to the solution of the acousto-aeroelasto dynamic interaction has been followed. The development of the foundation for the computational scheme for the calculation of the influence of the acoustic disturbance to the aeroelastic stability of the structure [56–60] starts from a rather simple and instructive model to a more elaborate FE-BE fluid–structure one. The formulation of the problem is summarized in Fig. 11.29.

[14] From Djojodihardjo et al. [3].

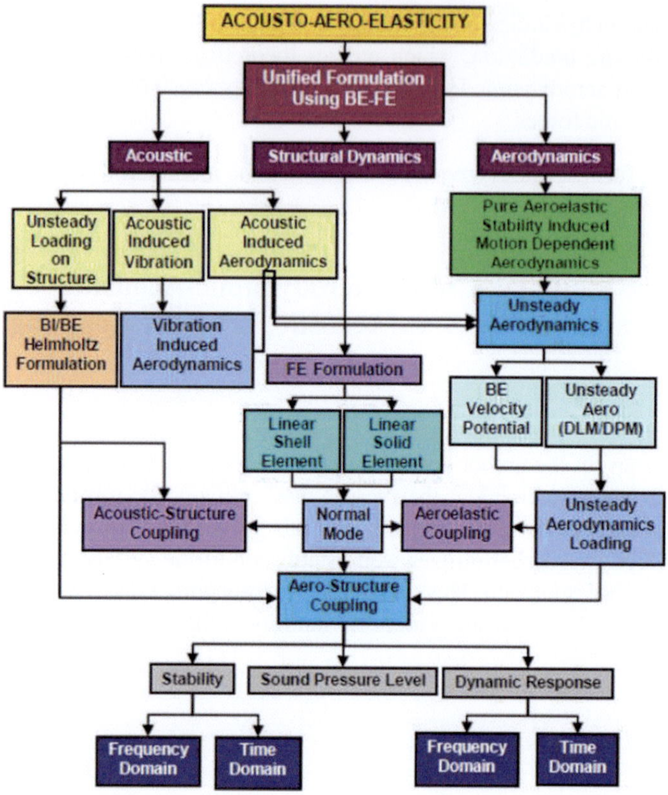

Fig. 11.29 Logical approach to treat the aeroacoustic effects on aeroelastic structure

It is of interest at this point to recall the concluding remarks made by Farassat [42] that linear unsteady aerodynamics should be viewed as part of the problem of aeroacoustics. There are still many problems in aeroacoustics as well as in unsteady aerodynamics to be solved by combined analytic and numerical methods. The efforts devoted here are only a small attempt in that direction.

11.7.2 Governing Equation for Acousto-Aeroelastic Problem

Referring to the development already carried out in earlier work, the governing equations for the acousto-aeroelasticity problem for an elastic wing submerged in a fluid and subject to acoustic pressure disturbance can be represented by the structural dynamic equation of motion with a general form of

$$\left[[K] - \omega^2([M] + [A_{\text{Aero}}] + [P_{\text{Acou}}])\right]\{x\} = \{F_{\text{Ext}}\} \tag{11.156}$$

where with appropriate considerations on their actual form K—stiffness-related terms

M—inertial mass-related term.

$F_{\text{Aero}}, A_{\text{Aero}}$—aerodynamic-related term.

$F_{\text{Acou}}, P_{\text{Acou}}$—acoustic-related term.

F—external forces other than those generated aerodynamically or acoustically.

The acoustic pressure is governed by Helmholtz equation, in discretized form can be written as

$$[H]\{p_{Acou}\} = i\rho\omega[G]\{v\} + \{p_{inc}\} \tag{11.157}$$

where with appropriate considerations on their actual form

H—acoustic pressure influence coefficient.

G—acoustic velocity influence coefficient.

v—acoustic velocity.

p_{Acou}—acoustic pressure.

The unified treatment which was the focus of the earlier work capitalizes on the following:

a. The set of Eqs. (11.156) and (11.157) represents the unified state of affairs of acousto-aeroelastic problem being considered, which in principle can be solved simultaneously. However, prevailing situations may allow the solution of the two equations independently.
b. The acoustic term incorporated in Eq. (11.156), as can be seen in subsequent development, is considered to consist of three components to be elaborated below and is also treated in unified fashion.
c. Analogous to the aerodynamic terms, the total acoustic pressure in the acoustic term incorporated in Eq. (11.156) will be separated into elastic structural motion dependent and independent parts, which is referred to here as acoustic-aerodynamic analogy. Such approach allows the incorporation of the structural motion dependent part to the aerodynamic term in Eq. (11.156).

In the work of Huang [63], by introducing acoustic pressure field using vibrating membrane of loud-speaker, his theoretical prediction of the direct influence of the acoustic pressure field to the aerodynamically induced structural vibration has been demonstrated to agree with his experiments. On the other hand, Lu and Huang [64] and Nagai et al. [65] have introduced the influence of the acoustic pressure field to the aerodynamic flowfield, recognizing the influence of trailing edge receptivity not accounted for in the earlier work.

With such perspective in mind, the unsteady aerodynamic loading can be considered to consist of three components, i.e.:

1. The unsteady aerodynamic load component induced by the elastic structural motion in the absence of acoustic excitation;
2. The unsteady aerodynamic load due to structural vibration which is induced by the ensuing acoustic pressure loading (Pinc);
3. The unsteady aerodynamic flow-field induced by acoustic pressure disturbance.

The combined acoustic and aerodynamic loading on the structure is then synthesized by linear superposition principle of small oscillation for the acoustic pressure disturbance to the aeroelastic problem, the latter due to aerodynamic–structure interaction. To facilitate the solution of the acousto-aeroelastic stability equation solved by V-g method, a novel approach is undertaken. Analogous to the treatment of dynamic aeroelastic stability problem of structure, in which the aerodynamic effects can be distinguished into motion independent and motion induced aerodynamic forces, the effect of acoustic pressure disturbance to the aeroelastic structure (acousto-aeroelastic problem) is considered to consist of structural motion independent incident acoustic pressure and structural motion dependent acoustic pressure, which is the scattering pressure. This is referred to in this work as the acoustic aerodynamic analogy. The governing equation for the acousto-aeroelastic problem is then formulated incorporating the total acoustic pressure (incident plus scattering pressure), and the acoustic aerodynamic analogy. A generic approach to solve the governing equation as a stability or dynamic response equation is formulated allowing a unified treatment of the problem.

11.7.3 *Review of Linearized Unsteady Aerodynamics and Acoustics Differential Equations*

The present work dwells further into the unified formulation of the unsteady aerodynamics and the acoustic terms. The unsteady velocity potential Φ satisfies the gas-dynamic equation [37]:

$$\nabla^2\Phi - \frac{1}{a^2}\left[\frac{\partial^2\Phi}{\partial t^2} + \frac{\partial}{\partial t}(\nabla\Phi\cdot\nabla\Phi) + (\nabla\Phi)\cdot\nabla\left(\frac{(\nabla\Phi\cdot\nabla\Phi)}{2}\right)\right] = 0 \quad (11.158)$$

For flows having free-stream velocity U_∞ in the direction of the positive x-axis, and assuming φ to be the perturbation velocity potential which satisfies:

$$\Phi = \phi + U_\infty x \quad (11.159)$$

$$u = \frac{\partial\phi}{\partial x} \text{ and assuming } u \ll U_\infty \quad (11.160)$$

then one obtains [45, 66].

$$\nabla^2\Phi - \frac{1}{a^2}\left[\frac{\partial}{\partial t} + U_\infty\frac{\partial}{\partial x}\right]^2\phi = F \cong 0 \quad (11.161)$$

where F is the contribution of nonlinear terms, which for practical purposes consistent with perturbation linearization assumption, can be neglected. Furthermore, it will be convenient to consider harmonic motion as an element of any unsteady motion which can be built up using the principle of superposition. Then one may assume:

$$\phi(x,t) = \overline{\phi}(x)e^{i\omega t} \tag{11.162}$$

Substitution into Eq. (11.161) gives

$$\frac{\partial^2\phi}{\partial x^2} - \frac{1}{a^2}\left[i\omega + U_\infty\frac{\partial}{\partial x}\right]^2\phi = 0 \tag{11.163a}$$

Defining $M \equiv \frac{U_\infty}{a}$, Eq. (11.163a) can be recast into:

$$\left(1 - M^2\right)\frac{\partial^2\overline{\phi}}{\partial x^2} + \frac{\partial^2\overline{\phi}}{\partial z^2} - 2\frac{i\omega M}{a}\frac{\partial\overline{\phi}}{\partial x} - \frac{\omega^2\overline{\phi}}{a} = 0 \tag{11.163b}$$

Define a new variable Φ^* following Prandtl–Glauert transformation as:

$$\overline{\phi}(x) = \Phi * \exp\left[i\frac{M^2}{\left(1-M^2\right)}\frac{\omega x}{U_\infty}\right] \tag{11.164}$$

and defines:

$$K^2 = \frac{M^2}{\left(1-M^2\right)}\frac{\omega^2}{U_\infty^2} \tag{11.165}$$

and after some algebra, the linearized unsteady aerodynamic Eq. (11.9) can be recast as

$$\nabla^2\Phi * + K^2\Phi* = 0 \tag{11.166}$$

or

$$\left\{\nabla^2 + K^2\right\}\Phi* = 0 \tag{11.167}$$

which has the wave-equation form similar to the Helmholtz equation for the acoustic pressure disturbance [56–60, 62].

$$\left\{\nabla^2 + k^2\right\}p = 0 \tag{11.168}$$

11.7.4 Boundary Element Formulation of the Solution of the Linearized Unsteady Aerodynamics Equations

To solve for the differential Eqs. (11.166) or (11.167), boundary integral formulation can be obtained by the application of the Green's theorem, similar to the approach taken for the solution of the acoustic pressure Eq. (11.14a, 11.14b). Applying the Green's theorem to Eq. (11.13) [31, 45, 49], then one obtains

$$\iiint_D \nabla\Phi_1^* \cdot \nabla\Phi_2^* dV = \iint_S \nabla\Phi_1^* (\nabla\Phi_2^* \cdot \mathbf{n}) dS = \iint_S \nabla\Phi_1^* \frac{\partial \Phi_2^*}{\partial n} dS \tag{11.169}$$

in which Φ_1^* is the transformed potential of interest and Φ_2^* the corresponding free-space Green's function, i.e.

$$\{\nabla^2 + K^2\}\Phi_1^* = 0 \tag{11.170}$$

$$\{\nabla^2 + K^2\}\Phi_2^* = \delta(R - R_1) \tag{11.171}$$

Then it follows that [31, 37, 49]

$$\Phi_1^* = \frac{1}{4\pi} \iint_S \Phi_1^* \left(\frac{\partial}{\partial v} \left(\frac{e^{-iK|R-R_1|}}{R - R_1} \right) \right) dS \tag{11.172}$$

Here the free-space Green's function with the general form

$$G(|R - R_1|) = \frac{e^{-iK|R-R_1|}}{4\pi |R - R_1|} \tag{11.173}$$

is used, where Φ_1^* refers to the R coordinate of the field point of interest and R_1 the coordinate of the source element [58–60, 67], and v (or u, as appropriate) is the unit outward normal vector of a surface element on S. The schematic of the acoustic domain is depicted in Fig. 11.30.

In the development that follows, attention is focused on lifting surfaces, for which the wing (or airfoil) thickness is assumed to be reasonably small so that many nonlinear terms can be neglected [66].[15] Then Eq. (11.172) reduces to

$$\Phi_1 * (x, y, z, \omega) = \frac{1}{4\pi} \iint_S \Delta\Phi_1 * \left(\frac{\partial}{\partial z_1} \left(\frac{e^{-iK|x-x_1|}}{x - x_1} \right) \right) dS \tag{11.174}$$

[15] Such assumptions can be relieved, and similar expression can be obtained by superposing Eq. (11.17) with the internal problem inside the body and wing surface, as carried out in Djojodihardjo, RH and Widnall, SE, [1].

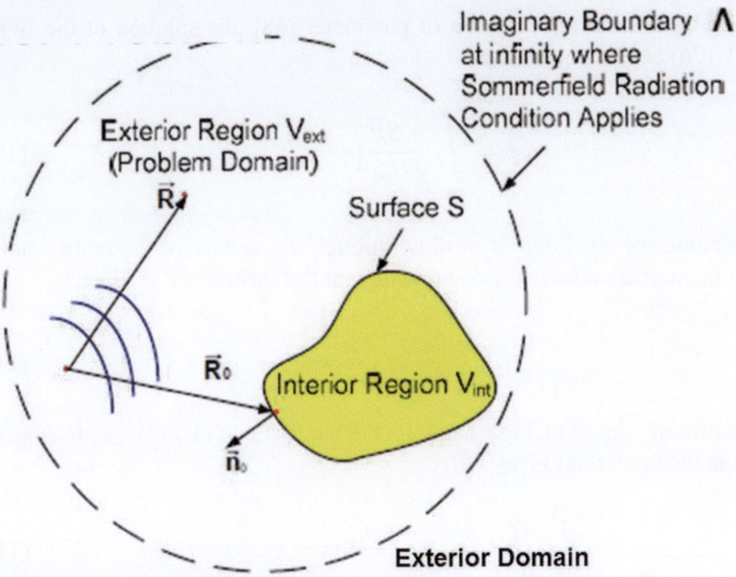

Fig. 11.30 Schematic of acoustic domain

where now $\Delta\Phi_1^*$ refers to the difference between the values of Φ_1^* at the upper and lower surface at the same coordinate location. Inverse transformation of 11.173 gives:

$$\bar{\phi} = \left[\begin{array}{l}\dfrac{1}{4\pi}\displaystyle\iint\limits_{S_{B-W}} \Delta\bar{\phi}_1 \exp\left[\dfrac{M^2}{(1-M^2)}\dfrac{i\omega}{U_\infty}x\right] \\ \left(\dfrac{\partial}{\partial z_1}\left(\dfrac{e^{-iKr}}{r_1}\right)\right)dS\end{array}\right]\exp\left[\dfrac{M^2}{(1-M^2)}\dfrac{i\omega}{U_\infty}x\right] \tag{11.175}$$

where $\overline{\phi}_1$ stands for the stationary part of the perturbation velocity potential of the flow field.

Integral equation for Pressure-Downwash relationship.

The velocity potential φ is related to the harmonically varying pressure difference ϕ by the Bernoulli equation:

$$\Delta\overline{P} = -\rho\left[i\omega + U_\infty\frac{\partial}{\partial x}\right]^2\overline{\phi} \tag{11.176}$$

where

$$\Delta P = \Delta\overline{P}e^{i\omega t} \tag{11.177}$$

Using the method of variation of parameter [68], the solution of the Bernoulli Eq. (11.176) can be obtained as

$$\overline{\phi} = \int_{-\infty}^{x} \frac{-\Delta\overline{P}}{\rho U_\infty}\left(e^{\frac{i\omega}{U_\infty}(\lambda - x)}\right)d\lambda \tag{11.178}$$

The boundary condition of no-flow through the surface of the body (and wing) defines the normal velocity or downwash w at the surface and is given by

$$\frac{\partial\overline{\phi}}{\partial n} = \frac{\partial\overline{\phi}}{\partial z} = w \tag{11.179}$$

Substituting Eqs. (11.178) and (11.179) into Eq. (11.175) (with appropriate algebraic manipulation) gives:

$$\frac{w}{U_\infty} = -\frac{1}{8\pi}\iint\limits_{S} \frac{-\Delta\overline{P}(x_1, y_1)}{\rho U_\infty} K(x - x_1, y - y_1)dS \tag{11.180}$$

where

$$K(x, y) = \lim_{z\to 0}\left[-\frac{i\omega}{U_\infty}x\right]\left[\left[\int_{-\infty}^{x-x_1} \exp\left[\frac{M^2}{\left(1 - M^2\right)}\frac{i\omega}{U_\infty}\xi\right]\right]\left(\frac{\partial^2}{\partial z^2}\left(\frac{e^{-iKr}}{r}\right)\right)d\xi\right] \tag{11.181}$$

which is the well-known standard Kernel function representation in lifting surface or panel method [43] and leads to further utilization of DPM (Doublet Point Method) or DLM (Doublet Lattice Method) in the unsteady aerodynamic problem as solution of 11.167 with appropriate boundary conditions. The detail of the formulation is elaborated elaborated in [3, 44].

Unified Treatment of Aerodynamic-Acoustic Forcing in the Acousto-Aeroelastic Boundary Element Formulation

The treatment of the aerodynamic problem elaborated above motivates further elaboration of the unified approach of the aerodynamic and acoustic part of the forcing function on the elastic structure, which leads to the governing Eq. (11.158) to be rewritten as[16]:

[16] Further detail can be found in Djojodihardjo and Safari [69], Djojodihardjo [70], Djojodihardjo [71].

$$\left[[K*] - \omega^2\left([M*] + \rho_0[F(ik_w)] + \frac{\rho}{2}\left(\frac{b_R}{k}\right)^2[A*(ik)]\right)\right]\{q\} = \{P*_{Inc_a}\} \tag{11.182}$$

and

$$\begin{aligned}&\rho_0\omega^2[G_{11}][T][\Xi]\{q\} + [H_{11}]\{p_a\} + [H_{12}]\{p_b\}\\&= i\rho\omega[G_{12}]\{v_b\} + \{p_{Inc_a}\}\end{aligned} \tag{11.183}$$

$$\begin{aligned}&\rho_0\omega^2[G_{21}][T][\Xi]\{q\} + [H_{21}]\{p_a\} + [H_{22}]\{p_b\}\\&= i\rho\omega[G_{22}]\{v_b\} + \{p_{Inc_a}\}\end{aligned} \tag{11.184}$$

where

$$[K*] = [\Xi]^T[K][\Xi] \tag{11.185a}$$

$$[M*] = [\Xi]^T[M][\Xi] \tag{11.185b}$$

$$[F_{ac}(ik_w)] = [\Xi]^T[L][P_{ac}(ik_w)][\Xi] \tag{11.185c}$$

$$[A*(ik)] = [\Xi]^T[G_{ips}]^T[A*(ik)][G_{ips}][\Xi] \tag{11.185d}$$

where Ξ is the eigenmodes obtained from modal analysis of the purely elastic structure governed by

$$[\mathbf{K} - \omega^2\mathbf{M}]\{\mathbf{x}\} = 0 \tag{11.186}$$

and where **L** is a coupling matrix of size **M** × **N**, where **M** is the number of FE degrees of freedom and **N** is the number of BE nodes on the coupled boundary a, depicted in Fig. 11.31. For each coinciding boundary and finite elements, the acoustic pressure elemental coupling term is given by Djojodihardjo [60], Holström [61] and Marquez et al. [62].

Here

$$L_e = \iint_S N_F n N_B dS \tag{11.187}$$

where N_F and N_B are the shape function of the coinciding finite element and boundary element, respectively, and n is the normal vector to the corresponding coinciding surface. Here $\{x\}$ is the structural deformation and transformed to the generalized coordinates $\{q\}$ using the following relationship:

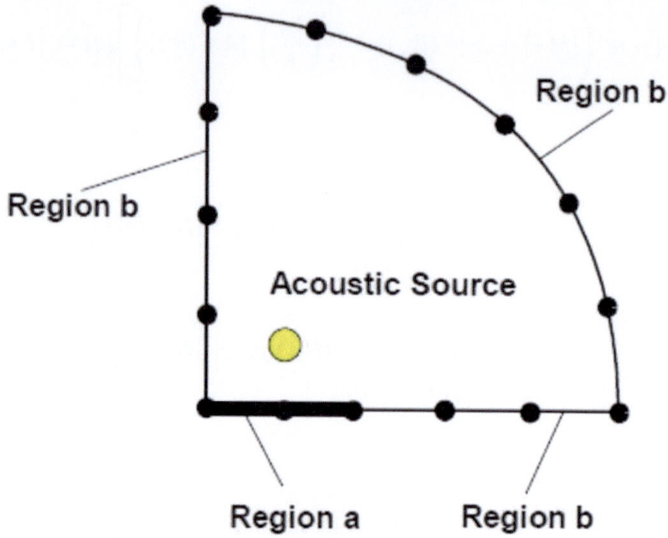

Fig. 11.31 Schematic of BE-FE fluid–structure interaction domain represented as a quarter space with defined boundary conditions on the boundaries

$$\{x\} = [\Xi]\{q\} \tag{11.188}$$

where Ξ is the modal matrix as solution to the eigenvalue problem corresponding to the free vibration of the structure.

By appropriate mathematical derivation, it was readily shown by Djojodihardjo [58, 60] that

$$\{p_a\} = -\rho_0\omega^2[P(ik_w)]\{x\} \tag{11.189}$$

With such formulation, similar procedure as elaborated in Djojodihardjo [58–60] is then followed.

11.7.5 *Acoustic-Aerodynamic Analogy*

The Acoustic-Aerodynamic Analogy procedure, which is very essential in solving the problems, is elaborated in Chap. 24 in this book[17] and will also be followed here. One significant equation point that is taken up here is that the solution of Eq. (11.182) will be facilitated by the use of modal approach, i.e. transforming $\{x\}$ into the generalized coordinate $\{q\}$ following the relationship $x = \Phi q$, where Φ is the modal matrix. In the examples worked out, a selected lower order natural modes will be employed

[17] Which originated from Djojodihardjo [71] and Djjodihardjo and Safari [69].

in Φ. Pre-multiplying Eq. (11.182) by ΦT and converting the dynamic pressure q_∞ into reduced frequency (k) as elaborated in Djojodihardjo [58–60], Eq. (11.182) can then be written as:

$$\bar{\phi} = \begin{bmatrix} \frac{1}{4\pi} \iint\limits_{S_{B-W}} \Delta\bar{\phi}_1 \exp\left[\frac{M^2}{(1-M^2)}\frac{i\omega}{U_\infty}x\right] \\ \left(\frac{\partial}{\partial z_1}\left(\frac{e^{-iKr}}{r_1}\right)\right) dS \end{bmatrix} \exp\left[\frac{M^2}{(1-M^2)}\frac{i\omega}{U_\infty}x\right]$$

$$\begin{aligned} &\left[[K*] - \omega^2\left([M*] + \rho_0[F(ik_w)] + \frac{\rho}{2}\left(\frac{b_R}{k}\right)^2 [A*(ik)]\right)\right] \\ &\{q\} = \{P*_{Inc_a}\} \\ &\Phi^T\left[\mathbf{K} - \omega^2\left(\mathbf{M} + \rho_0\left(\frac{L}{k}\right)^2 [\mathbf{A}^*(ik)] + \rho_0[\mathbf{F}_{ac_\infty}(k_w)]\right)\right]\Phi\{q\} \\ &= \Phi^T[\{\mathbf{F}_{ac_{inc}}(k_w)\} + \{\mathbf{F}\}] \end{aligned} \tag{11.190}$$

Since all of the acoustic terms are functions of wave number (k_w), Eq. (11.189) will be solved by utilizing iterative procedure. Incorporation of the scattering acoustic component along with the aerodynamic component in the second term of Eq. (11.189) can be regarded as one manifestation of what is referred to here as the acoustic-aerodynamic analogy followed in this approach.

11.7.6 Case Study and Numerical Results

Generic case of a fixed flat plate with a pulsating monopole acoustic source located at certain point on its upper surface has been considered for benchmarking in earlier work [58–60]. The present work will elaborate the application of the in-house Doublet Point Method code written in MATLAB©, which has been rederived using boundary integral formulation in Sects. 11.3 and 11.4. Further formulation of the acoustic-aerodynamic analogy as implied by Eq. (11.189) will account for the change in the flow boundary conditions due to the presence of the acoustic pulsating monopole source as was carried out in preceding work [60]. The use of modified BAH wing [52] as carried out previously will be utilized for benchmarking for the three-dimensional case. Results from this approach, along with the previous results for the direct acoustic effect on the acoustic influence on aeroelastic stability [37, 45, 72], will be discussed. For this purpose, the in-house MATLAB program was developed following the Doublet Point (DPM) Scheme of Ueda and Dowell [2] for incorporation into the combined acousto-aeroelastic solution of flutter stability problem of concern. The validation of the in-house DPM program by the benchmarking application of the in-house DPM code the equivalent BAH wing considered in Djojodihardjo [58–60]

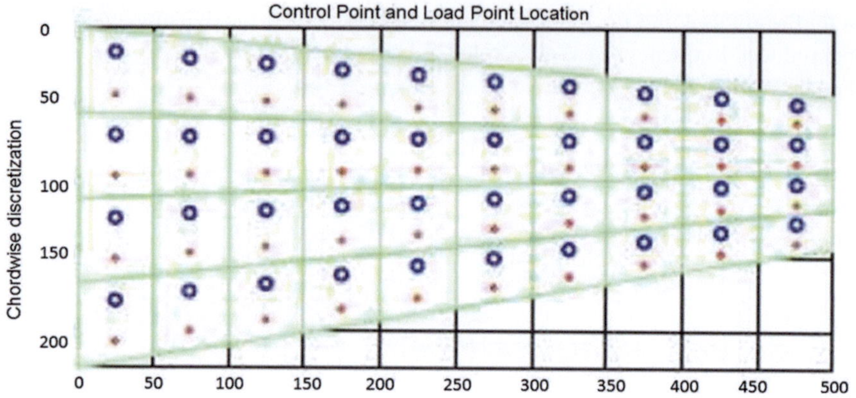

Fig. 11.32 Equivalent BAH wing discretization for unsteady aerodynamic load calculation

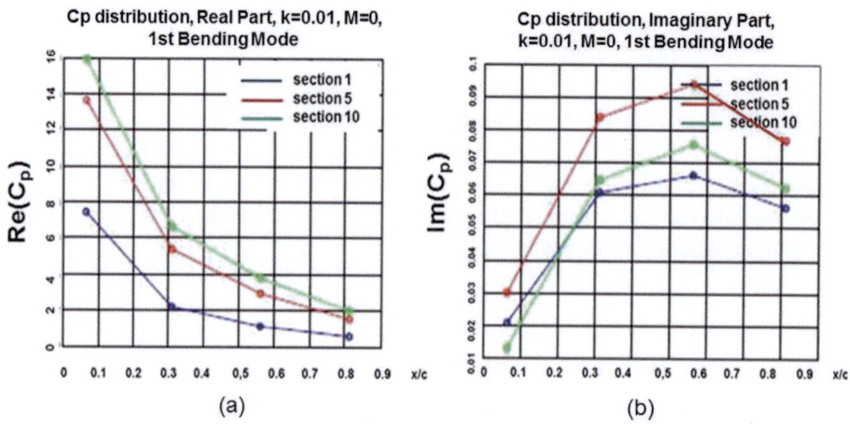

Fig. 11.33 **a** Real part of C_p distribution and **b** imaginary part of C_p distribution

is elaborated in Akbar et al. [44] and Djojodihardjo et al. [3], and presented here. For the purpose of this study, the equivalent BAH wing planform was discretized for 40 aerodynamic boxes with 10 elements span-wise and 4 elements chord-wise as depicted in Fig. 11.32.

To validate the method, calculations were carried out for several values of k (k = 0, 0.3, 0.6, 0.8). Figure 11.33 shows C_p distribution at station 1, 5 and 10 for those k values. These results have been assessed to compare well with the typical C_p distribution from Zwaan [4],[18] which is shown in Fig. 11.33.

[18] Djojodihardjo et al. [3].

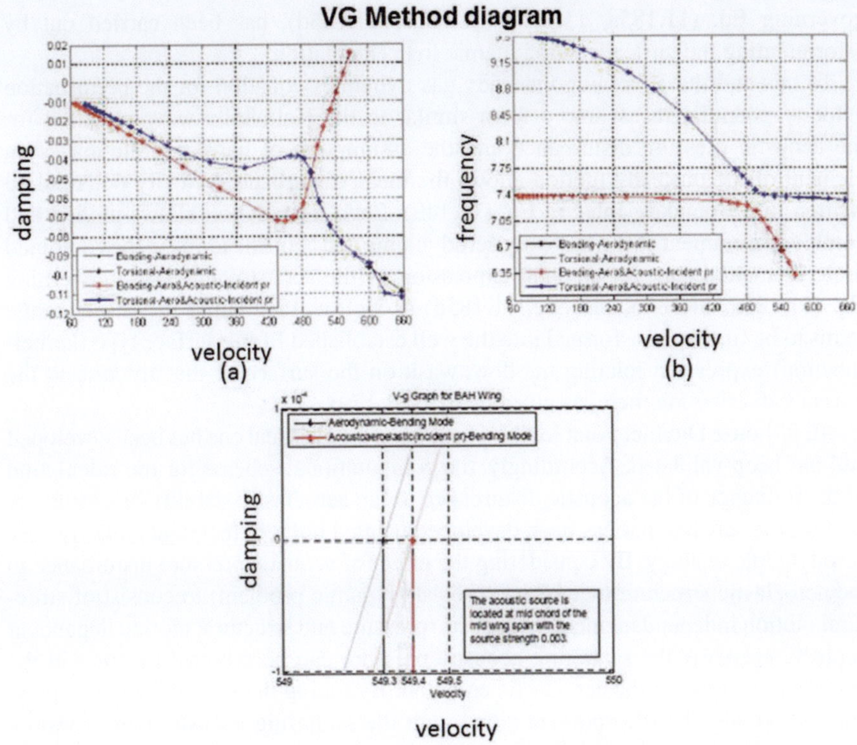

Fig. 11.34 **a** Damping and **b** frequency diagram for BAH wing calculated using V-g method written in MATLAB® for the acousto-aeroelastic problem (the total acoustic pressure already incorporates the scattering pressure). **c** enhanced V-g curve at flutter

Figure 11.34 exhibits the application of the method using the reformulation of the aerodynamic problem for the flutter stability problem of modified BAH Wing; (a) Damping and (b) frequency diagram for BAH wing calculated using V-g method written in MATLAB® for the acousto-aeroelastic problem. The V-g diagrams are drawn for pure aeroelastic as well as for acousto-aeroelastic cases. The figure illustrates the influence of the acoustic disturbance on the aeroelastic behavior of the structure considered.

11.7.7 Concluding Remarks for Case-Study IV

The objective of the present work to establish a unified approach for the acoustic and aerodynamic component of the acousto-aeroelastic problem using boundary integral approach, which will facilitate the unified treatment of the acousto-aeroelastic governing Eq. (11.185a, 11.185b, 11.185c, 11.185d), has been carried out by reformulating the unsteady aerodynamic load calculation.

To this end the governing unsteady gas dynamics equation for the perturbation velocity potential recast into a form similar to the Helmholtz wave equation for the acoustic pressure disturbance and the assumption of harmonic motion as an element of the unsteady motion allows the mean disturbance velocity potential be written in a Boundary Integral Eq. (11.149). Such formulation allows unified and combined treatment of the unsteady aerodynamic and acoustic terms in the combined finite element–boundary element expression of the acousto-aeroelastic governing Eq. (11.185a, 11.185b, 11.185c, 11.185d) as well as facilitating the aerodynamic terms to be further transformed into the well established lifting surface type (kernel-function) expression relating the downwash on the surface of the structure to the pressure distribution there, as given by Eq. (11.179).

An in-house Doublet Point Method following the classical one has been developed and has been validated. Accordingly, the computational scheme for the calculation of the influence of the acoustic disturbance to the aeroelastic stability of a structure follows the scheme that has been developed using a unified treatment and acoustic-aerodynamic analogy. By considering the effect of acoustic pressure disturbance to the aeroelastic structure (acousto-aero-elastodynamic problem) to consist of structural motion independent incident acoustic pressure and structural motion dependent acoustic pressure, the scattering acoustic pressure can be grouped together in the aerodynamic term of the aeroelastic equation. By tuning the incident acoustic pressure, it can also be incorporated along with the scattering acoustic term, forming the acousto-aeroelastic stability equation. For this purpose the problem domain has been defined to consist of those subjected to acoustic pressure only and that subject to acoustic structural coupling, which is treated as acousto-aeroelastic equation. Using BE and FE as appropriate, an integrated formulation is then obtained as given by the governing Eq. (11.189), which relates all the combined forces acting on the structure to the displacement vector of the structure. The solution of Eq. (11.186)—and after using modal approach in structural dynamics, Eq. (11.189)—can be obtained by solving it as a stability equation in a"unified treatment". The disturbance acoustic pressure already incorporates the total pressure (incident plus scattering pressure), which has been "tuned" to behave like the aerodynamic terms in the modal Eq. (11.189). Such approach allows the application of the solution of the acousto-aeroelastic stability equation in the frequency domain using V-g method.

Appendix: For Case Study II—Calculation of 3D Unsteady Subsonic Flow with Separation Bubble Using Singularity Method

The Singular Part of Kernel Functions[19]

Noting the complexity of the kernel functions, it is intended to integrate them using numerical approach. It is obvious that the singularity contained in the kernel functions will tend to spoil the numerical integration if they are not carefully treated. The usual technique to take care of this problem is to isolate the region in the vicinity of the singular point, and to integrate it in an analytical manner applying a suitable limiting process. The remaining regular kernel can then be easily and accurately integrated using a standard numerical quadrature. The limiting processes for the second and third kernels are well documented in many literatures. Such procedure leads to Eqs. (11.98) and (11.99). Afterward, the singularities of the first and fourth kernel are presented.

The integral in the chordwise direction will contain singularities if η and η 'take the same values. This situation is closely related to the integration technique applied for the spanwise direction. The singular part of the first kernel can be examined to consist of some poles and a logarithmic function, since the Bessel function has a logarithmic behaviour for small arguments. This singular part is [42]

$$K_{1s}\left(\xi,\eta;\xi^{'},\eta^{'}\right)=\frac{e^{-ik\left(\xi-\xi^{'}\right)}}{s^{2}}\left(\begin{array}{c}-\dfrac{1}{\left(\eta-\eta^{'}\right)^{2}}\left|1+\dfrac{\xi-\xi^{'}}{r^{'}}\right|\\+\dfrac{ik}{r^{'}}-\dfrac{k^{2}}{2}\ln\left|r^{'}-\left(\xi-\xi^{'}\right)\right|\end{array}\right) \tag{11.191}$$

The singular part of the fourth kernel has a less complexity since the singularity is caused only by the *r* '. Consequently it produces some poles, which can be taken separately as

$$K_{1s}\left(\xi,\eta;\xi^{'},\eta^{'}\right)\approx\frac{1}{2\pi}\left[-\frac{ik}{\beta^{2}\left|\xi-\xi^{'}\right|}\left(1-\frac{M\left(\xi-\xi^{'}\right)}{\left|\xi-\xi^{'}\right|}\right)\right] \tag{11.192}$$

If these two singular parts obtained are subtracted from the original kernels, the rest will become regular functions which facilitate the numerical treatment of them. Proceeding along this line, then the non-singular of the first and fourth kernel are

$$K_{1ns}\left(\xi,\eta;\xi^{'},\eta^{'}\right)=K_{1}\left(\xi,\eta;\xi^{'},\eta^{'}\right)-K_{1s}\left(\xi,\eta;\xi^{'},\eta^{'}\right) \tag{11.193}$$

[19] Not all the appendices and the equations elaborated in the referenced paper are reproduced here. Further details are given in Djojodihardjo, Sekar, Prananta [2].

Other Appendices and related equations can be found in the referred document.

$$K_{4ns}\left(\xi, \eta; \xi', \eta'\right) = K_4\left(\xi, \eta; \xi', \eta'\right) - K_{4s}\left(\xi, \eta; \xi', \eta'\right) \tag{11.194}$$

Integration Along the Chordwise Direction

The integration of the first kernel is worked out exactly in the same manner as the one in Laschka [42], where the singular parts are treated analytically and the rest numerically. Proceeding along Laschka's approach the singular integral is divided into three parts[20]

$$f_{1r}\left(\xi, \eta; \xi', \eta'\right) = \frac{1}{2\pi}\sum_{r=0}^{R}\left\{\lim_{\varepsilon\to 0}\left[\begin{array}{l}\frac{2}{\varepsilon}a_r^*(\eta)f_{11rs}\left(\xi, \eta; \eta'\right) \\ -\underset{\substack{\text{singular}\\ \text{integral}}}{\int_{-1}^{1}}\frac{a*_r\left(\eta'\right)}{\left(\eta-\eta'\right)}f_{11r\,ns}\left(\xi, \eta; \eta'\right)d\eta'\end{array}\right]\right\}$$

$$+\int_{-1}^{1}a_r^*\left(\eta'\right)\ln\left|\eta-\eta'\right|f_{12r}\left(\xi, \eta; \eta'\right)d\eta' + \int_{-1}^{1}a_r^*\left(\eta'\right)f_{13r}\left(\xi, \eta; \eta'\right)d\eta' \tag{11.195}$$

The Gauss-Jacobian Quadrature is employed for the integration of the non-singular kernel times the series expansion. The quadrature chosen takes the advantage of the series expansion, $h_r(X)$, as the weighting function,

$$L(r, X) = \int_0^1 h_r\left(X'\right)F\left(X, \eta; X', \eta'\right)dX' = \sum_{j=0}^{J}B_{rj}F\left(X, \eta; X_j, \eta'\right) \tag{11.196}$$

where

$$B_{rJ} = \frac{\cos\left(r\phi_j\right)\cos\left((r+1)\phi_j\right)}{J+1} \tag{11.197}$$

The components of $f_{1r}(\xi, \eta; \eta)$ are then evaluated one by one using this formula as

[20] The integration $\underset{\substack{\text{singular}\\ \text{integral}}}{\int_{-1}^{1}}$ signifies singular integral that should be dealt with following the established procedure.

$$f_{11rs}\left(\xi,\eta;\eta'\right)=\sum_{j=0}^{J}B_{rj}e^{-ik\frac{c(\eta)}{s}(X-X_j)}\left(1+\frac{X-X_j}{R_j}\right) \tag{11.198}$$

$$\begin{aligned} f_{12rvs}\left(\xi,\eta;\eta'\right)=&f_{1rs}\left(\xi,\eta;\eta'\right)\\ &+\left(\frac{dh_r(X)}{dX'}+ik\frac{c(\eta)}{s}h_r\left(X'\right)\right)\\ &\left(\frac{\beta s}{c(\eta)}\right)^2\left(\eta-\eta'\right)^2\ln\left|\eta-\eta'\right| \end{aligned} \tag{11.199}$$

$$\begin{aligned} f_{12r}\left(\xi,\eta;\eta'\right)=&-k^2\sum_{j=0}^{J}B_{rj}e^{-ik\frac{c(\eta)}{s}(X-X_j)}\\ &+\left(\frac{\beta s}{c(\eta)}\right)^2\left[\frac{dh_r(X)}{dX'}+ik\frac{c(\eta)}{s}h_r(X)\right]-2ik\frac{s}{c(\eta')}h_r(X) \end{aligned} \tag{11.200}$$

and

$$\begin{aligned} f_{13r}\left(\xi,\eta;\eta'\right)=&\frac{f_{1rvs}\left(\xi,\eta;\eta'\right)}{\left(\eta-\eta'\right)^2}-\ln\left|\eta-\eta'\right|f_{2r}\left(\xi,\eta;\eta'\right)\\ &-\frac{1}{\left(\eta-\eta'\right)^2}\sum_{j=0}^{J}B_{rj}e^{-ik\frac{c(\eta)}{s}(X-X_j)}\\ &\left\{\begin{array}{l} k\left|\eta-\eta'\right|\left[\begin{array}{l} K_1\left(k\left|\eta-\eta'\right|\right)-i+\frac{\pi i}{2}\left(I_1\left(k\left|\eta-\eta'\right|\right)-L_1k\left|\eta-\eta'\right|\right)\\ +i\int_0^{U\cdot Br}\frac{\tau}{\left(1+\tau^2\right)^{\frac{1}{2}}}e^{2ik(\eta-\eta')\tau}d\tau \end{array}\right]\\ +\frac{X-X_j}{R}e^{-ik\frac{c(\eta)}{s\beta^2}(X-X_j-MR_j)} \end{array}\right\} \end{aligned} \tag{11.201}$$

where the upper boundary of the integral becomes

$$U\cdot Br=\frac{\left(X-X'-MR\right)}{\beta^2\left|\eta-\eta'\right|}\frac{c\left(\eta'\right)}{s} \tag{11.202}$$

and

$$R=\left[\left(X-X'\right)^2+\beta^2\frac{s^2}{c^2(\eta)}\left(\eta-\eta'\right)^2\right]^{\frac{1}{2}} \tag{11.203}$$

$$X'=\frac{X'-X_{\mathrm{LE}}}{c(\eta)}=\frac{1}{2}\left(1-\cos\left(\phi'\right)\right) \tag{11.204}$$

$$\phi' = \frac{j\pi}{J+1} \tag{11.205}$$

$$j = 0, 1, 2, \ldots j \tag{11.206}$$

Following Watkins et al.[54], the approximation of the integrand is carried out by expansion like,

$$\frac{\tau}{\left(1+\tau^2\right)^{\frac{1}{2}}} = 1 - 1.101e^{-0.329\tau} - 0.899e^{-1.4067\tau} - 0.09480933e^{-0.329\tau}\sin\pi\tau \tag{11.207}$$

Using this expression the integral appeared in Eq. (11.201) could be worked out.

References

1. Djojodihardjo, R.H., and Sheila E. Widnall. 1969. A numerical method for the calculation of nonlinear unsteady potential flow problems. *AIAA Journal* 7 (10), October.
2. Djojodihardjo, H., W.K. Sekar, and B.B. Prananta. 1992. Calculation of 3D unsteady subsonic flow with separation bubble using singularity method. In *18th, ICAS Congress*, Beijing.
3. Djojodihardjo, M. Akbar, L. Gunawan, and M. Kusni. 2010. Unified aerodynamic-acoustic formulation for aero-acoustic structure coupling. In *27th Congress of the International Council of the Aeronautical Sciences 2010, ICAS 2010*, vol. 3, issue: Nice, France.
4. Haryanto, I., A preliminary study on buffeting problem utilizing dynamic response approach. In *ICAS94-10.7.3, and 4*.
5. Greidanus, J.H., A.I. van de Vooren, and H. Bergh. 1952. Experimental determination of the aerodynamic coefficients of an oscillating wing in incompressible two-dimensional flow. Repts, F-101 to F-104, National Luchtvaart Laboratorium.
6. Ransleben, G.E., and H.N. Abramson. 1962. Experimental determination of oscillatory lift and moment distributions of fully submerged flexible hydrofoils. Rept. 2, Contract Nonr-3335(00), Southwest Research Institute, San Antonio, Texas.
7. Spurck, J. 1963. Messungen der Aerodynamischen Beiwerte Schwingender Flugelprofil in Windkanal und Vergleich met der Theorie. NR. 29, Mitteilungen Max-Planck Institute, Gottingen.
8. Küssner, H.G. and G. von Gorup. 1960. Instationare Lineasierte Theorie der Flugelprofile endlicher Dicke in Inkompressibler Stromung. NR. 26, Mitteilungen Max-Planck Institute, Gottingen.
9. Hewson-Browne, R.C. 1963. The oscillation of a thick aerofoil in an incompressible flow. *Quarterly Journal of Mechanics and Applied Mathematics* 16: 79–92.
10. van de Vooren, A.I., and H. van de Vel. 1963. Unsteady profile theory in incompressible Flow. Rept. TW-17, Mathematics Institute, Rijksuniversiteit, Groningen, The Netherlands.
11. Giesing, J.P. 1968. Nonlinear two-dimensional unsteady potential flow with lift. *Journal of Aircraft* 5(2): 135–143.
12. Chen, C.F., and R.A. Wirtz. 1968. Second-order theory for flow past oscillating foils. *AIAA Journal* 6(8): 1556–1562.
13. Ashley, H., S.E. Widnall, and M.T. Landahl. 1965. New directions in lifting surface theory. *AIAA Journal* 3(1): 3–15.

14. Landahl, M.T., and V.J.E. Stark. 1968. Numerical lifting surface theory—Problems and progress. In *AIAA Paper 68–72.*
15. Miller, R.H. 1964a. Rotor blade harmonic loading. *AIAA Journal* 2(7): 1254–1269.
16. Miller, R.H. 1964b. Unsteady air loads on helicopter rotor blades. *Journal of the Royal Aeronautical Society* 68: 640.
17. Ashley, H. 1966. *Machine Computation of Aerodynamic Loads in Linear and Nonlinear Situations.* Rept. 66–5, AFOSR 66–1440, Fluid Dynamics Research Lab., Massachusetts Institute of Technology.
18. Hess, J.L., and A.M.O. Smith. 1966. *Calculation of Potential Flow About Arbitrary Bodies. Progress in Aeronautical Sciences*, vol. 8. Pergamon.
19. Rubbert, P.E. et al. 1967. A general method for determining the aerodynamic characteristics of fan-in-wing configuration. TR 67–61A, U. S. Army Aviation Materiel Lab.
20. Djojodihardjo, R.H. 1969. *A Numerical Method for the Calculation of Nonlinear Unsteady Lifting Potential Flow Problems.* Sc.D. thesis, Feb. Dept. of Aeronautics and Astronauics, Massachusetts Institute of Technology.
21. Hadamard, J. 1952. *The Finite Part of an Infinite Simple Integral. Lectures on Cauchy's Problem in Linear Partial Differential Equations*, 133–141. Dover.
22. Mangier, K.W. 1952. Improper integrals in theoretical aerodynamics. C.P. 94, Aeronautical Research Council, Great Britain.
23. Timoshenko, S., D.H. Young, and W. Weaver. 1974. *Vibration Problems in Engineering.* New York: Wiley.
24. Watkins, C.E., H.L. Runyan, and D.S. Woolston. 1995. On the kernel function of the integral equations relating the lift and downwash distributions of oscillating finite wing in subsonic flow, NACA Rep. 1234.
25. Wagner, H. 1925. Uber die Entstehung des Dynamischen Auftriebes von Traflugeln. *Zeitscrift fur Angewandte Maihematik und Mechanik* 5(1), Feb.
26. Chen, X.Z., M. Matsumoto, and A. Kareem. 2000. *Time domain flutter and buffeting response analysis of bridges.* https://www3.nd.edu/~nathaz/journals/. Time Domain Flutter and Buffeting Response Analysis of Bridges.
27. Jones, R.T. 1939. The unsteady lift of a finite wing. TN 682, NACA.
28. Garrick, I.E. 1957. *Nonsteady Wing Characteristics.* VII. Princeton, NJ: Princeton University Press.
29. Bratt, J.B. 1953. Flow patterns in the wake of oscillating airfoil. R & M 2773, Royal Aeronautical Establishment, Great Britain.
30. Jones, W.P. 1945. Aerodynamic forces on wings in nonuniform motion. R & M 2117, Aeronautical Research Council, Great Britain.
31. Chi, R.M. 1985. Separated flow unsteady aerodynamic theory. *Journal of Aircraft* 22 (11)
32. Ericsson, L.E., and J.P. Reding. 1981. Unsteady airfoil stall, Review and extension. *AIAA Journal* 8: 609–616.
33. Perumal, P.V.K., and F. Sisto. 1974. Lift and moment prediction for an oscillating airfoil with moving separation point. *ASME Transaction, Journal of Engineering of Power* 96: 372–378.
34. Dowell, E.H. 1977. A simplified theory of oscillating airfoil in transonic flow: Review and extention, *AIAA Paper*, 77–445, March.
35. Huber, T., B. Rochholz, and B. Laschka. 1991. Modelling of separated flow around oscillating cascade by classical singularity method. International Forum on Aerolasticity and Structural Dynamics, Aachen, June 3–6.
36. Djojodihardjo, H., W.K. Sekar, and B. Laschka. 1991. Calculation of pressure distribution on two dimensional wing in unsteady subsonic flow with separation bubble using kernel function method. International Forum on Aeroelasticity and Structural Dynamics, Aachen, Germany.
37. Bendat, J.S., and A.G. Piersol. 1971. *Random Data: Analysis and Measrument Procedure.* Canada: Wiley.
38. Laschka, B.1963. Zur Theorie der Harmonischschwingenden, Tragenden Flaeche bel Unterschallanstroemung, Zeitschrift fuer Flugwissen schaften, Braunschweig.
39. Küssner, H.G. 1940. Allgemeine Tragflachentheorie. *Luftfahrtforschung 17*, 370–378.

40. Watkins, C.E., D.S. Woolston, and H.J. Cunningham. 1960. A systematic kernel function procedures for determining aerodynamics forces on oscillating or steady finite wing at subsonic speed, NASA TRR-48.
41. Wrobel, L.C. 2003. *The Boundary Element Method, Volume 1, Applications in Thermo-Fluids and Acoustics*. Wiley.
42. Boyden, R.P., and W.G. Johson. 1982. Result of buffet tests in a cryogenic wind tunnel. NASA-TM 84520, September.
43. Djojodihardjo, H., and W.K. Sekar. 1991. Calculation of unsteady aerodynamics load on a finite span wing with separated region using strip theory approach. *IPTN Internal Report*.
44. Akbar, M., L. Gunawan, and M. Kusni. 2010. Calculation of unsteady aerodynamic loads on oscillating lifting surface using doublet point method. Paper 128, Proceedings, Regional Conference on Mechanical and Aerospace Technology Bali, February 9–10.
45. Ashley, H., and M.T. Landahl. 1965. *Aerodynamics of Wings and Bodies*. Reading, MA: Addison & Wesley.
46. Djojodihardjo, H., and W.K. Sekar. 1993. Flutter calculation of two and three dimensional subsonic wing with separation bubble as preliminary estimation of buffeting problem. In *The International Forum on Aeroelasticity and Structural Dynamics, Strasbourg, France*, May 24–26.
47. Djojodihardjo, H., W.K. Sekar, and E. Suleman. 1993. Flutter and dynamic response calculation of unsteady separated flow using singularity method or preliminary estimation of buffeting problem. In *Symposium of The Advances in Aerospace Sciences*, Stanford, CA, USA, September 23–24.
48. Försching, H. 1981. *Aeroelastic Buffeting Prediction Techniques—A General Review*. DFVLR.
49. Chyu, W.J., and M.K. Au-Yang. 1972. Random response of rectanguler panels to the pressure field beneath a turbulent boundary layor in subsonic flows. NASA TN D-6970, October.
50. Ueda, T., and E.H. Dowell. 1982. A new solution method for lifting surfaces in subsonic flow. *AIAA Journal* 20: 348–355.
51. Zwaan, R.J. 1981. Aeroelasticity of aircraft. Lecture Notes, Special Lecture, Short Course offered at Institut Teknologi Bandung, Indonesia, Lecture Notes delivered at Special Meeting at ITB in August 1981 (private collection).
52. Bisplinghoff, R.L., H. Ashley, and Halfman. 1995. *Aerolasticity*. Reading, MA: Addison Wesley.
53. Blevins, R.D. 1977. *Flow-Induced Vibration*. van Nostrand Reinhold Co.
54. Bull, M.K. 1967. Wall-pressure fluctuations associated with subsonic turbulent boundary layer flow. *Journal of Fluid Mechanics* 28(4): 719–754.
55. Djojodihardjo, H., and E. Tendean. 2004. A computational technique for the dynamics of structure subject to acoustic excitation. In *Paper IAC-04-3I.2.09. Presented at the 55th International Astronautical Congress*, 4–8 October, Vancouver, Canada.
56. Djojodihardjo, H., and I. Safari. 2005. Further development of the computational technique for dynamics of structure subjected to acoustic excitation. In *Paper IAC-05-C2.2.08, presented at the 56th International Astronautical Congress*, 17–21 October, Fukuoka Japan.
57. Djojodihardjo, H., and I. Safari. 2006. Unified computational scheme for acoustic aeroelastomechanic interaction. In *Paper IAC-06-C2.3.09, presented at the 57th International Astronautical Congress 2006*, 1–6 October, Valencia Spain.
58. Djojodihardjo, H. 2007a. BE-FE coupling computational scheme for acoustic effects on aeroelastic structures. In *Paper IF-086, Proceedings Of The International Forum on Aeroelasticity and Structural Dynamics*, held at the Royal Institute of Technology (KTH), Stockholm, Sweden, 18–20 June.
59. Djojodihardjo, H. 2007b. BEM-FEM acoustic-structural coupling for spacecraft structure incorporating treatment of irregular frequencies. In *Paper IAC.07-C2.3.4, IAC Proceedings, presented at the 58th International Astronautical Congress*, 23–28 September 2007, Hyderabad, India.
60. Djojodihardjo, H. 2008. Unified BE-FE aerodynamic-acoustic-structure coupling scheme for acoustic effects on aeroelastic structures. In *Paper I-586, 26th International Congress of the Aeronautical Sciences*, 14–19 September, Anchorage, Alaska, USA.

61. Holström, F. 2001. Structure-acoustic analysis using BEM/FEM; Implementation in MATLAB®. Master's Thesis, Copyright © 2001 by Structural Mechanics & Engineering Acoustics, LTH, Sweden, Printed by KFS i Lund AB, Lund, Sweden, May, Homepage: http://www.akustik.lth.
62. Marquez, A., S. Meddahi, S., and V. Selgas. 2004. A new BEM–FEM coupling strategy for two-dimensional fluid–solid interaction problems. *Journal of Computational Physics*, 199: 205–220.
63. Huang, X.Y. 1987. Active control of aerofoil flutter. *AIAA Journal* 25 (8).
64. Lu, P.J., and L.J. Huang. 1992. Flutter suppression of thin airfoils using active acoustic excitations. *AIAA Journal* 30 (12), December.
65. Nagai, K., K. Toshimitsu, and M. Namba. 1996. Active control of cascade flutter by means of sound waves from duct wall sources. *JSME International Journal, Series B* 39 (3): 608–614.
66. Morino, L., and C.C. Kuo. 1974. Subsonic potential aerodynamics for complex configurations: A general theory. *AIAA Journal* 12 (2), February.
67. Zingel, H. 1986. *Zur Bestimmung des aeroelastischen Verhaltens von Auftriebssystemen infolge von stromungsablosung*. DFVLR.
68. Kreyszig, E. 2006. *Advanced Engineering Mathematics*, 9th ed. Wiley.
69. Djojodihardjo, H., and I. Safari. 2013. BEM-FEM coupling for acoustic effects on aeroelastic stability of structures. *Computer Modeling in Engineering and Sciences* 91 (3): 205–234.
70. Djojodihardjo, H. 2013. Unified aerodynamic-acoustic formulation for aero-acoustic structure coupling. *JMEA* 3 (4)
71. Djojodihardjo, H. 2015. Vibro-acoustic analysis of the acoustic-structure interaction of flexible structure due to acoustic excitation. *Acta Astronautica* 108: 129–145.
72. Becker, J., and A. Gravelle. 1985. Some results of experimental and analytical buffeting investigations on a delta wing. In *International Symposium on Aeroelasticity*, Aachen, April 1–3.

Chapter 12
Introduction to Aeroservoelasticity with Case Studies

Abstract Aeroservoelasticity (ASE) extends the concept of aeroelasticity to address the aeroelastic interactions between aerodynamic forces and a flexible structure, which may include a control system. Then the classic Collar aeroelastic triangle can be extended to form the aeroservoelastic pyramid, where there are now forces resulting from the control system as well as the aerodynamic, elastic and inertial forces. Considerations of the aeroservoelastic interactions increase the importance of engineering efforts to incorporate lightweight and flexible structures, as well as high-gain digital flight control systems. Such considerations are important to account for aeroservoelastic instability. The dynamics of the guidance and control system may significantly affect aeroelastic problems, or vice versa, hence the term aeroservoelasticity. The FEM-based structural analysis is also essential for static aeroelastic studies in the nascent field of compliant blade performance modification. By way of introduction, mathematical modeling of a simple aeroelastic system with a control surface is elaborated. As a basic aeroservoelastic System that can shed some light on its state of affairs, the binary aeroelastic model will be helpful in analyzing flutter behavior as a baseline. Consideration is given to the effect of gusts. A particular case study on Aeroelastic Analysis of an Aircraft with Stand-By Actuator Using State-Space Approach is also presented for detailed treatment of the problem. Other examples discussed are the Design and Optimization of an Aeroservoelastic Wind Tunnel Model and Aeroservoelastic Modelling and Analysis of a Highly Flexible Flutter Demonstrator.

Keywords Aeroelasticity · Aeroservoelasticity · Controls surface · Control system · Physical and mathematical modeling · Aircraft system modeling

12.1 Introduction

Aeroservoelasticity (ASE) extends the concept of aeroelasticity to address the aeroelastic interactions between aerodynamic forces and a flexible structure, which may include a control system [1]. Then the classic Collar aeroelastic triangle can be

© The Author(s), under exclusive license to Springer Nature Singapore Pte Ltd. 2023

H. Djojodihardjo, *Introduction to Aeroelasticity*,
https://doi.org/10.1007/978-981-16-8078-6_12

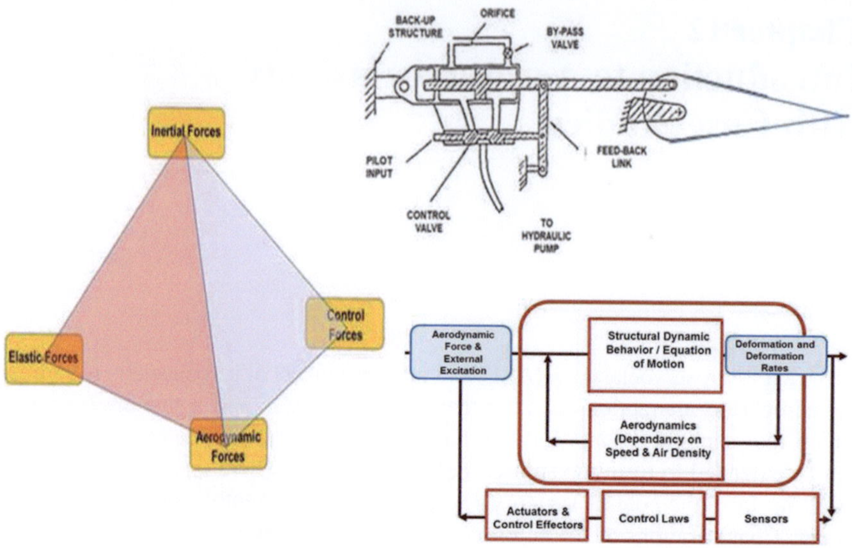

Fig. 12.1 Aeroservoelasticity pyramid, sample block diagram and sample system

extended to form the aeroservoelastic pyramid shown in Fig. 12.1, which has incorporated below forces resulting from the control system as well as the aerodynamic, elastic and inertial forces. This concept can be extended to aeroservothermoelasticity, if the state of affairs incorporates changes due to thermal effects. ASE effects become increasingly important in modern aircraft design as many applications employ some form of flight control system (FCS) to improve the handling and stability, flight performance and ride quality throughout the flight envelope, and also to reduce loads and improve service life, as depicted in the lower half of Fig. 12.1. For commercial aircraft, the FCS might include a gust and/or maneuver load alleviation system in addition to a control system that meets the basic handling requirements.

Aeroservoelasticity

Aeroservoelasticity (ASE) is multidisciplinary in nature and incorporates scientific considerations of the interactions between the aircraft steady and unsteady aerodynamic forces, the inertial forces, the elasticity of the structure and the flight control system of the aircraft. ASE interactions can affect the flight performance, structural integrity and handling qualities of the aircraft, such as what is occurring on a modified F-4 aircraft during a sideslip maneuver. The instability, excited by the control surfaces, occurred because of an overlap in frequency between a structural mode and the pitching motion of the aircraft.

In other applications, the dynamics of the guidance and control system may significantly affect aeroelastic problems, or the other way around, hence the term aeroservoelasticity. The FEM-based structural analysis is also essential for static aeroelastic studies in the nascent field of compliant blade performance modification.

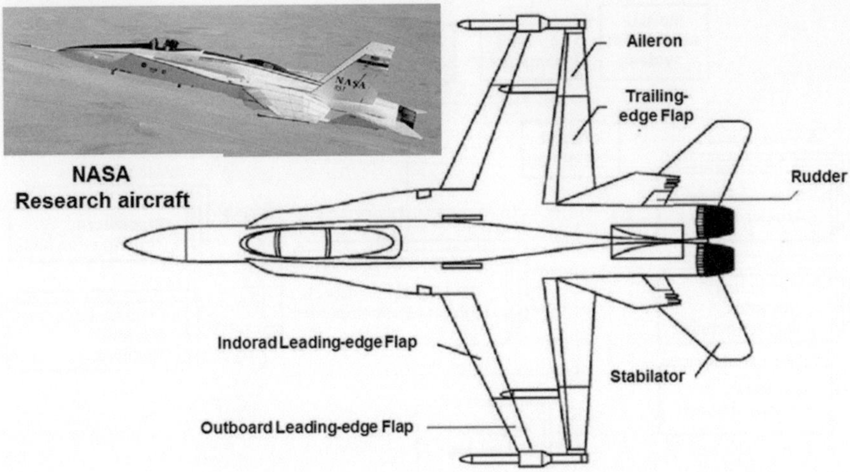

Fig. 12.2 An example of an aeroservoelastic system—NASA active aeroelastic wing aerodynamic model-modified F-A-18a-F (courtesy of NASA [3])

The compliance of the blade in an annular cascade represents a passive means of controlling the aerothermodynamic performance of the turbomachine by aeroelastic tailoring. This topic comes under the overarching subject of aeroservoelasticity, the application of automatic control theory to fundamental aeroelastic problems. In the blading of turbomachinery, the enhanced compliance and its chordwise distribution are introduced intentionally by design [2].

The concept of performance "map" may be introduced and will have to be extended to include the parametric dependence of performance on a representative value of a new dimensionless quantity of interest.

In Fig. 12.1, a Block Diagram of an Aeroservoelastic System is exemplified, while a physical example of such system is depicted on the right-hand side. Figure 12.2 depicts an example of an aeroservoelastic system—F/A-18A Aircraft Modeled as an Active Aeroelastic Wing research aircraft. Figure 12.3 depicts a Modeling example of an Aeroservoelastic System—F/A-18 control system component (courtesy of NASA [3]), while Fig. 12.4 shows some examples of Aeroservoelastic problems represented by Information flowchart for traditional and modern methods to compute flutter margins with possible extension of Aeroelasticity as first proposed by Collar to incorporate Control and Thermal Effects is depicted as Aeroservothermo Elasticity in Fig. 12.5. These illustrative examples are given here to provide an overview on the extent of aeroservoelastic system and solution problem directions.

It is feasible to develop flutter suppression systems that enable aircraft to fly beyond the flutter speed; however, such an approach has a very high risk and so far has only been demonstrated on wind tunnel models. Modern military aircraft is often designed for carefree handling and the ability to fly with reduced, or unstable, open-loop static stability so as to improve their maneuverability; however, they can only stay airborne through the use of the Flight Control System (FCS). All control

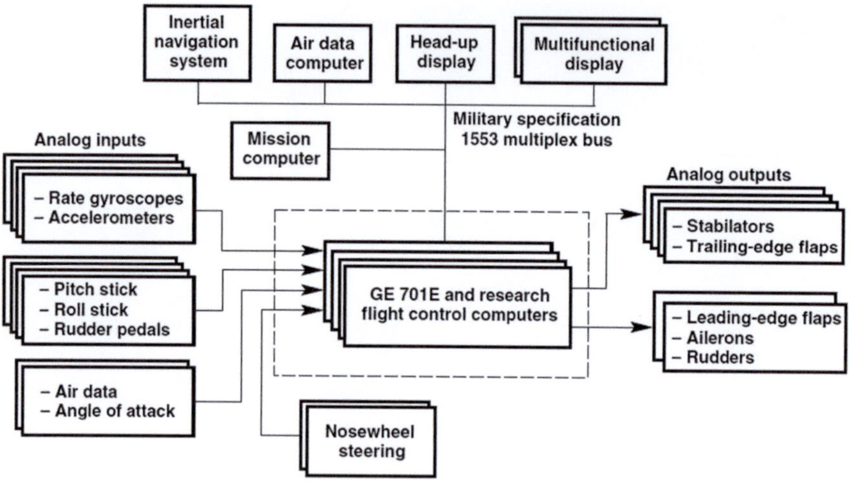

Fig. 12.3 Modeling Example of an aeroservoelastic system—F/A-18 control system component (courtesy of NASA [4])

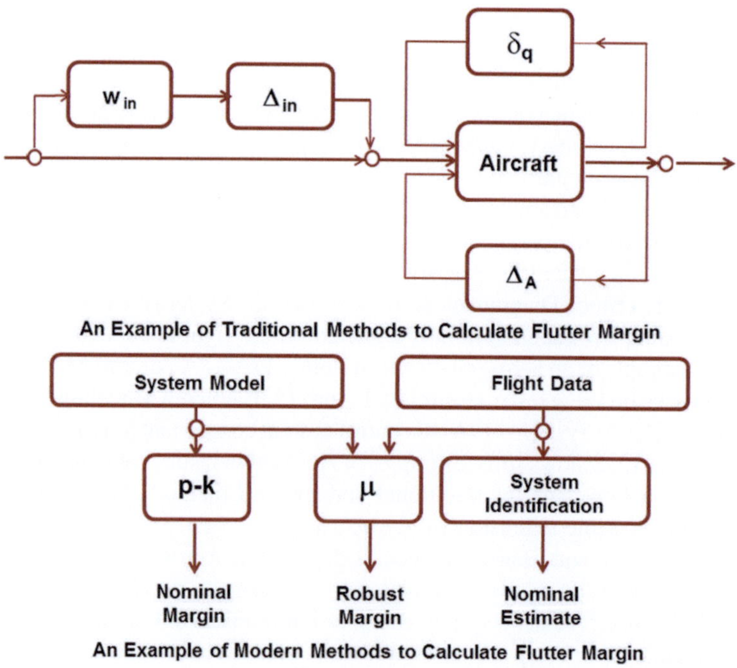

Fig. 12.4 Some examples of aeroservoelastic problems represented by information flowchart for traditional and modern methods to compute flutter margins (courtesy of NASA [5])

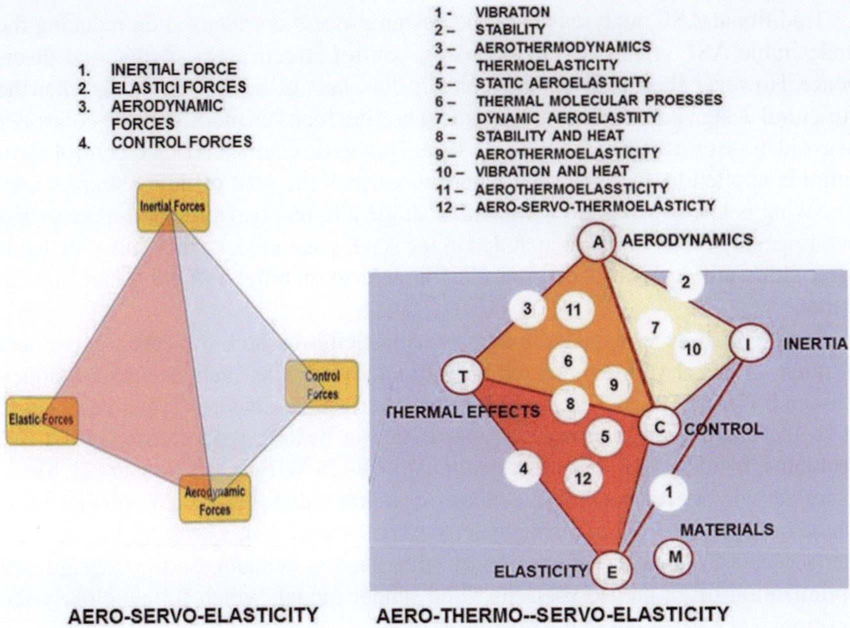

Fig. 12.5 Aeroservothermo elasticity

implementations involve the use of sensors, usually accelerometers and rate gyros placed at the aircraft center of mass and air data sensors (e.g. angle of incidence, air speed). Some form of control input could then be applied via the control surfaces.

Aeroservoelastic (ASE) effects, sometimes referred to as “structural coupling,” can potentially cause a major structural failure due to flutter involving coupling of the aeroelastic and control systems. However, there is also the possibility of causing fatigue damage and reducing control surface actuator performance. Most structural coupling problems occur when the motion sensors detect not only the aircraft rigid-body motion but also the motion in the flexible modes, and these vibrations are fed back into the FCS. In this case, the movement of the control surfaces is then likely to excite the flexible modes, so causing the aircraft to vibrate further. Notch filters are often used to remedy this problem by introducing significant attenuation of the response in the region of critical frequencies.

In the subsequent development, the use of feedback control on a simple binary aeroelastic system with a control surface is considered and the effects of the control law on the stability and response investigated. Following Wright and Cooper, [2], a simple PI controller is used to demonstrate how the gust response can be reduced and the flutter speed increased. Modeling of ASE systems in the time or frequency domain is also considered, including representation of reduced frequency-dependent aerodynamics and control effects along with the use of state space models.

Traditional ASE analyses [1] in the design process are focused on reducing the undesirable ASE effects on aircraft loads, control effectiveness, flutter, and divergence. However, these analyses are typically done later in the design process when the structural design is somewhat mature and therefore focus on modifying the controller to avoid these effects. For example, a Linear Quadratic Gaussian (LQG) control algorithm is applied to an aeroelastic wing model with the goal of flutter suppression. The wing is low-aspect ratio wing with a single aileron. Two structural modes (first bending and first torsion) are included in the state space model. Once full-state feedback gains are designed, the controller is able to stabilize the wing and prevent flutter.

Some recent developments in aeroservoelasticity research projects along a host of main thrusts at various research institutions worldwide, such as NASA Langley Research Center [6] and DLR, the German Aerospace Center [7], can be resorted to in the elaboration of analytical approaches in dealing with aeroservoelasticity problems, in addition to excellent treatment given in Wright and Cooper [2]. Three examples will be elaborated to illustrate problems associated with aeroservoelasticity. These are the modeling of a simple aeroservoelastic system with a control surface, aeroservoelastic analysis of an aircraft with standby actuator, and the design and optimization of an aeroservoelastic wind tunnel model, which follow closely the development carried out in the references cited.

12.2 Mathematical Modeling of a Simple Aeroelastic System with a Control Surface

As a basic aeroservoelastic System that can shed some light on its state of affairs, the present development follows closely the binary aeroelastic model elaborated by Wright and Cooper [2], as depicted in Fig. 12.6, which represents the classical typical section, to be analyzed for its flutter behavior as a baseline. The binary aeroelastic model consists of a uniform rigid rectangular wing with pitch θ and flap κ degrees of freedom introduced via two springs at the root. As further depicted in Figs. 12.6 and 12.7, a full span rigid flap as a control surface is incorporated. It has an infinite stiffness attachment to the wing but can be moved to any angle β as required. To gain a general understanding, the inertial effects of the control surface will be ignored. Therefore the control surface will be considered as an excitation device and not be considered in the basic dynamics of the wing.

The lift and pitching moment acting upon an elemental strip of wing, with the $M\theta$ term being included to allow for simple unsteady aerodynamic effects on the airfoil, and considering additional lift and moment components due to the application of the control surface through angle β, can be written as

$$dL = \frac{1}{2}\rho V^2 c\, dy \left[a_w \left(\frac{y^2}{s^2 V} \dot{q}_b + \frac{y}{s} q_t \right) + a_c \beta \right] \tag{12.1}$$

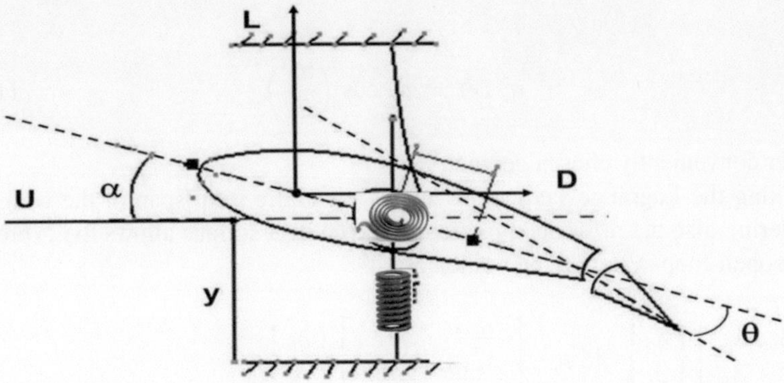

Fig. 12.6 A Typical section representation of Binary flutter system with a control surface

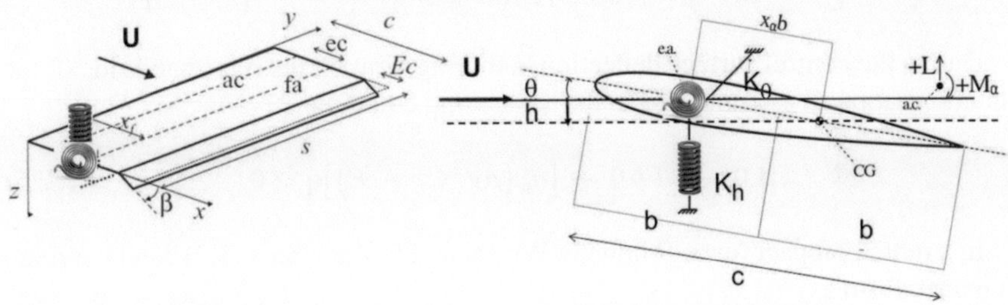

Fig. 12.7 A Binary flutter system with a control surface represented by Fig. 12.6

$$dM = \frac{1}{2}\rho V^2 c^2 dy \left[e\, a_w \left(\frac{y^2}{s^2 V} \dot{q}_b + \frac{y}{s} q_t \right) + M_{\dot{\theta}} \frac{cy}{4sV} \dot{q}_t + b_c \beta \right] \quad (12.2)$$

where there are now lift and moment components present due to the application of the control surface through angle β and the $M_{\dot{\theta}}$ term being included to allow for simple unsteady aerodynamic effects on the airfoil with control surface. For wings without a significant taper, then strip theory may be modified to account for the reduction toward the tip. In this case, the lift curve slope may be varied across the span, either by assuming functions of the form[1]

$$a_w(y) = a_1 \left[1 - \left(\frac{y}{s}\right)^2 \right] \quad (12.3a)$$

or

$$a_w(y) = a_1 \left[1 - \left(\frac{y}{s}\right) \right]^2 \quad (12.3b)$$

[1] See Wright and Cooper Chap. 4 for further detail.

or

$$a_w(y) = a_1 \cos\left(\frac{\pi y}{2s}\right) \tag{12.3c}$$

or other conveniently chosen geometry.

Writing the Lagrange's equations across the entire semi-span of the wing after considering also the prescribed motion of the control surface allows the expression for the open-loop system to be written as

$$\begin{aligned} &\begin{bmatrix} A_{bb} & A_{bt} \\ A_{tb} & A_{tt} \end{bmatrix} \begin{Bmatrix} \ddot{q}_b \\ \ddot{q}_t \end{Bmatrix} + \rho V \begin{bmatrix} \frac{c a_w s}{6} & 0 \\ \frac{c^2 e\, a_w s}{8} & \frac{-c^3 M_{\dot{\theta}} s}{24} \end{bmatrix} \begin{Bmatrix} \dot{q}_b \\ \dot{q}_t \end{Bmatrix} \\ &+ \left(\rho V^2 \begin{bmatrix} 0 & \frac{c a_w s}{8} \\ 0 & -\frac{c^2 e a_w s}{6} \end{bmatrix} + \begin{bmatrix} \frac{4EI}{s^3} & 0 \\ 0 & \frac{GJ}{s} \end{bmatrix} \right) \begin{Bmatrix} q_b \\ q_t \end{Bmatrix} = \rho V^2 \begin{Bmatrix} -\frac{c a_c s}{6} \\ \frac{c^2 b_c s}{4} \end{Bmatrix} \beta \end{aligned} \tag{12.4}$$

Due to the control surface deflection, a forcing term on the right-hand side of the equations appears as compared to the basic wing equation.

$$A\ddot{\mathbf{q}} + \rho V b\dot{\mathbf{q}} + \left[a_w\left(\rho V^2 C + E\right)\right]\mathbf{q} = 0 \tag{12.5}$$

In a more compact form, following Wright and Cooper Eq. (12.4) can be written in matrix form as

$$A\ddot{\mathbf{q}} + \rho V b\dot{\mathbf{q}} + \left[a_w\left(\rho V^2 C + E\right)\right]\mathbf{q} = \mathbf{g}\beta \tag{12.6}$$

As compared to the basic wing equation.

where for convenience the term ρ*V*2 is embedded in the excitation vector **g**. Then, a harmonic excitation β = β0e*iωt* and response **q** = **q**0e*iωt* lead to the frequency response function between the response degrees of freedom and the control surface rotation. A frequency domain approach may then be used to determine the response of the system due to the input of the control system at any point in the flight envelope. Alternatively, an approach based on time domain numerical integration could be employed for any general excitation input.

12.3 Incorporation of the Gust Terms

In what follows, consideration is given to the effect of gusts only. Similar approach can later be applied to include the effects of turbulence. Such elaboration is intended to exemplify the analytical approach for aircraft with control surfaces and for illustrative purposes. Now consider the effect of a uniform vertical gust of velocity wg encountered along the whole span of the wing. This uniform vertical gust serves as a disturbance to the wing. Such a simplified approach actually contains a number of assumptions, which could be used as a creative problem for the reader.

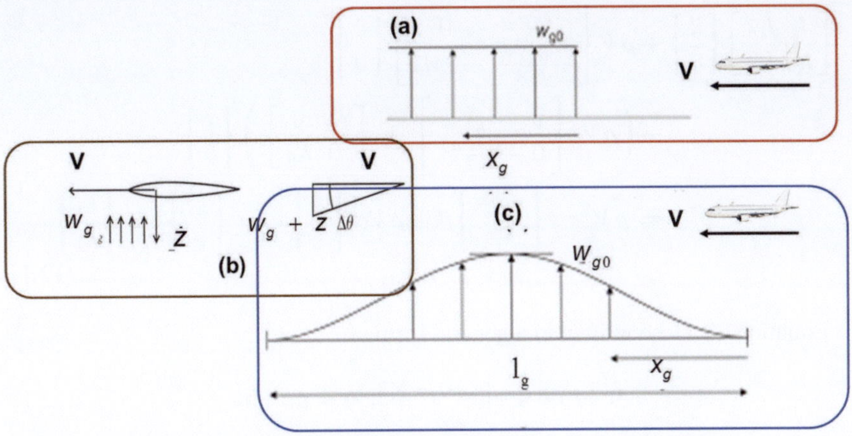

Fig. 12.8 Effective angle of incidence due to the vertical gust

The gust depicted in Fig. 12.8, for the uniform gust, shows that it gives rise to an effective instantaneous change of incidence Δθ of

$$\Delta\theta = \frac{w_g + \dot{z}}{V} \tag{12.7}$$

Therefore an extra lift term due to the gust velocity is generated, together with that from the vertical velocity term and the incidence due to pitch that may already be considered.

Then the lift and pitching moment of an elemental streamwise strip on the wing is now given by

$$dL = \frac{1}{2}\rho V^2 c dy\left[\alpha_w \left(\theta + \frac{w_g}{V} + \frac{y\dot{\kappa}}{V}\right) + \alpha_c \beta\right] \tag{12.8}$$

and

$$dM = \frac{1}{2}\rho V^2 c^2 dy\left[e\alpha_w \left(\theta + \frac{w_g}{V} + \frac{y\dot{\kappa}}{V}\right) + M_\theta \frac{\dot{\theta}_c}{4V} + b_c \beta\right] \tag{12.9}$$

Then the open-loop equations of motion are given as

$$\begin{aligned} &\begin{bmatrix} I_\kappa & I_{\kappa\theta} \\ I_{\kappa\theta} & I_{\kappa\theta} \end{bmatrix}\begin{Bmatrix} \ddot{\kappa} \\ \ddot{\theta} \end{Bmatrix} + \rho V \begin{bmatrix} \frac{cs^3\alpha_w}{6} & 0 \\ \frac{ec^2 s^2 \alpha_w}{4} & \frac{c^3 s}{8} M_\theta \end{bmatrix}\begin{Bmatrix} \dot{\kappa} \\ \dot{\theta} \end{Bmatrix} \\ &+ \left(\rho V^2 \begin{bmatrix} 0 & \frac{cs^3\alpha_w}{6} \\ 0 & \frac{ec^2 s^2\alpha_w}{2} \end{bmatrix} + \begin{bmatrix} K_\kappa & 0 \\ 0 & K_\theta \end{bmatrix}\right)\begin{Bmatrix} \kappa \\ \theta \end{Bmatrix} = \rho V^2 c s \begin{Bmatrix} \frac{s\alpha_c}{4} \\ \frac{c b_c}{2} \end{Bmatrix} \end{aligned} \tag{12.10}$$

$$\begin{bmatrix} I_\kappa & I_{\kappa\theta} \\ I_{\kappa\theta} & I_{\kappa\theta} \end{bmatrix} \begin{Bmatrix} \ddot{\kappa} \\ \ddot{\theta} \end{Bmatrix} + \rho V \begin{bmatrix} \frac{cs^3 a_w}{6} & 0 \\ -\frac{ec^2 s^2 a_w}{4} & \frac{c^3 s}{8} M_\theta \end{bmatrix} \begin{Bmatrix} \dot{\kappa} \\ \dot{\theta} \end{Bmatrix}$$
$$+ \left(\rho V^2 \begin{bmatrix} 0 & \frac{cs^3 a_w}{6} \\ 0 & -\frac{ec^2 s^2 a_w}{2} \end{bmatrix} + \begin{bmatrix} K_\kappa & 0 \\ 0 & K_\theta \end{bmatrix} \right) \begin{Bmatrix} \kappa \\ \theta \end{Bmatrix}$$
$$= \rho V^2 c s \begin{Bmatrix} -\frac{s a_c}{4} \\ \frac{c b_c}{2} \end{Bmatrix} \beta + \rho V c s \begin{Bmatrix} \frac{s}{4} \\ \frac{c}{2} \end{Bmatrix} w_g = \begin{Bmatrix} g_1 \\ g_2 \end{Bmatrix} \beta + \begin{Bmatrix} h_1 \\ h_2 \end{Bmatrix} w_g \tag{12.11}$$

Equation 12.11 be written in a general form as

$$A\ddot{\mathbf{q}} + \rho V B \dot{\mathbf{q}} + \left[\rho V^2 C + E\right]\mathbf{q} = \mathbf{g}\beta + \mathbf{h} w_g \tag{12.12}$$

The term ρV is embedded in the gust excitation vector h for convenience in further analysis. Considering the gust disturbance term as an excitation force, it is here written on the right-hand side of the equations together with the control surface input. Then the response due to a known gust time history can now be readily calculated.

12.4 Application of a Control System

One of the simplest forms of control system is the PI approach, and when implemented here the control surface demand angle is linearly proportional to the velocity and displacement of the system. For simplicity, consider that there is a transducer at the wing leading edge a distance $s0$ from the root, and the control surface deflection is taken as being proportional to its displacement and velocity such that

$$\beta = K_v \dot{z}_{wing} + K_d z_{wing} = K_v \left(\dot{\kappa} s_0 - \dot{\theta} \frac{c}{2} \right) + K_d \left(\kappa s_0 - \theta \frac{c}{2} \right)$$
$$= K_v \left(s_0 \; - \tfrac{c}{2} \right) \begin{Bmatrix} \dot{\kappa} \\ \dot{\theta} \end{Bmatrix} + K_d \left(s_0 \; - \tfrac{c}{2} \right) \begin{Bmatrix} \kappa \\ \theta \end{Bmatrix} \tag{12.13}$$

where K_v and K_d are weightings (commonly called feedback "gains") applied to the velocity and displacement terms, respectively.

Thus the aircraft response is fed back via the control surface to modify the aircraft characteristics. This mathematical model of the wing with gust and control input, modified by the feedback law, can be represented by the block diagram shown in Fig. 12.9 of a generalized Aeroservoelastic system of an aircraft (not just a wing).

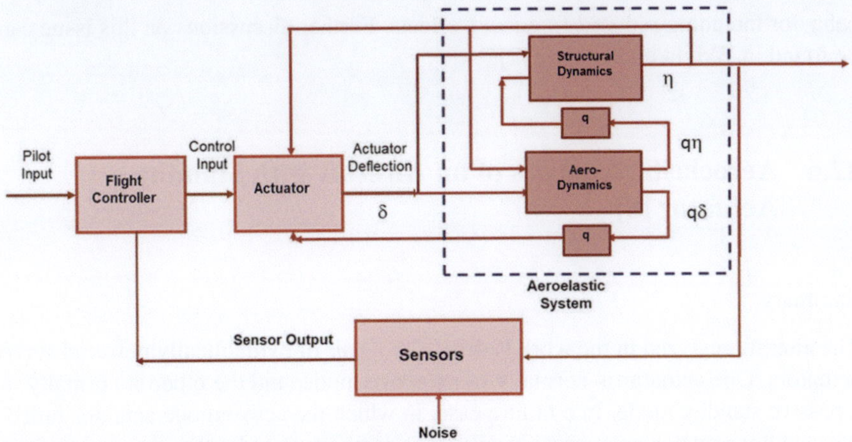

Fig. 12.9 An example of the block diagram of an aeroservoelastic system

12.5 Determination of Closed-Loop System Stability

In order to examine the stability of the closed-loop system, the feedback law in Eq. (12.8) must be combined with the basic wing equations in 5. Thus, the closed-loop equations of motion 12.11 of the wing plus control surface (but in the absence of a gust disturbance) becomes

$$
\begin{aligned}
A\ddot{\mathbf{q}} + \rho VB\dot{\mathbf{q}} + [\rho V^2 C + E]\mathbf{q} &= \mathbf{g}\beta \\
&= K_v \begin{pmatrix} g_1 \\ g_2 \end{pmatrix} \left(S_0 - \frac{c}{2} \right) \begin{Bmatrix} \dot{\kappa} \\ \dot{\theta} \end{Bmatrix} + K_d \begin{pmatrix} g_1 \\ g_2 \end{pmatrix} \left(S_0 - \frac{c}{2} \right) \begin{Bmatrix} \kappa \\ \theta \end{Bmatrix} \\
&= K_v \begin{bmatrix} g_1 s_0 & -g_1 \frac{c}{2} \\ g_2 s_0 & -g_2 \frac{c}{2} \end{bmatrix} \begin{Bmatrix} \dot{\kappa} \\ \dot{\theta} \end{Bmatrix} + K_d \begin{bmatrix} g_1 s_0 & -g_1 \frac{c}{2} \\ g_2 s_0 & -g_2 \frac{c}{2} \end{bmatrix} \begin{Bmatrix} \kappa \\ \theta \end{Bmatrix}
\end{aligned}
\tag{12.14}
$$

where F and G are composite feedback matrices with coefficients that are a function of the control gains, density and air speed. This equation can be rearranged as

$$A\ddot{\mathbf{q}} + (\rho V\mathbf{B} - \mathbf{F})\mathbf{q} + [\rho V^2 \mathbf{C} + \mathbf{E} - \mathbf{G}]\mathbf{q} = 0 \tag{12.15}$$

and the closed-loop system can now be solved to examine its stability. Clearly, the dynamics of the system have now been altered since there are extra stiffness and damping matrices present due to the control system, and this will affect the aeroelastic behavior including the flutter speed. Unsteady aerodynamic effects also reduce the control surface effectiveness as the application frequency increases. In practice, more sophisticated models need to be developed to allow for the unsteady aerodynamic behavior including reduced frequency effects. Further discussions on this issue can be found in Wright and Cooper [2].

12.6 Aeroelastic Analysis of an Aircraft with Standby Actuator [8]

Summary

The aircraft analyzed in the work is driven by a pair of hydraulically powered servo actuators. One actuator is normally in an active mode, and the other is normally in a passive standby mode. In a failure case, in which the active mode actuator failed, the standby actuator must provide sufficient damping in order that the airplane can still maintain flutter free condition. For this purpose, a method for aeroservoelastic analysis of an airplane with a standby actuator using state space approach has been developed. The actuator used in this analysis was assumed to exhibit linear transfer function and viscous damping behavior. The eigenmode, eigenfrequencies and generalized mass were calculated with MSC/NASTRAN software package. The generalized unsteady aerodynamic forces were calculated using Doublet Lattice Method. All of these data then will be used to set up the equations of motion of the airplane with standby actuator. This set of equations will be formulated as an eigenvalue problem in a state space form using MATLAB package software program. The generalized aerodynamic forces were expressed in Laplace domain using commonly used Pade approximation technique. Flutter analysis was carried out for several combination of altitudes and mass configurations. The results of this analysis are the impedance curve boundary that will be used as a design criteria of the actuator in the standby mode. Some of the results were also presented as examples.

12.6.1 Introduction

In the present example, taken from reference, considerations are given to a high wing aircraft with a T-tail configuration (as exhibited in Fig. 12.10, powered by two six-bladed Propellers. The aircraft has an Electrically Controlled Hydraulically Powered flight control system, where each control surface (Rudder, Elevators and Ailerons) is equipped with two actuators. The use of electrically controlled and hydraulically powered flight control system requires particular attention in its aeroservoelastic behavior.

In the present section, only the formulation of the problem and outline of the approach will be elaborated, to provide some impression of the state of affair and solution procedure.

Fig. 12.10 N-250 aircraft considered in the modeling of the aeroservoelastic system[2]

An important aspect in the aeroservoelastic analysis to meet certification requirements for this aircraft is to addresses failure cases. There are several failure cases in the flight control system that must be taken into account in order to meet the regulation stipulated in FAR 25.629 (Amendment 25–77 dated 29 July 1992); these are:

(a) Any single failure, or malfunction, or combinations thereof, in the flight control system, and any single failure in any flutter damper system.
(b) Any single failure of the stability augmentation system, or any other automatic or power operated system.
(c) Control surfaces, including tabs, should be investigated for nominal conditions and for failure modes that include single structural failures (such as actuator disconnects, hinge failures or, in the case of aerodynamic balance panels, failed seals), single and dual hydraulic system failures and any other combination of failure not shown to be extremely improbable. Where other structural components contribute to the flutter stability of the system, failures of those components should be considered for possible adverse effects.

The flight control system is considered to be operating in one of two modes, normal flight or failure condition.

1. In normal flight, one actuator has Power On and the other one is in standby. In standby a bypass valve is actuated allowing fluid to flow to each side of the actuator piston through an orifice sized to give damping required to prevent flutter, in case the Power On actuator fails. The schematic of the system is depicted on the right-hand side of Fig. 12.11. Therefore in the Finite Element Model, under normal flight conditions, each control surface has one actuator which is

[2] From the author's collection/repository.

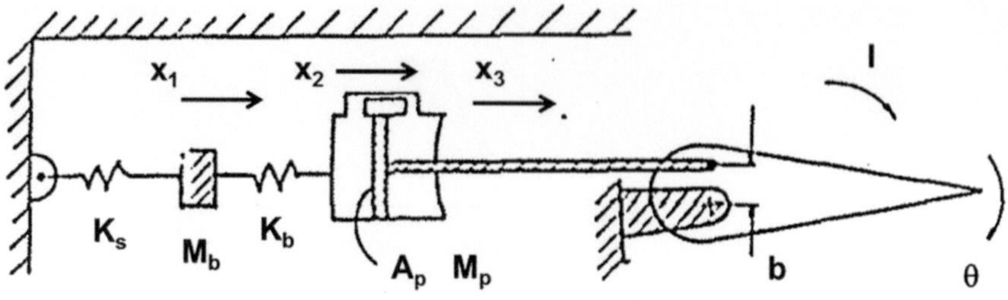

Fig. 12.11 Detail of the control surface sub-system for the aeroservoelastic system

represented as stiffness only and another one which is represented as a stiffness and damper in series. The stiffness is represented by the stiffness of the back-up structure and the actuator in series.
2. In failure condition, the Federal Aviation Regulation (FAR) of USA requires the flight control surfaces to be flutter free with adequate damping up to the Design Dive Speed (Vd) for any single failure, two hydraulic failures or a combined hydraulic and structural failure.

In case the active actuator fails, in the modeling idealization one hydraulic and one structural failure took place. In this condition the standby actuator will take part and is represented in the FE Model by stiffness and damping in series. The natural frequencies, mode shapes and the generalized unsteady aerodynamics were calculated by MSC/NASTRAN package, and the results then transferred to the MATLAB software. The equation of motion of the airplane without standby actuator is set up in S-domain. For this purpose, the generalized unsteady aerodynamics must be transformed into Laplace domain using commonly used Pade approximation technique.

The actuator transfer function is derived as presented in subsequent section below. The assumption made is that the actuator can be represented as spring and damper in series. The same procedure is followed here to set up the equation of motion of the aircraft with standby actuator.

The flutter analysis is performed for several "*a*" values. With these analyses, the minimum damping coefficient required, *C*, can be obtained from the most critical V-G diagram.

In what follow, the State-Space Form of the Aeroservoelastic equation is described, identifying the significant variables considered and salient feature of the governing equation. Then a simple approximation of the Unsteady Aerodynamic Equation is described. The detail of the viscous damper and spring modeling of the aeroservoelastic system can be followed in that particular chapter. The Linear State Space Approach to the problem will be summarized to allow general understanding of the approach, and the results are discussed, with the illustration of the V-G diagram for particular cases.

12.6.2 A State Space Form for an Aeroservoelastic System of an Aircraft with Stand-By Actuator

In the following, the set of equations that represent the airplane with unsteady aerodynamics written in a state space or first-order form will be summarized. Unsteady aerodynamic computer codes such as the doublet lattice codes in MSC/NASTRAN compute the aerodynamics in the reduced frequency domain. This is ideally suited for a k type flutter solution but does not lend itself to a time domain or s domain system of equations. Recently, an approximation using a form of Pade approximation has become popular in the aeroelastic and aeroservoelastic analyses. With the unsteady aerodynamics written in this form a state space formulation is possible. In this state space form, s domain, time domain and z domain transformations are not only possible but simple. At this point the airplane can be treated as any other control block and combined with other control blocks using common block diagram algebra.

The equations of motion of an airplane with unsteady aerodynamics can be expressed as a set of generalized coordinates in the s domain.[3] This expression is given by

$$s^2\left[\widehat{\mathbf{M}}\right]\mathbf{q} + s\left[\widehat{\mathbf{C}}\right]\mathbf{q} + \left[\widehat{\mathbf{K}}\right]\mathbf{q} = \mathbf{q}\left[\widehat{\mathbf{Q}}\right]\mathbf{q} + \left\{\widehat{\mathbf{F}}\right\} \tag{12.16}$$

where the $\{\mathbf{F}\}$ are the other forces (internal or external) acting on the airplane and may or may not be needed in more extensive equation development. These forces could be gust forces for a gust loads analysis, control surface forces caused by inputs of the flight control system, landing gear forces for taxi analysis, discrete velocity square damping terms, or many others.

The generalized aerodynamic forces of Eq. (12.16) can be approximated using the Aerodynamic Term results from the Unsteady Aerodynamics in the time domain, as expressed by Eq. (12.22), to be elaborated subsequently. This is an approximation and some caution is needed for its use. With this approximation the aircraft equations of motion become:

$$\begin{aligned} s^2\left[\widehat{\mathbf{M}}\right]\{\mathbf{q}\} + s\left[\widehat{\mathbf{C}}\right]\{\mathbf{q}\} + \left[\widehat{\mathbf{K}}\right]\{\mathbf{q}\} &= \overline{\mathbf{q}}[\mathbf{A}_0]\{\mathbf{q}\} + s\overline{\mathbf{q}}[\mathbf{A}_1]\{\mathbf{q}\} + s^2\overline{\mathbf{q}}[\mathbf{A}_0]\{\mathbf{q}\} \\ &\quad + \overline{\mathbf{q}}[\mathbf{N}_0]\{D\}^{-1}\{\mathbf{q}\} + s\overline{\mathbf{q}}[\mathbf{N}_1]\{D\}^{-1} + \left\{\widehat{\mathbf{F}}\right\} \end{aligned} \tag{12.17}$$

where the elements of **A**0, **A**1, **A**2, **N**0 and **N**1 matrices (i integration mode and j pressure mode) are made up of the corresponding terms of the **A**0, **A**1, **A**2, **N**0 and **N**1 scalars which were obtained in the corresponding aerodynamics.

[3] The s domain is the Laplace transform of the aeroelastic or aeroservoelastic equation like Eq. (12.11) which is in the time domain into the s domain.

Using inverse Laplace Transform as elaborated in the referenced paper, after considerable algebra, there is obtained in the standard control equation form as a first-order ordinary differential equation in time domain of the aeroservoelastic equation, given by

$$\left\{\dot{\mathbf{X}}\right\} = [\mathbf{A}]\{\mathbf{X}\} + [\mathbf{B}]\{\mathbf{U}\} \tag{12.18}$$

where [**A**] is called the plant matrix and contains all the information needed to compute the stability of the aeroservoelastic system as a linear one.

It should be noted that the approximation of the unsteady aerodynamics carried out here is one approximation adopted for convenience and quick impression of the state of affairs.

12.6.3 A Simple Approximation for the Unsteady Aerodynamics in the Time Domain

The generalized aerodynamic force coefficients are computed in the k (reduced frequency) domain and cannot be directly transformed into the s domain. This approach is a variation of a technique that is commonly known as Pade approximation. This technique takes advantage of

$$ik = \frac{b}{\nu}s \tag{12.19}$$

The unsteady aerodynamic force matrices as a function of k can be examined one term at a time and the fit obtained in this fashion. The coefficients of the fit can then be assembled in matrix form. The detail of the derivation can be followed in the paper [8]. Then after considerable algebra, using the relationship of ik and s in Eq. (12.19), the unsteady aerodynamic force in s domain can be expressed as

$$A(s) + iB(s) = a_0 + \left(\frac{b}{\nu}\right)sa_1 + \left(\frac{b}{\nu}\right)^2 s^2 a_2 + \frac{\left(\frac{b}{\nu}\right)sa_3}{\left(\frac{b}{\nu}\right)s + b_3} + \frac{\left(\frac{b}{\nu}\right)sa_4}{\left(\frac{b}{\nu}\right)s + b_4} \tag{12.20}$$

Finally this expression can be reduced to

$$A(s) + iB(s) = A_0 + sA_1 + s^2 A_2 + \frac{sN_1 + N_0}{s^2 + sR_1 + R_0} \tag{12.21}$$

where the coefficients can readily be interpreted, as also elaborated in [8]

The matrix equivalent form of Eq. (12.21) is

$$[\mathbf{Q}] = [\mathbf{A}_0] + s\{\mathbf{A}_1\} + s^2[\mathbf{A}_2] + [\mathbf{N}_0][\mathbf{D}]^{-1} + s[\mathbf{N}_1][\mathbf{D}]^{-1} \tag{12.22}$$

12.6.4 Linear State Space Approach to the Aeroservoelastic System of an Aircraft with Stand-By Actuator

Next we will derive the equations of motion of the aircraft equipped with a standby linearized actuator. These equations are put in a suitable form for the determination of stability. This set of equations will be formulated as an eigenvalue problem using state space methods so that the existing software, MATLAB, can be used for the solution and stability determination.

The need for a model of an actuator in the standby mode comes from the way the flight control is integrated into the airplane. The actuator can be "on," "standby" or (failed) "disconnected." In the "on" condition the actuator is modeled with stiffness and damping in series. In the "disconnect" condition the actuator is in a "free" condition with no stiffness or damping (both structural and hydraulic failures).

Previous section formulates the airplane in a state space form with a generalized force as the input. In this section the equations of the actuator force are written and combined with the airplane equations. The actuator moment (output) will be written in terms of the aileron rotation (input). The airplane aileron response (output) will be written in terms of actuator moments (input).

These two blocks will be combined into a single system of equations from which the stability will be seen from the eigensolution. The actuator in the standby mode looks like a velocity squared damper in series with a linear spring. The model for this formulation of the actuator was selected as simply a mathematical curve, because it was desired to have an impedance curve boundary to be used as design criteria of the actuator in the standby mode. The curve selected as the mathematical model of the actuator can be expressed as

$$f = \frac{Ks}{s+a}\Delta x \tag{12.23}$$

Though less important, this curve does have some basis in a physical system. This physical system is a viscous damper in series with a spring which is not like the actual actuator. The detail of the derivation of the impedance of a viscous damper in series with a spring is elaborated in the paper. The actuator force acts over an arm (which is at a distance to the hinge line) to create a moment on the control surface. This actuator moment is

$$m = hf \tag{12.24}$$

A similar relationship exists that relates the actuator displacement and the control surface relative angle (detailed given in the paper).

$$\Delta x = h\Delta\theta \tag{12.25}$$

A substitution of these equations into the actuator force Eq. (12.23) gives a relationship between the moment and the control surface relative angle, which is given by

$$m = \frac{h^2 Ks}{s+a}\Delta\theta \tag{12.26}$$

Transforming this into the time domain yields a linear first-order differential equation

$$\dot{m} + am = h^2 K\left(\Delta\dot{\theta}\right) \tag{12.27}$$

Solving for the higher order differential equation

$$\dot{m} = -am + h^2 K\left(\Delta\dot{\theta}\right) \tag{12.28}$$

In state space form where the input is the relative angular velocity and the output is the generalized moment, Eq. (12.28) becomes

$$\dot{m} = A_m m + B_m \Delta\dot{\theta} \tag{12.29a}$$

$$\dot{m} = C_m m + D_m \Delta\dot{\theta} \tag{12.29b}$$

where $A_m = -a$, $B_m = h^2 K$, $C_m = 1$ and $D_m = 0$.

The relative angular velocity, $\Delta\theta$, can be made as the angle on the wing side of the actuator and angle at the control surface side

$$\Delta\dot{\theta} = \dot{\theta}_c - \dot{\theta}_a \tag{12.30}$$

Since the mode shapes and generalized velocity is available, these angles can be computed and the expression becomes

$$\Delta\dot{\theta} = ([\Phi_c] - [\Phi_a])\{q_2\} \tag{12.31a}$$

or

$$\Delta\dot{\theta} = [\Delta\Phi]\{q_2\} \tag{12.31b}$$

The airplane was put into a first-order differential form in Eq. (12.3), which can then be elaborated into

$$\begin{Bmatrix} \{\dot{q}_1\} \\ \{\dot{q}_2\} \\ \{\dot{q}_3\} \\ \{\dot{q}_4\} \end{Bmatrix} = \begin{bmatrix} 0 & [I] & 0 & 0 \\ [\hat{a}_1] & [\hat{a}_2] & [\hat{a}_3] & [\hat{a}_4] \\ 0 & 0 & 0 & [I] \\ [I] & 0 & -[R_0] & -[R_1] \end{bmatrix} \begin{Bmatrix} \{q_1\} \\ \{q_2\} \\ \{q_3\} \\ \{q_4\} \end{Bmatrix} + \begin{bmatrix} 0 \\ \left(\left[\hat{M}\right] - [A_2]\right)^{-1} \\ 0 \\ 0 \end{bmatrix} \left\{\hat{f}\right\} \tag{12.32}$$

Again the detail of the derivation is elaborated in the paper.

The generalized force can be expressed in terms of the actuator moment, which can be written as

$$\left\{\hat{f}\right\} = [\Phi_a]^T \{m\} \tag{12.33}$$

Combining Eqs. (12.19, 12.20, 12.21, 12.22 and 12.23) yields

$$\begin{Bmatrix} \{\dot{m}\} \\ \{\dot{q}_1\} \\ \{\dot{q}_2\} \\ \{\dot{q}_3\} \\ \{\dot{q}_4\} \end{Bmatrix} = \begin{bmatrix} -a & 0 & h^2K(\Delta\Phi) & 0 & 0 \\ 0 & 0 & [I] & 0 & 0 \\ \left(\left[\hat{M}\right] - [A_2]\right)^{-1} [\Phi_A]^T & [\hat{a}_1] & [\hat{a}_2] & [\hat{a}_3] & [\hat{a}_4] \\ 0 & 0 & 0 & 0 & [I] \\ 0 & [I] & 0 & -[R_0] & -[R_1] \end{bmatrix} \begin{Bmatrix} \{m\} \\ \{q_1\} \\ \{q_2\} \\ \{q_3\} \\ \{q_4\} \end{Bmatrix} \tag{12.34a}$$

or in short

$$\left\{\dot{\mathbf{X}}\right\} = [\mathbf{A}]\{\mathbf{X}\} \tag{12.34b}$$

The eigenvalues can be obtained by considering the standard form

$$|[\mathbf{A}] - \lambda[\mathbf{I}]| = \{0\} \tag{12.35}$$

These eigenvalues of the A matrix look are the roots of the characteristic equation. For the complex conjugate (oscillatory) roots

$$\zeta = \frac{\Re(\lambda)}{|\lambda|}; \quad \omega = |\lambda| \tag{12.36}$$

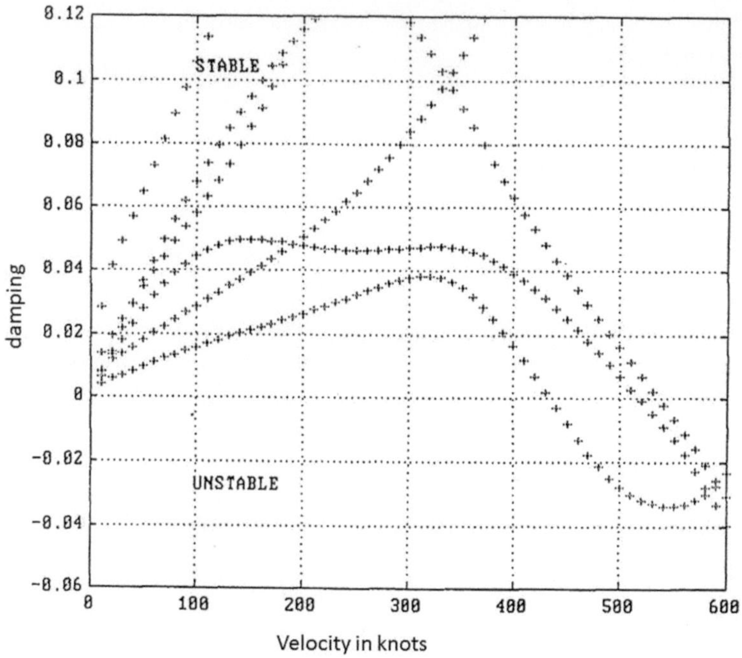

Fig. 12.12 Velocity-damping diagram for $a = 75$ (MATLAB)

Plots of the velocity versus damping and velocity versus frequency and/or root locus give an understanding of the stability boundary and stability characteristics.

12.6.5 Results and Discussions

Using the equations developed in preceding sections, computations have been performed for all the failure cases. However, here and in the referenced paper, results obtained using the method developed are only exhibited for the most critical case 3.

Some of these results, indicating the flutter solution plotted as frequency-velocity and damping-velocity diagram (V-G diagram), are exhibited in Figs. 12.12 and 12.13 for the value of stiffness to viscous damping ratio a. The results as exemplified here and elaborated in the referenced paper show that all cases being considered meet the flutter free envelope requirements.

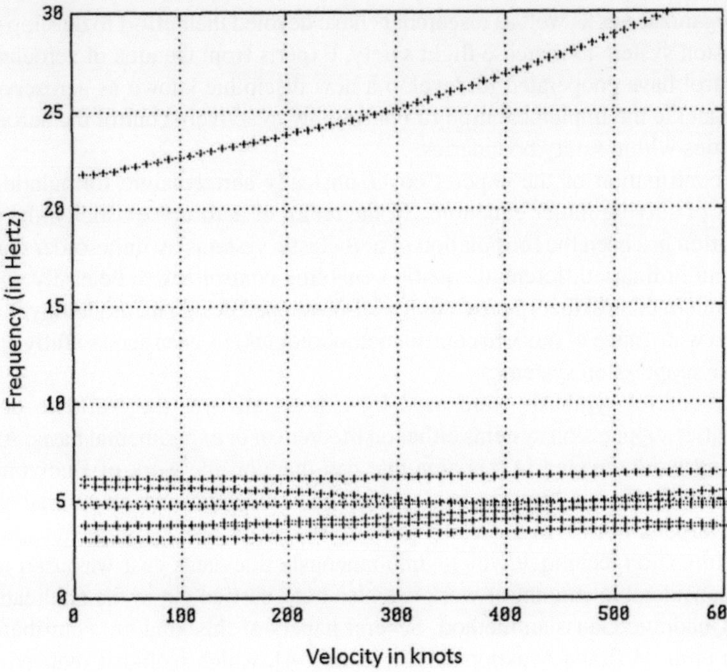

Fig. 12.13 Velocity-frequency diagram for $a = 75$

12.7 The Application of Artificial Neural Networks on Flutter Suppression System [9]

Summary

The study highlights the benefits of the implementations of artificial neural networks in aeroservoelastic problem. The neural networks are investigated as an alternative approach to flutter suppression system. Their use in both modeling the system identification and controller is studied for feasibility and to identify advantages and disadvantages. The connectionist neural networks approach implemented in an indirect adaptive control is demonstrated. The results from simulation were discussed, and the performance of indirect adaptive control using neural networks is critically evaluated.

12.7.1 Introduction

With the progress of aircraft design optimization which results in the increase in speed and flexibility of modern aircraft, flutter has become a key feature in aircraft design.

Many organizations as well as researchers have devoted their effort to develop flutter suppression system to increase flight safety. Experts from the area of aeroelasticity and control have cooperated to develop a new discipline known as aeroservoelasticity to tackle the implementation of control law to actively control the aeroelastic instabilities within safety boundaries.

The contribution of the experts cover unsteady aerodynamic formulation and methods of solving flutter equations. In the realm of aeroservoelasticity, the major contribution has been the formulation of aeroelastic systems by finite-order constant coefficient ordinary differential equation enabling control law to be easily applied. Since modern control theory provides a well-developed design methodology for such system, a wide range of modern control methodologies has been successfully applied on flutter suppression systems.

Many control synthesis were used by experts all over the world to develop active flutter suppression systems either on theoretical or experimental basis. Among pioneering work devoted in this area one can mention the work of Budiyono [9], Edwards et al. [10] and Newsom et al. [11], who based their work on optimal control theory. Karpel's work [12] especially presents the use of pole assignment technique. The optimization technique which simultaneously alleviates gust was used in that study. Considerable amount of work has also been carried out in the application of Linear Quadratic Gaussian method. Several papers of this kind are contributed by Mahesh et al. [13] and Mukhopadhyay et al. [14], which included reduced order model design. Another approach was made by Garrard et al. using robust Kalman Filter in Nissim [15–19] gave another contribution by introducing aerodynamic energy method as an alternative of classical control theory application. The work based on this approach which studied the active external store flutter suppression in the YF-17 flutter model discussed in the work of Nissim just mentioned. Abel [20] reviewed the results of classical control theory and those of aerodynamic energy concept. A recent paper by Lu and Huang 171 suggests a new technique using active acoustic excitation of thin airfoil.

An interesting contribution was made by Harris et al. [21] who presents adaptive flutter suppression system which automatically adjusts system parameters based on changes in flight condition and store configuration. With specific respect to control technique, a considerable attempt has been made to develop an intelligent controller which can adapt to its changing environment. In pace with the unfolding of the control technique, the recent decades have witnessed a great deal of time and effort spent in developing a new branch of computing and mathematics, the so-called artificial neural networks. This new approach is basically inspired by the physical working principles of the human central neural systems and operate in a very different manner to conventional engineering tools. Its implementation has been investigated in diverse areas including, but not limited to flight control systems, instrument landing systems, large flexible space structure control systems, signal processing and pattern recognition. Several considerations below motivate the use of artificial neural networks in the area of aeroservoelasticity.

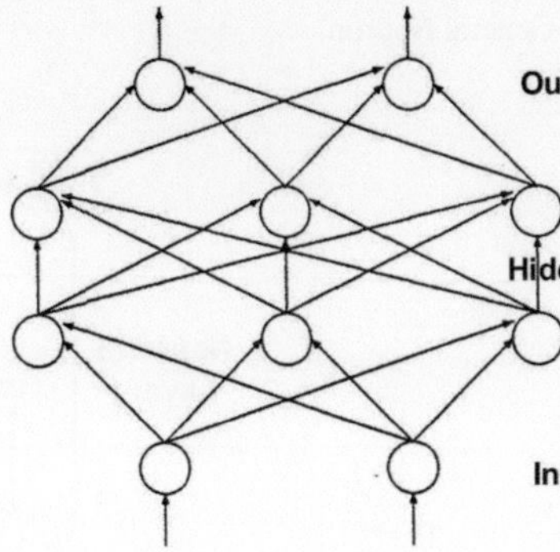

Fig. 12.14 Typical neural network structure

In real operation, several parameters may be uncertain and not of constant value. This necessitates a learning-type controller. Component failure which may happen in aeroservoelastic systems establishes the requirement of an adaptive-type controller.

12.7.2 *The Basic Concept of Artificial Neural Network (ANN)*

12.7.2.1 The Architecture of ANN

The basic feature of artificial neural networks approach in solving problems is that one need not have well-defined process for algorithmically transforming an input to an output.

What one need for most neural networks is a set of representative examples of the desired transformation. The trained artificial neural networks then reproduce the desired outputs when presented to the example input. The power of an artificial neural networks approach lies not necessarily in the elegance of the particular solution, but rather in the generality of the networks to find its own solution to particular problems given only examples of the desired behavior [22–24]. The diagram in Fig. 12.14 displays a typical neural networks. The network architecture is defined by the basic elements and the way in which they are interconnected. The basic processing element of the connectionist (ANN) architecture is often called a neuron, but other names such as unit [17], node [25] or perceptron [26] are also used. The neurons are connected by weights, also referred to as connections, or synapses, which conveys the information. Neurons are also often collected into group called layers within which the neurons have a similar function or structure. Depending on their function in the net, one can distinguish three types of layers. The layers whose activations are problem input to the net are called input layers, the layers whose output represent the output of the net output layers. The remaining layers are called hidden layers, because they are invisible from outside.

In general, there can be any number of layers, and neurons in any layer can be connected to neurons in any other layers. Some neurons have no inputs from other

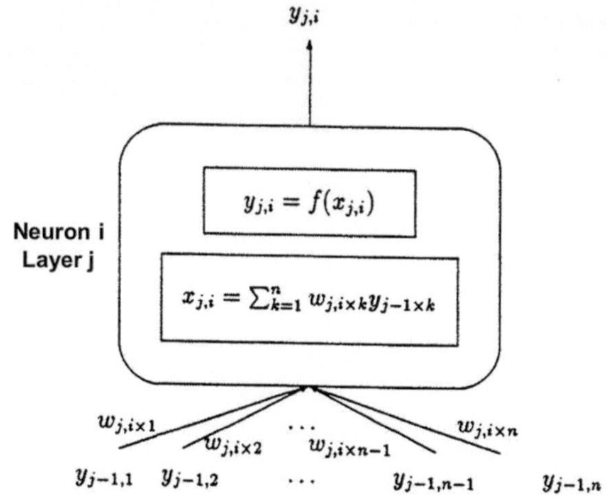

Fig. 12.15 General Neuron Model

neurons, and known as bias or threshold units. The memory content or signal strength of neuron is referred to as activation level. Usually, the activation levels of the input and output units are scaled such that the activation levels are appropriate to the neuron functions and the observable values are given a physical meaning.

12.7.2.2 The ANN Model

Let us review a single neuron from the networks structure which is depicted in Fig. 12.15.

An individual neuron has many inputs depending on the number of connections. Each connection to the neuron has a weight associated with it. After the total neuron input signal is calculated, that is given the sum of the product of inputs and weights:

$$x_{j,i} = \sum_{k=i}^{n} w_{j,i\times k} w_{j-1,k} \tag{12.37}$$

it is converted into an activation value through a functional relationship. The power of ANN lies within this transfer function.

$$y_{j,i} = f\left(x_{j,i}\right) \tag{12.38}$$

The sigmoidal function is the most commonly used in neural networks since it accommodates unlimited degree of nonlinearity. This function is differentiable, step-like and positive (bounded by (0,1)) function. In addition to sigmoidal function, hyperbolic tangent, threshold and ramp functions are occasionally used.

The output of the neuron is obtained from the activation value also through functional relationship. Identity function is usually used in this step. In general, the selection of the function used in neural networks depend on the type of patterns (input–output value pair). At present, the selection of an activation function is more an art than science and subject to much research. The important point to remember is that any nonlinear function will provide the networks with the ability of representing any nonlinear mappings.

12.7.2.3 Learning Methods

Various learning rules in neural networks training has been the subject of investigation of many studies. In the following paragraphs we will review briefly supervised learning (role-model following) known as backpropagation, which is probably one of the most widely used learning methods in training neural networks. The equations used follows those given by Werbos [27].

For the sake of clarity, layered networks shown in Fig. 12.14 will be taken as an example. In basic backpropagation, the output of a neural network depend on its inputs and weights is defined in the following algorithm:

$$x_i = X_j \quad 1 \le i \le m \tag{12.39}$$

$$\text{net}_i = \sum_{j=1}^{i-1} w_{j,i} x_1 \quad m \le i \le N + n \tag{12.40}$$

$$x_1 = s(\text{net}_i) \quad m \le i \le N + n \tag{12.41}$$

$$Y_i = x_{1+N} \quad 1 \le i \le n \tag{12.42}$$

where m is the number of input neurons in the networks, n the number of output neurons and N any integer larger than m. The value $N + n$ determines the number of neurons in the network.

The main idea of backpropagation is somewhat similar to that of the famous least-squares method in which one minimizes square error shown below:

$$E = \sum_{t=1}^{T} E(t) = \sum_{t=1}^{T} \sum_{i=1}^{n} \frac{1}{2} \left(\hat{Y}_i(t) - Y_i(t) \right)^2 \tag{12.43}$$

The way by which the error is minimized is the essence of the backpropagation method. The process begins with computing partial derivatives of the error with respect to the output:

$$\frac{\partial E}{\partial \hat{Y}_i(t)} = \hat{Y}_i(t) - Y_i(t) \quad 1 \le i \le n \tag{12.44}$$

Using the definitions of ordered derivatives [8], one can write:

$$\frac{\partial^+ E}{\partial X_i(t)} = \frac{\partial E}{\partial X_i(t)} + \sum_{j=i+1}^{N+n} \frac{\partial^+ E}{\partial X_j(t)} \times \frac{\partial X_i(t)}{\partial X_i(t)} \quad N+n \geq i \geq 1 \tag{12.45}$$

where

$$\frac{\partial E}{\partial X_i(t)} = \begin{cases} 0 & 1 \leq i \leq m \\ \frac{\partial E}{\partial \widehat{Y}_{1-N}(t)} & N+n \geq i \geq m+1 \end{cases} \tag{12.46}$$

and

$$\frac{\partial X_i(t)}{\partial X_i(t)} = s'\left(\text{net}_j\right) \times w_{jt} \text{ with } s'() = \frac{\partial s()}{\partial()} \tag{12.47}$$

Now the ordered derivatives for the weights can be calculated as

$$\frac{\partial^+ E}{\partial X_i(t)} = \frac{\partial E}{\partial X_i(t)} \times s'(\text{net}_i) \times X_j(t) \tag{12.48}$$

Finally, the weights are updated using

$$\text{new } w_{ij} = w_{ij}\big|_{\text{new}} = w_{ij} - \varepsilon \times \frac{\partial' E}{\partial w_{ij}} \begin{cases} 1 \leq i \leq N+n \\ 1 \leq j \leq N+n \end{cases} \tag{12.49}$$

where ε is defined as the learning rate which is some small constant chosen on adhoc basis.

12.7.3 Aeroservoelastic System

The aeroelastic system depicted in Fig. 12.16, showing definitions of coordinates and system properties, is taken as the plant under study. The parameters of the flutter model are elaborated and exemplified in Harris et al. [21], Hunt et al. [28] and Karpel [12]. The oscillation of typical section in three degree of freedom: heaving motion (bending), pitching motion (torsion) and control surface (aileron) deflection, representing the wing flexibility.

The motion of the plant can be described by the following equation:

$$[M]\left\{\ddot{\mathbf{q}}\right\} + [D]\left\{\dot{\mathbf{q}}\right\} + [\mathbf{K}]\{\mathbf{q}\} = \frac{1}{2}\rho U^2[\mathbf{A}]\{\mathbf{q}\} \tag{12.50}$$

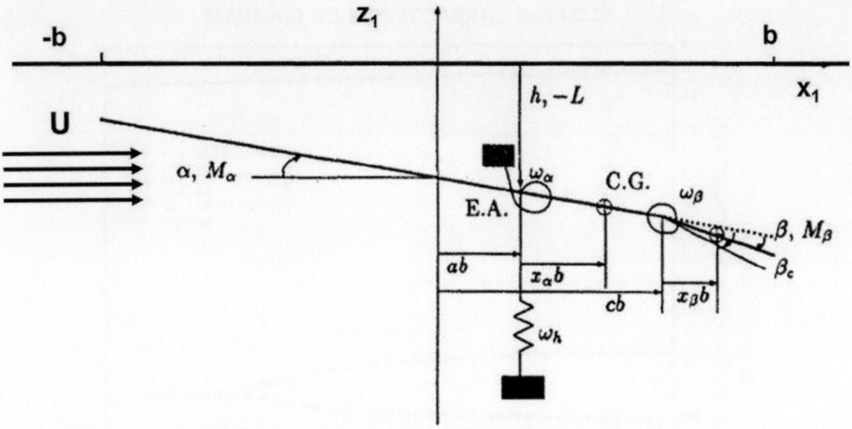

Fig. 12.16 Aeroservoelastic typical section model studied

where [**M**], [**D**], [**K**] and [**A**] are the matrices of mas, damping, stiffness and unsteady and operator, respectively. {**q**} is the generalized coordinate vector which represents the vibration modes. Equation (12.50) can be rewritten in Laplace domain by substituting

$$\{\mathbf{q}\} = \{\hat{\mathbf{q}}\}e^{st} \tag{12.51a}$$

where

$$s = \sigma + i\omega \tag{12.51b}$$

Through the use of Roger's approximation, the unsteady aerodynamic forces can be expressed in the form of rational functions in finite-order polynomials so that the equation of motion of the plant can be written in the state space form.

Figures 12.17 and 12.18 depict the flutter characteristic diagrams of the system under study. The diagrams show that the flutter which consists of two coupled modes (bending and pitching) prevails at the speed of $\frac{U}{b\omega} = 3.02$.

The control equations are then obtained by adding a single-input-multi-output equation which reads[4]

$$\{\dot{\mathbf{x}}\} = [\mathbf{A}]\{\mathbf{x}\} + [\mathbf{B}]\{\mathbf{u}\} \tag{12.52a}$$

$$\{\dot{\mathbf{y}}\} = [\mathbf{C}]\{\mathbf{x}\} + [\mathbf{D}]\{\mathbf{u}\} \tag{12.52b}$$

[4] By referring to and understnding the context, the control variables [**A**] and [**D**] notations apparently have been utilized conventionally in most literature.

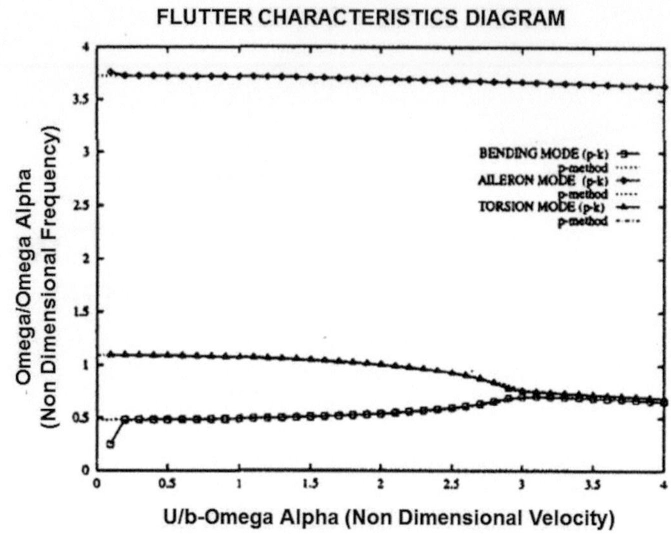

Fig. 12.17 Flutter characteristics diagram (frequency)

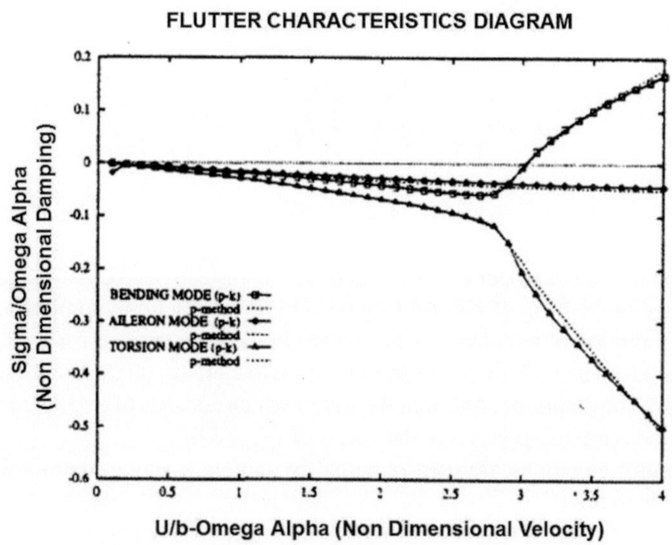

Fig. 12.18 Flutter characteristics diagram (damping)

where each of the variables {**x**} and [**A**] is state vector and dynamic matrix, respectively, u is the control variable and {**y**} is the measurement vector. Each of the variables {**B**}, [**C**] and {**D**} is input, calibration and input–output relation matrix, respectively. The parameters of Eq. (12.52) are

$$\{\mathbf{x}\}^T = \left\{\frac{h}{b}, \alpha, \beta, \frac{\dot{h}}{b}, \dot{\alpha}, \dot{\beta}, \mathbf{B}_1, \mathbf{B}_2, \mathbf{B}_3, \mathbf{B}_4\right\} \tag{12.53a}$$

and

$$\{\mathbf{u}\} = \beta_c \tag{12.53b}$$

For simplicity [**C**] is taken to be identity and {**D**} zero in this present work. The design target is for constant coefficient control law,

$$\{\mathbf{u}\} = [\mathbf{G}]\{\mathbf{y}\} \tag{12.54}$$

which suppresses flutter throughout the desired flight envelope, i.e. $U/1x0\,a = 0$–3.5.

12.7.4 Neural Adaptive Control

The application of neural networks for adaptive-type controllers have been studied by several authors (Narendra [29], Garcia-Padilla and Morant-Anglada [30], Wharington [31], Rickard and Bartholomew [32], Krisnakumar et al. [33]) and has shown to be reliable thus far. Its use in control area is justified by the following aspect: the need of controlling systems which are difficult to model; and design requirements which must be met by this type of systems and finally, to achieve the objective of control with less precise knowledge of the plant and its environment. The application of artificial neural networks for control will allow us to obtain benefits such as undertaking faster computations since this will be carried out in parallel and tolerance to failure, and moreover artificial neural networks present a natural robustness to calculate the parameters without previous modeling due to its generalization properties [7].

12.7.4.1 Adaptive Neuro-Control Concepts

The implementation of neural networks in control problems can be divided into two general categories: direct and indirect control. Under the first category one can distinguish two general approaches. These are copying an existing controller and reinforcement learning. The second category also consists of two fundamental approaches, namely inverse control and supervised control using error propagation through a model. The discussion that follows will be restricted to the supervised control since it is chosen as the control technique in this work.

Supervised neuro-control is similar to indirect control used in adaptive one; a linear system model is identified and using this model the control is adapted. Similarly, in supervised neuro-control, the first system is available for the sole purpose of training a neuro-controller. Neuro-model which is often called neural network model of the

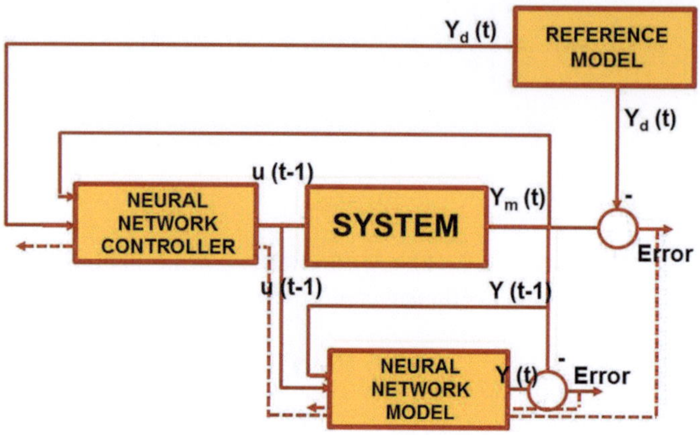

Fig. 12.19 Indirect adaptive control scheme

plant is used to generate partial derivatives of the error between the identified model and the desired (reference model) outputs with respect to the control inputs.

The neuro-controller can then be directly updated using backpropagation of error through the neural network model. While updating the neuro-controller, the neural network model weights are held constant. Again, the role of neuro-model is to relate the output error to the controller error. It should be emphasized, though, that the error definition for the model and the controller are different as shown in Fig. 12.19. The error in case of the controller is the error between the actual system output and the desired output generated by the desired model.

12.7.5 Indirect Adaptive Control for Flutter Suppression

The previously mentioned indirect adaptive control scheme is implemented as a tool for flutter suppression system. For the sake of clarity, time delays and other possible sources of nonlinearities such as back-lash in the aileron hinge are not taken into account in this present work. The supervised control using error propagation through a model is demonstrated. First, the aeroelastic plant is copied using artificial neural networks and supervised learning techniques.

Neural network controller is then designed using the supervised control technique and a fixed neural network model. It should be noted here that the controller acts as a regulator functioning as futter suppression means. In this case, there is no model to be tracked. Instead, the output of the plant (of the neural networks model) is compared with a certain set point which is zero position. The scheme of aeroservoelastic plant using neuro-controller is depicted in Fig. 12.20.

The logic of the indirect adaptive control system is as follows:

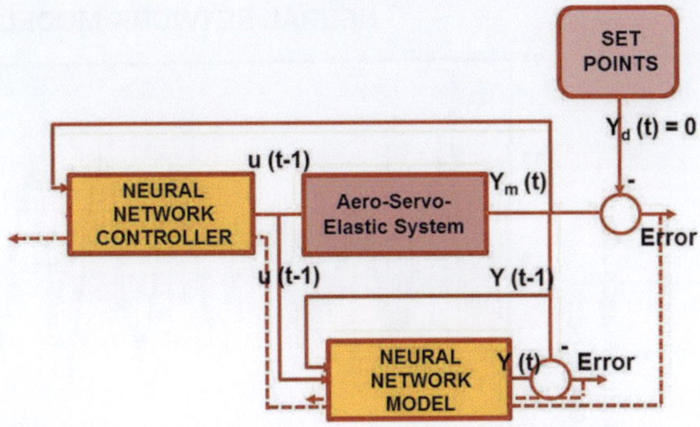

Fig. 12.20 Aeroservoelastic system using indirect adaptive control scheme

1. The neural networks model (NNM) is first trained to imitate the dynamic of the system.
2. Apply a control command to the plant and NNM.
3. Calculate the error between plant output and setting point.
4. Inversely propagate the error through the networks model to find the control command which minimizes the error.
5. Train the neural network controller (NNC) with the target of control command found in the previous step.
6. Execute control command resulted from trained NNC.

12.7.6 Results and Discussion

The actual system was simulated for one second to generate patterns data (input and output value pair) which was used to train the neural networks model. Two architectures of neural networks model are studied. The first network model is one with interconnections, and second architecture is partly connected between neurons in adjacent networks (fully connected layered networks).

The networks, i.e. the first model in which some interconnections, are truncated. Figure 12.21 shows the performance of the first networks model and Fig. 12.22 the second model. The figures display that the second architecture outperforms the first one. The results indicate that tracking performance can be improved by reducing interconnections between neurons to some extent.

With the second architecture as the identification systems, the indirect adaptive control for flutter suppression system was performed at $U/\mathrm{bco} = 3.5$ (16% above the flutter speed). The results are shown in Figs. 12.23 and 12.24. The results show that the neural control system can suppress the violent bending-torsion flutter at the targeted velocity above the flutter speed. However, small low-frequency oscillations still remains. The study to improve the suppression quality is the subject of the current work.

The results displays the promising benefits of implementing indirect adaptive control technique as it involves a learning-type controller and identification system.

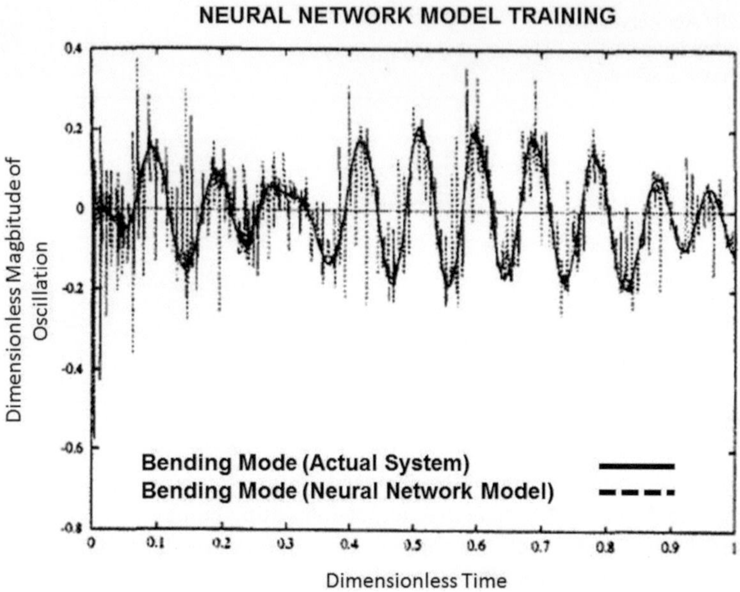

Fig. 12.21 Neural networks model performance

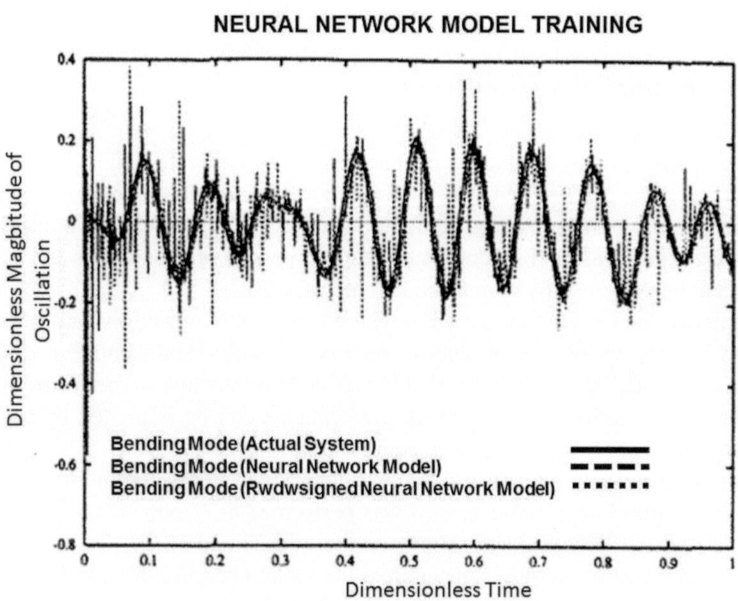

Fig. 12.22 Redesign neural networks model performance

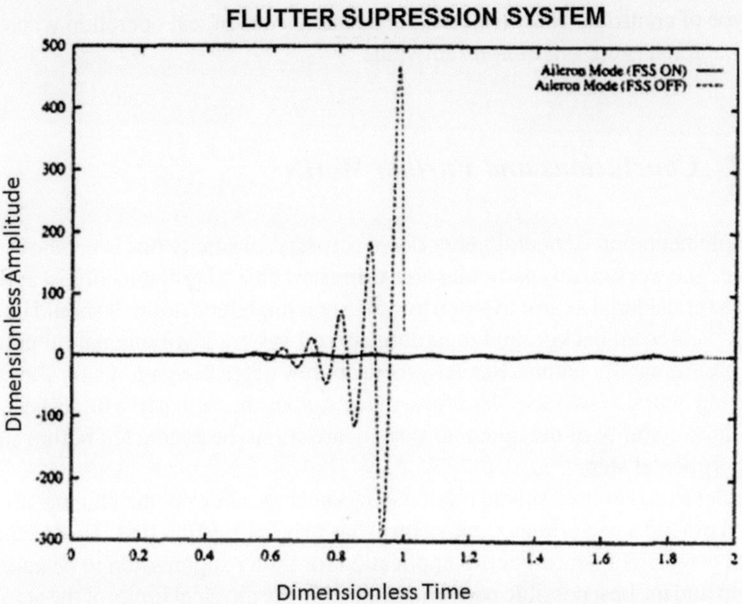

Fig. 12.23 Flutter suppression using indirect adaptive control (Aileron mode)

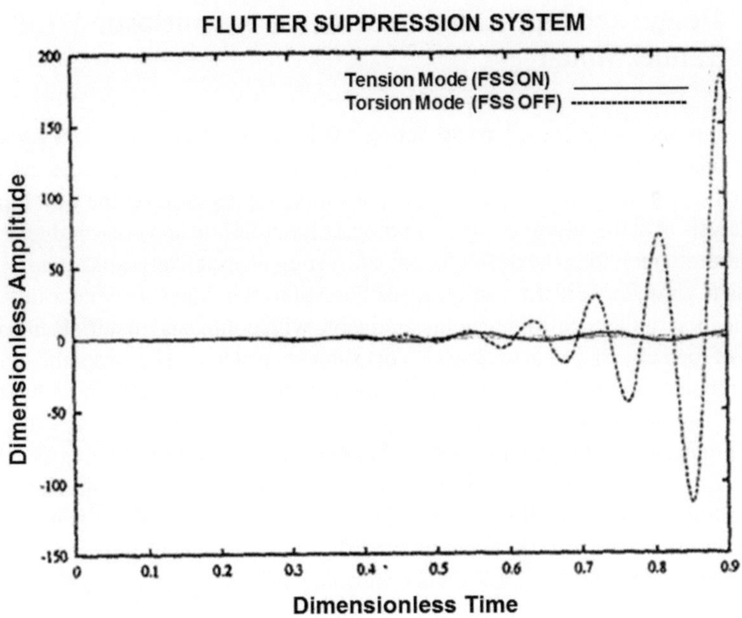

Fig. 12.24 Flutter suppression using indirect adaptive control (Torsion mode)

This type of controller will be especially advantageous in real operation where most of the parameters are not of constant value.

12.7.7 Conclusions and Further Works

The implementation of neural networks in aeroservoelasticity has been shown to be possible. The work on this particular area using new emerging connectionist approach should be considered as an early step toward a real implementation. It should be noted that the theoretical background regarding control system involving neural networks is still preliminary in nature. Besides, there are few other design tools for the system employing neural networks. Therefore, many studies, particularly with respect to the closed-loop stability of the dynamic system, are still to be conducted before starting the experimental step.

In order to arrive at a realistic implementation of the neuro-controller, the trade-off between practical performance and control has to be carried out first. This means that the next problem of neuro-control application for flutter suppression to be addressed is how to find the best possible performance within the physical limits of the actuators.

12.8 Design and Optimization of an Aeroservoelastic Wind Tunnel Model

The design and optimization of an aeroservoelastic wind tunnel model presented here is taken from the work carried out at DLR, the German Aerospace Center, as elaborated in the reference cited [7]. The work was focused on the development and analysis of configurations and technologies for emission and noise reduced short range aircraft, including the development and testing of an actively controlled flexible composite wing for validating active load alleviation techniques. In what follows, the identification of the manifold benefits attainable with composite materials in passive means of load alleviation is reviewed. This subject has led to a considerable amount of research dealing with weight minimization subject to combinations of buckling, strength, displacement and twist responses.

Experience gained in previous works has ranged from the structural layout, static aeroelastic optimization for various objective functions, model manufacturing and eventually to a wind tunnel test series. These tests may serve as test bench for the optimization process and the manufacturability of the stacking sequences of non-conventional laminate. The DLR owned subsonic Side Wind Tunnel Facility (SWG) in Göttingen which can produce 60 m/sec flow velocity has been utilized for this purpose.

The first model utilized in the research comprised a non-swept wing of 1.0 m span, 0.2 m constant chord and a symmetric airfoil, built by DLR in collaboration

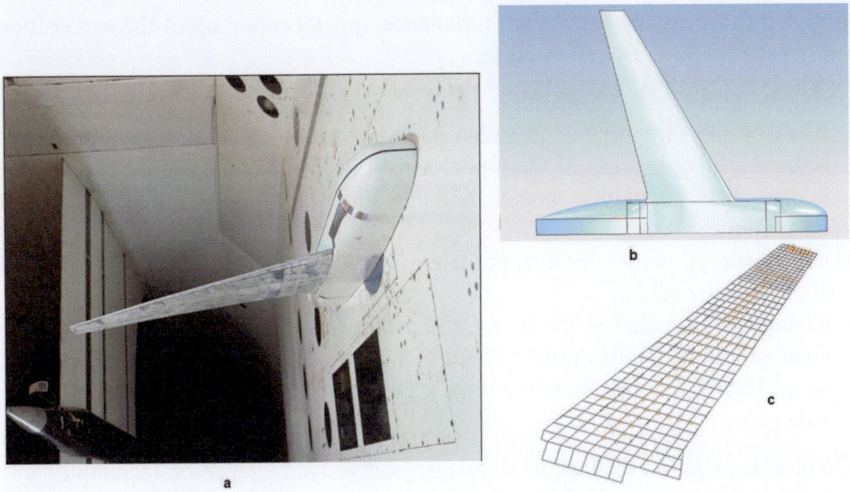

Fig. 12.25 An overview of various aeroelastic model for wind tunnel testing along with CAD clean wing configuration and FEM modeling. **a** NASA Langley's Models for Aeroelastic Validation Research Involving Computations (MAVRIC) program, model installed in the TDT (NASA Langley Transonic Dynamic Tunnel); **b** Computer Aided Design model of the MAVRIC clean-wing configuration; **c** The MAVRIC clean-wing FEM. (courtesy of NASA [34])

with the Technical University of Delft, designed and fabricated to produce optimized maximum static tip deflection. To obtain an overview of an Aeroelastic and Aeroservoelastic Experimental Model Wind-Tunnel Testing, a Comprehensive and similar impression is depicted by Fig. 12.25, which exhibits an aeroelastic model for Wind Tunnel Testing along with CAD Clean wing Configuration and FEM modeling. (a) NASA Langley's Models for Aeroelastic Validation Research Involving Computations (MAVRIC) program, model installed in the TDT (NASA Langley Transonic Dynamic Tunnel); (b) Computer Aided Design model of the MAVRIC clean-wing configuration; (c) The MAVRIC clean-wing FEM.

General lessons learned from the research and the use and validity of measurement results from the previous experiments can be summarized as follows:

- Aeroelastic experiments combine the uncertainties of structural, aerodynamic and measurement disciplines.
- Eliminating uncertainties is key to finding realistic explanations for aeroelastic effects and elimination of uncertainties. These can be achieved by:
  - Separating the disciplines in pretests where ideally only one discipline at a time is involved which allows for the identification of effects that can only be generated by the discipline investigated.
  - Keeping disciplines as simple and predictable as possible.

It was observed that the behavior of the wind tunnel with freestream velocities larger than 40 to 50 m/s became somewhat unpredictable. Possible errors may be attributed to increasing freestream turbulence, quick heating up of the test section and freestream flow, as well as noise.

Associated with structural layout the following has been observed:

- It is of utmost importance to gather and document as much information on the actual building process of the model as possible which allow for a meaningful update of the simulation models, and thus the elimination of a major source of uncertainties.
- Composite properties have to be determined, ideally, with every new material applied in the model.
- Clamping the model at the root and attaching it to the balance can be a large source for uncertainties which relates to the internal structure of the model itself, as well as the external structure required to transfer loads from the model to the balance.

The consideration of flaps to investigate active load alleviation techniques was one of the most prominent advancements to be addressed. The additional complexity introduced by this again emphasized the need to minimize uncertainties in the related aerodynamic and structure disciplines.

12.8.1 Model Specification

A relatively simple rectangular, non-swept wing layout was chosen over a complex shape, which was chosen on the basis of the findings in the research and to reduce modeling complexity. The work emphasizes the attainment of an adequate finite element and aerodynamic representation. The span of the forward swept was set to the previously realized 1.6 m wing, representing a viable compromise among aspect ratio, wing flexibility and the wind tunnel cross-sectional dimensions.

Two standard symmetric airfoils, NACA 0012 and NACA 0015, with a relative thickness of 12% and 15%, respectively, were preselected for application in the model. Considering a wing chord of 0.25 m, two-dimensional (2D) lift and drag polars for a Reynolds number range of 0.1×106 to 1.0×106 were computed with XFOIL, a two-dimensional interactive flow simulation software based on a 2D panel method coupled with a viscous boundary layer calculation for both NACA 0012 and NACA 0015 airfoils. For the lower Reynolds number, both airfoils indicate a noticeably nonlinear lift curve slope, as a result of laminar separation bubbles on the upper surface and a forward running transition location. As expected, on the one hand, the drag coefficients of the thicker NACA 0015 airfoil, in most areas, surpasses the ones for the thinner NACA 0012 airfoil. Figure 12.26 schematic illustrates the aeroelastic analysis similar to that carried out by Dillinger et al. for the wind tunnel test of their aeroservoelastic system.

Table 12.1 lists the different flap deflection cases investigated in the following. The flap deflection is again achieved by the application of appropriate airfoil coordinates, as generated with XFOIL.

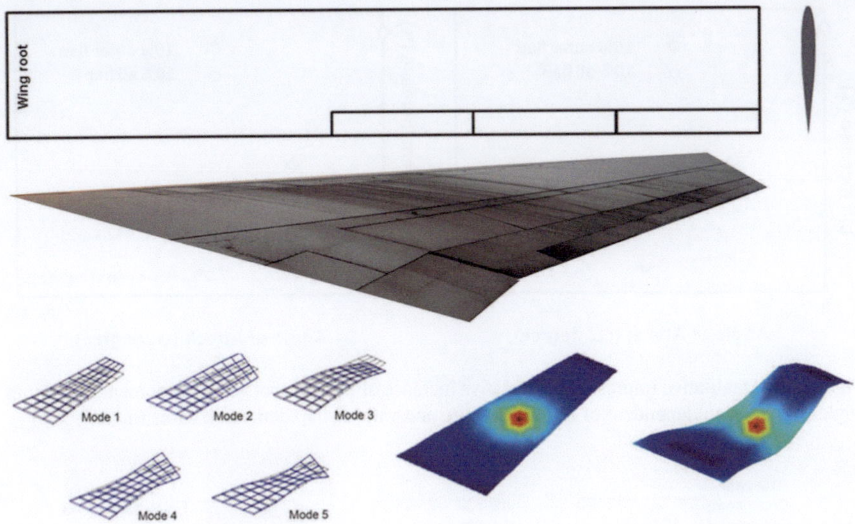

Fig. 12.26 A schematic illustrating the aeroelastic analysis similar to that carried out by Dillinger et al. for the wind tunnel test of their aeroservoelastic system[5,6]

Table 12.1 Case definition for different flap combinations[7]

Flap	Spanwise boundaries (m)	Case					
		1 (°)	2 (°)	3 (°)	4 (°)	5 (°)	6 (°)
1	0.6–0.9	0	0	0	10	0	10
2	0.9–1.2	0	0	10	0	10	10
3	1.2–1.5	0	10	0	0	10	10

As in the single flapped case, static calculations for an alpha range of 0°–10° and velocities from 10 to 50 m/s were investigated. The relative change of lift force and root bending moment with respect to Case 1 (no flap deflection) is plotted in Fig. 12.27.

Closely investigating the relative lift forces, the plot on the left, on the one hand, reveals that a flap situated more toward the root is able to generate a larger lift force increase for a defined flap deflection than a flap on the outer wing. This can be explained by means of the lift (circulation) distribution. While circulation, and thus lift force as a result of pressure equalization, tend toward zero at the tip, the effectiveness in increasing overall circulation by deflecting a flap is the highest in the

[5] Illustration created and adapted as an illustrative example based on information extracted from Dillinger et al. [7].

[6] Wing photo courtesy of stocksnap (copyright free); https://stocksnap.io/photo/airplane-wing-U2IMAIJ2.

[7] Created and adapted using information and data from Dillinger et al. [7].

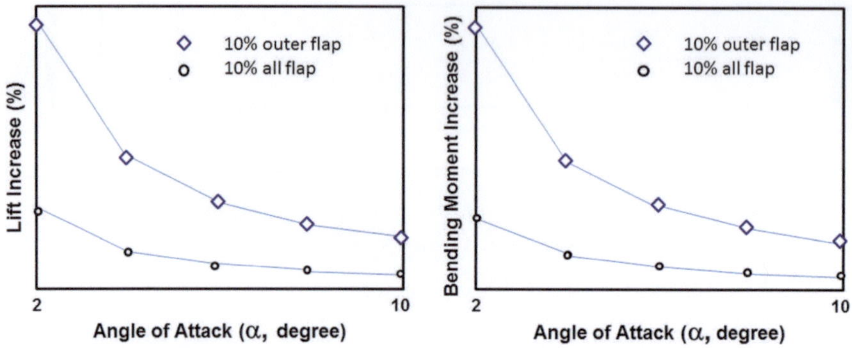

Fig. 12.27 Qualitative impression of Relative increase of lift and root bending moment for various angles of attack (independent of flight velocity), and various flap deflection cases (not to scale)[8]

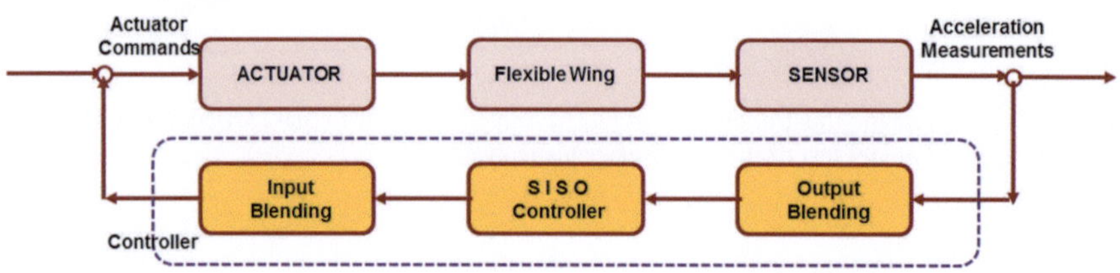

Fig. 12.28 Control loop interconnection[9]

area of highest circulation. In the case of a rectangular, aerodynamically untwisted wing the circulation is the highest in the root. Thus, the closer a flap is positioned toward the root, the more effective it becomes in increasing lift.

The root bending moment, on the other hand, shows the opposite trend, the further outside the flap is located, the more additional root bending moment it can generate. Evidently, the increased lever arm dominates over the diminished lift force in the outer wing, meaning that a flap further outboard generates more root bending moment than an inboard flap, see the plot on the right in Fig. 12.25.

The control approach isolates the target mode(s) by blending control inputs and measurement outputs in an $\mathscr{H}_2$ optimal way in order to enable a subsequent single-input single-output (SISO) controller design. In other words, the measurement outputs were weighted and summed up such that the resulting scalar signal best represents the response of target mode. Similarly, the control commands computed by the SISO controller were distributed to the given flap actuators in a way that the target mode was damped efficiently without exciting other modes. The resulting closed-loop interconnection is depicted in Fig. 12.28.

---

[8] Created and adapted using information and data from Dillinger et al. [7].

[9] Created and adapted based on data and information from Dillinger et al. [7].

12.9 Aeroservoelastic Modeling and Analysis of a Highly Flexible Flutter Demonstrator [35]

An example of an the aeroservoelastic modeling of a highly flexible flutter demonstrator is elaborated subsequently the referenced paper. A finite element model of the demonstrator is generated and condensed to a reduced number of degrees of freedom to represent the structural dynamics. The unsteady aerodynamics are captured by the doublet lattice method based on potential theory. By interconnection of structural dynamics and unsteady aerodynamics an aeroelastic model is derived, which provides a basis for the design of a utter suppression controller. In order to enable an efficient utter suppression a clear separation of the occurring utter mechanisms in speed and frequency is desired. To achieve this, the positions of the actuators controlling the flaps are varied within the scope of the aircraft design process.

The heart of the active utter suppression system is a flutter controller, which controls actuators connected to the ailerons. The development of an active flutter control system is based on aeroelastic models. It is therefore essential to derive sophisticated aeroelastic models beforehand.

The aerodynamic model is based on a set of trapezoidal aerodynamic boxes, which represent the lifting surfaces. By applying either vortices or doublets to all boxes, the flow field of the demonstrator is approximated using the potential theory. With the application of vortices this approach is called Vortex Lattice Method (VLM), while with doublets it is known as the doublet lattice method (DLM), respectively.

The interconnection between the structural dynamics and aerodynamics is carried out through splining, which results in an aeroelastic model. To prevent flutter and guarantee stability within the entire flight envelope, active flutter suppression can be used. The active flutter suppression methods dubbed as Flutter Free FLight Envelope eXpansion for ecOnomic Performance improvement (FLEXOP) are developed in the project elaborated in the paper referenced and were tested on the demonstrator aircraft shown in Fig. 12.29.

The modeling process within the FLEXOP project is described in Fig. 12.30. A finite element (FE) model is used for representing the structural dynamics of the demonstrator. The heart of the active flutter suppression system is a flutter controller, which controls actuators connected to the ailerons. The development of an active flutter control system is based on aeroelastic models. It is therefore essential to derive sophisticated aeroelastic models beforehand.

The structural dynamics of the flexible aircraft can be divided into a rigid-body and an elastic-body motion. The rigid-body motion basically describes the maneuver characteristics of the aircraft. In contrast, the elastic-body motion represents the dynamics due to the flexibility of the aircraft structure. In the approach taken by the referenced paper, the rigid-body motion is described in nonlinear form, but the equation of the elastic-body motion is considered to be linear. The detailed FE model serves as basis for the structural model of the aircraft.

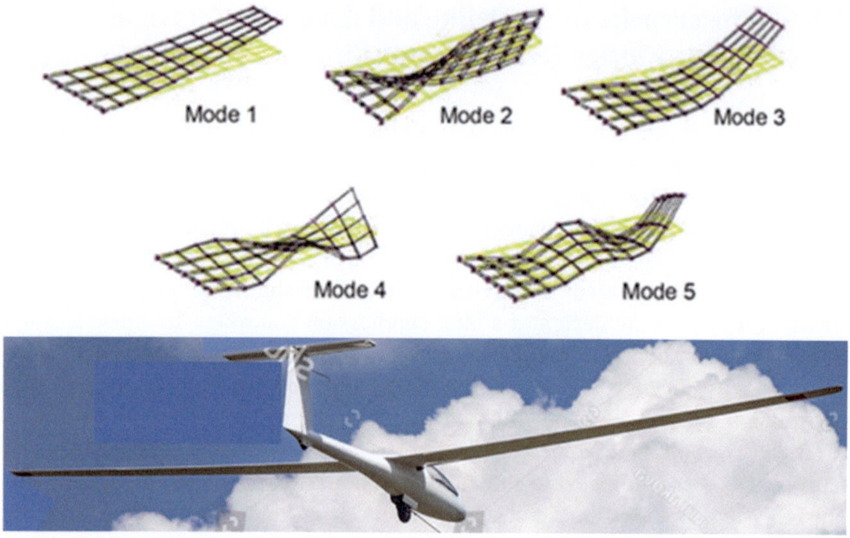

Fig. 12.29 Impression of finite element structural discretization that can be applied to a Glider. Boundary element aerodynamic boxes can also be generated. The FE deformation images are taken from Chap. 17 to illustrate similar procedure carried out for FLEXOP [36] demonstrator aircraft Full FE model and aerodynamic boxes [19][10,11]

12.9.1 Finite Element Model

Using MSC NASTRAN software, the aircraft structural FE model comprising the wing, fuselage and empennage can be generated. The wing is represented by a high-fidelity FE model comprising beam, surface and solid elements. Rigid-body interpolation elements have been added at predefined locations throughout the wing to facilitate the required model reduction. Furthermore, the wing mass model is density based as opposed to a lumped mass model. The fuselage structure is modeled using beam elements. The equivalent beam stiffness is obtained utilizing the cross-sections of the fuselage at different sections. The mass is then lumped at the two beam nodes. The V-tail empennage FE model is shell-element based comprising of the main structural load-bearing entities, the upper and lower skins, structural rubs, spars and the non-structural masses. Similar to the wing FE model, a density-based mass representation is used for the empennage as well.

---

[10] Created and adapted based on data and information from Dillinger et al. [7] and Rozov et al. [37].

[11] Glider image courtesy of stocksnap; https://stocksnap.io/photo/airplane-glider.

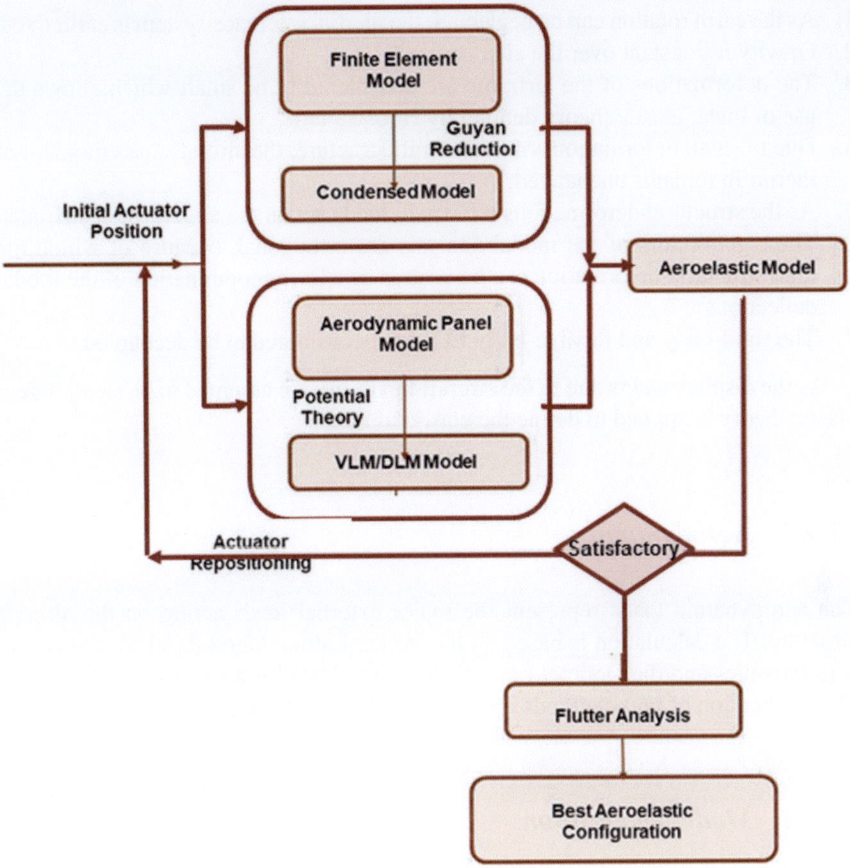

Fig. 12.30 Modeling process of the aeroelastic part of the demonstrator aircraft (information utilized as the courtesy of DLR/TUM[12])

12.9.2 Equations of Motion

For the derivation of the nonlinear flight mechanical EOM, the aircraft is considered as a rigid body with a constant mass mb and constant mass moment of inertia Jb. The Equation of Motion (EOM) representing the rigid and elastic-body motion are defined for the condensed model. The condensed model features rigid-body and flexible modes, which are described by the Equation of Motion (EOM). The EOM are based on an equilibrium of forces and moments, describing the behavior of the aircraft due to external loads originating from the aerodynamics and thrust. The assumptions below were made for simplification:

[12] The diagram is generated from information contained in Wüstenhagen et al. [35].

1. As the earth rotation can be neglected, the inertial reference system is earth fixed.
2. Gravity is constant over the airframe.
3. The deformations of the airframe are considered to be small which allows the use of linear elastic theory defined by Hooke's law.
4. Due to small deformations of the aircraft structure, the aircraft mass moment of inertia Jb remains unchanged.
5. As the structural deformations are small, loads act on the undeformed airframe.
6. The eigenvectors of the modal analysis are orthogonal, because of which the total structural deformation can be written as a linear combination of the modal deflections.
7. The rigid-body and flexible-body EOM are considered to be decoupled.

As the displacements due to the aircraft flexibility are assumed to be small, linear elastic theory is applied to define the elastic motion.

12.9.3 Aerodynamics

The aerodynamic loads represent the major external loads acting on the aircraft structure. The calculation is based on the Vortex Lattice Method (VLM) for steady aerodynamics and the Doublet Lattice Method (DLM) for unsteady aerodynamics. The application of both methods is elaborated in the referenced paper.

12.9.4 Model Integration

The overall aeroservoelastic model is built of several sub-models, as depicted in Fig. 12.30, and elaborated in the referenced paper [35]. The core forms the aeroelastic system, which represents the coupling of aerodynamics and structural dynamics.

The aerodynamic loads P_g^{aero} directly correlate with the structural motion and can be expressed in terms of Δc_{pj} as

$$\mathbf{P}_g^{aero} = q_\infty T_{kg}^T S_{kj} \Delta c_{pj} \tag{12.55}$$

where S_{kj} is an integration matrix relating the pressure in the aerodynamic boxes at point j with the forces at the aerodynamic grid points k.

The forces at the aerodynamic grid points k are then interpolated onto the structural grid points via the transpose of the spline matrix T_{kg}. The structural nodes used for the splining comprise of the condensation nodes as well as two nodes per condensation node, extended to the leading and trailing edge at that point using rigid elements. A similar splining model is used for the empennage as well. Multiplying with the dynamic pressure q_∞ leads then to the aerodynamic loads acting on the structure. The aircraft structure reacts on the aerodynamic loads by performing rigid-body and flexible motions, which directly affect the aircraft aerodynamics. Therefore the aeroelastic model is considered as a loop between structural dynamics and aerodynamics.

An aeroservoelastic model is derived by adding an active utter control system to the aeroelastic model. The active utter control system comprises actuators for detecting the relevant control surfaces, sensors for observing the dynamic behavior of the aircraft and the utter controller. As the actuators introduce additional inertial loads to the aircraft structure, they need to be considered during the aeroelastic design process of the aircraft. In the following section the utter characteristics of the aeroelastic model are evaluated with respect to the actuator position.

12.9.5 Model Analysis

The main goal of the FLEXOP project is to increase the flutter speed by means of active flutter suppression techniques. To demonstrate active utter suppression, the open-loop aeroelastic system needs to be designed in a way that the occurring flutter can be controlled. Thus, the aircraft design process is tailored to achieve a separation of the flutter phenomena in terms of frequencies and speeds. With increasing speed, two aeroelastic modes representing symmetric and asymmetric flutter cross the imaginary axis and become unstable. The flutter speed of symmetric flutter is determined to be 48:51 m = s at a frequency of 8:08 Hz. In comparison asymmetric utter becomes unstable at 50:58 m = s and 6:83 Hz. This demonstrates that the flutter characteristics are very close to each other in frequency and speed. Therefore, the idea is to separate the utter phenomena for an easier flutter controller design by using the position of the actuators for flutter control as design freedom. The separation of the flutter frequencies allows designing decoupled utter controllers for each flutter phenomenon. The separation in the flutter speeds helps to examine the stabilization of one flutter mode without having an interaction of the second flutter mode.

12.10 Concluding Remarks

The discussions elaborated in previous sections cover some selected topics in Aeroservoelasticity (ASE), which have demonstrated the extent of the concept of aeroelasticity to address the aeroelastic interactions between aerodynamic forces, flexible structures and a control system. Several examples drawn from recent research works from selected researcher have been referred and their salient and relevant parts are elaborated to gain a comprehensive coverage of their objective and scopes. Various equations utilized can be traced back to the basic fundamentals elaborated in the first part of the present book. Hence the case studies that serve as examples of Aeroservoelasticity (ASE) can be realized to be multidisciplinary that combine first principles basic approach, scientific considerations of the interactions between the aircraft steady and unsteady aerodynamic forces, the inertial forces, the elasticity of the structure and the flight control system of the aircraft. These have been shown to affect the flight performance, structural integrity and handling qualities of the aircraft as conceived by the designers and researchers.

References

1. Babcock, J.T. 2013. *Aeroservoelastic Design for Closed-Loop Flight Dynamics of an MAV*. University of Florida.
2. Wright, J.R., and J.E. Cooper. 2015. *Introduction to Aircraft Aeroelasticity and Loads*, 2nd ed. Wiley. https://doi.org/10.1002/9781118700440.
3. Clarke, R., M.J. Allen, and R.P. Dibley. Flight test of the F-A-18 active aeroelastic wing airplane. In *NASA-TM-2005–213664*, NASA Dryden Flight Research Center.
4. Cumming, S.B., and C.G. Diebler, Active aeroelastic wing aerodynamic model development and validation for a modified F-A-18A. In *NTRS-NASA TRS 20050204039*.
5. Lind, R., and M.J. Brenner. 1998. Robust flutter margin analysis that incorporates flight data. In *NASA/TP-1998-206543*.
6. Noll, T.E. 1990. Aeroservoelasticity. In *NASA-TM-102620*, Mar.
7. Dillinger, J.K.S., J.M. Meddaikar, J. Lübker, M. Pusch, and T. Kier. 2020. Design and optimization of an aeroservoelastic. *Fluids* 5 (1): 35.
8. Mursal, Y., H. Djojodihardjo, and F. Fadil. State-space aeroelastic analysis of an aircraft with stand-by sctuator. In *Paper ICAS-94-5.6.1*.
9. Budiyono, A. and H. Djojodihardjo. 1994. The application of artificial neural networks on flutter suppression system. In *Paper ICAS-94-7.8.2*, copyright by the author.
10. Edwards, J.W. et al. 1978. Active flutter control using generalized unsteady aerodynamic theory. *Journal of Guidance and Control*, Jan–Feb.
11. Newsom, Jerry R. et al. 1979. Control law synthesis for active flutter suppression using optimal control theory. *Journal of Guidance and Control*, Sept–Oct.
12. Karpel, Mordechay. 1982. Design for active flutter suppression and gust alleviation using state-space aeroelastic modelling. *Journal of Aircraft*, Mar.
13. Mahesh, J.K. et al. 1981. Control law synthesis for flutter suppression using linear Gaussian Theory. *Journal of Guidance and Control*, July–Aug.
14. Mukhopadhyay, V. et al. 1982. Reduced-order optimal control feedback control law synthesis for flutter suppression. *Journal of Guidance and Control*, Aug.
15. Nissim, E. 1971. Flutter suppression using controls based on concept of aerodynamic energy. In *Technical Report NASA TN D-6199*, NASA.
16. Nissim, E. 1977. Recent advances in aerodynamic energy concept for flutter suppression and gust alleviation using active controls. In *Technical Report NASA TN D-8519*, NASA, Sept 26.
17. Nissim, E., and I. Lottati. 1979. Active external store flutter suppression in the YF-17 flutter model. *Journal of Guidance and Control*, Sept–Oct.
18. Nissim, E., and I. Lottati. 1979. On single-degree-of-freedom flutter induced by activated controls. *Journal of Guidance and Control*, Sept–Oct.
19. Nissim, E., and Irving Abel. 1978. Development and application of an optimization procedure for flutter suppression using aerodynamic energy concept. In *Technical Report NASA TP 1137*, NASA.
20. Abel, Irving et al. 1978. Two synthesis techniques applied to flutter suppression on flight research wing. *Journal of Guidance and Control*, Sept–Oct.
21. Harris, T.M. et al. 1985. A review of aeroelastic research at flight dynamics laboratory. In *2nd International Symposium of Aeroelasticity and Structural Dynamic*.

22. Budiyono, Agus. 1992. *Study of Flutter Suppression System on Typical Section with Control Surface*. Institute of Technology Bandung, B-Eng. Thesis (in Indonesian).
23. Budiyono, Agus. 1993. Design of an active flutter suppression system using artificial neural networks. In *Technical Report*, RMIT (author's repository).
24. Alanis, A., N. Arana-Daniel, and C. Lopez-Franco. 2019. *Artificial Neural Networks for Engineering Applications*, 1st ed., February 7.
25. Linden, A., et al. 1993. *SESAME: Software Environment for Simulation of Adaptive Modular Systems*. GMD Schloss Birlinghoven.
26. Minsky, M.L. 1969. *Perceptrons: An Introduction to Computational Geometry*. MIT Press.
27. Werbos, Paul J. 1990. Backpropagation through time: What it does and how to do it. In *Proceeding of the IEEE*, Vol. 78, No. 10.
28. Hunt, K.J. et al. 1992. *Neural Networks for Control Systems-A Survey*. International Federation of Automatic Control.
29. Narendra, K.S. 1990. *Adaptive Control Using Neural Networks*. MIT Press, USA.
30. Garcia-Padilla, Felix, and Fransisco Morant-Anglada. 1991. A supervisory technique to apply neural networks in control. In *Artificial Neural Networks*. Springer-Verlag, Sept.
31. Wharington, John M. 1992. *Neural Networks for Automatic Flight Control Systems*. Royal Melbourne Institute of Technology, B-Eng. thesis.
32. Rickard, J. Susan, R. Susan Bartholomew, et al. 1993. Neurocontrollers for complex system. *31st Aerospace Sciences Meeting & Exhibit*.
33. Krisnakumar, K. et al. 1993. Adaptive neuro-control for the space station freedom. In *31st Aerospace Sciences Meeting & Exhibit*.
34. Florance, J.P., Pawel Chwalowski, and Carol D. Wieseman. 2010. Aeroelasticity benchmark assessment subsonic fixed wing program interim report, March. https://ntrs.nasa.gov/api/citations/20100016316/downloads/20100016316.pdf.
35. Wüstenhagen, M., T. Kier, M. Pusch, D. Ossmann, M.Y. Meddaikar, and A. Hermanutz. 2018. Aeroservoelastic stability of a 2D airfoil section equipped with a trailing edge Flap, In *2018 Atmospheric Flight Mechanics Conference*, 3150 AIAA-1471524.
36. AGARD CP No.46, Aeroelastic effects from a flight mechanics standpoint.
37. Rozov, Vladyslav, Andreas Hermanutz, Christian Breitsamter and Mirko Hornung. 2017. Aeroelastic analysis of a flutter demonstrator with a very flexible high-aspect-ratio Swept Wing, IFASD 2017, Italy.
38. Akmeşe, A., M.D. Cömert, and B.E. Platin. 2004. Aeroservoelastic analysis of missile control surfaces via robust control methods. In *16th IFAC Symposium on Automatic Control in Aerospace*, 14–18 June, St. Petersburg, Russia
39. Djojodihardjo, Harijono and Agus Budiyono. 1993. State space aeroelastic modelling for active suppression system. In *Proceedings, National Symposium on Aeronautical Science and Technology*, Bandung, 27–29 July.
40. Djojodihardjo, Harijono, S. Hakim Ad-Dairi, and A. Suwandi Ahmad. 1991. *Expanding Horizon of Aerospace Vehicle Design: Psychological Engineering as Part of Man Machine System with Case Study on Plant Identification Using Artificial Neural Network*. International Scientific Meeting on Aviation Psychology and Psychiatry, Jakarta, February 15–16.
41. Dowell, Earl H. et al, *A Modern Course in Aeroelasticity*, 5th ed.
42. Freeman, James A., and David M. Skapura. 1991. *Neural Networks Algorithm, Applications and Programming Techniques*. Addison-Wesley Publishing Company.
43. Garrard, W.L. et al. 1982. Robust Kalman filter design for active flutter suppression systems. *Journal of Guidance and Control*, July–Aug.
44. Kosko, Bart. 1992. *Neural Networks and Fuzzy Systems. A Dynamical Systems Approach to Machine Intelligence*. Prentice Hall, Englewood Cliffs.
45. Lu, Pong-Jeu, and Li-Deng Huang. 1993. Flutter suppression of thin airfoils using active acoustic excitation. *AIAA Journal*.
46. Miller, et al. (eds.). 1990. *Neural Networks for Control*. MIT Press.
47. Moreno, C.P., P.J. Seiler, and G.J. Balas. 2014. Model reduction for aeroservoelastic systems. *Journal of Aircraft* 5 (11).

48. Abel, Irving. 1979. An analytical technique for predicting the characteristic of a flexible wing equipped with an active flutter-suppression systems and comparison with wind-tunnel data. In *Technical Report TN D-6199*, NASA.
49. Scott, R.C., M. A. Castelluccio, D.A. Coulson, and J. Heeg. 2011. Aeroservoelastic wind-tunnel tests of a free-flying, joined-wing SensorCraft model for gust load alleviation. NASA
50. Tang, Deman, and Earl H. Dowell. 2016. Experimental aeroelastic models design and wind tunnel testing for correlation with new theory. Aerospace.
51. Whitcomb. 1976. *A Design Approach and Selected Wind-Tunnel Results at High subsonic Mounted Speeds for Winglets*. NADA TN D-3280.
52. Zell, Andreas et al. 1993. *User Manual Stuttgart Neural Network Simulator. v 3.0*. University of Stuttgart.

Chapter 13
Introduction to Aircraft Loads

Abstract The basic understanding of the significance of aircraft loads and their relationship to aeroelasticity will be elaborated comprehensively. The estimation of loads acting on an aircraft structure is an indispensable task ranging from conceptual, preliminary and detailed design to loads flight testing and the manufacturing and operation of aircrafts, when an aircraft is already in service, there are a large range of topics that need to be elaborated and proven, analytically as well as by experiments. Much of these problems can be understood and mastered, by looking into selected topic which can be classified into those related to (a) stability (e.g. flutter), (b) static deformation (e.g. static aeroelastic effects, steady flight maneuvers) and (c) dynamic response (e.g. maneuvers, gusts, turbulence). Once the response deformations and accelerations are obtained, the loads and stresses generated in the aircraft must also be determined so that the strength and fatigue/damage tolerance behavior may be assessed. Structural loads are the major reason for degradation of structures. Loads is a general term that incorporates both forces and moments. Various basic concepts relevant to loads in general will be introduced, including Newton's laws of motion for a particle and their generalization to a body, as well as D'Alembert's principle as sources of externally applied/reactive loads, which will further generate internal load within a structure. Basic analysis using the principles of mechanics, strength of materials, aerodynamics and flight mechanics will be elaborated. The determination of loads together with the qualification for static strength and fatigue by calculation and test for all important structural components is a main prerequisite for successful design and safe operation of any aircraft.

Keywords Aeroelasticity · Aircraft loads · Loads process · Loads and fatigue · Load factor · V-n diagram

13.1 Introduction

The estimation of loads acting on an aircraft structure is an indispensable task ranging from conceptual, preliminary and detailed design to loads flight testing and the manufacturing and operation of aircrafts, when an aircraft is already in service, there are a

© The Author(s), under exclusive license to Springer Nature Singapore Pte Ltd. 2023

H. Djojodihardjo, *Introduction to Aeroelasticity*,
https://doi.org/10.1007/978-981-16-8078-6_13

large range of topics that need to be elaborated and proven, analytically as well as by experiments. Physics-based loads and aerodynamic performance analysis methods in predesign can be carried out using a parameterized and automatized process nested in an overall aircraft design loop. Such process can be used for conventional and unconventional configurations. However, the basis of treating aircraft loads problems can be based on basic and fundamental principles, which are the fundamental laws of mechanics, including structural analysis and strength of materials, and thermo-sciences conservation principles which are the foundation of aerodynamics. However, much of these problems can be understood and mastered, by looking into selected topic which can be classified into those related to (a) stability (e.g. flutter), (b) static deformation (e.g. static aeroelastic effects, steady flight maneuvers) and (c) dynamic response (e.g. maneuvers, gusts, turbulence). The aircraft dynamic response needs to be calculated for ground maneuvers (e.g. taxiing, takeoff and landing), flight maneuvers (e.g. response to control movements) and gust/turbulence encounters. However, once the response deformations and accelerations are obtained, the loads and stresses generated in the aircraft must also be determined so that the strength and fatigue/damage tolerance behavior may be assessed.

Structural loads, leading to fatigue as well as corrosion, depending on the usage environment, are the major reason for degradation of structures [1]. Loads is a general term that incorporates both forces and moments, discrete and distributed, and external and internal. Various basic concepts relevant to loads in general will be introduced, including Newton's laws of motion for a particle and their generalization to a body, D'Alembert's principle (leading to discrete inertia forces/couples and distributed inertia forces), externally applied/reactive loads, Free-Body diagrams (FBDs), loads generated internally within a structure (i.e. internal loads) and inter-component loads.

Therefore, the present chapter will provide information to the students and practicing engineers from selected analysis from topics identified above, the logical foundation of the analysis, and some detailed examples. To obtain a comprehensive impression of these, the discussion in the chapter will resort to selected modern articles presented in recent scientific conferences and publications.

A general understanding of the various types of airframe loads, their generation and application during the design process, the transfer processes from "external loads" into "structural loads," and load qualification during ground and flight testing is therefore of importance in the process of monitoring, derivation, testing and inspection associated with structural loads. The estimation of loads acting on an aircraft structure is an indispensable task ranging from conceptual, preliminary and detailed design to loads flight testing when an aircraft is already in service. Aircraft load analysis is an important engineering discipline playing a major role from the first sketch of a new aircraft until the decommissioning.

As an example of basic elements of load analysis, the following advisory circular of the US Department of Transportation FAA (Federal Aviation Administration) Advisory Circular on Dynamic Gust Loads AC No: 25.341–1 [2][1] which addresses

[1] US Department of Transportation FAA Advisory Circular on Dynamic Gust Loads AC No: 25.341–1

discrete gust and continuous turbulence (or continuous gust) requirements is adopted to have an impression of the basic elements of the analysis requirements.

13.2 Basic Elements of Analysis for Aircraft Loads

This advisory circular provides some of the acceptable methods of modeling airplanes, airplane components and configurations, and the validation of those modeling methods for the purpose of determining the response of the airplane to encounters with gusts.

The dynamic analysis utilized to model various airplane parameters can have a significant influence on the design load levels. Figure 13.1 schematically depicts the basic elements to be modeled in the analysis comprising the elastic, inertial, aerodynamic and control system characteristics of the complete airplane. The degree of sophistication and detail required in the modeling is dictated by the complexity of the airplane and its systems. The figure also depicts loads problem in the perspectives of Aeroservoelasticity.

In particular, Fig. 13.1 depicts Aircraft Loads problem associated with Gust. Design loads for encounters with gusts are a combination of the steady, level 1 g flight loads and the just incremental loads including the dynamic response of the airplane. The steady 1 g flight loads can be realistically defined by the basic external

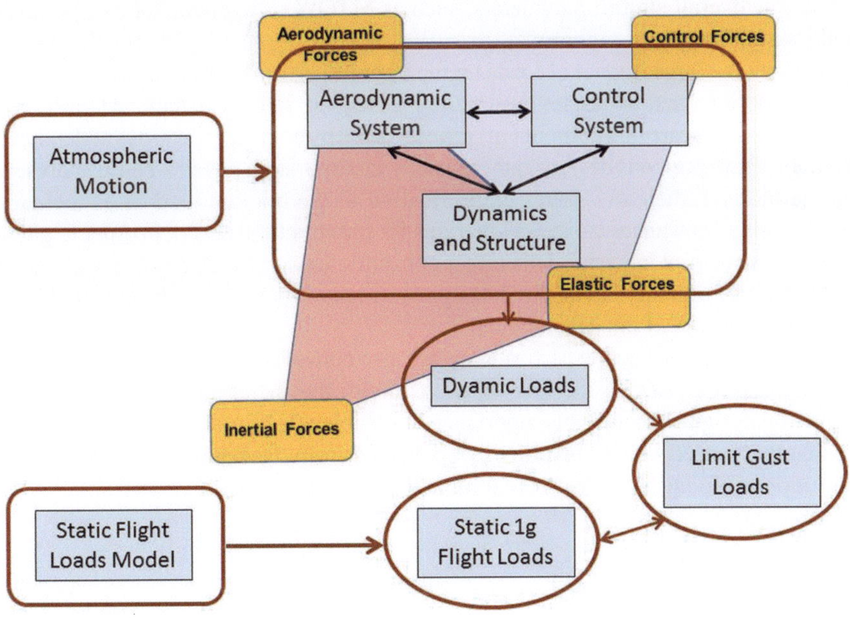

Fig. 13.1 Basic elements of analysis for loads due to gust response and aeroservoelasticity

parameters such as speed, altitude, weight and fuel load. They can be determined using static aeroelastic methods.

The gust incremental loads are the results of the interaction of atmospheric turbulence and airplane rigid body and elastic motions. These may be calculated using linear analysis methods when the airplane and its flight control systems can be reasonably (or conservatively) approximated by linear analysis.

Nonlinear solution methods are necessary for airplane and flight control systems that are not reasonably or conservatively represented by linear analysis models. Nonlinear features generally raise the level of complexity, particularly for the continuous turbulence analysis, because they often require that the solutions be carried out in the time domain.

13.3 Loads Analysis in Overall Aircraft Predesign

The early aircraft development stages focus on the feasibility and on the assessment of the overall performance of the investigated configurations. During the predesign stages, design details are typically not available, and the overall aircraft synthesis relies on certain specifications.

A schematic of a basic framework of load calculation is depicted in Fig. 13.2.[2] For example, this was specified as Top Level Aircraft Requirements, which specifies those such as transportation mission and operational constraints, and provides as output the overall aircraft parameters, such as MTOW (Maximum Takeoff Weight) and mission fuel consumption.

Despite the large impact of loads on the aircraft performance, conceptual overall aircraft design methodologies are mainly based on statistical data. At best, the airframe's loads carrying components are estimated by empirical relations or database available from previous designs rather than by incorporating physics-based analyses. Furthermore, at the early stages the estimation of the relevant load cases includes conservative allowances and safety margins for the structural design in order to guarantee the structural integrity. This rather conventional design approach usually adds an "aeroelastic penalty" to the final designed structures, which may be unveiled only at the later phases of the development.

Flexibility effects on the performance are also conventionally not included during the predesign phases of the investigations. With the increasing efficiency of the structural design concepts, the importance of flexible effects is significantly growing, so that these effects could be properly accounted from the beginning of the aircraft development to meet the design requirements and objectives.

---

[2] Many refined schemes has been introduced recently in a Multi Fidelity, Multi Disciplinary Analysis and Optimization Framework, such as those by Ciarella et al [3], Klimmek et al [4], Kafkas et al [5] and Karpuk et al [6]

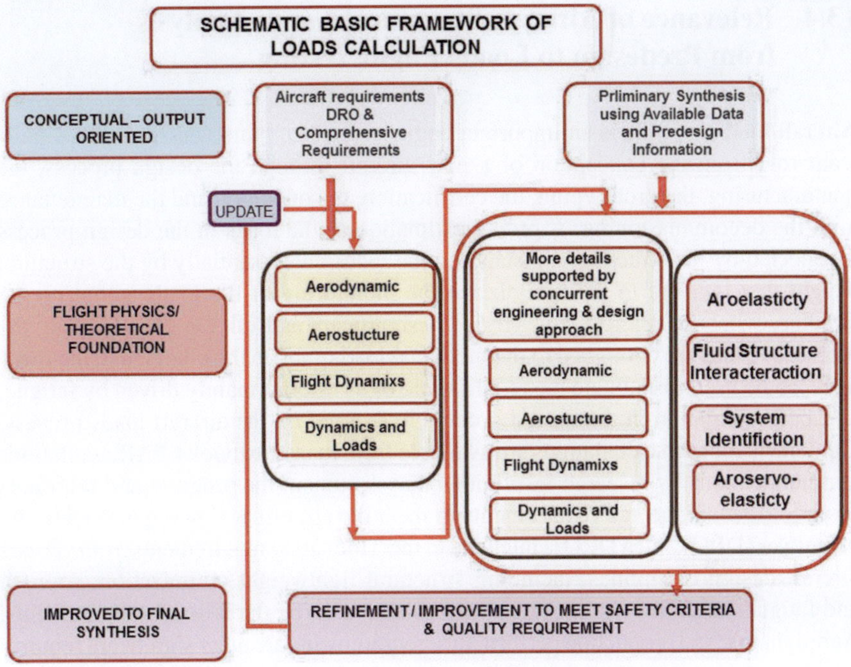

Fig. 13.2 Schematic basic framework of loads calculation

13.3.1 Multi-fidelity Loads Process

A fully automated multi-fidelity process, as shown in Fig. 13.2, is characterized by 3 phases, with focus on the aerostructural design modules which are used to investigate how a physics loads analysis process may impact the overall aircraft synthesis and the designed performance. The first phase is the initialization of the aircraft, labeled in the scheme as "1st Conceptual." The initialization step provides the first full aircraft synthesis for a given set of Top Level Aircraft Requirements. The synthesis is provided by a conceptual aircraft synthesis module, whose formulations are based on typical aircraft design methodologies.

The output is a representation of synthesized aircraft, and its performance, such as the aircraft design masses and mission fuel, is also shown in Fig. 13.2. Further, the solution is augmented by additional features in order to enable to progress the design solution to the next phase which is using low fidelity physics-based analysis modules. Examples of enhanced details are the initialization of a representative wing internal structural topology and the three-dimensional shape of the aircraft components.

The focus is then placed on taking into account a low fidelity loads analysis process and investigate the impact it has on the early design stages, so that expensive redesign activities or the degradation of the prospected performance are minimized.

13.4 Relevance of Aircraft Structural Loads Analysis from Predesign to Loads Flight Testing

Aircraft loads analysis is an important engineering discipline which plays a significant role from the conception of a new aircraft through the design process, the manufacturing, the prototyping, the certification, the operation and the maintenance until the decommissioning. A proper estimation of the loads in the design process is reflected by the structure withstanding the loads and essentially by the structural weight that is about to be minimized. The monitoring of the loads acting on the structure when the aircraft is in service is significant as well. On the one hand, the measured loads are compared to the predicted loads, and on the other hand, the loads data can be used to estimate the real lifetime of an aircraft mainly driven by fatigue.

Load calculation in the design process is focused on the aircraft loads process. Guidelines for such calculations are available from many textbooks, FAR regulations and publications by research institutions that document the progress and relevance of various technique and requirements to meet the aircraft design requirements and objectives (DRO). Such DRO is intended to meet the safety requirements from various aircraft design disciplines, including structural lightweight optimization, strength and fatigue, and most significantly, what is known to be the fail-safe requirements. Aerodynamic and propulsion performance requirements are other significant requirements, leading to new aircraft configuration, and in recent years and in the near future, fuel efficient or "green" requirements dictated by the global climate and environmental concerns. Focusing on aircraft loads analysis, few publications are available where a comprehensive survey regarding loads analysis is presented. Few publications are available where a comprehensive survey describing various load analysis methods for almost all load types that have to be considered in aircraft designs, such as those elaborated in Bisplinghoff, Ashley and Halfman [7], Taylor [8], Lomax [9], Dowell et al. [10], and Wright and Cooper [1].

To include as much as possible (or specified) of these loading scenarios in the early process of designing a new type of aircraft is the responsibility of the loads engineering department, while ensuring that these loads can be safely endured throughout the specified life is the task of the design and stress engineers. Further to the description of the scope and extent of Aircraft Load Calculation as depicted previously in Figs. 13.1 and 13.2, Figure 13.3 gives an idea how loads are initially generated and how they are used throughout the design-, qualification- and usage process.

13.4.1 Loads and Fatigue

The determination of loads together with the qualification for static strength and fatigue by calculation and test for all important structural components is a main prerequisite for successful design and safe operation of any aircraft. Whereas for transport aircraft with their rather limited range of operational maneuvers and high

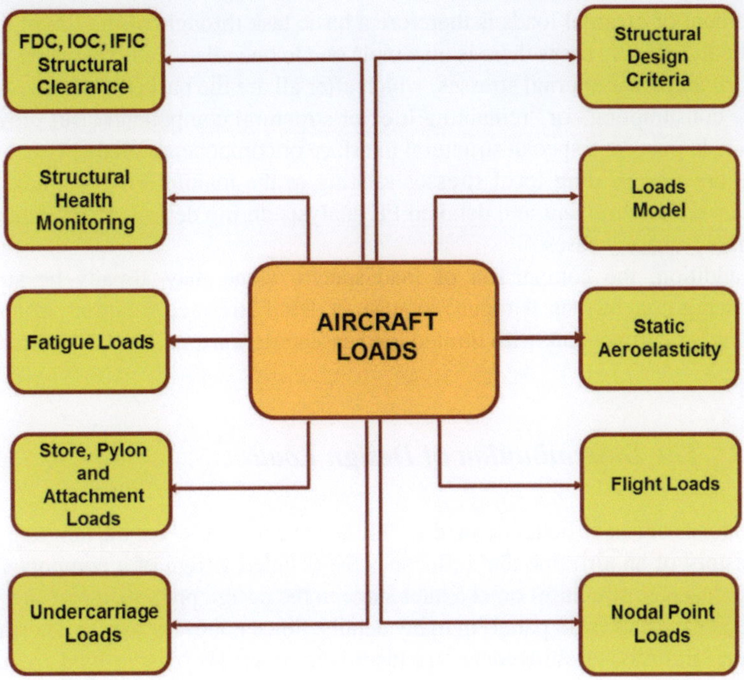

Fig. 13.3 Loads main tasks [12]

number of flight hours cycles fatigue is the main design driver for the airframe, fighter aircraft are predominantly designed to (static) limit load cases for the "corners" of the envisaged flight envelope (or V-n Diagram), which in general cover high strength requirements for fatigue of their comparatively short life.

This is only true as long as fighter life does not exceed the originally planned lifetime and the roles, missions and so forth, which are compatible with the design criteria at the beginning. Aging aircraft in both cases does not only mean that an aircraft is getting older in terms of flight hours and flight cycles, and it also means that some of the reference data for the basic design criteria have changed during time, i.e.:

- Airframe and equipment mass growth
- Enhancement of systems performance, especially engine thrust
- New configurations (stores)
- Update of flight control systems (FCS) (electronically or hardware changes like added slats or enlarged ailerons)
- Mission profiles and additional/changed roles
- Actual usage spectrum.

Most of these changes have an immediate impact on aircraft load scenarios, others will not change load levels but may change underlying statistic, e.g. fatigue spectra.

Assessment of external loads is therefore a basic task throughout the life of a fleet. Admittedly in many cases there is no simple one to one relationship between "external" load and local internal stresses, which after all are the basis for the assessment of "life consumption" or "remaining life" of structural components. But providing loads are known for a special structural interface or component, reliable conclusions can be drawn regarding local stresses relating to the manifold of load case from experience, measurement and detailed FE analysis during design, qualification and test phases in many cases.

In addition, the comparison of load spectra alone may already be suitable for drawing conclusions without recourse to detail stress calculations of specific locations for components with limited load case variations, i.e. landing gears.

13.4.2 The Determination of Design Loads

Design loads or can be better defined as "Initial Design Loads" are the first step in the load history of an airframe that influences the detailed design of a component (i.e. wing or fuselage structure) or, at a later stage in the design process, a part (i.e. wing spar cap or fuselage skin panel) in many details. Since not every load is determining these design tasks, establishment and identification of the "design load cases" are important [12].

In the typical example "loads loop" depicted in Fig. 13.4, the methods of calculating design load cases are determined. The loop considered incorporate special attention to points where an immediate context with fatigue calculations exists. This loop is usually repeated several times in the different phases of the aircraft design. First of all the Structural Design Criteria (SDC*)* are prepared as a basis for design, specifying the basic performance and flight parameters, and then a Loads Model (*LM*) is built, based on the SDC's, the aerodynamic, flight mechanic and weight and balance data of the aircraft. The loads module ensures that load cases selected for design are analyzed for an overall balanced aircraft (mass, inertia and aerodynamic forces) for all maneuvers and the loads analysis is performed in a time history sequence, thus providing load information on structural interfaces for every time step of the chosen maneuver.

The *loads* module which is incorporated in the Global Finite Element Analysis considers external load distribution for any component (such as bending, torque and shear force distribution along fuselage stations for all load cases) or a distribution of loads on the Finite Element grid nodal points.

Thus, starting with the SDC the load loop ends with the preparation of external loads for stress analysis of components. Usually an improved or changed data basis results in an update of the LM and consequently in more accurate and more detailed design load cases. Typical improvements are a better aerodynamic data basis (i.e. via extensive wind tunnel testing) or a refined FE-model because of an advanced design status. Modifications in the mass and balance status, control laws, etc., may also result

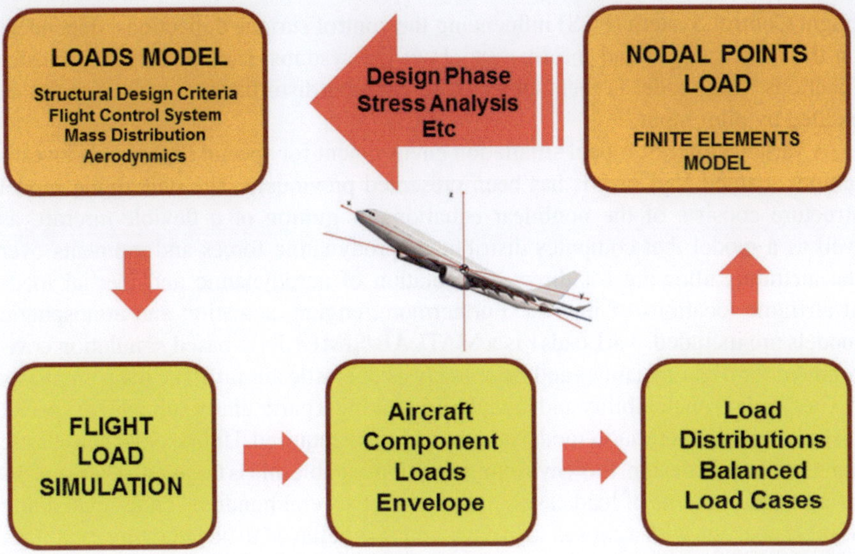

Fig. 13.4 Loads loop [12]

in substantial changes of the loads model, especially in advanced computer-controlled flight vehicles.

The importance of the link between knowledge of external loads and structural stress distributions for the assessment of fatigue life cannot be underestimated. With the progress of computational techniques and computer technology, today there are many commercial software that can be utilized for that purpose. Computers power plays an important role in obtaining better and refined results in the assessment of loads. Nevertheless, the correct selection of the critical maneuvers for the fatigue spectrum and their loads analysis still influences the fatigue performance of a structure during the design phase.

13.5 Load and Certification of Aircraft

For certification of an aircraft, it has to be demonstrated that its structure can withstand the loads acting on it without damage. In order to design the structure accordingly, a so-called loads envelope has to be computed [13]. This loads envelope is comprised critical combinations of flight conditions, (e.g. altitude, Mach number), mass configurations and excitations (e.g. maneuvers, gusts, dynamic landing). These so-called load cases are specified for commercial aircraft in the FAR/JAR Part 25 Regulations.

The inputs for loads analysis models are generally a Finite Element Model (FEM) containing the stiffness and mass data for each mass case, the distributed steady and/or unsteady aerodynamic loading at the specified Mach number, control laws of the

Flight Control System (FCS) influencing the control surface deflections, dependent on the current flight and information about the systems, such as actuator transfer functions. The model is then subjected to external disturbances, such as gusts, or excited by pilot input.

A variable aircraft model simulation environment for special flight loads investigations, named VarLoads1, has been presented previously. The underlying model structure consists of the nonlinear equations of motion of a flexible aircraft, as well as a model that computes distributed aerodynamic forces and moments over the airframe, allowing for direct computation of aerodynamic and inertial loads at airframe locations of interest. Furthermore, engine, actuation and atmospheric models are included. VarLoads1 is a MATLAB/SIMULINK-based simulation environment for flight dynamics and loads analysis of elastic aircraft. The main emphasis of CarLoads is on flexibility and modularity, making it particularly suitable for special investigations, where short model update cycles are required. Unlike other aeroelastic simulations in redesign, this environment is loop capable and is targeted toward calculating large amounts of load cases in the order of several hundreds rather than doing a single analysis with a chosen load case, which is believed to be the sizing case. This approach is closer to the actual design process of commercial aircraft and ensures a smooth transition from the predesign stage to the actual design load loop calculation, preventing unpleasant surprises (e.g. the assumed sizing load case turns out not to be the actual sizing load case).

The presented loads analysis is one part of a predesign process, illustrated in Fig. 13.5. Various version may be offered in the literature, such as by Kier et al. [13].

The geometry and a subsequently derived finite element model (FEM) are provided by a parameterized Multi Model Generator (MMG). A FEM containing all structural information is usually very large and is therefore statically condensed with the Guyan3 reduction.

The parametric MMG also provides the geometry for aerodynamic calculations. The aerodynamics for the loads analysis are generated by projection of the lifting surfaces. With the given input data, a dynamic aircraft model is constructed and used in a loads loop calculation. The resulting loads are used for the sizing of the structure. This process is repeated until a converged solution is reached. The sized design is then assessed and subsequently optimized to meet the demanded requirements.

The scope of this work is the development of dynamic aircraft models with VarLoads and the associated aerodynamic models for simulation. The versatility of VarLoads is demonstrated by merging existing with newly calculated aerodynamic models and the application to not only loads analysis, but also stability and control related problems.

It is quite common that, for an existing aircraft, a single component has to be optimized rather than the whole aircraft design. The present work describes an example, where the geometry of the vertical tail-plane (VTP) is altered, while the rest of the aircraft remains unchanged. A model of a transport aircraft based on an aerodynamic database is available from previous studies. Therefore, a substantial part of already available data can be used, which is supplemented by calculations with simple theories. First, the geometry of the baseline configuration is computed with

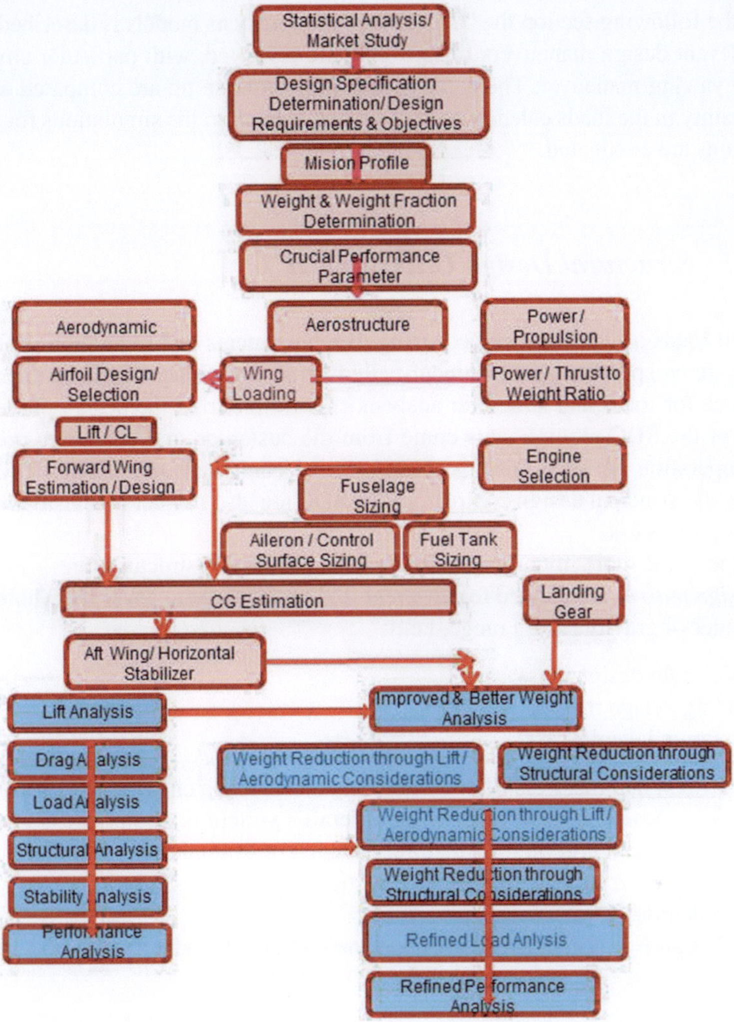

Fig. 13.5 Design process [14]

these simple theories and compared to the available database values. The differences in the resulting loads serve as uncertainty estimations for the changed geometries. As a typical critical load case for the VTP, a yawing maneuver according to JAR/FAR 25.351 is chosen.

The investigated vertical tail planes vary in their span from -10% to $+10\%$, leaving their sweep angle unchanged. The present work concentrates on the calculation of loads. Obviously, the loads are not the only driver for the geometry of the fin. Therefore, as an example, a stability and control study of the changed geometry on the minimum control speed *VMC* is carried out.

In the following section the basis of the loads analysis models is described. Next the relevant design maneuvers for the VTP are reviewed, with particular emphasis on the yawing maneuver. The aerodynamics for baseline fin are compared and the uncertainty in the loads calculation is assessed. Hereafter, the simulations for the fin variations are conducted.

13.5.1 Structural Design Criteria (SDC)

Aircraft loads are determined according to requirements and regulations collected in a systems specification document called Structural Design Criteria, the major reference for loads and structural analysis engineers during the design phase [12]. Many of the SDC requirements come from the customer, and others are prepared in co-operation between customer and original equipment manufacturer (OEM), usually the principal design contractor. The SDC are also subject to revisions during the design process.

Some of the more important items regarding loads and structures are:

Design masses are defined for different flight conditions to cover the whole mass and center of gravity (C.G.) range, i.e.:

- Basic flight design mass
- Landing design mass
- Maximum takeoff mass.

Total mass and mass distribution not only affect loads on wing as is sometimes believed but loads on most parts of the aircraft's structure. Design mass is one of the most important criteria for structural design. For example the basic flight design mass is coupled to the max/min allowed vertical load factor $\underline{Nz}$, for increased masses through the rule:

Nz·Weight = const. to avoid overloads or assessing the effects of over-g's.

13.6 Free-Body Diagrams for Loads Analysis

A Free-Body diagram (FBD) is a diagrammatic representation of the forces acting on a whole body, or part of it. The body is isolated from its supports, and all the applied and reactive forces are drawn on a diagram of the body, positioned at their effective points of action (e.g. center of mass for inertia forces, center of pressure for aerodynamic forces). This is because only their total effect is required for *overall* considerations; however, when the internal loads are required (see later), then the *distributed* nature of the loads will need to be retained.

Once an FBD is available, then generalized Newton's laws of motion may be applied to yield relationships between the forces and any resulting accelerations; alternatively, D'Alembert's principle may be applied. As an example of an FBD, consider an aircraft supported against vertical motion at the nose and main landing gear **positions** but free to roll horizontally, as shown in Fig. 13.5; it is accelerating forward under thrust loading with friction effects ignored.

The mass is m, weight $W = $ mg, thrust T and total support reactions R_N and R_M. The acceleration at the instant of interest is $a = T/m$. The dimensions are shown in the figure with the center of mass being a distance d above the thrust line and h above the ground. The forces acting are shown on an FBD in the lower part of the figure. The arrows are the forces acting upon the aircraft if the generalized Newton's law were to be employed. On the other hand, where D'Alembert's principle to be applied, an additional horizontal inertia force ma would need to be present (not shown here).

Figure 13.6a exhibits the Free-Body Diagram of an aircraft in flight, where Newton's Second Law can be applied. Figure 13.6b exhibits the Free-Body diagram of an aircraft on ground, when the Equilibrium of Forces applies.

Figure 13.6c shows the body diagram of an aircraft on Ground Run when Lift not yet generated (can be ignored); there is no pitch acceleration or inertia moment because both landing gears are assumed to remain in contact with the ground. By referring to this Figure, consider determining the nose gear reaction using Newton's law; the equations for translation and rotation about the center of mass are

$$\mathbf{T} = m\mathbf{a} \tag{13.1}$$

$$\mathbf{R_N} + \mathbf{R_M} - \mathbf{W} = 0 \tag{13.2}$$

$$\mathbf{R_N}b - \mathbf{R_M}c + \mathbf{T}d = \mathbf{I}_c\,\ddot{\alpha} = \mathbf{0} \tag{13.3}$$

Solving for the nose gear reaction yields

$$\mathbf{R_N} = \frac{\mathbf{W}c - \mathbf{T}d}{b + c} \tag{13.4}$$

However, if moments are taken about the point of contact of the main landing gear with the ground instead of about the center of mass, then the resulting reaction would be incorrect because the basis of the generalized Newton's law is that moments must be taken about the center of mass for a dynamic problem. If D'Alembert's principle were to be employed, the inertia force $m\,\boldsymbol{a}$ is included in the FBD and the problem is a static one; then where moments are taken is immaterial.

13.6.1 V–n Diagram

V–n Diagrams define the regime of speeds in combination with max/min allowable load factor Nz including gust conditions, see Fig. 13.7. For low speed regimes the attainable limit N_z depends on the maximum lift and dynamic pressure for the wing

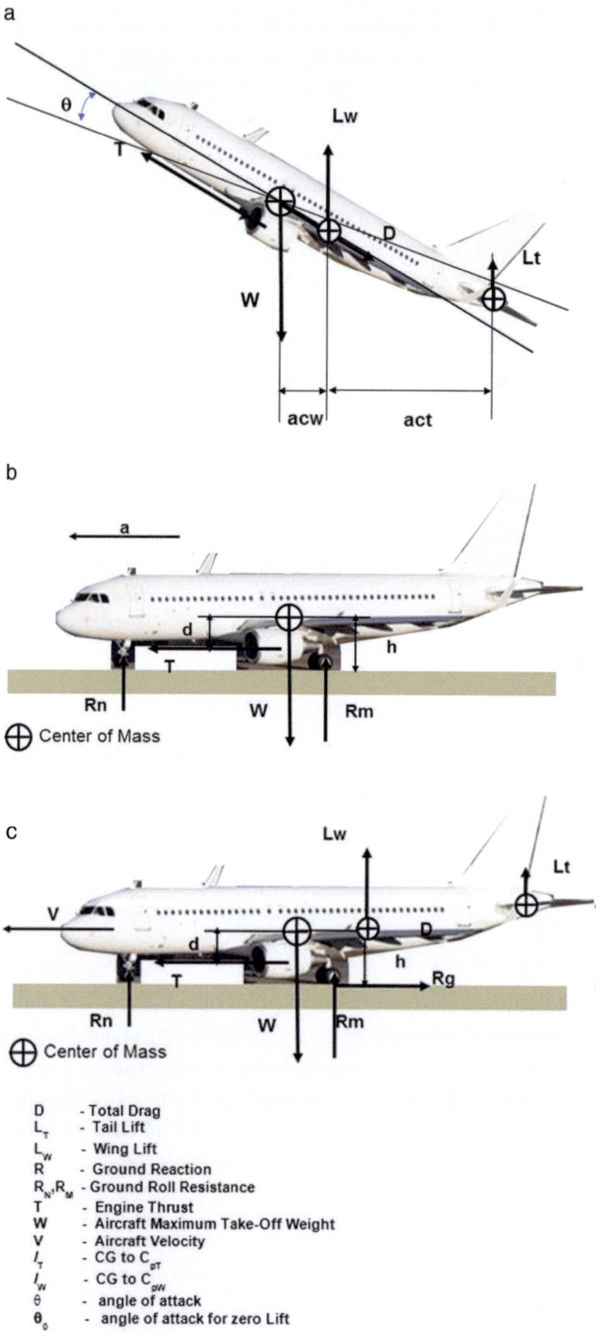

Fig. 13.6 **a** The free-body diagram of an aircraft in flight[3] **b** free-body diagram of an aircraft on ground run when lift not yet generated (can be ignored) **c** free-body diagram of an aircraft on ground run

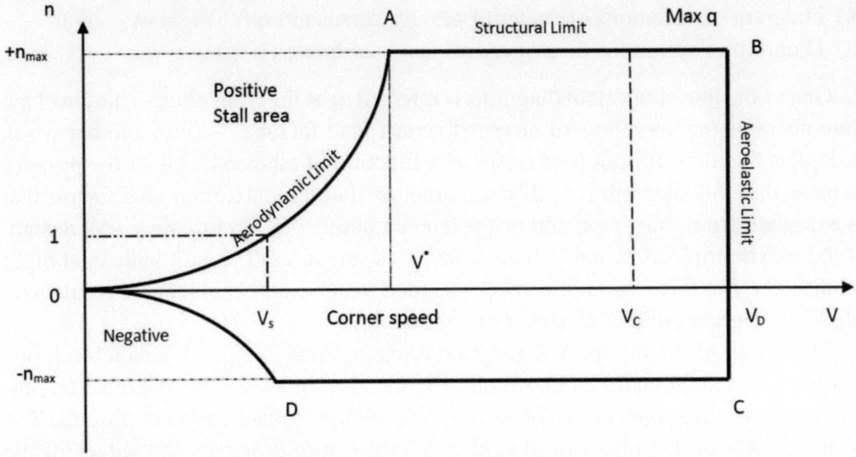

Fig. 13.7 A typical V–n diagram for a General Aviation aircraft

whereas for higher speed N_z is limited by the structural strength of the aircraft. The V–n diagram is referenced to a specific mass and store configuration, i.e. clean wing and design mass.

Flight regime of any aircraft includes all permissible combinations of speeds, altitudes, weights, centers of gravity and configurations. This regime is shaped by aerodynamics, propulsion, structure and dynamics of aircraft [10]. The borders of this flight regime are called flight envelope or maneuvering envelope.

The safety of human onboard is guaranteed by aircraft designer and manufacturer. Pilots are always trained and warned through flight instruction manual not to fly out of flight envelope, since the aircraft is not stable, or not controllable or not structurally strong enough outside the boundaries of flight envelope. A mishap or crash is expected, if an aircraft is flown outside flight envelope.

The flight envelope has various types; each of which is usually the allowable variations of one flight parameter versus another parameter. These envelopes are calculated and plotted by flight mechanics engineers and employed by pilots and flight crews. For instance, the load masters of a cargo aircraft must pay extra caution to the center of gravity location whenever they distribute various loads on the aircraft. There are several crashes and mishaps that safety board's report indicated that load master are responsible, since they deployed more loads than allowed, or misplaced the load before takeoff. Nose heavy and tail heavy are two flight concepts that pilots are familiar and experienced with, and are trained to deal with them safely.

Pilots are using several graphs and charts in their flight operations. Four important envelopes are as follows [11, 15]:

1. Diagram of variations of aircraft lift coefficient versus Mach number (C_L–M)
2. Diagram of variations of airspeed versus altitude (V–h)

---

[3] Aircraft model generated from copyright free site, https://stocksnap.io/search/

3. Diagram of variations of center of gravity versus aircraft weight (X_{cg}–W)
4. Diagram of variations of airspeed versus load factor (V–n).

One of the most important diagrams is referred to as *flight envelope*. This envelope demonstrates the variations of airspeed versus load factor (V–n). In another word, it depicts the aircraft limit load factor as a function of airspeed. One of the primary reasons that this diagram is highly important is that the maximum load factor; that is extracted from this graph and is a reference number in aircraft structural design. If the maximum load factor is under-calculated, the aircraft cannot withstand flight load safely. For this reason, it is recommended to structural engineers to recalculate the V–n diagram on their own as a safety factor.

The technique to plot the V–n diagram is exemplified in Fig. 13.7, which exhibits a typical V–n diagram for a GA aircraft. Figure 13.8 shows similar, but it is a combination of two diagrams: (1) The V–n diagram without consideration of gust, (2) The V–n diagram on the effect of gust. The V–n diagram describes the values of the load factor as experienced by the aircraft during its mission, which defined the flight envelope. The load factor is defined as the ratio between lift and weight

$$n = \frac{\mathbf{L}}{\mathbf{W}} \tag{13.5}$$

The load factor n associated with the phenomena of gust and gust load will then be elaborated and identified in the flight envelope or the V–n diagram.

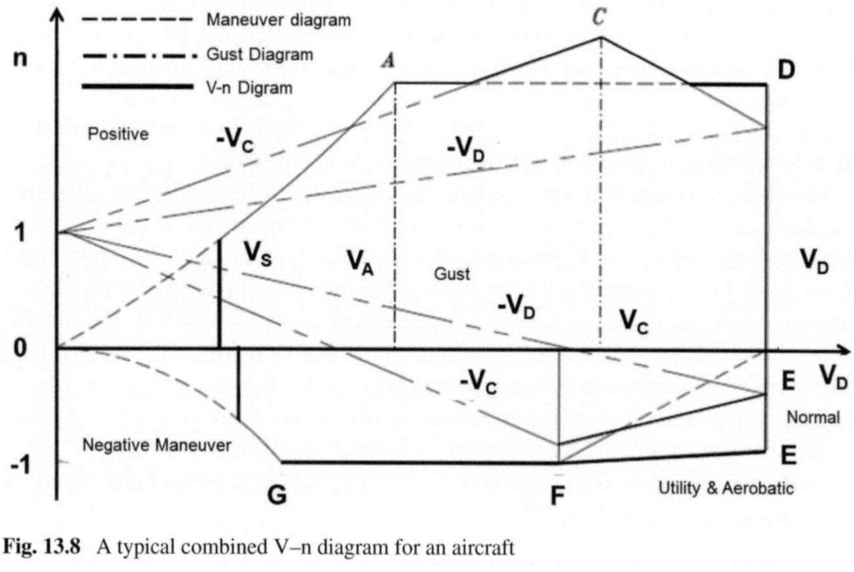

Fig. 13.8 A typical combined V–n diagram for an aircraft

Load Factor

The load to the aircraft on the ground is naturally produced by the gravity (i.e. 1 times g). But there are other sources of load to the aircraft during flight; one of which is the acceleration load. This load is usually normalized through load factor (i.e. "*n*" times g). In another word, aircraft load is expressed as a multiple of the standard acceleration due to gravity (g = 9.81 m/sec^2 = 32.17 ft/sec^2).

In some flight instances of such as turn and pull-up, the aircraft must generate a lift force such that it is larger than the weight. For instance, load factor in a pull-up from Eq. (13.5) can be re-written as:

$$n = \frac{\mathbf{a}}{\mathbf{g}} + 1 \tag{13.6}$$

where "***a***" is the centrifugal acceleration (V^2/R). This acceleration increases if, for instance, the airspeed or velocity increases or the turning turn decreases.

Accordingly, as implied by Eq. (13.6), the load factor will increase also. For other flight operations, similar expressions can be drawn. In some instances; especially for missiles; this load factor may reach as high as 30. Therefore, the structure is required to be capable of carrying such load with high factor of safety safely. Accordingly, the aircraft structure must be strong enough to carry other loads including acceleration load such that aircraft is able to perform its mission safely. Figure 13.8 Exhibits a typical combined V–n diagram for an aircraft, while Table 13.1 demonstrates real values of load factor for several aircrafts.

If the load is more than allowable design value, the structure will lose its integrity and may disintegrate during flight. Load factor is usually positive, but in some instances including pull-down, or when encountering a gust; it may become negative. In general, the absolute value of maximum negative load factor must not exceed 0.4 times maximum positive load factor. Past Experiences forced Federal Aviation Administration to regulate load factor on aircraft.

Table 13.1 Load factor for various types of aircraft

No.	Aircraft type	Maximum positive load factor	Maximum negative load factor
1	Normal (non-aerobatic)	2.5–3.8	−1 to −1.5
2	Utility (semi-aerobatic)	4.4	−1.8
3	Aerobatic	6	−3
4	Homebuilt	5	−2
5	Transport	3–4	−1 to −2
6	Highly maneuverable	6.5–12	−3 to −6
7	Bomber	2–4	−1 to −2

13.6.2 V–n Diagram Without Gust Effect

As Fig. 13.7 shows, V–n diagram is an envelope that indicates the limits of load factor and speed for a safe flight. It is usually composed of two curves plus few lines. The two curves on the left-hand side represent the aerodynamic limit on load factor imposed by stall ($C_{L_{max}}$). The expression for the top curve is extracted from stall equation, which can readily be derived as:

$$\mathbf{V}_{s_t} = \sqrt{\frac{2nm\mathbf{g}}{\rho SC_L}} \tag{13.7}$$

Hence,

$$n_{\max} = \frac{|\mathbf{V}|^2 \rho SC_{L_{\max}}}{2m\mathbf{g}} = \frac{\mathbf{V}^2 \rho SC_{L_{\max}}}{2m\mathbf{g}} \tag{13.8}$$

The top curve is literally a plot of Eq. (13.8). The region above this curve in the V–n diagram is the stall area. Since no aircraft can fly continuously at a flight condition above this curve, then this curve defines one of the limits on the aircraft maneuverability.

This is due to the fact that at this condition, the aircraft angle of attack will be above stalled angle. Based on the Eq. (13.8), as the airspeed increases, the maximum load factor will increase proportionally to V^2. However, n_{max} cannot be allowed to increase indefinitely. It is constrained by the structural strength (structural limit load factor). The top horizontal line denotes the positive limit load factor in the V–n diagram.

The flight velocity corresponding to the intersection between the left curve and top horizontal line (Point A) is referred to as *corner velocity* and designated as **V*** (**V** star). The corner velocity can be obtained by solving Eq. (13.8) for velocity, yielding:

$$\mathbf{V}' = \sqrt{\frac{2n_{\max}m\mathbf{g}}{\rho SC_{L_{\max}}}} \tag{13.9}$$

where the value of n_{max} corresponds to that at point A in Fig. 13.8.

This speed is sometimes referred to as *maneuvering speed* ($\mathbf{V}_A$) and is summarized as

$$\mathbf{V}_A = \sqrt{n_{\max}}\mathbf{V}_s \tag{13.10}$$

The point A is then called the maneuver point. At this point, both lift coefficient and load factor are simultaneously at their highest possible values. The corner velocity is an interesting velocity for fighter pilots. At speeds less than **V***, it is not possible to structurally damage the aircraft due to generation of load factor less than n_{max}.

However, the bank angle is not high enough for a tight turn. In contrast, at speeds greater than **V***, maneuverability decreases, since the speed is too high. Thus, fighter pilots are recommended to select this speed for much of their maneuvering missions. The right-hand side of the V–n diagram, vertical line BC, is a high speed limit. This speed is usually selected to be the dive speed. At flight speeds higher than this limit, the dynamic pressure (q) is higher than the design value for the aircraft. At the speed above dive speed, destructive phenomena such as flutter, aileron reversal and wing divergence may happen that leads to structural damage, or failure, or disintegration. This speed limit (dive speed) is a red-line speed for the aircraft; it should never be exceeded. The dive speed (VD) is usually higher than aircraft maximum speed (V_{max}), and the aircraft maximum speed (V_{max}) is often higher than aircraft cruising speed (VC). From FAR Part 23, the following regulations have been directly copied:

$$\mathbf{V}_D \geq 1.4\ \mathbf{V}_c \quad \text{(Normal aircraft)} \tag{13.11}$$

$$\mathbf{V}_D \geq 1.5\ \mathbf{V}_c \quad \text{(Utility aircraft)} \tag{13.12}$$

$$\mathbf{V}_D \geq 1.55\ \mathbf{V}_c \quad \text{(Acrobatic aircraft)} \tag{13.13}$$

Further discussions on the techniques to calculate maximum and cruising speeds can be followed in Bisplinghoff, Ashley and Halfman [7], Dowell [10] and Wright and Cooper [1] and Sadraey [11, 15]. The bottom line of the V–n diagram, given by horizontal line CD, corresponds with maximum negative limit load factor that is a structural limit when the aircraft is in a situation such as inverted flight. The bottom left curve corresponds to negative stall angle of attack. Since most wing airfoils have positive camber, their positive stall angles are often much higher than the absolute values of their negative stall angles. This curve defines the negative stall area.

13.6.3 Gust V–n Diagram

The atmosphere is a dynamic system that encompasses variety of phenomena. Some of these phenomena include turbulence, gust, wind shear, jet stream, mountain wave and thermal flow. In this section, we concentrate on only gust, since it is not predictable, but is happening during most high altitude flights. When an aircraft experiences a gust, the immediate effect is an increase or decrease in the angle of attack.

Figure 13.9 shows the geometry of an upward gust. When an upward gust with a velocity of Vg hits under the nose of an aircraft with the velocity of V, the instantaneous change (increase) in the angle of attack (Δα) is determined through:

$$\Delta\alpha = \tan^{-1}\left(\frac{\mathbf{V}_g}{\mathbf{V}}\right) \approx \left(\frac{\mathbf{V}_g}{\mathbf{V}}\right) \tag{13.14}$$

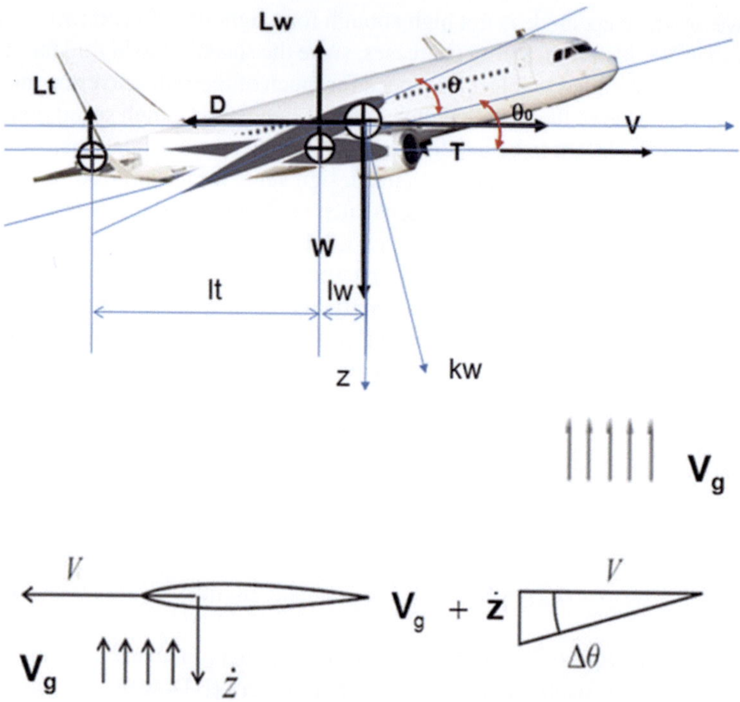

Fig. 13.9 The geometry of an upward gust in airplane gust encounter

Any sudden change (increase) in the angle of attack will produce a sudden change (increase) in the aircraft lift coefficient (ΔC_L):

$$\Delta C_L = C_{L_\alpha} \alpha \tag{13.15}$$

This in turn will generate a sudden change (increase) in lift (ΔL) as:

$$\Delta \mathbf{L} = q S \Delta C_L \tag{13.16}$$

Recall the definition of load factor. This change in lift will create a change in load factor:

$$\Delta n = \frac{\Delta \mathbf{L}}{\mathbf{W}} \tag{13.17}$$

This indicates that gust will change load factor and will generate a gust load. The loads experienced when an aircraft encounters a strong gust may sometimes exceed the maneuver load. Therefore when plotting V–n diagram the gust load should be given consideration. The gust load can be determined when the gust velocity is known. The gust velocity is not easy to measure, since it happens suddenly. The design requirements for gust velocities are extracted from flight test data.

There are various models for gust prepared by various researchers. Here, we refer to FAR for the gust modeling. According to FAR 232, a General Aviation aircraft must be able to withstand gust with a velocity of 50 ft/sec from sea level up to 20,000 ft. From 20,000 ft to 50,000 ft the gust velocity decreases linearly to 25 ft/sec. An aircraft must be able to fly safely at maneuver speed when encountering a gust with the velocity of 66 ft/sec. The aircraft must carry gust load during dive speed, if the gust speed is 25 ft/sec. These data are employed to plot the gust V–n diagram, as exhibited in Figs. 13.7 and 13.8. The following equation for modeling the "gust-induced load factor" as a function of gust speed is recommended by FAR:

$$n = 1 + \frac{k_g\, V_{gE}\, V_E\, a\, \rho\, S}{2\, W} \tag{13.18}$$

where k_g is a coefficient determined by:

$$k_g = \frac{0.88\, \mu_g}{5.3 + \mu_g} \tag{13.19}$$

Here μ_g is the aircraft mass ratio, which is defined as:

$$\mu_g = \frac{2\, m}{\rho\, \overline{C}\, a\, S} \tag{13.20}$$

In the above equations, m is aircraft mass, ρ is air density, $\overline{C}$ is wing mean aerodynamic chord, S is wing area, V_E is aircraft equivalent speed, V_{gE} is gust equivalent speed, and a is wing lift curve slope during gust encounter. Please note that the unit system in these equations is metric (or SI system). The gust V–n diagram is plotted using lines based on the Eq. (13.18) for various speeds (i.e. 7.62, 15.24 and 20.12 m/sec). Then the intersections between these three lines, respectively, with maneuver speed (V_A), cruising speed (V_C) and dive speed (V_D) must be marked. The gust V–n diagram is plotted for several altitudes to determine the highest load factor. Figures 13.7 and 13.8 show typical gust V–n diagrams. This diagram is finally combined; in a special technique; with the basic V–n diagram, to obtain the final applicable V–n diagram. Note that the limiting curves in the V–n diagram or Flight Envelope are determined by the aerodynamic, structural and aeroelastic limits, as indicated in Fig. 13.7.

13.7 Loads Analysis Models

13.7.1 Equations of Motion

The following results follow closely and are based on Kier et al. [13]. One of the main motivations for the development of the simulation environment Scheme[4] is to have one common model that can treat the highly nonlinear motions of a maneuver, as well as the small perturbations that account for the flexibility of the structure. Therefore, the equations of motion have been derived using "mean axes" as body reference systems, which is then expressed as

$$\begin{bmatrix} M_b\left(\dot{V}_b + \Omega_b \times V_b\right) \\ I_b\Omega_b + \Omega_b \times (I_b\Omega_b) \end{bmatrix} = \Phi_{ba}^T P_a^{\text{ext}}$$
$$M_{ff}\,\underset{f}{\ddot{u}} + D_{ff}\,\underset{f}{\dot{u}}\,K_{ff}\,\underset{f}{\ddot{u}} = \Phi_{fa}^T P_a^{\text{ext}} \tag{13.21a}$$

or alternatively

$$\begin{bmatrix} M_b\left(\dot{V}_b + \Omega_b \times V_b + T_{\text{bE}}\{0, 0, g\}\right) \\ I_b\Omega_b + \Omega_b \times (I_b\Omega_b) \end{bmatrix} = \Phi_{ba}^T P_a^{\text{ext}}$$
$$M_{ff}\,\underset{f}{\ddot{u}} + D_{ff}\,\underset{f}{\dot{u}}\,K_{ff}\,\underset{f}{\ddot{u}} = \Phi_{fa}^T P_a^{\text{ext}} \tag{13.21b}$$

where

$\mathbf{M}_b$—matrix with aircraft mass on main diagonal
$\mathbf{I}_b$—inertia tensor
$\mathbf{V}_b$—velocity vector
$\boldsymbol{\Omega}_b$—rotation vector
$\mathbf{M}_{\text{ff}}$—generalized mass matrix
$\mathbf{D}_{\text{ff}}$—generalized damping matrix
$\mathbf{K}_{\text{ff}}$—generalized stiffness matrix
$\mathbf{U}_{\text{ff}}$—generalized coordinates
$\boldsymbol{\Phi}_{\text{fa}}$—modal matrix for flexible modes
$\boldsymbol{\Phi}_{\text{ba}}$—modal matrix for rigid-body modes
P_g^{ext}—external force.
$\mathbf{T}_{bE}$—transformation from inertial to body fixed reference coordinate system, if applied.

These equations consist of the nonlinear rigid-body equations of motion, as used by the °flight mechanics community, and the linear elastic equations of motion used in structural dynamics. The derivation is based on the Lagrange equations. These equations consist of the first of Eq. (13.20), the nonlinear rigid-body equations of

[4] Referred to as VarLoads.

motion commonly used in flight mechanics, and the second, the linear elastic equations of motion in structural dynamics. The coupling of the rigid and flexible parts is driven solely by the external forces and moments

$$P_g^{\text{ext}} = P_g^{\text{aero}} + P_a^{\text{prop}} \tag{13.22}$$

which originated from the aerodynamics and the propulsion.

1. The structure is composed of concentrated masses.
2. Hooke's law is valid, i.e. deformations are small.
3. Eigenvectors are available from a modal analysis and are orthogonal w.r.t. the mass matrix deformations and rate of deformations are small and co-linear, i.e. their cross product can be neglected.
4. The tensor of inertia is assumed constant.
5. An earth fixed inertial reference system is assumed.
6. Gravity is uniform within this reference system.

The detailed derivation of the equations of motion and the underlying assumptions can be followed in the paper by Wesak and Schmidt [17, 18].

13.7.2 Aerodynamics

The calculation of aerodynamic forces acting on the structure is one of the key aspects of the loads analysis. For this purpose, resort can be made to a host of available theories, ranging from simple lifting line to more sophisticated and high fidelity Navier–Stokes Computational Fluid Dynamic Methods. However, other considerations are also given to composite effectiveness of computational time, cost and accuracy. Therefore, it is customary to utilize the fundamental classical methods from potential theory, which should and can be easily validated using first principles.

The results of these linear methods can be refined with small number of CFD calculations at points in the flight envelope where nonlinearities are expected. These aerodynamic nonlinearities are usually found in the high Mach number region.

When gust loads are to be calculated, unsteady aerodynamics have to be considered. The standard method is the Doublet Lattice Method (DLM) [19], which solves the acceleration potential equations in the frequency domain. In order to use the results in the time domain, a rational function approximation (RFA) has to be carried out, as described by Buttrill [18], Roger [20] or Carpel [21].

For predesign applications, the unsteadiness of the flow field can also be modeled with transfer functions. The use of Wagner function for sudden changes in angle of attack and the Küssner function for intrusion in sharp edged gusts is well known [7, 10]. The use of these functions will considerably reduce the modeling effort. Wagner $\Phi(\tau)$ and Küssner $\psi(\tau)$ functions, as exhibited in Fig. 13.10, can be calculated using Computational Fluid Dynamics Method, as exemplified by one such early attempts

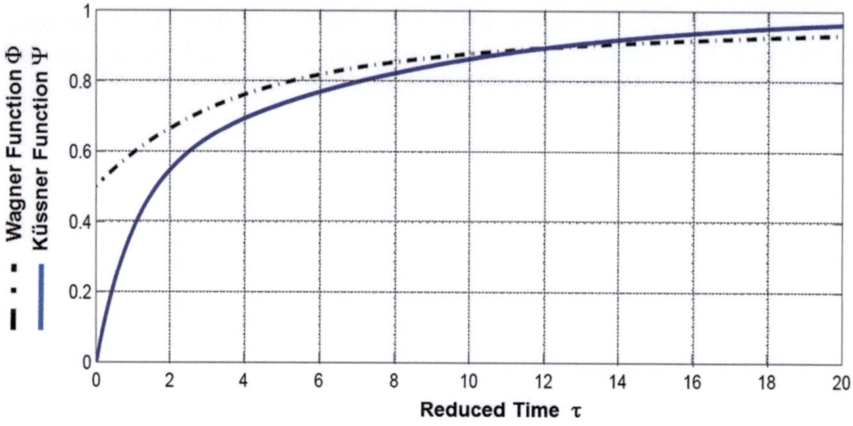

Fig. 13.10 Wagner $\Phi(\tau)$ and Küssner $\psi(\tau)$ function vs. reduced time τ[5]

by Djojodihardjo and Widnall [21] for a gust onset velocity equal to 0.1 times the free-stream velocity U_0 striking a two-dimensional flat plate.

The following papers (Kier [22]; Kier, Looye and Hoftee [13]; and Kier [23]) are refer erred here to elaborate instructive examples of comprehensive maneuver loads analysis. For this load analysis it is considered instructive and sufficient to utilize quasi-steady aerodynamics. A well-established method, which is available as a preprocessing module, is the so-called Vortex Lattice Method (VLM).

13.7.3 Vortex Lattice Method

The Vortex Lattice Method (VLM)[6] is a method solving the velocity potential equations, resulting in the *aerodynamic influence coefficient* (AIC) matrix [7]. The Aerodynamic Influence Coefficient (AIC) matrix relates the velocity at a control point to forces at an acting point (also known as source point).

This (quasi-)steady AIC matrix is particularly suitable for time domain simulations, since it has to be computed only once for each Mach number during the preprocessing. The varying forces during a dynamic simulation can then be calculated by multiplication with the velocity vectors at the control point that can be computed from the rigid body and elastic motions. The computation of the aerodynamic loading reduces to matrix multiplication. The Vortex Lattice Method (VLM)

---

[5] Synthesized from Djojodihardjo and Widnall [14]; further detail of Wagner $\Phi(\tau)$ and Küssner $\psi(\tau)$ functions can be found in R. L. Bisplinghoff, H. Ashley, R. L. Halfman [7].

[6] The Vortex Lattice Method is the extension of Prandtl's *lifting-line theory*, where the wing of an aircraft is modeled as an infinite number of *Horseshoe vortices*. The name was coined by V.M. Falkner in his *Aeronautical Research Council* paper of 1946. The method has since then been developed and refined further by W.P. Jones, H. Schlichting, G.N. Ward and others.

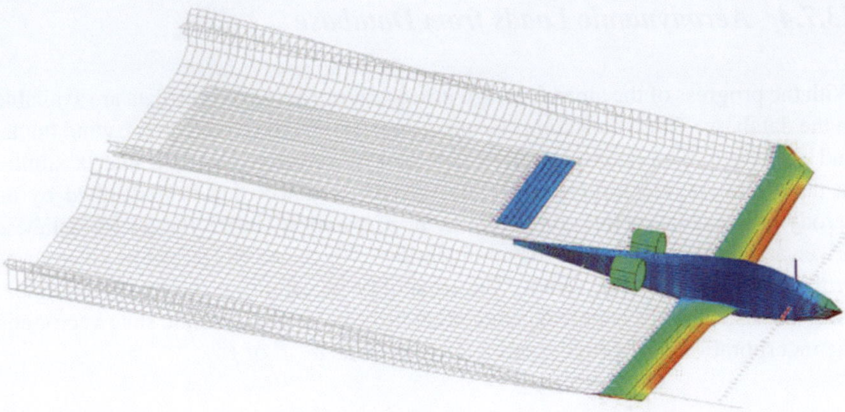

Fig. 13.11 A schematic of an example of the application **a** vortex lattice method simulation of an aircraft model [24]

models lifting surfaces. Therefore, to represent the fuselage and other cylindrical shaped or volumetric bodies, they can be approximated by a cruciform shape. Two examples of a Vortex Lattice grid for predesign studies are shown in Fig. 13.11.

The Vortex Lattice Method (VLM) discretizes lifting surfaces as small elementary wings (panels), as also exemplified by Fig. 11a, which consist of a horseshoe-shaped vortex line at the quarter chord and a control point. At this control point, which is located at the 75% chord, the normal component of the velocity has to be zero, since there is no flow through the solid surface. In order to achieve this, the circulation strengths of the vortex lines has to be set accordingly, so that their induced velocity compensates the free-stream velocity.

When the circulation strengths are known, the resulting aerodynamic forces can be computed with the Kutta-Joukowski theorem:

$$\mathbf{P}_g^{\mathrm{aero}} = \mathbf{n}_k \rho V_{TAS} \Gamma_k b_k \tag{13.23}$$

where

$\mathbf{P}_g^{\mathrm{aero}}$—aerodynamic lift acting on panel
$\mathbf{n}_k$—normal vector of panel
ρ—air density
V_{TAS}—true airspeed
Γ_k—circulation of panel
b_k—width of panel

13.7.4 Aerodynamic Loads from Database

With the progress of the aircraft design process, more aerodynamic data are available in the database. These data comprise the results from CFD calculations, wind tunnel and flight test. The database values can then be reformulated in a matrix similar to the Aerodynamic Influence Coefficient (AIC) matrix that is multiplied by an aerodynamic state vector, containing α, β, p, q and r. Unlike the VLM, database driven aerodynamics can include nonlinearities, which can be applied by cross terms in the aerodynamic state vector. The capability of capturing the highly nonlinear pressure distributions is only dependent on the size of aerodynamic state vector and the incorporation of the higher order terms such as α , β or β^2.

13.7.5 Finite Element Model of the Structural

The structural model for loads analyses is usually a Finite Element Model (FEM) to which the masses (systems, payload, fuel and structural) can be added, which a much simpler task than developing a separate model of the elements with density properties. Such an approach can be utilized to decouple the mass estimation method from the FEM model. The model is then statically condensed using Guyan reduction.

13.7.6 Structure and Aerodynamics Interaction

Since usually the aerodynamic and structural grid do not coincide, a methodology for the transfer of loads and displacements between the grids has to be employed. This is referred to as splining. A method-based beam shape functions, which is capable of handling kinks in the beam axis to accomplish the splining, which has been developed [25], can be applied. Such fluid structure interaction has also been treated for acoustic and structure interaction by Djojodihardjo and Safari [26].

13.8 Load Recovery

To recover the nodal loads acting on the condensed structural grid, Kier [13] employed the mode acceleration method, which is a Force Summation Method. Kier et al. utilized Force Summation Method that requires the external forces to be available on the structural grid; hence the Aerodynamic Influence Matrix (AIC matrix) has to be available in a half generalized form. Alternatively, the Mode Displacement Method can be used to recover the nodal loads, which only requires a fully generalized AIC matrix and a half generalized stiffness matrix.

Since the mode displacement method has a poor convergence behavior [27], the Force Summation Method is preferred. This is defined as

$$\mathbf{P}_a = \mathbf{P}_a^{\text{gust}} = \mathbf{M}_{ah} \begin{bmatrix} \mathbf{T}_{bE}\{0, 0, g\} + \boldsymbol{\Omega} \times \mathbf{V_b} + \mathbf{V_b} \\ \dot{\boldsymbol{\Omega}}_b \\ \ddot{\mathbf{u}} \end{bmatrix} \tag{13.24}$$

$\mathbf{P}_a$—nodal loads
$\mathbf{T}_{bE}$—transformation from inertial to Body fixed
$\mathbf{M}_{ah}$—half generalized mass matrix
g—gravity

The integrated loads are the outboard loads, that are summed up at monitoring stations, located between the structural nodes. The loads envelope is formed by sorting the integrated loads, yielding the sizing load cases for every grid point degree of freedom (DOF), as well as the correlated loads.

Flight Control System

A further very important factor influencing the loads is the presence of a Flight Control System (FCS). The dynamic simulations were calculated with an active yaw damper system, affecting the aircraft response and in turn the magnitude of the occurring loads. Here pilot model initiates the yawing maneuver by applying a rudder input, while simultaneously trying to counteract the roll motion and keeping the altitude constant.

The commanded control surface defections are passed through an actuation system, which is modeled with second-order transfer functions. The results obtained from the Vortex Lattice Method (VLM) aerodynamics of the baseline geometry are compared to the database values, and sources of differences can be identified. Then the yawing maneuver is simulated for both aerodynamic methods. After assessing the accuracy of the VLM method, geometry variations are examined.

The span of the Vertical Tail Plane (VTP) is varied, while keeping the sweep angle unchanged. The resulting maximum loads at the root of the VTP are then compared to the baseline configuration in order to recognize trends. Additionally, the dynamic pressures q_{dyn} at V_{MC} for the variations are compared.

13.9 Concluding Remarks

Discussions elaborated in this chapter have been addressed for the estimation of loads acting on an aircraft structure, which is an indispensable task ranging from conceptual, preliminary and detailed design to loads flight testing and the manufacturing and operation of aircrafts.

The basis of treating aircraft loads problems has been demonstrated to be based on basic and fundamental principles, which are the fundamental laws of mechanics, including structural analysis and strength of materials, thermo-sciences conservation principles which are the foundation of aerodynamics. To a large extent, the considerations, analysis and computational work associated with aircraft loads are addressed for the certification of an aircraft, which should be demonstrated that its structure can withstand the loads acting on it without damage, following official certification standards. The requirements cover loads envelope comprising critical combinations of flight conditions, such as FAR/JAR Part 25 Regulations for commercial aircrafts.

Several research works have been referred in the chapter to exemplify the objectives, scope and extent of the analysis and computational efforts that have been devoted to these cases. Flight regime of any aircraft includes all permissible combinations of speeds, altitudes, weights, centers of gravity and configurations. This regime is shaped by aerodynamics, propulsion, structure and dynamics of aircraft. *V–n Diagrams* define the regime of speeds in combination with max/min allowable load factor Nz including gust conditions. The aircraft loads environment may incorporate a large amount of load cases which is believed to be relevant for sizing. Such series of approaches may be followed in the actual design process of commercial aircraft and ensures a smooth transition from the predesign stage to the actual design load loop calculation, preventing unpleasant situations.

The inputs for loads analysis models in these examples are generally a Finite Element Model (FEM) associated with the stiffness and mass of the aircrafts or parts of them, the steady and/or unsteady aerodynamic loading, control laws of the incorporated Flight Control System (FCS) influencing the deflections of the control surface, and information about the systems. The model generated based on valid assumptions are then being examined for their responses to various loadings and external disturbances, such as gusts, or excited by pilot input, and their structural strength. One significant requirement is the fail-safe conditions that should be met by these aircraft.

References

1. Wright, J.R., and J.E. Cooper. 2015. *Introduction to Aircraft Aeroelasticity and Loads*, 2nd ed. Wiley. https://doi.org/10.1002/9781118700440.
2. Anonymous 2014. US Department of Transportation FAA Advisory Circular on Dynamic Gust Loads AC No: 25.341–1
3. Ciarella, A., C. Tsotkas, M. Hahn, N.P.M. Werter, R. De Breuke, C.S. Beaverstock, M.I. Friswell, Y. Yang, S. Özgen, A. Antoniadis, P. Tsoutsanis, and D. Drikakis. 2015. A multi fidelity, multi disciplinary analysis and optimization framework for the design of morphing UAVWings. In *16th AIAA/ISSMO Multidisciplinary Analysis and Optimization Conference*, Dallas, Texas.
4. Klimmek, T., P. Ohme, P.D. Ciampa, and V. Handojo. *Aircraft loads—An important task from pre-design to loads flight testing*. https://www.dglr.de/publikationen/2017/420223.pdf. Accessed June 6, 2023.
5. Kafkas, A., S. Kilimtzidis, A. Kotzakolios, V. Kostopoulos, and G. Lampeas. 2021. Multi-fidelity optimization of a composite airliner wing subject to structural and aeroelastic constraints. *Aerospace* 8: 398. https://doi.org/10.3390/aerospace8120398.

6. Karpuk, S, Mosca‡, V., Liu, S. and Elham, A. 2022. Development of a Multi-fidelity Design, Analysis, and Optimization Environment for Future Transport Aircraft, *AIAA SciTech Forum*. https://doi.org/10.2514/6.2022-0686.
7. Bisplinghoff, R.L., H. Ashley, and R. Halfman. 1955. *Aeroelasticity*. Dover Publications Inc.
8. Taylor, J. *Manual on Aircraft Loads*. AGARD-AG-83. https://apps.dtic.mil/sti/pdfs/AD0458256.pdf.
9. Lomax, T. 1996. *Structural Loads Analysis for Commercial Transport Aircraft: Theory and Practice*. AIAA Education Series.
10. Dowell, Earl, et al. 2015. *A Modern Course in Aeroelasticity*, 5th ed.
11. Sadraey, Mohammad H. 2017. *Aircraft Performance: An Engineering Approach*. CRC Press. ISBN 9781498776585.
12. Anonymouys 2000. *NATO, Aging Aircraft Fleets-Structural and RTO-EN-015*. https://www.sto.nato.int. Accessed November 2, 2017.
13. Kier, T., G. Looye, and J. Hofstee. 2005. *Development of Aircraft Flight Loads Analysis Models with Uncertainties for Pre-design Studies*, IFASD 2005, Munich, Germany.
14. Djojodihardjo, Harijono and Kim Ern Foong. 2013. Conceptual design and aerodynamic study of joined-wing business jet air. *Journal of Mechanics Engineering and Automation* 3: 263–278.
15. Sadraey, M. (2009). *Aircraft Performance Analysis*. Saarbrücken VDM Verlag.
16. Wesak, M. R., and D.K. Schmidt. 1988. Flight dynamics of aeroelastic vehicles. *Journal of Aircraft, 25* (6): 563–571.
17. Buttrill, C.S., M.R. Waszak, and D.K. Schmidt. 1992. Modelling and model simplification of aeroelastic vehicles: an overview. In *Technical Report NASA TM-107691*, NASA LARC.
18. Albano, E., and W.P. Rodden. 1969. A doublet lattice method for calculating lifting disturbances of oscillating surfaces in subsonic flows. *Journal of Aircraft* 7 (2): 279–285.
19. Roger, K.L. 1977. Airplane math modelling methods for active control design. In *AGARD Structures and Materials Panel*, CP-228, AGARD.
20. Carpel, M., B. Moulin, and P.C. Chen. 2003. Dynamic response of aeroservoelastic systems to gust excitation. In *International Forum on Aeroelasticity and Structural Dynamics*.
21. Djojodihardjo, H., and S.E. Widnall. 1969. Nonlinear, unsteady lifting potential flow problems. *AIAA Journal*.
22. Kier, T. 2005. Comparison of unsteady aerodynamic modelling methodologies with respect to flight loads analysis. In *Proceedings of AIAA Atmospheric Flight Mechanics Conference*.
23. Kier, T. 2017. An integrated model for lateral gust loads analysis and Dutch roll flight dynamics using a 3-D panel method. In *International Forum on Aeroelasticity and Structural Dynamics*, IFASD 25–28 June Como, Italy.
24. Guillermo, H. 2016. Open VOGEL, an open source framework for aerodynamic simulations based in the unsteady vortex lattice method (UVLM).
25. Mauermann, T. 2005. The femspline-method for connecting condensed structural models and aerodynamic models. In *Technical Report EGLG31 RP0500909, Airbus Loads and Aeroelastics*.
26. Djojodihardjo, H., and I. Safari. 2013. BEM-FEM coupling for acoustic effects on aeroelastic stability of structures. *CMES: Computer Modelling in Engineering and Sciences* 91: 205–234.
27. Livne, E., and G. Blando. 2000. Reduced-order design-oriented stress analysis using combined direct and adjoint solutions. *Journal of Aircraft* 38 (5): 898–909.

Chapter 14
Aeroelastic Experiments-Ground Vibration and Flight Flutter Test Case Studies

Abstract Aircraft operations demand that every aircraft should be safe to fly, and to that end it is mandatory that the conception, design, manufacturing and operations for every aircraft follow the fail-safe paradigm. Therefore, in addition that the design of aircraft could guarantee such mandatory requirements, physical experiments along the design, manufacturing and operation of aircraft need to be proved and validated by physical experiments, which is indeed highly challenging. Important aspects of carrying out experiments have many useful purposes, such as to assess the accuracy and validity of theoretical models, to study phenomena beyond the current reach of theory, and in the field of aeroelasticity, to verify the safety and integrity of aeroelastic systems through various test, from ground vibration tests to wind tunnel tests and to flight tests. In the present chapter, some of the prevailing fundamental aspects of experimental aeroelasticity are discussed. To provide hands-on impression, a case study on flight flutter test will be elaborated.

Keywords Aeroelasticity · Flight flutter test · Flutter · Modal analysis · Parameter identification · Vibration testing · Wind tunnel experiments

14.1 Introduction

The clearance process relating to aeroelastic and load issues during the design and certification phase of a new aircraft is a combination of numerical modeling backed up by testing. An outline of the main tests used to validate various elements of the aircraft mathematical model for aeroelastic issues is shown in Fig. 14.1. It is essential that there is an opportunity to update the numerical models of each element based on the test results. A similar series of tests are undertaken to certify aircraft for ground and flight loads. In this chapter, only a very brief outline of the methods will be given.

Figure 14.2 also illustrates the relevance of aeroelastic tests and experiment such as ground vibration test, various model test and accompanying analytical verification and validation to meet aircraft design requirements.

It must be emphasized that no one test is capable of providing the information for full validation of the mathematical models used for certification. That must inevitably

© The Author(s), under exclusive license to Springer Nature Singapore Pte Ltd. 2023
H. Djojodihardjo, *Introduction to Aeroelasticity*,
https://doi.org/10.1007/978-981-16-8078-6_14

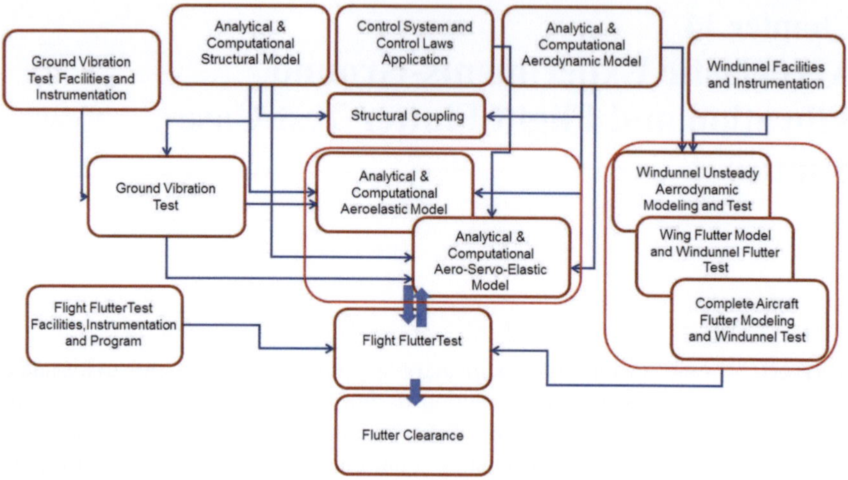

Fig. 14.1 Analysis and test certification process for aeroelasticity

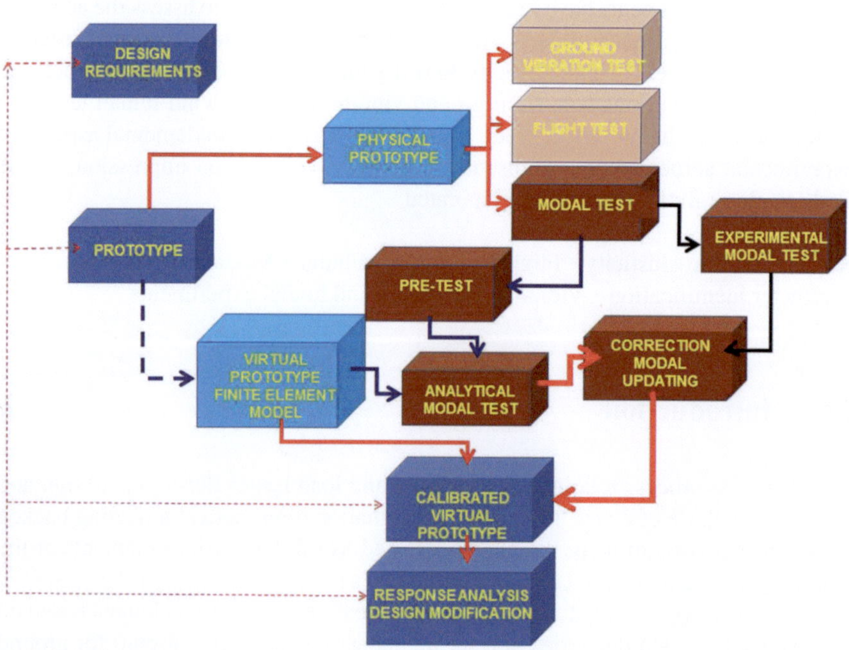

Fig. 14.2 Experimental test and analysis to meet aircraft design requirements

come by building up a range of test results on different aspects (e.g. structural stiffness, mass, mass distribution, center of mass, wind tunnel tests, systems tests, etc.). The ground tests can be quite accurate and are backed up by checks that show, when assembled together, that the structural dynamic and the flight response properties are both reasonable. However, tests performed in-flight to demonstrate aeroelastic stability and validate flight loads are subject to a number of uncertainties. The test set-up for flight flutter testing is much less ideal (e.g. noisy environment, inadequate excitation) than that for ground vibration testing, whereas flight loads have to be determined from strain gauge readings and these calibrations can be problematic.

14.2 Ground Vibration Testing

The ground vibration test (or GVT, sometimes called a modal test (Gasparetto et al [1], Giclais et al [2], Pickrel et al [3] and Salehi and Ziaei-Rad [4])) is performed on the prototype aircraft to obtain estimates of the whole aircraft normal modes (natural frequencies, damping ratios, mode shapes and modal masses). These modal data may then be used to confirm or adjust the calculated normal modes used in the critical flutter calculations, as well as providing substantiated estimates of damping. Different fuel and hydraulics configurations are studied.

An impression of a very sophisticated vibration test facility is exhibited in Fig. 14.3 which shows a spacecraft vibration test at the Naval Research Laboratory.

Fig. 14.3 Spacecraft vibration test at the Naval Research Laboratory [5]

Fig. 14.4 Ground vibration test of N-219 (a 19 Passenger regional aircraft designed and built by PT Dirgantara Indonesia) in 2019 (courtesy of Mr. Andi Eriawan).[1] **a** general set-up. **b** tire decompression as suspension means. **c** one location of the shaker (exciter)

This facility qualifies and acceptance tests spacecraft and spaceflight components by simulating the various vibration loading environments present during flight operations and demonstrating compliance to design specifications. Using the facility's electrodynamic shakers, an assortment of quasi-static, vibratory and shock loads can be generated, and test article characteristics can be quantified.

Figure 14.4 exhibits ground vibration test of N-219 (a 19 Passenger regional aircraft designed and built by PT Dirgantara Indonesia) in 2019. Figure 14.4a exhibits the entire set-up and GVT event, Fig. 14.4b shows one of the locations of the Shakers, and 4c shows the landing gear reduced tire pressure as a method of aircraft suspension during GVT. Figure 14.5 shows one GVT event of Ariane IV sections in 1977.

[1] All figures related to N-219, GVT and Flight Flutter Test facilities in this chapter (and book) are unpublished work (courtesy of PT Dirgantara Indonesia/PT Indonesian Aerospace) obtained through the good offices of/private communication with Mr.Andi Eriawan, a colleague at PT Dirgantara

Fig. 14.5 Ground vibration testing at NASA Armstrong, emphasizing on passive aeroelastic tailored wing in 2019 (courtesy of NASA [6])

These images are exhibited here to provide a comprehensive impression on ground vibration test of various aerospace objects, carried out in aerospace and aircraft industries.

14.3 Modal Analysis

With vibration being an essential element of aeroelasticity, structural vibration testing and analysis, modal analysis is an important tool for understanding the vibration characteristics of aerospace as well as other engineering structures. Modal Analysis is an essential tool to interpret vibration experiments that henceforth can be utilized as an elegant and convenient yet plausible physical and mathematical model for further analysis and synthesis of various engineering objects or problems (Fig. 14.6).

Modal analysis converts the vibration signals of excitation and responses measured on a complex structure that is difficult to perceive, into a set of modal parameters which can be straightforward to foresee. While the mathematics and physics that govern this subject are not simple, the basic concept and application can be easily understood through the examples and concepts elaborated and discussed

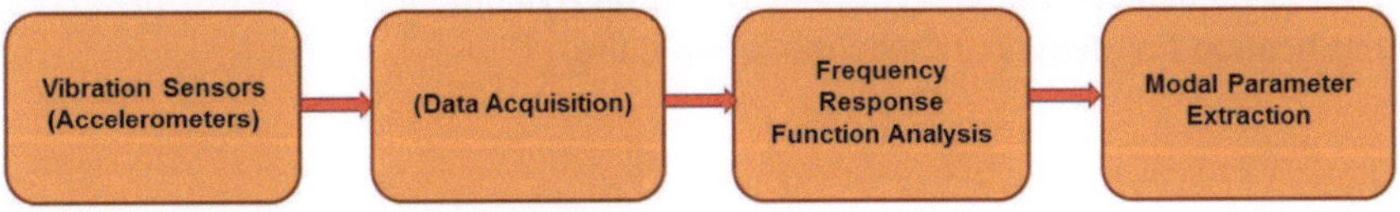

Fig. 14.6 Experimental modal analysis

Indonesia (PT.DI), as part of his work and responsibilities as a specialist and director in charge of Load, Dynamics and Aeroelasticity, including the work of his team.

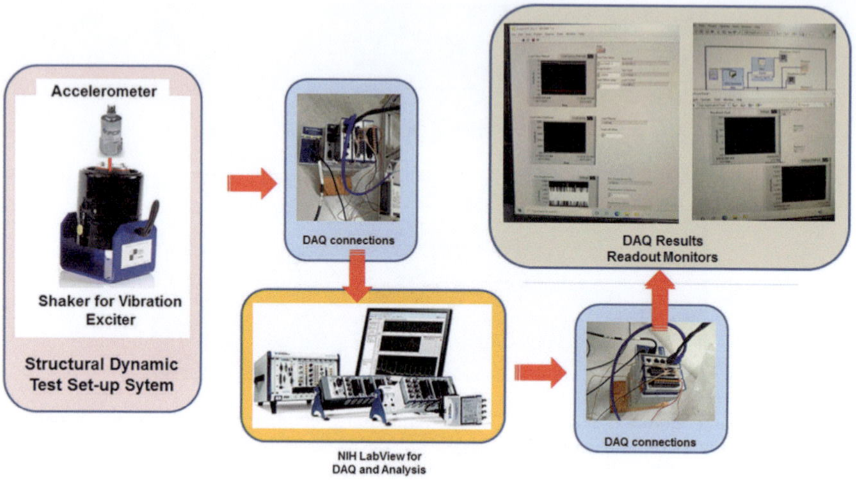

Fig. 14.7 Schematic of structural dynamic test set-up utilizing NIH LabView Modal Analysis Equipments at Universiti Putra Malaysia (courtesy of Dr. Moh Faisal A. Hamid,[2] Aerospace Department, Universiti Putra Malaysia)

subsequently. Assuming that the reader has a basic understanding of vibration terminology and theory, the concept of modal analysis, its applications where modal analysis is useful, data acquisition and visualization techniques will be discussed only in introductory fashion, capitalizing on the vibration principles.

14.4 Structural Dynamic Testing Equipments

14.4.1 Sample Hardware for Structural Dynamic Modal Testing

Figures 14.7 and 14.8 are examples of vibration experiment measurement and analyzer system and of laboratory equipment for university structural dynamic laboratory and some structural dynamics laboratory facility for instructions, respectively, which is fundamental in carrying out structural dynamic test as part of parameter identification for analytical modeling of structure.

[2] Personal Communication.

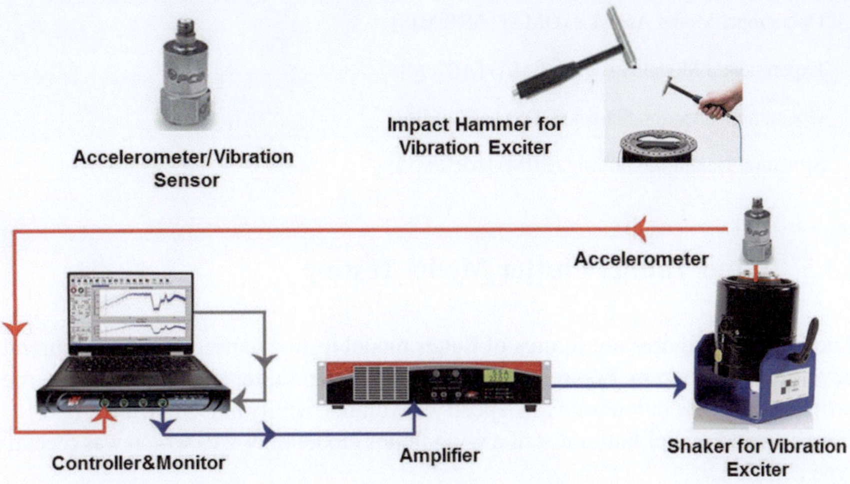

Fig. 14.8 Commercially off-the-shelf structural dynamic laboratory facilities set-up conceived and partially utilized by the author at Universiti Sains Malaysia School of Aerospace Engineering in 2007 for instruction and academic research

14.4.2 Some Softwares That Can Be Utilized for Structural Dynamics Modal Testing Analysis

To have an overall impression on softwares that are available commercially, the following are some commercial ones that can be utilized for modal analysis and other related analysis[3]:

FFT Analysis Software for Pulse Systems [B&K]

Pulse Reflex Advanced Modal Analysis Software [B&K]

MIMO Analysis Software [B&K]

Nastran—ANSYS Interface Software [B&K]

Operational Modal Analysis (OMA) Pro Software [B&K]

Rotating Machinery Analysis Software [B&K]

Modal Updating Software [FEMTools]

ANSYS Finite Element Modeling and Analysis Package Program

(Structural Analysis using ANSYS Workbench)

[3] Not exhaustive. Without any conflict of interest.

Operational Modal Analysis (OMA) [ARTeMIS]

Experimental Modal Analysis (EMA) [ARTeMIS]

Operating Deflection Shapes (ODS) [ARTeMIS]

Structural Health Monitoring (SHM) [ARTeMIS].

14.5 Wind Tunnel Flutter Model Testing

The following figures are figures of flutter model testing carried out during aircraft development program. Figure 14.9 shows wind tunnel flutter test of an N-250 aeroelastic model at the Indonesian low-speed wind tunnel facility in 1996, while Fig. 14.10 shows a wind tunnel flutter test of a wing flutter model of N-219 which was recently type certified.

Fig. 14.9 N-250 wind tunnel flutter model and experimental set-up at the Indonesian low-speed tunnel ILST-LAGG, BPPT, Serpong, Indonesia [7]

Fig. 14.10 A half-wing flutter model being tested at the PT IPTN low-speed tunnel, Bandung, Indonesia[4]

14.6 Aircraft Ground Vibration Test of N-219 Prototype for Certification—A Case Study[5]

The present elaboration is excerpted from and drawn from the technical document of PT Dirgantara Indonesia on the ground vibration test (GVT) vibration evaluation for N-219 aircraft flaps. Detailed explanation of the equipment, test installation and set-up are shown in "Ground Vibration Test Procedure of N-219" document provided by ground test group of PT Dirgantara Indonesia.

14.6.1 Objective

The objective of the GVT is to measure a set of modal parameters: mode shapes, natural frequencies and damping ratios of tested aircraft flap system. The vibration of the flap system will also be evaluated.

[4] Author and colleagues unpublished work at IPTN, 1992–1996, and from Djojodihardjo, N-250 in the framework of Aircraft Science, Technology and Industry Development (in Indonesian: 'N-250 Dalam Bingkai Pengembangan Sains, Teknologi Dan Industri Pesawat Terbang Di Indonesia'), to appear in a forthcoming book by Rahardi Ramelan, 2021.

[5] This work is based on and partial reproduction of the Ground Vibration Test of N-219 Prototype for Certification, as carried out at PT Dirgantara Indonesia and made available for the present book through the good offices of Mr.Andi Eriawan and colleagues of the Aeroelastic and Structures group.

The results will be used for:

1. Verification of the calculated dynamic characteristics (analytical model) and for updating the analytical model. Subsequent aeroelastic calculations are based on the updated analytical model.
2. Determining the vibration course from flaps surface to establish the proper solution for reducing the vibration level.

14.6.2 Scope

The technical document is intended to provide a set of vibration characteristics of the vibration of flap system as required by carrying out GVT on N-219 aircraft flap system. This includes the aircraft configuration of the flap system (such as the description of the test set: locations and direction of accelerometer and exciter required to obtain mode shapes and frequencies) and to evaluate the vibration characteristics of the flaps.

14.6.3 Applicable Requirements

The ground vibration test is performed to comply with the requirements of CASR 23.629, paragraph (a) (2).[6] The vibration evaluation test is conducted to comply with the requirements of CASR 23.251, paragraph (a).[7]

14.6.4 Description of Test Article

14.6.4.1 Test Article

N-219 Aircraft will be tested with flight test mass configurations with interior panel removed and without passengers. No other activity is allowed to be conducted on or in the aircraft during GVT. The three-view drawings of the N-219 aircraft are shown in Fig. 14.11.

[6] "The natural frequencies of main structural components must be determined by vibration tests or other approved methods".

[7] "There must be no vibration or buffeting severe enough to result in structural damage, and each part of the airplane must be free from excessive vibration, under any appropriate speed and power conditions up to VD/MD".

Fig. 14.11 Three view drawing of N 219[8]

14.6.4.2 Flap System

During GVT execution, the main structure of the wing is supported by trestle and all control surfaces in the wing and the horizontal and vertical tail are left in mid-position and locked. Flap system should be unlocked throughout the period of test to test out different flap deflections. The flap is going to be tested in various test conditions: Flap 0° (Up), Flap 10°, Flap 20°, Flap 30° and Flap 40°.

14.6.5 Description of Test Set-Up

14.6.5.1 Test Method

Vibration will be applied to the aircraft using phase separation method with multiple-input-multiple-output (MIMO) during the test. The dynamic response of the aircraft due to the excitation signal is measured using accelerometer transducers mounted on the aircraft. Both excitation signal and responses are fed to the data acquisition system to obtain the frequency response function (FRF). The procedure is repeated for other exciter configurations.

---

[8] All figures related to N-219, GVT and Flight Flutter Test facilities in this chapter (and book) are unpublished work obtained through the good offices of/private communication with Mr.Andi Eriawan, a colleague at PT Dirgantara Indonesia (PT.DI), as part of his work and responsibilities as a specialist and director in charge of Load, Dynamics and Aeroelasticity, including the work of his team.

Table 14.1 List of shaker location and direction

No. Shaker configuration	Shaker location	Direction
1	Inner LH Flap—Trailing Edge	**Z**
2	Inner RH flap—trailing edge	**Z**
3	Mid-LH flap—trailing edge (near pivot B)	**Z**
4	Mid-RH flap—trailing edge (near pivot B)	**Z**
5	Outer LH flap—trailing edge (near pivot C)	**Z**
6	Outer RH flap—trailing edge (near pivot C	**Z**
7	Inner LH and RH flap—trailing edge	**Z**
8	Mid-LH and RH flap—trailing edge	**Z**
9	Outer LH and RH flap—trailing edge	**Z**

14.6.5.2 Response Measurements

The dynamic responses of the aircraft due to the excitation signal shall be measured by accelerometer transducers.

The shakers locations are listed in Table 14.1. The number of accelerometers used should be adequate to measure the complete mode shapes and obtain a good representation. The accelerometers should be located on the hard-points of the structure, such as on the spar and rib intersection, for example. This is done to avoid measuring local vibration of the skin panel which is not the primary concern of the test. The detailed accelerometer locations and direction of the control surfaces are shown in Fig. 14.12 and listed in Table 14.2. Figure 14.12 illustrates a qualitative impression of the accelerometer location on the N-219 aircraft prototype LH RH wing, LH RH flap and for LH ad RH flap pivot clevis.[9] Figure 14.13 exhibits the flap bell-crank bracket, while Figure is 14 the schematic of accelerometer location for flap system rod (LH, RH).

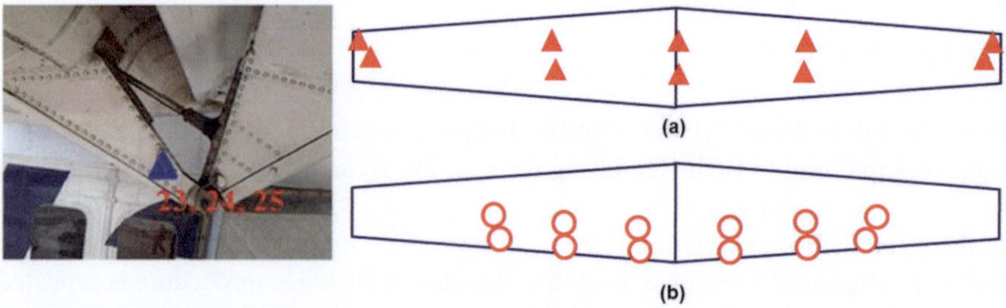

Fig. 14.12 Qualitative schematic of accelerometer location on **a** left-hand and right-hand wing, and **b** on left-hand and right-hand flap, and for left-hand and right-hand flap pivot clevis[10]

[9] LH – Left Hand, RH – Right Hand.

[10] Reproduced qualitatively for general impression from original drawing provided by PT Dirgantara Indonesia through good offices of Mr. Andi Eriawan.

Table 14.2 Accelerometer location and direction of wing and flaps

No.	Accelerometer position	Direction	No.	Accelerometer position	Direction
1	LH tip wing front	z	27	RH flap pivot clevis	y
2	LH tip wing rear	z	28	RH flap pivot clevis	z
3	LH mid wing front	z	29	LH flap bell-crank bracket lower (inner)	x
4	LH mid wing rear	z	30	LH flap bell-crank bracket lower (inner)	y
5	Center wing front	z	31	LH flap bell-crank bracket lower (inner)	z
6	Center wing rear	z	32	RH flap bell-crank bracket lower (inner)	x
7	RH mid wing front	z	33	RH flap bell-crank bracket lower (inner)	y
8	RH mid wing rear	z	34	RH flap bell-crank bracket lower (inner)	x
9	RH tip wing front	z	35	LH flap system rod	x
10	RR tip wing rear	z	36	LH flap system rod	y
11	LH flap outer front	z	37	LH flap system rod	z
12	LH flap outer rear	z	38	RH flap system rod	x
13	LH flap mid front	z	39	RH flap system rod	y
14	LH flap mid rear	z	40	RH flap system rod	z
15	LH flap inner front	z	41	Flap motor support	x
16	LH flap inner rear	z	42	Flap motor support	y
17	RH flap inner front	z	43	Flap motor support	z
18	RH flap inner rear	z	44	Flap motor support	x
19	RH flap mid front	z	45	Flap motor support	y
20	RH flap mid rear	z	46	Flap motor support	z
21	RH flap outer front	z	47	Fuselage frame 7751	x
22	RH flap outer rear	z	48	Fuselage frame 7751	y
23	LH flap pivot clevis	x	49	Fuselage frame 7751	z
24	LH flap pivot clevis	y	50	Fuselage frame 4290	x
25	LH flap pivot clevis	z	51	Fuselage frame 4290	y
26	RH flap pivot clevis	x	52	Fuselage frame 4290	z

All accelerometers installed to follow their orientations in X (STA), Y (BL) and Z (WL) axes directions of the aircraft basic coordinate system. The installation location of the accelerometers can be placed approximately according to its designed and calculated locations (Fig. 14.14).

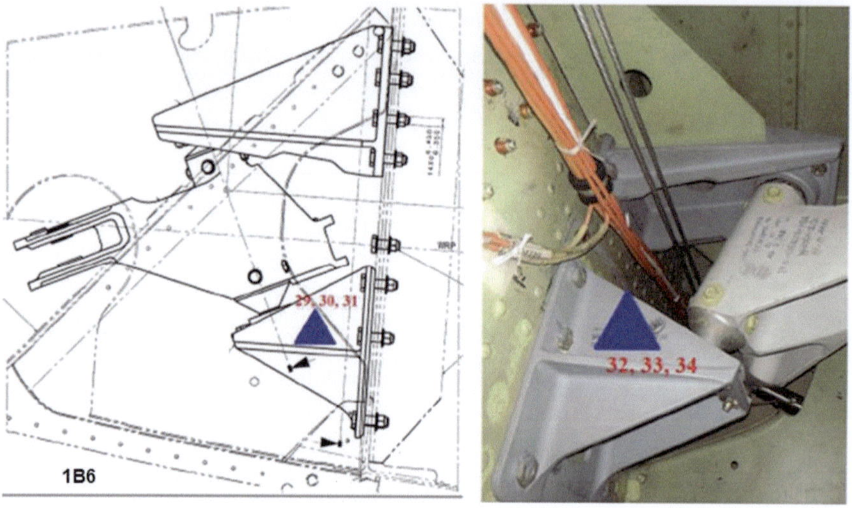

Fig. 14.13 Flap bell-crank bracket where accelerometers are located for test[11]

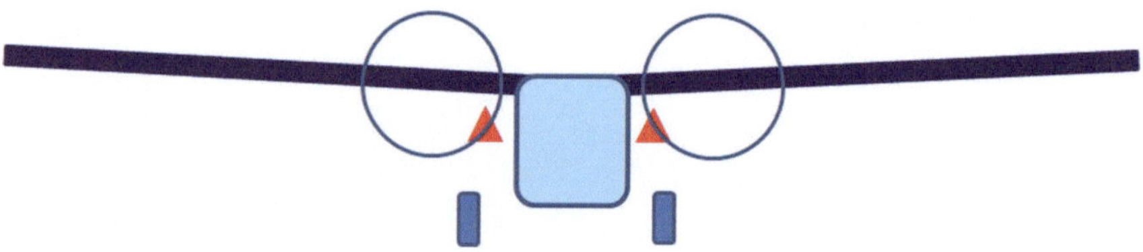

Fig. 14.14 Schematic of accelerometer location for flap system rod (LH, RH)

14.6.5.3 Excitation System

Multi-point excitation technique with sweeping or random burst of vibration will apply to excite the aircraft prototype structure. The frequency range of excitation is 1 Hz to the 3rd harmonic blade passage frequency (BPF) at a range of 480–533.3 Hz. That range is chosen since it has been estimated it will simulate the frequency range based on the propeller rpm (1800–2000 rpm with 4 blades propeller). The shakers will be supported in such a way that the exciter system will have a minimum influence on the dynamic characteristics of the aircraft. The shaker supports should allow excitation in the proper direction.

The shaker attachment should be such that it does not introduce any local stiffening or other distortions to the aircraft. The flexibility of the mechanical attachment of the shakers should be minimized.

[11] For detailed schematic drawing the reader should refer to Table 14.1 which has been provided by PT Dirgantara Indonesia, but cannot be reproduced here for graphical issues.

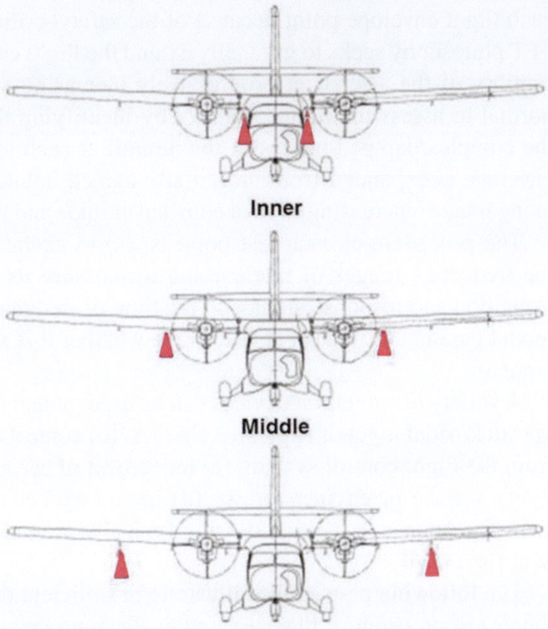

Fig. 14.15 Front view of shaker location (schematic)–(top to bottom) inner, middle, outer flap

The maximum number of the shakers used simultaneously is two (2). Shaker configuration is going to be at LH side only, RH side only or LH and RH at the same time. The location and configuration for the shakers are partially shown in Fig. 14.15. The configuration of the shaker utilized depends on the expected quality of the test results. Some adjustments can be made to tune for the most appropriate configuration prior and during the test. Further detail of the report on the technical document on the ground vibration test (GVT) vibration evaluation of N-219 can be found in the technical documentation report produced for the particular purpose by PT Dirgantara Indonesia.

14.7 Flight Flutter Test

The uncertainty in the aeroelastic model used for flutter calculations, and especially the unsteady aerodynamics, means that calculated flutter speeds will probably be inaccurate to some extent, due to the limitations in the theory that has been developed thus far. It is therefore a requirement of the certification process to validate the flutter behavior and demonstrate freedom from aeroelastic instability over the flight envelope in a flight flutter test (FFT).

On the basis of calculations, a nominal flight envelope is cleared to permit a first flight to take place. Thereafter, the FFT program precedes every other flight test at each flight envelope point because of the safety–critical nature of flutter. The basic FFT philosophy seeks to gradually extend the flight envelope by assessing the flutter stability of the aircraft at progressively increasing speed and Mach number. It is normal to assess the flutter stability by identifying the frequency and damping of the complex/damped modes of the aircraft at each test point. The allowable flight envelope is expanded from an initially agreed boundary by examining the results along lines of increasing EAS at constant altitude and lines of constant Mach number.

The procedure at each test point is: (a) to excite vibration of the aircraft over the frequency ranges of interest and to measure its response, (b) to curve-fit the excitation and response signals in the time or frequency domain and to identify the model parameters and (c) to determine whether it is safe to proceed to the next test point.

A variety of excitation devices can be used, namely (a) control surface movement via stick/pedal input or explosive charges, (b) control surface movement via a signal from the flight control system, (c) movement of aerodynamic vane fitted to aircraft flying surface or engine/store or (d) inertia exciter mounted in the fuselage. The aircraft response is measured using typically around 20 accelerometers, far fewer than for a GVT.

The following case studies illustrate in sufficient detail the ground vibration and flutter test procedures; the case studies illustrate these tests, which were carried out in 2020 for N-219 and 1996 for N-250 aircraft prototypes.

14.8 Aircraft Prototype Flight Flutter Testing—A Case Study [8]

Summary

Modern flight flutter testing of aircraft requires application of artificial excitation. The work reported here presents the development of N250 Prototype 1 flight flutter testing using an aerodynamic vane. An important consideration in aircraft flight flutter test is the use of controllable aerodynamics vanes to generate a well-defined shaking force. The force has to be high enough to excite the aircraft structure beyond its normal flight vibration level. The response of the aircraft structure to the exciting force is analyzed to determine the vibrational stability parameters.

The best exciter locations are chosen to excite the symmetrical and antisymmetrical bending and torsion modes of the lifting surface beyond its normal flight vibration level. Sine-sweep and sine-dwell excitation types yield higher structural response levels as compared to random or impulse step excitation. The former is less complicated to analyze.

The instrumentation to measure the vane forces and responses are located at positions where the aircraft responds best to the excitation. The flight test procedure starts with test points from a safe initial condition and proceeds with small speed increments up to the ultimate speed, Mach condition, VD/MD. Theoretical predictions in combination with the resonance frequencies and the corresponding damping and transmissibility factors, as a function of speed, provide an estimation of the critical speed beyond which vibrational instability can occur.

The work reported here concludes that the vane, developed by Dynamic Engineering Inc. (DEI), is easily operated for in-flight excitation. Flight flutter testing of the aircraft is accomplished in a quick, accurate and safe way. Higher excitation frequencies are reached than were possible with stick raps.

14.8.1 Introduction

Modem flight flutter testing of aircraft requires artificial excitation. The work reported here presents the development of N250 Prototype 1 flight flutter testing excitation using an aerodynamic vane. An important element in aircraft flight flutter testing is the use of controllable aerodynamic vanes to generate a well-defined shaking force. The force has to be high enough to excite the aircraft structure beyond its normal flight vibration level. The response of the aircraft structure due to the excitation has to be analyzed to determine the vibrational stability parameters.

The flight flutter testing is carried out for a series of test points starting from a safe initial condition and progressing with small speed increments up to the ultimate speed, Mach condition, VI/M1. A brief overview of the new vane application for the N250-prototype 1 is presented, and some test results are highlighted. The flutter clearance testing should be completed early in order that:

a. Other flight testing may proceed safely.
b. Maximum reaction time is available to correct problems.

The work reported here consists of five sections. Section 14.1 presents a brief description of the aircraft, flutter test objectives, and scope of tests. Section 14.2 describes general methods and test procedures as well as those specifically used for the N250 Prototype 1 flight flutter testing. Section 14.3 covers methods to analyze flight flutter test data. Finally, Sects. 14.4 and 14.5 present, respectively, the test results and conclusions.

14.8.2 General Description of Aircraft

N250 Prototype 1 aircraft is a turboprop commuter short-haul aircraft for 50 passengers, high wing and T-tail (See Fig. 14.16). The aircraft is powered by Allison. AE-2100C turboprop engine with six-bladed Dowty-Rotol propellers that are mounted on the inboard wing structure.

The retractable tri-cycle Messier Bugatti that is mounted in the fuselage supports the aircraft. The flight control system is a fly-by-wire design. Each control surface is actuated by two actuators: one is active while the other is standby. The design of

Fig. 14.16 N250-PA1 three-view drawing[12]

the flight control system insures that the stiffness and damping of the actuators are adequate for normal and failure conditions. This implies that mass balance is not allowed on any control surface for flutter prevention. The active actuator provides the stiffness while the standby actuator generates the damping force.

The general technical data of N250 Prototype 1 Aircraft are as follows:
Wingspan—28,000 mm.
Total Length—26,629 mm.
Height—8787 mm.
Max. Takeoff Weight—22,000 kg.
Max. Landing Weight—21,800 kg.
Max. Cruise Speed at Sea Level—250 knots.
Max. Dive Speed at Sea Level—300 knots.
Max. Operational Ceiling—25,000 ft.

14.8.3 Objectives of Flight Flutter Tests

14.8.3.1 The Main Objectives Are

1. To confirm and demonstrate the aircraft freedom from flutter in its flight envelope.
2. To validate the methodology of theoretical flutter analysis.

[12] Image of N-250 is personal collection of the author from his work at PT IPTN.

3. To meet the airworthiness requirements. FAR/JAR/CASR Part 25–629.

To successfully attain these objectives, the following are essential:

a. reliability of the theoretical flutter analysis predictions
b. verification of the dynamic characteristics by ground vibration test
c. an efficient and reliable excitation system
d. online and post-flight flutter test data evaluation
e. good correlation between predicted values and test results data.

The remainder of the work reported here shows that the use of controllable aerodynamic vane provides definite advantages over conventional excitation methods. In particular, the use of the DEI vane for the N250 Prototype 1 flight flutter testing accomplishes all of the above objectives.

14.8.3.2 Scope

The flight flutter test of N250 Prototype 1 using flutter exciter is performed up to VD at altitudes of: 25,000 ft., 10,000 ft. and 17,000 ft. Two mass configurations are considered, namely.

a. Max. Weight—most forward CG
b. Light Weight—most aft CG.

14.8.4 N250 Prototype 1 Flight Flutter Testing Development

14.8.4.1 Vibrational Stability Parameters

In the frequency range of interest, the vibrational stability parameters of the aircraft are:

1. the resonance frequencies
2. the accompanying damping value for each resonant frequency
3. the accompanying complex transmissibility factor for each resonant frequency. This value represents the ratio of the vibration mode response and the modal excitation force. For each test point, the vibrational stability parameters are determined from the excitation forces and the corresponding responses.

14.8.5 Excitation System

14.8.5.1 General Methods of Excitation

The acceptable approach to demonstrate flutter stability requires that the structural modes important for flutter can be excited in flight.

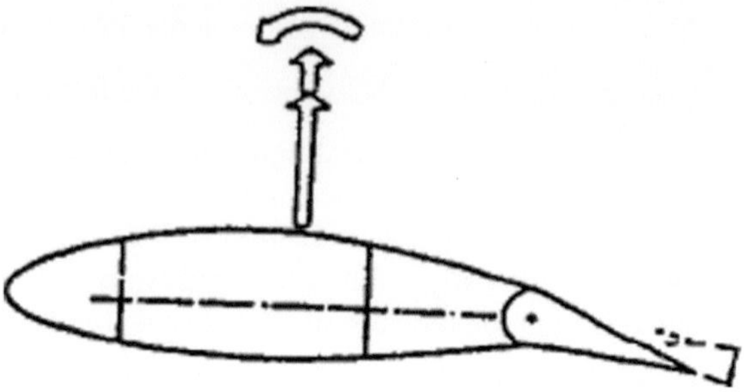

Fig. 14.17 Schematic of stick rap

The best exciter locations have to be chosen so that both the symmetrical and antisymmetrical bending and torsion modes of the lifting surfaces can be excited beyond their normal flight vibration level. Sine-sweep and sine-dwell excitation types yield higher structural response levels as compared with random, impulse or step excitation. The former is also less complicated to analyze.

The following is a brief summary of various conventional excitation methods and their limitations.

(1) Control pulses (Stick Raps and Servo Control Input) can be used to shake the aircraft by its own control surfaces and thereby avoid the installation of special devices shown in Fig. 14.17. However, the frequency range is usually limited to low frequency (maximum about 10 Hz) and the quantification of the excitation forces generated by the control surface deflections is not possible. There is, therefore, a poor repeatability in generation of forces.
(2) The low-quality impulse excitation with Pyrotechnics (Bonkers) is applicable to supersonic airplane flight flutter testing and control surface testing, where the installation of vanes or inertia shakers is not allowed. This type of excitation is very good for higher frequencies (about 65 Hz). However, this excitation is poor for lower frequencies and, furthermore, the excitation forces are not measurable.
(3) The excitation by air turbulence is still currently in use. The determination of the vibrational stability of the aircraft for the entire flight envelope using air turbulence is not accurate. The method requires the adoption of certain assumptions on the spectral contents of the natural source. This cheapest excitation source is not always available wherever and whenever needed and is, therefore, limited in practical application.
(4) The advantages of the use of excitation by the controllable aerodynamic vanes are: the lower and upper frequency range of the excitation can be enlarged and accurate measurements of the excitation forces are possible.

14.8.5.2 Types of Aerodynamic Vanes Are Applied

For the flight flutter testing carried out, three types of aerodynamic vanes are utilized:

(a) Rotating vane with adjustable width (Lockheed application, Fig. 14.18a, b,

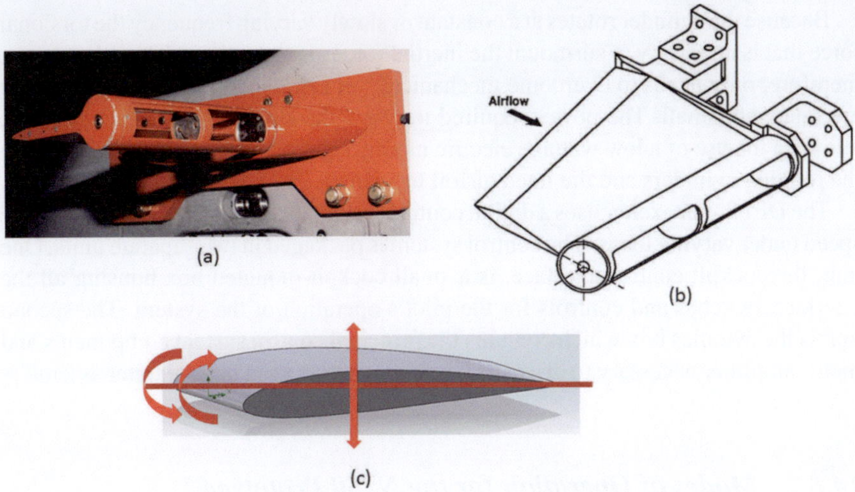

Fig. 14.18 **a** and **b** General impression of rotating vane; **c** Schematic of pitching vane

(b) Pitching vane, Fig. 14.18c,
(c) Fixed vane with rotating slotted cylinder along trailing edge (not exhibited).

14.8.6 Excitation Methods Used for N250 Prototype 1

Two methods of excitation were used in sequence at each test point in the flight flutter testing of the N250 Prototype 1.

(1) The first method is the use of control surface pulses by stick raps and rudder kicks. This method has the benefit of being able to excite the low frequencies and to quickly anticipate the occurrence of the low decay rate of the aircraft structure. After the aircraft structural vibrations due to the control surface pulses are damped out, the second method of excitation is activated.
(2) The second method of excitation uses a flutter exciter system with slotted rotation cylinders. (See Fig. 14.18c). It was developed by Wilmer H. Reed III and is distributed by Dynamic Engineering Inc. (DEI). The exciters are mounted on the wing tip and horizontal stabilizer tip, both left- and right-hand sides. The advantage of using this flutter exciter is that it can sweep all modes in the frequency range of interest. In this case the structural modes that are important for flutter can be excited.

The basic flutter exciter unit is a pair of rotatable concentric cylinders. Each cylinder has a slot that allows the air to pass through. As the cylinder rotates about its axis the airstream is alternately deflected upward and downward at a frequency twice that of the cylinder's rotation frequency. The amplitude of the excitation force depends upon the dynamic pressure and the degree of slot opening.

Because the cylinder rotates at a constant or slowly varying frequency the torsional force that is necessary to surmount the inertial forces is not present. The drive motor, therefore, only needs to overcome mechanical and aerodynamic friction forces that are relatively small. The power required to rotate the slotted cylinders is minimal, allowing the use of a low wattage electric motor. This can be mounted together with the rotating cylinders and the fixed airfoil to a simple.

The DEI flutter exciter uses a digital control system to maintain the proper cylinder speed under varying loads. The control system is packaged in two separate units. One unit, the cockpit control interface, is a small cockpit-mounted box housing all the interface switches and controls for the pilot's operation of the system. The second unit is the avionics box which contains the principal control system components and motor amplifier necessary to drive the brushless motors used in the exciter assembly.

14.8.7 Modes of Operation for the N250 Prototype 1

The system allows three principal modes of operation which can be selected by the pilot:

(1) Sine dwell
(2) Linear sweeps
(3) Logarithmic.

Linear sweeps in the range of 25 Hz are used with 60 s duration time for the N250 Prototype 1 flight flutter testing.

Shorter duration sweeps and logarithmic sweeps did not always sufficiently excite the structural modes above 20 Hz. The system also provides the choice of two force measurements outputs: high or low level. These levels correspond to the opening and closing of the outboard cylinder. Closing the outboard cylinder results in a 50% reduction in force. An additional blockage plug is included which allows the inboard slot to be half-closed for an additional 25% reduction in force. The inboard plug is attached on the ground prior to flight. Thus, it cannot be modified while in flight.

The N-250 Prototype 1 flight flutter testing used 75% of maximum force (the slotted cylinder is 75% open) in order to obtain adequate excitation for important flutter modes in flight. The control system maintains phase control over the two exciters.

The pilot can choose to operate the units either in-phase or 180 out-of-phase, allowing symmetric or antisymmetric excitation, respectively.

The induced sinusoidal forces and bending moments are calculated from measurements by strain gauges mounted at the root of the exciter vane. The amplitude and frequency of excitation is displayed on the operator's control panel and is also available as analogue voltage inputs for data recording systems similarly to the control panel switch positions.

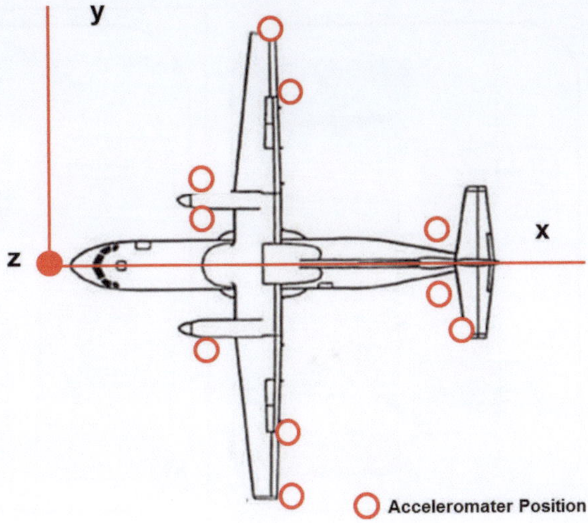

Fig. 14.19 Schematic of some accelerometers locations

14.8.8 *Instrumentation*

The essential instrumentation for flight flutter tests is the accelerometers distributed over the aircraft for the determination of structural vibrations, and the induced forces of the flutter exciters.

The vane forces and the responses have to be measured. For that purpose, the aircraft has to be instrumented at locations where the aircraft responds best to the excitation. In general, this will be at the extremities of the lifting surfaces of the aircraft, namely the tips of the wing, the stabilizer and the fin.

Thirty-five accelerometers and potentiometers are distributed over the N250 Prototype 1 aircraft and 4 strain gauges are located on the four flutter exciters as shown in Fig. 14.19.

Also important is the instrumentation to determine speed, Mach number, altitude and fuel condition. All these signals were recorded on an on-board pulse code magnetic (PCM) tape recorder. In addition, they were sent via telemetry to the mission control room for real-time quick-look monitoring on stripcharts and for a quasi-online analysis of frequency and damping values of dominant modes (Fig. 14.20).

14.8.9 *Flight Test Procedure*

14.8.9.1 General Aspects of Flight Test Procedure

The flight test procedure starts with test points from a safe initial condition and proceeds with small speed increments up to the ultimate speed, Mach condition,

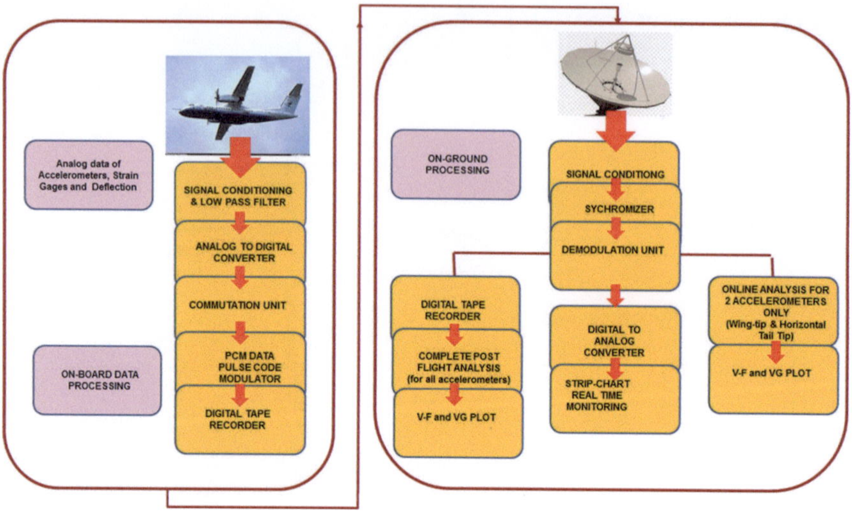

Fig. 14.20 N-250 flight flutter test telemetry and data preprocessing

VD/MD. To reduce the complexity of the analysis, the symmetrical and the antisymmetrical behavior are considered separately. At each test point, the sweep excitation is performed for symmetric and anti symmetric excitation by which the vanes acts, respectively, in phase and in counterphase sequence. The length and the number of repetitions of the sweep excitation can be regarded as an acceptable compromise to simulating the non-steady flight test conditions at the high-speed descending flight path.

The progress of a flight flutter test depends strongly on the test procedure, on previous experience with similar aircraft configurations, and on the results of reliable flutter calculations.

The general philosophy of testing is that such test points are scheduled for each flight which allowed expansion of the flight envelope to higher speeds only after extrapolation of the damping trends. If the analysis of the flight records at lower speeds showed no evidence of the flutter onset, a clearance for the next airspeed increment was given. Maximum use is made of real-time analysis techniques to reduce as much as possible the number of flight trials required.

The basis for the test is the availability of a comprehensive flutter analysis of the aircraft in all possible configurations. Starting from an initial flight envelope, cleared from a flutter point of view by analysis the flight envelope expansion is made in progressive steps. At each progressive step the in-flight damping and frequency of all important structural modes of the aircraft and its control surfaces are determined.

14.8.9.2 Implementation of Test Procedures to N250 Prototype 1

An efficient flight flutter test program of the N250 Prototype 1 of clean configuration (retracted flaps) for maximum weight—most forward CG and lightweight—most Aft CG were performed up to Vd fur 25,000 ft, 17,000 ft and 10,000 ft and up to 170 KEAS with flaps down.

The flight testing was conducted at three altitude levels, 25,000 ft, 10,000 ft and 17,000 ft with a velocity range of 160 KEAS to 250 KEAS and a Mach number range of 0.28–0.63 (see Fig. 14.21), test points of N250-PA1 Flight Flutter Test). For each test point, control surface pulses and flutter exciter were both used to obtain frequency and damping for the elevator and rudder control surfaces. This operation was used to clear each test point qualitatively and to obtain a rough approximation of decay rate. The 8 selected parameters were displayed on stripchart for "Quick-look" monitoring.

After completing the application of the impulse forces described above the flutter exciters were activated to obtain better frequency and damping calculation. This is done by activating the symmetric and antisymmetric wing flutter exciters in sequence. When this is completed, the symmetric and antisymmetric tail flutter exciters are activated similarly. Because of the limitation of altitude tolerance, only control surface impulses were done at dive speeds.

The linear frequency sweeps were performed from I to 25 Hz at 60 s. duration. This frequency range covers the primary modes of interest for the N250-PAl. The accelerometer response data were telemetered to a mission control room for real-time data collection and monitoring. Thirty-nine (39) parameters were telemetered and 2 parameters (one on wingtip and one on horizontal tip) were selected for a quasi-online analysis of frequencies and damping values for dominant modes to clear next test point. The other 8 selected parameters were displayed also on stripchart for

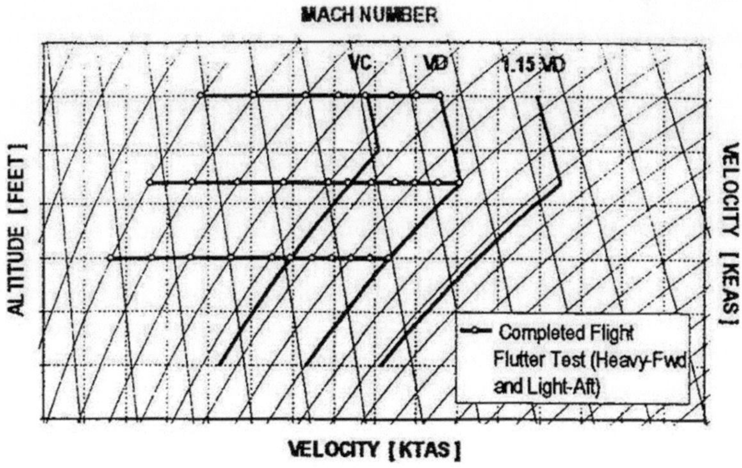

Fig. 14.21 Flight flutter test point

"Quick-look" monitoring. These 39 parameters were recorded and kept in storage. The data sampling rate is 125 samples/sec. For back-up data all these signals were recorded on-board by PCM tape recorder.

14.8.10 Aircraft Configurations and Test Points

14.8.10.1 Mass Configurations

Flight flutter testing of N250-PA1 using flutter exciter has been done up to VD at altitudes of 25,000 ft., 10,000 ft. and 17,000 ft. Two mass configurations were performed namely: Max. Weight—most fwd CG and Light Weight—most aft CG.

Max. Weight—most fwd CG

Max. Weight = 22 Ton

- Most fwd CG = 17% MAC
- UB tank was burnt first.

Test was stopped at approximately 1000 kg fuel used (tolerance consideration), at which:

Light Weight—most aft CG

Light Weight = 20 Ton

- Most aft CG = 41% MAC
- **I/B** tank was burnt first.

14.8.10.2 Test Procedure

Test was stopped at approximately 600 kg fuel remain (safety consideration). All altitude levels were tested with tolerance of 2000 ft. All test points were tested with "yaw damper-off" up to VD and several test points with "yaw damper-on" up to V. Also both "autopilot–on" and "yaw damper-on" were tested for several test points.

Test configurations completed according to the theoretical prediction, the "autopilot-off" and "yaw damper-on" condition is more critical than the "autopilot-on" and "yaw damper-off". Since the rudder movement of "yaw damper-on" can excite the fin torsion which is the flutter mode, it is clear that "yaw damper-on" is more critical than "autopilot-on". Therefore, the flight flutter testing for the "autopilot-on" and "yaw damper-on" condition need not be performed.

14.8.11 Analysis Methods of Flight Flutter Test Data

14.8.11.1 Quick-Look Analysis

In conjunction with theoretical predictions, the sequence of resonance frequencies, the damping values and the transmissibility factors, as a function of the speed, provide a good estimation of the critical speed beyond which vibrational instability can occur. In deciding the safe continuation of a flutter test, the damping value of the least damped vibration mode at each test point must be immediately available. This requires quick-look analysis of a number of transmitted responses and excitation signals at each test point.

Two ways of quick-look analysis are used at PT. IPTN. The most important and direct way in making the decision to continue flutter testing to the next test point based on real-time monitoring of the time traces of a limited number of response signals. The label of the vibration and the shape of the response due to sweep excitation contain important information about the vibrational damping, especially in relation to the previous test point. If a low damped mode is present, the comparison of the shape on the measured sweep response with precalculated sweep responses of a single-degree-of-freedom system for a series of damping values gives a good indication of the amount of damping remaining in that low damped mode.

The second way of quick-look analysis is to analyze the least damped mode with a dedicated software package. The methods of analysis are based on curve-fitting one or a limited number of resonance peaks in either the power spectrum or the frequency response function (FRF) or its complex plane representation.

14.8.11.2 Post-Flight Analysis

Post-flight analysis was performed by an accurate, multiple degrees of-freedom curve-fitting technique on all relevant power spectra or FRF's on all on-board registered signals.

14.8.12 Analysis of Measured Vibration Signals

14.8.12.1 Methods

To analyze the measured vibration signals, several methods are available. For all methods it is assumed that the flying aircraft can be represented by a linear system. The analysis of this linear system can be done in time domain, or more commonly, in the frequency domain. The analysis is carried out in the time domain if the stick-rap excitation is used. It is possible to use the impulsive response of the aircraft structure to predict flutter onset. In most cases, the impulsive response is of short duration and, therefore, the accuracy of the measurement is reduced. However, a brief analysis directly from the stripchart recording can provide an initial guess of flutter onset. If the flutter exciter is used, the analysis is usually done in the frequency domain. In this case, the transfer function between excitation and response is calculated. From the averaged input and cross-power spectrum, the transfer function is computed by using fast Fourier transform. This represents the complete system parameter description available.

The following two methods are used to analyze measured vibration signals:

- Levy Complex-Curve-Fitting Method.
- Zimmerman Flutter Criterion.

14.8.12.2 Levy Complex-Curve-Fitting Method

For the system identification and parameter estimation, the Levy Complex-Curve-Fitting method is used.

During the flight test the above methods are used to extract frequencies and damping from two accelerometer data that are located on the right wingtip and on the right horizontal tail plane (HTP). Therefore, the evaluation is done only for the critical modes. In addition to the extraction of frequencies and damping data, these methods can also predict the flutter onset [9–11] based on the two critical modes. This prediction provides clearance information to proceed to the next test point. After the flight flutter test a more detailed evaluation of all accelerometers data was also done by using Levy complex-curve-fitting techniques.

14.8.12.3 Zimmerman Flutter Criterion

One difficult aspect of flutter testing is that a flutter mode can suddenly become unstable with only a small increase in dynamic pressure. Furthermore, there may be little or no indication of approaching instability in a plot of modal damping against dynamic pressure.

The Zimmerman flutter criterion [10, 11] is based on considering the value of measured frequency and damping data for the characteristic equation of a two-degrees-of-freedom system as a flutter margin parameter with increasing speed. Using this information, it is possible to predict the behavior of the flutter margin at speeds not yet flown and to predict the speed of flutter onset especially for online analysis. It can be expressed in terms of the measured frequency and damping of the two modes which are input as parameters.

14.8.12.4 Flight Flutter Test Results

In addition to online analysis of wingtip and HTP during the flight test with telemetry data, other accelerometer data were used for offline analysis by playing back the

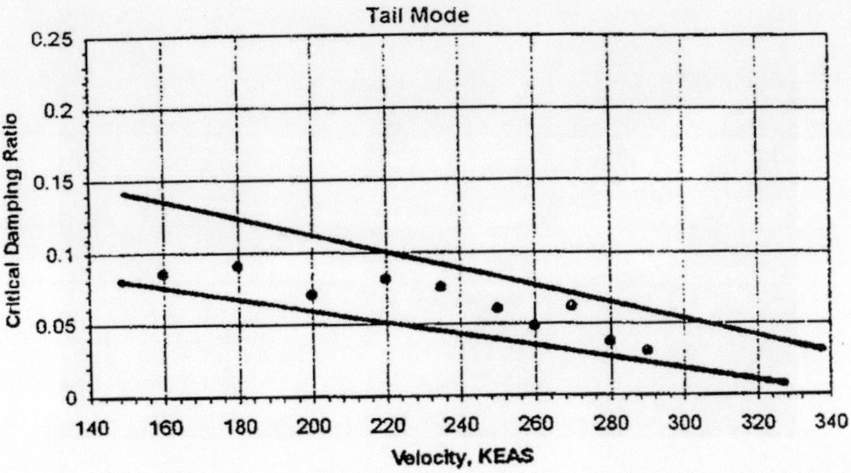

Fig. 14.22 Critical damping ratio evaluation

telemetry data. The time histories of accelerometer and reference excitation signals were first plotted to determine the position of sweep and the parameters which contained the modes of interest.

All frequency response function (FRF) parameters and curve fits described previously were performed with "Levy' Curve-Fitting". After determining the frequencies and damping values, the V-g plots were supplemented with due consideration of the aircraft configurations. The fuel condition did not have a big influence on the frequencies and damping, since the critical mode is fin mode.

V-g and V-f results of the antisymmetric modes of flight flutter test results for max. weight-most fwd CG were also recorded [8][13]. The damping results of the testing were presented as critical damping ratio. For small values of damping ratio, the structural damping (g) is equal to two times of critical damping ratio. Because of the limitation of altitude tolerance, only control surfaces impulses were done at dive speeds. Sample of the critical damping ratio evaluation is illustrated in Fig. 14.22.

The plots of the measured frequency and damping trends versus speed at all altitudes show that the N250-PAI is free from flutter up to VD, and that damping values for all modes exceed g-0.03 under test conditions. The damping trends also show that there is no large and rapid reduction in damping near VD. In general, the theoretical predicted trends of suspected critical modes for frequency and damping versus airspeed and altitude showed moderate compliance, with test results. The damping extrapolation based on the test data for higher speeds do not show trends for inadequate flutter stability within the 1.15 VD range.

Several test points were measured using "yaw damper-on". It can be seen that the frequencies and damping with "yaw damper-on" were in close agreement to the

[13] Sample of Data and Figures related to such measurement can be found in [8].

Fig. 14.23 Flight flutter test of Pt Dirgantara Indonesia 19 Passenger N-219 Prototype

results of the "yaw damper-off". It is concluded that there is no feedback from "yaw damper-on" (control system) to the flutter mode (Fig. 14.23).

14.8.12.5 Appendix[14]

Some figures associated with the flight flutter test of PT DIRGANTARA Indonesia N-219 are here displayed to illustrate some of the missing illustrations of the previous N-250 flight flutter test as well as to indicate some progress that has been made in the technique and facilities of flight flutter test procedure (Figs. 14.24, 14.25, and 14.26).

14.9 Conclusions

The following conclusions are reached on the basis of the flutter test results:

a. The fuel condition does not have a significant influence on the frequencies and damping, since the critical mode is fin mode.
b. The theoretically predicted trends of suspected critical modes for frequency and damping versus airspeed and altitude show moderate compliance with test results. Extrapolating the above test data for higher speeds indicate no flutter instability within the 1.15 VD range.

[14] Images related to N-219, GVT and Flight Flutter Test are courtesy of PT Dirgantara Indonesia/PT Indonesian Aerospace through the good offices of Mr. Andi Eriawan (personal contact) at PT Dirgantara Indonesia, and are greatly acknowledged.

Fig. 14.24 Flutter exciter for N-219 flight flutter test

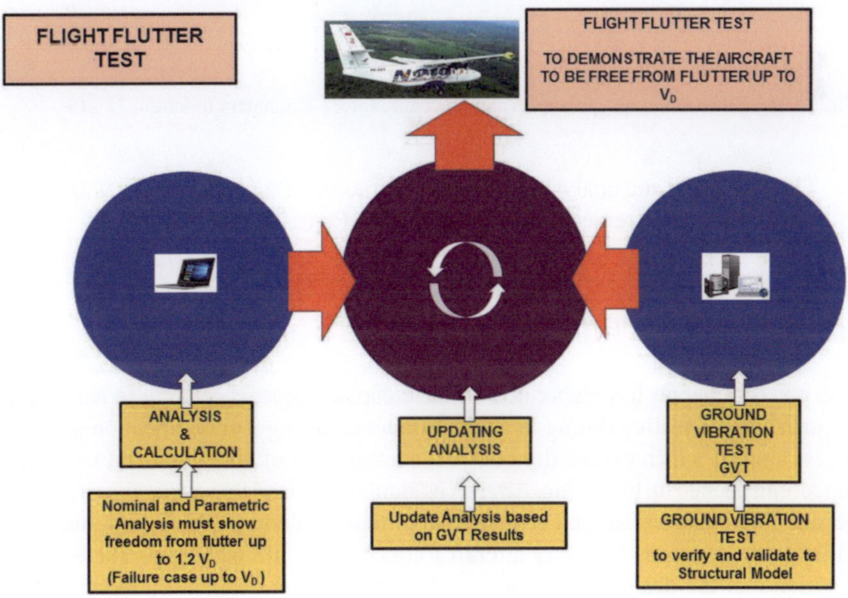

Fig. 14.25 N-219 Flight flutter test integrated approach

c. The tests show that the frequencies and damping with "yaw damper-on" are in very close agreement to the results of the "yaw damper-off." It is concluded that there is no feedback from "yaw damper-on" (control system) to the flutter mode.
d. Based on comprehensive post-flight analysis after flight testing up to VD, it is concluded that no flutter behavior is predicted within the VD flight and CG envelopes of N250-PA1 including flaps speed placard.

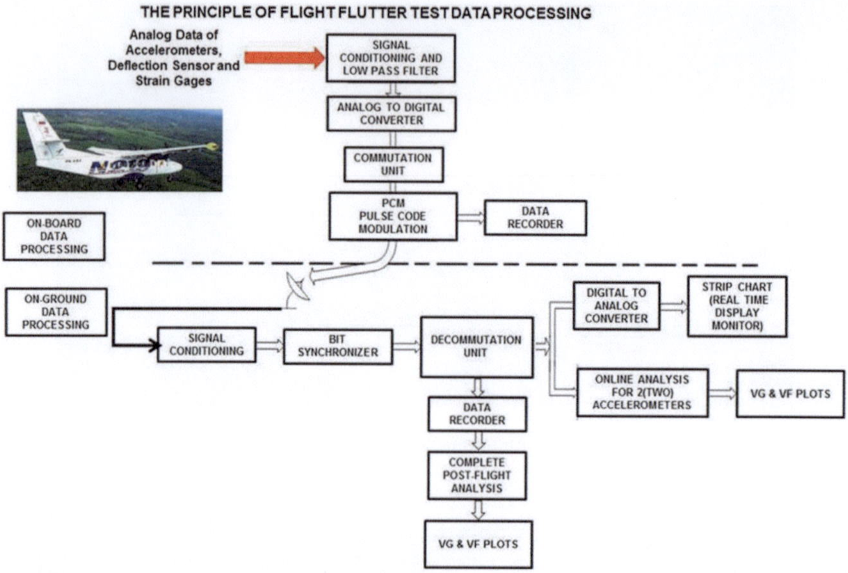

Fig. 14.26 Flight flutter test principles and procedure for PT Dirgantara Indonesia N-219

e. The flutter test and analysis of the N250 Prototype 1 aircraft comply with all the certification requirements of AR/JAR/CASR Part 25–269.

14.10 Concluding Remarks

The present chapter has elaborated on the clearance process relating to aeroelastic experiment and testing during the design, manufacturing and certification phase of a new aircraft which essentially covers a combination of analytical and numerical modelling backed up by testing. Aircraft operations demand that every aircraft should be safe to fly, and to that end it is mandatory that the conception, design, manufacturing and operations for every aircraft follows a fail-safe paradigm. Therefore, in addition that the design of aircraft could guarantee such mandatory requirements, physical experiments along the design, manufacturing and operation of aircraft need to be proved and validated by physical experiments, which is indeed highly challenging.

In this conjunction, some principles have been elaborated in introductory level, accompanied by some illustrations drawn from the industry, research institutions and universities. To provide hands-on impression, some case studies on ground vibration and flight flutter tests have been elaborated to a certain extent.

References

1. Gasparetto, V.E.L., M.R. Machado, S.H.S. Carneiro. 2020. *Experimental modal analysis of an aircraft wing prototype for SAE aerodesign competition.* https://doi.org/10.15446/dyna.v87n214.75361, https://www.redalyc.org/journal/49649666177012/html. Accessed August 15, 2022.
2. Giclais, G., P. Lubrina, C. Stephan. 2016. *Aircraft ground vibration testing at ONERA.* Aerospace Lab, Alain Appriou, pp. 1–18, https://doi.org/10.12762/2016.AL12-05 hal-01511792.
3. Pickrel, C.R. 2002. *Airplane ground vibration testing—Nominal modal model correlation, sound and vibration.* November. http://www.sandv.com/downloads/0211pick.pdf.
4. Salehi, M., and S. Ziaei-Rad. 2007. Ground vibration test (GVT) and correlation analysis of an aircraft structure model. *Iranian Journal of Science & Technology, Transaction B, Engineering* 31 (B10): 65–80.
5. Anonymous, Naval Research Laboratory Facilities. 2008. https://www.nrl.navy.mil/content_images/2008Facilities_revised.pdf. Accessed April 4, 2020.
6. Spivey, Natalie. 2019. *Ground vibration testing at NASA Armstrong, emphasizing on passive aeroelastic tailored wing ground vibration test using fixed base correction method.* NASA Document ID 20,190,030,841. https://ntrs.nasa.gov/citations/20190030841. Accessed on April 4, 2020.
7. Djojodihardjo, H. 2002. *Aeroelastics and Unsteady Separated Aerodynamics of Aircrafts: Review and Outlook, in the Memorial Volume on the Occasion of the Emeritierung of Professor Boris Laschka.* Technische Universität München.
8. Fadil, R., Indrawanto, and H. Djojodihardjo. 1998. N_250_Prototype_1_Flight_Flutter_Testing. In *Paper ICAS-98-4.5.3, 19th Congress of the International Council of the Aeronautical Sciences*, Anaheim, California, U.S.A., 18–23 September 1994.
9. Levy, E.C. 1958. *Complex-Curve Fitting.* Los Angeles, CA: Space Technology Labs, November 3.
10. Bennett, R.M. 1982. *Application of Zimmerman Flutter-Margin Criterion to a Wind-Tunnel Model.* NASA Technical Memorandum 84545.
11. König, K. 1983. Flight vibration test analysis methods, theory and application. MBB GmbH, Bremen, Germany; also König, K. 1983. Problems of system identification in flight. In *AGARD Report No. 720, ADA142556.*

Chapter 15
Piezoaeroelastic Wing Section for Energy Harvester

Abstract Piezoaeroelastic energy harvesters convert airflow-induced vibrations into electrical energy, while the availability and affordability of piezoelectric transducers offer a class of flapping foil energy harvesters mostly in micro- to milliwatts scale which need to be tuned to match the characteristic frequencies. The present work presents a brief review of aeroelastic instability of a generic typical wing section due to the free-stream flow field which is utilized as an oscillating foil energy converter. For propaedeutic analysis a generic piezoaeroelastic cantilevered beam is defined and treated as a typical section. The basic governing equation of this generic structure is treated as a three degree-of-freedom electrodynamic system, with the first two degree of freedom comprising the standard binary aeroelastic system with additional relevant terms to represent the influence of a piezoelectric embedded element on the cantilevered wing. Following the philosophical approach of binary aeroelastic system, the problem is mathematically formulated and solved for the range of solutions that can be obtained depending on the prevailing physical properties of the system, focusing on the stability characteristics of the generic system. The characteristics of the unsteady aerodynamics of the oscillating system associated with favorable energy harvesting capabilities are assessed.

Keywords Aeroelasticity and flapping mechanism · Bio-inspired · Flapping piezoelectric energy harvester · Flow energy harvesting · Oscillatory foil

15.1 Introduction and Brief Review

The progress of energy harvesting technology and self-powered systems that has been prompted by the development of low-power electronics offers autonomous systems that require minimum maintenance that are essential for their deployment in previously inaccessible locations [1]. Energy harvesting can be defined as the process of capturing energy, which is relatively minute, from the naturally occurring

---

The present chapter is based on a paper by H. Djojodihardjo, Aeroelastic and Performance Baseline Analysis of Piezoaeroelastic Wing Section, presented at First International Symposium on Flutter and its Application 2016 (ISFA), Tokyo, Japan.

© The Author(s), under exclusive license to Springer Nature Singapore Pte Ltd. 2023

H. Djojodihardjo, *Introduction to Aeroelasticity*,
https://doi.org/10.1007/978-981-16-8078-6_15

("Green") energy sources and then accumulating them and storing them for later use. Autonomous system and smart material technology that are now widely utilized take advantage of the thermal energy that can be converted to electrical energy by the Seebeck-Peltier effect or mechanical energy extraction in piezoelectric system (first discovered by Jacques and Pierre Curie). These make extracting energy from mechanical energy an attractive approach for powering electronic systems, as in piezoelectric system.

The piezoelectric effect was discovered by the brothers Pierre Curie and Jacques Curie in 1880, while the converse effect was later discovered by Gabriel Lippmann in 1881 through the mathematical aspect of the theory. These behaviors were labeled the piezoelectric effect and the inverse piezoelectric effect, respectively [2]. New approaches have been developed using first-principles computations, based on fundamental physics. First-principles theory laid the framework for a basic understanding of the origins of piezoelectric behavior and properties.

The principle of the composite energy harvesting system relies on the fact that the material, such as polyvinylidene fluoride (PVDF), has natural piezoelectric properties. The conversion of mechanical energy to electricity by the piezoelectric PVDF is characterized by the coupling coefficient k, which can be as high as 15%.

A generic model for energy to electrical energy conversion using vibrating wing section and piezoelectric effects for energy harvesters is illustrated in Fig. 15.1. The conversion of flow energy to aeroelastic vibrational energy consists of two parts, in conformity to the concept of aeroelasticity, as illustrated in Fig. 15.2, consisting of the mechanical vibration part (a) and the aerodynamic part (b).

The damping consists of two parts, the mechanical and structural damping incorporated in part (a) and the aerodynamic damping incorporated in the aerodynamic forcing of part (b), as shown in Fig. 15.2. The piezoelectric element then establishes the piezoaeroelastic vibration energy harvester, which harvests the flow energy to be delivered to the electrical loads [1].

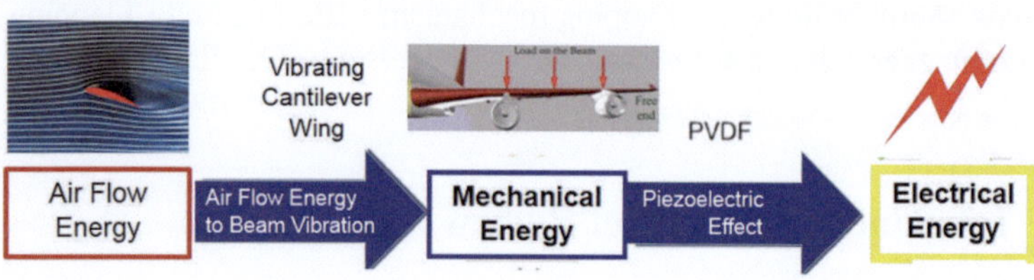

Fig. 15.1 Schematic estimation of flow energy to electrical energy conversion using vibrating wing section and piezoelectric effects

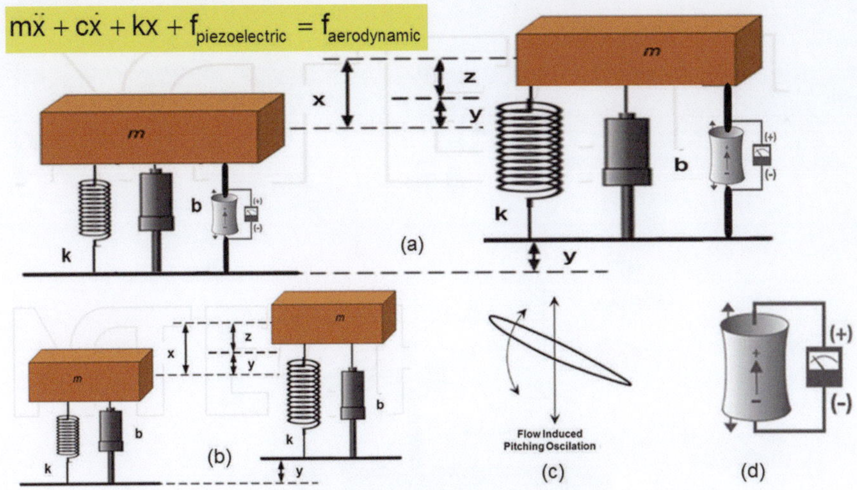

Fig. 15.2 Generic model of piezoaeroelastic vibration energy harvester comprising the mechanical and structural **a**, aerodynamic **b** and piezoelectric **c** parts

15.2 Mechanics of Piezoelectric-Patched Cantilevered Beam as Energy Harvester

For the purposes of propaedeutic analysis a generic piezoaeroelastic cantilevered beam is defined and treated as a binary aeroelastic cantilevered wing section as the base line, with piezoelectric element embedded at the wing root, as further simplified implementation of the generic piezoaeroelastic vibration energy harvester.

The schematic of the piezoaeroelastic energy harvesting system from airflow excitation is depicted in Fig. 15.3 as the baseline of the present analysis. Aeroelastic vibrations of the cantilever wing bending (plunge or heaving motion, represented by bending angular deflection) strain the piezoelectric patches dynamically and produce the electrical output. Therefore, if piezoelectric coupling is added to the plunge DOF of the typical section as schematically depicted in Figs. 15.2 and 15.3, the resultant force on the piezoelectric element will result in a resultant voltage.

The governing linear aeroelastic equations to be utilized in the piezoaeroelastic energy harvesting system can be derived from the binary aeroelastic system as given in the classical literature (e.g. Bisplinghoff et al. [3]; Wright and Cooper [4]; Djojodihardjo and Yee [5]; Djojodihardjo [6]):

$$(m + m_e)\ddot{h} + mbx_\alpha\ddot{\alpha} + d_h\dot{h} + K_h h = -L \tag{15.1}$$

$$mbx_\alpha\ddot{h} + I_\alpha\ddot{\alpha} + d_h\dot{\alpha} + K_\alpha\alpha = M \tag{15.2}$$

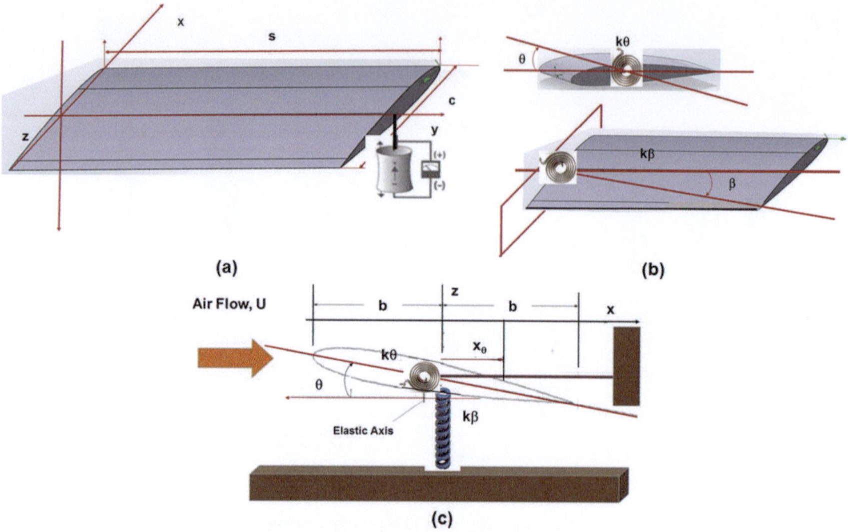

Fig. 15.3 Generic model of linear piezoaeroelastic energy harvester

where m is the airfoil mass per length (in the span direction), m_e is the fixture mass (connecting the airfoil to the plunge springs) per length, M is the aerodynamic moment, L is the aerodynamic lift, and the over-dot represents differentiation with respect to time (t).

Incorporating piezoelectric element in the model, the following set of piezoaeroelastic model is obtained (similar to that obtained by Marqui and Erturk [7]):

$$(m + m_e)\ddot{h} + mbx_\alpha\ddot{\alpha} + d_h\dot{h} + K_h h - \frac{\Theta}{l} = -L \tag{15.3}$$

$$mbx_\alpha\ddot{h} + I_\alpha\ddot{\alpha} + d_h\dot{\alpha} + K_\alpha\alpha = M \tag{15.4}$$

$$c_{\text{piezo}}\dot{V} + \frac{V}{R_l} + \Theta\dot{h} = 0 \tag{15.5}$$

Here the fixture mass (m_e) is defined for the case when the system slightly deviates from the ideal typical section depicted in Fig. 15.5 due to the masses of the shaft, spring mass and other attachments in real experiments [8], while it is zero in the ideal representation of Fig. 15.3, which is assumed here for the baseline solution.

After incorporating all these factors, further simplification, modification and redefinition of the relevant and prevailing piezoaeroelastic parameters, Eq. (15.5) can be reduced to:

$$I_\beta\ddot{\beta} + k_\beta\beta + I_{\beta\theta}\ddot{\theta} + q\mathrm{S}C_{L\alpha}\theta - \frac{k_\beta}{s}\mathrm{DV} = 0 \tag{15.6}$$

$$I_{\beta\theta} s\ddot{\beta} + I_\theta \ddot{\theta} + k_\theta \theta - 2q\mathrm{Seb}C_{L\alpha}\theta = 0 \tag{15.7}$$

$$+k_\beta D\dot{\beta} + C_{\mathrm{piezo}}\dot{V} + \frac{1}{R_l}V = 0 \tag{15.8}$$

which will be utilized as the baseline for the solution procedure developed subsequently. All the symbols are given in the nomenclature. It should also be noted that for the present case the structural and aerodynamic damping have been ignored; further analysis which incorporates these ignored terms will follow after the baseline case is established.

15.3 Synthesis of Baseline Solution Procedure

The problem posed by the set of Eqs. (15.6–15.8) will be formulated in the following sequence of problems, which is considered to follow an appropriate rationale:

1. Formulate the equation as stability equation, which may be met by certain behavior of piezoelectric terms.
2. Establish the relationships between the piezoelectric output in terms of the aeroelastic parameters setting and flow parameters and determine the related setting of the piezoelectric system to meet certain objectives.

Based on the insight gained in the solution of the problem, preliminary design specifications for a piezoaeroelastic system can be defined.

Following the technique introduced by Wright and Cooper [4], the general form of the baseline aeroelastic Eqs. (15.1) and (15.2) is in the classical second-order form for N degrees-of-freedom system:

$$\mathbf{A}\ddot{\mathbf{q}} + (\rho_0 U\mathbf{B} + \mathbf{D})\dot{\mathbf{q}} + (\rho U^2\mathbf{C} + \mathbf{E})\mathbf{q} = 0 \tag{15.9}$$

where $\mathbf{A}, \mathbf{B}, \mathbf{C}, \mathbf{D}$ *and* $\mathbf{E}$ are the structural inertia, aerodynamic damping, aerodynamic stiffness, structural damping and structural stiffness matrices respectively, and $\boldsymbol{q}$ is the generalized coordinates (typically modal coordinates). It is important to note that the $\mathbf{B}$ *and* $\mathbf{C}$ matrices only apply for the reduced frequency for which they are defined.

Equation (15.9) can be transformed to eigenvalue formulation of flutter equations by adding an identity equation of the form

$$\boldsymbol{I}\dot{\boldsymbol{q}} - \boldsymbol{I}\dot{\boldsymbol{q}} = 0 \tag{15.10}$$

Upon combining Eq. (15.9) with Eq. (15.10), the following Eq. (15.11) is obtained:

$$\begin{bmatrix} \mathbf{I} & 0 \\ 0 & \mathbf{A} \end{bmatrix} \begin{Bmatrix} \dot{\mathbf{q}} \\ \ddot{\mathbf{q}} \end{Bmatrix} - \begin{bmatrix} 0 & \mathbf{I} \\ -(\rho U^2 \mathbf{C} + \mathbf{E}) & -(\rho U \mathbf{B} + \mathbf{D}) \end{bmatrix} \begin{Bmatrix} \mathbf{q} \\ \dot{\mathbf{q}} \end{Bmatrix} = \begin{Bmatrix} 0 \\ 0 \end{Bmatrix} \tag{15.11}$$

Equation (15.11) can further be reduced to as

$$\begin{Bmatrix} \dot{\mathbf{q}} \\ \ddot{\mathbf{q}} \end{Bmatrix} - \begin{bmatrix} 0 & \mathbf{I} \\ -\mathbf{A}^{-1}(\rho U^2 \mathbf{C} + \mathbf{E}) & -\mathbf{A}^{-1}(\rho U \mathbf{B} + \mathbf{D}) \end{bmatrix} \begin{Bmatrix} \mathbf{q} \\ \dot{\mathbf{q}} \end{Bmatrix}$$
$$= \begin{Bmatrix} 0 \\ 0 \end{Bmatrix} \text{ or } \dot{\mathbf{x}} - \mathbf{Q}\mathbf{x} = 0 \tag{15.12}$$

Equation (15.12) is now in the first-order form although the Q matrix has the size of 2N × 2N, twice the size of the matrices in the aeroelastic Eqs. (15.6–15.9).

For convenience, following the practice in aeroelasticity, the analysis of flutter stability can be obtained by assuming oscillatory solution of the form:

$$\{\mathbf{x}\} = \{\mathbf{x}_0\}e^{\lambda t}; \ \ \lambda = \sigma + i\omega \tag{15.13}$$

where σ is the real (damping) part and ω is the imaginary (oscillatory) part.

Hence by assuming $\mathbf{x} = \mathbf{x}_0 e^{\lambda t}$ Eq. (15.12) can be reduced to the form

$$(\mathbf{I}\lambda - \mathbf{Q})\mathbf{x}_0 = 0 \text{ or } (\mathbf{Q} - \mathbf{I}\lambda)\mathbf{x}_0 = 0 \tag{15.14}$$

Two approaches can be taken to solve Eqs. (15.12) and/or (15.14), namely the frequency-domain method and time-integration method in the time domain.

a. **Frequency-Domain Method**

The frequency-domain method is used to solve the eigenvalue problem as defined by Eqs. (15.12) or (15.14). The eigenvalue solutions are computed for a range of reasonable flow velocity range, calculated for any airspeed. The appearance of positive real part of the eigenvalues indicates the occurrence of instability or flutter.

b. **The time-integration method**

The time-integration method is used to integrate Eqs. (15.12) or (15.14) numerically and calculate the time response for a specific airflow velocity using a conveniently chosen numerical integration method, such as the Runge–Kutta method.

15.4 Fundamental Solution Procedure—Decoupled Linear Approach

The baseline aeroelastic Eq. (15.14) can be decoupled into two sets of equations, in order that each set has similar solution vector.

Hence one can write:

$$\begin{bmatrix} I_\beta & I_{\beta\theta} \\ I_{\beta\theta} & I_\theta \end{bmatrix}\begin{Bmatrix} \ddot{\beta} \\ \ddot{\theta} \end{Bmatrix} + \begin{bmatrix} 0 & 0 & 0 \\ 0 & 0 & 0 \\ k_\beta D & 0 & C_{piezo} \end{bmatrix}\begin{Bmatrix} \dot{\beta} \\ \dot{\theta} \\ \dot{V} \end{Bmatrix} + \begin{bmatrix} k_\beta & 0 & -\frac{k_\beta}{s}D \\ 0 & k_\alpha & 0 \\ 0 & 0 & \frac{1}{R_l} \end{bmatrix}\begin{Bmatrix} \beta \\ \theta \\ V \end{Bmatrix} = \begin{bmatrix} 0 & -qSC_{L\alpha} \\ 0 & qSebC_{L\alpha} \end{bmatrix}\begin{Bmatrix} \beta \\ \theta \end{Bmatrix} \tag{15.15}$$

After substituting Eq. (15.13) into Eq. (15.15), one obtains

$$\lambda^2\begin{Bmatrix} \beta \\ \theta \end{Bmatrix} - \begin{bmatrix} I_\beta & I_{\beta\theta} \\ I_{\beta\theta} & I_\theta \end{bmatrix}^{-1}\begin{bmatrix} 0 & -qSC_{L\alpha} \\ 0 & q\mathrm{Seb}C_{L\alpha} \end{bmatrix}\begin{Bmatrix} \beta \\ \theta \end{Bmatrix} = -\lambda\begin{bmatrix} I_\beta & I_{\beta\theta} \\ I_{\beta\theta} & I_\theta \end{bmatrix}^{-1}\begin{bmatrix} 0 & 0 & 0 \\ 0 & 0 & 0 \\ k_\beta D & 0 & C_{\mathrm{piezo}} \end{bmatrix}\begin{Bmatrix} \beta \\ \theta \\ V \end{Bmatrix} - \begin{bmatrix} I_\beta & I_{\beta\theta} \\ I_{\beta\theta} & I_\theta \end{bmatrix}^{-1}\begin{bmatrix} k_\beta & 0 & -\frac{k_\beta}{s}D \\ 0 & k_\alpha & 0 \\ 0 & 0 & \frac{1}{R_l} \end{bmatrix}\begin{Bmatrix} \beta \\ \theta \\ V \end{Bmatrix} \tag{15.16}$$

In the simplified analysis, Eq. (15.16) is split into two sets of equations. These are

$$\lambda^2\{Q\} + \lambda[M_d]\{Q\} + [M_k]\{Q\} - [M_A]\{Q\} = 0 \tag{15.17}$$

and

$$\lambda[M_d]\{V\} + [M_k]\{V\} = 0 \tag{15.18}$$

The latter equation can be reduced to

$$V\left(\lambda C_{\mathrm{piezo}} + \frac{1}{R_l}\right) = \lambda k_\beta D\beta \tag{15.19}$$

or

$$V = \frac{\lambda k_\beta D\beta}{\left(\lambda C_{\mathrm{piezo}} + \frac{1}{R_l}\right)} \tag{15.20}$$

A significant loss of information is noted here, since the simplified piezoaeroelastic instability formulation failed to include any piezoelectric variable. D disappears from the eigenvalue problem, since the second row of the "simplified" piezoaeroelastic problem does not contain a term incorporating β. Nevertheless, for the interest of developing the solution further, the piezoaeroelastic eigenvalue problem will be treated. The piezoelectric variables, however, are determined by the eigenvalues and the flow dynamic pressure $q = \rho U^2/2$ in $V = \frac{\lambda k_\beta \beta D}{\lambda C_{\text{piezo}} + \frac{1}{R_l}}$.

The solution procedure then can be formulated as follows:

1. Find the four eigenvalues $\lambda = \rho = \sigma + i\omega$ of

$$\lambda^4\left(\mathbf{I}_\beta \mathbf{I}_\theta - \mathbf{I}_{\beta\theta}^2\right) + \left(k_\beta k_\theta - \rho U^2 \frac{cs^2 C_{L\alpha}}{4} k_\beta\right) = 0 \tag{15.21}$$

2. For each set of eigenvalue, find the harvested voltage

$$V = \frac{\lambda k_\beta \beta D}{\lambda C_{\text{piezo}} + \frac{1}{R_l}} \tag{15.22a}$$

which consists of

$$V = V_{\text{real}}(\lambda) + V_{\text{imag}}(\lambda) \tag{15.22b}$$

The sustained fluctuating values of voltage V will be considered as plausible solutions, dictated by the prevailing aeroelastic properties to be further analyzed.

15.5 Principal Piezoaeroelastic Stability Equation

Following the approach similar to the transformation of Eq. (15.10) to eigenvalue (flutter equations) formulation Eq. (15.20) by adding an identity equation,

$$\boldsymbol{I}\dot{\boldsymbol{Q}} - \boldsymbol{I}\dot{\boldsymbol{Q}} = 0 \tag{15.23a}$$

where

$$\mathbf{Q} \equiv \begin{Bmatrix} \beta \\ \theta \\ V \end{Bmatrix} \tag{15.23b}$$

to Eqs. (15.5) and (15.6) which have been written in a form similar to Eq. (15.9), then one gets

$$\begin{bmatrix} 1 & 0 & 0 & 0 & 0 & 0 \\ 0 & 1 & 0 & 0 & 0 & 0 \\ 0 & 0 & 1 & 0 & 0 & 0 \\ 0 & 0 & 0 & I_\beta & I_{\beta\theta} & 0 \\ 0 & 0 & 0 & I_{\beta\theta} & I_\theta & 0 \\ 0 & 0 & 0 & 0 & 0 & 0 \end{bmatrix} \begin{Bmatrix} \dot{\beta} \\ \dot{\theta} \\ \dot{V} \\ \ddot{\beta} \\ \ddot{\theta} \\ \ddot{V} \end{Bmatrix} -$$

$$\begin{bmatrix} 0 & 0 & 0 & 1 & 0 & 0 \\ 0 & 0 & 0 & 0 & 1 & 0 \\ 0 & 0 & 0 & 0 & 0 & 1 \\ k_\beta & qSC_{L\alpha} & -\frac{k_\beta}{s}D & 0 & 0 & 0 \\ 0 & k_\theta - 2qSebC_{L\alpha} & 0 & 0 & 0 & 0 \\ 0 & 0 & \frac{1}{R_l} & k_\beta D & 0 & C_{\text{piezo}} \end{bmatrix} \begin{Bmatrix} \beta \\ \theta \\ V \\ \dot{\beta} \\ \dot{\theta} \\ \dot{V} \end{Bmatrix} = \begin{Bmatrix} 0 \\ 0 \end{Bmatrix} \tag{15.24}$$

or

$$\begin{bmatrix} \mathbf{I} & 0 \\ 0 & \mathbf{A}' \end{bmatrix} \begin{Bmatrix} \dot{\mathbf{Q}} \\ \ddot{\mathbf{Q}} \end{Bmatrix} - \begin{bmatrix} 0 & \mathbf{I} \\ \mathbf{C}' & \mathbf{B} \end{bmatrix} \begin{Bmatrix} \mathbf{Q} \\ \dot{\mathbf{Q}} \end{Bmatrix} = \begin{Bmatrix} 0 \\ 0 \end{Bmatrix} \tag{15.25}$$

where

$$\left[A'\right] \equiv \begin{bmatrix} I_\beta & I_{\beta\theta} & 0 \\ I_{\beta\theta} & I_\theta & 0 \\ 0 & 0 & 0 \end{bmatrix} \tag{15.26a}$$

$$[B] \equiv \begin{bmatrix} 0 & 0 & 0 \\ 0 & 0 & 0 \\ k_\beta D & 0 & C_{\text{piezo}} \end{bmatrix} \tag{15.26b}$$

$$\left[C'\right] \equiv \begin{bmatrix} k_\beta & qSC_{L\alpha} & -\frac{k_\beta}{s}D \\ 0 & k_\theta - 2q\text{Seb}C_{L\alpha} & 0 \\ 0 & 0 & \frac{1}{R_l} \end{bmatrix} \tag{15.26c}$$

and

$$[\mathbf{I}] \equiv \begin{bmatrix} 1 & 0 & 0 \\ 0 & 1 & 0 \\ 0 & 0 & 1 \end{bmatrix} \tag{15.26d}$$

Equation (15.24) or (15.25) can be written as.

15.5.1 Eigenvalue Problem in Frequency Domain

$$\begin{Bmatrix} \dot{\mathbf{Q}} \\ \ddot{\mathbf{Q}} \end{Bmatrix} - \begin{bmatrix} \mathbf{I} & 0 \\ 0 & \mathbf{A}' \end{bmatrix}^{-1} \begin{bmatrix} 0 & \mathbf{I} \\ \mathbf{C}' & \mathbf{B} \end{bmatrix} \begin{Bmatrix} \mathbf{Q} \\ \dot{\mathbf{Q}} \end{Bmatrix} = \begin{Bmatrix} 0 \\ 0 \end{Bmatrix} \tag{15.27}$$

or

$$\left\{\dot{\mathbf{G}}\right\} - \left[\mathbf{A}''\right]\left[\mathbf{B}''\right]\{\mathbf{G}\} = \{\mathbf{0}\} \text{ or } \left\{\dot{\mathbf{G}}\right\} - [\mathbf{M}_\mathbf{G}]\{\mathbf{G}\} = \{\mathbf{0}\} \tag{15.28}$$

where

$$\{\mathbf{G}\} = \begin{Bmatrix} \mathbf{Q} \\ \dot{\mathbf{Q}} \end{Bmatrix} = \begin{Bmatrix} \beta \\ \theta \\ V \\ \dot{\beta} \\ \dot{\theta} \\ \dot{V} \end{Bmatrix} \tag{15.29}$$

$$\left[\mathbf{A}''\right] \equiv \begin{bmatrix} \mathbf{I} & 0 \\ 0 & \mathbf{A}' \end{bmatrix}^{-1}; \quad \left[\mathbf{B}''\right] \equiv \begin{bmatrix} 0 & \mathbf{I} \\ \mathbf{C}' & \mathbf{B} \end{bmatrix} \tag{15.30}$$

$$[\mathbf{M}_\mathbf{G}] \equiv \left[\mathbf{A}''\right]\left[\mathbf{B}''\right] \tag{15.31}$$

For further elaboration, different from the simplified split equations approach, one should again look at the oscillating solution assumption to obtain the solution of the extended matrix solution of Eq. (15.28).

Assuming periodic solution of $\{\mathbf{G}\}$ in the form:

$$\{\boldsymbol{G}\} = \{\boldsymbol{G}_0\}e^{\lambda t} = \{\boldsymbol{G}_0\}e^{(\sigma + i\omega)t} \tag{15.32}$$

where $\lambda = (\sigma + i\omega)t$

then Eq. (15.8) can be converted to

$$\lambda\{\boldsymbol{G}\} = [\boldsymbol{M}_G]\{\boldsymbol{G}\} \rightarrow \{\boldsymbol{G}\} = \frac{1}{\lambda}[\boldsymbol{M}_G]\{\boldsymbol{G}\} \ or \ [\lambda - [\boldsymbol{M}_G]] = 0 \tag{15.33}$$

which leads to finding the roots λ that satisfy

$$\{\boldsymbol{\lambda} - [\boldsymbol{M}_G]\} = 0 \tag{15.34}$$

or

$$
\begin{bmatrix} \lambda & 0 & 0 & 0 & 0 & 0 \\ 0 & \lambda & 0 & 0 & 0 & 0 \\ 0 & 0 & \lambda & 0 & 0 & 0 \\ 0 & 0 & 0 & \lambda & 0 & 0 \\ 0 & 0 & 0 & 0 & \lambda & 0 \\ 0 & 0 & 0 & 0 & 0 & \lambda \end{bmatrix} - \left(\begin{bmatrix} 1 & 0 & 0 & 0 & 0 & 0 \\ 0 & 1 & 0 & 0 & 0 & 0 \\ 0 & 0 & 1 & 0 & 0 & 0 \\ 0 & 0 & 0 & I_\beta & I_{\beta\theta} & 0 \\ 0 & 0 & 0 & I_{\beta\theta} & I_\theta & 0 \\ 0 & 0 & 0 & 0 & 0 & 0 \end{bmatrix}^{-1} \begin{bmatrix} 0 & 0 & 0 & 1 & 0 & 0 \\ 0 & 0 & 0 & 0 & 1 & 0 \\ 0 & 0 & 0 & 0 & 0 & 1 \\ k_\beta & qSC_{L\alpha} & -\frac{k_\beta}{s}D & 0 & 0 & 0 \\ 0 & k_\theta - 2q\text{Seb}C_{L\alpha} & 0 & 0 & 0 & 0 \\ 0 & 0 & \frac{1}{R_l} & k_\beta D & 0 & C_{\text{piezo}} \end{bmatrix} \right) = 0 \quad (15.35)
$$

Then, for each set of piezoaeroelastic parameters and wind velocity, represented by q, there are six roots of λ and consequently six each of the corresponding σ and ω values. The voltage V of the piezoelectric elements will be the solution vector. The latter notion will be the subject of further investigation.

15.5.2 Time-Integration Problem in the Time Domain/State Space

$$
\begin{bmatrix} \mathbf{I} & 0 \\ 0 & \mathbf{A}' \end{bmatrix} \begin{Bmatrix} \dot{\mathbf{Q}} \\ \ddot{\mathbf{Q}} \end{Bmatrix} = \begin{bmatrix} 0 & \mathbf{I} \\ \mathbf{C}' & \mathbf{B} \end{bmatrix} \begin{Bmatrix} \mathbf{Q} \\ \dot{\mathbf{Q}} \end{Bmatrix} \rightarrow \begin{Bmatrix} \dot{\mathbf{Q}} \\ \ddot{\mathbf{Q}} \end{Bmatrix} = \begin{bmatrix} \mathbf{I} & 0 \\ 0 & \mathbf{A}' \end{bmatrix}^{-1} \begin{bmatrix} 0 & \mathbf{I} \\ \mathbf{C}' & \mathbf{B} \end{bmatrix} \begin{Bmatrix} \mathbf{Q} \\ \dot{\mathbf{Q}} \end{Bmatrix} \quad (15.36a)
$$

or

$$
\{\dot{G}\} = \left[A''\right]\left[B''\right]\{G\} \quad (15.36b)
$$

or

$$
\{\dot{G}\} = [M_G]\{G\} \quad (15.36c)
$$

15.6 Results

15.6.1 Baseline Aeroelastic Stability Results

For the purpose of comparison and assessment, the baseline flutter stability results for a typical section as a binary aeroelastic system representing a cantilever Isogai

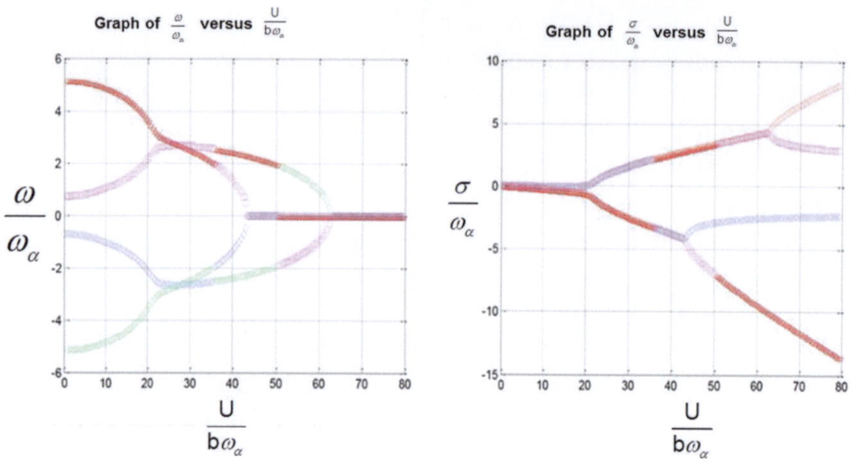

Fig. 15.4 Upper-left: frequency-reduced velocity diagram for low-frequency aerodynamic model. Upper-right: damping-reduced velocity diagram for low-frequency aerodynamic model. Lower images: corresponding diagram reproduced from Zwaan [9] for similar analysis

Wing with low-frequency aerodynamics following Done-Zwaan aeroelastic stability analysis are exhibited in Fig. 15.4. The favorable agreement validates the aeroelastic stability analysis algorithm used in developing the extended piezoaeroelastic stability analysis.

The baseline flutter stability results for a typical section as a binary aeroelastic system representing a cantilever wing section utilizing quasi-steady aerodynamics following the aeroelastic stability algorithm for a typical section as a binary aeroelastic system representing a cantilever wing section utilizing Theodorsen unsteady aerodynamics are reproduced in Fig. 15.5. The data utilized in the present study is presented in Table 15.1.

Figure 15.6 exhibits the flutter stability diagram for the pure wing section treated as a binary system, solved using quartic root approach to solve the eigenvalues of Eq. (15.15) using steady aerodynamic approximation. The computation of the corresponding solutions using Theodorsen unsteady aerodynamics is in progress.

15.6.2 Decoupled Linear Equations Approach for the Binary Aeroelasticity Based Piezoaeroelastic System for Solving the System Output Voltage

The parameters utilized for the cantilevered wing section are those given in Table 15.1. Table 15.2 shows the reference value of bending angle β of the cantilever wing section and various reference parameters for the piezoelectric element utilized in the piezoaeroelastic system.

The resulting frequency and damping of the wing section represented as typical section as a function of the airflow velocity (represented by its normalized values

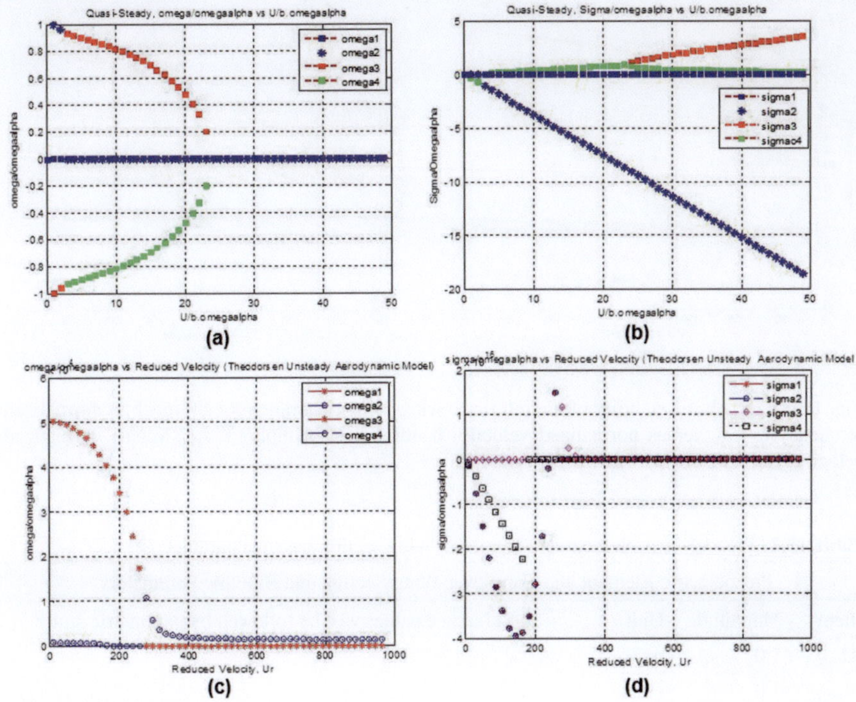

Fig. 15.5 **a** Frequency-reduced velocity diagram low-frequency aerodynamic model; **b** damping-reduced velocity diagram low frequency aerodynamic model; **c** frequency-reduced velocity diagram Theodorsen unsteady aerodynamic model; **d** damping-reduced velocity diagram Theodorsen unsteady aerodynamic model

Table 15.1 Cantilever wing section parameters

Parameter	Magnitude	Information
s	7.5	m, semi-span
c	2.0	m
x_f	0.48c	$X = 0$ at the leading edge
e	$(x_f$–c/4) –1.0	
a	$2x_f$/c – 1.0	
b	c/2	m, half-chord
th	0.12c	m, wing section thickness
I_θ	$(10 \times 2\pi)^2$	Nm/rad
x_α	π x c x th	m
Area	π × c × th	m^2, Wing Section Area, Aross-Sectional Area, Approximated by an Ellipse

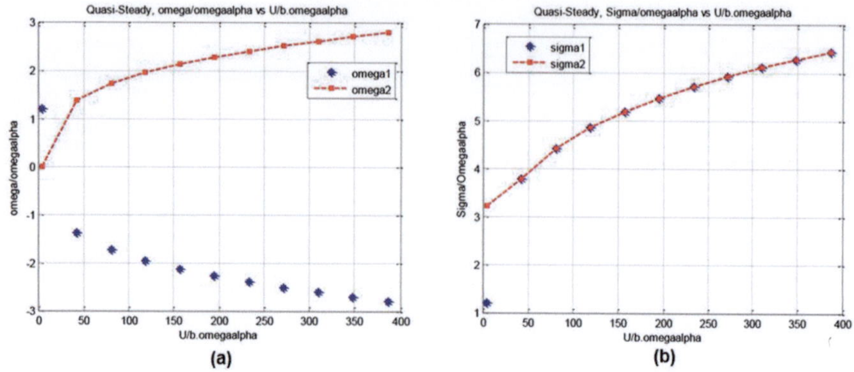

Fig. 15.6 Aeroelastic stability of the reference wing using quadratic root approach; **a** normalized frequency ω/ω_α versus normalized velocity; **b** normalized damping σ/ω_α versus normalized velocity

Table 15.2 Piezoelectric element and cantilever wing section baseline parameters

Piezoelectric Element and Cantilever Wing Section and Baseline Parameters			
Item	Magnitude	Unit	Reference values will be followed by parametric study
D_{piezo}	1.0	m/Volt	
R_{piezo}	1.0	Ohm	
C_{piezo}	1.0	Coulomb/Volt	
Beta	1.0	radian	The Angular Bending Deflection of the Cantilever Beam for initialization, if required

$\omega U/b$ are shown in Figs. 15.7a, b. The corresponding voltage generated by the piezoelectric element for various values of the cantilever wing section bending angle β as functions of the airflow velocity is shown in Figs. 15.8a–d.

15.7 Discussions and Concluding Remarks

Following the simplified aeroelastic binary system-based piezoaerolastic energy harvester depicted in Fig. 15.3, the aeroelastic stability problem associated with the system using a baseline data has been formulated, elaborated and solved using various approaches. In addition, the baseline solutions obtained and presented in the figures are based on quasi-steady aerodynamics. The formulation and computation of the solution using more elaborate Theodorsen aerodynamics are in progress. The formulation of two fundamental solution approaches for the piezoaerolastic energy harvester system has also been formulated, which incorporates systematic solution procedures using frequency-domain binary aeroelastic-based stability approach and

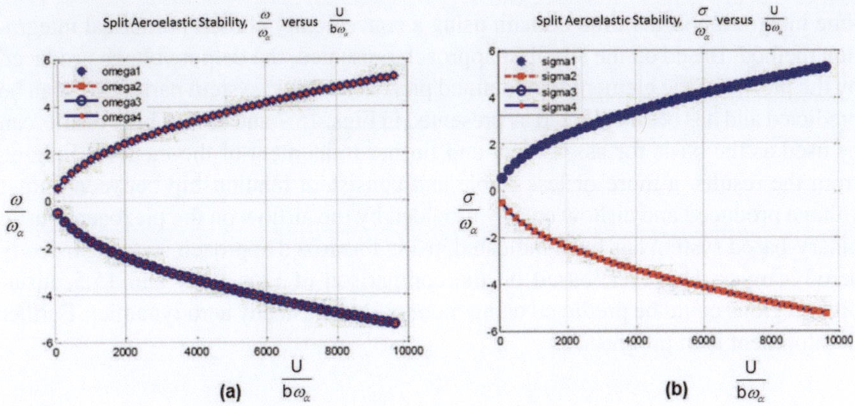

Fig. 15.7 Resulting frequency and damping of the wing section represented as typical section as a function of the airflow velocity (represented by its normalized values $\omega U/b$; **a** variation of the normalized frequency ω/ω_α versus normalized airflow velocity $\omega U/b$; **b** variation of the normalized damping σ/ω_α versus normalized airflow velocity $\omega U/b$. Results are obtained for $\beta = 0.1$

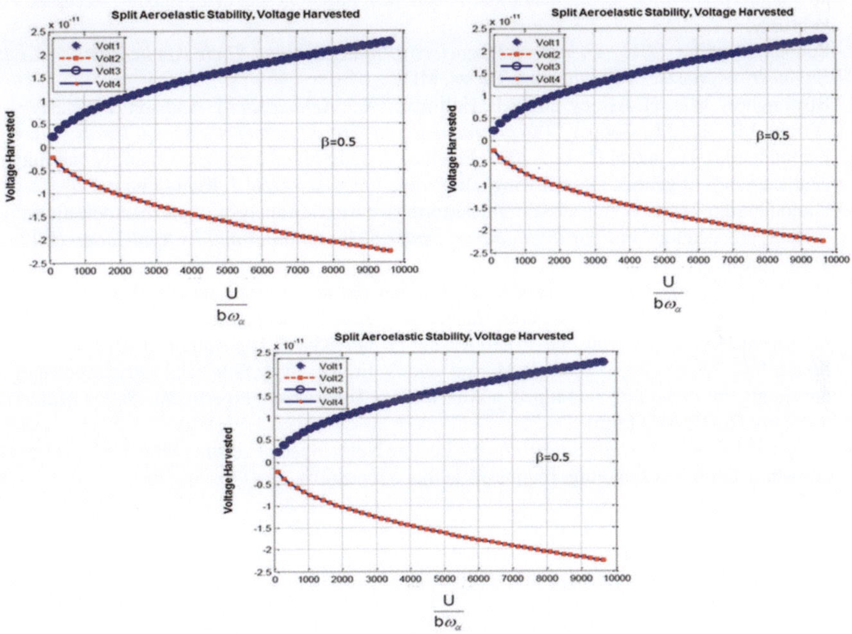

Fig. 15.8 Output voltage produced by the piezoaeroelastic wing section energy harvester represented as typical section as a function of the airflow velocity (represented by its normalized values $\beta U/b$ and various values of β. **a** For $\beta = 0.1$; **b** for $\beta = 0.5$; and **c** for $\beta = 1.0$

time integration in the time domain using a conveniently chosen numerical integration method. Based on the simplest approach presented, the output voltage produced by the piezoelectric element for a defined piezoaeroelastic system parameters can be predicted and has been obtained as presented in Figs. 15.7 and 15.8. These results can be used as the basis for assessment and further refinement of the method. Judging from the results, a more or less stable and consistent relationship between output voltage produced and airflow energy provided by the airflow on the piezoaeroelastic binary-based system has been indicated, using linearized approach and quasi-steady aerodynamics. However, based on the comparison of Figs. 15.4 and 15.5, instability regions could be predicted using more refined unsteady aerodynamics. Further development is in progress.

References

1. Zhu, Dubin. 2016. *Vibration energy harvesting: Machinery vibration, human movement and flow induced vibration.* U Southampton, Intechopen. http://citeseerx.ist.psu.edu/. Accessed 20 February.
2. Ledoux, Antoine. 2011. *Theory of Piezoelectric Materials and Their Applications in Civil Engineering*, Master of Engineering Thesis, MIT.
3. Bisplinghoff, R.L., H. Ashley, and R.L. Halfman. 1955. *Aeroelasticity*. Addison & Welesly.
4. Wright, J.R., and J.E. Cooper. 2007. *Introduction to Aircraft Aeroelasticity and Loads*. Wiley.
5. Djojodihardjo, H., and H.H. Yee. 2007. Parametric study of the aeroelastic stability and flutter characteristics of generic aircraft wings. In *Proceedings, Aerotech II*, Kuala Lumpur.
6. Djojodihardjo, H. 2016. Aeroelastic and performance baseline analysis of piezoaeroelastic wing section, proceedings. In *First International Symposium on Flutter and its Application* (ISFA), Tokyo, Japan.
7. De Marqui, Carlos, and Alper Erturk. 2012. Electroaeroelastic analysis of airfoil based wind energy harvesting using piezoelectric transduction and electromagnetic induction. *Journal of Intelligent Material Systems and Structures* 24 (7): 846–854, © The Author(s) 2012.
8. Sousa, V.C., M. de Anicezio, C. de Marqui, and A. Erturk. 2011. Enhanced aeroelastic energy harvesting by exploiting combined nonlinearities: Theory and experiment. *Smart Material Structure* 20: 094007 (8pp).
9. Zwaan, R.J. 1981. *Aeroelasticity of aircraft*. Lecture Notes, Special Lecture, Short Course offered at Institut Teknologi Bandung, Indonesia, author's repository.

Chapter 16
Introduction and Selected Case Studies in Hydroelasticity

Abstract As a generic definition, hydroelasticity can further be defined as a branch of science concerned with the motion and distortion of deformable bodies responding to environmental excitations in the sea, as it evolves and is modified from the original Collar triangle. The discipline is concerned with phenomena involving interaction between inertial, hydrodynamic, i.e. the fluid pressure acting on the structure, and elastic forces on the structure which modifies its dynamic state and, in return, the motion and distortion of the structure. Considerations on hydroelasticity relevant to offshore oil production platforms, low-speed conventional ships and high-speed monohull or multihull vessels, which are affected by several types of dynamic loads including environmental actions, such as wind and waves, will be discussed to provide comprehensive understanding of the state of affairs. Then engineering analyses for the prediction of induced dynamic responses of such engineering systems will be elaborated in terms of a formulation of fluid–structure interaction via integration of hydrodynamics, structural mechanics and use of novel modeling techniques. Specific example in this subject is the author and colleague work on numerical boundary element computation of submerged body-surface wave interaction, which will be elaborated in detail. To provide some introductory examples in hydroelasticity, attention is given to the state of affairs and equation of motion of hydrofoils moving in incompressible and inviscid or viscous flow, and discussions on methods of solutions for stability and dynamic response. Inviscid fluid–structure coupling modeling and solution scheme will also be discussed, employing finite element method and representing the hydrofoil by a typical section. Many figures utilized in this chapter have been adopted or adapted from recent publications, for which the author would like to thank the corresponding authors whole-heartedly.

Keywords Fluid-structure interaction · Hydrodynamics · Hydroelasticity · Off-shore structures · Physical and mathematical modeling ships and high-speed vessels · Submerged body-surface wave interaction

© The Author(s), under exclusive license to Springer Nature Singapore Pte Ltd. 2023

H. Djojodihardjo, *Introduction to Aeroelasticity*,
https://doi.org/10.1007/978-981-16-8078-6_16

16.1 Review and Introduction to Hydroelasticity of Ships

Offshore oil production platforms, low-speed conventional ships and high-speed monohull or multihull vessels are floating structures that are affected by several types of dynamic loads including environmental actions, such as wind and waves. Hydroelasticity considers the latter as the most important source of motions and structural responses in floating as well as submerged structures [1–3]. Then engineering analyses for the prediction of induced dynamic responses of such engineering systems could be based on a realistic formulation of the fluid–structure interaction (FSI) domain via integration of hydrodynamics, structural mechanics and use of novel modeling techniques. The degree of interaction between the structural response and hydrodynamic loads of floating structures and seagoing vessels determines the practical solution in each of the structural (Ω_s) and fluid domains (Ω_f) respectively and provides valuable information with regard to global strength analysis and assessment, as exhibited in Fig. 16.1a. According to the degree of structural response, problem areas can be divided between those where the ship can be treated as a rigid body and those as a flexible body for which the degree of dynamic response is significantly affected by its inherent flexibility.

The treatment of ships as rigid bodies responding to waves has continued to be significant, and the prediction of seaway-induced dynamic loads has been approached by two distinct procedures, the hydrodynamic analysis and the quasi-static structural safety assessment. In both these approaches, bodily motions imply that the structure is a rigid body experiencing neither strains nor stresses. Concepts such as mode shapes, natural and resonance frequencies, fatigue, however, are not incorporated due to the rigidity restriction. Although the traditional methodologies have been successful to a

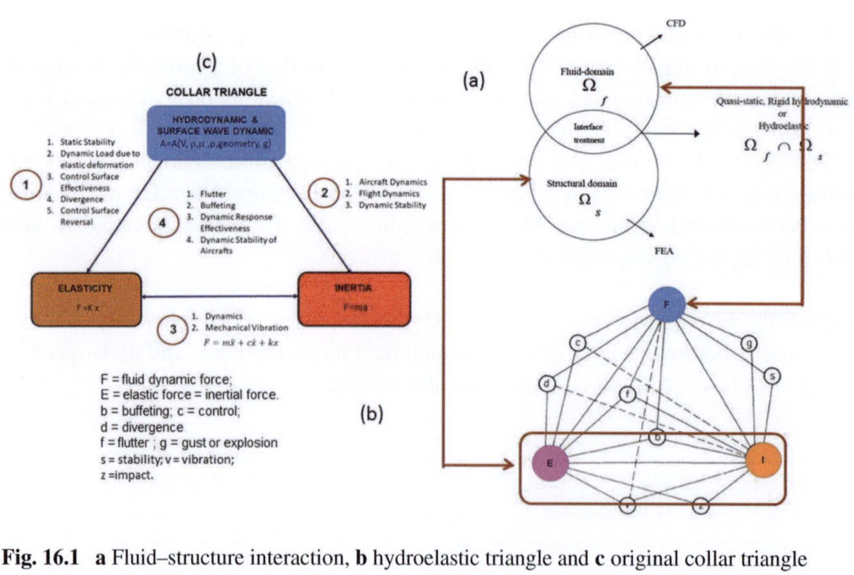

Fig. 16.1 **a** Fluid–structure interaction, **b** hydroelastic triangle and **c** original collar triangle

certain extent and are continuously being improved, catastrophes cannot be prevented from happening as a result of excessive wave excited hull responses of ships in rough seas.

Methods based on existing rules applied for the prediction of wave-induced loads have been proven to be reasonably adequate in dealing with problems arising and relying on empirical rules supported by service experience. With the increasing market demands for more slender, higher speed open deck ocean going carriers and the continuously updated unconventional multihull designs, the advantage of assessing the usefulness and applicability of alternative load prediction tools in ship design cannot be overemphasized.

The degree of interaction between structural response and hydrodynamic loads, the ship function and structural design characteristics undoubtedly determines the practicality and complexity of alternative engineering design assessment procedures. Sophisticated alternatives suggest that the system may be assessed within frequency or time domains, where the structure can be treated either as a rigid or flexible body by means of the finite element method and structural dynamic techniques. Along these lines, hydroelasticity could provide a preferred and valid option since it has arisen from the need to assess the flexible symmetric and antisymmetric dynamic behavior of floating, flexible structures in regular waves and irregular seaways using a fluid–structure interaction model[1] encompassing the physical properties of both fluid and structure.

Many recent papers, without underestimating the effectiveness of current techniques, utilize the theoretical background and applicability of hydroelasticity for the identification of global dynamic loads of long slender monohull and high-speed multihull vessels. State-of-the-art technological developments and fundamental theoretical formulations are still essential in understanding the general state of affairs. However, emphasis should be given to the description of general, flexible, fluid–structure interaction modeling procedures in an attempt to highlight the advantages and limitations of alternative idealizations that could be employed in preliminary or detailed design stages. Hence, improved communication between academics and practicing specialists concerned with hydrodynamics and global strength assessment of marine structures is relevant.

[1] Such as elaborated in Chap. 24, BEM-FEM Coupling for Acoustic Effects on Aeroelastic Stability of Structures.

16.2 Definition and Scope of Hydroelasticity

As a generic definition, hydroelasticity can be defined to be the branch of science concerned with the interactions of deformable bodies with the water environment in which they operate, which is also depicted in Fig. 16.1.

Hydroelasticity can further be defined as a branch of science concerned with the motion and distortion of deformable bodies responding to environmental excitations in the sea [4]. Figure 16.1 depicts hydroelasticity phenomena as it evolves and is modified from the original Collar's triangle of forces representing types of aeroelastic effects. The discipline is concerned with phenomena involving interaction between inertial, hydrodynamic and elastic forces [5]. By way of introduction and to have a comprehensive impression of hydroelasticity of ship, the work of Senjanovic et al. [3] is chosen, and some pertinent points and examples are presented.

Hence hydroelasticity implies, among others:

- A branch of science concerned with the motion and distortion of deformable bodies responding to environmental excitations in the sea [4].
- A discipline concerned with phenomena involving interaction between inertial, hydrodynamic and elastic forces [5].
- According to Heller and Abramson [5]: the naval counterpart to aeroelasticity—the fluid pressure acting on the structure modifies its dynamic state and, in return, the motion and distortion of the structure disturb the pressure field around it.[2]

Hydroelasticity of ships was brought to the attention of the Naval Architecture community in the 1970s through the work of Bishop and Price [6], culminating with the publication of the synonymous book in 1979 which elaborates comprehensive reviews of advances in ship hydroelasticity. Chen et al. [4], Hirdaris and Temarel [2] are helpful references in this field. ISSC reports regularly review advances in numerical approaches, model tests and full-scale measurements with hydroelastic effects included.

Figure 16.2 depicts a schematic of Naval Hydrofoil Ships and Fish hydrodynamics to provide a preliminary impression of some practice in Naval architecture and ocean engineering and problems associated with this field, which are closely related to mankind culture, environment and effort to learn from nature, such as fish hydrodynamics. Figure 16.3 depicts a nutshell summary of the scope of hydroelasticity of ships, as elaboration and interpretation of Fig. 16.1, while Fig. 16.4 summarizes model testing for ship and a schematic classification of ship hydroelasticity testing as assessment and learning process in ships hydroelasticity. One also may note that ships hydroelasticity, in addition to its relationships to aeroelasticity, may be related

[2] As also stipulated by Spyros Hirdaris in [1] as a generic definition, hydroelasticity is the branch of science concerned with the interactions of deformable bodies with the water environment in which they operate. Hydroelasticity as the naval counterpart to aeroelasticity recognizes that at fluid–structure interaction level significant differences may exist between the hydrodynamic, inertia, and elastic forces experienced by a floating marine structure. In other words, the fluid pressure acting on the structure modifies its dynamic state and, in return, the motion and distortion of the structure disturb the pressure field around it.

Fig. 16.2 Schematic of hydrofoil ship,[3] reflecting hydroelasticity and fish hydrodynamics

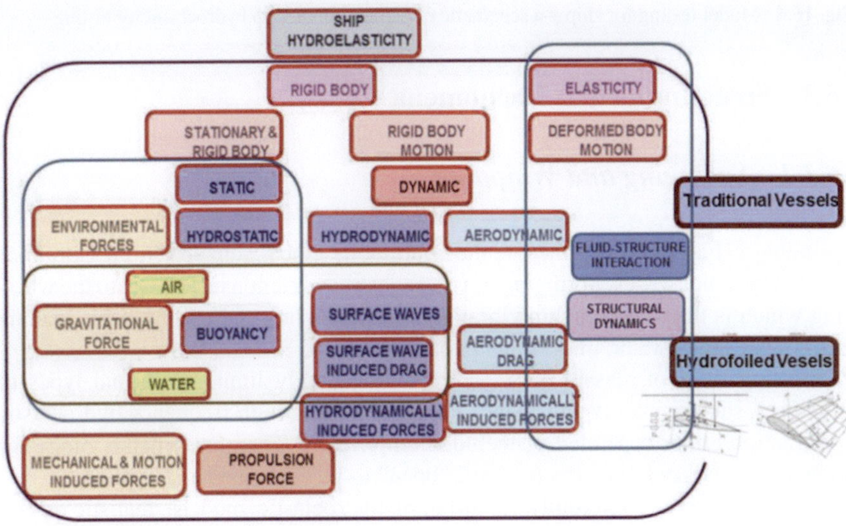

Fig. 16.3 Scope of hydroelasticity of ships

also to fluid–structure interaction, taking advantage of the progress in computational fluid dynamics and computational technology. With such introduction and elaboration on hydroelasticity, the topics in the present chapter will be elaborated in some colorful fashion.

---

[3] Hydrofoil Ship Image courtesy of unsplash.com. https://unsplash.com/photos/vijlBhahDk8.

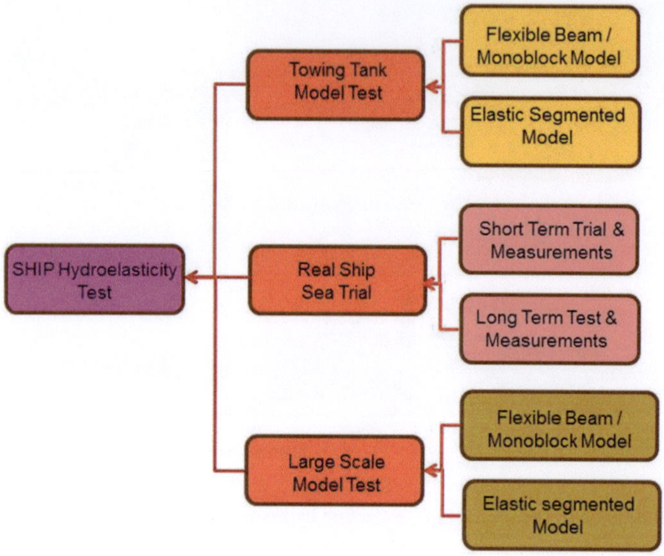

Fig. 16.4 Model testing for ship—a schematic classification of ship hydroelasticity testing

16.3 Some Important Phenomena

16.3.1 Springing and Whipping

Springing [3] is usually defined as the continuous global ship structural vibrations induced by water waves. Springing is a resonant phenomenon in contrast to the whipping which is the transient ship vibrational response induced by impulsive loading (slamming, green water, underwater explosion, etc.).

Existing rules of classification societies cover only limited size and types of structures, such as mainly quasi-static approach, where high-frequency hydroelastic contribution either neglected or included empirically. Therefore, methodology for inclusion of hydroelastic effects is still "open" to be developed and applied.

These may cover reliability of different hydroelastic models, realistic operational profile and statistical post-processing of analytical approaches, which should also include extreme loading and fatigue cases. Design methodology within direct calculation approach should not contradict the existing rule values for existing ships.

16.3.2 Research Techniques

As in other engineering fields, the following research techniques and methods can be resorted to in hydroelasticity.

- **Model tests**
  - Expensive
  - Limited number of cases
  - Problem of similitude (hydroelasticity, viscosity…)
- **Numerical simulations**
  - Numerical modeling difficulties
  - Lack of full validation
  - CPU time
- **Full-scale measurements**
  - Limited number of operating conditions
  - Difficulties related to the measurement of the sea states
- **Overall**
  - Selection of the representative conditions (ship speed, loading conditions, scatter diagram, probability levels).

Figure 16.2 depicts a typical ship with hydrofoils, which can exhibit observable hydroelastic phenomena. Exemplifying some relevant phenomena in hydroelasticity, Fig. 16.4 summarizes the scope of the model testing for ship as elaborated in the referenced paper, Fig. 16.5 exhibits numerical model and simulation of hydrostructure interactions while Fig. 16.6a schematic of progressive solving of hydroelastic problem at different levels of complexity, accuracy, problem solving efforts and delivery time. These figures serve to exemplify the scope and depth of ship hydroelasticity.

16.4 Global Hydroelastic Response of LNG Ships

To illustrate the analytical and experimental treatment of hydroelastic problem for LNG, ship, the following selected graphical results are taken from Senjanovic work [3].

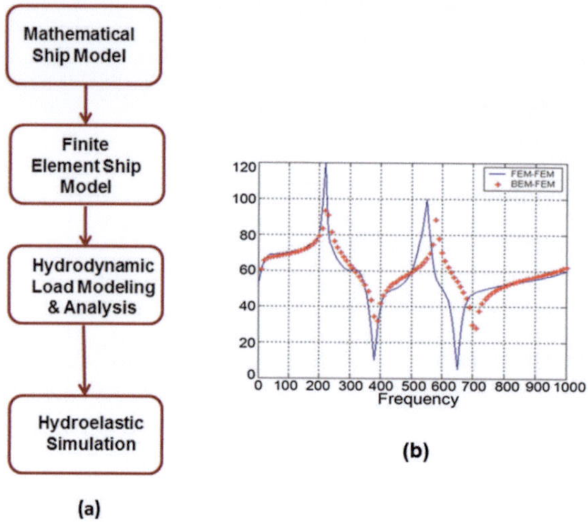

Fig. 16.5 Numerical model and simulation of hydrostructure interactions; **a** modeling sequence; **b** qualitative impression of hydroelastic simulation[4]

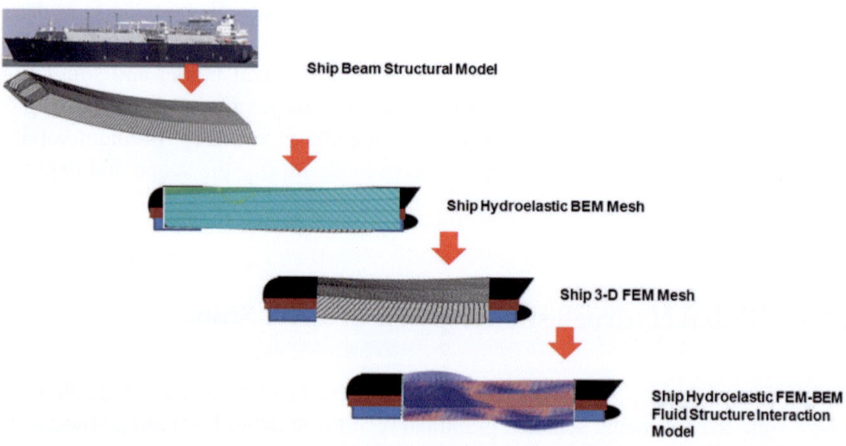

Fig. 16.6 Schematic of progressive solving of hydroelastic problem at different levels of complexity, accuracy, problem solving efforts and delivery time[5,6]

[4] Created and adapted from data and information from other resources, among others Senjanovic et al. [6] (with the authors' permission) and Malenica and Derbanne [7].

[5] Created and adapted from data and information from various resources, among others Senjanovic et al. [3] (with the authors' permission) and other considerations.

[6] Oceangoing LNG Vessel Picture Courtesy of stocksnap; https://stocksnap.io/search/Oceangoing+and+LNG+Vessels; shutterstock.com 1142981348.

The work elaborated was a joint development project with the goal to develop hydroelastic model for ships with internal liquid (LNGC, Tankers, etc.). The LNG ship was modeled as a beam structural model and a 3D-finite element (3DFE) structural model. Some impressions of results of the work elaborated in reference obtained using the said example ship, using beam hydroelastic model and 3D FEM hydroelastic model and describe semi-analytical solution modeling and validation procedures as illustrated in Figs. 16.6, 16.7 and 16.8. Detailed elaboration of the analysis, for example, can be found in Senjanovic et al. [3] and Hirdaris et al. [8]

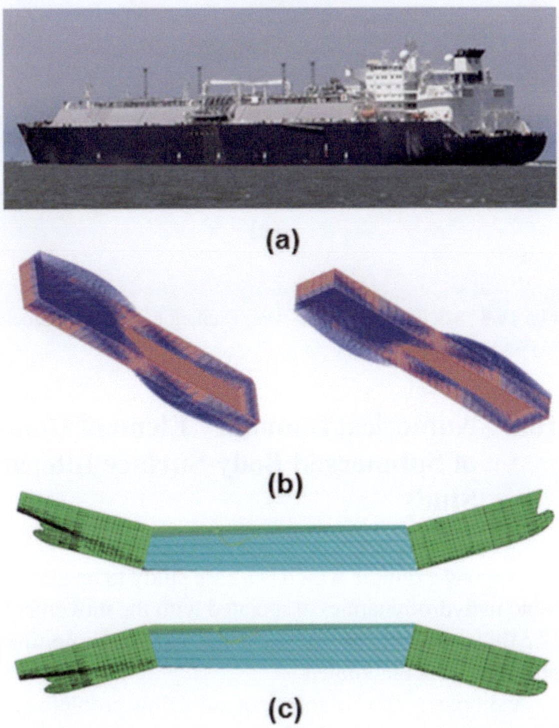

Fig. 16.7 Impression of **a** a typical oceangoing or LNG vessels, **b** and **c** schematic illustrative example of its simulated model dynamics[7,8]

[7] The figure has been created and adapted using data and information from various online resources, among others Spyros Hirdaris et al. [8] and (a) and (b) from https://stocksnap.io/search/Oceangoing+and+LNG+Vessel/.

[8] Oceangoing LNG Vessel Picture Courtesy of stocksnap; https://stocksnap.io/search/Oceangoing+and+LNG+Vessels; shutterstock.com 114298134.

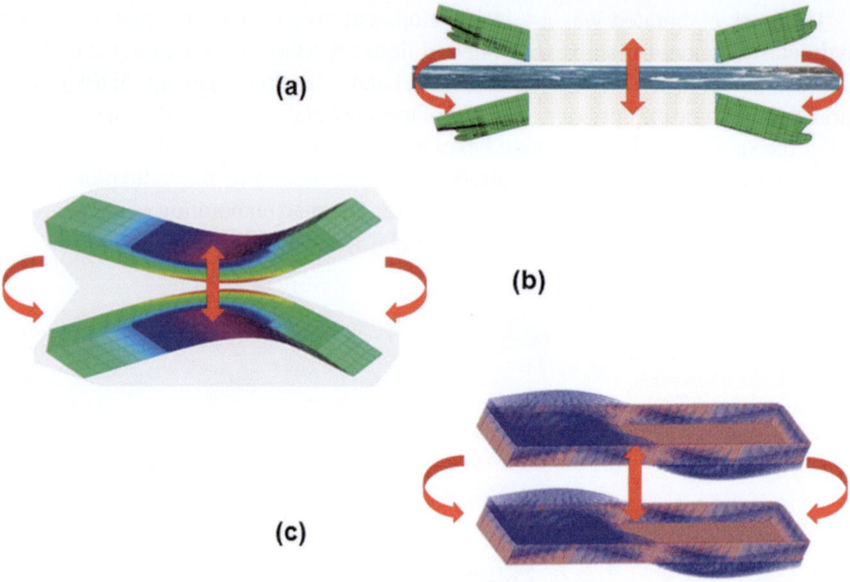

Fig. 16.8 **a** Schematic of ship-wave excitation, **b** beam hydroelastic BEM model, **c** beam structural hydrodynamic FEM

16.5 Numerical Boundary Element Computation of Submerged Body-Surface Interaction—A Case Study

The second example which is a case study to be elaborated will address fundamental topic in hydrodynamics associated with the movement of a sphere under the surface. For this purpose, the work carried out by Djojodihardjo et al. [9] concerning the above topic is elaborated.

A simple method to solve potential flow problem of submerged body-surface wave interaction is presented. The equation governing flow below the surface is Laplace equation. The boundary condition on the body is of Neumann type, while on the free surface the nonlinear dynamic free surface condition has to be satisfied. In addition, far-field and radiation condition, which determine the behavior of wave going to infinity, have to be satisfied. Taking advantage of potential flow model, then Green Identity is used to transform the problem of Laplace equation with nonlinear boundary condition into a nonlinear problem in its boundary. To satisfy the nonlinear dynamic free surface condition, Newton iteration is employed. The scheme for relaxation is obtained by expanding the nonlinear dynamic free surface condition to obtain correction terms to the velocity potential.

Truncation up to linear terms results in a simple scheme. Since the dynamic condition is valid on the unknown location of the free surface, in each iteration step the position of the free surface where the equation is applied is corrected using

Bernoulli equation. In this manner, dynamic free surface condition is satisfied exactly at the collocation points. Results for two- and three-dimensional non-lifting problem are presented as examples.

The problem of submerged body-surface wave interaction is relevant in many practical applications, such as in the prediction of the forces when the body is subject to certain flow conditions in the design of off-shore structures and ship hulls. Of great interest is the drag component attributed by the free surface wave induced by the motion of the submerged structure.

The resistance of a body experiencing relative motion with respect to the otherwise undisturbed flow comprises two components, viscous and wave resistances, which may be assumed to be independent of each other [10]. Proceeding along this assumption the viscous resistance depends only on Reynolds number and the wave resistance is governed only by Froude number. Consequently, the calculation of two components can be carried out separately. Furthermore, if the occurrence of wave braking, bow vortices and other flow details are not taken into account, the irrotational condition of the flow is preserved and a velocity potential field may be defined. Therefore in the present work attention is focused on the reduction of the problem of the determination of wave-induced force or wave resistance to the solution of incompressible potential flow around moving submerged body with free surface condition [10].

This paper deals with the solution of potential flow around a submerged body which generates waves on the free surface. This effort is motivated by the need to have a simple accurate computational tool to calculate wave resistance on moving submerged hulls. The problem of potential flow around submerged body represents a class of problems in fluid mechanics where the governing equation is the Laplace equation and in addition to the boundary conditions on known surface of the body and in the far-field, the presence of free surface above the body needs to be resolved. Particular attention should be devoted to the notion that the geometry of the free surface is known. In addition to the kinematic boundary condition, the pressure must be a known constant at the free surface. These two conditions constitute a nonlinear boundary condition known as the dynamic free surface condition. The dynamic free surface condition must be satisfied at the water surface whose location is unknown and is a part of the solution.

The lifting case would be the most general problem of practical interest. However, since the non-lifting case would be a simpler one, this problem will first be worked out. The result will be useful for validation purposes and to gain more insight into the numerical complexities of the problem. This will hopefully assist further effort to tackle the lifting cases.

The application of non-lifting method will be restricted to a symmetric case without slip stream. Similar approach may be followed in the hydrodynamic interaction between off-shore structures and surface waves. To proceed toward this end, the linear method of Dawson [11] is extended to satisfy complete nonlinear boundary condition on the exact free surface location. Dawson uses double model potential as the base potential and takes undisturbed free surface as the approximate free surface. In this work, extension of Green identity approach to potential problem [12]. Jensen and Soding [13, 14] worked out this problem, and a simple relaxation scheme· is used to solve the correct potential and surface position iteratively.

16.5.1 Basic Problem Analyzed

As a basic problem, steadily moving submerged body, with constant velocity U and zero slip stream, in an inviscid undisturbed deep water is considered. A coordinate system moving with the body is used. In this case, steady-incompressible flow will be used for analysis. Since no irrotationality is present, a scalar field can be defined to represent velocity vector as

$$V = \nabla\Phi \tag{16.1}$$

The mass conservation equation for steady potential flow can then be written as

$$\nabla^2\Phi = 0 \tag{16.2}$$

This Laplace equation is the governing equation for the flow below free surface. It can also be applied for the perturbation potential Φ.

The Neumann condition of zero normal velocity on the surface boundary of the body can be written as

$$\nabla\Phi \cdot \mathbf{n} = 0 \tag{16.3}$$

On the free surface, two conditions, no flow through the moving free surface and constant pressure condition, have to be satisfied. Taking into account the velocity of free surface, the first condition reads

$$\frac{\mathrm{D}\zeta}{\mathrm{D}t} = \Phi_z \tag{16.4}$$

where ς is surface elevation. For steady flow this expression is

$$\nabla\Phi \cdot \nabla\zeta = \Phi_z \tag{16.5}$$

Applying Bernoulli equation on the free surface and requiring that the pressure on the free surface is atmospheric, one obtains

$$\frac{1}{2}\left[(\nabla\Phi)^2 - U^2\right] + gz = 0 \tag{16.6}$$

This equation and the kinematic condition may be recast into one equation, eliminating z in the last equation.

$$\frac{1}{2}\nabla\Phi \cdot \nabla(\nabla\Phi)^2 + g\Phi_z = 0 \tag{16.7}$$

The last equation is known as the nonlinear dynamic free surface condition. This equation is valid only on the correct location of the free surface. The radiation condition dictates that the waves behave properly, namely propagate only downstream. In three-dimensional case, the waves propagate in a sector. At a distance far away from the body, disturbances vanish:

$$\nabla\Phi = (U, 0, 0) \tag{16.8}$$

The problem now becomes a boundary value problem of Laplace equation with a nonlinear boundary condition, Eq. 16.7, and other boundary conditions, Eqs. 16.3 and 16.8 on the internal and external boundaries, respectively.

16.5.2 Green Identity Formulation

The notion that the governing equation has the form of Laplace equation makes possible the use of Green Identity transformation. For the particular problem considered, Green identity formulation for three-dimensional case in perturbation potential ϕ can be utilized [12, 15]; hence:

$$4\pi\phi = \iint \left[\frac{\partial\phi}{\partial n}\frac{1}{r} - \phi\frac{\partial}{\partial n}\left(\frac{1}{r}\right)\right] \mathrm{d}S \tag{16.9}$$

where x is in domain V and S is its boundary. This transformation turns the field problem into problem on its boundary. The physical interpretation is that potential in the field can be represented by potential of distributed sources and doublets on the boundary. The normal derivative of potential in the first term and potential in the second term can be treated as the strength of the singularities. It should be noted that source and doublet satisfy far-field boundary condition.

Computational aerodynamics code capitalizing on this approach has been developed by the first author for nonlinear as well as linear cases [12, 16]. This technique will now be used in this particular problem; it should be noted that the boundary condition on the free surface is implicit, and hence making the problem highly nonlinear.

Several possibilities can be followed in the use of Eq. (16.9). For non-lifting problem, one can use sources singularities only. It has been shown [12] that doublet singularities can be used in either thickness or lifting problems, thus possessing more advantageous characteristics than sources or vortices.

In this particular study, source singularities will be used to represent the non-lifting case. It is considered to be simple and instructive and serves to demonstrate the plausibility of the approach. Later work will be devoted to the treatment of general approach.

Let us define the source strength as u, then

$$\phi(x) = \frac{1}{4\pi} \iint \sigma \frac{1}{r} \mathrm{d}S \tag{16.10}$$

Now consider the domain of free surface problem. From the physical consideration, distributing source at the boundary seems to be able to satisfy only the body boundary condition. Dawson [11] uses double model to be able to satisfy the linearized free surface condition. The drawback using this approach for a more complex free surface condition is the difficulty found in isolating singularities on the free surface. Another way to overcome the problem is by keeping the source distribution a distance above the free surface [13, 14]. It is noted that this technique inconsistent with the Green Identity approach. On the other hand, Jensen found that the use of only point sources above the free surfaces satisfactory accuracy. Therefore similar approach will be adopted. Our computational study also indicates that the distance between source points and control points turned out not to affect the free surface height,

The source integral over the boundary is difficult to analyze since the shape of boundary can be arbitrary. Following the approach of panel method, this integral is divided into elements and u is assumed to be constant on each elements.

$$\phi(x) = \sum_i \sigma_i \quad \varphi_i(x) \tag{16.11a}$$

where

$$\varphi_i(x) = \frac{1}{4\pi} \iint\limits_{S_i} \frac{1}{r} \mathrm{d}S_i \tag{16.11b}$$

The integral on the right-hand side of the equation is taken over each surface element (panel) and depends only on geometrical data. In this study, the shape of the element is taken to be planar quadrilateral, but better approximation can readily be produced.

Defining the body boundary and nonlinear dynamic free surface conditions as $N(\Phi) = 0$ and $F(\Phi) = 0$ respectively, the substitution of the potential by the expression due to source distribution yields

$$N(\sigma) = 0 \tag{16.12a}$$

and

$$F(\sigma) = 0 \tag{16.12b}$$

These two equations comprise a new boundary value problem with source strength as the unknown. Since $F(u)$ is nonlinear, proper treatment should be used to solve it.

16.5.3 Solution of the Nonlinear. Free Surface Condition

Standard method to solve a nonlinear equation is by using relaxation. The residual of the equation is relaxed step by step using linearized equation. The linearized equation does not influence the accuracy of the solution, but it determines the convergence of the iteration. Here, Newton method is used to find the linearized equation.

First, consider a condition in an iteration step. Approximation to Φ is obtained from previous iteration step, or from other solution:

$$F(\Phi) = \frac{1}{2}\nabla\Phi \cdot \nabla(\nabla\Phi)^2 + g\Phi_z \tag{16.13}$$

A better approximation to Φ is found by substituting $\Phi + \mathrm{d}\Phi$ into $F(\Phi)$, truncating up to first order, and forcing the equation to be zero:

$$\frac{1}{2}\nabla \mathrm{d}\Phi \cdot \nabla(\nabla\Phi)^2 + \nabla\Phi \cdot \nabla(\nabla\Phi \cdot \nabla \mathrm{d}\Phi) + g\mathrm{d}\Phi_z = -F(\Phi) \tag{16.14}$$

If the new solution is Φ', so as $\mathrm{d}\Phi = \Phi' - \Phi$, a linearized equation for a better solution is obtained as:

$$\frac{1}{2}\nabla\Phi' \cdot \nabla(\nabla\Phi)^2 + \nabla\Phi \cdot \nabla\left(\nabla\Phi \cdot \nabla\Phi'\right) + g\Phi'_z = \nabla\Phi \cdot \nabla(\nabla\Phi)^2 \tag{16.15}$$

This equation is solved in every iteration step together with equation on the body surface. The equation of dynamic free surface condition is valid on the unknown free surface location. Since this location is unknown, an approximation must be taken, and every iteration step is ·corrected using

$$z = \frac{1}{2g}\left[U^2 - (\nabla\Phi)^2\right] \tag{16.16}$$

The initial solution of if > can be chosen as potential for double model, free-stream potential or potential from linear solution. The initial location of free surface is usually taken as its undisturbed position.

It can be noted here that due to the movement of the position where Eq. 16.15 or Eq. 16.14 is applied in the next iteration, the correction to potential ϕ is not an accurate one since the expansion was performed only in the potential variable. In Jensen's approach, the variation of the potential is also dependent on dz. This results in a much more complex expression and higher derivatives. In this particular study, no expansion in dz was carried out to keep the method simple. The result is found to be accurate since the surface where the dynamic free surface equation applied is corrected in every iteration step.

16.5.4 Computational Detail

In this section, the detail of computation is presented. Two-dimensional case is considered here for the sake of clearness. Three-dimensional case will only need minor modifications. Equation 16.15 can be appreciated better by writing it in expanded form:

$$\begin{aligned}&2(\Phi_z\Phi_{xx}+\Phi_x\Phi_z\Phi_{xx})\Phi'+2\Phi_x\Phi_z\Phi'_{zz}+2\Phi_x^2\Phi'_{xx}\\&+2\left(\Phi_x\Phi_{xz}+\Phi_z\Phi_{zz}+\frac{g}{3}\right)\Phi'_z+\Phi_z\Phi'_{zz}=\mathrm{RHS}(\Phi)\end{aligned}\tag{16.17}$$

To simplify the notation, Eq. 16.17 is rewritten in a more convenient form.

$$A(\Phi)\Phi'_x+B(\Phi)\Phi'_{xz}+C(\Phi)\Phi'_{xx}+D(\Phi)\Phi'_z+E(\Phi)\Phi'_{zz}=\mathrm{RHS}(\Phi)\tag{16.18}$$

Now, using expression of perturbation potential due to source distribution on the body and source points above the free surface, one can write

$$\Phi'(x)=Ux+\sum_i\sigma_i\quad\varphi_i(x)\tag{16.19}$$

Equation 16.18 is applied at the same number of points on the body and on the current location of free surface:

$$\sum_i\sigma_i\left[A(\Phi)\Phi'_x+B(\Phi)\Phi'_{xz}+C(\Phi)\Phi'_{xx}+D(\Phi)\Phi'_z+E(\Phi)\Phi'_{zz}\right]=\mathrm{RHS}(\Phi)\tag{16.20}$$

In each iteration step, this equation is solved together with equation on the body surface for slip condition:

$$\sum_i\sigma_i\left[n_x\varphi_x+n_z\varphi_z\right]=-U\cdot n_j\tag{16.21}$$

to obtain source strength. Using source strength data, the potential, residual and new approximation for surface location can be calculated using Eqs. 16.19, 16.13 and 16.16, respectively. Note that in some cases, under-relaxation the iteration can be stabilized. This is carried out as

$$\sigma^{n+1}=(1-\omega)\sigma^n+\omega\sigma^{\mathrm{new}}\tag{16.22}$$

where ω is the relaxation factor; $\omega < 1$ means under-relaxation and $\omega > 1$ means over-relaxation. The iteration is stopped after certain tolerance has been satisfied by the residual or change in surface elevation.

16.5.5 Two-Dimensional Results

Two-dimensional cases considered here are coded using FORTRAN on scalar machines such as HP 9000 and IBM R6000. The most CPU-consuming part of the method is the calculation of panel integral. In this case the integral on a line segment is worked out using five points Gauss Legendre quadrature.

Submerged vortex was used to compare this method with the others. This case was also used by Salvesen and Kerczek [17], Dawson [11], Jensen and Soding [13, 14]. Two cases of $\gamma = -2.70$ ft/s and $\gamma = 2.70$ ft/s are considered. The negative circulation problem, Fig. 16.9a, showed a nice convergence rate, and the iteration were stopped after the residual reached −4.00 on log-scale. The elevation changes also converged in the same manner. This means that the exact nonlinear dynamic free surface condition is satisfied at the correct position of the free surface. This proved that the idea of having a simple relaxation scheme, instead of the accurate one, is plausible. The case of positive circulation, Fig. 16.9, on the other hand, showed some instability on the relaxation.

The surface change had to be damped with under-relaxation to keep the iteration converging. This reconfirmed the problem which was also faced by Jensen and Soding [13, 14]. The comparisons were made for both problems with nonlinear results of Jensen and Soding [13, 14]. Comparisons with Dawson's linear method [11] are shown in Fig. 16.10.

16.5.6 Three-Dimensional Case

The three-dimensional case poses a more difficult problem in fulfilling the radiation condition. This is carried out by misaligning the control points on the free surface and the source points above them. The code for three-dimensional cases was developed on CRA YY-MP 8/64 machine. This machine has 8 processors, and each processor has a 64-array vector register. This method has promising possibility to be vectorized or processed in parallel fashion, since the most CPU-consuming part is the integral on each surface element.

The integral on each element is a process which can be carried out separately for each element, hence it can be treated by parallel processors. Careful treatment of the loop in the element integral and the use of numerical integration made possible the gain execution time to a factor of 80.

The results in Figs. 16.11a and b show the cases where radiation condition was not satisfied and satisfied, respectively. This was a problem of submerged doublet.

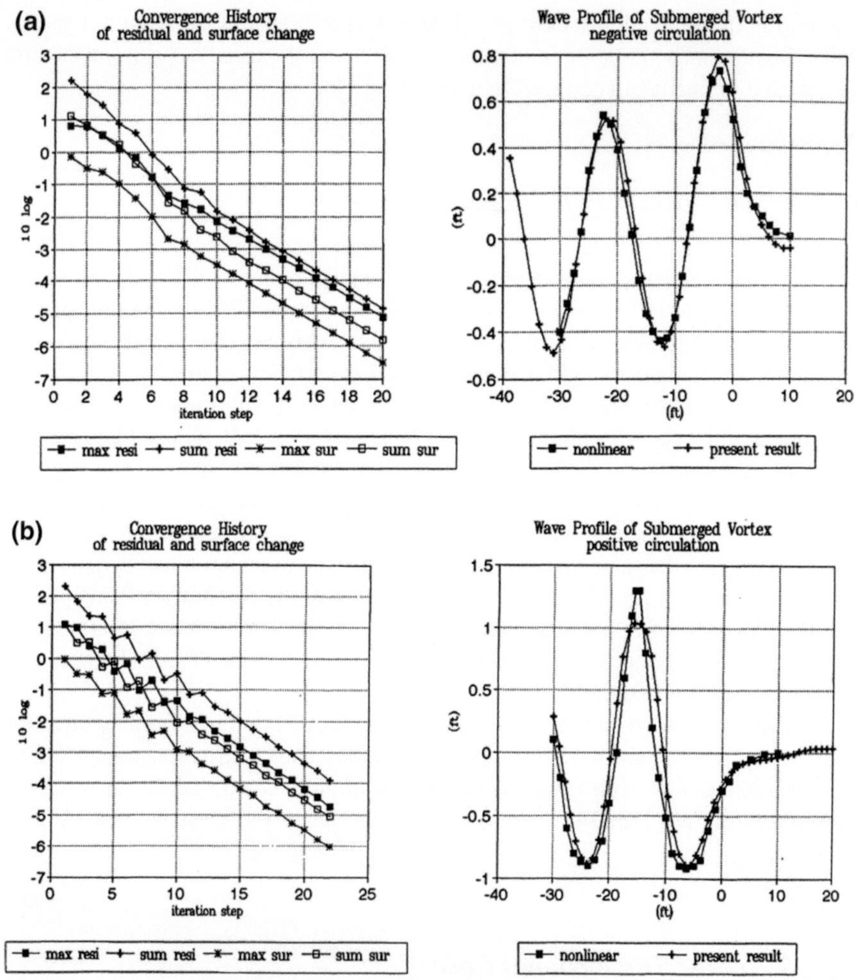

Fig. 16.9 **a** Convergence history and wave profile of submerged vortex located 4.5 ft below the surface. $\gamma = -2.70$ ft/s. **b** Convergence history and wave profile of submerged vortex located 4.5 ft below the surface. $\gamma = 2.70$ ft/s

Figure 16.11a, b and Fig. 16.12a, b proved the usefulness of the simple procedure by misaligning the source points and control points.

Other results for submerged sphere and ellipsoid are shown in Fig. 16.13a and b. The velocity vectors on the surface of an ellipsoid are shown in Fig. 16.14. Figures 16.15 and 16.16 show two cases of submerged ellipsoid at two different Froude numbers. The sector angle where the wave exists seems to be the same; hence it is believed that this is a correct tendency.

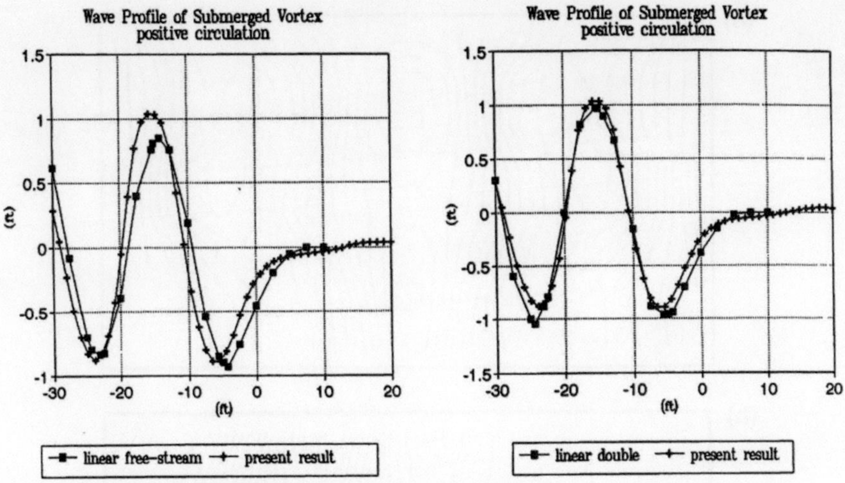

Fig. 16.10 Comparison with linear method of Dawson, left figure is free-stream linearized and right double model approximation

16.5.7 Closing Remarks

What has been elaborated above exemplifies the development of a simple method to solve nonlinear free surface problem for submerged body. The method uses a simple relaxation scheme, which results in expansion only in the velocity potential, to solve the nonlinear dynamic free surface condition. In two-dimensional case this method has worked out satisfactory. It should be noted that for the three-dimensional case further work must still be carried out to investigate the stability of the method. The stability boundary of the method seems to have lower values than Jensen's. This may be attributed to the fact that the gradient was not expanded in elevation.

16.6 Hydroelastic Equation of Motion, Dynamic Response and Stability

Governing equations in hydroelasticity in essence are similar to those in aeroelasticity, but other specific forcing function and their relevant sources may need special treatment. For example, the problem of cavitation is not addressed in aeroelasticity. Some of these specific topics will not be discussed in the present chapter, and resort should be made to relevant publications in the literature.

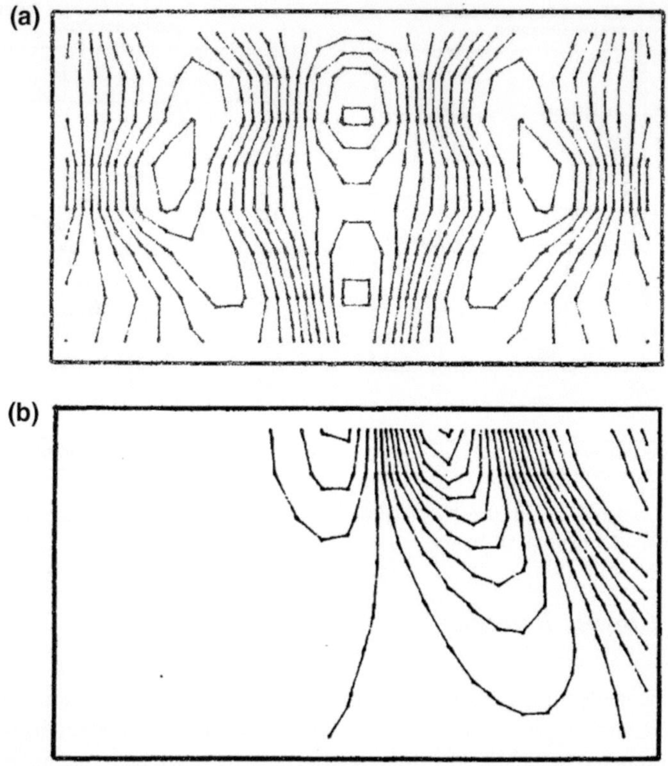

Fig. 16.11 **a** Incorrect radiation condition for a submerged dipole, the waves travel upstream instead of only downstream. **b** Correct radiation condition for a submerged dipole obtained by misaligning the free surface collocation and source points

In the present section, to provide some introductory examples in hydroelasticity, attention is given to the state of affairs and equation of motion of hydrofoils moving in incompressible and inviscid or viscous flow, by referring to Djojodihardjo [18], Chae [19] and Ducoin and Young [20], as well as solutions for stability and dynamic response. Most importantly, attention is given to the dynamics governing equations, fluid–structure interactions, basic equations for typical hydrofoil sections and some elaboration into more complex multidegree-of-freedom systems that require discretization in order to arrive at numerical solutions utilizing commercially available computational methods, such as ANSYS, CFD, CSD and CFX softwares [21].

The baseline governing linear hydroelastic equations to be utilized in the state of affairs and equation of motion of hydrofoils moving in incompressible and inviscid or viscous flow can be derived from the binary aeroelastic system (e.g. Bisplinghoff et al. 22; Zwaan 23; Wright and Cooper 24; Djojodihardjo and Yee 25):

$$m\ddot{\mathbf{h}} + mbx_\alpha\ddot{\alpha} + d_h\ddot{\mathbf{h}} + K_h\mathbf{h} = -\mathbf{L} \tag{16.23}$$

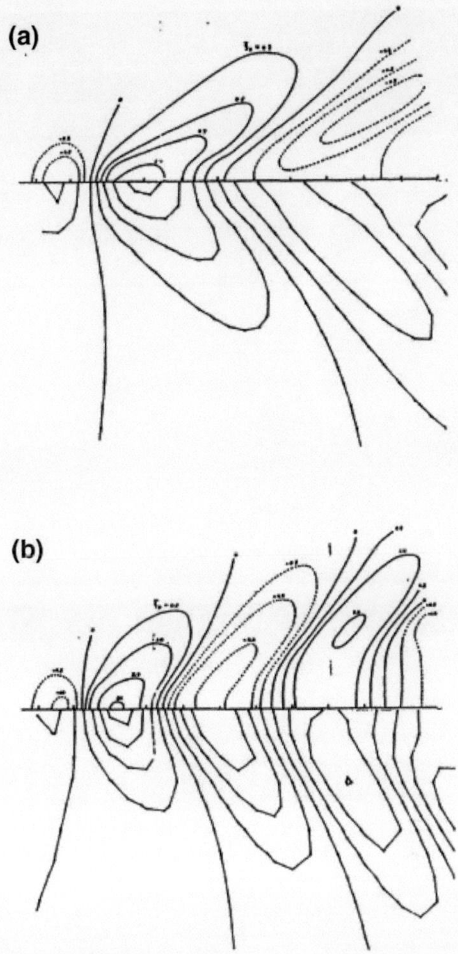

Fig. 16.12 **a** Comparison of linear result (lower half) with analytical result presented in Jensen and Soding [13]. $F = gd/U^2 = 0.50$. **b** Comparison of linear result (lower half) with analytical result presented in Jensen and Soding [13]. $F = gd/U^2 = 1.00$

$$mbx_\alpha \ddot{\mathbf{h}} + I_\alpha \ddot{\alpha} + d_h \dot{\alpha} + K_\alpha \dot{\alpha} = \mathbf{M} \tag{16.24}$$

where m is the hydrofoil mass per length (in the span direction), m_e is the fixture mass (connecting the hydrofoil to the plunge springs) per length, M is the hydrodynamic moment, L is the hydrodynamic lift and the over-dot represents differentiation with time.

Here the fixture mass (m_e) is defined for the case when the system slightly deviates from the ideal typical section depicted in Fig. 16.17 due to the masses of the shaft, spring mass and other attachments in real experiments (Sousa et al. 2011), while it is

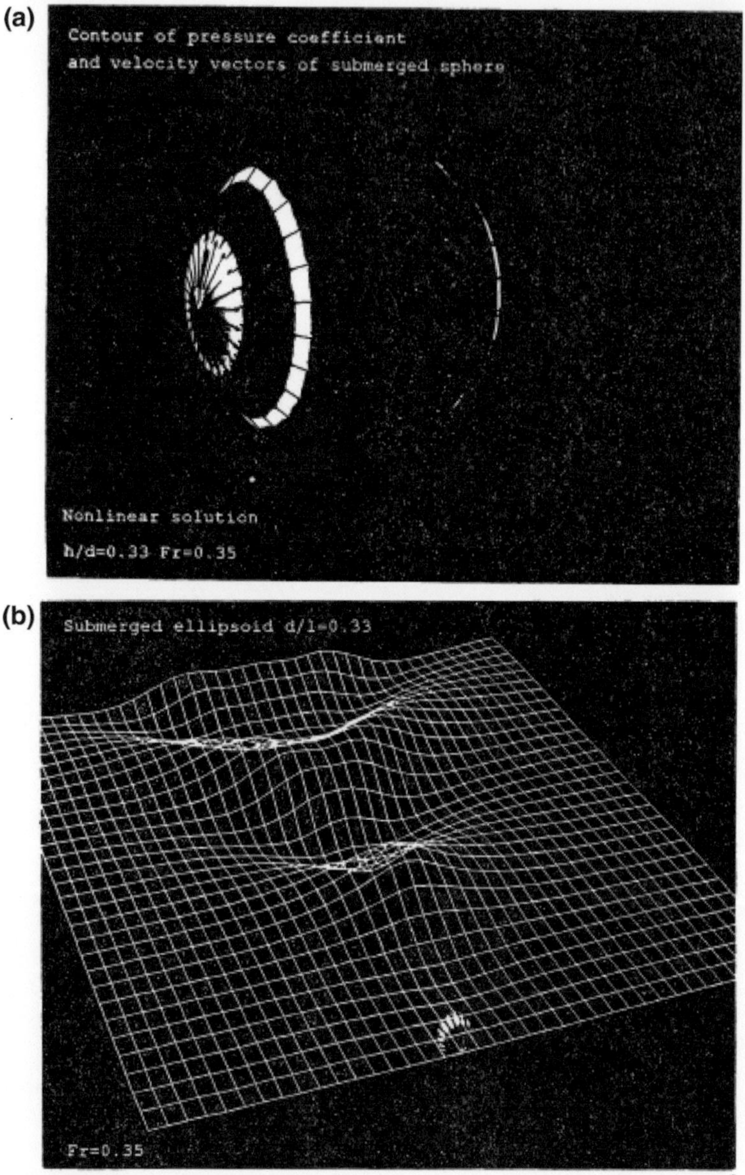

Fig. 16.13 **a** Wave induced by a submerged sphere at Froude number 0.35. **b** Wave induced by a submerged ellipsoid at Froude number 0.35

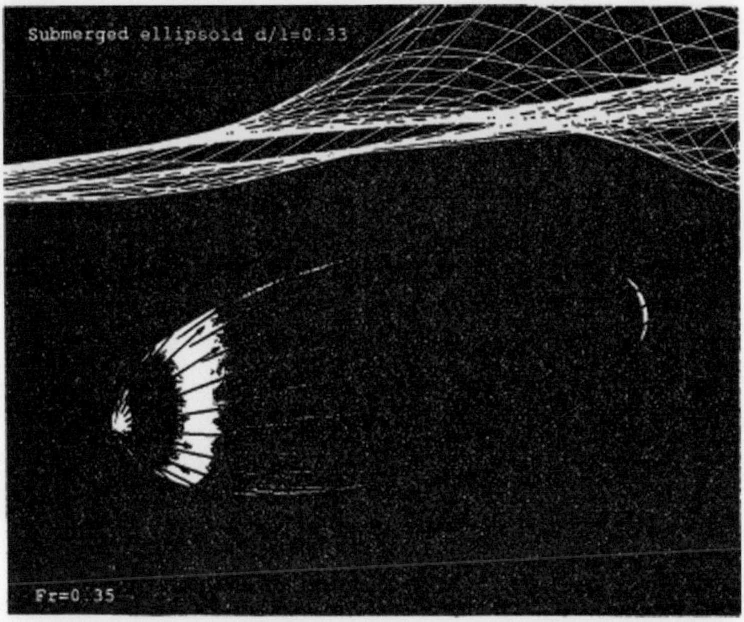

Fig. 16.14 Velocity vectors on the surface of the ellipsoid showing correct conditions

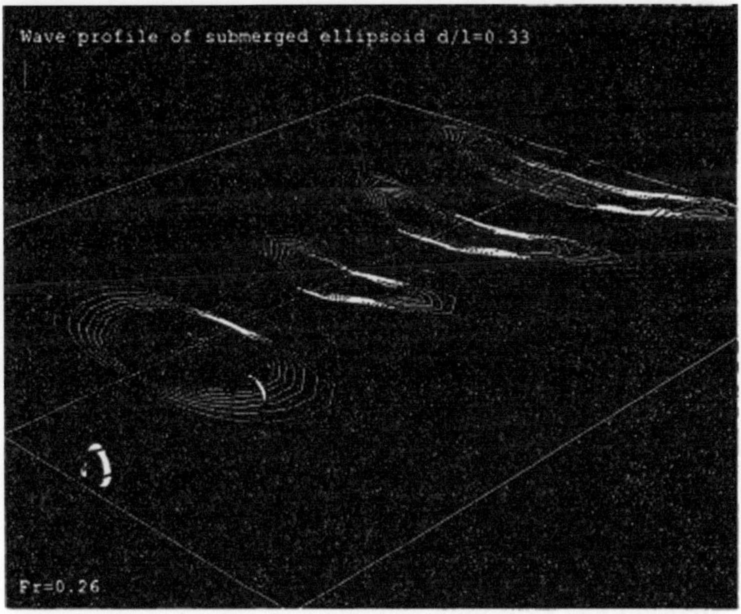

Fig. 16.15 Wave induced by submerged ellipsoid at Froude number 0.26

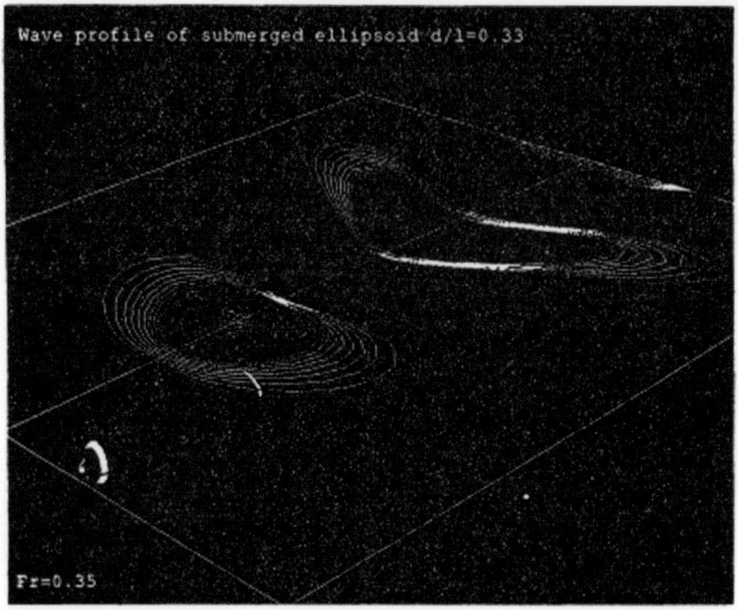

Fig. 16.16 Wave induced by submerged ellipsoid at Froude number 0.35

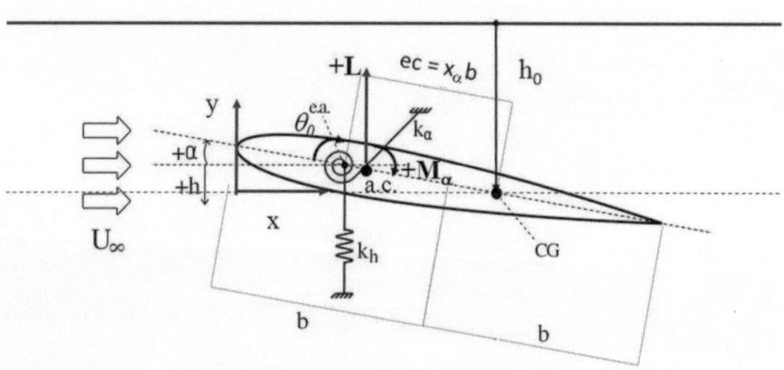

Fig. 16.17 2-DOF model representing plunging (h_0) and pitching (θ_0) degrees of freedom at the tip of a 3D, cantilevered, rectangular hydrofoil

zero in the ideal representation of Fig. 16.18, which is assumed here for the baseline solution.

16.7 Inviscid FSI Coupling Model

Using inviscid fluid–structure coupling model, Chae [19] develops the following approach. The inviscid FSI coupling model employs a time domain (TD) fully

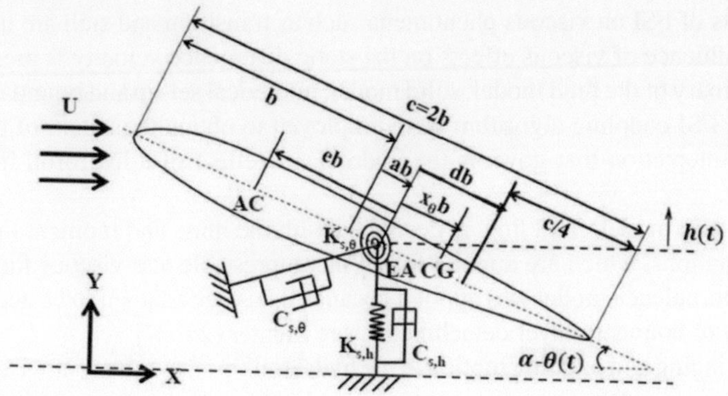

Fig. 16.18 2-DOF solid model of a cantilevered hydrofoil which spanwise bending and twisting

coupled (FC) method. The FC method is possible because the analytical representation of both the solid and the inviscid fluid forces are known. Employing finite element method, the discretized non-dimensional form can be recast from that written for a hydrofoil represented by a typical section as illustrated in Fig. 16.17 into a fluid–structure coupling for the hydrofoil represented by a set of mass (M), damping (d) and stiffness (K) matrices in a non-dimensional, fully coupled equation of motion (EoM)

$$\left(M_s + M^T\right)\ddot{\mathbf{x}}_n + \left(C_s + C^T\right)\dot{\mathbf{x}}_n + \left(C_s + C^T\right)\mathbf{x}_n = \mathbf{F}_{\text{static}} \tag{16.25}$$

where the subscript n is the time step index in an iterative computational approach that has been employed for his purpose.

The investigations of the effects of fluid–structure interaction on the lift and drag coefficients of a hydrofoil in viscous flow, as well as changes in laminar to turbulent transition, static stall in lift and divergence speed studied by Ducoin and Young [20] will be elaborated to some extent. These may be formulated as the translation of the basic 2D typical section hydroelastic governing equations for inviscid flow to inviscid flow [20] and to finite element model (FEM) for more general hydroelastic fluid–structure interaction (FSI) [19].

In their work, the hydrodynamic response of a flexible hydrofoil is simulated using the commercial CFD solver CFX the plunging and pitching motion at the foil tip due to spanwise bending and twisting deformation of a 3D, rectangular, cantilevered hydrofoil is simulated using a 2-DOF model. The fluid and solid solvers are coupled via a user-defined subroutine in CFX, which exchanges the fluid pressure field and foil motions between the solvers.

In particular, the flow over a NACA66 hydrofoil was studied for initial angles of attack range from $\alpha_0 = 0°$ to $8°$ and free-stream velocity range from $V_\infty = 5$ to 20 m/s. The fluid, solid and FSI coupling models are presented, followed by a brief description of the experimental set-up of the rigid and flexible hydrofoil studies. Comparisons of numerical predictions with experimental measurements are shown.

The effects of FSI on viscous phenomena such as transition and stall are discussed, and the influence of viscous effects on the static divergence velocity is presented.

A summary of the fluid model, solid model, numerical set-up and boundary conditions, and FSI coupling algorithm were employed to obtain the extent of the fluid–structure interaction that governs the hydroelastic effect of a hydrofoil in viscous flow.

In the fluid model, fluid flow is described with the mass and momentum conservation equations, which are applied for an incompressible and viscous fluid. The $k - \omega$ *SST* turbulence model was applied because it has been shown to be accurate for prediction of boundary layer detachments, see Menter [26, 27].

The plunging and pitching motion at the hydrofoil tip caused by spanwise bending and twisting of a 3D, rectangular, cantilevered hydrofoil is represented using a 2-DOF solid model, as shown in Fig. 16.18. The hydrofoil is assumed to be rigid in the chordwise direction; the spanwise bending, $h(z)$, and twisting, $\theta(z)$, deformations along the z-direction are assumed to be captured by the shape functions $f(z)$ and $g(z)$, respectively:

$$\mathbf{h} = \mathbf{h}_0 f(z) \tag{16.26a}$$

$$\boldsymbol{\theta} = \boldsymbol{\theta}_0 f(z) \tag{16.26b}$$

where h_0 is the vertical translation about the elastic axis. The hydrofoil tip motion is defined positive downward, and θ_0 is the rotation about the elastic axis at the hydrofoil tip defined positive in the clockwise direction, as shown in Fig. 16.18.

It should be noted that $f(z)$ and $g(z)$ are obtained by applying the hydrodynamic pressure distributions (computed via CFX) on to a finite element model of the 3D, cantilevered, rectangular hydrofoil and are found to be approximately the same for the range of angle of attacks of interest.

A partitioned FSI coupling method can be employed to faciliate the use of modular CFD and CSD solvers. Classically, partitioned FSI methods couple the two different CFD and CSD solvers via either the loosely coupled (LC) or tightly coupled (TC) techniques. The LC technique uses the CFD solution at the previous time step as an input to the CSD solution for the new time step, while the TC technique iterates between the solutions of the CFD and CSD solvers per each new time step. The discretized Equation of Motion non-dimensional viscous LHC method in a time domain is given by:

$$\left(M_s + M^T\right)\ddot{\mathbf{x}}_n + \left(C_s + C^T\right)\dot{\mathbf{x}}_n + \left(C_s + C^T\right)\mathbf{x}_n = (\mathbf{F}_{\mathrm{CFD}})_n - \left(\mathbf{F}_{\mathrm{FSI}}^T\right)_n \tag{16.27}$$

F_{CFD} is the non-dimensional viscous fluid force on the hydrofoil that is computed by the CFD solver, which includes both static and dynamic force components. The loose hybrid coupled (LHC) method subtracts the potential ow estimate of the hydroelastic force that depends on the ow-induced deformations, FT.

The 2-DOF solid model of a cantilevered hydrofoil which spanwise bending and twisting is exhibited in Fig. 16.18. The bending deformation, *h*, and the twisting deformation, θ, are defined as positive upwards and counterclockwise, respectively, with regard to the elastic axis (EA) of the flexible hydrofoil. The flow velocity is defined as U, the geometric angle of attack as α_0, and effective angle of attack as $\alpha_{\text{effective}} = \alpha_0 - \theta$. EA is located at a distance ab in the downstream direction from the mid-chord, where $b = c/2$ is semi-chord length. The aerodynamic center (AC), at which the pitch moment coefficient does not vary with the lift coefficient, is located at a distance *eb* in the upstream direction from the EA. The actual center of pressure, where the real lift acts, changes with operating conditions, particularly for thick and rounded nose sections common to many maritime applications.

The center of gravity (CG) is located at distance $x_\theta b$ in the downstream direction from the EA. To validate the solvers, numerical predictions are compared with experimental measurements of the rigid and flexible hydrofoil described above. The upper curves (red) in Fig. 16.19 show the comparison of lift coefficients (C_L) for the rigid hydrofoil, and the lower curves (blue) in Fig. 16.19 shows the comparison of tip section displacements (y) measured at the leading edge for the flexible hydrofoil. Good general agreements have been reported for the lift coefficient. Details about the flexible hydrofoil experimental set-up and results are elaborated in Ducoin et al. [28].

The aim of this section is to investigate the viscous FSI response of the flexible hydrofoil, including laminar to turbulent transition, stall, hydrofoil performances and deformations. To avoid confinement effects, the numerical predictions are obtained using infinite domain mesh.

The FSI solver was validated by comparing numerical predictions with experimental measurements of a rigid and a flexible hydrofoil conducted at the French Naval Academy, France. The results show that the flexible hydrofoil undergoes a clockwise rotation about the elastic axis because the center of pressure is to the left of the elastic axis, which in turn increases the effective angle of attack. The increase in effective angle of attack caused by the elastic deformation accelerates transition

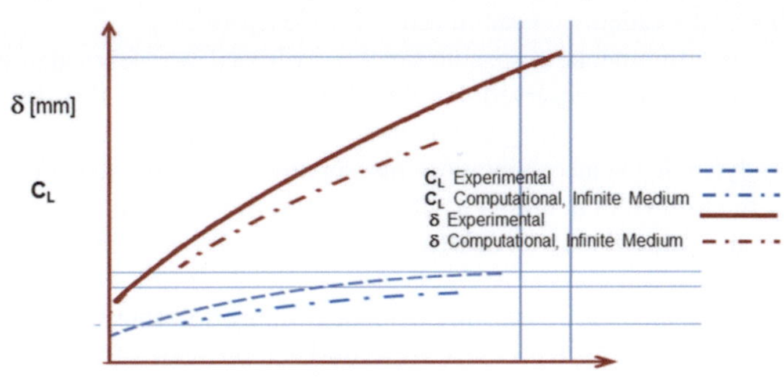

Fig. 16.19 Comparisons of the measured versus predicted lift coefficient for the rigid hydrofoil and deflection for flexible hydrofoil (qualitative) [28].

and stall, and changes the location of the center of pressure, all of which result in nonlinearities that have strong effects on the FSI response and static divergence.

16.8 Experimental Analysis of Hydroelastic Response of Flexible Hydrofoils

The next example to be elaborated will address the topic of experimental analysis of hydroelastic response of flexible hydrofoils, extracted from Astolfi et al. [29].

With the recent emphasis on high-speed foiling yachts, the interest of the sailing community for highly deformable loaded hydrofoils is enhanced further. Consequently, basic studies of flexible hydrofoils in a well-controlled flow situation as it can be done in a hydrodynamic tunnel are essential. The aim of this paper is to present an original experimental set-up developed at the French Naval Academy to analyze the hydroelastic response of flexible homogeneous hydrofoils for various flow conditions including cavitating flow.

It is based on the measurements of hydrodynamic forces using a hydrodynamic balance, structural stress by means of strain gauges embedded in the hydrofoil structure and vibration spectrum by vibrometry laser. Several physical observations are reported that should be very beneficial for sailing yacht designers.

The experiments were carried out in the cavitation tunnel of the French Naval Academy Research Institute (IRENav). A comprehensive impression of hydrofoil hydroelastic test set-up and facilities for hydroelastic response of flexible hydrofoils is illustrated in Fig. 16.20.

The flow velocity ranges between 3 and 12 m s^{-1} and the pressure between 100 mbar and 3 bar to control cavitation inception and development. The turbulence intensity in the middle of the test section is 2%. The experiment was carried out by Astolfi et al. in order to analyze the behavior of flexible hydrofoils, and it was based on the measurements of forces, structural strain measurements and vibration response together with high-speed visualization performed in a cavitation tunnel at the French Naval Academy on cantilevered flexible hydrofoils.

For force measurements, a specific flexible hydrofoil was fabricated using 3D printing techniques. It is observed that the lift tends to be lower than for the rigid blade and that the change of drag and moment coefficients fell into the uncertainties of the hydrodynamic balance. Vibrations measurements were carried out using laser vibrometry in subcavitating and in cavitating flows. Several features are observed. First the bending and twisting mode's frequencies have been identified in still air and in still water and added mass effect is clearly quantified.

Then measurements were performed in non-cavitating flow for various flow velocities and angle of incidence. It is observed that the bending mode's frequency is nearly constant when the angle of attack or the flow velocity increases, whereas the twisting mode's frequency increases with the flow velocity and the angle of incidence.

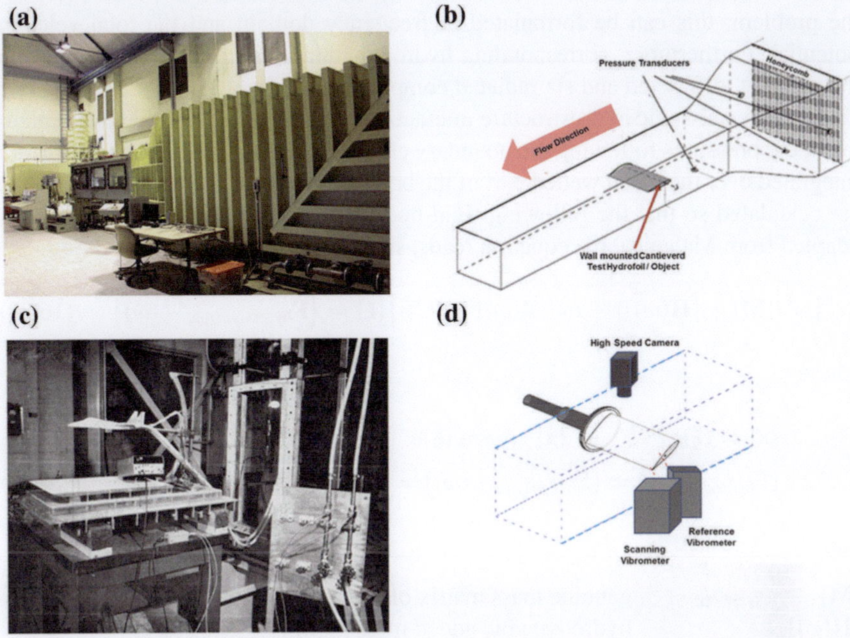

Fig. 16.20 Comprehensive impression of hydrofoil hydroelastic test set-up and facilities. **a** The Cavitation Tunnel at Hydrodynamics Laboratory of BPPT, the Indonesia Agency for the Development of Technology Indonesia [30], **b** schematic of the hydrofoil model in the test section of a cavitation tunnel; **c** view showing an aircraft model installed on Angle of Attack Mechanism at the Langley 16 b6 24-inch water tunnel [30]; **d** impression of an example of data acquisition sensors for aeroelastic simulation of hydroelastic testing

16.9 Hydro Structure Interaction Models

The different hydro-structural issues can schematically be separated with respect to the nature of the hydrodynamic loading and the nature of the structural response. As far as the hydrodynamic loading is concerned, the usual practice is to classify it into three different categories:

- Linear hydrodynamic loading
- Weakly nonlinear, non-impulsive loading
- Impulsive hydrodynamic loading.

Within the potential flow hydrodynamic models, which would be an instructive approach, the linear hydrodynamic loading means the classical method of solution which has been the classical seakeeping hydrodynamic analysis. The advantage in the case of the quasi-static structural responses is that the hydrodynamic and structural calculations can be performed separately.

The usual procedure passes through the solution of the rigid-body diffraction radiation problem using the boundary element method (BEM). Due to the linearity the problem, this can be formulated in frequency domain and the total velocity potential. Furthermore, corresponding hydrodynamic pressure is decomposed into the incident, diffracted and six radiated components.[9]

Linear quasi-static hydrostructure interaction model [7]. implies that once calculated, the pressures following the boundary element method (BEM) formulation are integrated over the mean wetted part of the body, and the hydrodynamic coefficients are calculated so that the following rigid-body motion equation can be written, as adapted from Malenica, the equation reads, similar to Eq. 16.25:

$$\left\{\omega^2([\mathbf{M}] + [\mathbf{H}(\omega)]) - i\omega([\mathbf{B}(\omega)] + [\mathbf{C}])\right\}\{\boldsymbol{\xi}\} = \left\{\mathbf{F}_{\text{hydrodynamic}}{}^{DI}(\omega)\right\} \tag{16.28}$$

where

$$\begin{gathered}\{\mathbf{x}\} = (\boldsymbol{\xi})e^{-i\omega t} \qquad \{\dot{\mathbf{x}}\} = -i\omega(\boldsymbol{\xi})e^{-i\omega t} \qquad \{\ddot{\mathbf{x}}\} = \omega^2(\boldsymbol{\xi})e^{-i\omega t} \\ \left(\mathbf{F}_{\text{hydrodynamic}}\right) = \left(\mathbf{F}_{\text{hydrodynamic}}(\omega)\right) = \left(\overline{\mathbf{F}}_{\text{hydrodynamic}}(\omega)\right)e^{-i\omega t}\end{gathered} \tag{16.29}$$

and

$[\mathbf{M}]$	genuine mass matrix of the body
$[\mathbf{H}(\omega)]$	hydrodynamic added mass matrix
$[\mathbf{B}(\omega)]$	hydrodynamic damping matrix
$[\mathbf{C}]$	hydrostatic restoring matrix
$\left\{\mathbf{F}_{\text{hydrodynamic}}{}^{DI}(\omega)\right\}$	hydrodynamic excitation vector
$\{\boldsymbol{\xi}\}$	body motions vector.

The solution of the motion equation gives the body motions, and the seakeeping problem is formally solved.

The next step consists in transferring the loading from hydrodynamic model to the structural finite element model. This is the critical step in the procedure and should be done with greatest care in order to build fully consistent loading cases which perfectly balance the rigid-body inertia and the hydrodynamic pressure loads. As far as the rigid-body inertia is concerned, the situation is simple, and we should just make sure that the rigid-body mass matrix is evaluated using the mass distribution from the FE model.

16.10 Influence of Waves in the Hydroelasticity of Ships

Further considerations are here given to relevant topics in hydroelasticity, in particular associated with the unique presence of waves in ocean going vehicles and off-shore structures. To this end reference will be made on the work of Hirdaris and Temarel [2].

[9] Further detail canbe found in Malenica [7].

Noting that hydroelasticity is the branch of science concerned with the interactions of deformable bodies with the water environment in which they operate, one may intuitively assume in their operations, ocean going ships and off-shore structures are experiencing strains and stresses, hence their structural flexibility, due to ocean environment, waves and wind. However, the influence of waves as exciting force in the motion and dynamics of ocean going ships and off-shore structures is unique in hydroelasticity. Nevertheless, the evolution of naval architecture, coupled with developments in computational technology, resulted in examining the dynamic behavior of ocean going ships and off-shore structures within distinct subject areas, each based on its individual set of assumptions, namely seakeeping of rigid bodies in waves, maneuvering of rigid bodies in calm water, and strength of ocean going ships and off-shore structures. The concept that ocean going ships and off-shore structures are flexible structures that can be modeled as elastic beams was the most straight forward engineering approach at hand. However, the term hydroelasticity was coined for the first time by Heller and Abramson [5]. They defined hydroelasticity as the naval counterpart to aeroelasticity and recognized that at fluid–structure interaction level significant differences may exist between the hydrodynamic, inertia and elastic forces experienced by a floating marine and off-shore structure. Consequently, the fluid pressure acting on the structure modifies its dynamic state and, in accordingly, the motion and distortion of the structure modify the pressure field around it.

The equations of motion for a ship traveling with forward speed U in regular, deep water waves of frequency v encountered at arbitrary heading x can be written as

$$\left[-\omega^2(\mathbf{a}+\mathbf{A})+i\omega(\mathbf{b}+\mathbf{B})+(\mathbf{c}+\mathbf{C})\right]\mathbf{p}=\boldsymbol{\Xi} \tag{16.30}$$

Here a, b and c denote the $n \times n$ generalized mass, structural damping (assumed diagonal), and stiffness (diagonal) matrices; a and c are obtained from the "dry" or "in vacuo" analysis where the free–free ship structure is considered, excluding external forces or internal damping.

The natural frequencies ω_r and corresponding mode shapes (e.g. vertical and horizontal displacements w_r and v_r and twist ϕ_r) and internal actions (e.g. vertical and horizontal bending and torsional moments; prying and yaw-splitting moments for multihulled vessels; direct and shear stresses, etc.) are obtained during the dry analysis, idealizing the ship's structure either as a beam or using three-dimensional FE modeling. $A(\omega_e)$, $\mathrm{B}(\omega_e)$, and C are the $n \times n$ generalized added inertia, fluid damping, and fluid restoring matrices, respectively. $\Xi(\omega_e, \chi)$ denotes the $n \times 1$ generalized wave excitation vector, containing both incident wave and wave diffraction contributions. These comprise the "wet" analysis, whereby the ship is divided into a number of strips (2D) or a number of panels over the mean wetted surface for the distribution of pulsating sources (3D), and subsequent determination of the hydrodynamic coefficients and wave excitation takes place. In the 2D case the underwater sections can be represented using Lewis or multiparameter conformal mapping. Equation 16.28 represents the wet analysis, and its solution is the $n \times 1$ principal coordinate vector expressed as

$$\mathbf{p}\{t\} = \overline{\mathbf{p}}^{-i\omega t} \tag{16.31}$$

where p represents the complex principal coordinate amplitudes, and ω_e the wave encounter frequency.

It can be seen that this system of equations of motion is unified in the sense that it allows for both rigid-body motions and distortions that are coupled through the effects of the fluid actions. This can be seen in the make-up of the generalized matrices of structural origin, as shown in Eq. (16.32)

$$a = \begin{bmatrix} a_R & 0 \\ 0 & a_D \end{bmatrix} \quad b = \begin{bmatrix} 0 & 0 \\ 0 & b_D \end{bmatrix} \quad a = \begin{bmatrix} 0 & 0 \\ 0 & c_D \end{bmatrix} \tag{16.32}$$

Here the indices R and D denote rigid and distortional modes, respectively. The generalized structural damping and stiffness matrices have no contributions from the rigid-body modes and are diagonal. Further details can be followed in the referenced paper by Hidari and Temarel.

Following the solution of Eq. 16.30 obtained in the principal coordinates in regular waves, distortions and internal actions, such as bending moments and stresses, are obtained. The evaluation of global wave-induced loads is carried out through modal summation. As an example, the vertical bending moment M_y at a cross section at distance x, measured from the stern, and the direct stress at a position s (x, y, z) are derived as

$$M_y(x,t) = e^{i\omega t} \sum_{r=7}^{n} M_{yr}(x) p_r \tag{16.33}$$

and

$$\sigma_x(s,t) = e^{i\omega t} \sum_{r=7}^{n} \sigma_{yr}(s) p_r \tag{16.34}$$

In Eqs. 16.30 and 16.31, M_{yr} and σ_{xr} denote the modal vertical bending moment and direct stress, respectively, for the rth mode shape, where the summation commences from the first distortion mode. The rigid-body modes are assumed to have zero internal actions. Equation 16.31 applies when 3D modeling of the structure is used, while Eq. 16.30 is more suitable for beam modeling of monohulled vessels.

In the time domain a long-crested irregular seaway can be modeled as a combination of R regular waves of amplitude a_j, $j = 1, 2, \ldots, R$, corresponding to wave frequency ω_j, obtained from a prescribed wave spectrum, and using random phase angles. The vertical bending moment as the steady-state response in the time domain, excluding transients, is obtained by employing the above consideration in Eq. 16.30:

$$\overline{M}_y(x,t) = \sum_{j=1}^{R} \sum_{r=7}^{n} a_j M_{yr}(x) p_r(t) \tag{16.35}$$

In the dynamics of floating ocean going and off-shore structure, slamming may occur and is a nonlinear phenomenon. The transient response due to slamming, however, can be evaluated through the use of the linear equations of motion by treating it as an arbitrary excitation and using the convolution integral. Then the transient principal coordinate, in regular or irregular waves, can be evaluated as

$$\breve{p}(x,t) = \int_0^t h(\tau) \breve{F}(t-\tau) \mathrm{d}\tau \tag{16.36}$$

where $h(t)$ is the matrix of impulse response functions, and $\breve{F}(t)$ is the transient generalized force vector. In the case of 2D hydroelasticity, bottom slamming in head waves is accounted for, including the influence of flare if required, by assuming that slamming may take place over a specified length of the hull's forefoot. Here slamming length is divided into a number of sections. In this case the transient generalized force can be expressed as

$$\breve{F}(t) = \int_0^{l_s} \breve{F}(x,t) w(x) \mathrm{d}x \tag{16.37}$$

where $w(x)$ denotes the vector of modal deflections of the dry hull and $F(x, t)$ denotes the total impulsive force per unit length applied to hull on impact. The evaluation of $F(x, t)$ can be carried out using empirical pressure that is available in the literature. In the case of 3D hydroelasticity a similar approach can also be used by identifying a slamming area S, divided into a number of panels, comprising the hull bottom and/or side and/or flare for monohulled vessels, as well as under the deck of the cross beam for multihulled vessels. The transient generalized force vector is obtained as

$$\breve{F}(t) = \iint_S n(x,y,z) u_r(x,y,z) p(x,y,z,t) \mathrm{d}S \tag{16.38}$$

where n denotes the normal vector and $u_r = \{u_r, v_r, w_r\}$, the principal mode shape vector. The coordinates (x, y, z) denote the center of the panel in the slamming area, which has been detected to have impacted with the water based on the attitude of the hull (heave and pitch, as well as roll motions in non-head waves) obtained from the linear equations of motion 16.30. The impact pressure p is evaluated from the empirical formulae. Then the vertical bending moment as the responses excited by slamming (whipping) in irregular unidirectional waves is given by

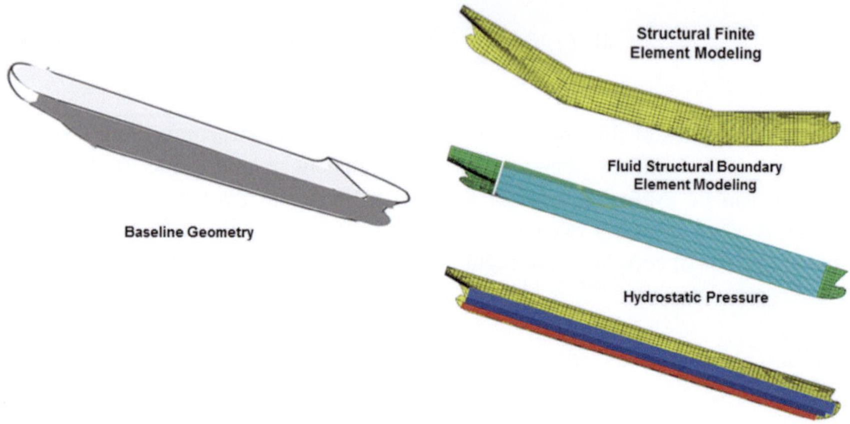

Fig. 16.21 Comprehensive schematic of three-dimensional FE structural idealization and fluid–structure interaction modeling of various hull shapes of ships and bulk carriers[10]

$$\breve{M}_y(x,t) = \sum_{r=7}^{n} M_{yr}(x)\breve{p}_r(t) \tag{16.39}$$

and the total response by

$$M_y(x,t) = \overline{M}_y(x,t) + \breve{M}_y(x, t - t_s), \tag{16.40}$$

where t represents the time elapsed from the beginning of the response record to the start of a slam.

16.11 Hydroelastic Modeling of a Bulk Carrier in Regular Waves

A crude 3D finite element (FE) model can be generated for a bulk carrier, an example of which is carried out using both 2D and 3D hydroelasticity analyses accounting for major structural components such as deck, side, inner/outer bottom, hopper spaces, bulkheads and major longitudinals.

Figure 16.21 comprehensively illustrates a three-dimensional FE structural idealization and fluid–structure interaction modeling of various hull shapes of ships and bulk carriers typical of such modeling. In some work, to avoid localized distortions during the dry analysis transverse frames and longitudinal stiffeners were omitted, and fictitious bulkheads of negligible mass and thickness were incorporated.

[10] Extracted from various data and information, among others Ni et al. [31].
Harding et al. [32].

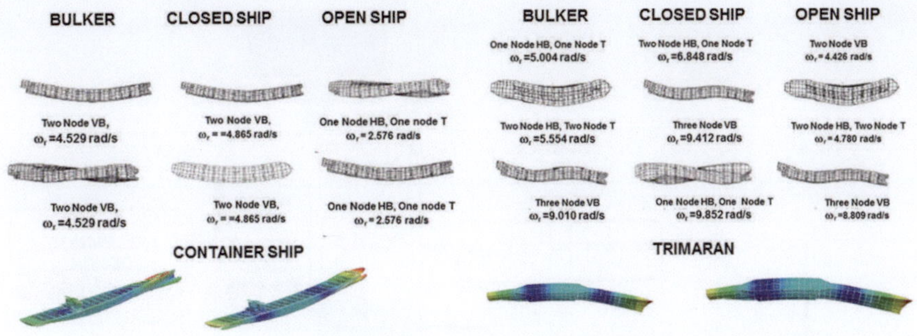

Fig. 16.22 Principal mode shapes and natural frequencies (rad/s) for all shell idealizations (VB, vertical bending; HB, horizontal bending; T, torsion, HB, T, HB,T dominant)[11]

16.12 Design Applications for Hydroelasticity

Design experience suggests that as a result of the imposed loading a slender ship throughout its normal life will bend and twist as a non-uniform elastic beam. Examples of principal mode shapes and natural frequencies (rad/s) for all shell idealizations for the finite element model of a slender ship investigated by Harding and Temarel [33] are exhibited in Fig. 16.22.

Taking bending in waves as an example, it will take place in three distinct frequency regimes, namely: (a) the ultra-low frequency (b) the low frequency, and (c) the high frequency.

From a design and standardization perspective, practical rules and state-of-the-art design procedures should be backed up by in-service experience and suitably validated software and take proper account of all still water and low-frequency wave effects. Non-slender monohulled vessels and multihulled vessels also experience distortions in frequency regimes exemplified above. The historical developments in design rules and procedures primarily put emphasis on bending, but not exclusively.

Traditionally, bending in still water is assessed by a simple computation accounting for the hull girder properties, the cargo and the buoyancy distribution. Low-frequency bending occurs at frequencies primarily associated with the natural heaving and pitching periods when the ship is in waves. This bending is primarily influenced by the time-dependent differential distribution of wave, buoyancy and hydrodynamic effects.

The transfer functions are used for stochastic analysis to identify the maximum wave loads occurring during the vessel's lifetime. The short-term analyses are performed for each irregular wave condition, namely the modal period and significant wave height.

[11] Created and adapted using data and information (with great thanks to) Harding and Temarel [2], the online version of this article can be found at http://pim.sagepub.com/content/223/3/305.

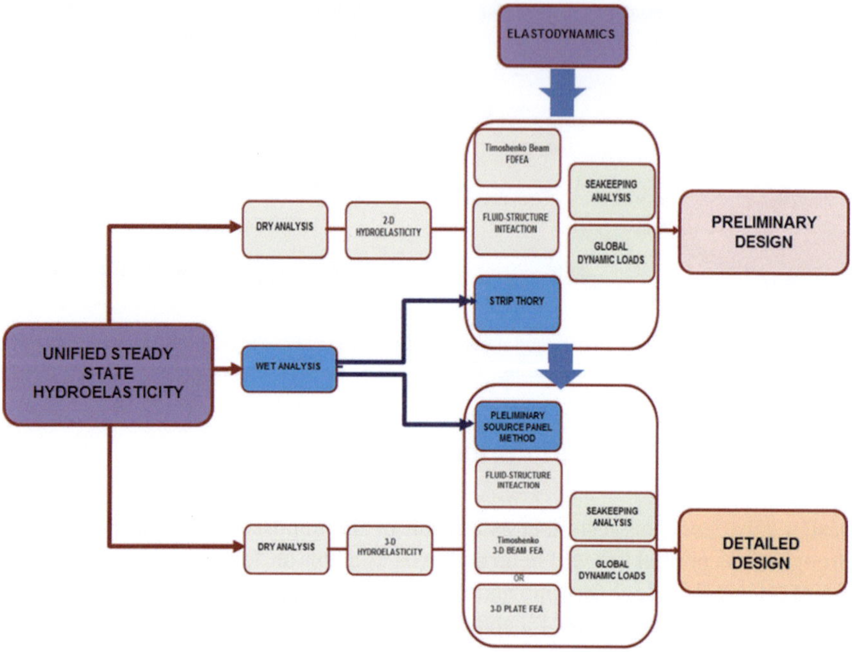

Fig. 16.23 Example illustrating the use of hydroelasticity theory in the design process of monohulled vessels[12]

Hydroelasticity theory in its 2D or 3D form is unified in the sense that it subsumes the principles of structural theory and marine hydrodynamics (conventional seakeeping and strength) by studying the behavior of a flexible body moving through a liquid.

When applied to a ship hull it may be used, within the concepts of linearity, to determine the inherent motions, distortions and stresses under the actions of external loading arising from the seaway, as well as other dynamic sources of excitation if required. A typical example for incorporating hydroelasticity in the design process of monohulled vessels is shown in Fig. 16.23. The relatively simpler two-dimensional analysis, comprising beam structural idealization and strip theory for the fluid forces and fluid-structure interactions, can be used during preliminary design.

On the other hand, a more detailed FE structural idealization combined with source distribution over the mean wetted surface can be used for the detailed design, where structural FE models are usually available. From a design point of view, it is always of interest to estimate the ship responses in certain sea states as well as the extreme responses a ship may experience in her whole life. Statistical analysis is then required by post-processing the responses from regular waves with wave spectra and wave data of the service area within the interested time scale. Although throughout the

[12] Created using data and information from various sources, among others (with great thanks to) Harding et al. [32] (with permission).

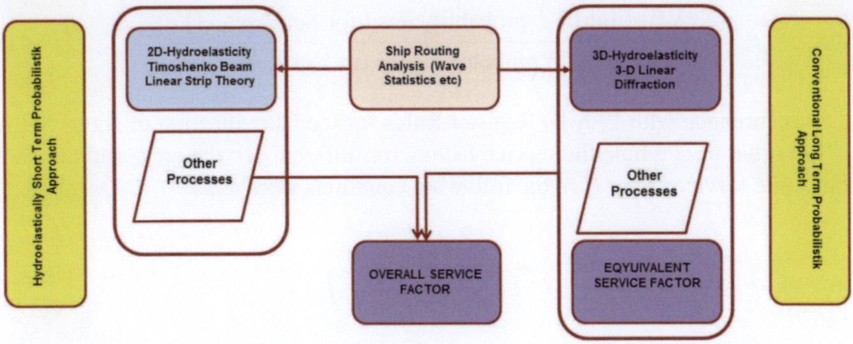

Fig. 16.24 Service factor assessment procedure[13]

years, the recommended wave spectrum functions or wave data might have changed, the general procedure remains almost the same.

16.13 Service Factor Assessment

The effects of hull flexibility when deriving an equivalent service factor for a single passage in certain waterways and lakes carried out by Hidaris and Temarel [2] produced an assessment procedure.

The assessment procedure derived for the service factor assessment is outlined in Fig. 16.24. For the 2D frequency domain symmetric hydroelastic analysis, the hull was discretized into 20 sections of equal length, and relevant properties were obtained for the Timoshenko beam model [34–36]. The longitudinal mass distribution was derived on the basis of combining lightweight and deadweight mass distributions for the ballast loading condition.

The transient hydroelastic analysis (whipping) loads were derived in the time domain, based on Eqs. 16.35, 16.36, 16.37, 16.39 and 16.40. A 3D frequency domain hydrodynamic (rigid body) analysis was carried out in regular waves and irregular waves.

Hidaris and Temarel [2] comparisons showed that achieving good agreement between predictions and measurements, both for ship-wave matching and springing, depends on the parameters of the wave spectra as well as the estimation of structural damping for the latter and any uncertainties involved in measuring such data [33].

By combining the long-term "rigid-body" wave bending moment with the effects of hydroelasticity, a suitable service factor (f_{si}) was derived as shown by

[13] Created and adapted using data and information from various sources, among others (with great thanks to) Hirdaris et al. [33] (with permission). Details are elaborated in this reference.

$$f_{\mathrm{si}} = \frac{\mathrm{VBM}\left(\text{at } 10^{-8} \text{ probabiity level for Sea Area, } i\right)}{\mathrm{VBM}\left(\text{at } 10^{-8} \text{ probabiity level for North Atlantic}\right)} \tag{16.41}$$

In accordance with Lloyd's Register Rules for the Classification of Naval Ships [37] in order to combine the service factors for different service areas and derive a composite service factor (f_{s}), the following equations were used:

$$f_{\mathrm{s}} = \ln\left(\sum_{i=1}^{m} P_i e^{f_{\mathrm{si}}}\right) \tag{16.42}$$

where P_i represents the probability of the ship operating in a particular sea area, i is the number of sea areas and $e = 2.7183$. The final service factor incorporating the 37.7% enhancement due to hydroelastic effects was

$$F_s = (A + B + 1) f_s \tag{16.43}$$

where A and B represent enhancement factors due to the effects of springing (from wave excitation) and whipping (from bottom slamming due to forefoot emergence) respectively. The service factors for Eq. 16.43 are described by Hirdaris et al. [33]. The example serves to demonstrate the use of hydroelasticity in a complementing role to more established procedures.

16.14 Concluding Remarks—Progress and Future Developments

Hydroelasticity, like aeroelasticity, has benefitted from the progress on fluid–structure interactions and wave-induced loads. As such, these are significant aspect of hydroelasticity, and along with experimental endeavor and operational observations, contribute to progressive understanding of the importance of developing analytical tools and design solutions that simulate more accurately the physics of the ship and off-shore structure hydroelasticity.

The still relevant design philosophy for the prediction of motions and wave-induced loads is based upon prescriptive rule requirements and first principles design procedures backed up by simulation-based design. The use of direct calculation methods that account simultaneously for the effects of dynamic wave environment using fully nonlinear hydrodynamics, Reynolds averaged Navier–Stokes equations and related softwares, as well as linear and, especially, nonlinear finite element method (FEM) and boundary element method (BEM) will be very useful and is expected to evolve further.

References

1. Hirdaris, S., Introduction to Hydroelasticity of Ships, 2004-P8_LRTA.
2. Hirdaris, S.E., and P. Temarel. 2009. Hydroelasticity of ships: Recent advances and future trends. *Proceedings of the Institution of Mechanical Engineers, Part M: Journal of Engineering for the Maritime Environment*. https://doi.org/10.1243/14750902JEME160
3. Senjanovic, I., N. Vladimir, and S. Malenica. 2019. An overview of ship hydroelasticity. In *7th International Conference on Marine Structures*, 8 May 2019, Dubrovnik, Croatia.
4. Chen, X.J., Y.S. Wu, W.C. Cui, and J.J. Jensen. 2006. Review of hydroelasticity theories for global response of marine structures. *Ocean Engineering* 33: 439–457.
5. Heller, S.R., and H.N. Abramson. 1959. Hydroelasticity—A new naval science. *Journal of the American Society of Naval Engineers* 71: 205–209.
6. Bishop, R.E.D., W.G. Price, and P.K.Y. Tam. 1977. A unified dynamic analysis of ship response to waves. *Transactions of Royal Institution of Naval Architects* 119: 363–390.
7. Malenica. 2014. Hydro-structural issues in the design of ultra large container ships. *International Journal of Naval Architecture and Ocean Engineering* 6: 983–999. https://doi.org/10.2478/IJNAOE-2013-0226
8. Hirdaris, S.E., W. Bai, D. Dessi, A. Ergin, X. Gu, O.A. Hermundstad, R. Huijsmans, K. Iijima, U.D. Nielsen, J. Parunov, N. Fonseca, A. Papanikolaou, K. Argyriadis, and A. Incecik. 2014. Loads for use in the design of ships and offshore structures. *Ocean Engineering* 78: 131–174. https://doi.org/10.1016/j.oceaneng.2013.09.012
9. Djojodihardjo, H., B.B. Prananta, and S.B. Aman. 1993. Numerical boundary element computation of submerged body-surface wave interaction. In *Proceedings of the Third International Offshore and Polar Engineering Conference*, Singapore, 6–11 June.
10. Van den Berg, W., H.C. Raven, and H.H. Valkhof. 1990. Free surface potential flow calculation for merchant vessels. In *CFD and CAD in ship design*, ed. G. van Oortmensen. Elsevier Science Publisher.
11. Dawson, C.W. 1977. A practical computer method for solving ship-wave problems. In *Proceedings of the 2nd Seminar on Numerical Ship Hydrodynamics*, UCLA, Berkeley.
12. Djojodihardjo, R.H., and S.E. Widnall. 1969. A numerical method for the calculation of nonlinear unsteady lifting potential flow problems. *AIAA Journal* 7: 2001–2009.
13. Jensen, G., and H. Soding. 1989. Ship wave resistance computation. In *Seminar on Marine Technology*, Bandung Institute of Technology, Bandung.
14. Jensen, G., Bertram, V., and Soding, H. 1991. Ship wave resistance computation. In *Proceedings of Fifth International Conference on Numerical Ship Hydrodynamics*. Washington, DC: National Academic Press.
15. Djojodihardjo, H., H. Suhartono, and M. Karnadi. 1986. The use of green identity in panel method for the computation of three dimensional subsonic aerodynamics (in Indonesian). *Indonesian Institute of Aeronautics and Astronautics Journal* 1(1).
16. Djojodihardjo, H. 1991. Computational study of the aerodynamics of passenger vehicle. In *Proceedings of the 6th International Conference on Automotive Engineering*, Seoul, Korea.
17. Salvesen, N., and V.C. Kerczek. 1976. Comparison of numerical and perturbation solutions of two-dimensional nonlinear water-wave problems. *Journal of Ship Research* 20: 160–170.
18. Djojodihardjo, H. 2016. Aeroelastic and performance baseline analysis of piezoaeroelastic wing section for energy harvester. In *First IntSymp Flutter and Application*, JAXA, Tokyo.
19. Chae, E.J. 2015. *Dynamic response and stability of flexible in incompressible viscous flow*. Ph.D., University of Michigan.
20. Ducoin, Antoine, and Yin L. Young. 2011. Hydroelastic response and stability of a hydrofoil in viscous flow. In *Second International Symposium on Marine Propulsors, SMP'11*, Hamburg, Germany.
21. Anonymous. ANSYS Fluent Tutorial Guide (ANSYS, ANSYS Workbench, Ansoft, AUTODYN, EKM, Engineering Knowledge Manager, CFX, FLUENT). http://users.abo.fi/rzevenho/ansys%20fluent%2018%20tutorial%20guide.pdf

22. Bisplinghoff, R.L., H. Ashley, and R.L. Halfman. 1955. *Aeroelasticity*. Addison-Wesley.
23. Zwaan, R.J. 1981. *Aeroelasticity of Aircraft, Lecture Notes, Special Lecture, Short Course offered at Institut Teknologi Bandung, Indonesia* (Author's reposutory).
24. Wright, J.R., and J.E. Cooper. 2015. *Introduction to aircraft aeroelasticity and loads*, 2nd ed. Wiley. https://doi.org/10.1002/9781118700440
25. Djojodihardjo, H., and H.H. Yee. 2007. Parametric study of the flutter characteristics of transport aircraft wings. In *Proceedings of AEROTECH-II, Conference on Aerospace Technology of XXI Century*, Kuala Lumpur.
26. Menter, F.R. 1993. *Improved two-equation k-turbulence models for aerodynamic flows*. NASA Technical Memorandum 103975 34. 38.
27. Menter, F.R. 1994. Two-equation eddy-viscosity turbulence models for engineering applications. *AIAA Journal* 32 (8): 1598–1605. https://doi.org/10.2514/3.12149
28. Ducoin, Antoine, and Yin L. Young. 2013. Hydroelastic response and stability of a hydrofoil in viscous flow. *Journal of Fluids and Structures* 38: 40–57.
29. Astolfi, J.A., A. Lelong, P. Bot, and J.-B. Marchand. 2015. Experimental analysis of hydroelastic response of flexible hydrofoils. In *Conference: 5th High Performance Yacht Design Conference*, Auckland.
30. Pendergraft, Jr., et al. 1992. *A user's guide to the Langley 16-by 24-inch water tunnel*. NASA TM 104200.
31. Ni, Zao, M. Dhanak, and T. C. Su. 2021. Performance of a hydrofoil operating close to a free surface over arange of angles of attack. *International Journal of Naval Architecture and Ocean Engineering* 13: 1–11.
32. Harding, R.D., S.E. Hirdaris, S.H. Miao, M. Pittilo, and P. Temarel. 2006. Use of hydroelasticity analysis in design. In *Proceedings of the 4th International Conference on Hydroelasticity in Marine Technology*, 1–12, China.
33. Hirdaris, S.E., N. Bakkers, N. White, and P. Temarel. 2009. Service factor assessment of a great lakes bulk carrier incorporating the effects of hydroelasticity. *Marine Technology and SNAME News* 46 (2): 116–221.
34. Timoshenko, S.P., and D.H. Young. 1968. *Theory of structures*, 2nd ed. USA: McGraw-Hill Inc.
35. Haque, Aamer. 2019. *Timoshenko beam theory*. Independently Published.
36. Öchsner, Andreas. 2021. *Classical beam theories of structural mechanics*. Springer. ISBN 978-3-030-76034-2 (ISBN 978-3-030-76035-9 (eBook). https://doi.org/10.1007/978-3-030-76035-9. Accessed January 2022.
37. Anonymous. Lloyd Register, 2018, Rules and Regulations for the Classification of Naval Ships, vol. 1.

Chapter 17
Introduction to Envaeroelasticity with Vibro-acoustics as a Case Study

Abstract The application of BE-FE acoustic-structure interaction on a structure subject to acoustic load is elaborated using the boundary element–finite element acoustic-structural coupling and the utilization of the computational scheme developed earlier. The plausibility of the numerical treatment is investigated and validated through application to generic cases. The analysis carried out in the work is intended to serve as a baseline in the analysis of acoustic-structure interaction for lightweight structures. Results obtained thus far exhibit the robustness of the method developed.

Keywords Acoustic-structure interaction · BEM-FEM coupling · Boundary element method · Envaeroelasticity · Vibro-acoustics

List of Acronyms and Symbols

Acronyms

[AIC]	Aerodynamics influence coefficient
$[A(ik)]$	Unsteady aerodynamics matrix

Symbols

$[A(t^*)]$	Unsteady aerodynamics matrix
b	Wing chord/span chosen for convenience
$[\mathbf{C}]$	Viscous damping
c	Constant for BEM equation, chord, or speed of sound
$[\mathbf{D}]$	Viscous damping
$[\mathbf{F}]$	External forces
H_{ij}	Influence coefficient matrices
Gij	Influence coefficient matrices
g	Free-space Green function

© The Author(s), under exclusive license to Springer Nature Singapore Pte Ltd. 2023
H. Djojodihardjo, *Introduction to Aeroelasticity*,
https://doi.org/10.1007/978-981-16-8078-6_17

$[\mathbf{K}]$	Stiffness matrix
k	Reduced frequency $\frac{\omega b}{U}$
kw	Wave number, in the Helmholtz equation
$[\mathbf{L}]$	Fluid–structure coupling matrix
L	Aerodynamic lift
M	Aerodynamic moment in flutter calculation
$[\mathbf{M}]$	Mass matrix
N	Shape function, as implied by the context
$\hat{n}_0$	Surface normal vector
$n0$	Surface normal vector
P	Acoustic pressure
p_{inc}	Incident acoustic pressure
psc	Scattering acoustic pressure
q	Generalized coordinates
$\{\mathrm{q}\}$	Vector of generalized coordinates
q_∞	Dynamic pressure of the fluid surrounding the structure
r	Radius of the acoustic monopole
$R0$	A point in boundary surface
S	Bounding surface
$v\mathrm{f}$	Flutter speed

Greek Symbols

Γ	Boundary at infinity
Γ_{U}	Closed boundary
δ	Kronecker's delta function
λ	Wavelength
ν	Normal velocity vector
ρ	Air density

17.1 Aerospace Vibro-acoustics as a Case Study on Envaeroelasticity

Fluid–structure interaction involving environmental excitation forces like aerodynamic and acoustics can be found in the space flights, In particular during the launch of a space vehicle.

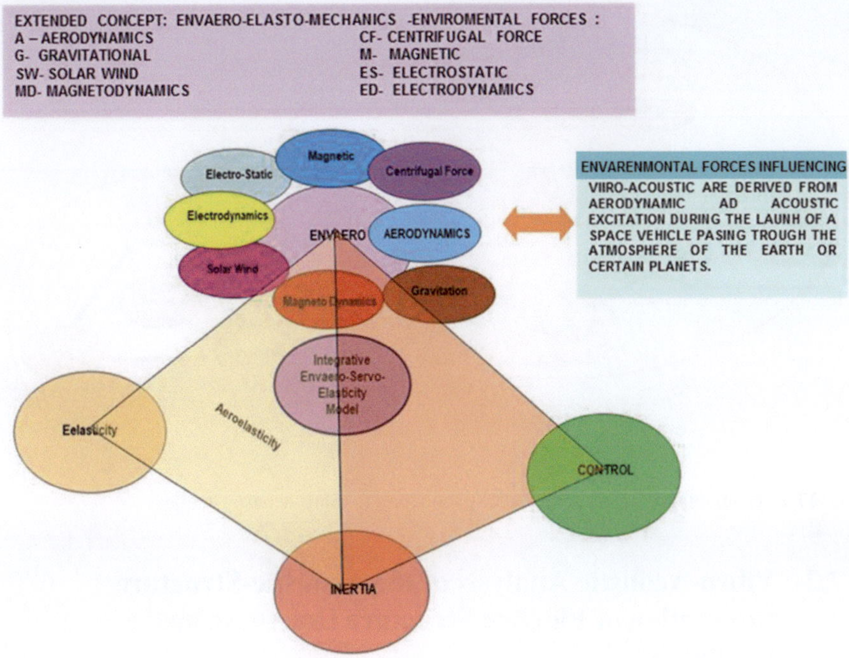

Fig. 17.1 Envaero-servo-elasto-mechanics

The application of BE-FE acoustic-structure interaction on a structure subject to acoustic load is elaborated using the boundary element–finite element acoustic structural coupling and the utilization of the computational scheme developed earlier. The plausibility of the numerical treatment is investigated and validated through application to generic cases. The analysis carried out in the work is intended to serve as a baseline in the analysis of acoustic-structure interaction for lightweight structures. Results obtained thus far exhibit the robustness of the method developed.

Figures 17.1 and 17.2 are reproduced from Figs. 1.8 and 1.9 in Chap. 1 which are displayed to obtain a comprehensive impression of the example being considered here as a case study of envaeroelasticity. The present chapter will exclusively discuss vibro-acoustic analysis of the acoustic–structure interaction of flexible structure due to acoustic excitation as an envaeroelasticity case study.[1]

[1] Based on the work of the author which is published in *Acta Astronautica, Volume 108*, March–April 2015, Pages 129–145 [1].

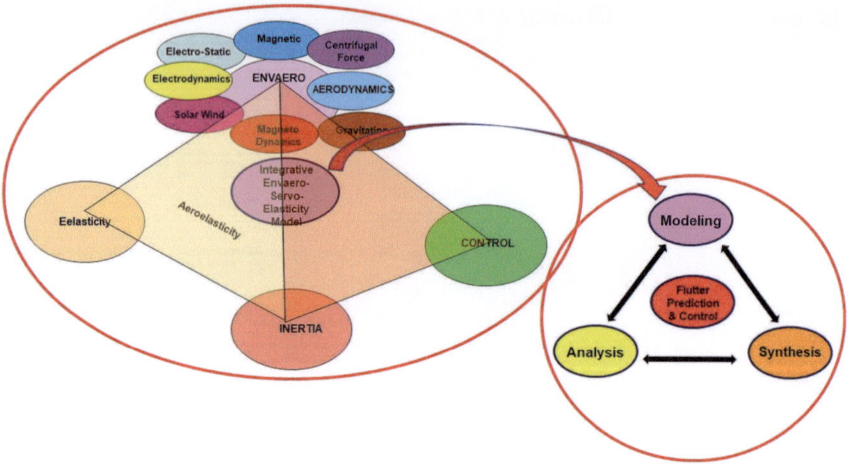

Fig. 17.2 From integrative envaeroelasticity modeling to vibro-acoustic analysis

17.2 Vibro-Acoustic Analysis of the Acoustic-Structure Interaction of Flexible Structure Due to Acoustic Excitation

17.2.1 Introductory Remarks

During flight missions, space vehicles, such as illustrated in Fig. 17.3 [2], are subjected to a severe dynamic pressure loading and broadband, aeroacoustics and structure-borne excitations of various circumstances, which can endanger the survivability of the payload and the vehicles electronic equipment, and consequently the success of the mission. Aerospace structures are generally characterized by the use of exotic composite materials of various configurations and thicknesses, as well as by their extensively complex geometries and connections between different subsystems. It is therefore of crucial importance for the modern aerospace industry, the development of analytical and numerical tools that can accurately predict the vibro-acoustic response of large, composite structures of various geometries and subject to a combination of aeroacoustics excitations.

Assisted with computing capability and user-friendly computer-aided analysis software, the analyst is challenged to ensure that the analysis includes all the relevant physical phenomena. Simple fundamental principles are mandatory, in order not to lose insight on the interrelationships between relevant elements and to devise simple methods that are robust to address various problems. Space-borne structure must be able to resist the loads induced by the launch environment and meet all the functional performances required on orbit such as dimensional stability and structural integrity. Noise and vibration should also be taken as critical consideration in the design of aerospace vehicles for fatigue of components arising from interior structural

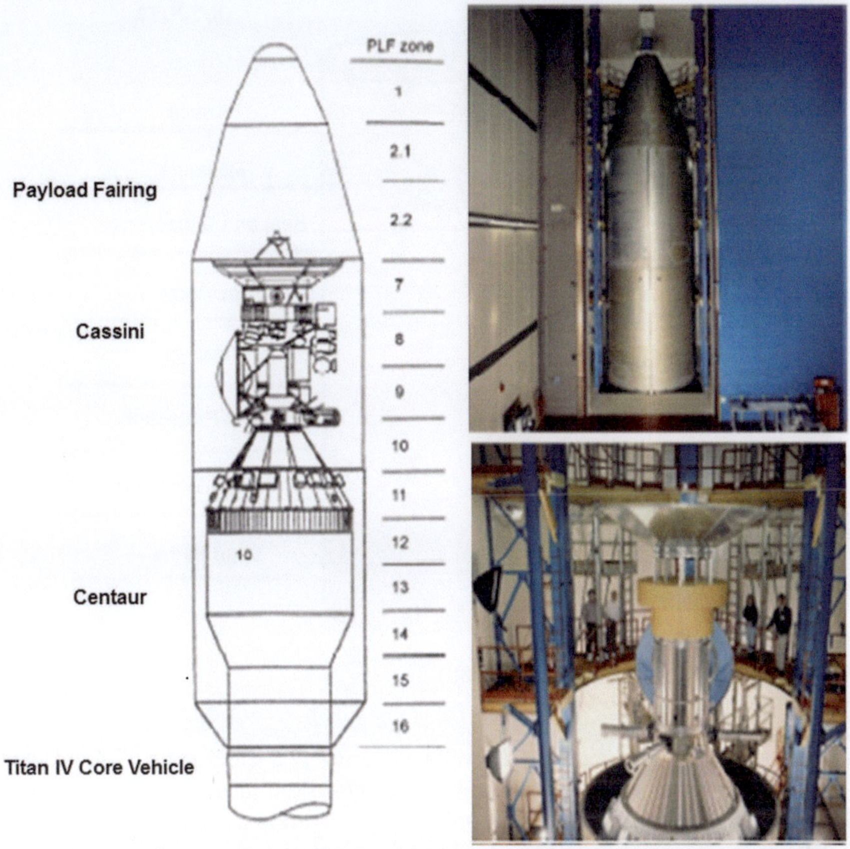

Fig. 17.3 Illustration of a spacecraft being inspected for vibro-acoustic specification; **a** Cassini spacecraft configured inside the Titan IV payload fairing; **b** Titan IV payload fairing in acoustic test chamber; **c** Cassini spacecraft mockup used in acoustic test chamber. Courtesy of NASA[2]

and acoustic pressure fluctuations due to external structural or acoustic loading [4–12]. Other related works are reported in references [13–18]. Pappa et al. [19–21] summarize modal testing activities at the NASA Langley Research Center for generic aircraft fuselage structures. Citarella et al. [15] set up an integrated approach for an automobile vibro-acoustic analysis to assess, visualize and compare vibro-acoustic performance to predetermined design targets and identifying and quantifying the forces and sound sources responsible for their prevailing behavior.

The great number of design variables allows to synergistically fulfill high stiffness and acoustic standards. Hence the objective of the present paper is to describe the application of BE-FE fluid–structure interaction on a structure subject to acoustic load and to elaborate FE formulation of the computational scheme for unified approach on

[2] William O. Hughes, Application of the Bootstrap Statistical Method in Deriving Vibroacoustic Specifications, NASA/TM—2006–214446 [3].

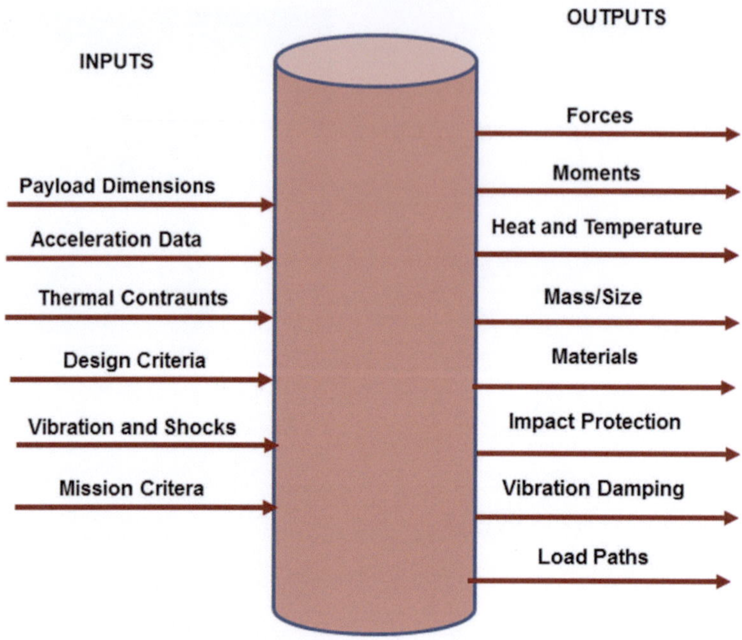

Fig. 17.4 Structures subsystem interface [28]

acoustic-aeroelastic interaction as developed earlier [22–27]. Figure 17.4, adapted from [21] shows schematically the structures subsystem interface. This is divided into two separate categories, inputs and outputs, to allow for easy referencing. A technique to estimate driving force spectra of equipment packages attached to cylindrical structures subjected to broadband random acoustic excitations is summarized in Figs. 17.4 and 17.5, adapted from [28]. This procedure is considered indicative of the present state of the art and will provide satisfactory predictions of the vibratory environment.

17.2.2 Problem Formulation

To address the problem associated with fluid–structure interaction, in particular the vibration of structures due to sound waves, aerodynamics and their combined effects, a generic approach to the solution of the elasto-acousto-fluid-dynamic interaction will be followed in order to develop the foundation for the computational scheme for the calculation of the influence of the acoustic disturbance to the aeroelastic stability of the structure, starting from a rather simple and instructive model to a more elaborate FE-BE fluid–structure one. The generic approach consists of two

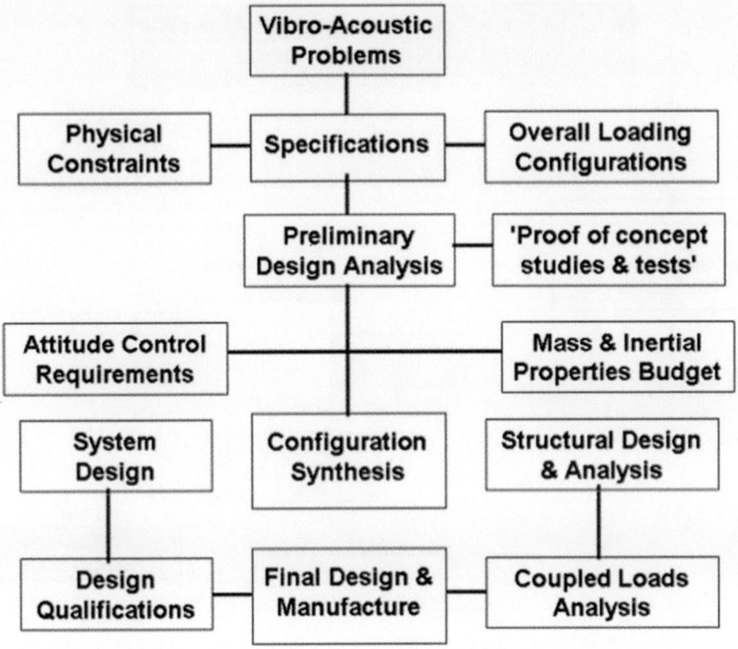

Fig. 17.5 Technique to estimate driving force spectra of equipment packages subjected to broadband random acoustic excitations [28]

parts, as schematically summarized in Fig. 17.6 which exhibits the computational strategy to treat the acoustic-structure interaction.

The first is the formulation of the acoustic wave propagation governed by the Helmholtz equation by using boundary element approach, which then allows the calculation of the acoustic pressure on the acoustic-structure boundaries. The influence of the acoustic excitation field has been given rigorous consideration by taking into account both the incident and scattering acoustic pressure, following the governing equations described by Dowling and Ffowcs-Williams [29] and Norton [30]. The second part addresses the structural dynamic problem using finite element approach. In the formulation of BE-FE coupling to treat the fluid–structure interaction, reference is made to the solution procedure for structural–acoustic interaction problems described by Chuh-Mei and Pates [14], Holström [31] and Marquez, Meddahi and Selgas [32]. A third part could be added to incorporate other environment effect, such as aerodynamics, as elaborated in references [22–26], to be utilized in the aeroacoustoelastic problem.

Following the scheme depicted in Fig. 17.6, the present work is a synthesis and adaptation of a host of methods that have been suggested by various authors as appropriate and elaborated subsequently for its effectiveness. The method has been applied to generic problems in the acoustic as well as structural dynamic domain, followed by its application for acoustic-structure coupling. It is then applied to address the stability

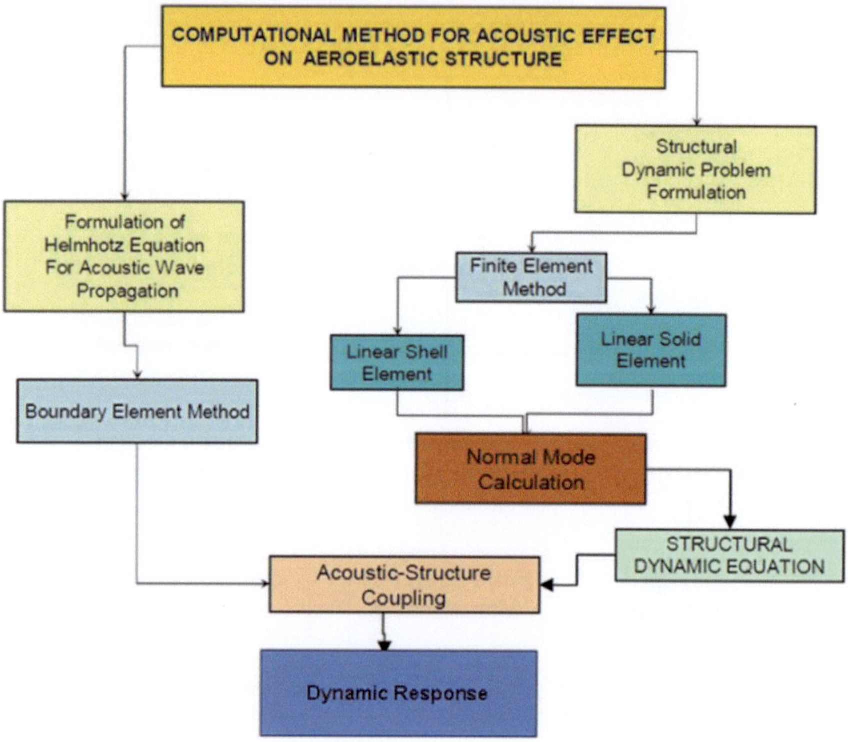

Fig. 17.6 Introduction of FEM-BEM approach in the integration strategy for computational method in aeroacoustics effect on aeroelastic structures [18, 23–27]

problem in aeroacoustoelasticity, by careful development of various elements of methods: BEM for acoustic domain, FEM for structural domain and BEM-FEM for the acoustic-structural coupling or acousto-aeroelasticity generic problems. Attention is focused on instructive method suitable for desk-top computers. Computations are carried out using in-house MATLAB-based program as well as NASTRAN for validation.

A novel approach is presented in addressing acoustic-aeroelastic problem, by the introduction of acoustic-aerodynamic analogy in treating acousto-aeroelastic stability problem. To avoid spurious results, CHIEF method (as elaborated subsequently) is utilized and assessed.

17.2.3 Discretization and Treatment of Helmholtz Integral Equation for the Acoustic Field Following Conventional BEM Formulation

For an exterior acoustic problem, the problem domain V is the free-space V_{ext} outside the closed surface S. V is considered enclosed between the surface S and an imaginary surface Γ at a sufficiently large distance from the acoustic sources and the surface S.

External radiation and scattering problems can be reduced to integral equations over a surface in various ways. The utilization of Helmholtz integral equation for the external acoustic field and their implementation to boundary element method (BEM) has been the subject of continued interest. One should note, however that there may be a discrete set of wave numbers or frequencies for which the solution to these integral equations either does not exist or is not unique. Benthien and Schenck [5] review many of these methods for handling these non-existence or non-uniqueness problems and have carried out comparative study on their accuracy and efficiency. Both problems stem from the rank deficiency of the matrix. The Fredholm alternative theorem and singular value decomposition (SVD) updating techniques have been utilized to study the rank-deficiency matrices by juhl [33]. Juhl also employed the SVD technique to study the numerical instability due to the irregular frequency. To deal with this problem, Burton and Miller [6] solved the problem by combining singular and hypersingular equations with an imaginary number. However, the calculation for the hypersingular integral is required. To avoid this computation, an alternative method was proposed by Schenck [34]. He proposed the Combined Helmholtz Interior Integral Equation Formulation (CHIEF) method by collocating the point outside the domain as an auxiliary constraint to promote the rank of influence matrices. Wu and Seybert [35] proposed a CHIEF-block idea to enforce the constraints in a weighted residual sense over a small region. If the CHIEF point locates on or near the nodal line of its corresponding mode, it may not provide a valid constraint [10–12].

With such preliminary note, one could proceed by considering the propagation of time harmonic acoustic waves in a homogeneous isotropic acoustic medium (which can be either finite or infinite) as described by the Helmholtz equation:

$$\nabla^2\varphi(x) + k^2\varphi(x) = 0 \quad \forall x \in V \tag{17.1}$$

where φ is the pressure field, $k = \omega/c$ is the wave number, ϖ is the angular frequency and c is the wave speed in the acoustic medium.

In the present case considered as illustrated in Fig. 17.7, the boundary condition for the Helmholtz equation is given by:

$$\varphi(x) = \tilde{\varphi}(x) \quad \forall x \in V \tag{17.2a}$$

$$\frac{\partial\varphi(x)}{\partial n} \equiv q(x) \equiv \tilde{q}(x) \equiv i\omega\rho v(x) \quad \forall x \in S \text{ or } \Gamma \tag{17.2b}$$

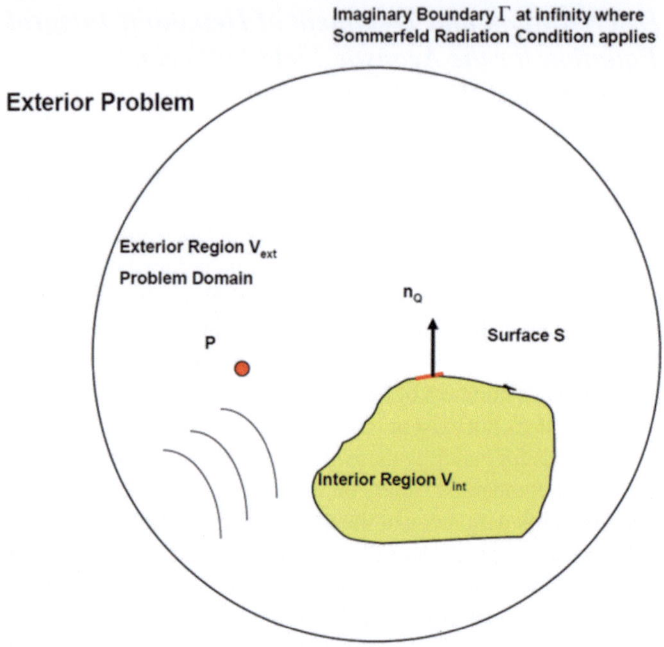

Fig. 17.7 Exterior problem for homogeneous Helmholtz equation

The barred quantities indicate known values on the boundary.

The corresponding boundary integral representation of the solution to the Helmholtz[3] integral equation [8, 22–27, 28–32] is given by

$$\begin{aligned} C(x)\varphi(x) &= \int_S \left(G(x,y)\frac{\partial\varphi(y)}{\partial n(y)} - \frac{\partial G(x,y)}{\partial n(y)}\varphi(y) \right) \mathrm{d}S + \varphi^i \\ &= \int_S \left(G(x,y)q(y) - \frac{\partial G(x,y)}{\partial n(y)}\varphi(y) \right) \mathrm{d}S + \varphi^i \quad \forall x \in V \end{aligned} \tag{17.3}$$

where **x** is the collocation point, **y** the source point, $n(y)$ the surface unit outward normal vector at y. Frequently one seeks to express the unknown exterior potential $\varphi\ (p)$ as the potential arising from some particular form of source distribution $\sigma(q)$ on the boundary, for example as a *single-layer potential*

$$\varphi(p) = \int_B \sigma(q)G(p,q)dS_q \equiv L[\sigma] \tag{17.4a}$$

or as a *double-layer potential*

[3] Also known as Kirchhoff-Helmholtz Integral Equation, A. D. Pierce, Acoustics, 1981 [39].

$$\varphi(p) = \int_B \mu(q) \frac{\partial}{\partial n_q} G(p, q) \mathrm{d}S_q \equiv M[\sigma] \tag{17.4b}$$

where $\partial/\partial n_q$ denotes differentiation along the outward normal at q. In the latter case $\mu(q)$ represents a distribution of dipoles or doublets on S. Equations 17.4a and 17.4b define $\varphi(p)$ explicitly at all points of S.

The incident wave $\phi^{\mathrm{i}}(\mathbf{x})$ in Eq. 17.3 will not be present for radiation problems. The value of C($\mathbf{x}$) depends on the location of $R(x)$ in the fluid domain, R_0 denote a point located on the boundary S and G the free-space Green's function. For 3D problems G is given by:

$$G(x, y) = \frac{e^{ikr}}{4\pi r}; \ r = |x - y| \rightarrow G(x, y) = \frac{e^{ik|x-y|}}{4\pi |x - y|} \tag{17.5}$$

where $i = \sqrt{(-1)}$.

The boundary condition on Γ satisfies Sommerfeld acoustic radiation condition as the distance approaches infinity. At the infinite boundary Γ, the Sommerfeld radiation condition in three dimensions can be written as [14, 29–31]:

$$\lim_{|R-R_0|\rightarrow\infty} R\left(\frac{\partial g}{\partial R} - ikg\right) \Rightarrow 0 \text{ as } R \Rightarrow \infty \tag{17.6}$$

which is satisfied by the fundamental solution. Letting point $\mathbf{x}$ approach the boundary leads to the following conventional boundary integral equation (CBIE):

$$C(x)\varphi(x) = \int_S \left(G(x, y)q(y) - \frac{\partial G(x, y)}{\partial n(y)}\varphi(y)\right)\mathrm{d}S(y) + \varphi^I(x) \quad \forall x \in S \tag{17.7}$$

where $C(x) = 1/2$ if S is smooth around x.

Equation 17.7 can be rearranged as:

$$C(x)\varphi(x) - \varphi^I(x) + \int_S \frac{\partial G(x, y)}{\partial n(y)}\varphi(y)\mathrm{d}S(y) = \int_S G(x, y)q(y)\mathrm{d}S(y) \quad \forall x \in S \tag{17.8a}$$

or alternatively:

$$C(x)\varphi(x) + \int_S \frac{\partial G(x, y)}{\partial n(y)}\varphi(y)\mathrm{d}S(y)$$

$$= \int_S G(x, y)q(y)\mathrm{d}S(y) + \varphi^I(x) \quad \forall x \in S \tag{17.8b}$$

where $\varphi^I(x)$ is the incident wave. The right-hand side of Eq. 17.8 represent the boundary condition on S, *and in the cases considered here, will be assumed to be known*. In discretized form, Eq. 17.8 can be written as

$$(E + B) \cdot \varphi = i\omega\rho A \cdot v \tag{17.9a}$$

or

$$(E + B) \cdot \varphi = A \cdot \overline{q} \tag{17.9b}$$

where the double-layer potential φ is the sound pressure vector (unknown), v is the velocity vector (given), and the entries of the influence coefficient matrices are represented, respectively, by

$$E_{ij} = -\frac{1}{2}\delta_{ij};\ E_{ij} = -\frac{1}{2} \text{ when } i = j \tag{17.10}$$

$$A_{ij} = a_j(\mathbf{R}_i) = \int_{\Gamma_1} N_j(\mathbf{R}_Q)G(\mathbf{R}_i, \mathbf{R}_Q)\mathrm{d}S_Q \tag{17.11}$$

$$B_{ij} = b_j(\mathbf{R}_i) = \int_{\Gamma} N_j(\mathbf{R}_Q)\frac{\partial G(\mathbf{R}_i, \mathbf{R}_Q)}{\partial n_Q}\mathrm{d}S_Q \tag{17.12}$$

Equation 17.9a or 17.9b can also be written as

$$\sum_{i=1}^{N} f_{ij}\varphi_i = \sum_{j=1}^{N} g_{ij}q_j + \hat{b}_i \tag{17.13a}$$

for node $i = 1, 2, \ldots N$ (12) or

$$\sum_{i=1}^{N} f_{ij}\varphi_i = \sum_{j=1}^{N} g_{ij}q_j + \varphi_i^I \tag{17.13b}$$

where, in general, N_j is the interpolation function of the j-th node and $\hat{b}$ is from the incident wave for scattering problems as represented in Eq. 17.8a, and the LHS of Eq. of 17.13b

$$f_{ij}\varphi_i = \int_{\Gamma_1, i\neq j} \frac{\partial G(x_i, y)}{\partial n(y)}\varphi_j\mathrm{d}S(y) + \delta_{ij}\varphi_j \tag{17.14}$$

while the RHS of Eq. 17.13b

$$g_{ij}q_j = \int_{\Gamma_1} G(x, y)q_j \mathrm{d}S(y) \tag{17.15}$$

Further detail related to the discretization of Helmholtz integral equation is given in Appendix 1.

17.2.4 Acoustic-Structure Coupling

For the purpose of establishing acoustic-structure coupling formulation, both the BE-Acoustic and the FE structure domain are schematically illustrated in Fig. 17.8. Learning from the aeroelastic case, aeroelastic modes are those that exist when the structural and aerodynamic modes are fully coupled; that is, oscillations of a fluid mode excite all structural modes and vice versa. In general, these aeroelastic modes also have complex eigenvalues and eigenvectors. At low speeds (well below the flutter speed, for example) one may usually identify the structural and aerodynamic eigenvalues separately, because structural/aerodynamic coupling is weak [36]. However, as stability issues are approached, the eigenvalues and eigenvectors may change substantially, and the fluid and structural modes become more strongly coupled.

The utilization of FE for the structural dynamic equation of motion in the structural domain leads to a system of simultaneous equations which relate the displacements at

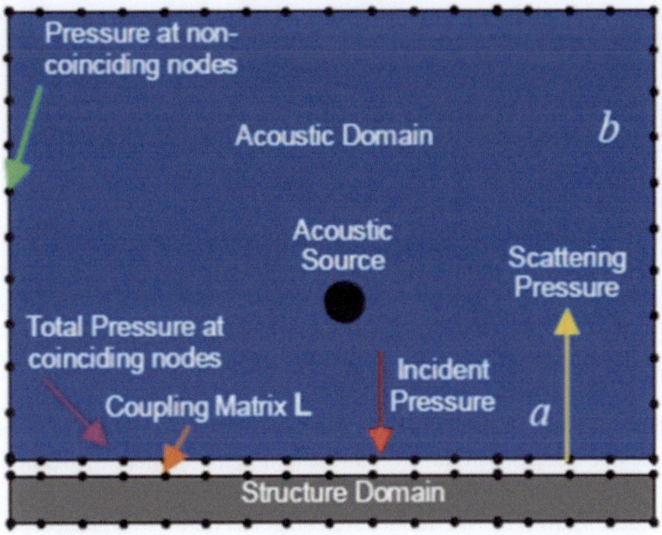

Fig. 17.8 Schematic of fluid–structure interaction domain

all the nodes to the *nodal forces*. In the BEM, on the other hand, a relationship between nodal displacements and *nodal tractions* should be established. Representing the elastic structure by FE model, the structural dynamic equation of motion is given by [23–27]:

$$[\mathbf{M}]\{\ddot{x}\} + [\mathbf{C}]\{\dot{x}\} + [\mathbf{K}]\{x\} = \{\mathbf{F}\} \tag{17.16}$$

where **M**, **C** and **K** are structural mass, damping and stiffness, respectively, which are expressed as matrices in a FE model, while **F** is the given external forcing function vector, and $\{x\}$ is the structural displacement vector. The elastic structure can be represented by FE model.

Let divide the BE nodes into two parts; part a, which is related to the finite elements that represents the surface of the acoustic medium in contact with the elastic structure, and part b, which is the remaining acoustic domain. Through the use of well-established modal analysis, the acoustic-structure problem as the BEM/FEM coupling represented by Eq. 17.15 can be recast into equivalent one using generalized coordinates as before, and yields

$$[\mathbf{M}]\{\ddot{q}\} + [\mathbf{C}]\{\dot{q}\} + [\mathbf{K}]\{q\} = \{\mathbf{F}\} \tag{17.17}$$

where **M**, **C** and **K** are now the generalized structural mass, damping and stiffness, respectively which are expressed as matrices in a FE model, while **F** is the given external forcing function vector, and q the generalized coordinate vector, while L represent the generalized acoustic-structure coupling.

For the normal fluid velocities and the normal translational displacements on the shell elements at the fluid–structure coupling interface, the coupling is governed by a relationship, which takes into account the velocity continuity over the coinciding nodes. This relationship is given by:

$$\mathrm{v} = i\omega(\mathbf{T} \cdot x) \tag{17.18}$$

Similar to **L**, $\mathbf{T}(n \times m)$ is also a global coupling matrix that connects the normal velocity of a BE node with the translational displacements of FE nodes obtained by taking the transpose of the boundary surface normal vector *n*. *The strength of the acoustic-structure coupling depends on various aspect of the computational procedure.* For weakly coupled problems a stable and efficient algorithm is obtained using one stage and a sufficiently accurate predictor. For strongly coupled problems, stability is enhanced by increasing the number of stages in the computational iteration.

L is the acoustic-structure coupling matrix of size $m \times n$, where m is the number of FE degrees of freedom on the structure and n is the number of BE nodes on the coupled boundary representing part a of the acoustic domain. Accordingly, the node pressure values, p, from the BE model are categorized into two groups, p_a and p_b.

Similarly, in the use of FMM-BEM scheme, following an elaborate procedure described in earlier work [22–27], the governing equation for the acoustic-aerodynamic-elastic structure coupling is also given by

$$\begin{bmatrix} [\mathbf{K} + i\omega\mathbf{C} - \omega^2\mathbf{M}] - q_\infty[\mathbf{A}(ik)]\{x\} & \mathbf{L} & 0 & 0 \\ \rho_0\omega^2\mathbf{G}_{11}T & \mathbf{H}_{11} & -i\rho_0\omega\mathbf{G}_{12} & \mathbf{H}_{13} \\ \rho_0\omega^2\mathbf{G}_{21}T & \mathbf{H}_{21} & -i\rho_0\omega\mathbf{G}_{22} & \mathbf{H}_{23} \\ \rho_0\omega^2\mathbf{G}_{31}T & \mathbf{H}_{31} & -i\rho_0\omega\mathbf{G}_{32} & \mathbf{H}_{33} \end{bmatrix} \begin{Bmatrix} x \\ p_a \\ v_b \\ p_c \end{Bmatrix} = \begin{Bmatrix} \mathbf{F} \\ -\mathbf{H}_{12}p_b + i\rho_0\omega\mathbf{G}_{13}v_c + p_{inc_a} \\ -\mathbf{H}_{22}p_b + i\rho_0\omega\mathbf{G}_{23}v_c + p_{inc_b} \\ -\mathbf{H}_{32}p_b + i\rho_0\omega\mathbf{G}_{33}v_c + p_{inc_c} \end{Bmatrix} \quad (17.19)$$

This equation forms the basis for the treatment of the acoustic-fluid–structure interaction in a unified fashion. The solution vector consisting of the displacement vector of the structure and total acoustic pressure on the boundaries of the acoustic domain, including the acoustic-structure interface, can be obtained by solving Eq. 17.18 as a dynamic response problems. Further detail related to the development of the Acoustic-Aeroelastic Coupling is given in Appendices 17.2 and 17.3.

17.2.5 Finite Element Numerical Simulation of a Rectangular Plate Under Pressure

To verify to the correctness of the developed finite element formulation, they are used to analyze the normal mode response of wing-like plate, incorporating an algorithm developed following the generic scheme of Fig. 17.6 and written in MATLAB®; results are then validated using commercial software package NASTRAN®.

The results of 5 normal mode analysis of Equivalent BAH wing using developed finite element program MATLAB® and validation using commercial package NASTRAN®, are illustrated in Figs. 17.10 and 17.11. The Stiffness properties of the equivalent BAH wing is illustrated in Fig. 17.9.

The natural frequencies associated with these five modes for solid and shell elements in the Finite Element Analysis are exhibited in Tables 17.1 and 17.2, respectively. Looking into these tables carefully, one may conclude that the differences are quite acceptable, and these may justify the plausibility as well as the accuracy of the method, in particular as proof of concept.

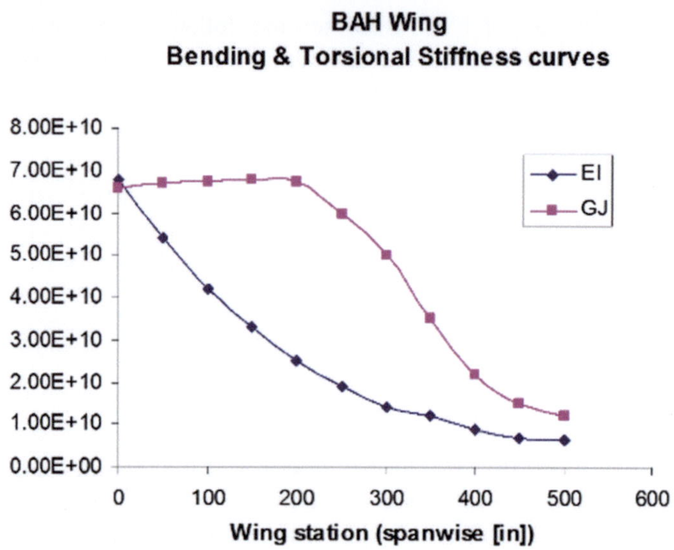

Fig. 17.9 Bending and torsional rigidity of Wing-like plate, here represented by equivalent BAH wing[4]

17.2.6 Flexible Structure Subject to Harmonic External Forces in Acoustic Medium

Several cases are considered to assess and validate the computational scheme as well as to evaluate its performance. An example is carried out for the wing-like a flat plate with flexural properties as elaborated in Figs. 17.10 and 17.11 subject to monopole acoustic loading. The incident pressure distribution is exhibited in Fig. 17.12 while its deformation and total acoustic pressure response are shown in Fig. 17.13.

Another generic example is exemplified by reproducing solution considered in [31], i.e. a flexible vibrating membrane on the top of a box whose all other walls are rigid subjected to harmonic external forces in acoustic medium as shown in Fig. 17.14.

The structure consists of five hard walls and one flexible membrane on the top and has the dimension $a \times b \times c = 305.8 \times 152.4 \times 152.4$ mm. The flexible structure, which consists of a 0.064 mm thick undamped aluminum plate, is modeled with coupled boundary and finite elements.

The frequency response due to the application of arbitrary harmonic excitation forces to the membrane following the present method using BEM-FEM Coupling is shown in Fig. 17.15, and is compared to results obtained using FEM-FEM approach [23–26, 31]. Comparable agreement has been indicated and serves to validate the present method.

[4] BAH wing is a wing developed by Bisplinghoff, Ashley and Halfman in [4] and has been used extensively for benchmarking.

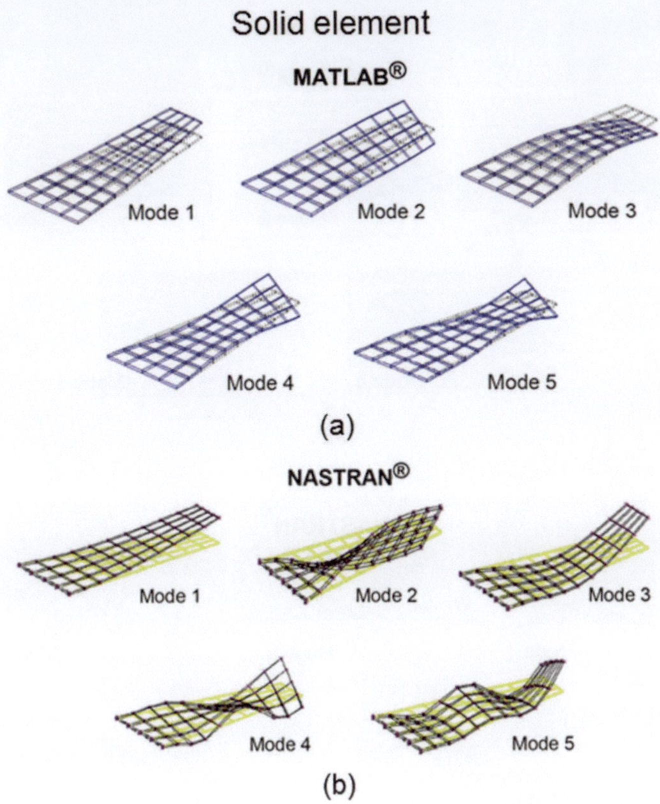

Fig. 17.10 Five normal mode analysis for equivalent BAH wing modeling with solid element, using **a** MATLAB®, **b** NASTRAN®

The convergence trend of the sound pressure level frequency response of a vibrating top membrane of an otherwise rigid box due to monopole source at the center of the box as studied in [31] is also reproduced and the results are exhibited in Fig. 17.17.

These results lend support to the extension of the present technique to tackle more involved cases, such as how the internal acoustics influences the response of an externally excited box. For the latter case, the governing Eq. 17.8a or 17.8b should be rewritten, such as, following Herrin, Wu and Seybert [40] as

$$p(P) = \int_S \left(G(r)\delta \mathrm{d}p - \frac{\partial G(r)}{\partial n}\delta p \right) \mathrm{d}S \tag{17.20}$$

where the primary variables are the single-layer (δ *dp*) and double-layer (δ*p*) potentials. When using the direct BEM, there is a distinction between an interior and exterior problem. Both sides of the boundary are considered simultaneously even

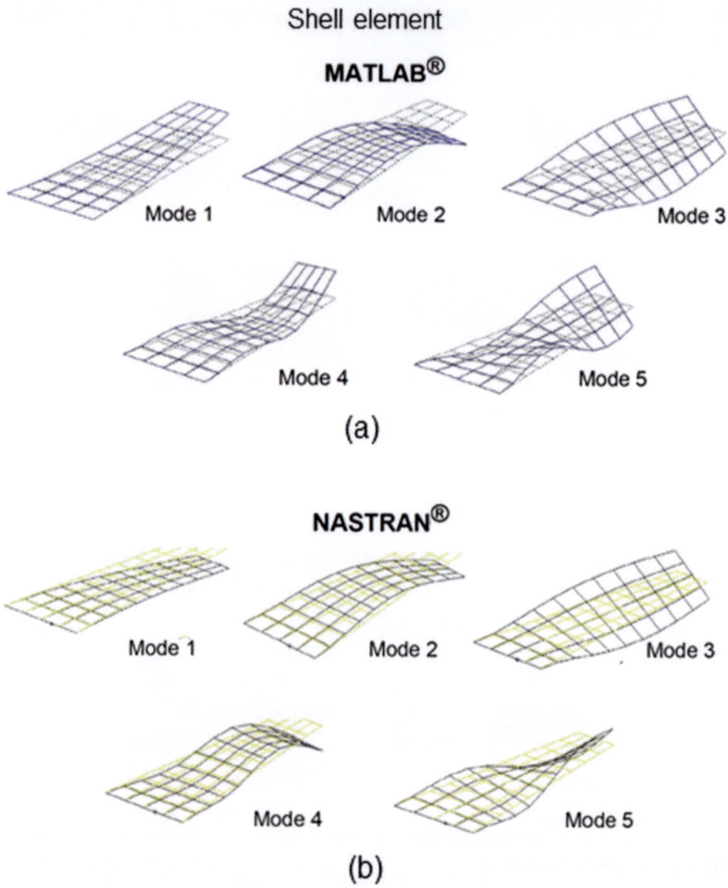

Fig. 17.11 Five normal mode analysis for equivalent BAH wing modeling with shell element

Table 17.1 Five normal mode analysis for equivalent BAH wing modeling with solid element

Mode	Natural frequency (Hz)	
	MATLAB®	NASTRAN®
1	1.8071	1.8758
2	3.1465	3.2851
3	7.7370	7.3132
4	9.4129	8.9200
5	24.4352	20.9072

Table 17.2 Five natural frequency for equivalent BAH wing modeling with shell element

Mode	Natural frequency (Hz)	
	MATLAB®	NASTRAN®
1	1.8071	1.8758
2	3.1465	3.2851
3	7.7370	7.3132
4	9.4129	8.9200
5	24.4352	20.9072

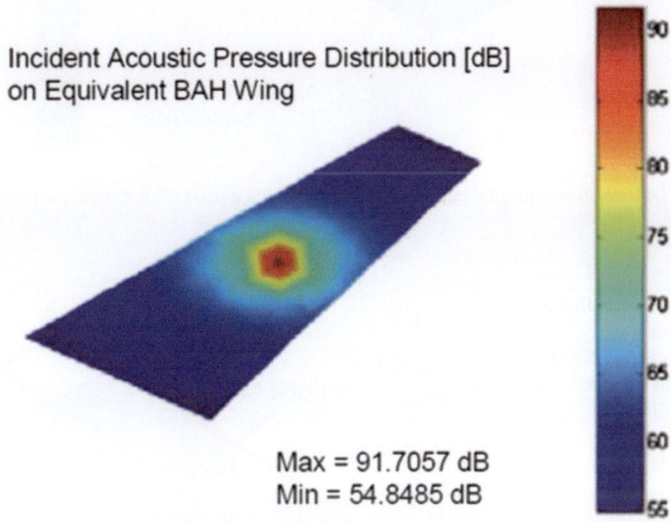

Fig. 17.12 Incident pressure distribution [dB] due to monopole acoustic source

though only one side of the boundary may be in contact with the fluid. In this case, the boundary consists of the inside and outside surfaces of the surface S shown in Fig. 17.7, and both sides should be analyzed at the same time.

17.3 Flexible Structure Subject to Acoustic Excitation in a Confined Medium

Application of the method to another example is carried out for a box shown in Fig. 17.17 with a dimension of $a \times b \times c = 450 \times 450 \times 270$ mm. Five walls of the structure are modeled as shell elements and the bottom of the box is modeled as a rigid wall. Each of the flexible walls is assumed to be made of 1 mm aluminum plate, and is modeled as coupled boundary and finite elements.

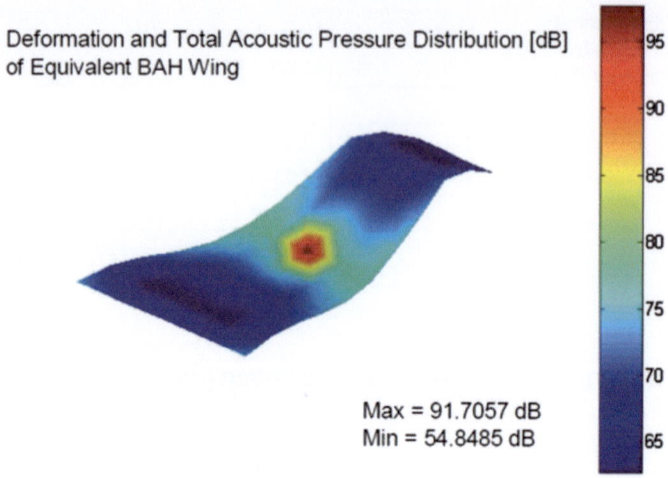

Fig. 17.13 Deformation and total acoustic pressure response [dB] on symmetric equivalent BAH wing

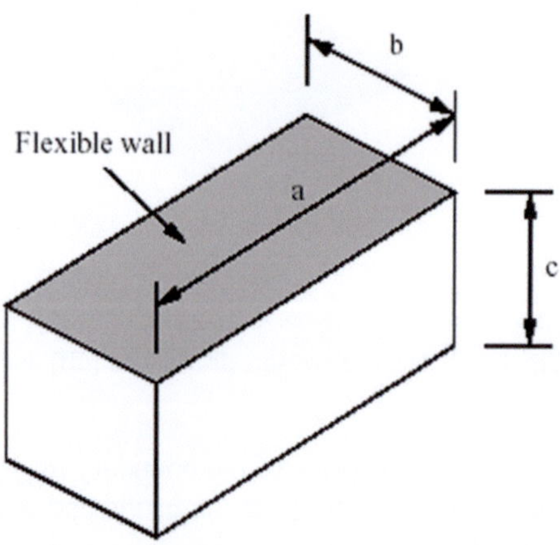

Fig. 17.14 Generic flexible structure typical of space-structure subjected to harmonic external forces in acoustic medium

This box structure is subjected to an acoustic monopole source at the center of the box and the acoustic medium is air with the following properties: density, $\rho = 1.21$ kg/m^3 and sound velocity $c = 340$ m/s. The discretization of the box is also depicted in Fig. 17.17.

17.3.1 Normal Mode Analysis

The result of the modal analysis of the box to obtain the first three normal modes using finite element program developed in MATLAB® is shown in Fig. 17.18. The first

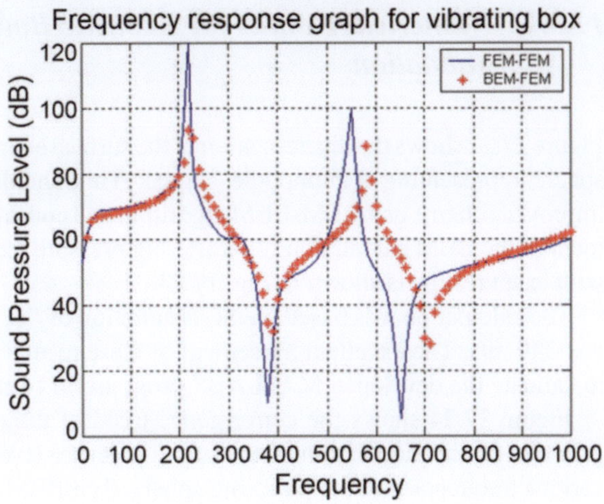

Fig. 17.15 Comparison between FEM-FEM and BEM-FEM approach for acoustic pressure [dB] at the center of the plate

Table 17.3 First five eigenfrequencies for box modeling with shell element

Mode	Natural frequency (Hz)	
	MATLAB®	NASTRAN®
1	17.644	16.190
2	36.026	35.986
3	41.072	41.520
4	45.585	45.787
5	51.231	51.689

five eigenfrequencies for the shell elements obtained using the present routine are compared to those obtained using commercial package NASTRAN®. As exhibited in Table 17.3, good agreement is obtained.

a. **Acoustic Excitation**

Acoustic excitation due to an acoustic monopole source with initial frequency, $f =$ 10 Hz, is applied at the center of the box.

No other external forces are applied. The resulting distribution of the incident acoustic pressure is shown in Fig. 17.19. The total pressures on the surface of the box obtained from the computational results are shown in Fig. 17.20. The frequency response for the incident pressure and total pressure on the center top surface of the box computed using Eq. 17.17 is shown in Fig. 17.21.

17.3.2 *Numerical Results for Acoustic Boundary Element Validation*

Figure 17.22 shows the discretization of the surface elements of an acoustics pulsating sphere representing a monopole source. This should be later compared to the improved scheme using FM-BEM algorithm and constant elements. Excellent agreement of the BEM calculation for scattering pressure from acoustic monopole source with exact results is shown in Fig. 17.23.

The calculation was based on the assumption of $f = 10$ Hz, $\rho = 1.225$ kg/m^3, and c = 340 m/s. The excellent agreement of these results with exact calculation serves to validate the developed MATLAB® program for further utilization.

Figure 17.24 shows the convergence trend of the computational scheme to the grid size on the calculation of the sound pressure level on the pulsating sphere for various frequencies of the pulsating sphere. The BEM-FEM simulation results on a spherical shell (Fig. 17.25a and b) subject to acoustic pressure is shown in Figs. 17.26 and 17.27. Analytical values are given by Junger and Feit [38].

The boundary integral equation of Eq. (17.4) fails at frequencies coincident with the interior cavity frequencies of homogeneous Dirichlet boundary conditions [17]. In the formulation of the exterior problem, these frequencies correspond to the natural frequencies of acoustic resonances in the interior region. When the interior region resonates, the pressure field inside the interior region has non-trivial solution. Since the interior problem and the exterior problem shares similar integral operators, the exterior integral equation may also break down. The discretized equation of the [**H**] matrix in Eq. (17.19) becomes ill-conditioned when the exciting frequency is close to the interior frequencies, thus providing an erroneous acoustic loading matrix.

This problem could be overcome by using the CHIEF [14, 23–25] or Burton–Miller method [6], a recent technique utilizing SVD and Fredholm alternative theorem [32] or others as discussed by Benthien and Schenk [5]. To avoid non-uniqueness problem, special treatment is carried out to inspect whether the H matrix is ill-behaved or not by utilizing SVD updating technique [11]. If it is ill-behaved, the present method resorts to the utilization of CHIEF method.

Without loss of generality, as exhibited in Figs. 17.26, 17.27 and 17.28, the application of CHIEF method for one and three CHIEF points proved to be successful, which verified Benthien and Schenck [5] observation that CHIEF continues to be a viable and efficient approach. As also pointed out by Wu and Seybert [35], the choice of CHIEF points can readily be made without losing accuracies. Following the reviewer's suggestions, the application of Burton–Miller reformulated method by Chen et al. [12] is also shown in Fig. 17.29.

17.4 Closing Remarks

Vibro-acoustic analysis of random vibration response of a flexible structure due to acoustic forcing has been carried out on typical aerospace generic structure. A brief survey was carried out to review and assess dynamic pressure loading and broadband, acoustic and structure-borne excitations of various situations designated not to endanger the survivability of the payload and the success of the mission. The analysis and numerical simulations for accurate prediction of the vibro-acoustic response of generic structures of various geometries subject to acoustic excitations are carried out using the unified BEM-FEM Acoustic-structural coupling developed in the author and colleague's earlier work. Results of FM-BEM are presented to show that the computational procedure for solving combined excitation due to acoustic and external forces on structural problem formulated as coupled FEM-BEM equation has given good results.

By using quarter space modeling for acoustic domain and varying the radius from 5 to 20 times the half wingspan and the number of elements from 400 to 1200, convergent acoustic pressure response on symmetric equivalent BAH wing have been obtained. Hence such numerical configuration can be used for the calculation of total acoustic pressure response in coupled BEM-FEM problem due to acoustic excitation.

Using BE and FE as appropriate, an integrated formulation is then obtained as given by the governing Eq. (17.17), which relates all the combined forces acting on the structure to the displacement vector of the structure. By analyzing the structural dynamic response due to acoustic excitation, the work has served as examples and baseline approach in the viability of the analysis of acoustic-structure interaction of aerospace structures. The work carried out thus far is focused on the formulation of the basic problem of acoustic excitation and vibration of elastic structure in a coupled fluid-elastic-structure interaction.

Judging from the results of the case studies, the computational procedure to solve the combined excitation due to acoustic and external forces on structural problem formulated as coupled FEM-BEM equation has shown to be satisfactory. The synthesized methods developed, including the in-house MATLAB code, proved to be effective. The effectiveness and efficiency as well as rapid convergence are exhibited in the Figs. 17.16, 17.17, 17.24 and 17.29, among others. Computations carried out using in-house MATLAB based program as well as NASTRAN have produced results with close accuracy. In retrospect, the method effectiveness, including the effectiveness of the coupling method, has been demonstrated. The efficiency of the method has been evaluated through comparison of the individual elements of the three components outlined in the introduction and summarized in Fig. 17.6 in qualitative way to particular method and exemplified in the generic examples mentioned. In retrospect, the method effectiveness, including the effectiveness of the coupling method, has been demonstrated. The efficiency of the method has been evaluated through comparison of the individual elements of the three components outlined in the introduction and summarized in Fig. 17.6 in qualitative way to particular method and exemplified in the generic examples mentioned. A simple solution using CHIEF regularization has been employed to address the fundamental issue of spurious solution. The solution of Eq. (17.18) as a result of the effect of acoustic disturbance to a structure is given as the pressure loading response on the structure as well as the pressure field in the fluid medium and is given as total pressure. This allows the calculation of the scattering pressure due to the incident pressure.

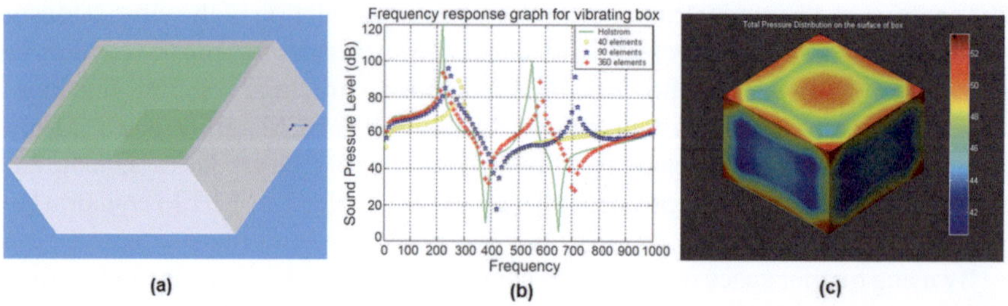

Fig. 17.16 Convergence trend of the sound pressure level frequency response of a vibrating top membrane of an otherwise rigid box due to monopole source at the center of the box (**a**) studied in [21] as a function of frequency calculated using present computational scheme for various grid size is shown in (**b**); **c** Total pressure distribution on the surfaces of the same box. Top surface is modeled as BE-FE, others as BE

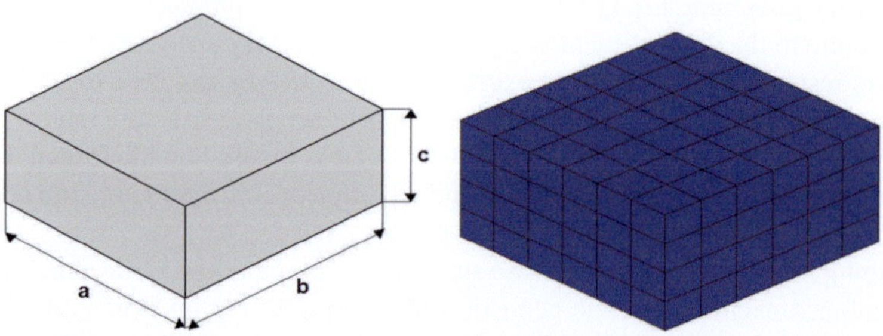

Fig. 17.17 Generic flexible structure typical of space-structure subjected to monopole acoustic source in an acoustic medium

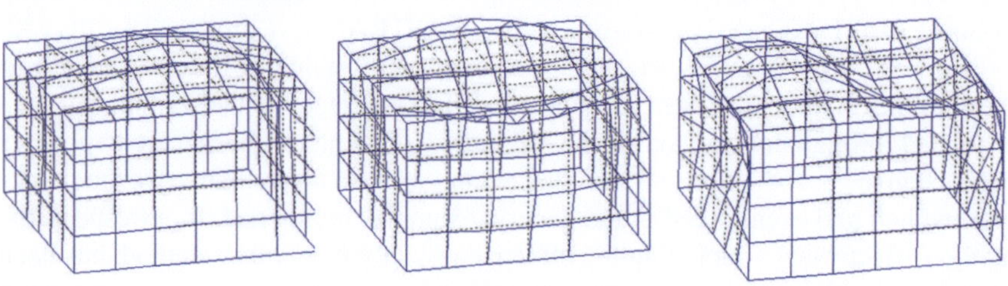

Fig. 17.18 First three normal modes for box modeled with five flexible walls using shell elements

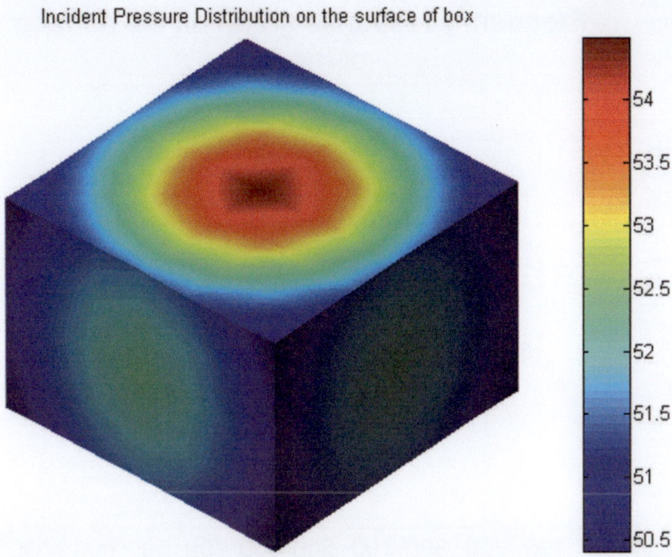

Fig. 17.19 Incident acoustic pressure distribution on the surface of box due to monopole acoustic excitation

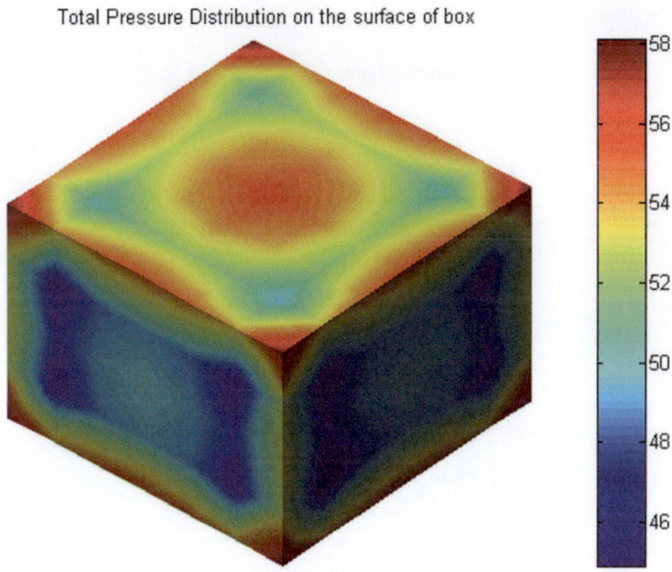

Fig. 17.20 Total acoustic pressure distribution on the surface of box due to monopole acoustic excitation

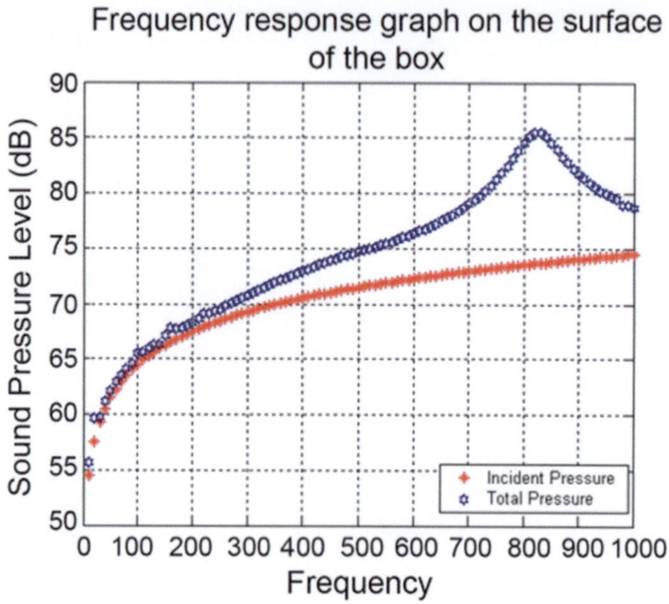

Fig. 17.21 Incident and total acoustic pressure distribution on the center top surface of box as a function of frequency

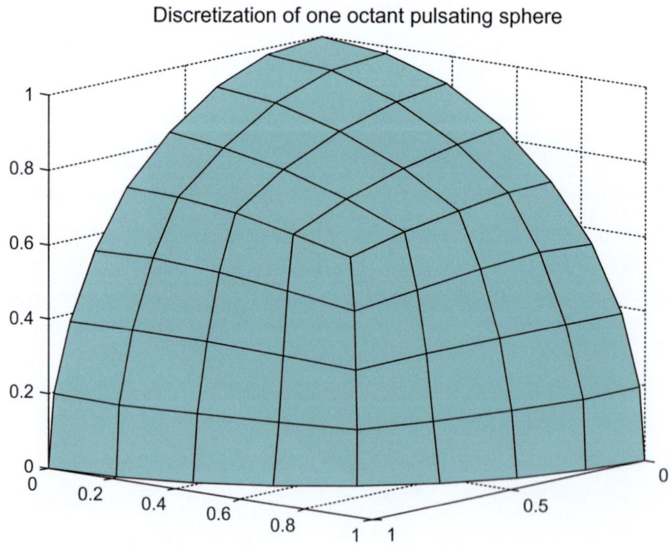

Fig. 17.22 Discretization of one octant pulsating sphere

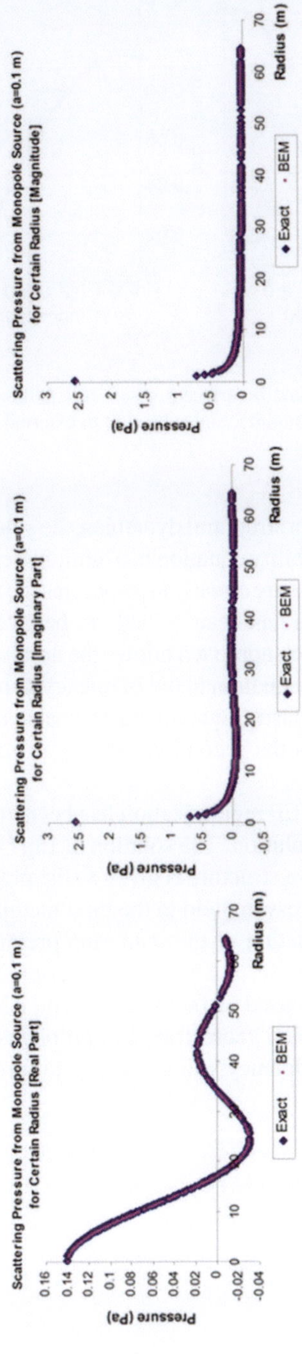

Fig. 17.23 Comparison of monopole source exact and BE scattering pressure results

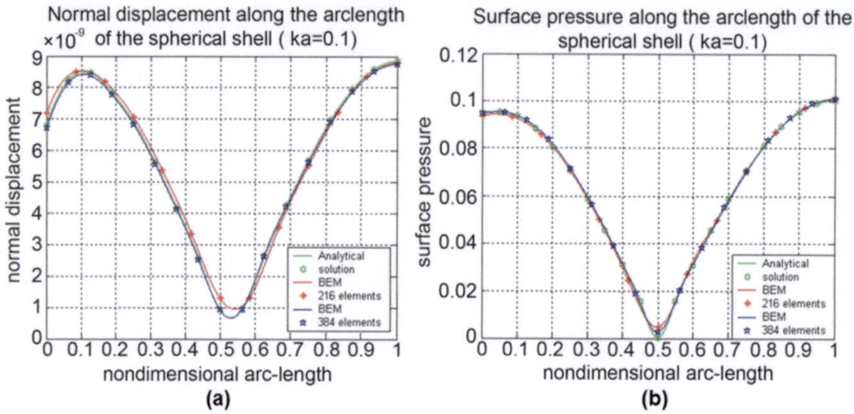

Fig. 17.24 **a** Normal displacement and **b** surface pressure distribution, on a completely flexible sphere modeled using BEM-FEM Coupling, and subjected to external normal force at a point on the surface

After using modal approach in structural dynamics, the solution of Eq. (17.18) can be obtained by solving it as a stability equation in a"unified treatment" (see Appendix 3). The disturbance acoustic pressure already incorporates the total pressure (incident plus scattering pressure), which has been "tuned" to behave like the aerodynamic terms in the modal equation. Such approach allows the application of the solution of the acousto-aeroelastic stability equation in the frequency domain using V-g method. Alternatively, the acousto-aeroelastic equation part can also be treated as a dynamic response problem, which forms the second generic approach and which has been dealt with in [23–25].

A simple solution using CHIEF regularization has been employed to address the fundamental issue of spurious solution. The solution of Eq. (17.17) as a result of the effect of acoustic disturbance to a structure is given as the pressure loading response on the structure as well as the pressure field in the fluid medium and is given as total pressure. This allows the calculation of the scattering pressure due to the incident pressure.

Since the method outlined focused on proof of concept, it has not been elaborated in its application in more complex geometries and problems. The improvement of the computational speed and efficiency will be subject to further separate work.

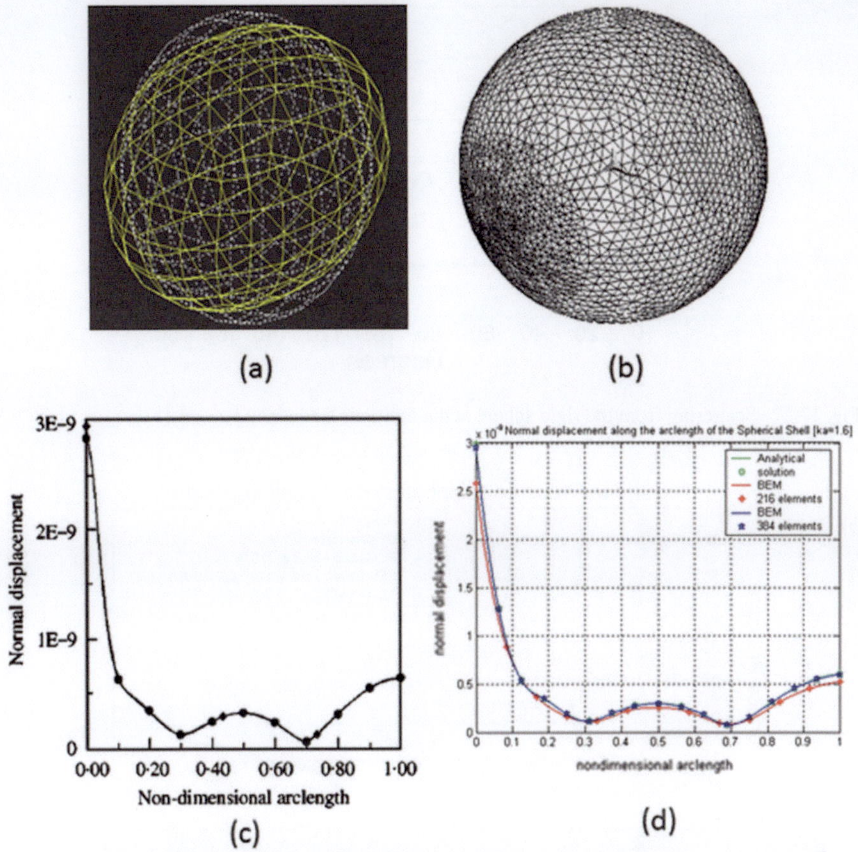

Fig. 17.25 **a** Typical mode shape, showing the triangular FEM Elements on the spherical shell; **b** Triangular FEM Shell elements adopted by Chen et al. [11] which is also used here for bench marking. **c** Normal displacement along a meridional arc for $k = 1.6$; **d** Normal displacement along a meridional arc for $k = 1.6$ obtained by Chen et al. [11]

Fig. 17.26 Frequency sweep plot for the pulsating sphere model [15]

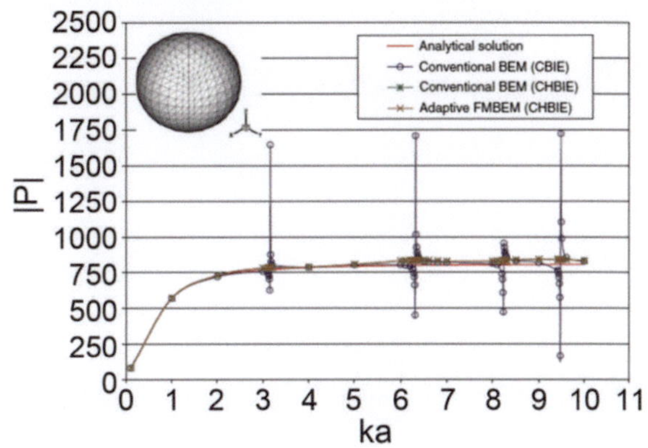

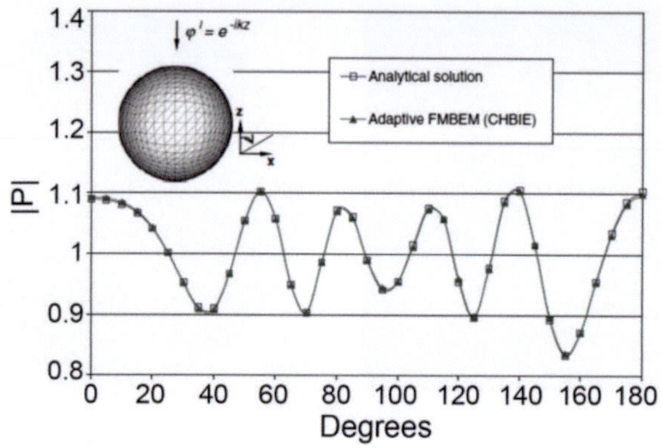

Fig. 17.27 Scattering from the rigid sphere at the fictitious frequency $k_a = \pi$ [11]

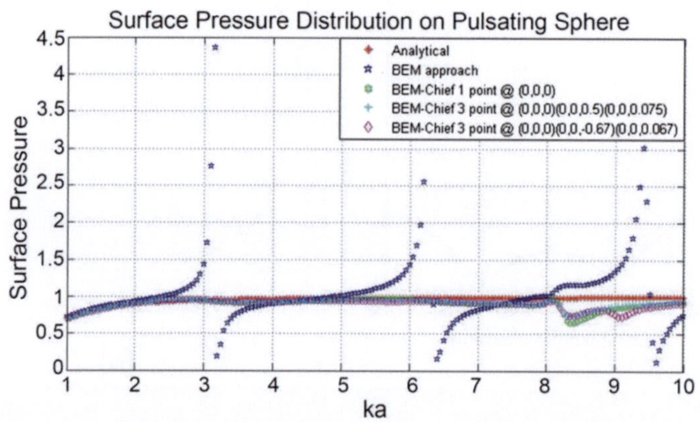

Fig. 17.28 Surface pressure distribution on pulsating sphere for analytical, BEM and BEM-CHIEF solution for one and two CHIEF point [9, 10]

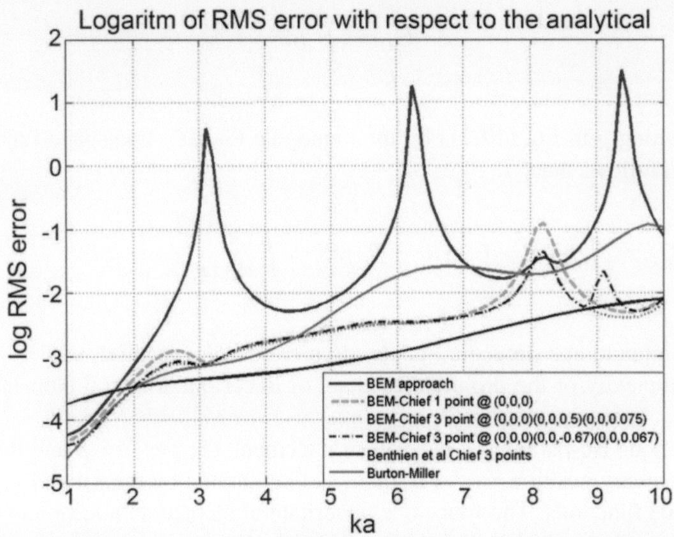

Fig. 17.29 Comparison of the logarithm of RMS error of surface pressure distribution of pulsating sphere using CHIEF method (present work), Benthiem–Schenck CHIEF results and Chen, Cheng & Harris Burton–Miller method [4]. 384 isometric surface elements are utilized, and the use of one (1) CHIEF point has demonstrated its ability to eliminate the spurious solution with reasonably good accuracy

Appendix 1: Discretization of Helmholtz Integral Equation

The Helmholtz equation is discretized by dividing the boundary surface S into N elements. The discretized boundary integral equation as given by Eq. (17.9a) and the following equations can be written as,

$$cp_i - p_{inc} - \sum_{j=1}^{N} \int_S p\overline{g}\mathrm{d}S = i\rho_0\omega \sum_{j=1}^{N} \int_S gv\mathrm{d}S \tag{17.21}$$

where i indicates field point, j source point and S_j surface element j, and for convenience, $\overline{g}$ is defined as

$$\overline{g} \equiv \frac{\partial g}{\partial n} \tag{17.22}$$

Let

$$\overline{H}_{ij} = \int_{S_j} \overline{g}\mathrm{d}S \tag{17.23a}$$

$$G_{ij} = \int\limits_{S_j} g\mathrm{d}S \tag{17.23b}$$

Substituting g in Eq. (17.21) by the monopole Green's free-space fundamental solution, it follows that:

$$G_{ij} = \int\limits_{S_j} g\mathrm{d}S = \int\limits_{S_j} g\left(R_i - R_j\right)\mathrm{d}S = \int\limits_{S_j} \frac{e^{-ik\left|R_i - R_j\right|}}{4\pi\left|R_i - R_j\right|}\mathrm{d}S \tag{17.24}$$

The computational procedure is based on collocation method, while depending on the complexity of the problem, constant or linear quadrilateral element will be employed.

A four-node iso-parametric quadrilateral element, the pressure p and the normal velocity v at any position on the element can be defined by their nodal values and linear shape functions. The four-node quadrilateral element can have any arbitrary orientation in the three-dimensional space. Using Cartesian coordinate system, and going through the algebra and using linear shape functions, the integral on the left-hand side of Eq. (17.20), considered over one element j, can be written as:

$$\begin{aligned}\int\limits_{S_j} p\overline{g}_i\mathrm{d}s &= \int\limits_{S_j} \left[N_1\ N_2\ N_3\ N_4\right]\overline{g}_i\mathrm{d}S \begin{bmatrix} p_1 \\ p_2 \\ p_3 \\ p_4 \end{bmatrix}_j \\ &= \left[\overline{h}_{ij}^1\ \overline{h}_{ij}^2\ \overline{h}_{ij}^3\ \overline{h}_{ij}^4\right] \begin{bmatrix} p_1 \\ p_2 \\ p_3 \\ p_4 \end{bmatrix}_j \end{aligned} \tag{17.25}$$

while that on the right-hand side

$$\begin{aligned}\int\limits_{S_j} gv\mathrm{d}S &= \int\limits_{S_j} \left[N_1\ N_2\ N_3\ N_4\right]g_i\mathrm{d}S \begin{bmatrix} v_1 \\ v_2 \\ v_3 \\ v_4 \end{bmatrix}_j \\ &= \left[g_{ij}^1\ g_{ij}^2\ g_{ij}^3\ g_{ij}{}^n\right] \begin{bmatrix} v_1 \\ v_2 \\ v_3 \\ v_4 \end{bmatrix}_i \end{aligned} \tag{17.26}$$

where

$$\overline{h}_{ij}^{k} = \int_{S_j} N_k \overline{g}_j \mathrm{d}S \quad k = 1, 2, 3, 4 \tag{17.27}$$

$$g_{ij}^{k} = \int_{S_j} N_k g_j \mathrm{d}S \quad k = 1, 2, 3, 4 \tag{17.28}$$

The integration in Eq. (17.27) and (17.28) can be carried out using Gauss integration [18, 31]. Going through the algebra, Eq. (17.21) can be recast into a discretized set of simultaneous linear equations, which relates the pressure p_i at field point i due to the boundary conditions p to v at source surface S_i of element i and the incident pressure p_{inc}, given by:

$$\sum_{j=1}^{N} \left[h_{ij}^{1}\ h_{ij}^{2}\ h_{ij}^{3}\ h_{ij}^{4} \right] \begin{bmatrix} p_1 \\ p_2 \\ p_3 \\ p_4 \end{bmatrix}_j = i\rho\omega_0 \sum_{j=1}^{N} \left[g_{ij}^{1}\ g_{ij}^{2}\ g_{ij}^{3}\ g_{ij}^{4} \right] \begin{bmatrix} v_{n1} \\ v_{n2} \\ v_{n3} \\ v_{n4} \end{bmatrix}_j + p_{\text{inc}} \tag{17.29}$$

or in matrix form:

$$[\mathbf{H}]\{p\} = i\rho_0\omega[\mathbf{G}]\{v\} + \{p_{\text{inc}}\} \tag{17.30}$$

where $\mathbf{H}$ and $\mathbf{G}$ are two $N \times N$ matrices of influence coefficients, while p and v are vectors of dimension N representing total pressure and normal velocity on the boundary elements. This matrix equation can be solved if the boundary condition $v = \partial p/\partial n$ and the incident acoustic pressure field p_{inc} are known.

At this point, a few remarks are necessary. Proper interpretation should be given to the diagonal terms of [H] in Eq. (17.30) as implied by the original boundary integral (7), since these terms concern the evaluation of influence coefficient for which the field point is located at the source element. Accordingly, [H] should be written as

$$[\mathbf{H}] = [\mathbf{H}]^{D} + [\mathbf{H}]^{OD} \tag{17.31}$$

i.e. the diagonal and the off-diagonal part. The matrix $[\mathbf{H}]^D$ as implied in Eq. (17.30) can be written as $[\mathbf{H}]^D = [\overline{H}]^D + [\mathbf{C}]$ where $\mathbf{C}$ is space angle constant which is the quotient of $\Omega/4\pi$ and $\overline{H}$ is the matrix implied by the second term of Eq. (17.25). For a node coinciding with three or four corner elements, Ω is the space angle toward the acoustic medium, and the space angle for a sphere is 4π. For a smooth surface the space angle is 2π, and $\mathbf{C} = ½$.

Appendix 2: BEM-FEM Acoustic-Aeroelastic Coupling (AAC)

The BE region is treated as a super finite element and its stiffness matrix is computed and assembled into the global stiffness matrix and identified as the coupling to finite elements. The state of affairs is schematically depicted in Fig. 17.8.

The utilization of FEM on the structural domain leads to a system of simultaneous equations which relate the displacements at all the nodes to the *nodal forces*. In the BEM, on the other hand, a relationship between nodal displacements and *nodal tractions* is established. Representing the elastic structure by FE model, the structural dynamic equation of motion is given by Bisplinghoff, Ashley and Halfman [4]

$$[\mathbf{M}]\{\ddot{x}\} + [\mathbf{C}]\{\dot{x}\} + [\mathbf{K}]\{x\} = \{\mathbf{F}\} \tag{17.32}$$

where **M**, **C** and **K** are structural mass, damping and stiffness, respectively, which are expressed as matrices in a FE model, while **F** is the given external forcing function vector, and $\{x\}$ is the structural displacement vector. The incorporation of the self-excited aerodynamic effects to the structural dynamic's equation can be written as [24–26]:

$$[\mathbf{M}]\{\ddot{\mathbf{x}}\} + [\mathbf{C}]\{\dot{\mathbf{x}}\} + [\mathbf{K}]\{\mathbf{x}\} - \mathbf{q}_{\infty}[\mathbf{A}(ik)]\{\mathbf{x}\} = \{\mathbf{0}\} \tag{17.33}$$

where $\mathbf{A}(ik)$ is an aerodynamic influence coefficient after applying aerostructure coupling from the control points of aerodynamic boxes to the structural finite element grid points as elaborated by Djojodihardjo and Safari [23].

Taking into account the acoustic pressure p on the structure at the fluid–structure interface as a separate excitation force, the acoustic-structure problem can be obtained from Eq. (17.33) by introducing a fluid–structure coupling term given by $[\mathbf{L}]\{p\}$. It follows that

$$[\mathbf{M}]\{\ddot{x}\} + [\mathbf{C}]\{\dot{x}\} + [\mathbf{K}]\{x\} - q_{\infty}[\mathbf{A}(ik)]\{x\} + [\mathbf{L}]\{p\} = \{\mathbf{F}\} \tag{17.34}$$

where **L** is a coupling matrix of size $M \times N$ in the BEM/FEM coupling thus formulated. M is the number of FE degrees of freedom and N is the number of BE nodes on the coupled boundary. For the BE part of the surface at the fluid–structure interface a, Eq. (17.30) can be utilized.

The global coupling matrix **L** transforms the acoustic fluid pressure acting on the nodes of boundary elements on the entire fluid–structure interface surface a, to nodal forces on the finite elements of the structure. Hence **L** consists of n assembled local transformation matrices L_e, given by

$$L_e = \int\limits_{S_e} N_F^T n N_B \mathrm{d}S \tag{17.35}$$

in which N_F is the shape function matrix for the finite element, and N_B is the shape function matrix for the boundary element. It can be shown that:

$$N_F = \begin{bmatrix} 1\,0\,0\,0\,0 \\ 0\,1\,0\,0\,0 \\ 0\,0\,1\,0\,0 \end{bmatrix} [N_i] \tag{17.36}$$

The rotational parts in N_F are neglected since these are considered to be small in comparison with the translational ones in the BE-FE coupling, consistent with the assumptions in structural dynamics as, for example, stipulated in [4].

For the normal fluid velocities and the normal translational displacements on the shell elements at the fluid–structure coupling interface, a relationship, which takes into account the velocity continuity over the coinciding nodes, should be established. This relationship is given by Eq. (17.18)

$$\mathrm{v} = i\omega(\mathbf{T} \cdot x) \tag{17.37}$$

Similar to **L**, **T**$(n \times m)$ is also a global coupling matrix that connects the normal velocity of a BE node with the translational displacements of FE nodes obtained by taking the transpose of the boundary surface normal vector n. The local transformation vector T_e can then be written as:

$$T_e = n^T \tag{17.38}$$

The presence of an acoustic source can further be depicted by Fig. 17.30. Three regions are considered, i.e. a, b and c; region a is the fluid–structure interface region, where FEM mesh and BEM mesh coincide and region b and c is the region where all of the boundary conditions (pressure or velocity) are known. For the coupled FEM-BEM regions, BEM equation can now be written as:

$$\begin{bmatrix} \mathbf{H}_{11} & \mathbf{H}_{12} & \mathbf{H}_{13} \\ \mathbf{H}_{21} & \mathbf{H}_{22} & \mathbf{H}_{23} \\ \mathbf{H}_{31} & \mathbf{H}_{32} & \mathbf{H}_{33} \end{bmatrix} \begin{Bmatrix} p_a \\ p_b \\ p_c \end{Bmatrix} = i\rho_0\omega \begin{bmatrix} \mathbf{G}_{11} & \mathbf{G}_{12} & \mathbf{G}_{13} \\ \mathbf{G}_{21} & \mathbf{G}_{22} & \mathbf{G}_{23} \\ \mathbf{G}_{31} & \mathbf{G}_{32} & \mathbf{G}_{33} \end{bmatrix} \begin{Bmatrix} v_a \\ v_b \\ v_c \end{Bmatrix} + \begin{Bmatrix} p_{\mathrm{inc}_a} \\ p_{\mathrm{inc}_b} \\ p_{\mathrm{inc}_c} \end{Bmatrix} \tag{17.39}$$

Considering $\mathbf{v}_a = i\omega(\mathbf{T} \cdot x)$, BEM equation can be modified as:

$$\begin{bmatrix} \mathbf{H}_{11} & \mathbf{H}_{12} & \mathbf{H}_{13} \\ \mathbf{H}_{21} & \mathbf{H}_{22} & \mathbf{H}_{23} \\ \mathbf{H}_{31} & \mathbf{H}_{32} & \mathbf{H}_{33} \end{bmatrix} \begin{Bmatrix} p_a \\ p_b \\ p_c \end{Bmatrix} = i\rho_0\omega \begin{bmatrix} \mathbf{G}_{11} & \mathbf{G}_{12} & \mathbf{G}_{13} \\ \mathbf{G}_{21} & \mathbf{G}_{22} & \mathbf{G}_{23} \\ \mathbf{G}_{31} & \mathbf{G}_{32} & \mathbf{G}_{33} \end{bmatrix} \begin{Bmatrix} i\omega[\mathbf{T}]x \\ v_b \\ v_c \end{Bmatrix} + \begin{Bmatrix} p_{\mathrm{inc}_a} \\ p_{\mathrm{inc}_b} \\ p_{\mathrm{inc}_c} \end{Bmatrix} \tag{17.40}$$

or:

$$\mathbf{H}_{11} p_a + \mathbf{H}_{12} p_b + \mathbf{H}_{13} p_c = -\rho_0\omega^2 \mathbf{G}_{11}\mathbf{T}x + i\rho_0\omega\mathbf{G}_{12} v_b + i\rho_0\omega\mathbf{G}_{13} v_c + \{p_{\mathrm{inc}_a}\}$$

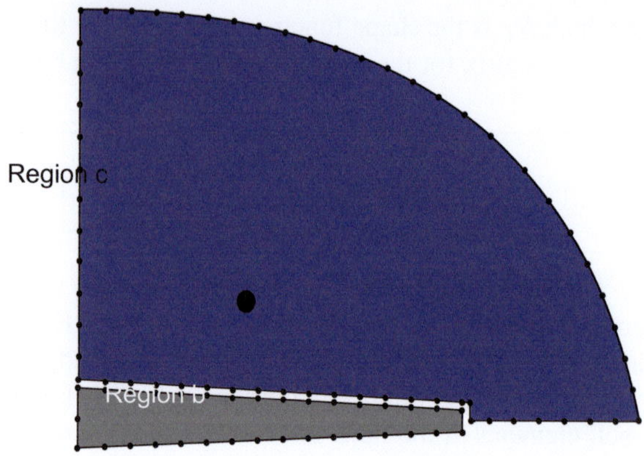

Fig. 17.30 Schematic of FE-BE problem representing quarter space problem domains for half wing

$$\begin{aligned}
\mathbf{H}_{21}p_a + \mathbf{H}_{22}p_b + \mathbf{H}_{23}p_c &= -\rho_0\omega^2\mathbf{G}_{21}\mathbf{T}x + i\rho_0\omega\mathbf{G}_{22}v_b + i\rho_0\omega\mathbf{G}_{23}v_c + \left\{p_{\text{inc}_b}\right\} \\
\mathbf{H}_{31}p_a + \mathbf{H}_{32}p_b + \mathbf{H}_{33}p_c &= -\rho_0\omega^2\mathbf{G}_{31}\mathbf{T}x + i\rho_0\omega\mathbf{G}_{32}v_b + i\rho_0\omega\mathbf{G}_{33}v_c + \left\{p_{\text{inc}_c}\right\}
\end{aligned} \tag{17.41}$$

If the pressure boundary condition on b (p_b), velocity boundary condition on c (v_c), and the incident pressure on a, b and c are known, by taking to the left side all the unknown the above equation can be written as:

$$\begin{aligned}
&\rho_0\omega^2\mathbf{G}_{11}\mathbf{T}x + \mathbf{H}_{11}p_a - i\rho_0\omega\mathbf{G}_{12}v_b + \mathbf{H}_{13}p_c \\
&\quad = -\mathbf{H}_{12}p_b + i\rho_0\omega\mathbf{G}_{13}v_c + \left\{p_{\text{inc}_a}\right\} \\
&\rho_0\omega^2\mathbf{G}_{21}\mathbf{T}x + \mathbf{H}_{21}p_a - i\rho_0\omega\mathbf{G}_{22}v_b + \mathbf{H}_{23}p_c \\
&\quad = -\mathbf{H}_{22}p_b + i\rho_0\omega\mathbf{G}_{23}v_c + \left\{p_{\text{inc}_b}\right\} \\
&\rho_0\omega^2\mathbf{G}_{31}\mathbf{T}x + \mathbf{H}_{31}p_a - i\rho_0\omega\mathbf{G}_{32}v_b + \mathbf{H}_{33}p_c \\
&\quad = -\mathbf{H}_{32}p_b + i\rho_0\omega\mathbf{G}_{33}v_c + \left\{p_{\text{inc}_c}\right\}
\end{aligned} \tag{17.42}$$

Since the pressure p on FEM equation lies in region a, Eq. (17.34) can be written as

$$[\mathbf{M}]\{\ddot{x}\} + [\mathbf{C}]\{\dot{x}\} + [\mathbf{K}]\{x\} - q_\infty[\mathbf{A}(ik)]\{x\} + [\mathbf{L}]\{p_a\} = \{\mathbf{F}\} \tag{17.43}$$

where p_a is the total acoustic pressure resulting from the application of acoustic disturbance force to the structure, which consists of the incident acoustic pressure p_{inc} and scattering acoustic pressure p_{sc}. The scattering acoustic pressure will be dependent on the dynamic response of the structure due to the incident acoustic pressure. Following the general practice in structural dynamics, solutions of Eq. (17.43) are sought by considering synchronous motion with harmonic frequency ω. Correspondingly, Eq. (17.43) reduces to:

$$\left[\mathbf{K} + i\omega\mathbf{C} - \omega^2\mathbf{M}\right]\{\overline{x}\} - q_\infty[\mathbf{A}(ik)]\{x\} + [\mathbf{L}]\{\overline{p}_a\} = \{\overline{\mathbf{F}}\} \tag{17.44}$$

where

$$x = \overline{x}e^{i\omega t}, \tag{17.45a}$$

$$p_a = \overline{p}_a e^{i\omega t} \tag{17.45b}$$

or, dropping the bar sign for convenience, but keeping the meaning in mind, Eq. (17.43) can be written as

$$\left[\mathbf{K} + i\omega\mathbf{C} - \omega^2\mathbf{M}\right]\{x\} - q_\infty[\mathbf{A}(ik)]\{x\} + [\mathbf{L}]\{p_a\} = \{\mathbf{F}\} \tag{17.46}$$

Combination of Eqs. (17.42) and (17.46) yields the coupled BEM-FEM Equation as given by Eq. (17.19).

Appendix 3: Further Treatment for Acoustic-Aeroelastic Coupling; Acoustic-Aerodynamic Analogy

At this point, the solution approach philosophy is in order. Analogous to the treatment of dynamic aeroelastic stability problem of structure, in which the aerodynamic effects can be distinguished into motion independent (self-excited) and motion-induced aerodynamic forces, the effect of acoustic pressure disturbance to the aeroelastic structure (acousto-aeroelastic problem) can be viewed to consist of structural motion independent incident acoustic pressure (excitation acoustic pressure) and structural motion dependent acoustic pressure, which is known as the scattering pressure. However the scattering acoustic pressure is also dependent on the incident acoustic pressure. The consequence of such treatment has been adopted in the above section and will further be implemented in the subsequent development.

For aeroelastic calculation purposes, further treatment to simplify Eq. (17.46) will be carried out. Since the pressure boundary condition on b ($p_b = 0$) and velocity boundary condition on c ($v_c = 0$), Eq. (17.42) can be written as:

$$\rho_0\omega^2\mathbf{G}_{11}\mathbf{T}x + \mathbf{H}_{11}p_a - i\rho_0\omega\mathbf{G}_{12}v_b + \mathbf{H}_{13}p_c = p_{\text{inc}_a} \tag{17.47a}$$

$$\rho_0\omega^2\mathbf{G}_{21}\mathbf{T}x + \mathbf{H}_{21}p_a - i\rho_0\omega\mathbf{G}_{22}v_b + \mathbf{H}_{23}p_c = p_{\text{inc}_b} \tag{17.47b}$$

$$\rho_0\omega^2\mathbf{G}_{31}\mathbf{T}x + \mathbf{H}_{31}p_a - i\rho_0\omega\mathbf{G}_{32}v_b + \mathbf{H}_{33}p_c = p_{\text{inc}_c} \tag{17.47c}$$

Since $\mathbf{G}_{22}$ and $\mathbf{H}_{33}$ is square matrix, Eqs. (17.47b) and (17.47c) can be written as

$$v_b = -\frac{1}{\rho_0\omega}[\mathbf{G}_{33}]^{-1}\left(\rho_0\omega^2\mathbf{G}_{21}\mathbf{T}x + \mathbf{H}_{21}p_a + \mathbf{H}_{23}p_c - p_{\mathrm{inc}_c}\right) \tag{17.48a}$$

$$p_c = -[\mathbf{H}_{33}]^{-1}\left(\rho_0\omega^2\mathbf{G}_{31}\mathbf{T}x + \mathbf{H}_{31}p_a - i\rho_0\omega\mathbf{G}_{32}v_b - p_{\mathrm{inc}_c}\right) \tag{17.48b}$$

Substituting Eq. (17.48b) into Eq. (17.47b)

$$\rho_0\omega^2\mathbf{A}_{21}\mathbf{T}x + \mathbf{B}_{21}p_a - i\rho_0\omega\mathbf{A}_{22}v_b + \mathbf{B}_{23}p_{\mathrm{inc}_c} - p_{\mathrm{inc}_b} = \{0\} \tag{17.49}$$

where

$$\begin{aligned}
\mathbf{A}_{21} &= \left(\mathbf{G}_{21} - \mathbf{H}_{23}[\mathbf{H}_{33}]^{-1}\mathbf{G}_{31}\right)\\
\mathbf{B}_{21} &= \left(\mathbf{H}_{21} - \mathbf{H}_{23}[\mathbf{H}_{33}]^{-1}\mathbf{H}_{31}\right)\\
\mathbf{B}_{23} &= \left(\mathbf{H}_{23}[\mathbf{H}_{33}]^{-1}\right)\\
\mathbf{A}_{22} &= \left(\mathbf{G}_{22} - \mathbf{H}_{23}[\mathbf{H}_{33}]^{-1}\mathbf{G}_{32}\right)
\end{aligned} \tag{17.50}$$

Since $\mathbf{A}_{22}$ is square matrix, Eq. (17.50) can be written as:

$$v_b = -\frac{i}{\rho_0\omega}[\mathbf{A}_{22}]^{-1}\left(\rho_0\omega^2\mathbf{A}_{21}\mathbf{T}x + \mathbf{B}_{21}p_a + \mathbf{B}_{23}p_{\mathrm{inc}_c} - p_{\mathrm{inc}_b}\right) \tag{17.51}$$

Substituting Eq. (17.48a) into Eq. (17.47c)

$$\rho_0\omega^2\mathbf{C}_{31}\mathbf{T}x + \mathbf{A}_{31}p_a + \mathbf{A}_{33}p_c + \mathbf{B}_{32}p_{\mathrm{inc}_b} - p_{\mathrm{inc}_c} = \{0\} \tag{17.52}$$

where

$$\begin{aligned}
\mathbf{C}_{31} &= \left(\mathbf{G}_{31} - \mathbf{G}_{32}[\mathbf{G}_{22}]^{-1}\mathbf{G}_{21}\right)\\
\mathbf{A}_{31} &= \left(\mathbf{H}_{31} - \mathbf{G}_{32}[\mathbf{G}_{22}]^{-1}\mathbf{H}_{21}\right)\\
\mathbf{A}_{33} &= \left(\mathbf{H}_{33} - \mathbf{G}_{32}[\mathbf{G}_{22}]^{-1}\mathbf{H}_{23}\right)\\
\mathbf{B}_{32} &= \left(\mathbf{G}_{32}[\mathbf{G}_{22}]^{-1}\right)
\end{aligned} \tag{17.53}$$

Since $\mathbf{A}_{33}$ is square matrix, Eq. (17.52) can be written as:

$$p_c = -[\mathbf{A}_{33}]^{-1}\left(\rho_0\omega^2\mathbf{C}_{31}\mathbf{T}x + \mathbf{A}_{31}p_a + \mathbf{B}_{32}p_{\mathrm{inc}_b} - p_{\mathrm{inc}_c}\right) \tag{17.54}$$

Substituting Eq. (17.48b) and (17.54) into Eq. (17.47a)

$$\rho_0\omega^2\mathbf{D}_{11}\mathbf{T}x + \mathbf{E}_{11}p_a + \mathbf{F}_{12}p_{\mathrm{inc}_b} + \mathbf{F}_{13}p_{\mathrm{inc}_c} = \left\{p_{\mathrm{inc}_a}\right\} \tag{17.55}$$

where

$$\mathbf{D}_{11} = \left(\mathbf{G}_{11} - \mathbf{G}_{12}[\mathbf{A}_{22}]^{-1}\mathbf{A}_{21} - \mathbf{H}_{13}[\mathbf{A}_{33}]^{-1}\mathbf{C}_{31}\right)$$

$$\begin{aligned}\mathbf{E}_{11} &= \left(\mathbf{H}_{11} - \mathbf{G}_{12}[\mathbf{A}_{22}]^{-1}\mathbf{B}_{21} - \mathbf{H}_{13}[\mathbf{A}_{33}]^{-1}\mathbf{A}_{31}\right)\\ \mathbf{F}_{12} &= \mathbf{G}_{12}[\mathbf{A}_{22}]^{-1} - \mathbf{H}_{13}[\mathbf{A}_{33}]^{-1}\mathbf{B}_{32}\\ \mathbf{F}_{13} &= \mathbf{H}_{13}[\mathbf{A}_{33}]^{-1} - \mathbf{G}_{12}[\mathbf{A}_{22}]^{-1}\mathbf{B}_{23}\end{aligned} \tag{17.56}$$

Since $\mathbf{E}_{11}$ is square matrix

$$p_a = -[\mathbf{E}_{11}]^{-1}\left(\rho_0\omega^2\mathbf{D}_{11}\mathbf{T}\{x\} - p_{\text{inc}_a} + \mathbf{F}_{12}p_{\text{inc}_b} + \mathbf{F}_{13}p_{\text{inc}_c}\right) \tag{17.57}$$

Matrix $\mathbf{E}_{11}$ and $\mathbf{D}_{11}$ are also a square matrix; finally by substituting Eq. (17.57) into Eq. (17.46), the BEM-FEM aeroacoustic-structure coupling can be obtained as:

$$\begin{aligned}&\left[\mathbf{K} + i\omega\mathbf{C} - \omega^2\mathbf{M}\right]\{x\} - q_\infty[\mathbf{A}(ik)]\{x\}\\ &\quad + [\mathbf{L}]\left(-[\mathbf{E}_{11}]^{-1}\left(\rho_0\omega^2\mathbf{D}_{11}\mathbf{T}\{x\} - p_{\text{inc}_a} + \mathbf{F}_{12}p_{\text{inc}_b} + \mathbf{F}_{13}p_{\text{inc}_c}\right)\right) = \{\mathbf{F}\}\end{aligned} \tag{17.58}$$

Incident pressure on region b and c will not influence the stability problem associated with the structures and may at this point be disregarded. Hence, without considering damping matrix $\mathbf{C}$ Eq. (17.58) simplifies to:

$$\begin{aligned}&\left[\mathbf{K} - \omega^2\mathbf{M}\right]\{x\} - q_\infty[\mathbf{A}(ik)]\{x\} - \rho_0\omega^2[\mathbf{L}][\mathbf{E}_{11}]^{-1}[\mathbf{D}_{11}][\mathbf{T}]\{x\}\\ &\quad = -[\mathbf{L}][\mathbf{E}_{11}]^{-1}\left\{p_{\text{inc}_a}\right\} + \{\mathbf{F}\}\end{aligned} \tag{17.59}$$

or

$$\left[\mathbf{K} - \omega^2\mathbf{M}\right]\{x\} - q_\infty[\mathbf{A}(ik)]\{x\} - \rho_0\omega^2\left[\mathbf{F}_{ac_{sc}}(k_w)\right]\{x\} = \left\{\mathbf{F}_{ac_{inc}}(k_w)\right\} + \{\mathbf{F}\} \tag{17.60}$$

where

$$\begin{aligned}\left[\mathbf{F}_{ac_{sc}}(k_w)\right] &= [\mathbf{L}][\mathbf{E}_{11}]^{-1}[\mathbf{D}_{11}][\mathbf{T}]\\ \left\{\mathbf{F}_{ac_{inc}}(k_w)\right\} &= -[\mathbf{L}][\mathbf{E}_{11}]^{-1}\left\{p_{inc_a}\right\}\end{aligned} \tag{17.61}$$

Equation (17.60) will not be solved directly since the size of the mass and stiffness matrices of the aircraft model are very large. Instead one uses the modal approach where the structural deformation $\{x\}$ is transformed to the generalized coordinate $\{q\}$ given by the following relation:

$$x = \Phi q \tag{17.62}$$

where Φ is the modal matrix whose columns contain the lower-order natural modes. Premultiplying by Φ^{T} and converting dynamic pressure q_∞ into reduced frequency (k) as elaborated by Djojodihardjo and Safari [23, 26], Eq. (17.60) can then be written as:

$$\Phi^T\left[K-\omega^2\left(M+\frac{\rho}{2}\left(\frac{L}{k}\right)^2[A(ik)]+\rho_0\left[F_{ac_{sc}}(k_w)\right]\right)\right]\Phi\{q\}$$
$$=\Phi^T\left\{F_{ac_{inc}}(k_w)\right\}+\Phi^T\{F\} \quad (17.63)$$

since all of the acoustic terms are functions of wave number (k_w). Equation (17.60) can be solved by utilizing iterative procedure.

Incorporation of the scattering acoustic term along with the aerodynamic term in the second term of Eq. (17.63) can be regarded as one manifestation of the acoustic-aerodynamic analogy followed in this approach.

Further method of approach for the solution of the acousto-aeroelastic problem is then dealt with. Following the same procedure as developed in earlier work [24, 25], the acoustic excitation is incorporated by coupling it to the unsteady aerodynamic load in the flutter stability formulation. Linearity and principle of superposition has been assumed. Hence the acoustic loading can be superposed to the aerodynamic loading on the structure and form the modified aeroelastic equation (acousto-aeroelastic equation) of the structural dynamic problem associated with acoustic and aerodynamic excitation. Detailed elaboration can be found in [24] and [27].

References

1. Djojodihardjo, H. 2015. Vibro-acoustic analysis of the acoustic-structure interaction of flexible structure due to acoustic excitation. *Acta Astronautica* 108: 129–145.
2. Annarella, C. 1991. Spacecraft Structures. www.thirdwave.de/3w/tech/spacecraft/spacecraftstructures.pdf.
3. Hughes, William O. 2006. *Application of the bootstrap statistical method in deriving vibroacoustic specifications*. NASA-TM—2006–214446.
4. Bisplinghoff, R.L., H. Ashley, and R.L. Halfman. 1955. *Aeroelasticity*. Addison-Wesley Publishing Company.
5. Benthien, G.W., and H.A. Schenck. 1991. Structural-acoustic coupling. In *Boundary element methods in acoustics*. London: Elsevier Applied Science.
6. Burton, A.J., and G.F. Miller. 1971. The application of the integral equation method to the numerical solution of some exterior boundary value problems. *Proceedings of the Royal Society of London Series A* 323: 201–210.
7. Calvi, A. 2010. Spacecraft structural dynamics and loads, an overview. www.ltas-vis.ulg.ac.be/uploads/File/CALVI---LIEGE_2010.
8. Chang, K.Y., J.C. Cockburn, and G.C. Kao. 1973. *Prediction of vibro-acoustic loading criteria for space vehicle components*. NASA-CR-171019, Wyle Laboratories—Research Staff, Report WR 73-9, September.
9. Chen, I.L., J.T. Chen, S.R. Kuo, and M.T. Liang. 2001. A new method for true and spurious eigensolutions of arbitrary cavities using the combined Helmholtz exterior integral equation formulation method. *The Journal of the Acoustical Society of America*.
10. Chen, I.L., J.T. Chen, and M.T. Liang. 2001. Analytical study and numerical experiments for radiation and scattering problems using the chief method. *Journal of Sound and Vibration* 248 (5): 809–828.
11. Chen, J.T., I.L. Chen, and K.H. Chen. 2006. Treatment of rank deficiency in acoustics using SVD. *Journal of Computational Acoustics*.

12. Chen, K., J. Cheng, and P.J. Harris. 2009. A new study of the Burton and Miller method for the solution of a 3D Helmholtz problem. *IMA Journal of Applied Mathematics* 74: 163–177.
13. Chronopoulos, D. 2012. *Prediction of the vibroacoustic response of aerospace composite structures in a broadband frequency range*. Ph.D. Thesis, Ecole Centrale De Lyon.
14. Chuh-Mei, and C.S. Pates III. 1995. *Analysis of random structure-acoustic interaction problems using coupled boundary element and finite element methods*. NASA-CR-195931.
15. Citarella, R., L. Federico, and A. Cicatiello. 2007. Modal acoustic transfer vector approach in a FEM-BEM vibroacoustic analysis. *Engineering Analysis with Boundary Elements* 31: 248–258.
16. Cohan, L.E., and D.W. Miller. 2011. Vibroacoustic launch analysis and alleviation of lightweight, active mirrors. *Optical Engineering* 50: 013002.
17. Coyette, J.-P. 2003. Sources of uncertainties in vibro-acoustic. www.infobruit.com/revues/-8-11098.
18. Coyette, J.P., Y. Detandt, G. Lielens, and B. Van den Nieuwenhof. 2008. *Vibro-acoustic simulation of mechanical components subjected to distributed pressure excitations*. Free Field Technologies.
19. R.S. Pappa, J.I. Pritchard, and R.D. Buehrle. 1999. *Vibro-acoustics modal testing at NASA Langley Research Center*. NASA TM-1999-209319.
20. Pappa, R.S., J.I. Pritchard, and R.D. Buehrle. 2001. *Structural dynamics experimental activities in ultra-lightweight and inflatable space structures*. NASA/TM-2001-210857, May.
21. Pritchard, J.I., R.D. Buehrle, R.S. Pappa, and R. Grosveld. 2001. Comparison of modal analysis methods applied to a vibro-acoustic test article. In *Proceedings, IMAC-XX Conference*, pp. 1144–1152. http://sem-proceedings.com/20i/sem.org-IMAC-XX-Conf-S34P01-Comparison-Modal-Analysis-Methods-Applied-Vibro-acoustic-Test-Article-pdf. Accessed January 8, 2013.
22. Djojodihardjo, H., and E. Tendean. 2005. A computational technique for the dynamics of structure subject to acoustic excitation. In *ICAS 2004*, Vancouver, October.
23. Djojodihardjo, H., and I. Safari. 2006. Unified computational scheme for acoustic aeroelastomechanic interaction. In *Paper IAC-06-C2.5.09, Presented at the 57th International Astronautical Congress*, 1–6 October, Valencia, Spain.
24. Djojodihardjo, H. 2007. BE-FE coupling computational scheme for acoustic effects on aeroelastic structures. In *Paper IF-086, Proceedings, The International Forum on Aeroelasticity and Structural Dynamics*, held at the Royal Institute of Technology (KTH), Stockholm, Sweden, 18–20 June.
25. Djojodihardjo, H. 2007. BEM-FEM acoustic-structural coupling for spacecraft structure incorporating treatment of irregular frequencies. In *Paper IAC.07-C2.5.4, IAC Proceedings, Presented at the 58th International Astronautical Congress*, 23–28 September/Hyderabad, India.
26. Djojodihardjo, H. 2013. Unified aerodynamic-acoustic formulation for aero-acoustic structure coupling. *Journal of Mechanics Engineering and Automation* 3: 209–220. https://www.google.com/webhp?source=search_app.
27. Djojodihardjo, H., and I. Safari. 2015. BEM-FEM coupling for acoustic effects on aeroelastic stability of structures. *Computer Modeling in Engineering & Sciences* 91 (3): 205–234.
28. Frazer, David, Hank Kleespies, and Cliff Vasicek. 1991. Chapter 9, Spacecraft_Structures, Edited and revised by Cyril Anarella. https://cmapspublic3.ihmc.us/rid=1NG0Q5LSC-24DZTG5-17B8/9_SpacecraftStructures.pdf. Accessed January 7, 2013.
29. Dowling, A.P., and J.E. Ffowcs-Williams. 1983. *Sound and sources of sound*. New York, Brisbane, Chichester, Toronto: Ellis Horwood Limited, Chichester–Wiley. ©1983, A.P. Dowling and J.E. Ffowcs Williams.
30. Norton, M.P. 1989. *Fundamentals of noise and vibration analysis for engineers*. Cambridge, New York, Melbourne, Sydney: Cambridge University Press.
31. Holström, F. 2001. Structure-acoustic analysis using BEM/FEM; Implementation In MATLAB®. Master's Thesis, Copyright © 2001 by Structural Mechanics and Engineering Acoustics, LTH, Sweden, Printed by KFS i Lund AB, Lund, Sweden, May 2001. Homepage: http://www.akustik.lth.

32. Marquez, A., S. Meddahi, and V. Selgas. 2005. A new BEM-FEM coupling strategy for two-dimensional fluid–solid interaction problems. *Journal of Computational Physics* 199: 205–220.
33. Juhl, P. 1994. A numerical study of the coefficient matrix of the boundary element method near characteristic frequencies. *Journal of Sound and Vibration.*
34. Schenck, H.A. 1968. Improved integral formulation for acoustic radiation problems. *Journal of the Acoustical Society of America* 44: 41–58.
35. Wu, T.W., and A.F. Seybert. 1991. A weighted residual formulation for the CHIEF method in acoustics. *Journal of the Acoustical Society of America* 90: 1608–1614.
36. Dowell, E.H., and K.C. Hall. 2001. Modeling of fluid-structure interaction. *Annual Review of Fluid Mechanics* 33: 445–490.
37. Herrin, D.W., T.W. Wu, and A.F. Seybert. 2015. Boundary element modeling. Published online: 8 April 2008. https://doi.org/10.1002/9780470209707.ch8. Accessed 15 April.
38. Junger, M.C., and D. Feit. 1986. *Sound, structures and their interaction.* MIT Press.
39. Pierce, A.D. 1981. *Acoustics: An introduction to its physical principles and applications.* New York: McGraw-Hill.

Chapter 18
Introduction and Case Studies in Aeroelasticity of Bridges and Tall Structures

Abstract Design of structures, including buildings and bridges, is dictated by safety and economy, which, due to progress in technology, materials and computational methods, has led to the utilization of light structures with optimum strength. In this regard, aeroelastic phenomenon that has been observed with great interest and concern by aerodynamicists, aeroelasticians and engineers is observed in civil structures, most spectacular of which are the aeroelastic flutter phenomena on a relatively slender bridge structure, due to vortex shedding. Tall buildings, now approaching the frontier of 1000 m height, have enormously spread worldwide in recent years and led to new challenging problems facing the international engineering community. Wind-induced vibration results in one issue that may be of concern for the serviceability design of tall buildings and which produces discomfort. The wind action on slender structures with low natural frequencies can introduce uncomfortable vibrations which could affect general well-being and interfere with the daily activities of the occupants. Consequently, the risk of exceeding acceptable vibration limits and of causing discomfort should be estimated and eliminated. These topics are here illustrated and discussed, accompanied by selected examples that should provide a comprehensive understanding. Analytical and experimental means of incorporating wind-induced vibration in bridges and structures will be discussed to gain a comprehensive, although introductory understanding. Method used for the study of flutter stability analysis of the structure during motion will be exemplified. Some illustrations are presented without elaborate discussions, since pictures implicitly contain a large body of information that may need lengthy elaboration, but most importantly, pictures will incite creativity and curiosity for further active learning efforts by the readers.

Keywords Aeroelasticity · Building aerodynamics · Bluff body aerodynamic oscillations · Civil structure aeroelasticity · Vortex shedding · Wind-induced vibration

© The Author(s), under exclusive license to Springer Nature Singapore Pte Ltd. 2023
H. Djojodihardjo, *Introduction to Aeroelasticity*,
https://doi.org/10.1007/978-981-16-8078-6_18

18.1 Aeroelastic Problems in Bridges and Tall Structures

18.1.1 Bridges

Aeroelastic phenomenon that has been observed with great interest and concern of aerodynamicists, aeroelasticians and engineers occurs in civil structures, most spectacular of which are the aeroelastic flutter phenomena on a relatively slender bridge structure, due to vortex shedding. Following historical observations, such phenomena have taken place many times, but the well-known one and which drew the attention of engineers was the collapse of Tacoma Narrows bridge, near Seattle, in the early 1940's.

The Tacoma Narrows Bridge is the historical name given to the twin suspension bridge—originally built in 1940—that spanned the Tacoma Narrows strait. It collapsed just four months afterward due to aeroelastic flutter. This event has then become very popular and has given rise to various studies analyzing and seeking solutions to the failure phenomenon of cable suspension bridges, and for that matter, long-span bridges.

The first Tacoma Narrows Bridge was located in the state of Washington, USA. The construction was completed and opened to the traffic on July 1, 1940. Tacoma Narrows Bridge was the very first bridge to incorporate a series of plate girders as roadbed support and the first bridge of the cable suspension type. At that time, it was also the third largest suspension bridge having a 853.44 m central span and two side spans of 335.28 m each.

The bridge's collapse has been described as "spectacular" and in subsequent decades "has attracted the attention of engineers, physicists and mathematicians." Throughout its short existence, it was the world's third-longest suspension bridge by main span, behind the Golden Gate Bridge and the George Washington Bridge.

Tacoma Narrows Bridge discussed (also known as the first Tacoma Narrows Bridge) is depicted in Fig. 18.1. One of the Physical Model Rationale of the First Tacoma Bridge Wind-Induced Oscillations is illustrated in Fig. 18.2. The pictures of bridge cross section are depicted in Fig. 18.3a, while for more general impression of suspension and long-span bridges, Fig. 18.3, while Fig. 18.4a and b shows the cross section of the Golden Gate and the Great Belt East Bridge, respectively. Construction of the first Tacoma Narrows Bridge began in September 1938. From the time the deck was built, it began to move vertically in windy conditions, so construction workers nicknamed the bridge **Galloping Gertie**. The motion continued after the bridge opened to the public, despite several damping measures. The bridge's main span finally collapsed in 64 km/h (40-mile-per-hour) winds on the morning of November 7, 1940, as the deck oscillated in an alternating twisting motion that gradually increased in amplitude until the deck tore apart, as depicted in Fig. 18.1.

The bridge's collapse had a lasting effect on science and engineering. The bridge collapsed because moderate winds produced aeroelastic flutter that was self-exciting and unbounded: for any constant sustained wind speed above about 56 km/h (35 mph) that produced torsional flutter oscillation, with observed amplitude to continuously

Fig. 18.1 1940 Tacoma Narrows Bridge, the first Tacoma Narrows Bridge, a suspension bridge in the US state of Washington that spanned the Tacoma Narrows strait of Puget Sound between Tacoma and the Kitsap Peninsula[1]

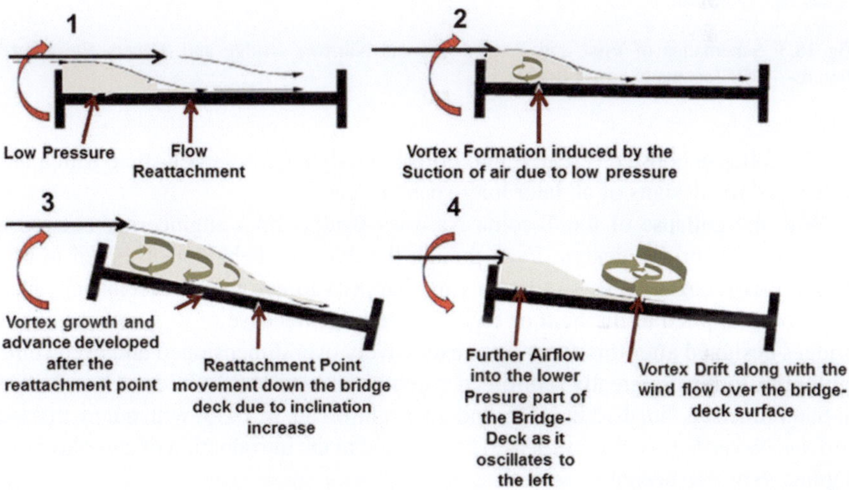

Fig. 18.2 One of the qualitative schematic of the physical model rationale of the First Tacoma Bridge Wind-Induced Oscillations[2]

increase, accompanied by a negative damping factor (i.e. a reinforcing effect, opposite to damping), as schematically depicted in Fig. 18.2. Figure 18.3 exhibits the schematic of this 1940 Tacoma Narrows Bridge known as "Galloping Gertie" and the 1950 replacement Bridge that have introduced several modifications and improvements to take into account the identified problems, as well as to allow denser traffic.

[1] Created by the author from information obtained in https://www.britannica.com/topic/Tacoma-Narrows-Bridge; also https://www.simscale.com/blog2018/07/tacoma-narrows-bridge-collapse/.

[2] Adapted from Green and Unruh [8].

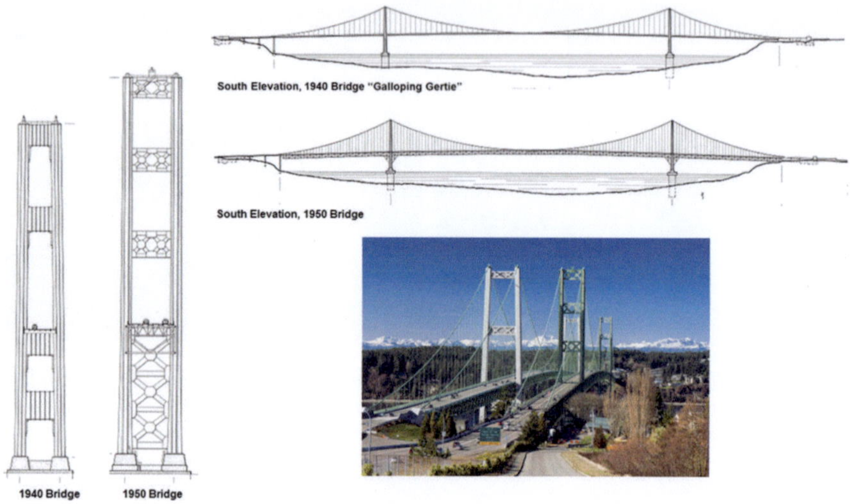

Fig. 18.3 Schematics of First and Present Tacoma Narrows Bridge and Towers Elevation.[3] Picture—1950 Tacoma Narrows Bridge[4]

The collapse boosted research into bridge aerodynamics-aeroelastics, which has influenced the designs of all later long-span bridges.

With the collapse of the Tacoma Narrows Bridge, civil engineering entered a new period of bridge design. Even though the dynamic behavior observed in this bridge was recognized as similar to what happens to aerofoils, aeroelasticity had never been applied to the field of civil engineering. Because of this, several of the bridges designed after this event were excessively over dimensioned and even some built prior to it were greatly reinforced. Such is the case of the Golden Gate Bridge in San Francisco, finished in 1937, and later reinforced, in 1955, with a new trussed surface. Nevertheless, this event also contributed to the introduction of aeroelasticity applied to bridge design.

Various impression of the complexities of Suspension Bridge Design and Load Analysis is exhibited schematically in Fig. 18.4.

To this day, several of the theories applied in aeronautical engineering have already been adapted to bridges and, even though the computation is still somewhat exhausting and demanding, hybrid theories have been proven to yield fairly approximate results when compared to fully experimental testing. These theories are mostly analytical but are based on experimental results.

[3] Tower elevations and roadway axonometrics (1940, 1950), 1940 bridge failure diagrams and illustrations - Tacoma Narrows Bridge, Spanning Narrows at State Route 16, Tacoma, Pierce County, WA Drawings from Survey HAER WA-99, https://www.loc.gov/resource/hhh.wa0453.sheet/?sp=1 and https://www.loc.gov/resource/hhh.wa0453.sheet/?sp=2.

[4] https://unsplash.com/s/photos/first-tacoma-narrows-bridge.

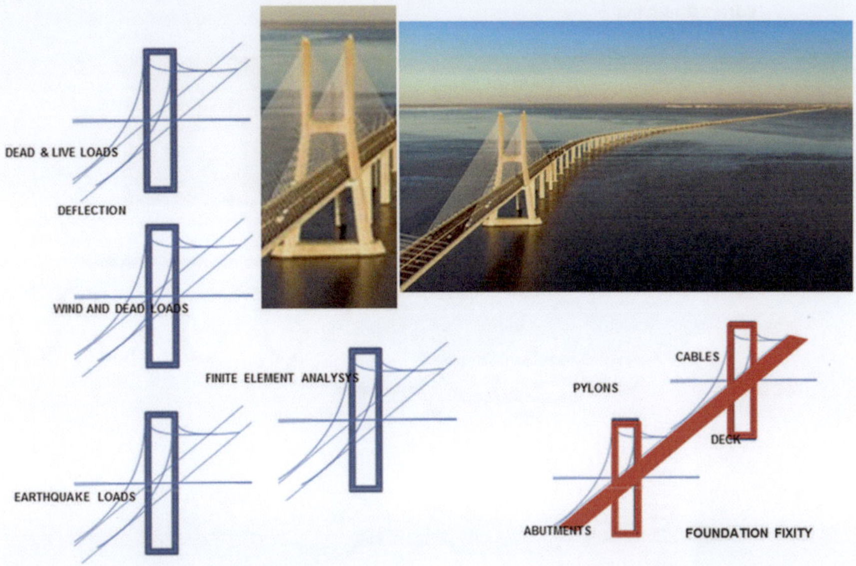

Fig. 18.4 Various impression of the complexities of suspension bridge design and load analysis[5]

18.1.2 *Vortex Shedding*

Tacoma Narrows suspension bridge collapse in 1942 has been attributed to vortex shedding. The separation of flow occurring around the body produces force on the body, i.e. a pressure force on the windward side and a suction force on the leeward side.

The pressure and suction forces result in the shedding of vorticity in the wake region causing structural deflections on the body, which in turn produces the change of fluid momentum along the entire body surface [19]. The shed vortices are convected downwind by local mean wind speed and viscous diffusion and will also form large-scale coherent patterns. The frequency of the shed vortices, which is characterized by the Strouhal number St, and is dependent on the Reynolds number Re, and the geometry and elasticity of the structure, and will dictate the structural response. Qualitative impressions of the flow pattern and corresponding vortices excitation of a bridge deck girder are exhibited in Fig. 18.5. The top figure shows a generic CFD visualization of vortex shedding pattern downstream a circular cylinder forming von Karman vortex street. The center figure illustrates possible flow pattern around and downstream a bridge deck girder, while the bottom figure illustrates CFD analysis visualization image of the vortex shedding flow pattern of a twin-box deck cable-stayed bridge girders.

[5] Picture of Vasco da Gama Bridge, Protugal downloaded from copyright-free site https://stocksnap.io/search/longest+bridge+in+world; shutterstock 12394471300. Schematics of Bridge Pylon is adapted from this picture.

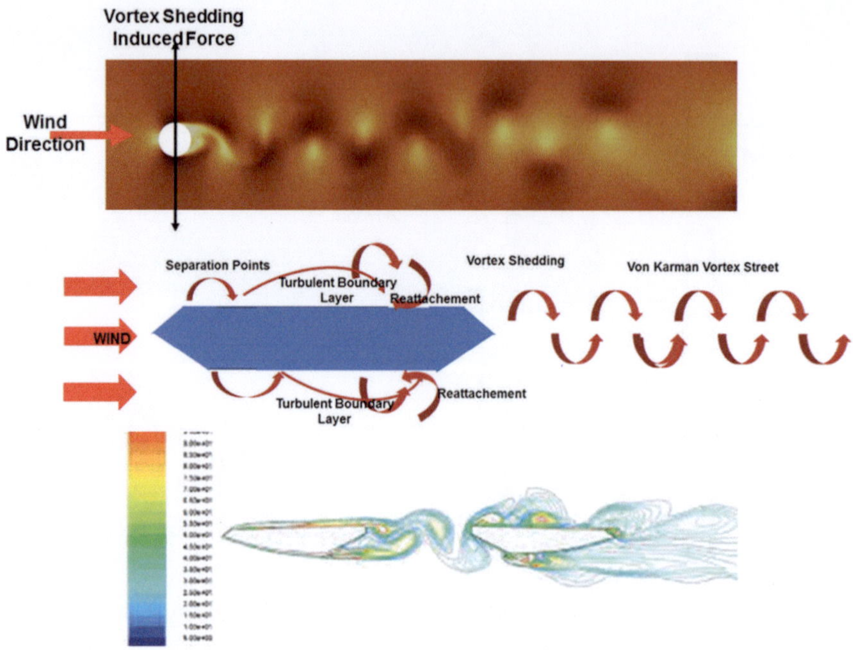

Fig. 18.5 Qualitative impression of flow pattern downstream a bridge deck girder; top: CFD visualization of flow downstream a cylinder[6]; center: a schematic of vortex shedding downstream a flow over a bridge deck; bottom: a typical CFD simulation image of vortex shedding downstream a twin bridge deck girders[7]

18.1.3 Tall Buildings

With the advances in structural design and progress in light and high strength materials, modern tall buildings and structures are built higher and higher. However, along with such progress, efficient structural systems, high strength materials, and increased height, consistent with light structure philosophy, result in the decrease in building weight and damping, and increase in slenderness. However, as is the long and slender bridges, as the height and slenderness increase, the structures and buildings flexibility also increase, which make them prone to wind-induced vibration. Flexible structures are affected by vibration under the action of wind which causes structural motion and plays an important role in the structural and architectural designs. Wind safe tall building design then has to account for the influence of the aerodynamics of wind on the structures from the very beginning, which

[6] M. A. Indira & Djojodihardjo CFD Simulation behind stationary cylinder (Chap. 8).

[7] Ftp.atdd.noaa.gov, CFD analysis of the vortex shedding response of a twin-box deck cable-stayed bridge.

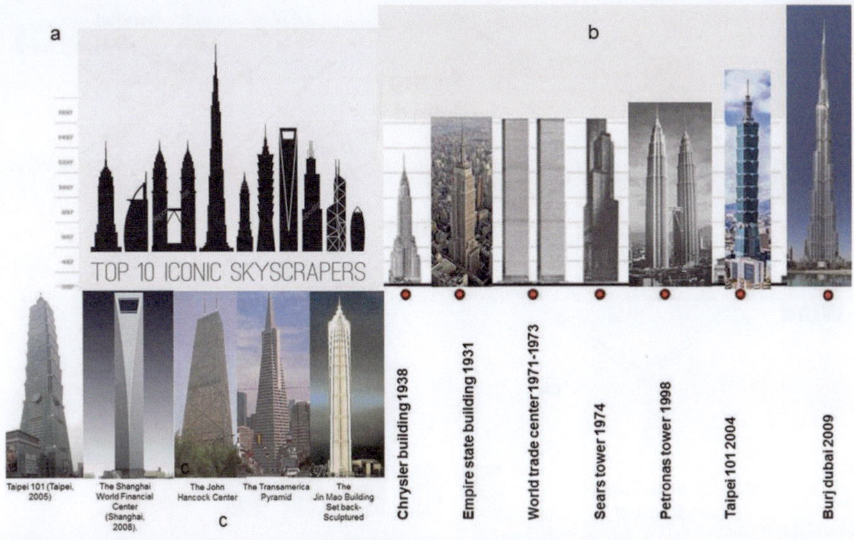

Fig. 18.6 Some selected world's tallest towers

then results in various combined concept of architecture along with building aerodynamics. Hence tall structure shaping strategies for aerodynamic wind excitation response modifications come into play.

Some selected world's tallest towers are depicted in Fig. 18.6. This modern phenomenon has now led to new challenging problems facing the international engineering community. One issue that has started to dominate the serviceability design of tall buildings is wind-induced vibration, which also fall within the domain of aeroelasticity, and which produces discomfort. In the case of slender structures with low natural frequencies, the wind action can introduce uncomfortable vibrations which could affect general well-being and interfere with the daily activities of the occupants.

An illustration of the effects of wind flow on tall buildings and structures is depicted in Fig. 18.7a. When a building is subjected to a wind flow, the originally parallel wind streamlines are displaced on both transverse sides of the building, and as illustrated in Fig. 18.5, produced vortices that eventually form wake reminiscent of von Karman vortex street. At low wind speeds, the vortices are shed somewhat symmetrically at the same instant on either transverse side of the building or structure, and the structure or building may not experience appreciable cross-flow vibration in the across-wind direction.

However, at higher wind speeds beyond certain critical values, as elaborated subsequently in Sects. 18.4 and 18.6, the vortices are shed alternately as illustrated in Figs. 18.14, 18.15 and 18.16. Such flow pattern of vortex shedding may produce appreciable across-wind structural vibrations as depicted in Fig. 18.7b. Figure 18.7c–f depicts some of tall structure shaping strategies to account for wind effects. Although many literature works have been dedicated to the perception of tall

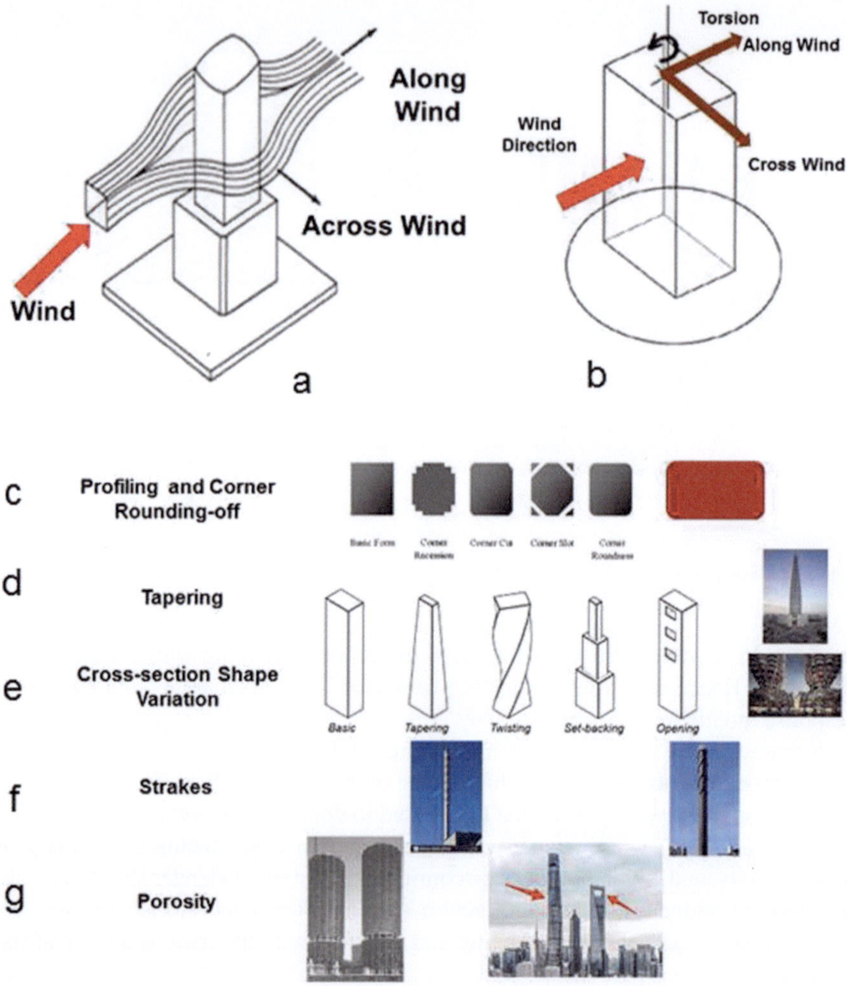

Fig. 18.7 **a** and **b** Interaction of cross and along wind with tall building and its response; **c**–**g** some of tall structure shaping strategies to account for wind effects

building vibrations, to the best of the author's knowledge internationally accepted design standard for satisfactory levels of wind-induced vibrations in tall buildings still needs to be established. If a modern building is prone to experience wind-induced vibrations, the risk of exceeding acceptable vibration limits and of causing discomfort should be estimated. Therefore, management and treatment of wind risk related to motion perception in tall buildings is of major relevance.

It is widely accepted that the perception of vibration is closely related to the acceleration response of buildings. The vibration of tall buildings due to the across-wind direction, i.e. perpendicular to the incoming wind direction, is usually greater than otherwise. The wind-induced response of tall buildings to satisfy comfort criteria has to be accurately estimated.

To that end, many experimental and numerical studies have underlined the complexity of the wind flow and of the wind action around prismatic objects modeling possible tall buildings geometries (see Figs. 18.6 and 18.7b). Aeroelastic or motion-induced effects can occur and lead to significantly different responses of the vibrating structures. Progressive improvements in computation fluid dynamics techniques are very promising, although the most reliable means for estimating wind-induced responses of tall buildings experiencing aeroelastic phenomena are still the wind tunnels experiments.

One of the possible approaches used to identify aeroelastic effects of tall buildings is that of aerodynamic damping. The aerodynamic damping in the direction of the wind is usually positive and small; it can be negative and relevant in the across-wind direction. Hence the accurate design of tall buildings in serviceability conditions is a challenging task, which incorporate several subjects such as bluff body aerodynamics, aeroelasticity, human perception of vibrations, wind tunnel measurements as well as risk management and treatment.

Figure 18.8a and b depicts another fundamental phenomena that have been observed experimentally. These are phenomena that are relevant in the consideration of wind-induced or aeroelastic oscillations on civil structures, particularly long-span structures and bridges, tall structures and chimneys. Figure 18.8a illustrates vortex shedding prevailing on a slender structure with cylindrical or rectangular cross-sectional subject to a uniform flow perpendicular to it. The flow pattern downstream the cylinder shown in the Fig. 18.8a is the well-known von Karman vortices. The two pictures on Fig. 18.8b exhibit two relatively tall chimneys with helical strakes on their upper part to suppress the cross-flow oscillation due to these von Karman vortices, that have been “locked-in.” The effect of Vortex shedding on the response of structures in the cross-flow direction is depicted schematically in Fig. 18.9. One of the first principle approaches to understand and initiate analysis has been elaborated in Chap. 15, with a case study on simple method to calculate the oscillating lift on a circular cylinder in potential flow.

18.1.4 Tall Structure Shaping Strategies for Aerodynamic Wind Excitation Response Modifications

Improvements on tall structures and buildings design have been carried out to account for the wind-induced vibration as elaborated, often accompanied by increased flexibility and may require better inherent damping. Although the structure still carries satisfactorily all the lateral loads, the serviceability requirements as occupants’ discomfort avoidance should be satisfied. Many researches and studies have been carried out in order to mitigate undesirable wind-induced excitation and to improve the performance of tall buildings against wind loads. Different design methods and

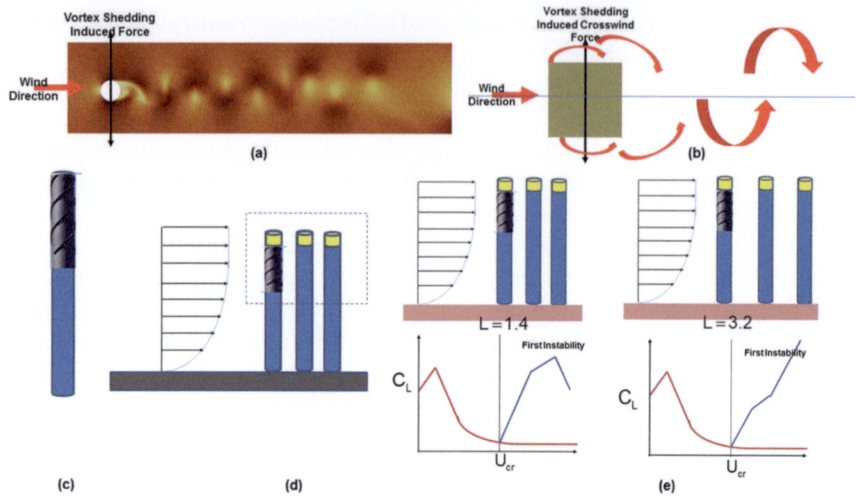

Fig. 18.8 Vortex shedding and across-flow-induced force before and after introducing helical strakes to induce locked-in vortices[8]; **a** and **b** vortex shedding flow pattern and forces-along cylindrical object; **c** typical cylindrical tower or chimney with helical strakes to suppress across-flow oscillation since the helical strakes induces "locked-in vortices"; **d** and **e** qualitative adaptation of the atmospheric boundary layer flow pattern and forces-along cylinders with locked-in vortices due to first cylinder having helical strakes

Fig. 18.9 Effect of vortex shedding on response

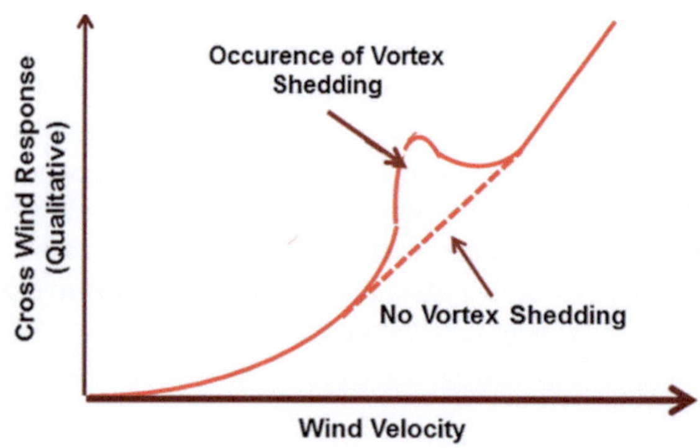

modifications are possible, such as some alternative structural systems and additional damping systems in order to ensure the functional performance of flexible tall structures in terms of wind-induced motion control should be introduced. Some of these efforts are illustrated in Fig. 18.7, and the buildings that have incorporated such efforts are illustrated in Fig. 18.6.

[8] See, for example, Fundamentals Of Vortex-Induced Vibration, https://www.bsee.gov/sites/bsee.gov/files/tap-technical-assessment-program/485ab.pdf.
Halse [11].
Simantiras and Willis [31].

Research by Ilgin and Günel,[9] for example, proposed the classification of tall structure shaping strategies for aerodynamic wind excitation response modifications as

1. Major architectural modifications: Modifications having effect on the architectural concept such as tapering, setbacks, sculptured building tops, varying the shape, openings.
2. Minor architectural modifications: Architectural modifications having no effect on architectural concept such as corner modifications and orientation of building in relation to the most frequent strong wind direction.

Some modifications on cross-sectional shape such as slotted, chamfered, rounded corners, and corner cuts on a rectangular building are depicted in Fig. 18.7c–g. and can have significant effects on both along-wind and across-wind responses of the building to wind. Addition of openings completely through the building, particularly near the top, is another very useful way of improving the aerodynamic response of that structure against wind by reducing the effect of vortex shedding forces which cause across-wind motion. Corner modifications may provide 25% reduction in base moment when compared to the original square section. Chamfers of the order of 10% of the building width make 40% reduction in the along-wind response and 30% reduction in the across-wind response when compared to the rectangular cross-sectional shape without corner cuts. Further details of these techniques can be found in Ilgin and Günel.

18.2 A Semi-analytical Approach Based of Wind Tunnel Tests on Rigid Models to Account for Across-Wind Aeroelastic Response of Square Tall Buildings

An overall impression of the aeroelasticity or the effect of wind-induced vibration on tall buildings will be exemplified by referring to the work of Venanzi and Materazzi[10] on the prediction of the across-wind aeroelastic response of square tall buildings. In particular, a semi-analytical procedure has been followed which is based on the assumption that square tall buildings do not experience vortex shedding resonance or galloping and fall in the range of positive aerodynamic damping for reduced velocities corresponding to operational conditions. Capitalizing on these conditions, aeroelastic wind tunnel tests on the elastic model may not be necessary, and the response can be well evaluated using wind tunnel tests on rigid models and analytical modeling of the aerodynamic damping.

To this end, a procedure consisting of two phases may be followed. First, in order to obtain the aerodynamic forces, simultaneous measurements of the pressure time histories in the wind tunnel on rigid models should be carried out. Then, aeroelastic

---

[9] Ilgin and Gunel [14].

[10] Venanzi and Materazzi [39].

forces are analyzed and calculated, and the structural response is computed through direct integration of the equations of motion considering the contributions of both the aerodynamic and aeroelastic forces. The procedure gives a conservative estimate of the aeroelastic response and has the advantage that aeroelastic wind tunnel tests are avoided.

Special attention should be given to the prediction of the wind response of tall buildings during the design stage. A simple illustration of the response of tall buildings due to wind is illustrated schematically in Fig. 18.6a, while some of tall structure shaping strategies to account for wind effects are depicted in Fig. 18.6b.

Although the along-wind aerodynamic damping that may arise is always positive and almost linearly increases with the reduced velocity, the across-wind aerodynamic damping often shows a monotonic growth for low reduced velocities. It can even become negative beyond a certain reduced velocity. Such conditions depend on many factors, such as the structural shape, the wind speed, the building's natural frequency, the turbulence intensity and the structural damping ratio. Negative aerodynamic damping in flexible structures is associated with the occurrence of vortex excited oscillation and galloping.

The vortex shedding resonance can theoretically occur at a reduced frequency corresponding to the Strouhal number. Alternatively, the response can be governed by galloping, an instability phenomenon typical of non-circular slender structures, due to zero or negative damping [26].[11]

In the recent years also the interesting case of buildings whose section varies with the height through taper and set-back, such as those illustrated in Fig. 18.5 and 18.6b, began to be addressed [16].[12] Ignoring aeroelastic effects can in general leads to slightly conservative results except under certain conditions like vortex shedding or galloping.

An aeroelastic magnification ratio (AMF) can be defined to represent the amplification of the aeroelastic across-wind response with respect to the response obtained neglecting the motion-induced forces. It can be estimated with reference to a square tall building with aspect ratio of 8. Results showed that the AMF does not exceed unity for reduced wind velocities lower than 10.

The effects of the building shape and the structural damping on the aerodynamic damping ratios were derived in the literature. The results showed that in the across-wind direction the aerodynamic damping ratio has first positive and increasing values and then becomes negative as the reduced velocity increases. The aerodynamic damping was found to become negative for reduced velocities greater than 10.

Across-wind aeroelastic response of square tall buildings literature references showing the results of aeroelastic tests on models of square tall buildings agree that in the across-wind direction the structural response is amplified when a critical reduced velocity corresponding to the onset of vortex shedding resonance or galloping is exceeded, although the galloping of tall buildings has never been reported in literature.

---

[11] Novak [26].

[12] Kim and Kanda [16].

The vortex shedding resonance occurs when the reduced velocity $U = V/fD$, where V is the top wind speed, f is the structural frequency in the across-wind direction and D is the side length, is close to the critical value $U_{\text{cr,v}}$ corresponding to the reciprocal of the Strouhal number St

$$C_{\text{cr,v}} = \frac{1}{\text{St}} = \frac{V}{f_s D} \tag{18.1}$$

where f_s is the vortex shedding frequency. The Strouhal number for buildings with square cross section is almost independent on the Reynolds number and is about 0.10 [10].[13]

The critical velocity for galloping onset is influenced by many variables: the side length D, the damping ratio ξ, the natural frequency in the across-wind direction f and the mass density per unit height m. To evaluate the critical velocity of galloping, the Glauert-Den Hartog criterion[14,15] can be applied, assuming that the quasi-steady theory is valid for every reduced velocity. Following this criterion, the critical galloping velocity is given by:

$$V_{\text{cr,v}} = \frac{8\pi m f \xi}{\rho D\left(\frac{\partial C_L}{\partial \alpha}\big|_0 + C_D|_0\right)} \tag{18.2}$$

where $\left(\frac{\partial C_L}{\partial \alpha}\big|_0 + C_D|_0\right)$ is the galloping instability factor [17].[16] Figure 18.10 shows the ratio of the critical galloping velocity and the wind speed at the top of a square building as a function of the reduced velocity, while the mass-damping coefficient is defined by

$$M_d = \frac{\int_0^H m(z)\Phi^2(z)\mathrm{d}z}{\int_0^H \Phi^2(z)\mathrm{d}z} \frac{\xi}{\rho D^2} \tag{18.3}$$

where m is the mass of the building per unit height, Φ is the modal shape in the across-wind direction and ρ is the air density.

The reduced critical galloping velocity for any fixed value of the mass-damping coefficient can be determined at the intersection between the surface and the plane $V_{\text{cr,g}}/V_{\text{top}} = 1$. The ratio values lower than unity correspond to the domain that is unsafe for galloping instability. If the mass-damping coefficient is higher, the critical reduced velocity for galloping $U_{\text{cr,g}}$, will also be higher, which corresponds to the intersection between the surface and the plane $V_{\text{cr,g}}/V_{\text{top}} = 1$. If the galloping instability factor is assumed to be constant with the Reynolds number, the exposure conditions do not affect significantly the critical velocity for galloping.

---

[13] Gu and Quan [10].

[14] Nikitas and Macdonald [25].

[15] Stickland and Scanlon [36].

[16] Kwok [17].

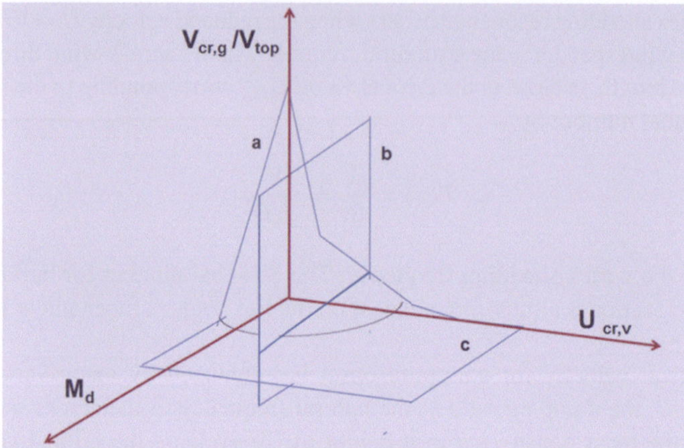

Fig. 18.10 Qualitative impression of the ratio of the critical galloping velocity and the wind speed at the top of the square building $V_{\mathrm{cr,g}}/V_{\mathrm{top}}$ as a function of the reduced velocity U and the mass-damping coefficient M_d (**a**), plane $U_{\mathrm{cr,v}} = 10$ (**b**), plane $V_{\mathrm{cr,g}}\,/V_{\mathrm{top}} = 1$[17]

As tall buildings are subjected to turbulent wind in which both the mean wind velocity and the turbulence intensity vary with height, a lock-in mechanism is activated and the excitation due to vortex shedding displays a broad band in the frequency domain. This modifies the structural response also at frequencies in the neighborhood of the frequency corresponding to the vortex resonance [18].[18]

For this reason most wind tunnel tests show that the across-wind aerodynamic damping begins to decrease for reduced velocities U lower than $U_{\mathrm{cr,v}}$, typically beyond 8. Figure 18.11 shows the across-wind aerodynamic damping ratio as a function of the reduced velocity for square tall buildings obtained from aeroelastic tests by Steckley [41], Marukawa et al. [22] and Gu and Quan [10]. The segment of positive aerodynamic damping is followed by a segment where the aerodynamic damping decreases until negative values are reached. In particular, the across-wind aerodynamic damping has positive and increasing values for reduced velocities between 8 and 10. Beyond that range, the across-wind response of tall buildings is amplified due to lock-in excitation.

Further detail of the semi-analytical computational approach based on wind tunnel tests on rigid models and the results for across-wind aeroelastic response of square tall buildings are elaborated in this work. The procedure followed has been applied in the range of positive aerodynamic damping that corresponds to the operational reduced velocities of most tall building lower than about 250 m. In that range, the aerodynamic damping computed using the proposed procedure seems to correctly represent the one evaluated using wind tunnel tests on flexible models reported in the literature.

---

[17] Adapted from information in Venanzi and Materazzi [39].

[18] Kwok and Melbourne [18].

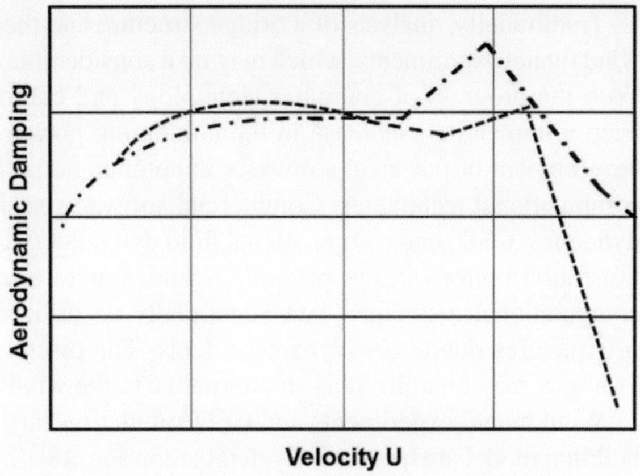

Fig. 18.11 Qualitative behavior of across-wind aerodynamic damping ratio for square tall buildings (no quantitative accuracy), as obtained by some researchers[19]

18.3 Computer Modeling Example of Aeroelastic Analysis of Bridge Girder Section

In each example from a selected recent works in the literature, emphasis is given to basic governing equations that can be derived from first principles. First principles are the physical laws and basic mathematical identities as implied in basic subjects such as basic physic, thermodynamics, fluid mechanics that can be easily understood using basic reasoning. Then by meticulous development of these principles elaborated in great detail using mathematical relationships, valid governing equations can be derived, which can be used to model the problem considered with close accuracy. More involved numerical procedures may be employed, some of which are available as commercial computational software.[20]

For the particular problem considered here, a numerical simulation of wind flow around bridges using the finite element method (FEM) and the principles of computational fluid dynamics (CFD) and computational structural dynamics (CSD) following the work of Selvam and Govindaswamy will be extracted to provide a computer modeling example of an aeroelastic analysis of bridge girder section. Long-span bridges are prone to wind-induced aerodynamics, the classical example of which is the Tacoma's narrows bridge failure in State of Washington. Capitalizing on this and many examples of suspension bridge failures in the past, the example elaborated below will be instructive in the analysis and design of girders of long-span bridges, in particular in dealing with wind-induced instabilities which should be checked during the design. The aerodynamic effect of wind on bridges, known as aeroelastic Instability, is primarily vortex shedding (as illustrated in Figs. 18.1 and 18.2), galloping, torsional divergence, flutter and buffeting. If the wind velocity exceeds the critical velocity for flutter that the bridge can withstand, then the bridge fails due to the phenomenon of flutter.

---

[19] Qualitative image based on information and data from Venanzi and Materazzi [38].

[20] Selvam and Govindaswamy [30].

Traditionally, analysis of a bridge structure and the wind effects is studied using wind tunnel experiments, which may take considerable time and could be very costly. With the progress of computer technology and computational methods, there has been a tremendous increase in the computing power and speed, which allows the development of powerful softwares in computational fluid dynamics and structural computational techniques. Commercial softwares such as computational structural dynamics (CSD) and computational fluid dynamics (CFD) are available for plausible utilization options in the prevailing fluid–structure interaction (FSI). Many other computational techniques can beneficially be utilized in the analysis and design of structures due to aerodynamic effects. The discrete vortex method (DVM), for example, may be utilized as an alternative to the wind tunnel procedures.[21,22,23]

Wind tunnel experiments and CFD simulation carried out by Shirzadi et al. on a building model are qualitatively depicted in Fig. 18.12. A power law boundary-layer velocity wind profile with an exponent $a = 0.18$ and a mean speed at the building height 8.6 m/s was assumed. The wind flow had an angle of attack 90° with the reference to the longer wall of the object. The length to depth ratio of the building was equal 1.5. The problem was simplified and reduced to a two-dimensional case without aeroelastic effects. The numerical simulation was performed for Reynolds' number $\mathrm{Re} = 2.3 \times 10^6$. The number of vortex particles varies from 12,000 to 24,000 during the computational process. Numerical results were compared with experimental data, as exhibited there. Figure 18.12b and c refers to the same moment of time. For a model of a bridge deck girder, Fig. 18.13 shows a flow visualization results of simulated von Karman vortex street developing in the wake of the object investigated.

The examples quoted by Nowicki [27] exhibit good conformity between numerical simulation results and experimental data. The most important achievement is the shape of the line obtained from computer simulations. It shows that DVM algorithms are able to model the phenomena of turbulence that occurs near edges of a body like the place where wall connect with the roof. The discrepancy between experiment and calculations, that can be noticed in the Fig. 18.12c, is derived from the fact that the building was not long enough to treat the flow as two dimensional. The influences of the walls parallel to the flow should not have been neglected, and a three-dimensional analysis of the flow may produce better agreement. However, the points being emphasized here are that first to be presented here is that to a good extent the discrete vortex method (DVM) has been exemplified to be able to serve as an alternative to the wind tunnel procedures, and second, the differences between the two approaches can be considered to be small and do not disqualify the results from engineering (civil) point of view.

A large eddy simulation (LES) turbulence model can be used and the rigid-body grid movement technique can be adopted. The critical velocity for flutter can be calculated directly using the free oscillation procedure. The influence of grid on

[21] Nowicki [27].

[22] Larsen and Walther [19].

[23] Wang [41].

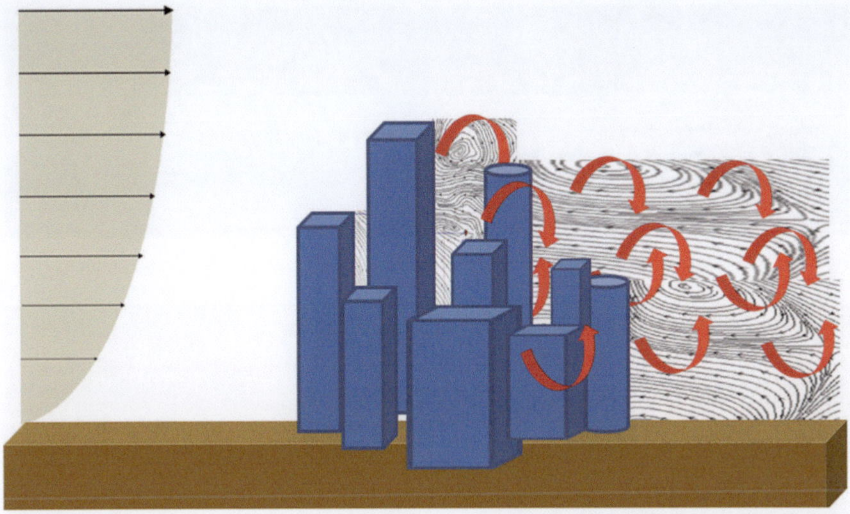

Fig. 18.12 Qualitative schematic visualization of complex vortices structure downstream of the flow over building[24]

critical velocity should be studied. The computed critical velocity for flutter could be compared with the wind tunnel measurements for validation. The criteria for the design of long-spanned suspension bridges are concerned with the static and dynamic responses of the bridge under wind loading. A basic knowledge of the wind forces is required to understand the issues involved in the design.

18.4 Aeroelastic Effects and Phenomena

18.4.1 Dynamic Behavior

The dynamic behavior of a long-span bridge girder section can be analyzed using first principles. Hence for this particular case, it is governed by Newton's second law. Then the motion of mass of the bridge girder section is described by the differential equation

$$m\ddot{x} + c\dot{x} + kx = F(t) \tag{18.4}$$

where $F(t)$ is the time-dependent load acting on the mass, k is the stiffness coefficient and c is the coefficient of damping. This equation can be rewritten in the form

[24] The flow visualization detail of this figure can be followed in many sources, such as from Shirzadi et al. [30], Adepalli and Pardyjal [1], and Mohan et al. [23].

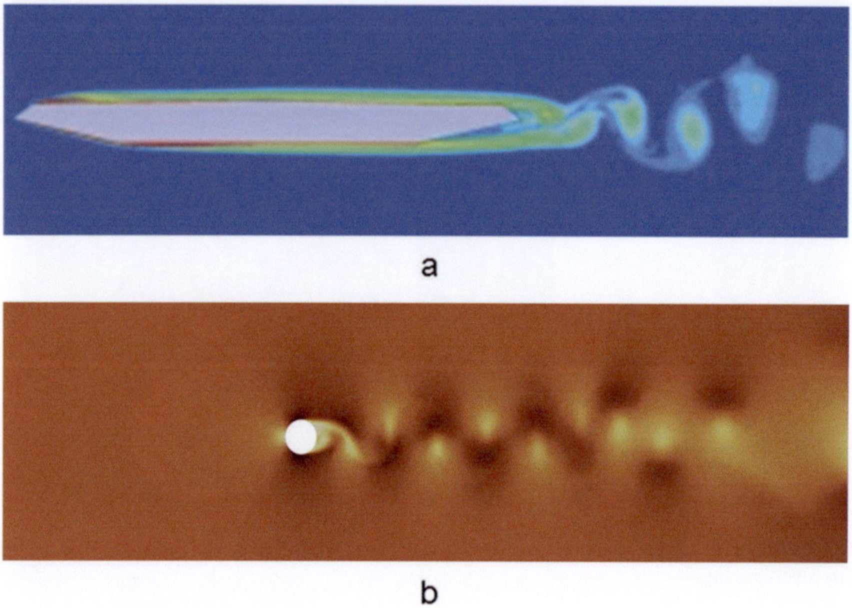

Fig. 18.13 **a** Typical flow visualization of simulated von Karman vortex street developing in the wake of a bridge girder section[25]; **b** von Karman vortex street animation by the author

$$\ddot{x} + 2\zeta\dot{x} + \omega^2 x = \frac{F(t)}{m} \tag{18.5}$$

where

$$\omega = \sqrt{\frac{k}{m}} \tag{18.6a}$$

and

$$\zeta = \frac{c}{2m\omega} \tag{18.6b}$$

or

$$\zeta = \frac{c}{2\sqrt{km}} \tag{18.6c}$$

Here ω is the natural circular frequency, and $2m\omega$ is the critical damping coefficient. Three cases can appear based on the value of ζ. If it is equal to unity, then it refers

[25] Image downloaded from a public domain site, wikiwand, to compare and discuss related isssues incorporated by Zhu [44].

to critically damped case, Otherwise, if ζ is less or greater than unity, the case is associated with under-damped, or over-damped response respectively.

Dynamic behavior includes the responses caused by vortex shedding excitation, self-excited oscillations and buffeting due to wind turbulence. Suspension bridges may oscillate in two natural modes, vertical and torsional. In the vertical mode, all joints at any cross section move with the same distance in the vertical plane. In the torsional mode, every cross section rotates about a longitudinal axis parallel to the roadway.

18.4.2 Aeroelastic Instability

When a structure is subjected to wind flow, it may vibrate or suddenly deflect in the airflow du to aeroelastic effects. Such structural motion produces flow pattern changes around the structure. If these wind pattern changes around the structure introduce aerodynamic forces that increase rather than decrease the resulting wind-induced vibration, then succeeding deflections of oscillatory and/or divergent character, which is known as aeroelastic instability, may result. In wind engineering of civil structures, such aeroelastic instability phenomena that are of great concern are those due to vortex shedding, which may result in torsional divergence, galloping, flutter and buffeting of the civil structure, analogous to those prevailing in aeronautical structure.

18.4.3 Flow Pattern

Very important flow properties that influence the flow pattern and hence the resulting aeroelastic forces on structures are the dynamics of the flow, the geometry of the structure and the properties of the fluid, including density ρ and dynamic viscosity μ of the fluid. The state of affairs is represented by the Reynolds number, which is defined as[26]

$$\mathrm{Re} = \frac{\rho V D}{\mu} \tag{18.7}$$

where Re is the Reynolds number, ρ is the density of the air, V is the velocity of the wind relative to the structure, D is the diameter or characteristic dimension of the structure, which can be taken as the height of the obstructing structure perpendicular to the wind direction.

[26] Note that U and V are used alternatively to represent velocity (its absolute value, a scalar).

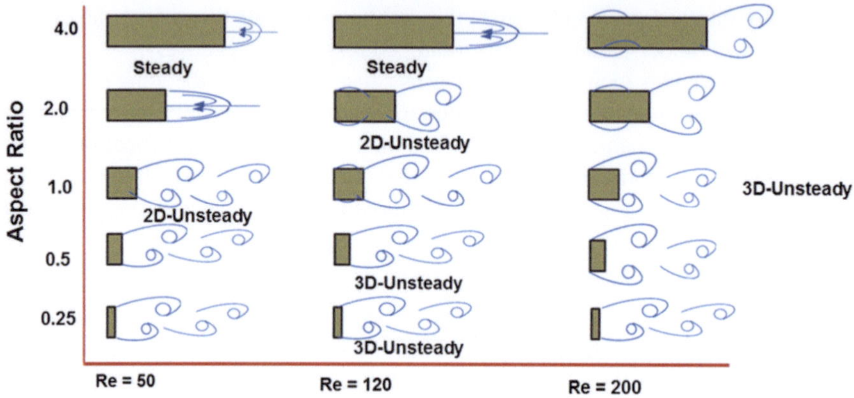

Fig. 18.14 Flow over various shapes of 2D cylinders—qualitative impression of the effects of cylinder aspect ratio AR and Reynolds number Re[27]

An insight into the understanding of the nature and extent of the vortex shedding phenomenon for different ranges of Reynolds number for two different cross-sections, a plate and a cylinder can be obtained from Figs. 18.14, 18.15 and 18.16.

The vortex shedding phenomena is describable in terms of a non-dimensional number St, which is defined as

$$\mathrm{St} = \frac{N_S D}{V} \tag{18.8}$$

where St is the Strouhal number, N_S is the frequency of the full cycles of the vortex shedding, D is the characteristic dimension of the structure projected normal on a plane perpendicular to the mean flow velocity and V is the relative velocity of the wind with respect to the structure. The Strouhal number St takes on different characteristic constant values depending upon the cross-sectional shape of the prism being enveloped by the flow. From the figures, as illustrated in Fig. 18.15, it is seen that for a very low Reynolds number, the flow remains the same, just circumventing the obstruction on its way. For higher Reynolds numbers, the flow starts to separate around the edges of the obstruction, and vortices are generated in the immediate wake of the obstruction. The flow pattern which is reminiscent of on Karman vortex street is schematically and qualitatively exhibited in Fig. 18.15. Thereafter further increase in the Reynolds number causes the creation of cyclically alternating vortices and they are carried over with the flow downstream. From there on, the inertial effects become dominant over the viscous effects and turbulence sets in, resulting in shear of the flow. So, this reasonably illustrates the vorticity phenomenon starting from a smooth and low speed flow to a turbulent and high-speed flow.

[27] Extracted from data and information from Yoon et al. [42].

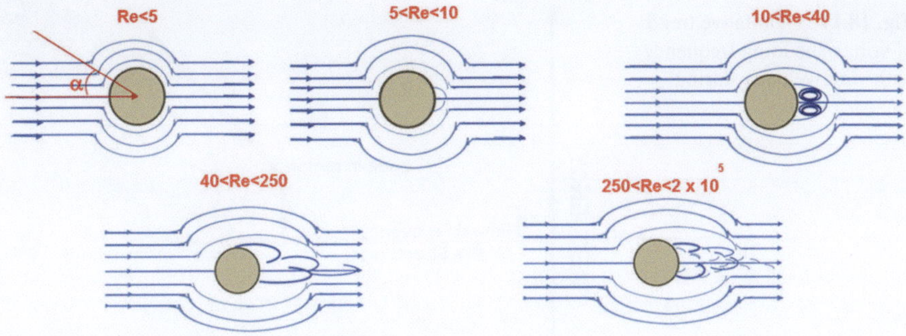

Fig. 18.15 Qualitative impression of flow past a circular cylinder at various Reynolds number[28]

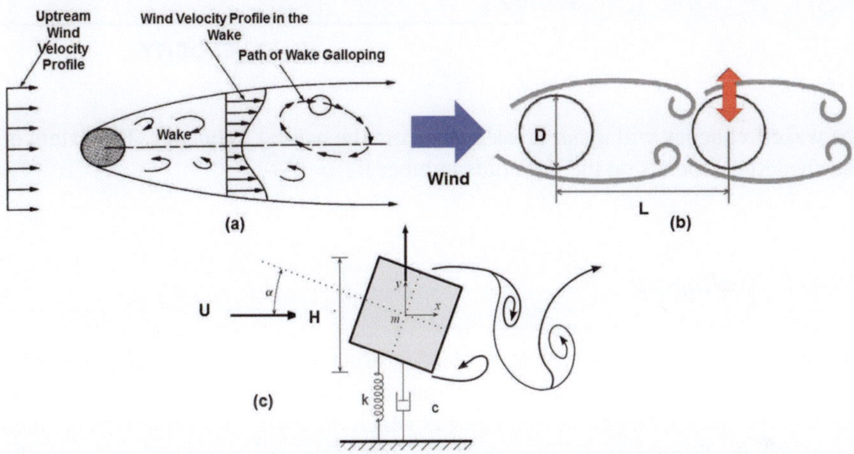

Fig. 18.16 **a** Flow pattern of wake galloping phenomenon ([33], with permission), **b** wake galloping of two circular cylinders and **c** a sketch of wake galloping of a square cross-sectional cylinder experimental set-up with variable angle of attack oscillating across the stream

The structural member acts as if rigidly fixed, when the frequency of vortex shedding (also called wake frequency) is not close to the natural frequency of the member. On the other hand, when the vortex-induced and the natural frequencies coincide, the resulting condition is called lock-in. During lock-in condition, the structural member oscillates with increased amplitude but rarely exceeding half of the across-wind dimension of the body [33]. The lock-in condition is illustrated in Fig. 18.16.

Figures 18.16 and 18.17 show that the wake frequency remains locked to that of natural frequency for a range of wind velocities. As the velocity further increases,

---

[28] Extracted from data and information from various sources, among others Thangadurai [24] and Simiu and Scanlan [33].

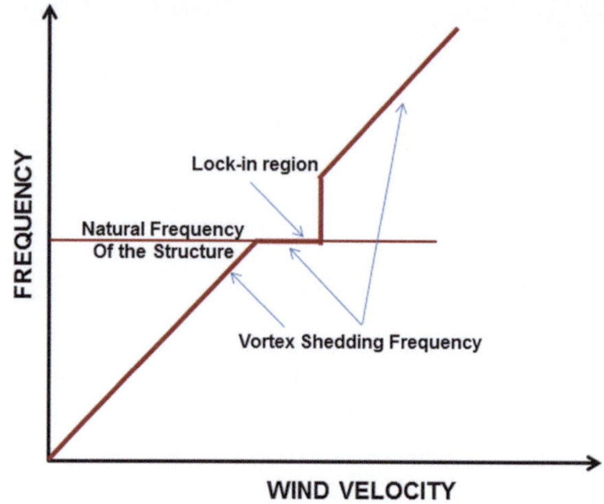

Fig. 18.17 Qualitative trend of vortex shedding frequency with wind velocity during lock-in[29]

the wake frequency will again break away from the natural frequency. The extent of the shedding depends on the Reynolds number $\text{Re} = \frac{\rho V D}{\mu}$.

18.4.4 Galloping

Galloping is an instability typical of slender structures.[30] This is a relatively low-frequency oscillatory phenomenon of elongated, bluff bodies acted upon by a wind stream. The natural structural frequency at which the bluff object responds is much lower than the frequency of vortex shedding. It is in this sense that galloping may be considered a low-frequency phenomenon. There are two types of galloping: wake and across-wind.

A related phenomenon, ***wake galloping*** can be observed by experimenting with two cylinders, one windward, producing a wake and one leeward, as depicted in Fig. 18.16b. In the wake downstream of the upstream cylinder, the downstream cylinder is subjected to galloping oscillations induced by the turbulent wake of the upstream cylinder. Due to this forcing, the upstream cylinder tends to rotate clockwise, and the downstream cylinder tends to rotate anticlockwise, thus inducing torsional oscillations.

The ***across-wind galloping*** in a bridge is an instability that is triggered by a turbulent wind blowing transversely across the deck. As the section of the bridge vibrates crosswise in a steady wind velocity U, the relative velocity changes, thus resulting in the changing of the angle of attack α. Consequently, an increase or decrease of the lift force of the cylinder is produced. Two possibilities may prevail.

[29] Generated qualitatively using data and information from Simiu and Scanlan [33].

[30] Zhao et al. [43].

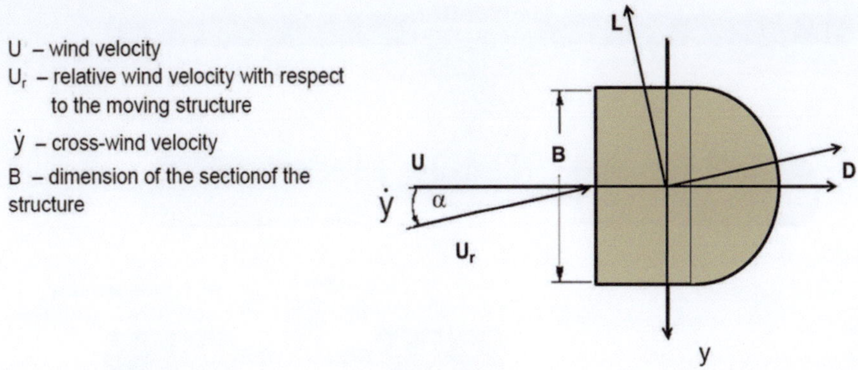

Fig. 18.18 Across-wind galloping: Wind and motion components, with resultant lift and drag, on a bluff cross section[31]

If due to the increase of α an increase in the lift force in the opposite direction of motion is produced, the associated situation is stable. Otherwise, if an increase of α produces a decrease in lift force, then the situation is unstable, and galloping occurs. Such situation is illustrated in Fig. 18.18.

18.5 Torsional Divergence or Quasi-static Divergence

A classic example of this phenomenon can be observed in ice covered power transmission lines. Galloping is reduced in these power transmission lines by reducing the distance between the supporting towers and increasing the tension of the lines, and/or by introducing mechanical damper known as Stockbridge damper, depicted in Fig. 18.19.

The wind flowing against a structure exerts a pressure in proportional to the square of the wind velocity. In general, wind pressure induces both forces and moments in a structure. At a critical wind velocity, the edge-loaded bridge may buckle "out of plane" under the action of a drag force or torsionally diverge under a wind-induced moment that increases with a geometric twist angle (or angle of attack) α. In reality, the divergence involves an inseparable combination of lateral buckling and torsional divergence.

The flow-induced moment, divergence angle and the critical divergence velocity $U_{\mathrm{cr,div}}$ can be derived to have the following relationships. For this purpose, consider a small rotation angle α as shown in Figs. 18.20 and 18.21. The pitch moment resulting from the aerodynamic excitation per unit span i is given by:

[31] Created qualitatively by modifying and adapting information and data from Simiu and Scanlan [33].

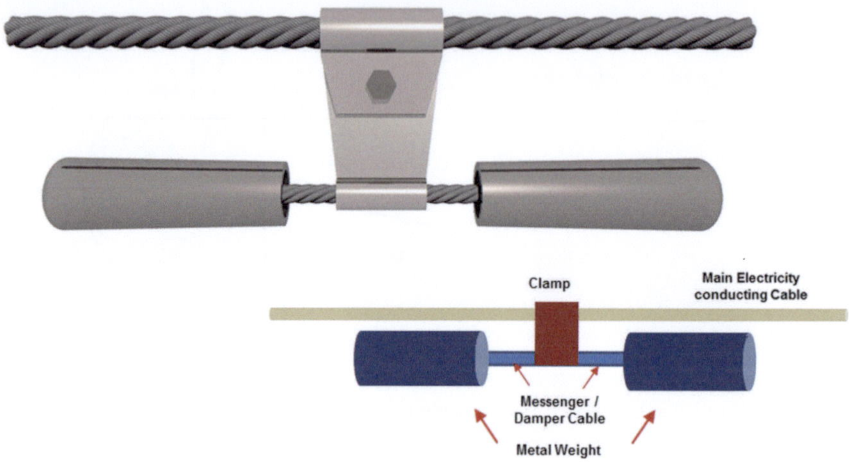

Fig. 18.19 Modern design Stockbridge damper with metal weights

$$M_\alpha = \frac{1}{2}\rho U^2 B^2 C_M(\alpha) \tag{18.9}$$

where ρ is density, U is the mean wind velocity, B is the deck width, α is the angle of twist and C_M is the aerodynamic moment coefficient about the twisting axis. At zero angle of attack the value of this moment is

$$M_\alpha(0) = \frac{1}{2}\rho U^2 B^2 C_{M0}(\alpha) \tag{18.10}$$

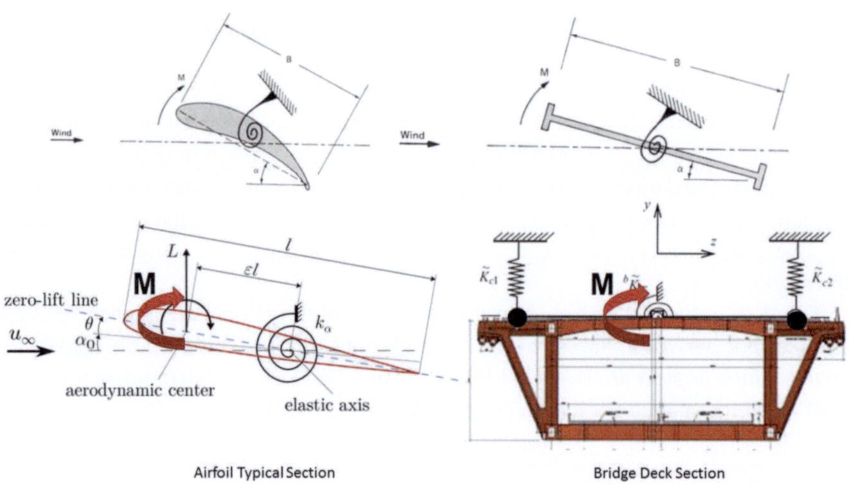

Fig. 18.20 Schematics of torsional divergence in typical section and bridge deck section

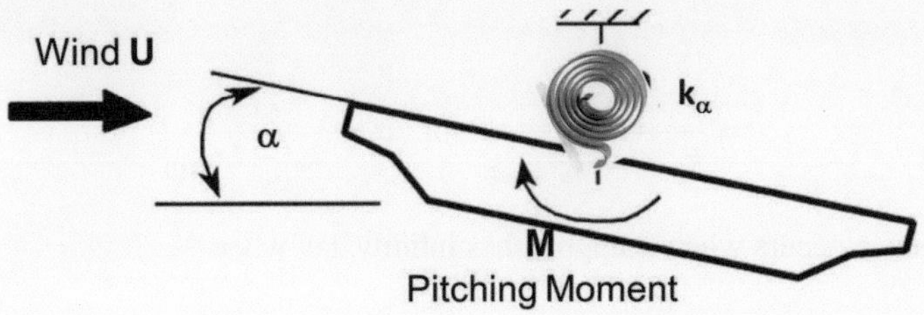

Fig. 18.21 Simplification of Fig. 18.20 to depict torsional divergence and divergence critical velocity derivation

where

$$C_{M0} = C_M(0)$$

For relatively small angle α M_α in Eq. 18.10 can be approximated by

$$M_\alpha = \frac{1}{2}\rho U^2 B^2 \left(C_{M0} + \left.\frac{\partial C_M}{\partial \alpha}\right|_{\alpha=0} \alpha \right) \tag{18.11}$$

When the pitch moment caused by wind exceeds the resisting torsional capacity, then the bridge displacement diverges.

Now equating the aerodynamic moment to the structural resisting moment gives

$$\frac{1}{2}\rho U^2 B^2 \left(C_{M0} + \left.\frac{\partial C_M}{\partial \alpha}\right|_{\alpha=0} \alpha \right) = k_\alpha \alpha \tag{18.12}$$

Equating $M(\alpha)$ to the internal torsional moment k_α α yields[32]

$$\alpha = \frac{\lambda C_{M0}}{k_\alpha - \lambda \left.\frac{\partial C_M}{\partial \alpha}\right|_{\alpha=0}} \tag{18.13}$$

where k_α is the constant of spring torsion.

Setting

$$\lambda = \frac{1}{2}\rho U^2 B^2 \tag{18.14}$$

in the above equation, we get

$$\left(k_\alpha - \lambda \frac{\partial C_{M0}}{\partial \alpha} \right)\alpha = \lambda C_{M0} \quad or \quad \left(k_\alpha - \lambda C'_{M0} \right)\alpha = \lambda C_{M0} \tag{18.15}$$

[32] Further discussions on torsional divergece and detail of the derivation can be found in Simiu and Yeo [34].

Hence

$$\alpha = \frac{\lambda C_{M0}}{k_\alpha - \lambda \frac{\partial C_M}{\partial \alpha}\Big|_{\alpha=0}} \quad \text{or} \quad \alpha = \frac{\lambda C_{M0}}{k_\alpha - \lambda C'_{M0}} \tag{18.16}$$

Divergence occurs when α approaches infinity, i.e. when

$$\lambda = \frac{k_\alpha}{\left(\frac{\partial C_M}{\partial \alpha}\Big|_{\alpha=0}\right)} \tag{18.17}$$

Hence, if due to the wind-induced moment, the torsional (or twist) angle α becomes infinite (hence the bridge girder cross-sectional diverges), the wind velocity at which such situation prevails is the critical divergence wind velocity for torsional divergence, and is given by:

$$U_{cr} = \sqrt{\frac{2k_\alpha}{\rho B^2 \frac{\partial C_M}{\partial \alpha}\Big|_{\alpha=0}}} \tag{18.18}$$

Torsional divergence is an instance of a static response of a structure. Torsional divergence was first associated with aircraft wings due to their susceptibility to twisting off at excessive air speeds [33].

The increase in α gives rise to higher torsional moment as the wind velocity increases. If the structure does not have sufficient torsional stiffness to resist this increasing moment, the structure becomes unstable and will be twisted to failure.

18.6 Flutter and Forced Oscillation

18.6.1 Flutter

The phenomenon of flutter is a very serious concern in the design of bridges. The failure of the Tacoma's narrows bridge was due to the flutter. In the later part of this chapter, a review of the Tacoma's Narrows bridge failure is reported to give a better insight into the flutter-induced instability that resulted in failure. The term flutter has been variously used to describe different types of wind-induced behavior. Flutter can be defined as a condition of negative aerodynamic damping wherein the deflection in the structure grows to enormous levels till failure once started. It is also known as classical flutter. The other types of flutter reported by Simiu and Scanlan [33] are stall flutter and panel flutter.

Stall flutter is a single-degree-of-freedom oscillation of airfoils in torsion due to the nonlinear characteristics of the lift [33]. The stall flutter phenomenon can also occur with structures having broad surfaces depending on the angle of approaching wind. The torsional oscillation of a traffic stop sign about its post is an example of this phenomenon.

Panel flutter is a sustained oscillation of panels typically the sides of large rockets, caused by the high-speed passage of air along the panel. The most prominent cases have been in supersonic flow regimes and so have not appeared in the wind engineering context. Flag flutter is closely related to panel flutter.

18.6.2 Free Oscillation Solution Procedure

This method was used in this work for the study of flutter stability analysis of the structure during motion. In this method the structure is elastically suspended and is given an initial perturbation in terms of heave or pitch and the structure is left to oscillate freely. The lift, drag and moment generated due to the applied displacement is then measured, and thus a time history data is generated. The governing equations of motion for translation and rotation are, from Eqs. 18.4 and 18.5, can be rewritten as

$$m\ddot{h} + c_h\dot{h} + k_h h = -L \tag{18.19}$$

$$I_\alpha\ddot{\alpha} + c_\alpha\dot{\alpha} + k_\alpha = M \tag{18.20}$$

Here m, I_α, h, α, L and M represents mass, moment of Inertia, heave, pitch, lift and moment, respectively. c and k represents damping and stiffness coefficients with the subscripts h and α meaning heave and rotation, respectively.

These two equations can be rewritten as

$$\ddot{h}' + \left(\frac{\omega_h}{\omega_\alpha}\right)\left(\frac{1}{u'}\right)^2 h = \frac{-C_L}{2R_m} \tag{18.21}$$

where

$$R_m = \frac{m}{\rho B^2} \tag{18.21a}$$

$$C_L = \frac{L}{\frac{1}{2}\rho U^2 B} \tag{18.21b}$$

and

$$\ddot{\alpha}^* + \left(\frac{1}{u^*}\right)^2 \alpha^* = \frac{C_m}{2R_L} \tag{18.22}$$

where

$$R_L = \frac{I_\alpha}{\rho B^4} \tag{18.22a}$$

$$C_M = \frac{M}{\frac{1}{2}\rho U^2 B^2} \tag{18.22b}$$

Furthermore, when it appears, C_D is defined as

$$C_D = \frac{D}{\frac{1}{2}\rho U^2 B} \tag{18.22c}$$

The derivation of (18.21) and (18.22) will be left as an exercise.

18.6.3 Forced Oscillation Procedure

In this method, the structure is forced in a torsional or heave sinusoidal motion relative to the flow with a prescribed frequency and amplitude.The lift and moment generated due to this applied force is measured and used for the calculation of the aerodynamic derivatives. The calculated aerodynamic derivatives are then used for the computation of the critical velocity for flutter.

The lift and moment loads exerted on an oscillating bridge section with 2 degrees of freedom namely the vertical or heave motion (h) and rotational or pitch motion (α) are given by the following equations [33].

$$L = \frac{1}{2}\rho U^2(2B)\left[KH_1^*\frac{\dot{h}}{U} + KH_2^*\frac{B\dot{\alpha}}{U} + K^2H_3^*\alpha + K^2H_4^*\frac{h}{B}\right] \tag{18.23}$$

$$M = \frac{1}{2}\rho U^2(2B^2)\left[KA_1^*\frac{\dot{h}}{U} + KA_2^*\frac{B\dot{\alpha}}{U} + K^2A_3^*\alpha + K^2A_4^*\frac{h}{B}\right] \tag{18.24}$$

where

$K = \frac{B\omega}{U}$	is the reduced non-dimensional frequency.
H_i^* and A_i^* ($i = 1,2,3,4$)	are the aerodynamic derivatives.
U	wind velocity.
B	chord deck width of the bridge.

For the pure heave motion, Eqs. 18.23 and 18.24 become

$$L = \frac{1}{2}\rho U^2(2B)\left[KH_1^*\frac{\dot{h}}{U} + K^2H_4^*\frac{h}{B}\right] \tag{18.25}$$

$$M = \frac{1}{2}\rho U^2(2B^2)\left[KA_1^*\frac{\dot{h}}{U} + K^2A_4^*\frac{h}{B}\right] \tag{18.26}$$

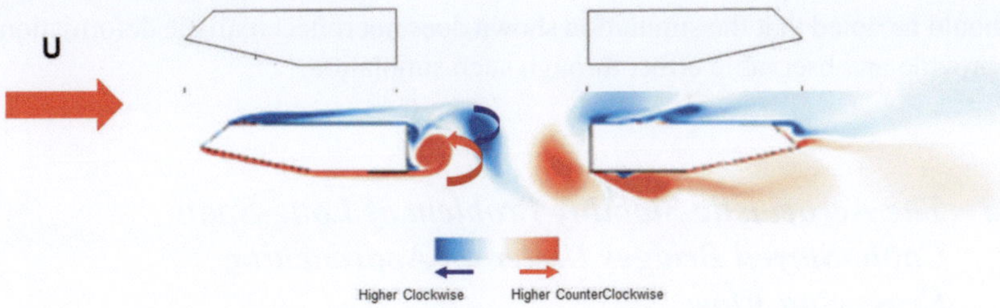

Fig. 18.22 Illustrative and qualitative example of flow field (vortex pattern) around a twin-deck bridge. Verification nd/oe CTUl configuration can be obtained from CFD simulation or wind-tunnel test. The results will be dependent on flow and geometric parameters

For *Sn* samples, $(L, M, h_i, \dot{h})$, $i = 1, 2, \ldots Sn$, the Eqs. 18.25 and 18.26 constitute two sets of over determined equations, which can be solved in the least squares sense, whose details can be followed in Walther [40].[33]

The detailed results of engineering interest can be obtained using more elaborate numerical computational method. Such procedure has been followed in the reference papers. The analysis for the bridge cross section by Selvam and Govindraswamy is performed in two broad categories, fixed and moving. Different grids varying in terms of the number of elements, spacing and density are used in the computations. The effect of grid in the accuracy of the prediction of vortex-induced response and critical velocity computation is studied by running the models for different grids. Since the flow is very complicated and highly nonlinear, the mathematical integration techniques and solution strategies used to solve the equations of structure and fluid are sensitive to errors.

Even if a very small error is induced in the solution process, over several iterations of computation, the error gets carried over and grows in magnitude during each step in the time marching solution procedure and finally blows up to enormous magnitudes. The artificial viscosity or numerical diffusion when added to the fluid flow helps to have a better control of the numerical stability. But when it is more, the flow becomes diffusive, and the results do not represent the actual behavior of the fluid. Hence several trials were made to find the lowest value of diffusion coefficient θ, as explained in earlier section that yields stable results. The coefficient of diffusion θ, is varied from 0.1 to 0.5 in the computations, with value of θ being similar to the first-order upwind procedure in finite difference method.

Figure 18.22 depicts an example of vortex pattern behind bridge girder cross CFD simulation, to demonstrate similar simulations that have been carried out by Selvam and Govindaswamy, aeroelastic analysis of bridge girder section using computer modeling,[34] Their results also show the case where the bridge model considered undergoes significant deflections after the onset of flutter occurred.

---

[33] Walther [40] and Larsen and Walther [19].

[34] Selvam and Govindaswamy [30].

It should be noted that the simulation shown does not reflect realistic deformation but to provide an observable effect through such simulation.

18.6.4 The Aeroelastic Stability Problem of Long-Span Cable-Stayed Bridges Under an Approaching Crosswind Flow

The aeroelastic stability problem of long-span cable-stayed bridges under an approaching crosswind flow can be analyzed starting from a continuous model of the fan-shaped bridge scheme with both H- or A-shaped towers, critical states of the coupled wind-structure system are identified by means of a variational formulation, accounting for torsional and flexural (vertical and lateral) bridge oscillations. The overall bridge dynamics is described by introducing simple mechanical systems with equivalent stiffness properties and, under the assumption of a prevailing truss-like bridge behavior, analytical estimates for dominant stiffness contributions are proposed. Several case studies are discussed, and comparisons with experimental evidences as well as with available analytical and numerical results are presented. The proposed simplified approach proves to be consistent and effective for successfully capturing the main wind-bridge interaction mechanisms, and it could be considered as a useful engineering tool for the aeroelastic stability analysis of long-span cable-stayed bridges. Once the flutter is initiated, the bridge undergoes enormous deflections until it finally fails.

18.7 Aeroelastic Equilibrium of the Bridge

The aeroelastic stability problem of long-span cable-stayed bridges under an approaching crosswind flow[35] can be analyzed starting from a continuous model of the fan-shaped bridge scheme with both H-or A-shaped towers; critical states of the coupled wind-structure system are identified by means of a variational formulation, accounting for torsional and flexural (vertical and lateral) bridge oscillations. The bridge scheme herein examined can be exemplified by those shown in Fig. 18.23. The side and central span lengths can be denoted by l_s and l, respectively The overall bridge dynamics is described by introducing simple mechanical systems with equivalent stiffness properties and, under the assumption of a prevailing truss-like bridge behavior, analytical estimates for dominant stiffness contributions are proposed. Several case studies are discussed and comparisons with experimental evidences as well as with available analytical and numerical results are presented. The proposed simplified approach proves to be consistent and effective for successfully

---

[35] Vairo [38].

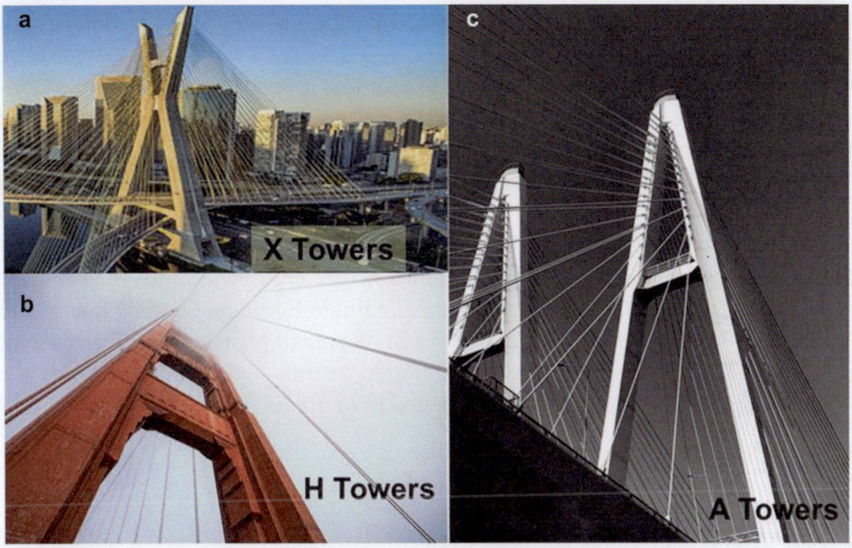

Fig. 18.23 Pictures of bridges pylon structure—**a** X towers; **b** H towers; **c** A towers[36]

capturing the main wind-bridge interaction mechanisms, and it could be considered as a useful engineering tool for the aeroelastic stability analysis of long-span cable-stayed bridges. Once the flutter is initiated, the bridge undergoes enormous deflections until it finally fails.

18.7.1 Aeroelastic Equilibrium of the Bridge: A Continuous Model

In Fig. 18.23, a beam-like girder, axially (i.e. along the z-direction) unconstrained, is hung to the tops of two piles, whose height with respect to the deck level is h, by means of four-plane fan-shaped stay curtains, with a constant stay spacing Δ.

Girder cross section is assumed to be constant with z and the bridge is symmetric with respect to both the vertical plane through the z-axis and the plane orthogonal to z through the bridge mid-span.

Due to the structural symmetry of the scheme, the twin reference system shown in Fig. 18.23 which displays H- and X-shaped towers (denoted in the following as HST and AST, respectively) can be assumed to incorporate pylons that are not joined with the girder.

Anchor cables are connected to the deck and at every point of the bridge's ends vertical displacements are prevented, so that torsional rotations are also fully

[36] Images downloaded from public domain sites, among others https://stocksnap.io/search/ and wikiwand (public domain).

restrained. Moreover, along-x girder displacement component is assumed to be zero at the bridge's ends and at the towers' locations. The girder's width B is assumed to be small in comparison with l and, in order to take into account usually employed efficient aerodynamic cross-sectional designs, is different from the distance $2b$ between the stay curtains at the girder level.

Since in modern long-span cable-stayed bridges, Δ is very small compared to l, and an equivalent diffused stay arrangement along the deck can be conveniently considered. Axial and shear deformability of towers and girder as well as flexural deformability of pylons in the plane (x, y) are neglected. Accordingly, the behavior of the deck can be described by employing the Euler–Bernoulli bending theory and the De Saint Venant torsional theory of torsion–flexure of prismatic members,[37] and, under general time-dependent loads, the bridge deformation is represented by the following displacement functions (elaborated in Vairo's article[38]):

- $s(z, t)$, horizontal deflection (in the plane (x, z)) of the girder;
- $v(z, t)$, vertical deflection (in the plane (y, z)) of the girder;
- $\theta(z, t)$, torsional rotation of the girder;
- $w(t)$, axial (along-z) displacement of the girder;
- $u(t)$, mean along-z displacement at the tower's tops;
- $\psi(t)$, rotation of the tower top-section around the y-axis.

Where the dependency on the time t has been emphasized.

Accordingly, equations governing the linearized flexural–torsional problem of the bridge under unsteady crosswind flow result in:

$$E_d I_y s_{\wp}^{iv}(z,t) = q_{cx}(z,t) - m_{cy}^{i}(z,t) - m\ddot{s}(z,t) + D(z,t) \quad (18.27)$$

$$E_d I_x s_{\wp}^{iv}(z,t) = q_{cy}(z,t) - m\ddot{v}(z,t) + L(z,t) \quad (18.28)$$

$$c_t \theta_{\wp}^{ii}(z,t) = -m_{cz}(z,t) - I_\theta \ddot{\theta}(z,t) - M(z,t) \quad (18.29)$$

$$\int_P -q_{cz}(z,t)\mathrm{d}z \mp k_p u(t) - S_{O\wp}(t) = 0 \quad (18.30)$$

$$\int_P -m_{cy}(z,t)\mathrm{d}z \mp k_{p\theta} \Psi_P(t) - M_{O\wp}(t) = 0 \quad (18.31)$$

where

$$v_{\wp}^{iv} = \frac{\partial^4 v_{\wp}}{\partial z^4} \quad (18.32a)$$

[37] Connor [4].

[38] Vairo [38].

$$\theta_{\wp}^{ii} = \frac{\partial^2 \theta_{\wp}}{\partial z^2} \tag{18.32b}$$

$$m_{cy}^{i} = \frac{\partial m_{cy}}{\partial z} \tag{18.32c}$$

and $\dot{F}$ is the first-time derivative of F.

Equations 18.27–18.29 denote flexural and torsional equilibria of the girder, while Eqs. 18.30 and 18.31 represent translation equilibrium along-z and rotation equilibrium around y for each tower, respectively. In Eqs. 18.30 and 18.31, the integration is performed on the stay curtains belonging to the left ($\wp = L$) or right ($\wp = R$) side of the bridge, applying sign "–" when $\wp = L$ and "+" when $\wp = R$.

Moreover, $E_d I_x$ ($E_d I_y$) and C_t are the bending and torsional stiffnesses of the deck, respectively, E_d is the Young's modulus of the girder; m and I_θ are the unit length mass and the girder's polar mass moment of inertia; q_{cj} (z, t) denotes the density of stays girder interaction forces along the j-axis ($j = x,y,z$): $m_{cz}(z, t)$ and m_{cy} (z, t) are the couple densities about z and y, induced on the girder by the stay curtains; $S_{o\wp}(t)$ and $M_{o\wp}(t)$ indicate the along-z force and the along-y couple produced by anchor stays at the tower tops; k_p and $k_{p\theta}$ are the flexural and torsional stiffnesses, respectively, at the tower top sections. For H-shaped towers made of pylons connected at the top, it can be assumed $k_{p\theta} = k_p b^2$, whereas for A-shaped towers, $k_{p\theta}$ can be considered approaching infinity, so that in this case, $\psi_L = \psi_R = 0$. Further details of the analysis could be followed in Vairo's paper referenced.

18.8 Closing Remarks

The aeroelastic stability of a selected civil structures, in particular tall structure and long-span bridges, has been addressed, and a simplified variational formulation for the dynamic problem of the wind-structure coupled system has been elaborated to reveal the relevant physical characteristics and the governing equations based on first principles. Some examples are selected from classical and recent literature, to reveal the detailed approaches, analytical elaboration and some wind tunnel experiments that can provide a comprehensive impression for dealing with similar or more involved problems.

In particular, to cite an example, starting from a continuous model of the fan-shaped bridge scheme with both H- and A-shaped towers, stability limit states, with regard to both torsional divergence and flutter, are identified by singularity conditions of an integral wind-structure impedance matrix. This latter is defined considering a general representation of the aeroelastic non-steady wind loads and introducing integral stiffness properties, which allow to describe the overall dynamic behavior of the bridge by means of simple lumped parameter mechanical systems. Under the assumption of a prevailing truss-like bridge behavior, integral stiffnesses have been analytically estimated considering damping-free torsional and flexural (vertical and lateral) bridge oscillations in still air. Moreover, proposed closed-form relationships prove that cable-stayed bridges with A-shaped towers exhibit torsional stiffness (deeply related with the bridge sensitivity to wind effects) greater than that of bridges based on H-shaped towers. Several wind-structure stability analyses have been elaborated.

Analytical procedures and analysis described should serve as a guideline on the ability of the described approach to successfully capture the physics of wind-induced motion and instabilities introduced to tall structures and bridges.

References

1. Adepalli, B., and E.R. Pardyjal. 2014. A study of flow fields in step-down street canyons. *Environmental Fluid Mechanics*.
2. Chai, Sun S., and Sergei Monten. 2000. Wind effects on long-span bridges. In *Wind effects on long-span bridges. Bridge engineering handbook*, ed. Wai-Fah Chen and Lian Duan. Boca Raton: CRC Press.
3. Chan, C.M., and J.K.L. Chui. 2006. Wind-induced response and serviceability design optimization of tall steel buildings. *Engineering Structures* 28 (4): 503–513.
4. Connor, St. Venant. Theory of where PQ is a force applied at Q in the direction of the displacement measure, dQ. Torsion-Flexure of Prismatic Members. https://ocw.mit.edu/courses/civil-and-environmental-engineering/1-571-structural-analysis-and-control-spring-2004/readings/connor_ch11.pdf.
5. Djojodihardjo, R.H., and S.E. Widnall. 1969. Numerical method for the calculation nonlinear unsteady lifting potential flow problem. *AIAA* 7 (10): 19.
6. Fundamentals of Vortex-Induced Vibration. https://www.bsee.gov/sites/bsee.gov/files/tap-technical-assessment-program/485ab.pdf.
7. Giesing, J.P. 1968. Nonlinear two-dimensional unsteady potential flow with lift. *Journal of Aircraft* 5 (2): 135–143 as quoted in Djojodihardjo, and Widnall. 1969. Numerical method for the calculation nonlinear unsteady lifting potential flow problem. *AIAA* 7 (10).
8. Green, Daniel, and William G. Unruh. 2006. The failure of the Tacoma Bridge—A physical model. *American Journal of Physics* 74: 706. https://doi.org/10.1119/1.2201854.
9. Grünbaum, Catrina. 2008. Structures of tall buildings, Lunds Tekniska Högskola-Rapport_TVBK-5156.
10. Gu, M., and Y. Quan. 2004. Across-wind loads of typical tall buildings. *Journal of Wind Engineering and Industrial Aerodynamics* 92 (13): 1147–1165.
11. Halse, K.H. 1997. *On vortex shedding and prediction of vortex-induced vibrations of circular cylinders*. Dr.-Ing.Thesis, Norwegian University of Science and Technology, Trondheim.
12. Howarth, H. 2015. Human exposure to wind-induced motion in tall buildings and assessment of guidance in ISO 6897 and ISO 10137.
13. Ilgin, H.E. 2006. *A study on tall buildings and aerodynamic modifications. Against wind excitation*. M.S. Dissertation, Department of Architecture, Room No: 69, Middle East Technical University.
14. Ilgin, H. Emre, and M. Halis Günel. 2007. The role of aerodynamic modifications in the form of tall buildings against wind excitation. *METU Journal of Faculty of Architecture*.
15. Kim, H.-J. 2009. *Mechanism of wake galloping of two circular cylinders*. Ph.D. thesis, Nagoya University.

16. Kim, Y., and J. Kanda. 2010. Effects of taper and set back on wind force and wind-induced response of tall buildings. *Wind and Structures* 13 (6): 499–517.
17. Kwok, K.C.S. 1977. *Cross-wind response of structures due to displacement dependent excitations*. Ph.D. Thesis, Monash University, Victoria, Australia.
18. Kwok, K.C.S., and W.H. Melbourne. 1981. Wind induced lock-in excitation of tall structures. *Journal of the Structural Division* 107 (1): 57–72.
19. Larsen, A., and J.H. Walther. 1996. *A new Computational Method for assessment of the Aeroelastic stability of long span bridges*. Copenhagen: IABSE.
20. Larsen, Allan, and Jens H. Walther. 1998. Discrete vortex simulation of flow around five generic bridge deck sections. *Journal of Wind Engineering and Industrial Aerodynamics* 77&78: 591–602.
21. Le, Thai-Hoa, and Luca Caracoglia. 2019. Modeling vortex-shedding effects for the stochastic response of 2 tall buildings in non-synoptic winds. https://www.sciencedirect.com/science/article/pii/S0889974615002893.
22. Marukawa, H., N. Kato, K. Fujii, and Y. Tamura. 1996. Experimental evaluation of aerodynamic damping of tall buildings. *Journal of Wind Engineering and Industrial Aerodynamics* 59 (2–3): 177–190.
23. Mohan, R., S. Sundararaj, and K.B. Thiagarajan. 2019. Numerical simulation of flow over buildings. In *The 11th National Conference on Mathematical Techniques and Applications, AIP Conference Proceedings*, vol. 2112, 020149-1-020149-7. https://doi.org/10.1063/1.5112334. Published by AIP Publishing. 978-0-7354-1844-8/$30.00.
24. Murugan, Thangadurai. 2015. *Effect of free stream turbulence on flow past a circular cylinder*. Central Mechanical Engineering Research Institute, Technical Report, October 2015. https://doi.org/10.13140/RG.2.1.1683.0162.
25. Nikitas, R.N., and J.H.G. Macdonald. 2014. Misconceptions and generalizations of the Den Hartog galloping criterion. *Journal of Engineering Mechanics* 140 (4).
26. Novak, M. 1972. Galloping oscillations of prismatic structures. *Journal of the Engineering Mechanics Division* 98 (EM1): 27–46.
27. Nowicki, Tomasz. 2009. *Survey of applications of discrete vortex method in civil engineering. Budownictwo i Architektura* 5.
28. Pozzuoli, Chiara. 2012. *Aeroelastic effects on tall buildings-performance-based*. Dr.-Ing.Thesis, U Firenze.
29. Da Silva Rebelo de Campos, A.M. 2014. *Bridge aerodynamic stability*. M.Sc. Thesis, Tecnoco Lisboa.
30. Selvam, R.P., and S. Govindaswamy. 2001. *Aeroelastic analysis of bridge girder section using computer modeling*. University of Arkansas.
31. Shirzadi, Mohammad Reza, Parham A. Mirzaei, and Mohammad Naghashzadegan. 2017. Improvement of k-epsilon turbulence model for CFD simulation of atmospheric boundary layer around a high-rise building using stochastic optimization and Monte Carlo Sampling technique. *Journal of Wind Engineering and Industrial Aerodynamics* 171: 366–379.
32. Simantiras, Paolo, and Neil Willis. 1999. Investigation on vortex induced oscillations and helical strakes effectiveness at very high incidence angles. In *The Ninth International Offshore and Polar Engineering Conference*, Brest, France, May 30–June 4.
33. Simiu, E., and R.H. Scanlan. 1996. *Wind effects on structures*, 3rd ed. New York: Wiley Interscience.
34. Simiu, and Yeo. 2019. *Wind effects on structures modern structural design for wind*, 4th ed. Wiley Blackwell.
35. Steckley, A. 1989. *Motion-induced wind forces on chimneys and tall buildings*. Ph.D Thesis, University of Western Ontario.
36. Stickland, M., and T.J. Scanlon. 2001. An investigation into the aerodynamic characteristics of catenary contact wires in a cross-wind. *Proceedings of the Institution of Mechanical Engineers Part F: Journal of Rail and Rapid Transit* 215 (4): 311–318.
37. Turkiyyah, G., D. Reed, J. Yang. 1995. Fast vortex methods for predicting wind-induced pressures on buildings. *Journal of Wind Engineering and Industrial Aerodynamics* 58.

38. Vairo, Giuseppe. 2010. A simple analytical approach to the aeroelastic stability problem of long-span cable-stayed bridges. *International Journal for Computational Methods in Engineering Science and Mechanics* 11 (1): 1–19.
39. Venanzi, I., and A.L. Materazzi. 2012. Across-wind aeroelastic response of square tall buildings. *Wind and Structures* 15 (6).
40. Walther, J.H. 1994. *Discrete vortex method for 2D flow past bodies of arbitrary shape undergoing prescribed rotary and translational motion.* Ph.D. Thesis, Department of fluid mechanics, Technical University of Denmark, Denmark.
41. Wang, Lu. 2016. *High-performance discrete-vortex algorithms for unsteady viscous-fluid flows near moving boundaries.* Ph.D., UC Berkeley.
42. Yoon, Dong-Hyeog, Kyung-Soo Yang, and Choon-Bum Choi. 2010. Flow past a square cylinder with an angle of incidence. *Physics of Fluids* 22: 04360.
43. Zhao, Jisheng, Justin S. Leontini, David Lo Jacono, and John Sheridan. 2014. Fluid–structure interaction of a square cylinder at different angles of attack. *Journal of Fluid Mechanics* 747: 688–721. ISSN 0022-1120.
44. Zhu, Zhiwen. 2015. LES prediction of aerodynamics and coherence analysis of fluctuating pressure on box girders of long-span bridges. *Computers & Fluids* 110 (30): 169–180.

Part III
Case Studies on Application Examples

Chapter 19
Aeroelastic Optimization of Tapered Wing Structure

Abstract A methodology for including maximum flutter speed requirement in the preliminary structural wing design is developed. The problem of minimizing structural weight while satisfying static strength, dynamic characteristics and aeroelastic behavioral constraints is stated in a nonlinear mathematical form, with beam width and thickness taken as design variables, and solved using gradient-based optimization technique. Dynamic characteristics of the structure are calculated using finite element model. Laplace form of the unsteady aerodynamics forces is obtained from Fourier transform of unit pulse aerodynamics response. The frequency-domain p-k method is applied for the calculation of aeroelastic stability boundaries. Based on constraint values and the required gradients, a first-order Taylor series approximation is used to develop an approximation linear programming for weight minimization. A modified feasible direction method is, then applied iteratively to solve the optimization problem. Validation of the method is carried out in the design of cantilever straight wing structure with 6% hyperbolic airfoil. It will be shown that the optimized wing design can significantly differ from those obtained without optimization process.

Keywords Aeroelasticity · Frequency-domain P-K method · Optimization · Structural dynamics

19.1 Introduction

With advancing design process in which input data for all calculations become more precise and easy to generate, an interdisciplinary design concept, that takes into account all important interdisciplinary mutual effects including aeroelastic stability boundary, must be initiated to provide a basis for design decisions on time. This

---

Originally presented at the 22nd International Congress of the Aeronautical Sciences, September 2000, Harrogate UK, as "Optimization of Tapered Wing Structure With Aeroelastic Constraint", by I Wayan Tjatra, Harijono Djojodihardjo and Ismojo Harjanto, Computational Aeroelastic Group, Aerospace Engineering Department, Institute of Technology Bandung, Jl. Ganesha 10, Bandung 40132, Indonesia.

© The Author(s), under exclusive license to Springer Nature Singapore Pte Ltd. 2023
H. Djojodihardjo, *Introduction to Aeroelasticity*,
https://doi.org/10.1007/978-981-16-8078-6_19

design concept should have capability in providing the effects of change on each of the design parameter on the behavior of the structure (stress level, damping, natural frequency, aeroelastic boundary, etc.). Furthermore this design concept should also be used to find optimal design solution by means of mathematical optimization tools. One such methodology that based upon this interdisciplinary design concept was proposed by Sobieszczanski-Sobieski [6], which explicitly includes the important aeroelastic stability constraints, i.e. maximum divergence and flutter speed. Since then, there were significant number of research works have been done in the field of optimal design of aircraft structures subject to aeroelastic stability constraints, the so-called aeroelastic tailoring. One of the important early contribution into this fields was made by Haftka and Yates [2] in which an optimization algorithm was developed for the used in repetitive aeroelastic calculation during the structural design process. The used of advanced composite/smart materials has further improved the possibilities in tailoring the dynamic response characteristics of the structure under unsteady aerodynamic loads and the availability of powerful computer opened the possibilities for lengthy optimization calculation. Because aeroelastic tailoring in the design process of an aircraft structure is true multidisciplinary analysis and optimization, design results cannot be achieved independently of the requirements of certain other disciplines.

With regard to the structure modeling, it is important to understand that there is a risk of multiplication in modeling error in each discipline when different models are combined during the optimization process. Aeroelastic optimization includes evaluation of nonlinear aeroelastic stability constraints, which consist of the solution of eigenvalues of nonlinear unsymmetrical and complex matrices. Evaluation of such nonlinear and non-differentiable constraints leads to discontinuities in divergence or flutter speed design parameters.

The purpose of present paper is to demonstrate the accuracy of an aeroelastic optimization procedure, which was developed based upon a simple modified feasible direction method (MFD), used in the design of flexible wing structure which is linear but the external aerodynamic loads are nonlinear and are function of structure response, i.e. transonic unsteady aerodynamics. The unsteady aerodynamics theories used for the aeroelastic optimization in the past were linear such as doublet lattice and Theodorsen's strip theory for unsteady subsonic flows and Mach box theory for unsteady supersonic flows. Nonlinear dependency of the unsteady transonic aerodynamic forces on structural response makes it difficult for the integration of aerodynamic solution into the solution of the aeroelastic equation of motion and stability. Few works specifically associated with wing structure design subject to transonic aeroelastic constraints are available at present time [5, 8]. In this work, the optimal design of wing structure was obtained using nonlinear unsteady transonic aerodynamic theory by solving transonic small disturbance (TSD) flow equations. The unsteady aerodynamic forces were calculated from the unit pulse aerodynamic response in time domain [7]. The structural parameter considered is those found in the aeroelastic equation of motion: mass ratio, static unbalance, first bending and first torsional natural frequencies and free-stream Mach number. Finite difference method is applied in the calculation of design sensitivities. The accuracy of this optimization methodology is validated using the results obtained using other methods and experimental data.

19.2 General Approach

The problem of minimizing wing structural weight while satisfying static strength and aeroelastic stability constraint of the structure is stated in nonlinear mathematical programming form and solved using linearized gradient-based optimization technique. Wing skin, spars and the other wing structural components are assumed to be built-up and modeled as straight stiffness beam, positioned along the elastic axis of the wing, with linear variation of thickness and width in the span-wise directions and concentrated masses at the beam center of mass. The dynamic characteristics of the wing are calculated using finite element modeling in which the stiffness beam is modeled using beam elements with three degree-of-freedom at each of its nodal point. The associated design variables consist of beam width and thickness.

It is assumed that the amplitude of bending and torsional deflection is small that the aerodynamic response of the wing geometry is considered linear with the change in the wing angle of attack. This assumption implies that the Pade approximation method can be used for the generalized unsteady aerodynamic force coefficients curve fitting in Laplace domain. The generalized unsteady aerodynamics forces coefficients in Laplace domain themselves are calculated from the Fourier transform of the time-domain aerodynamic response of the wing geometry due to a unit pulse displacement. Unsteady transonic flow fields are represented using transonic small disturbance (TSD) equations. This method is considered more accurate compared with the indicial response method where a jump in the indicial displacement, even with moderately small amplitude, can generate divergent aerodynamic response. Meanwhile, the continuously unit pulse displacement can avoid inaccuracy in the aerodynamic response.

Aeroelastic stability boundary of the structure is calculated using the p-k method and a straightforward finite difference method is used for the approximation of flutter constraints sensitivities, which are represented by the rate of change of the damping coefficients. The optimization problem itself is solved iteratively by linear modified feasible direction method.

19.3 Minimum Weight Optimization

The problem of minimizing the optimum structure total weight can be represented in the form of nonlinear mathematical programming as follows:

Minimize the objective function: $F(x)$.

Subject to the constraint conditions of:

$$\begin{aligned} &g_j(x) \leq 0 \quad && j = 1, 2, \ldots n_g \\ &h_k(x) \leq 0 \quad && k = 1, 2, \ldots n_h \\ &x_i^l \leq x_i \leq x_i^u \; && j = 1, 2, \ldots n \end{aligned} \tag{19.1}$$

where $F(x)$ is the total weight of the structure, and x is the vector of design variables which contain all structural physical properties which are changing during the optimization process, such as dimension of the cross section, skin thickness and the concentrated masses. The g_j and h_k functions contain all of the in-equality and equality constraint, respectively, such as the allowable structural component stresses and strains and aeroelastic damping or rate of change of the aeroelastic damping. In addition to those two constraints, there is a side constraint x_i which specify the upper bounds values, x_i^u , and the lower bounds values, x_i^l , of each of the design variables.

For the present optimization problem of wing structural design with aeroelastic constraint using a flexible beam model divided into (numel)-elements, the objective function is the total structural weight of the wing and is represented by

$$\begin{aligned} W(x) &= \sum_{i=1}^{\text{num.el}} \rho_{mi} L_i A_i \\ &= \sum_{i=1}^{\text{num.el}} \left[m_0 L \frac{x_1 x_2}{x_1^0 x_2^0} \right] \Bigg|_i \end{aligned} \tag{19.2}$$

where $W(x)$ is the total design weight, ρ_{mi}, L_i and A_i are the mass density, length and cross-sectional area of the ith beam elements, respectively. Two design variables are the flexible beam width, x_1, and the beam thickness, x_2. Zero subscript or superscript represents values of the variables at the wing root.

19.4 Aeroelastic Constraint

Since variables that come in the aeroelastic constraints function depend on the solution method used in the flutter analysis of the structure, before the aeroelastic constraint can be defined, the method used in flutter analysis of the wing structure has to be decided first. Considering all the available flutter solution method, the *p-k* iterative method which is based upon the eigenvalues solution of the stability equation, with flow free-stream velocity as the input and motion damping coefficients and frequencies as output, is the most ideal to be used in this optimization problem. Using the *p-k* method for the flutter analysis, the fundamental wing structure equation of motion can be expressed as [4, 7]:

$$\left[M_{hh}p^2 + \left(-\frac{1}{4k}\rho c V Q_{hh}^I\right)p + \left(K_{hh} - \frac{1}{2}\rho V^2 Q_{hh}^R\right)\right]\{u_h\} \tag{19.3}$$

where M_{hh} and K_{hh} are the generalized mass and stiffness of the structure, respectively. The real and imaginary part of the generalized aerodynamic forces is denoted, respectively, by Q_{hh}^R and Q_{hh}^I which are function of flow free-stream velocity, V, and motion reduced frequency $k = (\omega\, b/2V)$. Meanwhile $\{u_h\}$ represents the structural generalized coordinate which contains nodal displacements. Eigenvalues of the system are given by complex variable p which is defined as

$$p = \omega(\gamma + 1) \tag{19.4}$$

where ω is the motion frequency (Hz) and γ is the transient damping coefficient.

The unsteady generalized aerodynamic force coefficients $[Q_{hh}]$ in the above equation are defined as

$$[Q_m] = [\phi]^T[\text{AFC}(ik)][\phi] \tag{19.5}$$

in which $[\varphi]$ and $[\text{AFC}(ik)]$ represent the structure natural mode shape and matrix of the aerodynamic response coefficient, respectively. This aerodynamic response coefficient is calculated by solving, in time domain, the transonic small disturbance flow equations around the wing structure having a unit pulse displacement. Then, by making used of Fourier transform along with Pade approximation function, the aerodynamic response coefficients are calculated from this transformation and can be written in terms of p—variable (Laplace variable) as [1]

$$Q_{ij}(p) = Q_0 + Q_1 p + Q_2 p^2 + \sum_{m=3}^{6} \frac{Q_m p}{(p + \beta_{m-2})} \tag{19.6}$$

where β_{m-2} represents phase-lag parameter. The approximating function coefficients $Q_o, Q_1, Q_2 \ldots$ are evaluated by least square curve fitting using complex values of Q_{ij} at discrete number of k or p—values. Before solution of the aeroelastic stability equation, Eq. 19.3, as an eigenvalue problem can be carried out, this equation should be written in a state-space form as:

$$[A - pI]\{U_h\} \tag{19.7}$$

where I is a unit matrix, A is a real matrix defined by matrix equation

$$[A] = \begin{bmatrix} 0 & I \\ -M_{hh}^{-1}\left[K_{hh} - \frac{1}{2}\rho V^2 Q_{hh}^R\right] & M_{hh}^{-1}\left[\frac{1}{4k}\rho c Q_{hh}^I\right] \end{bmatrix} \tag{19.8}$$

and the vector of generalized coordinate $\{U_h\}$ contains not only nodal point displacement but also velocity. Because A is not a symmetric matrix, conventional numerical technique can not be applied in the calculation of the eigenvalues and eigenmodes of this equation.

Dynamic characteristics of the wing structure required for the analysis, the structural natural frequency, ω_i, and mode shape, φ, are obtained using a finite element method. The stiffness beam is divided into several beam element with three degree-of-freedom (d.o.f) at each nodal point, which are one transverse displacement and two torsional/rotation d.o.f.

The flutter constraint is defined by satisfying requirements on modal damping at a series of velocities, rather than defined straight on the actual flutter speed. Aeroelastic constraints, therefore, can be expressed as [3]

$$\begin{aligned} \gamma_{ij} &\leq \gamma_{j\,\mathrm{req}} \\ g_j(x_j) &= \frac{\gamma_{ij} - \gamma_{j\,\mathrm{req}}}{\mathrm{GFACT}} \leq 0 \end{aligned} \tag{19.9}$$

in which γ_{ij} is the calculated damping coefficient for the *i*-th mode at the *j*-th velocity and $\gamma_{j\,\mathrm{req}}$ is the required damping level at the *j*-th velocity. Both of those constraints have to be satisfied for all mode shape (i) and velocities (j) used in the analysis. The general normalization factor (GFACT) is used to normalize the constraint values (which cannot be normalized with respect to $\gamma_{j\,\mathrm{req}}$ because this variable could take a value of zero). As described in Ref. [7], the GFACT value used for all calculation in this study is 0.1.

In addition to the aeroelastic constraints described above, two side—constraints are also applied in this optimization process to take into account the maximum and minimum values the design variables x_1 and x_2 (i.e. the width and thickness of the elastic beam) can take on.

19.5 Sensitivity of Aeroelastic Constraint

Derivative of the constraint function, $g_j(x_i)$, with respect to design variable, x_i, is defined from Eq. 19.9 as

$$\frac{\partial g_j(x_i)}{\partial x_i} = \frac{1}{\mathrm{GFACT}} \frac{\partial \gamma_{ij}(x_i)}{\partial x_i} \tag{19.10}$$

Since this sensitivity derivative is represented in terms of γ_{ij} which is obtained from the nonlinear eigenvalue solution of the stability equation, Eq. 19.7, solution of the constraint sensitivity as given in the above equation will be efficiently obtained using numerical approximation. In this study, calculation of the right-hand side of Eq. 19.10 are carried out using forward difference formulae. The discrepancy of this approach is that, as in any other finite difference approximation, it has a computational error which may be large that the approximation for the sensitivity become inaccurate. In order avoid this problem, each step interval have to be carefully defined.

19.6 Optimization Procedure

After the engineering and sensitivity analysis is completed, the structure is then optimized by solving the nonlinear mathematical programming problem stated in Eq. 19.1. In this study, the method of modified feasible direction (MFD) is applied to solve the optimization problem [8]. The main task in this method is to find an accurate usable-feasible search direction, S^q, which will define the direction with maximum gradient in the objectives function, $F(x_i)$ but still lies in the feasible domain (no constraints are violated). Once this search direction is defined, new vector for design variables is composed as

$$x^q = x^{q-1} + \alpha S^q \tag{19.11}$$

where

$q =$ iteration number and

α is the scalar displacement parameter.

Values of this scalar parameter are estimated at the beginning of each iteration based upon the gradient of the objective function and constraints. Linear Taylor series approximation is applied in the calculation of the required values of the objective function and also constraints at every iteration step

$$F\left(x^q\right) = F\left(x^{q-1}\right) + \left[\frac{dF\left(x^{-1}\right)}{d\alpha}\right]\alpha$$

$$g_j\left(x^q\right) = g_j\left(x^{q-1}\right) + \left[\frac{dg_j\left(x^{-1}\right)}{d\alpha}\right]\alpha \tag{19.12}$$

The procedure to perform the optimization process is as follows:

a. Select a free-stream Mach number and the corresponding velocity, altitude and wing configuration.
b. Define the initial values for design variables and initial dynamic characteristics of the wing structure.
c. Calculate static aerodynamic pressure distribution (by solving the steady TSD equation) and calculate static aeroelastic stability of the structure
d. Based upon the static aeroelastic deformation of the structure, calculate the unit pulse response of the wing structure (solve the unsteady TSD flow equations).
e. Once the unit pulse response is forces in Laplace domain can be determined and the flutter damping constraint and sensitivities evaluated.
f. With the results from step e, an optimization process can be started, yielding a new set of design variables. Calculate a new dynamic characteristic of the structure.
g. With the new design variables, repeat step c to f until a converged optimum results is obtained. At the optimum, the normalized values of the constraint must

not larger than the defined error of EMIN and the objective function cannot have moved by more than FMIN % from the previous iteration.

19.7 Example Problem

To demonstrate the preceding optimization derivation, a straight wing model is considered at various flutter speed target and flight conditions. The wing has a rectangular, unswept, untapered planform that uses a stiffness beam representation for the structure's flexibility and concentrated masses for the distributed mass representation. This same rectangular wing was selected to demonstrate transonic flutter prediction in Ref. [5]. The wing has a moderate aspect ratio of 5.34 with a 6% hyperbolic airfoil, taper ratio of 0.7 and a tapered cantilever stiffness beam that represents the flexibility of the wing. As shown in Fig. 19.1, the elastic axis of the stiffness beam is placed at 33% chord length from the leading edge, meanwhile its beam mass is positioned at 43% chord length from the leading edge.

Elastic beam is made of aluminum with $E = 1.5E + 09$ lb/ft^2 and $G = 5.5E + 08$ lb/ft^2. The structural properties are given in Table 19.1. For the determination of structural dynamic characteristics the stiffness beam is divided into beam elements of equal length with the same number of discrete mass points. Initial values of the design variables are evaluate based upon static load requirements and taken to be x_1, the stiffness beam width equal to 0.2950 ft and the beam thickness x_2 equal to 0.9221 ft. With this beam initial dimension, the first ten natural frequencies of the structure are in Table 19.2, and comparison is made with respect to the uniform model. It can be seen that, in general, the prediction of the bending frequencies using uniform structural model is more accurate compared to the torsional frequencies.

For aeroelastic analysis it is assumed that the beam stiffness gives no contribution to the 6% hyperbolic airfoil aerodynamics. A cubical spline is applied to accomplished transformation of structural mesh into the aerodynamic mesh required for the solution of the Transonic Small Disturbance (TSD) flow equation. The aerodynamic response at certain free-stream Mach number is calculated using a mesh system which consists of 100 × 23 × 40 mesh points.

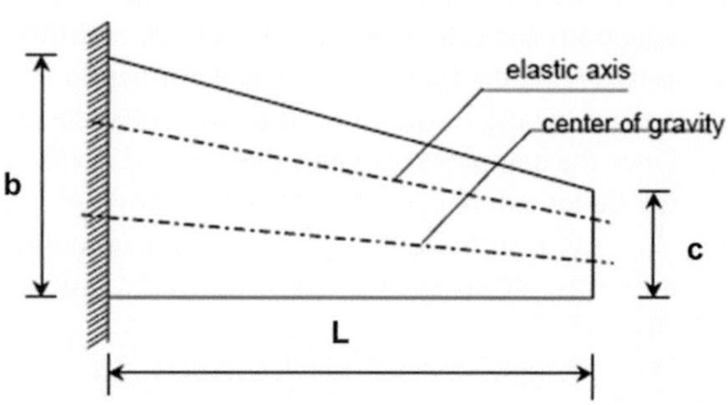

Fig. 19.1 Rectangular wing planform with taper ratio of 0.7

Table 19.1 Structural properties of tapered wing

Parameter	Values
Chord length	72.0 in
Span	240.0 in
EI	$23.65\ \mathrm{E}+06$ lb ft^2
GJ	$2.39\ \mathrm{E}+06$ lb ft^2
Mass/wingspan	0.746 slug/ft
S_α	0.447 slug ft/ft
I_α	1.943 slug ft^2/ft

Table 19.2 First ten natural frequencies of tapered wing structure

No.	Modes	Natural frequency (Hz)		% error
		Uniform	Tapered	
1	1st bending	7.872	8.877	12.8
2	1st torsion	12.995	14.667	12.9
3	2nd torsion	38.989	40.199	3.1
4	2nd bending	49.038	51.686	5.4
5	3rd torsion	64.973	66.713	2.7
6	4th torsion	90.962	93.775	3.1
7	5th torsion	116.951	121.480	3.9
8	3rd bending	138.051	141.650	2.6
9	6th torsion	142.940	149.970	4.9
10	7th torsion	168.929	179.400	6.2

Aeroelastic analysis of the initial structure, carried out using p-k method, shows that the finite element model gives the highest flutter speed for the structure at almost the same flutter frequency, as shown in Table 19.3. The damping versus velocity curve will show that the initial wing structure undergoes a mild flutter in bending mode.

Wing structure optimization is studied for several flutter speed target, which are : 720, 840, 960, 1080 and 1200 ft/s, or an increase of 20, 40, 60, 80 and 100% in flutter speed from the initial configuration, with maximum error in the constraint and objective function of 0.0001 and 0.001, respectively. The side constraints of the problem are defined as

$$0.01\ \text{ft} < x_1 < 0.5\ \text{ft} \quad \text{and}$$
$$0.70\ \text{ft} < x_2 < 1.0\ \text{ft}$$

The change in design variables from its initial values to the optimum values at various flutter speed target is given in Table 19.4.

Table 19.3 Flutter properties of the tapered wing

Flutter Parameter	Values		
	Exact	Assumed modes	FEM
U_F (ft/s)	576.53	564.785	600.00
f_F (Hz)	10.54	11.73	11.04
k_F	0.344	0.358	0.347

Table 19.4 Comparison between the initial and optimum value of the design variables for tapered wing

u_f	Initial condition			Optimum values		
	Objective Function	x_1	x_1	Objective Function	x_1	x_1
720	15.78	0.295	0.922	15.67	0.321	0.899
840	15.78	0.295	0.922	17.65	0.351	0.927
960	15.78	0.295	0.922	19.45	0.383	0.934
1080	15.78	0.295	0.922	20.52	0.416	0.952
1200	15.78	0.295	0.922	22.55	0.470	0.883

Time history of the objective function and the aeroelastic constraints during the iteration process are given in Fig. 19.2 up to Fig. 19.5. Meanwhile, comparison of the variation of damping values with respect to the free stream velocity for initial and optimal design variables is shown in Fig 19.5. From Fig. 19.2 it is shown that both the objective function and aeroelastic constraints converged in less than 5 iterations regardless of the flutter speed target. The increase in flutter speed from the initial to the optimal design can be seen in Fig. 19.4, along with the fact that the flutter mechanism does not change from the initial design.

The percentage of change in the design variables from their initial values at several flutter seed targets can be summarized in Table 19.5. This table shows that the change in x_1 variables to the optimum design is larger compared to the change of x_2 variable, which indicate that the optimization problem is more sensitive to the x_1 design variable. This change in design variables for tapered wing structure is smaller compared to that for uniform wing structure.

The change in structural natural frequency at the optimal design compared to the initial design values for various flutter speed target is shown in Table 19.6. It is found that the change in the torsional frequency is much higher compared to the change in the bending frequency. This can be explained as the change of the x_1 design variable from its initial value is higher compared to the change in the x_2 variable. This causes the change in torsional stiffness of the beam will also be higher compared to the change in bending stiffness. This in turn will cause a higher change in the torsional frequency of the beam (Fig. 19.2 and Table 19.5).

The variation of the change in the structure total mass with respect to the percentage of change in the flutter speed target is given in Fig. 19.5. Relation between these two parameters is represented as an S curve (Figs. 19.3 and 19.4; Table 19.6).

19.8 Concluding Remarks

A methodology for including flutter speed requirements in the design of a wing structure is developed and tested. The problem of minimizing structural weight while

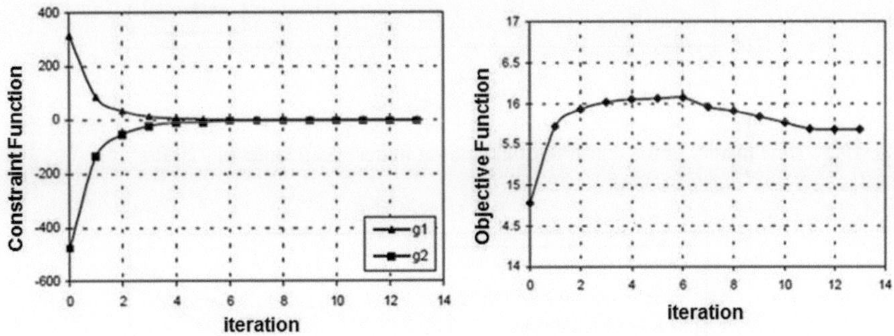

Fig. 19.2 Time history of the objective function at flutter speed target of 720 ft/s

Table 19.5 Changes in values of the design variables at several flutter speed targets

u_f	% change	
	x_1	x_2
720	8.81	−2.49
840	18.98	0.55
960	29.83	1.30
1080	41.02	3.25
1200	59.32	−4.23

Table 19.6 Comparison between the initial and optimum natural frequency at several flutter speed targets

u_f	Initial Frequency (Hz)		Optimal Values (Hz)		% Change	
	$\omega_{bending}$	$\omega_{torsion}$	$\omega_{bending}$	$\omega_{torsion}$	$\omega_{bending}$	$\omega_{torsion}$
720	8.88	14.67	8.65	16.42	2.52	11.94
840	8.88	14.67	8.92	18.17	0.55	22.89
960	8.88	14.67	9.00	20.20	1.34	37.71
1080	8.88	14.67	9.16	22.18	3.23	51.22
1200	8.88	14.67	8.50	22.69	4.28	75.16

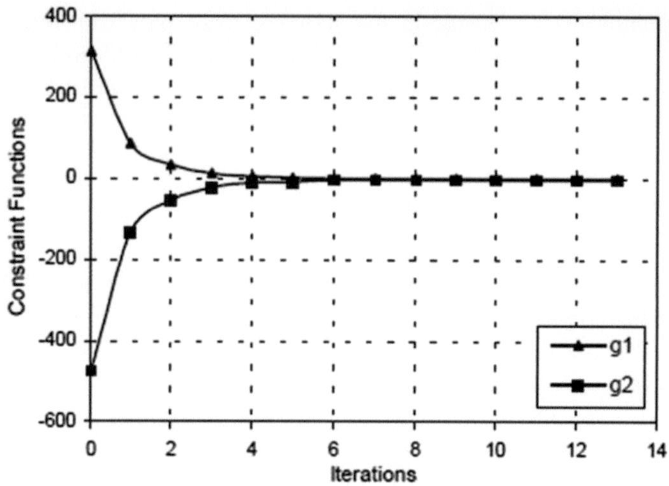

Fig. 19.3 Time history of the constraint functions at flutter speed target of 720 ft/s

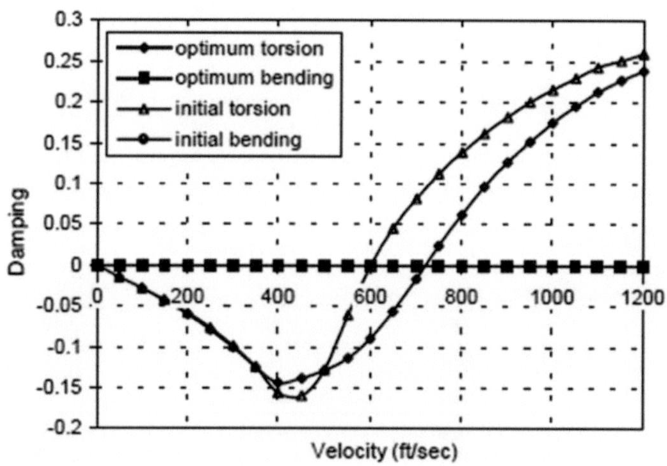

Fig. 19.4 Variation of the damping coefficients with velocity for rectangular wing at flutter speed target of 720 ft/s

satisfying static behavioral constraints is stated as a nonlinear programming which is solved using a modified feasible direction optimization procedure. The wing structure is modeled as a stiffness beam with discrete masses using finite element method and the associated design variables consist of beam width and thickness. The unsteady aerodynamic generalized forces are calculated based upon TSD flow solution using unit pulse response technique.

For the straight wing test case, the optimization problem converged in less than 10 iterations. A higher flutter speed constraint/target will give an optimum design

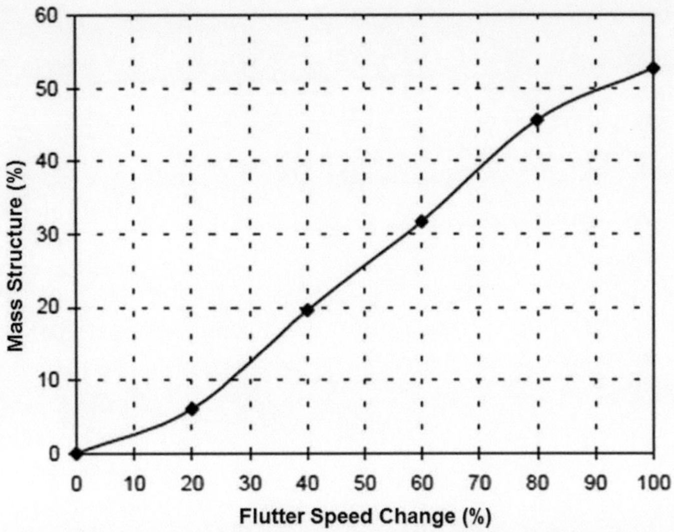

Fig. 19.5 Variation of the total mass with respect to changes in flutter speed target

with a larger difference between the first two bending and torsion natural frequencies and a larger total mass.

References

1. Baker, G.A., Jr. 1974. *Essential of Pade approximates*. New York: Academic Press.
2. Haftka, R.T., and E.C. Yates. 1976. Repetitive flutter calculations in structural design. *Journal of Aircraft* 13 (7).
3. Hajela, P. 1985. A root locus based flutter synthesis procedure. AIAA paper 83-0063.
4. NASTRAN user manual. 1997. The MacNeal-Schwendler Corporation, Version 69+.
5. Shirk, M.H., et al. 1986. Aeroelastic tailoring—Theory, practice and promise. *Journal of Aircraft* 23 (1)
6. Sobieszczanski-Sobieski, J. 1989. Multidisciplinary optimization for engineering systems: Achievements and potential, optimization: Methods and applications, possibilities and limitations. In *Proceeding of an International Seminar organized by DLR*, 42–62, Bonn, June 1989.
7. Tjatra, I.W. 1998. Calculation of three-dimensional unit impulse response using non-linear transonic flow equations. In *AIAA Atmospheric Flight Mechanics Conference*, Boston, USA.
8. Wilkinson, K., et al. 1977. FASTOP a flutter and strength optimization program for lifting surface structures. *Journal of Aircraft* 14 (6).

Chapter 20
Acoustic Effects on Binary Aeroelastic Model

Abstract Acoustics is the science concerning the study of sound. The effects of sound on structures attract overwhelming interests and numerous studies were carried out in this particular area. Many of the preliminary investigations show that acoustic pressure produces significant influences on structures such as thin plate, membrane and also high impedance medium like water (and other similar fluids). Thus, it is useful to investigate structural response to acoustics on aircraft, especially on aircraft wings, tails and control surfaces which are vulnerable to flutter phenomena. The present paper describes the modeling of structure-acoustic interaction to simulate the external acoustic effect on binary flutter model. Here, the model is illustrated as a rectangular wing where the aerodynamic wing model is constructed using strip theory with simplified unsteady aerodynamics involving the terms for flap and pitch degree of freedom. The external acoustic excitation, on the other hand, is modeled using a four-node quadrilateral isoparametric element via finite element approach. Both equations are then carefully coupled and solved using eigenvalue solution. Next the mentioned approach is implemented in MATLAB, and the outcome of the simulated results is later described, analyzed and illustrated.

Keywords Aeroelasticity · Binary model · Flutter · Structural-acoustic-aerodynamic coupling

Nomenclature

$[c,]$	Structural damping matrix
$[D_c]$	Proportional structural damping matrix
$[K_a]$	Acoustic stiffness matrix
$[K,]$	Structural stiffness matrix
$[M_{\mathrm{a}}]$	Acoustic mass matrix
$[M_{\mathrm{s}}]$	Structural mass matrix
$[R_{\mathrm{as}}]$	Structural-acoustic coupling matrix

This chapter is reproduced from Kok Hwa Yu, Harijono Djojodihardjo, and A Halim Kadarman, "*Acoustic Effects on Binary Aeroelasticity Model*", IIUM Engineering Journal, Vol. 12, No. 2, 2011.

© The Author(s), under exclusive license to Springer Nature Singapore Pte Ltd. 2023
H. Djojodihardjo, *Introduction to Aeroelasticity*,
https://doi.org/10.1007/978-981-16-8078-6_20

$\{p\}$	Vector of generalized pressures
$\{q,\}$	Vector of generalized structural displacements
$\{u\}$	Vector of generalized displacements-pressures
A	Combined structural-acoustic mass matrix
B	Combined structural-acoustic damping matrix
C	Combined structural-acoustic stiffness matrix
D	Strain displacement matrix
G	Transformation matrix
1	Identity matrix
I_κ	Moment inertia for flap, kgm2
I_θ	Moment inertia for pitch, kgm2
K_κ	Flap stiffness, Nm/rad
K_θ	Pitch stiffness, Nm/rad
L	Lift
M	Pitching moment, Nm
M_B	Non-dimensional pitch damping derivative
N_f	Shape function of acoustic fluid
N	Shape function of wing structure
V	Freestream velocity, m/s
a	Speed of sound, m/s
a	Two-dimensional lift curve slope
c	Wing chord
m	Mass per unit area, kg/m^2
ec	Distance between aerodynamic center with Flexural axis wingspan
t	Thickness of acoustic medium
$x_{,,,}$	Mass axis
x_f	Flexural axis

Greek Letters

ρ	Air density, kg/m^3
κ	Flap degree of freedom
θ	Pitch degree of freedom

20.1 Introduction

Aeroelasticity, a study on structure stability in response to aerodynamic loads, is regarded as a major aspect in designing an aircraft. In aeroelastic phenomena, the interaction of aerodynamic, elastic and inertia forces on elastic structures causes undesired distortions in deformed mode shape and could even lead to a destructive vibration known as aeroelastic flutter. This type of instability produces unstable oscillation which may trigger catastrophic damages to the whole structure. For an aircraft, slender bodies such as aircraft wings, tails and control surfaces are typically vulnerable to this unexpected threat and each aeroelastic factor needs to be taken into consideration upon the design and flight performance of the aircraft. Due to their significant influences, these aeroelastic problems have been widely addressed in classical and standard textbooks [2, 5] which also discuss the theory and basic principles toward the understanding of aeroelasticity. In taking steps to reduce this catastrophic risk, heavier structures were purposely designed for flutter prevention. However, this approach creates a major drawback in reducing the aircraft efficiency in terms of mission performance and operation cost. An improved approach was later proposed by employing an active control system on a lifting surface called active flutter suppression [11] to stabilize the vibration of airframe structures and also overcome the weight penalty caused by the former approach. However, cheaper alternatives are being considered to replace the current flutter control system.

One of the alternative solutions, which is presently being investigated, comprises the use of external acoustic excitation. To our best knowledge, the initial studies on structural analysis with the presence of acoustic excitation can be found in the work of Fahy and Wee [4] and also Rama Bhat et al. [1]. Both studies were carried out due to the concern of aircraft structural integrity when dealing with intense acoustic environments. For an aircraft, the sound, or frequently referred to as noise is generated from the propeller, exhaust, engine vibration and airflow around aircraft structure. For example, the sound pressure level produced by multiengine of a typical aircraft is approximately 130 dB and can reach 150 dB under supersonic condition. For the past few decades, many of the preliminary investigations show that acoustic pressure produces significant influences on structures such as thin plate, membrane and also high-impedance medium like water (and other similar fluids).

The aeroelastic flutter analysis on rotating disk in an unbounded acoustic medium [6] for instance is one of the latest studies conducted in this specific research area. On the other hand, the previous works of Djojodihardjo [7–9] demonstrate that the acousto-aeroelastic problem using BE-FE approach leads to significant influence on the performance of aeroelastic structure. It is thus useful to investigate the acoustic effect on aircraft wing structure using a different method in which the acoustic is formulated using FEM approach [12].

20.2 Computational Method

20.2.1 Binary Aeroelastic Model

Due to the complexity of aircraft structures, it is often crucial to take account of simplifying assumptions in this methodology to allow computational of the elastic properties. Here, a simple model—a two-degree-of-freedom system (bending and

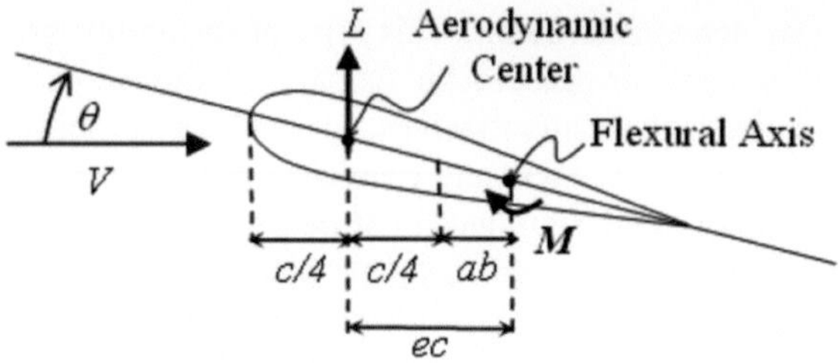

Fig. 20.1 A two-dimensional airfoil with notations

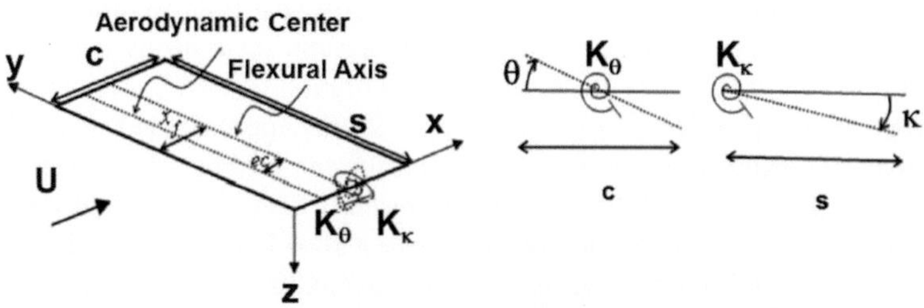

Fig. 20.2 Schematic layout of binary aeroelastic model (Hancock wing model)

torsion) consisting of a rigid wing with constant chord is adopted. Considering the two-dimensional airfoil, the airfoil with the chord length c is visualized in the flight condition with a uniform free stream of velocity, V, shown in Fig. 20.1.

Using the notation given in Fig. 20.1, the binary model is constructed based on the binary concept used in Hankock and Wright [10]. Illustrated in Fig. 20.2, the rectangular wing of span s and chord c is assumed to be rigid with two rotational springs at the root to provide flap (κ) and pitch (θ) degrees of freedom. In addition, the aerodynamics is modeled using a modified strip theory which allows calculations for unsteady effects.

According to Wright and Cooper [13], the full equations of motion can be written in the form of

$$[\mathbf{M}_s]\{\ddot{\mathbf{q}}_s\} + [\mathbf{C}_s]\{\dot{\mathbf{q}}_s\} + [\mathbf{K}_s]\{q_s\} = \{0\} \tag{20.1}$$

where mass matrix $[M_s]$, damping matrix $[C,]$ and stiffness matrix $[K]$ for the wing structure can be expressed as

$$[\mathbf{M}_s] = \begin{bmatrix} \frac{1}{3}ms^3c & \frac{1}{2}ms^2\left(\frac{1}{2}c^2 - cx_f\right) \\ \frac{1}{2}ms^2\left(\frac{1}{2}c^2 - cx_f\right) & ms\left(\frac{1}{3}c^3 - c^2x_f - cx_f^2\right) \end{bmatrix} \tag{20.1a}$$

$$[\mathbf{C}_s] = \rho V\begin{bmatrix} \frac{1}{6}cs^3a_w & 0 \\ -\frac{1}{4}ec^2s^2a_w & -\frac{1}{8}c^3sM_{\dot{\theta}} \end{bmatrix} + [\mathbf{D}_s] \tag{20.1b}$$

$$[\mathbf{K}_\mathrm{s}] = \rho V^2 \begin{bmatrix} 0 & \frac{1}{4}cs^2 a_w \\ 0 & \frac{1}{2}ec^2 s a_w \end{bmatrix} + \begin{bmatrix} K_k & 0 \\ 0 & K_\theta \end{bmatrix} \tag{20.1c}$$

20.2.2 Structural-Acoustic Coupling

To predict the acoustic effect on binary aeroelastic model, a proper coupling mechanism involving acoustic and structural interaction is included [3]. Taking this into consideration and by referring to Eq. (20.1), the equations for a flexible structure with an acoustic enclosure can now be written as

$$[\mathbf{M}_\mathrm{s}]\{\ddot{\mathbf{q}}_\mathrm{s}\} + [\mathbf{C}_\mathrm{s}]\{\dot{\mathbf{q}}_\mathrm{s}\} + [\mathbf{K}_\mathrm{s}]\{\mathbf{q}_\mathrm{s}\} - [\mathbf{R}_\mathrm{as}]^T\{\mathbf{p}\} = \{0\} \tag{20.2}$$

while the equation of motion for an acoustic enclosure coupled to a flexible structure is given by

$$[\mathbf{M}_\mathrm{a}]\{\ddot{\mathbf{p}}\} + [\mathbf{K}_\mathrm{a}]\{\mathbf{p}\} - \rho[\mathbf{R}_\mathrm{as}]^T\{\ddot{\mathbf{q}}_\mathrm{s}\} = \{0\} \tag{20.3}$$

where $[M_\mathrm{a}]$ and $[K_\mathrm{a}]$ are the acoustic mass and stiffness matrices and $[R_\mathrm{as}]$ is the structural-acoustic coupling matrix. They can be expressed as

$$\mathbf{M}_\mathrm{a} = t \int_A \mathbf{N}_f^T \mathbf{N}_f \mathrm{d}A \tag{20.4}$$

$$\mathbf{K}_\mathrm{a} = a^t t \int_A \mathbf{D}^T \mathbf{D} \mathrm{d}A \tag{20.5}$$

$$\mathbf{R}_\mathrm{as} = t\ \mathbf{G}^T \int_0^L \mathbf{N}_s^T \mathbf{N}_f \mathrm{d}\overline{x} \tag{20.6}$$

where a is the speed of sound and t is the thickness of the fluid medium, while N_f and N_S is the shape function for acoustic fluid and wing structure. The matrices G and D are the transformation matrix and strain displacement matrix.

Combining both Eqs. (20.2) and (20.3), the coupled system can be written as

$$\begin{bmatrix} \mathbf{M}_\mathrm{s} & 0 \\ -\rho_0 \mathbf{R}_\mathrm{as} & \mathbf{M}_\mathrm{a} \end{bmatrix} \begin{Bmatrix} \ddot{\mathbf{q}}_\mathrm{s} \\ \ddot{\mathbf{p}} \end{Bmatrix} + \begin{bmatrix} \mathbf{C}_\mathrm{s} & 0 \\ 0 & 0 \end{bmatrix} \begin{Bmatrix} \dot{\mathbf{q}}_\mathrm{s} \\ \mathbf{p} \end{Bmatrix} + \begin{bmatrix} \mathbf{K}_\mathrm{s} & -\mathbf{R}_\mathrm{as}^T \\ 0 & \mathbf{K}_\mathrm{a} \end{bmatrix} \begin{Bmatrix} \mathbf{q}_\mathrm{s} \\ \mathbf{p} \end{Bmatrix} = \begin{Bmatrix} 0 \\ 0 \end{Bmatrix} \tag{20.7}$$

This can be expressed in simpler form as

$$[\mathbf{A}]\{\ddot{\mathbf{u}}\} + [\mathbf{B}]\{\dot{\mathbf{u}}\} + [\mathbf{C}]\{\mathbf{u}\} = \{0\} \tag{20.8}$$

20.2.3 Flutter Solution

The acoustic-aeroelastic system in frequency domain is then solved by the use of solutions of eigenvalues and eigenvectors in a state-space form. Thus, the corresponding equation can be written as

$$\begin{bmatrix} \mathbf{I} & 0 \\ 0 & \mathbf{A} \end{bmatrix} \begin{Bmatrix} \dot{\mathbf{u}} \\ \ddot{\mathbf{u}} \end{Bmatrix} + \begin{bmatrix} 0 & \mathbf{I} \\ \mathbf{C} & -\mathbf{B} \end{bmatrix} \begin{Bmatrix} \mathbf{u} \\ \dot{\mathbf{u}} \end{Bmatrix} = \begin{Bmatrix} 0 \\ 0 \end{Bmatrix} \tag{20.9}$$

20.3 Results and Discussion

Before analyzing the acoustic influence on wing structure, the rectangular wing model with semi-span $s = 7.5$ m and chord $c = 2$ m was first modeled using finite element method (FEM) in order to determine more precise estimation on flutter occurrence condition in terms of the natural frequencies involved. Using the in-house FEM code written in MATLAB, the mode shapes of the flexible wing made of Aluminum 6061 with Young's modulus, $E = 69$ GPa and Poisson's ratio, $\nu = 0.33$ are illustrated in Fig. 20.3 for low-frequency modes and Fig. 20.4 for high-frequency modes.

Adopting values from [12], the wing structure is assumed to have a uniform mass distribution of 100 kg/m^2. The mass axis is placed at the semi-chord $xm = 0.5c$ and the flexural axis is at $x = 0.48c$. In addition, other specified parameters like the lift curve

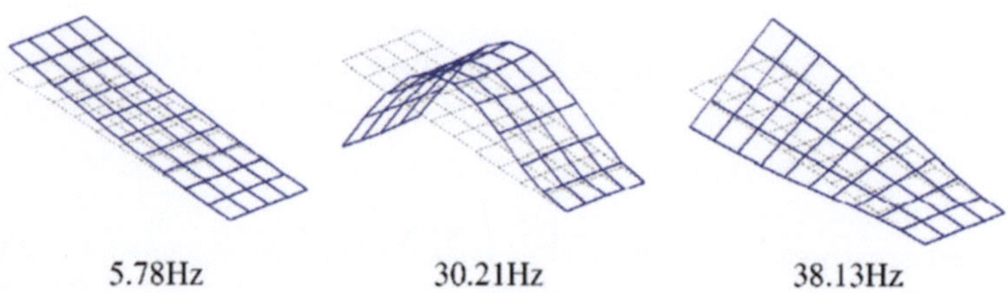

Fig. 20.3 Mode shapes of binary aeroelastic model at low-frequency modes

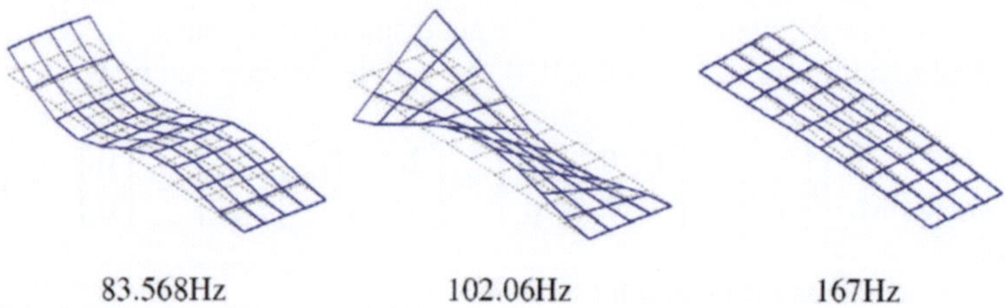

Fig. 20.4 Mode shapes of binary aeroelastic model at high-frequency modes

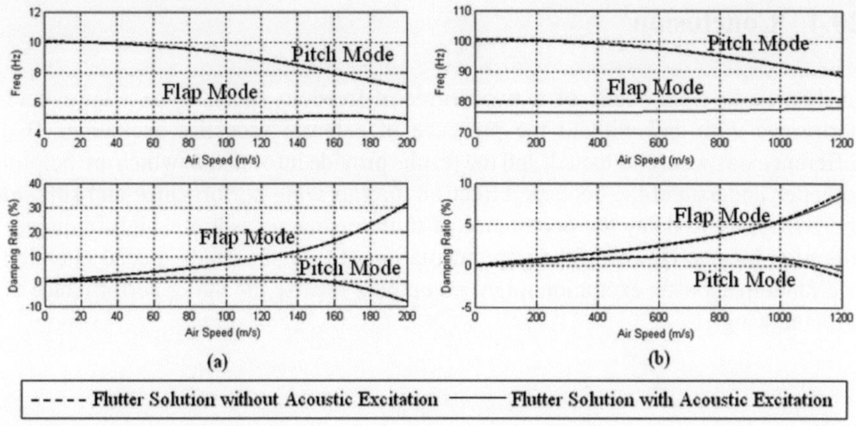

Fig. 20.5 Frequency and damping plots for binary aeroelastic model with: **a** low-frequency vibration, **b** high-frequency vibration

slope $a_{,,,} = 27$ T, air density $p = 1.225$ kg/m^3 and non-dimensional pitch damping derivative which is assumed to be -1.2 were included. Based on the information obtained from FEM simulation, the flutter analysis for two cases (low-frequency vibration and high-frequency vibration) was carried out. The detailed specifications for both cases are listed below:

a. Low-frequency vibration ($K_k = I_k\ (5 \times 2\pi)^2$ Nm/rad and $K_\theta = I_\theta(10 \times 2\pi)^2$ Nm/rad)
b. High-frequency vibration ($K_k = I_k\ (80 \times 2\pi)^2$ Nm/rad and $K_\theta = I_\theta(100 \times 2\pi)^2$ Nm/rad).

Using the parameters mentioned, the acoustic-aeroelastic problem was solved using MATLAB. Then the results of the analysis are shown in Fig. 20.5.

As shown in Fig. 20.5a, the acoustic excitation has no significant influence on flutter performance at low-frequency mode as both flutter solution results (with and without acoustic source) are the same. Meanwhile, for the case of high-frequency mode in Fig. 20.5b, the result obtained presents small changes for flutter solution with the inclusion of acoustic excitation compared with pure flutter solution. By observing the pitch mode in Fig. 20.5b, the flutter speed (damping ratio equal to zero) for acoustic-aeroelastic problem has increased. From the result, the flutter speed for binary wing model under acoustic influence has increased to 1145 m/s from 1080 m/s. This indicates that the flutter suppression involving external acoustics source has the potential which can be implemented in order to delay flutter condition from occurring.

20.4 Conclusion[1]

In this work, simulations of a two-degree-of-freedom flutter system have been performed with and without the presence of external acoustics excitation. Two different cases were conducted, and the results provide information which are helpful to better understand the acoustic effect on aircraft wing performance and support the possibility to delay the occurrence of flutter using acoustic for high-frequency vibration modes. However, the implementation of acoustic needs special attention and random acoustic excitation might potentially reverse the flutter performance of airplane wing.

References

1. Bhat, R., B.V.A. Rao, and H. Wagner. 1973. Structural response to random acoustic excitation. *Earthquake Engineering and Structural Dynamics* 2: 117–132.
2. Bisplinghoff, R.L., H. Ashley, and R.L. Halfman. 1955. *Aeroelasticity*. USA: Addison-Wesley Publishing Company Inc.
3. Coskuner, J. 2004. *Combined direct-adjoint approximations for large-scale design-oriented structural-acoustics finite-element analysis*. Master Thesis, University of Washington.
4. Fahy, F.J., and R.B.S. Wee. 1968. Some experiments with stiffened plates under acoustic excitation. *Journal of Sound and Vibration* 7 (3): 431–436.
5. Fung, Y.C. 1968. *An introduction to the theory of aeroelasticity*. Dover Publications.
6. Jana, A. Raman. 2006. Aeroelastic flutter of a disk rotating in an unbounded acoustic medium. *Journal of Sound and Vibration* 289: 612–631.
7. Djojodihardjo, H. 2007a. BEM-FEM fluid-structure coupling due to acoustic or structural loading disturbance. In *Proceedings of Aerotech-II 2007 Conference on Aerospace Technology of XXI Century*, 20–21 June 2007, Kuala Lumpur.
8. Djojodihardjo, H. 2007b. BE-FE coupling computational scheme for acoustic effects on aeroelastic structures. In *Proceedings of the International Forum on Aeroelasticity and Structural Dynamics*, 18–19, Paper IF-086, held at the Royal Institute of Technology (KTH), Stockholm, Sweden, June.
9. Djojodihardjo, H. 2008. Unified BE-FE aerodynamic-acoustic-structure coupling scheme for acoustic effects on aeroelastic structures. In *26th International Congress of the Aeronautical Sciences*, Paper ICAS2008 7.7.7.5, Anchorage, Alaska, September.
10. Hancock, G.J., A. Simpson, and J.R. Wright. 1985. On the teaching of the principles of wing flexure/torsion flutter. *Aeronautical Journal* 89: 285–305.
11. Nissim, E. 1971. Flutter suppression using active controls based on the concept of aerodynamic energy. NASA TN D-6199.

---

[1] The present chapter is originally published as "Acoustic Effects on Binary Aeroelasticity Model", co-authored by Kok Hwa Yu[a] (Corresponding Author, Email: yukokhwa@yahoo.com), Harijono Djojodihardjo[b] and A.Halim Kadarman[a]

[a] School of Aerospace Engineering, Engineering Campus, Universiti Sains Malaysia, 14300 Penang, Malaysia.

[b] At the time of publication at Aerospace Engineering, Faculty of Engineering, Universiti Putra Malaysia, 43400 Serdang, Selangor, Malaysia, now Chairman of the Board, The Institute for the Advancement of Aerospace Science and Technology "Persada Kriyareka Dirgantara", Jakarta, Indonesia, Email: harijono.djojodihardjo@gmail.com.

12. Sandberg, Göran, Per-Anders Wernberg, and Peter Davidsson. 2008. Fundamentals of fluid-structure interaction. In *Computational aspects of structural acoustics and vibration*, 23–101 (CISM, vol. 505).
13. Wright, J.R., and J.E. Cooper. 2007. *Introduction to aircraft aeroelasticity and loads*. Wiley.

Chapter 21
Application of a Multipole Secondary Source for Propeller Active Noise Control

Abstract The commercial feasibility of active noise control (ANC) is very promising due to its capability beyond passive noise control (PNC). To some extent ANC becomes a complement of PNC. The active noise reduction is also capable and beneficial in reducing noise selectively. However, the active noise reduction using a conventional secondary source can become very complicated if a significant noise level reduction is required, since a large number of secondary sources will be needed. The active noise reduction is also less effective for reducing high-frequency noise. With such perspectives, a novel approach has been developed using a multipole secondary source to address the problems mentioned. In addition, the multipole secondary source will be used for numerical simulation of noise reduction in of propeller noise source in a free field.

Keywords Acousto-aeroelasticity · Active noise control · Multipole source · Vibration suppression

21.1 Introduction

Active noise control is a significant demand for human comfort and environmental requirements, as also established by aviation authorities (such as FAA). Efforts for noise control have been carried out in many literatures, among others in the work by Kempton [1], Germain [2], Johansson [3], Pabst et al. [4] and Griffin [5]. Establishing very high-frequency bound is a challenging problem in active noise reduction. In addition, the distance of the secondary source with respect to the primary source is another factor that can seriously inhibit effective active noise reduction. Such problem may arise in the use of monopole secondary source. Such issue motivates the development and utilization of the method of multipole secondary source. The advantage of using the multipole secondary source is in the reduction of higher frequency noise and the number of secondary sources, which could lead further to a more compact active noise reduction system.

Originally appeared as Kusni et al. [6].

© The Author(s), under exclusive license to Springer Nature Singapore Pte Ltd. 2023
H. Djojodihardjo, *Introduction to Aeroelasticity*,
https://doi.org/10.1007/978-981-16-8078-6_21

Although the concept of the multipole secondary source is a widely known technique, the use of multipole secondary source for noise elimination has a limitation, since it is impossible to use the multipole secondary source of infinite order. To solve the problem, an optimization of secondary source strength will be performed so that the radiation of the sound power of the sound field can be minimized. Such effort will allow efficient sound reduction using limited multipole order. Such attempt has been made by Kusni and Soenarko [7–9]. The concept of sound reduction with multipole source has been employed in the application of the boundary integral method to sound radiation problems Seybert et al. [10]. In aerodynamics, the use of mutipoles has been utilized by Hess and Smith [11], Djojodihardjo and Widnall [12], as also comprehensively described by Ashley and Rodden [13], among others. Utilization of monopole and multipole acoustic sources has also been discussed by Djojodihardjo [14], Yu et al. [15] and Djojodihardjo [16]. Techniques on the use of multipole sources for description and cancelation of sound fields have been discussed by Nelson et al. [17], Ruijgrok [18]. Koopman et al. [19], Bolton and Beauvilain [20, 21], Beauvilain and Bolton [21] and Bolton et al. [22]. These techniques have inspired the development of the present method. Kempton [1] develops the concept of sound reduction with a multipole source, by expanding the monopole source by multipole sources using Taylor series. Each analytic function, g(x), can be expressed in the Taylor series as follows:

$$g(x + h) = \sum_{n=0}^{\infty} \frac{h^n}{n!} \frac{\partial^n}{\partial x^n} g(x). \tag{21.1}$$

Here h is an incremental rise of variable x, in which the formulation above will be convergent to the area $x > h$. Kempton uses the result to indicate that the sound pressure field of monopole source located at—h on the x-axis can be expressed as a multipole point field of infinite order situated at the center of the coordinate system. Figure 21.1 exhibits the geometry and coordinate system used. The monopole primary source located along the x-axis at – h will generate sound pressure at any point M (at the O-xyz frame of reference), which can be expressed as,

$$P_{\text{pm}}(x, y, z) = \frac{j\omega\rho}{4\pi} Q_{\text{pm}} \frac{e^{-jkr_h}}{r_h} \tag{21.2}$$

where

$$r_h = \sqrt{(x + h)^2 + y^2 + z^2} \tag{21.2a}$$

and where Q_{pm} is the strength of the monopole primary source, r_h is given as Eq. 21.2a and ρ Is the air density, $\omega = 2\pi f$ where f is the prevailing frequency, $k = \omega/c$ where c is the sound velocity, and $j = \sqrt{(-1)}$. Expanding the right-hand side of Eq. 21.2 in Taylor series, then it will become,

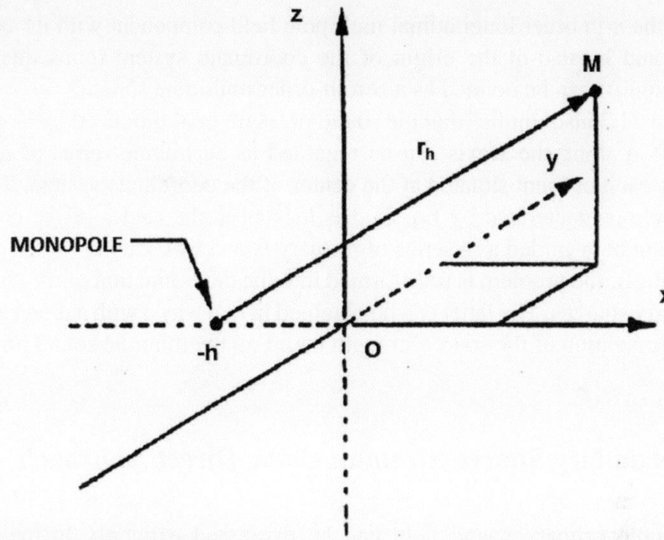

Fig. 21.1 Geometry and coordinate system

$$P_{\mathrm{pm}}(x, y, z) = \frac{j\omega\rho}{4\pi} \sum_{n=0}^{\alpha} \frac{\partial^n}{\partial x^n}\left(\frac{e^{-jkr}}{r}\right) \tag{21.3}$$

where α is a suitably chosen value, and

$$r = \sqrt{x^2 + y^2 + z^2} \tag{21.3a}$$

It can be shown that for $n = 0$,

$$\frac{j\omega\rho}{4\pi} Q_{\mathrm{pm}}\left(\frac{e^{-jkr}}{r}\right) \tag{21.3b}$$

is the sound pressure field formed by a monopole point located at the center of the coordinate system. Similarly, for $n = 1$,

$$\frac{j\omega\rho}{4\pi} Q_{\mathrm{pm}} h \frac{\partial}{\partial x}\left(\frac{e^{-jkr}}{r}\right) \tag{21.3c}$$

represents the sound pressure field formed by a dipole point located at the origin of the coordinate system with the dipole axis along the x-axis. Hence the general form

$$\frac{j\omega\rho}{4\pi} Q_{\mathrm{pm}}\left(\frac{h^n}{n!}\right) \frac{\partial}{\partial x}\left(\frac{e^{-jkr}}{r}\right) \tag{21.3d}$$

represents the n-th order longitudinal multipole field component with its axis along the x-axis and located at the origin of the coordinate system (consequently, the monopole source can be defined as a zeroth-order multipole source).

Equation 21.3 also implies that the sound pressure field produced by a monopole located at – h along the x-axis can be regarded as an infinite series of multipole points with each of them situated at the center of the coordinate system. Therefore if a multipole characterized by Eq. 21.3 is located at the center of the coordinate system, it can be regarded as a series of primary sources of sound.

Accordingly, the problem is transformed into the determination of the strength of the secondary sources. The latter can be obtained in two ways: with a direct approach and the optimization of the source strength based on the minimal sound power.

21.2 Secondary Source Strength Using Direct Approach

The monopole primary sound field can be expressed explicitly in the form of multipole source strength

$$P_{\text{pm}}(x, y, z) = \frac{j\omega\rho}{4\pi} \sum_{n=0}^{\alpha} Q_{sn} \frac{\partial^n}{\partial x^n} \left(\frac{e^{-jkr}}{r} \right) \tag{21.4}$$

The multipole source strength in Eq. 21.4, Q_{sn}, can be determined by comparing Eq. 21.4 with Eq. 21.3, if the primary monopole source Q_{pm}, and the distance of the secondary source, h, have been determined. Clearly the monopole in the series expansion has the same source strength as the primary monopole, i.e., $Q_{s0} = Q_{\text{pm}}$. The dipole secondary source has a source strength $Q_{s1} = hQ_{\text{sn}}$, and, in general, the nth-order multipole component has a source strength of $Q_{\text{sn}} = \frac{h^n}{n!} Q_{\text{pm}}$. This approach is used to define a secondary resource strength; thus this characterizes a direct approach.

Q_{sn}'s are the source strengths required to produce exactly a monopole primary sound pressure field when the Taylor series expansions of Eqs. 21.3 and 21.4 are carried out to an infinite number of terms. In practice, it is not possible to generate an infinite series of multipole sources; hence one has to truncate the series up to a certain point. Consequently, with a limited number of multipole components, Q_{sn} obtained with the direct approach will not result in an optimal monopole equivalent primary field. Therefore, an alternative procedure is suggested, as outlined below, to obtain a minimal composite sound strength by a combination of primary and secondary source radiation.

21.3 Secondary Source Strength Using Optimized Approach

The optimal secondary source strength can be established by resorting to the following procedure. Firstly the sound power has to be determined. The sound power, W, radiated by an acoustic source, can be expressed as

$$W = \iint_S \mathbf{I} \cdot \mathbf{n} \mathrm{d}S \tag{21.5}$$

where $\mathbf{I}$ is an acoustic intensity vector; $\mathbf{n}$ is an outward normal vector to the surface S enclosing the source and dS is an incremental element of that surface. Assuming harmonic motion the average intensity is given by

$$\mathbf{I} = \mathrm{Re}(p * \mathbf{v}) \tag{21.6}$$

where p is the sound pressure, and v is the velocity vector defined by Euler's equation,

$$\mathbf{v} = -\left(\frac{1}{j\omega\rho}\right)\nabla P \tag{21.7}$$

Re{.} expresses the real part of the argument, superscript * expresses a complex conjugate and ∇ is a gradient operator. Integration of Eq. 21.5 over the surface of the sphere S yields the sound strength which can be expressed as

$$W = \int_0^{2\pi}\int_0^{2\pi} I_r r^2 \sin\theta \, \mathrm{d}\theta \mathrm{d}\phi \tag{21.8}$$

Here

$$I_r = \frac{1}{2}\mathrm{Re}(P * V_r) \tag{21.9}$$

is a radial component of the sound intensity, V_r is a radial component of the particle velocity and the angles θ and ϕ are shown in Fig. 21.2.

For the various cases discussed in here, the distribution of sound pressure is defined to be spherically symmetric to the x-axis, so that there is no pressure variation with respect to ϕ. Hence Eq. 21.8 can be further reduced to

$$W = 2\pi r^3 \int_0^{\pi} I_r \sin\theta \, \mathrm{d}\theta \tag{21.10}$$

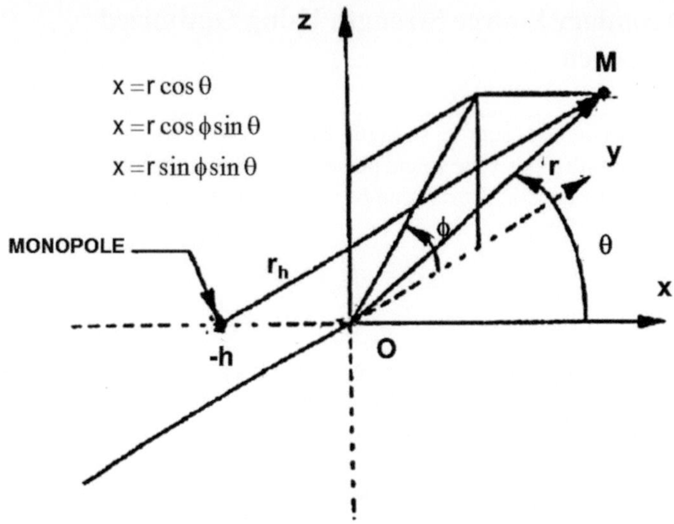

Fig. 21.2 Coordinate transformation

Prior to the use of Eq. 21.10 for calculating the sound power, it is desirable to utilize a combination of monopole primary source and multipole secondary source. To that end it is convenient to utilize spherical coordinate system. The sound field produced by a monopole primary source having a source strength Q_{pm} and located at – h along the x-axis can be expressed as

$$P_{\text{pm}}(r, \theta) = \frac{j\omega\rho}{4\pi} Q_{\text{pm}} \frac{e^{-jkr_h}}{r_h} \tag{21.11}$$

where r_h is distance of monopole primary source to the field point.

By reference to Fig. 21.2, the following coordinate transformation can be carried out, by utilizing the relationship

$$r_h = \sqrt{x^2 + h^2 + 2r\,h\,\cos\theta} \tag{21.12}$$

The sound field formed by secondary monopole, dipole, longitudinal quadrupole and longitudinal octupole sources is successively defined as follows:

$$P_{\text{sm}}(r, \theta) = \frac{j\omega\rho}{4\pi} Q_{\text{pm}} \frac{e^{-jkr}}{r} \tag{21.13}$$

$$P_{sd}(r, \theta) = \frac{\omega\rho}{4\pi} k\,\cos\theta \left(1 + \frac{1}{jkr}\right) Q_{sd} \frac{e^{-jkr}}{r} \tag{21.14}$$

$$P_{sq}(r,\theta) = -\frac{j\omega\rho}{4\pi}k^2\left[\left(1+\frac{3}{jkr}+\frac{3}{(jkr)^2}\right)\cos^2\theta - \left(\frac{1}{jkr}+\frac{1}{(jkr)^2}\right)\right]Q_{sq}\frac{e^{-jkr}}{r} \tag{21.15}$$

$$P_{sO}(r,\theta) = -\frac{j\omega\rho}{4\pi}k^3\cos\theta\left[\begin{array}{c}\left(1+\frac{6}{jkr}+\frac{15}{(jkr)^2}+\frac{15}{(jkr)^3}\right)\cos^2\theta \\ -\left(\frac{3}{jkr}+\frac{9}{(jkr)^2}+\frac{9}{(jkr)^3}\right)\end{array}\right] Q_{sO}\frac{e^{-jkr}}{r} \tag{21.16}$$

The total sound power of primary and secondary source consisting of monopole, dipole, quadrupole and octupole secondary sources can be obtained from Eqs. 21.13, 21.14, 21.15 and 21.16.[1] Then the secondary source power is optimized for the total minimal sound power.

21.4 Applications

This section describes a simulation of noise reduction in several cases occurring in the vicinity of our environment. The first case uses a simple-multiple frequency primary sources generated by the blower of an aircraft air-conditioning system, fluorescent lamp, transformer, etc. The second case uses a primary source in the form of noise issued by an aircraft propeller. The third case uses a broadband primary source in the form of white noise.

21.4.1 The Case of Simple-Multiple Frequency Noise Reduction of Aircraft Air-Conditioning Blower

This subsection elaborates the results of the reduction of noise pressure consisting of several single frequencies. This noise sound is in the form of monotonous sound. This droning sound occurred in the aircraft electronic instruments, for example: noise sound of fluorescent lamp, transformer, cooling machine, etc.

Figure 21.3 show that the multipole method can reduce successfully the droning noise, since the droning sound usually consists of several single fairly low frequency.

[1] Subscript m, d, q and o in the variables in these equations stand for monopole, doublet, quadrupole and octupole,rspectively.

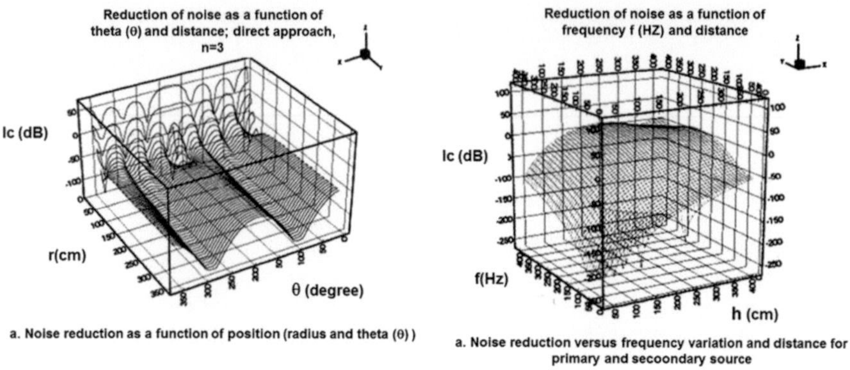

Fig. 21.3 Results of droning sound reduction consisting of a single sound frequency

21.4.2 Noise Reduction of Cessna 150 Propeller

The section describes the result of noise reduction derived from the aircraft engine of Cessna 150. The Cessna 150 aircraft belongs to a small aircraft category using propeller engine. Figure 21.4 is graph of sound pressure in the time domain obtained by measuring the aircraft noise.

The measurement is carried out while the aircraft is on ground. Figure 21.5 shows the response of noise frequency of the aircraft engine.

The Cessna 150 aircraft has a single engine and a two-bladed propeller, driven by a four-cylinder piston engine. The peaks in the frequency response shown in Fig. 21.5 are related to the frequency as a result of the rotation of propeller blades and the combustion cycles in the piston engine. The magnitude of the frequency as a result of the rotation of the propeller blade, f_1, is expressed by:

$$f_1 = B\frac{n_p}{60}\ \text{Hz} \tag{21.17}$$

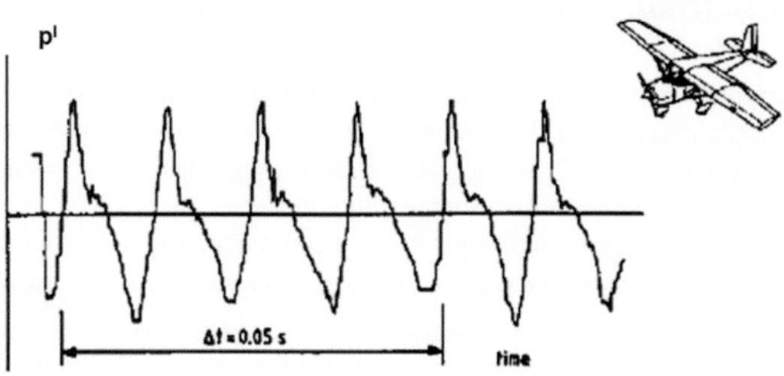

Fig. 21.4 Picture of C-150 aircraft and diagram of time response of aircraft noise

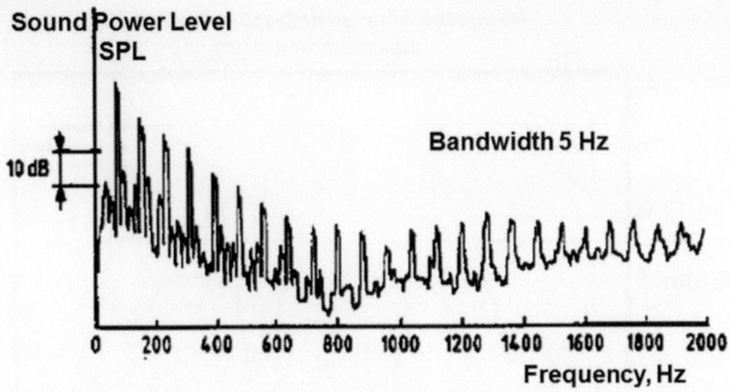

Fig. 21.5 Spectrum of frequencies of Cessna 150 aircraft noise

where B is the number of propeller blades, and n_p is rotational frequency of propeller in rpm.

For the aircraft considered, the number of blades $B = 2$ and $n_p = 2400$ rpm. Hence

$$f_1 = 2\frac{2400}{60}\ \text{Hz} = 80\ \text{Hz} \tag{21.17a}$$

Since the aircraft engine is a four-cycle one, the frequency due to the combustion cycles, f_e, can be expressed as

$$f_e = N\frac{n/2}{60}\ \text{Hz} \tag{21.17b}$$

where N is the number of cylinders and n is a velocity of engine rotation in rpm. With $N = 4$ and $n = np = 2400$ rpm, then $f_e = f_1$. Hence the contribution of propeller and engine noise cannot be separated in the frequency spectra (Fig. 21.6).

21.4.3 Noise Reduction of Broadband White Noise

This section describes noise reduction of broadband white noise. There are three kinds of variation carried out as related to this broadband noise reduction. The first is the variation of the distance between the primary and the secondary sources. Figure 21.7 shows that the farther the distance of primary and secondary sources, the reduction of the broadband noise (1–20.000 Hz) is the more difficult. The second is the variation of number of frequency contents. Figure 21.8 shows that the number of frequency content does not influence the magnitude of the noise reduction.

The third case illustrates the capability of the noise reduction technique to identify the degree of white noise. As exhibited in Fig. 21.9, the results show that the

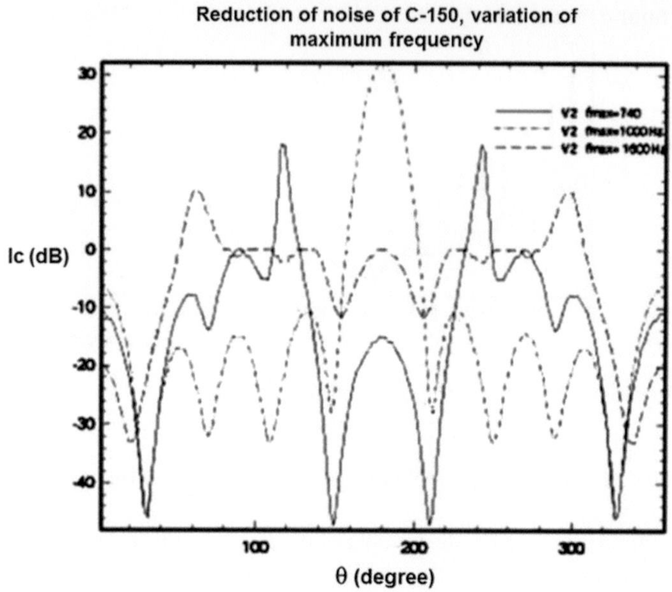

Fig. 21.6 Result of the reduction of propeller noise of C-150 aircraft for various frequencies

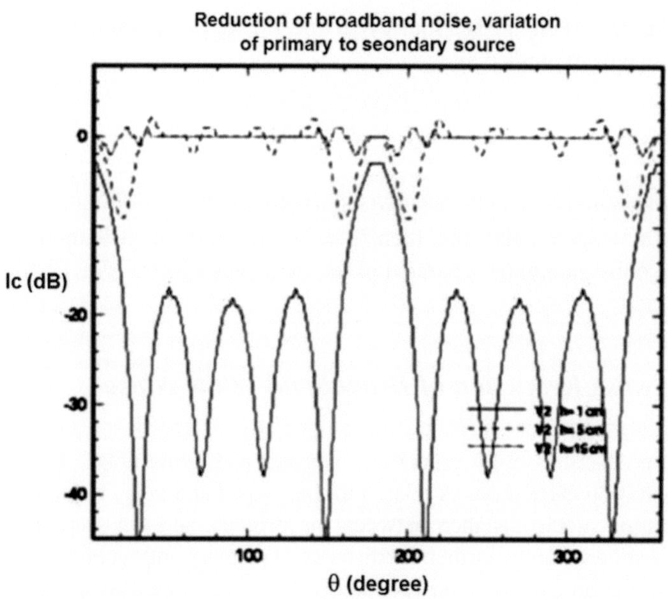

Fig. 21.7 Variation of the distance of primary to secondary sources

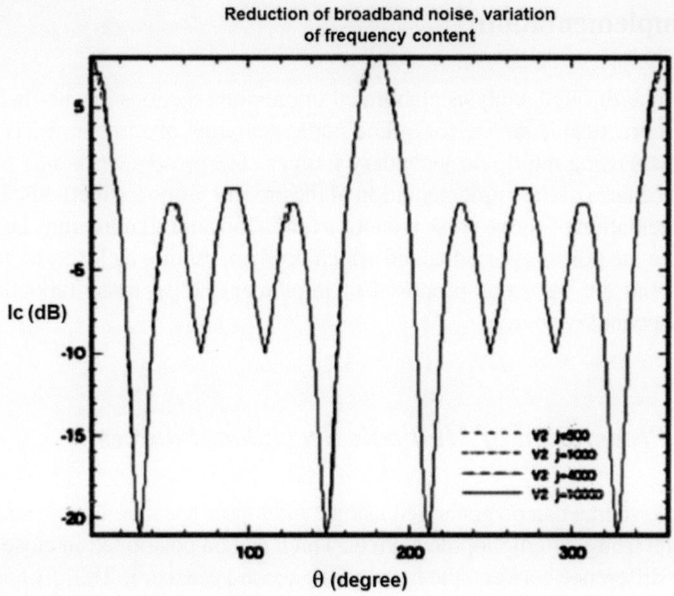

Fig. 21.8 Reduction of broadband white noise results for case 2

broadband noise is not really a white noise, since the maximum frequency used is limited in value (below 20.000 Hz).

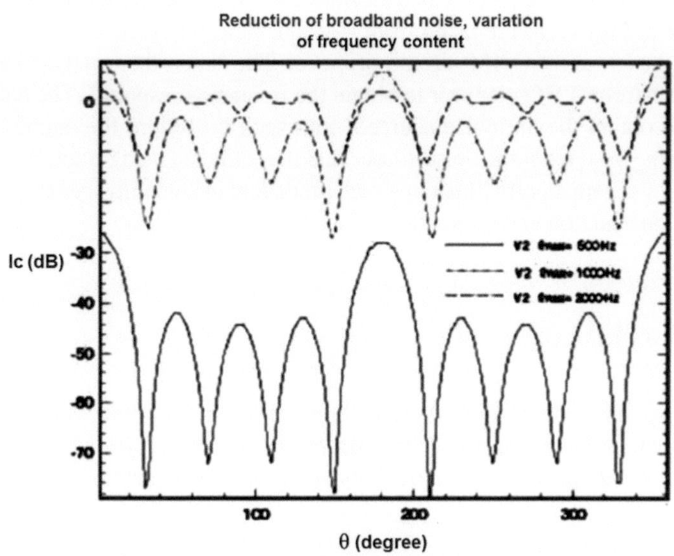

Fig. 21.9 Reduction of broadband white noise results for case 3

21.5 Implementation

The technique through analysis elaborated in previous sections has not been developed into a practicable device for online implementation of active noise control (of aircraft noise) using multipole secondary sources. The present work only presents a projected scenario of the implementation of the present proposed methods. However, the implementation of active noise reduction for C-150 aircraft noise may be advantageous, since the noise is periodic, and which does not require a closed-loop control. The following are the steps proposed to implement active noise reduction using multipole secondary source.

21.5.1 Preparation of Multipole Secondary Sources

A monopole source can be generated using a monopole speaker. Dipole sources can be produced using two monopole sources which can be positioned in close vicinity. The phase difference between the first and the second speaker is 180°. A quadrupole source can be produced using two dipole sources. Microphone or accelerometer can be utilized as noise sensors. The advantage of using accelerometer is that one can directly identify the source of the noise. If microphone is used, one has to invert the signal to identify the source of noise.

Figure 21.10 shows the reconstruction of C-150 noise. The data is taken at $t =$ 0.00 until $t = 1.65$ s. Comparison between the reconstruction and measurement results show good agreement. The data shown in Fig. 21.4 and those reconstructed and shown at Fig. 21.10 are taken from a distance of 20 m at an elevation of 1.2 m exactly at the left of the C-150 aircraft propeller. The data is then analyzed using fast Fourier transform (FFT) analyzer to obtain the frequency response. The data is then inverted to obtain the antinoise source. From such procedure the antinoise signal can be synthesized from the noise source and that radiated using multipole speaker. Figure 21.11 schematically illustrates such procedure using multi speaker system for antinoise introduction and interaction.

21.6 Conclusion

Active noise reduction of aircraft propeller noise using multipole secondary source has been meticulously elaborated. The technique and results presented indicate that the method is very promising, although its practical implementation requires further work.

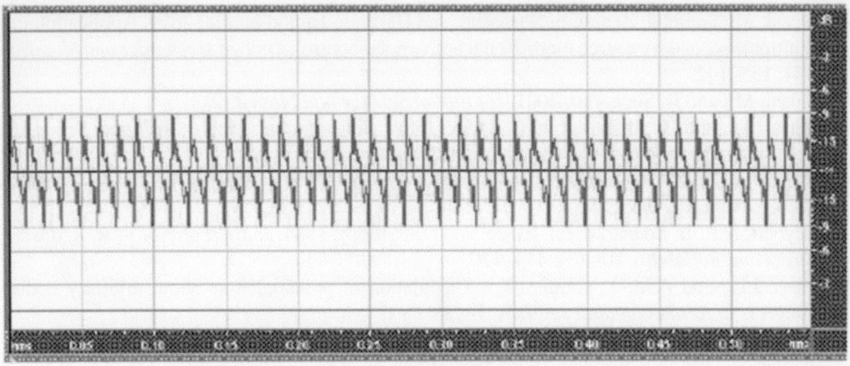

Fig. 21.10 Sample of C-150 noise data reconstruction taken at $t = 0.0$ until $t = 1.65$ s

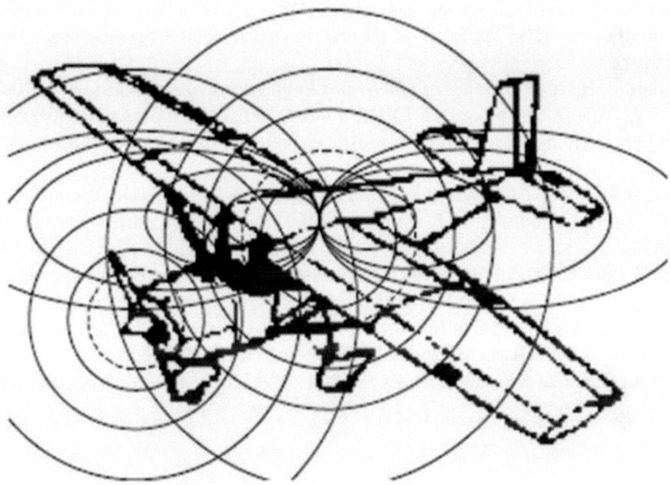

Fig. 21.11 Schematic of antinoise signal introduction and interaction

References

1. Kempton, J. 1976. *Journal of Sound and Vibration* 48: 475–483.
2. Germain, Pierre. 2000. *Active Control of Run-Up Noise from Propeller Aircraft*. M. Appl. Sci. Thesis, Uni. British Columbia, May 2000.
3. Johansson, Sven. 2000. *Active Control of Propeller Induced Noise in Aircraft—Algorithms & Methods*. Ph.D. Thesis, Department of Telecommunications and Signal Processing, Blekinge Institute of Technology, Sweden, December 2000.
4. Pabst, Oliver, Thomas Kletschkowski, and Delf Sachau, 2009. Combined active noise control and audio in a light jet. In *The 16th International Congress on Sound and Vibration*. Krakow, 5–9 July 2009.
5. Griffin. 2009. *The Control of Interior Cabin Noise Due to a Turbulent*. M.Sc. Thesis, VPISU.

6. Kusni, Muhammad, Benjamin Soenarko, and Harijono Djojodihardjo. 2014. Application of a multipole secondary source for propeller active noise control. *Applied Mechanics and Materials* 564: 135–142.
7. Kusni, M., and B. Soenarko. 2001. *Journal of Science and Math* 2 (2).
8. Kusni, M., and B. Soenarko. 2001. *Dipole Wave Cancellation Using a Multipole Secondary Source, Symposium on Wave Propagation*, September 2001.
9. Kusni, M. 2002. *The Active Noise Control of Propeller Noise Using a Multipole Secondary Source*. Paper ICAS 2002-852.
10. Seybert, A.F., B. Soenarko, F.J. Rizzo, and D.J. Shippy. 1984. *Journal of Vibration, Acoustics, Stress and Reliability* 106 (3): 414–420.
11. Hess, J.L., and A.M.O. Smith. 1967. Calculation of potential flow about arbitrary bodies. *Progress in Aeronautical Sciences* 8: 1–138.
12. Harijono Djojodihardjo, R., and Sheila E. Widnall. 1969. *AIAA Journal* 7 (10).
13. Ashley, H., and W.P. Rodden. 1972. *Annual Review of Fluid Mechanics* 4: 431–472.
14. Djojodihardjo, Harijono. 2009. Application of fast multipole boundary element method for unified FMBEM-FEM acoustic-structural coupling. In *Proceedings, International Astronautical Congress 2009*. Paper number IAC-09.C2.3.5.
15. Kok Hwa, Yu, A. Halim Kadarman, and Harijono Djojodihardjo. 2010. *Development and Implementation of Some BEM Variants—A Critical Review, Engineering Analysis with Boundary Elements*.
16. Djojodihardjo, H. 2013. *Journal of Mechanics Engineering and Automation* 3: 209–220.
17. Nelson, P.A., A.R.D. Curtis, and S.J. Elliott. 1986. Optimal multipole source distribution for the active suppression and absorption of acoustic radiation. *Proceedings of Euromech Colloquia* 213.
18. Ruijgrok, G.J.J. 1993. *Elements of Aviation Acoustics*. Delft University Press.
19. Koopman, G.H., L. Song, and J.B. Fahnline. 1989. *Journal of Acoustics Society America* 86: 2433–2438.
20. Bolton, J.S., and T.A Beauvilain. 1992. *Journal of Acoustics Society America* 91: 2349 (A).
21. Beauvilain, T.A., and J.S. Bolton. 1993. Cancellation of radiated sound fields by the use of multipole secondary sources. In *Proceedings of the Second Conference on Recent Advances in Active Control of Sound and Vibration*, 957–968. Lancaster, PA: Technomic.
22. Bolton, J.S., B.K. Gardner, and T.A. Beauvilain. 1995. *Journal of Acoustics Society America* 98 (A).

Chapter 22
Kinematic and Unsteady Aerodynamic Study of Bi- and Quad-Wing Ornithopter

Abstract The potential of Flapping-Wing Micro-Air Vehicles (MAVs) for sensing and information gathering relevant for environmental and disaster monitoring and security surveillance leads to the identification and modeling the salient features and functional significance of the various components in the flying reasonably sized biosystems. The dynamics, kinematics and aerodynamics of their wing systems and the production of mechanical power output for lift and thrust will be synthesized following a simplified and generic, but meticulous, model for a flapping-wing ornithopter. Basic unsteady aerodynamic approach incorporating viscous effect and leading-edge suction is utilized. The first part of the study is focused on a bi-wing ornithopter. Later, parametric study is carried out to obtain the lift and thrust physical characteristics in a complete cycle for evaluating the plausibility of the aerodynamic model and for the synthesis of an ornithopter model with simplified mechanism. Further analysis is carried out by differentiating the pitching and flapping motion phase lag and studying its respective contribution to the flight forces. A similar procedure is then applied to flapping quad-wing ornithopter model. Results are discussed in comparison with various selected simple models in the literature, with a view to develop a practical ornithopter model.

Keywords Bi-wing ornithopter · Flapping-wing aerodynamics · Micro-air vehicle · Quad-wing ornithopter · Unsteady aerodynamics

Nomenclature

AR	Aspect ratio
B	Semi-wingspan
c	Chord
$C(k)$	Theodorsen function
$C(k)_{\text{jones}}$	Modified Theodorsen function
C_{df}	Drag coefficient due to skin friction

Published in ASD Journal (2016), Vol. 4, No. 1, pp. 1–23, as Kinematic and Unsteady Aerodynamic Study on Bi- and Quad-Wing Ornithopter.

© The Author(s), under exclusive license to Springer Nature Singapore Pte Ltd. 2023
H. Djojodihardjo, *Introduction to Aeroelasticity*,
https://doi.org/10.1007/978-981-16-8078-6_22

dD_{camber}	Sectional force due to camber
dD_f	Sectional friction drag
dF_x	Sectional chordwise force
dL	Sectional lift
dy	Width of sectional strip under consideration
dN	Sectional total normal force
dN_c	Sectional circulatory normal force
dN_{nc}	Sectional apparent mass effect
dT	Sectional thrust
dT_s	Leading edge suction force
V_x	Flow speed tangential to section
V_{rel}	Relative velocity
V_i	Induced velocity
w_0	Downwash velocity at ¾-chord point
Γ	Circulation
ρ	Air density
β	Flapping angle
β_0	Maximum flapping angle
θ	Pitching angle
$\dot{\theta}$	Pitching rate
$\ddot{\theta}$	Pitching acceleration
θ_0	Maximum pitch angle
$\theta_{hindwing}$	Effective pitching angle of hind-wing
θ_f	Angle of flapping axis with respect to flight velocity (incidence angle)
θ_p	Mean pitch angle of chord with respect to flapping axis
ϕ	Lag angle between pitching and flapping angle
δ	Incidence angle
α	Relative angle of attack

22.1 Introduction

Motivated by flying biosystems, flight engineering has been initiated since hundreds of years ago and has gradually grown from the time of Leonardo Da Vinci to Otto Lilienthal's gliders, to modern aircraft technologies and present flapping flight research. Recent interest in the latter has grown significantly particularly for small flight vehicles (or Micro-Air Vehicles) with very small payload-carrying capabilities to allow remote sensing missions in hazardous as well as confined areas. Some of these vehicles may have a typical wingspan of 15 cm, with a weight restriction of less than 100 g [1]. Perhaps the most comprehensive account of insect flight or entomopter to date is given by Weis-Fogh [2], Ellington [3–5], Shyy et al. [6, 7], Dickinson et al. [8], Żbikowski [9] and Ansari et al. [10], while one of the first successful attempts to develop bird-like flapping flight was made by DeLaurier [11]. In a recent paper,

Wang [12, 13] has elaborated the peculiar nature of insects hovering, which has been efficiently acquired not by the dominant aerodynamic lift, but by the drag in a paddling-like motion.

Although our interest in developing a mathematical and experimental model is on more or less rigid bi- and quad-wing ornithopter, it should also take note on other relevant lightweight, flexible wings characteristics of insects and hummingbirds, that undergo large deformations while flapping, which can increase the lift of flapping wings (Rosenfeld [14]), as applicable.

In the past, flapping-wing designs have been created with varied success, for forward or hover mode, but not both, based on observations of hummingbirds and bats (Nicholson et al. [15]). According to Maybury and Lehmann [16], the dragonfly has the capability to shift flight modes simply by varying the phase lag between its fore and hind-wings. With that observation, a quad-winged flapping system could be conceived as the simplest mechanism that has the capabilities to shift between flight modes [17].

In one of the recent works in developing quad flapping-wing micro-air vehicle, Ratti [17] has theoretically shown that a flight vehicle with four flapping wings has 50% higher efficiency than that with two flapping wings. Inspired by the flight of a dragonfly, Prosser [18] analyzed, developed and demonstrated a Quad-Wing Micro-Air Vehicle (QW-MAV) which can produce higher aerodynamic performance and energy efficiency, and increased payload capacity compared to a conventional flapping-wing MAV (FW-MAV). In developing a generic model of flapping-wing ornithopter, the bi-wing ornithopter will first be reviewed and developed, and then extended to quad-wing ornithopter. In addition, the present approach is aimed at finding the simplest ornithopter configuration which can be used as the baseline for progressive and continued development. Also, by analyzing and synthesizing simple ornithopter configuration, the latter can be built into a simple mechanized one that can be used for experimental studies and further development. For this purpose, the sequence of Figs. 22.1, 22.2 and 22.3 is presented, which are ordered according to the wing structure characteristics. Figure 22.1 exhibits wing geometries of a Pterosaur, an eagle and a dragonfly, which could inspire the development of the geometrical and aerodynamic modeling of an ornithopter.

(a) (b) (c)

Fig. 22.1 **a** Pterosaur [19]; **b** soaring eagle; **c** a dragonfly exhibiting its wing geometry and structural detail

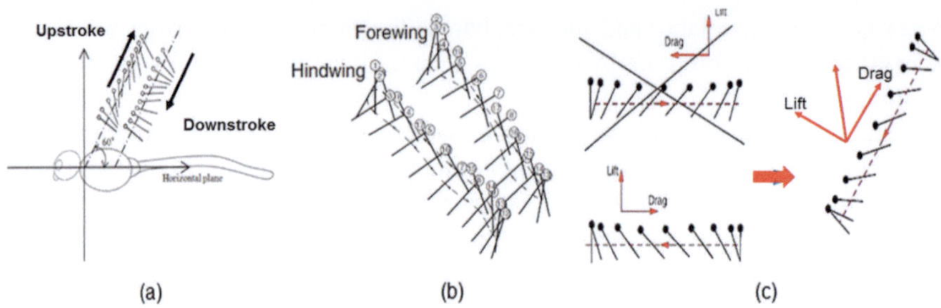

Fig. 22.2 **a**, **b** Upstroke and downstroke motion of dragonfly (Adapted from [20]); **c** saving power by eliminating a half stroke in normal hovering [12, 13]

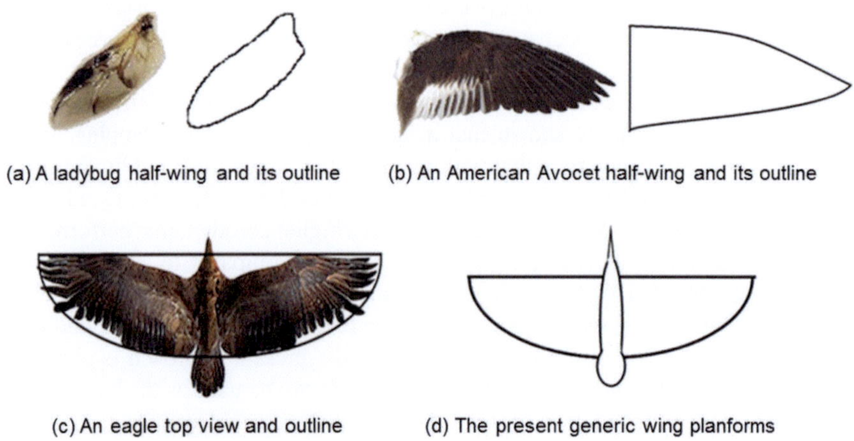

Fig. 22.3 Ornithopter flapping-wing aerodynamics computation

The image displayed in Fig. 22.2 exhibits a dragonfly, which will later be imitated to take advantage of the quad-wing kinematic and aerodynamic interactions, in the effort of improving the performance of the ornithopter to be developed. Figure 22.2 also schematically exhibits the flapping motion of the quad-wing dragonfly, as studied by Wang and Russell [20]. It is of interest to note that a systematic study by Wang [12, 13] shows that in the case of insects with dominant hovering movement, due to comparable lift and drag components of the aerodynamic force, the sustainment of flight is attributed to the drag, like in paddling movement. Figure 22.2c, adapted from Wang [12, 13], shows how by eliminating a half stroke in hovering, like in dragonfly, power efficiency can be achieved.

Taha et al. [21] made a thorough review of the significant work done so far in the area of flight dynamics and control of flapping-wing micro-air vehicles (MAVs), covering the flapping kinematics, the aerodynamic modeling and the body dynamics. They identified the missing gap between hover and forward speed movement, where $k > 0.1$, flapping frequency ω in the order of the body natural frequency, and relative flow angle $\alpha > 25°$ or dynamic stall, where there is dominant LEV contribution and coupling between the aerodynamic forces and the body modes.

Addressing this gap and in dealing with LEV, Taha et al. [22] embarked upon a novel approach of using a state-space formulation for the aerodynamics of flapping flight by extending Duhamel's principle in the linear unsteady flows to non-conventional lift curves to capture the LEV contribution. Their proposed model has been validated through a comparison with direct numerical simulations of Navier–Stokes on hovering insects.

It has also been observed that the flapping frequency tends to decrease with body mass increase [6]. In view of these findings, the classification tabulated in Table 22.1 could summarize some of the relevant features of flapping biosystems that may give us an overview for developing flapping ornithopter MAV. Whereas crane-flies, mosquitoes and other Nematocera as well as many larger Brachycera and Cyclorrhapha undoubtedly use normal hovering in most cases [23]. Birds habitually perform aerial maneuvers that exceed the capabilities of best anthropogenic aircraft control systems (Tedrake et al. [24]).

The complexity and variability of the aerodynamics during these maneuvers are difficult, with dominant flow structures (e.g. vortices) that are difficult to predict robustly from first-principles (or Navier–Stokes) models. In this conjunction, machine learning will play an important role in the control design process for responsive flight by building data-driven approximate models of the aerodynamics and by synthesizing high-performance nonlinear feedback policies based on these approximate models and trial-and-error experience.

Biosystem flapping flights are characterized by a relatively low Reynolds number, flexible wing, highly unsteady flow, laminar separation bubble, non-symmetrical upstroke and downstroke and for entomopters, the presence and significant role of the leading-edge vortex and wake vortices capture, among others.

Hence the objective of the present work are as follows: first, to understand and mimic the kinematics and unsteady aerodynamics of biosystems that can be adopted in the present bi-wing FW-MAV and quad-wing QW-MAV. Second, following our previous attempt to develop pterosaur-like ornithopter to produce lift and thrust for forward flight as a simple and workable ornithopter flight model [25, 26], the present work will simulate and analyze the kinematics and aerodynamics of bird-like rigid Bi- and Quad-Wing Ornithopter.

At the present stage, which is addressed on bi-wing ornithopter mimicking bird's forward flight, the work does not incorporate leading-edge vortex effects. In modeling and simulating the influence of the leading-edge vortex in our future work, information gained from many recent approaches such as those of Ansari et al. [10], Wang [13] and Taha et al. [21, 22] will be taken into consideration. A simplistic and heuristic leading-edge vortex modeling which associate the shed vortices with rapid pronation of the wing is presented in a companion paper [28].

Table 22.1 Overview of some relevant characteristics of flapping biosystems[1] (extension based on [25, 26])

Items	Insects	Humming bird	Bat	Small birds	Large birds	Flapping MAV	Small low speed airplanes
1. Types	Beetles, bumblebees, butterflies, dragonflies	Amazilia	Plecotus Auritus	Sparrows, swifts, robins	Eagle, hawk, vulture, falcon, skua gull	DARPA DRO [1]	Cessna 210
2. Weight typical (gf)[1]	25×10^{-5}–12.8	5.1	9.0	35–82	952–4300	≤ 50	1,045,000
3. Wing semi-span (cm)[1]	0.062–7.7	5.9	11.5	20–48	58–102	< 7.5	5600
Wing-loading (g/cm^2)	10^{-3}–10^{-1}	0.4	0.072	0.029–0.152	0.35–0.67	10^{-2}–1	11.18
4. Typical power (gf cm s^{-1} per gf)	5.3–238	130	83	93–110	42–57	≈ 39	$\approx 1.3 \times 10^4$
5. Dominant wing movement	Hover	Hover and fly	Fly	Fly	Fly	Hover and fly	Fly
6. Motion elements of wing kinematics	Pronation and supination in stroke plane	Pronation and supination in stroke plane	Flapping and pitching	Flapping and pitching	Flapping and pitching	Flapping and pitching	–
7. Flight speed (m/s)	1.05–9	15	10–14	6–10	10–20	3–10	99 m/s (cruise at 6100 m altitude)

(continued)

[1] Power functions of wing dimensions and flight parameters against body mass m, following Shyy [7] and Norberg [23]. The exponent of correlation is for (Mass) exponent.

Table 22.1 (continued)

Items	Insects	Humming bird	Bat	Small birds	Large birds	Flapping MAV	Small low speed airplanes
8. Reynolds No	10–1000	7500	14,000	10^3–10^4	10^4–10^5	10^4–10^5	10,000,000
9. Leading-edge vortex/LEV	LEV by swept wing at Re = 5 × 10^3	Yes	Yes	Yes	Yes	Yes	No
10. Entering its own TEV/ Wake capture	Yes	Yes	No	No	No	No	No
11. Laminar separation bubble/LSB	Yes	Yes	Yes	Yes	Yes	Yes	No
12. Leading-edge flap	–	–	Has been observed on bats	–	E.g. Mallard, at Re = 6 × 104 (Jones [27]) 		
13. Self-activated flaps at TE					E.g. Skua Gull [6, 7] 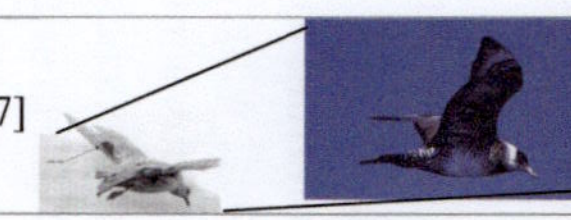		

22.2 Theoretical Development of the Generic Aerodynamics of Flapping Wings

Following the frame of thought elaborated in the previous section, several generic flying biosystem wing planforms are chosen as baseline geometries for the ornithopter. Referring to the eagle wing and for convenience of baseline analysis, the semi-elliptical wing (shown in Fig. 22.3) is selected for current study with the backdrop of various wing-planform geometries utilized by various researchers.

Analytical approaches of quasi-steady and unsteady model are carefully evaluated in the present work in order to deal with the aerodynamic problem. In agreement with the quasi-steady model, based on observation on flying birds, it can be assumed that the flapping frequencies are sufficiently slow that shed wake effects are negligible, as in pterosaur and medium- to large-sized birds. The unsteady approach attempts to model the wake like hummingbird and insects will be deferred to succeeding work. The present unsteady aerodynamic approach is synthesized using basic foundations that may exhibit the generic contributions of the motion elements of the bio-inspired bi-wing ad quad-wing air vehicle characteristics. To account for the unsteady effects, Theodorsen unsteady aerodynamics [29] and its three-dimensional version by Jones [30] have been incorporated. The computation of lift and thrust generated by pitching and flapping motion of three-dimensional rigid wing is conducted in a structured approach using strip theory and Jones' modified Theodorsen approach (DeLaurier [11]; Jones [30]) for a wing without camber. Furthermore, the Polhamus leading-edge suction [31, 32] is also incorporated. The total lift and thrust for the wing are calculated by the summation of the contributions from each strip for a whole flapping cycle. Fully unsteady lifting surface theory [33–36] may later be incorporated.

At the present stage, which will be assessed a posteriori based on the results, DeLaurier's [11] unsteady aerodynamics and modified strip theory approach for the flapping wing is utilized with post-stall behavior. The computational logic in the present work is summarized in the flowchart exhibited in Fig. 22.4.

To obtain insight into the mechanism of lift and thrust generation of flapping and pitching motion, Djojodihardjo et al. [37, 38] analyzed the wing flapping motion by looking into the individual contribution of the pitching, flapping and coupled pitching-flapping to the generation of the aerodynamic forces. Also the influence of the variation of the forward speed, flapping frequency and pitch-flap phase lag has been analyzed. Such approach will also be followed here through further scrutiny of the motion elements. The generic procedure is synthesized from the pitching-flapping motion of rigid wing developed by DeLaurier [11] and Harmon [39]. The flapping motion of the wing is distinguished into three distinct motions with respect to the three axes: these are: (a) *Flapping*, which is up and down plunging motion of the wing; (b) *Feathering* is the pitching motion of wing and can vary along the span; and (c) *Lead-lag,* which is in-plane lateral movement of wing, as incorporated in Fig. 22.5. For further reference to the present work, the lead-lag motion could be interpreted to apply to the phase lag between pitching and flapping motion, while the fore-and-aft movement can be associated with the orientation of the stroke plane. The

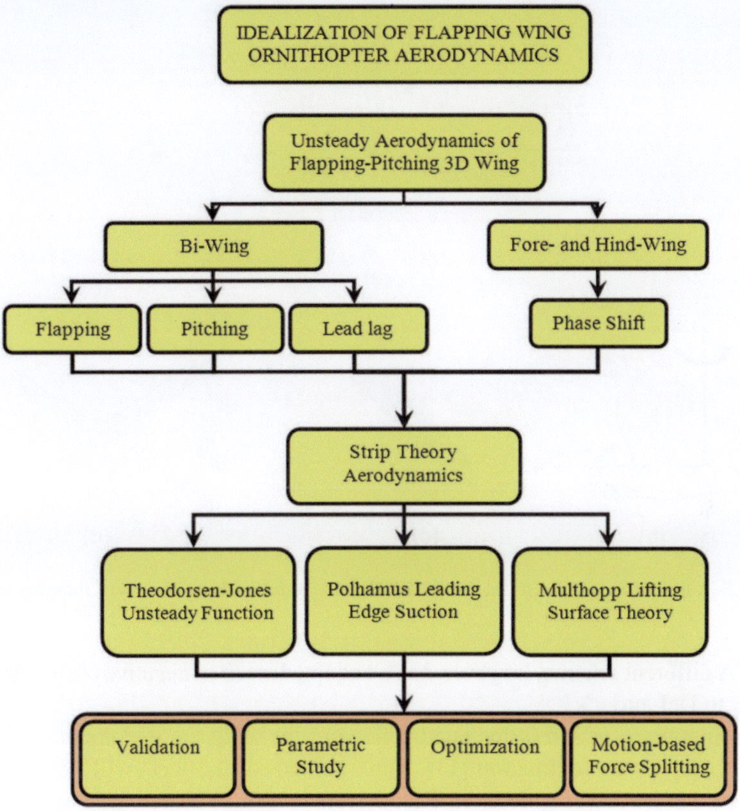

Fig. 22.4 Ornithopter flapping-wing aerodynamics computational scheme

degree of freedom of the motion is depicted in Fig. 22.5. The flapping angle β varies as a sinusoidal function and pitching angle θ are given by the following equations.

$$\beta(t) = \beta_0 \cos \omega tt \tag{22.1}$$

$$\theta(t) = \theta_0 \cos(\omega t + \phi) + \theta_{\text{fp}} \tag{22.2}$$

where θ_0 and β_0 indicate maximum value for each variable, ϕ is the lag between pitching and flapping angle and y is the distance along the span of the wing, and θ_{fp} is the sum of the flapping axis angle with respect to flight velocity (incidence angle) and the mean angle of the chord line with respect to the flapping axis. The present method is exemplified by the use of elliptical planform wing.

As a baseline, by referring to Eqs. 22.1 and 22.2, β and θ is considered to oscillate following a sine function; such scheme indicates that these motions start from zero

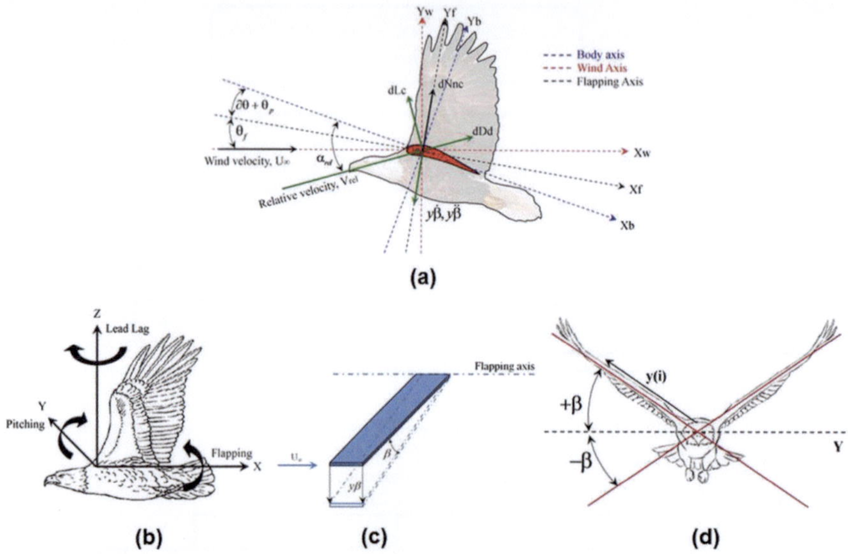

Fig. 22.5 **a** Forces on section of the wing. **b–d** Flapping and pitching motion of flapping wing

value. A different scheme, however, can be adopted, such as negative cosine function similar to DeLaurier's.

Leading-edge suction is included following the analysis of Polhamus [31, 32] and DeLaurier's approximation [11]. Three-dimensional effects will later be introduced by using Scherer's modified Theodorsen-Jones Lift Deficiency Factor [40], in addition to the Theodorsen unsteady aerodynamics [29] and its three-dimensional version by Jones [30]. Further refinement is made to improve accuracy. Following Multhopp approach (Multhopp [41]), simplified physical approach to the general aerodynamics of arbitrary planform is adopted, i.e. a lifting line in the quarter-chord line for calculating the downwash on the three-quarter-chord line for each strip.

In the present analysis, no linear variation of the wing's dynamic twist is assumed for simplification and instructiveness. However, in principle, such additional requirements can easily be added due to its linearity.

The total normal force is acting perpendicularly to the chord line and is given by

$$\mathrm{d}N = \mathrm{d}N_{\mathrm{c}} + \mathrm{d}N_{\mathrm{nc}} \tag{22.3}$$

The circulatory normal force for each section acts at the quarter chord and also perpendicular to the chord line is given by

$$\mathrm{d}N_{\mathrm{c}} = \frac{\rho U V}{2} C_n(y) c \mathrm{d}y \tag{22.4}$$

$$\mathrm{d}N_{\mathrm{nc}} = \frac{\rho \pi c^2}{4} \dot{V}_{\mathrm{mid\text{ - }chord}} \mathrm{d}y \tag{22.5}$$

where

$$\dot{V}_{\text{mid - chord}} = U\dot{\alpha} - \frac{1}{4}c\ddot{\theta} \tag{22.6}$$

Using these relationships, the relative velocity at three-quarter-chord point which is used for the calculation of the aerodynamic forces can be established. The relative angle of attack at three-quarter chord, α, is then given by

$$\alpha = \frac{\left(\dot{h}\cos(\theta - \theta_{\mathrm{f}}) + \frac{3}{4}c\dot{\theta} + U\left(\theta - \theta_{\mathrm{fp}}\right)\right)}{U} \tag{22.7}$$

$$\alpha = Ae^{i\omega t} \tag{22.8}$$

which is schematically elaborated in Fig. 22.6.

The modified Theodorsen Lift Deficiency function for finite aspect ratio wing is given by Jones [30]. Another derivation for unsteady forces for finite aspect ratio wing carried out by Scherer [40] arrived at a similar form to the Theodorsen two-dimensional case. It is utilized here for convenience and takes the following form

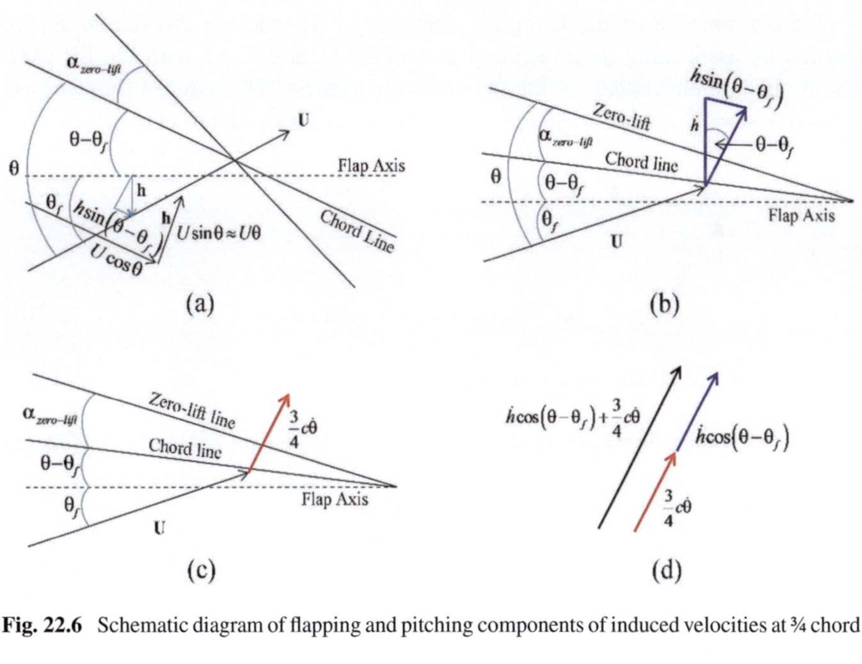

Fig. 22.6 Schematic diagram of flapping and pitching components of induced velocities at ¾ chord

$$C(k)_{\text{Jones}} = \frac{\overbrace{\text{AR}}^{\text{Aspect Ratio}} \quad C(k)}{2 + \overbrace{\text{AR}}^{\text{Aspect Ratio}}} \tag{22.9}$$

where

$$C(k) = F(k) + iG(k) \tag{22.10}$$

$C(k)$, $F(k)$ and $G(k)$ relate to the well-known Theodorsen function [29, 30] which are functions of reduced frequency, k. Following the methodological philosophy of Theodorsen [29] and Garrick [42, 43] and the classical unsteady aerodynamics, the unsteady lift is expressed as [43]:

$$L = \pi c \rho U C(k) Q \tag{22.11}$$

where Q is given by $Q = \omega e^{i\omega t}$. Then, substitution Q into Eq. 22.11 gives

$$L = \pi c \rho U C(k) \left(\omega e^{i\omega t}\right) \tag{22.12}$$

The convenience of the Complex Analysis of Theodorsen is exemplified by Garrick by associating the imaginary part of 22.11 and 22.12 with the lift [43]. The details are elaborated for the sake of completeness. The reduced frequency is defined as $k = \frac{\omega c}{2U}$, or $\omega t = \frac{\omega c}{2U} . \frac{2Ut}{c} = ks$. Assuming sinusoidal motion

$$\omega e^{i\omega t} = \omega(\cos\,\omega t + i\,\sin\,\omega t) \tag{22.13}$$

or

$$\omega e^{i\omega t} = \omega(\cos\,ks + i\,\sin\,ks) \tag{22.14}$$

Combining 22.10 and 22.13, one obtains:

$$L = \pi c \rho U \omega [F(k) + iG(k)](\cos\,ks + i\,\sin\,ks) \tag{22.15}$$

Note that

$$|C(k)| = \mathbf{C}(k) = |F(k) + iG(k)| = \left|\left((F(k))^2 + (G(k))^2\right)^{\frac{1}{2}}\right| \tag{22.16}$$

where the imaginary value of Eq. 22.15 is the lift:

$$\alpha_{\text{Theodorsen}} = \tan^{-1} \frac{G(k)}{F(k)} \tag{22.17}$$

$$F(k) = |C(k)| \cos \alpha_{\text{Theodorsen}} = \mathbf{C}(k) \cos \alpha_{\text{Theodorsen}} \tag{22.18}$$

$$G(k) = |C(k)| \sin \alpha_{\text{Theodorsen}} = \mathbf{C}(k) \sin \alpha_{\text{Theodorsen}} \tag{22.19}$$

After some algebraic manipulation, Eq. 22.15 reduces to

$$L = \pi c \rho U \omega \cdot I.P. \begin{bmatrix} \mathbf{C}(k) \cos \alpha_{\text{Theodorsen}} - \mathbf{C}(k) \sin \alpha_{\text{Theodorsen}} \\ +i\mathbf{C}(k)(\cos(ks) \sin \alpha_{\text{Theodorsen}} + \sin(ks) \cos \alpha_{\text{Theodorsen}}) \end{bmatrix} \tag{22.20}$$

and the imaginary parts (I.P.) of the above equation is

$$\mathbf{C}(k)(\cos(ks) \sin \alpha_{\text{Theodorsen}} + \sin(ks) \cos \alpha_{\text{Theodorsen}}) \tag{22.21a}$$

or

$$\mathbf{C}(k) \sin(ks + \alpha_{\text{Theodorsen}}) \tag{22.21b}$$

Therefore:

$$L = \pi c \rho U \omega \left[\left((F(k))^2 + (G(k))^2 \right)^{\frac{1}{2}} \sin\left(ks + \tan^{-1} \frac{G(k)}{F(k)} \right) \right] \tag{22.22}$$

Consistent with the strip theory, the downwash for untwisted planform wing is given by [44, 45]

$$\frac{w_0}{U} = \frac{2(\alpha_0 + \theta_{\text{fp}})}{2 + \overbrace{\text{AR}}^{\text{Aspect Ratio}}} \tag{22.23}$$

Considering all of these basic fundamentals, the relative angle of attack at three-quarter-chord point α' is given by

$$\alpha' = \frac{\overbrace{\text{AR}}^{\text{Aspect Ratio}}}{2 + \underbrace{\text{AR}}_{\text{Aspect Ratio}}} \left[F(k)\alpha + \frac{c}{2U} \frac{G(k)}{k} \dot{\alpha} \right] - \frac{w_0}{U} \tag{22.24}$$

which has taken into account the three dimensionality of the wing.

From Fig. 22.6c, the flow velocity which includes the downwash and the wing motion relative to free-stream velocity, V, can be formulated as

$$V = \left[\left(U\cos\theta - \dot{h}\sin(\theta - \theta_{\mathrm{f}})\right)^2 + \left(U\left(\alpha' - \theta_{\mathrm{fp}}\right) - \frac{1}{2}\mathrm{c}\dot{\theta}\right)^2 \right]^{\frac{1}{2}} \tag{22.25}$$

where the third and fourth terms are acting at the three-quarter-chord point. The apparent mass effect for the section is perpendicular to the wing, acts at mid-chord, and can be calculated as

$$\mathrm{d}N_{\mathrm{nc}} = \frac{\rho\pi c^2}{4}\left(U\dot{\alpha} - \frac{1}{4}\mathrm{c}\ddot{\theta}\right)\mathrm{d}yy \tag{22.26}$$

The term $U\dot{\alpha} - \frac{1}{4}\mathrm{c}\ddot{\theta}$ is the normal velocity's time rate of change at mid-chord due to the motion of the wing.

The total chordwise force, $\mathrm{d}F_{x,}$ is accumulated by three force components; these are the leading-edge suction, force due to camber and chordwise friction drag due to viscosity effect. All of these forces are acting along and parallel to the chord line.

$$\mathrm{d}F_x = \mathrm{d}T_{\mathrm{s}} - \mathrm{d}D_{\mathrm{camber}} - \mathrm{d}D_{\mathrm{f}} \tag{22.27}$$

The leading-edge suction, $\mathrm{d}T_{\mathrm{s}}$, following Garrick [42, 43], is given by.

$$\mathrm{d}T_{\mathrm{s}} = 2\pi\eta_{\mathrm{s}}\left(\alpha' + \theta_{\mathrm{fp}} - \frac{1}{4}\mathrm{c}\dot{\theta}\right)\frac{\rho U V}{2}\mathrm{c}\mathrm{d}yy \tag{22.28}$$

while following DeLaurier [11] the chordwise force due to camber and friction is, respectively, given by.

$$\mathrm{d}D_{\mathrm{camber}} = -2\pi\alpha_0\left(\alpha' + \theta_{\mathrm{fp}}\right)\frac{\rho U V}{2}\mathrm{c}\mathrm{d}yy \tag{22.29}$$

$$\mathrm{d}D_{\mathrm{f}} = \frac{1}{2}\rho V_x^2 C_{d_{\mathrm{f}}}\mathrm{c}\mathrm{d}yy \tag{22.30}$$

The efficiency term η_{s} is introduced for the leading-edge suction $\mathrm{d}T_{\mathrm{s}}$ to account for viscosity effects. The vertical force dN and the horizontal force $\mathrm{d}F_x$ at each strip dy will be resolved into those perpendicular and parallel to the free-stream velocity, respectively. The resulting vertical and horizontal components of the forces are then given by

$$\mathrm{d}L = \mathrm{d}N\cos\theta + \mathrm{d}F_x\sin\theta \tag{22.31}$$

$$\mathrm{d}T = \mathrm{d}F_x\cos\theta - \mathrm{d}N\sin\theta \tag{22.32}$$

To obtain a three-dimensional lift for each wing, these expressions should be integrated along the span, b; hence

$$L = \int_{0}^{b} \mathrm{d}L\mathrm{d}x \tag{22.33}$$

$$T = \int_{0}^{b} \mathrm{d}T\mathrm{d}x \tag{22.34}$$

22.3 Results

22.3.1 Results for Bi-Wing

For later comparison with appropriate results from the literature, numerical computations are performed using the following wing geometry and parameters: the wingspan of 40 cm, aspect ratio of 6.36, flapping frequency of 7 Hz, total flapping angle of 60°, forward speed of 6 m/s, maximum pitching angle of 20° and incidence angle of 6° and there is no wing dihedral angle. In the calculation, both the pitching and flapping motions are in cosine function by default, which is subject to parametric study, and the upstroke and downstroke have equal time duration. The wake capture has not been accounted for in the current computational procedure. The computational scheme developed has been validated satisfactorily, starting from the verification of present work with other works by Zakaria et al. [46] and DeLaurier [11] which is shown in Fig. 22.7, where it uses the pterosaur's wing model.

The result of bi-wing calculated using chosen wing geometry and parameters is shown qualitatively in Fig. 22.8 (left), and the average values of both lift and thrust forces are shown in Table 22.2. Also in Fig. 22.8 (right), the criterion for post-stall

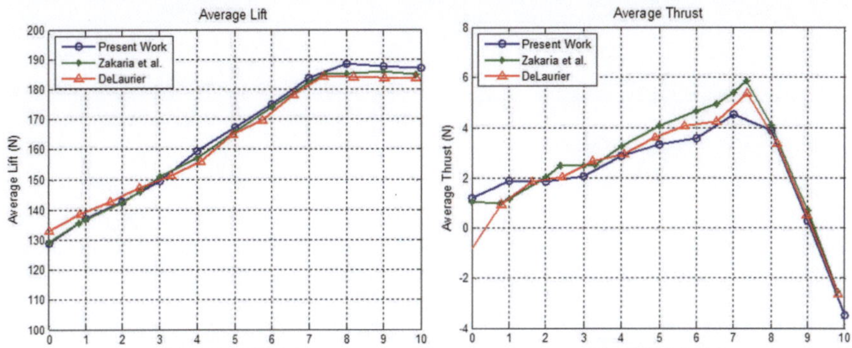

Fig. 22.7 Verification of aerodynamic modeling of present work with work by Zakaria et al. [46] and DeLaurier [11]

behavior is shown to emphasize that at certain angle of attack, it is exceeding the limit of maximum stall angle to enter the region of post-stall condition, even though the angle is only accounted for the upper (positive) limit, following DeLaurier's assumption in his work.

In Fig. 22.9, the present work uses Yu et al.'s [47] parameter to produce comparable agreements qualitatively and quantitatively with the behavior of Yu et al.'s [47] results. For the thrust force per cycle, during downstroke, the force is not really pronounced and low due to the stall condition, causing such trend (Table 22.3).

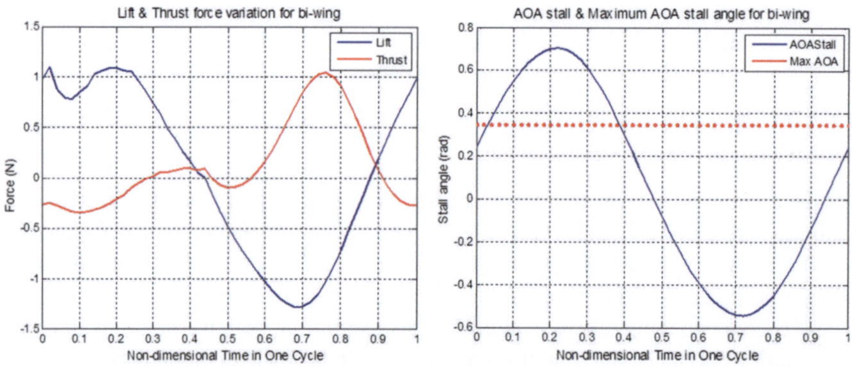

Fig. 22.8 Upper left: Lift and thrust for bi-wing ornithopter; upper right: stall angle criterion to show post-stall behavior within the present modeling

Table 22.2 Average lift and thrust of present work (bi-wing)

Forces	Present work
Average lift (N)	0.0662
Average thrust (N)	0.1110

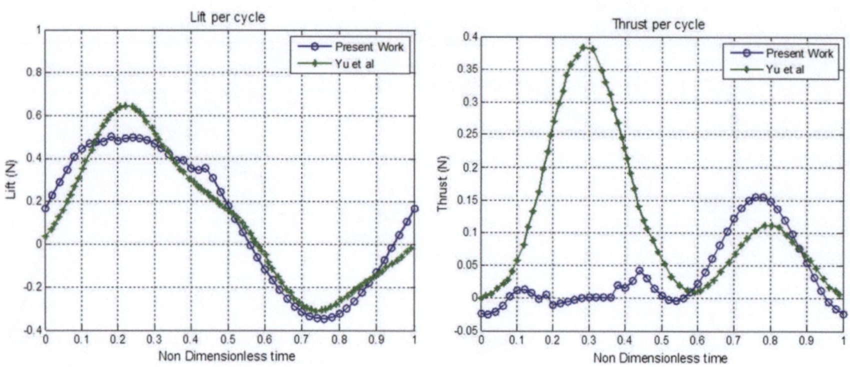

Fig. 22.9 Validation with Yu et al. [47]

Table 22.3 Average lift and thrust of present work (bi-wing, modified) and Yu et al. [47]

Forces	Present work (modified)	Yu et al. [47]
Average lift (N)	0.124339349	0.121
Average thrust (N)	0.033662	0.119

22.3.1.1 Variation of Oscillatory Articulation of the Bi-Wing

In modeling the pitching and flapping motion of the ornithopter wing, one may learn from the biosystems as summarized in Table 22.1, but could also attempt to introduce variations. Based on a close observation to selected avians, such as soaring eagles, one can observe that before taking off, they expand (flap) their wings up to a maximum position and stretch their legs simultaneously. It follows that the oscillatory motion can be modeled as a cosine function. It is noted, however, that DeLauriers [11] uses negative sinusoidal motion. Motivated by these meticulous observations, various possible models can be defined and utilized accordingly to account for every possible flapping kinematics. The results are shown in Fig. 22.10 and Table 22.4.

What can be seen in Figs. 22.10 and 22.11 is that, in conformity with our observation and those researchers like DeLaurier [11], Fujiwara et al. [48] and Chen et al. [49], flapping motion should be in cosine function. Interestingly, as observed by Chen et al. and assumed by DeLaurier, the pitching motion is prominent in negative sine function and exhibited by our calculation. Table 22.5 lists the average lift and thrust for various pitch articulation. Judging from these results, at least within

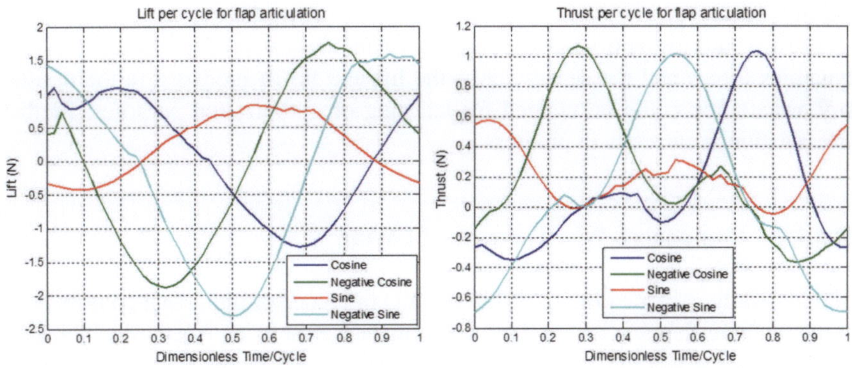

Fig. 22.10 Lift and thrust for bi-wing ornithopter for each kinematics definition (flap articulation)

Table 22.4 Average lift and thrust (bi-wing) for flap articulation

Forces	Flap articulation (pitch-cosine function)			
	Cosine	Negative cosine	Sine	Negative sine
Average lift (N)	0.06621196	0.025954553	0.219303352	– 0.093042201
Average thrust (N)	0.111028054	0.195295971	0.213896502	0.057148544

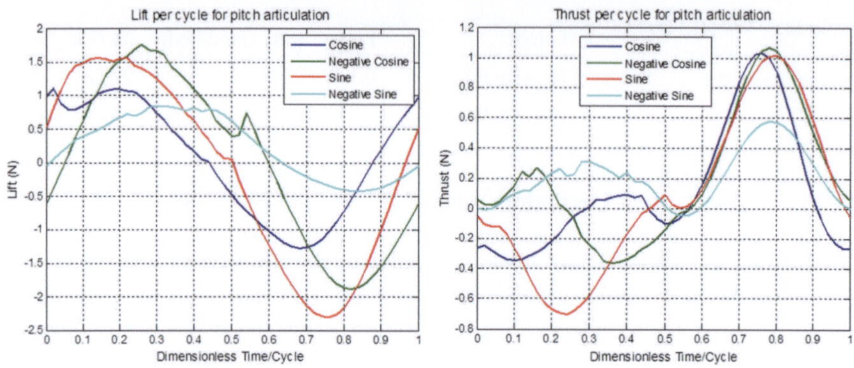

Fig. 22.11 Lift and thrust for bi-wing ornithopter for each kinematics definition (pitch articulation)

Table 22.5 Average lift and thrust (bi-wing) for pitch articulation

Forces	Pitch articulation (flap-cosine function)			
	Cosine	Negative cosine	Sine	Negative sine
Average lift (N)	0.06621196	0.006560787	– 0.109378927	0.228908605
Average thrust (N)	0.111028054	0.200125707	0.070766743	0.206238465

the assumptions adopted in the present work, one can obtain an impression, which combination of cosine-flap and negative sine-pitch produces the highest value of lift and thrust forces.

These results exemplify that the flapping kinematics can produce significant aerodynamics forces and the sensitivity of the lift and thrust produced to the oscillatory articulation could be utilized for tailoring or optimization purposes. Further investigation is currently in progress.

22.3.1.2 Component-Wise Forces for Bi-Wing

Another study is carried out to investigate the influence of individual contributions of the pitching-flapping on the flight performance. Results obtained as exhibited in Fig. 22.12 show that the lift is dominated by the incidence angle while the thrust is dominated by the flapping angle (other parameters remaining constant).

From the above component-wise force analysis, it can be deduced that also an appropriate combination of these force elements can be obtained to produce optimum lift and thrust. The optimization of this problem is also currently under study (Table 22.6).

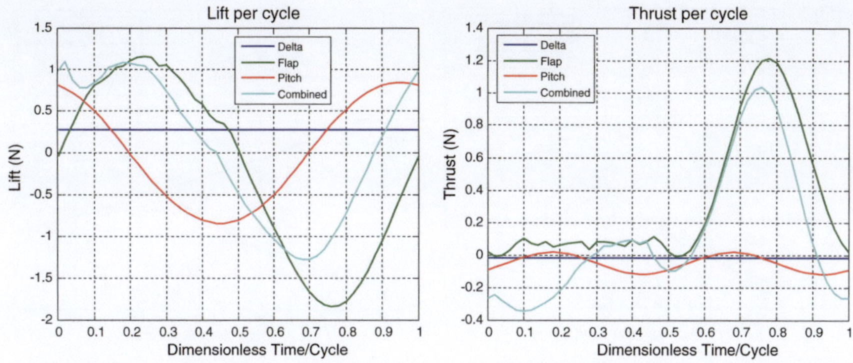

Fig. 22.12 The influence of individual contributions of the pitching-flapping motion and incidence angle on the flight performance

Table 22.6 Average lift and thrust (bi-wing) for each individual contribution

Forces	Individual contribution for bi-wing			
	Incidence only	Flap only	Pitch only	Combined
Average lift (N)	0.277616461	– 0.186354266	0.01614587	0.06621196
Average thrust (N)	– 0.018025915	0.312341058	– 0.051815978	0.111028054

22.3.1.3 Parametric Study for Bi-Wing

A parametric study is carried out to assess the influence of some flapping-wing motion parameters to the flight performance desired. The study considers the following parameters: the effect of flapping frequency and total flapping angle. The results are exhibited in Fig. 22.13. From this study, in general, the lift (at higher degree of frequency) and thrust forces always increase as the flapping frequency increases. As for the flapping amplitude, the thrust increases but the lift decreases over the cycle.

By referring to Pennycuick's [50] and Tucker's formula [51] to associate wing aspect ratio and wing area of birds, the present ornithopter model operating frequencies as anticipated in Fig. 22.13 are close to the operating flapping frequency values of selected birds shown in Table 22.18.

The phase lag angle between the pitching and flapping motion should be in such way that when the relative air velocity is at its peak, the pitch angle should be maximum. With such condition, this can be achieved only if the phase angle is $\pi/2$ (90°). Table 22.9 shows the value for both average values of lift and thrust forces, and it is in agreement with the statement above (Figs. 22.14 and 22.15; Tables 22.7, 22.8).

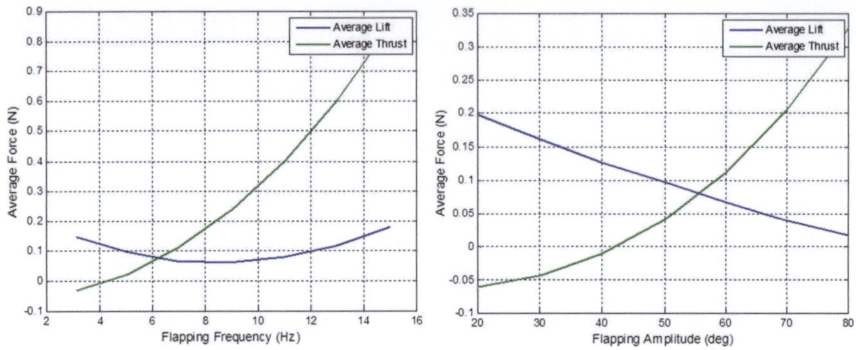

Fig. 22.13 Parametric study on the influence of flapping frequency and flapping amplitude on cyclic lift and thrust (bi-wing, semi-elliptical planform)

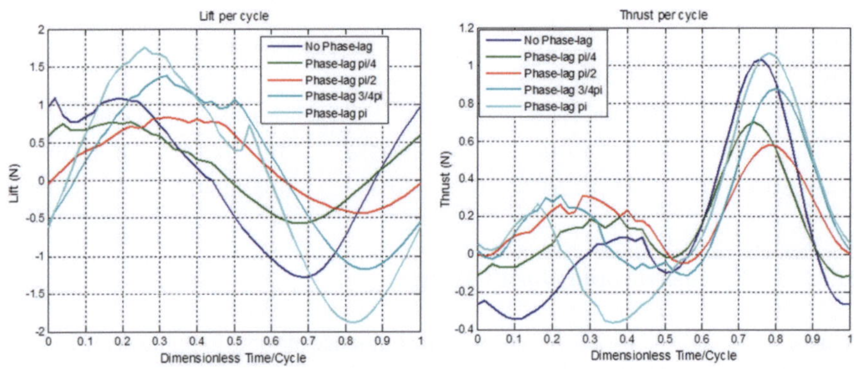

Fig. 22.14 Lift and thrust variation with phase lag angle

22.3.2 *Analysis and Results for Quad-Wing*

For the quad-wing kinematics and aerodynamics, the present work takes into account the influence of the fore-wing-induced downwash on the hind-wing effective angle of attack. This effect is modeled by assuming that, at any instant, the circulation Γ of the fore-wing acts at its quarter-chord point, and the induced downwash is calculated at the three-quarter-chord point of the hind-wing, as depicted in Fig. 22.16.

Following Kutta-Joukowski Law, the instantaneous equivalent circulation generated by the fore-wing is given by

$$\Gamma = \frac{L_{\text{force}}}{\rho U_\infty} \tag{22.35}$$

and the induced velocity V_i, following Biot-Savart law is given by

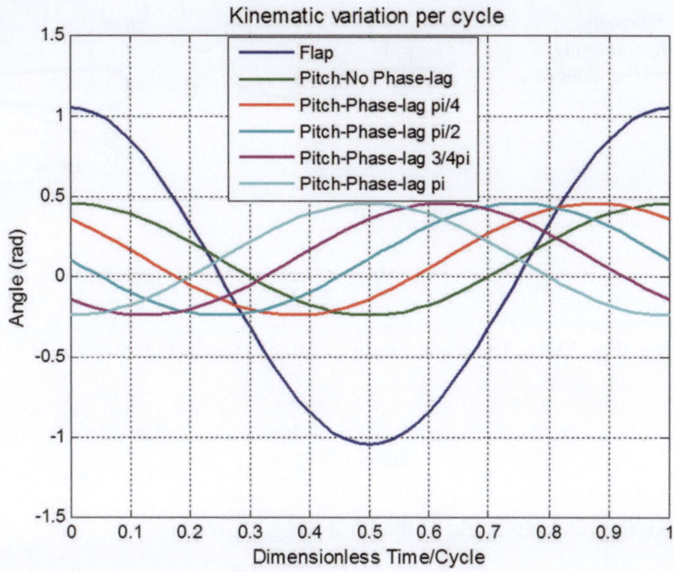

Fig. 22.15 Kinematics variation with phase lag angle

Table 22.7 Average lift and thrust with variation of flapping frequency

Forces	Frequency, f (Hz)						
	3.14	5	7	9	11	13	15
Average lift (N)	0.14807	0.09848	0.06621	0.06189	0.07878	0.11861	0.17849
Average thrust (N)	– 0.03212	0.02167	0.11103	0.23660	0.39895	0.60309	0.85149

Table 22.8 Average lift and thrust with variation of flapping amplitude

Forces	Flapping amplitude, β (°)						
	20	30	40	50	60	70	80
Average lift (N)	0.19759	0.16056	0.12520	0.09623	0.06621	0.03899	0.01668
Average thrust (N)	– 0.06182	– 0.04379	– 0.01122	0.04039	0.11103	0.20526	0.32870

Table 22.9 Average lift and thrust variation with lag of phase angle

Forces	Pitch and flap phase lag				
	0	$\pi/4$	$\pi/2$	$3\pi/4$	π
Average lift (N)	0.06621196	0.187714154	0.228901395	0.161863869	0.005574254
Average thrust (N)	0.111028054	0.157128996	0.206268504	0.226252777	0.199126131

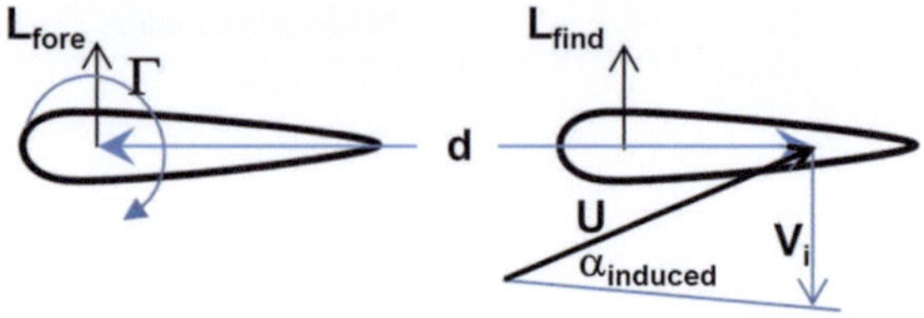

Fig. 22.16 Schematic diagram of the fore-wing downwash and the induced angle of attack on the hind-wing

$$V_i = \frac{\Gamma}{2\pi d} \tag{22.36}$$

Following Fig. 22.16, for small angle of attack, the induced angle is formulated as

$$\alpha_{\text{induced}} \approx \frac{V_i}{U_\infty} \tag{22.37}$$

Therefore the pitching angle of the hind-wing is given by

$$\theta(t) = \frac{\mathrm{d}y}{B}\theta_0 \sin(\omega t + \phi) + \frac{V_i}{U} + \theta_{\text{fp}} \tag{22.38}$$

The analysis is then carried out for the quad-wing with similar wing geometry as for the bi-wing. Initial initiative was done with an assumption that the fore- and hind-wings are closely attached, meaning of inexistence of gap between the leading edge of the hind-wing and the trailing edge of the fore-wing. The results for quad-wing configuration below are obtained using the same wing geometry and parameters used in bi-wing case, for fore- (front) wing and hind- (latter) wings. The results are compared and analyzed to appreciate the influence of physical refinements in the computational procedure and for validation purposes. This analysis also accounts for the induced angle of attack on the hind-wing due to downwash of the fore-wing. The results are presented in Figs. 22.17 and 22.18, and Table 22.10.

Figure 22.17 compares the lift computed using the present model to Wang and Russell's more elaborate model calculation [20]. This comparison is very qualitative, for proof of concept considerations. Figure 22.18 shows the default motions of flapping and pitching for both fore-wing and hind-wing, which are in cosine function.

22.3.2.1 Variation of Oscillatory Articulation of the Quad-Wing

Following the procedure and parametric study carried out for bi-wing ornithopter, the present study also addresses the flapping kinematics accordingly, by taking into considerations what has been learned from bi-wing parametric study.

The fore-wing and hind-wing are arranged in tandem without gap, so that the leading edge of the hind-wing touches the trailing edge of the fore-wing, and they are moving simultaneously. The pitching motion of both fore-wing and hind-wing moves

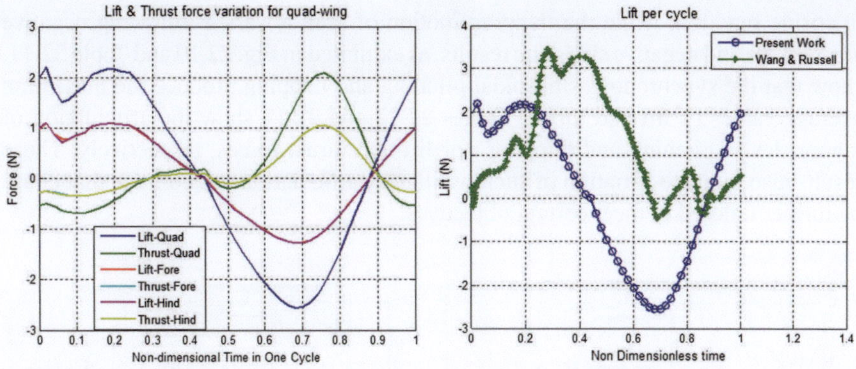

Fig. 22.17 Left: Lift and thrust for quad-wing ornithopter; right: qualitative comparison with Wang and Russell results [20]

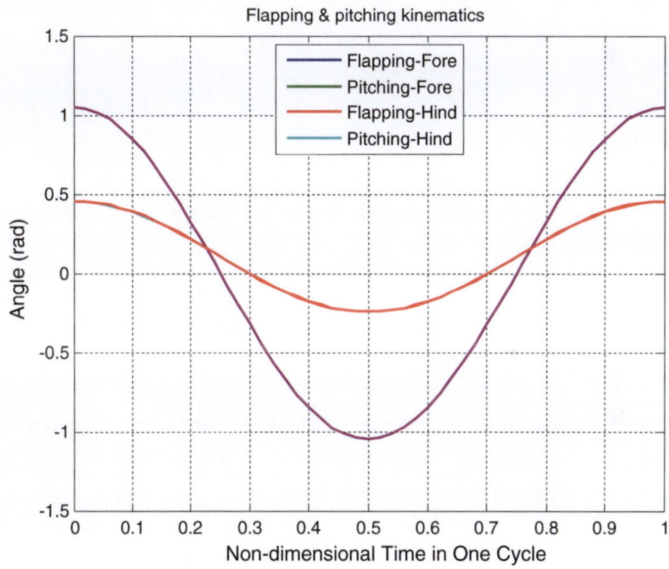

Fig. 22.18 Flapping and pitching kinematics for fore-wing and hind-wing

Table 22.10 Average lift and thrust for present work

Forces	Present work	Fore-wing	Hind-wing	Wang and Russell [20]
Average lift (N)	0.1306	0.0662	0.0644	1.136567
Average thrust (N)	0.2212	0.1110	0.1102	–

in cosine function, while the flapping motion of both is varied following negative cosine, sine and negative sine. The results, as exhibited in Fig. 22.19 and Table 22.11, show that the synchronous sinusoidal pitching and flapping produce the maximum average values of lift and thrust. Tables 22.12 and 22.13 show the articulation of kinematics in pitching and flapping, for lift and thrust forces, respectively. These results also indicate variation of such oscillatory articulation possibilities that could be further tailored to meet certain objectives.

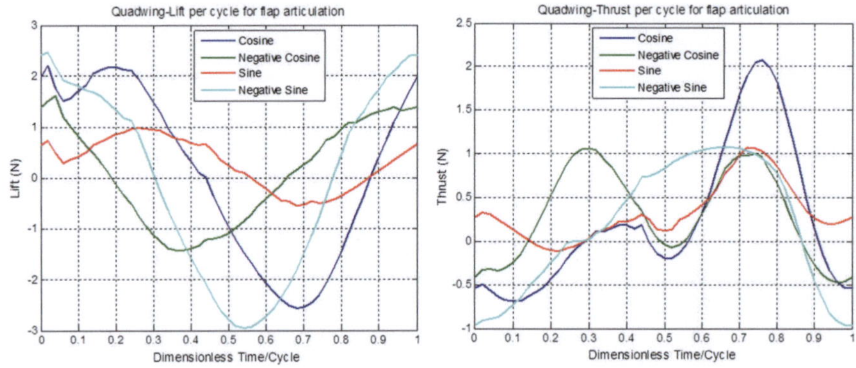

Fig. 22.19 Lift and thrust for quad-wing ornithopter for each flapping kinematics definition (pitching motion in sine function)

Table 22.11 Average lift and thrust for quad-wing ornithopter for each flapping kinematics definition (pitching motion in sine function)

Forces	Quad-wing (fore-wing articulation) (hind-wing in cosine function)			
	Cosine	Negative cosine	Sine	Negative sine
Average lift (N)	0.13057331	0.091443385	0.285027383	– 0.028624158
Average thrust (N)	0.221245293	0.305811394	0.325031883	0.166480894

Table 22.12 Average lift for quad-wing ornithopter for each pitching and flapping kinematics definition

LIFT	FLAP (fore-wing)				
		Cosine	Negative cosine	Sine	Negative sine
FLAP (hind-wing)	Cosine	0.13057331	0.091443385	0.285027383	– 0.028624158
	Negative cosine	0.093518449	0.049941679	0.244836351	– 0.067288766
	Sine	0.285901213	0.246693043	0.439445182	0.13004561
	Negative sine	– 0.028557342	– 0.068838928	0.12563759	– 0.191357656

Table 22.13 Average thrust for quad-wing ornithopter for each pitching and flapping kinematics definition

THRUST	FLAP (fore-wing)				
		Cosine	Negative cosine	Sine	Negative sine
FLAP (Hind-wing)	Cosine	0.221245293	0.305811394	0.325031883	0.166480894
	Negative cosine	0.307394289	0.388278097	0.407754701	0.252427872
	Sine	0.324244072	0.410206416	0.428256265	0.274096829
	Negative sine	0.167752521	0.251135277	0.27130442	0.111445429

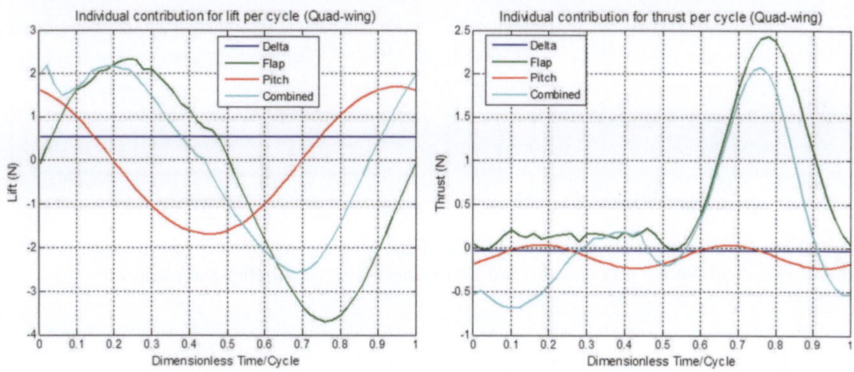

Fig. 22.20 Component-wise contribution for quad-wing

22.3.2.2 Component-Wise Forces for Quad-Wing

Individual contributions of the pitching-flapping motion on the flight performance are assessed. The calculation is performed on semi-elliptical wing.

Results obtained as exhibited in Fig. 22.20 depict similar behavior to the bi-wing cases that the lift is dominated by the incidence angle. For the thrust, the flapping motion has a very dominant influence over the force.

From the above component-wise force analysis, it can be summarized that also a suitable combination of these force elements can be attained in order to generate.

22.3.2.3 Parametric Study for Quad-Wing

A parametric study is carried out to assess the influence of certain flapping-wing motion parameters to the flight performance desired. The study considers the following parameters: the effect of flapping frequency and the gap distance between fore-wing and hind-wing. The results are exhibited in Fig. 22.21, Tables 22.15 and

22.16. It can be seen that the effect of flapping frequency exhibits similar trend as bi-wing's. For the gap distance effect, the lift and thrust increase as the distance increases but in very small magnitude, which is almost negligible (Table 22.14).

Results obtained as exhibited in Fig. 22.22 show the lift produced for various scenarios involving phase combinations between flapping and pitching motions of the individual fore- and hind-wings. Table 22.17 summarizes the average forces per cycle for the selected scenarios.

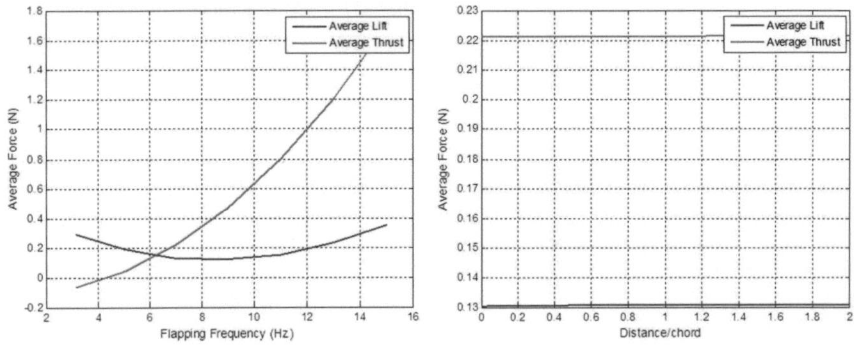

Fig. 22.21 Parametric study on the influence of flapping frequency and gap distance between fore-wing and hind-wing on cyclic lift and thrust (quad-wing, semi-elliptical planform)

Table 22.14 Average lift and thrust (quad-wing) for each individual contribution

Forces	Individual contribution for quad-wing			
	Incidence only	Flap only	Pitch only	Combined
Average lift (N)	0.556235687	– 0.374071356	0.032346548	0.13057331
Average thrust (N)	– 0.036124137	0.62456649	– 0.104059056	0.221245293

Table 22.15 Average lift and thrust with variation of flapping frequency

Forces	Flapping frequency, *f* (Hz)						
	3.14	5	7	9	11	13	15
Average lift (N)	0.29656	0.19699	0.13057	0.12290	0.15600	0.23465	0.35296
Average thrust (N)	– 0.06460	0.04293	0.22125	0.47201	0.79564	1.20212	1.69614

Table 22.16 Average lift and thrust with variation of gap distance between fore-wing and hind-wing

Forces	Distance, *d*			
	No gap	Half-chord	One chord	Two chord
Average lift (N)	0.13057331	0.130673458	0.130733521	0.130773553
Average thrust (N)	0.221245293	0.221402859	0.221497164	0.221559936

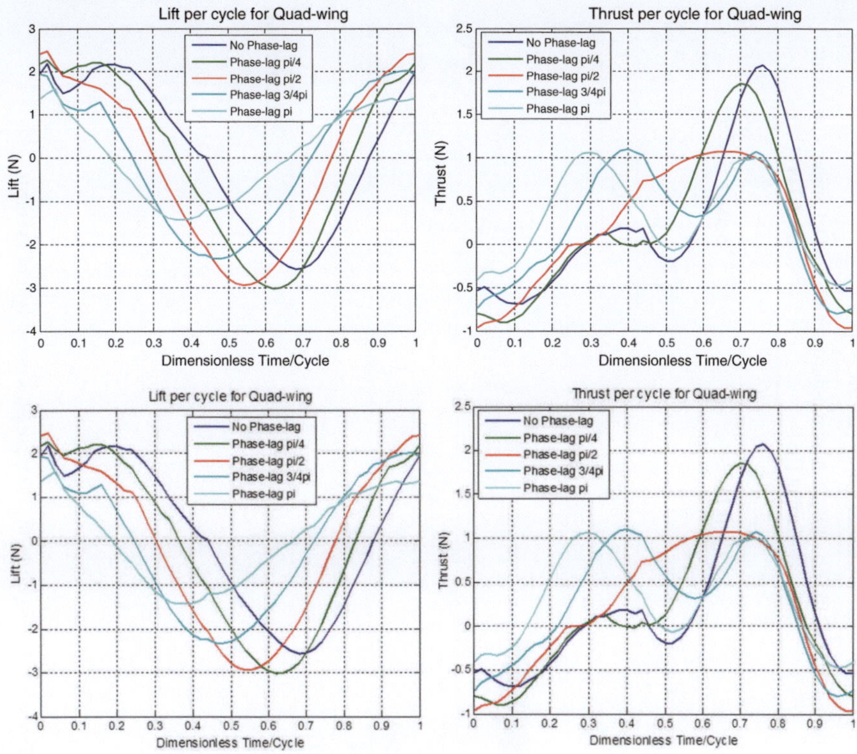

Fig. 22.22 Fore flapping phase lag variation for quad-wing

Table 22.17 Average lift and thrust with variation of flapping phase lag for fore-wing

Forces	Phase lag of fore-wing flapping motion (fore- and hind-wing in cosine function)				
	0	$\pi/4$	$\pi/2$	$3\pi/4$	π
Average lift (N)	0.13057331	0.021852034	– 0.028624573	– 0.010914094	0.09011779
Average thrust (N)	0.221245293	0.175313837	0.166509991	0.227573236	0.30449

A deduction can be made from the results from Table 22.17 that having no phase lag produces the maximum lift and thrust, among the others. In conformity with the observation by Deng and Hu [52] and Alexander [53] in their study, the present study also indicates that when quad-wing insect-like dragonfly performs aggressive maneuvers, they will employ in-phase flight to generate larger aerodynamic forces. However further analysis to optimize the combination of these parameters is still under progress.

Table 22.18 Comparison of basic performance of ornithopter models and birds (extended from earlier work[2] [25, 26, 54])

Ornithopter model	Present work (semi-elliptical)	Previous work (semi-elliptical) [26]	Malik and Ahmad [55]	Byl's hummingbird-scale robot [56]	DeLaurier's pterosaur model [11]
Length	–	–	–	–	–
Wingspan (m)	0.4	0.4	0.4	0.16	5.4864
Aspect ratio	6.36	6.36	6.2	–	–
Wing area (m^2)	0.0251	0.0251	0.0251	–	–
Lift/cycle (N)	0.0662	0.1403	0.1705	0.2000	177.923
Thrust/cycle (N)	0.1110	1.3163	0.0201	–	–
		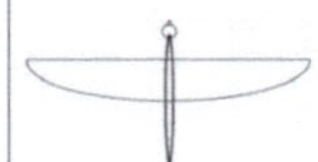	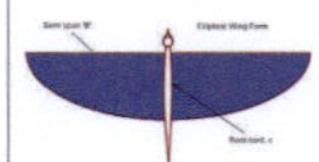	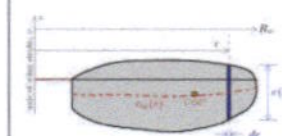	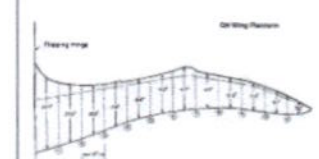
$\frac{\text{Lift(N)}}{\text{Wingspan(m)}}$ (N/m)	0.1655	0.351	0.4263	1.25	32.431
$\frac{\text{Thrust(N)}}{\text{Wingspan(m)}}$ (N/m)	0.2775	3.291	0.0503	–	–
Wing loading (N/m^2)	2.64	5.592	6.79	–	–
$\frac{\text{Lift(N)}}{\text{Aspect Ratio(m)}}$ (N)	0.025	0.022	0.027	–	–

(continued)

[2] For the Ornithopter models, the lift used in the calculation is the Lift/Cycle whereas for the birds, the lift represents the weight of the birds.

Table 22.18 (continued)

BIRD	Wandering albatross	Turkey vulture	Red-tailed Hawk	Bald eagle	Peregrine falcon
Length	110 cm	67 cm (26 in.)	49 cm (19 in.)	79 cm (31 in.)	46 cm (18 in.)
Wingspan	300 cm	171 cm (67 in.)	125 cm (49 in.)	203 cm (80 in.)	116 cm (46 in.)
Aspect ratio	15	7.0	7.1	6.6	8.91
Wing area (m^2)	0.643	0.612	0.188	0.703	0.151
Weight	9.00 kg	1.8 kg (4 lb)	1.082 kg (2.4 lb)	4.3 kg (9.5 lb)	0.952 kg (2.1 lb)

$\frac{\text{Lift(N)}}{\text{Wingspan(m)}}$ (N/m)	29.4	10.5	8.656	21.182	8.206
Wing loading (N/m^2)	137.31	28.84	57.41	60.00	61.80
$\frac{\text{Lift(N)}}{\text{Aspect Ratio(m)}}$ (N)	5.87	2.57	1.52	6.52	1.07

22.4 Comprehensive Assessment of Modeling Result

Better understanding of the production of lift and thrust is intended for current simplified modeling of both bi-wing and quad-wing ornithopters. It is also meant to build a comprehensive foundation and act as a guideline to develop a simple experimental model ornithopter. A more sophisticated computational and experimental prototype can be built in a progressive manner by superposing other significant characteristics. To gain better comprehension into the kinematic and aerodynamic modeling of bi-wing and quad-wing ornithopters, comparison will be made on the basic characteristics and performance of selected ornithopter models with those of selected real birds and insects.

The use of CFD computation to simulate the vorticity field for quad-wing is a complex study as reported by Wang and Russell [20]. Although on average, an upward net force is generated on the wing due to the downward flow created by the wing motion, the computation is not readily related to the computational results for lift and thrust. For future progress, such result could be the basis platform to the present aerodynamics and kinematics modeling of non-deforming quad-wing ornithopter, which can extensively and progressively be further redeveloped and refined to approach the genuine living biosystem flight features, such as those of dragonfly and other related entomopters.

For this purpose, Table 22.18 has been prepared as an extension of the earlier table presented in [26], to obtain an insight of the flight characteristics and basic performance of ornithopter and entomopter models, and birds and insects. Table 22.18 exhibits the ratio of the lift per cycle, thrust per cycle, lift per aspect ratio and the wing loading calculated using the present simplified computational model and those obtained by other investigators; for comparison, the weight per wingspan of a selected sample of birds is also exhibited. Although the comparison is by no means rigorous, it may shed some light on how the geometrical modeling and the flapping motion considered in the computational model may contribute to the total lift produced and how further refinement could be synthesized.

The development carried out in this work is addressed to biomimicry of biosystem flying in the Reynolds number range of 1.0×10^4–1.0×10^5 which is turbulent. The projected ornithopter and MAV will be operating in this range of Reynolds number also. The aerodynamics that has been adopted in the present work takes into account viscous correction appropriately (DeLaurier [11]; Shyy et al. [7]). Shyy et al. [7] show that for all airfoils, the CL/CD ratio exhibits a clear Reynolds number dependency. For Re varying between 7.5×10^4 and 2.0×10^6, CL/CD changes by a factor of 2 to 3 for the airfoils tested.

In the present work, viscosity effects are taken into account following the approach and results of DeLaurier [11], using the computational formulation as given in the present paper as a simplified approach to the problem, but validated through comparison with comparable experimental results range. Such approach can be justified as a preliminary step toward more accurate approach and to develop simple flapping ornithopter MAV.

22.5 Conclusions

The present work has been performed to assess the effect of flapping-pitching motion with pitch-flap phase lag in the flight of ornithopter. In this conjunction, a computational model has been considered, and a generic computational method has been adopted, utilizing strip theory and two-dimensional unsteady aerodynamic theory of Theodorsen with modifications to account for three-dimensional and viscous effects and leading-edge suction. The study is carried out on semi-elliptical wing planforms. The results have been compared and validated with other literatures within similar unsteady aerodynamic approach and general physical data, and within the physical assumptions limitations; encouraging qualitative agreements or better have been indicated, which meet the proof of concept objectives of the present work. For the bi-wing flapping ornithopter, judging from lift per unit span, the present flapping-wing model performance is comparable to those studied by Yu et al. [47]. The analysis and simulation by splitting the flapping and pitching motion show that: (a) the lift is dominantly produced by the incidence angle; (b) the thrust is dominated by flapping motion; and (c) phase lag could be utilized to obtain optimum lift and thrust for each wing configurations.

For the quad-wing ornithopter, at the present stage, the simplified computational model adopted verified the gain in lift obtained as compared to bi-wing flapping ornithopter, in particular by the possibility of varying the phase lag between the flapping and pitching motion of individual wing as well as between the fore- and hind-wings. A structured approach has been followed to assess the effect of different design parameters on lift and thrust of an ornithopter, as well as the individual contribution of the component of motion. These results lend support to the utilization of the generic modeling adopted in the synthesis of a flight model, although more refined approach should be developed. Various physical elements could be considered to develop ornithopter kinematic and aerodynamic modeling, as well as using more refined aerodynamic computation, such as CFD or lifting surface methods. In retrospect, a generic physical and computational model based on simple kinematics and basic aerodynamics of a flapping-wing ornithopter has been demonstrated to be capable of revealing its basic characteristics and can be utilized for further development of a flapping-wing MAV. Application of the present kinematic, aerodynamic and computational approaches shed some light on some of the salient aerodynamic performance of the quad-wing ornithopter.

References

1. Ho, S., H. Nassef, N. Pornsinsirirak, Y.-C. Tai, and C.-M. Ho. 2003. Unsteady aerodynamics and flow control for flapping wing flyers. *Progress in Aerospace Sciences* 39: 635–681.
2. Weis-Fogh, T. 1973. Quick estimates of flight fitness in hovering animals, including novel mechanisms for lift production. *Journal of Experimental Biology* 5: 169–230.

3. Ellington, C.P. 1984. The aerodynamics of hovering insect flight, III. Kinematics. *Philosophical Transactions of the Royal Society of London B* 305: 41–78.
4. Ellington, C.P. 1999. The novel aerodynamics of insect flight: Applications to micro-air vehicles. *The Journal of Experimental Biology* 202: 3439–3448.
5. Ellington, C.P. 1984. The aerodynamics of hovering insect flight, I, quasi-steady analysis. *Philosophical Transactions of the Royal Society of London B* 305: 1–15.
6. Shyy, W., M. Berg, and D. Ljungqvist. 1999. Flapping and flexible wings for biological and micro air vehicles. *Progress in Aerospace Sciences* 35: 455–505.
7. Shyy, W., H. Aono, S.K. Chimakurthi, P. Trizila, C.-K. Kang, C.E.S. Cesnik, and H. Li. 2010. Recent progress in flapping wing aerodynamics and aeroelasticity. *Progress in Aerospace Science* 46 (7): 284–327.
8. Dickinson, M.H., F.O. Lehmann, and S.P. Sane. 1999. Wing rotation and the aerodynamic basis of insect flight. *Science* 284 (5422): 1954–1960.
9. Zbikowski, R. 2002. On aerodynamic modelling of an insect-like flapping wing in hover for micro air vehicles. *Philosophical Transactions of the Royal Society of London A* 360: 273–390.
10. Ansari, S.A., R. Zbikowski, and K. Knowles. 2006. Aerodynamic modelling of insect-like flapping flight for micro air vehicles. *Progress in Aerospace Sciences* 42: 129–172.
11. DeLaurier, J.D. 1993. An aerodynamic model for flapping wing flight. *Aeronautical Journal of the Royal Aeronautical Society* 97: 125–130.
12. Wang, Z.J. 2004. The role of drag in insect hovering. *Journal of Experimental Biology* 207: 4147–4155.
13. Wang, Z.J. 2005. Dissecting insect flight. *Annual Review of Fluid Mechanics* 37: 183–210.
14. Rosenfeld, N.C. 2011. *An Analytical Investigation of Flapping Wing Structures for MAV*. Ph.D. Thesis, University of Maryland.
15. Nicholson, B., S. Page, H. Dong, and J. Slater. 2007. *Design of a Flapping Quad-Winged Micro Air Vehicle*. AIAA-4337.
16. Maybury, W.J., and F.-O. Lehmann. 2004. The fluid dynamics of flight control by kinematic phase lag variation between two robotic insect wings. *Journal of Experimental Biology* 207: 4707–4726.
17. Ratti, J. 2011. *QV-The Quad Winged, Energy Efficient, Six Degree of Freedom Capable Micro Air Vehicle*. Ph.D. Thesis, Georgia Institute of Technology.
18. Prosser, D.T. 2011. *Flapping Wing Design for a Dragon-Fly like MAV*. M.Sc. Thesis, Rochester Institute of Technology.
19. Strang, K.A. 2009. *Efficient Flapping Flight of Pterosaurs*. Ph.D. Thesis, Stanford University.
20. Wang, Z.J., and D. Russell. 2007. Effect of forewing and hindwing interactions on aerodynamic forces and power in hovering dragonfly flight. *Physical Review Letters* 99 (148101).
21. Taha, H.E., M.R. Hajj, and A.H. Nayfeh. 2012. Flight dynamics and control of flapping wing MAV—A review. *Nonlinear Dynamics* 70: 907–939.
22. Taha, H.E., M.R. Hajj, and P.S. Beran. 2014. State-space representation of the unsteady aerodynamics of flapping flight. *Aerospace Science & Technology* 34: 1–11.
23. Norberg, U.M. 1970. Hovering Flight of *Plecotusauritus. L. Bijdr. Dierk* 40: 62–66; *Proceedings of 2nd International Bat Research Conference.*
24. Tedrake, R., Z. Jackowski, R. Cory, J.W. Roberts, and W. Hoburg. 2009. *Learning to Fly Like a Bird, Communications of the ACM*.
25. Djojodihardjo, H., and A.S.S. Ramli. 2012. Kinematic and aerodynamic modelling of flapping wing ornithopter. *Procedia Engineering* 50: 848–863.
26. Djojodihardjo, H., and A.S.S. Ramli. 2013. Kinematic and unsteady aerodynamic modelling, numerical simulation and parametric study of flapping wing ornithopter. In *Proceedings, International Forum on Aeroelasticity and Structural Dynamics*. Bristol.
27. Jones, A.R., N.A. Bakhtian, and H. Babinsky. 2008. Low Reynolds number aerodynamics of leading-edge flaps. *Journal of Aircraft* 45 (1): 342–345.
28. Djojodihardjo, H., and M.A.A. Bari. 2014. *Kinematic and Unsteady Aerodynamic Modelling of Flapping Bi- and Quad-Wing Ornithopter*. Proceedings-ICAS 2014, Paper 2014-0433.

29. Theodorsen, T. 1949. *General Theory of Aerodynamic Instability and the Mechanism of Flutter*. NACA Report No. 496.
30. Jones, R.T. 1940. *The Unsteady Lift of a Wing of Finite Aspect Ratio*. NACA Report No. 681.
31. Polhamus, E.C. 1966. *A Concept of the Vortex Lift of Sharp-Edge Delta Wings Based on a Leading-Edge-Suction Analogy*. NASA TN D-3767.
32. Polhamus, E.C. 1968. *Application of the Leading-Edge-Suction Analogy of Vortex Lift to the Drag Due to Lift of Sharp-Edge Delta Wings*. NASA TN D-4739.
33. Ashley, H., M.T. Landahl, and S.E. Widnall. 1965. New directions in lifting surface theory. *AIAA Journal* 3 (1).
34. Albano, A., and W.P. Rodden. 1969. A doublet-lattice method for calculating lift distributions on oscillating surfaces in subsonic flows. *AIAA Journal* 7 (2): 279–285.
35. Djojodihardjo, R.H., and S.E. Widnall. 1969. A numerical method for the calculation of nonlinear, unsteady lifting potential flow problems. *AIAA Journal* 7 (10): 2001–2009.
36. Murua, J., R. Palacios, and J.M.R. Graham. 2012. Applications of the unsteady vortex-lattice method in aircraft aeroelasticity and flight dynamics. *Progress in Aerospace Sciences* 55: 46–72.
37. Djojodihardjo, H., and A.S.S. Ramli. 2012. Generic and parametric study of the aerodynamic characteristics of flapping wing micro-air-vehicle. *Applied Mechanics and Materials* 225: 18–25.
38. Djojodihardjo, H., M.A.A. Bari, A.S.M. Rafie, and S. Wiriadidjaja. 2014. Further development of the kinematic and aerodynamic modelling and analysis of flapping wing ornithopter from basic principles. *Applied Mechanics and Materials* 629: 9–17.
39. Harmon, R.L. 2008. *Aerodynamic Modelling of a Flapping Membrane Wing Using Motion Tracking Experiments*. M.Sc. Thesis, University of Maryland.
40. Scherer, J.O. 1968. *Experimental and Theoretical Investigation of Large Amplitude Oscillating Foil Propulsion Systems*. Hydronautics, Laurel, Md.
41. Multhopp, H. 1955. *Methods for Calculating the Lift Distribution of Wings (Subsonic Lifting-Surface Theory)*. ARC R&M No. 2884.
42. Garrick, I.E. 1936. *Propulsion of a Flapping and Oscillating Aerofoil*. NACA Report No. 567.
43. Garrick, I.E. 1938. *On Some Reciprocal Relations in the Theory of Nonstationary Flows*. NACA Report No. 629.
44. Kuethe, A.M., and C.-Y. Chow. 1986. *The Finite Wing, Foundations of Aerodynamics*, 4th ed., 145–164. New York: John Wiley.
45. Anderson, J.D. *Fundamentals of Aerodynamics*, 4th ed. New York: McGraw-Hill.
46. Zakaria, M.Y., H.E. Taha, and M.R. Hajj. 2014. *Shape and Kinematic Design Optimization of Pterosaur Replica*. AIAA 2014-2869.
47. Yu, C., D. Kim, and Y. Zhao. 2014. Lift and thrust characteristics of flapping wing aerial vehicle with pitching and flapping motion. *Journal of Applied Mathematics and Physics* 2: 1031–1038.
48. Fujiwara, T., K. Hirakawa, S. Okuma, T. Udagawa, S. Nakano, and K. Kikuchi. 2008. Development of a small flapping robot: Motion analysis during takeoff by numerical simulation and experiment. *Mechanical Systems and Signal Processing* 22 (6): 1304–1315.
49. Chen, M.W., Y.L. Zhang, and M. Sun. 2013. Wing and body motion and aerodynamic and leg forces during take-off in droneflies. *Journal of the Royal Society Interface* 10.
50. Pennycuick, C.J. 1990. Predicting wingbeat frequency and wavelength of birds. *Journal of Experimental Biology* 150: 171–185.
51. Tucker, V.A. 1987. Gliding birds: The effect of variable wing span. *Journal of Experimental Biology* 133.
52. Deng, X., and Z. Hu. 2008. *Wing-Wing Interactions in Dragonfly Flight*. Ine-Web.Org. https://doi.org/10.2417/1200811.1269.
53. Alexander, D.E. 1984. Unusual phase relationships between the forewings and hindwings in flying dragonflies. *Journal of Experimental Biology* 109: 379–383.
54. Djojodihardjo, H., A.S.S. Ramli, M.S.A. Aziz, and K.A. Ahmad. 2013. Numerical modelling, simulation and visualization of flapping wing ornithopter, keynote address. In *International Conference on Engineering Materials and Processes (ICEMAP)*. Chennai.

55. Malik, M.A., and F. Ahmad. 2010. Effect of different design parameters on lift, thrust, and drag of an ornithopter. In *Proceedings of the World Congress on Engineering 2010*, vol. II. London, U.K.
56. Byl, K. 2010. A passive dynamic approach for flapping wing micro aerial vehicle control. In *ASME Dynamic Systems and Controls Conference*.

Chapter 23
Analysis and Computational Study of the Aerodynamics, Aeroelasticity and Flight Dynamics of Flapping-Wing Ornithopter Using Linear Approximation

Abstract The present work addresses the aerodynamics, aeroelasticity and flight dynamics of birdlike bio-inspired bi-wing flapping-wing ornithopter in forward flight. The main interest is to compare the dynamics of rigid and flexible flapping wing in forward flight. First, a generic approach is followed to model the geometry, kinematics and aerodynamics of flapping-wing ornithopter by considering a three-dimensional rigid and thin wing in flapping and pitching motion with and without phase lag. The unsteady aerodynamic approach incorporates viscous effects and leading-edge suction. Next a fundamental representation of unsteady air loads and structural flexibility interaction is developed for the analysis and numerical simulation based on a generic linear aeroelastic analysis using forward speed and oscillatory flapping motion as disturbances, to find out the influence of wing flexibility on its aeroelastic stability and aerodynamic performance. Further, a simplified and generic model flight dynamic model is presented based on using the same aerodynamic approach, by only considering the equation of motion in the plane of symmetry to gain insight for further development of refined model. Parametric studies are carried out both for the aeroelastic and flight dynamic problems to assess the plausibility of the present approach. The elaboration in this chapter follows Djojodihardjo in (Analysis and computational study of the aerodynamics, aeroelasticity and flight dynamics of flapping wing ornithopter using linear approximation, 2016, [6]).

Keywords Aerodynamics · Aeroelasticity · Computational study · Flight dynamics · Flapping wing · Ornithopter

Nomenclature

AR	Aspect ratio
B	Semi-wingspan
c	Chord
$C(k)$	Theodorsen function
$C(k)_{\text{jones}}$	Modified Theodorsen function
C_{df}	Drag coefficient due to skin friction

© The Author(s), under exclusive license to Springer Nature Singapore Pte Ltd. 2023

H. Djojodihardjo, *Introduction to Aeroelasticity*,
https://doi.org/10.1007/978-981-16-8078-6_23

$\mathrm{d}D_{\mathrm{camber}}$	Sectional force due to camber
$\mathrm{d}D_{\mathrm{f}}$	Sectional friction drag
$\mathrm{d}F_x$	Sectional chordwise force
Fx	X (Horizontal) component of the resultant pressure force acting on the vehicle
Fz	Z (Vertical) component of the resultant pressure force acting on the vehicle
f, g	Generic functions
h	Height
i	Time index during navigation
j	Waypoint index
$\mathrm{d}L$	Sectional lift
$\mathrm{d}y$	Width of sectional strip under consideration
$\mathrm{d}N$	Sectional total normal force
$\mathrm{d}N_{\mathrm{c}}$	Sectional circulatory normal force
$\mathrm{d}N_{\mathrm{nc}}$	Sectional apparent mass effect
$\mathrm{d}t$	Time step
$\mathrm{d}T$	Sectional thrust
$\mathrm{d}T_{\mathrm{s}}$	Leading-edge suction force
$F(k)$	Theodorsen function real component
$G(k)$	Theodorsen function imaginary component
$\dot{h} \equiv \frac{\mathrm{d}h}{\mathrm{d}t}$	Plunging rate
k	Reduced frequency
L	Total lift
L_{fore}	Lift force of fore-wing
t	Time
T	Total thrust
U	Flight velocity
V	Relative velocity at quarter-chord point
V_x	Flow speed tangential to section
V_{rel}	Relative velocity
V_i	Induced velocity
w_0	Downwash velocity at ¾-chord point
Γ	Circulation
ρ	Air density
β	Flapping angle
β_0	Maximum flapping angle
θ	Pitching angle
$\dot{\theta}$	Pitching rate
$\ddot{\theta}$	Pitching acceleration
θ_0	Maximum pitch angle
$\theta_{\mathrm{hindwing}}$	Effective pitching angle of hind-wing
θ_{f}	Angle of flapping axis with respect to flight velocity (incidence angle)
θ_{p}	Mean pitch angle of chord with respect to flapping axis
ϕ	Lag angle between pitching and flapping angle

δ	Incidence angle
α	Relative angle of attack
α'	Flow's relative angle of attack at three-quarter-chord point (DeLaurier)
α_0	Zero-lift angle
$\alpha_{\text{Theodorsen}}$	Phase angle of Theodorsen function
η_s	Efficiency coefficient
ω	Flapping frequency

23.1 Introduction

The development of a flapping-wing ornithopter mimicking flapping biosystems can be carried out in a progressive manner, by first looking into the geometric, kinematic and aerodynamic characteristics, and may be followed by the incorporation of the flexibility features. In view of the complexity of the fluid-structure interaction of a flexible flapping wing, earlier work has been focused on rigid wings, deferring the aeroelastic analysis after a comprehensive understanding and model has been gained on the geometric, kinematic and aerodynamic model of the rigid flapping ornithopter. The biological flapping flyers have flexible wings with anisotropic flexibility in both spanwise and chordwise directions, utilizing complicated wing motions of flapping, twisting, folding, rotating motions or area expansion and contraction [1–8]. The passive or active deformation of the wing contributes to the generation of appropriate aerodynamic performances according to various flight modes. The artificial flapping flyers inspired from the biological flappers also have thin and flexible passive wings structurally similar to those of insects. Therefore, for the optimal design and the real-time control of flapping-wing flight, an efficient aerodynamic model applicable to general flapping wings is necessary, and an efficient aeroelastic analysis method should be also developed. Wing flexibility may be desirable for improving the flapping-wing ornithopter aerodynamic performance, and passive and active stability. Hence, following earlier studies carried out [9–13], a generic approach is first followed to model the geometry, kinematics and aerodynamics of flapping-wing ornithopter; considerations are given to the motion of a three-dimensional rigid and thin wing in flapping and pitching motion with and without phase lag. Basic Unsteady Aerodynamic Approach incorporating viscous effect and leading-edge suction is utilized as baseline approach. The study is focused on a Bi-Wing ornithopter. Parametric study is carried out to reveal the flapping Bi-Wing ornithopter aerodynamic characteristics and for comparative analysis with various selected simple models in the literature. Further analysis is carried out by differentiating the pitching and flapping motion and studying its respective contribution to the flight forces.

Chimakurti [6] mentioned in his work that wing flexibility is found to have a favorable effect on lift generation. However, Zhang et al. [14], in their experimental investigation on the effects of flapping-wing aspect ratio and flexibility on its aerodynamic performance, show that the effect of flexibility reduces the lift. Dai et al. [4], in their study on the aerodynamics and aeroelasticity of flapping wings, give results that indicate that the lift decreases significantly with flexibility. It is then of great interest to understand the overriding flexibility factors that influence the flight performance of the flapping-wing biosystem or ornithopter.

Starting with basic fundamentals and simplified approaches, a fundamental representation of unsteady air loads and structural flexibility interaction is developed for the analysis and numerical simulation based on a generic linear aeroelastic analysis using forward speed and oscillatory flapping motion as disturbances. The aeroelastic characteristics of the flapping-wing ornithopter are investigated following an approach elaborated by Djojodihardjo and Yee [7]. By representing the wing as its typical section, the influence of the flexibility of the wing on its aeroelastic stability characteristics as well as on the aerodynamic performance is investigated. The series of work already carried out in references [9–13] will be used as the baseline for carrying out the aerodynamic performance of a flexible ornithopter wing.

Based on the results obtained in the first part of the work, the second part of the work will look at the influence of the flexibility of the wing. For this purpose, the flexibility studies carried out in the first part of the study are used to generate a heuristic model of the influence of the aeroelastic characteristics of the wing on its aerodynamic characteristics, in particular lift and thrust. The results will then be assessed based on physical framework and other available results and approaches in the literature. Further elaborate work will be discussed. Incorporating the tail within the model of flying biosystem can assist in generating aerodynamic force that enhances the flight control of the ornithopter, but greatly reduce or eliminate the ornithopter ability to hover. Such notion grows with smaller size of vehicle, as the control surfaces shrink and significantly reduce the aerodynamic efficiency of the control surfaces, and limiting their ability to generate adequate control forces and moments [1]. However, since fixed tails may be suitable to control the bird-sized ornithopter, it will be considered in the present study. It will be of interest, however, to assess the relative capabilities of flapping wing and fixed tail control. With such motivation, using results from previous studies [11, 13], the present paper addresses the flight dynamics of a generic bi-wing ornithopter, first without and afterward with a tail. The objective of the study is to design a simplified strategy for the articulation of the wing, in a simplified but basic geometry and characteristics of the body, wing and tail of the ornithopter, and evaluate the lift and thrust at various pitching and flapping phase lag values. After establishing the flapping-wing ornithopter configuration and kinematics, the detailed equations of motion derived in [11, 13] are applied in the flight mechanics of the bird-wing flapping-wing ornithopter. In the present work, preliminary development is carried out by only considering the flight mechanics of the bird-wing system flight in the plane of symmetry in the local horizon coordinate frame of reference. Based on the developed model, parametric study is carried out and flapping strategies are elaborated and numerically simulated, followed by assessment on the design and articulation strategy of some generic flapping ornithopter. At the present stage, which is addressed on bi-wing ornithopter mimicking bird's forward flight, the present work does not incorporate leading-edge vortex and stall effects.

23.2 Theoretical Development of the Aerodynamic, Aeroelastic and Flight Dynamic Modeling of Flapping Wings

23.2.1 Aerodynamic Model

The present aerodynamic approach is synthesized using basic foundations that may exhibit the generic contributions of the motion elements of the bio-inspired bi-wing and quad-wing air vehicle characteristics. These are the strip theory and thin wing aerodynamic approach [15–19], and Jones' modified Theodorsen approach [20] which incorporates Garrick's leading-edge suction [15]. The computation of lift and thrust generated by pitching and flapping motion of three-dimensional rigid wing is carried out in a structured approach. Later, the computational model will take into account certain physical parameters that can be identified via observations and established results of various researchers.

To obtain an insight into the mechanism of lift and thrust generation, Djojodihardjo and Ramli [10, 11] and Djojodihardjo and Bari [12, 13] analyzed the wing flapping motion by looking into the individual contribution of the pitching, flapping and coupled pitching-flapping to the generation of the aerodynamic forces. Also the influence of the variation of the forward speed, flapping frequency and pitch-flap phase lag has been analyzed. Such approach will also be followed here through further scrutiny of the motion elements.

The flapping motion of the wing is distinguished into three distinct motions with respect to the three axes; these are (a) *Flapping*, which is up and down plunging motion of the wing; (b) *Feathering* is the pitching motion of wing and can vary along the span; and (c) *Lead-lag*, which is in-plane lateral movement of wing, as incorporated in Fig. 23.1. For further reference to the present work, the lead-lag motion could be interpreted to apply to the phase lag between pitching and flapping motion, while the fore-and-aft movement can be associated with the orientation of the stroke plane. The degree of freedom of the motion is also depicted in Fig. 23.1. The flapping angle β varies as a sinusoidal function and pitching angle θ are given by the following equations.

$$\beta(t) = \beta_0 \cos \omega t \tag{23.1}$$

$$\theta(t) = \theta_0 \cos (\omega t + \phi) + \theta_{fp} \tag{23.2}$$

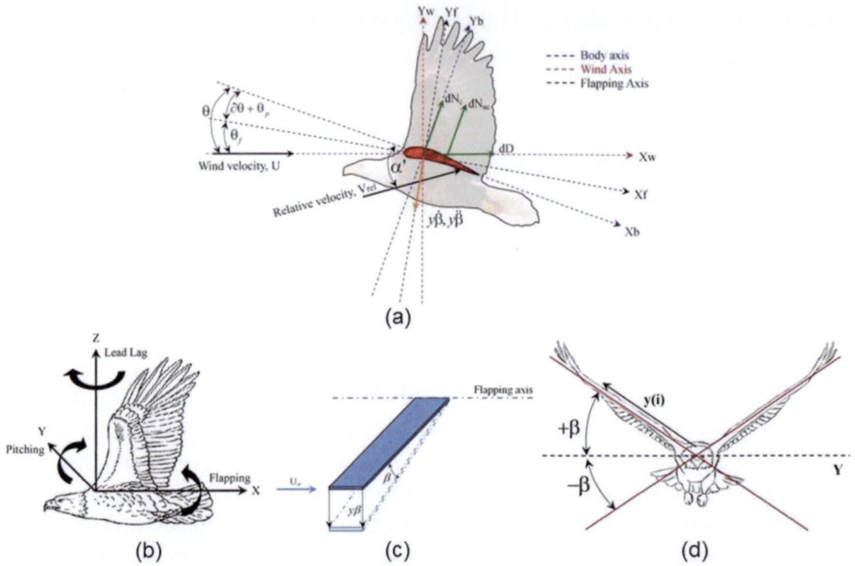

Fig. 23.1 **a** Forces on section of the wing. **b–d** Flapping and pitching motion of flapping wing

where θ_0 and β_0 indicate maximum value for each variables, ϕ is the phase lag between pitching and flapping angle and y is the distance along the span of the wing, and θ_{fp} is the sum of the flapping axis angle with respect to flight velocity (incidence angle) and the mean angle of the chord line with respect to the flapping axis, as exhibited in Fig. 23.1.

The present method is exemplified by the use of elliptical planform wing. As a baseline, by referring to Eqs. 23.1 and 23.2, β and θ are considered to oscillate following a cosine function; such scheme indicates that these motions start from specified values. A different scheme, however, can be adopted.

Leading-edge suction is included following the analysis of Polhamus [21] and DeLaurier's approximation [5]. Three-dimensional effects will later be introduced by using Scherer's modified Theodorsen-Jones Lift Deficiency Factor [22], in addition to the Theodorsen unsteady aerodynamics [23] and its three-dimensional version by Jones [20]. Further refinement is made to improve accuracy. Following Multhopp approach [24] simplified physical approach to the general aerodynamics of arbitrary planform is adopted, i.e. a lifting line in the quarter-chord line for calculating the downwash on the three-quarter-chord line for each strip.

In the present analysis no linear variation of the wing's dynamic twist is assumed for simplification and instructiveness. However, in principle, such additional requirements can easily be added due to its linearity. The total normal force acting is perpendicularly to the chord line and is given by

$$\mathrm{d}N = \mathrm{d}N_{\mathrm{c}} + \mathrm{d}N_{\mathrm{nc}} \tag{23.3}$$

which consists of the circulatory normal force for each section, acts at the quarter chord and is also perpendicular to the chord line, given by

$$dN_c = \frac{\rho U V}{2} C_n(y) c dy \tag{23.4}$$

and the apparent mass effect that is perpendicular to the wing, acts at mid-chord, and can be calculated as

$$dN_{nc} = \frac{\rho \pi c^2}{4} \left(U\dot{\alpha} - \frac{1}{4} c\ddot{\theta} \right) dy \tag{23.5}$$

The total chordwise force, dF_x, is accumulated by three force components; these are the leading-edge suction, force due to camber and chordwise friction drag due to viscosity effect. All of these forces are acting along and parallel to the chord line.

$$dF_x = dT_s - dD_{camber} - dD_f \tag{23.6}$$

The leading-edge suction, dT_s, following Garrick [15], is given by

$$dT_s = 2\pi \eta_s \left(\alpha' + \theta_{fp} - \frac{1}{4} \frac{c\dot{\theta}}{U} \right) \frac{\rho U V}{2} c dy \tag{23.7}$$

while following DeLaurier [5] the chordwise force due to camber and friction is, respectively, given by

$$dD_{camber} = 2\pi \left(\alpha' + \theta_{fp} \right) \frac{\rho U V}{2} c dy \tag{23.8}$$

$$dD_f = \frac{1}{2} \rho V_x^2 C_{d_f} c dy \tag{23.9}$$

The efficiency term η_s is introduced for the leading-edge suction dT_s to account for viscosity effects. The vertical force dN and the horizontal force dF_x at each strip dy will be resolved into those perpendicular and parallel to the free-stream velocity, respectively. The resulting vertical and horizontal components of the forces is then given by

$$dL = dN \cos(\theta) + dF_x \sin(\theta) \tag{23.10}$$

$$dT = dF_x \cos(\theta) - dN \sin(\theta) \tag{23.11}$$

To obtain a three-dimensional lift for each wing, these expressions should be integrated along the span, b; hence

$$L = \int_{0}^{b} \mathrm{d}L\mathrm{d}x \tag{23.12}$$

$$T = \int_{0}^{b} \mathrm{d}T\mathrm{d}x \tag{23.13}$$

For later comparison with appropriate results from the literature, numerical computations are performed using the following wing geometry and parameters: the wingspan of 40 cm, aspect ratio of 6.36, flapping frequency of 7 Hz, total flapping angle of 60°, forward speed of 6 m/s, maximum pitching angle of 20° and incidence angle of 6° and there is no wing dihedral angle.

In the calculation, both the pitching and flapping motions are in cosine function by default, which is subject to parametric study, and the upstroke and downstroke have equal time duration. The wake capture has not been accounted for in the current computational procedure. The computational scheme developed and the aerodynamic forces for bi-wing have been validated and verified satisfactorily in previous work (Djojodihardjo et al. [11, 13]).

To simplify the modelling scheme of flight dynamics, rectangular shape is chosen for this study for convenience of baseline analysis. In agreement with the quasi-steady model, based on observation on flying birds, it can be assumed that the flapping frequencies are sufficiently slow that shed wake effects are negligible, as in medium-to large-sized birds. The present unsteady aerodynamic approach is synthesized using basic foundations that may exhibit the generic contributions of the motion elements of the bio-inspired bi-wing air vehicle characteristics.

The computational logic in the present work is summarized in the flowchart exhibited in Fig. 23.2. To obtain insight into the mechanism of lift and thrust generation of flapping and pitching motion, Djojodihardjo et al. [12, 13] analyzed the wing flapping motion by looking into the individual contribution of the pitching, flapping and coupled pitching-flapping to the generation of the aerodynamic forces, as well as the influence of the variation of the forward speed, flapping frequency and pitch-flap phase difference. Such approach will also be followed here through further scrutiny of the motion elements. The generic procedure is synthesized from the pitching-flapping motion of rigid wing developed by DeLaurier [5].

23.3 Synthesis of Aeroelastic Approach

In the consideration of the aeroelastic effect of flapping ornithopter, a simplified two-dimensional approach for the typical section of the flapping-wing ornithopter will be utilized following earlier approach elaborated in [13].

Following the Unsteady Aerodynamics elaborated in the previous section, the lift L and moment M of the wing as given by Eqs. 23.12 and 23.13 could be taken to be

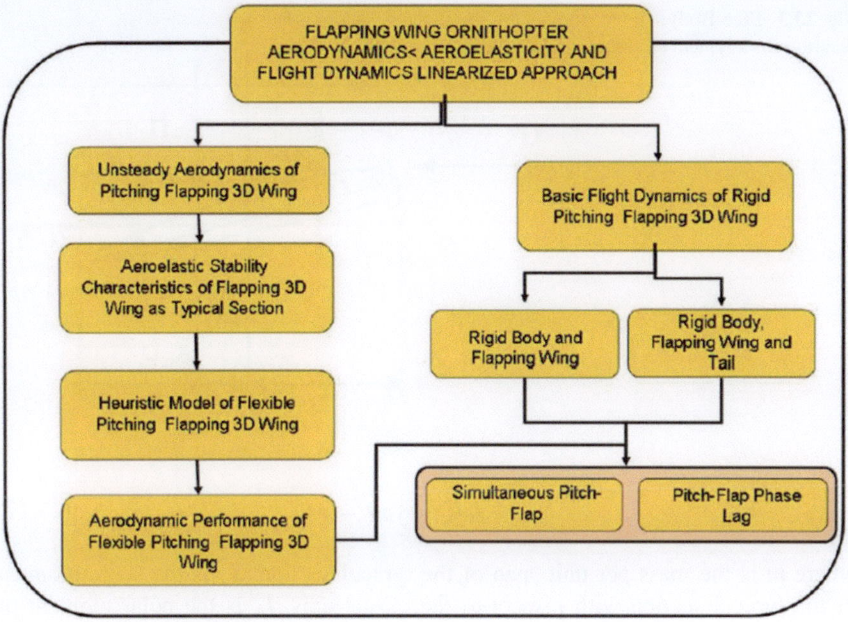

Fig. 23.2 Ornithopter flapping-wing aerodynamics computational scheme

acting on the typical section. To start the analysis, one should look at the dynamics of the wing represented by the typical section following the Lagrange equation (by referring to Fig. 23.3 [7, 25, 26]):

$$-\frac{\mathrm{d}}{\mathrm{d}t}\left(m\ddot{h} + S_\alpha\ddot{\alpha} + K_\alpha\alpha = -L\right) - K_\alpha\alpha + M_y = 0I_\alpha\dot{\alpha}$$
$$m\ddot{h} + S_\alpha\ddot{\alpha} + K_\alpha h = -L$$
$$S_\alpha\ddot{h} + I_\alpha\ddot{\alpha} + K_\alpha\alpha = M_y$$
$$I_\alpha\ddot{\alpha} + c_\alpha\dot{\alpha} + k_\alpha = M$$

$$-\frac{\mathrm{d}}{\mathrm{d}t}\left(m\dot{h} + S_\alpha\dot{\alpha}\right) - K_\mathrm{h}h - L = 0 \quad (23.14)$$

$$-\frac{\mathrm{d}}{\mathrm{d}t}\left(S_\alpha\dot{h} + I_\alpha\dot{\alpha}\right) - K_\alpha\alpha + M_y = 0 \quad (23.15)$$

which can be reduced to the general governing equation for such two-dimensional pitching and flapping (heaving) aerodynamic section given by:

$$m\ddot{h} + S_\alpha\ddot{\alpha} + K_\alpha h = -L \quad (23.16)$$

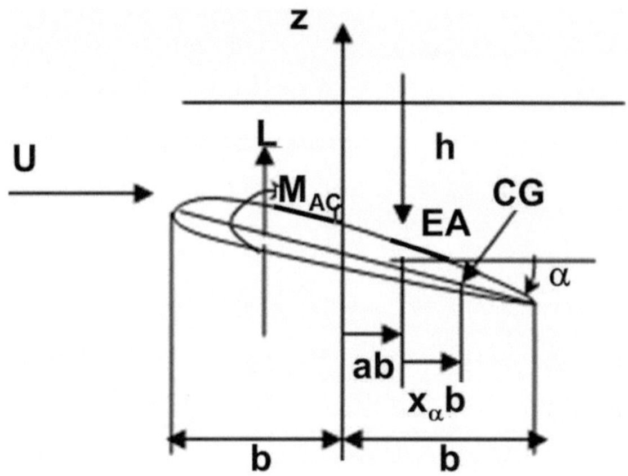

Fig. 23.3 Free-Body diagram for a typical section

$$S_\alpha \ddot{h} + I_\alpha \ddot{\alpha} + K_\alpha \alpha = M_y \tag{23.17}$$

where m is the mass per unit span of the typical section, S_α is the static moment of the typical section with respect to the elastic axis, I_α is the polar moment of inertia of the typical section with respect to the elastic axis, and K_{h} and K_α are the bending (heaving) and torsional spring stiffness, respectively, of the typical section. The typical section experiences movement in two degrees of freedom, i.e. h, heaving (bending) displacement in the vertical direction (positive downward), and α, pitching angular displacement (positive nose-up). L and M_{AC} are the aerodynamic Lift and Moment, respectively; both L and M_{AC} are acting on the aerodynamic center (L positive upward, M_{AC} positive nose-up).

Since the Eqs. 23.16 and 23.17 are written for a two degree-of-freedom pitching and heaving typical section, one may consider that the ornithopter wing is represented by a typical section. This implies that the inertial and other related properties of the wing are considered to be "collapsed" at the typical section.

Noting that the wing is essentially following pitching and flapping motion for its aerodynamic performance, then the elasticity of the wing could be considered to modify the aerodynamic pitching and flapping. In other words, each of the pitching and flapping motions has two components, the motion based and the elasticity based.

Hence the dynamics of the flexible pitching and flapping wing can be identified with the parameters:

$$\alpha_{\text{flex - wing}} = \alpha_{\text{rigid}} + \alpha_{\text{el}} \tag{23.18}$$

$$h_{\text{flex - wing}} = h_{\text{rigid}} + h_{\text{el}} \tag{23.19}$$

The rigid part can be evaluated using earlier approach [8–13]:

$$\alpha_{\text{rigid}} = \alpha \tag{23.20}$$

$$\alpha_{\text{rigid}} = \alpha \tag{23.21}$$

Oscillatory motion assumes that

$$\alpha_{\text{el}} = (\alpha_{\text{el}})_0 e^{i\omega t} \tag{23.22}$$

$$h_{\text{el}} = (h_{\text{el}})_0 e^{i\omega t} \tag{23.23}$$

where ω_{el} is the harmonic frequency, due to the flexibility of the wing structure (and not to be confused with the flapping or pitching frequency). Substituting into the dynamic equation of motion,

$$m\ddot{h}_{\text{flex}} + S_\alpha \ddot{\alpha}_{\text{flex}} + K_\alpha h_{\text{flex}} = -L(\alpha_{\text{flex}}, h_{\text{flex}}) \tag{23.24}$$

$$S_\alpha \ddot{h}_{\text{flex}} + I_\alpha \ddot{\alpha}_{\text{flex}} + K_\alpha \alpha_{\text{flex}} = M_y(\alpha_{\text{flex}}, h_{\text{flex}}) \tag{23.25}$$

or

$$\begin{bmatrix} S_\alpha \omega^2 & m\omega^2 + K_h \\ I_\alpha \omega^2 + K_\alpha & S_\alpha \omega^2 \end{bmatrix} \begin{Bmatrix} \alpha_{\text{flex}} \\ h_{\text{flex}} \end{Bmatrix} = \begin{Bmatrix} -L(\alpha_{\text{flex}}, h_{\text{flex}}) \\ My(\alpha_{\text{flex}}, h_{\text{flex}}) \end{Bmatrix} \tag{23.26}$$

The terms on the left-hand side of the equation lead to an eigenvalue problem, which can be solved to yield the eigenfrequencies and eigenmodes:

$$\begin{bmatrix} S_\alpha \omega^2 & m\omega^2 + K_h \\ I_\alpha \omega^2 + K_\alpha & S_\alpha \omega^2 \end{bmatrix} = \begin{Bmatrix} 0 \\ 0 \end{Bmatrix} \tag{23.27}$$

Using the dynamic response relationship of the flexible wing due to aerodynamic and other exciting forces, one will be able to evaluate the total prevailing aerodynamic angle of the flapping and pitching motion.

23.4 Typical Section Representation of Flapping Wing for Aeroelastic Analysis

For aeroelastic analysis, two different aerodynamic approximations, with increasing complexity, will be utilized; these are the Quasi-Steady and classical unsteady (harmonic) aerodynamics of Theodorsen [23]. First, to gain insight into the problem, the simplest approach, the Quasi-Steady Aerodynamic Model, will be followed. For the quasi-steady aerodynamic model, the aerodynamic Lift L and Moment M_{AC}, as well known in steady linearized aerodynamics, are given by:

$$L(t) = qSC_{L_\alpha}\alpha(t) \tag{23.28}$$

and

$$M_{\mathrm{AC}} = 0 \tag{23.29}$$

Since linearized aerodynamics is used, the airfoil essentially is regarded as a flat plate. From Fig. 23.3, the aerodynamic moment with respect to the elastic axis is given by

$$M_{\mathrm{EA}} = 2Leb + M_{\mathrm{AC}} = 2qSebC_{L_\alpha}\alpha(t) \tag{23.30}$$

Hence Eqs. 23.16 and 23.17 can be rewritten as

$$m\ddot{h} + S_\alpha\ddot{\alpha} + K_h h + qSC_{L_\alpha}\alpha = 0 \tag{23.31}$$

$$S_\alpha\ddot{h} + I_\alpha\ddot{\alpha} + K_\alpha\alpha - 2qScbC_{L_\alpha}\alpha = 0 \tag{23.32}$$

which is known as the flutter equation (since in the form given by Eqs. 23.31 and 23.32, or the following equation, Eqs. 23.33, 23.34, 23.35 and 23.36, this is an eigenvalue or stability equation). In matrix notation, this is given by:

$$[\mathbf{M}]\{\mathbf{R}\} + ([\mathbf{K}] - q[\mathbf{A}_0])\{\mathbf{x}\} = \{0\} \tag{23.33}$$

where

$$[\mathbf{M}] = \begin{bmatrix} m & S_\alpha \\ S_\alpha & I_\alpha \end{bmatrix} \tag{23.34}$$

$$[\mathbf{K}] = \begin{bmatrix} K_h & 0 \\ 0 & K_\alpha \end{bmatrix} \tag{23.35}$$

$$[\mathbf{A}_0] = \begin{bmatrix} 0 & SC_{L_\alpha} \\ 0 & 2SebC_{L_\alpha} \end{bmatrix} \tag{23.36}$$

For convenience, following the practice in aeroelasticity, the analysis of flutter stability can be obtained by assuming a solution of the form:

$$h = \hat{h}^{pt} \tag{23.37a}$$

$$\alpha = \hat{\alpha}^{pt} \tag{23.37b}$$

and

$$\{\mathbf{x}\} = \{\hat{\mathbf{x}}\}e^{pt} \tag{23.38}$$

Solving as eigenvalue problem, going through the algebra will result in the flutter stability characteristic equation given by:

$$(I_\alpha m - S_\alpha)p^4 + \left(K_\alpha m - 2qSebC_{L_\alpha}m + K_\mathrm{h}I_\alpha - qSC_{L_\alpha}S_\alpha\right)p^2 + \left(K_\alpha K_\mathrm{h} - 2qSebC_{L_\alpha}K_\mathrm{h}\right) = 0 \tag{23.39}$$

which has a general form of:

$$a_4p^4 + a_2p^2 + a_0 = 0 \tag{23.40}$$

where

$$a_4 = I_\alpha m - S_\alpha^2 \tag{23.41a}$$

$$\begin{aligned} a_2 &= m\left(K_\alpha - 2qebSC_{L_\alpha}\right) + I_\alpha K_\mathrm{h} - S_\alpha qSC_{L_\alpha} \\ &= mK_\alpha + I_\alpha K_\mathrm{h} - (2meb - S_\alpha)qSC_{L_\alpha} \end{aligned} \tag{23.42b}$$

$$a_0 = K_\mathrm{h}\left(K_\alpha - 2qSebC_{L_\alpha}\right) \tag{23.43c}$$

The characteristic equation is a fourth-order polynomial which has four roots;

$$\begin{aligned} p_{1,2,3,4} &= (\sigma + i\omega) \\ &= \pm\sqrt{\frac{1}{2a}\left(-a_2 \pm \sqrt{a_2^2 - 4a_4a_0}\right)} \end{aligned} \tag{23.44}$$

and the solution is given in the form

$$\{\mathbf{x}\} = \{\hat{\mathbf{x}}\}e^{\sigma t}e^{i\omega t} \tag{23.45}$$

where σ is damping, ω is frequency and $\{\hat{x}\}$ Vibration mode represents the displacement vector.

Following Done [25, 26], as utilized by Djojodihardjo [13], the solution can be conveniently and comprehensively represented by damping and frequency diagrams as functions of either dynamic pressure q, or reduced frequency (or reduced velocity U_R) $k_\mathrm{R} = U/(b.\omega)$, or velocity, as illustrated subsequently, or summarized in a table. The table allows the classification of solution according to the values of the coefficients a_0, a_2, a_4 and the stability categories that result. The stability of motion depends on the value of σ (aerodynamic damping). As can be concluded from Eq. (23.45), and summarized in Table 23.1, if $\sigma > 0$, then the displacement vector $\{x\}$ will oscillate with increasing amplitude in time, and the resulting motion will be

Table 23.1 Flutter stability solution categories[1]

$a_2^2 - 4a_4a_0$	> 0			< 0
a_0	> 0		< 0	>< 0
a_2	> 0	< 0	>< 0	>< 0
p2	$-\omega_1^2, -\omega_2^2$	$-\sigma_1^2, -\sigma_2^2$	$\sigma^2, -\omega^2$	$-g + ih$
p	$\pm i\omega_1, \pm i\omega_2$	$\pm i\sigma_1, \pm i\sigma_2$	$\pm \sigma, \pm i\omega$	$\pm \sigma, \pm i\omega$
Type of motion	Harmonic	Aperiodic	Aperiodic	Oscillatory
	2 pos freq	2 diverging	1 diverging	1 div.pos.freq
	2 neg freq	2 converging	1 converging	1 conv.pos.freq
			Harmonic	1 div.neg.freq
			1 pos freq	1 conv.neg.frq
			1 neg freq	
Type of instability	**Neutral**	**Divergent**	**Divergent**	**Flutter**
Category	**I**	**III**	**IV**	**II**

unstable. If $\sigma = 0$, a neutrally stable oscillation will result. Only when $\sigma > 0$ the oscillation will subside in time.

23.5 Aeroelastic Analysis of Flapping-Wing Ornithopter Represented as Typical Section with Low-Frequency Aerodynamics

For the purpose of aeroelastic modeling and assessment using Low-Frequency Aerodynamic model approach, the Ornithopter Flapping-Wing Model elaborated in References [9–14] is represented by a typical section.

Based on the data utilized, baseline sectional properties of the typical section model have been evaluated and tabulated in Table 23.2. Based on the data given there, some parametric study can be carried out to obtain a favorable configuration and aeroelastic configuration.

[1] Adapted from Done G. T. S., The Flutter and Stability of Undamped Systems, School of Engineering Science, University of Edinburgh, Report and Memoranda No. 3553, November 1966, and Done, G. T. S., A study of binary flutter roots using a method of system synthesis, Aeronautical Research Council, 1967, R&M No. 3554.

Table 23.2 Flapping-wing typical section characteristic of the ornithopter flapping-wing model

Parameter	Unit	Values
Mass	**kg**	2.5480 E − 04
Span	**m**	0.4
c	**m**	0.08
b	**m**	0.04
e	**–**	− 0.02
Omega-alpha (ω_α)	**rad/s**	5.9178 E + 04
Omega-h (ω_h)	**rad/s**	5.9178 E + 04
ρ	**kg/m^3**	1.225
$C_{L\text{-alpha}}$ ($C_{L\alpha}$)	–	6.2832
y	**M**	0.2
E	**GPa**	2.9000 E + 09
I	**m^4**	8.3333 E + 10
G	**Gpa**	1.1154 E + 09

23.6 Computational Results

The computations are performed using the following wing geometry and parameters: the wingspan of 40 cm, chord length of 8 cm and the wing shape is rectangular. The data for the typical section representing the flapping ornithopter wing is tabulated in Table 23.2. Several simplifying assumptions have been made in order to obtain some insight into the flexibility characteristics of the biomimicking ornithopter flapping wing.

The elastic properties listed there are based on keratin [27]. The results as shown in Fig. 23.4 and Table 23.3 indicate characteristics typical of the second column of Table 23.1, which will not lead to aeroelastic instability. In addition, the results also show that the prevailing eigenfrequencies estimated using the quasi-steady aerodynamics in the operational range of the flapping ornithopter are much smaller than the pitching and flapping frequency of the ornithopter wing.

Such conclusion is considered reasonable and in confirmation with observation on biosystem. In addition, the flexural property as represented by K_α shows that, if quasi-steady aerodynamics is assumed, the elastic deflection due to the prevailing aerodynamic force as calculated using the unsteady aerodynamics elaborated in Sect. 23.2 will produce at most 5% change in θ or α'. Such situation is taken into consideration in establishing a heuristic model as elaborated in succeeding section.

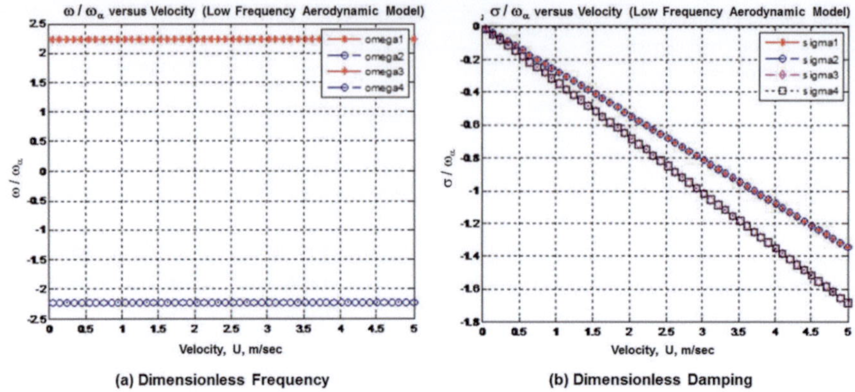

Fig. 23.4 Numerical computation to determine the flutter stability of the low-frequency model of ornithopter wing typical section. Dimensionless frequency and damping are plotted against velocity

Table 23.3 Computational results of aeroelastic stability characteristics—low-frequency aerodynamic model

x_α		0.5000	r_α	0.0015	
Mass		2.5480 E − 4		3.4106 E − 07	
S_α		5.0960 E − 06		1.1944 E = 03	
K_n		906.2500		1.1593 e + 06	
a_1		3.4122 E + 05		0.0561	
a_3		9.5106 E − 05		6.0933 e − 11	
Determinant		0.0029	V	− 0.9972	− 0.9972
				0.0747	− 0.0747
D	1.0 E + 29 multiplied to				
	− 0.1079	0			
	− 0	5.5821			
Matrix A	1.0 E + 30 multiplied to				
	0.2737	3.800			
	0.0213	0.2737			

23.6.1 *Theodorsen Unsteady Aerodynamic Aeroelastic Analysis of Flapping-Wing Ornithopter Model Represented as Typical Section*

Theodorsen unsteady aerodynamic model

The unsteady aerodynamic forces for harmonic motions can be represented by [2, 29]

$$\frac{\hat{L}}{\pi\rho b^3} = \omega^2\left\{\frac{k_a}{k}\left(\frac{\hat{h}}{b} - (0.5+a)\hat{\alpha}\right) + \frac{k_b}{k^2}\hat{\alpha}\right\} \tag{23.46a}$$

$$\frac{\hat{M}_{AC}}{\pi\rho b^4} = -\omega^2\left\{\frac{m_a}{k^2}\left(\frac{\hat{h}}{b} - (0.5+a)\hat{\alpha}\right) + \frac{m_b}{k^2}\hat{\alpha}\right\} \tag{23.46b}$$

where

$$\begin{aligned}
k_\alpha &= k'_\alpha + ik''_\alpha \\
k'_\alpha &= -2kG(k) \\
k''_\alpha &= -2kF(k) \\
k_b &= k'_b + ik''_b \\
k'_b &= -2F(k) - 2kG(k) - \frac{k^2}{2} \\
k''_b &= 2kF(k) + 2G(k) + k \\
m_\alpha &= m'_\alpha + im''_\alpha \\
m'_\alpha &= -\frac{1}{2}k^2 \\
m''_\alpha &= 0 \\
m_b &= m'_b + im''_b \\
m'_b &= -\frac{3}{8}k^2 \\
m''_b &= k
\end{aligned} \tag{23.47a–l}$$

were defined by Küssner [30], while $F(k)$ and $G(k)$ are the real and imaginary parts of the Theodorsen function $C(k)$. Substituting Eq. (23.23) into the right-hand side of Eq. (23.1) using Eq. 23.47(a–l), then the eigenvalue or flutter stability equation equivalence of Eq. 23.6 becomes

$$[\mathbf{A}(k)]\{\hat{\mathbf{q}}\} = \lambda\{\hat{\mathbf{q}}\} \tag{23.48}$$

where A is a 2×2 matrix whose elements are given by

$$A_{11} = \left(\frac{\omega_\alpha}{\omega}\right)^2\left(\mu - \frac{k_a}{k^2}\right) \tag{23.49a}$$

$$A_{12} = \left(\frac{\omega_\alpha}{\omega_h}\right)\left(\mu x_\alpha - \frac{k_D}{k^2} + (0.5+a)\frac{k_a}{k^2}\right) \tag{23.49b}$$

$$A_{21} = \mu x_\alpha - \frac{m_a}{k^2} + (0.5+a)\frac{k_a}{k^2} \tag{23.49c}$$

$$A_{22} = \mu r_\alpha^2 \frac{m_b}{k^2} + (0.5 + a)\left(\frac{m_a}{k^2} + \frac{k_b}{k^2}\right) - (0.5 + a)^2 \frac{k_a}{k^2} \tag{23.49d}$$

Equation 23.48 is another polynomial of the eigenvalue λ. At this point, the following remarks are in order.

First, following the flutter stability analysis as carried out by Done and elaborated in paragraphs 2.1 and 2.2, similar procedure for solving Eqs. 23.49a–d can be followed to obtain the roots λ, which in general are complex. The square roots of λ are also complex numbers, consisting of the real parts, which are indicative of the damping terms and imaginary parts, which are indicative of the frequencies of the flutter solution. A diagram indicating the values of the frequencies and damping terms of the flutter solution can be drawn, from which one may obtain significant information on the behavior of the flutter solution as a function of flow properties, represented by the airspeed U, dynamic pressure q or reduced frequency

$$k_R = U/\omega b.$$

Second, following the classical procedure known as VG [2] or K-Method [29], the flutter equation can be solved for the prevailing modal frequencies (bending and torsion for the particular binary flutter problem) and the associated damping, by assuming a solution of the form

$$\lambda = \left(\frac{\omega_\alpha}{\omega_h}\right)^2 (1 + ig) \tag{23.50}$$

Then the frequency and damping of the flutter solution are given by

$$\lambda = \frac{(1 + ig)}{\omega^2} = \mathrm{Re}(\lambda) + \mathrm{Im}(\lambda) = \frac{1}{\omega^2} + \mathrm{Im}\left(\frac{g}{\omega^2}\right) \tag{23.51a}$$

$$\omega = \sqrt{\frac{1}{\mathrm{Re}(\lambda)}} \tag{23.51b}$$

$$g = \omega^2 \mathrm{Im}(\lambda) \tag{23.51c}$$

The onset of flutter, if it prevails, is given by the value of U (then here is the flutter velocity UF), for which one of the damping (g) curves crosses the zero-axis. At this point, the two modal frequencies (bending and torsion) will approach each other [2, 29].

23.6.2 Computational Results

The results as shown in Fig. 23.4 also indicate characteristics typical of the second column of Table 23.1, which will not lead to aeroelastic instability.

In addition, the results also show that the prevailing eigenfrequencies estimated using the quasi-steady aerodynamics in the operational range of the flapping ornithopter are much smaller than the pitching and flapping frequency of the ornithopter wing. Such conclusion is considered reasonable and in confirmation with observation on biosystem. The flexural property as represented by K_α shows that, if quasi-steady aerodynamics is assumed, the elastic deflection due to the prevailing aerodynamic force as calculated using the unsteady aerodynamics elaborated in Sect. 23.2 will produce at most 5% change in of θ or α'. Such situation is taken into consideration in establishing a heuristic model as elaborated in succeeding section.

23.6.3 Incorporation of Quasi-Steady Aerodynamics Flexibility in a Heuristic Model for Aerodynamic Performance Estimation

Based on the findings obtained in previous section, a heuristic aeroelastic model can be established. The simplest one is to incorporate the influence of the aeroelastic properties being reduced to the static flexibility properties. It is also assumed that the flexibility effect acted instantly. Following such rationale, then the effect of aeroelasticity, hence flexibility, is to modify the pitching and heaving angle linearly to a small percentage. The results of such heuristic model assumption to the aerodynamic performance of the flapping-wing ornithopter can be calculated using the procedure already outlined in section one. Essentially, a constant Flexibility Coefficient γ_f is introduced to account for flexibility of the wing in pitching and flapping motion (Fig. 23.5).

Then the pitching and heaving motion will be modified as follows

$$\theta(t) = \gamma_\mathrm{f}(\theta_0 \cos(\omega t + \phi)) + \theta_\mathrm{fp} \tag{23.52a}$$

$$\dot{\theta}(t) = \gamma_\mathrm{f}(-\omega\theta_0 \sin(\omega t + \phi)) \tag{23.52b}$$

$$\ddot{\theta}(t) = \gamma_\mathrm{f}\big(-\omega^2\theta_0 \cos(\omega t + \phi)\big) \tag{23.52c}$$

$$h(t) = \gamma_\mathrm{f}(-y\beta_0 \cos(\omega t)) \tag{23.52d}$$

$$\dot{h}(t) = \gamma_\mathrm{f}(y\omega\beta_0 \sin(\omega t)) \tag{23.52e}$$

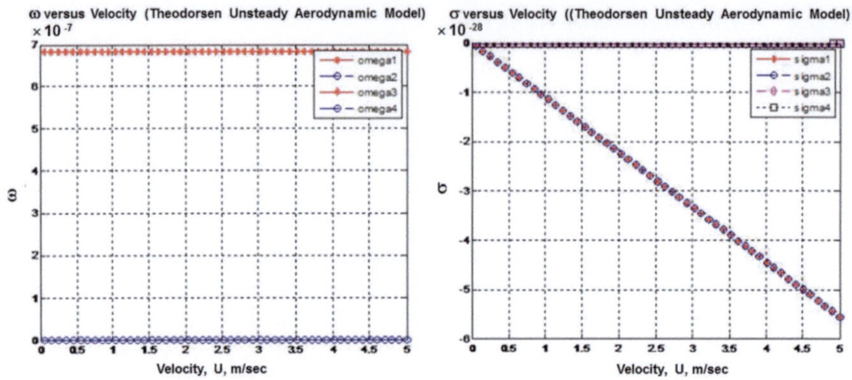

(a) Dimensionless Frequency **(b) Dimensionless Damping**

Fig. 23.5 Numerical computational results of the flutter stability of the model of ornithopter wing typical section using Theodorsen unsteady aerodynamic aeroelastic analysis. Dimensionless frequency and damping are plotted against velocity

$$\ddot{h}(t) = \gamma_{\mathrm{f}}\big(y\omega^2\beta_0\cos(\omega t)\big) \tag{23.52f}$$

The results are exhibited in Fig. 23.4a, b, which describe the influence of the flexibility on the lift and thrust produced by the flapping wing, if the flexibility effects is introduced on θ and h. All the results are computed by considering the dynamic stall criterion for attached flow similar to that utilized by DeLaurier [5].

Figure 23.6a, b and Table 23.4 show the influence of introducing 5 and 10% flexibility as a representation of the aeroelastic effect using quasi-steady aerodynamics. If the flexibility factor γ_{f} is introduced into the apparent angle of attack α', the prevailing equation will be modified as:

$$\alpha' = \gamma_{\mathrm{f}}\left[\frac{\overbrace{\mathrm{AR}}^{\text{Aspect Ratio}}}{2+\overbrace{\mathrm{AR}}^{\text{Aspect Ratio}}}\left[F(k)\alpha + \frac{c}{2U}\frac{G(k)}{k}\alpha\right] - \frac{w_0}{U}\right] \tag{23.53}$$

The results are exhibited in Fig. 23.7a, b, and Table 23.5. These results show that the effect of static aeroelasticity tends to reduce the lift and increase the thrust. In addition, the introduction of the static aeroelasticity introduced to the primary variables θ and h will produce slightly different values than if the aeroelastic effect is introduced in the derived variable α'. Noting that the heuristic model is a first approximation to the actual state of affairs, such difference may be attributed to many simplifying assumptions, such as the three dimensionality of the flow as represented by α', among others.

Proceeding to the investigation on the static aeroelasticity effects on the individual contribution of pitching and flapping motion components, the results are shown in

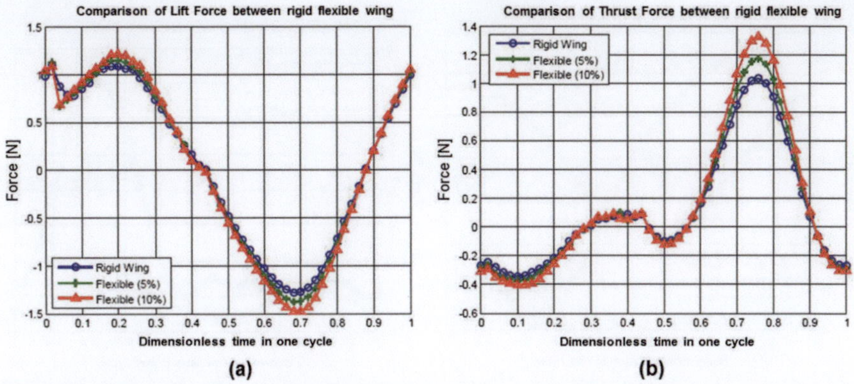

Fig. 23.6 **a** Lift and **b** thrust variation with rigid wing and flexible wing of 5 and 10% using heuristic model

Table 23.4 Comparison of the average lift and thrust of rigid and flexible ornithopter wing using heuristic model

	Rigid wing	Flexible wing (5%)	Flexible wing (10%)
Average lift [N]	0.0662	0.0498	0.0386
Average thrust [N]	0.1110	0.1272	0.1457

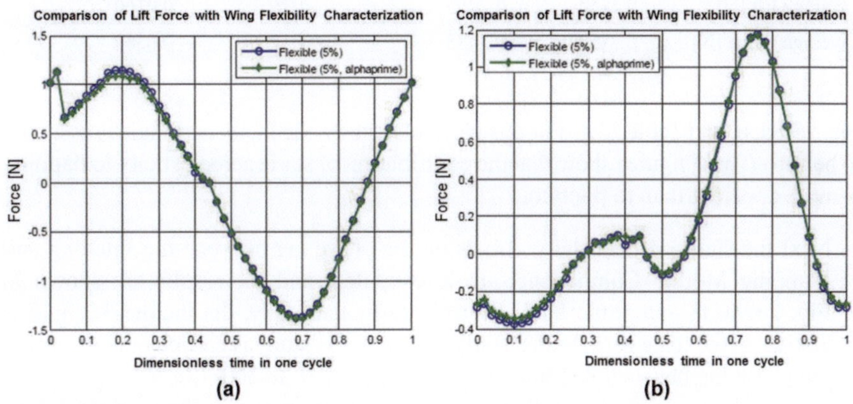

Fig. 23.7 **a** Lift and **b** thrust variation for 5% flexible wing, and 5% flexible wing and incorporating alphaprime (α')

Table 23.5 Comparison of the average lift and thrust of rigid and flexible ornithopter wing using heuristic model using α' (alphaprime) as basis of flexible deformation

	Flexible wing (5%)	Flexible wing (5% alphaprime, α')
Average lift [N]	0.0498	0.0306
Average thrust [N]	0.1272	0.1395

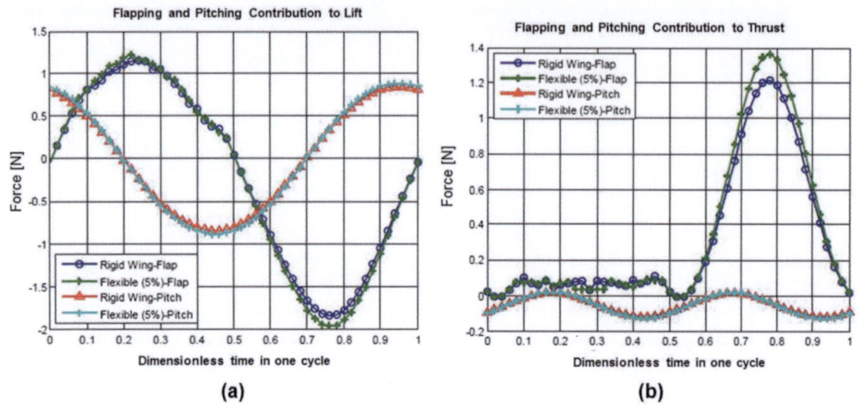

Fig. 23.8 Contribution of flapping and pitching motion individually on **a** lift and **b** thrust forces for rigid wing and flexible wing of 5%

Table 23.6 Comparison of the average lift and thrust of rigid and flexible ornithopter wing using heuristic model contributed by pitching and flapping motion

	Flapping only		Pitching only	
	Rigid wing	Flexible wing (5%)	Rigid wing	Flexible wing (5%)
Average lift [N]	– 0.1864	– 0.2065	0.0161	0.0160
Average thrust [N]	0.3123	0.3435	– 0.0518	– 0.0562

Fig. 23.8a, b, and Table 23.6. For this particular study, the incidence angle is assumed to be zero. These figures show that the contribution of static aeroelasticity to flapping is more apparent than to pitching.

- Next the static aeroelasticity effects on the phase lag between the Pitching and Flapping Motion Components are investigated, and the results are shown in Fig. 23.9a, b, and Table 23.7. For this particular study, the incidence angle is also assumed to be zero. In this study, a parametric study is carried out by varying the phase lag between flapping and pitching from 0° to 360° (2π).
- The results as exhibited by these figures show the extent of the contribution of static aeroelasticity to the influence of the phase lag between the pitching and flapping motion on the lift and thrust generated by the flapping-wing ornithopter.

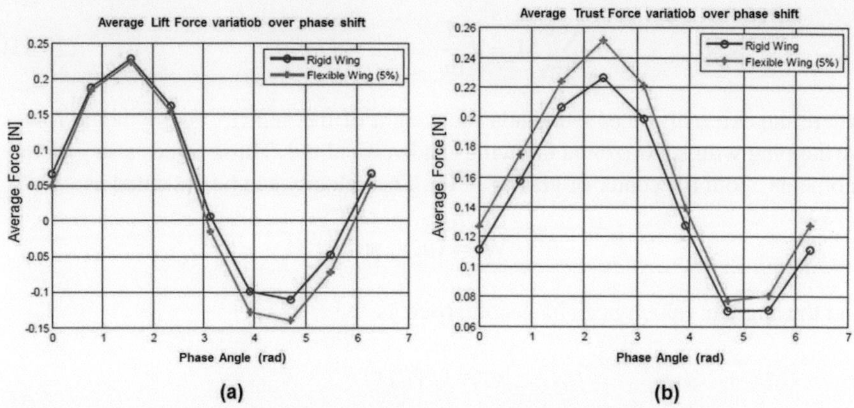

Fig. 23.9 Phase shift influence on **a** lift and **b** thrust forces for rigid wing and flexible wing of 5%

Table 23.7 Comparison of the average lift and thrust of rigid and flexible

	Rigid wing		Flexible wing (5%)	
Phase	Average lift [N]	Average thrust [N]	Average lift [N]	Average thrust [N]
0	0.0662	0.1110	0.0498	0.1272
0.25π	0.1877	0.1571	0.1823	0.1748
0.5π	0.2289	0.2063	0.2240	0.2238
0.75π	0.1619	0.2263	0.1540	0.2534
π	0.0056	0.1991	– 0.0148	0.2212
1.25π	– 0.0987	0.1274	– 0.1288	0.1399
1.5π	– 0.1111	0.0700	– 0.1402	0.0775
1.75π	– 0.0482	0.0798	– 0.0723	0.0808
2π	0.0665	0.1111	0.0502	0.1273

23.7 Flight Dynamics Considerations

23.7.1 Flight Dynamic Model

The assumptions are made for the derivation of the equation of motion (EOM); Rotation of the earth is negligible, vehicle mass is constant and mass distribution is constant with time.

The EOM of the flapping-wing MAV can then be obtained by applying Newton's second laws, given by

$$\sum \mathbf{F} = m\frac{\mathrm{d}}{\mathrm{d}t}\mathbf{V}_b + \omega \times (m\mathbf{V}_b) \tag{23.54}$$

$$\sum \mathbf{M} = I\frac{\mathrm{d}}{\mathrm{d}t}\omega + \omega \times (I\,\omega) \tag{23.55}$$

where the external forces F include the weight of the vehicle, aerodynamic forces by flapping wings, horizontal tail wings and vertical tail. Those forces also generate moments about the center of gravity (CG). The velocity can be indicated as

$$\mathbf{V}_b = (u, v, w)$$

and the angular velocity can be decomposed as

$$\omega = (p, q, r)$$

The methodology to analyze a fixed-wing vehicle was applied to a flapping-wing vehicle.

The formulation of tail wing forces and moments is simpler since it is a fixed wing. Treated as it is quad-wing configuration, the tail is in rectangular shape with the same span length and calculated using the present modeling developed, as illustrated schematically in Fig. 23.10.

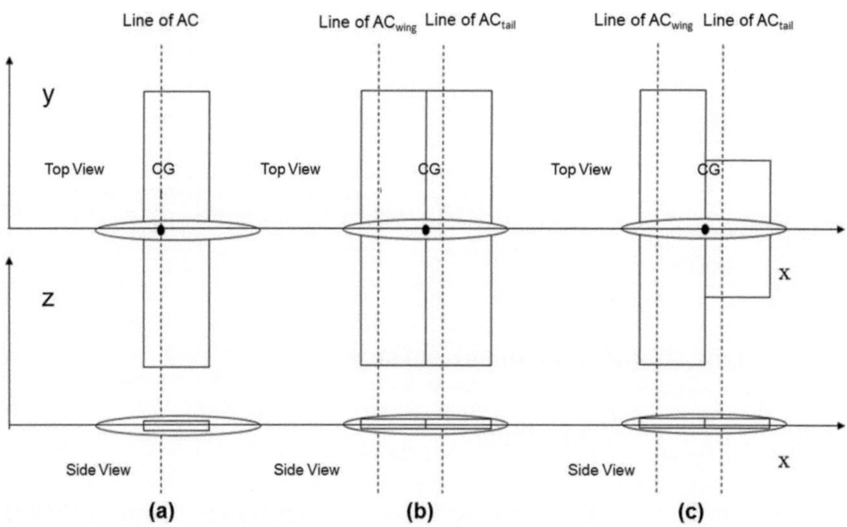

Fig. 23.10 Proposed configuration baseline for simplified generic modeling calculation; **a** single wing without tail, **b** wing with tail and **c** wing with tail (half-length of wingspan)

23.7.2 Formulation of Overall Force and Moment

By summing all calculated values from proposed aerodynamic modeling using specified parameters, the overall forces and moments are attainable. The forces are

$$\mathbf{F}_x = \mathbf{F}_{x_{\text{wing}}} + \mathbf{F}_{x_{\text{tail}}} \tag{23.56a}$$

$$\mathbf{F}_z = \mathbf{F}_{z_{\text{wing}}} + \mathbf{F}_{z_{\text{tail}}} \tag{23.56b}$$

$$\mathbf{F}_y = 0 \tag{23.56c}$$

and the moments are

$$\mathbf{M}_x = 0 \tag{23.57a}$$

$$\mathbf{M}_z = 0 \tag{23.57b}$$

$$\mathbf{M}_y = \mathbf{F}_{z_{\text{wing}}} l_w + \mathbf{F}_{x_{\text{wing}}} h_w - \mathbf{F}_{z_{\text{tail}}} l_t - \mathbf{F}_{x_{\text{tail}}} h_{\text{tail}} \tag{23.57c}$$

Here l_{wing} and l_{tail} are the distance between the aerodynamic center of the wing and the tail with respect to the ornithopter center of gravity (CG), respectively. In the present case, the corresponding vertical distances h_{w} and h_{t} are zero. The resultant force along the y-direction is approximately zero because the fore-wing (wing) and the hind-wing (tail) are symmetric. The detailed trajectory and attitude are attainable by expanding and solving Eqs. 23.54 and 23.55.

23.8 Results and Analysis

The results computed from the present modeling are shown in Fig. 23.10a–c and Table 23.8. The baseline model is being configured as a bi-wing without the tail. The next step is to include more details and to be optimized so that the model will be more enriched and sophisticated.

Table 23.8 Total lift and thrust of main wing (without tail)

	Wing
Average lift [N]	0.4278
Average thrust [N]	0.3179

23.8.1 Parametric Study

In previous work, the phase angle variation between pitching and flapping is carried out for optimization study, and it is found that the best aerodynamic performance is gained during phase shift of 90° [10]. With such variation, the trajectory of each case is calculated to see the influence of it. The result is shown in Fig. 23.12. More detailed flight dynamic model and more degree of freedom will be accounted for the flight dynamic model considered in forthcoming work (Fig. 23.11).

In solving the flight dynamic equation, an approximate approach is utilized by step-wise linear integration of the equations of motion (23.54) and (23.55). For reasonably small time steps, this should be adequate and assessed a posteriori by comparison with more exact scheme or results from other researchers, as qualitatively exhibited in Fig. 23.13b. In addition to a tailless ornithopter, the ornithopter with wing and tail is also considered. The computation of the flight velocities and trajectories

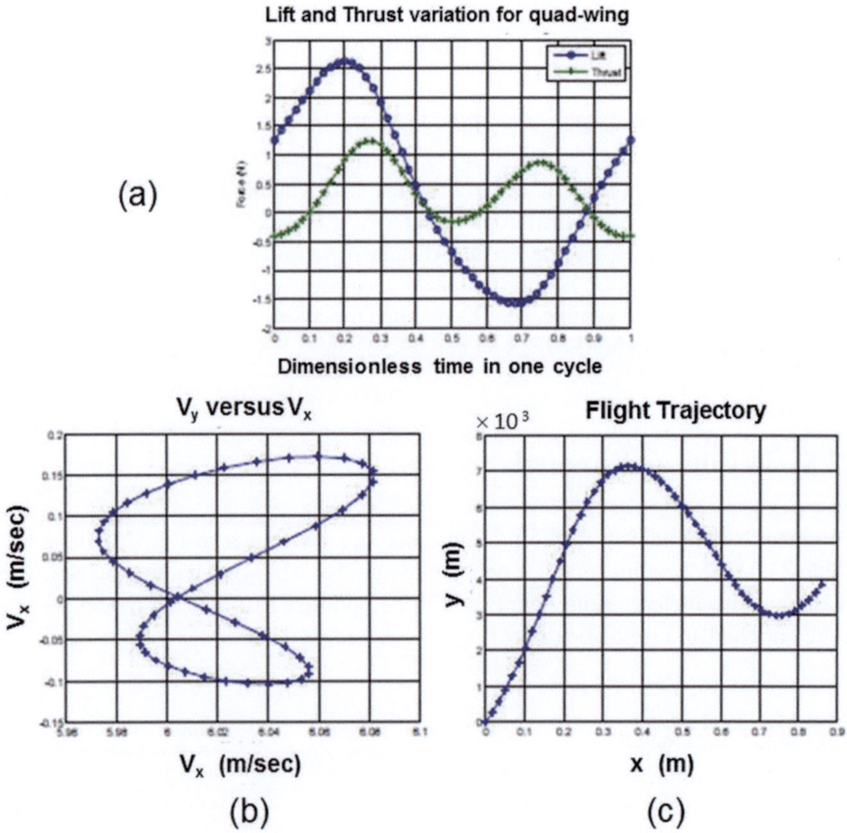

Fig. 23.11 **a** Total lift and thrust of single wing without tail; **b** the graph of V_y versus V_x; and the calculated flight trajectory over one cycle (without dynamic stall consideration)

of these models has been initiated from the mechanically equilibrium states, taken into account the static stability at the beginning of the flight cycle. The results are exhibited in Fig. 23.13a, b.

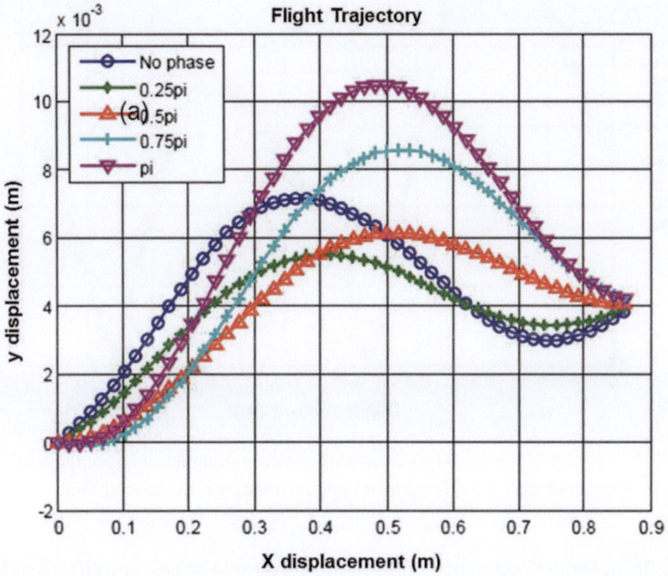

Fig. 23.12 Flight trajectory in the plane of symmetry for rectangular wing without tail as a baseline investigation in flight dynamic study considering the variation of phase lag between pitching and flapping motion

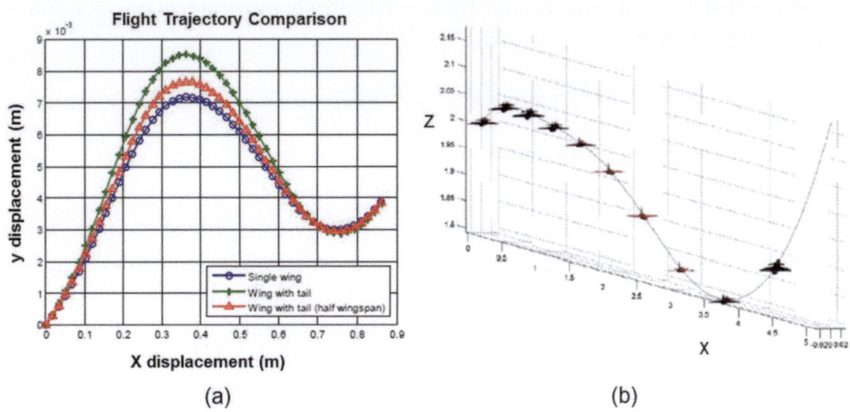

Fig. 23.13 **a** Flight trajectory in the plane of symmetry for rectangular wing without tail (blue) and wing with tail (green and red), calculated using approximation step-wise method. **b** The pattern of trajectory developed qualitatively resembles the motion of "rectilinear flight" of Tamkang Golden **Snitch** [19]

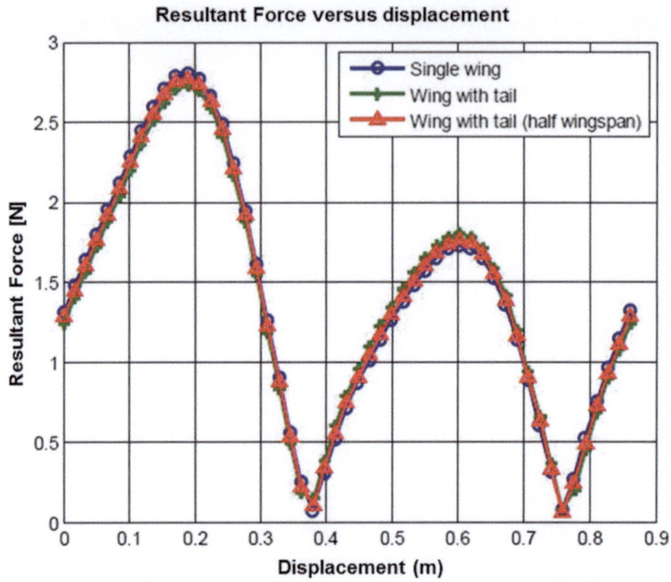

Fig. 23.14 Resultant force versus displacement trend for rectangular wing without tail (blue) and wing with tail (green and red), calculated using approximation step-wise method

The resultant force versus displacement graphs have been computed for three wing and tail configurations, as exhibited in Fig. 23.14. Some remarks are in order. First, the computational results show that the results for two other more developed models differ slightly from the tailless flapping-wing configuration flight characteristics, represented by the aerodynamic forces and flight trajectories. Another interesting outcome is that the total aerodynamic force and the trajectories are all positive within the cyclic time frame.

23.9 Concluding Remarks[2]

The philosophical approach and computation presented in the present paper are based on the utilization of quasi-steady aerodynamics a typical section approximation of the flexible flapping-wing ornithopter, based on model utilized in [8–13]. In addition, based on the Aeroelastic stability characteristics obtained using such approach, a heuristic flexibility model has been assumed in estimating the influence of flexibility on the aerodynamic performance of the flexible flapping-wing ornithopter.

[2] The present work was carried out under Universiti Putra Malaysia (UPM) Research University Grant Scheme (RUGS) No. 9378200, the Ministry of Higher Education Research Grants ERGS: 5527088 and FRGS:5524250. The author would also like to thank the assistance of Messrs. ASS Ramli and MAA Bari in contributing to the present work.

These approaches are considered to be educational. With the introduction of all these simplification, one may expect to obtain a qualitative impression of the influence of flexibility using zeroth-order approximation, but yet may gain some insight of the issue using lower cost effort. Further approximation may have to be judged on the cost of the effort compared to the more sophisticated approach using refined model and computational scheme, such as exemplified by [5, 6, 8]. Nevertheless, comparing the present results with those of Zhang et al. [14] and Dai et al. [4], the present results exhibit some similar trend, in the sense that the flapping-wing low flexibility exhibits minor influence on the aerodynamic performance. The present approach and model, however, indicate that the flexibility. The present approach and model, however, indicate that the influence of flexibility of the flapping wing improves its capability to produce thrust rather than lift.

The present work has been performed to investigate the two-dimensional flight dynamic modeling of flapping-wing ornithopter with pitch-flap phase lag. In this conjunction, a computational model has been developed, and a generic computational method has been adopted, utilizing strip theory and two-dimensional unsteady aerodynamic theory of Theodorsen with modifications to account for three-dimensional and viscous effects and leading-edge suction. The present work is based on basic approach, for instructiveness and to gain insight into the problem, and can readily be computed without external software. In the flight mechanic modeling, the computational results show that the results for two other more developed models differ slightly from the tailless flapping-wing configuration flight characteristics, as represented by the aerodynamic forces and flight trajectories. The total aerodynamic force and the trajectories are all positive within the cyclic time frame. Further refinement of the model with more complexity will be carried out progressively.

References

1. Anderson, M.L. 2011. *Design and Control of Flapping Wing Micro Air Vehicles*. Ph.D. Thesis, AFIT, dayton, USA.
2. Bisplinghoff, R.L., H. Ashley, and R.L. Halfman. 1955. *Aeroelasticity*. Dover Publications, Inc., copyright © 1955 by Addison-Wesley Publishing Co., Inc., copyright © renewed 1983.
3. Chikamurti, S.K. 2009. *A Computational Aeroelasticity Framework for Analyzing Flapping Wings*. Ph.D. Thesis, University of Michigan.
4. Dai, H., F.-B. Tian, J. Song, and H. Luo. 2012. *Aerodynamics and Aeroelasticity of Flapping Wings*. Nashville, Tennessee, USA: Department of Mechanical Engineering, Vanderbilt University, ICTAM.
5. DeLaurier, J.D. 1993. An aerodynamic model for flapping wing flight. *Aeronautical Journal of the Royal Aeronautical Society* 97: 125–130.
6. Djojodihardjo, H. 2016. *Analysis and Computational Study of the Aerodynamics, Aeroelasticity and Flight Dynamics of Flapping Wing Ornithopter Using Linear Approximation*. AIAA paper AIAA 2016-2027, AIAA SciTech Forum.

7. Djojodihardjo, H., and Y.H. Hoe. 2007. *Parametric Study of the Flutter Characteristics of Transport Aircraft Wings*. Kuala Lumpur, Malaysia: AEROTECH II.
8. Djojodihardjo, H., and A.S.S. Ramli. 2012. Kinematic and aerodynamic modelling of flapping wing ornithopter. *Procedia Engineering* 50: 848–863.
9. Djojodihardjo, H., and A.S.S. Ramli. 2013. Kinematic and unsteady aerodynamic modelling, numerical simulation and parametric study of flapping wing ornithopter. In *Paper IFASD 2013 S3B, International Forum on Aeroelasticity and Structural Dynamics (IFASD)*. Bristol.
10. Djojodihardjo, H., A.S.S. Ramli, M.S.A. Aziz, and K.A. Ahmad. 2013. Numerical modelling, simulation and visualization of flapping wing ornithopter. In *International Conference on Engineering Materials and Processes (ICEMAP 2013)*. Chennai.
11. Djojodihardjo, H., and A.S.S. Ramli. 2014. Modelling studies of bi- and quad-wing flapping ornithopter kinematics and aerodynamics. *International Journal of Automotive and Mechanical Engineering*. ISSN-2180-1606 (online).
12. Djojodihardjo, H., and M.A.A. Bari. 2014. Kinematic and unsteady aerodynamic modelling of flapping bi- and quad-wing ornithopter. In *Paper International Council of Aeronautical Sciences(ICAS) Congress*. St. Petersburg, ICAS 2014-043.
13. Djojodihardjo, H., A.S.S. Ramli, and M.A.A. Bari. 2015. Kinematic and unsteady aerodynamic study on bi- and quad- wing ornithopter. *Journal of Aeroelasticity and Structural Dynamics*.
14. Zhang, C., Z.A. Khan, and S.K. Agrawal. 2010. Experimental investigation of effects of flapping wing aspect ratio and flexibility on aerodynamic performance. In *IEEE International Conference on Robotics and Automation*. Anchorage Convention District, Anchorage, Alaska.
15. Garrick I. E., 1936. *Propulsion of a Flapping and Oscillating Aerofoil*, NACA Report No. 567.
16. Friedmann, P., and P. Ifju. 2007. Computational aerodynamics of low Reynolds number plunging, pitching and flexible wings for MAV applications. In *46th AIAA Aerospace Sciences Meeting and Exhibition, Reno. on AIAA paper 2008-522*.
17. Gogulapati, A., P.P. Friedmann, E. Kheng, and W. Shyy. 2010. Approximate aeroelastic modelling of flapping wings: Comparison with CFD and experimental data. In *51st AIAA/ASME/ASCE/AHS/ASC Structures, Structural Dynamics, and Materials Conference on Paper AIAA 2010-2707*. Orlando, Florida.
18. Shyy, W., Y. Lian, J. Tang, H. Liu, P. Trizila, B. Stanford, L. Bernal, C. Cesnik, P. Friedmann, and P. Ifju. 2007. Computational aerodynamics of low Reynolds number plunging, pitching and flexible wings for MAV applications. In *46th AIAA Aerospace Sciences Meeting and Exhibition, Reno. on Paper AIAA 2008-522*.
19. Hsiao, F.Y., T.M. Yang, and W.C. Lu. 2012. Dynamics of flapping-wing MAVs: Application to the Tamkang golden snitch. *Journal of Applied Science and Engineering* 15(3): 227–238.
20. Jones, R.T. 1940. *The Unsteady Lift of a Wing of Finite Aspect Ratio*. NACA Report No. 681.
21. Polhamus, E.C. 1968. *Application of the Leading-Edge-Suction Analogy of Vortex Lift to the Drag Due to Lift of Sharp-Edge Delta Wings*. NASA TN D-4739.
22. Scherer, J.O. 1968. *Experimental and Theoretical Investigation of Large Amplitude Oscillating Foil Propulsion Systems*. Hydronautics, Laurel, Md.
23. Theodorsen, T. 1935. *General Theory of Aerodynamic Instability and the Mechanism of Flutter*. NACA Report No. 496.
24. Multhopp, H. 1955. *Methods for Calculating the Lift Distribution of Wings (Subsonic Lifting-Surface Theory)*. ARC R&M No. 2884.
25. Done, G.T.S. 1966. *The Flutter and Stability of Undamped Systems*. School of Engineering Science, University of Edinburgh, Report and Memoranda No. 3553.
26. Done, G.T.S. 1967. *A Study of Binary Flutter Roots Using a Method of System Synthesis*. Aeronautical Research Council, R&M No. 3554.
27. Kock, J. 2006. *Physical and Mechanical Properties of Chicken Feather Materials*. M.Sc. Thesis, Georgia Institute of Technology.

28. Shyy, W., Berg, M., and Ljungqvist, D. (1999). Flapping and flexible wings for biological and micro air vehicles. *Progress in Aerospace Sciences* 35: 455–505.
29. Zwaan, R.J. 1981. Aeroelasticity of Aircraft, Lecture Notes, ITB, Bandung, Indonesia (Djojodihardjo's Repository).
30. Küssner, H.G. 1963. Zusammenfassender Bericht über den instationären Auftrieb von Flügeln, Luftfahrtforschung, Bd. 13, Nr. 12, December.

Chapter 24
BEM–FEM Coupling for Acoustic Effects on Aeroelastic Stability of Structures

Abstract A series of work has been carried out to develop the foundation for the computational scheme for the calculation of the influence of the acoustic disturbance to the aeroelastic stability of the structure. The generic approach consists of three parts. The first is the formulation of the acoustic wave propagation governed by the Helmholtz equation by using boundary element approach, which then allows the calculation of the acoustic pressure on the acoustic-structure boundaries. The structural dynamic problem is formulated using finite element approach. The third part involves the calculation of the unsteady aerodynamics loading on the structure using generic unsteady aerodynamics computational method. Analogous to the treatment of dynamic aeroelastic stability problem of structure, the effect of acoustic pressure disturbance to the aeroelastic structure is considered to consist of structural motion-independent incident acoustic pressure and structural motion-dependent acoustic pressure, referred to as the acoustic-aerodynamic analogy. Results are presented and compared to those obtained in earlier work.

Keywords Boundary element method · Finite element method · Fluid–structure coupling · Computational mechanics

List of Symbols

$[AIC]$	Aerodynamics Influence Coefficient
$[A(ik)]$	Unsteady Aerodynamics Matrix
b	wing chord/span chosen for convenience
$[C]$	Viscous Damping
c	constant for BEM equation, or speed of sound
$[\mathbf{F}]$	External Forces
G_{ij}	influence coefficient matrices

The present chapter is reproduced from an article *co-authored by* Harijono Djojodihardjo and Irtan Safari which appeared in *CMES: Computer Modeling in Engineering & Sciences, Vol. 91, No. 3, pp. 205–234, 2013* At the time of writing, Irtan Safari is at the Garuda Maintenance Facility, Jakarta, Indonesia.

© The Author(s), under exclusive license to Springer Nature Singapore Pte Ltd. 2023
H. Djojodihardjo, *Introduction to Aeroelasticity*,
https://doi.org/10.1007/978-981-16-8078-6_24

g	free-space green function
H_{ij}	influence coefficient matrices
$[K]$	Stiffness Matrix
k	reduced frequency
kw	wave number, in the Helmholtz equation
$[\mathbf{L}]$	fluid–structure coupling matrix
$[\mathbf{M}]$	Mass Matrix
N	shape function, as implied by the context
n_0	surface normal vector
p	acoustic pressure
p_{inc}	incident acoustic pressure
p_{sc}	scattering acoustic pressure
g	generalized coordinates
q	dynamic pressure of the fluid surrounding the structure
R_0	a point in boundary surface
S	bounding surface
V_f	flutter speed
ν	normal velocity vector
δ	Kronecker's delta function
λ	wave-length
ρ	air density

24.1 Introduction

The foundation for the computational scheme for the calculation of the influence of the acoustic disturbance to the aeroelastic stability of the structure has been developed in earlier work (Djojodihardjo and Tendean 2004) [1], Djojodihardjo and Safari [2, 3]). Analogous to the treatment of dynamic aeroelastic stability problem of structure, in which the aerodynamic effects can be distinguished into motion-independent and motion-induced aerodynamic forces, the effect of acoustic pressure disturbance to the aeroelastic structure (acousto-aero-elastic problem) can be viewed to consist of structural motion-independent incident acoustic pressure and structural motion-dependent acoustic pressure, which is known as the scattering pressure. This can be referred to as the acoustic-aerodynamic analogy. Proceeding to the formulation of BE-FE coupling to treat the fluid–structure interaction, reference is made on the solution of structural–acoustic interaction problems using BEM–FEM coupling given by Holström [4] and Meddahi et al. [5]. Applying similar approach to solve the acousto-aeroelastic problem, the present work consists of three parts. The first part involves the formulation of the acoustic wave propagation governed by the Helmholtz equation by using boundary element approach, which then allows the calculation of the acoustic pressure on the acoustic-structure boundaries. The governing Helmholtz equation will be solved using Boundary Element method, following the procedure

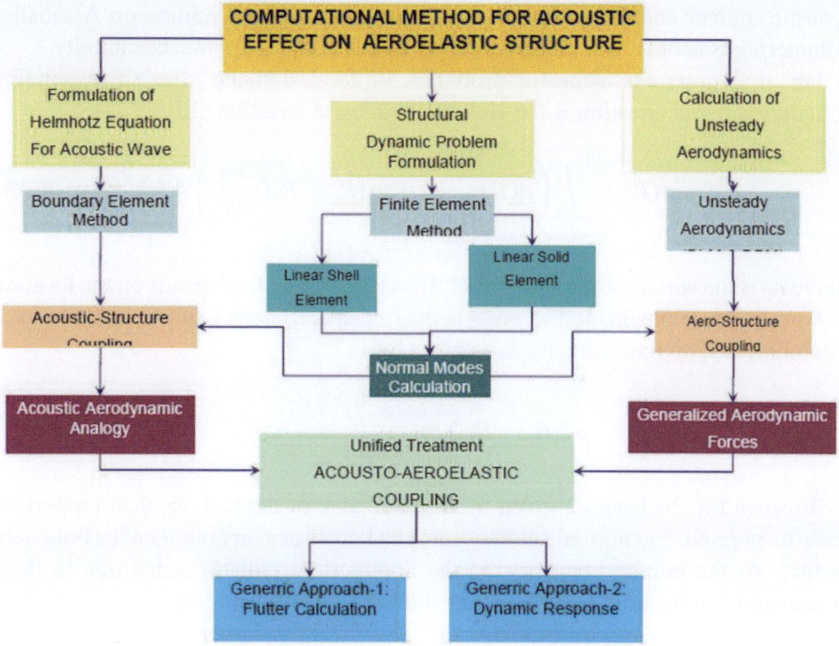

Fig. 24.1 Computational strategy for the calculation of acoustic effects on aeroelastic structures

elaborated by [6] and taking into consideration various techniques and development described in many recent literature, such as those elaborated by [4, 7–9]. The second part addresses the structural dynamic problem using finite element approach. The acoustic-structure interaction is then given special attention to formulate the BEM–FEM fluid–structure coupling. The third part involves the calculation of the unsteady aerodynamic loading on the structure using a conveniently chosen unsteady aerodynamics computational method, to be utilized in the aeroelastic problem. The acoustic pressure disturbance is then superposed to the aeroelastic problem, following the acoustic-aerodynamic analogy. Solution procedure can then be readily formulated. Figure 24.1 shows the computational strategy to treat the aeroacoustic effects on aeroelastic structure.

24.2 Discretization of the Helmholtz Integral Equation for the Acoustic Field

For an exterior acoustic problem, as depicted in Fig. 24.2, the problem domain V is the free-space V_{ext} outside the closed surface S. V is considered enclosed between the surface S and an imaginary surface Λ at a sufficiently large distance from the acoustic sources and the surface S such that the boundary condition on Λ satisfies Sommerfeld's acoustic radiation condition as the distance approaches infinity.

For time-harmonic acoustic problems in fluid domains, the corresponding boundary integral equation is the Helmholtz integral equation [10]

$$cp(R) = \int\limits_S \left(p(R)\frac{\partial g}{\partial n_0} - g(|R - R_0|)\frac{\partial p}{\partial n_0} \right) \mathrm{d}S \tag{24.1}$$

where n_0 is the surface unit normal vector, and the value of c depends on the location of R in the fluid domain, and where g is the free-space Green's function. R_0 denotes a point located on the boundary S, as given by

$$g\left(|R - R_0|\right) = \frac{e^{-ik|R - R_0|}}{4\pi|R - R_0|} \tag{24.2}$$

To solve Eq. 24.1 with g given by Eq. 24.2, one of the two physical properties, acoustic pressure and normal velocity, must be known at every point on the boundary surface. At the infinite boundary Λ, the Sommerfeld radiation condition in three dimensions can be written as (Dowling and Ffowcs-Williams 1983):

$$\lim_{|R - R_0|\to\infty} r\left(\frac{\partial g}{\partial r} + ikg\right) \Rightarrow 0\,\text{as}\,t \Rightarrow \infty,\; r = |R - R_0| \tag{24.3}$$

which is satisfied by the fundamental solution.

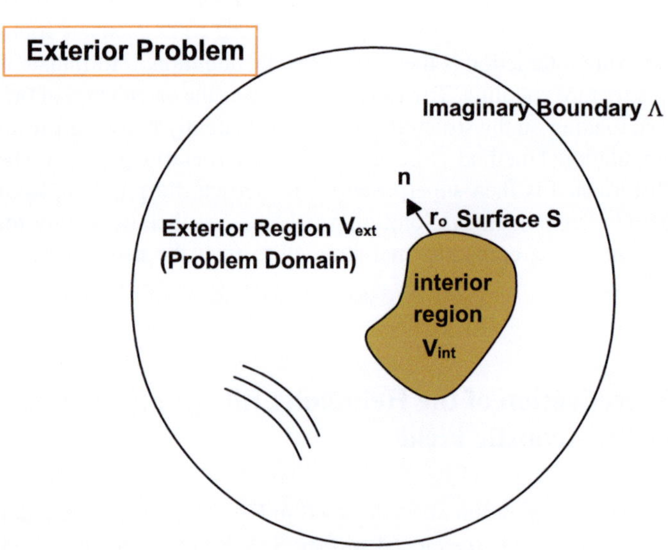

Fig. 24.2 Exterior problem for homogeneous Helmholtz equation

The total pressure, which consists of incident and scattering pressure, serves as an acoustic excitation on the structure. The integral equation for the total wave is given by

$$cp(R) - p_{inc}(R) = \int\limits_{S} \left[p(R)\frac{\partial g(R - R_0)}{\partial n_0} - \frac{\partial p(r)}{\partial n_0} g(R - R_0) \right] \mathrm{d}S \tag{24.4}$$

where $p = p_{inc} + p_{sc}$, and where

$$c = \begin{cases} 1, & \mathrm{R} \in \mathrm{V_{ext}} \\ 1/2, & \mathrm{R} \in \mathrm{S} \\ \Omega/4\pi, & \mathrm{R} \in \mathrm{S} \quad \text{(non smooth surface)} \\ 0, & \mathrm{R} \in \mathrm{V_{int}} \end{cases} \tag{24.5}$$

The Helmholtz equation is then discretized by dividing the boundary surface S into N elements. The discretized boundary integral equation becomes,

$$cp_i - p_{inc} - \sum_{j=1}^{N} \int\limits_{S} p\overline{g}\mathrm{d}S = i\rho_0\omega \sum_{j=1}^{N} \int\limits_{S} gv\mathrm{d}S \tag{24.6}$$

where i indicates field point, j source point and S_j surface element j, and for convenience, $\overline{g}$ is defined as

$$\overline{g} \equiv \frac{\partial g}{\partial n} \tag{24.7}$$

Let

$$\overline{H}_{ij} = \int\limits_{S_j} \overline{g}\mathrm{d}S \tag{24.8}$$

$$G_{ij} = \int\limits_{S_j} g\mathrm{d}S \tag{24.9}$$

Substituting g in Eq. (24.2) to be the monopole Green's free-space fundamental solution, it follows that:

$$G_{ij} = \int\limits_{S_j} g\mathrm{d}S = \int\limits_{S_j} g(|R_j - R_i|)\mathrm{d}S = \int\limits_{S_j} \frac{e^{ik|R_j - R_i|}}{4\pi|R_j - R_i|}\mathrm{d}S \tag{24.10}$$

or, in Cartesian coordinate system,

$$G_{ij} = \int\limits_{S_j} \frac{e^{-ik\sqrt{(x_j-x_i)^2+(y_j-y_i)^2+(z_j-z_i)^2}}}{4\pi\sqrt{(x_j-x_i)^2+(y_j-y_i)^2+(z_j-z_i)^2}} \mathrm{d}S \tag{24.11}$$

where R_j is the coordinate vector of the midpoint of element j and R_i is the coordinate vector of the node i. In the development that follows, four-node iso-parametric quadrilateral elements are used throughout. To calculate $\overline{H}_{ij}$, the derivative $\overline{g}$ has to be evaluated

$$\overline{H}_{ij} = \int\limits_{S_j} \overline{g}\mathrm{d}S = \int\limits_{S_j} \frac{\partial g}{\partial \hat{n}} \mathrm{d}S = \int\limits_{S_j} (\nabla g)^T \hat{n} \mathrm{d}S \tag{24.12}$$

where

$$\hat{n} = \begin{bmatrix} n_x \\ n_y \\ n_z \end{bmatrix} \tag{24.13}$$

and

$$\nabla g = \begin{bmatrix} -\frac{xe^{-ik\sqrt{(x_i-x_j)^2+(y_i-y_j)^2+(z_i-z_j)^2}}}{4\pi\sqrt{(x_i-x_j)^2+(y_i-y_j)^2+(z_i-z_j)^2}}\left(ik + \frac{1}{\sqrt{(x_i-x_j)^2+(y_i-y_j)^2+(z_i-z_j)^2}}\right) \\ -\frac{ye^{-ik\sqrt{(x_i-x_j)^2+(y_i-y_j)^2+(z_i-z_j)^2}}}{4\pi\sqrt{(x_i-x_j)^2+(y_i-y_j)^2+(z_i-z_j)^2}}\left(ik + \frac{1}{\sqrt{(x_i-x_j)^2+(y_i-y_j)^2+(z_i-z_j)^2}}\right) \\ -\frac{ze^{-ik\sqrt{(x_i-x_j)^2+(y_i-y_j)^2+(z_i-z_j)^2}}}{4\pi\sqrt{(x_i-x_j)^2+(y_i-y_j)^2+(z_i-z_j)^2}}\left(ik + \frac{1}{\sqrt{(x_i-x_j)^2+(y_i-y_j)^2+(z_i-z_j)^2}}\right) \end{bmatrix} \tag{24.14}$$

For a four-node iso-parametric quadrilateral element, the pressure p and the normal velocity v at any position on the element can be defined by their nodal values and linear shape functions, i.e.

$$v(\xi, \eta) = N_1 v_1 + N_2 v_2 + N_3 v_3 + N_4 v_4 = \begin{bmatrix} N_1 & N_2 & N_3 & N_4 \end{bmatrix} \begin{bmatrix} v_1 \\ v_2 \\ v_3 \\ v_4 \end{bmatrix} \tag{24.15}$$

$$p(\xi, \eta) = N_1 p_1 + N_2 p_2 + N_3 p_3 + N_4 p_4 = \begin{bmatrix} N_1 & N_2 & N_3 & N_4 \end{bmatrix} \begin{bmatrix} p_1 \\ p_2 \\ p_3 \\ p_4 \end{bmatrix} \tag{24.16}$$

where the shape functions in the element coordinate system are,

$$N_1 = \frac{1}{4}(\xi - 1)(\eta - 1) \quad N_2 = -\frac{1}{4}(\xi + 1)(\eta - 1)$$
$$N_3 = \frac{1}{4}(\xi + 1)(\eta + 1) \quad N_2 = -\frac{1}{4}(\xi - 1)(\eta + 1) \tag{24.17}$$

The four-node quadrilateral element can have any arbitrary orientation in the three-dimensional space. Using the shape functions (24.20), the integral on the left-hand side of Eq. 24.6, considered over one element j, can be written as:

$$\int_{S_j} p\overline{g}_i \mathrm{d}s = \int_{S_j} \left[N_1 \; N_2 \; N_3 \; N_4 \right] \overline{g}_i \mathrm{d}S \begin{bmatrix} p_1 \\ p_2 \\ p_3 \\ p_4 \end{bmatrix}_j = \left[\overline{h}_{ij}^1 \; \overline{h}_{ij}^2 \; \overline{h}_{ij}^3 \; \overline{h}_{ij}^4 \right] \begin{bmatrix} p_1 \\ p_2 \\ p_3 \\ p_4 \end{bmatrix}_j \tag{24.18}$$

while that on the right-hand side

$$\int_{S_j} gv\mathrm{d}S = \int_{S_j} \left[N_1 \; N_2 \; N_3 \; N_4 \right] g_i \mathrm{d}S \begin{bmatrix} v_1 \\ v_2 \\ v_3 \\ v_4 \end{bmatrix}_j = \left[g_{ij}^1 \; g_{ij}^2 \; g_{ij}^3 \; g_{ij}^n \right] \begin{bmatrix} v_1 \\ v_2 \\ v_3 \\ v_4 \end{bmatrix}_i \tag{24.19}$$

where

$$\overline{h}_{ij}^k = \int_{S_j} N_k \overline{h}_j^k dS, \quad \mathrm{k} = 1, 2, 3, 4 \tag{24.20}$$

$$\overline{g}_{ij}^k = \int_{S_j} N_k \overline{g}_j dS, \quad \mathrm{k} = 1, 2, 3, 4 \tag{24.21}$$

The integration in Eqs. 24.18 and 24.19 can be carried out using Gauss points [4, 11]. These Gauss points in the iso-parametric system are defined as:

$$(\xi_1, \eta_1) = \frac{1}{\sqrt{3}}(1, -1) \quad (\xi_2, \eta_2) = \frac{1}{\sqrt{3}}(1, 1)$$
$$(\xi_3, \eta_3) = \frac{1}{\sqrt{3}}(-1, 1) \quad (\xi_4, \eta_4) = \frac{1}{\sqrt{3}}(-1, -1) \tag{24.22}$$

Substituting Eqs. 24.18 and 24.19 into Eq. 24.6 for all elements j, there is obtained

$$c_i p_i - p_{inc} - \sum_{j=1, j\neq i}^{N} \left[\overline{h}_{ij}^1 \ \overline{h}_{ij}^2 \ \overline{h}_{ij}^3 \ \overline{h}_{ij}^4 \right] \begin{bmatrix} p_1 \\ p_2 \\ p_3 \\ p_4 \end{bmatrix}_j = i\rho\omega_0 \sum_{j=1}^{N} \left[g_{ij}^1 \ g_{ij}^2 \ g_{ij}^3 \ g_{ij}^4 \right] \begin{bmatrix} v_{n1} \\ v_{n2} \\ v_{n3} \\ v_{n4} \end{bmatrix}_j \tag{24.23}$$

Equation 24.23 can be rewritten as

$$\sum_{j=1}^{N} \left[h_{ij}^1 \ h_{ij}^2 \ h_{ij}^3 \ h_{ij}^4 \right] \begin{bmatrix} p_1 \\ p_2 \\ p_3 \\ p_4 \end{bmatrix}_j = i\rho\omega_0 \sum_{j=1}^{N} \left[g_{ij}^1 \ g_{ij}^2 \ g_{ij}^3 \ g_{ij}^4 \right] \begin{bmatrix} v_{n1} \\ v_{n2} \\ v_{n3} \\ v_{n4} \end{bmatrix}_j + p_{inc} \tag{24.24}$$

Hence the discretized equation forms a set of simultaneous linear equations, which relates the pressure p_i at field point i due to the boundary conditions p to v at source surface S_i of element i and the incident pressure p_{inc}.

In matrix form:

$$[\boldsymbol{H}]\{p\} = i\rho_0\omega[\boldsymbol{G}]\{v\} + \{p_{inc}\} \tag{24.25}$$

where, **H** and **G** are two $N \times N$ matrices of influence coefficients, while p and v are vectors of dimension N representing total pressure and normal velocity on the boundary elements. This matrix equation can be solved if the boundary condition $v = \partial p/\partial n$ and the incident acoustic pressure field p_{inc} are known.

At this point, a few remarks are necessary. Proper interpretation should be given to the diagonal terms of [H] in Eq. 24.25 as implied by the original boundary integral (24.4) since these terms concern the evaluation of influence coefficient for which the field point is located at the source element. Accordingly, [H] should be written as

$$[\boldsymbol{H}] = [\boldsymbol{H}]^D + [\boldsymbol{H}]^{OD} \tag{24.26}$$

i.e. the diagonal and the off-diagonal part.

The matrix $[\text{H}]^{\text{D}}$ as implied in (24.26) can be written as $[\text{H}]^{\text{D}} = [\overline{H}]^{\text{D}} + [\text{C}]$ where **C** is space angle constant as implied in Eq. 24.4 which is the quotient of $\Omega/4\pi$ and $\overline{H}$ is the matrix implied by the second term of Eq. 24.23. For a node coinciding with three or four element corners, Ω is the space angle toward the acoustic medium, and the space angle for a sphere is 4π. For a smooth surface the space angle is 2π, and **C** = ½.

The boundary integral equation of Eq. 24.4 fails at frequencies coincident with the interior cavity frequencies of homogeneous Dirichlet boundary conditions [6]. In the case of the formulation of the exterior problem, these frequencies correspond to the natural frequencies of acoustic resonances in the interior region. When the interior region resonates, the pressure field inside the interior region has non-trivial solution.

Since the interior problem and the exterior problem share similar integral operators, the exterior integral equation may also break down. The discretized equation of the [H] matrix in Eq. 24.25 becomes ill-conditioned when the exciting frequency is close to the interior frequencies, thus providing an erroneous acoustic loading matrix. This problem could be overcome by using the CHIEF [7, 12] or Burton-Miller method [13, 14] or a recent technique utilizing Singular Value Decomposition SVD [15] and Fredholm alternative theorem [7, 14, 16]. To avoid non-uniqueness problem, reference [17] describes special treatment to be carried out to inspect whether the **H** matrix is ill-behaved or not by utilizing SVD updating technique. The present method, however, resorts to the utilization of CHIEF method, if Eq. 24.25 ill-behaved. Such approach applied to the present method has proved to be successful, as indicated in Fig. 24.3.

24.3 BEM–FEM Acoustic-Aeroelastic Coupling (AAC)

Following [11], the BE region is treated as a super finite element and its stiffness matrix is computed and assembled into the global stiffness matrix, and identified as the coupling to finite elements. The state of affairs is schematically depicted in Fig. 24.4a.

The utilization of FEM on the structural domain leads to a system of simultaneous equations which relate the displacements at all the nodes to the *nodal forces*. In the BEM, on the other hand, a relationship between nodal displacements and *nodal tractions* is established. Representing the elastic structure by FE model, the structural dynamic equation of motion is given by [18]

$$[\boldsymbol{M}]\{\ddot{x}\} + [\boldsymbol{C}]\{\dot{x}\} + [\boldsymbol{K}]\{x\} = \{\boldsymbol{F}\} \tag{24.27a}$$

where **M**, **C** and **K** are structural mass, damping and stiffness, respectively, which are expressed as matrices in a FE model, while **F** is the given external forcing function vector, and $\{x\}$ is the structural displacement vector. The incorporation of the self-excited aerodynamic effects to the structural dynamics equation can be written as [18, 19] and Zona-Tech (1992) [20]:

$$[\boldsymbol{M}]\{\ddot{x}\} + [\boldsymbol{C}]\{\dot{x}\} + [\boldsymbol{K}]\{x\} - q_\infty[\boldsymbol{A}(ik)]\{x\} = \{0\} \tag{24.27b}$$

where **A**(ik) is an aerodynamic influence coefficient after applying aero structure coupling from the control points of aerodynamic boxes to the structural finite element grid points as elaborated in [3].

Taking into account the acoustic pressure p on the structure at the fluid–structure interface as a separate excitation force, the acoustic-structure problem can be obtained from Eqs. 24.27a, 24.27b by introducing a fluid–structure coupling term given by $[\mathbf{L}]\{p\}$. It follows *that*

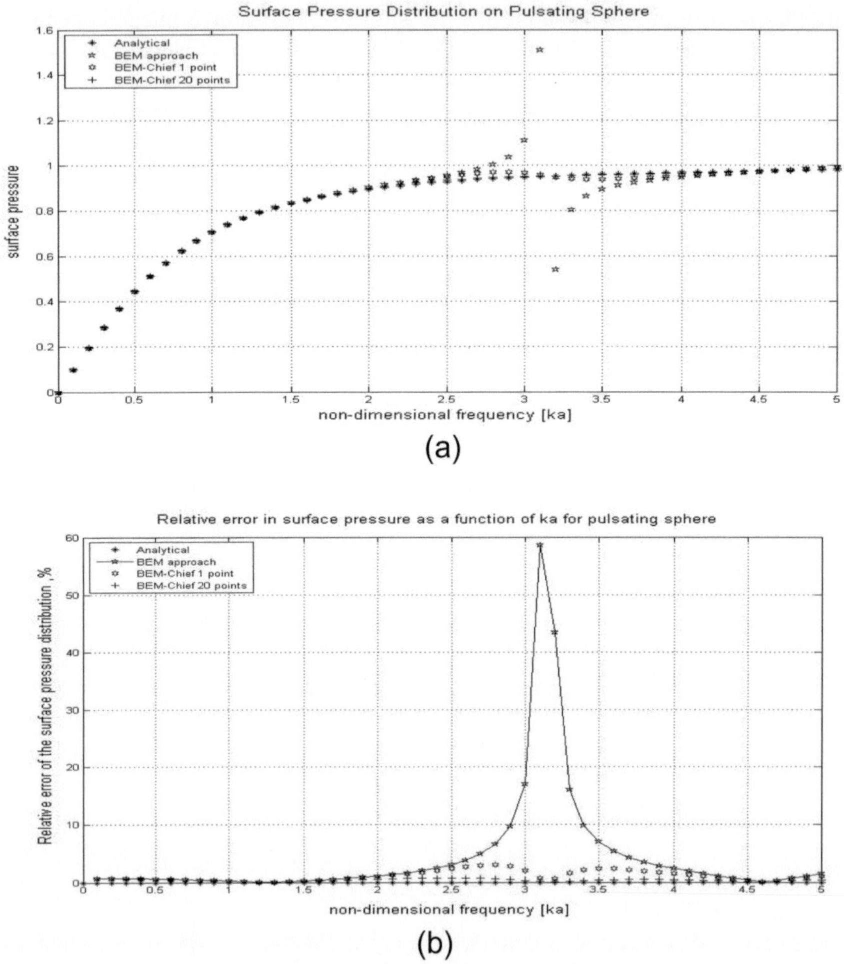

Fig. 24.3 **a** Surface pressure distribution on pulsating sphere for analytical, BEM, and BEM-CHIEF solution for one and two CHIEF point; **b** error of the surface pressure distribution in (**a**) with respect to the exact solution. 384 isometric surface elements are utilized, and the use of one (1) CHIEF point has been able to eliminate the spurious solution with reasonably good accuracy

$$[\boldsymbol{M}]\{\ddot{x}\} + [\boldsymbol{C}]\{\dot{x}\} + [\boldsymbol{K}]\{x\} - q_\infty[\boldsymbol{A}(ik)]\{x\} + [\boldsymbol{L}]\{p\} = \{\boldsymbol{F}\} \tag{24.28}$$

where **L** is a coupling matrix of size $M \times N$ in the BEM/FEM coupling thus formulated. M is the number of FE degrees of freedom and N is the number of BE nodes on the coupled boundary. For the BE part of the surface at the fluid–structure interface a, Eq. 24.25 can be utilized.

The global coupling matrix **L** transforms the acoustic fluid pressure acting on the nodes of boundary elements on the entire fluid–structure interface surface a, to nodal forces on the finite elements of the structure. Hence **L** consists of n assembled local

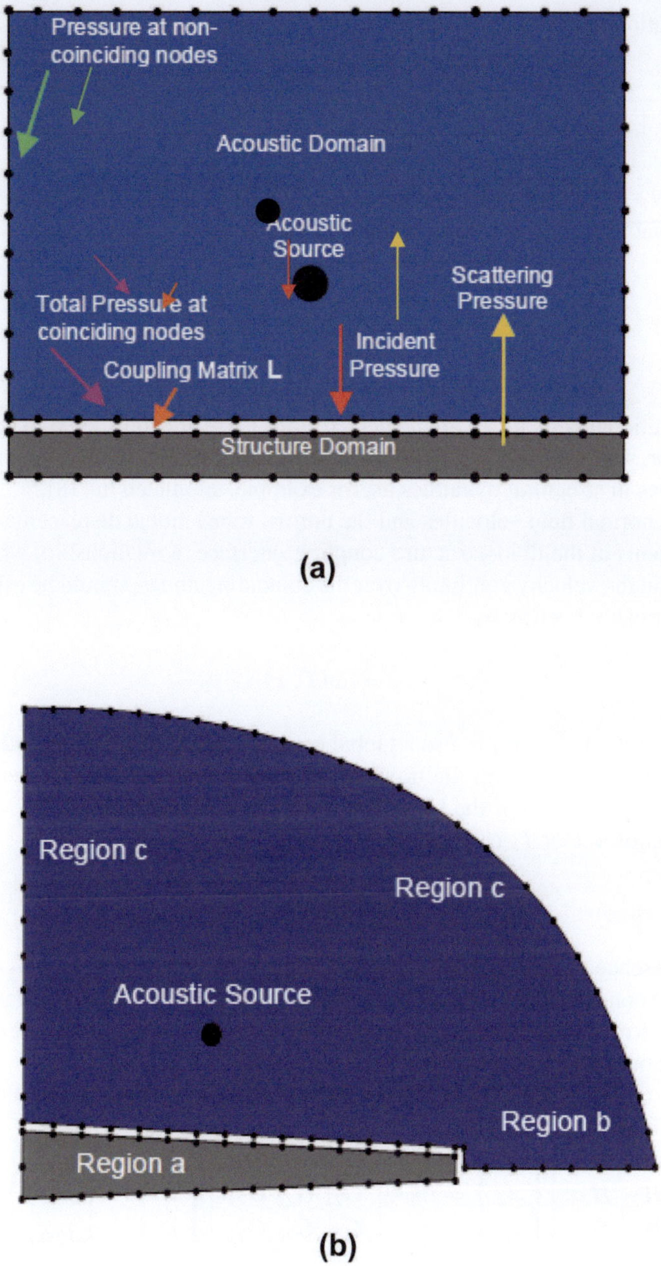

Fig. 24.4 **a** Schematic of fluid–structure interaction domain. **b** Schematic of FE-BE problem representing quarter space problem domains for half wing

transformation matrices L_e, given by

$$\boldsymbol{L}_e = \int\limits_{S_e} \boldsymbol{N}_F^T \boldsymbol{n} \boldsymbol{N}_B \mathrm{d}S \tag{24.29}$$

in which N_F is the shape function matrix for the finite element and N_B is the shape function matrix for the boundary element. It can be shown that:

$$\boldsymbol{N}_F = \begin{bmatrix} 1\,0\,0\,0\,0 \\ 0\,1\,0\,0\,0 \\ 0\,0\,1\,0\,0 \end{bmatrix} [\boldsymbol{N}_i] \tag{24.30}$$

The rotational parts in N_F are neglected since these are considered to be small in comparison with the translational ones in the BE-FE coupling, consistent with the assumptions in structural dynamics as, for example, stipulated in [18].

For the normal fluid velocities and the normal translational displacements on the shell elements at the fluid–structure coupling interface, a relationship, which takes into account the velocity continuity over the coinciding nodes, should be established. This relationship is given by

$$\mathrm{v} = i\omega(\boldsymbol{T}, x) \tag{24.31}$$

Similar to **L**, **T** $(n \times m)$ is also a global coupling matrix that connects the normal velocity of a BE node with the translational displacements of FE nodes obtained by taking the transpose of the boundary surface normal vector n [4, 21]. The local transformation vector T_e can then be written as:

$$T_e = n^T \tag{24.32}$$

The presence of an acoustic source can further be depicted by Fig. 24.4b. Three regions are considered, i.e. *a*, *b* and c; region *a* is the fluid–structure interface region, where FEM mesh and BEM mesh coincide and region *b* and *c* is the region where all of the boundary conditions (pressure or velocity) are known.

For the coupled FEM–BEM regions, BEM equation can now be written as:

$$\begin{bmatrix} \boldsymbol{H}_{11} & \boldsymbol{H}_{12} & \boldsymbol{H}_{13} \\ \boldsymbol{H}_{21} & \boldsymbol{H}_{22} & \boldsymbol{H}_{23} \\ \boldsymbol{H}_{31} & \boldsymbol{H}_{32} & \boldsymbol{H}_{33} \end{bmatrix} \begin{Bmatrix} p_a \\ p_b \\ p_c \end{Bmatrix} = i\rho_0\omega \begin{bmatrix} \mathbf{G}_{11} & \mathbf{G}_{12} & \mathbf{G}_{13} \\ \mathbf{G}_{21} & \mathbf{G}_{22} & \mathbf{G}_{23} \\ \mathbf{G}_{31} & \mathbf{G}_{32} & \mathbf{G}_{33} \end{bmatrix} \begin{Bmatrix} v_a \\ v_b \\ v_c \end{Bmatrix} + \begin{Bmatrix} p_{inc_a} \\ p_{inc_b} \\ p_{inc_c} \end{Bmatrix} \tag{24.33}$$

Considering $\boldsymbol{v}_a = i\omega(\boldsymbol{T}.x)$, BEM equation can be modified as:

$$\begin{bmatrix} \boldsymbol{H}_{11} & \boldsymbol{H}_{12} & \boldsymbol{H}_{13} \\ \boldsymbol{H}_{21} & \boldsymbol{H}_{22} & \boldsymbol{H}_{23} \\ \boldsymbol{H}_{31} & \boldsymbol{H}_{32} & \boldsymbol{H}_{33} \end{bmatrix} \begin{Bmatrix} p_a \\ p_b \\ p_c \end{Bmatrix} = i\rho_0\omega \begin{bmatrix} \boldsymbol{G}_{11} & \boldsymbol{G}_{12} & \boldsymbol{G}_{13} \\ \boldsymbol{G}_{21} & \boldsymbol{G}_{22} & \boldsymbol{G}_{23} \\ \boldsymbol{G}_{31} & \boldsymbol{G}_{32} & \boldsymbol{G}_{33} \end{bmatrix} \begin{Bmatrix} i\omega[\boldsymbol{T}]x \\ v_b \\ v_c \end{Bmatrix} + \begin{Bmatrix} p_{inc_a} \\ p_{inc_b} \\ p_{inc_c} \end{Bmatrix} \tag{24.34}$$

or:

$$\begin{aligned} \boldsymbol{H}_{11}p_a + \boldsymbol{H}_{12}p_b + \boldsymbol{H}_{13}p_c &= -\rho_0\omega^2\boldsymbol{G}_{11}\boldsymbol{T}x + i\rho_0\omega\boldsymbol{G}_{12}v_b + i\rho_0\omega\boldsymbol{G}_{13}v_c + \{p_{inc_a}\} \\ \boldsymbol{H}_{21}p_a + \boldsymbol{H}_{22}p_b + \boldsymbol{H}_{23}p_c &= -\rho_0\omega^2\boldsymbol{G}_{21}\boldsymbol{T}x + i\rho_0\omega\boldsymbol{G}_{22}v_b + i\rho_0\omega\boldsymbol{G}_{23}v_c + \{p_{inc_b}\} \\ \boldsymbol{H}_{31}p_a + \boldsymbol{H}_{32}p_b + \boldsymbol{H}_{33}p_c &= -\rho_0\omega^2\boldsymbol{G}_{31}\boldsymbol{T}x + i\rho_0\omega\boldsymbol{G}_{32}v_b + i\rho_0\omega\boldsymbol{G}_{33}v_c + \{p_{inc_c}\} \end{aligned} \tag{24.35}$$

If the pressure boundary condition on $b(p_b)$, velocity boundary condition on $c(v_c)$ and the incident pressure on a, b and c are known, by taking to the left side all the unknown the above equation can be written as:

$$\begin{aligned} \rho_0\omega^2\boldsymbol{G}_{11}\boldsymbol{T}x + \boldsymbol{H}_{11}p_a - i\rho_0\omega\boldsymbol{G}_{12}v_b + \boldsymbol{H}_{13}p_c &= -\boldsymbol{H}_{12}p_b + i\rho_0\omega\boldsymbol{G}_{13}v_c + \{p_{inc_a}\} \\ \rho_0\omega^2\boldsymbol{G}_{21}\boldsymbol{T}x + \boldsymbol{H}_{21}p_a - i\rho_0\omega\boldsymbol{G}_{22}v_b + \boldsymbol{H}_{23}p_c &= -\boldsymbol{H}_{22}p_b + i\rho_0\omega\boldsymbol{G}_{23}v_c + \{p_{inc_b}\} \\ \rho_0\omega^2\boldsymbol{G}_{31}\boldsymbol{T}x + \boldsymbol{H}_{31}p_a - i\rho_0\omega\boldsymbol{G}_{32}v_b + \boldsymbol{H}_{33}p_c &= -\boldsymbol{H}_{32}p_b + i\rho_0\omega\boldsymbol{G}_{33}v_c + \{p_{inc_c}\} \end{aligned} \tag{24.36}$$

Since the pressure p on FEM equation lies in region a, Eq. 24.28 can be written as

$$[\boldsymbol{M}]\{\ddot{x}\} + [\boldsymbol{C}]\{\dot{x}\} + [\boldsymbol{K}]\{x\} - q_\infty[\boldsymbol{A}(ik)]\{x\} + [\boldsymbol{L}]\{p_a\} = \{\boldsymbol{F}\} \tag{24.37}$$

where p_a is the total acoustic pressure resulting from the application of acoustic disturbance force to the structure, which consists of the incident acoustic pressure p_{inc} and scattering acoustic pressure p_{sc}. The scattering acoustic pressure will be dependent on the dynamic response of the structure due to the incident acoustic pressure. Following the general practice in structural dynamics, solutions of Eq. 24.37 are sought by considering synchronous motion with harmonic frequency ω. Correspondingly, Eq. 24.37 reduces to:

$$\left[\boldsymbol{K} + i\omega\boldsymbol{C} - \omega^2\boldsymbol{M}\right]\{\overline{x}\} - q_\infty[\boldsymbol{A}(ik)]\{x\} + [\boldsymbol{L}]\{\overline{p}_a\} = \{\overline{\boldsymbol{F}}\} \tag{24.38}$$

where

$$x = \overline{x}e^{i\omega t};\ p_a = \overline{p}_a e^{i\omega t} \tag{24.39}$$

or, dropping the bar sign for convenience, but keeping the meaning in mind, Eq. 24.38 can be written as

$$\left[\boldsymbol{K} + i\omega\boldsymbol{C} - \omega^2\boldsymbol{M}\right]\{x\} - q_\infty[\boldsymbol{A}(ik)]\{x\} + [\boldsymbol{L}]\{p_a\} = \{\boldsymbol{F}\} \tag{24.40}$$

Combining Eqs. 24.36 and 24.40, the coupled BEM–FEM equation can then be written as:

$$\begin{bmatrix} \left[\boldsymbol{K} + i\omega\boldsymbol{C} - \omega^2\boldsymbol{M}\right] - q_\infty[\boldsymbol{A}(ik)]\{x\} & \boldsymbol{L} & 0 & 0 \\ \rho_0\omega^2\boldsymbol{G}_{11}T & \boldsymbol{H}_{11} & -i\rho_0\omega\boldsymbol{G}_{12} & \boldsymbol{H}_{13} \\ \rho_0\omega^2\boldsymbol{G}_{21}T & \boldsymbol{H}_{21} & -i\rho_0\omega\boldsymbol{G}_{22} & \boldsymbol{H}_{23} \\ \rho_0\omega^2\boldsymbol{G}_{31}T & \boldsymbol{H}_{31} & -i\rho_0\omega\boldsymbol{G}_{32} & \boldsymbol{H}_{33} \end{bmatrix} \begin{Bmatrix} x \\ p_a \\ v_b \\ p_c \end{Bmatrix} = \begin{Bmatrix} \boldsymbol{F} \\ -\boldsymbol{H}_{12}p_b + i\rho_0\omega\boldsymbol{G}_{13}v_c + p_{inc_a} \\ -\boldsymbol{H}_{22}p_b + i\rho_0\omega\boldsymbol{G}_{23}v_c + p_{inc_b} \\ -\boldsymbol{H}_{32}p_b + i\rho_0\omega\boldsymbol{G}_{33}v_c + p_{inc_c} \end{Bmatrix} \tag{24.41}$$

This equation forms the basis for the treatment of the fluid–structure interaction in a unified fashion. The solution vector consisting of the displacement vector of the structure and total acoustic pressure on the boundaries of the acoustic domain, including the acoustic-structure interface, can be obtained by solving Eq. 24.41 as a dynamic response.

24.4 Further Treatment for AAC; Acoustic-Aerodynamic Analogy

At this point, the solution approach philosophy is in order. Analogous to the treatment of dynamic aeroelastic stability problem of structure, in which the aerodynamic effects can be distinguished into motion-independent (self-excited) and motion-induced aerodynamic forces, the effect of acoustic pressure disturbance to the aeroelastic structure (acousto-aero-elastic problem) can be viewed to consist of structural motion-independent incident acoustic pressure (excitation acoustic pressure) and structural motion-dependent acoustic pressure, which is known as the scattering pressure. However the scattering acoustic pressure is also dependent on the incident acoustic pressure. The consequence of such treatment has been adopted in the above section and will further be implemented in the subsequent development.

For aeroelastic calculation purposes, further treatment to simplify Eq. 24.41 will be carried out. Since the pressure boundary condition on b ($p_b = 0$) and velocity boundary condition on c ($v_c = 0$), Eq. 24.36 can be written as:

$$\rho_0\omega^2\boldsymbol{G}_{11}\boldsymbol{T}x + \boldsymbol{H}_{11}p_a - i\rho_0\omega\boldsymbol{G}_{12}v_b + \boldsymbol{H}_{13}p_c = p_{inc_a} \tag{24.42a}$$

$$\rho_0\omega^2\boldsymbol{G}_{21}\boldsymbol{T}x + \boldsymbol{H}_{21}p_a - i\rho_0\omega\boldsymbol{G}_{22}v_b + \boldsymbol{H}_{23}p_c = p_{inc_b} \tag{24.42b}$$

$$\rho_0\omega^2 \boldsymbol{G}_{31}\boldsymbol{T}x + \boldsymbol{H}_{31}p_a - i\rho_0\omega\boldsymbol{G}_{32}v_b + \boldsymbol{H}_{33}p_c = p_{inc_c} \quad (24.42c)$$

Since $\mathbf{G}_{22}$ and $\mathbf{H}_{33}$ is square matrix, Eqs. 24.42b and 24.42c can be written as

$$v_b = -\frac{i}{\rho_0\omega}[\boldsymbol{G}_{22}]^{-1}\left(\rho_0\omega^2\boldsymbol{G}_{21}\boldsymbol{T}x + \boldsymbol{H}_{21}p_a + \boldsymbol{H}_{23}p_c - p_{inc_b}\right) \quad (24.43a)$$

$$p_c = -[\boldsymbol{H}_{33}]^{-1}\left(\rho_0\omega^2\boldsymbol{G}_{31}\boldsymbol{T}x + \boldsymbol{H}_{31}p_a - i\rho_0\omega\boldsymbol{G}_{32}v_b - p_{inc_c}\right) \quad (24.43b)$$

Substituting Eq. 24.43b into Eq. 24.42b

$$\rho_0\omega^2\boldsymbol{A}_{21}\boldsymbol{T}x + \boldsymbol{B}_{21}p_a - i\rho_0\omega\boldsymbol{A}_{22}v_b + \boldsymbol{B}_{23}p_{inc_c} - p_{inc_b} = \{0\} \quad (24.44)$$

where

$$\begin{aligned}
\boldsymbol{A}_{21} &= \left(\boldsymbol{G}_{21} - \boldsymbol{H}_{23}[\boldsymbol{H}_{33}]^{-1}\boldsymbol{G}_{31}\right)\\
\boldsymbol{B}_{21} &= \left(\boldsymbol{H}_{21} - \boldsymbol{H}_{23}[\boldsymbol{H}_{33}]^{-1}\boldsymbol{H}_{31}\right)\\
\boldsymbol{B}_{23} &= \left(\boldsymbol{H}_{23}[\boldsymbol{H}_{33}]^{-1}\right)\\
\boldsymbol{A}_{22} &= \left(\boldsymbol{G}_{22} - \boldsymbol{H}_{23}[\boldsymbol{H}_{33}]^{-1}\boldsymbol{G}_{32}\right)
\end{aligned} \quad (24.45)$$

Since $\mathbf{A}_{22}$ is square matrix, Eq. 24.44 can be written as:

$$v_b = -\frac{i}{\rho_0\omega}[\boldsymbol{A}_{22}]^{-1}\left(\rho_0\omega^2\boldsymbol{A}_{21}\boldsymbol{T}x + \boldsymbol{B}_{21}p_a + \boldsymbol{B}_{23}p_{inc_c} - p_{inc_b}\right) \quad (24.46)$$

Substituting Eq. 24.43a into Eq. 24.42c

$$\rho_0\omega^2\boldsymbol{C}_{\mathbf{31}}\boldsymbol{T}x + \boldsymbol{A}_{\mathbf{31}}p_a + \boldsymbol{A}_{33}p_c + \boldsymbol{B}_{32}p_{inc_b} - p_{inc_c} = \{0\} \quad (24.47)$$

where

$$\begin{aligned}
\boldsymbol{C}_{\mathbf{31}} &= \left(\boldsymbol{G}_{31} - \boldsymbol{G}_{32}[\boldsymbol{G}_{22}]^{-1}\boldsymbol{G}_{21}\right)\\
\boldsymbol{A}_{\mathbf{31}} &= \left(\boldsymbol{H}_{31} - \boldsymbol{G}_{32}[\boldsymbol{G}_{22}]^{-1}\boldsymbol{H}_{21}\right)\\
\boldsymbol{A}_{33} &= \left(\boldsymbol{H}_{33} - \boldsymbol{G}_{32}[\boldsymbol{G}_{22}]^{-1}\boldsymbol{H}_{23}\right)\\
\boldsymbol{B}_{32} &= \left(\boldsymbol{G}_{32}[\boldsymbol{G}_{22}]^{-1}\right)
\end{aligned} \quad (24.48)$$

Since $\mathbf{A}_{33}$ is square matrix, Eq. 24.47 can be written as:

$$p_c = -[\boldsymbol{A}_{33}]^{-1}\left(\rho_0\omega^2\boldsymbol{C}_{\mathbf{31}}\boldsymbol{T}x + \boldsymbol{A}_{\mathbf{31}}p_a + \boldsymbol{B}_{32}p_{inc_b} - p_{inc_c}\right) \quad (24.49)$$

Substituting Eqs. 24.46 and 24.49 into Eq. 24.42a

$$\rho_0\omega^2\boldsymbol{D}_{11}\boldsymbol{T}x + \boldsymbol{E}_{11}p_a + \boldsymbol{F}_{12}p_{inc_b} + \boldsymbol{F}_{13}p_{inc_c} = \left\{p_{inc_a}\right\} \quad (24.50)$$

where

$$\begin{aligned}
\boldsymbol{D}_{11} &= \left(\boldsymbol{G}_{11} - \boldsymbol{G}_{12}[\boldsymbol{A}_{22}]^{-1}\boldsymbol{A}_{21} - \boldsymbol{H}_{13}[\boldsymbol{A}_{33}]^{-1}\boldsymbol{C}_{31}\right)\\
\boldsymbol{E}_{11} &= \left(\boldsymbol{H}_{11} - \boldsymbol{G}_{12}[\boldsymbol{A}_{22}]^{-1}\boldsymbol{B}_{21} - \boldsymbol{H}_{13}[\boldsymbol{A}_{33}]^{-1}\boldsymbol{A}_{31}\right)\\
\boldsymbol{F}_{12} &= \boldsymbol{G}_{12}[\boldsymbol{A}_{22}]^{-1} - \boldsymbol{H}_{13}[\boldsymbol{A}_{33}]^{-1}\boldsymbol{B}_{32}\\
\boldsymbol{F}_{13} &= \boldsymbol{H}_{13}[\boldsymbol{A}_{33}]^{-1} - \boldsymbol{G}_{12}[\boldsymbol{A}_{22}]^{-1}\boldsymbol{B}_{23}
\end{aligned} \tag{24.51}$$

Since $\mathbf{E}_{11}$ is square matrix

$$p_a = -[\boldsymbol{E}_{11}]^{-1}\left(\rho_0\omega^2\boldsymbol{D}_{11}\boldsymbol{T}\{x\} - p_{inc_a} + \boldsymbol{F}_{12}p_{inc_b} + \boldsymbol{F}_{13}p_{inc_c}\right) \tag{24.52}$$

Matrix $\mathbf{E}_{11}$ and $\mathbf{D}_{11}$ are also a square matrix, finally by substituting Eq. 24.52 into Eq. 24.40 BEM–FEM aero-acoustic-structure coupling can be obtained as:

$$\begin{aligned}
&\left[\boldsymbol{K} + i\omega\boldsymbol{C} - \omega^2\boldsymbol{M}\right]\{x\} - q_\infty[\boldsymbol{A}(ik)]\{x\}\\
&+[\boldsymbol{L}]\left(-[\boldsymbol{E}_{11}]^{-1}\left(\rho_0\omega^2\boldsymbol{D}_{11}\boldsymbol{T}\{x\} - p_{inc_a} + \boldsymbol{F}_{12}p_{inc_b} + \boldsymbol{F}_{13}p_{inc_c}\right)\right) = \{\boldsymbol{F}\}
\end{aligned} \tag{24.53}$$

Incident pressure on region b and c will not influence the stability problem associated with the structures, and may at this point be disregarded. Hence, without considering damping matrix $\mathbf{C}$ Eq. 24.53 simplifies to:

$$\begin{aligned}
&\left[\boldsymbol{K} - \omega^2\boldsymbol{M}\right]\{x\} - q_\infty[\boldsymbol{A}(ik)]\{x\}\\
&\qquad -\rho_0\omega^2[\boldsymbol{L}][\boldsymbol{E}_{11}]^{-1}[\boldsymbol{D}_{11}][\boldsymbol{T}]\{x\} = -[\boldsymbol{L}][\boldsymbol{E}_{11}]^{-1}\left\{p_{inc_a}\right\} + \{\boldsymbol{F}\}
\end{aligned} \tag{24.54}$$

or

$$\left[\boldsymbol{K} - \omega^2\boldsymbol{M}\right]\{x\} - q_\infty[\boldsymbol{A}(ik)]\{x\} - \rho_0\omega^2\left[\boldsymbol{F}_{ac_{sc}}(k_w)\right]\{x\} = \left\{\boldsymbol{F}_{ac_{inc}}(k_w)\right\} + \{\boldsymbol{F}\} \tag{24.55}$$

where

$$\begin{aligned}
\left[\boldsymbol{F}_{ac_{sc}}(k_w)\right] &= [\boldsymbol{L}][\boldsymbol{E}_{11}]^{-1}[\boldsymbol{D}_{11}][\boldsymbol{T}]\\
\left\{\boldsymbol{F}_{ac_{inc}}(k_w)\right\} &= -[\boldsymbol{L}][\boldsymbol{E}_{11}]^{-1}\left\{p_{inc_a}\right\}
\end{aligned} \tag{24.56}$$

Equation 24.55 will not be solved directly since the size of the mass and stiffness matrices of the aircraft model are very large. Instead one uses the modal approach where the structural deformation $\{x\}$ is transformed to the generalized coordinate $\{q\}$ given by the following relation:

$$x = \Phi q \tag{24.57}$$

where Φ is the modal matrix whose columns contain the lower order natural modes. Pre-multiplying by Φ^{T} and converting dynamic pressure q_∞ into reduced frequency (k) as elaborated in [3], Eq. 24.56 can then be written as:

$$\begin{aligned}&\Phi^T\left[\boldsymbol{K}-\omega^2\left(\boldsymbol{M}+\frac{\rho}{2}\left(\frac{L}{k}\right)^2[\boldsymbol{A}(ik)]+\rho_0\left[\boldsymbol{F}_{ac_{sc}}(k_w)\right]\right)\right]\Phi\{q\}\\&=\Phi^T\left\{\boldsymbol{F}_{ac_{inc}}(k_w)\right\}+\Phi^T\{\boldsymbol{F}\}\end{aligned} \tag{24.58}$$

since all of the acoustic terms are functions of wave number (k_w) Eq. 24.58 can be solved by utilizing iterative procedure.

Incorporation of the scattering acoustic term along with the aerodynamic term in the second term of Eq. 24.58 can be regarded as one manifestation of the acoustic-aerodynamic analogy followed in this approach.

Further method of approach for the solution of the acousto-aeroelastic problem is then dealt with. Following the same procedure as developed in earlier work [1, 2], the acoustic excitation is incorporated by coupling it to the unsteady aerodynamic load in the flutter stability formulation. Linearity and principle of superposition have been assumed. Hence the acoustic loading can be superposed to the aerodynamic loading on the structure, and form the modified aeroelastic equation (acousto-aeroelastic equation) of the structural dynamic problem associated with acoustic and aerodynamic excitation. In the earlier work, tacit consideration is only given to the incident acoustic pressure as the acoustic excitation, without considering the acoustic scattering effects, and without considering **L**. The incident pressure p_{inc} was also assumed to belong to a certain class that allows its incorporation in the aerodynamics term. Thus Eq. 24.41 was treated in a decoupled fashion. Such approach has given instructive results.

In the present development, rigorous consideration has been devoted to the acoustic scattering problem. Two generic approaches to solve Eq. 24.41 can be followed. The first is to solve Eq. 24.41 as a stability equation in a "unified treatment," and the disturbance acoustic pressure already incorporates the total pressure, i.e. the incident plus the scattering acoustic pressure. The treatment of the incident acoustic pressure p_{inc} in Eq. 24.58 follows similar approach adopted for the scattering acoustic pressure by tuning p_{inc} so that it behaves like the aerodynamic terms in the modal Eq. 24.58. This is considered logical for the problem considered since the intention is to look at its enhancement effect to the aerodynamic one, thus only a class of p_{inc} will meet the eigenvalue requirements of Eq. 24.58. This approach will be elaborated further in the following section. The second generic approach is to solve Eq. 24.41 as a dynamic response problem due to acoustic excitation. The left-hand side of Eq. 24.41 incorporates the scattering acoustic pressure term. Appropriate algebraic manipulation is carried out to allow modal approach of p_{inc} [17, 22].

24.5 Acoustic Modified Flutter Formulation (Stability Problem Using k-Method)

For the calculation of the influence of acoustic effect on aeroelastic stability problem, Eq. 24.58 can be further formulated by making special treatment to the acoustic incident force and without other external forces ($\mathbf{F} = 0$) in the right-hand side. This treatment can be made by "tuning" that term to behave like the aerodynamic terms in generalized variables, in addition to the treatment of the scattering acoustic pressure. This assumption has been made by assuming linearity and principle of superposition. This assumption allows the superposition of the acoustic loading to the aerodynamic loading on the structure and forms the modified aeroelastic equation (acousto-aeroelastic equation) of the structural dynamic problem associated with acoustic and aerodynamic excitation. Let define

$$\left\{\boldsymbol{F}^*_{ac_{inc}}(k_w)\right\} = \Phi^T\left\{\boldsymbol{F}_{ac_{inc}}(k_w)\right\} \tag{24.59}$$

Then Eq. 24.58 can be written as:

$$\left[\boldsymbol{K}^* - \omega^2\left(\boldsymbol{M}^* + \frac{\rho}{2}\left(\frac{L}{k}\right)^2\left[\boldsymbol{A}^*(ik)\right] + \rho_0\left[\boldsymbol{F}^*_{ac_{sc}}(k_w)\right]\right)\right]\{q\} = \omega^2\left[\boldsymbol{F}^{**}_{ac_{inc}}(k_w)\right]\{q\} \tag{24.60}$$

where

$$\boldsymbol{M}^* = \Phi^T\boldsymbol{M}\Phi = \text{ generalized mass matrix} \tag{24.61a}$$

$$\boldsymbol{K}^* = \Phi^T\boldsymbol{K}\Phi = \text{ generalized stiffness matrix} \tag{24.61b}$$

$$\boldsymbol{A}^*(ik) = \Phi^T\boldsymbol{A}(ik)\Phi = \text{ generalized aero matrix} \tag{24.61c}$$

$$\boldsymbol{F}^*_{ac_{sc}}(k_w) = \Phi^T\left[\boldsymbol{F}_{ac_{sc}}(k_w)\right]\Phi \tag{24.61d}$$

$$\left(\boldsymbol{F}^{**}_{ac_{inc}}\right)_i = \frac{\left(\boldsymbol{F}^*_{ac_{inc}}\right)_i}{\omega_i^2 q_i} \tag{24.61e}$$

As modified aeroelastic stability problem, Eq. 24.60 can be written as

$$\left[\boldsymbol{K}^* - \omega^2\left(\boldsymbol{M}^* + \frac{\rho}{2}\left(\frac{L}{k}\right)^2\left[\boldsymbol{A}^*(ik)\right] + \rho_0\left[\boldsymbol{F}^*_{ac_{sc}}(k_w)\right] + \left[\boldsymbol{F}^{**}_{ac_{inc}}(k_w)\right]\right)\right]\{q\} = \{0\} \tag{24.62}$$

which can be simplified as:

$$[\boldsymbol{M}^{**} - \lambda \boldsymbol{K}]q = 0 \tag{24.63}$$

where

$$[\boldsymbol{M}^{**}] = \left[\boldsymbol{K}^{*} - \omega^{2}\left(\boldsymbol{M}^{*} + \frac{\rho}{2}\left(\frac{L}{k}\right)^{2}[\boldsymbol{A}^{*}(ik)] + \rho_{0}[\boldsymbol{F}^{*}_{ac_{sc}}(k_{w})] + [\boldsymbol{F}^{**}_{ac_{inc}}(k_{w})]\right)\right] \tag{24.64}$$

and

$$\lambda = \frac{1 + ig}{\omega^{2}} \tag{24.65}$$

Equation 24.64 is solved as an eigenvalue problem for a series of values for parameters k and ρ. Since $\mathbf{M}^{**}$ is in general a complex matrix, the eigenvalues λ are also complex numbers. For n structural modes, there are n eigenvalues corresponding to n modes at each k. The airspeed, frequency and structural damping are related to the eigenvalue λ as follows:

$$\omega_{f} = \frac{1}{\sqrt{Re(\lambda)}} 24.66\text{a},\ U_{f} = \frac{\omega_{f} b}{k} 24.66\text{b and } g = \frac{\text{Im}(\lambda)}{\text{Re}(\lambda)} \tag{24.66c}$$

To evaluate the flutter speed, *V*–*g* and *V*–*f* diagrams are constructed [18, 23]. The V-g diagram plots the structural damping as a function of velocity, and the V-f diagram plots the frequency as a function of velocity. The flutter critical speeds are indicated in the *V*–*g* diagram as the lowest velocity V at which the g curve crosses the V axis from its negative (stable region) to its positive value (unstable region), i.e. when $g = 0$.

24.6 Numerical Results

24.6.1 Acoustic Boundary Element Simulation

In order to verify the validity of the boundary elements acoustic models, a numerical test case is conducted to test the validity of the method. To avoid complexity, the acoustic source is assumed to be a monopole source which creates the acoustic pressure. This acoustic pressure is interacting with the unsteady aerodynamic forces.

For a pulsating sphere an exact solution for acoustic pressure a at a distance r from the center of a sphere with radius a pulsating with uniform radial velocity U_a is

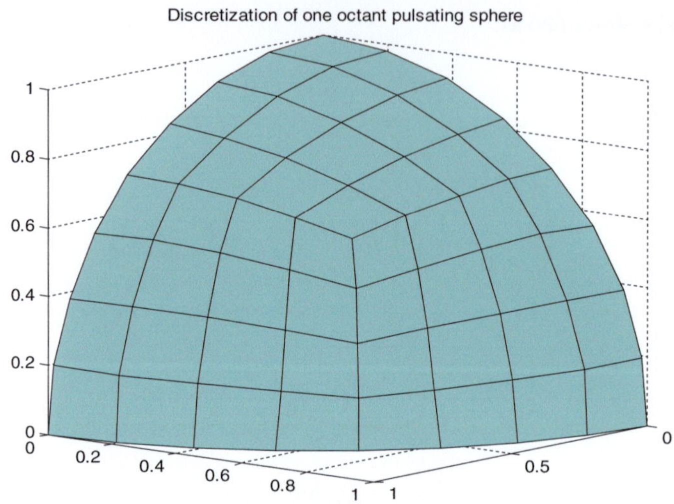

Fig. 24.5 Discretization of one octant pulsating sphere

$$p(r) = \frac{a}{r} U_a \frac{i z_0 ka}{1 + ika} e^{-ik(r-a)} \tag{24.67}$$

where z_0 is the acoustic characteristic impedance of the medium and k is the wave number.

Figure 24.5 shows the discretization of the surface elements of an acoustics pulsating sphere representing a monopole source. BEM calculation for scattering pressure from acoustic monopole source will be compared with exact results. Figure 24.6 shows the excellent agreement between the computational procedure developed and the exact one.

The calculation depicted in Fig. 24.6 was based on the assumption of $f = 10$ Hz, $\rho = 1.225$ kg/m^3 and $c = 340$ m/s. The excellent agreement of these results with exact calculation serves to validate the developed MATLAB® program for further utilization.

24.6.2 Coupled BEM–FEM Numerical Simulation

The BAH wing structure [18] and the surrounding boundary representing quarter space of the problem are discretized as shown in Fig. 24.7. The problem domain is divided into two parts. First the near field region is a quarter space with radius of two times the BAH wing span and is relatively more densely discretized; second, the intermediate to far-field region is a quarter space with radius ten times the BAH wing span and is less densely discretized compared to the near field region, and is modeled with BEM only.

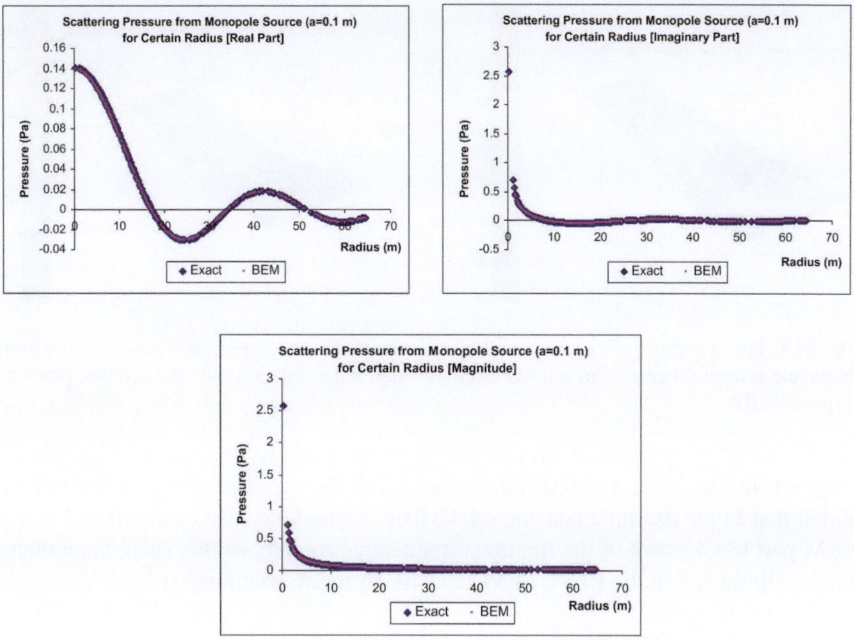

Fig. 24.6 Comparison of monopole source exact and BEM scattering pressure results

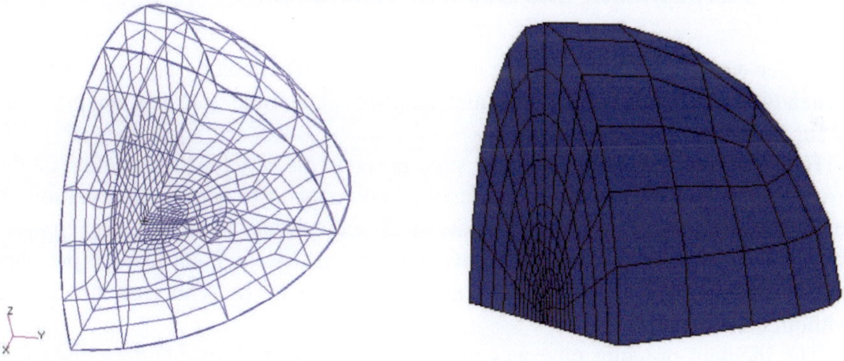

Fig. 24.7 D3-D domain representing of BAH wing structure and its surrounding boundary

The BAH wing which is modeled as FEM and BEM is subjected to an excitation due to an acoustic monopole source; the acoustic medium is air with density $\rho = 1.225\ \text{kg/m}^3$ and the sound velocity is $c = 340$ m/s. The monopole acoustic source is placed at the intersecting line of the half span and half chord planes of the BAH wing structure, and at about 0.1 m above the wing surface. The monopole source has the frequency of 10 Hz, radius of $a = 0.1$ m. The result of applying Eq. 24.61 for $F = 0$ is presented as the incident, total and scattering pressures drawn as color-coded

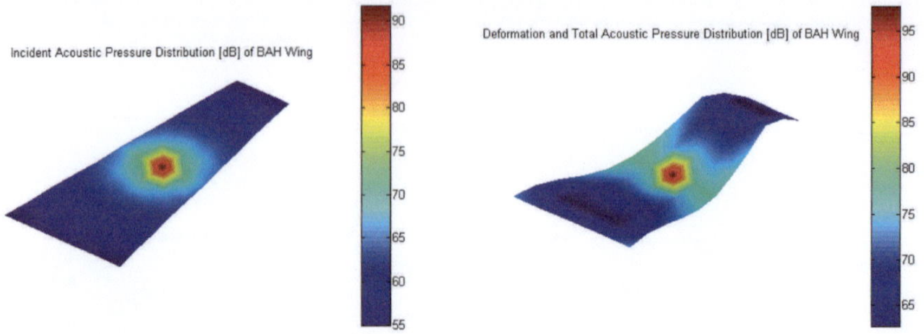

Fig. 24.8 Pressure distribution on symmetric equivalent BAH wing; **a** incident pressure (dB) from monopole acoustic source as an acoustic excitation and **b** deformation and total acoustic pressure response (dB)

diagrams in Fig. 24.8, which qualitatively exhibit the expected behavior. It could be added that in the example considered, CHIEF method has also been utilized in the BEM part to take care of the fictitious frequency problem, so that such phenomena can be eliminated in the BEM–FEM acoustic-structure coupling.

24.6.3 Flutter Calculation for Coupled Unsteady Aerodynamic and Acoustic Excitations

In earlier work [1, 2], the acoustic excitation is incorporated by coupling it to the unsteady aerodynamic load in the flutter stability calculation. Linearity and principle of superposition have been assumed.

Hence the acoustic loading can be superposed to the aerodynamic loading on the structure and form the modified aeroelastic equation (acousto-aeroelastic equation) of the structural dynamic problem associated with acoustic and aerodynamic excitation. In the earlier work, tacit consideration was only given to the incident acoustic pressure as the acoustic excitation, without considering the acoustic scattering effects and without considering *L*.

The incident pressure pinc was also assumed to belong to a certain class that allows its incorporation in the aerodynamics term. Thus Eq. 24.61 was treated in a decoupled fashion Such approach has resulted in the delay of the inception of flutter. Figure 24.9 shows the unsteady aerodynamics pressure (*Cp*) and mode shape of the structure when flutter occurs.

In the present development, rigorous consideration has been devoted to the acoustic scattering problem. by solving Eq. 24.61 as a stability equation in a "unified treatment," and the disturbance acoustic pressure already incorporates the total pressure, which has been "tuned" to behave like the aerodynamic terms in the modal Eq. 24.61. The solution is exhibited as Fig. 24.10, which shows that the flutter inception is delayed at a higher speed also.

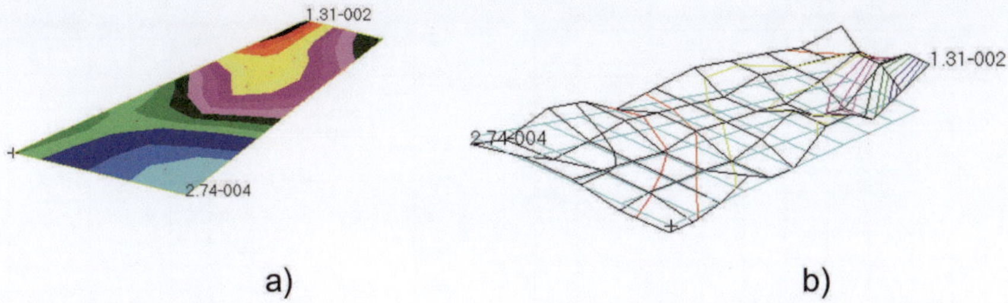

Fig. 24.9 **a** Unsteady aerodynamics pressure distribution (C_p), and **b** mode shape of the wing structure when flutter occurs

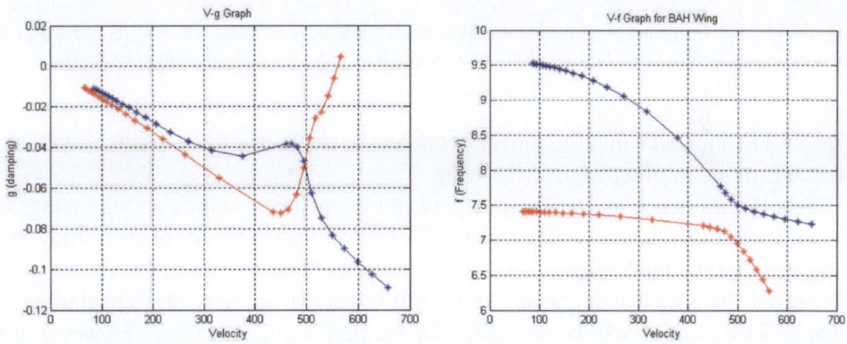

Fig. 24.10 Damping and frequency diagram for BAH wing calculated using V–g method written in MATLAB® for the acousto-aeroelastic problem (the total acoustic pressure already incorporates the scattering pressure)

24.7 AAC (Acoustic-Aeroelastic Coupling) Parametric Study

The computational scheme for the distribution of acoustic pressure on the surface of the pulsating sphere, the total acoustic pressure for coupled BEM–FEM problem, and the influence of placing an acoustic monopole above a three-dimensional wing (a BAH wing) to the flutter velocity, by using coupled BEM–FEM formulation for the acoustic incident pressure induced by the monopole source have been validated using NASTRAN® [19] and ZAERO® [20]. The calculation of the unsteady aerodynamic terms in Eq. 24.60 is carried out using Doublet Point Method as elaborated in [24, 25], and developed into a routine written in MATLAB®, as elaborated in [2].

It is of interest to look into some simple applications to obtain the usefulness of the method. Along this line, several parametric studies are carried out. The first study looks into the influence of the intensity of the acoustic source on the flutter stability by varying its location above the wing. Figure 24.11a indicates that the most effective way in placing the acoustic monopole source is on the tip of the wing, and

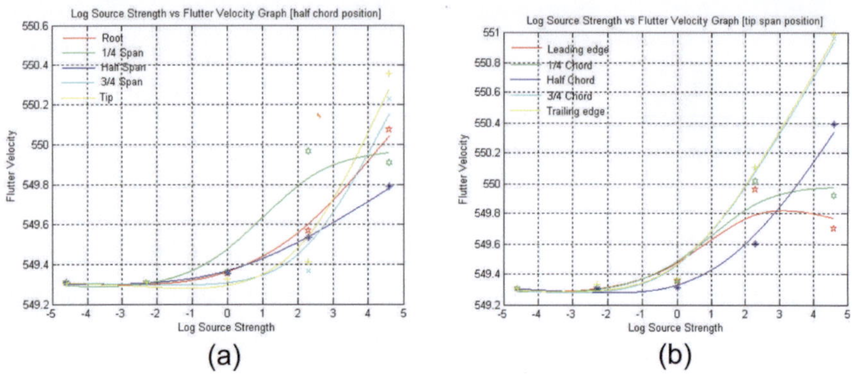

Fig. 24.11 The Influence of acoustic monopole source Intensity on flutter velocity as a function of monopole position; **a** at mid-chord along wing span, **b** at wing-tip section along the chord

Fig. 24.11b indicates that the most effective way in placing the acoustic monopole source is on the trailing edge of the wing.

Next the influence of the distance between the acoustic source and the wing on the flutter stability is investigated. Figure 24.12 exhibits the results of such study and indicates that the most effective way in placing the acoustic monopole source is on the nearest distance from the wing. These results serve to indicate the logical trend of such problem, which will be useful for further practical applications. However, the favorable effect of the introduction of a monopole source closer to the wing should be accompanied by the increase of its strength.

24.8 Concluding Remarks

The computational scheme for the calculation of the influence of the acoustic disturbance to the aeroelastic stability of a structure has been developed using a unified treatment by applying acoustic-aerodynamic analogy.

By considering the effect of acoustic pressure disturbance to the aeroelastic structure (acousto-aero-elastic problem) to consist of structural motion-independent incident acoustic pressure and structural motion-dependent acoustic pressure, the scattering acoustic pressure can be grouped together in the aerodynamic term of the aeroelastic equation. By tuning the incident acoustic pressure, it can also be incorporated along with the scattering acoustic term, forming the acousto-aeroelastic stability equation. For this purpose the topology of the problem domain has been defined to consist of those subjected to acoustic pressure only and that subject to acoustic structural coupling, which is treated as acousto-aeroelastic equation. Using BE and FE as appropriate, an integrated formulation is then obtained as given by the governing Eq. 24.41, which relates all the combined forces acting on the structure to the displacement vector of the structure. The solution of Eq. 24.41—and after using modal

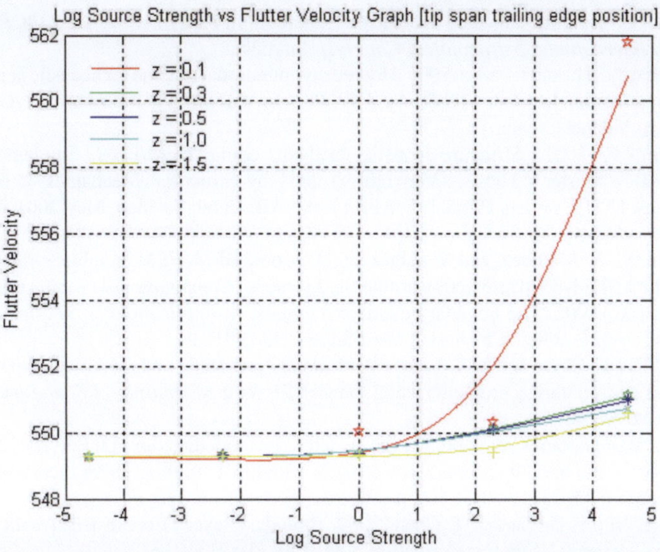

Fig. 24.12 The Influence of acoustic monopole source strength on flutter velocity as a function of the distance of the monopole source above the wing from the tip-chord point of the wing-tip section

approach in structural dynamics, Eq. 24.58—can be obtained by solving Eq. 24.62 as a stability equation in a"unified treatment", and the disturbance acoustic pressure already incorporates the total pressure (incident plus scattering pressure), which has been "tuned" to behave like the aerodynamic terms in the modal Eq. 24.60. Such approach allows the application of the solution of the acousto-aeroelastic stability equation in the frequency domain using V–g method. Such technique forms the first generic approach to solve Eq. 24.41. Alternatively, the acousto-aeroelastic equation part can also be treated as a dynamic response problem, which forms the second generic approach and which has been dealt with in [2, 17].

The method developed has been demonstrated to be capable of solving the acoustic-aero-elastomechanic coupling problem. Specifically, the results of both generic approaches to the example worked out show that the presence of acoustic excitation at a frequency near the original flutter frequency can delay the flutter inception, thus confirming our expectation. Further improvements in the computational technique based on efficient algorithm and specific numerical behavior may take advantage of the work of [26–28].

References

1. Djojodihardjo, H., and E. Tendean. 2004. A computational technique for the dynamics of structure subject to acoustic excitation. In *ICAS 2004*. Vancouver.
2. Djojodihardjo, H., and I. Safari. 2005. Further development of the computational technique for

dynamics of structure subjected to acoustic excitation. In *Paper IAC-05-C2.2.08, Presented at the 56th International Astronautical Congress*, Fukuoka.
3. Djojodihardjo, H., and I. Safari. 2006. Unified computational scheme for acoustic aeroelastomechanic interaction. In *Paper IAC-06-C2.3.09, Presented at the 57th International Astronautical Congress*. Valencia Spain.
4. Holström, F. 2001. Structure-acoustic analysis using BEM/FEM; implementation in MATLAB®. Master's Thesis, Copyright © 2001 by Structural Mechanics & Engineering Acoustics, LTH, Sweden, Printed by KFS i Lund AB, Lund, Sweden, May 2001, Homepage: http://www.akustik.lth
5. Meddahi, S., A. Marquez, and V. Selgas. 2004. A new BEM–FEM coupling strategy for two-dimensional fluid–solid interaction problems. *Journal of Computational Physics* 199: 205–220.
6. Wrobel, L.C. 2002. The boundary element methods. In *Applications in Thermo-Fluids and Acoustics*, vol. 1. Wiley, Chichester, West Sussex PO19 IUD.
7. Chen, J.T., I.L. Chen, and M.T. Liang. 2001. Analytical study and numerical experiments for radiation and scattering problems using the CHIEF method. *Journal of Sound and Vibration* 248 (5): 809–828.
8. Papacharalampopoulos, A., G.F. Karlis, A. Charalambopoulos, and D. Polyzos. 2010. BEM solutions for 2D and 3d dynamic problems in Mindlin's strain gradient theory of elasticity. *CMES* 58 (1): 45–73.
9. Zhang, Y. M., Y. Gu, and J. T. Chen. 2009. Boundary layer effect in BEM with high order geometry elements using transformation. *CMES* 45 (3): 227–247.
10. Dowling, A. P., and J. E. Ffowcs Williams. 1983. *Sound and Sources of Sound*. Chichester: Ellis Horwood Limited—New York Brisbane Chichester Toronto: Wiley, ©1983, A.P. Dowling and J.E. Ffowcs Williams.
11. Weaver, W., and P.R. Johnston. 1987. *Structural Dynamics by Finite Elements*. Prentice-Hall, Inc.
12. Schanz, M. 1976. Improved integral formulation for acoustic radiation problems. *Journal of the Acoustical Society of America* 44 (1): 41–58.
13. Burton, J., and G.F. Miller. 1971. The application of integral equation methods to numerical solutions of some exterior boundary value problem. *Proceedings of Royal Society of London Series A* 323: 201–210.
14. Chen, J.T., I.L. Chen, S.R. Kuo, and M.T. Liang. 2001. A new method for true and spurious eigensolutions of arbitrary cavities using the CHEEF method. *Journal of Acoustic Society of America* 109 (3). https://doi.org/10.1121/1.1349187.
15. Chen, J.T., I.L. Chen, and K.H. Chen. 2006. Treatment of rank deficiency in acoustic using SVD. *Journal of Composite Acoustic* 14 (2): 157–183.
16. Chen, J.-T., I.-L. Chen, Y.-T. Lee, and J.-W. Lee. 2012. Non-uniqueness and treatment in dual BIEM/BEM using SVD. In *The Cross-Straits Workshop on Applied Mathematics*.
17. Djojodihardjo, H. 2007. BEM-FEM acoustic-structural coupling for spacecraft structure incorporating treatment of irregular frequencies. In *Paper IAC-07-C2.3.4, 58th International Astronautical Congress*, Hyderabad.
18. Bisplinghoff, R.L., H. Ashley, and R. Halfman. 1955. *Aeroelasticity*. Addison Wesley.
19. Rodden, W.P., and E.H. Johnson. 1994. MSC/NASTRAN aeroelastic analysis user's manual. *Version* 68: 237–257.
20. Anonymous, Zona Tech. 2004. *ZAERO Theoretical Manual ZONA Technology, Inc*. 13th ed, 12-04 © 2004.
21. Beer, G., and J.O. Watson. 1992. *Introduction to Finite and Boundary Element Methods for Engineers*. Baffins Lane, Chichester, England: Wiley Ltd.
22. Djojodihardjo, H. 2008. Unified BE-FE aerodynamic-acoustic-structure coupling scheme for acoustic effects on aeroelastic structures paper ICAS-2008–585. In *26th International Congress of the Aeronautical Sciences*, Anchorage.
23. Dowell, E.H. (ed.), H.C. Curtiss, R.H. Scanlan, and F. Sisto. 1980. *A Modern Course in Aeroelasticity*. Alphen aan den Rijn, The Netherlands: Sijthoff & Noordhoff.
24. Houbolt, J.C. 1969. Some new concepts in oscillatory lifting surface theory. *AFF DLTR-68-2*.

25. Ueda, T., and E.H. Dowell. 1982. A new subsonic method for lifting surfaces in subsonic flow. *AIAA Journal* 20 (3): 348–355. Technical Comment, Rodden, W.P. (vol. 22, no. 1, p. 160); and Reply by Ueda, T., Dowell, E.H. (vol. 2, no. 4, p. 575, 1983).
26. Schanz, M. 2010. On a reformulated convolution quadrature based boundary element method. *CMES* 58 (2): 109–128.
27. Wu, H. J., Y. J. Liu, and W. K. Jiang. 2012. Analytical integration of the moments in the diagonal form fastmultipole boundary element method for 3-D acoustic wave problems. *Engineering Analysis with Boundary Elements* 36: 248–254.
28. Zhang, Y., Y. Gu, and J.-T. Chen. 2009. Analysis of 2D thinwalled structures in BEM with high-order geometry elements using exact integration. *CMES* 50 (1): 1–20.

Chapter 25
Active Vibration Suppression of a Generic Smart Composite Structure

Abstract An efficient analytical method for vibration analysis of a Euler–Bernoulli beam with spring loading at the tip has been developed as a baseline for treating flexible beam attached to central-body space structure, followed by the development of MATLAB© finite element method computational routine. Extension of this work is carried out for the generic problem of active vibration suppression of a cantilevered Euler–Bernoulli beam with piezoelectric sensor and actuator attached as appropriate along the beam. Such generic example can be further extended for tackling lightweight structures in space applications, such as antennas, robot's arms and solar panels. For comparative study, three generic configurations of the combined beam and piezoelectric elements are solved. The equation of motion of the beam is expressed using Hamilton's principle, and the baseline problem is solved using Galerkin-based finite element method. The robustness of the approach is assessed.

Keywords Aero-servo-elasticity · Finite element method · Hamiltonian mechanics · Piezoelectric material · Active vibration control · Structural dynamics

25.1 Introduction

Vibration control of lightweight structures is of great interest of many studies and investigations [1–3]. The high cost of sending heavy masses and large volumes into space has prompted the wide utilization of lightweight structures in space applications, such as antennas, robot's arms and solar panels. A model of such set-up is exemplified in Fig. 25.1 [2]. These kinds of structures are largely flexible, which results in lightly damped vibration, instability and fatigue. To suppress the adverse effect of vibration, sophisticated controller is needed.

Active control approaches are widely reported in the literatures for the vibration control of structures [4–10]. The active control approach makes use of actuators

(Originally published in Composite Structure as Active Vibration Suppression of an Elastic Piezoelectric Sensor and Actuator fitted Cantilevered Beam Configurations as a Generic Smart Composite Structure co-authored by H. Djojodihardjo, M. Jafari, S. Wiriadidjaja, K. A. Ahmad, Composite Structures · Volume 132, 15 November 2015, Pages 848–863).

© The Author(s), under exclusive license to Springer Nature Singapore Pte Ltd. 2023

H. Djojodihardjo, *Introduction to Aeroelasticity*,
https://doi.org/10.1007/978-981-16-8078-6_25

Fig. 25.1 Solar panel on a typical satellite (left) and an experimental model of free-floating platform with two flexible appendages designed by Gasbarri et al. [2] (right)

and sensors to find out some essential variables of the structure and suppress its vibration through minimizing the settling time and the maximum amplitude of the undesirable oscillation. This method requires a specific level of understanding about the dynamic behavior of continuous structures via mathematical modeling [4, 5]. Selecting adequate sensor and actuator is essential in active vibration control [11, 12]. The conventional form of sensor and actuator, such as electrohydraulic or electromagnetic actuator, is not applicable to implement on the lightweight space structures. Thus, in recent years, a new form of sensor and actuator has been studied using smart materials, such as shape memory alloys and piezoelectric materials. The definition of smart material may be expressed as a material which adapts itself in response to environmental changes. Among smart materials, piezoelectric materials are widely studied in literatures, since they have many advantageous such as adequate accuracy in sensing and actuating, applicable in the wide frequency range of operations, applicable in distributed or discrete manner and available in different size, shape and arrangement.

Space structures can be simplified using beam and plate. The present investigation is based on the vibration analysis of simple beam as a generic structure. Without loss of generalities, the theoretical development utilizes Euler–Bernoulli beam approximation, which can readily be extended to other refined models. Euler–Bernoulli beam theory is applicable to thin and long span, for which plane sections can be assumed to remain plane and perpendicular to the beam axis, and shear stress and rotational inertia of the cross section can be neglected. Solar panel and antenna are very flexible and slender, so that Euler–Bernoulli beam theory can be considered.

The equation of motion of the beam will be developed using Hamilton's method and Lagrange method [13, 14]. Hamiltonian mechanics is an elegant and convenient approach, since scalar equation of motion of the beam and boundary conditions are obtained simultaneously. The partial differential equation of motion of the beam can be solved by analytical methods such as separation of variables or numerical methods such as finite element method.

Since these structures are flexible, there is a need for vibration control in these structures which is required in order not to disturb the functionality of the space structure as a whole as well as to facilitate maneuvering and attitude control for well-behaved space structural dynamics. There are several ways to control the vibration [15, 16]. Then the effort is aimed for devising a simple and effective controller to manipulate the vibration of a flexible structure. One of the adequate and simple controllers is proportional-integral-derivative (PID) controller, which is classified as classic and linear controller [16]. PID controller minimizes the steady state error of the system. Linear quadratic regulator (LQR) controller is another convenient method. LQR is expressed as optimal and modern controller, which is based on minimizing the cost function of a dynamic system [16]. To develop a successful operation, most controllers have been developed for a finite number of natural modes where the controllability and observability conditions are met. It is also noted that various recent literature addresses similar issues which are utilized for comparative purposes.

25.2 Formulation of Generic Problems

Following a series of previous investigation on the analysis of impact resilient structure [17–20], and vibration analysis of an elastically clamped cantilever beam [21, 22], the main aim of this investigation is to design a straightforward and convenient controller for suppressing the transverse vibration in a cantilever aluminum flexible beam through the use of sensing and actuating transducers. The Euler–Bernoulli beam theory is utilized to model the flexible beam with piezoelectric patches. The equation of motion and boundary conditions of the beam is derived by using Hamilton's principle. To validate and assess the analytical as well as the finite element computational schemes, the solution for a cantilevered beam with spring loading at the tip as illustrated in Fig. 25.2, which has the potential to be expanded in many variations, will be demonstrated. Next, considerations will be given for the active vibration control for Cantilevered Bernoulli Beam. Three different piezoelectric material configurations on the aluminum beam are considered for comparative study. Finite element method is utilized to achieve the natural frequencies and natural modes. Case study one is validated by analytical solution. Previous experimental result is used for validation of case study two.

To design the controller, two first major natural modes of the beam vibration are considered, since other natural modes has insignificant effect [4, 12, 23, 24]. The

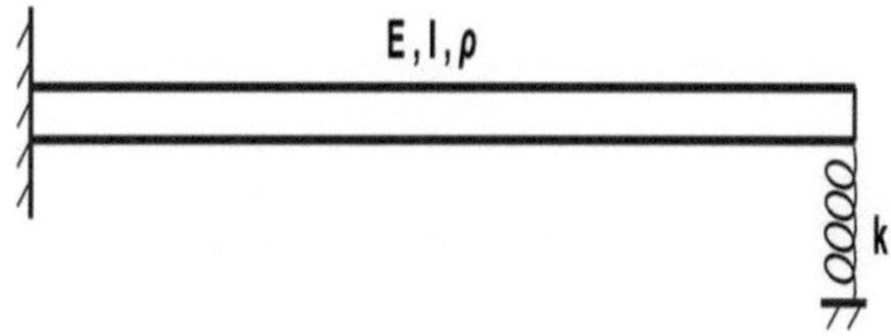

Fig. 25.2 Assumed model for the beam

dynamic equation of the beam is transferred to state-space form in order to design controllers. Two controllers are designed for each case study: PID controller and LQR controller with observer. These controllers are easy to perform and effective to suppress the vibration of the beam. The systematic of the problem formulation and the methods of approach including the objective of this study is summarized in Fig. 25.3.

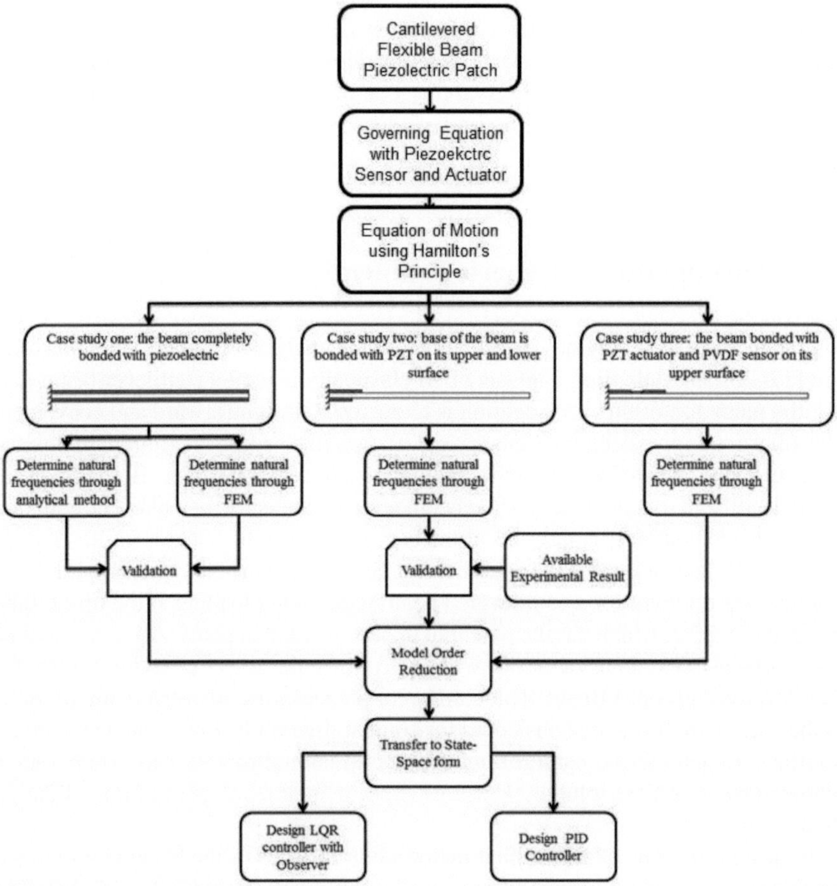

Fig. 25.3 Problem formulation and solution flow chart

25.3 Solution Scheme

25.3.1 *Equation of Motion of the Euler–Bernoulli Beam Using Hamilton's Principle*

Figure 25.4 depicts the cantilevered Euler–Bernoulli beam with spring loading at the free end. Following Euler–Bernoulli beam model, the shear displacement and the rotation of cross sections are neglected in comparison to the translation; hence the cross sections remain constant after deformation [14, 25]. From Fig. 25.4

$$u = -z\frac{\partial w\,(x,t)}{\partial x},\ \varepsilon_{xx} = \frac{\partial u}{\partial x} = -z\frac{\partial^2 w}{\partial x^2} \tag{25.1}$$

where the deflection w is a function of space and time, $w = w(x, t)$. The strain energy (potential energy) of the beam due to the linear spring can be expressed as [14, 25–27]

$$\Pi = \frac{1}{2}\iiint\limits_V \sigma_{xx}\varepsilon_{xx}\mathrm{d}V + \left(\frac{1}{2}\mathrm{kw}^2(l,t)\right) = \frac{1}{2}\int_0^l \mathrm{EI}\left(\frac{\partial^2 w}{\partial x^2}\right)\mathrm{d}x + \left(\frac{1}{2}\mathrm{kw}^2(l,t)\right) \tag{25.2}$$

where I and E are the area of the moment of inertia and elasticity modulus, respectively. k is the stiffness of linear spring. T is the kinetic energy of the system, which is given by

$$T = \frac{1}{2}\int_0^l \iint\limits_A \rho\left(\frac{\partial W}{\partial X}\right)^2 \mathrm{d}A\mathrm{d}x = \frac{1}{2}\int_0^l \rho\left(\frac{\partial W}{\partial X}\right)^2 \mathrm{d}x \tag{25.3}$$

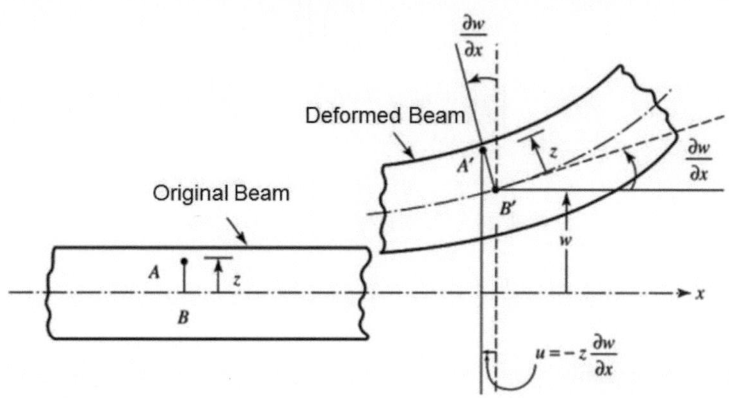

Fig. 25.4 Beam bending [25]

The work is done by the force on beam is given by

$$W = \int_0^l f\,w\,\mathrm{d}x \tag{25.4}$$

Following Hamilton general principle [14, 25–27], the equation of motion of the system can be obtained as [12–14]

$$\delta \int_0^l L(x(t), \dot{x}(t), t) = 0 \quad \text{where} \quad L = \Pi - T - W \tag{25.5}$$

and where L is the Lagrangian. Π, T and W are potential energy, kinetic energy and work, respectively. Substitution of Eqs. (25.2–25.4) into Eq. (25.1) yields

$$\delta \int_{t_1}^{t_2} \left\{ \frac{1}{2} \int_0^l EI \left(\frac{\partial^2 w}{\partial x^2} \right)^2 \mathrm{d}x + \left(\frac{1}{2} k w^2(l,t) \right) - \frac{1}{2} \int_0^l \rho A \left(\frac{\partial w}{\partial x} \right)^2 dx - \int_0^l f w\,\mathrm{d}x \right\} = 0 \tag{25.6}$$

Integrating Eq. 25.6 by parts, dynamic equation of the Euler–Bernoulli beam can be written as (25.2).

$$\frac{\partial^2}{\partial x^2} \left(EI \frac{\partial^2 w}{\partial x^2} \right) + \rho A \frac{\partial^2 w}{\partial t^2} = f(x,t) \tag{25.7}$$

For the transverse vibration of the beam, the boundary conditions for $x = 0, l$, are given by

$$\left[-EI \frac{\partial^2 w}{\partial x^2} \delta \left(\frac{\partial w}{\partial x} \right) \right] \bigg|_{x=0} = 0 \tag{25.8}$$

$$\left[EI \frac{\partial^2 w}{\partial x^2} \delta \left(\frac{\partial w}{\partial x} \right) \right] \bigg|_{x=l} = 0 \tag{25.9}$$

$$\left[\frac{\partial}{\partial x} \left(EI \frac{\partial^2 w}{\partial x^2} \right) \delta w \right] \bigg|_{x=0} = 0 \tag{25.10}$$

$$\left[\left(-\frac{\partial}{\partial x} \left(EI \frac{\partial^2 w}{\partial x^2} \right) + kw \right) \delta w \right]_{x=l} = 0 \tag{25.11}$$

For convenience, define

$$c = \sqrt{\frac{EI}{\rho A}} \tag{25.12}$$

Then Eq. 25.7 can be written as

$$c^2 \frac{\partial^4 w}{\partial x^4}(x, t) + \frac{\partial^2 w}{\partial t^2}(x, t) = 0 \tag{25.13}$$

Using the methods of separation of variables [28],

$$w(x, t) = W(x)T(t) \tag{25.14}$$

Substituting Eq. 25.14 into Eq. 25.13 can be split into two equations

$$\frac{\partial^4 W(x)T(t)}{\partial x^4} - \alpha^2 W(x) = 0 \tag{25.15}$$

and

$$\frac{\partial^2 T(t)}{\partial x^2} + \omega^2 T(t) = 0 \tag{25.16}$$

Their solutions are given by:

$$\begin{aligned} W(x) = {} & A(\cos\alpha x + \cosh \alpha x) + B(\cos\alpha x - \cosh \alpha x) \\ & + C(\sin\alpha x + \sinh \alpha x) + D(\sin\alpha x - \sinh \alpha x) \end{aligned} \tag{25.17}$$

and

$$\omega = \alpha^2 \sqrt{\frac{EI}{\rho A}} \tag{25.18}$$

where

$$\alpha^4 = \frac{\omega^2}{c^2} = \frac{\rho A \omega^2}{EI} \tag{25.19}$$

Applying boundary conditions Eqs. 25.8–25.11 for a cantilever beam with spring loading at the free end as appropriate [21, 22], the natural frequencies and vibration modes of the system may be obtained.

25.3.2 Finite Element Approach

The fourth-order differential Eq. 25.7 can be solved using the weighted residual method. Among the numerous techniques for applying the weighted residual method, the Galerkin approach will be applied [26, 29]. After separating the differential equation Eq. 25.7, the technique is applied to Eq. 25.14. For this purpose, this equation can then be multiplied by a trial function *$\phi(x)$*, and the approximate solution is a linear combination of trial functions. The product of the trial function and the differential equation is integrated over the domain and set to zero. Following the procedure, one obtains

$$\int \phi(x)\left\{EI\frac{\partial^4 w}{\partial x^4} - \rho A\omega^2 W(x)\right\}\mathrm{d}x = 0 \tag{25.20}$$

$W(x)$ is assumed as

$$W(x) = L_1 W_{j-1} + L_2\theta_{j-1} + L_3 W_j + L_4\theta_j \quad (j-1)h \le x \le h \tag{25.22}$$

where W is displacement and θ is the rotation of the nodes. A non-dimensional coordinate ξ is now introduced, as depicted in Fig. 25.5. The essence of the Galerkin method is that the trial function is chosen to be [14, 25–27, 29]

$$\phi(x) = W(x) \tag{25.23}$$

For a system of n elements, the local stiffness matrix can be written as

$$K_j - \omega^2 M_j = 0 \quad j = 1, 2, \ldots, n \tag{25.24}$$

where

$$K_j = \frac{EI}{h^3}\begin{bmatrix} 12 & 6h & -12 & 6h \\ 6h & 4h^2 & -6h & 2h^2 \\ -12 & -6h & 12 & -6h \\ 6h & 2h^2 & -6h & 4h^2 \end{bmatrix} \tag{25.25}$$

$$M_j = \frac{h\rho A}{420}\begin{bmatrix} 156 & 22h & 54 & -13h \\ 22h & 4h^2 & 13h & -3h^2 \\ 54 & 13h & 156 & -22h \\ -13h & -3h^2 & -22h & 4h^2 \end{bmatrix} \tag{25.26}$$

The natural frequencies of the beam can be obtained by determining the eigenvalue of Eq. 25.25. The eigenvector corresponds to the natural mode of the system [14].

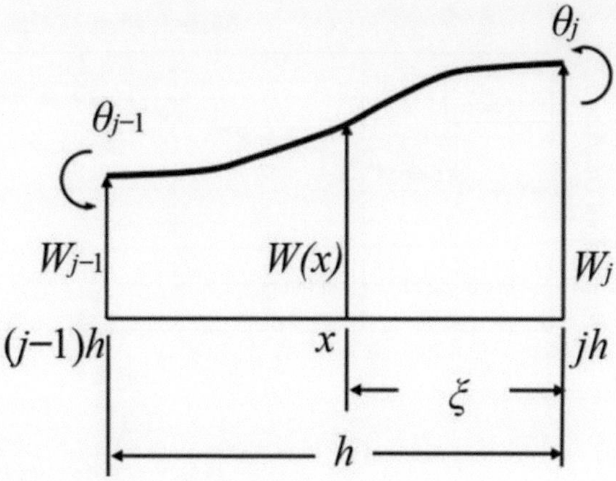

Fig. 25.5 Local coordinate system

25.3.3 Results and Discussions of the Baseline Problem

In the example, the material of the beam is considered to be made of stainless steel. The properties and dimension of the beam are shown in Table 25.1.

The analytical solution of Eq. 25.13 for the eigenfrequencies and eigenmodes for the beam considered is elaborated in [21, 22]. The natural frequencies of the numerical approach are determined by Eq. 25.24. The first three natural frequencies determined by both analytical and finite element approaches are shown in Table 25.2 and illustrated in Fig. 25.6. These results indicate that both the analytical and finite element approaches are agreeable in that the errors of numerical results are significantly small in comparison to the analytical ones.

The first natural mode of the numerical result has an excellent agreement with exact answer and its error is almost zero. Similarly, the next two modes exhibit similar

Table 25.1 Properties of the system

Stainless steel	Young's Modulus	Density	Poisson ratio	Length	Width	Thickness	Spring stiffness
	210 Gpa	7850 kg/m^2	0.3	450 mm	20 mm	3 mm	10,000 N/m

Table 25.2 Natural frequencies of exact and numerical solution

Natural modes	Eigen solution α	Frequency from analytical results (Hz)	Frequency from numerical results (Hz)	Numerical results error (%)
1st	8.06796	46.4038220296	46.4038189479	$1.1491.10^{-3}$
2nd	12.4237	110.035527965	110.032263817	$2.9665.10^{-3}$
3rd	17.9451	229.571454472	229.569470466	$8.6423.10^{-3}$

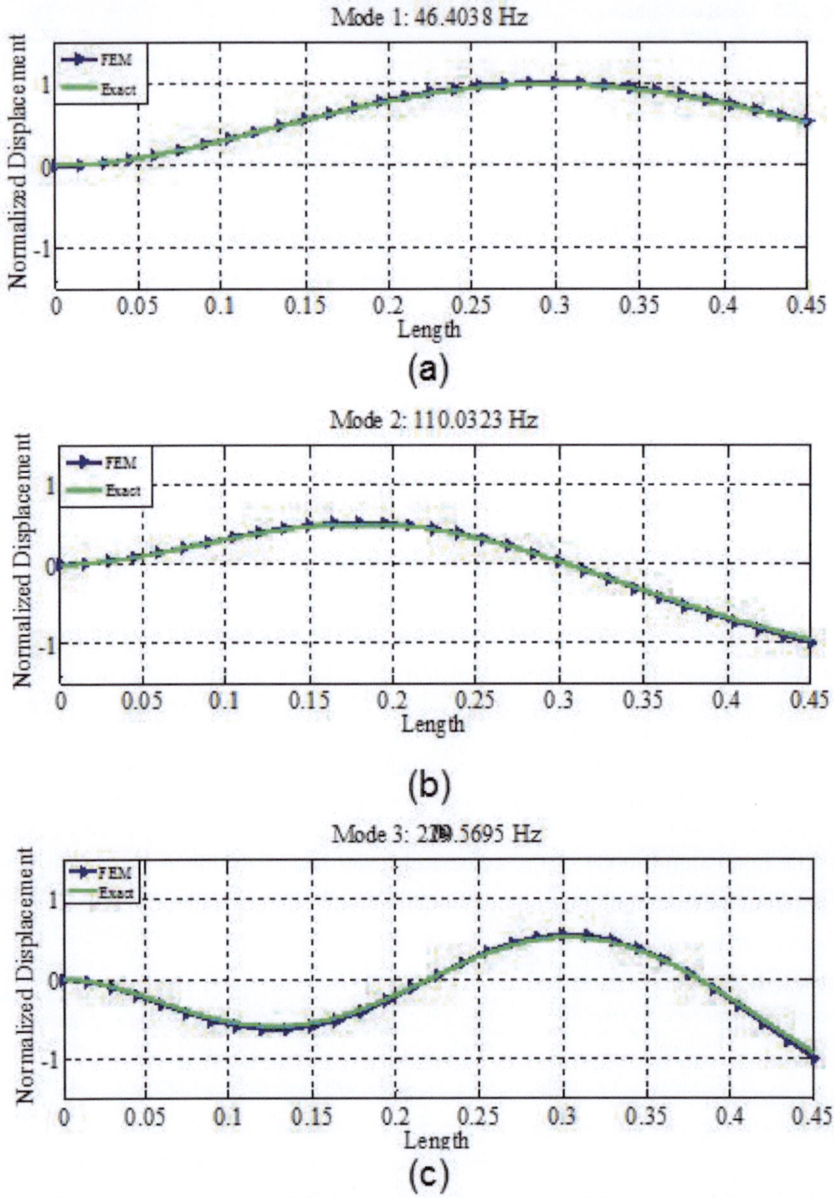

Fig. 25.6 **a** First natural mode, **b** second natural mode, **c** third mode; 1.Green line: analytical approach 2. Blue line: numerical approach

accuracy. The result thus obtained agrees with the work of Maurizi, Rossi and Reyes [30] for a degenerate case, as elaborated in references [21, 22].

25.4 The Utilization of Piezoelectric Sensors and Actuators

Two common piezoelectric materials are widely used in many recent studies [31–36], which are piezoceramics, lead zirconium titanate (PZT) and piezopolymer, PVDF (polyvinylidene fluoride) [11, 37, 38]. Piezoceramic PZTS can be utilized both as sensor and actuator because they are applicable in the wide range of frequency and required lower actuating voltage; piezopolymer pvdfs are usually used as sensor [23, 24]. These materials may be analyzed for the configuration in which the piezosensor and/or piezoactuator is attached to a base structure, and then the dynamic behavior of the total composite structure is modeled accordingly [24].

Figure 25.7 illustrates a beam bonded with piezoelectric material. The constitutive equation of piezoelectric material can be linearly described as Eqs. 25.27 and 25.28 [11, 12, 37, 38].

$$\varepsilon_p = S^E_{pq}\sigma_q + d_{pi}E^e_i \tag{25.27}$$

where indices $i, j = 1, 2, 3$ are directions of the polarization of the piezoelectric material and indices p, $q = 1, \ldots, 6$ indicate stress or strain direction on the perpendicular planes, which are illustrated in Fig. 25.8. S_{pq} (m^2/N) is the elastic compliance, which presents the inverse of the elastic modulus. d_{pi} (m/volt) is the piezoelectric strain constant, which is the strain developed along p-direction over electrical field applied along i-direction. Equation 25.27 is applicable for the actuator.

The dynamic equation of sensor, Eq. 25.28, is given by.

$$D_i = d_{ip}\sigma_q + \xi^{\sigma}_{ij}E^e_i \tag{25.28}$$

where ξ is the dielectric permittivity of the material in the units of Farads per meter (F/m). D_i is electrical displacement, which is described as charge per unit area (Coulomb/m^2).

Fig. 25.7 Beam bonded with piezoelectric material

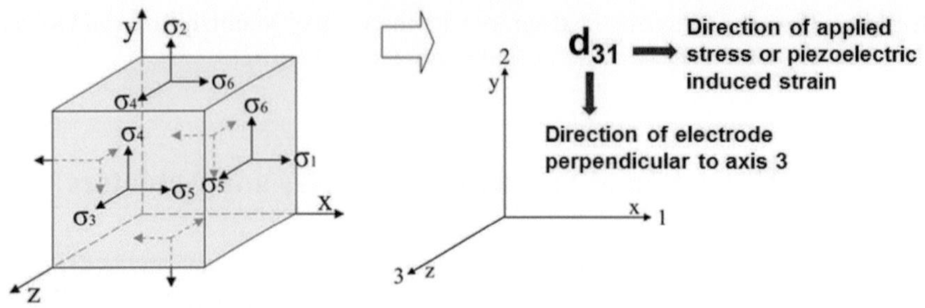

Fig. 25.8 Notation convention adopted for stress tensor and piezoelectric constants (on the left) and the direction of polarization (on the right) [23, 24]

25.4.1 Actuator

To determine the piezoelectric actuator moment for one-dimensional motion, the piezoelectric layer is assumed to be very thin and perfectly attached to the beam. The strains only arise in *x*-direction. Figure 25.9 illustrates the one-dimensional beam with actuator system. If the piezoelectric layer is assumed to be free and not bonded with the beam, Eq. 25.27 can be reduced to:

$$\varepsilon_p = d_{31}\frac{V}{h_{ac}} \tag{25.29}$$

where V is the applied control voltage (Volt), and h_{ac} (m) is the thickness of the piezoelectric actuator. The laminated beam is assumed to be uniform along its length. Since the piezoelectric layer is bonded to the beam, it is not free to move, and the effect of the voltage is to induce a negative prestrain in the piezoelectric layer. The prestrain can be assumed to be negative, since it is in the opposite direction of the strain of the free piezoelectric layer. If the free piezoelectric layer strain is ε_p, a prestrain of $-\varepsilon_p$ is needed to bring the piezoelectric layer back into position to be bonded to the beam [38].

The moment, M, is produced because the prestrain, ε_p results in a force in the beam which acts through the moment arm to the neutral axis. The moment can be calculated by taking the integral of the longitudinal force multiplied by the moment

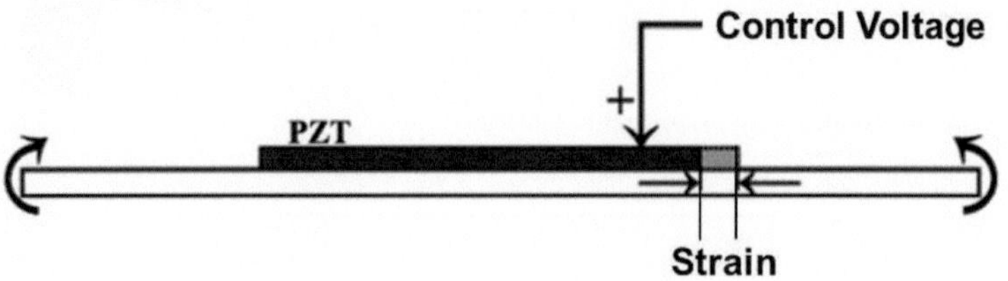

Fig. 25.9 Schematic of piezoelectric actuator

arm to the neutral axis and integrated over the cross section of the beam.

$$M(x,t) = \int_{-\frac{b}{2}}^{\frac{b}{2}} \left\{ \begin{array}{l} \int_{-\left(\frac{h_{bm}}{2}+h_{ac}\right)}^{\frac{h_{bm}}{2}} E_{ac}(\varepsilon_l + \varepsilon_p) z \, dz + \int_{-\frac{h_{bm}}{2}}^{\frac{h_{bm}}{2}} E_{bm} \varepsilon_l z \, dz \\ \int_{\frac{h_{bm}}{2}}^{\frac{h_{bm}}{2}+h_{ac}} E_{ac}(\varepsilon_l - \varepsilon_p) z \, dz \end{array} \right\} \tag{25.30}$$

Since ε_p at the bottom of the beam has an opposite polarization with upper one, it has an opposite sign. By considering a symmetric beam and the same properties for actuators, Eq. 25.30 reduces to

$$M(x,t) = -2bE_{ac}h_{ac}\left(\frac{h_{bm} + h_{ac}}{2}\right)\varepsilon_p \tag{25.31}$$

Substituting Eq. 25.29 into Eq. 25.31 results in

$$M(x,t) = -2bE_{ac}\left(\frac{h_{bm} + h_{ac}}{2}\right)d_{31}V(x,t) \tag{25.32}$$

It follows that for a symmetric beam bonded with piezoelectric material, the moment produced on the beam is equal to.

$$M(x,t) = -bE_{ac}\left(\frac{h_{bm} + h_{ac}}{2}\right)d_{31}V(x,t) \equiv C_{ac} \cdot V(x,t) \tag{25.33}$$

where C_{ac} is a constant that expresses the moment produced per unit control voltage. In the antisymmetric cases, the moment induced by piezoactuator around neutral axis can be determined by the following integral [23]:

$$M(x,t) = \int_{-\frac{b}{2}}^{\frac{b}{2}} \int_{h_{bm}-h_{NA}}^{h_{bm}+h_{ac}-h_{NA}} E_{ac}\varepsilon_p z \, dz \, dy \tag{25.34}$$

where h_{NA} is the location of neutral axis. By solving the integral, Eq. 25.34, the governing equation of piezoactuator with antisymmetric configuration can be obtained.

$$M(x,t) = bE_{ac}\left(h_{bm} + \frac{h_{ac}}{2} - h_{NA}\right)d_{31}V(x,t) \tag{25.35}$$

25.4.2 Sensor

When the moment is applied to the beam, the stress is produced in the piezoelectric material layer as expressed schematically in Fig. 25.10. If no electrical fields are applied to the beam, Eq. 25.28 is reduced to Eq. 25.36.

$$\begin{aligned} D_1 &= d_{ip}\sigma_p + \xi_{ij}^{\sigma} E_j^e \\ D_3 &= d_{31}\sigma_1 \end{aligned} \tag{25.36}$$

where D_3 is electrical displacement in z-direction. Here$\sigma_1$is the stress in the piezoelectric material in x-direction, which is expressed as

$$\begin{aligned} \sigma_1 &= E_{sn}\varepsilon_{sn} \\ \varepsilon_{sn} &= -\left(\frac{h_{bm} + h_{sn}}{2}\right)\frac{\partial^2 w}{\partial x^2} \end{aligned} \tag{25.37}$$

where ε_s is the strain in the mid-plane of piezoelectric material in x-direction, and E_{sn} is the Young's modulus of the sensor. Substitution of Eq. 25.37 into Eq. 25.36, results in

$$D_3 = -d_{31} E_{sn}\left(\frac{h_{bm} + h_{sn}}{2}\right)\frac{\partial^2 w}{\partial x^2} \tag{25.38}$$

By integrating Eq. 25.38 over the area of the piezoelectric patches, the generating charge, q, can be obtained as

$$q = \iint\limits_{A_s} D_3 \, \mathrm{d}x \, \mathrm{d}y \tag{25.39}$$

Substituting Eq. 25.38 into Eq. 25.39 gives

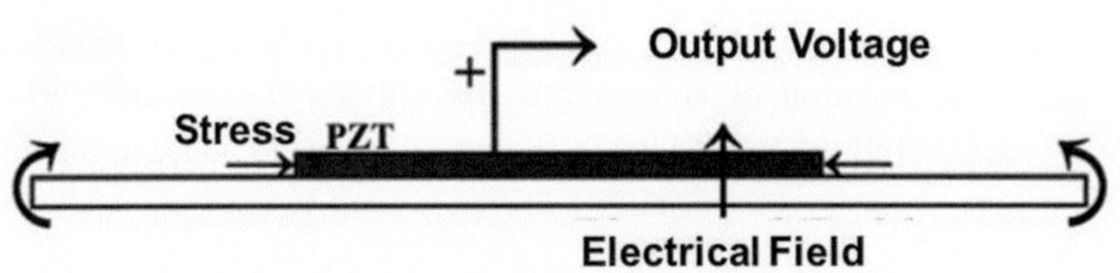

Fig. 25.10 Piezoelectric sensor scheme

$$q = -d_{31} E_{sn} \left(\frac{h_{bm} + h_{sn}}{2} \right) \int_{-\frac{b}{2}}^{\frac{b}{2}} \int_{x_{s1}}^{x_{s2}} \frac{\partial^2 w}{\partial x^2} \, \mathrm{d}x \, \mathrm{d}y$$

$$q = -d_{31} E_{sn} b \left(\frac{h_{bm} + h_{sn}}{2} \right) \left(\left. \frac{\partial w}{\partial x} \right|_{x_{sn,1}}^{x_{sn,2}} \right) \tag{25.40}$$

The generating voltage of sensor can be obtained from the generating charge divided by capacitance, C;

$$V_{sn} = \frac{q}{C} \tag{25.41}$$

where the capacitance can be expressed as

$$C = \frac{\varepsilon_{33}^T A_{sn}}{h_{sn}} = \frac{\varepsilon_{33}^T b(x_{sn,2} - x_{sn,1})}{h_{sn}} \tag{25.42}$$

and where ${\varepsilon^T}_{33}$ is the permittivity constant of piezoelectric material in the z-direction.

Figure 25.11 summarizes the sensing and actuating procedure.

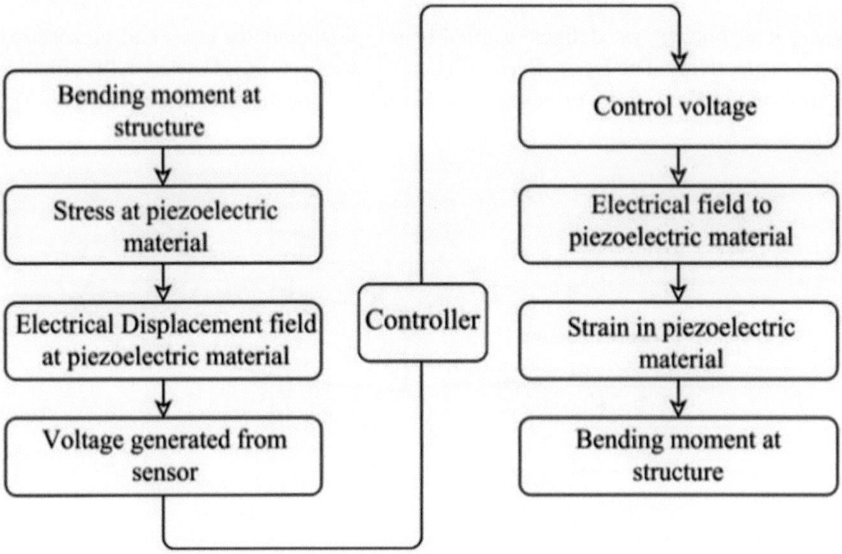

Fig. 25.11 Basic mechanism of sensor and actuator

25.5 Equation of Motion of Euler–Bernoulli Beam with Piezoelectric Patches

A symmetric beam with piezoelectric patches on the top and bottom will be considered for the vibration-controlled Euler–Bernoulli beam. Displacements of the beam under the bending around y-axis are considered [14, 25, 26] as illustrated in Fig. 25.12.

$$u = -z\frac{\partial w(x,t)}{\partial x}, v = 0, w = w(x,t) \tag{25.43}$$

Here u, v and w are displacements in x-, y- and z-directions, respectively. Strain, $\varepsilon_{b,xx}$, and stress, $\sigma_{b,xx}$, result from beam bending in the x-direction which are described as.

$$\begin{aligned} \varepsilon_{b,xx} &= \frac{\partial u}{\partial x} = -z\frac{\partial^2 w}{\partial x^2}, \\ \varepsilon_{yy} &= \varepsilon_{zz} = \varepsilon_{xy} = \varepsilon_{yz} = \varepsilon_{zx} = 0 \end{aligned} \tag{25.44}$$

$$\begin{aligned} \sigma_{b,xx} &= E_n\varepsilon_{b,xx} = -zE_n\frac{\partial^2 w}{\partial x^2}, \\ \sigma_{yy} &= \sigma_{zz} = \sigma_{xy} = \sigma_{yz} = \sigma_{zx} = 0 \end{aligned} \tag{25.45}$$

where $n = bm, ac, sn$, defines original beam, piezoactuator layer and piezosensor layer, respectively. The Euler–Bernoulli beam theory is utilized to derive the equation of motion via Hamilton's principle. The Hamilton's principle is described as [13].

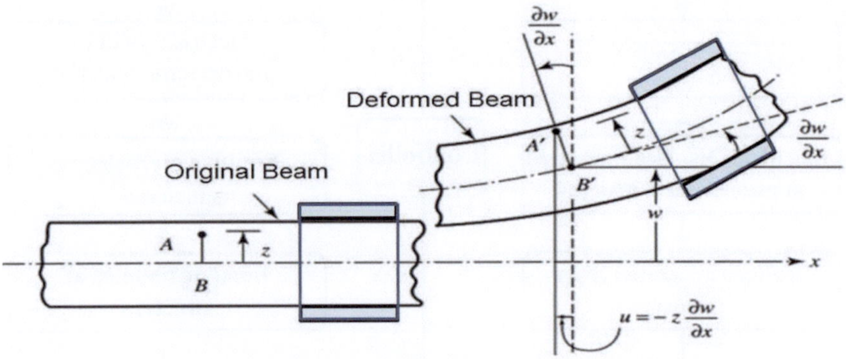

Fig. 25.12 Schematic of the bending of beam with piezoelectric patches on the top and bottom

$$\delta \int_{t_1}^{t_2} (L)\,\mathrm{d}t = 0$$
$$L = T - \Pi \tag{25.46}$$

where Π and T are potential energy (strain energy) and kinetic energy, respectively. Here a linear stress distribution is assumed in the beam with symmetric piezoelectric material patches. The stress in x-direction can be represented as summation of bending and longitudinal stress, $\sigma_{xx} = \sigma_{b,xx} + \sigma_l \sigma_{xx} = \sigma_{b,xx} + \sigma_l$. In the actuator layer, the stress induced by the prestrain is added to the total stress, $\sigma_{xx} = \sigma_{b,xx} + \sigma_l + \sigma_p$. The expression for the potential energy of the original beam is given by

$$\Pi = \frac{1}{2} \iiint_V (\sigma_{xx}\varepsilon_{xx} + \sigma_{yy}\varepsilon_{yy} + \sigma_{zz}\varepsilon_{zz} + \sigma_{xy}\varepsilon_{xy} + \sigma_{yz}\varepsilon_{yz} + \sigma_{zx}\varepsilon_{zx})\,\mathrm{d}V \tag{25.47}$$

The potential energy of the system is obtained by the summation of the sensor, the original beam and the actuator potential energy. The potential energy of the piezoelectric layer can be written as

$$\Pi = \frac{1}{2}\int_0^l \int_0^b \left\{ \int_{-\left(\frac{h_{bm}}{2}+h_{sn}\right)}^{-\frac{h_{bm}}{2}} E_{sn}K_{sn}(x)(\varepsilon_b + \varepsilon_l)^2 \mathrm{d}z + \int_{-\frac{h_{bm}}{2}}^{\frac{h_{bm}}{2}} E_{bm}(\varepsilon_b + \varepsilon_l)^2 \mathrm{d}z \right.$$
$$\left. + \int_{\frac{h_{bm}}{2}}^{\frac{h_{bm}}{2}+h_{ac}} E_{ac}K_{ac}(x)(\varepsilon_b + \varepsilon_l - \varepsilon_p)^2 \mathrm{d}z \right\} \mathrm{d}y\mathrm{d}x \tag{25.48}$$

where $K(x)$ represents the location of the piezoelectric material on the beam [11].

By considering similar properties and thickness for upper and lower layer and symmetric beam segment element, and substituting Eq. 25.43 into Eq. 25.48, and integrating, there is obtained.

$$\Pi = \int_0^l \left\{ \frac{1}{2}[E_{sn}I_{sn}K_{sn}(x) + E_{bm}I_{bm} + E_{ac}I_{ac}K_{ac}(x)]\left(\frac{\partial^2 w}{\partial x^2}\right)^2 \right.$$
$$+ \frac{1}{2}b\left[E_{sn}K_{sn}(x)\varepsilon_l^2 h_{sn} + E_{bm}\varepsilon_l^2 h_{bm} + E_{ac}K_{ac}(x)h_{ac}(\varepsilon_l - \varepsilon_p)^2\right]$$
$$\left. - b\left[E_{ac}K_{ac}(x)\varepsilon_p h_{ac}\left(\frac{h_{bm}+h_{ac}}{2}\right)\right]\frac{\partial^2 w}{\partial x^2} \right\} dx \tag{25.49}$$

The first expression in brackets is the strain energy due to bending. The second expression represents the strain energy due to longitudinal strain, and the last one illustrates the work done by the moment. Taking the variations of the potential energy with respect to transverse displacement, w, maintaining the first and the third expression, and noting that the second expression is eliminated due to variation, since it does not contain w term, then the variation with respect to w from Eq. 25.49 results in

$$\Pi = \int_0^l \left\{ \overline{I}(x) \frac{\partial^2 w}{\partial x^2} \delta\left(\frac{\partial^2 w}{\partial x^2}\right) - M(x,t)\delta\left(\frac{\partial^2 w}{\partial x^2}\right) \right\} \mathrm{d}x \tag{25.50}$$

where $\overline{I}(x) = E_{bm} I_{bm} + E_{sn} I_{sn} K_{sn}(x) + E_{ac} I_{ac} K_{ac}(x)$. The kinetic energy, T, of the system can be shown as [25, 26]

$$T = \int_0^l \frac{1}{2} (\rho_{sn} A_{sn} K_{sn}(x) + \rho_{bm} A_{bm} + \rho_{ac} A_{ac} K_{ac}(x)) \left(\frac{\partial w}{\partial t}\right)^2 \mathrm{d}x \tag{25.51}$$

The variation of kinetic energy with respect to transverse displacement, $w(x, t)$, is

$$T = \int_0^1 \overline{\rho} \frac{\partial w}{\partial t} \delta\left(\frac{\partial w}{\partial t}\right) \mathrm{d}x$$

$$\overline{\rho}(x) = \rho_{bm} A_{bm} + \rho_{sn} A_{sn} K_{sn}(x) + \rho_{ac} A_{ac} K_{ac}(x) \tag{25.52}$$

Substituting Eqs. 25.50 and Eq. 25.52 into Eq. 25.46, one obtains

$$\int_{t_1}^{t_2} \left\{ \left[\int_0^l \left[\left(\overline{\rho}(x) \frac{\partial w}{\partial t} \delta\left(\frac{\partial w}{\partial t}\right) \mathrm{d}x \right) \right] - \int_0^l \left[\overline{I}(x) \frac{\partial^2 w}{\partial x^2} \delta\left(\frac{\partial^2 w}{\partial x^2}\right) \right.\right.\right.$$
$$\left.\left.\left. - \delta\left(M(x,t) \frac{\partial^2 w}{\partial x^2}\right) \mathrm{d}x \right] \right] \right\} \mathrm{d}t = 0 \tag{25.53}$$

Integrating the first expression in brackets of Eq. 25.53 with respect to time and the second expression with respect to x by parts, there is obtained

$$\int_{t_1}^{t_2} \left\{ \left[\left(\overline{I}(x) \frac{\partial^2 w}{\partial x^2} - M(x,t) \right) \delta\left(\frac{\partial w}{\partial x}\right) \Big|_0^l \right] - \left[\left(\frac{\partial}{\partial x} \left(\overline{I}(x) \frac{\partial^2 w}{\partial x^2} \right) - \frac{\partial M(x,t)}{\partial x} \right) \delta w \Big|_0^l \right] \right.$$
$$\left. + \left[\int_0^l \left(\overline{\rho}(x) \frac{\partial^2 w}{\partial t^2} + \frac{\partial^2}{\partial x^2} \left(\overline{I}(x) \frac{\partial^2 w}{\partial x^2} \right) - \frac{\partial^2 M(x,t)}{\partial x^2} \right) \delta w \mathrm{d}x \right] \right\} \mathrm{d}t = 0 \tag{25.54}$$

where the first and second expressions in brackets represent the boundary conditions, and the third expression in bracket represents the equation of motion of the beam. By setting each term equal to zero, the conditions and the equation of motion can be obtained.

$$\overline{\rho}(x)\frac{\partial^2 w}{\partial t^2} + \frac{\partial^2}{\partial x^2}\left(\overline{I}(x)\frac{\partial^2 w}{\partial x^2}\right) = \frac{\partial^2 M(x,t)}{\partial x^2} \tag{25.55}$$

The boundary conditions for the clamped end are obtained from the geometry of the beam, which gives

$$w(0,t) = 0 \tag{25.56}$$

$$\frac{\partial w(0,t)}{\partial x} = 0 \tag{25.57}$$

For the free end, the boundary conditions are achieved from the first and second term of Eq. 25.33 at the point $x = l$.

$$\overline{I}(l)\frac{\partial^2 w(l,t)}{\partial x^2} - M(l,t) = 0 \tag{25.58}$$

$$-\frac{\partial}{\partial x}\left(\overline{I}(l)\frac{\partial^2 w(l,t)}{\partial x^2}\right) + \frac{\partial M(l,t)}{\partial x} = 0 \tag{25.59}$$

25.6 Solution of the Free Vibration of Beam with Piezoelectric Patches Using Finite Element Method

The Galerkin finite element method is utilized to obtain natural frequencies and natural modes of a beam bonded with piezoelectric layer [26]. Beam and patch elements are combined to acquire finite element model for the beam with piezoelectric layer.

Patch elements are modification of beam elements, in order to account for the added mass and stiffness of the piezoelectric layer patched on the beam. Standard beam element is applied on the element without piezoelectric patches. A beam model with six elements considered is illustrated in Fig. 25.13. The beam element has two nodes, where each node has two degrees of freedom. Each node has a displacement, W, and a rotational, θ, degree of freedom. To implement Galerkin method, a test function $\phi(x)$ is multiplied by Eq. 25.7 and integrated with respect to x over the domain as shown in Eq. 25.60.

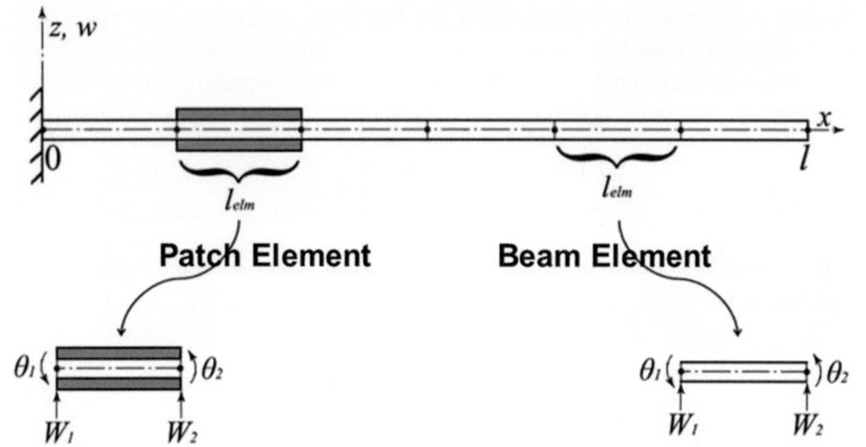

Fig. 25.13 Finite element model of the beam patched with piezoelectric layer

$$\int_0^l \phi(x)\left\{\overline{I}\frac{\mathrm{d}^4 W(x)}{\mathrm{d}x^4} - \overline{\rho}\omega^2 W(x)\right\}\mathrm{d}x = 0 \tag{25.60}$$

After integrating by parts and applying the boundary conditions, there is obtained.

$$\int_0^l \frac{\mathrm{d}^2\phi(x)}{\mathrm{d}x^2}\left(\overline{I}\frac{\mathrm{d}^2 W(x)}{\mathrm{d}x^2}\right)\mathrm{d}x - \int_0^l \phi(x)\left(\overline{\rho}\omega^2 W(x)\right)\mathrm{d}x = 0 \tag{25.61}$$

which can be rewritten in following form

$$\sum_{j=1}^{n}\int_{(j-1)h}^{jh} \overline{I}\left[\frac{\mathrm{d}^2\phi(x)}{\mathrm{d}x^2}\right]\left[\frac{\mathrm{d}^2 W(x)}{\mathrm{d}x^2}\right]\mathrm{d}x - \omega^2\sum_{j=1}^{n}\int_{(j-1)h}^{jh} \overline{\rho}\phi(x)W(x)\mathrm{d}x = 0 \tag{25.62}$$

where j represents the number of element. Each integral equation can be used for each element to determine the stiffness and mass matrices. The transverse deflection, $W(x)$ is assumed as

$$\begin{aligned}
W(x) &= \mathbf{L}^T\bar{w}_j,\\
\frac{\mathrm{d}}{\mathrm{d}x}W(x) &= \left(\frac{1}{l_{elm}}\right)\mathbf{L}'^T\bar{w}_j\\
\frac{\mathrm{d}^2}{\mathrm{d}x^2}W(x) &= \left(\frac{1}{l_{elm}}\right)^2\mathbf{L}''^T\bar{w}_j
\end{aligned} \tag{25.63}$$

where **L** and **w** are the shape function and nodal vector, respectively, and can be expressed as

$$\boldsymbol{L} = \left[3\xi^2 - 2\xi^3 \; l_{elm}\left(\xi^2 - \xi^3\right) \; 1 - 3\xi^2 + 2\xi^3 \; l_{elm}\left(-\xi + 2\xi^2 - \xi^3\right) \right]^T \quad (25.64)$$

$$\overline{\boldsymbol{w}}_j = \left[W_{j-1}\, \theta_{j-1}\, W_j\, \theta_j \right]^T \quad (25.65)$$

$$\begin{aligned} &(j-1)l_{elm} \le x \le j l_{elm},\ \xi = j - \frac{x}{l_{elm}}, \\ &0 \le \xi \le 1 \end{aligned} \quad (25.66)$$

By setting the test function equal to $W(x)$ for each element, and using the relationships Eqs. 25.64–25.66, 25.62 can be written as

$$\begin{aligned} &\sum_{j=1}^{n} \overline{\boldsymbol{w}}^T k_j \overline{\boldsymbol{w}} - \omega^2 \sum_{j=1}^{n} \overline{\boldsymbol{w}}^T m_j \overline{\boldsymbol{w}} = 0 \\ &k_j = \frac{\overline{I}}{h^3} \int_0^1 \boldsymbol{L}'' L''^T d\xi \\ &m_j = h\overline{\rho} \int_0^1 \boldsymbol{L}\boldsymbol{L}^T d\xi \end{aligned} \quad (25.67)$$

The k_j represents the stiffness matrix and m_j describes the mass matrix for one beam element. By determining the integrals of Eq. 25.67, these matrices reduce to

$$k_j = \frac{\overline{I}}{l_{elm}^3} \begin{bmatrix} 12 & 6l_{elm} & -12 & 6l_{elm} \\ 6l_{elm} & 4l_{elm}^2 & -6l_{elm} & 2l_{elm}^2 \\ -12 & -6l_{elm} & 12 & -6l_{elm} \\ 6l_{elm} & 2l_{elm}^2 & -6l_{elm} & 4l_{elm}^2 \end{bmatrix} \quad (25.68)$$

$$m_j = \frac{l_{elm}\overline{\rho}}{420} \begin{bmatrix} 156 & 22l_{elm} & 54 & -13l_{elm} \\ 22l_{elm} & 4l_{elm}^2 & 13l_{elm} & -3l_{elm}^2 \\ 54 & 13l_{elm} & 156 & -22l_{elm} \\ -13l_{elm} & -3l_{elm}^2 & -22l_{elm} & 4l_{elm}^2 \end{bmatrix} \quad (25.69)$$

where $\overline{I} = E_{bm} I_{bm}$ and $\overline{\rho} = \rho_{bm} A_{bm}$ are considered for the original the beam element and $\overline{I} = E_{bm} I_{bm} + E_{sn} I_{sn} + E_{ac} I_{ac}$ and $\overline{\rho} = \rho_{bm} A_{bm} + \rho_{sn} A_{sn} + \rho_{ac} A_{ac}$ for the patch element. By considering the number of nodes, the stiffness and the mass matrices for each element can be assembled together and synthesized into the global stiffness and mass matrices. Thus, the eigenvalue problem of Eq. 25.67 can be represented by [25, 26].

$$K\overline{\mathbf{w}} = \omega^2 M\overline{\mathbf{w}} \tag{25.70}$$

where K and M are global stiffness and mass matrices for an arbitrary beam, respectively. Finite element formulation is utilized to write the in-house MATLAB program. The actuator distributed moment on the beam element can be obtain through the virtual work. The virtual work done by moment is expressed as follows:

$$\delta W = \int\limits_{(j-1)l_{elm}}^{jl_{elm}} M_{ac}(x,t)\cdot\delta\left(\frac{\partial^2 w}{\partial x^2}\right)\mathrm{d}x \tag{25.71}$$

By substituting Eq. 25.35 for actuator moment into Eq. 25.71, there is obtained

$$\delta W = -bE_{ac}\left(\frac{h_{bm}+h_{ac}}{2}\right)d_{31}V(t)\cdot\delta\left(\frac{\partial w}{\partial x}\right)\Bigg|_{(j-1)l_{elm}}^{jl_{elm}} \tag{25.72}$$

Substituting Eq. 25.63 into Eq. 25.72 and changing the integration boundaries commensurate with the local coordinates Eqs. 25.66 and 25.72 can be rewritten as

$$\begin{aligned} \delta W &= -bE_{ac}\left(\frac{h_{bm}+h_{ac}}{2}\right)d_{31}V(t)\cdot\delta\left(\bar{w}^T\right)\mathbf{L}'\Big|_0^1 \\ \delta W &= -\delta\left(\bar{w}^T\right)bE_{ac}\left(\frac{h_{bm}+h_{ac}}{2}\right)d_{31}V(t)\left[0\ 1\ 0\ -1\right]^T \end{aligned} \tag{25.73}$$

where the first term after the equality sign shows the transverse variation and, thus, the actuation force as expressed by

$$\begin{aligned} \{P_{ac}\} &= bE_{ac}\left(\frac{h_{bm}+h_{ac}}{2}\right)d_{31}V(t)\left[0\ 1\ 0\ -1\right]^T \\ \{P_{ac}\} &= \{f_{ac}\}V(t) \end{aligned} \tag{25.74}$$

f_{ac} is the force vector of piezoactuator, which maps the control voltage to the structure.

At this point, some remarks are in order regarding the computational modeling and computational method adopted in this work, to provide legitimate support to further work on vibration control optimization. To this end, comparison of the present baseline results to similar work is elaborated in Appendix I.

25.7 Control and Control Performance

25.7.1 System Response

For a multidegree-of-freedom (MDOF) system, the time response of the system can be expressed in matrix form [16, 25, 38] as

$$[m]\{\ddot{\eta}(t)\} + [c]\{\dot{\eta}(t)\} + [k]\{\eta(t)\} = \{P_{ac}\} + \{P_{ex}\} \tag{25.75}$$

where m is the mass matrix, k is the stiffness matrix, P_{ac} is the actuation force and P_{ex} is the external force. c is the damping matrix, which can be expressed as proportional damping, which is typically mentioned as Rayleigh damping. For damped structure, the closed forms of the solutions are not generally feasible. However, the idealized solution of damped motion can be assumed by utilizing classical damping. Classical damping is usually divided into two categories, Rayleigh damping and Cauchy damping. Here, Rayleigh damping method is utilized to represent a linear combination of mass and stiffness matrices [30].

$$[c] = \alpha[m] + \beta[k] \tag{25.76}$$

where α and β are known proportional Rayleigh coefficients respect to mass and stiffness matrices, respectively. α and β are defined as

$$\alpha = 2\xi_i\left(\frac{\omega_i \times \omega_j}{\omega_i + \omega_j}\right),\ \beta = 2\xi_i\left(\frac{1}{\omega_i + \omega_j}\right) \tag{25.77a,b}$$

ξ is the modal viscous damping coefficient, which correspond to undamped natural frequency, ω_i. The damping coefficient is the dynamic property of material, which cannot be determined theoretically. In this study, a uniform damping coefficient of 0.5% is assumed taking into account the results obtained from experimental investigations [4, 23, 24, 38] and dynamic properties of metals [4, 23, 24, 38].

25.7.2 Modal Order Reduction

To facilitate the solution of the dynamical system Eq. 25.75, which involves very large matrices, resort is made to order reduction method. The concept is to estimate the high-dimensional state space by using an appropriate low-dimensional subspace to obtain a smaller system with approximately similar properties. To design the linear controller, first and second natural modes of the beam vibration are considered, since the other natural modes is expected to exhibit insignificant effect in comparison to the first two modes. In this regard, modal order reduction technique is utilized to reduce the large number of order of the system, which is obtained by finite element solution. Accordingly, first the coordinate of the system is reduced by considering the first two modes.

$$\{\eta(t)\} = [T]_{n\times 2}\{g(t)\}_{2\times 1}$$
$$\{g(t)\} = \begin{Bmatrix} g_1 \\ g_2 \end{Bmatrix} \tag{25.78a,b}$$

where T is the reduced matrix of eigenvectors based on the first two modes, and g is the reduced coordinates. By substituting Eq. 25.78a, b into Eq. 25.75 and multiplying by $[T]^T$, the reduced order transfer function of the system can be obtained.

$$[T]^T[m][T]\{\ddot{g}(t)\} + [T]^T[c][T]\{\dot{g}(t)\} + [T]^T[k][T]\{g(t)\} = [T]^T\{f_{ac}\} + [T]^T\{f_{ex}\} \tag{25.79}$$

where the mass, damping and stiffness matrices and the force vectors can be defined as follows:

$$\begin{aligned} \left[\widehat{m}\right]_{2\times 2} &= [T]^T_{2\times n}[m]_{n\times n}[T]_{n\times 2}, \\ \left[\widehat{c}\right]_{2\times 2} &= [T]^T_{2\times n}[c]_{n\times n}[T]_{n\times 2}, \\ \left[\widehat{k}\right]_{2\times 2} &= [T]^T_{2\times n}[k]_{n\times n}[T]_{n\times 2}, \\ \left\{\widehat{f}_{ac}\right\}_{2\times 1} &= [T]^T_{2\times n}\{f_{ac}\}_{n\times 1}, \\ \left\{\widehat{f}_{ex}\right\}_{2\times 1} &= [T]^T_{2\times n}\{f_{ex}\}_{n\times 1}, \end{aligned} \tag{25.80a,b,c,d,e,f}$$

Thus, Eq. 25.75 can be rewritten as

$$\left[\widehat{m}\right]\{\ddot{g}(t)\} + \left[\widehat{c}\right]\{\dot{g}(t)\} + \left[\widehat{k}\right]\{g(t)\} = \left\{\widehat{f}_{ac}\right\} + \left\{\widehat{f}_{ex}\right\} \tag{25.81}$$

25.7.3 State-Space Representation

Equation 25.81 is transformed to a state-space vector dynamic equation for designing the state feedback control system, using g as the solution vector. Hence

$$\{\dot{g}(t)\} = \{\dot{g}(t)\} \tag{25.82}$$

$$\{\ddot{g}(t)\} = -\left[\widehat{m}\right]^{-1}\left[\widehat{c}\right]\{\dot{g}(t)\} - \left[\widehat{m}\right]^{-1}\left[\widehat{k}\right]\{g(t)\} + \left[\widehat{m}\right]^{-1}\left\{\widehat{f}\right\}_{ac} + \left[\widehat{m}\right]^{-1}\left\{\widehat{f}\right\}_{ex} \quad (25.83)$$

Then, **X** vector is introduced in order to reduce the order of Eq. 25.75 as follows:

$$\{g(t)\} = \begin{bmatrix} g_1 \\ g_2 \end{bmatrix} = \begin{bmatrix} X_1 \\ X_2 \end{bmatrix},$$
$$\{\dot{g}(t)\} = \begin{bmatrix} \dot{g}_1 \\ \dot{g}_2 \end{bmatrix} = \begin{bmatrix} \dot{X}_1 \\ \dot{X}_2 \end{bmatrix} = \begin{bmatrix} X_3 \\ X_4 \end{bmatrix}, \quad (25.85)$$
$$\{\ddot{g}(t)\} = \begin{bmatrix} \ddot{g}_1 \\ \ddot{g}_2 \end{bmatrix} = \begin{bmatrix} \ddot{X}_1 \\ \ddot{X}_2 \end{bmatrix} = \begin{bmatrix} \dot{X}_3 \\ \dot{X}_4 \end{bmatrix}$$

Thus, Eq. 25.82 and Eq. 25.83 can be represented, respectively, as

$$\begin{bmatrix} \dot{X}_1 \\ \dot{X}_2 \end{bmatrix} = \begin{bmatrix} X_3 \\ X_4 \end{bmatrix} \quad (25.86)$$

$$\begin{bmatrix} \dot{X}_3 \\ \dot{X}_4 \end{bmatrix} = -\left[\widehat{m}\right]^{-1}\left[\widehat{c}\right]\begin{bmatrix} X_3 \\ X_4 \end{bmatrix} - \left[\widehat{m}\right]^{-1}\left[\widehat{k}\right]\begin{bmatrix} X_1 \\ X_2 \end{bmatrix} + \left[\widehat{m}\right]^{-1}\left\{\widehat{f}\right\}_{ac} + \left[\widehat{m}\right]^{-1}\left\{\widehat{f}\right\}_{ex} \quad (25.87)$$

which can be rewritten as

$$\begin{bmatrix} \dot{X}_1 \\ \dot{X}_2 \\ \dot{X}_3 \\ \dot{X}_4 \end{bmatrix} = \begin{bmatrix} [0]_{2\times 2} & I_{2\times 2} \\ -\left[\widehat{m}\right]^{-1}\left[\widehat{k}\right] & -\left[\widehat{m}\right]^{-1}\left[\widehat{c}\right] \end{bmatrix}\begin{bmatrix} X_1 \\ X_2 \\ X_3 \\ X_4 \end{bmatrix} + \begin{bmatrix} [0]_{2\times 1} \\ \left[\widehat{m}\right]^{-1}\left\{\widehat{f}\right\}_{ac} \end{bmatrix} + \begin{bmatrix} [0]_{2\times 1} \\ \left[\widehat{m}\right]^{-1}\{f_{ex}\} \end{bmatrix} \quad (25.88)$$

or

$$\left\{\dot{X}\right\} = [A]\{X\} + [B] + \left[B^*\right] \quad (25.89)$$

where A, B and B^* are termed as the state matrix, the input matrix corresponding to the actuator and the input matrix corresponding to the external force, respectively. The output of the system is expressed as Y. In this study, the output of the system is sensor voltage derived in previous section. Thus, the output can be represented as

$$Y = [C]\{X\} \quad (25.90)$$

where C is the output matrix, and is given by

$$[C] = C_{sn}\big[\theta_1(x_{s2}) - \theta_1(x_{s1})\ \theta_2(x_{s2}) - \theta_2(x_{s1})\ 0\ 0\big] \tag{25.91}$$

where θ is the derivative of the displacement and x_s is the location of the piezoelectric sensor on the beam. The benefits of state-space approach are in the formulation of the appropriate control to obtain the desired output.

25.7.4 Control Strategy Formulation

Two linear control methods are applied in the present investigation to suppress the vibration of the beam. The state-space model consisting of the first two natural modes of the system is utilized to design the controller. First, PID control method is considered, which is a well-known classical control method. Then, LQR modern control is utilized to design an optimal controller in order to compare with PID control.

25.7.4.1 PID Control

The proportional-integral-derivative (PID) control, which is well-known control tool due to its robustness and simplicity, will be considered for comparative purposes. The proportional feedback constant, *P*, controls the natural frequency of the system, and hence the amplitude of vibration. The integral constant, *I*, sets the necessary adjustment for the damping, or energy dissipation of the system. The combination of proportional and integral control action gives the controller a way to minimize steady state error, while having the ability to minimize the effects of disturbances to the system. Proportional and integral constants manipulate past control error and cannot predict the future control error. Thus, the derivative constant, *D*, is proportional to the change in the error. In other words, it manipulates the speed or response of the controller. The convenient selections of the PID constants are a key aspect in the success of executing the PID controller. The transfer function of the PID controller is given by [16]:

$$u(t) = K_p e(t) + K_i \int e(t)\,\mathrm{d}t + K_d \frac{\mathrm{d}e(t)}{\mathrm{d}t} \tag{25.92}$$

where K_p is the proportional gain, K_i is the integral gain and K_d is the derivative gain. Figure 25.14 shows the block diagram of PID control system.

25.7.4.2 LQR Control with Observer

Linear quadratic regulator (LQR) is then considered for optimal control, to provide a symmetric way to determine the state feedback control matrix [16]. It is noticed that

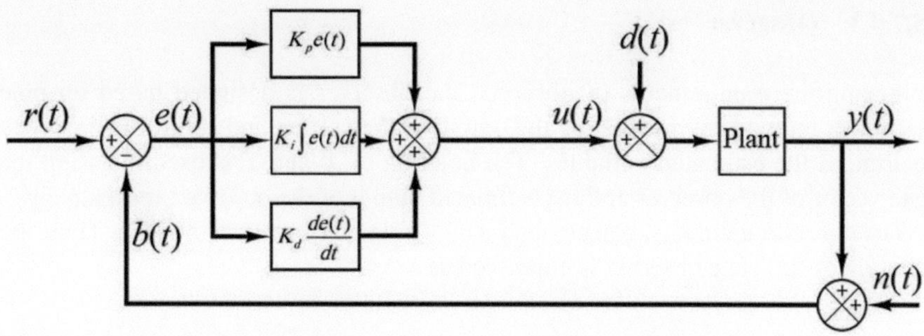

Fig. 25.14 Closed-loop system with parallel PID control block diagram

for controlling the beam vibration, all variables for LQR control are not available. An observer is then necessary to design an estimator for the unavailable feedback values for its ability to estimate the unavailable variables. In order to control the system with observer, it should be controllable and observable. The definition of system controllability, system observability, observer and LQR is briefly elaborated subsequently.

A system is called controllable, if a system, with regard to unconstrained input control vector, can be transferred from any initial value $\mathbf{X}(t_0)$ at time t_0 to the any specified value in a specific time $t_0 \leq t \leq t_1$. The controllability matrix is expressed as [16].

$$\left[\boldsymbol{B} \,\middle|\, \boldsymbol{AB} \,\middle|\, \ldots \,\middle|\, \boldsymbol{A}^{n-2}\boldsymbol{B} \,\middle|\, \boldsymbol{A}^{n-1}\boldsymbol{B} \right]_{n \times n} \tag{25.93}$$

The system is controllable, if and only if the controllability matrix Eq. 25.93 is a full rank matrix (rank of n); in other words, each vector of the matrix Eq. 25.93 should be linearly independent. In this regard, MATLAB program can be utilized to determine the controllability matrix.

One system is called observable, if every state vector $\mathbf{X}(t_0)$ can be obtained from the observation output in a specific time, $t_0 \leq t \leq t_1$. The observability matrix of a system is given by [16]

$$\left[\boldsymbol{C} \,\middle|\, \boldsymbol{CA} \,\middle|\, \ldots \,\middle|\, \boldsymbol{CA}^{n-2} \,\middle|\, \boldsymbol{CA}^{n-1} \right]^T \tag{25.94}$$

One system is observable, if and only if the observability matrix has n linearly independent vectors, or it is full rank matrix. Same as controllability, MATLAB program is applicable to determine observability matrix of a system and check its rank.

25.7.4.3 Observer

Following the requirements of observer, the observer is designed based on pole placement method. By measuring the output of the system and control variables, it determines the estimated variables. The notation of $\hat{X}$ and $\hat{Y}$ is used to assign the state vector of the observer and the estimated output of the observer, respectively.

The observer gain, K_{ob}, can be obtained by pole placement method [7]. Thus, the equation of full state observer is expressed as

$$\dot{\hat{X}} = [A]\hat{X} + [B]f + K_{ob}(Y - \hat{Y}) \tag{25.95}$$

where A and B are state matrix and actuation matrix, which are same as the state-space Eq. 25.89. f and Y are input and output of the system, respectively, which are expressed as

$$\begin{aligned} Y &= \boldsymbol{CX} \\ \hat{Y} &= \boldsymbol{C}\hat{\boldsymbol{X}} \\ f &= -K_{lqr}\hat{\boldsymbol{X}} \end{aligned} \tag{25.96}$$

Equation 25.97 is illustrated in the block diagram form exhibited in Fig. 25.15. By substituting Eq. 25.96 into Eq. 25.95, they can be rewritten as

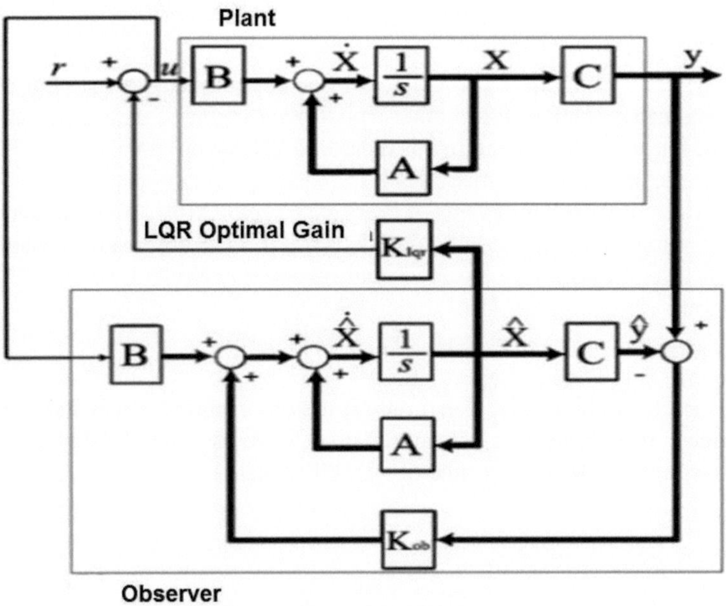

Fig. 25.15 Block diagram of the closed-loop system with observer

$$\dot{\hat{X}} = [A - K_{ob}C]\hat{X} + [B]f + K_{ob}Y$$
$$\dot{\hat{X}} = [A_{ob}]\hat{X} + \begin{bmatrix} B & K_{ob} \end{bmatrix}\begin{Bmatrix} f \\ Y \end{Bmatrix} \tag{25.97}$$
$$\dot{\hat{X}} = [A_{ob}]\hat{X} + [B_{ob}]\begin{Bmatrix} f \\ Y \end{Bmatrix}$$

25.7.4.4 LQR Gain

LQR control provides an approach to calculate the state feedback gain of the control system [16]. The system equation is given as

$$\dot{X} = [A]X + [B]f \tag{25.98}$$

The optimal actuator input (control vector) can be determined as

$$\boldsymbol{f}(t) = -K_{lqr}\boldsymbol{X}(t) \tag{25.99}$$

The state feedback gain K_{lqr} is optimized to minimize the following

$$J = \int_0^\infty (\boldsymbol{X}^T\boldsymbol{Q}\boldsymbol{X} + \boldsymbol{f}^T\boldsymbol{R}\boldsymbol{f})\mathrm{d}t \tag{25.100}$$

where **Q** and **R** are the constant weighting matrices, which are real symmetric and positive-definite matrices. **Q** and **R** can be estimated by experiments or trial and error; however, assigning **Q** large with respect to **R** means that the response attenuation has more weight than the control effort and conversely. The minimized control gain is expressed as

$$K_{lqr} = \boldsymbol{R}^{-1}\mathbf{B}^T\boldsymbol{P} \tag{25.101}$$

where **P** is the unique and positive solution of the well-known *Riccati* equation [16].

$$\boldsymbol{A}^T\boldsymbol{P} + \boldsymbol{P}\boldsymbol{A} - \boldsymbol{P}\boldsymbol{B}\boldsymbol{R}^{-1}\boldsymbol{B}^T\boldsymbol{P} + \boldsymbol{Q} = \boldsymbol{0} \tag{25.102}$$

For multidegree of freedom system, the solution of Eq. 25.102 for P is difficult to achieve; however, it can be solved numerically. MATLAB software has a built-in command to solve *Riccati* Eq. 25.102 and therefore obtain K_{LQR} and *P*. LQR controller with observer is performed in MATLAB/SIMULINK©. Detailed elaboration is given in [39].

25.8 Results and Discussions

An aluminum beam with three different configurations is considered. The properties of aluminum beam can be obtained from established references and experimental work [23, 24, 38]. These are tabulated in Table 25.3. The natural frequencies and natural modes for each case study are obtained through the procedure discussed in previous section.

The PID controller and LQR controller are utilized to manipulate the vibration of the beam, where controllers are designed using MATLAB/SIMULINK©. The results are obtained by assuming 1 *N* impulse for the duration of 0.001 s at the tip of the beam. If the actuator voltage exceeds the maximum operating voltage of the piezoelectric material, the latter may lose its polarization and piezoelectricity property. In this regard, the input control voltage is limited to ±90 V, which is much less than maximum operating voltage of PZT and PVDF. Since the classical method for determining PID coefficients is not applicable here, these coefficients are obtained from available experiments [23, 40] and tabulated in Table 25.4 for each case study.

In the present work, the LQR controller with observer is designed and simulated to control the vibration of the system. The controllability and observability of systems were checked in first step, where state-space forms of the assumed beams were all controllable and observable. The observer control method is based on pole placement; thus, new poles should be chosen with respect to the poles of the system [16]. The most conventional approach is to obtain new poles based on experience.

Table 25.3 Properties of aluminum and piezoelectric materials [23, 24]

	Aluminium	PZT5−H	PVDF
Modulus of elasticity	71	61	2
Density (Kg/m^3)	2710	7500	1780
Width (m)	0.014	0.014	0.014
Thickness (mm)	0.66	0.75	0.11
Length (m)	0.319	–	–
Strain constant d_{31}	–	-171×10^{-12}	23×10^{-12}
Voltage constant g_{31}	–	0.0114	0.216

Table 25.4 PID coefficients that are obtained by experiment for each case study

PID coefficients	Case study I	Case study II	Case study III
Proportional K_p	700	25	15
Integral K_I	675	22.5	13.5
Derivative K_D	900	30	18

Table 25.5 LQR parameters that obtained by experiment for each case study

LQR parameter	Case study I	Case study II	Case study III
Q	50×10^7	10^6	10^8
R	10^{-4}	10^{-3}	10^{-2}

To determine the LQR optimal gain, first, Q and R should be specified. Q and R are obtained for the best settling time by experiment or trial and error method. In this study, R is only a scalar number, and Q is a square diagonal matrix in which the entries of the main diagonal are all one and multiplied by a coefficient, q. R and q for each case study are presented in Table 25.5. The results of each case study are given and discussed in the following paragraphs.

The first configuration is considered as a beam completely bonded with PVDF on the top and the bottom. The sensor is located at the tip of the beam, which is 15 mm long. Other PDVF patches are utilized as actuator. The length of PVDF on the upper surface is same as the length of the beam, and the lower one is 304 mm. Case study one is schematically illustrated in Figure 25.16. The free vibration analysis of this case study is carried out by both analytical and numerical approaches. Properties of PVDF material are listed in Table 25.3.

For analytical approach, the natural frequencies of the beam considered can be obtained by substituting β using the following relation:

$$\omega = \beta^2 \sqrt{\frac{1}{\rho}} \tag{25.103}$$

Similar to Eqs. 25.18 and 25.70 is utilized to obtain the eigenvalue and, consequently, the natural frequencies, via FEM.

The results from the analytical approach and finite element analysis are given in Table 25.6. Again, the results of the FEM solution show very small error in comparison to the analytical solution.

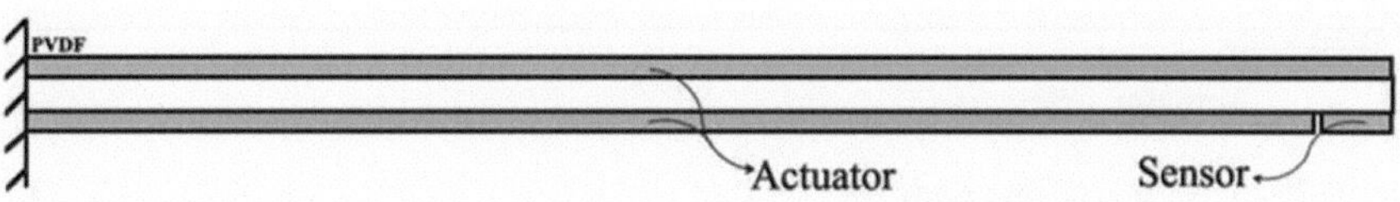

Fig. 25.16 Case study one: beam with complete boned piezoelectric patch

Table 25.6 Natural frequencies of case study one

Natural frequencies	Analytical (Hz)	FEM (Hz)	Error %
1st mode	4.9501	4.9502	2.02×10^{-3}
2nd mode	31.0197	31.0222	8.05×10^{-3}
3rd mode	86.8618	86.8636	2.07×10^{-3}

Controlled and uncontrolled responses of the case study one are shown in Fig. 25.17. Time response of PID and LQR controller is independently compared with uncontrolled system. In this case, LQR shows a little better settling time in comparison to PID. Table 25.9 lists the calculated results for settling time of these controllers.

In case study two, two PZT are bonded at the base of the beam, which are 38 mm long. PZT sensor is located right after the piezoactuator on the top of the beam. The length of the sensor is 15 mm, as shown in Fig. 25.18. The natural frequencies of this case study are obtained by finite element methods and validated by experimental results [23] and expressed in Table 25.7.

The result of PID and LQR controllers is given in Fig. 25.19, which exhibits significant performance. Both of them can effectively control the vibration of the system; however, the results of LQR control are a little bit better than PID.

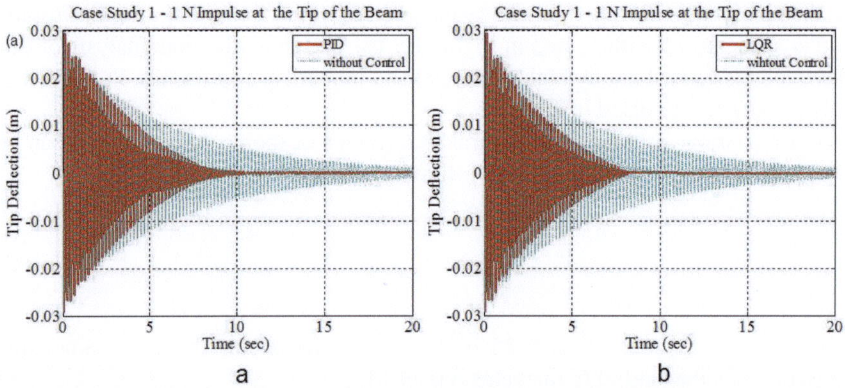

Fig. 25.17 Time response of the beam, case study one **a** PID control, **b** LQR control

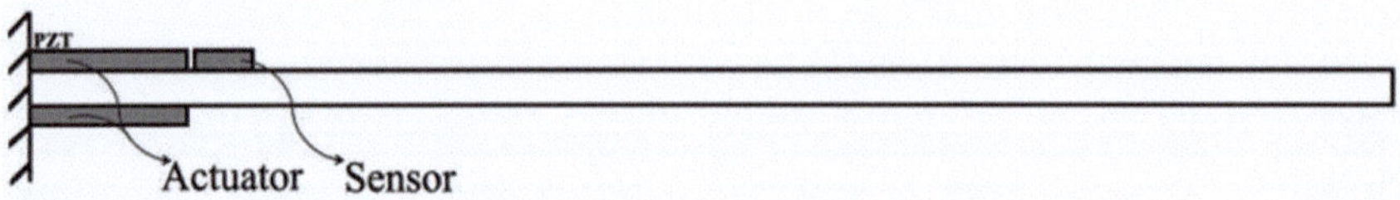

Fig. 25.18 Schematic of case study two

Table 25.7 Natural frequencies of case study two compared with available experimental study from Hong [23]

Natural frequencies	Experimental study (Hz)	FEM (Hz)	Error %
1st mode	6.3	6.44	2.22
2nd mode	38	39.45	3.81
3rd mode	99	107.65	8.73

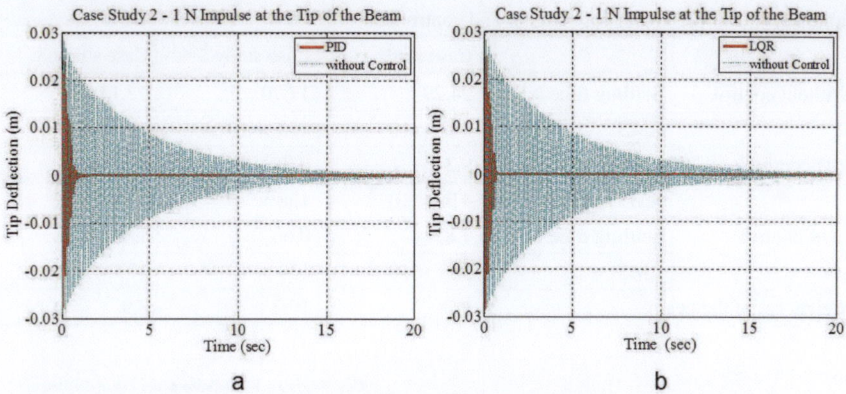

Fig. 25.19 Time response of the beam, case study two **a** PID control, **b** LQR control

One PZT actuator and one PVDF sensor are considered in case study three. The length actuator is 38 mm and laid on the upper surface of the beam. The sensor is located next to the piezoactuator with 2 mm distance from it. The length of piezosensor is 15 mm. Figure 25.20 shows the location of actuators and sensor on the beam. In this case, natural frequencies and natural modes are determined by finite element method, which are shown in Table 25.8.

The controlled and uncontrolled system responses of case study three are illustrated in Fig. 25.21. The performance of LQR is also shown to be better than that of PID.

Three case studies have been considered to evaluate which one has better performance in controlling the system. Settling time and root-mean-square (RMS) value of all of these case studies are exhibited in Table 25.9. The settling time is defined as the time taken for the response of the system to reach and stay in a range around the desired value, which is usually considered to be 2–5% of the final value. The

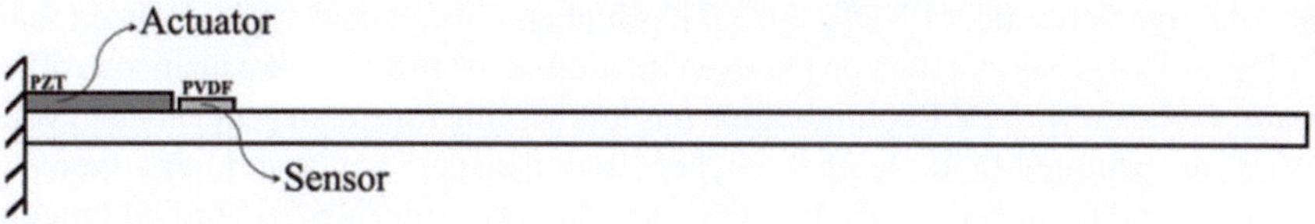

Fig. 25.20 Schematic of case study three

Table 25.8 Natural frequencies of case study three determined by finite element method

Natural frequencies	FEM
1st mode	6.22
2nd mode	37.82
3rd mode	102.92

Table 25.9 Comparison of case studies and controllers

		Case study 1	Case study 2	Case study 3
Without control	Settling time (s)	24.29	17.70	17.13
	RMS	483×10^{-5}	400×10^{-5}	427×10^{-5}
PID control	Settling time (s)	9.34	0.88	1.47
	RMS	414×10^{-5}	138×10^{-5}	192×10^{-5}
LQR control	Settling time (s)	7.87	0.66	1.13
	RMS	385×10^{-5}	134×10^{-5}	177×10^{-5}
Weight (g) of the beam		9.7	10.2	8.9

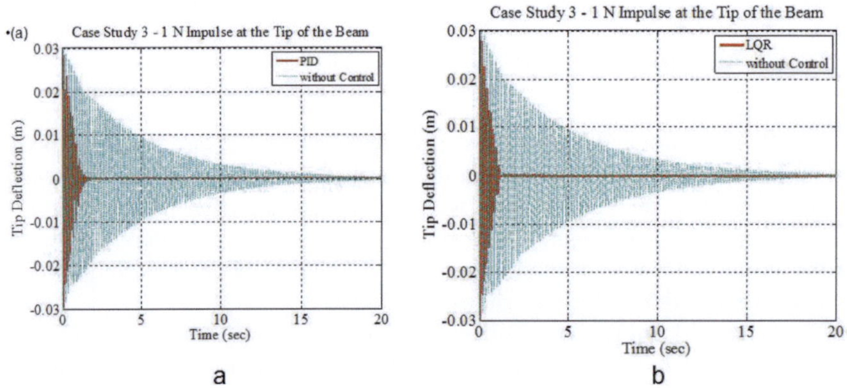

Fig. 25.21 Time response of the beam, case study three **a** PID control, **b** LQR control

RMS value is the square root of the arithmetic mean of the squared magnitudes of the waveform, which is a measure of the wave amplitudes.

As shown in the figures, it is obvious that the beam with PZT actuator has much better response in comparison to the beam bonded with PVDF. PZT can surpass the vibration better than PVDF, since it has higher stiffness and strain constant than PVDF. Stiffness has an effect on passive vibration of the beam, where in uncontrolled beam, the beam with PZT configuration has less settling time than beam bonded with PVDF, as exhibited in Table 25.9. Higher strain constant results in higher moment on the beam due to Eq. 25.35. In this regard, the system can be controlled quickly with PZT actuator. The settling time and RMS of case study two and three are close; however, case study two can surpass the vibration slightly better than case study three. By utilizing the PVDF sensor in the case study three, the weight of the system reduces in comparison to case study two.

Both controllers provide significant vibration suppression. For all of these case studies, LQR shows a slightly better settling time performance than PID. Case study two with LQR controller shows the best result in this study. Settling time and RMS value of it are less than the others, and the weight of the beam is acceptable.

Two other issues will be addressed following the reviewer comments. The first issue concerns the spill–over effect, which is exemplified in Appendix II. In the worked-out example, the spill–over effect gives rise to overshoots in higher modes; however, it does not affect the settling time of the system, which is the objective of the present study. Two ways can be suggested to reduce this effect in this structure: first, to provide some controllers to avoid the spill–over effect, and, second, to provide some passive damping on the structure. The second issue concerns the uncertainties, which is elaborated to some detail in Appendix III. These considerations show that the control considered in this work for the particular system is robust.

25.9 Conclusions

In this investigation, active vibration suppression of an elastic structure using piezoelectric sensor and actuator has been elaborated by solving generic beam problem and three configurations of the beam-piezoelectric elements composite beams. The active control of the vibration of an aluminum flexible beam has been studied through the bonding of the piezoelectric elements onto a beam as a smart material. The governing equation of piezoelectric sensor and actuator is derived for the beam structure. The general equation of motion of smart beam structure bonded with piezoelectric layer is derived through Hamilton's principle based on Euler–Bernoulli beam theory. For baseline considerations, a cantilevered Euler–Bernoulli beam with spring loading at the tip has been solved analytically and numerically using finite element method. Then three different configurations of cantilevered Euler–Bernoulli beams with piezoelectric patches are considered as case studies. In the first configuration, the beam is completely integrated with PVDF on both sides. The free vibration of this configuration is solved using analytical method and Galerkin finite element method. The beam is patched with PZT at the base of the beam in the second case study. The second case study is solved by finite element and compared to an available experimental work. The free vibration solution of third one, which is the beam with PZT actuator and PVDF senor, is determined by finite element only. The state-space representation of the dynamic equation of the beam is formed in order to design the controllers and to obtain the response of the beam vibration. Then, the full-order state observer LQR and PID are presented and designed to control the vibration of the beam. These controllers are utilized in these case studies for comparative purposes and to gain insight into the problems addressed. The results thus obtained exhibit the effectiveness of the various controllers chosen and give a beneficial insight on the utilization of the stability and control of flexible beam structure, which will be beneficial in addressing more involved problems in the future.

Appendix 1: Comparative Study of the Baseline Results with Other Works

For comparative study, FEM and analytical solution of this study is compared with two other works; the first is based on Timoshenko beam theory [4] and the second on theoretical and experimental study [36]. Table 25.10 compares the results of Narayanan [4] Timoshenko beam theory with the results of the present model, which uses both analytical and finite element computational schemes, and using the same thickness to length to ratio of 0.004.

It can be observed that both works exhibit excellent agreement, which serve to validate and indicate the robustness of the present method in obtaining the baseline solution.

The second comparison serves to verify that for larger thickness to length ratio, the Euler–Bernoulli beam theory may fail to show good agreement with experiments. Baillargeon and Vel [36] work deals with thick sandwich beam, where the thickness to length ratio of the beam is 0.1. However, it is interesting to note that the first natural frequency calculated using our method and Baillargeon and Vel's ABAQUS-based solution and experimental results indicates an error of 19.7% for the first mode, as exhibited in Table 25.11. Nevertheless, such qualitative agreement is encouraging.

Table 25.10 Comparison of the present FEM results with analytical and Timoshenko beam theoretical results [4]

Number of modes	FEM	Analytical Euler-Bernoulli beam theory	Error between FEM and analytical	Timoshenko beam theory [4]	Error between FEM and Timoshenko
1st	6.89	6.89	0.0	6.89	0.0
2nd	43.17	43.17	0.0	43.285	2.6×10^{-3}
3rd	120.89	120.89	0.0	121.22	2.7×10^{-3}
4th	236.89	236.90	4.2×10^{-5}	237.72	3.5×10^{-3}
5th	391.57	391.60	7.7×10^{-5}	393.58	4.6×10^{-3}
6th	584.98	585.00	3.4×10^{-5}	589.63	7.9×10^{-3}

Table 25.11 Comparison the first natural frequency of the present FEM results with experimental results

Number of modes	FEM	Analytical Euler-Bernoulli beam theory	Error between FEM and analytical	Experimental work [36]	Error between FEM and experimental
1st	147.4	147.39	0.0	123.06	19.7%

Appendix 2: The Spill–Over Effect on the Vibration Control of Structures

For vibration control of a smart structure, controllers are normally designed based on a reduced order model (ROM) of the system. When such a ROM-based controller operates in closed loop with the actual structure, spill–over phenomenon occurs because higher mode which are not accounted for in the model, hence in ROM, will be excited [15].

For that purpose, case study three is utilized to exemplify the spill–over effect. Figure 25.22 illustrates the uncontrolled vibration of the beam if three different combinations of the natural modes of the beam are considered; first with two modes, the second with three modes, and the third with six modes. The results show that

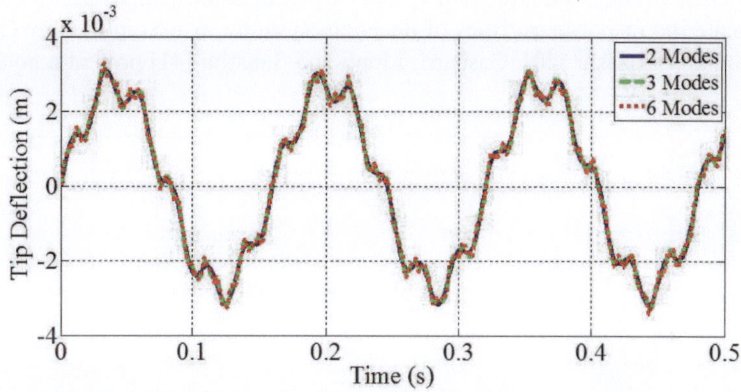

Fig. 25.22 Time response of uncontrolled vibration of the beam

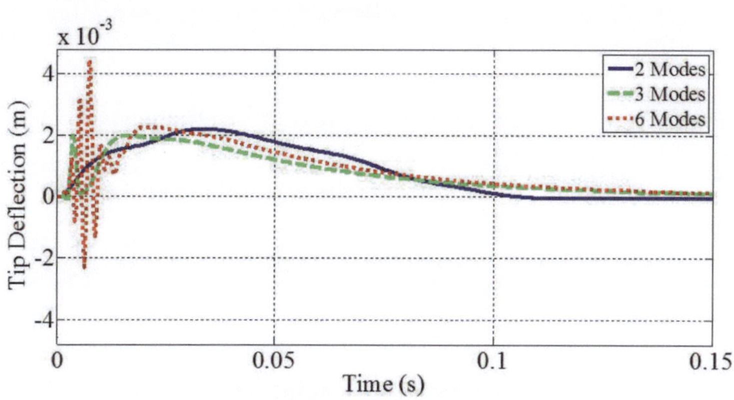

Fig. 25.23 Time response of the LQR controlled vibration of the beam

with the incorporation of higher modes, more fluctuation is exhibited. Figure 25.23 depicts the LQR controlled vibration of the beam by considering the higher modes. The spill–over effect can be observed in the beam vibration that incorporates the first six (6) major modes. The controller may excite the mode 5 or 6, and it results in overshoots. However, the settling time of the controlled vibration, which is desirable in this investigation, is not significantly changed.

Appendix 3: Uncertainty Analysis of the Current Model

The uncertainty analysis for structural dynamics can be carried out using several approaches, such as deterministic and stochastic approaches, such as deterministic and stochastic approaches (Fig. 25.24).

In the field of stochastic structural dynamics, perturbation methods are widely used to estimate the response statistics of uncertain systems, as exemplified by Cortez, Ferguson and Bhaskar [40]. Gasbarri, Monti and Sabatini [41] present a nonlinear

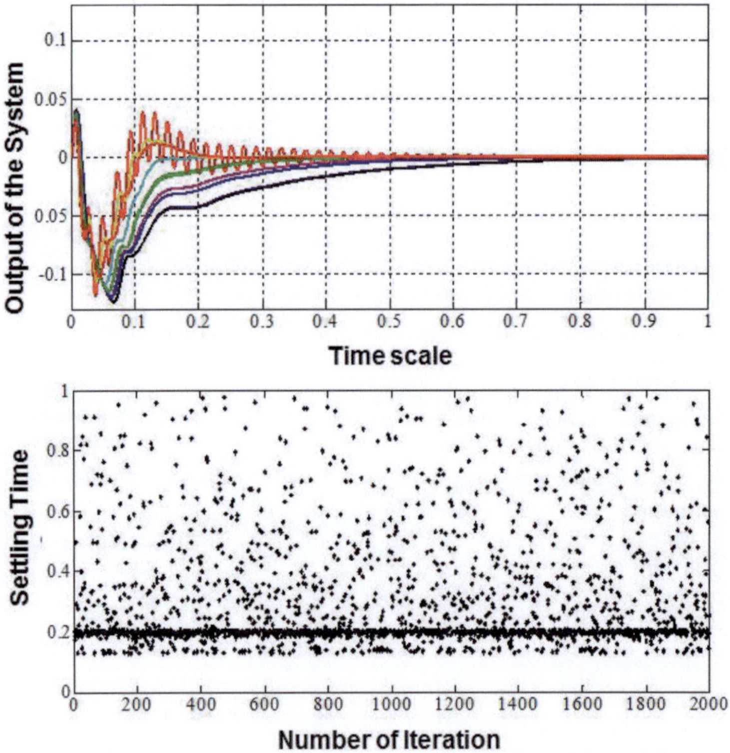

Fig. 25.24 Simulations of the system with uncertainties in stiffness of the structure

algorithm to control the attitude motion of a large flexible satellite using command shaping technique to reduce the mutual interaction between the attitude motion and the flexibility that has been shown to be robust to uncertainties on the controlled plant. A robustness analysis on the uncertainties of elastic properties of the spacecraft has been demonstrated using a Monte Carlo analysis.

In these examples, the statistical source may contain the stiffness or damping fluctuations caused by random variations in material properties, randomness in boundary conditions, and variations caused by manufacturing and assembly techniques. The inaccuracies and assumptions, which have been introduced in the mathematical modeling of the structure, are the non-statistical sources.

Ibrahim [42] stipulated that continuous systems may involve uncertainties from the geometry, the material properties and the support mechanism of the system. They can be determined using various techniques, such as perturbation method, variational methods, asymptotic estimate methods and integration methods. On the other hand, the statistics of the eigenvalue problem are directly determined by performing averaging analysis to the system's partial differential equation and its associated

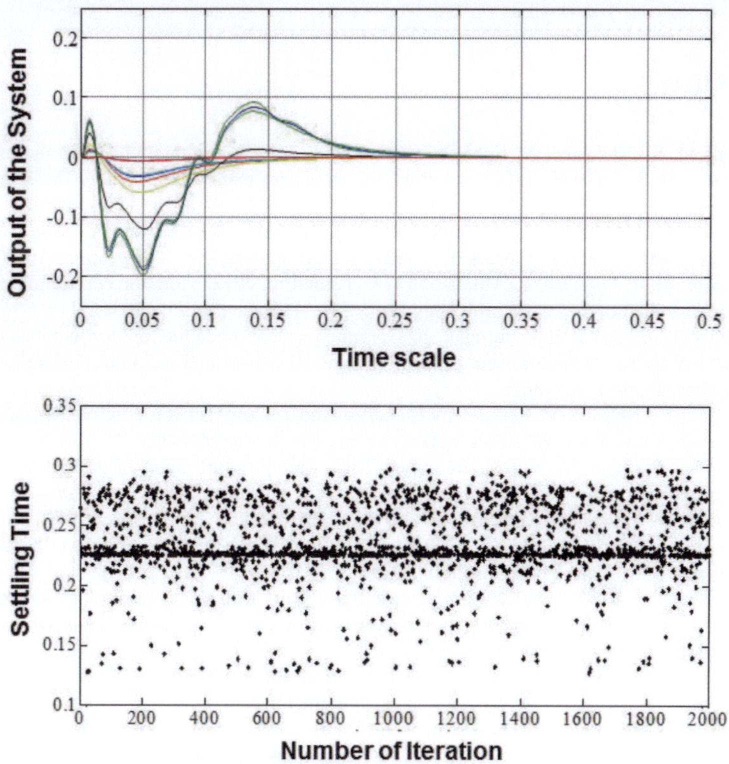

Fig. 25.25 Simulations of the system with uncertainties in the force on the system

boundary conditions. The statistics can be evaluated by using either iteration methods or hierarchy methods.

To account for uncertainties in the present study, which deals with relatively generic model, random stiffness and random magnitude of the force are individually considered as an uncertainty in the system with full-state observer LQR controller. For the stiffness, an uncertainty parameter is multiplied with stiffness matrices in Eq. 25.79. Thus, the stiffness randomly varies between $\pm 35\%$ of its original value. As demonstrated in Fig. 25.25a, the system is controlled for ten uncertain and random stiffnesses. To ensure that the results are reliable, Monte Carlo simulation is performed for two thousand times, and the settling times of each simulation is measured. The results, which are shown in Fig. 25.25b, demonstrate that the settling time of the system with uncertainties in stiffness is within a certain bound, and, thus, acceptable.

This simulation is repeated for another uncertainty associated with the magnitude of force on the structure. The uncertain parameter is multiplied with the force in Eq. 25.79. Figure 25.25 again demonstrates that this system is robust for different amount of uncertainties in the force.

References

1. Azadi, M., S. Fazelzadeh, M. Eghtesad, and E. Azadi. 2011. Vibration suppression and adaptive-robust control of a smart flexible satellite with three axes maneuvering. *Acta Astronautica* 69: 307–322.
2. Gasbarri, P., M. Sabatini, N. Leonangeli, and G.B. Palmerini. 2014. Flexibility issues in discrete on–off actuated spacecraft: Numerical and experimental tests. *Acta Astronautica* 101: 81–97.
3. Sabatini, M., P. Gasbarri, R. Monti, and G.B. Palmerini. 2012. Vibration control of a flexible space manipulator during on orbit operations. *Acta Astronautica* 73: 109–121.
4. Narayanan, S., and V. Balamurugan. 2003. Finite element modelling of piezolaminated smart structures for active vibration control with distributed sensors and actuators. *Journal of Sound and Vibration* 262: 529–562.
5. Xu, S., and T. Koko. 2004. Finite element analysis and design of actively controlled piezoelectric smart structures. *Finite Elements in Analysis and Design* 40: 241–262.
6. Pommier-Budinger, V., and M. Budinger. 2014. Sizing optimization of piezoelectric smart structures with meta-modelling techniques for dynamic applications. *International Journal of Applied Electromagnetics and Mechanics*. ISSN 1875-8800.
7. Kumar, S., R. Srivastava, and R.K. Srivastava. 2014. Active vibration control of smart piezo cantilever beam using PID controller. *IJRET*.
8. Vasques, C.M.A. 2012. Modal piezoelectric transducers with shaped electrodes for improved passive shunt vibration control of smart piezo-elastic beams. In *6th European Workshop on Structural Health Monitoring—Tu.3.C.4.*
9. Basdogan, I., U. Boz, S. Kulah, and M.U. Aridogan. 2012. *Active Control of Plate-Like Structures for Vibration and Sound Suppression*. Intechopen.
10. Yayli, M.Ö., M. Aras, and S. Aksoy. 2014. *An Efficient Analytical Method for Vibration Analysis of a Beam on Elastic Foundation with Elastically Restrained Ends, Shock & Vibration.*
11. Fuller, C.C., S. Elliott, and P.A. Nelson. 1996. *Active Control of Vibration*. Academic Press.
12. Jalili, N. 2010. Piezoelectric-based vibration control. In *From Macro to Micro/nano Scale Systems*. Springer.
13. Baruh, H. 1999. *Analytical Dynamics*. WCB/McGraw-Hill Boston.

14. Meirovitch, L. 1997. *Principles and Techniques of Vibrations*. Prentice Hall.
15. Alkhatib, R., and M. Golnaraghi. 2003. Active structural vibration control: A review. *Shock and Vibration Digest* 35: 367.
16. Ogata, K., and Y. Yang. 1970. *Modern Control Engineering*.
17. Djojodihardjo, H. 2013. Computational simulation for analysis and synthesis of impact resilient structure. *Acta Astronautica* 91: 283–301.
18. Djojodihardjo, H., and A. Shokrani. 2010. Generic study and finite element analysis of impact loading on elastic panel structure, Paper IAC-10.C2.6.2. In *Presented at the 61th International Astronautical Congress*. Prague, The Czech Republic.
19. Djojodihardjo, H., P. M. Ng, and L. K. Soo. 2009. Analysis and simulation of impact loading on elastic beam structure with case studies, Paper IAC-09-C2.6.01. In *Presented at the 60th International Astronautical Congress*. Daejeon, Republic of Korea.
20. Djojodihardjo, H., and I. Safari. 2013. BEM-FEM coupling for acoustic effects on aeroelastic stability of structures. *CMES: Computer Modelling in Engineering & Sciences* 91: 205–234.
21. Jafari, M., H. Djojodihardjo, and K.A. Ahmad. 2014. Vibration analysis of a cantilevered beam with spring loading at the tip as a generic elastic structure. *Applied Mechanics and Materials* 629: 407–413.
22. Jafari, M., and H. Djojodihardjo. 2014. Vibration control of an elastic structure using piezoelectric sensor and actuator with cantilevered beam as a case study. In *Scientific Cooperations International Workshops on Electrical and Computer Engineering Subfields*, 22–23 August, Koc University, Istanbul/Turkey.
23. Hong, S.-Y. 1992. *Active Vibration Control of Adaptive Flexible Structures Using Piezoelectric Smart Sensors and Actuators*.
24. Yousefi-Koma. 1997. *Active Vibration Control of Smart Structures Using Piezoelements*. Carleton University.
25. Rao, S.S. 2007. *Vibration of Continuous Systems*. Wiley.
26. Meirovitch, L. 1980. *Computational Methods in Structural Dynamics*. Springer.
27. Chopra, A.K. 1995. *Dynamics of Structures*, vol. 3. Prentice Hall New Jersey.
28. Kreyszig, E. 2007. *Advanced Engineering Mathematics*. Wiley.
29. Brebbia, C.A., and J.J. Connor. 1973. *Fundamentals of Finite eElement Techniques for Structural Engineers*. Butterworths.
30. Maurizi, M.J., R.E. Rossi, and J.A. Reyes. 1976. Vibration frequencies for a uniform beam with one end spring-hinged and subjected to a translational restraint at the other end. *Journal of Sound & Vibration*.
31. Chhabra, D., K. Narwal, and P. Singh. 2012. Design and analysis of piezoelectric smart beam for active vibration control. *IJART*.
32. Wrona, S., and M. Pawełczyk. 2013. Controllability-oriented placement of actuators for active noise-vibration control of rectangular plates using a memetic algorithm. *Archives of Acoustics*.
33. Safizadeh, M.R., I.Z. Mat Darus, and M. Mailah. 2010. *Optimal Placement of Piezoelectric Actuator for Active Vibration Control of Flexible Plate*, academia edu.
34. Shoushtari, N.D. 2013. *Optimal Active Control of Flexible Structures Applying Piezoelectric Actuators*. Ph.D., Waterloo.
35. Padoin, M., and F. Perondi. 2013. Modelling and LQR/LQG control of a cantilever beam using piezoelectric material. In *22nd International Congress of Mechanical Engineering (COBEM 2013)*. Ribeirão Preto, SP, Brazil.
36. Baillargeon, B.P., and B.S. Vel. 2005. Active vibration suppression of sandwich beams using piezoelectric shear actuators: Experiments and numerical simulations. *Journal of Intelligent Material Systems and Structures* 16.
37. Meitzler, H.T., A. Warner, D. Berlincourt, G. Couqin, and F. Welsh III. 1988. *IEEE Standard on Piezoelectricity, Society*.
38. Bailey, T., and J. Ubbard. 1985. Distributed piezoelectric-polymer active vibration control of a cantilever beam. *Journal of Guidance, Control, and Dynamics* 8: 605–611.
39. Djojodihardjo, H., and M. Jafari. 2014. Vibration analysis of a cantilevered beam with piezoelectric actuator as a controllable elastic structure. In *Paper IAC-14,C2,3,8, Proceedings, The International Astronautical Congress*. Toronto.

40. Cortes, L., N.S. Ferguson, and A. Bhaskar. 2012. A perturbation method for locally damped dynamic systems. In *19th International Congress on Sound and Vibration*, Vilnius, Lithuania. http://eprints.soton.ac.uk/338022/1/ICSV19LluisCortes.pdf
41. Gasbarri, P., R. Monti, and M. Sabatini. 2014. Very large space structures: Non-linear control and robustness to structural uncertainties. *Acta Astronautica* 93: 252–265.
42. Ibrahim, R.A. 1987. Structural dynamics with parameter uncertainties. *Applied Mechanics Reviews* 40 (3): 309–328.

Chapter 26
Transonic Flow Computation of Slender Body of Revolution Using Transonic Small Disturbance and Navier–Stokes Equations

Abstract For drag minimization of slender body of revolution in transonic flow purposes, computational schemes are developed. Selected methods are reviewed and adapted to obtain relatively simple and fast procedures to facilitate parametric studies for drag optimization. The slender body integral approach was first resorted to for its elegance and its simplicity to facilitate parametric studies for fast, analytical and structured search procedure in the optimization scheme, such as that provided by MATLAB code. The scheme was incorporated in the preliminary step of an optimization cycle that will have refined search using more accurate codes at later stages. The finite difference transonic small disturbance schemes will then be used to assess the aerodynamic characteristics of the candidate geometries in better detail. Two different finite difference computational schemes have been pursued. The finite difference schemes follow the well-known transonic small disturbance computational techniques. Resort is also made to commercially available Navier–Stokes flow solvers, to validate computational schemes developed as well as an instrument for numerical experimental studies.

Keywords Aerodynamis · Navier–Stokes flow solvers · Transonic aerodynamics · Slender body · Transonic small disturbance

26.1 Introduction

The analytical method, which is based on the well-known transonic small perturbation slender body theory elaborated in the classical and elegant exposition of Ashley and Landahl [1], is considered to be convenient and can reveal the generic contributions of geometrical elements to the drag.

Two finite difference type schemes have been developed. Similar to the approach of Murman and Cole [2], Bailey [3], and Krupp and Murman [4], the first one has

Originally presented as KEYNOTE LECTURE, delivered at the International Conference on Computational and Experimental Engineering Sciences ICCES03, Corfu, Greece, 2003. Co-authored by H. Djojodihardjo, A. F. Widodo and E. Priyono, W. K. Sekar, C. Weishäupl and B. Laschka.

© The Author(s), under exclusive license to Springer Nature Singapore Pte Ltd. 2023

H. Djojodihardjo, *Introduction to Aeroelasticity*,
https://doi.org/10.1007/978-981-16-8078-6_26

been worked out [5] while the second one is the extension of the two-dimensional scheme developed by Sekar, Weishäupl and Laschka [6].

Computational studies using commercial Navier–Stokes flow solvers have also been carried out [7, 8]. The computational results are validated by using experimental data at MBB [9].

26.2 Governing Equation for Transonic Flow About Slender Bodies of Revolution

Divergence form of transonic small perturbation for axisymmetrical slender body

$$\left[(1 - M_\infty^2)\phi_x - \frac{(\gamma + 1)}{2} M_\infty^2 \phi_x^2\right]_x + \frac{1}{r}[r\phi_r]_r = 0 \tag{26.1}$$

Quasi-linear form of transonic small perturbation for axisymmetrical slender body

$$\left[(1 - M_\infty^2) - M_\infty^2(\gamma + 1)\frac{\partial\phi}{\partial x}\right]\frac{\partial^2\phi}{\partial x^2} + \frac{1}{r}\frac{\partial}{\partial r}(r\frac{\partial\phi}{\partial r}) = 0 \tag{26.1a}$$

In Fig. 26.1:

$$\varphi = V_\infty x + \phi$$
$$u = \varphi_x = V_\infty + \phi_x$$
$$v = \varphi_y = \phi_y$$

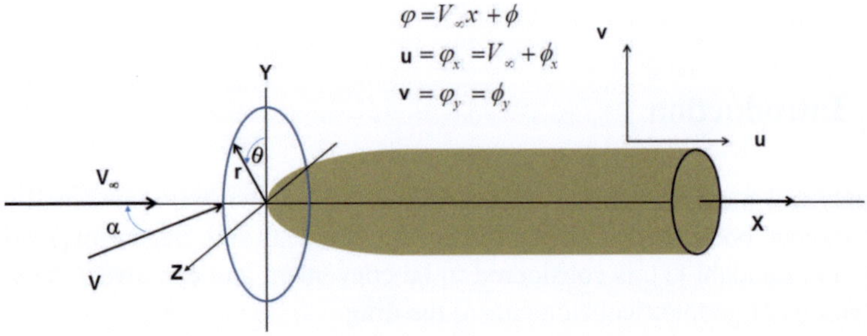

Fig. 26.1 Schematics of axisymmetric slender body and coordinate system

26.3 Transonic Small Disturbance Approach with Boundary Condition Derived from Transonic Small Perturbation Integral Equation (1st TSD Method)

The governing equation for axisymmetrical body moving in transonic flow is given by the transonic small disturbance equation

$$\left[(1 - M_\infty^2) - M_\infty^2(\gamma + 1)\frac{\partial \phi}{\partial x}\right]\frac{\partial^2 \phi}{\partial x^2} + \frac{1}{r}\frac{\partial}{\partial r}(r\frac{\partial \phi}{\partial r}) = 0 \tag{26.2}$$

(Quasi-Linear form)or

$$\lambda \varphi_{\mathrm{xx}} + \frac{1}{r}(\mathrm{r}\varphi_{\mathrm{r}})_r = 0 \tag{26.2a}$$

where

$$\lambda = \left\{(1 - M_\infty^2) - (\gamma + 1)\frac{\varphi_x}{U_\infty}M_\infty^2\right\} \tag{26.2b}$$

The flow tangency boundary condition at the body surface is given by the first-order slender body approximation[1]:

$$\lim_{\mathrm{r}->0}(r\varphi_r) = R\frac{\mathrm{d}R}{\mathrm{d}x} \tag{26.3}$$

To avoid singularities in the application of the transonic small perturbation approach at the x-axis along the length of the body, the solution obtained using integral approach for axisymmetrical slender bodies (Ashley and Landahl [1]), which has been developed using inner and outer expansions of small parameter approach to the transonic small perturbation differential equation, has been utilized. Using such approach, the disturbance potential in the vicinity of the singularity axis can be given by:

$$\varphi = \frac{S'(x)}{2\pi}\ln r + g(x) \tag{26.4}$$

The first term on the right-hand side of Eq. (26.4) can be calculated readily at the particular point near the x-axis, and the value of φ next depends on the complete solution of the function $g(x)$.

Since $g(x)$ does not depend on r, a relatively simple numerical scheme to solve for $g(x)$ can be constructed, which can be superposed to the first term.

[1] Frank R. Bailey, Numerical Calculation of Transonic Flow About Slender Bodies of Revolution, NASA TN D-6582".

26.3.1 Boundary Conditions

Following Ashley and Landahl [1], the boundary condition at the surface of the axisymmetric slender body was derived by matching the inner solution due to the singular nature of the flow as $r \to 0$ and yields

$$\varphi_r = \lim_{r \to 0} \frac{1}{r} \frac{S'(x)}{2\pi} \tag{26.4b}$$

So that

$$\varphi = \frac{S'(x)}{2\pi} \ln r + g(x) \tag{26.5}$$

from the geometrical relationship: $S'(x) = 2\pi\ R\ (x) \frac{\mathrm{d}R\ (x)}{\mathrm{d}x}$ so that

$$r\ \varphi_r = R\ (x) \frac{dR\ (x)}{dx} \tag{26.6}$$

or

$$\varphi = R(x) \ln r\ \mathrm{d}R(x)/\mathrm{d}x\ +\ g(x) \tag{26.7}$$

Integrating (26.6) between r_1 and r_2 yields

$$\begin{aligned} \varphi_{i,1} &= R'(x)\ R(x)\ \ln r_1 + \varphi_{i,2} - R'(x)\ R(x) \\ \ln r_2 &= \varphi_{i,2} + R'(x)\ R(x)[\ln r_1 - \ln r_2] \end{aligned} \tag{26.8}$$

The disturbance velocity along the x-direction on the surface is then given by:

$$v = \frac{\partial \phi}{\partial x} = \frac{\left[\frac{(S'(x)_{i+1} - S'(x)_{i-1})}{2\pi} \ln\ R(x) + g(x)_{i+1,0} - g(x)_{i-1,0}\right]}{2\Delta\ x} \tag{26.9}$$

where $g(x)_{i,0} = g(x)_{s,1}$ from previous step.

26.3.1.1 Far Field Boundary Condition

$$\varphi_x = 0 \text{ and } \varphi_r = 0 \tag{26.10}$$

26.3.1.2 Boundary Condition on the Shock Surfaces

To take into account the discontinuous change in velocities that occur at shock surfaces, using the divergence form Eq. (26.1a):

$$\left[(1 - M_\infty^2)\phi_x - \frac{(\gamma + 1)}{2} M_\infty^2 \phi_x^2\right]_x + \frac{1}{r}\,[r\phi_r]_r = 0 \tag{26.1a}$$

Obtain the jump condition (Lax [10])

$$\left[(1 - M_\infty^2) - M_\infty^2(\gamma + 1)\left(\frac{u_1 + u_2}{2}\right)\right](u_1 - u_2)^2 + (v_1 - v_2) = 0 \tag{26.11}$$

$$\varphi_1 = \varphi_2 \tag{26.12}$$

26.3.2 *Grid Generation and Solution in the Vicinity of the Surface*

Various grid configurations have been utilized; the simplest one is the uniformly spaced grid. The distance between consecutive points in the grid is Δx (along the axial direction) and Δr along the radial direction (Fig. 26.2).

To avoid using excessive number of nodes, use gradually, geometrically increasing grid spacing both in the *x*- and *y*-directions away from the slender body surface, i.e. use

$$\begin{aligned}\Delta x_{\text{next}} &= K_1 \Delta x_{\text{previous}} \\ \Delta y_{\text{next}} &= K_2 \Delta y_{\text{previous}}\end{aligned} \tag{26.13}$$

where K_1 and K_2 are stretching factors, between 1.05 and 1.15 (Sankar [11]).

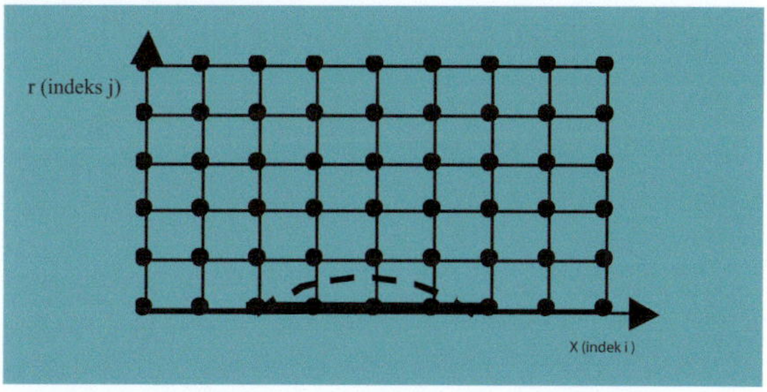

Fig. 26.2 Grid generation around the axisymmetric slender body

At grid points along $j = 1$, the velocity potential is calculated following the inner solution for slender axisymmetric body [1, 5] as given by:

$$\varphi = \frac{S'(x)}{2\pi} \ln r + g(x) \tag{26.5}$$

$g(x)$ is the part of the velocity potential that is invariant to r and is dependent only on x. It is given by the following expression:

$$g(x)_{i,j} = \varphi_{i,j} - \frac{S'(x)}{2\pi} \ln r_j \tag{26.14}$$

26.4 Finite Difference Computation

26.4.1 Disturbance Velocity Along x Direction on the Surface

$$v = \frac{\partial \varphi}{\partial x} = \frac{\partial \left[\frac{S'(x)}{2\pi} \ln r + g(x) \right]}{\partial x} \tag{26.15}$$

in finite difference form

$$v = \frac{\left[\frac{(S'(x)_{i+1} - S'(x)_{i-1})}{2\pi} \ln R(x) + g(x)_{i+1,0} - g(x)_{i-1,0} \right]}{2\Delta x} \tag{26.16}$$

where $g(x)_{i,0} = g(x)_{i,1}$ that was calculated in previous step.

26.4.2 Discretization of the Governing Equation for the Interior Points

(a) Central finite difference scheme for the elliptical (subsonic) regimes

$$\varphi_{i,j} = \frac{\frac{\lambda_{i,j}}{(\Delta x)^2} \left(\varphi_{i+1,j} + \varphi_{i-1,j} \right) + A_k \varphi_{i,j-1} + C_k \varphi_{i,j+1}}{\left[B_k + \frac{2\lambda_{i,j}}{\Delta x^2} \right]} \tag{26.17a}$$

$$\lambda_{i,j} = \left[\beta^2 - M_\infty^2 (\gamma + 1) \left(\frac{\varphi_{i+1,j} - \varphi_{i-1,j}}{\Delta x} \right) \right] \tag{26.17b}$$

$$A_k = \frac{1}{\Delta r^2} \left[1 - \frac{\Delta r}{2} \left(\frac{1}{r_j} \right) \right] \tag{26.17c}$$

$$B_k = \frac{2}{\Delta r^2} \tag{26.17d}$$

$$C_k = \frac{1}{\Delta r^2}\left[1 + \frac{\Delta r}{2}\left(\frac{1}{r_j}\right)\right] \tag{26.17e}$$

(b) Backward finite difference scheme for the hyperbolical (supersonic) regimes:

$$\varphi_{i,j} = \frac{\frac{\lambda_{i,j}}{(\Delta x)^2}\left(-2\varphi_{i-1,j} + \varphi_{i-2,j}\right) + A_k\varphi_{i,j-1} + C_k\varphi_{i,j+1}}{\left[B_k - \frac{\lambda_{i,j}}{(\Delta x)^2}\right]} \tag{26.18a}$$

$$\lambda_{i,j} = \left[\beta^2 - M_\infty^2(\gamma + 1)\left(\frac{\varphi_{i,j} - \varphi_{i-1,j}}{\Delta x}\right)\right] \tag{26.18b}$$

$$A_k = \frac{1}{\Delta r^2}\left[1 - \frac{\Delta r}{2}\left(\frac{1}{r_j}\right)\right] \tag{26.18c}$$

$$B_k = \frac{2}{\Delta r^2} \tag{26.18d}$$

$$C_k = \frac{1}{\Delta r^2}\left[1 + \frac{\Delta r}{2}\left(\frac{1}{r_j}\right)\right] \tag{26.18e}$$

26.4.2.1 Iterative Matrix Solution

The flow field values are obtained by solving the finite difference algebraic equations in an iterative fashion, starting with assumed initial values.

To facilitate rapid convergence, the successive over-relaxation scheme based on Gauss–Seidel method is employed. At iteration stage n, the flow field is calculated by solving directly the finite difference equations, to yield field values between n and $n + 1$ iterations (here regarded as the $(n + 1/2)$ iteration level). The final solution associated with $(n + 1)$s iteration is obtained using corrector scheme:

$$\varphi_{i,j}^{n+1} = \varphi_{i,j}^{n} + \omega_j(\varphi_j^{n+1/2} - \varphi_j^{n}) \tag{26.24}$$

ω_j is a diagonal matrix of relaxation parameter [3].

The direct solution of the algebraic equation obtained at the nth iteration is used as the value for the $(n + 1/2)$th iteration. For the interior points, since both elliptical and hyperbolical regimes may prevail, the numerical computation should be carried out following different schemes for each case. The type of the differential equation for each case can be characterized by $\lambda_{i,j}$, which is defined as

$$\lambda_{i,j} = \left[\beta^2 - M_\infty^2(\gamma+1)\left(\frac{\varphi_{i+1,j} - \varphi_{i-1,j}}{2\Delta x}\right)\right] \tag{26.25}$$

which is calculated using central difference scheme.

For the interior points, since both elliptical and hyperbolical regimes may prevail, the numerical computation should be carried out following different schemes for each case. The type of the differential equation relevant for each case can be characterized by the value of $\lambda_{i,j}$. Therefore it is essential to calculate first the value of $\lambda_{i,j}$ at each point following central finite difference scheme.

If the value of $\lambda_{i,j}$ obtained is positive, proceed with central finite difference scheme.

If the value of $\lambda_{i,j}$ obtained is negative, recalculate $\lambda_{i,j}$ using backward finite difference scheme. If $\lambda_{i,j}$ so calculated turns out to be negative, proceed with backward finite difference scheme.

If $\lambda_{i,j}$ calculated using central finite difference scheme was found to be negative, whereas if calculated using backward finite difference scheme turned out to be positive, then $\lambda_{i,j}$ is set to be zero and the calculation then may proceed using either backward finite difference- or central finite difference- scheme.

If upon completion of the computational sequence the solution converged, one then proceeds by calculating the difference between the values of φ obtained after the nth and $(n+1)$th iteration levels, defined as the residual:

$$\text{RESIDUAL} = \sum_{i=0}^{i_{max}} \sum_{j=0}^{j_{max}} \left| \phi_{i,j}^{n+1} - \phi_{i,j}^{n} \right| \tag{26.26}$$

26.5 TSD Equation for Axisymmetric Body (2nd TSD Method)[2]

A more rigorous version of the transonic small disturbance (TSD) equation is derived from the full potential equation by adopting the small disturbance assumption but retaining the nonlinear term in the flow direction to capture the shock. From the axisymmetric conservative form [6]:

$$\frac{\partial f_0}{\partial t} + \frac{\partial f_1}{\partial x} + \frac{\partial f_2}{\partial r} + \frac{f_2}{r} = 0 \tag{26.27}$$

where the fluxes are defined as:

$$f_0 = -(A\varphi_t + B\varphi_x) \tag{26.28a}$$

[2] Computational program developed by W. K. Sekar.

$$f_1 = E\varphi_x + F\varphi_x^2 \tag{26.28b}$$

$$f_2 = \varphi_r \tag{26.28c}$$

$$A = M_\infty^2, \quad B = 2M_\infty^2, \tag{26.28d}$$

$$E = 1 - M_\infty^2, \quad F = -\frac{1}{2}(\gamma + 1)M_\infty^2 \tag{26.28e}$$

In the computational coordinates, the TSD equation in the computational coordinates can be described as:

$$-\frac{\partial}{\partial t}\left(A\frac{\varphi_t}{\xi_x} + B\varphi_\xi\right) + \frac{\partial}{\partial \xi}\left[E\varphi_\xi\xi_x + F(\varphi_\xi\xi_x)^2\right] + \frac{\partial}{\partial \zeta}\left(\frac{\varphi_\zeta}{\xi_x}\right) + \frac{\varphi_\zeta}{\zeta} = 0 \tag{26.29}$$

where

$$\xi(x) = \frac{x}{L} \tag{26.29a}$$

$$\zeta = \frac{z}{L} \tag{26.29b}$$

Equation (26.29) can be rewritten as

$$\overline{R}(\varphi^{n+1}) = 0 \tag{26.30}$$

where $\overline{R}$ is a differential operator.

In the solution of the TSD equation, the disturbance velocity potential ϕ at specified time level $n + 1$ is given by:

$$\varphi^{n+1} = \varphi^* + \Delta\varphi \tag{26.31}$$

with ϕ^* as an assumed value of ϕ^{n+1}.

By applying a Taylor approximation then the governing Eq. (26.30) can be written as:

$$\left(\frac{\partial \overline{R}}{\partial \varphi}\right)_{\varphi=\varphi^*} \Delta\varphi = -\overline{R}(\varphi^*) \tag{26.32}$$

Equation (26.32) will be solved by an approximate factorization to facilitate independency of ξ and ζ directions that will accelerate the iteration process. The LHS of Eq. (26.32) is approximately factored into a product of independent operator L_ξ and L_ζ as:

$$L_{\xi} L_{\zeta} \Delta\varphi = -\overline{R}\left(\varphi^{*}, \varphi^{n}, \varphi^{n-1}, \varphi^{n-2}\right) \tag{26.33}$$

where

$$L_{\xi} = 1 + \left(\frac{3B}{4A}\xi_x \Delta t\right)\frac{\partial}{\partial \xi} - \left(\frac{\xi_x \Delta t^2}{2A}\right)\frac{\partial}{\partial \xi} F_1 \frac{\partial}{\partial \xi} \tag{26.34}$$

$$L_{\zeta} = 1 - \left(\frac{\xi_x \Delta t^2}{2A}\right)\frac{\partial}{\partial \zeta}\frac{1}{\xi_x}\frac{\partial}{\partial \zeta} - \left(\frac{\Delta t^2}{2A}\right)\frac{1}{\varsigma}\frac{\partial}{\partial \zeta} \tag{26.35}$$

$$F_1 = E\xi_x + 2F\varphi_{\xi}^{*}\xi_x^2 \tag{26.36}$$

$$\begin{aligned} R\left(\varphi^{*}, \varphi^{n}, \varphi^{n-1}, \varphi^{n-2}\right) &= -\frac{1}{2}\left(2\varphi^{*} - 5\varphi^{n} + 4\varphi^{n-1} - \varphi^{n-2}\right) \\ &- \frac{B}{4A}\Delta t \xi_x \left(3\varphi_x^{*} - 4\varphi_x^{n} + \varphi_x^{n-1}\right) \\ &+ \frac{(\Delta t)^2}{2A}\xi_x \frac{\partial}{\partial \xi}\left(E\xi_x \varphi_{\xi}^{*} + F\xi_x^2 \varphi_{\xi}^{*2}\right) + \frac{(\Delta t)^2}{2A}\left(\frac{\partial}{\partial \zeta} + \frac{1}{\varsigma}\right)\varphi_{\zeta}^{*} \end{aligned} \tag{26.37}$$

a. Wake boundary conditions:

$$[\phi_x + \phi_t] = 0 \tag{26.38a}$$

$$\left[\varphi_z\right] = \delta_x^{*+} - \delta_x^{*-} \tag{26.38b}$$

[] denotes jump of property inside the bracket across the wake.

26.5.1 Entropy Correction

An entropy correction introduced by Fuglsang [12] due to the discontinuity of pressure across a shock wave is implemented in this TSD equation, and its influence will be added to the value of the pressure coefficient.

The total pressure distribution consists of isentropic part and non-isentropic part as follows:

$$C_p = C_{p_i} + C_{p_s} \tag{26.39}$$

with the pressure distribution due to the entropy jump:

$$C_{p_s} = \frac{2(s - s_{\infty})/c_v}{\gamma(\gamma - 1)M_{\infty}^2} \tag{26.40}$$

and the isentropic pressure distribution part:

$$C_{p_i} = \frac{2}{\gamma M_\infty^2}\left\{\left[1 + \frac{\gamma - 1}{2} M_\infty^2 \left(1 - \nabla\varphi \circ \nabla\varphi + 2\frac{\partial \varphi}{\partial t}\right)\right]^{\frac{\gamma}{\gamma-1}} - 1\right\} \tag{26.41}$$

26.6 The Navier–Stokes Equations for Axisymmetric Body

The Navier–Stokes Equations, which involve the mass, momentum and energy conservation equations, can be written in the conservation form and can be recast into:

$$\frac{\overline{Q}^{n+1} - \overline{Q}^{n}}{\Delta\tau} + \left(\frac{\partial \overline{E}}{\partial \xi}\right)^{n+1} + \left(\frac{\partial \overline{F}}{\partial \varsigma}\right)^{n+1} + \alpha\left(\overline{H}\right)^{n+1} = 0 \tag{26.42}$$

To facilitate numerical computation, the above equation is written in finite difference formulation. The time derivative is approximated by a first-order backward difference quotient and the remaining terms are evaluated at time level $n + 1$. Then Eq. (26.42) can be written in finite difference formulation as:

$$\frac{\overline{Q}^{n+1} - \overline{Q}^{n}}{\Delta\tau} + \left(\frac{\partial \overline{E}}{\partial \xi}\right)^{n+1} + \left(\frac{\partial \overline{F}}{\partial \varsigma}\right)^{n+1} + \alpha\left(\overline{H}\right)^{n+1} = 0 \tag{26.43}$$

Equation (26.48) is nonlinear. Further simplification is afforded by linearized approach, which then yields:

$$\overline{E}^{n+1} = \overline{E}^{n} + \left(\frac{\partial \overline{E}}{\partial \overline{Q}}\right)\Delta\overline{Q} + O(\Delta\tau)^2 \tag{26.44}$$

$$\overline{F}^{n+1} = \overline{F}^{n} + \left(\frac{\partial \overline{F}}{\partial \overline{Q}}\right)\Delta\overline{Q} + O(\Delta\tau)^2 \tag{26.45}$$

$$\overline{H}^{n+1} = \overline{H}^{n} + \left(\frac{\partial \overline{H}}{\partial \overline{Q}}\right)\Delta\overline{Q} + O(\Delta\tau)^2 \tag{26.46}$$

where

$$\frac{\partial \overline{E}}{\partial \overline{Q}} = A; \quad \frac{\partial \overline{F}}{\partial \overline{Q}} = B; \quad \frac{\partial \overline{H}}{\partial \overline{Q}} = C \tag{26.47}$$

are Jacobian matrices.

The numerical computation then is carried out by solving the following equation in the computational grid.

$$\Delta\overline{Q} + \Delta\tau\{\frac{\partial}{\partial\xi}[\overline{E}^n + A^n\overline{Q}] + \frac{\partial}{\partial\varsigma}[\overline{F}^n + B^n\Delta\overline{Q}] + \alpha[\overline{H}^n + C^n\Delta\overline{Q}] = 0 \quad (26.48)$$

26.7 Results

Computational results of the first TSD method have been validated using experimental results obtained at NASA Ames [4]. The general agreement for the most part of the flow regimes is very good, despite the coarseness and simple nature of the computational grid chosen, as exemplified in Fig. 26.1a–e for $M = 1$.

The numerical computation for validation purposes for both 1st and the 2nd TSD schemes has been carried out for MBB-3 Bodies of Revolution[3] at zero angle of attack, at $M = 1.2$. The experiments were conducted at Reynolds number of 10 million based on the length of the body.

For validation purposes, the results of the 2nd TSD scheme for MBB-5 and MBB-3 Body of revolution are compared with the experimental data, as shown in Figs. 26.3a–e and 26.4a–e, for $M = 0.8$ (a) and $M = 1.2$, (b), respectively.

Comparison of TSD1 with NASA AMES Results:

See Fig. 26.3b–e.

26.8 MBB Bodies of Revolution

For comparative purposes, present study is also carried out on the MBB Bodies of revolution [9], with fore- and aft-body geometry following third-order polynomials.

Further study will utilize the present results to generate other slender axisymmetric bodies with other aft-body geometries for optimum drag considerations.

Since experimental data are available, the geometries of MBB Body of Revolution 1, 3 and 5 [9] will be utilized both for validation and a reference for aerodynamic optimization search to be pursued later. The geometries of these bodies are described below.

The forebody geometry is represented as a third-order polynomial in x-direction as follows:

$$r(x) = a_3x^3 + a_2x^2 + a_1x$$

[3] Lorenz-Meyer, W. and Aulehla, F., "MBB - Body of Revolution No. 3", AGARD AR 138, London, May 1979, Editor J. Barche.

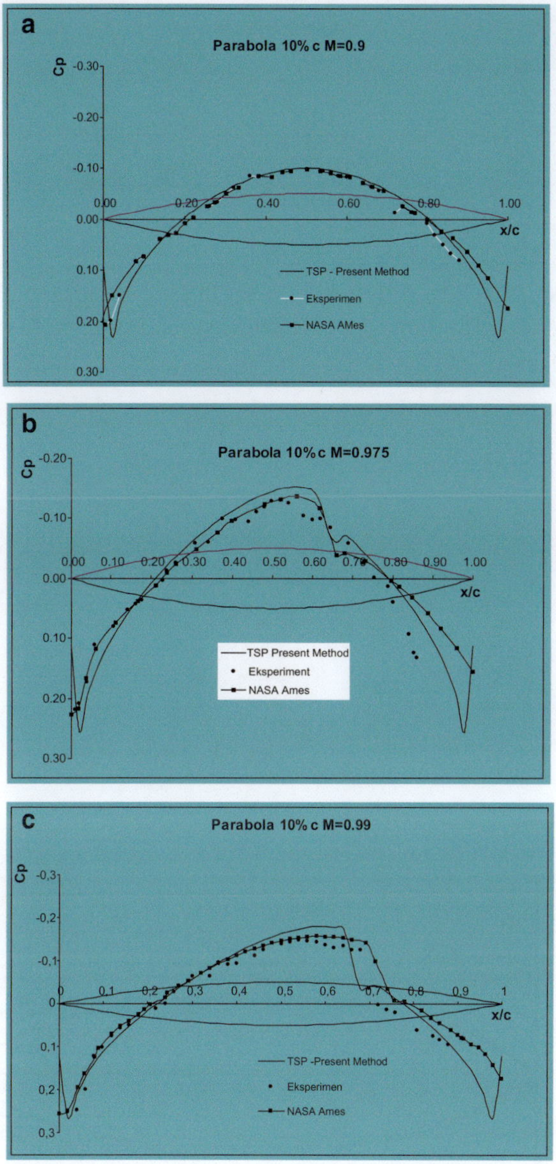

Fig. 26.3 **a** Computational results for $M = 0.9$ and comparison with computational and experimental results at NASA Ames. **b** Computational results for $M = 0.975$ and comparison with computational and experimental results at NASA Ames. **c** Computational results for $M = 0.99$ and comparison with computational and experimental results at NASA Ames. **d** Computational results for $M = 1.0$ and comparison with computational and experimental results at NASA Ames. **e** Computational results for $M = 1.2$ and comparison with computational and experimental results at NASA Ames

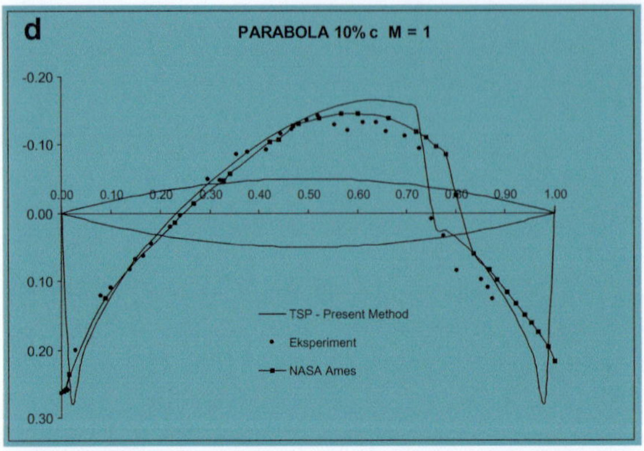

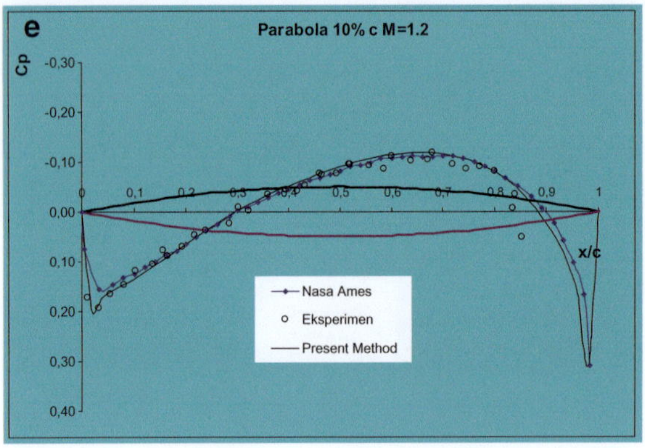

Fig. 26.3 (continued)

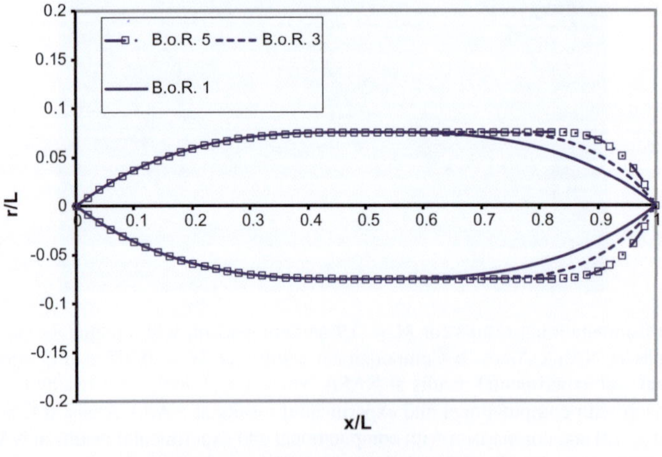

Fig. 26.4 The geometries of MBB body of revolution 1, 3, and 5

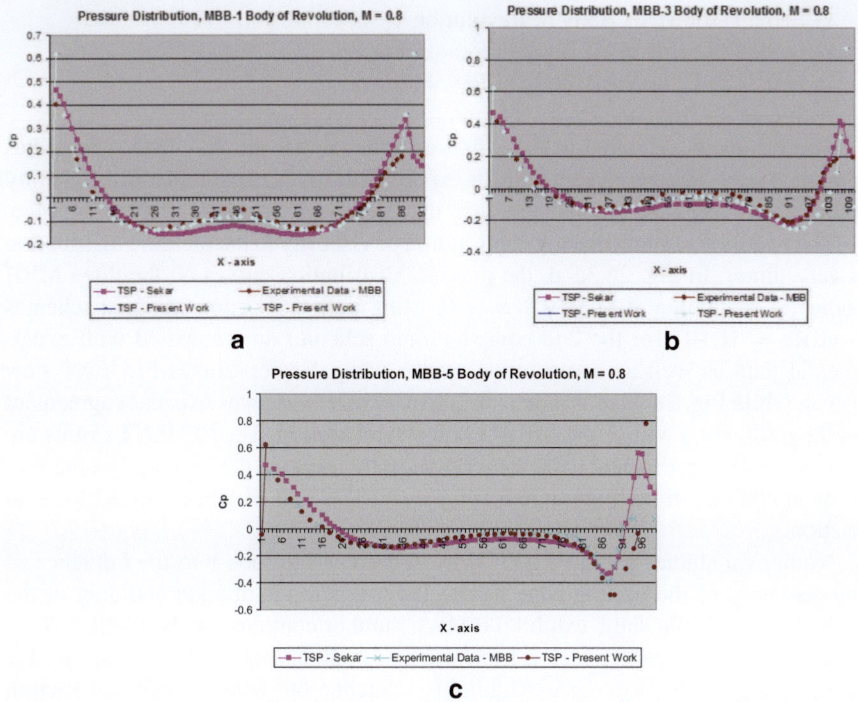

Fig. 26.5 **a**, **b** Pressure distribution on MBB body of revolution. **c** Pressure distribution on MBB body of revolution

with:

$$a_3 = r_{\max}/x_1^3$$
$$a_2 = -3r_{\max}/x_1^2$$
$$a_1 = 3r_{\max}/x_1$$
$$x_1 = 0.5\,\mathrm{L}$$

Figure 26.4 the afterbody geometry represented as third-order polynomial in x: $r(x) = b_3\,x^3 + b_2\,x^2 + b_1\,x + b_0$.

The afterbody geometry represented as third-order polynomial in x: $r(x) = b_3\,x^3 + b_2\,x^2 + b_1\,x + b_0$.

with:

$$b_3 = r_{\max}/(x_2 - L)^3,\ b_2 = -3x_2 r_{\max}/(x_2 - L)^3,\ b_1 = 3x_2^2 r_{\max}/(x_2 - L)^3$$
$$b_0 = \left(L\,r_{\max}/(x_2 - L)^3\right)\left(3x_2 L - L^2 - 3x_2^2\right)$$

where

$x_2 = 0.5$ L for MBB Body of Revolution 1.
$x_2 = 0.6875$ L for MBB Body of Revolution 3.
$x_2 = 0.8125$ L for MBB Body of Revolution 5.
with $L = 800$ mm and $r_{max} = 60$ mm.

The results as exhibited in these figures indicate that the 2nd TSD scheme has produced very close agreement with the experimental data, in particular in the vicinity of the adverse pressure gradient region of the afterbody, except near the trailing edge.

The results also show that the contribution of viscosity to the pressure distribution is very minor. In Fig. 26.3a–d, the pressure distribution curves on the three MBB bodies of revolution at $M = 0.8$, $\alpha = 0$, using both TSD computational schemes (and Re = 1E+07 for the 2nd computational scheme) are compared with experimental data as well as computation results using Navier–Stokes FLUENT flow solver, while Fig. 26.4a–e, for $M = 1.2$. FLUENT result gives excellent agreement on the midsection part of the MBB-3 body of revolution. The FLUENT results are very close to experimental data. Better grids may improve the accuracy, at the cost of computational time. Figure 26.6 compares TSD2 and FLUENT results for skin friction.

Numerical studies using FLUENT are carried out to look into the influence of the geometry of the trailing edge part to the pressure distribution and drag of the body. Figure 26.10a and b exhibit the Mach number contours for the MBB-3 Body of revolution at $M = 0.8$ which is truncated at a small distance from its trailing edge (boat tailing), while Fig. 26.9 exhibits its influence on pressure and skin friction distribution. Computational results for various aft-body geometries are shown in Table 26.1 (Figs. 26.10, 26.11 and 26.12)

26.9 Conclusions

Performance of two finite difference type TSD schemes following the well-known transonic small disturbance computational techniques has been shown to give efficient and accurate results. Following the approach of Murman and Cole [2], Bailey [3] and Krupp and Murman [4], and combining with slender body Integral formulation for the boundary condition at the body surface [1], the 1st TSD scheme has been developed. The more accurate second TSD scheme has been developed and extended to axisymmetric case.

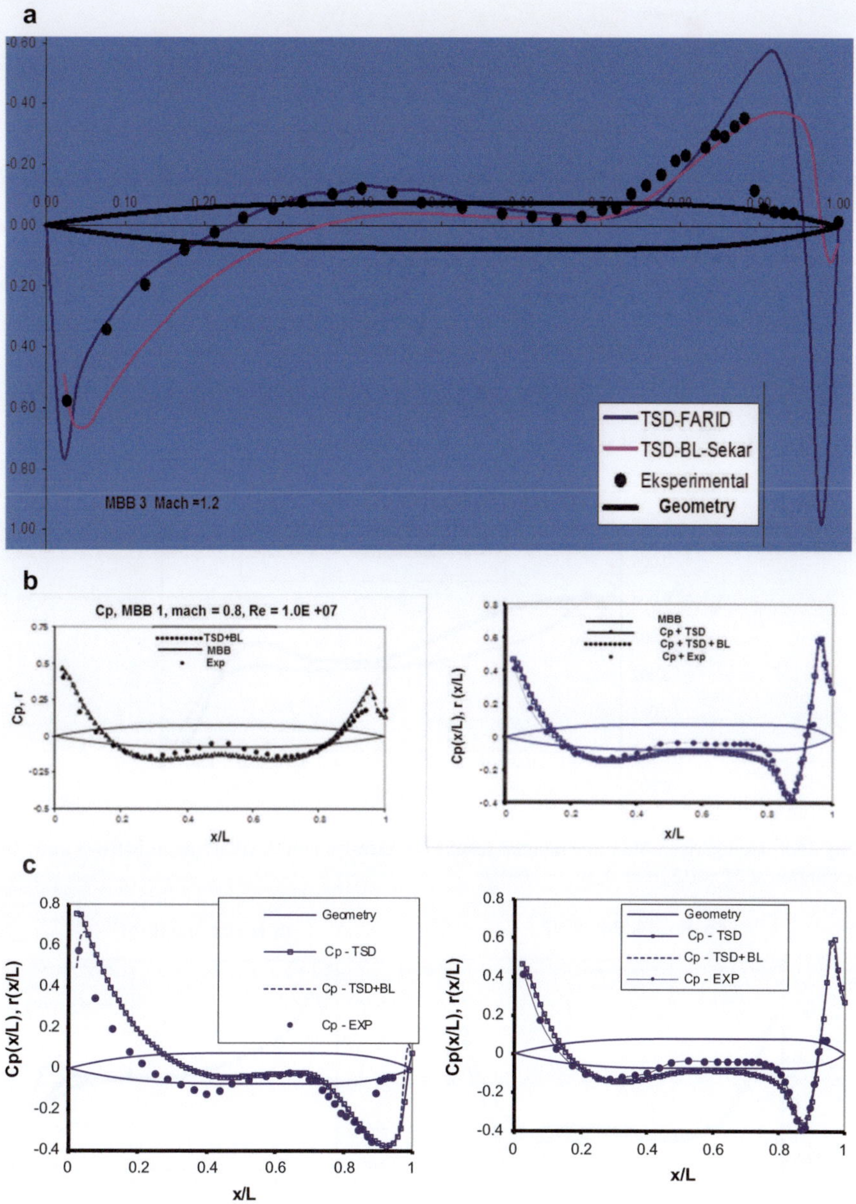

Fig. 26.6 **a** Comparison of the results of the 1st and 2nd TSD Computational schemes with MBB-3 at $M = 1.2$. **b** (left) MBB-1 at $M = 0.8$, **c** (right) MBB-5 at M = 0.8 2nd TSD with and without viscosity. **d** (left) MBB-3 at $M = 1.2$ 2nd TSD with and without viscosity-e. (right) Comparison of the results of the 2nd TSD Computational schemes (without and with viscosity) for MBB-5 at M = 1.2

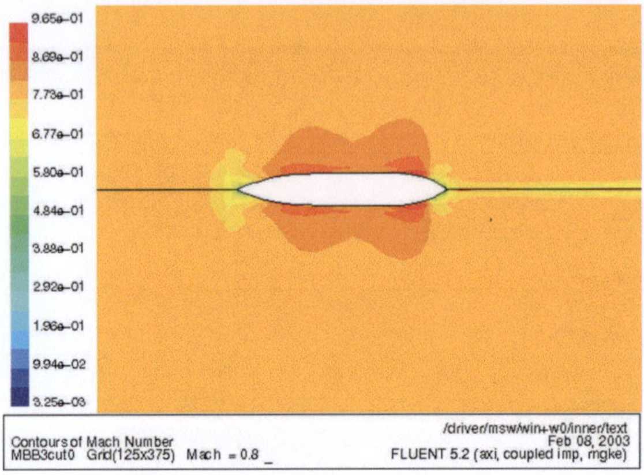

Fig. 26.7 Mach contour of MBB-3 body of revolution at $M = 0.8$, $\alpha = 0$

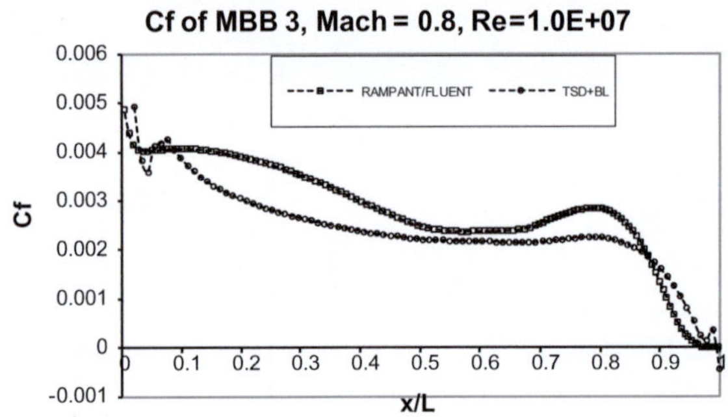

Fig. 26.8 Comparison of computational results for skin-friction distribution on MBB-3 body of revolution at $M = 0.8$, $\alpha = 0$, Re = 1E+07

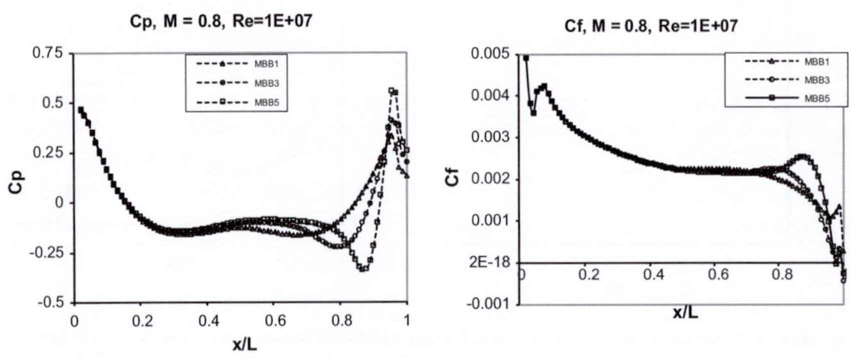

Fig. 26.9 Influence of the aft-geometrical shape of MBB Bodies of revolution on the distribution of pressure and friction coefficients, at $M = 0.8$, $\alpha = 0$, Re = 1E+07

Table 26.1 Computational results for various aft-body geometries

Mach. No.	Flat base	11th power	11th power (TSD)	Tanh	Tanh (TSD)	1/cos	MBB-3	5th power
1.2	41.107933E−2	1.088242E−2	1.8E−2	1.324629E−2	9.38E−2	3.61E02	8.93E−2	
Integral method			0.961009E−2					10.5381E−2

Note TSD—TSD2; MBB—experiment by MBB; Power-11: $R(x) = 1 - x^{11}$; 1/cos: $R(x) = 1/\cos(px/l)$

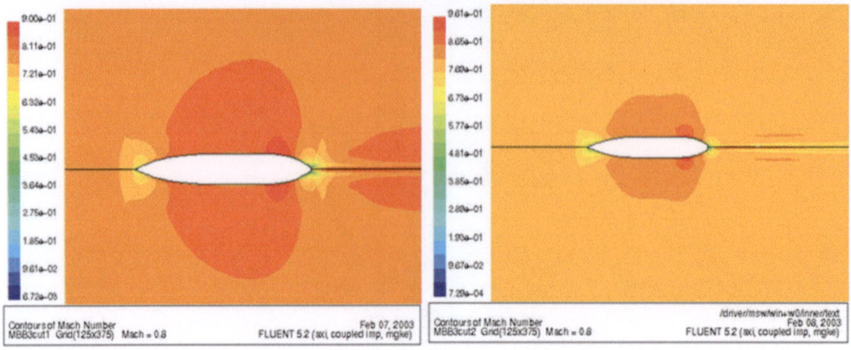

Fig. 26.10 **a** and **b** Mach contour of MBB-3 body of revolution at $M = 0.8$, $\alpha = 0$, Re = 1E+07, truncated at a small distance e from its trailing edge (boat tailing)

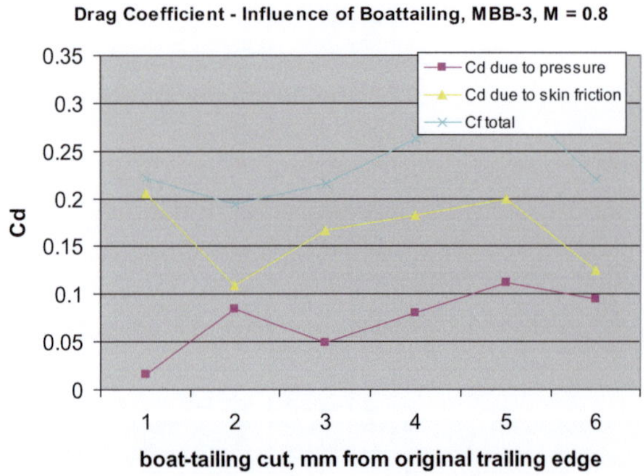

Fig. 26.11 Computational results of drag of various afterbody geometries at $M = 1.2$, showing the influence on boat tailing, calculated using FLUENT

The 1st TSD computational scheme outlined can be utilized for preliminary estimation of the transonic aerodynamic characteristics of shock-free slender axisymmetric body geometries. The 2nd TSD-viscous computational scheme can be utilized for detailed and relatively rapid computation of the transonic aerodynamic characteristics of slender axisymmetric body geometries, including the effect of viscosity.

Future work will look at the use of other numerical methods for the divergence form of the transonic equation.

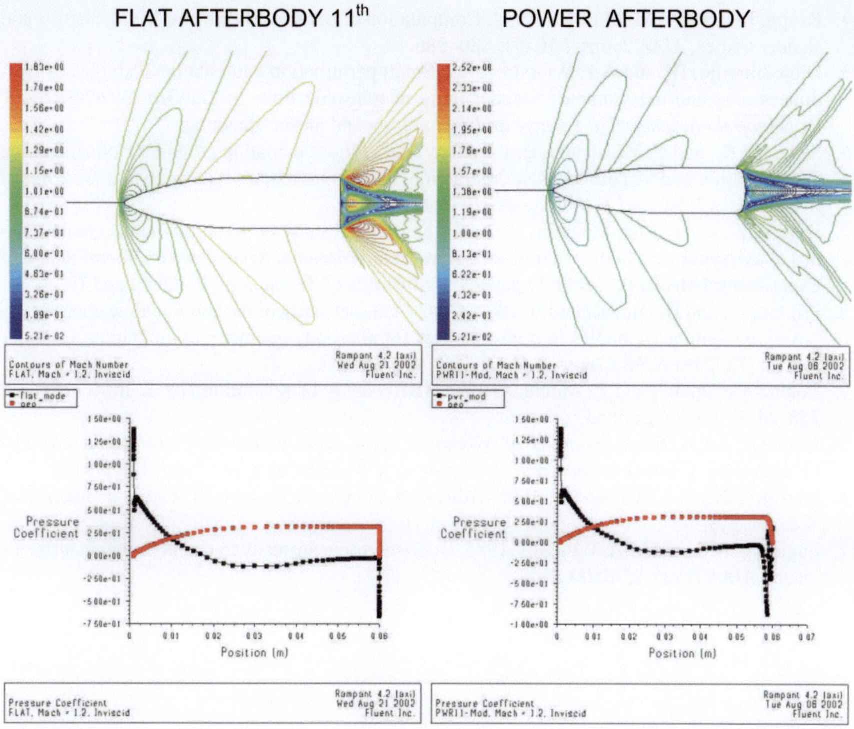

Fig. 26.12 Flow field and pressure distribution indicating the influence of boat tailing, calculated using FLUENT

Acknowledgements The organization of the work reported here was made possible partly during the first author's stay during his sabbatical leave from Institut Teknologi Bandung as a Visiting Professor at the Lehrstuhl für Fluidmechanik, then chaired by University Prof.Dr.-Ing.B.Laschka at Technische Universität München, to whom he extends his great appreciation, and sponsored by DAAD, and partly during his stay as a Visiting Professor at the International Cooperation Center for Engineering Education Development at Toyohashi University of Technology, for which he extends his thanks to Toyohashi University of Technology, and in particular, to Professor Hiroomi Homma. The use of various computational routines such as FLUENT was made possible through the courtesy of PT Dirgantara Indonesia.

References

1. Ashley, H., and M.T. Landahl. 1965. *Aerodynamics of Wings and Bodies.* Addison-Wesley Publishing Co.
2. Murman, E.M., and J.D. Cole. 1971. Calculation of plane steady transonic flows. *AIAA Journal* 9 (1): 114–121.
3. Frank, R.B. 1971. Numerical calculation of transonic flow about slender bodies of revolution. *NASA TN D-6582.*

4. Krupp, J.A., and E.M. Murman. 1972. Computation of transonic flows past lifting airfoils and slender bodies. *AIAA Journal* 10 (7): 880–886.
5. Djojodihardjo, H., and A.F. Widodo. 2002. Small perturbation computational studies of two-dimensional and axisymmetric slender body in transonic flow. In *The 3rd INDO-TAIWAN Workshop on Aeronautical Science and Technology*. Indonesia: Bandung.
6. Sekar, W.K., and C. Weishäupl, B. Laschka. 2002. Flow calculation around airfoils using the TSD equation and viscous-inviscid interaction. In *Aeronautika dan Astronautika (Journal of Aerospace Science and Technologies)*. Indonesia.
7. Prijono, E., and H. Djojodihardjo. 2002. Computational study of the aerodynamic characteristics of axi-symmetric bodies in transonic flow. In *International Symposium on Theoretical and Experimental Mechanics*. Bali: Organized by Institute of Technology Bandung and JICA.
8. Prijono, E., and H. Djojodihardjo. 2002. Computational study of the aerodynamic characteristics of axi-symmetric bodies in transonic flow for afterbody geometry optimization. In *Paper ICAS 0375, 23rd ICAS Congress*, 8–13. Toronto, Canada.
9. Lorenz-Meyer, W., and F. Aulehla. 1979. MBB—Body of revolution No. 3. In *AGARD AR 138*, ed. J. Barche. London.
10. Lax, P.D. 1954. *Weak Solutions of Nonlinear Hyperbolic Equations and Their Numerical Computation, Communication of Pure and Applied Mathematics*, vol. 7, no. 1, 159–193.
11. Sankar, L.N., and M.J. Smith. 1995. *Advanced Compressible Flow II*. Georgia Institute of Technology.
12. Fuglsang, D.F., and M.H. Williams. 1985. Non-isentropic unsteady transonic small disturbance theory. *AIAA Paper* 85-0800.

Chapter 27
Computational Modeling, Simulation and Tailoring of Non-penetrating and Impact Resilient Generic Structure

Abstract A computational modeling and simulation study is carried out to gain insight and formulate strategy for the design and tailoring of panel-like space structure that can withstand space debris impact without penetration. To represent a generic engineering structure, the impacted panel structure is modeled as a set of bonded Mindlin plates. The analysis is based on fundamental principles which are elaborated and numerically simulated. The objective is to identify optimum configuration in terms of loading, structural dimensions, material properties and composite layup. The analyses are based on dynamic response with emphasis on the elastic region. The direct numerical simulation is carried out in parallel for the analysis, synthesis, parametric study and optimization. Simulation results of impact loading by a spherical rigid body at certain velocity perpendicular to the panel show how fiber-metal laminates can be structurally tailored to achieve a non-penetrating impact.

Keywords Impact analysis · Composites · Engineering analysis · Engineering design · Finite element analysis · Structural analysis · Structural dynamics

27.1 Introduction

Impact resilient structures and non-penetrating impact to structures are of great interest in many engineering applications for protective purposes, varying from civil, land vehicle, aircraft and space structures, and personal utilization, to mention a few examples. Non-penetrating impacts on aerospace structures such debris collision and similar loading events may pose a serious threat to the structural integrity by inducing hidden damages, and for composites, intra-ply matrix cracking and interply delamination may ensue. Such damage will grow progressively during the subsequent loading–unloading cycles and may cause catastrophic failure of the structures [1].

The present chapter is based on two articles published in ACTA ASTRONAUTICA, Djojodihardjo. Computational simulation for analysis and synthesis of impact resilient structure. Acta Astronautica, 2013 and ASCE Journal of AerospaceEngineering, Djojodihardjo and Mahmud, Computational Modeling, Simulation, and Tailoring of Nonpenetrating Impact, Journal of Aerospace Engineering,2015, DOI: 10.1061/(ASCE)AS.1943-5525.0000531.

© The Author(s), under exclusive license to Springer Nature Singapore Pte Ltd. 2023
H. Djojodihardjo, *Introduction to Aeroelasticity*,
https://doi.org/10.1007/978-981-16-8078-6_27

To design impact resilient structures, it will be instructive and convenient to resort to fundamental principles and take into account the progress in analytical and computational approaches as well as in material science and technologies.

There is a need for resiliency analysis tools; in the early design stage, system engineers need to assess the threats and the reliability requirements of their design by employing resiliency analysis tools that first gauge the robustness of the bare unprotected design to check if it meets the specified reliability target. The process of accurately assessing the robustness of a bare unprotected design, or evaluating the effectiveness of candidate fault-tolerant techniques places the requirements on the resiliency analysis. The impact velocity of a projectile will affect the performance of compliant and resilient laminates.

In addition, spacecraft designers and operators will have to deal with a debris hazard far into the future, and a non-impact-penetrated structure would be of great interest. An analysis of generic problems will be instrumental and gratifying. To this end, research problem conceived can be formulated as the loading and response characteristics of a set of rectangular isotropic and homogeneous plates but of different material properties, bonded together and placed under the impact of a solid and rigid spherical impactor. In particular, as an example, the meteorite impact on spacecraft structure can be idealized by considering the generic case of a plate structure modeled as Mindlin plate [2] subject to impact loading, and then analysis, numerical simulation and parametric study can be carried out. The first objective of the work is to develop a computational algorithm to analyze flat plate as a generic structure subjected to impact loading. The analysis will be based on dynamic response analysis. Consideration is given to elastic region. The second objective is to utilize the computational algorithm for direct numerical simulation, and as a parallel scheme, commercial off-the-shelf numerical code is utilized for parametric study, optimization and synthesis. Through such analysis and numerical simulation, effort is devoted to arrive at optimum configuration in terms of loading, structural dimensions, material properties and composite layup, among others. Results will be discussed in view of practical applications.

One critical issue could be the resilience and structural integrity of the generic structure subject to impact, such as for the wall of microsatellites, in the elastic range (or at most within the plastic range), prior to its collapse. The problem can be idealized as the loading and response characteristics of a set of bonded rectangular isotropic and homogeneous plates with different material properties, subjected under the impact of a solid and rigid spherical impactor, as depicted in Fig. 27.1. The use of composites, in particular the bimetallic composite, is investigated for structural tailoring the plate materials.

There would be a prohibitive weight penalty if the structure was designed for all possible impacts, and thus the design is usually based on the assumption that the spacecraft will not be impacted by debris larger than a predetermined size. A size that has been used for space structures designed for use on or around the moon is a 1-g micrometeorite with a 1.56 cm diameter. In the present simulation studies, spherical impactor of a smaller size is considered [3, 4]. There are a few different structural methods for debris protection, two of which are sacrificial bumpers and multilayered

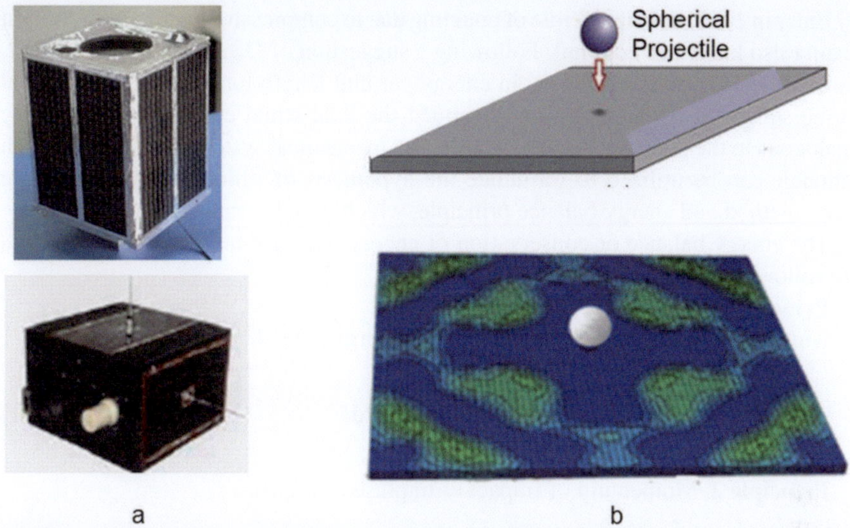

Fig. 27.1 **a** An example of a class of microsatellite structures with flat plate panels; **b** baseline model of a plate and spherical impactor used for impact analysis

protection. The theory behind sacrificial bumpers is that a bumper can be designed to vaporize both the micrometeorite of known mass and the local bumper wall upon impact.

27.2 Philosophical Approach in the Modeling and Simulation of Non-penetrating Impact

For later validation purpose, following the considerations outlined in the introduction, the present work resort to fundamental and basic principles, as well as simplified problems. These principles or hypotheses are [5]:

1. Algorithm and results for more involved plate impact problem should be justifiable and can be validated using the degenerate case of beam impact problem, by using some simplifying approaches.
2. Impact problem can be assessed and justified from a simpler static loading case by using energy conservation principle or hypothesis; i.e. the strain energy of impact should be proportional to the strain energy of static loading, and using quasi-steady assumption, should be equivalent.

Such principle is utilized here and could be traced back to 1705, when James Bernoulli proposed that the curvature of a beam was proportional to the bending moment. Using this theory, in 1744 Euler and in 1751 Daniel Bernoulli (1700–1782, John's son) proposed the transverse vibrations of beams, which was also suggested by Euler in 1757, in the principle of buckling due to compressive loading on a straight beam (also known as column). Following a suggestion of Daniel Bernoulli in 1742, Euler in 1744 introduced the strain energy per unit length for a beam, proportional to the square of its curvature and regarded the total strain energy as the quantity analogous to the potential energy of a discrete mechanical system [5–7]. Hence such rationale can be utilized to formulate the hypothesis of validation by quasi-static static method and energy balance principle, which is elaborated subsequently.

The energy balance or conservation of energy principle utilized here comprises the following principles (see Fig. 27.2):

Principle 1:

Kinetic Energy of Impact = Total Strain Energy

$$\frac{1}{2}\,\mathrm{mV}^2 = \int P\mathrm{d}\delta \tag{27.1}$$

Principle 2: Momentum of Impact = Impluse of Impact

$$mV = F\Delta t \tag{27.2}$$

Principle 3:

$$\Delta i_{mpulse} = \text{time elapsed starting from the state when the impactor hasthe initial kinetic energy} \frac{1}{2}\mathrm{mV}^2 \text{ to the instant when it comes to a stop.} \tag{27.3}$$

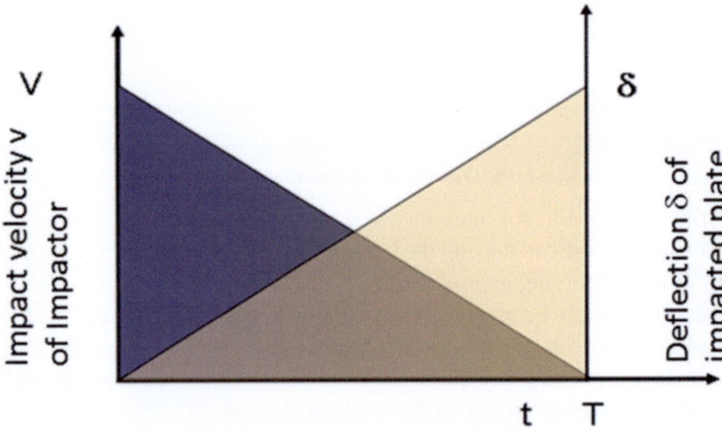

Fig. 27.2 State of affairs of impact and deflection for energy balance consideration

At each instant, the "spring force" due to elastic deformation develops until the plate reached a state of maximum deformation $\delta_{\max}$, and due to the "spring force," the beam or plate exerts an equivalent resisting force $P_{\max}$, when the impactor comes to a stop. For elastic deformation:

$$\int P\mathrm{d}\delta = \tfrac{1}{2}P_{\max}\delta_{\max} \tag{27.4}$$

Therefore

$$\frac{1}{2}mV^2 = \tfrac{1}{2}P_{\max}\delta_{\max} \tag{27.5}$$

This equation, which is derived from conservation of energy or energy balance principle, can be used for comparing and validating more involved numerical computational results by analytical means, by resorting to a degenerate case. However, if one is concerned with uncertainties and accuracy, other means of validation or approach have to be adopted. In addition, Eq. (27.5) can be used to obtain the concentrated force P acting at the center of the beam (or plate, as appropriate) for the equivalence between static and "quasi-static" impact, for validation purpose using first principles. In reality, all kinetic energy applied to deform the structure or system is transferred when the deformation reaches its maximum. At this maximum deflection, the impactor velocity reaches zero. Then the energy balance equation for the impact is [6],

$$\frac{1}{2}mv^2 = E_b + E_s + E_m + E_c \tag{27.6}$$

where E_{b} = bending energy, E_{s} = shear energy, E_{m} = energy due to deformation of the membrane, E_{c} = stored energy at contact region during impact, M = mass, V = Velocity before impact, k = contact stiffness and T_{c} = duration of contact. For the simplified case, all except E_{b} (bending energy) can reasonably be assumed to be zero.

27.3 Method of Approach in the Modeling and Simulation of Non-penetrating Impact

For the computational simulation of impact resilient structure, the study is focused on the numerical finite element simulation of impact of a rigid sphere on an elastic panel. The general logical framework for the analysis and numerical simulation of impact on panel structure is comprehensively summarized as the dotted part of Fig. 27.3, which excludes surface impact effect. The problem will be studied analytically and numerically and simulated using finite element codes, such as NASTRAN™ and/or ABAQUS™. The deformation and stress distribution on the impacted structure due

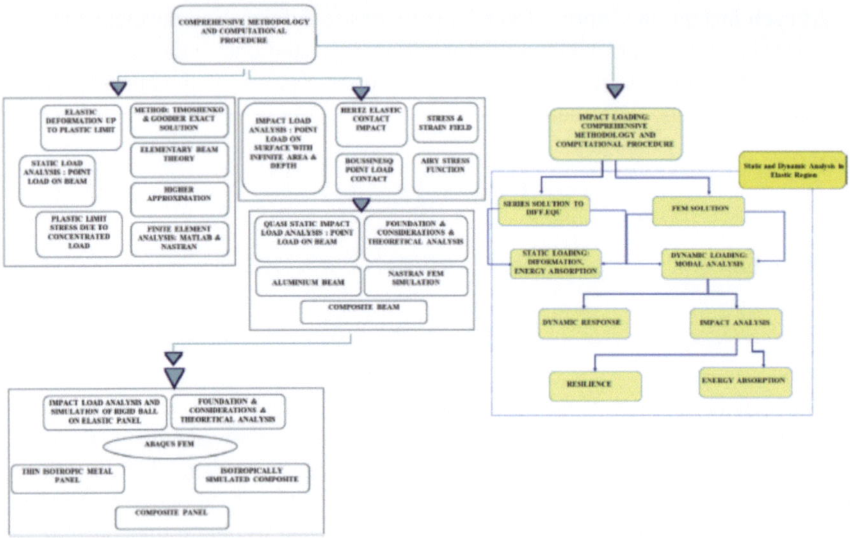

Fig. 27.3 Comprehensive scheme for analytical and computational impact study

to the impact loading will be determined for structural integrity and dynamic stability purposes.

As a further elaboration of earlier work, for the plate impact analysis, the present work utilizes Mindlin plate model and finite element numerical simulation using commercial software. The beam impact problem which will be used for validation in the degenerate case is taken to be the simplest, i.e. the Euler–Bernoulli beam. A procedure for such validation will be rigorously elaborated subsequently. Within such a roadmap, examples have been and will be elaborated on: Analysis of a Euler–Bernoulli and/or Timoshenko beam subject to impact, analysis of laminated side impact protection bar using finite element analysis and simulation, simulation of isotropic Mindlin plate subject to impact, simulation of bonded two-component plates subject to impact and simulation of laminated metal composites Mindlin plates. Such methodological approach is depicted in Sect. 27.4. A procedure for such validation will be rigorously elaborated subsequently.

27.4 Cross-Validation of Analytical and Numerical Simulation

As a further elaboration of earlier work, for the plate impact analysis, several means of cross-validation of analytical and numerical simulation have been carried out. Three approaches were taken. The first is by simulating a degenerate case in order that it can be validated using a simple analytical one. The second is by carrying out analytical approach for the same case simulated by numerical software and comparing the results to see that the numerical simulation is satisfactory. The third is by developing in-house computational routine using finite element approach and MATLAB™ (Fig. 27.4).

The first validation approach has been carried out in Refs. [7–11] as well as in the present work. This is exhibited in Table 27.1.

For the latter particular approach, the validation is more qualitative in nature. For validation using second approach, the results given by Navier's series solution and finite element method written in MATLAB™ (in-house) computational code have been validated using the exact solution as well as ANSYS™ code results [10, 11].

Validation of the results using the third approach is carried out using in-house MATLAB™ series solution and ANSYS™ FEM, which has been carried out in accuracy and plausibility of both in-house MATLAB™ codes and numerical simulation results.

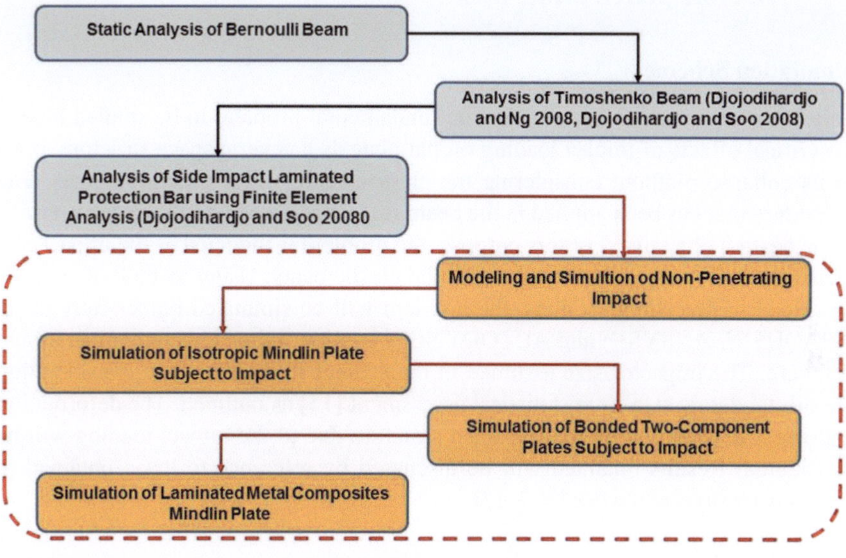

Fig. 27.4 Schematic of the methodological approach followed in the present work; two related works are depicted by three upper boxes [7, 8] boxes within the dashed line box are addressed in the present work

Table 27.1 Qualitative validation of Euler beam and degenerate plate under impact adapted from [12]

Deformation of the steel rectangular plate 100 × 100 mm, $E = 2e09$ N/m^2, $p = 0.3$ N/m^2, $\rho = 8760$ kg/m^3, Thickness = 2 mm		
	Maximum deformation δ [M]	% Error
Exact[a]	5.6784e−4	
ANSYS™ (FE)[a]	5.6509e−4	0.48
MATLAB™ series solution[a]	5.6793e−4	0.01
MATLAB™ FEM[a]	5.6780e−4	0.00
Plate treated as Euler–Bernoulli Beam—Exact (present)	1.5625e−5	Reasonable qualitative agreement

27.5 Parametric Study of Plates Under Impact by Finite Element Simulation for Structural Tailoring of Non-penetrating Case

Simulation Scheme

Following earlier studies [3, 7–12], the fundamental problem to be studied here is the critical effects of impact loading on flat plate as a generic space structure, prior to its collapse, without considering the elasto-plastic surface impact effects. The procedure that has been applied in the beam impact case can be followed to arrive at some desired objective. For this purpose, the problem is modeled as the impact of a relatively rigid body of small size to a solid elastic panel. Using method of analysis described in preceding sections, the problem will be simulated numerically using ABAQUS™, ANSYS™ and AUTODYNE™ explicit finite element computational software. The impactor size assumed in the present study takes into consideration the orbital debris size from statistical assessment [15] as outlined. The deformation and stress distribution on the impacted structure due to the impact loading will be considered. Results obtained will be discussed by reference to those obtained in previous sections and work [3, 7–12].

27.5.1 von Mises Stress Evaluation

In the present study, tacit consideration is given to the elastic range of loading, which incorporates the fundamental assumptions of linear elasticity including small deformations and linear relationship between stress and strain components. To emphasize the range of loading should remain in elastic region, the von Mises yield criterion that should be satisfied by the stress field is reproduced below. This criterion proposes that the material yielding commences when a specific parameter pertinent to stress tensor reaches a critical value.

$$\sigma_c = \sqrt{\frac{1}{2}[(\sigma_1 - \sigma_2)^2 + (\sigma_2 - \sigma_3)^2 + (\sigma_3 - \sigma_1)^2]} \tag{27.7}$$

where σ_c is equivalent stress (von Mises) and σ_1, σ_2 and σ_3 are the principal stresses. As far as the calculated value for σ_c is less than σ_y, the yield stress, the material response can be assumed as elastic.

The maximum equivalent strain (von Mises strain) is calculated by equation below,

$$\varepsilon_e = \frac{1}{1+\nu'}\left[\frac{(\varepsilon_1 - \varepsilon_2)^2 + (\varepsilon_2 - \varepsilon_3)^2 + (\varepsilon_3 - \varepsilon_1)^2}{2}\right]^{1/2} \tag{27.8}$$

Here, σ_i = stress at the ith direction, ε_i = strain at the ith direction and ν' = effective Poisson's ratio for elastic and thermal strains for a given body reference temperature. It is well known that the von Mises stress can be used to predict the elastic response of material under any loading condition based on the results of simple uniaxial tensile tests [15]. In addition, the von Mises stress criterion leads to similar values for two different states of stresses with equal distortion energy. Since von Mises criterion is independent of the first stress invariant, it is applicable for the analysis of plastic deformation relevant to ductile materials as the onset of yield for these materials does not depend on the hydrostatic component of the stress tensor. The associated total deformation U is given by

$$U = \sqrt{U_x^2 + U_y^2 + U_z^2} \tag{27.9}$$

27.5.2 Selection of Laminated Metal Composites for the Present Study

Based on encouraging results from previous studies [3–5, 7–9, 12], the present study is focused on laminated metal composites, such as aluminum-titanium composites, also known as Alpolic [16], GLARE and other glass/epoxy composites. For this purpose, attention is given on GLARE [17], and it has superior properties suited for the present study.

Material properties for various materials such as steel is readily available online (such as Engineer's Handbook, retrieved 6 February 2015) and a suitable source for 2024 Aluminum is used and indicated in the reference (High Strength Glass Fibres 2014). However the epoxy for the GLARE is sourced for the manufactures of the compound.

Here S2-Glass fiber is used (such as AGY, 2014). This glass fiber is common and used for Airbus A380 skin panes. However, only one direction Young's modulus data and relevant perpendicular plane shear modulus are available. Therefore, in this particular work assumption is made that the properties of this Epoxy for other directions are assumed using Poisson's ratio multiplied by the maximum Young's

Table 27.2 Deformation of the rectangular plate

Deformation of the rectangular plate 100 × 100 mm, $E = 2e11$, $p = 0.3$, $\rho = 8760$ kg/m^3, Thickness = 2 mm		
	Maximum deformation [M]	% Error
Exact	5.6784e−004	
ANSYS™ (FE)	5.6509e−004	0.48
MATLAB™ series solution	5.6793e−004	0.01
MATLAB™ FEM	5.6780e−004	0.00

Table 27.3 First natural frequency for rectangular plate

First natural frequency for rectangular plate 200 × 200 mm, $E = 2e11$, $\nu = 0.3$, $\rho = 8760$ kg/m^3, Thickness = 2 mm		
	Frequency [Hz] (first mode)	% Error
Exact	437.09	0.00
ANSYS™ (FE)	438.59	**0.34**
MATLAB™ (first term of spatial component)	441.5393	1.01

Table 27.4 Aluminum-titanium composite properties

Tensile strength	ASTM E8	MPa, N/mm^2	69
0.2% proof stress	ASTM E8	MPa, N/mm^2	60
Elongation	ASTM E8	%	11.1
Flexural elasticity, E	ASTM C393	GPa, kN/mm^2	49.0
Flexural rigidity, E × I	ASTM C393	kN mm^2/mm	256
Punching shear resistance			
Shear resistance	ASTM D732	MPa, N/mm^2	48

modulus and shear ratio. Detail of the properties is given in Tables 27.4, 27.5, 27.6 27.7, 27.8 and 27.9. For the purpose of this study a hypothetical metal laminate will be used based on Alpolic metal laminates (TiAl) which will be referred to as HML.

27.6 Analysis of Surface Impact Using Hertz Elastic Contact Impact Theory

Following the theory, the sphere ball is assumed to be perfectly rigid, the flat slab surface subject to impact is non-rigid, ductile and deformable, and the contact surface is frictionless. Accordingly, the shape of the surface of contact satisfies certain boundary conditions, as depicted in Fig. 27.5, namely.

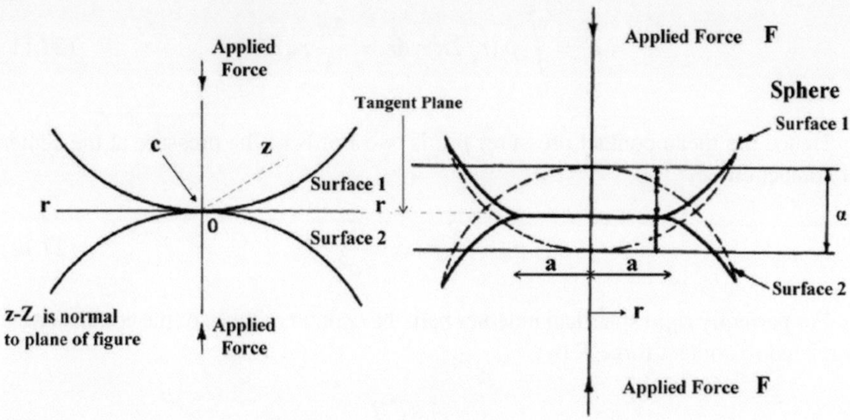

Fig. 27.5 Compression and indentation of two spherical contact surfaces

i. The radii of curvature of contacting bodies are large compared with radius of the circle of contact. Thus, each surface is treated as an elastic half-space.
ii. The bodies are in frictionless contact. It means that only a normal pressure is transmitted between the ball and the specimen.
iii. Dimensions of each body are large compared with radius of circle of contact.
iv. The distance between the surfaces of the two bodies is zero inside and greater than zero outside the circle of contact.
v. The integral of pressure distribution within the circle of contact with respect to the area of the circle of contact gives the force acting between the two bodies.
vi. The surfaces are continuous and non-conforming.
vii. The strains are small.

For non-conforming elastic bodies B and B' coming into contact at a point 0, the surfaces of the bodies in the vicinity have radii of curvature R_1 and R_2. If F is the compressive force in the normal direction, then the contact region spreads to radius a and within the contact area there is an elliptical distribution of contact pressure

$$p\,(r,t) = p_0(t)\left[1 - \left(\frac{r}{a\,(t)}\right)^2\right]^{1/2}, \; r \leq a \tag{27.10}$$

where r is radial coordinate originating at the center and p_0 is the pressure at the center of the contact area as schematically illustrated in Fig. 27.5.

Outside the contact circle, the normal stress pressure $p(r)$ is zero, and it is a free surface. This pressure distribution results in a compressive reaction force F on each body, given by

$$F = \int_{0}^{a} p(r)\, 2\pi\, r\, \mathrm{d}r = \frac{2\pi}{3} p_0\, a^2 \tag{27.11}$$

Hence the mean contact pressure, p_m, is two-thirds of the pressure at the center of contact circle,

$$p_m = \frac{2}{3} p_0 = \frac{F}{\pi a^2} \tag{27.12}$$

For perfectly rigid spherical indenter ball, the contact radius a of the contact circle is related to contact force F by:

$$a^3 = \frac{4}{3} \frac{kFR_1}{E_*} \tag{27.13}$$

where, following Fischer-Cripps [18], k is an elastic mismatch factor given by

$$k = \frac{9}{16}\left[\left(1 - \upsilon_2^2\right) + \frac{E_2}{E_1}\left(1 - \upsilon_1^2\right)\right] \tag{27.14}$$

where E_2, υ_2 and E_1, υ_1 denote Young's modulus and Poisson's ratio of the specimen and the spherical indenter ball, respectively. Here E_* is the equivalent effective Young's modulus,

$$\frac{1}{E_*} = \frac{\left(1 - \upsilon_1^2\right)}{E_1} + \frac{\left(1 - \upsilon_2^2\right)}{E_2} \tag{27.15}$$

where υ = Poisson's ratio.

Since the spherical indenter ball is assumed perfectly rigid, $E_1 = \infty$, so that Eq. (27.14) becomes:

$$k = \frac{9}{16}\left(1 - \upsilon_2^2\right) \tag{27.16}$$

With no load applied, and with the indenter just touching the specimen, the distance d between a point on the periphery of the spherical indenter ball to the specimen surface as a function of radial distance r is given by:

$$d = \frac{r^2}{2R} \tag{27.17}$$

Consider a load applied to the ball in contact with a flat surface, as shown in Fig. 27.6, such that the point at which load is applied moves a vertical distance δ.

This distance is called the "load-point displacement" and may be considered to be the distance of mutual approach between the spherical ball and the flat surface.

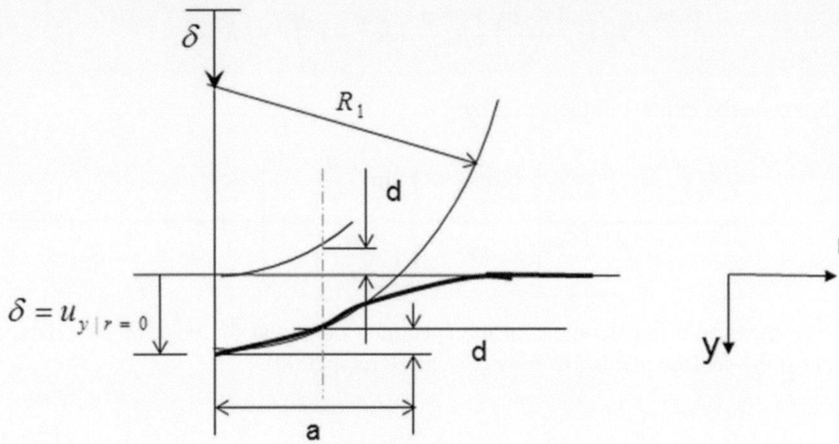

Fig. 27.6 Schematic of the contact between a rigid, spherical indenter ball and a flat, non-rigid slab

The distance of mutual approach δ (when $h = 0$ at $r = 0$) between contacting bodies is given by [19] as:

$$\delta = \frac{a^2}{R_*} = \left(\frac{3}{4E_*}\right)^{2/3}\left(\frac{F^2}{R_*}\right)^{1/3} \tag{27.18}$$

where R_* is the characteristic radius,

$$\frac{1}{R_*} = \frac{1}{R_1} + \frac{1}{R_2} \tag{27.19}$$

Here, the sphere with radius R_1 is impacted on a flat slab surface. Thus $R_2 = \infty$ for flat slab surface.

Since the spherical ball is assumed to be perfectly rigid, $E_1 = \infty$ and

$$\frac{1}{E_*} = \frac{\left(1 - \upsilon_2^2\right)}{E_2} \tag{27.20}$$

27.6.1 Surface Stresses and Distribution of Displacement Induced by a Spherical Indentation Ball

For rigid spherical ball impact on non-rigid flat slab, following Hertz theory [18, 20] the distribution of pressure, Eq. (27.1) produces displacement of the slab specimen surface inside the contact circle given by:

$$u_y = \frac{1-\upsilon_2^2}{E_2}\frac{3}{2}p_m\frac{\pi}{4a}\left(2a^2 - r^2\right) r \le a \tag{27.21}$$

and outside the contact circle given by:

$$u_y = \frac{1-\upsilon_2^2}{E_2}\frac{3}{2}p_m\frac{1}{2a}\left[\begin{array}{c}\left(2a^2 - r^2\right)\sin^{-1}\frac{a}{r}\\ +r^2\frac{a}{r}\left(1-\frac{a^2}{r^2}\right)^{1/2}\end{array}\right], \quad r \ge a \tag{27.22}$$

The maximum tensile stress in the specimen occurs at the edge of the contact circle at the surface and is given by (Fischer-Crisp, 2000):

$$\sigma_{max} = (1 - 2\,\upsilon_2)\frac{F}{2\,\pi\,a^2} \tag{27.23}$$

This stress, acting in a radial direction on the surface outside the spherical ball, decreases as the inverse square of the distance away from the center of contact. Combining Eqs. 27.12 and 27.13, the maximum tensile stress outside the spherical ball can be expressed in terms of the spherical ball radius R_1:

$$\sigma_{max} = \left[\frac{(1-2\,\upsilon_2)F}{2\,\pi}\right]\left(\frac{3E_2}{4k}\right)^{2/3} F^{1/3} R_1^{-2/3} \tag{27.24}$$

Inside the contact circle, the radial stress distribution at the surface is:

$$\frac{\sigma_r}{p_m} = \frac{1-2\,\upsilon_2}{2}\frac{a^2}{r^2}\left[1-\left(1-\frac{r^2}{a^2}\right)^{3/2}\right] - \frac{3}{2}\left(1-\frac{r^2}{a^2}\right)^{1/2}; \quad r \le a \tag{27.25a}$$

and on the surface outside the contact circle:

$$\frac{\sigma_r}{p_m} = \frac{1-2\,\upsilon_2}{2}\frac{a^2}{r^2} \quad r > a \tag{27.25b}$$

The maximum value of σ_r occurs at $r = a$. In-plane displacements on the surface beneath the spherical ball in the radial direction are given by, for $r \le a$:

$$u_r = -\frac{(1-2\,\upsilon_2)(1+\upsilon_2)}{2E_2}\frac{a^2}{r}p_m\left[1-\left(1-\frac{r^2}{a^2}\right)^{3/2}\right] \tag{27.26}$$

Note that for all values of $r < a$, the radial displacement of points on the surface is directed inward toward the center of contact. Outside the contact area, the radial displacement is given by:

$$u_r = -\frac{(1-2\upsilon_2)(1+\upsilon_2)}{3E_2}\frac{a^2}{r}\frac{3}{2}p_m \quad r > a \tag{27.27}$$

The hoop stress, on the surface, is always a principal stress and outside the contact circle is equal in magnitude to the radial stress:

$$\sigma_\theta = -\sigma_r \quad r > a \tag{27.28}$$

Within the interior of the specimen, the stresses are calculated from [15]:

$$\begin{aligned}\frac{\sigma_r}{p_m} = \frac{3}{2}\Bigg\{&\frac{(1-2\upsilon_2)}{3}\frac{a^2}{r^2}\left[1-\left(\frac{y}{u^{1/2}}\right)^3\right]+\left(\frac{y}{u^{1/2}}\right)^3\frac{a^2u}{u^2+a^2y^2}\\ &+\frac{y}{u^{1/2}}\left[u\frac{1-\upsilon_2}{a^2+u}+(1+\upsilon_2)\frac{u^{1/2}}{a}\tan^{-1}\left(\frac{a}{u^{1/2}}\right)-2\right]\Bigg\}\end{aligned} \tag{27.29}$$

$$\begin{aligned}\frac{\sigma_\theta}{p_m} = -\frac{3}{2}\Bigg\{&\frac{(1-2\upsilon_2)}{3}\frac{a^2}{r^2}\left[1-\left(\frac{y}{u^{1/2}}\right)^3\right]\\ &+\frac{y}{u^{1/2}}\left[\begin{array}{c}2\upsilon_2+u\dfrac{1-\upsilon_2}{a^2+u}\\ -(1+\upsilon_2)\dfrac{u^{1/2}}{a}\tan^{-1}\left(\dfrac{a}{u^{1/2}}\right)\end{array}\right]\Bigg\}\end{aligned} \tag{27.30}$$

$$\frac{\sigma_y}{p_m} = -\frac{3}{2}\left(\frac{y}{u^{1/2}}\right)^3\frac{a^2u}{u^2+a^2y^2} \tag{27.31}$$

$$\frac{\tau_{ry}}{p_m} = -\frac{3}{2}\left(\frac{ry^2}{u^2+a^2y^2}\right)\left(\frac{a^2u^{1/2}}{a^2+u}\right) \tag{27.32}$$

where

$$u = \frac{1}{2}\left\{\begin{array}{l}\left(r^2+y^2-a^2\right)\\ +\left[\left(r^2+y^2-a^2\right)^2+4a^2y^2\right]^{1/2}\end{array}\right\} \tag{27.33}$$

An example of the computed vertical normal stress around point load due to contact point force on the surface of a beam is shown in Fig. 27.7. Boussinesq and Hertz methods [18, 20] are used.

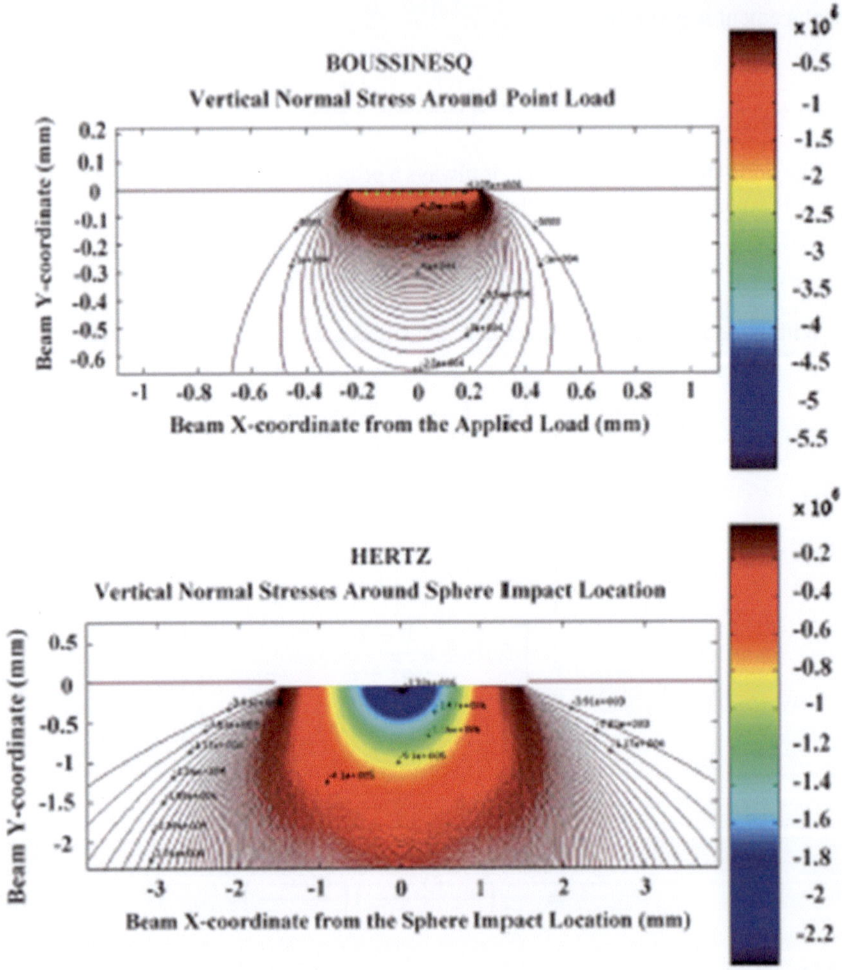

Fig. 27.7 Vertical normal stress around point load

27.6.2 Indentation Response of Materials

The indentation stress–strain response of specimen material can be generally divided into three regimes [18]:

a. $p_{\mathrm{m}} <\approx \sigma_y$; fully elastic response where σ_y is the uniaxial yield stress of the specimen material.
b. $\sigma_y \approx< p_{\mathrm{m}} < C\sigma_y$; elastic–plastic stage happens where plastic deformation occurs beneath the surface but the plastic zone is completely surrounded by elastically strained material. C is the constraint factor whose value depends upon the material and the spherical ball geometry.

c. $p_m = C\sigma_y$; plastic zone continues to grow in size such that the indentation contact area increases at a rate which gives little or no increase in the mean contact pressure for further increases in spherical ball load. The ratio between the indentation mean contact pressure and the uniaxial compressive yield stress is called the constraint factor and is given the symbol C.

In this study, the rigid-perfectly plastic material model is used. It assumes that the elastic deformation is small enough to be negligible and the material flows plastically at a constant stress σ_y in tension.

During the initial application of load, the response is elastic and can be predicted from the Hertz relation, by combining Eqs. (27.34) and (27.35):

$$p_m = \left(\frac{3E_*}{4\pi k}\right)\frac{a}{R_1} \tag{27.34}$$

Equation (27.34) assumes linear elasticity and makes no allowance for yield within the specimen material.

For a fully elastic response of static Hertzian contact, the principal shear stress for indentation with a spherical ball is a maximum at ≈ 0.47 pm at a depth of ≈ 0.5 a beneath the specimen surface (directly beneath the ball impactor) or point of contact.

27.7 Analytical and Computational Results

Following the analytical procedures outlined in previous sections, computational routine has been developed, both for the analytical cases and using finite element approach. Results obtained are used to obtain insight into the stages and situations for the impact load on an elastic structure, modeled first as the generic case of a pointed load on a beam, and second as a flat elastic surface of a semi-infinite structure subjected to impact loading due to a spherical indenter ball.

27.7.1 Collapse Load Applied at Midspan of Beam: $L_1 = L/2$

The resulting moment, stress and strain due to a pointed load applied at the midspan of a beam are illustrated in Fig. 27.8. These results show that yield and plastic moment, and yield and plastic stress are independent of applied load's location. The weakest part of the beam is at its midspan. Higher collapse load is required to cause beam failure as the applied load distance from beam midspan is increasing. Plastic hinge length and shape are predictable from bending moment diagram. Maximum bending stress occurs under the point load. As the distance of the fibers from neutral axis is decreasing, bending stresses will also decrease progressively.

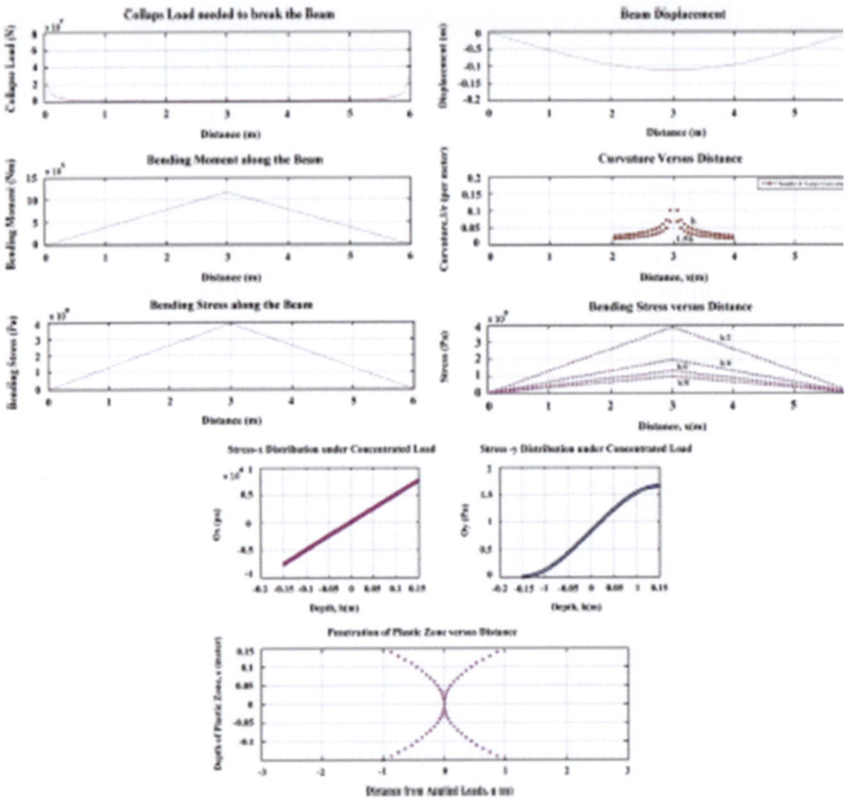

Fig. 27.8 Collapse load applied at midspan of beam

27.7.2 Static and Dynamic Analysis of Isotropic Flat Plate Bending

Problem Formulation

Specific problem is then addressed on the loading response of a flat plate under prescribed boundary conditions subject to dynamic external loading; the static loading case is also considered for comparative and validation purposes. The state of affairs is illustrated in Fig. 27.9.

To gain some insight into the impact resilient structure in elastic region, related parametric study and numerical simulation will be carried out for a generic example.

Addressing static loading as a baseline and future reference, the governing differential equation for the deflections for rectangular thin plate bending analysis based on Kirchhoff's assumptions [7, 21, 22] is the biharmonic equation is given by

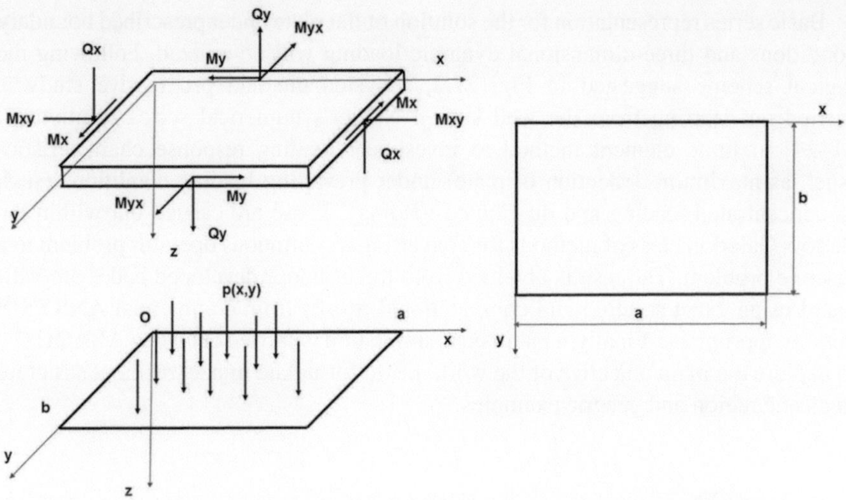

Fig. 27.9 Generic problem of loading on a flat plate under prescribed boundary condition

$$\frac{\partial^4 w}{\partial x^4} + 2\frac{\partial^4 w}{\partial x^2 \partial y^2} + \frac{\partial^4 w}{\partial y^4} = \frac{p}{D} \tag{27.35}$$

For flat plate simply supported at its four edges, the boundary condition is given by:

$$\mathrm{w} = 0 \quad \frac{\partial^2 w}{\partial x^2} = 0 \text{ at x} = 0 \text{ and x} = \text{a} \tag{27.36}$$

$$\mathrm{w} = 0 \quad \frac{\partial^2 w}{\partial y^2} = 0 \text{ at y} = 0 \text{ and y} = \text{a} \tag{27.37}$$

where

$$D = \frac{Et^2}{12\left(1 - \nu^2\right)} \tag{27.38}$$

t = thickness of plate, ν = Poisson ratio of material and E is the modulus of elasticity.

The problem to be solved is to find the distribution of deflection and stresses of the plate subject to a vertical point load at various positions across the plate, which is the fallback situation for static loading case, and the distribution of deflection and stresses of the plate under dynamic loading. The first part is a classical problem of flat plate deformation and stresses while the second part requires modal analysis of the dynamic load of the flat plate. As a generic case, the plate will be assumed to be isotropic and homogeneous.

Basic series representation for the solution of flat plate under prescribed boundary conditions and three-dimensional dynamic loading will be utilized. Following the logical scheme suggested in Fig. 27.2, a systematic and progressive study is carried out starting from the well-known Navier's numerical series solution [7, 21–24] to finite element method to investigate loading response characteristics (such as maximum deflection of plate) under prevailing loading conditions (such as concentrated-loading and distributed-loading). These are carried out within the Petrov–Galerkin class of methods for converting a continuous operator problem to a discrete problem. The results obtained from the in-house developed codes are validated using exact solution and computational results from commercial ANSYS™ code as appropriate. Finally, a parametric simulation is carried out using ABAQUS™ to explore the main objective of the work, i.e. to formulate impact resilient structure in elastic region and generic examples.

27.7.3 *Methodological Scheme*

To develop the computational code for solving the problem as stated above, the scheme of approach depicted in Fig. 27.4, which outlined the methodological scheme for a comprehensive impact study including Hertz elastic contact impact, is used as a framework. The present stage, however, is contained within the dotted rectangle in the methodological approach summarized in Fig. 27.4. The generic methods to address impact load on elastic plate developed are:

1. Series solution of equation (Navier's method) using analytical technique
2. Finite element solution, which will be solved by using MATLAB™.

The computational scheme developed is validated using known exact solutions [7] as well as using commercial code ANSYS™. Using such framework, the in-house computational code can be further extended to cover more involved cases.

The next scheme is to utilize the developed code as well as ANSYS™ for parametric study of flat plate as a generic element of structure subject to dynamic and impact loading and define some performance measure related to load characteristics, geometrical characteristics and constraints, as well as material properties. These performance measures will be defined in terms of resilience, energy absorption, and flexibility, among others.

27.7.4 *Theoretical Foundation and Generic Analysis*

The basic approach undertaken in modeling and analysis of flat plate under impact load follows the approach taken in [8, 9] for a beam subject to impact load. For the baseline analysis, the load is considered to be applied at the center of the beam or panel, and parametric studies with respect to relevant variables such as the variation of the location of the application point for the beam, and variation of the geometry of the beam or panel.

Computational finite element simulations using MATLAB™-based in-house program and NASTRAN™ for the beam and ABAQUS™ for the panel are carried out for particular examples. For the case of the beam, purely analytical approach is first resorted to as the baseline as well as for validation purposes, since the solution to analytical approach is limited to certain class of geometries only. Such approach may also be utilized to project possible behavior of panel and designing the numerical simulation.

The work reported here is a continuation of previous papers which attempts to build up a comprehensive account on the development of impact resistant structure through theoretical analysis, numerical simulation for parametric studies and eventually multi-variables optimization.

a. **Series Solution using Navier's Method**

Navier's method or double series solution is used to determine unknowns for a simply supported rectangular thin plate when a concentrated and distributed force is applied. Solution of the bending problem employs the following Fourier series for load p and deflection *w* [7, 21–23] is given by

$$p(x, y) = \sum_{m=1}^{\infty}\sum_{n=1}^{\infty} p_{mn} \sin\frac{m\pi x}{a} \sin\frac{n\pi y}{b} \tag{27.39}$$

$$w(x, y) = \sum_{m=1}^{\infty}\sum_{n=1}^{\infty} a_{mn} \sin\frac{m\pi x}{a} \sin\frac{n\pi y}{b} \tag{27.40}$$

where p_{mn} and a_{mn} are the coefficients to be determined. The deflection must satisfy the differential Eq. (27.25) and the prescribed boundary conditions (27.36)–(27.37). The detail of the solution procedure is well known [7, 21–24] and is given by:

$$w(x, y) = \frac{1}{\pi^4 D}\sum_{m=1}^{\infty}\sum_{n=1}^{\infty} \frac{p_{mn}}{\left[\left(^m/_a\right)^2 + \left(^n/_b\right)^2\right]^2} \sin\frac{m\pi x}{a} \sin\frac{n\pi y}{b} \tag{27.41}$$

where

$$p_{mn} = \frac{4}{ab}\int_0^b\int_0^a p(x, y) \sin\frac{m\pi x}{a} \sin\frac{n\pi y}{b} dx dy \tag{27.42}$$

Evaluation of a_{mn} in Eq. (27.40) requires substitution of Eqs. (27.41) and (27.42) into (27.35),

$$\sum_{m=1}^{\infty}\sum_{n=1}^{\infty}\left\{\left[\left(\frac{m\pi}{a}\right)^4+2\left(\frac{m\pi}{a}\right)^2\left(\frac{n\pi}{b}\right)^2+\left(\frac{n\pi}{b}\right)^4\right]\right\}$$
$$\times a_{mn}\sin\frac{m\pi x}{a}\sin\frac{n\pi y}{b}=0 \tag{27.43}$$

This equation must apply for all x and y, and therefore it can be concluded that

$$a_{mn}=\frac{1}{\pi^4 D}\frac{p_{mn}}{\left[\left(\frac{m}{a}\right)^2+\left(\frac{n}{b}\right)^2\right]^2} \tag{27.44}$$

Substituting Eq. (27.42) into Eq. (27.41) produces the equation of the deflection surface of the plate:

$$w(x,y)=\frac{1}{\pi^4 D}\sum_{m=1}^{\infty}\sum_{n=1}^{\infty}\frac{p_{mn}}{\left[\left(\frac{m}{a}\right)^2+\left(\frac{n}{b}\right)^2\right]^2}\sin\frac{m\pi x}{a}\sin\frac{n\pi y}{b} \tag{27.45}$$

When a rectangular plate is subjected to a uniformly distributed load $p(x, y) = P_0$, the results of the previous section are simplified considerably. After integration, Eq. (27.42) becomes

$$p_{mn}=\frac{4P_0}{\pi^2 mn}(1-\cos m\pi)(1-\cos n\pi)=\frac{4P_0}{\pi^2 mn}\left[1-(-1)^m\right]\left[1-(-1)^n\right] \tag{27.46}$$

Introducing p_{mn} into Eq. (27.45), one obtains (Fig. 27.10)

$$w(x,y)=\frac{16P_0}{\pi^6 mn}\sum_{m=1}^{\infty}\sum_{n=1}^{\infty}\frac{\sin\frac{m\pi x}{a}\sin\frac{n\pi y}{b}}{mn\left[\left(\frac{m}{a}\right)^2+\left(\frac{n}{b}\right)^2\right]^2} \tag{27.47}$$

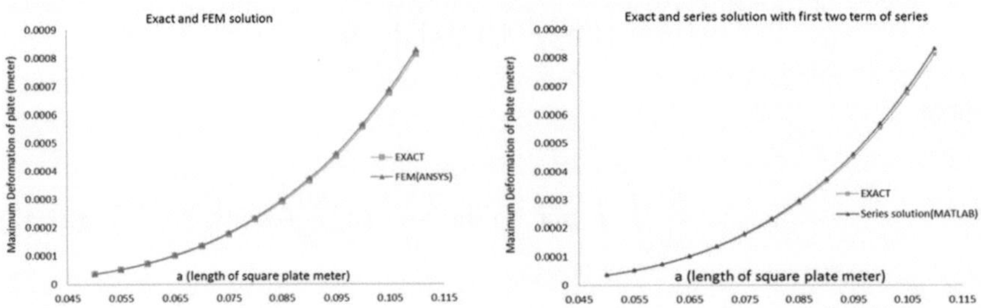

Fig. 27.10 Comparison of various computational results for the maximum deflection of square plate

where m, n = 1, 3, ... The maximum deflection occurs at the center of the plate ($x = a/2$, $y = b/2$) and its value, from Eq. (27.45) is

$$w_{\max} = \frac{16P_0}{\pi^6 mn} \sum_{m=1}^{\infty} \sum_{n=1}^{\infty} \frac{(-1)^{\frac{m+n}{2}-1}}{mn\left[\left(\frac{m}{a}\right)^2 + \left(\frac{n}{b}\right)^2\right]^2} \tag{27.48}$$

b. **Validation of the MATLAB TM computational codes**

For this validation, a square plate (0.1 m × 0.1 m) with thickness of 0.002 m and steel is utilized. The results obtained using series solution and finite element method in the MATLAB™ and that obtained using ANSYS™ finite element codes are compared with the exact solution.

The results given by Navier's series solution and finite element method written in MATLAB™ (in-house) computational code are validated using the exact solution as well as ANSYS™ code results, as exhibited in Fig. 27.7. Validation of the results obtained using in-house method is shown in Table 27.1 (in percentile error with respect to the exact solution) and Figs. 27.8 and 27.9. These results serve to validate the accuracy and plausibility of both in-house MATLAB™ codes.

27.7.5 Static and Dynamic Analysis

Static Analysis

In the first part of this section, parametric study is carried out for rectangular plate with various dimensions for equivalent energy absorption analysis. The deformation of a rectangular plate is investigated by using series solution (MATLAB™) and finite element method (ANSYS™) is exhibited in Table 27.2.

Dynamic Analysis (Modal Analysis)

In the second part of this section, modal analysis is carried out for rectangular plate with various dimensions for later use in impact analysis. Variables of interest are as maximum deformation and impact energy of elastic square plate subject to impact by a solid and rigid impactor, which will be calculated using ANSYS™ (FE-modeling). For generic cases, the computation is also carried out using in-house MATLAB™ code. Sample modal analysis simulation results of a square plate are illustrated in Figs. 27.11, 27.12 and 27.13. The first natural frequency obtained for a rectangular plate using in-house MATLAB™ code and ANSYS ™ is exhibited in Table 27.3.

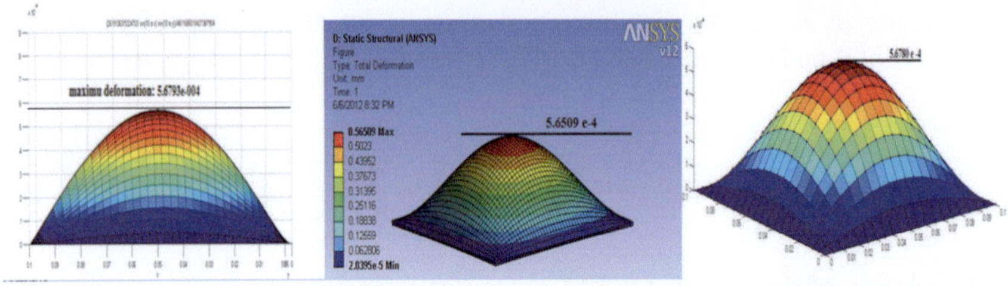

Fig. 27.11 Maximum deformation of plate: **a** MATLAB™ in-house code, series solution; **b** ANSYS™; **c** MATLAB™ in-house code, FE method

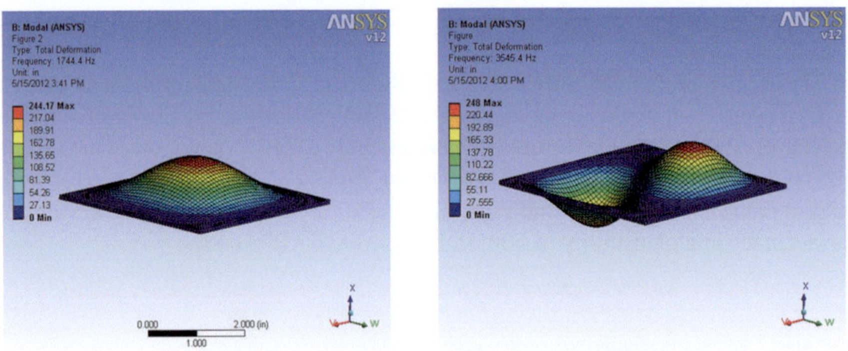

Fig. 27.12 Preliminary modal analysis simulation results (first two modes of square plate)

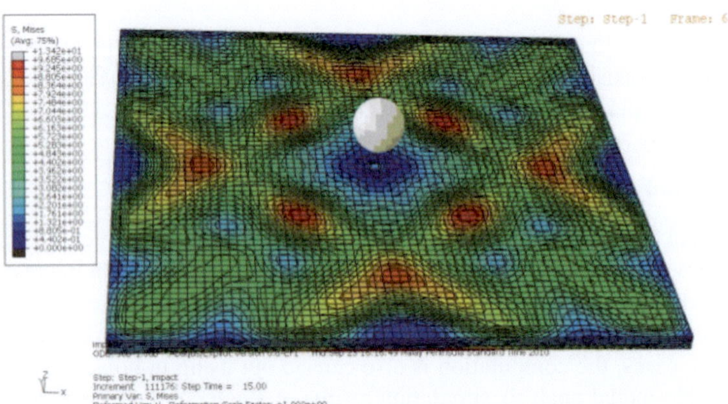

Fig. 27.13 Distribution of von Mises stresses, of a 200 × 200 mm aluminum alloy panel with 5 mm thickness subject to impact by rigid sphere impactor of 20 mm diameter at a velocity of 2 m/s at 0.015 s. after impact

27.8 Finite Element Impact Simulation of Flat Plate Subject to Impact for Exploring Resilient Structure

Simulation Scheme

The fundamental problem to be studied here is the critical effects of impact loading on flat plate as a generic space structure, prior to its collapse, without considering the elasto-plastic surface impact effects. For this purpose, the problem is modeled as the impact of a relatively rigid body of small size to a solid elastic panel. Using method of analysis described in preceding sections, the problem will be simulated numerically using ABAQUS™ finite element computational software. The deformation and stress distribution on the impacted structure due to the impact loading will be considered. Results obtained will be discussed by reference to those obtained in previous sections and work [10].

Two types of panels will be considered, i.e. an elastic aluminum alloy panel and a composite aluminum-titanium panel. Without losing generalities, aluminum alloy and aluminum-titanium composite properties considered for the present study are summarized in Table 27.4.

Simulation Results

To gain insight on how a particular panel structure responds to impact load due to spherical rigid body at a certain velocity normal to it, numerical simulation was performed for panel specifications as given in Figs. 27.14 and 27.15.

By carrying out parametric study, one may obtain useful information for design and operational purposes.

Sample results from the finite element computational simulation using ABAQUS™ are exhibited in Figs. 27.16, 27.17, 27.18, 27.19, and 27.20.

Fig. 27.14 Distribution of maximum principal strain of a 200 × 200 mm steel plate subject to impact by rigid sphere impactor of 20 mm at a velocity of 200 mm/s at 0.15 s after impact

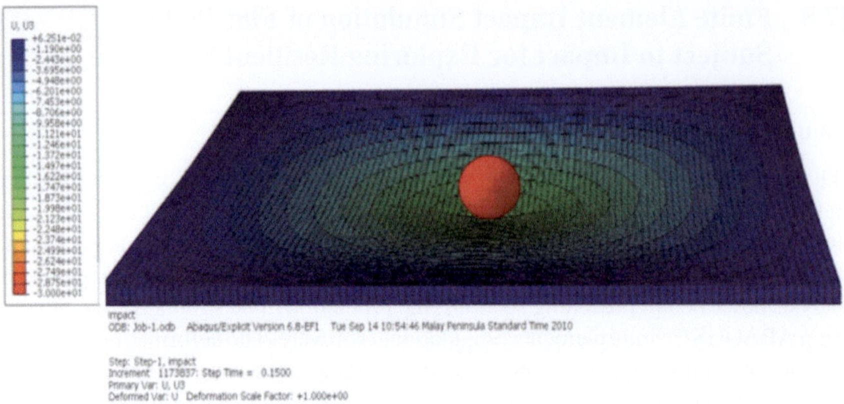

Fig. 27.15 Displacement of a 200 × 200 mm steel plate subject to impact by rigid sphere impactor of 20 mm at a velocity of 200 mm/s at 0.15 s after impact

Figures 27.17, 27.18, 27.19 and 27.20 exhibit the progression of the kinetic energy, strain energy and total energy of the whole system for the aluminum alloy case.

These figures can be validated using the principle that the energy should not exceed the initial energy of the impacting ball which for the present case is equal to ½ $mV^2 = 1/2 * 35 * 2^2 = 70$ mJ.

By carrying out such simulation for different values of relevant parameters, parametric assessment can be made for the search or choice of favorable design specification. The procedure that has been applied in the beam impact case with specific examples can be followed to arrive at some desired objective.

These figures exhibit the desired expected results corresponding to the materials utilized in the numerical simulation:

1. Aluminum alloy has lower maximum von Mises Stress contour compared to steel, and the aluminum-titanium alloy has lower maximum von Mises Stress contour compared to aluminum alloy, for the specifications, conditions and elastic range considered.
2. The strain energy of the whole model in the aluminum-titanium composite case is much lower than that exhibited by aluminum alloy.

These are some of the meaningful results that can be gained from these numerical simulations. Much more significant conclusions can be drawn by further analysis and optimum refinement of the meshes.

27.9 Case Studies Simulation Rationale

To gain insight on how a particular panel structure responds to impact load due to spherical rigid body at a certain velocity normal to it, numerical simulation was

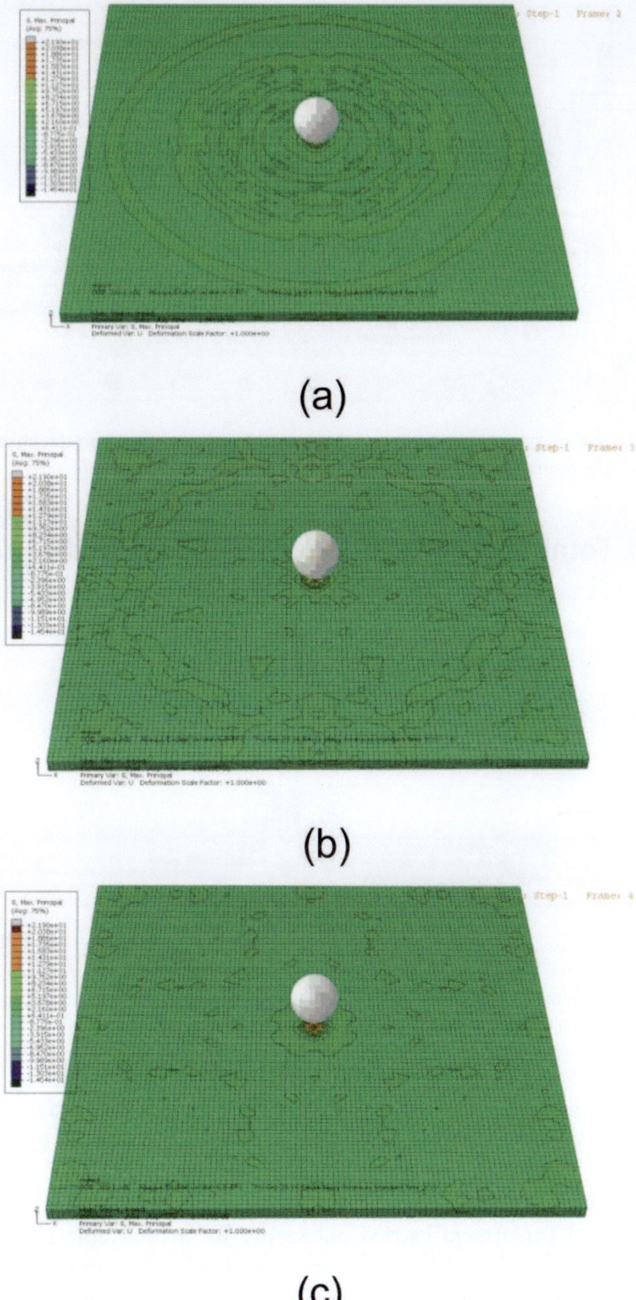

Fig. 27.16 Distribution of maximum principal stress of a 200 × 200 mm aluminum-titanium composite panel with 5 mm thickness subject to impact by rigid sphere impactor of 20 mm diameter at a velocity of 2 m/s, at **a** 0.01 s, **b** 0.015 s and **c** 0.02 s after impact

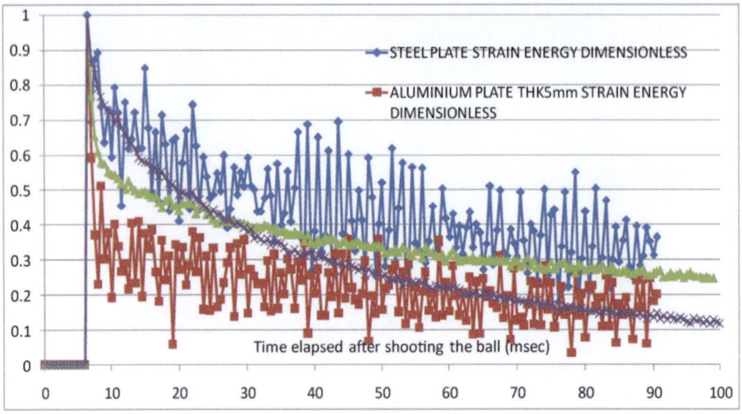

Fig. 27.17 Progression of strain energy of various metal panels with 5 mm thickness subject to impact by rigid sphere impactor of 20 mm diameter at a velocity of 200 mm/s

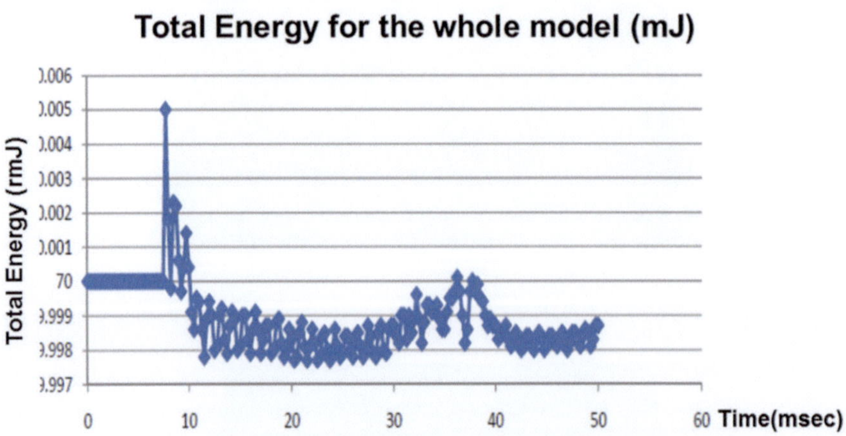

Fig. 27.18 Progression of kinetic energy of aluminum alloy panel with 5 mm thickness subject to impact by rigid sphere impactor of 20 mm diameter at a velocity of 2.0 m/s

performed for some cases. By carrying out parametric study, one may obtain useful information for design and operational purposes. Sample results from the finite element computational simulation using ABAQUS™ and ANSYS™ exemplify the results of certain choice of criteria [11, 12].

In a numerical experiment carried out earlier for a bimetallic aluminum-steel composite with similar dimensions as the other panel, with the more ductile aluminum panel (and having less yield stress limit than steel) facing the impact load, shows that by imposing bonding boundary conditions, the integrity of the composite structure is insured. It seems by using bonding the strength of the composite structure which

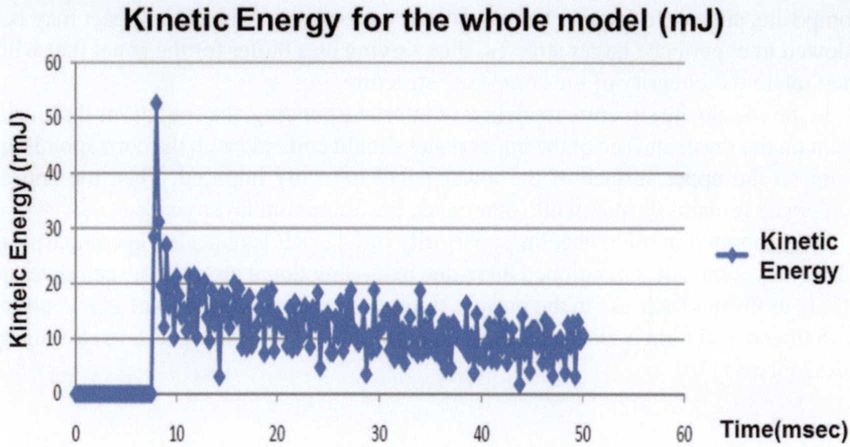

Fig. 27.19 Progression of strain energy of aluminum alloy panel with 5 mm thickness subject to impact by rigid sphere impactor of 20 mm diameter at a velocity of 2.0 m/s

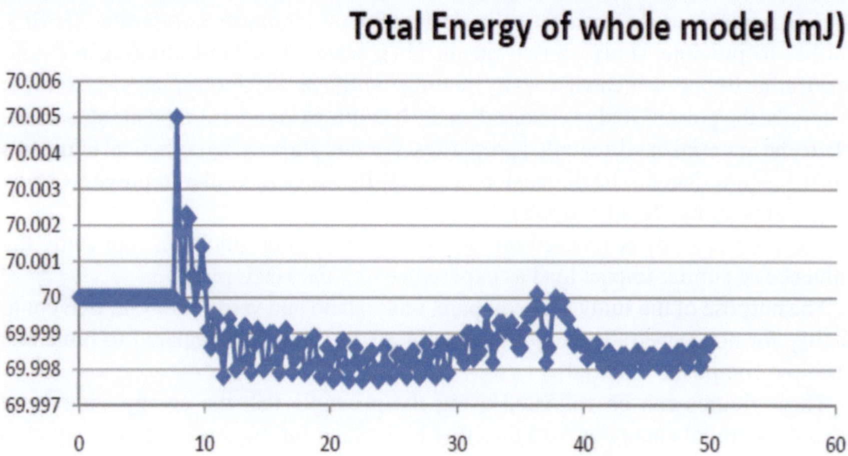

Fig. 27.20 Progression of total energy of the whole model for aluminum alloy panel with 5 mm thickness subject to impact by rigid sphere impactor of 20 mm diameter at a velocity of 2.0 m/s

could be improved. Therefore, the present study is focused on the use of composite materials to device structures which will produce non-penetrating impact.

For this purpose, the following materials will be investigated: Hypothetical metal laminate (HML) based on Alpolic and GLARE 3 3/2. Alpolic was chosen since it was utilized in previous work [11, 12, 23, 24], so that comparison can be made to other materials, in particular GLARE 3.

To produce impact resilient structure, for an isotropic metallic panel, the impact loading should produce stresses that remain in the elastic region. For a bimetallic composite, such criterion can be relaxed; the panel that is subject to impact may be allowed to experience larger stresses, thus serving as a buffer for the panel that will then retain the integrity of the composite structure.

In the composite structure modeling of interface bonding, the conditions that each point on the lower surface of the upper panel should coincide with the corresponding point on the upper surface of the lower panel is tacitly imposed. Then the entire composite remains elastic, while otherwise, the aluminum layer yields.

As elaborated in reference, the superiority of GLARE is indicated in comparison with other composites mentioned there due to its directional properties; such state of affairs motivates their use in the present study. Directional properties of glass/epoxy with fiber orientation is clearly indicated with GLARE to be superior (High Strength Glass Fibers) [16].

27.9.1 Simulation Studies

In what follows, two simulation studies are carried out. The first simulation deals with AlTi (Alpolic) and its hypothetically derived metal laminates composites (HML). Further to previous study carried out in [11], where structural tailoring for non-penetrating impact was carried out by treating multilayered Alpolic plates as Isotropic materials, the present study is focused on the benefits of layered composite structures, where the material has directional properties. For this purpose, hypothetical laminates (HML) is introduced. The dimension of the HML plates is similar to previous study [11], that is 20 by 20 cm and total thickness of 10 mm.

The second study is focused on GLARE with 1.1 mm thickness and study the influence of similar impact load as experienced by the HML plate.

The purpose of the study is to compare, understand and verify how GLARE could qualify for non-penetrating requirements in a superior way compared to bimetalic isotropic composite material.

These figures can be validated using the principle that the energy should not exceed the initial energy of the impacting ball which for the present case is equal to $\frac{1}{2}\,mV^2 = (\frac{1}{2}) \times 35 \times 2^2 = 70$ mJ.

27.9.2 Simulation Studies on Hypothetical Metal Laminate (HML) by Assumed Directional Properties

The plates are bonded together and the solver is set to consider that the bonding is frictionless and unbreakable, with maximum bond offset to be 1e−7 m along the bond. For the mesh, the minimum edge length is 5e−3 m, transition ratio of 0.272 and a growth rate of 1.2. The structural analysis uses AUTODYN explicit dynamics solver. There is no prestress applied to the model. The plates are clamped individually along the four sides of each plate. The impactor ball is placed perpendicularly at the center of the top plates with a small gap such that it will impact the top plate at a velocity of 2 m/s. The impact contact is set to be frictionless. The analysis iteration time is 5e−3 s, with time step safety factor of 0.9 (90%) and the maximum allowable energy error is 0.1 (10%). The solver uses a double precision floating point number for programming side precision control.

Figure 27.21 exhibits the progression of kinetic energy of aluminum alloy panel with 5 mm thickness subject to impact by rigid sphere impactor of 20 mm diameter at a velocity of 2.0 m/s, while Figs. 27.22 and 27.23 related to strain energy and total energy, respectively. Figure 27.24 shows the comparisons of composite materials total energy, internal energy and kinetic energy for HML material which has been tailored into a set of different composite structures. 9 (nine) Alpolic-based HML composite structures and 2 (two) isotropic metal layers have been investigated. The two other plate structures are pure Ti-alloy, and pure Alpolic plate, respectively. The other nine composite structures are composed of multilayered HML plates, each having different polytrophic properties orientations. Here, the Alpolic material is hypothetically treated as having unidirectional properties and the elastic properties in the *x*-direction, while the *y*-direction has been assumed to follow certain rules.

One rule is that the property in the *y*-direction to be proportional to the *x*-direction as reflected in Poisson ratio fashion, i.e. using the following relationship

$$E_{\text{minor}} = \nu E_{\text{major}} \tag{27.49}$$

The other rule follows the following relationship:

$$\frac{1}{E_{\text{minor}}} = \frac{2}{E_{\text{major}}} + \frac{1}{G_{\text{major}}} \tag{27.50}$$

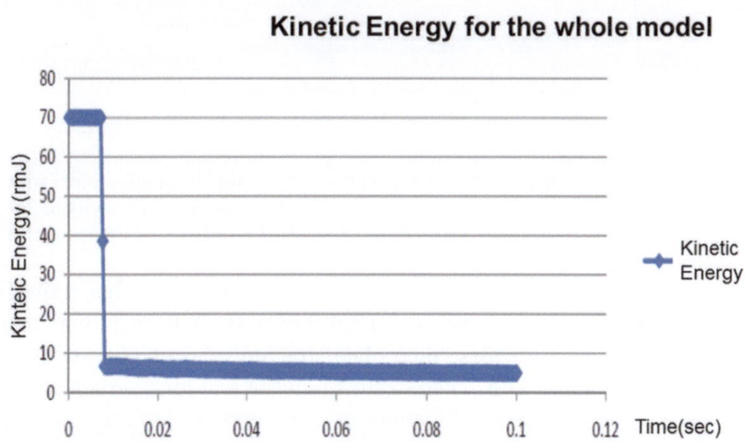

Fig. 27.21 Progression of kinetic energy of aluminum alloy panel with 5 mm thickness subject to impact by rigid sphere impactor of 20 mm diameter at a velocity of 2.0 m/s

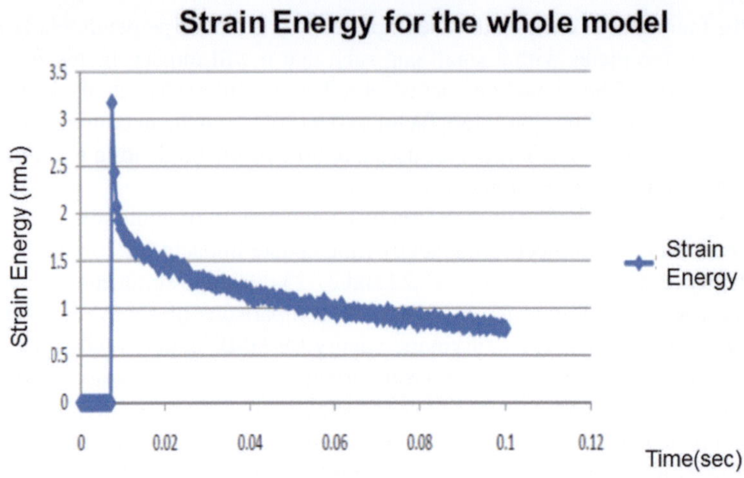

Fig. 27.22 Progression of strain energy of aluminum alloy panel with 5 mm thickness subject to impact by rigid sphere impactor of 20 mm diameter at a velocity of 2.0 m/s

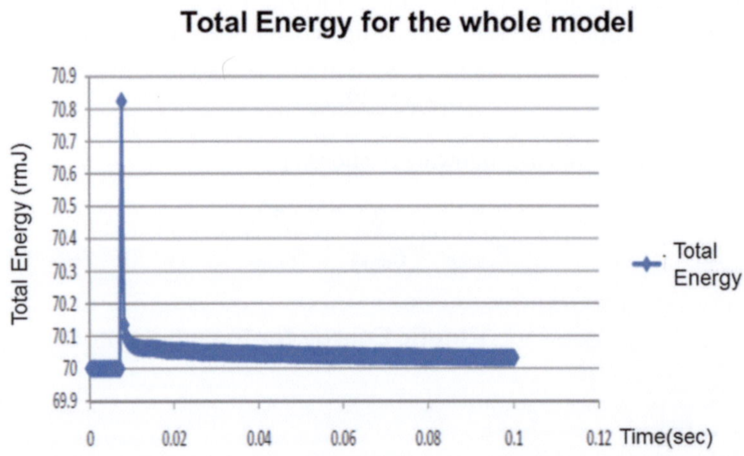

Fig. 27.23 Progression of total energy of the whole model for aluminum alloy panel with 5 mm thickness subject to impact by rigid sphere impactor of 20 mm diameter at a velocity of 2.0 m/s

Both estimates arrive at close value, i.e. 21.574 Gpa and 21.0222 Gpa, respectively.

All these composite plates have same total thickness. With such hypothetically tailored composites, each of these nine plate structures has unique orientation as reflected in Figs. 27.4 and 27.5. These results indicate that for a certain orientation one composite composition exhibits certain superior property. For example, Alpolic-based HML composite 3 layers 0-90-0 has the least maximum stress and Alpolic-based HML composite 3 layers 45-90-135 has the least maximum strain.

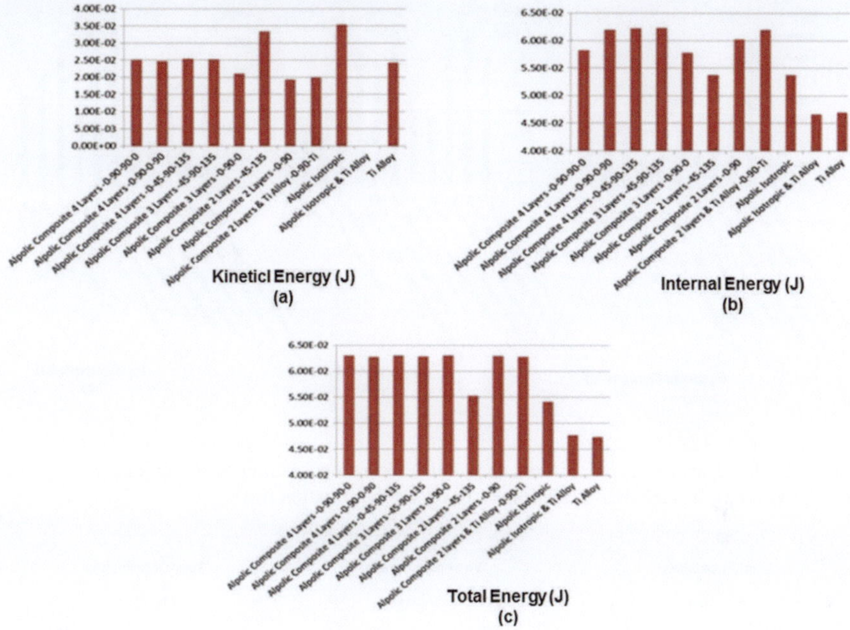

Fig. 27.24 Energy comparison for different composites combinations for a total thickness of 1 cm: **a** kinetic energy (J); **b** internal energy (J); **c** total energy (J)

Figure 27.25 exhibits a comparison of stress, strain, and total deformation for composites combinations: (a) von Mises stress (Pa); (b) von Mises strain (m = m); (c) total deformation (m). This figure also illustrates the maximum von Mises stress, von Mises strain and total deformation for these cases for impact velocity of 10 m/s.

In addition, the maximum von Mises stress, von Mises strain and total deformation are illustrated in Fig. 27.26, using similar calculations for impact velocity of 100 m/s to explore the structural limitations. It was found that the plastic limit occurs between these two cases. Figure 27.27 exhibits the worst and best von Mises stress conditions obtained by numerical simulation for flat and curved plates, which has been carried out to identify the inception of non-penetrating boundary. Other cases have also been worked out. For impact velocities larger than the investigated cases, appropriate design considerations (such as resizing, reinforcement and the like) should be resorted.

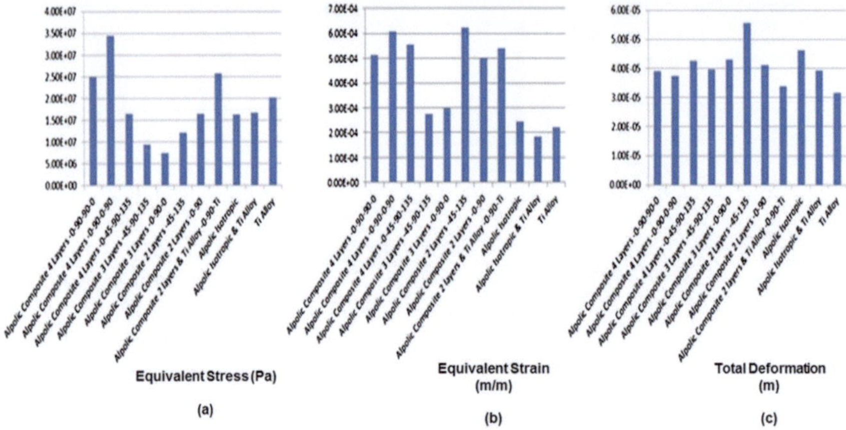

Fig. 27.25 Comparison of stress, strain, and total deformation for composites combinations: **a** von Mises stress (Pa); **b** von Mises strain (m = m); **c** total deformation (m)

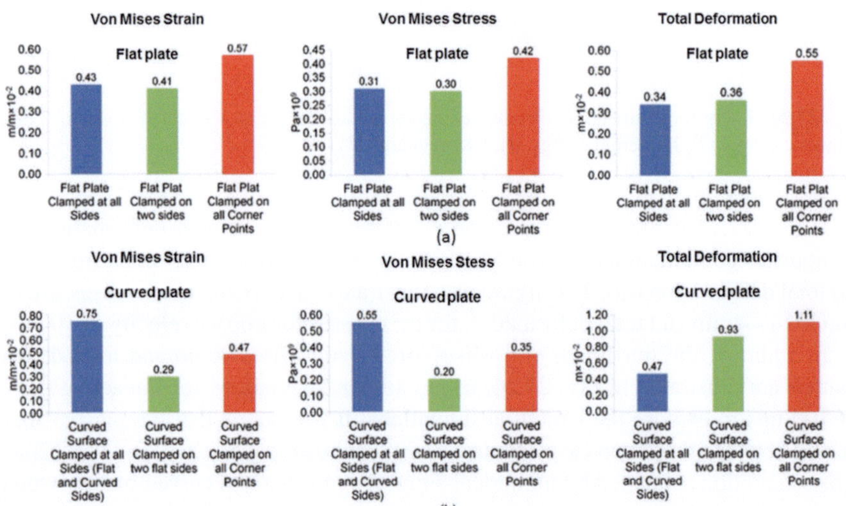

Fig. 27.26 Maximum von Mises stress, maximum von Mises strain, maximum total deformation for plate for all four sides clamped (left bar), two sides clamped (center bar), and four corners clamped (right bar) **a** for flat composite plates; **b** for curved plate (curvature $l = R$ ¼ 0.01, $l =$ width of plate, $R =$ radius of curvature of the plate; the curved plate is curved only in y-direction); impact velocity is 10 m = s

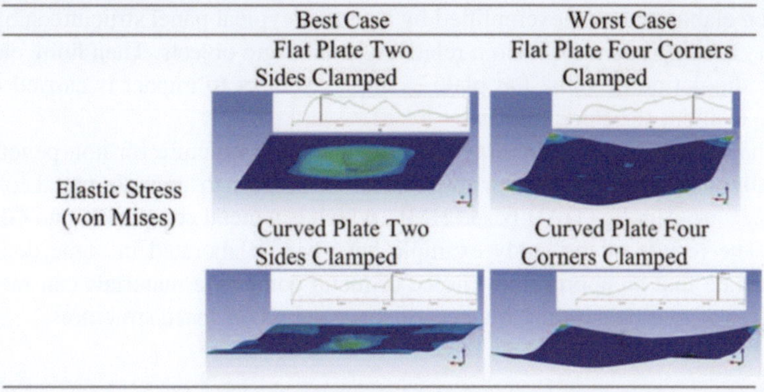

Fig. 27.27 Exhibits the worst and best von Mises stress conditions obtained by numerical simulation for flat and curved plates, which has been carried out to identify the inception of non-penetrating boundary

27.10 Concluding Remarks[1]

A systematic analysis has been elaborated to exemplify the utilization of static and dynamic analysis for exploration of impact load on generic structure, starting from fundamental to more practical cases. A comprehensive scheme for analytical and computational impact study has been summarized in Fig. 27.3 and the logical methodology as summarized in Fig. 27.4 has been utilized to illustrate a comprehensive look on impact problem to a structure for design and operational purposes.

The comprehensive analysis and numerical procedure incorporate component of surface impact analysis of using Hertz elastic contact impact theory and analysis of elastic deformation of material up to plastic limit. Focusing attention on impact loading on elastic panel structure, a computational finite element simulation has been developed and exemplified by a generic typical panel structure subject to arbitrary loading and impact from relatively small rigid objects.

Then finite element impact simulation of some flat plate examples subject to impact is carried out to explore resilient structure tailoring. The simulation exhibited the panel structure response in terms of its deflection, stress distribution and energy absorption which are relevant aspects for design and operational purposes.

As an example of structural tailoring of space structure subject to Impact for non-penetrating case, an analysis has been elaborated using the principles of energy balance (or equivalence) of impact to allow validation and guidelines for the direction of numerical simulation. The example is focused on the impact load on generic structure, which can readily be extended to more practical cases. Focusing attention on impact loading on elastic panel structure, a computational finite element simulation

[1] The author would like to thank Universiti Putra Malaysia (UPM) for granting Research University Grant Scheme (RUGS) No. 9378200, and the Ministry of Higher Education for granting ERGS 5527088 and FRGS 5524250 research funding, under which the present research is carried out.

has been elaborated, and exemplified by a generic typical panel structure subject to arbitrary loading and impact from relatively small rigid objects. Then finite element impact simulation of some flat plate examples subject to impact is carried out to explore resilient structure tailoring.

In the search for and structural tailoring of a plate structure for non-penetrating impact by space debris, particular attention has been given to laminate-metal composites; these are Alpolic (TiAl)-based HML, which is a metal composite, and GLARE 3 3/2. The results of the study example have been elaborated in some details to demonstrate that an appropriate choice of metal composite materials can meet the objectives of non-penetrating impacts of space debris on space structures.

Appendix

Tables related to the properties of materials used in the work of Djojodihardjo and Mahmud [11] are exhibited as Tables 27.4, 27.5, 27.6, 27.7, 27.8, 27.9, 27.10 while computational results on cases considered in [11] on the influence of the plate boundary conditions and curvature on von Mises stress, strain and total deformation is exhibited in Table 27.11.

Table 27.5 Structural steel impactor, treated to be rigid

Density, kg m^{-3}	Young's modulus, Pa	Poisson's ratio	Bulk modulus, Pa	Shear modulus, Pa
7850	2.e+11	0.30	1.6667e+11	7.6923e+10

Table 27.6 2024 Aluminum T3, isotropic elasticity

Density kg m^{-3}	Young's modulus, Pa	Poisson's ratio	Bulk modulus, Pa	Shear modulus, Pa
2780	7.31e+10	0.33	7.1667e+10	2.7481e+10

Table 27.7 The isotropic plate material utilised in the study; for comparative purposes and for simplicity, the Alpolic (TiAl) composite is treated in one study as isotropic

Descriptions	Young's modulus, Pa	Shear modulus	Poisson's ratio
Pure aluminum plate	2.9e8	1.09e8	0.33
Titanium alloy plate	9.6e10	3.53e10	0.36
Alpolic isotropic composite	7.06e10	2.65e10	0.33

Table 27.8 Alpolic composite treated within its hypothetical orthotropic properties. The Poisson's Ratio of 0.33 applied for XY;YZ;XZ plane utilized data from reference (Alpolic, Mitsubishi 2014). (a) Young modulus; (b) Shear modulus

(a)

Fiber orientation hypothetical metal laminate (HML)	Young's modulus *X*-direction, Pa	Young's modulus *Y*-direction, Pa	Young's modulus *Z*-direction, Pa
0°	7.06e10	1.0e10	1.0e10
90°	1.0e10	7.06e+010	1.0e10

(b)

	Shear modulus *XY*, Pa	Shear modulus *YZ*, Pa	Shear modulus *XZ*, Pa
0°	5.0e9	2.65e10	5.0e9
90°	5.0e9	5.0e9	2.65e10

The Poisson's Ratio of 0.33 applied for *XY*;*YZ*;*XZ* planes utilized data from Ref. [3]. (a) Young's modulus; (b) Shear modulus

Table 27.9 Epoxy for GLARE 0°, Orthotropic Elasticity

Density, kg m^{-3}	Young's modulus *X*-direction, Pa	Young's modulus *Y*-direction, Pa	Young's modulus *Z*-direction, Pa
2488	2.1e+10	9.38e+010	2.1e+10
Poisson's ratio *XY-YZ-XZ*	Shear modulus *XY*, Pa	Shear modulus *YZ*, Pa	Shear modulus *XZ*, Pa
0.23	8.76e+9	3.81e+10	8.76e+9

Table 27.10 Epoxy for GLARE 90°, orthotropic elasticity

Density kg m^{-3}	Young's modulus *X*-direction, Pa	Young's modulus *Y*-direction, Pa	Young's modulus *Z*-direction, Pa
2488	2.1e+10	9.38e+010	2.1e+10
Poisson's ratio *XY-YZ-XZ*	Shear modulus *XY*, Pa	Shear modulus *YZ*, Pa	Shear modulus *XZ*, Pa
0.23	8.76e+9	8.76e+9	3.81e+10

Table 27.11 Influence of plate boundary conditions and curvature on von Mises stress, strain and total deformation

Plate impact velocity: 100 m/s elapsed time of computation: 0.001 s	Illustration	von Mises stress (Pa)	von Mises strain (mm)	Total deformation (m)
		Maximum during the computed elapsed time		
Flat plate clamped at all sides	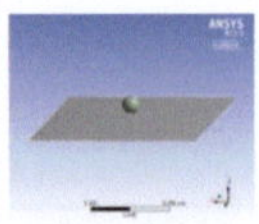	2.505e9	3.5602e−2	1.1593e−2
Flat plat clamped on two sides	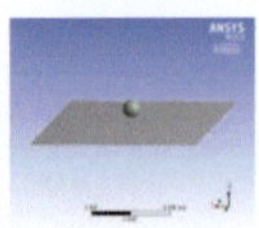	2.4894e9	3.4438e−2	1.2349e−2
Flat plate clamped on all corner points	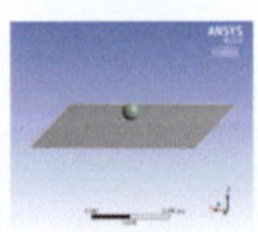	3.9016e9	5.3374e−2	1.9134e−2
Curved plate clamped at all sides (flat and curved sides)	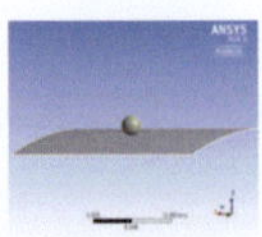	3.2925e9	4.5042e−2	1.3407e−2
Curved plate clamped on two flat sides	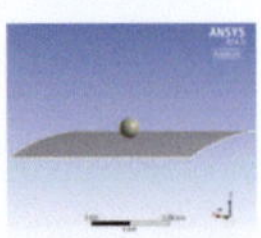	2.2381e9	3.2406e−2	1.9637e−2
Curved plate clamped on all corner points	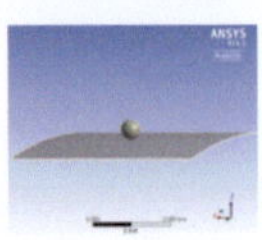	6.6063e9	8.874e−2	2.2372e−2

References

1. Roy, S., and I. Mueller, V. Janapati, S. Das, F.K. Chang. 2012. Real-time prediction of impact-induced damage for composite structures based on failure analysis and efficient database methods. *SPIE Smart Structures and Materials Nondestructive Evaluation and Health Monitoring* 83481L.
2. Mindlin, R.D. 1951. Influence of rotary inertia and shear on flexural motions of isotropic, elastic plates. *Journal of Applied Mechanics* 18: 31–38.
3. Djojodihardjo, H. 2013. Computational simulation for analysis and synthesis of impact resilient structure. *Acta Astronautica* 91: 283–301.
4. Djojodihardjo, H., and A. Shokrani. 2010. Generic study and finite element analysis of impact loading on elastic panel structure, Paper IAC-10.C2.6.2. In *Proceedings, 61st International Astronautical Congress*, 26 September–1 October 2010. Prague: The Czech Republic.
5. Djojodihardjo, H., and A.S. Mahmud. 2014. Finite element computational modeling and simulation studies of non-penetrating impact using fundamental principles. *Applied Mechanics and Materials* 629: 62–70.
6. Abrate, S. 2001. Modeling of impacts on composite structures. *Composite Structures* 51 (2): 129–138.
7. Djojodihardjo, H., and S.L. Kai, IJSE. 2008. Modeling and analysis of FRP car side door impact beam. In *Proceeding of the 2008 International Joint Conference in Engineering—IJSE2008*, August 4–5. Jakarta, Indonesia.
8. Djojodihardjo, H., and P. -M. Ng. 2008. Numerical simulation of impact loading on elastic structure and case studies, Paper IAC-08-C2.6.1. In *Proceedings, 59th International Astronautical Congress*, 29 September–3 October 2008. Glasgow, United Kingdom.
9. Djojodihardjo, H., P. -M. Ng, and S. Lin-Khai. 2009. Analysis and simulation of impact loading on elastic beam structure with case studies. Paper IAC-09-C2.6.1. In *60th International Astronautical Congress*. Daejon: Republic of Korea.
10. Djojodihardjo, H., and H. Jamali, A. Shokrani, F. Mustapha, R. Zahari, S. Wiriadidjaja. 2012. Computational simulation for static and dynamic load of rectangular plate in elastic region for analysis of impact resilient structure. *Applied Mechanics and Materials* 225: 150–157.
11. Rice, J.R. 2010. *Solid Mechanics, School of Engineering and Applied Sciences, and Department Earth and Planetary Sciences*. Cambridge, MA: Harvard University. http://esag.harvard.edu/rice/e0_Solid_Mechanics_94_10.pdf.
12. Anonymous. 2013. High-speed particle impacts suspected in two spacecraft anomalies. *NASA Orbital Debris Quarterly News* 17 (3).
13. Anonymous. (2014, June 10). *Alpolic, Mitsubishi*. http://www.alpolic.com.
14. Vogelesang, L.B., and A. Vlot. 2000. Development of fibre metal laminates for advanced aerospace structures. *Journal of Materials Processing Technology* 103 (1): 1–5.
15. Fisher-Cripps, A.C. 2000. *Introduction to Contact Mechanics*. Springer.
16. Stronge, W.J. 2000. *Impact Mechanics*. Cambridge University Press.
17. Johnson, K.L. 2000. *Contact Mechanics*. Cambridge University Press, August 28,1987.
18. Care, G., and A.C. Fischer-Cripps. 1997. Elastic-plastic indentation stress fields using the finite-element method. *Journal of Materials Science* 32: 5653–5659.
19. Swanson, S.R. 1997. *Introduction to Design Analysis with Advance Composite Materials*. Prentice Hall.
20. Craig, R.R., and A.J. Kurdila. 2005. *Fundamentals of Structural Dynamics*. Wiley Publishers.
21. Weaver, W., and P.R. Johnston. 1987. *Structural Dynamics by Finite Elements*. Prentice-Hall, Inc.
22. Beer, G., and J.O. Watson. 1992. *Introduction to Finite and Boundary Element Methods for Engineers*. Baffins Lane, Chichester, England: John Wiley & Sons Ltd..
23. Felippa, C.A., and R.W. Clough. 1969. *The Finite Element Method in Solid Mechanics*. American Mathematical Society.
24. Botelho, E.C. (2006). Review on the development and prop continuous fibre/epoxy/aluminum hybrid composites for aircraft structures. *Materials Research* 9 (3).

Appendix A
MATLAB Program for Chap. 9 Flutter Calculation Method

```
%%%%%%%%%%%%%%%%%%%%%%%%%%%%%%%%%%%%%%%%%%%%
% VG Method applied to quasi-steady
%%%%%%%%%%%%%%%%%%%%%%%%%%%%%%%%%%%%%%%%%%%%
Pp = [a4,a2,a0];
psq = roots(Pp);
   lambda = psq;
% omega1 = omegaalpha/sqrt(real(lambda(1)));
if real(lambda(1))>0
omega1 = omegaalpha/sqrt(real(lambda(1)));
end
g1 = imag(lambda(1))*omega1*omega1/omegaalpha/omegaalpha;
if real(lambda(2))>0
omega2 = omegaalpha/sqrt(real(lambda(2)));
end
g2 = imag(lambda(2))*omega2*omega2/omegaalpha/omegaalpha;
figure(8),plot(U,omega1,'*',U,omega2,'o');
   xlabel(' U, m/sec ');
   ylabel('frequency omega');
   title('frequency-velocity diagram, K-Steady ');
   grid on
   hold on
   figure(9),plot(U,g1,'s',U,g2,'d');
   xlabel(' U, m/sec ');
   ylabel('non-dimensional damping g');
   title('damping-velocity diagram, K-Steady ');
   grid on
   hold on
   yo1 = omega1/omegaalpha;
   yo2 = omega2/omegaalpha;
   figure(10); plot(yo1,g1,'*',yo2,g2,'o');
  title('Polar Diagram of Imaginary vs Real root (Steady & K-Method,
Polynomial)');
   xlabel('frequency/omegaalpha');
   ylabel('non-dimensional damping g');
   grid on
   hold on
```

© The Editor(s) (if applicable) and The Author(s), under exclusive license to Springer Nature Singapore Pte Ltd. 2023

H. Djojodihardjo, *Introduction to Aeroelasticity*,
https://doi.org/10.1007/978-981-16-8078-6

```
%%%%%%%%%%%%%%%%%%%%%%%%%%%%%%%%%%%%%%%%%%%%%%%%%%%
end
toc
  r = roots(P);
  lambda = roots(P);
% r is lambda
  y1=imag(r(1))/omegaalpha;
  y2=imag(r(2))/omegaalpha;
  y3=imag(r(3))/omegaalpha;
  y4=imag(r(4))/omegaalpha;
  y5=real(r(1))/omegaalpha;
  y6=real(r(2))/omegaalpha;
  y7=real(r(3))/omegaalpha;
  y8=real(r(4))/omegaalpha;
omega1 = omegaalpha/sqrt(real(lambda(1)));
g1 = imag(lambda(1))*omega1*omega1/omegaalpha/omegaalpha;
omega2 = omegaalpha/sqrt(real(lambda(2)));
g2 = imag(lambda(2))*omega2*omega2/omegaalpha/omegaalpha;
omega3 = omegaalpha/sqrt(real(lambda(3)));
g3 = imag(lambda(3))*omega1*omega3/omegaalpha/omegaalpha;
omega4 = omegaalpha/sqrt(real(lambda(4)));
g4 = imag(lambda(4))*omega2*omega4/omegaalpha/omegaalpha;
     if r(1) < 0
       omega1 = 0;
       g1 = 0;
     end
     if r(2) < 0
       omega2 = 0;
       g2 = 0;
     end
     if r(3) < 0
       omega3 = 0;
       g3 = 0;
     end
     if r(4) < 0
       omega4 = 0;
       g4 = 0;
     end
%  FLUTTER  STABILITY-TYPICAL  SECTION-steady-low  frequency-
Theodorsen-AGARD-K
% -for all-23 June 07-BAH-Variation-2
%PROGRAM TO DETERMINE THE FLUTTER STABILITY OF THE TYPICAL SECTION:
%STRAIGHT WING CASE:
% Data adapted from YHH Boeing 747-like wing modified using
Zwaan/Isogai and BAH Wing in the absence
% of data
% Configurations:
% VG/K-Method applied to all for flutter solution
% part one : for steady case
% part two : for low frequency refinement
% part three : Theodorsen aerodynamics
% part four : solution of flutter equation using K-Method - polyno-
mial solution
% part five : solution of flutter equation using
```

```
% K-Method - eigenvalue problem
% Can further be developed for Done's Flutter Stability Analysis
% Developed based on Done's Flutter Stability Analysis
% Data based on Zwaan nomenclature
% Data taken from BAH Wing
% mass is based on m (which is defined), and myu is a derived quantity
% Developed based on Done's Flutter Stability Analysis
% Data based on Zwaan nomenclature
% Data taken from BAH Wing
% mass is based on m (which is defined), and myu is a derived quantity
tic;
% To determine the range of the velocity in m/sec in the flutter
stability
% doiagram
% Vinit - origin of diagram absis
% Vfinal - end of diagram absis
% interval - increment of velocity
Vinit = 100;
Vfinal = 600;
interval = 25;
% m = 0.0161; % BAH;
myu = 60; % NACA 0012
%%%%%%%%%%%%%%%%%%%%%%%%%%%%%%%%%%%%%%%%%%%
x_alpha = 0.2; %BAH
r_alpha = 0.5; % BAH
c = 5.715;%BAH;
% c = 2.0;
S = 12.7;
b = c/2;
% Br = c/2;
Br = c/2;
% e = 0.25;% case study 1;
e = 0.75;%case study 2;
% e = -0.2;%case study 3;
x_AC = -0.5;
a = e-x_AC;
rho = 1.225;
% myu = m/pi/rho/b/b; % BAH;
m = myu*pi*rho*b*b;
% Cl_alpha = 0.1788;
Cl_alpha = 2*pi;
y = 8.89;
EI = 1.94e+008;
GJ = 2.69e+008;
K_alpha = GJ/y;
Kh = 3*EI/(y*y*y);
omegaalpha = 22.3570;
omegah =12.8000;
% to get the value of S_alpha and I_alpha:
S_alpha = m*x_alpha*Br;
I_alpha = m*r_alpha*r_alpha*Br*Br;
A=zeros(2);
%%%%%%%%%%%%%%%%%%%%%%%%%%%%%%%%%%%%%%%%%%%
omega1 = 0.0;
```

```
omega2 = 0.0;
% to determine the flutter roots:
% ******************************************************
% Part One
% Steady / Quasi Steady Aerodynamic Model
% ******************************************************
% for k =0.1:0.02:2
% V = omegaalpha*b/k;
for V=Vinit:interval:Vfinal
   q = 0.5*rho*V*V;
   Ur = V/omegaalpha/b;
   k = omegaalpha*b/V;
   a4 = (I_alpha/S_alpha/b)-(S_alpha/(myu*pi*rho*b^3));
        a2  =  (I_alpha*Kh/(myu*pi*rho*b^2)+K_alpha)/b/S_alpha-
q*S*Cl_alpha*(1+2*e/x_alpha)/(myu*pi*rho*b^3);
   a0 = Kh*(K_alpha-(2*q*S*e*b*Cl_alpha))/m/b/S_alpha;
   P = [a4,0,a2,0,a0];
   r = roots(P);
   U = V;
   y1=imag(r(1))/omegaalpha;
   y2=imag(r(2))/omegaalpha;
   y3=imag(r(3))/omegaalpha;
   y4=imag(r(4))/omegaalpha;
   y5=real(r(1))/omegaalpha;
   y6=real(r(2))/omegaalpha;
   y7=real(r(3))/omegaalpha;
   y8=real(r(4))/omegaalpha;
      if y1 < 0
         y1 = 0;
         y5 = 0;
      end
      if y2 < 0
         y2 = 0;
         y6 = 0;
      end
      if y3 < 0
         y3 = 0;
         y7 = 0;
      end
      if y4 < 0
         y4 = 0;
         y8 = 0;
      end
   figure(1);
   plot(q,y1,'*',q,y2,'o',q,y3,'d',q,y4,'s');
    title('Imaginary root vs dynamic pressure (Steady Aerodynamic
Model)');
   xlabel('Dynamic pressure, p, N-m/m2');
   ylabel('omega/omegaalpha');
   grid on
   hold on
   figure(2);
   plot(q,y5,'*',q,y6,'o',q,y7,'d',q,y8,'s');
```

```
      title('Real root vs dynamic pressure (Steady Aerodynamic
Model)');
   xlabel('Dynamic pressure, p, N-m/m2');
   ylabel('sigma/omegaalpha');
   grid on
   hold on
   figure(3);
   plot(Ur,y1,'*',Ur,y2,'o',Ur,y3,'d',Ur,y4,'s');
    title('Imaginary root vs Reduced Velocity (Steady Aerodynamic
Model)');
   xlabel('Reduced Velocity, Ur');
   ylabel('omega/omegaalpha');
   grid on
   hold on
   figure(4);
   plot(Ur,y5,'*',Ur,y6,'o',Ur,y7,'d',Ur,y8,'s');
      title('Real root vs Reduced Velocity (Steady Aerodynamic
Model)');
   xlabel('Reduced Velocity, Ur');
   ylabel('sigma/omegaalpha');
   grid on
   hold on
   figure(5);
   plot(U,y1,'*',U,y2,'o',U,y3,'d',U,y4,'s');
   title('Imaginary root vs Velocity (Steady Aerodynamic Model)');
   xlabel('Velocity, U m/sec');
   ylabel('omega/omegaalpha');
   grid on
   hold on
   figure(6);
   plot(U,y5,'*',U,y6,'o',U,y7,'d',U,y8,'s');
      title('Real root vs Reduced Velocity (Steady Aerodynamic
Model)');
   xlabel('Velocity, U m/sec');
   ylabel('sigma/omegaalpha');
   grid on
   hold on
   figure(7); plot(y1,y5,'*',y2,y6,'o',y3,y7,'d',y4,y8,'s');
  title('Polar Diagram of Imaginary vs Real root (Steady Aerodynamic
Model)');
   xlabel('frequency/omegaalpha');
   ylabel('sigma/omegaalpha');
   grid on
   hold on
%%%%%%%%%%%%%%%%%%%%%%%%%%%%%%%%%%%%%%%%%%%%%%%%%%%%%%%%%%%%%%%%%%%%%%%%
% VG Method applied to quasi-steady
%%%%%%%%%%%%%%%%%%%%%%%%%%%%%%%%%%%%%%%%%%%%%%%%%%%%%%%%%%%%%%%%%%%%%%%%
Pp = [a4,a2,a0];
psq = roots(Pp);
   lambda = psq;
% omega1 = omegaalpha/sqrt(real(lambda(1)));
if real(lambda(1))>0
omega1 = omegaalpha/sqrt(real(lambda(1)));
end
```

```
g1 = imag(lambda(1))*omega1*omega1/omegaalpha/omegaalpha;
if real(lambda(2))>0
omega2 = omegaalpha/sqrt(real(lambda(2)));
end
g2 = imag(lambda(2))*omega2*omega2/omegaalpha/omegaalpha;
figure(8),plot(U,omega1,'*',U,omega2,'o');
   xlabel(' U, m/sec ');
   ylabel('frequency omega');
   title('frequency-velocity diagram, K-Steady ');
   grid on
   hold on
   figure(9),plot(U,g1,'s',U,g2,'d');
   xlabel(' U, m/sec ');
   ylabel('non-dimensional damping g');
   title('damping-velocity diagram, K-Steady ');
   grid on
   hold on
   yo1 = omega1/omegaalpha;
   yo2 = omega2/omegaalpha;
   figure(10); plot(yo1,g1,'*',yo2,g2,'o');
  title('Polar Diagram of Imaginary vs Real root (Steady & K-Method,
Polynomial)');
   xlabel('frequency/omegaalpha');
   ylabel('non-dimensional damping g');
   grid on
   hold on
%%%%%%%%%%%%%%%%%%%%%%%%%%%%%%%%%%%%%%%%%%%%%
end
toc
%PROGRAM TO DETERMINE THE FLUTTER STABILITY OF THE TYPICAL SECTION:
%STRAIGHT WING CASE:
% Data adapted from YHH Boeing 747-like wing modified using
Zwaan/Isogai and BAH Wing in the absence
% of data
% Configurations:
% VG/K-Method applied to all for flutter solution
% part one: for steady case
% part two: for low frequency refinement
% part three: Theodorsen aerodynamics
% part four: solution of flutter equation using K-Method - polyno-
mial solution
% part five: solution of flutter equation using
% K-Method - eigenvalue problem
% Can further be developed for Done's Flutter Stability Analysis
% Developed based on Done's Flutter Stability Analysis
% Data based on Zwaan nomenclature
% Data taken from BAH Wing
% mass is based on m (which is defined), and myu is a derived quantity
% Developed based on Done's Flutter Stability Analysis
% Data based on Zwaan nomenclature
% Data taken from BAH Wing
% mass is based on m (which is defined), and myu is a derived quantity
tic;
```

```
% To determine the range of the velocity in m/sec in the flutter
stability
% doiagram
% Vinit - origin of diagram absis
% Vfinal - end of diagram absis
% interval - increment of velocity
Vinit = 50;
Vfinal = 600;
interval = 50;
myu= 100;
%m = 0.0161;
x_alpha = 0.00000974;
r_alpha = 0.21461;
% c is taken from 0.7 span BAH wing
c = 4.92;
b = c/2;
a = -2.0;
%BAH span
Span = 500*0.0254;
% Span = 45.0;
S = c;
e = 0.51;
rho = 1.225;
m=myu*22/7*rho*b^2;
Cl_alpha =2*22/7;
% omegaalpha is taken from BAH wing 1st torsional mode
omegaalpha=22.357;
omegah=12.8;
y =0.75*Span;
EI = 154494478;
GJ = 151565673;
% K_alpha = GJ/y;
I_alpha=811.34131;
K_alpha= I_alpha*omegaalpha^2;
% Kh = 3*EI/(y*y*y);
% Kh = 125746;
Kh=m*omegah^2;
% to get the value of S_alpha and I_alpha:
S_alpha = m*x_alpha*b;
% I_alpha = m*r_alpha*r_alpha*Br*Br;
% to determine the flutter roots:
A=zeros(2);
% to determine the flutter roots:
% ***********************************************************
% Part One
% Steady / Quasi Steady Aerodynamic Model
% ***********************************************************
% for k =0.1:0.02:2
% V = omegaalpha*b/k;
for V=Vinit:interval:Vfinal
   q = 0.5*rho*V*V;
   Ur = V/omegaalpha/b;
   k = omegaalpha*b/V;
   a4 = (I_alpha*m)-(S_alpha*S_alpha);
```

```
a2 = (m*K_alpha)+(I_alpha*Kh)-((2*m*e*b)+S_alpha)*q*S*Cl_alpha;
 a0 = Kh*(K_alpha-(2*q*S*e*b*Cl_alpha));
 % z = (a2*a2)-4*a4*a0;
 P = [a4,0,a2,0,a0];
 r = roots(P);
 U = V;
 y1=imag(r(1))/omegaalpha;
 y2=imag(r(2))/omegaalpha;
 y3=imag(r(3))/omegaalpha;
 y4=imag(r(4))/omegaalpha;
 y5=real(r(1))/omegaalpha;
 y6=real(r(2))/omegaalpha;
 y7=real(r(3))/omegaalpha;
 y8=real(r(4))/omegaalpha;
    if y1 < 0
       y1 = 0;
       y5 = 0;
    end
    if y2 < 0
       y2 = 0;
       y6 = 0;
    end
    if y3 < 0
       y3 = 0;
       y7 = 0;
    end
    if y4 < 0
       y4 = 0;
       y8 = 0;
    end
 figure(1);
 plot(q,y1,'*',q,y2,'o',q,y3,'d',q,y4,'s');
  title('Imaginary root vs dynamic pressure (Steady Aerodynamic
Model)');
 xlabel('Dynamic pressure, p, N-m/m2');
 ylabel('omega/omegaalpha');
 grid on
 hold on
 figure(2);
 plot(q,y5,'*',q,y6,'o',q,y7,'d',q,y8,'s');
    title('Real root vs dynamic pressure (Steady Aerodynamic
Model)');
 xlabel('Dynamic pressure, p, N-m/m2');
 ylabel('sigma/omegaalpha');
 grid on
 hold on
 figure(3);
 plot(Ur,y1,'*',Ur,y2,'o',Ur,y3,'d',Ur,y4,'s');
  title('Imaginary root vs Reduced Velocity (Steady Aerodynamic
Model)');
 xlabel('Reduced Frequency, kr');
 ylabel('omega/omegaalpha');
 grid on
 hold on
```

```
   figure(4);
   plot(Ur,y5,'*',Ur,y6,'o',Ur,y7,'d',Ur,y8,'s');
      title('Real root vs Reduced Velocity (Steady Aerodynamic
Model)');
   xlabel('Reduced Frequency, kr');
   ylabel('sigma/omegaalpha');
   grid on
   hold on
   figure(5);
   plot(U,y1,'*',U,y2,'o',U,y3,'d',U,y4,'s');
   title('Imaginary root vs Velocity (Steady Aerodynamic Model)');
   xlabel('Velocity, U m/sec');
   ylabel('omega/omegaalpha');
   grid on
   hold on
   figure(6);
   plot(U,y5,'*',U,y6,'o',U,y7,'d',U,y8,'s');
      title('Real root vs Reduced Velocity (Steady Aerodynamic
Model)');
   xlabel('Velocity, U m/sec');
   ylabel('sigma/omegaalpha');
   grid on
   hold on
   figure(7); plot(y5,y1,'*',y6,y2,'o',y7,y3,'d',y8,y4,'s');
  title('Polar Diagram of Imaginary vs Real root (Steady Aerodynamic
Model)');
   ylabel('frequency/omegaalpha2');
   xlabel('sigma/omegaalpha');
   grid on
   hold on
%%%%%%%%%%%%%%%%%%%%%%%%%%%%%%%%%%%%%%%
% VG Method applied to quasi-steady
%%%%%%%%%%%%%%%%%%%%%%%%%%%%%%%%%%%%%%%
   psq = roots(P);
   lambda = psq;
omega1 = omegaalpha/sqrt(real(lambda(1)));
if real(lambda(1))>0
omega1 = omegaalpha/sqrt(real(lambda(1)));
end
g1 = imag(lambda(1))*omega1*omega1/omegaalpha/omegaalpha;
omega2 = omegaalpha/sqrt(real(lambda(2)));
if real(lambda(2))>0
omega2 = omegaalpha/sqrt(real(lambda(2)));
end
g2 = imag(lambda(2))*omega2*omega2/omegaalpha/omegaalpha;
figure(8),plot(U,omega1,'*',U,omega2,'o');
   xlabel(' U, m/sec ');
   ylabel('frequency omega');
   title('frequency-velocity diagram, K-Steady ');
   grid on
   hold on
   figure(9),plot(U,g1,'s',U,g2,'d');
   xlabel(' U, m/sec ');
   ylabel('damping g');
```

```
   title('damping-velocity diagram, K-Steady ');
   grid on
   hold on
   figure(10); plot(g1,omega1,'*',g2,omega2,'o');
  title('Polar Diagram of Imaginary vs Real root (Steady & K-Method,
Polynomial)');
   ylabel('frequency/omegaalpha2');
   xlabel('damping g');
   grid on
   hold on
%%%%%%%%%%%%%%%%%%%%%%%%%%%%%%%%%%%%%%%%%%%%%
end
toc
tic;
% for k =0.1:0.02:2
% V = omegaalpha*b/k;
for V = Vinit:interval:Vfinal
   q = 0.5*rho*V*V;
   Ur = V/omegaalpha/b;
   k = omegaalpha*b/V;
   U = V;
%**************************************************************
% Part Two
% low frequency refinement
%**************************************************************
   a4 = (I_alpha*m)-(S_alpha*S_alpha);
   a3 = (q/V)*S*Cl_alpha*((2*e*b*S_alpha)+I_alpha);
  a2 = (m*K_alpha)+(I_alpha*Kh)-((2*m*e*b)+S_alpha)*q*S*Cl_alpha;
   a1 = (q/V)*S*Cl_alpha*K_alpha;
   a0 = Kh*(K_alpha-(2*q*S*e*b*Cl_alpha));
   P = [a4,a3,a2,a1,a0];
      if y1 < 0
         y1 = 0;
         y5 = 0;
      end
      if y2 < 0
         y2 = 0;
         y6 = 0;
      end
      if y3 < 0
         y3 = 0;
         y7 = 0;
      end
      if y4 < 0
         y4 = 0;
y8 = 0;
      end
   figure(11);
   plot(q,y1,'*',q,y2,'o',q,y3,'d',q,y4,'s');
   title('Imaginary root vs dynamic pressure (Low Frequency Aerody-
namic Model)');
   xlabel('Dynamic pressure, p, N-m/m2');
   ylabel('omega/omegaalpha');
   grid on
```

```
   hold on
   figure(12);
   plot(q,y5,'*',q,y6,'o',q,y7,'d',q,y8,'s');
   title('Real root vs dynamic pressure (Low Frequency Aerodynamic
Model)');
   xlabel('Dynamic pressure, p, N-m/m2');
   ylabel('sigma/omegaalpha');
   grid on
   hold on
   figure(13);
   plot(Ur,y1,'*',Ur,y2,'o',Ur,y3,'d',Ur,y4,'s');
   title('Imaginary root vs Reduced Velocity (Low Frequency Aerody-
namic Model)');
   xlabel('Reduced Frequency, kr');
   ylabel('omega/omegaalpha');
   grid on
   hold on
   figure(14);
   plot(Ur,y5,'*',Ur,y6,'o',Ur,y7,'d',Ur,y8,'s');
   title('Real root vs Reduced Velocity (Low Frequency Aerodynamic
Model)');
   xlabel('Reduced Frequency, kr');
   ylabel('sigma/omegaalpha');
   grid on
   hold on
   figure(15);
   plot(U,y1,'*',U,y2,'o',U,y3,'d',U,y4,'s');
    title('Imaginary root vs Velocity (Low Frequency Aerodynamic
Model)');
   xlabel('Velocity, U m/sec');
   ylabel('omega/omegaalpha');
   grid on
   hold on
   figure(16);
   plot(U,y5,'*',U,y6,'o',U,y7,'d',U,y8,'s');
   title('Real root vs Reduced Velocity (Low Frequency Aerodynamic
Model)');
   xlabel('Velocity, U m/sec');
   ylabel('sigma/omegaalpha');
   grid on
   hold on
   figure(17); plot(y5,y1,'*',y6,y2,'o',y7,y3,'d',y8,y4,'s');
    title('Polar Diagram of Imaginary vs Real root (Low Frequency
Aerodynamic Model)');
   ylabel('frequency/omegaalpha2');
   xlabel('sigma/omegaalpha');
   grid on
   hold on
%%%%%%%%%%%%%%%%%%%%%%%%%%%%%%%%%%%%%%%%%%%%%%%
%% VG-Method applied to low Frequency
%% Refinement
%%%%%%%%%%%%%%%%%%%%%%%%%%%%%%%%%%%%%%%%%%%%%%%
   psq = roots(P);
   lambda = psq;
```

```
if real(lambda(1))>0
omega1 = omegaalpha/sqrt(real(lambda(1)));
end
g1 = imag(lambda(1))*omega1*omega1/omegaalpha/omegaalpha;
if real(lambda(2))>0
omega2 = omegaalpha/sqrt(real(lambda(2)));
end
g2 = imag(lambda(2))*omega2*omega2/omegaalpha/omegaalpha;
% U1 = omega1*b/k;
% U2 = omega2*b/k;
   figure(18),plot(U,omega1,'*',U,omega2,'o');
   xlabel(' U, m/sec ');
   ylabel('frequency omega');
   title('frequency-velocity diagram, K-Low Frequency ');
   grid on
   hold on
   figure(19),plot(U,g1,'s',U,g2,'d');
   xlabel(' U, m/sec ');
   ylabel('damping g');
   title('damping-velocity diagram, K-Low frequency ');
   grid on
   hold on
   figure(20); plot(g1,omega1,'*',g2,omega2,'o');
    title('Polar Diagram of Imaginary vs Real root (Low Frequency &
K-Method, Polynomial)');
   ylabel('frequency/omegaalpha2');
   xlabel('damping g');
   grid on
   hold on
%%%%%%%%%%%%%%%%%%%%%%%%%%%%%%%%%%%%%%%%%%%
end
toc
tic;
%**********************************************************
% Part Three
% Theodorsen Aerodynamics
%**********************************************************
A=zeros(2);
% for k =0.1:0.02:2
% V = omegaalpha*b/k;
for V = Vinit:interval:Vfinal
      q = 0.5*rho*V*V;
      Ur = V/omegaalpha/b;
k = omegaalpha*b/V;
% Solution of Flutter Stability Equation
% k = b*omegaalpha/U;
      kr = 1/k;
      im=sqrt(-1);
      U=kr*b*omegaalpha;
      q = 0.5*rho*V*V;
   Ur=V/omegaalpha/b;
%%%% THEODORSEN FUNCTION%%%%%%%%%%%%%%%%%%%%%%%%%%%
   H1 = besselh(1,2,k);
   H2 = besselh(0,2,k);
```

```
  Ck = H1/(H1+im*H2);
  Fk = real(Ck);
  Gk = imag(Ck);
%%%%%%%%%%%%%%%%%%%%%%%%%%%%%%%%%%%%%%%%%%%%%%
%Calculation of non-dimensional unsteady aerodynamic coefficients
due to
%heaving and pitching following Kuessner
%%%%%%%%%%%%%%%%%%%%%%%%%%%%%%%%%%%%%%%%%%%%%%
  ka = -2*k*Gk+im*2*k*Fk;
  kb = (-2*Fk-2*k*Gk-k*k/2)+im*(2*k*Fk+2*Gk+k);
  ma = -k*k/2;
  mb = -3*k*k/8+im*k;
%%%%%%%%%%%%%%%%%%%%%%%%%%%%%%%%%%%%%%%%%%%%%%
%%%%%Calculate each term of the Flutter Equations Matrix expressed
in
%%%%%generalized quantities
%%%%%%%%%%%%%%%%%%%%%%%%%%%%%%%%%%%%%%%%%%%%%%
  A(1,1) =(omegaalpha/omegah)^2*(myu-ka/k/k);
              A(1,2)    =(omegaalpha/omegah)^2*(myu*x_alpha-
kb/k/k+(0.5+a)*ka/k/k);
  A(2,1) = myu*x_alpha-ma/k/k+(0.5+a)*ka/k/k;
   A(2,2) = myu*r_alpha*r_alpha-mb/k/k+(0.5+a)*(ma/k/k+kb/k/k)-
(0.5+a)^2*ka/k/k;
%%%%%%%%%%%%%%%%%%%%%%%%%%%%%%%%%%%%%%%%%%%%%%
%%%%%%%%Solve  eigenvalue  problem  [A(k)]{x}='lambda'{x}  using
polynomial solution
%%%%%%%%calculate omega1 and U=omega*b/k
%%%%%%%%%%%%%%%%%%%%%%%%%%%%%%%%%%%%%%%%%%%%%%
% p=[1 -(A(1,1)+A(2,2)) A(1,1)*A(2,2)-A(1,2)*A(2,1)];
% lambda =roots(p);
  %P = [1,0,-(A(1,1)+A(2,2)),0,A(1,1)*A(2,2)-A(1,2)*A(2,1)];
  %r = roots(P);
  % lambda = r;
%*********************************************************
  P = [1,-(A(1,1)+A(2,2)),A(1,1)*A(2,2)-A(1,2)*A(2,1)];
  psq = roots(P);
  lambda = psq;
  psq1 =real(psq(1))+imag(psq(1));
  psq2 =real(psq(2))+imag(psq(2));
  p1 =sqrt(psq1);
  p2 =-sqrt(psq1);
  p3 =sqrt(psq2);
  p4 =-sqrt(psq2);
  y1=imag(p1);
  if y1<0
    y1=imag(p2);
  end
  y2=imag(p3);
  if y2<0
    y2=imag(p4);
  end
  sigma1=real(p1)/omegaalpha;
  sigma2=real(p2)/omegaalpha;
  sigma3=real(p3)/omegaalpha;
```

```
   sigma4=real(p4)/omegaalpha;
   omega1=y1/omegaalpha;
   omega2=y2/omegaalpha;
% omega3=imag(r(3))/omegaalpha;
% omega4=imag(r(4))/omegaalpha;
% y1=omega1;
% if y1<0
% y1=omega2;
% end
% y2=omega3;
% if y2<0
% y2=omega4;
% end
%***********************************************************
   %sigma1=real(r(1))/omegaalpha;
   %sigma2=real(r(2))/omegaalpha;
   %sigma3=real(r(3))/omegaalpha;
   %sigma4=real(r(4))/omegaalpha;
   %omega1=imag(r(1))/omegaalpha;
   %omega2=imag(r(2))/omegaalpha;
   %omega3=imag(r(3))/omegaalpha;
   %omega4=imag(r(4))/omegaalpha;
   %y1=omega1;
   %if y1<0
   % y1=omega2;
   %end
   %y2=omega3;
   %if y2<0
   % y2=omega4;
   %end
   y5=sigma1;
   y6=sigma2;
   y7=sigma3;
   y8=sigma4;
   figure(21);
   plot(q,y1,'*',q,y2,'o');
  title('omega/omegaalpha vs dynamic pressure (Theodorsen Unsteady
Aerodynamic Model)');
   xlabel('Dynamic pressure, p, N-m/m2');
   ylabel('omega/omegaalpha');
   grid on
   hold on
   figure(22);
   plot(q,y5,'*',q,y6,'o',q,y7,'d',q,y8,'s');
  title('sigma/omegaalpha vs dynamic pressure (Theodorsen Unsteady
Aerodynamic Model)');
   xlabel('Dynamic pressure, p, N-m/m2');
   ylabel('sigma/omegaalpha');
   grid on
   hold on
   figure(23);
   plot(Ur,y1,'*',Ur,y2,'o');
  title('omega/omegaalpha vs Reduced Velocity (Theodorsen Unsteady
Aerodynamic Model)');
```

```
   xlabel('Reduced Frequency, kr');
   ylabel('omega/omegaalpha');
   grid on
   hold on
   figure(24);
   plot(Ur,y5,'*',Ur,y6,'o',Ur,y7,'d',Ur,y8,'s');
  title('sigma/omegaalpha vs Reduced Velocity (Theodorsen Unsteady
Aerodynamic Model)');
   xlabel('Reduced Frequency, kr');
   ylabel('sigma/omegaalpha');
   grid on
   hold on
   figure(25);
   plot(U,y1,'*',U,y2,'o');
   title('omega/omegaalpha vs Velocity (Theodorsen Unsteady Aero-
dynamic Model)');
   xlabel('Velocity, U, m/sec');
   ylabel('omega/omegaalpha');
   grid on
   hold on
   figure(26);
   plot(U,y5,'*',U,y6,'o',U,y7,'d',U,y8,'s');
   title('sigma/omegaalpha vs Velocity (Theodorsen Unsteady Aero-
dynamic Model)');
   xlabel('Velocity, U, m/sec');
   ylabel('sigma/omegaalpha');
   grid on
   hold on
   figure(27); plot(y5,y1,'*',y6,y2,'o',y7,y3,'d',y8,y4,'s');
   title('Polar Diagram of Imaginary vs Real root (Theodorsen Aero-
dynamic Model)');
   ylabel('frequency/omegaalpha2');
   xlabel('sigma/omegaalpha');
   grid on
   hold on
end
toc
tic;
%**********************************************************
% Part four
% Solution of Flutter Equation using K-Method
% Zwaan: AGARD Procedure
%**********************************************************
%**********************************************************
% Part Three
% Theodorsen Aerodynamics
%**********************************************************
A=zeros(2);
% for k =0.1:0.02:2
% V = omegaalpha*b/k;
for V=Vinit:interval:Vfinal
   q = 0.5*rho*V*V;
   Ur = V/omegaalpha/b;
   k = omegaalpha*b/V;
```

```
% Solution of Flutter Stability Equation
% k = b*omegaalpha/U;
  kr = 1/k;
  im=sqrt(-1);
  U=kr*b*omegaalpha;
  q = 0.5*rho*V*V;
  Ur=V/omegaalpha/b;
%%%% THEODORSEN FUNCTION%%%%%%%%%%%%%%%%%%%%%%%%%%
  H1 = besselh(1,2,k);
  H2 = besselh(0,2,k);
  Ck = H1/(H1+im*H2);
  Fk = real(Ck);
  Gk = imag(Ck);
%%%%%%%%%%%%%%%%%%%%%%%%%%%%%%%%%%%%%%%%%%%%%%
%Calculation of non-dimensional unsteady aerodynamic coefficients
due to
%heaving and pitching following Kuessner
%%%%%%%%%%%%%%%%%%%%%%%%%%%%%%%%%%%%%%%%%%%%%%
  ka = -2*k*Gk+im*2*k*Fk;
  kb = (-2*Fk-2*k*Gk-k*k/2)+im*(2*k*Fk+2*Gk+k);
  ma = -k*k/2;
  mb = -3*k*k/8+im*k;
%%%%%%%%%%%%%%%%%%%%%%%%%%%%%%%%%%%%%%%%%%%%%%
%%%%%Calculate each term of the Flutter Equations Matrix expressed
in
%%%%%generalized quantities
%%%%%%%%%%%%%%%%%%%%%%%%%%%%%%%%%%%%%%%%%%%%%%
  A(1,1) =(omegaalpha/omegah)^2*(myu-ka/k/k);
                  A(1,2)     =(omegaalpha/omegah)^2*(myu*x_alpha-
kb/k/k+(0.5+a)*ka/k/k);
  A(2,1) = myu*x_alpha-ma/k/k+(0.5+a)*ka/k/k;
   A(2,2) = myu*r_alpha*r_alpha-mb/k/k+(0.5+a)*(ma/k/k+kb/k/k)-
(0.5+a)^2*ka/k/k;
%%%%%%%%%%%%%%%%%%%%%%%%%%%%%%%%%%%%%%%%%%%%%%
%%%%%%%%Solve  eigenvalue  problem  [A(k)]{x}='lambda'{x}  using
polynomial solution
%%%%%%%%calculate omega1 and U=omega*b/k
%%%%%%%%%%%%%%%%%%%%%%%%%%%%%%%%%%%%%%%%%%%%%%
     P = [1,-(A(1,1)+A(2,2)),A(1,1)*A(2,2)-A(1,2)*A(2,1)];
  psq = roots(P);
  lambda = psq;
if real(lambda(1))>0
omega1 = omegaalpha/sqrt(real(lambda(1)));
end
g1 = imag(lambda(1))*omega1*omega1/omegaalpha/omegaalpha;
if real(lambda(2))>0
omega2 = omegaalpha/sqrt(real(lambda(2)));
end
g2 = imag(lambda(2))*omega2*omega2/omegaalpha/omegaalpha;
% U1 = omega1*b/k;
% U2 = omega2*b/k;
   figure(28),plot(U,omega1,'*',U,omega2,'o');
   xlabel(' U, m/sec ');
   ylabel('frequency omega');
```

```
   title('frequency-velocity diagram, Theodorsen Unsteady ');
   grid on
   hold on
   figure(29),plot(U,g1,'s',U,g2,'d');
   xlabel(' U, m/sec ');
   ylabel('damping g');
   title('damping-velocity diagram, Theodorsen Unsteady ');
   grid on
   hold on
   figure(30); plot(g1,omega1,'*',g2,omega2,'o');
    title('Polar Diagram of Imaginary vs Real root (Theodorsen & K-
Method, Polynomial)');
   ylabel('frequency/omegaalpha2');
   xlabel('damping g');
   grid on
   hold on
%%%%%%%%%%%%%%%%%%%%%%%%%%%%%%%%%%%%%%%%%%%%%
% Part Five
% Solution of Flutter Equation using K-Method
% but
%%%%%%%%Solve eigenvalue problem [A(k)]{x}='lambda'{x}
%%%%%%%%for comparative purposes
%%%%%%%%%%%%%%%%%%%%%%%%%%%%%%%%%%%%%%%%%%%%%
     [V,D] = eig(A);
%%%%%%%%%%%%%%%%%%%%%%%%%%%%%%%%%%%%%%%%%%%%%
%%%%%%%%calculate omega1 and U=omega*b/k
%%%%%%%%%%%%%%%%%%%%%%%%%%%%%%%%%%%%%%%%%%%%%
if real(D(1,1))>0
omega1 = omegaalpha/sqrt(real(D(1,1)));
end
g1 = (imag(D(1,1)))*omega1*omega1/omegaalpha/omegaalpha;
if real(D(2,2))>0
omega2 = omegaalpha/sqrt(real(D(2,2)));
end
g2 = (imag(D(2,2)))*omega2/omega2/omegaalpha/omegaalpha;
%*****************************************************
   figure(31),plot(U,omega1,'*',U,omega2,'o');
   xlabel(' U, m/sec ');
   ylabel('frequency omega');
   title('frequency-velocity diagram, Theodorsen Unsteady ');
   grid on
   hold on
   figure(32),plot(U,g1,'s',U,g2,'d');
   xlabel(' U, m/sec ');
   ylabel('damping g');
   title('damping-velocity diagram, Theodorsen Unsteady ');
   grid on
   hold on
   figure(33); plot(g1,omega1,'*',g2,omega2,'o');
    title('Polar Diagram of Imaginary vs Real root (Theodorsen & K-
Method, Eigenvalue)');
   ylabel('frequency/omegaalpha2');
   xlabel('damping g');
   grid on
```

```
  hold on
end
toc

%PROGRAM TO DETERMINE THE FLUTTER STABILITY OF THE TYPICAL SECTION:
%STRAIGHT WING CASE:
% Data adapted from YHH Boeing 747-like wing modified using
Zwaan/Isogai and BAH Wing in the absence
% of data
% Configurations:
% VG/K-Method applied to all for flutter solution
% part one : for steady case
% part two : for low frequency refinement
% part three : Theodorsen aerodynamics
% part four : solution of flutter equation using K-Method - polyno-
mial solution
% part five : solution of flutter equation using
% K-Method - eigenvalue problem
% Can further be developed for Done's Flutter Stability Analysis
% Developed based on Done's Flutter Stability Analysis
% Data based on Zwaan nomenclature
% Data taken from BAH Wing
% mass is based on m (which is defined), and myu is a derived quantity
% Developed based on Done's Flutter Stability Analysis
% Data based on Zwaan nomenclature
% Data taken from BAH Wing
% mass is based on m (which is defined), and myu is a derived quantity
tic;
% To determine the range of the velocity in m/sec in the flutter
stability
% doiagram
% Vinit - origin of diagram absis
% Vfinal - end of diagram absis
% interval - increment of velocity
Vinit = 50;
Vfinal = 600;
interval = 50;
myu= 100;
%m = 0.0161;
x_alpha = 0.00000974;
r_alpha = 0.21461;
% c is taken from 0.7 span BAH wing
c = 4.92;
b = c/2;
a = -2.0;
%BAH span
Span = 500*0.0254;
% Span = 45.0;
S = c;
e = 0.51;
rho = 1.225;
m=myu*22/7*rho*b^2;
Cl_alpha =2*22/7;
% omegaalpha is taken from BAH wing 1st torsional mode
```

```
omegaalpha=22.357;
omegah=12.8;
y =0.75*Span;
EI = 154494478;
GJ = 151565673;
% K_alpha = GJ/y;
I_alpha=811.34131;
K_alpha= I_alpha*omegaalpha^2;
% Kh = 3*EI/(y*y*y);
% Kh = 125746;
Kh=m*omegah^2;
% to get the value of S_alpha and I_alpha:
S_alpha = m*x_alpha*b;
% I_alpha = m*r_alpha*r_alpha*Br*Br;
% to determine the flutter roots:
A=zeros(2);
% to determine the flutter roots:
% **********************************************************
% Part One
% Steady / Quasi Steady Aerodynamic Model
% **********************************************************
% for k =0.1:0.02:2
% V = omegaalpha*b/k;
for V=Vinit:interval:Vfinal
q = 0.5*rho*V*V;
Ur = V/omegaalpha/b;
k = omegaalpha*b/V;
a4 = (I_alpha*m)-(S_alpha*S_alpha);
a2 = (m*K_alpha)+(I_alpha*Kh)-((2*m*e*b)+S_alpha)*q*S*Cl_alpha;
a0 = Kh*(K_alpha-(2*q*S*e*b*Cl_alpha));
% z = (a2*a2)-4*a4*a0;
P = [a4,0,a2,0,a0];
r = roots(P);
U = V;
y1=imag(r(1))/omegaalpha;
y2=imag(r(2))/omegaalpha;
y3=imag(r(3))/omegaalpha;
y4=imag(r(4))/omegaalpha;
y5=real(r(1))/omegaalpha;
y6=real(r(2))/omegaalpha;
y7=real(r(3))/omegaalpha;
y8=real(r(4))/omegaalpha;
if y1 < 0
y1 = 0;
y5 = 0;
end
if y2 < 0
y2 = 0;
y6 = 0;
end
if y3 < 0
y3 = 0;
y7 = 0;
end
```

```
if y4 < 0
y4 = 0;
y8 = 0;
end
figure(1);
plot(q,y1,'*',q,y2,'o',q,y3,'d',q,y4,'s');
title('Imaginary root vs dynamic pressure (Steady Aerodynamic
Model)');
xlabel('Dynamic pressure, p, N-m/m2');
ylabel('omega/omegaalpha');
grid on
hold on
figure(2);
plot(q,y5,'*',q,y6,'o',q,y7,'d',q,y8,'s');
title('Real root vs dynamic pressure (Steady Aerodynamic Model)');
xlabel('Dynamic pressure, p, N-m/m2');
ylabel('sigma/omegaalpha');
grid on
hold on
figure(3);
plot(Ur,y1,'*',Ur,y2,'o',Ur,y3,'d',Ur,y4,'s');
title('Imaginary root vs Reduced Velocity (Steady Aerodynamic
Model)');
xlabel('Reduced Frequency, kr');
ylabel('omega/omegaalpha');
grid on
hold on
figure(4);
plot(Ur,y5,'*',Ur,y6,'o',Ur,y7,'d',Ur,y8,'s');
title('Real root vs Reduced Velocity (Steady Aerodynamic Model)');
xlabel('Reduced Frequency, kr');
ylabel('sigma/omegaalpha');
grid on
hold on
figure(5);
plot(U,y1,'*',U,y2,'o',U,y3,'d',U,y4,'s');
title('Imaginary root vs Velocity (Steady Aerodynamic Model)');
xlabel('Velocity, U m/sec');
ylabel('omega/omegaalpha');
grid on
hold on
figure(6);
plot(U,y5,'*',U,y6,'o',U,y7,'d',U,y8,'s');
title('Real root vs Reduced Velocity (Steady Aerodynamic Model)');
xlabel('Velocity, U m/sec');
ylabel('sigma/omegaalpha');
grid on
hold on
figure(7); plot(y5,y1,'*',y6,y2,'o',y7,y3,'d',y8,y4,'s');
title('Polar Diagram of Imaginary vs Real root (Steady Aerodynamic
Model)');
ylabel('frequency/omegaalpha2');
xlabel('sigma/omegaalpha');
grid on
```

```
hold on
%%%%%%%%%%%%%%%%%%%%%
% VG Method applied to quasi-steady
%%%%%%%%%%%%%%%%%%%%%
psq = roots(P);
lambda = psq;
omega1 = omegaalpha/sqrt(real(lambda(1)));
if real(lambda(1))>0
omega1 = omegaalpha/sqrt(real(lambda(1)));
end
g1 = imag(lambda(1))*omega1*omega1/omegaalpha/omegaalpha;
omega2 = omegaalpha/sqrt(real(lambda(2)));
if real(lambda(2))>0
omega2 = omegaalpha/sqrt(real(lambda(2)));
end
g2 = imag(lambda(2))*omega2*omega2/omegaalpha/omegaalpha;
figure(8),plot(U,omega1,'*',U,omega2,'o');
xlabel(' U, m/sec ');
ylabel('frequency omega');
title('frequency-velocity diagram, K-Steady ');
grid on
hold on
figure(9),plot(U,g1,'s',U,g2,'d');
xlabel(' U, m/sec ');
ylabel('damping g');
title('damping-velocity diagram, K-Steady ');
grid on
hold on
figure(10); plot(g1,omega1,'*',g2,omega2,'o');
title('Polar Diagram of Imaginary vs Real root (Steady & K-Method,
Polynomial)');
ylabel('frequency/omegaalpha2');
xlabel('damping g');
grid on
hold on
%%%%%%%%%%%%%%%%%%%%%
end
toc
tic;
% for k =0.1:0.02:2
% V = omegaalpha*b/k;
for V=Vinit:interval:Vfinal
q = 0.5*rho*V*V;
Ur = V/omegaalpha/b;
k = omegaalpha*b/V;
U = V;
% ************************************************************
% Part Two
% low frequency refinement
% ************************************************************
a4 = (I_alpha*m)-(S_alpha*S_alpha);
a3 = (q/V)*S*Cl_alpha*((2*e*b*S_alpha)+I_alpha);
a2 = (m*K_alpha)+(I_alpha*Kh)-((2*m*e*b)+S_alpha)*q*S*Cl_alpha;
a1 = (q/V)*S*Cl_alpha*K_alpha;
```

```
a0 = Kh*(K_alpha-(2*q*S*e*b*Cl_alpha));
P = [a4,a3,a2,a1,a0];
if y1 < 0
y1 = 0;
y5 = 0;
end
if y2 < 0
y2 = 0;
y6 = 0;
end
if y3 < 0
y3 = 0;
y7 = 0;
end
if y4 < 0
y4 = 0;
y8 = 0;
end
figure(11);
plot(q,y1,'*',q,y2,'o',q,y3,'d',q,y4,'s');
title('Imaginary root vs dynamic pressure (Low Frequency Aerody-
namic Model)');
xlabel('Dynamic pressure, p, N-m/m2');
ylabel('omega/omegaalpha');
grid on
hold on
figure(12);
plot(q,y5,'*',q,y6,'o',q,y7,'d',q,y8,'s');
title('Real root vs dynamic pressure (Low Frequency Aerodynamic
Model)');
xlabel('Dynamic pressure, p, N-m/m2');
ylabel('sigma/omegaalpha');
grid on
hold on
figure(13);
plot(Ur,y1,'*',Ur,y2,'o',Ur,y3,'d',Ur,y4,'s');
title('Imaginary root vs Reduced Velocity (Low Frequency Aerody-
namic Model)');
xlabel('Reduced Frequency, kr');
ylabel('omega/omegaalpha');
grid on
hold on
figure(14);
plot(Ur,y5,'*',Ur,y6,'o',Ur,y7,'d',Ur,y8,'s');
title('Real root vs Reduced Velocity (Low Frequency Aerodynamic
Model)');
xlabel('Reduced Frequency, kr');
ylabel('sigma/omegaalpha');
grid on
hold on
figure(15);
plot(U,y1,'*',U,y2,'o',U,y3,'d',U,y4,'s');
title('Imaginary root vs Velocity (Low Frequency Aerodynamic
Model)');
```

```
xlabel('Velocity, U m/sec');
ylabel('omega/omegaalpha');
grid on
hold on
figure(16);
plot(U,y5,'*',U,y6,'o',U,y7,'d',U,y8,'s');
title('Real root vs Reduced Velocity (Low Frequency Aerodynamic
Model)');
xlabel('Velocity, U m/sec');
ylabel('sigma/omegaalpha');
grid on
hold on
figure(17); plot(y5,y1,'*',y6,y2,'o',y7,y3,'d',y8,y4,'s');
title('Polar Diagram of Imaginary vs Real root (Low Frequency Aero-
dynamic Model)');
ylabel('frequency/omegaalpha2');
xlabel('sigma/omegaalpha');
grid on
hold on
%%%%%%%%%%%%%%%%%%%%%%%%%%%%%%%%%%%%%%%%%%%%%
%% VG-Method applied to low Frequency
%% Refinement
%%%%%%%%%%%%%%%%%%%%%%%%%%%%%%%%%%%%%%%%%%%%%
psq = roots(P);
lambda = psq;
if real(lambda(1))>0
omega1 = omegaalpha/sqrt(real(lambda(1)));
end
g1 = imag(lambda(1))*omega1*omega1/omegaalpha/omegaalpha;
if real(lambda(2))>0
omega2 = omegaalpha/sqrt(real(lambda(2)));
end
g2 = imag(lambda(2))*omega2*omega2/omegaalpha/omegaalpha;
% U1 = omega1*b/k;
% U2 = omega2*b/k;
figure(18),plot(U,omega1,'*',U,omega2,'o');
xlabel(' U, m/sec ');
ylabel('frequency omega');
title('frequency-velocity diagram, K-Low Frequency ');
grid on
hold on
figure(19),plot(U,g1,'s',U,g2,'d');
xlabel(' U, m/sec ');
ylabel('damping g');
title('damping-velocity diagram, K-Low frequency ');
grid on
hold on
figure(20); plot(g1,omega1,'*',g2,omega2,'o');
title('Polar Diagram of Imaginary vs Real root (Low Frequency & K-
Method, Polynomial)');
ylabel('frequency/omegaalpha2');
xlabel('damping g');
grid on
hold on
```

```
%%%%%%%%%%%%%%%%%%%%%%%%%%%
end
toc
tic;
% ************************************************************
% Part Three
% Theodorsen Aerodynamics
% ************************************************************
A=zeros(2);
% for k =0.1:0.02:2
% V = omegaalpha*b/k;
for V=Vinit:interval:Vfinal
q = 0.5*rho*V*V;
Ur = V/omegaalpha/b;
k = omegaalpha*b/V;
% Solution of Flutter Stability Equation
% k = b*omegaalpha/U;
kr = 1/k;
im=sqrt(-1);
U=kr*b*omegaalpha;
q = 0.5*rho*V*V;
Ur=V/omegaalpha/b;
%%%% THEODORSEN FUNCTION%%%%%%%%%%%%%%%%%%%%%%%%%%%%%%%
H1 = besselh(1,2,k);
H2 = besselh(0,2,k);
Ck = H1/(H1+im*H2);
Fk = real(Ck);
Gk = imag(Ck);
%%%%%%%%%%%%%%%%%%%%%%%%%%%%%%%%%%%%%%%%%%%%%%%%%%%%%%%%
%Calculation of non-dimensional unsteady aerodynamic coefficients
due to
%heaving and pitching following Kuessner
%%%%%%%%%%%%%%%%%%%%%%%%%%%%%%%%%%%%%%%%%%%%%%%%%%%%%%%%
ka = -2*k*Gk+im*2*k*Fk;
kb = (-2*Fk-2*k*Gk-k*k/2)+im*(2*k*Fk+2*Gk+k);
ma = -k*k/2;
mb = -3*k*k/8+im*k;
%%%%%%%%%%%%%%%%%%%%%%%%%%%%%%%%%%%%%%%%%%%%%%%%%%%%%%%%
%%%%%Calculate each term of the Flutter Equations Matrix expressed
in
%%%%%generalized quantities
%%%%%%%%%%%%%%%%%%%%%%%%%%%%%%%%%%%%%%%%%%%%%%%%%%%%%%%%
A(1,1) =(omegaalpha/omegah)^2*(myu-ka/k/k);
A(1,2)                    =(omegaalpha/omegah)^2*(myu*x_alpha-
kb/k/k+(0.5+a)*ka/k/k);
A(2,1) = myu*x_alpha-ma/k/k+(0.5+a)*ka/k/k;
A(2,2)  =  myu*r_alpha*r_alpha-mb/k/k+(0.5+a)*(ma/k/k+kb/k/k)-
(0.5+a)^2*ka/k/k;
%%%%%%%%%%%%%%%%%%%%%%%%%%%%%%%%%%%%%%%%%%%%%%%%%%%%%%%%
%%%%%%%%Solve  eigenvalue  problem  [A(k)]{x}='lambda'{x}  using
polynomial solution
%%%%%%%%calculate omega1 and U=omega*b/k
%%%%%%%%%%%%%%%%%%%%%%%%%%%%%%%%%%%%%%%%%%%%%%%%%%%%%%%%
% p=[1 -(A(1,1)+A(2,2)) A(1,1)*A(2,2)-A(1,2)*A(2,1)];
```

```
% lambda =roots(p);
%P = [1,0,-(A(1,1)+A(2,2)),0,A(1,1)*A(2,2)-A(1,2)*A(2,1)];
%r = roots(P);
% lambda = r;
% ********************************************************
P = [1,-(A(1,1)+A(2,2)),A(1,1)*A(2,2)-A(1,2)*A(2,1)];
psq = roots(P);
lambda = psq;
psq1 =real(psq(1))+imag(psq(1));
psq2 =real(psq(2))+imag(psq(2));
p1 =sqrt(psq1);
p2 =-sqrt(psq1);
p3 =sqrt(psq2);
p4 =-sqrt(psq2);
y1=imag(p1);
if y1<0
y1=imag(p2);
end
y2=imag(p3);
if y2<0
y2=imag(p4);
end
sigma1=real(p1)/omegaalpha;
sigma2=real(p2)/omegaalpha;
sigma3=real(p3)/omegaalpha;
sigma4=real(p4)/omegaalpha;
omega1=y1/omegaalpha;
omega2=y2/omegaalpha;
% omega3=imag(r(3))/omegaalpha;
% omega4=imag(r(4))/omegaalpha;
% y1=omega1;
% if y1<0
% y1=omega2;
% end
% y2=omega3;
% if y2<0
% y2=omega4;
% end
% ********************************************************
%sigma1=real(r(1))/omegaalpha;
%sigma2=real(r(2))/omegaalpha;
%sigma3=real(r(3))/omegaalpha;
%sigma4=real(r(4))/omegaalpha;
%omega1=imag(r(1))/omegaalpha;
%omega2=imag(r(2))/omegaalpha;
%omega3=imag(r(3))/omegaalpha;
%omega4=imag(r(4))/omegaalpha;
%y1=omega1;
%if y1<0
% y1=omega2;
%end
%y2=omega3;
%if y2<0
% y2=omega4;
```

```
%end
y5=sigma1;
y6=sigma2;
y7=sigma3;
y8=sigma4;
figure(21);
plot(q,y1,'*',q,y2,'o');
title('omega/omegaalpha vs dynamic pressure (Theodorsen Unsteady
Aerodynamic Model)');
xlabel('Dynamic pressure, p, N-m/m2');
ylabel('omega/omegaalpha');
grid on
hold on
figure(22);
plot(q,y5,'*',q,y6,'o',q,y7,'d',q,y8,'s');
title('sigma/omegaalpha vs dynamic pressure (Theodorsen Unsteady
Aerodynamic Model)');
xlabel('Dynamic pressure, p, N-m/m2');
ylabel('sigma/omegaalpha');
grid on
hold on
figure(23);
plot(Ur,y1,'*',Ur,y2,'o');
title('omega/omegaalpha vs Reduced Velocity (Theodorsen Unsteady
Aerodynamic Model)');
xlabel('Reduced Frequency, kr');
ylabel('omega/omegaalpha');
grid on
hold on
figure(24);
plot(Ur,y5,'*',Ur,y6,'o',Ur,y7,'d',Ur,y8,'s');
title('sigma/omegaalpha vs Reduced Velocity (Theodorsen Unsteady
Aerodynamic Model)');
xlabel('Reduced Frequency, kr');
ylabel('sigma/omegaalpha');
grid on
hold on
figure(25);
plot(U,y1,'*',U,y2,'o');
title('omega/omegaalpha vs Velocity (Theodorsen Unsteady Aerody-
namic Model)');
xlabel('Velocity, U, m/sec');
ylabel('omega/omegaalpha');
grid on
hold on
figure(26);
plot(U,y5,'*',U,y6,'o',U,y7,'d',U,y8,'s');
title('sigma/omegaalpha vs Velocity (Theodorsen Unsteady Aerody-
namic Model)');
xlabel('Velocity, U, m/sec');
ylabel('sigma/omegaalpha');
grid on
hold on
figure(27); plot(y5,y1,'*',y6,y2,'o',y7,y3,'d',y8,y4,'s');
```

```
title('Polar Diagram of Imaginary vs Real root (Theodorsen Aerody-
namic Model)');
ylabel('frequency/omegaalpha2');
xlabel('sigma/omegaalpha');
grid on
hold on
end
toc
tic;
%*********************************************************
% Part four
% Solution of Flutter Equation using K-Method
% Zwaan: AGARD Procedure
%*********************************************************
%*********************************************************
% Part Three
% Theodorsen Aerodynamics
%*********************************************************
A=zeros(2);
% for k =0.1:0.02:2
% V = omegaalpha*b/k;
for V=Vinit:interval:Vfinal
q = 0.5*rho*V*V;
Ur = V/omegaalpha/b;
k = omegaalpha*b/V;
% Solution of Flutter Stability Equation
% k = b*omegaalpha/U;
kr = 1/k;
im=sqrt(-1);
U=kr*b*omegaalpha;
q = 0.5*rho*V*V;
Ur=V/omegaalpha/b;
%%%% THEODORSEN FUNCTION%%%%%%%%%%%%%%%%%%%
H1 = besselh(1,2,k);
H2 = besselh(0,2,k);
Ck = H1/(H1+im*H2);
Fk = real(Ck);
Gk = imag(Ck);
%%%%%%%%%%%%%%%%%%%%%%%%%%%%%%%%%%%%%%%%%%
%Calculation of non-dimensional unsteady aerodynamic coefficients
due to
%heaving and pitching following Kuessner
%%%%%%%%%%%%%%%%%%%%%%%%%%%%%%%%%%%%%%%%%%
ka = -2*k*Gk+im*2*k*Fk;
kb = (-2*Fk-2*k*Gk-k*k/2)+im*(2*k*Fk+2*Gk+k);
ma = -k*k/2;
mb = -3*k*k/8+im*k;
%%%%%%%%%%%%%%%%%%%%%%%%%%%%%%%%%%%%%%%%%%
%%%%%Calculate each term of the Flutter Equations Matrix expressed
in
%%%%%generalized quantities
%%%%%%%%%%%%%%%%%%%%%%%%%%%%%%%%%%%%%%%%%%
A(1,1) =(omegaalpha/omegah)^2*(myu-ka/k/k);
```

```
A(1,2)                          =(omegaalpha/omegah)^2*(myu*x_alpha-
kb/k/k+(0.5+a)*ka/k/k);
A(2,1) = myu*x_alpha-ma/k/k+(0.5+a)*ka/k/k;
A(2,2)   =   myu*r_alpha*r_alpha-mb/k/k+(0.5+a)*(ma/k/k+kb/k/k)-
(0.5+a)^2*ka/k/k;
%%%%%%%%%%%%%%%%%%%%%%%%%%%%%%%%%%%%%%%%%%%%
%%%%%%%%Solve  eigenvalue  problem  [A(k)]{x}='lambda'{x}  using
polynomial solution
%%%%%%%%calculate omega1 and U=omega*b/k
%%%%%%%%%%%%%%%%%%%%%%%%%%%%%%%%%%%%%%%%%%%%
P = [1,-(A(1,1)+A(2,2)),A(1,1)*A(2,2)-A(1,2)*A(2,1)];
psq = roots(P);
lambda = psq;
if real(lambda(1))>0
omega1 = omegaalpha/sqrt(real(lambda(1)));
end
g1 = imag(lambda(1))*omega1*omega1/omegaalpha/omegaalpha;
if real(lambda(2))>0
omega2 = omegaalpha/sqrt(real(lambda(2)));
end
g2 = imag(lambda(2))*omega2*omega2/omegaalpha/omegaalpha;
% U1 = omega1*b/k;
% U2 = omega2*b/k;
figure(28),plot(U,omega1,'*',U,omega2,'o');
xlabel(' U, m/sec ');
ylabel('frequency omega');
title('frequency-velocity diagram, Theodorsen Unsteady ');
grid on
hold on
figure(29),plot(U,g1,'s',U,g2,'d');
xlabel(' U, m/sec ');
ylabel('damping g');
title('damping-velocity diagram, Theodorsen Unsteady ');
grid on
hold on
figure(30); plot(g1,omega1,'*',g2,omega2,'o');
title('Polar Diagram of Imaginary vs Real root (Theodorsen & K-
Method, Polynomial)');
ylabel('frequency/omegaalpha2');
xlabel('damping g');
grid on
hold on
%%%%%%%%%%%%%%%%%%%%%%%%%%%%%%%%%%%%%%%%%%%%
% Part Five
% Solution of Flutter Equation using K-Method
% but
%%%%%%%%Solve eigenvalue problem [A(k)]{x}='lambda'{x}
%%%%%%%%for comparative purposes
%%%%%%%%%%%%%%%%%%%%%%%%%%%%%%%%%%%%%%%%%%%%
[V,D] = eig(A);
%%%%%%%%%%%%%%%%%%%%%%%%%%%%%%%%%%%%%%%%%%%%
%%%%%%%%calculate omega1 and U=omega*b/k
%%%%%%%%%%%%%%%%%%%%%%%%%%%%%%%%%%%%%%%%%%%%
if real(D(1,1))>0
```

```
omega1 = omegaalpha/sqrt(real(D(1,1)));
end
g1 = (imag(D(1,1)))*omega1*omega1/omegaalpha/omegaalpha;
if real(D(2,2))>0
omega2 = omegaalpha/sqrt(real(D(2,2)));
end
g2 = (imag(D(2,2)))*omega2/omega2/omegaalpha/omegaalpha;
% ***********************************************************
figure(31),plot(U,omega1,'*',U,omega2,'o');
xlabel(' U, m/sec ');
ylabel('frequency omega');
title('frequency-velocity diagram, Theodorsen Unsteady ');
grid on
hold on
figure(32),plot(U,g1,'s',U,g2,'d');
xlabel(' U, m/sec ');
ylabel('damping g');
title('damping-velocity diagram, Theodorsen Unsteady ');
grid on
hold on
figure(33); plot(g1,omega1,'*',g2,omega2,'o');
title('Polar Diagram of Imaginary vs Real root (Theodorsen & K-
Method, Eigenvalue)');
ylabel('frequency/omegaalpha2');
xlabel('damping g');
grid on
hold on
end
toc
```

Appendix B
SI Units and Conversion Tables[1]

See Tables B.1, B.2, B.3, B.4 and B.5.

Accepted Units

For practical reasons a number of non-metric units are accepted for use. These include units of time (minute, hour, etc.), units of plane angle (degree, etc.), and a few units for special applications, such as the nautical mile, used in navigation. Section 5 includes accepted units and shows their areas of application. These units may be used in full compliance with the provisions of the Metric Conversion Law (15 U.S.C. 205(a)), Executive Order 12,770, and the Federal Register Notice, "Metric System

Table B.1 Common SI units

Measurement	Unit	Symbol	Equivalents
Length	1 milimeter 1 cm 1 m 1 km	mm cm m km	1000 μm (μm) 10 milimeters (mm) 100 cm (cm) 1000 m (m)
Area	1 square meter 1 square km	m^2 km^2	10,000 square centimeters (cm^2) 1000 000 square meters (m^2)
Volume	1 mililitre 1 L	mL L	1 cubic centimeter (cm^3 or cc) 1000 mililitres (mL)
Mass	1 g 1 kg 1 ton	g kg t	1000 miligrams (mg) 1000 g (g) 1000 kg (kg) 1 metric ton
Time	1 s	s	
Temperature	1 K	K	1 degree Celcius (°C)

[1] Most of our information are adopted from GOVPUB-NIST-The International System of Units (SI)-Conversion Factors, https://www.govinfo.gov/content/pkg/GOVPUB-C13-f10c2ff9e7af209 1314396a2d53213e4.pdf.

© The Editor(s) (if applicable) and The Author(s), under exclusive license to Springer Nature Singapore Pte Ltd. 2023

H. Djojodihardjo, *Introduction to Aeroelasticity*,
https://doi.org/10.1007/978-981-16-8078-6

Table B.2 Metric conversion units (1), (2) and (3)

B.2a

Unit	Multiply by	Obtain	Measurement	
		Unit	Multiply by	Obtain
Inches	2.54	Centimeters	0.394	Inches
Feet	0.3085	Meters	3.281	Feet
Yards	0.914	Meters	1.0936	Yards
Miles	1.609	Kilometers	0.62	Miles
Square inches	6.45	Square centimeters	0.155	Square inches
Square feet	0.093	Square meters	10.76	Square feet
Square yards	0.836	Square meters	1.196	Square yards
Acres	0.405	Hectares	2.471	Acres
Square miles	2.59	Square kilometerd	0.386	Square miles
Fluid ounces	29.57	Milliliters (mL)	0.0338	Fluid ounces
Quarts	0.946	Liters (L)	1.057	Quarts
Gallons	3.785	Liters (L)	0.264	Gallons
Ounces	28.35	Grams (g)	0.0353	Ounces
Pounds	0.4536	Kilograms (kg)	2.2046	Pounds
Tons	0.907	Metric tons (t)	1.102	Tons

B.2b

From unites	Conversion to	Units
Fahrenheit	Subtract 32; then divide by 1.8	To find Celsius
Celsius	Multiply by 1.8; then add 32	To find Fahrenheit

B.2c

To convert from	To	Multiply by
Foot	Meter	0.3045

Table B.3 CGS units not to be used

CGS units that shall not be used	Symbol equivalents
erg, dyne, gal	Used in mechanics
Poise, stikes	Used in fluid dynamics
stilb, phot, lambett	Used in photometry
emu, esu, gauss, oersted, maxwell, gilbert, biot, franklin, abampere, abvolt, statvolt, etc	Used electricity and magnetism

of Measurement; Interpretation of the International System of Units for the United States" (63 F.R. 40,334, July 28, 1998).

Deprecated Names or Symbols

Other units from older versions of the metric system, some terms not recommended for continued use, and jargon that shall not be used include.

Table B.4 Deprecated names and symbols

Deprecated term or symbol	Correct unit
Kilo	Kilogram
Calorie	Joule (J), if the value is used in physics Kilojoule (kJ), if the value is used in nutrition
Candle or candlepower	Candela
Centiliter	Milliliter or liter
Fermi	Femtometer
Gamma	Nanotesla
Micron	Micrometer
Millimicron	Nanometer
Mho	Siemens
γ (gamma)	Microgram
λ (lambda)	Cubic millimeter or microliter

Table B.5 Various conversions (1)–(8)

B.5.1 Quantities of space and time (1)		
Plane angle		
Radian	Degree arc	57.29578
Length		
Angstrom Fathom	Nanometer (nm) Meter (m)	0.1 1.828804
Foot	Meter (m)	0.3048006
Inch (in) Inch (in)	Centimeter (cm) Millimeter (mm)	2.54 25.4
Microinch (μin)	Micrometer (μm)	0.0254
Mil (0.001 in)	Millimeter (mm)	0.0254
Mil (0.001 in)	Micrometer (μm)	25.4
Yard (yd)	Meter (m)	0.9144
Mile, international (5280 ft) (mi) Nautical mile Point (pointer's)	Kilometer (km) Kilometer (km) Millimeter (mm)	1.609344 1.852 0.35146
Pica	Millimeter (mm)	4.2175

(continued)

Table B.5 (continued)

B.5.1 Quantities of space and time (2)		
Volume		
Fluid ounce	Milliliter (mL)	29.57353
Cubic inch	Cubic centimeter (cm^3)	16.387664
Velocity		
Foot per second	Meter per second (m/s)	0.3048
Mile per hour (mph) Knot (nautical mile per hour)	Kilometer per hour (km/h) Kilometer per hour (km/h)	1.609344 1.862
Acceleration		
Inch per second squared	Meter per second squared (m/s^2)	0.0254
Foot per second squared	Meter per second squared (m/s^2)	0.3048
Standard acceleration of gravity (g)	Meter per second squared (m/s^2)	9.80665
Flow rate		
Cubic foot per second	Cubicmeter per second (m^3/s)	0.02831685
Cubic foot per minute Cubic foot per minute	Cubicmeter per second (m^3/s) Liter per second (L/s)	0.0004719474 0.4719474
Cubic yard per minute Gallon per minute	Liter per second (L/s) Liter per second (L/s)	12.74258 0.0630902
Gallon per day	Liter per day (L/d)	3.785412
Fuel efficiency		
Mile per gallon	Kilometer per liter (km/L)	0.4251437
Quantities in mechanics		
Mass (weight)		
Ton (long) (2240 lbs)	Kilogram (kg)	1.016047
Ton (long) (2240 lbs)	Metric ton (t)	1.1061047
Ton (short) (2000 lbs)	Kilogram (kg)	907.18474
Ton (short) (2000 lbs)	Metric ton (t)	0.90718474

(continued)

Table B.5 (continued)

B.5.1 Quantities of space and time (2)		
Slug	Kilogram (kg)	14.5939
Pound (avoirdupois)	Kilogram (kg	0.45359237
Ounce (troy)	Gram (g)	31.10348
Ounce (avoirdupois)	Gram (g)	28.34932
Grain	Milligram (mg)	64.79891
B.5.1 Quantities of space and time (3)		
Moment of mass		
Pound foot	Kilogram meter (kg m)	0.138255
Density		
Ton (2000 lb [short]) per cubic yard Ton (2000 lb [short]) per cubic yard	Kilogram per cubic meter (kg/m^3) Metric ton per cubic meter (t/m^3)	1.186553 0.0011865
Pound per cubic foot	Kilogram per cubic meter (kg/m^3)	16.01846
Concentration (mass)		
Pound per gallon Ounce (avoirdupois) per gallon	Gram per liter (g/L) Gram per liter (g/L)	119.8264 7.489152
Momentum		
Pound foot per second	Kilogram meter per second (kg m/s)	0.1382550
Moment of inertia		
Pound square foot	Kilogram square meter (kg m^2)	0.04214011
Force		
Pound-force	Newton (N)	4.448222
Poundal	Newton (N)	0.1382550
Moment of force; torque		
Pound force foot	Newton meter (Nm)	1.355818
Pound force inch	Newton meter (Nm)	0.1129848
Pressure; stress		
Standard atmosphere	Kilopascal (kPa)	101.325
Bar	Kilopascal (kPa)	100
Millibar	Kilopascal (kPa)	0.1

(continued)

Table B.5 (continued)

B.5.1 Quantities of space and time (3)		
Pound force per square inch (psi) Kilopound force per square inch (psi) Pound force per square foot Inch of mercury (32 °F)	Kilopascal (kPa) Megapascal (MPa) Kilopascal (kPa) Kilopascal (kPa)	6.894757 6.894757 0.047b88026 3.386 38
Foot of water (39.2 °F) Inch of water (39.2 °F) Millimeter of mercury (32 °F) Torr (Torr)	Kilopascal (kPa) Kilopascal (kPa) Kilopascal (kPa) Pascal (Pa)	2.98898 0.249082 0.1333224 133.3224

B.5.1 Quantities of space and time (4)		
Viscosity (dynamic)		
Centipoise	Millipascal second (mPa s)	1
Viscosity (kinematic)		
Centistokes	Square millimeter per second (mm^2/s)	1
Energy, work, heat		
Kilowatthour Calorie (as used in physics) Calorie (as used in nutrition)	Megajoule (MJ) Joule (J) Kjoule (kJ)	3.6 4.184 4.184
Btu	Kjoule (kJ)	1.055056
Therm (US)	Megajoule (MJ)	105.4804
Horsepower hour	Megajoule (MJ)	2.684520
Foot pound-force	Joule (J)	1.355818

B.5.1 Quantities of space and time (5)		
Power		
Ton, refrigeration (12 000 Btu/h) Btu per second Btu per hour Horsepower (550 foot pound force per second)	Kilowatt (kW) Kilowatt (kW) Watt (W) Watt (W)	3.516853 1.055056 0.2930711 745.6999
Horsepower electric	Watt (W)	746
Foot pound force per second	Watt (W)	1.355818

(continued)

Table B.5 (continued)

B.5.1 Quantities of space and time (5)		
Quantities of heat		
Temperature		
Linear expansion coefficient		
Reciprocal degree Fahrenheit	Reciprocal kelvin (K^{-1})	1.8
Reciprocal degree Fahrenheit	Reciprocal kelvin ($°C^{-1}$)	1.8
Heat		
Heat flow rate		
Thermal conductivity		
Btu inch per hour square food degree Fahrenheit	Watt per meter kelvin [W/(m K)]	0.1442279
Coefficient of heat transfer		
Btu inch per hour square food degree Fahrenheit	Watt per square meter kelvin [W/(m^2 K)]	5.678263
Heat capacity		
Btu per degree Fahrenheit	Kilojoule per kelvin (kJ/K)	1.899101
Specific Heat capacity		
Btu per pound degree Fahrenheit	Kilojoule per kilogram kelvin (kJ/kg K)	4.1868
Entropy		
Btu per degree Rankine	Kilojoule per kelvin (kJ/K)	1.899101
Specific entropy		
Btu per pound degree Rankine	Kilojoule per kilogram kelvin [kJ/(kg K)]	4.1868
Specific internal energy		
Btu per pound	Kilojoule per kilogram (kJ/kg)	3.326

B.5.2 Quantities of electricity and magnetism (6)		
Magnetic field strength		
Oersted	Ampere per meter (A/m)	7,957,747
Magnetic flux		
Maxwell	Nanoweber (nWb)	10
Magnetic flux density		
Gauss	Millitesla (mT)	0.1
Electric charge		
Ampere hour	Coulomb (C)	3 600
Resistivity		

(continued)

Table B.5 (continued)

B.5.2 Quantities of electricity and magnetism (6)		
Ohm circular mil per foot	Nanoohm meter (nΩ m)	1.662426
Conductivity		
Mho per centimeter	Siemens per meter	100
B.5.3 Quantities of light and related electromagnetic radiation (7)		
Wavelength		
Ångström	Nanometer (nm)	0.1
Luminance		
Lambert (L)	Candela per square meter (cd/m^2)	3,183,099
Candela per square inch	Candela per square meter (cd/m^2)	1440.003
Footlambert	Candela per square meter (cd/m^2)	3.426259
Luminous exitance		
Lumen per square foot	lux (lx)	10.76391
Phot	lux (lx)	10,000
Illuminance		
Footcandle	lux (lx)	10.76391
B.5.4 Quantities of radiology (8)		
Activity		
Curie	Megabecquerel (MBq)	37,000
Absorbed dose		
Rad	Gray (Gy)	0.01
Rad	Centigray (cGy)	1
Dose equivalent		
Rem	Sievert (Sv)	0.01
Rem	Millisievert (mSv)	10
Millirem	Millisievert (mSv)	0.01
Millirem	Microsievert (μSv)	10
Exposure		
Roentgen	Coulomb per kilogram (C/kg)	0.000258